THERMODYNAMICS, KINETICS, AND MICROPHYSICS OF CLOUDS

Climate change has provided a new impetus for research on clouds and precipitation. One of the greatest uncertainties in current global climate models is cloud feedback, arising from uncertainties in the parameterization of cloud processes and their impact on the global radiation balance. In the past two decades, substantial progress has been made in the simulation of clouds using cloud resolving models. However, most of the parameterizations employed in these models have been empirically based. New theoretical descriptions of cloud processes are now being incorporated into cloud models, using spectral microphysics based on the kinetic equations for the drop and crystal size spectra along with the supersaturation equation, and newer parameterizations of drop activation and ice nucleation based on the further development of the classical nucleation theory. From these models, cloud microphysics parameterizations are being developed for use in global weather and climate models.

Thermodynamics, Kinetics, and Microphysics of Clouds reflects this shift to an increasingly theoretical basis for the simulation and parameterization of cloud processes. The book presents a unified theoretical foundation that provides the basis for incorporating cloud microphysical processes in cloud and climate models in a manner that represents interactions and feedback processes over the relevant range of environmental and parametric conditions. In particular, this book provides:

- the closed system of equations of spectral cloud microphysics that includes kinetic equations for the drop and crystal size spectra for regular and stochastic condensation/deposition and coagulation/accretion along with the supersaturation equations;
- the latest theories and theoretical parameterizations of aerosol hygroscopic growth, drop activation, and ice homogeneous and heterogeneous nucleation, derived from the general principles of thermodynamics and kinetics and suitable for cloud and climate models;
- a theoretical basis for understanding the processes of cloud particle formation, evolution, and precipitation, based on numerical cloud simulations and analytical solutions to the kinetic equations and supersaturation equation;
- a platform for advanced parameterizations of clouds in weather prediction and climate models using these solutions; and
- the scientific foundation for weather and climate modification by cloud seeding.

This book will be invaluable for researchers and advanced students engaged in cloud and aerosol physics, and air pollution and climate research.

Vitaly I. Khvorostyanov is Professor of Physics of the Atmosphere and Hydrosphere, Central Aerological Observatory (CAO), Russian Federation. His research interests are in cloud physics, cloud numerical modeling, atmospheric radiation, and cloud-aerosol and cloud-radiation interactions, with applications for climate studies and weather modification. He has served as Head of the Laboratory of Numerical Modeling of Cloud Seeding at CAO, Coordinator of the Cloud Modeling Programs on Weather Modification by Cloud Seeding in the USSR and Russia, Member of the International GEWEX Radiation Panel of the World Climate Research Program, and Member of the International Working Group on Cloud-Aerosol Interactions. Dr. Khvorostyanov has worked as a visiting scientist and Research Professor in the United States, United Kingdom, France, Germany, and Israel. He has co-authored nearly 200 journal articles and four books: *Numerical Simulation of Clouds* (1984), *Clouds and Climate* (1986), *Energy-Active Zones: Conceptual Foundations* (1989), and *Cirrus* (2002). Dr. Khvorostyanov is a member of the American Geophysical Union and the American Meteorological Society.

Judith A. Curry is Professor and Chair of the School of Earth and Atmospheric Sciences at the Georgia Institute of Technology. She previously held faculty positions at the University of Colorado, Penn State University, and Purdue University. Dr. Curry's research interests span a variety of topics in the atmospheric and climate sciences. Current interests include cloud microphysics, air and sea interactions, and climate feedback processes associated with clouds and sea ice. Dr. Curry is co-author of *Thermodynamics of Atmospheres and Oceans* (1999) and editor of the *Encyclopedia of Atmospheric Sciences* (2003). She has published more than 190 refereed journal articles. Dr. Curry is a Fellow of the American Meteorological Society, the American Association for the Advancement of Science, and the American Geophysical Union. In 1992, she received the Henry Houghton Award from the American Meteorological Society.

The Cloud
Percy Bysshe Shelley (1820)

I bring fresh showers for the thirsting flowers,
From the seas and the streams;
I bear light shade for the leaves when laid
In their noonday dreams.
From my wings are shaken the dews that waken
The sweet buds every one,
When rocked to rest on their mother's breast,
As she dances about the sun.
I wield the flail of the lashing hail,
And whiten the green plains under,
And then again I dissolve it in rain,
And laugh as I pass in thunder.

I am the daughter of Earth and Water,
And the nursling of the Sky;
I pass through the pores of the oceans and shores;
I change, but I cannot die.
For after the rain, when with never a stain
The pavilion of Heaven is bare
And the winds and sunbeams with their convex gleams
Build up the blue dome of air,
I silently laugh at my own cenotaph
And out of the caverns of rain,
Like a child from the womb, like a ghost from the tomb,
I arise and unbuild it again.

(Poetical Works of Shelley (Cambridge Editions),
by Percy Bysshe Shelley (Author), Newell F. Ford (Introduction).
Publisher: Houghton Mifflin; Revised edition,
January 1975, 704 pages, ISBN-10: 0395184614)

THERMODYNAMICS, KINETICS, AND MICROPHYSICS OF CLOUDS

VITALY I. KHVOROSTYANOV
Central Aerological Observatory, Russia

JUDITH A. CURRY
Georgia Institute of Technology, USA

CAMBRIDGE
UNIVERSITY PRESS

CAMBRIDGE
UNIVERSITY PRESS

Shaftesbury Road, Cambridge CB2 8EA, United Kingdom

One Liberty Plaza, 20th Floor, New York, NY 10006, USA

477 Williamstown Road, Port Melbourne, VIC 3207, Australia

314–321, 3rd Floor, Plot 3, Splendor Forum, Jasola District Centre, New Delhi – 110025, India

103 Penang Road, #05–06/07, Visioncrest Commercial, Singapore 238467

Cambridge University Press is part of Cambridge University Press & Assessment,
a department of the University of Cambridge.

We share the University's mission to contribute to society through the pursuit of
education, learning and research at the highest international levels of excellence.

www.cambridge.org
Information on this title: www.cambridge.org/9781107016033

First published 2014

A catalogue record for this publication is available from the British Library

Library of Congress Cataloging-in-Publication data
Khvorostyanov, Vitaly I.
Thermodynamics, kinetics, and microphysics of clouds / Vitaly I. Khvorostyanov,
Judith A. Curry. — First edition.
pages cm
Includes bibliographical references and index.
ISBN 978-1-107-01603-3
1. Precipitation (Meteorology)—Measurement. 2. Cloud forecasting.
3. Precipitation forecasting. 4. Atmospheric thermodynamics. I. Curry, Judith A. II. Title.
QC925.K47 2014
551.57'6—dc23 2014001806

ISBN 978-1-107-01603-3 Hardback

Contents

Preface

Cloud microphysics is a branch of cloud physics that studies initiation, growth, and dissipation of cloud and precipitation particles. Cloud microphysics is governed by the thermodynamic and kinetic processes in clouds. The field of cloud microphysics has been intensively developed since the 1940s when the first successful experiments on cloud seeding were performed. The field has received additional impetus in recent years from the challenges associated with forecasting precipitation and understanding aerosol-cloud interactions in the context of climate change and feedback processes. Several books on cloud microphysics are available, including Mason (1957), Fletcher (1962, 1970a), Dufour and Defay (1963), Sedunov (1974), Voloshchuk and Sedunov (1975), Voloshchuk (1984), Matveev (1984), Young (1993), Pruppacher and Klett (1997), and Cotton et al. (2011).

Thermodynamics, Kinetics, and Microphysics of Clouds extends the subject of cloud microphysics beyond these previous treatments. The goals and contents of this book are formulated to:

- Present in compact form the major thermodynamic relations and kinetic equations required for theoretical consideration of cloud microphysics;
- Review the currently known states of water in liquid, crystalline, and amorphous forms, and the conceptual modern theories of water and equations of state for water in various states;
- Formulate a closed system of equations that describe the kinetics of cloud microphysical processes and is suitable both for analytical studies and for inclusion in numerical models;
- Derive from theory generalized analytical parameterizations for aerosol deliquescence, hygroscopic growth, efflorescence, and drop activation and ice nucleation in various modes;
- Demonstrate that these theoretical parameterizations generalize and unify previous parameterizations and include them as particular cases; express previous empirical parameters via atmospheric and aerosol parameters and theoretical quantities;
- Derive the kinetic equations of stochastic condensation and coagulation and obtain their analytical solutions that reproduce the observed drop and crystal size spectra; express parameters of empirical distributions from theory; and
- Outline a path for future generalizations of the kinetic equations of cloud microphysics based on the Chapman–Kolmogorov and Fokker–Planck equations.

Using the general principles of thermodynamics and kinetics, a closed system of equations is formulated that includes kinetic equations for the drop and crystal size spectra along with the supersaturation equations. Using these equations and further developing classical nucleation theory, theories are developed of aerosol hygroscopic growth, drop activation, and ice homogeneous and heterogeneous nucleation. Analytical expressions are obtained for the particle concentration, critical radii and energies of nucleation, nucleation rates that are expressed as functions of temperature, saturation ratio, pressure, and aerosol concentration simultaneously and in factorized form. It is shown that the new theoretical expressions generalize previous empirical parameterizations, can reduce to them in some particular cases, and their empirical parameters are expressed via the aerosol parameters and physical constants. The validity of these new theoretical expressions is verified in comparison with experimental data, previous empirical and semi-empirical parameterizations, and parcel model simulations. A similar theory is developed for the aerosol deliquescence and efflorescence. This allows for the first time calculation from the theory of a unified phase diagram for solutions that are in agreement with experimental phase diagrams.

Various analytical solutions to the kinetic equations and supersaturation equations are obtained for adiabatic and non-adiabatic processes. These solutions are suitable both for analytical studies of condensation and for inclusion in the numerical models. This system of equations, including kinetic equations for drops and crystals and integral supersaturation equations, is generalized for the turbulent atmosphere and multidimensional models. A fast algorithm for a numerical solution based on the splitting method is described. Spectral bin microphysical method was applied for many years in various 1D, 2D, and 3D models for various cloud types, and its applicability in the models of various scales and dimensions is discussed.

The kinetic equations of stochastic condensation in a turbulent atmosphere are derived and generalized taking into account the coagulation and accretion processes. Various analytical solutions to these stochastic equations are obtained, whose functional forms are similar to the gamma distributions and to exponential and inverse power laws that have been observed in clouds and precipitation. The solution parameters are expressed via the atmospheric characteristics and physical constants, and the solutions are verified through comparison with experimentally observed size spectra. These solutions provide explanations of various empirical parameterizations and a platform for their refinement.

In addition to advancing our basic understanding of cloud microphysical processes, the theoretical approach employed in this book supports the explanation and interpretation of laboratory and field measurements in the context of instrument capabilities and limitations and motivates the design of future laboratory and field experiments. In the context of models that include cloud processes, ranging from small-scale models of clouds and atmospheric chemistry to global weather and climate models, the unified theoretical foundations presented here provide the basis for incorporating cloud microphysical processes in these models in a manner that represents the process interactions and feedback processes over the relevant range of environmental and parametric conditions. Further, the analytical solutions presented here provide the basis for computationally efficient parameterizations that include the relevant parametric dependencies. The methods of cloud simulation using spectral bin microphysics described here are especially suitable for modeling of weather modification by cloud seeding since these methods are almost always based on modification of cloud microstructure and phase state. These methods are also convenient for studies of inadvertent cloud modification by anthropogenic and natural pollutions and for studies of cloud-radiation interactions.

This book incorporates the heritage of Russian cloud physics that introduced and developed the kinetic equations for drop and crystal diffusion growth, the fast numerical algorithms for their solutions, and the stochastic approach to cloud microphysical processes. This Russian heritage is combined with the best knowledge of cloud microphysics acquired and described in the Western literature over the past several decades. A large amount of the material presented in this book is based on original work conducted jointly by the authors over almost two decades. Some of this research has been published previously in journal articles, but a large portion of this material is being published here in this book for the first time, notably the parameterization of heterogeneous ice nucleation and the theory of aerosol deliquescence and efflorescence.

Integration of Russian and Western perspectives on cloud physics was facilitated by the 1972 bilateral treaty between the U.S. and USSR on Agreement and Cooperation in the Field of Environmental Protection, specifically under Working Group VIII – The Influence of Environmental Change on Climate. Its regular meetings and exchanges of delegations and information promoted international collaboration, provided the foundation for long-term cooperation, and outlined proposals for joint research. With the advent of the World Climate Research Programme (WCRP) in 1980, both Khvorostyanov and Curry subsequently became members of the WCRP Working Group on Radiative Fluxes, which later became the Radiation Panel of the Global Water and Energy Exchange Experiment (GEWEX). The GEWEX Radiation Panel had regular annual meetings (where the authors participated and met), which initiated the collaboration that has lasted for almost two decades, resulted in more than 30 joint publications, and culminated in this book.

This book bridges Russian and Western perspectives of cloud physics. Khvorostyanov's involvement in the evolution of the Russian school of cloud physics includes development of cloud models with spectral bin microphysics and applications to cloud seeding and cloud-radiation interactions. Curry's early cloud microphysics research focused on aircraft observations of cloud microphysics and the development of parameterizations for cloud and climate models. Over the past 18 years, Khvorostyanov and Curry have collaborated on a range of cloud microphysical topics of relevance to understanding and parameterizing cloud processes for cloud and climate models, that integrate the Russian perspectives on cloud microphysics into the broader community, and that combine Eastern and Western approaches to cloud microphysics. In addition to summarizing and integrating these perspectives and the broad body of recent research in cloud microphysics, throughout the book a number of new results are included, as well as extensions and generalizations of existing ones.

This monograph is intended to provide a source of information for scientists engaged in teaching and research in cloud physics and dynamics, aerosol physics, air pollution, and weather modification. The book can be used as a textbook to provide graduate-level students with the theoretical foundations of cloud microphysics. Researchers and students should have a basic background in physics and thermodynamics and mathematical physics before using this book. Beyond this basic background, the authors have made every effort to make the book as self-inclusive as possible. Formal derivations and analytical solutions are emphasized, with every effort made to make the mathematical steps easy to follow, including additional details in the appendices. A comprehensive bibliography is provided that references seminal material in the primary literature and previous textbooks and monographs.

The authors gratefully acknowledge support from the DOE Atmospheric Radiation Measurement Program and numerous NASA projects. Many basic concepts and views described in this book were accumulated in multiyear fruitful collaboration with Prof. Mikhail Buikov, Prof. Kirill Kondratyev, and Prof. Kenneth Sassen, to whom the authors are thankful. The authors also greatly appreciate the numerous useful discussions over many years on multiple aspects of cloud and climate studies with Drs. Al Arking, Stefan Bakan, Neville Fletcher, Steve Ghan, Hartmut Grassl, Olaf Hellmuth, Peter Hobbs, Paul Mason, Hugh Morrison, Leonid Matveev, Anna Pirnach, Bill Rossow, Yuri Sedunov, Robert Schiffer, Victor Smirnov, Alexander Stepanov, Graeme Stephens, Vladimir Voloshchuk, and many of our colleagues and co-authors who helped clarify various aspects of cloud microphysics.

In addition, the authors would like to thank Dr. Vladimir Chukin, Dr. Paul DeMott, and Dr. Hitoshi Kanno for their useful discussions on ice nucleation and for providing experimental data. Dr. Osamu Mishima and Dr. Thomas Koop are also thanked for their permission to adapt and use their conceptual current schemes of the water states and phase diagrams. The authors are also grateful to Mrs. Sylvaine Ferrachat from the group of Prof. Ulrike Lohmann at the Institute for Atmospheric and Climate Science, Zurich, Switzerland, for preparing and providing the figure of the observed climatic global cloud field based on the ISCCP data. They would also like to thank Dr. Yuxin Yun from the group of Prof. Joyce Penner at the University of Michigan for providing the figure of the global cloud field simulated with the climate model. Useful discussions with, and help from, Prof. Stephen Warren and Dr. Ryan Eastman in preparing the figures of the global cloud field are greatly appreciated. Thanks also to Drs. Vladimir Chukin, Olaf Hellmuth, and Vladimir Nikulin for their help in preparing some of the figures.

This book was greatly facilitated and supported by the editors at Cambridge University Press: Dr. Matt Lloyd, Ms. Sarika Narula, and Mrs. Shari Chappell; Ms. Saradha Chandrahasan (Project Manager at S4Carlisle); and Mr. Michael McGee (Copy Editor). The authors greatly appreciate their excellent organizational and editorial work. The authors are grateful to the Art Team of the S4Carlisle Publishing Services who carefully and skillfully created the artwork, preparing the figures for publication. Lastly, the authors gratefully acknowledge the permission granted them for reproducing figures from published articles by the American Meteorological Society, the American Geophysical Union, John Wiley & Sons, and the American Chemical Society.

Vitaly I. Khvorostyanov, Moscow, Russia
Judith A. Curry, Atlanta, Georgia, USA

1
Introduction

1.1. Relations among Thermodynamics, Kinetics, and Cloud Microphysics

The most distinctive feature of the Earth when viewed from space is the presence of clouds covering approximately 60% of its surface area. Clouds are a major factor in determining the Earth's radiation budget, by reflecting shortwave radiation and emitting longwave radiation. Through their ability to precipitate, clouds provide virtually all of the fresh water on Earth. Clouds are associated with some of the most damaging weather in the world: torrential rains, severe winds and tornadoes, hail, thunder and lightning, and snowstorms. The latent heat released in clouds is an important source of energy for scales of motion ranging from the global atmospheric circulation, to hurricanes and mid-latitude cyclones, to individual storms. Clouds are also important in atmospheric chemistry because they play an active role in many chemical reactions and transport chemicals through updrafts and scavenging associated with precipitation.

A major challenge in understanding and modeling clouds is the broad range of spatial scales involved. The scales range from the micron scale of individual cloud drops, to the scale of an individual cloud (kilometers), up to the scale of the largest cloud systems (1000 km). Our present computational capability allows only a small range of spatial scales to be simulated explicitly in a single model; processes on the other scales are either specified or parameterized.

Cloud microphysics addresses processes at the smallest of these scales associated with the initiation, growth, and dissipation of cloud and precipitation particles. Some of the most challenging issues in cloud microphysics are associated with interactions between aerosol particles and cloud particles. These interactions include heterogeneous nucleation of liquid and ice particles, homogeneous ice nucleation, rates of water adsorption onto growing particles, the number concentration and size distribution of a population of cloud particles, and the likelihood that the cloud will form precipitation-sized particles. The influence of aerosols on cloud processes moulds cloud optical properties directly through determination of the size and phase of cloud particles, and they indirectly influence radiative transfer through determination of whether or not the cloud precipitates, and thereby determines cloud lifetime. Cloud microphysical processes are intimately connected with cloud dynamical processes associated with air motions on a range of scales, and through latent heat release and precipitation-induced downdrafts, cloud microphysical processes influence cloud dynamics. An understanding of cloud–aerosol interactions and ice nucleation has a direct application to weather modification through cloud seeding.

This book focuses on cloud microphysics processes in the context of providing a systematic and unified theoretical treatment of the thermodynamics and kinetics of cloud microphysical processes. Thermodynamics is a branch of theoretical physics that studies the properties of thermodynamic systems in states of thermodynamic equilibrium and during transitions from one equilibrium state to the other.

Physical kinetics is a branch of theoretical physics that considers the processes in systems that are brought to a non-equilibrium state by external forces. One of the major premises of physical kinetics is an assumption that local thermodynamic equilibrium is reached sufficiently fast in small subdomains of the system (even though they contain a large number of particles), while the overall system is in a non-equilibrium state. The microscopic method in physical kinetics describes the processes in non-equilibrium states using distribution functions for various variables and solving kinetic equations for them. The first kinetic equation was formulated by Ludwig Boltzmann in 1872 for the kinetic theory of gases, which has served as a prototype for other numerous kinetic equations for various processes, including the kinetic equations of coagulation and diffusion particles growth that are considered in this book.

A cloud consisting of many drops and/or crystals surrounded by water vapor and other gases comprises a thermodynamic system. Some microphysical processes in this system have small relaxation times for adjustment to an equilibrium state, and these subsystems can be considered in thermodynamic equilibrium. For example, the vapor and temperature fields around a growing or evaporating particle adjust almost instantaneously to the equilibrium state; nucleation of drops and crystals usually occurs at conditions close to thermodynamic equilibrium. Thus, the properties of nucleating particles can be evaluated from thermodynamic equilibrium equations.

Other microphysical processes (e.g., the diffusion and coagulation growth of drops and crystals) have longer durations, the systems remain in non-equilibrium, and the evolution of their microphysical properties can be considered using kinetic equations for drops and crystals distribution functions. Thus, microphysical processes in clouds can be considered to be a combination of thermodynamic equilibrium processes and kinetic non-equilibrium processes, depending on the system state.

1.2. The Correspondence Principle

There are many different parameterizations of various cloud microphysical processes and properties used in atmospheric models: deliquescence of cloud condensation nuclei (CCN) and hygroscopic growth; the activation of CCN into drops; homogeneous and heterogeneous ice nucleation; critical humidities for ice nucleation; terminal velocities; the size spectra of drops and crystals (e.g., gamma or exponential distributions) and their moments; and so on. Most of these parameterizations are based on fits to experimental data, and their parameters may depend on specific observations, air mass, or aerosol samples. Some of these empirical parameterizations extend and develop previous empirical parameterizations, but many suggest newer expressions or newer parameter values. The correspondence between the new and old parameterizations is often unclear. This creates uncertainty in the choice of the optimum parameterizations for atmospheric models, which may lead to uncertainty in the model simulation results. These uncertainties could be diminished if these parameterizations could be derived from the theory and their parameters expressed via the atmospheric and aerosol properties.

A framework for pursuing this strategy is the *correspondence principle*, which is one of the major principles in physics. The correspondence principle was formulated by Niels Bohr between 1913 and 1920 while developing his model of the atom and was later generalized in order to explain the correspondence to, and remove the contradictions between, the new quantum mechanics and the old classical physics. Subsequently, the correspondence principle has been generalized over several decades and extended to other phenomena in physics and other sciences. The correspondence principle states that a new theory or parameterization should not reject the previous correct theory or parameterization but rather generalize them, so that the old (previous) theory becomes a particular case of the new theory. The new theory or parameterization contains a new parameter absent in the previous theory; when its value tends to some limiting value, the new theory transforms into the old theory. While the formulation of the correspondence principle is simple, it is nevertheless a very powerful methodological tool in understanding natural phenomena and developing correct generalizations of the existing theories and parameterizations. An important consequence of the correspondence principle is that a newer theory should be able to express the empirical parameters of the previous theories or parameterizations via the physical constants. The historical applications of the correspondence principle are beyond the scope of this book. Here, we emphasize that: When developing a new theory or parameterization, one should attempt to generalize previous theories and express the empirical parameters via physical quantities. One of the major goals of this book is to describe and develop further the theories that derive and generalize the known parameterizations of cloud microphysics, and to express the empirical parameters via the parameters of the theory and fundamental atmospheric constants. The correspondence principle provides an integrating framework for this book, and many examples of correspondence between the older and newer theories and parameterizations are described.

1.3. Structure of the Book

Chapter 2 is devoted to the general description of the global cloudiness and its major properties. Cloud microphysical properties are defined and their characteristics are given. Mathematical formulations are reviewed for various size spectra and their moments used for parameterization of cloud and aerosol microphysical and optical properties: inverse power laws, generalized and ordinary gamma distributions, lognormal distributions and their equivalent—algebraic distributions. Analytical methods for evaluation of the moments of the distributions are described. A brief review is given of cloud optical properties, as well as various parameterizations of their dependence on cloud microstructure. Methods for evaluation of the extinction, scattering, and absorption coefficients are described for polydisperse ensembles of drops and crystals.

Chapter 3 is devoted to the general description of the thermodynamic potentials and relations that are used in the subsequent chapters. Five statistical energy distributions are described: the Gibbs distribution, and its four particular cases—the Maxwell, Boltzmann, Bose–Einstein, and Fermi–Dirac distributions. Phase rules are formulated for bulk phases including systems with curved interfaces. The equations of state for ideal and non-ideal gases are derived. Basic thermodynamic characteristics of solutions are introduced, and their thermodynamic relations are described. A general equilibrium equation (entropy equation) is derived, which serves as a basis for derivation of the following equations: the Clausius–Clapeyron equation for various interface systems (liquid and ice in bulk, water drop and vapor, ice crystal and vapor), the Kelvin equation, and the Köhler equation.

The following thermodynamic properties of gas mixtures and solutions are derived: partial gas pressures in a mixture of gases; equilibrium of two bulk phases around a phase transition point; Raoult's law; freezing point depression and boiling point elevation, and the relation of water activity and freezing point depression. Equations are developed for dry and wet adiabatic processes that serve as a platform for constructing Lagrangian parcel and Eulerian multidimensional cloud models.

The properties of water and aqueous solutions, and their parameterizations, are considered in Chapter 4. Various forms of water are described, including those that form at very low temperatures and high pressures (liquid water, crystalline ices, and amorphous forms of water and ice), along with possible transitions among them. Modern theories of water are reviewed, including stability limit conjecture, singularity free theory, and the theories based on a hypothesis of the second critical point. The temperature and pressure ranges in the clouds of Earth and on other planets of our solar system are described. The equivalence of pressure and solution effects in aqueous solutions is also discussed. A compact review is given of existing parameterizations of some basic water and ice thermodynamic properties (saturated vapor pressures, surface tensions, heat capacities, heats of phase transitions, etc.).

Various equations of state for ice and liquid water are reviewed, including the Einstein-Debye thermodynamic equations of state for ice, modern equations of van der Waals type, equations of state for ice based on the Gibbs function and for fluid water based on the Helmholtz function and based on the concept of the second critical point.

In Chapter 5, the diffusion and coagulation growth of drops and crystals is examined. The equations for the growth rates of individual drops and of crystals are derived. The equations account for the effective diffusion coefficients, kinetic corrections, psychrometric corrections due to latent heat release, and ventilation corrections, and are formulated in a convenient factorized form as a product of the terms describing these factors and are valid in a wide range of diffusion growth rates from diffusion to kinetic limits.

Calculations of drop and crystal diffusion growth require supersaturation equations. A detailed derivation of the general equations for the fractional water and ice supersaturations is given from the heat and vapor balance equations. The supersaturation absorption or relaxation times for drops and crystals are introduced, which are the major scaling times of supersaturation absorption at condensation in clouds or the supersaturation release at evaporation. General mathematical expressions are derived for the relaxation times, including their diffusion and kinetic limits. The supersaturation equations are presented in various forms that can be convenient for various models or parameterizations. In particular, the quasi-equilibrium supersaturations over water and ice are obtained for mixed-phase, liquid, and crystalline clouds that can be used in many cases in the models of all scales instead of the saturation adjustment method. Kinetic equations of condensation and deposition in the adiabatic process are derived, their analytical solutions in terms of the integral supersaturation are obtained for arbitrary values of the condensation coefficients in the wide range of particle growth rates, and the diffusion and kinetic limits of the solutions are found.

Chapter 6 examines aerosol hygroscopic growth and drop activation from cloud condensation nuclei (CCN). Reviews are given of the existing empirical and semi-empirical parameterizations of these processes and of the Köhler theory and its subsequent modifications. Hygroscopic growth of mixed aerosol particles and activation of CCN are considered based on a modification of the Köhler theory without the assumption of a dilute solution in CCN and using a newer parameterization of a soluble CCN fraction that can be proportional to the volume or surface of CCN. New analytical expressions are derived for the equilibrium radius of the wet aerosol in a cloudless atmosphere, and in clouds, and for the critical radii and supersaturations for CCN activation. These expressions generalize the known equations of the Köhler theory and are valid not only for diluted solutions but for arbitrary insoluble fractions and for both volume-distributed soluble fractions and soluble shells on the surface of an insoluble core (e.g., mineral dust particle). The accuracy and the regions of applicability of the classical expressions are clarified by the new solutions. Based on these new expressions for the wet radii, a general but simple method is derived for calculation of the wet aerosol size spectrum from the dry aerosol size spectrum. A general method is developed for transformation of the CCN size spectra into differential CCN activity spectra by supersaturations using the equations for the critical radii. Droplet concentrations are calculated by integration over supersaturations of the differential CCN activity spectra. A generalized power law for droplet concentration is derived that includes Twomey's power law as a limiting case, provides physically based expressions for Twomey's parameters, and ensures finite drop concentrations limited by CCN concentration at any large supersaturation.

The kinetics of drop formation in clouds is considered in Chapter 7. To study the kinetics of drop nucleation in clouds, the integro-differential equation for integral water supersaturation in a cloud is derived and analyzed. Solving the supersaturation equation using the algebraic form of the cloud condensation nuclei (CCN) activity spectrum yields analytical expressions for the time of the CCN activation process, the maximum supersaturation, and droplet concentration. All three quantities are expressed as functions of the vertical velocity and characteristics of the CCN size spectra (mean geometric radius, dispersion, and parameters of solubility). Analytical expressions for droplet concentration are found as algebraic functions of maximum supersaturation and as a generalized quasi-power law, and are limited by the total CCN concentration at high supersaturations.

The kinetics of drop activation and the effects of aerosol properties and dynamics on drop concentration are studied using two new methods based on numerical and analytical solutions. In the numerical method, a simple and fast numerical algorithm and solution were developed that allow users to obtain all these characteristics rapidly without running of extensive simulations using parcel models. Analytical expressions are obtained for the time of CCN activation, maximum supersaturation, and the concentration of activated droplets. Solutions are found for these parameters that are the products of power laws by 6 variables: CCN concentration, mean radius, soluble fraction, vertical velocities, surface tension, and condensation coefficient. The solution includes 4 limits: one is a generalization of Twomey's power law, and the other 3 limits are new.

In Chapter 8, the homogeneous nucleation of drops and crystals is considered. Using the concept of metastable states, the Fokker–Planck and Frenkel–Zeldovich kinetic equations are derived for the size distribution function of the germs of a new phase. Based on thermodynamic principles and extending further classical nucleation theory, general equations for the critical germ radius, free energy, and the nucleation rates are derived for homogeneous ice nucleation that express these properties as functions of temperature, water saturation ratio, pressure, and the finite size of freezing particles simultaneously. It is shown that the new expressions generalize and unify previously derived equations and empirical parameterizations, and include them as particular cases.

Using these equations of extended classical nucleation theory, critical freezing and melting temperatures of homogeneous freezing are derived as functions of saturation ratio, and critical saturation ratios over water and ice are derived as a function of temperature. These equations explain, generalize and unify existing empirical parameterizations, express their parameters via the atmospheric and aerosol properties, and outline the limits of their applicability. A simple nonlinear equation is obtained for the liquidus curves, and an analytical relation is found between the freezing and melting point depressions. A simple quantitative expression is derived that establishes equivalence of the solution and pressure effects on freezing, that is in good agreement with observed relations. The empirical water activity shift method is derived from the classical theory and its limitations are studied. The kinetics of homogeneous ice nucleation is evaluated at various temperatures, aerosol concentrations, and vertical velocities using parcel model simulations and the new equations for the critical energies and nucleation rates. Simultaneous account for the temperature and supersaturation effects on the nucleation rates produces very strong negative feedback that allows nucleation into crystals of only a very small fraction of deliquescent haze particles.

A new parameterization of homogeneous ice nucleation for the models is developed based on extended classical nucleation theory, whereby the critical energies and nucleation rates are presented in a separable form as a product of the functions of temperature, supersaturation, and vertical velocities. Limits of these equations are found for the diffusion and kinetic regimes of crystal growth, which allows its application for pure and polluted clouds.

Chapter 9 is devoted to heterogeneous nucleation of drops and ice crystals. Nucleation of drops by vapor deposition on water-insoluble particles is considered; the critical radii, energies, and nucleation rates are derived; and the shape factor is introduced. The four basic modes of heterogeneous ice nucleation on ice nuclei (deposition, deliquescence-freezing, immersion, and contact) are described, along with the properties of ice nuclei (IN). A review is given of existing empirical and semi-empirical parameterizations of heterogeneous ice nucleation. Nucleation of crystals in the deposition mode on water-insoluble particles is considered in detail. The major focus of this chapter is on the deliquescence-freezing and immersion-freezing modes. Through further developing classical nucleation theory, equations are derived for the ice germ critical radius, energy and nucleation rate of ice crystals as functions of temperature, water saturation ratio, droplet radius, external pressure, and misfit strain between ice and insoluble substrate crystalline lattices. These equations generalize and unify the previous expressions found for the critical germ radius and energy. The expressions for the critical energy and nucleation rate are presented in the separable form as a product of the function depending on temperature and supersaturation. This unification and separable representation of the temperature and supersaturation dependencies provide a physical basis for existing empirical parameterizations of crystal concentrations as functions of temperature or supersaturation or combined parameterizations and allow further a rigorous theoretical derivation of the combined parameterization.

The equations are derived from classical nucleation theory for the critical freezing and melting temperature as a function of saturation ratio, drops size, pressure, misfit strain, and active sites. Derivation is given for the cases of volume heterogeneous freezing, surface quasi-heterogeneous freezing, and surface quasi-heterogeneous melting. It is found that a simple nonlinear equation for liquidus curves, which describes the freezing and melting point depressions, is a particular limiting case of these equations for the infinitesimally small nucleation rates. Critical saturation ratios or water activities of heterogeneous freezing are derived as functions of temperature, drops size, pressure, misfit strain, and active sites, and the empirical water activity shift method for heterogeneous nucleation is derived from classical nucleation theory. The derived critical energies and nucleation rates of heterogeneous nucleation are applied and verified in parcel model simulation of a mixed-phase cloud in a wide range of temperatures, aerosol concentrations, and vertical velocities. The kinetics of ice nucleation and cloud glaciation are studied in detail. A semi-empirical parameterization of the final concentrations of nucleated crystals is found as a function of temperature and vertical velocities. The simulation results are compared with analogous simulations of homogeneous freezing described in Chapter 8, which reveal the similarities and differences between heterogeneous and homogeneous nucleation. The thermodynamic constrains on the empirical parameterizations are considered.

In Chapter 10, parameterizations of heterogeneous ice nucleation suitable for use in cloud and climate models with large time and space steps are derived from extended classical nucleation theory. Considering heterogeneous freezing of haze particles and drops, the critical energies and nucleation rates are presented in a separable form as a product of the functions of temperature, supersaturation, active site parameter, and vertical velocity. Integrating this nucleation rate over time and solving the integral supersaturation equation yields a new equation for the crystal concentrations as the function of temperature, critical or maximum supersaturation, vertical velocity, and a few fundamental constants. It is shown that the new equations generalize and unify several previous empirical and semi-empirical parameterizations, reduce them in some particular cases, and express their empirically determined parameters via the atmospheric and aerosol properties and fundamental constants. The limits of these equations are found for the diffusion and kinetic regimes of crystal growth, which allows its application for pure and polluted clouds. These parameterizations of heterogeneous freezing are compared with the analogous parameterizations of homogeneous freezing, and the differences are described quantitatively. Similar parameterizations are developed for deposition ice nucleation and for immersion drop freezing near water saturation. Contact ice nucleation is described: its general properties, the three major mechanisms of aerosol scavenging by drops that may lead to contact nucleation (Brownian diffusion, thermophoresis, and diffusiophoresis), the collection rates, probabilities of freezing, and scavenging in polydisperse ensembles of drops and aerosols.

Chapter 11 considers the phenomena of deliquescence and efflorescence in atmospheric aerosols. The general properties of these processes are described, and previous theories and models of deliquescence and efflorescence are reviewed. New models of deliquescence and efflorescence are developed based on extensions of classical nucleation theory that are applied to salt crystallization in solutions. Using the general equilibrium equation, expressions are derived for the critical radii, energies, and nucleation rates of liquid germs on the surface of a particle at deliquescence, and the critical radii, energies, and nucleation rates of the salt crystallization germs in supersaturated solutions. These models are applied to analytical derivations and calculations of the temperature dependencies of the dissolution heat and solubility, and to the temperature dependence of the deliquescence relative humidity. The results are compared with experimental data and good agreement is found. Application of the efflorescence model to salt crystallization, along with the ice nucleation model, allows calculation of the location of the eutectic point in solution and the right branch with salt crystallization on the phase diagram of solutions. These calculations are combined with calculations of the freezing temperatures and a unified phase diagram of solutions is plotted, where all four branches (deliquescence, efflorescence, freezing, and melting curves) are calculated based on a unified basis using extended classical nucleation theory.

Chapter 12 presents a unified treatment of cloud particle fall velocities for both liquid and crystalline cloud particles over the wide size range observed in the atmosphere. A review is given of previous theories and parameterizations of particle fall velocities. The representation in this book is formulated in terms of the Best (or Davies) number, X, and the Reynolds number, Re. The fall velocities are represented as generalized power laws. The coefficients of the power laws for the

Re-X relation and for the fall velocities are found as the continuous analytical functions of X or particle diameter over the entire hydrometeor size range, which makes this method convenient in applications. The turbulent corrections for the drag coefficients and terminal velocities are derived. Analytical asymptotic solutions are obtained for the coefficients of generalized power laws for the two regimes that represent large and small particles and correspond to potential and aerodynamic flows, respectively. The analytical temperature and pressure corrections for the wide size ranges are derived and compared with previous parameterizations. The expressions for Re-X relation and drag coefficients are applied to spherical and non-spherical drops, as well as several crystal habits, with special attention paid to the turbulent corrections.

Chapter 13 considers stochastic condensation in clouds and formation of the broad size spectra of drops and crystals. An extended review is given of various mechanisms that were suggested to explain existence of the broad size spectra in clouds at all stages of their development. A new version of the kinetic equation for stochastic condensation for the small-size fraction of drops and crystals spectra is derived that is valid for arbitrary relative values of Lagrangian turbulent time and supersaturation relaxation time. Analytical solutions are obtained under a number of assumptions. For this purpose, a model of the condensation process in a turbulent cloud (stochastic condensation) is developed and the Reynolds procedure is applied to the regular and fluctuation parts of all the quantities yielding an equation in terms of covariances, which are incorporated into a kinetic equation to yield the final kinetic equation of stochastic condensation. Particular cases of low-frequency and high-frequency approximations are examined.

Using a few basic assumptions and simplifications, various analytical solutions are obtained for the small-size fraction. Neglecting the diffusional growth of larger particles, solutions have the form similar to the gamma distributions. Their indices that determine the breadth of the spectra and the rates of coagulation and precipitation formation are expressed via atmospheric properties and fundamental constants. Other solutions are obtained in the form of the generalized gamma distributions for the cases, including the diffusional growth of large particles, sedimentation, and coagulation with the large fraction when these processes are important. Finally, based on the integral Chapman–Kolmogorov equation for stochastic processes, a general integral stochastic kinetic equation is formulated. Using this equation, the differential Fokker–Planck equation for stochastic condensation is derived, and various versions of the kinetic equations considered in this chapter are shown to be particular cases of this Fokker–Planck equation.

Chapter 14 is devoted to analytical solutions to the stochastic kinetic equation of coagulation and formation of the large-size fractions of the drop and crystal size spectra for precipitating clouds. The kinetic equation of coagulation in approximation of continuous collection is derived from the general coagulation equation, and the corresponding assumptions are discussed. Then, the basic stochastic integral kinetic equation is simplified and reduced to a differential equation. The analytical solutions to this equation are obtained for various cases in various forms that are similar to the gamma distributions, exponential Marshal–Palmer and inverse power law size spectra that have been determined empirically. Finally, it is shown that the coagulation equation represents a particular case of the integral Chapman–Kolmogorov equation and can be presented also as the differential Fokker–Planck equation. Possible ways of generalization of the kinetic equations are outlined.

2

Clouds and Their Properties

2.1. Cloud Classification

Clouds represent the large visible ensembles of drops or crystals suspended in the gaseous atmosphere. Clouds in the Earth's atmosphere result from the condensation or deposition of water vapor. Clouds on other planets may result from the condensation of the other gases. For example, clouds on Venus may consist of sulfuric acid drops, some clouds on Mars may consist of CO_2 ice crystals, and the clouds on Jupiter may consist of H_2O–NH_3 drops and CH_3 ice crystals (e.g., Curry and Webster, 1999). Hereafter, we consider the Earth's water clouds, although most of the equations in this book can be applied with appropriate modifications to the clouds of other substances on other planets.

Numerous studies of the condensation process have shown that condensation occurs when the water vapor pressure slightly exceeds the saturated vapor pressure, or is *supersaturated*—i.e., the relative humidity slightly exceeds 100%. At the same time, it is known that the relative humidity near the ground is most often 40–80%, decreases upward, and is 2–3 times smaller at the level of the tropopause. Therefore, cloud formation requires a mechanism that increases the relative humidity to $\approx$100%. Cloud formation mechanisms include cooling due to vertical uplift (*convection*) or radiative cooling, horizontal transport (*advection*) of heat and moisture, or the mixing of air masses. There is a great variety of conditions and motions in the atmosphere that lead to cloud formation, which results in a diversity of cloud properties and appearances.

Various classifications of clouds are used in meteorological literature. *Classification by levels* is based on sorting clouds by the heights of their locations in the troposphere, and clouds are classified as upper-, middle-, and low-layer clouds, and clouds of vertical development that have significant vertical extension. *Classification by temperature* defines a cloud as warm if the entire cloud is warmer than 0 °C; otherwise, the cloud is classified as cold. The similar *classification by phase state* characterizes clouds as liquid, mixed-phase, or ice clouds.

The most commonly used cloud classification is based on the morphology of the cloud as defined by its appearance as seen by a ground-based observer. This classification is adopted in the International Atlas of Clouds developed and published by the World Meteorological Organization (WMO, 1975, 1987). This classification is tree-like (branched), similar to biological classifications of animals and plants, and classifies the clouds using Latin names. The classification defines ten main groups called *genera*, which are subdivided into *species* determined by the cloud shapes or internal structure, and in turn, are subdivided into varieties determined by some specific features within a species—e.g., the transparency or shapes of the boundaries.

Upper-level clouds are often called *cirriform clouds*. These clouds form at very low temperatures, generally below –40 °C, are mostly pure crystalline, and contain ice crystals of numerous forms. Cirriform clouds often form in the front upper parts of atmospheric fronts just below the tropopause when a warm air mass lifts above the cold air mass. Therefore, cirriform clouds often serve as indicators of changing weather or approaching storms. Three of the 10 genera defined in the International Atlas of Clouds as cirriform clouds are:

Cirrus (*Ci*)—Detached clouds in the form of white, delicate filaments, or white or mostly white patches or narrow bands. These clouds have a fibrous (hairlike) appearance, or a silky sheen, or both. The height of cirrus cloud bases is typically 7–10 km in the midlatitudes, and reaches 18 km in the tropics. Precipitation from these clouds forms bands of small crystals but they do not reach the ground.

Cirrocumulus (***Cc***)—Thin white patches, sheets, or layers of cloud without shading, composed of very small elements in the form of grains or ripples, merged or separate, and more or less regularly arranged. The cloud base is 6–8 km in the midlatitudes, and the cloud thickness is typically 0.2–0.4 km.

Cirrostratus (***Cs***)—Transparent whitish cloud veils with a fibrous (hair-like) or smooth appearance that totally or partially cover the sky, and generally produce a halo phenomena. The cloud base is 6–8 km in the midlatitudes, and can be much lower in the polar regions; the cloud thickness ranges from 0.1 to 3 km. Their precipitation usually does not reach the ground.

A detailed description of various properties of cirriform clouds is given in Cirrus (2002).

The two genera classified as **middle-level clouds** are:

Altocumulus (***Ac***)—White or grey, or both white and grey, patches, sheets, or layers of cloud, generally with shading, and composed of laminae, rounded masses, or rolls, which are sometimes partially fibrous or diffuse and which may or may not be merged. The cloud base is 2–6 km in the midlatitudes, and the cloud thickness ranges from 0.2–0.7 km. Its precipitation may reach the ground as raindrops or snow crystals. There are 2 species of Ac: wave-like (*Ac undulatus, Ac und*) and convective-like (*Ac cumuliformis, Ac cu*).

Altostratus (***As***)—Greyish or bluish cloud sheet or layer with a striated, fibrous, or uniform appearance, totally or partially covering the sky, and having parts thin enough to reveal the sun at least dimly, as if through ground glass. The slightly fibrous structure may be caused by the bands of precipitation. Altostratus does not produce halos. The cloud base is 3–5 km in the midlatitudes, and the cloud thickness is about 1 km. Its precipitation often may reach the ground. Types of As include the fog-like cloud *As nebulosus*, *As neb*, and the wave-like *As undulatus, As und*.

A recent review of the properties of mid-level clouds is given by Sassen and Wang (2012).

Low-level clouds include the following genera:

Stratocumulus (*Sc*)—Grey or whitish, or both grey and whitish, patches, sheets, or layers of cloud which almost always have dark parts, composed of crenellations, rounded masses, or rolls, which are nonfibrous (except when virga-inclined trails of precipitation are present) and which may or may not be merged. The cloud base height is 0.5–1.5 km, and the cloud thickness is about 0.2–0.8 km. Weak short-lived precipitation may reach the ground as rain or snow. The species of this cloud genus include wave-like Sc (*Stratocumulus undulatus, Sc und*) and convective-like Sc (*Stratocumulus cumuliformis, Sc cuf*).

Stratus (***St***)—Generally grey clouds with a fairly uniform base, which may produce drizzle, ice prisms, or snow grains. If the sun is visible through the cloud, its outline is clearly discernible. Stratus clouds do not produce halo phenomena except, possibly, at very low temperatures. The height of the cloud base is 0.1–0.7 km, the cloud thickness is about 0.2–0.8 km. Sometimes precipitation may reach the ground as drizzle or small snow. The species of this cloud genus include fog-like clouds (*Stratus nebulosus, St neb*), wave-like clouds (*St undulatus, St und*), and broken *St* (*St fractus, St fr*).

The other 3 of the main 10 genera may fill thick layers in the troposphere and spread among several levels. They are the following.

Nimbostratus (*Ns*)—Grey cloud layer, often dark, sometimes with yellowish or bluish color, the appearance of which is rendered diffuse by more or less continuously falling rain or snow, which in most cases reaches the ground. The cloud layer is thick enough to completely obscure the sun. Low, ragged clouds *Frnb* frequently occur below the nimbostratus layer and may obscure the Ns layer above. The cloud base height is 0.1–1 km, cloud depth is 2–5 km and greater.

Cumulus (***Cu***)—Detached clouds, generally dense and with sharp outlines developing vertically in the form of rising mounds, domes, or towers, of which the bulging upper part often resembles a cauliflower. The sunlit parts of these clouds are mostly brilliant white. Their base is relatively dark, often grayish or bluish, and nearly horizontal. Sometimes cumulus clouds are ragged. The cloud base is usually at 0.8–1.5 km in the midlatitudes, but can vary over a wide range, depending on the relative humidity near the ground. Cloud depth may vary from a few hundred meters to several km. Precipitation is usually absent. This cloud genus includes three species. (a) *Cumulus humilis* (*Cu hum*)—cumulus clouds of only a slight vertical extent, usually not greater than 1 km. They generally appear flattened because their vertical extent is smaller than the horizontal sizes. (b) *Cumulus mediocris* (*Cu med*)—cumulus clouds of moderate vertical extent, the tops of which show fairly small protuberances. (c) *Cumulus congestus* (*Cu cong*)—cumulus clouds that exhibit marked vertical development; their bulging upper part frequently resembles a cauliflower.

Cumulonimbus (***Cb***)—Heavy, dense clouds, with a considerable vertical extent. At least part of their upper portion is usually smooth, fibrous, or striated and is nearly always flattened. This part often spreads out in the shape of an anvil or vast plume. Under the base of these clouds, which is generally very dark, there are frequently low ragged clouds and precipitation, sometimes in the form of virga, which is trails of precipitation (fall streaks) falling from the base but not reaching the earth's surface). Cb clouds produce heavy rain and hail in summer and heavy snowfall and graupel in winter. They are also the source of thunderstorms and lightning.

Fog is not considered in the Atlas of Clouds as a separate genus, but is defined as follows: ***Fog***— Composed of very small water droplets (sometimes ice crystals) in suspension in the atmosphere. It reduces the visibility at the earth's surface generally to less than 1,000 m. The vertical extent of fog ranges between a few meters and several hundred meters.

There are several types of fog, classified by the mechanism of formation. ***Radiative fog*** forms usually during calm nights due to radiative cooling of the underlying surface and the lower boundary layer of the atmosphere. ***Advective fog*** forms due to advection and cooling of a warmer air mass over a colder surface. ***Steam fog*** forms when water vapor evaporates from a warm surface into the cold air. Steam fogs frequently form above cold ocean currents, over non-frozen rivers and lakes in winter, and in polar regions over cracks in the sea ice and polynyas. In the cold polar regions or midlatitudes at low temperatures, so called "***diamond dust***" may occur, where ice crystals form in the clear sky, sparkling in the sunlight. These clouds may spread down to the ground forming ice fog.

A ***condensation trail*** or ***contrail*** forms behind the exhaust of an aircraft engine at high altitude as a visible white plume. Contrails are similar to cirrus clouds and may have a noticeable effect on weather and climate in regions with intensive air traffic (Liou et al., 1991; Schumann, 1994; Sassen, 1997). Another similar phenomenon is ***ship tracks*** that often form due to the warm moist air of ship exhausts and that appear on satellite images as linear features of enhanced albedo (brighter white bands) in the field of Sc clouds expelled by ship exhausts (Coakley et al., 1987). The effects of ship tracks on weather and climate have also been studied (e.g., Ackerman, et al., 2003, and references therein).

Orographic clouds and cloud systems formed above the mountains due to orographic uplift and waves are sometimes included into the above 10 genera, but are often considered a separate cloud genus.

Some typical cloud properties are given in Table 2.1, although the values can vary in the wide ranges. A major factor determining cloud properties is the characteristic value of vertical velocity w in updrafts, which influences their genesis. Low-level clouds are also strongly affected by the radiative heat exchange and surface-atmosphere, heat-moisture exchange.

Table 2.1. *The typical characteristics of various cloud types: depth H, vertical velocity w, life time τ_{lt} (hours or minutes), liquid water content LWC or ice water content IWC, the wet adiabatic cooling rate $(\partial T/\partial t)_{wa} = w(\gamma_a - \gamma_{wa})$ and the infrared cooling rate $(\partial T/\partial t)_{ir}$. The data are compiled from Borovikov et al. (1963), Mason (1971), Young (1993), Pruppacher and Klett (1997), and Cotton et al. (2011).*

Cloud type	Fog St, Sc	Ns, As, Ac	Cu hu Cu med	Cu cong	Cb	Orographic	Ci, Cs Cc
Depth H, km	0.05–0.6	0.5–3	1.5	5 km	6–12	1.5–6	0.5–3
w, m s^{-1}	±0.010	0.01–1	1–3	3–10	5–30	0.5–2.0	0.01–1.5
Life time τ_{lt}	2–6 h	6–12 h	10–40 min	20–45 min	45 min	3–10 h	0.2–12 h
LWC, IWC, g/m3	0.05–0.3	0.05–0.4	0.3–1.0	0.5–2.5	1.5–5.0	0.2–2.0	0.001–0.2
$(\partial T/\partial t)_{wa}$ °C/h	0.2	0.2–20	20–60	60–200	100–600	10–40	0.2–20
$(\partial T/\partial t)_{ir}$ °C/h	1–4	1–6	1–4	1–4	1–4	1–4	0.1–3

2.2. Cloud Regimes and Global Cloud Distribution

A close correlation of cloudiness with vertical motions and areas of thermal contrasts, baroclinicity and cyclogenesis, seen in Table 2.1, is supported by a comparison shown in Fig. 2.1 of observed climatological global pressure fields with climatic fronts and cyclones from Blüthgen (1966), shown in Fig. 2.2 observed global cloud amounts, and shown in Fig. 2.3 global cloud amounts simulated with a modern general circulation model (GCM). The methods of construction of observed cloud

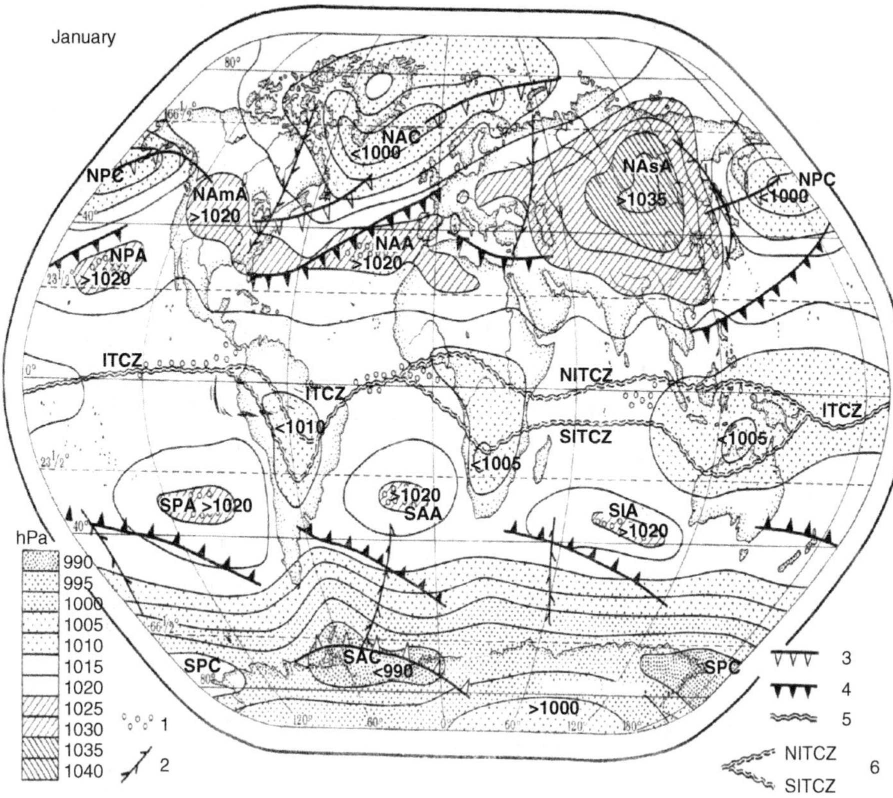

Figure 2.1. Global pressure field and dynamic structure of the atmosphere in January. 1) Regions of high pressure; 2) low-pressure troughs; 3) Arctic; and 4) Polar climatological fronts in the Northern and Southern Hemispheres; 5) Intertropical Convergence Zone (ITCZ); and 6) its North and South branches (NITCZ and SITCZ). The other notations mean the following. The Northern Hemisphere: NAmA is the North-American Anticyclone; NAC is the North-Atlantic Cyclone (Iceland Depression), NAsA is the North Asian Anticyclone (Siberian High), NPC is the North Pacific Cyclone (Aleutian Depression), NPA is the North Pacific Anticyclone (Honolulu High), and NAA is the North Atlantic Anticyclone (Azores High). The Southern Hemisphere: SPA is the South Pacific Anticyclone, SAA is the South Atlantic Anticyclone (Saint Helene Island High), SIA is the South Indian Anticyclone, SPC and SAC are the South Pacific and South Atlantic Cyclones over Antarctica. Adapted from Blüthgen (1966), with changes.

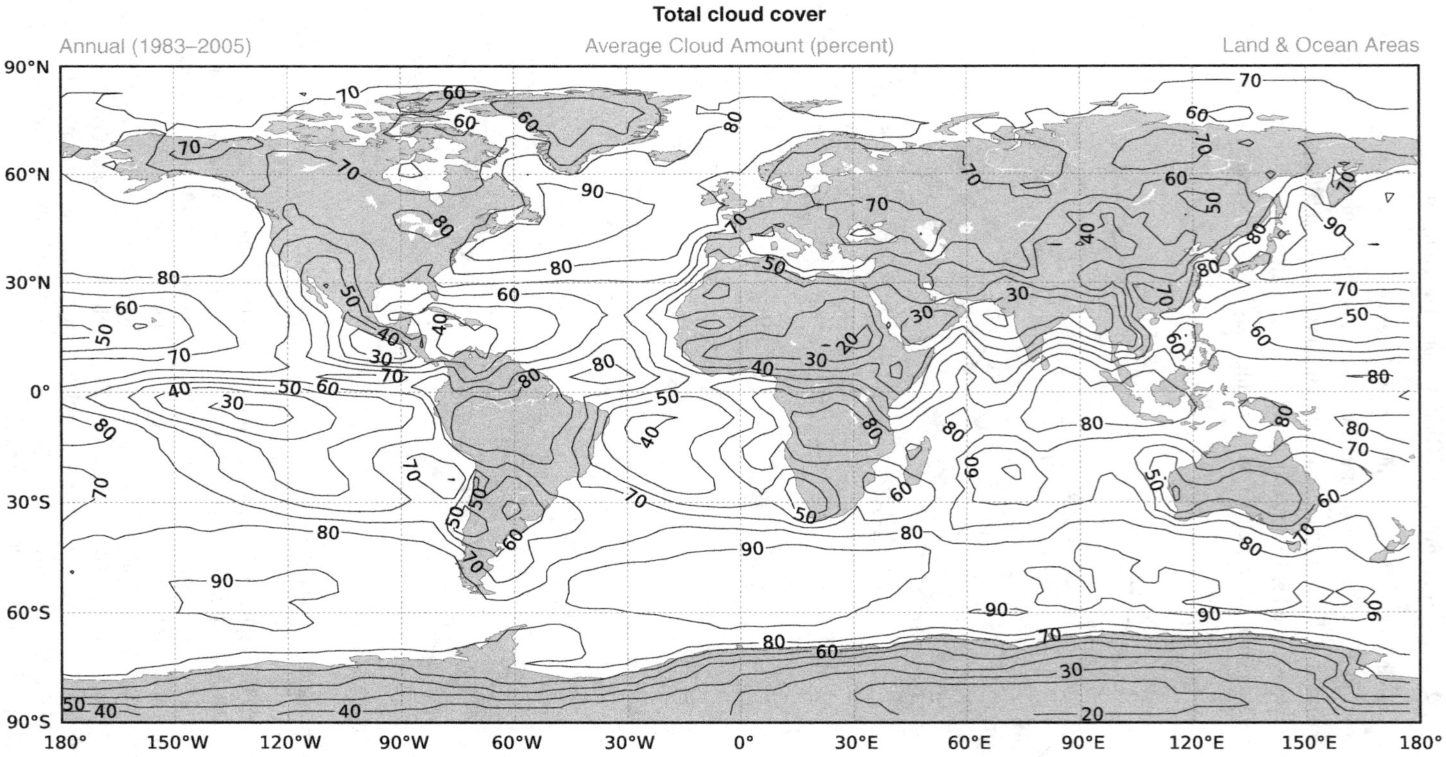

Figure 2.2. The observed cloud amount, annual average in percent over the years 1983–2005 from the International Satellite Cloud Climatology Project (ISCCP), (Schiffer and Rossow, 1983; Han, et al., 1988; Rossow and Schiffer, 1999) (http://isccp.giss.nasa.gov/products/browsed2.html). The figure was prepared using the D2 products of ISCCP and kindly provided by Mrs. Sylvaine Ferrachat from the group of Prof. Ulrike Lohmann at the Institute for Atmospheric and Climate Science, Zurich, Switzerland.

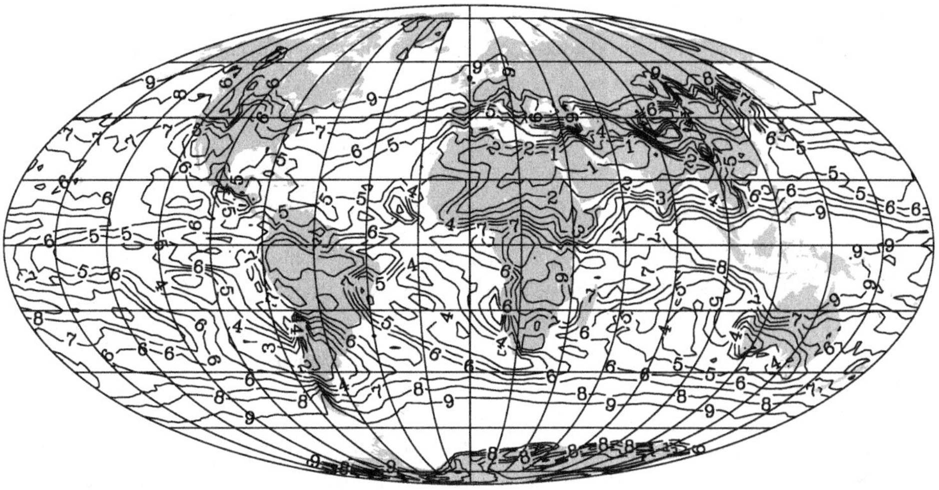

Figure 2.3. Five-years' averaged global cloud distribution for January simulated with the coupled GCM CAM-IMPACT model described in Yun and Penner (2012). The figure was prepared and kindly provided by Dr. Yuxin Yun from the group of Prof. Joyce Penner, University of Michigan.

climatologies are described by Warren et al. (2013). The observed cloud climatologies have been developed mostly from two kinds of data: surface-based and satellite data. (1) The surface-based data sets use visual observations of clouds from the Earth's surface, as coded in weather reports from stations on land and ships in the ocean (e.g., London, 1957; Berlyand and Strokina, 1980; Hahn et al., 1988; Hahn and Warren, 2007; Warren et al., 1985, 2013) (www.atmos.washington.edu/CloudMap). (2) The satellite data sets use radiances measured by satellites in polar and geostationary orbits. Satellites detect clouds principally at visible and thermal infrared wavelengths. The most comprehensive satellite data set was collected in the framework of the International Satellite Cloud Climatology Project (ISCCP). It originated in 1983 as the first project of the World Climate Research Program (WCRP) and has continued for more than 30 years (Schiffer and Rossow, 1983; Han et al., 1988; Rossow and Schiffer, 1999) (http://isccp.giss.nasa.gov/products/browsed2.html).

The observed cloud field is from ISCCP data during the years 1983–2005, prepared using the D2 products of ISCCP and kindly provided by Mrs. Sylvaine Ferrachat from the group of Prof. Ulrike Lohmann at the Institute for Atmospheric and Climate Science, Zurich, Switzerland. Fig. 2.3 (kindly provided by Yuxing Yun) shows global cloud distribution simulated with the use of the coupled CAM-IMPACT climate model described in Yun and Penner (2012). The model consists of two components: the NCAR Community Atmospheric Model (CAM3) (Collins et al., 2006), and the University of Michigan IMPACT aerosol model derived from the Lawrence Livermore National Laboratory (LLNL) chemical transport model (Rotman et al., 2004; Penner et al., 1998). A comparison of Figs. 2.1–2.3 shows that the features of global fields of pressure, winds, and climatic fronts (Fig. 2.1), and cloudiness (Fig. 2.2) at present are sufficiently well reproduced by the general circulation model (GCM) (Fig. 2.3).

Since maxima of ascending velocities and advection rates are located in the climatic cyclones and fronts (Fig. 2.1), these figures show a good correlation of clouds and dynamics—i.e., of maxima of the monthly cloud amount and the areas of maximum vertical velocity. In this section, the global distribution of cloudiness is analyzed with regard to cloud genesis.

2.2.1. *Large-Scale Condensation in Fronts and Cyclones*

The term "large scale condensation" is used in the general circulation model (GCM) to denote the extended areas of stratiform cloudiness that form in atmospheric fronts and cyclones. Its global distribution and genesis are illustrated in Figs. 2.1, 2.2, and 2.3. The midlatitudes exhibit three maxima of the cloud amount in Figs. 2.2 and 2.3, which appear due to the frontal and cyclonic activity shown in Fig. 2.1.

1. In the Atlantic region, there is a pressure trough (the Iceland minimum) and two climatic fronts, Arctic and Polar, in the northern part (Fig. 2.1). In winter, the Arctic front begins near the eastern Canadian coast (about 40°N), runs south of Newfoundland to the northeast, passes the Iceland depression and extends to the Novaya Zemlya islands, and sometimes further eastward, along the entire Siberian coast. This front runs parallel and north of the Gulf Stream, continuing as the warm North Atlantic Current and further as the Norwegian Current (Fig. 2.4).

   The front is formed by the temperature contrast between Arctic snow and ice fields in the north and the warm Gulf Stream in the south. The winter Polar front begins near the

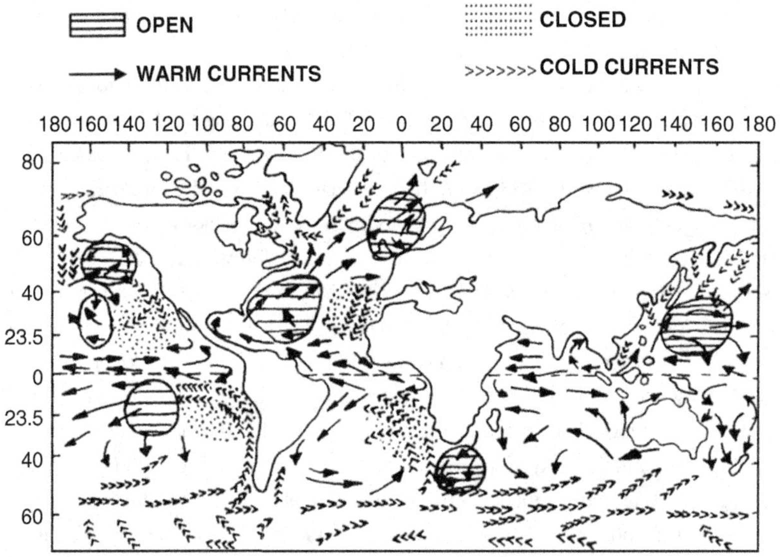

Figure 2.4. Warm and cold ocean currents, open and closed convective cells and global distribution of low cloudiness over the oceans (adapted from Agee, 1985, with changes). Horizontal axis is longitude; vertical axis is latitude (degrees).

coast of Florida and runs south of the Gulf Stream and along both it and the Arctic front (Figs. 2.1, 2.4).

Therefore, temperature contrasts between the warm ocean currents—the Gulf Stream and the North Atlantic Current—and colder areas to the northwest (the cold Labrador Current and the cold continents) and the south are sources of baroclinicity, which is responsible for the formation of climatological fronts. The western sections of these fronts, where baroclinicity is increased due to the temperature contrasts between the cold American continent and the warmer ocean, are the scene of the maximum cyclogenesis. Developing cyclones move to the northeast along both fronts (*stormtracks*) toward Europe. The cyclones are areas of permanent updrafts, which generate persistent cloudiness in this area.

Thus, the dynamic structure of the pressure field and fronts determine the cloud distribution. Figs. 2.2 and 2.3 show that a long band of the maximum cloud amount (up to 80–90%) is formed in January, stretching from Cape Hatteras to Greenland and further northeastward to the Novaya Zemlya Islands. This band coincides with the maximum of the large-scale updrafts along the climatic fronts and the Iceland pressure trough (Fig. 2.1). That is why satellite images exhibit persistent broad cloud bands along all of the Gulf Stream and North Atlantic Current. The cloud amount here can be higher in winter than the annual average due to increasing horizontal temperature gradients and baroclinicity in winter, which leads to the increase in cyclogenesis.

2. There is a kind of "translation symmetry" between the Atlantic and Pacific Oceans—that is, a "parallel transfer" of the pressure and cloud patterns from one ocean to another that gives a similar picture. In the Northern Pacific, similar to the Northern Atlantic, the maximum of the monthly cloud amount (up to 80–90%, Figs. 2.2 and 2.3) is present at the Arctic and Polar fronts and the Aleutian pressure minimum (Fig. 2.1), along the warm Kuroshio ocean current continuing as the North Pacific Current (Fig. 2.4). This cloudiness can be seen in satellite pictures as a broad band, stretching from Japan to the northeast, up to Alaska.
3. There is a kind of "mirror symmetry" between the Southern and Northern Hemispheres in the Atlantic and Pacific oceans—that is, a mirror reflection of the Northern ocean relative to the equator that gives a similar but mirror picture in the Southern ocean. This is caused by the change of the sign of the Coriolis force and the direction of cyclonic rotation between the hemispheres. In the Southern Hemisphere clouds, a maxima in amount up to 70–90% are observed in the three oceans at the Southern Polar fronts and in the pressure trough along the Antarctic Circumpolar Current (Figs. 2.1 and 2.4).

Satellite images usually show these cloud areas in the form of "white spirals or commas," which begin to the west of the southern parts of South America, Africa, and Australia, and then run to the southeast and merge with the cloud band formed in the pressure trough around the Antarctic Circumpolar Current. The counterclockwise cyclonic rotation in the Northern Hemisphere is changed to the opposite direction in the Southern Hemisphere due to the change of sign in the Coriolis force, which determines the shape of these cloud spirals and commas and "the mirror symmetry" between the Northern and Southern Hemispheres.

The clouds, linked to fronts and cyclones in these low-pressure patterns, often form three-level systems of the Ns-As-Cs type and deep convective Cb clouds, especially at the cold fronts, so that almost all 10 genera from the classifications described in Section 2.1 are met in these systems.

2.2.2. Sc-St Clouds and Types of Cloud-Topped Boundary Layer

An increase of cloud amount in the eastern and western parts of the oceans (Figs. 2.2, 2.3) is caused by the extended fields of Sc-St above the warm and cold ocean currents. The same western parts of climatological fronts above the warm currents are the areas of widespread Sc fields within the atmospheric *cloud-topped boundary layer* (CTBL), which consists mostly of open convective cells (Fig. 2.4).

The five extended bands of low stratus clouds St-Sc exist within ABL over the cold ocean currents near the western coasts of the continents (Figs. 2.2, 2.3, 2.4): near North America (cold California Current), South America (cold Peru Current), North Africa (cold Azores or Canary Current), South Africa (cold Benguala Current), West Australia (cold South Indian Current). The CTBL in eastern parts of the oceans consists usually of closed convective cells that form above cold ocean currents. Sc clouds in the ABL are formed under the influence of heat and moisture exchange with the surface and by the CTBL dynamic structure.

Six major types of cloud-topped boundary layer (CTBL), important for cloud-resolving models and climate modeling with GCMs, may be identified (CTBL, 1985; Agee, 1985, 1987; Brown, 1974; Brümmer et al., 1985, 1992; Bakan, 1984; Chlond, 1992; Cotton et al., 2011; Curry and Herman, 1985; Curry et al., 1986, 1988, 1996, 2000; Herman and Goody, 1976; Herman and Curry, 1984; Khvorostyanov, 1982; Klein et al., 2009; Lilly, 1968; Mason, 1985, 1989; Matveev, 1984; Moeng and Arakawa, 1980; Morrison et al., 2005a,b; Randall, 1980; Randall et al., 2003).

1. **Cold air outbreaks.** Clouds form when cold air moves from the east of continents or from the ice edge over warm ocean currents. The initial cloud formation is rather quick and is dominated by strong heating and moisture transport from below. The clouds are initially cumuliform (sometimes marine fog forms) and are organized in *cloud streets* along the wind, but eventually develop into stratocumulus in the form of closed or open convective cells.
2. **Marine stratocumulus boundary layers.** This type of CTBL forms over cold ocean currents generally to the west of continents. The dominant energy source for vertical mixing is radiative cooling of the cloud tops. Clouds are usually organized in the closed convective cells.
3. **Continental cloud-topped boundary layer.** Under the influence of high-pressure patterns, low cloud layers may form at night and persist for several days. These cloud layers are relatively thin and are dominated by radiative processes.
4. **Polar St-Sc clouds.** Stratus clouds cover about 70% of the central Arctic basin during summer. Radiative or advective cooling of relatively warm, moist air that is advected over the Arctic Ocean is the dominant formation mechanism.
5. **Stratocumuli associated with large-scale weather systems.** The dominant feature of these clouds is the strong large-scale vertical motions that are upward ahead of the low-pressure center and downward behind. These cloud patterns move with the weather systems.
6. **Trade-wind cloud-topped boundary layer.** Trade-wind clouds form due to the buoyancy caused by evaporation at the ocean surface. They usually form in the eastern parts of the oceans, where the surface water temperature is relatively low and mostly cold air advection and a rather low inversion (< 1,000 m) are observed. The inversion height increases and the stratiform structure breaks up toward the equator and toward the west.

The climatic distribution of the well-mixed cloud-topped boundary layers with convective cells is given in Fig. 2.4 according to Agee (1985). Cellular cloud patterns are often exhibited on satellite images of these areas, which correspond to the CTBL of types 1, 2, and 6 of the earlier classification. The figure shows that the open convective cells with ascending motion and clouds at their boundaries are formed over warm currents, and closed cells with upward motion and clouds at the cell centers emerge over the cold currents.

Several types of models have been developed to explain the formation of the CTBL with clouds. These are the bulk models, conceptual models, and models based on primitive equations. The last type, in turn, contains two large classes: large eddy simulation (LES) models and turbulence closure models. Some models are described and reviewed in Lilly (1968), Brown (1974), Moeng and Arakawa (1980), Randall (1980), CTBL (1985), Curry et al. (1988, 1996), Chlond (1992), Khvorostyanov (1995), Bretherton et al. (1999a,b), Randall et al. (2003), Khvorostyanov et al. (2003), and Cotton et al. (2011).

2.2.3. Convective Cloudiness in the Intertropical Convergence Zone

In the tropics, there are clearly pronounced maxima of cloud amounts (mostly due to convective clouds) which are linked to the *intertropical convergence zone* (ITCZ) consisting of one or two branches (Fig. 2.1) and moving seasonally around the equator after the sun. The cloud amount maxima are especially large (up to 70–90%) in the regions where the ITCZ intersects the coastlines of South America, South and Central Africa, Malaysia, and North Australia. Here, diurnal deep convection also makes considerable contribution to the total cloud amount and precipitation.

2.2.4. Orographic Cloudiness

In the mountains, orographic clouds make considerable contributions to the total cloud and precipitation amounts. The orographic clouds are most frequent in the zones of updrafts on the upwind slopes of the mountains, especially on the western slopes of the Cordillera, in the Rocky Mountains, in Scandinavia due to prevailing western zonal winds, on the southwestern slopes of the Himalayas, and on the Hida crest in Japan during the monsoon season (Figs. 2.1, 2.2). After passing a mountain crest in the westerly winds, the descending and heating air becomes dry and clouds dissipate on the eastern (downwind) slopes of the mountains, causing a lack of precipitation. If climatic westerly winds prevail, a sharp asymmetry in precipitation often occurs between the western and eastern slopes of the mountain ridges. For example, the annual precipitation sum may be 6000 mm on the western slopes of Cordillera and only 200 mm on the eastern slopes and downwind in Patagonia, causing droughts. This lack of rainwater or snow is a typical feature of many areas located to the east of the mountain crests. Under appropriate conditions, the situation may be improved and the amount of precipitation falling out on the eastern downwind slopes can be increased by artificial cloud seeding with ground-based aerosol generators or from aircraft located near the upwind slopes of the mountain ridge (e.g., Young, 1974a, 1993; Kachurin, 1978; Dennis, 1980; Cotton et al., 1986; Berjulev et al., 1989; Khvorostyanov and Khairoutdinov, 1990).

The preceding brief review shows that all major, global, monthly, and annual cloud maxima are closely linked to large-scale or mesoscale ascents due either to frontal and cyclonic circulations or to convective and orographic forcing. The exception are St-Sc clouds in CTBL that may exist for a long time in the areas of large-scale descent, such as the Azores, Hawaii and the Arctic, St. Helene, and other high-pressure centers. Mechanisms of their formation are more complicated and depend on surface fluxes and on the entrainment rate in CTBL from above (e.g., Lilly, 1968; Randall, 1980; Moeng and Arakawa, 1980; Agee, 1985, 1987). Cirrus clouds occur frequently as parts of the other cloud systems—frontal, convective, or orographic cloudiness—but sometimes they can appear separately in the areas of pressure troughs, wave disturbances, or jet streams in the upper troposphere (Cirrus, 2002).

2.3. Cloud Microphysical Properties

The microstructure of clouds is characterized by the phase, shape, and size spectra of their particles. The size spectra of cloud drops and ice crystals are described by the size distribution functions $f(r)$ of the particles by radii r, or by diameters D or by other characteristic dimensions like the crystal axes or maximum length. The size distribution function by radii $f(r)$ is the number of particles in the range from a radius r to $r + dr$ in some specified volume. The integral of $f(r)$ over all radii is equal to the number of particles N in this volume. If it is a unit volume (1 cm^3 or 1 L or 1 m^3), then N is equal to the particle concentration and it is said that $f(r)$ is normalized to the concentration N. Sometimes the size spectrum can be normalized to unity—i.e., the integral of $f(r)$ over r is equal to 1, then $f(r)$ represents a probability to find a particle in the range $(r, r + dr)$. Evaluation of all moments with such normalization requires multiplication by the particle concentration N. The size distribution function by masses $f(m)$ is the number of particles in the range of masses $(m, m + dm)$ in a unit volume.

The most important and commonly used characteristics of cloud microstructure are concentrations, mean radii, modal radii, effective radii, supersaturation absorption (relaxation) times, liquid or ice water contents, visibility, precipitation rate, lidar and radar reflectivities, and the extinction, scattering, and absorption coefficients.

Droplet and crystal concentrations or number densities, N_d and N_c, are the numbers of particles in a unit volume of a cloud; they are the zero moments of the size spectra; for drops we have

$$N_d = \int_0^\infty f_d(r_d)dr_d = M_d^{(0)}. \tag{2.3.1}$$

Hereafter, $M^{(n)}$ denotes the n-th moment of the size spectrum. The same definitions are valid for ice crystals with a change of the subscript from "d" to "c"; for common definitions, we omit the subscripts here. *The mean radius* $\overline{r_d}$ *of the droplets or the effective (mean) radius of the crystals* $\overline{r_c}$ is a radius of the particles averaged over their ensemble. It is proportional to the first moment of the size spectrum

$$\overline{r_d} = \frac{1}{N_d}\int_0^\infty r_d f_d(r_d)dr_d = \frac{M_d^{(1)}}{N_d}. \tag{2.3.2}$$

The modal radius, r_{dm} or r_{cm}, of a size spectrum is the radius where the size spectrum has the maximum, it is determined from the condition

$$\left.\frac{df}{dr}\right|_{r=r_m} = 0. \tag{2.3.3}$$

The effective radius, $r_{d,eff}$ or $r_{c,eff}$, is the ratio of the 3rd to the 2nd moments of the spectrum. It is used in calculations of radiative transfer in clouds:

$$r_{eff} = \frac{<r^3>}{<r^2>}. \tag{2.3.4}$$

The supersaturation absorption (relaxation) times, τ_{fd} and τ_{fc}, are the characteristic times of supersaturation absorption by the droplets and crystals. As will be shown in Chapters 5, 6, and 7, they are inversely proportional to the first moment of the size spectra in the diffusion regime of particle growth:

$$\tau_{fd} \approx (4\pi D_v N_d \bar{r}_d)^{-1}, \qquad \tau_{fc} \sim (4\pi D_v N_c \bar{r}_c)^{-1}. \tag{2.3.5}$$

The *liquid water content* q_l (LWC), or *ice water content* q_i (IWC), is the mass of drops or crystals in a unit cloud volume:

$$q_l = \frac{4\pi}{3}\rho_d \int_0^\infty r_d^3 f_d(r_d)\, dr_d = \frac{4\pi}{3}\rho_w N_d <r_d^3> = \frac{4\pi}{3}\rho_w N_d M_d^{(3)}, \tag{2.3.6}$$

$$q_i = \frac{4\pi}{3}\rho_i \int_0^\infty r_c^3 f_c(r_c)\, dr_c = \frac{4\pi}{3}\rho_i N_c <r_c^3> = \frac{4\pi}{3}\rho_i N_c M_c^{(3)}. \tag{2.3.7}$$

The liquid water path (LWP) and the ice water path (IWP) are defined as the integrals of LWC and IWC over the cloud layer from the cloud bottom z_b to the cloud top z_t:

$$LWP = \int_{z_b}^{z_t} q_l(z)dz, \qquad IWP = \int_{z_b}^{z_t} q_i(z)dz. \tag{2.3.7a}$$

They characterize the amount of condensate contained in a cloud and are used for parameterizing cloud optical properties.

The *volume extinction*, $\tilde{\sigma}_{ext,k}$, *scattering*, $\tilde{\sigma}_{sc,k}$, *and absorption*, $\tilde{\alpha}_{abs,k}$, *coefficients* of the k-th substance that characterize attenuation of radiation per unit length and are measured in cm^{-1} or in km^{-1} are defined as

$$\tilde{\sigma}_{ext,k} = \pi \int_0^\infty r_k^2 K_{ext}\left(\frac{2\pi r_k}{\lambda}, m_k\right) f_k(r_k)\, dr_k, \tag{2.3.8a}$$

$$\tilde{\sigma}_{sc,k} = \pi \int_0^\infty r_k^2 K_{sc}\left(\frac{2\pi r_k}{\lambda}, m_k\right) f_k(r_k)\, dr_k, \tag{2.3.8b}$$

$$\tilde{\alpha}_{abs,k} = \pi \int_0^\infty r_k^2 K_{abs}\left(\frac{2\pi r_k}{\lambda}, m_k\right) f_k(r_k)\, dr_k, \tag{2.3.8c}$$

$$\tilde{\sigma}_{sc,k} = \tilde{\sigma}_{ext,k} - \tilde{\alpha}_{abs,k}, \quad \omega_{\lambda,k} = \frac{\tilde{\sigma}_{sc,k}}{\tilde{\sigma}_{ext,k}} = \frac{\tilde{\sigma}_{sc,k}}{\tilde{\sigma}_{sc,k} + \tilde{\alpha}_{abs,k}}, \tag{2.3.8d}$$

where $f_k(r_k)$ is the size spectrum of the k-th substance, and λ is the wavelength of the radiation, $\omega_{\lambda,k}$ is the single-scattering albedo, K_{ext}, K_{sc}, and K_{abs} are the factors of extinction, scattering, and absorption, respectively. These factors can be calculated using *Mie theory* or various analytical approximations (e.g., van de Hulst, 1957; Liou, 1980, 1992). If a drop or crystal is much greater than the wavelength and the condition $2\pi r_k/\lambda >> 1$ is satisfied and absorption is absent, then

$K_{ext}(2\pi r_k/\lambda) \approx 2, K_{sc}(2\pi r_k/\lambda) \approx 2$. The corresponding mass coefficients that characterize scattering or absorption of radiation per unit mass and are measured in cm^2g^{-1} or m^2g^{-1} are defined as

$$\sigma_{ext,k} = \tilde{\sigma}_{ext,k}/q_{lk}, \qquad \sigma_{sc,k} = \tilde{\sigma}_{sc,k}/q_{lk}, \qquad \alpha_{abs,k} = \tilde{\alpha}_{abs,k}/q_{lk}, \tag{2.3.8e}$$

where q_{lk} is the LWC, IWC, or mass content of the other substance. These coefficients will be considered in more detail in Section 2.5. The optical thickness $\tau_{ext,k}$ of the k-th substance in a cloud is

$$\tau_{ext,k} = \int_{z_b}^{z_t} \sigma_{ext,k}(z)dz. \tag{2.3.8f}$$

It characterizes the total (integral) extinction of radiation by a k-th substance in the cloud (by drops, crystals, vapor, or aerosol). Similar optical thicknesses $\tau_{sc,k}$, $\tau_{abs,k}$ are considered sometimes with respect to the scattering and absorption coefficients.

The visibility or *visual range* in a fog or cloud, L_{vis}, is inversely proportional to the second moments of the size spectra. It is usually defined in meteorology as a distance from which an object is seen under the angle of the minimum threshold of 1/50. In liquid clouds, visibility is inversely proportional to the extinction coefficient:

$$L_{vis} = \frac{3.91}{\sigma_{ext}} = 3.91\left[\pi \int_0^\infty r_d^2 K_{ext}\left(\frac{2\pi r_d}{\lambda}\right) f_d(r_d)dr_d\right]^{-1}. \tag{2.3.9}$$

Ohtake and Huffman (1969) modified this equation for the visibility in a mixed-phase or crystalline fog or cloud by introducing corrections due to the presence of the ice phase. Assuming $K_{ext} \approx 2$, which is a reasonable approximation for the visible wavelength $\lambda \sim 0.5$ μm and drops or crystals of a few microns and larger, the expression of Ohtake and Huffman (1969) can be written as

$$L_{vis} = \frac{3.91}{2\pi}\left[\int_0^\infty r_d^2 f_d(r_d)\, dr_d + \left(\int_0^\infty r_c^2 f_c(r_c)\, dr_c\right)\left(\frac{2.5}{\bar{r}_c}\left(\frac{6q_i}{\pi\rho_i N_i}\right)^{1/3} - 3\right)^{-1}\right]^{-1}. \tag{2.3.10}$$

With $f_c = 0$, this converts into the previous (Eqn. 2.3.9). This expression can be used for evaluation of the visibility range in mixed and crystalline clouds and fogs.

The precipitation rate, P, is the intensity or mass flux of precipitation per unit area per unit time in various forms: liquid (rain), mixed (rain with snow), or ice (snow):

$$P = \frac{4\pi}{3}\sum_{k=1,2}\rho_k \int_0^\infty v_k(r_k) r_k^3 f_k(r_k)\, dr_k = \frac{4\pi}{3}\sum_{k=1,2}\rho_k a_{vk} M_k^{(3+b_k)}, \tag{2.3.11}$$

where $k = 1$ and $k = 2$ mean droplets and crystals, $M^{(3+b)}$ is the (3+b)-th moment, and it is assumed that the terminal velocity is described by the power law $v_k = a_{vk} r_k^{b_k}$. The more complicated dependencies of $v(r)$ are discussed in Chapter 12.

The radar reflectivity factor Z_R in terms of particle radii is proportional to the sixth moment of the size spectrum:

$$Z_R = \langle r^6 \rangle = \int_0^\infty r^6 f(r)\, dr = M^{(6)}. \tag{2.3.12}$$

The radar reflectivity factor in terms of particle diameters Z_D can be related to Z_R using the relation $f_r(r)dr = f_D(D)dD$, where subscripts "r" and "D" denote corresponding distributions. It then follows from (Eqn. 2.3.12) that

$$Z_D = \langle D^6 \rangle = \int_0^\infty D^6 f_D(D)dD = M_D^{(6)} = 64Z_R. \tag{2.3.13}$$

The corresponding backscatter cross-sections for spherical particles at radar wavelengths, $\lambda >> r_m$, in the Rayleigh scattering limit are

$$A_k = 64\alpha_{rad} Z_R, \qquad \alpha_{rad} = \frac{\pi^5}{\lambda^4}\left|\frac{n_{k\lambda}^2 - 1}{n_{k\lambda}^2 + 2}\right|^2, \tag{2.3.14}$$

where $n_{k\lambda} = m_{k\lambda} - ik_{k\lambda}$ is the complex refractive index for water ($k = 1$) and ice ($k = 2$), $m_{k\lambda}$ and $k_{k\lambda}$ are its real and imaginary parts. In the typical temperature range for not very cold clouds and the wavelengths of meteorological radars, $|(n_{k\lambda}^2 - 1)/(n_{k\lambda}^2 + 2)|^2 \approx 0.93$ for water and ≈ 0.21 for ice. That is, the contribution of ice into reflectivity is almost five times smaller for the same size spectra of drops and crystals.

2.4. Size Spectra and Moments

Various parameterizations of the drop and crystal size distribution functions $f(r)$ have been developed in cloud physics, including *gamma distributions*, *generalized or modified gamma distributions*, *log-normal distributions*, *exponential distributions*, and *inverse power laws*. They are usually constructed as empirical fits to measured size spectra. Size spectra can also be obtained as solutions of the corresponding kinetic equations that describe evolution of the size distribution functions and are considered in Chapters 5, 13, and 14. In this section, we present the major empirical parameterizations used in cloud physics for the aerosol, droplet, and crystal size spectra and the analytical evaluation of their moments used in various applications. The theoretical foundations for some of these expressions are given in Chapters 13 and 14. Aerosol size spectra play a key role in the formation of drops on aerosol particles (Chapters 6 and 7) and ice nucleation (Chapters 8, 9, and 10) and are considered here along with the drop and crystal size spectra.

2.4.1. *Inverse Power Laws*

The inverse power law is frequently used for parameterization of the aerosol size spectra or the size spectrum of large drops and crystals in the form

$$f_a(r_a) = c_{Na} r_a^{-\nu}. \tag{2.4.1}$$

The inverse power law was suggested by Junge in the 1950s (see Junge, 1963) as an empirical fit to the observed aerosol size spectra. Junge obtained the slope $\nu \approx 4$ as the most frequent value in many measurements and in a significant radii range. This empirical inverse power law fit for aerosols was later justified. It was derived from the solution of the coagulation equation, but the values of the slopes n appeared to be different in various ranges of the aerosol size spectrum. For example, $\nu \approx -5/2$ at radii $r_a \sim 0.1$ μm and $\nu \approx -9/2$, at radii $r_a \sim 5$ μm. That is, the slopes increase with increasing radius, and this power law was actually a "quasi power law" with variable slope (e.g., Junge, 1963; Lushnikov and Smirnov, 1975; Smirnov, 1978; Pruppacher and Klett, 1997; Seinfeld and Pandis, 1998). The inverse power law explains some aerosol properties, in particular the inverse power law for the aerosol extinction coefficient, $\sigma_{ext} \sim \lambda^{-a}$ (see Ångström, 1929, 1964, and Chapter 6 here), which is convenient in evaluating those moments over limited ranges of the spectrum. Deficiencies of the inverse power law are that the integrals over the entire spectrum diverge at small sizes, and it is necessary to use several different slopes at various radii to approximate a measured spectrum. Therefore, other distributions have been used to avoid these problems.

The inverse power law (Eqn. 2.4.1) with a negative index can approximate the spectrum of larger cloud drops over the radius range from 100 to between 300 and 800 μm with $\nu = -2$ to -12 (Okita, 1961; Borovikov et al., 1965; Nevzorov, 1967; Ludwig and Robinson, 1970). The inverse power law (Eqn. 2.4.1) has also been used to represent the ice crystal size spectra in cirrus and frontal clouds in the intermediate size region from ~20 to between 100 and 800 μm, with values of ν in the range -2 to -8 (Heymsfield and Platt, 1984; Platt, 1997; Poellot et al., 1999; Ryan, 2000).

2.4.2. Lognormal Distributions

Lognormal distributions are used for parameterization of the aerosol size spectra (e.g., Whitby, 1978; Jaenicke, 1988; Vignati et al., 2004) and for the drop and crystal size spectra in analyses of measured cloud size spectra and some cloud models (e.g., Levin, 1954; Clark, 1976; Feingold and Levin, 1986). The lognormal size spectrum of aerosol or cloud drops and crystals $f_a(r_a)$ can be presented in the form:

$$f_a(r_a) = \frac{N_a}{\sqrt{2\pi}(\ln\sigma_a)r_a} \exp\left[-\frac{\ln^2(r_a/r_{a0})}{2\ln^2\sigma_a}\right], \tag{2.4.2}$$

where N_a is the particle number concentration, σ_a is the dispersion of the spectrum and r_{a0} is the mean geometric radius that is related to the modal radius r_{am} as

$$r_{am} = r_{a0}\exp(-\ln^2\sigma_a). \tag{2.4.3}$$

The general n-th moment of the lognormal distribution is calculated in Appendix A.2 of Chapter 2 and is

$$M^{(n)} = \int_0^\infty r_a^n f_a(r_a)\,dr_a = \frac{N_a}{\sqrt{2\pi}(\ln\sigma_a)} \int_0^\infty r_a^{n-1} \exp\left[-\frac{\ln^2(r_a/r_{a0})}{2\ln^2\sigma_a}\right] dr_a$$
$$= N_a r_{a0}^n \exp\left(\frac{n^2\ln^2\sigma_a}{2}\right). \tag{2.4.4}$$

In particular, the concentration, mean radius $\bar{r}$, mean squared radius $\bar{r}_2$, and mean cubic radius $\bar{r}_3$ with lognormal distributions follow from (Eqn. 2.4.4) for $n = 0, 1, 2$, and 3:

$$N_a = M^{(0)}, \tag{2.4.5}$$

$$\bar{r}_1 = \frac{1}{N_a}\int_0^\infty r_a f_a(r_a)\,dr_a = \frac{M^{(1)}}{N_a} = r_{a0}\exp\left(\frac{\ln^2\sigma_a}{2}\right), \tag{2.4.6a}$$

$$\bar{r}_2 = \left(\frac{1}{N_a}\int_0^\infty r_a^2 f_a(r_a)\,dr_a\right)^{1/2} = \left(\frac{M^{(2)}}{N_a}\right)^{1/2} = r_{a0}\exp\left(\ln^2\sigma_a\right), \tag{2.4.6b}$$

$$\bar{r}_3 = \left(\frac{1}{N_a}\int_0^\infty r_a^3 f_a(r_a)\,dr_a\right)^{1/3} = \left(\frac{M^{(3)}}{N_a}\right)^{1/3} = r_{a0}\exp\left(\frac{3\ln^2\sigma_a}{2}\right). \tag{2.4.6c}$$

The aerosol mass content, liquid water content (LWC), or ice water content (IWC) in a unit volume of the air is

$$q_l = \frac{4}{3}\pi\rho_a\int_0^\infty r_a^3 f_a(r_a)\,dr_a = \frac{4}{3}\pi\rho_a M^{(3)} = \frac{4}{3}\pi\rho_a N_a r_{a0}^3\exp\left(\frac{9\ln^2\sigma_a}{2}\right), \tag{2.4.7}$$

where ρ_a is the density of the aerosol material, water, or ice.

Several models of aerosol size spectra have been suggested as a superposition of several lognormal modes as an approximation for more complex aerosols of various types: marine, continental, rural, and urban (e.g., Whitby, 1978; Jaenicke, 1988; Vignati et al., 2004; Ghan et al., 2011)

$$f_a(r_a) = \sum_{i=1}^{m}\frac{N_{ai}}{\sqrt{2\pi}(\ln\sigma_{ai})r_{ai}}\exp\left[-\frac{\ln^2(r_{ai}/r_{ai0})}{2\ln^2\sigma_{ai}}\right], \tag{2.4.8}$$

where the subscript i means the i-th mode, summation is performed by i over m modes; $m = 4$ in the Jaenicke's (1988) model, but can be smaller or greater in the other models. The value $m = 3$ is usually chosen in parameterizations for polymodal aerosols, which describe a superposition of the three aerosol modes: the smallest Aitken particles ($r_a < 0.01$ μm), the larger submicron fraction ($0.1 \le r_a \le 1$ μm), and a coarse fraction ($r_a \ge 1$ μm, up to a few tens of microns) (e.g., Abdul-Razzak et al., 2000). A more detailed model with $m = 7$ and the superposition of seven lognormal distributions was developed in the aerosol module M7 by Vignatti et al. (2004), which consisted of four internally mixed aerosols containing both soluble and insoluble fractions, and three insoluble modes. This model M7 was also used in mesoscale weather forecast and regional climate models (Zubler et al., 2011).

2.4.3. Algebraic Distributions

A lognormal size spectrum can be approximated by the superposition of the power laws with positive indices at $r_a < r_{a0}$ and negative indices at $r_a > r_{a0}$ (see Section 6.4.2). Another convenient approximation of the lognormal distribution was found by Khvorostyanov and Curry (2006, 2007) who introduced *the algebraic size spectra*

$$f_a(r_a) = \frac{k_{a0}N_a}{r_{a0}} \frac{(r_a/r_{a0})^{k_{a0}-1}}{[1+(r_a/r_{a0})^{k_{a0}}]^2}, \tag{2.4.9}$$

where the index k_{a0} is defined as that introduced by Ghan et al. (1993):

$$k_{a0} = \frac{4}{\sqrt{2\pi}\,\ln\sigma_a}. \tag{2.4.10}$$

Derivation of (Eqn. 2.4.9) from the lognormal spectrum (Eqn. 2.4.2) is given in Section 6.4.4. The distribution (Eqn. 2.4.9) well approximates the lognormal spectrum, tends to zero at small and large radii, and it allows easy analytical evaluation of the moments and various algebraic functions that cannot be evaluated analytically with the lognormal spectrum, as will be illustrated in Chapter 6.

2.4.4. Gamma Distributions

The size spectra $f(r)$ of cloud and fog drops and crystals can be conditionally separated into two major fractions: particles with radii smaller than 25–50 μm, called *small-size fraction*; and larger particles, called *large-size fraction*. The droplets and crystals of the small-size fraction with a modal radii of a few microns have been reasonably approximated by *generalized or modified gamma distributions* (Levin, 1954; Borovikov et al., 1963; Deirmendjan, 1969; Clark, 1974):

$$f(r) = c_N r^p \exp(-\beta r^\lambda). \tag{2.4.11}$$

We omit in this section the subscripts "*d*" for drops and "*c*" for crystals; these subscripts will appear later when necessary. Here, $p > 0$ is the spectral index or the shape parameter of the spectrum, β and λ determine its exponential tail, and c_N is a normalization constant. This size spectrum is usually normalized to the droplet or crystal concentration N, relating c_N to N.

Calculation of the n-th moment of the generalized gamma distribution (Eqn. 2.4.11) gives (Gradshteyn and Ryzhik, 1994)

$$\begin{aligned} M^{(n)} &= \int_0^\infty r^n f(r)\,dr = c_N \int_0^\infty r^{p+n} \exp(-\beta r^\lambda)\,dr \\ &= c_N \lambda^{-1} \beta^{-(p+n+1)/\lambda} \Gamma\left(\frac{p+n+1}{\lambda}\right), \end{aligned} \tag{2.4.12}$$

where $\Gamma(x)$ is Euler's gamma function. The concentration N is the 0-th moment and it follows from this equation that

$$\begin{aligned} N = M^{(0)} &= c_N \int_0^\infty r^p \exp(-\beta r^\lambda)\,dr \\ &= c_N \lambda^{-1} \beta^{-(p+1)/\lambda} \Gamma\left(\frac{p+1}{\lambda}\right). \end{aligned} \tag{2.4.13}$$

The constant c_N can be expressed via N using this relation

$$c_N = N\lambda\beta^{(p+1)/\lambda}\left[\Gamma\left(\frac{p+1}{\lambda}\right)\right]^{-1}, \tag{2.4.14a}$$

and the normalized size spectrum can be written as

$$f(r) = N\lambda\beta^{(p+1)/\lambda}\left[\Gamma\left(\frac{p+1}{\lambda}\right)\right]^{-1} r^p \exp(-\beta r^\lambda). \tag{2.4.14b}$$

The first moment determines the mean radius $\bar{r}$. Using (Eqns. 2.4.13 and 2.4.14a), we obtain

$$M^{(1)} = N\bar{r} = \int_0^\infty rf(r)dr = N\beta^{-(1/\lambda)}\Gamma\left(\frac{p+2}{\lambda}\right)\left[\Gamma\left(\frac{p+1}{\lambda}\right)\right]^{-1}. \tag{2.4.15}$$

The mean radius $\bar{r}$ of the distribution follows from this relation:

$$\bar{r} = \frac{M^{(1)}}{N} = \beta^{-(1/\lambda)}\Gamma\left(\frac{p+2}{\lambda}\right)\left[\Gamma\left(\frac{p+1}{\lambda}\right)\right]^{-1}. \tag{2.4.16}$$

The liquid (or ice) water content (LWC or IWC) is defined following equations from Section 2.3:

$$q_l = \frac{4\pi}{3}\rho_w < Nr^3 > = \frac{4\pi}{3}\rho_w\int_0^\infty r^3 f(r)\,dr = \frac{4\pi}{3}\rho_w M^{(3)}$$

$$= \frac{4\pi}{3}\rho_w N\beta^{-3/\lambda}\Gamma\left(\frac{p+4}{\lambda}\right)\left[\Gamma\left(\frac{p+1}{\lambda}\right)\right]^{-1}. \tag{2.4.17}$$

The radar reflectivity in terms of radii Z_R is defined as a sixth moment of the spectrum. It is obtained from (Eqn. 2.4.12) and by substituting c_N from (Eqn. 2.4.14a):

$$Z_R = < r^6 > = \int_0^\infty r^6 f(r)\,dr = M^{(6)} = N\beta^{-6/\lambda}\frac{\Gamma[(p+7)/\lambda]}{\Gamma[(p+1)/\lambda]}. \tag{2.4.18}$$

The radar reflectivity in terms of particle diameters Z_D can be related to Z_R using a relation $f_r(r)dr = f_D(D)dD$, where subscripts "r" and "D" denote the distributions by radius and diameter. Thus, it follows from (Eqn. 2.4.18) that

$$Z_D = < D^6 > = \int_0^\infty D^6 f_D(D)dD = M_D^{(6)} = 2^6 Z_R$$

$$= 2^6 N\beta^{-6/\lambda}\Gamma\left(\frac{p+7}{\lambda}\right)\left[\Gamma\left(\frac{p+1}{\lambda}\right)\right]^{-1}. \tag{2.4.18a}$$

The absolute spectral dispersion by radii is evaluated as

$$\sigma_{ra}^2 = \frac{1}{N}\int_0^\infty (r-\bar{r})^2 f(r)dr$$

$$= \frac{1}{N}(M^{(2)} - 2\bar{r}M^{(1)} + \bar{r}^2 M^{(0)}). \tag{2.4.19a}$$

Substituting here the moments from (Eqns. 2.4.12 and 2.4.13) yields

$$\sigma_{ra}^2 = \beta^{-2/\lambda}\Gamma^{-1}\left(\frac{p+1}{\lambda}\right)\left[\Gamma\left(\frac{p+3}{\lambda}\right) - \Gamma^2\left(\frac{p+2}{\lambda}\right)\Gamma^{-1}\left(\frac{p+1}{\lambda}\right)\right]. \tag{2.4.19b}$$

The relative dispersion from (Eqns. 2.4.19b to 2.4.16) is

$$\sigma_{rr} = \frac{\sigma_{ra}}{\bar{r}} = \left[\frac{\Gamma[(p+1)/\lambda)]\Gamma[(p+3)/\lambda]}{\Gamma^2[(p+2)/\lambda]} - 1\right]^{1/2}. \tag{2.4.19c}$$

A particular case of the generalized gamma distribution (Eqn. 2.4.11) with $p = \lambda - 1$ is called *the Weibull distribution*:

$$f(r) = c_N r^{\lambda-1}\exp(-\beta r^{\lambda}). \tag{2.4.20a}$$

The relative spectral dispersion for the Weibull distribution follows from (Eqn. 2.4.19c):

$$\sigma_{rr} = \frac{\sigma_{ra}}{\bar{r}} = \left[\frac{\Gamma(1+2/\lambda)}{\Gamma^2(1+1/\lambda)} - 1\right]^{1/2}. \tag{2.4.20b}$$

A particular case that occurs as an analytical solution to some kinetic equations is the Weibull distribution with $\lambda = 2$, $p = 1$—i.e., $f(r) = c_N r\exp(-\beta r^2)$ (see Chapter 13). The relative dispersion for this spectrum is

$$\sigma_{rr} = \left[\frac{\Gamma(2)}{\Gamma^2(3/2)} - 1\right]^{1/2} = \left(\frac{4}{\pi} - 1\right)^{1/2} = 0.523, \tag{2.4.20c}$$

that is, it does not depend on the spectral indices p and λ. We used here the recurrent property of the gamma function and its value at $x = ½$:

$$\Gamma(p+1) = p\Gamma(p), \quad \Gamma(1) = 1, \quad \Gamma(1/2) = \sqrt{\pi}. \tag{2.4.20d}$$

Another important particular case of (Eqn. 2.4.11) with $p = 0$ is the exponential distribution suggested by Marshall and Palmer (1948) for raindrops, and by Gann and Marshall (1958) for snowflakes:

$$f(r) = N_0\exp(-\beta r), \tag{2.4.21}$$

β and N_0 are often called the slope and the intercept of the spectrum.

A particular case of the generalized gamma distributions (Eqn. 2.4.11) with $\lambda = 1$,

$$f(r) = c_N r^p\exp(-\beta r), \tag{2.4.22}$$

is referred to as the *simple gamma distribution* or simply as the *gamma distribution*. Note that the inverse power law (Eqn. 2.4.1) is a particular case of the gamma distribution (Eqn. 2.4.22) with $\beta = 0$ and $p < 0$. All the moments of (Eqn. 2.4.22) can be obtained from the previous equations setting $\lambda = 1$, but these size spectra are often used without reference to the generalized gamma distributions and the relations of the moments is simplified due to the recurrent relation (Eqn. 2.4.20d) for the Euler gamma functions. Therefore, it is useful to provide a brief outline of evaluating moments of the gamma distribution. The general moment can be calculated using the definition of $\Gamma(x)$ (Gradshteyn and Ryzhik, 1994):

$$M^{(n)} = \int_0^\infty r^n f(r)dr = c_N \int_0^\infty r^{p+n} \exp(-\beta r)dr$$

$$= \frac{c_N}{\beta^{p+n+1}} \int_0^\infty x^{p+n} \exp(-x)dx = \frac{c_N}{\beta^{p+n+1}} \Gamma(p+n+1). \quad (2.4.23)$$

We assume that the size spectrum (Eqn. 2.4.22) is normalized to the concentration N:

$$N = \int_0^\infty f(r)dr = M^{(0)} = \frac{c_N}{\beta^{p+1}} \Gamma(p+1). \quad (2.4.23a)$$

The normalization constant c_N can be expressed via N as

$$c_N = N \frac{\beta^{p+1}}{\Gamma(p+1)}. \quad (2.4.24)$$

Substitution of c_N into (Eqn. 2.4.23) gives for n-th moment:

$$M^{(n)} = \frac{N}{\beta^n} \frac{\Gamma(p+n+1)}{\Gamma(p+1)} = \frac{N}{\beta^n}(p+n)(p+n-1)...(p+1), \quad (2.4.25)$$

where we used the recurrent relation (Eqn. 2.4.20d), and where the gamma functions vanish. The first three moments are

$$M^{(0)} = N, \quad M^{(1)} = \frac{N}{\beta}(p+1), \quad M^{(2)} = \frac{N}{\beta^2}(p+2)(p+1),$$

$$M^{(3)} = \frac{N}{\beta^3}(p+3)(p+2)(p+1). \quad (2.4.25a)$$

The mean radius is determined from the first moment:

$$\bar{r} = \frac{1}{N} \int_0^\infty r f(r)dr = \frac{M^{(1)}}{N} = \frac{(p+1)}{\beta}. \quad (2.4.26)$$

The modal radius r_m is determined from the condition $df(r_m)/dr = 0$, which yields from (Eqn. 2.4.22)

$$r_m = \frac{p}{\beta} = \bar{r} \frac{p}{p+1}. \quad (2.4.27)$$

The second equality relates the modal and mean radii. Using these definitions of c_N and the mean radius, the normalized size spectrum can be rewritten in the two forms convenient for calculations, via β, and via the mean radius expressing from (Eqn. 2.4.26) $\beta = (p+1)/\bar{r}$:

$$f(r) = N \frac{\beta^{p+1}}{\Gamma(p+1)} r^p \exp(-\beta r) \quad (2.4.28a)$$

$$= N \frac{(p+1)^{p+1}}{\Gamma(p+1)} \frac{1}{\bar{r}} \left(\frac{r}{\bar{r}}\right)^p e^{-(p+1)r/\bar{r}}. \quad (2.4.28b)$$

The mean squared radius $<r^2>$ needed in optical calculations is

$$<r^2> = \frac{1}{N}M^{(2)} = \frac{(p+2)(p+1)}{\beta^2} = \frac{p+2}{p+1}\bar{r}^2. \tag{2.4.29}$$

The effective radius r_{eff} is the ratio of the 3rd to the 2nd moment of the spectrum:

$$r_{eff} = \frac{<r^3>}{<r^2>} = \frac{M^{(3)}}{M^{(2)}} = \frac{p+3}{p+2}\bar{r} = \frac{p+3}{p}r_m. \tag{2.4.30}$$

The liquid water content q_L is evaluated via the 3rd moment:

$$q_L = \frac{4}{3}\pi\rho_w M^{(3)} = \frac{4}{3}\pi\rho_w N\frac{(p+3)(p+2)}{(p+1)^2}\bar{r}^3. \tag{2.4.31}$$

The ice water content q_I is calculated with the same formula, replacing ρ_w with ρ_i, and keeping in mind that N is then the crystal concentration. The absolute dispersion of the spectrum is expressed via the 0th, 1st, and 2nd moments as

$$\sigma_{ra}^2 = \frac{1}{N}\int_0^\infty (r-\bar{r})^2 f(r)dr = \frac{1}{N}\int_0^\infty (r^2 - 2r\bar{r} + \bar{r}^2)^2 f(r)dr$$

$$= \frac{1}{N}(M^{(2)} - 2\bar{r}M^{(1)} + \bar{r}^2 M^{(0)}) = \frac{\bar{r}^2}{p+1}. \tag{2.4.32}$$

The index p of the gamma distribution determines the relative spectral dispersion σ_{rr} of the gamma distribution (Eqn. 2.4.28a,b):

$$\sigma_{rr} = \frac{\sigma_{ra}}{\bar{r}} = \frac{1}{\bar{r}}\left(\frac{1}{N}\int_0^\infty (r-\bar{r})^2 f(r)dr\right)^{1/2} = (p+1)^{-1/2}. \tag{2.4.33}$$

Typical values for local volumes in liquid clouds of the index $p = 6 - 15$ were measured by Levin (1954) and confirmed in many subsequent experiments (e.g., Noonkester, 1984; Austin et al., 1985; Curry, 1986; see the review in Chapter 13). These values of p correspond to the relative dispersion $\sigma_r = 0.38 - 0.25$. Vice versa, the index p can be expressed from (Eqn. 2.4.33) via σ_{rr}:

$$p = 1/\sigma_{rr}^2 - 1. \tag{2.4.33a}$$

If the relative dispersion of the spectrum is known from measurements (see Chapter 13), and assuming that the spectrum is the gamma distribution, the index p can be evaluated along with the other moments, including cloud optical properties using the equations given earlier. The index p and spectral dispersions determine the rate of coagulation and accretion processes and therefore play an important role in precipitation formation. The smaller the p and the larger the spectral dispersions in a cloud, the faster this cloud may produce precipitation.

In calculations of the droplets and crystal growth rates and the evolution of the size spectrum, the minus first moment may be required (Chapter 5). It follows from the previous equations:

$$M^{(-1)} = N\frac{\Gamma(p)}{\Gamma(p+1)}\beta = \frac{N}{r_m} = \frac{N}{\bar{r}}\frac{p+1}{p}. \tag{2.4.34}$$

The radar reflectivity defined in Section 2.1 is evaluated via the 6-th moment (Eqn. 2.4.18) with $\lambda = 1$:

$$Z_R = \langle r^6 \rangle = \int_0^\infty r^6 f(r)dr = M^{(6)} = N\beta^{-6}\frac{\Gamma(p+7)}{\Gamma(p+1)}. \tag{2.4.35}$$

Using the recurrent relation $\Gamma(p+1) = p\Gamma(p)$ and the relation for $\beta = (p+1)/\bar{r}$, yields

$$Z_R = \langle r^6 \rangle = N\bar{r}^6\frac{(p+6)(p+5)(p+4)(p+3)(p+2)}{(p+1)^5}. \tag{2.4.36}$$

An example of the typical size spectra observed in liquid clouds is shown in Fig. 2.5.

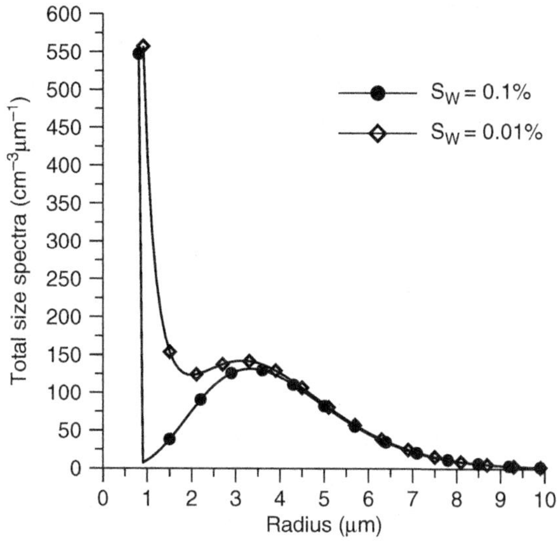

Figure 2.5. An example of the typical size spectra observed in liquid clouds consisting of the aerosol and droplet size spectra. The left branch with radii smaller than 1–2 μm is the size spectrum of the wet interstitial aerosol described by the inverse power law or lognormal spectrum and limited by the boundary radius $r_b \sim 0.5$–1 μm decreasing with increasing water supersaturation s_w. The right branch with modal radius near 3.5 μm is a droplet size spectrum with larger radii calculated as a gamma distribution with mean radius $\bar{r} = 4$ μm, index $p = 6$, and normalized to the droplet concentration $N = 500$ cm^{-3}. The minimum near 1–2 μm is caused by the superposition of these two spectra.

These size spectra consist of the superposition of the aerosol and droplet size spectra. The left branch with $r < 1$–2 μm is the size spectrum of the wet interstitial aerosol described by the inverse power law or lognormal spectrum and limited by the boundary radius, $r < r_b \sim 0.5$–1 μm, decreasing with increasing water supersaturation s_w as described in Chapter 6. The right branch with a modal radius near 3.5 μm is a droplet size spectrum with larger radii. It is calculated here as a gamma distribution with mean radius $\bar{r} = 4$ μm, index $p = 6$, and is normalized to the droplet concentration $N = 500$ cm^{-3}. The minimum near 1–2 μm is caused by the superposition of these two spectra; its position moves to smaller radii with increasing s_w.

2.5. Cloud Optical Properties

As described in Section 2.3, the mass absorption coefficients and scattering coefficients of droplets and crystals in a polydisperse cloud can be calculated as integrals of the absorption and scattering efficiencies with the size distribution function. The precise calculations of the absorption and scattering efficiencies for spherical particles are based on Mie theory, with expansions into power series by Legendre polynomials (e.g., van de Hulst, 1971; Liou, 1980), but they are too time-consuming for cloud or climate models. The nonsphericity of drops and crystals and the mixture of various crystal habits and shapes in clouds further complicate calculations. Therefore, various approximate methods have been developed for parameterizations of cloud optical properties. A detailed description of the subject is beyond the scope of this book, and in the following text we give only a brief introductory overview of some methods for calculation of optical parameters that are related to cloud microphysics.

One method of parameterization of the absorption and scattering coefficients α_{Li}, $\sigma^s_{\lambda Li}$, $\alpha^s_{\lambda Li}$, optical thickness τ_i, asymmetry parameter $<\cos\theta>$ and single-scattering albedo ω_λ represents them as linear functions of powers of effective radius r_e (Eqn. 2.3.4). Slingo and Schecker (1982) developed such parameterizations of shortwave ($\lambda = 0.25$–4 μm) optical properties for water clouds with 24-band spectral resolution. Slingo (1989) showed that this scheme can be averaged over as few as 4 bands and showed that this resolution allows reproduction with good accuracy the characteristics of solar radiation measured by Herman and Curry (1984) in arctic St clouds. Ebert and Curry (1992), using the results of calculations in Takano and Liou (1989) based on the ray-tracing technique, generalized such parameterizations for crystalline clouds and non-spherical crystals with 5-band spectral resolution. The optical parameters in such schemes are parameterized as

$$\tau_i = IWP(a_i + b_i / \bar{r}_e), \qquad 1 - \omega_i = c_i + d_i\bar{r}_e, \tag{2.5.1}$$

$$< \cos\theta_i > = e_i + f_i\bar{r}_e, \qquad \alpha_{Li} = a_{Li} + b_{Li}/\bar{r}_e, \tag{2.5.2}$$

where $\bar{r}_e$ is the effective crystal radius, i is the number of spectral interval or the transmission band, IWP denotes here the ice water path for crystalline clouds and the liquid water path for liquid clouds. The values of the parameters were fitted in Ebert and Curry (1992) for five intervals in the solar spectrum from 0.25 μm to 3.5 μm, and for five intervals in the infrared (longwave) spectrum from 4.0 to 200 μm. Liou (1992) and Fu and Liou (1993) described a parameterization for crystalline clouds with expansion terms up to the third order by the effective crystal size. Such parameterizations are suitable for both cloud and climate models.

Another method is based on approximations of the exact numerical Mie calculations performed for the particles with the size parameter $x = 2\pi r/\lambda$, where λ is the wavelength, and the complex refractive index $n_\lambda = m_{i\lambda} - ik_{i\lambda}$, where $m_{i\lambda}$ and $k_{i\lambda}$ are its real and imaginary parts. Shifrin (1955) calculated the absorption efficiency K_{abs} from Mie theory for spherical particles and found a good analytical approximation for a particle with radius r:

$$K_{abs}^{Sh}(2\pi r/\lambda) = 1 - \exp(-\psi), \qquad \psi = 8\pi r \kappa_\lambda / \lambda . \tag{2.5.3}$$

This formula is an interpolation between the small and large particle limits, is valid both for weak and strong absorption, and ensures good accuracy in calculations of K_{abs}. This expression should be slightly corrected in order to give the correct Rayleigh–Gans limit (van de Hulst, 1971) at small ψ as

$$K_{abs}^{cor}(2\pi r/\lambda) = 1 - \exp[-(2/3)\psi]. \tag{2.5.4}$$

In the bands with weak absorption or for small particles, $\psi << 1$, and in bands with strong absorption, $\psi \geq 1$, the limits of (Eqn. 2.5.4) are

$$K_{abs}^{cor} \approx \frac{2\psi}{3} = \frac{16\pi\kappa_\lambda r}{3\lambda}, \text{ at } \psi << 1, \quad \text{and} \quad K_{abs}^{cor} \approx 1 \text{ at } \psi >> 1. \tag{2.5.5}$$

The absorption cross-section is calculated as

$$\sigma_{abs} = \pi r^2 K_{abs}^{cor} = \pi r^2 \{1 - \exp[-(2/3)\psi]\}. \tag{2.5.6}$$

Using (Eqns. 2.5.6 with 2.5.5) yields the limits at weak and strong absorption:

$$\sigma_{abs} \approx \frac{4\pi\kappa_\lambda}{\lambda} V, \text{ at } \psi << 1, \quad \text{and} \quad \sigma_{abs} \approx A \text{ at } \psi \geq 1, \tag{2.5.7}$$

where $V = (4/3)\pi r^3$ and $A = \pi r^2$ are the volume and the projected surface area of a particle with radius r. These equations show that the efficiency factor K_{abs}^{cor} in (Eqn. 2.5.4) satisfies the requirement that the weak absorption is volume-dependent (the first equation in (Eqn. 2.5.7)) and the strong absorption is surface-dependent (second equation in (Eqn. 2.5.7)), as follows from the rigorous radiation theory (van de Hulst, 1957; Kondratyev, 1969; Paltridge and Platt, 1976; Liou, 1980, 1992; Goody, 1995).

Van de Hulst (1957) developed an approximate but sufficiently accurate method, called the *anomalous diffraction theory* (ADT), which considers the rays passing through and around the particle, and therefore have different phases behind the particle. The extinction K_{ext} and absorption K_{abs} factors in ADT result from the interference of the rays with different phases that pass through the particle with those that do not. Based on ADT, van de Hulst (1957) derived expressions for the scattering and absorption factors for spherical particles:

$$K_{ext}\left(\frac{2\pi r}{\lambda}, n\right) = 2 - 4\frac{\cos z}{\delta} e^{-y} \sin(\delta - z) + 4\left(\frac{\cos z}{\delta}\right)^2 [\cos 2z - e^{-y}\cos(\delta - 2z)], \tag{2.5.8a}$$

$$K_{abs}^{H}\left(\frac{2\pi r}{\lambda}, n\right) = 1 + 2\frac{\exp(-\psi)}{\psi} + 2\frac{\exp(-\psi) - 1}{(\psi)^2}, \tag{2.5.8b}$$

where $n = m_{i\lambda} - ik_{\lambda i}$ is the complex refractive index, and

$$\delta = \frac{4\pi r_i}{\lambda}(m_{\lambda i} - 1), \qquad z = \arctan\frac{\kappa_{\lambda i}}{m_{\lambda i} - 1},$$

$$y = \frac{\kappa_{\lambda i}\delta}{m_{\lambda i} - 1} = \frac{4\pi r_i \kappa_{\lambda i}}{\lambda}, \qquad \psi = \frac{8\pi r_i \kappa_{\lambda i}}{\lambda}. \tag{2.5.9}$$

The extinction and absorption factors (Eqn. 2.5.8a,b) were derived in the ADT by van de Hulst (1957) under the following two assumptions:

$$|n - 1| << 1, \quad \text{and} \quad x = 2\pi r/\lambda >> 1. \tag{2.5.10}$$

The first condition, referred to as *the soft-particle approximation*, means that the refraction index of the particle substance should be close to 1, $|n| \sim 1$. The second requirement, $x >> 1$, is typical for the geometric optics approximation. However, subsequent tests and comparisons with exact Mie calculations showed that these equations still describe the efficiency factors with reasonable accuracy up to $m_\lambda \sim 1.3$–1.5, and for $x = 2\pi r/\lambda \le 1$ (e.g., van de Hulst, 1957; Stephens, 1984; Chylek and Klett, 1991a,b; Chylek and Videen, 1994) and can be applied for drops, many types of aerosol particles, and for crystals with some modifications, taking into account their shapes. ADT has been a useful tool in remote sensing of clouds with radars for evaluation of the optical properties of clouds consisting of various crystal habits and shapes (Matrosov, 1997). By expanding into the power series, it is easily shown that the asymptotics of K^H_{abs} in (Eqn. 2.5.8b) at weak and strong absorptions are the same as that of the interpolation equation (Eqn. 2.5.5), and the limits of σ_{abs} for these cases also reduce to (Eqn. 2.5.7) with volume and surface absorption, respectively.

It is possible to integrate analytically formulae (Eqns. 2.5.3) – (2.5.9) for some types of the size spectra—e.g., gamma distributions, and hence obtain analytical representations of the mass absorption and scattering coefficients (e.g., van de Hulst, 1957; Khvorostyanov, 1982, 1995; Stephens, 1984; Ackerman and Stephens, 1987; Kondratyev and Khvorostyanov, 1989; Kondratyev et al., 1990; Stephens et al., 1990; Chylek and Klett, 1991a,b; Chylek and Videen, 1994; Mitchell and Arnott, 1994; Khvorostyanov and Sassen, 1998a,b). ADT is also convenient for numerical integration with the measured size spectra (Matrosov, 1997). The analytical approach for polydisperse clouds is described in the following text.

Size spectra $f_i(r_i)$ of drops and crystal are parameterized by the gamma distribution:

$$f_i(r_i) = c_N N_i r_i^p \exp[-(p_i+1)r_i / \overline{r}_i], \tag{2.5.11}$$

where p_i is the index of gamma-distribution, $\overline{r}_i$ is the mean radius, N_i is the number concentration, c_N is the normalizing factor, and the subscript i denotes drops or crystals. Substitution of K_{ext} from (Eqn. 2.5.8a) and $f_i(r_i)$ from (Eqn. 2.5.11) into expression (Eqn. 2.3.8a) for the extinction volume coefficient $\tilde{\sigma}_{ext,i}$ and integration yields an analytical expression for $\tilde{\sigma}_{ext,i}$:

$$\begin{aligned}\tilde{\sigma}_{ext,i} &= \pi\int_0^\infty r_i^2 K_{ext}\left(\frac{2\pi r_i}{\lambda}, n_i\right) f_i(r_i)\, dr_i \\ &= A_g(1 - D_{2,ext} + D_{3,ext} - D_{4,ext}),\end{aligned} \tag{2.5.12a}$$

$$A_g = 2\pi\overline{r}_i^2 N_i \frac{p_i + 2}{p_i + 1},$$

where A_g is the geometric cross-section πr^2 integrated over the size spectrum of the particles ensemble related to a unit volume, and

$$D_{2,ext} = \frac{2(p_i + 2)\cos\bar{z}}{\bar{\varphi}^{(p_i+2)/2}\bar{\delta}} \sin\left[(p_i + 2)\arctan\frac{\bar{\delta}}{p_i + \bar{y} + 1} - \bar{z}\right], \tag{2.5.12b}$$

$$D_{3,ext} = \frac{p_i + 1}{p_i + 2}\frac{(2\cos^2\bar{z}\cos 2\bar{z})}{\bar{\delta}^2}, \tag{2.5.12c}$$

$$D_{4,ext} = \frac{p_i + 1}{p_i + 2}\frac{2\cos^2\bar{z}}{\bar{\delta}^2\bar{\varphi}^{(p_i+1)/2}}\cos\left[(p_i + 1)\arctan\frac{\bar{\delta}}{p_i + \bar{y} + 1} - 2\bar{z}\right], \tag{2.5.12d}$$

$$\bar{\delta} = \frac{4\pi\bar{r}_i(m_{\lambda i} - 1)}{\lambda}, \qquad \bar{z} = \arctan\frac{\kappa_{\lambda i}}{m_{\lambda i} - 1}, \tag{2.5.13a}$$

$$\bar{y} = \frac{\bar{\delta}\kappa_{\lambda i}}{m_{\lambda i} - 1} = \frac{4\pi\bar{r}_i\kappa_{\lambda i}}{\lambda}, \qquad \bar{\varphi} = \left[1 + \frac{\bar{y}}{(p_i + 1)^2} + \frac{\bar{\delta}}{(p_i + 1)^2}\right]^{1/2}, \tag{2.5.13b}$$

and all the parameters with overbar are calculated with the mean radius $\bar{r}_i$. The last three terms in (Eqn. 2.5.12a) are the corrections due to diffraction wave effects. After integration over the size spectrum, the accuracy of ADT approximation may increase relative to its accuracy for a single particle since the phase shift between the rays passing through the particle and outside is smoothed due to differences in the particle sizes. Thus, (Eqn. 2.5.12a–d) can be applicable for many particle types including drops, crystals, and aerosols.

Equation (2.5.12a) is simplified for the case of solar radiation at $\lambda = 0.4 - 4$ μm and for characteristic sizes of drops and crystals of a few microns and greater. The expression (Eqn. 2.5.12a) for $\tilde{\sigma}_{ext,i}$ contains four terms in parentheses. The first term is 1 and the contribution of the other three terms have maxima at the minimum values of $\bar{\delta}$ and $\bar{\varphi}$, which are in the denominators here. If $\bar{r}_i$ is fixed, $\bar{\delta}$ and $\bar{\varphi}$ are minimum at maximum λ. An estimate for $\lambda = 3.8$ μm for the droplet size spectrum with $p_i = 6$ and $\bar{r}_i = 4.7$μm yields $\bar{\delta} \approx 4.6$ and $\bar{\varphi} \approx 1.6$, and the contribution of the third term $D_{3,ext}$ into (Eqn. 2.5.12a) is ~10%. The coefficients before sin(...) in the second term $D_{2,ext}$ and before cos(...) in the fourth term $D_{4,ext}$ are respectively 0.9 and 1.5%, and so these terms can be neglected. At smaller wavelengths, $\bar{\delta}$ and $\bar{\varphi}$ are greater, and the contribution of the terms $D_{2,ext}$ and $D_{4,ext}$ is even smaller. Thus, only the first and third terms, 1 and $D_{3,ext}$, are retained for the SW (shortwave or solar) radiation and for cloud particles of a few microns and greater. Note, however, that the contribution into $\tilde{\sigma}_{ext,i}$ in (Eqn. 2.5.12a) of the terms $D_{2,ext}$ and $D_{4,ext}$ can be comparable to the others for greater λ—e.g., for LW (longwave or infrared) or microwave radiation or for smaller particles sizes ~1–2 μm or submicron particles, when the contribution of these terms may be comparable to $D_{3,ext}$ and all four terms should be retained in (Eqn. 2.5.12a). Retaining in (Eqn. 2.5.12a) only the first and third terms (1 and $D_{3,ext}$) for cloud drops and spherically equivalent crystals in the SW spectrum, and expressing $\bar{z}$, $\cos\bar{z}$, and $\cos(2\bar{z})$ via $m_{\lambda i}$ and $k_{\lambda i}$ from (Eqn. 2.5.13a), we have

$$\tilde{\sigma}_{ext,i} = 2\pi N_i\bar{r}_i^2\frac{p_i + 2}{p_i + 1}\left[1 + \frac{p_i + 1}{p_i + 2}\left(\frac{\lambda^2}{8\pi^2\bar{r}_i^2}\right)\frac{(m_{\lambda i} - 1)^2 - \kappa_{\lambda i}^2}{[(m_{\lambda i} - 1)^2 + \kappa_{\lambda i}^2]^2}\right]. \tag{2.5.14}$$

This expression predicts an anomalous dispersion—i.e., an increase in the extinction coefficient with increasing wavelength, $\sim\lambda^2$, of a few percent in the SW region, in agreement with observations. A more universal quantity that does not depend on particle concentration and is used in the radiative transfer equations is the mass extinction coefficient $\sigma_{\lambda i}^{ext}$, which can be obtained by dividing (Eqn. 2.5.14) by the LWC or IWC. For the gamma distribution:

$$q_{Li} = \frac{4}{3}\pi\rho_w N_i \frac{(p_i+3)(p_i+2)}{(p_i+1)^2}\bar{r}_i^3, \tag{2.5.15}$$

where ρ_w denotes the density of water or ice, thus $\sigma_{\lambda i}^{ext} = \tilde{\sigma}_{ext,i}/q_{Li}$ and the mass coefficient is

$$\sigma_{\lambda i}^{ext} = \frac{3}{2\rho_i\bar{r}_i}\frac{p_i+1}{p_i+3}\left(1+\frac{p_i+1}{p_i+2}\frac{\lambda^2}{8\pi^2\bar{r}_i^2}\frac{(m_{\lambda i}-1)^2-\kappa_{\lambda i}^2}{[(m_{\lambda i}-1)^2+\kappa_{\lambda i}^2]^2}\right). \tag{2.5.16}$$

The volume absorption coefficient $\tilde{\alpha}_{abs,i}$ can be evaluated using either the interpolation formula (Eqn. 2.5.4) or van de Hulst's (1957) Equation (2.5.8b). Integrating (Eqn. 2.5.4) with the gamma distribution (Eqn. 2.5.11), we obtain for any λ and $\bar{r}_i$

$$\begin{aligned}\tilde{\alpha}_{abs,i} &= \int_0^\infty \pi r^2 K_{abs}^{cor}\left(\frac{2\pi r}{\lambda}\right) f_i(r_i)\,dr \\ &= \pi N_i\bar{r}_i^2\frac{p_i+2}{p_i+1}\left[1-\left(1+\frac{(16\pi/3)\kappa_{\lambda i}\bar{r}_i}{(p_i+1)\lambda}\right)^{-(p_i+3)}\right].\end{aligned} \tag{2.5.17}$$

The volume scattering coefficient can be evaluated as $\tilde{\sigma}_{sc,i} = \tilde{\sigma}_{ext,i} - \tilde{\sigma}_{abs,i}$. Using (Eqns. 2.5.14 and 2.5.17), we obtain

$$\tilde{\sigma}_{sc,i}=\pi N_i\bar{r}_i^2\frac{p_i+2}{p_i+1}\left\{1+\frac{p_i+1}{p_i+2}\frac{\lambda^2}{8\pi^2\bar{r}_i^2}\frac{(m_{\lambda i}-1)^2-\kappa_{\lambda i}^2}{[(m_{\lambda i}-1)^2+\kappa_{\lambda i}^2]^2}+\left(1+\frac{(16\pi/3)\kappa_{\lambda i}\bar{r}_i}{(p_i+1)\lambda}\right)^{-(p_i+3)}\right\}. \tag{2.5.18}$$

The mass scattering and absorption coefficients of drops and crystals in [cm^{2}g^{-1}] or [m^{2}g^{-1}] are obtained by dividing by q_{Li}. They can be expressed via the mean radius $\bar{r}_i$ or via the effective radii r_e often used in radiation calculations (given in (Eqn. 2.4.30)), $r_e = \overline{r^3}/\overline{r^2} = (p_i+3)/(p_i+2)\bar{r}_i$,

$$\begin{aligned}\sigma_{\lambda i}^{sc} = \frac{\tilde{\sigma}_{sc,i}}{q_{Li}} &= \frac{3}{4\rho_i\bar{r}_i}\frac{p_i+1}{p_i+3}\left\{1+\frac{p_i+1}{p_i+2}\frac{\lambda^2}{8\pi^2\bar{r}_i^2}\frac{(m_{\lambda i}-1)^2-\kappa_{\lambda i}^2}{[(m_{\lambda i}-1)^2+\kappa_{\lambda i}^2]^2}\right. \\ &\left.+\left(1+\frac{(16\pi/3)\kappa_{\lambda i}\bar{r}_i}{(p_i+1)\lambda}\right)^{-(p_i+3)}\right\},\end{aligned} \tag{2.5.19a}$$

$$\begin{aligned}\sigma_{\lambda i}^{sc} &= \frac{3}{4\rho_i r_e}\frac{p_i+1}{p_i+2}\left\{1+\frac{(p_i+1)(p_i+3)}{(p_i+2)^3}\frac{\lambda^2}{8\pi^2 r_e^2}\frac{(m_{\lambda i}-1)^2-\kappa_{\lambda i}^2}{[(m_{\lambda i}-1)^2+\kappa_{\lambda i}^2]^2}\right. \\ &\left.+\left(1+\frac{(p_i+2)}{(p_i+1)(p_i+3)}\frac{16\pi\kappa_{\lambda i}r_e}{3\lambda}\right)^{-(p_i+3)}\right\},\end{aligned} \tag{2.5.19b}$$

$$\alpha_{Li} = \frac{\tilde{\alpha}_{ext,i}}{q_{Li}} = \frac{3}{4\rho_i \bar{r}_i} \frac{p_i + 1}{p_i + 3} \left[1 - \left(1 + \frac{(16\pi/3)\kappa_{\lambda i} \bar{r}_i}{(p_i + 1)\lambda} \right)^{-(p_i+3)} \right], \tag{2.5.20a}$$

$$\alpha_{Li} = \frac{3}{4\rho_i r_e} \frac{p_i + 1}{p_i + 2} \left[1 - \left(1 + \frac{(p_i + 2)}{(p_i + 1)(p_i + 3)} \frac{16\pi\kappa_{\lambda i} r_e}{3\lambda} \right)^{-(p_i+3)} \right]. \tag{2.5.20b}$$

The single scattering albedo now can be calculated as $\omega_\lambda = \sigma_{sc}/\sigma_{ext}$, or $1 - \omega_\lambda = \sigma_{abs}/\sigma_{ext}$.

Shown in Fig. 2.6 is the wavelength dependence of the mass absorption coefficient α_{Li} calculated using (Eqn. 2.5.20a), $k_{\lambda i}$ from Liou (1992), the index of gamma distribution $p_i = 6$, and four mean radii of 5, 10, 20, and 50 μm. One can see that for the small particles of 5 μm, the difference between the maxima in the absorption bands near 4.5 μm, 6 μm, and 12–20 μm, and the minima in the transparency windows near 5.5 μm and 8–12 μm reaches a factor of 2–3. With increasing size, this difference becomes smaller—i.e., the spectral dependence weakens and almost vanishes for $\bar{r} = 50\,\mu\text{m}$.

Comparison of these mass scattering and absorption coefficients $\sigma_{\lambda i}^{sc}$ and α_{Li} obtained using ADT with the corresponding coefficients calculated with Mie theory and the values of n_λ, k_λ tabulated in Warren (1984) and Liou (1992) shows that the errors of the expressions (Eqns. 2.5.19a,b and 2.5.20a,b) for the SW spectrum mostly do not exceed 9–13%, and the errors at $\lambda < 2.4$ μm do not exceed 5%. These errors are much smaller than the errors that may occur in calculations of the cloud radiative properties with some prescribed scattering and absorption coefficients that are independent of temporal and spatial variations and cloud microstructure.

Some particular cases are useful for developing broadband parameterizations. For weak absorption, $X_\lambda = (16\pi/3)\bar{r}_i \kappa_{\lambda i}/\lambda << 1$ (i.e., $k_\lambda < 1.5 \times 10^{-2}$ in the SW region at $\lambda < 2.5$ μm for drops or

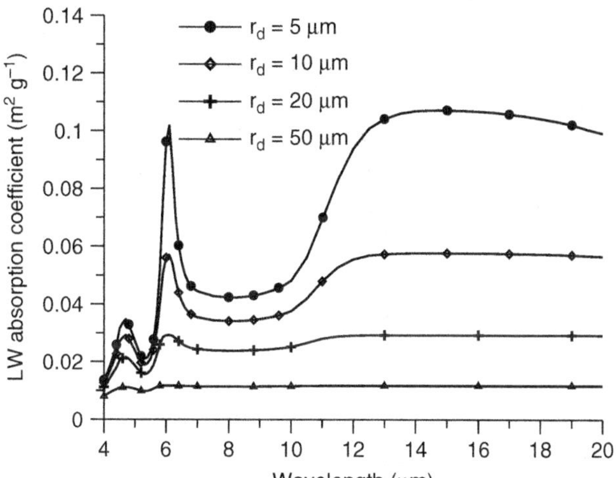

Figure 2.6. Spectral dependence on the wavelength λ of the LW absorption coefficient calculated with anomalous diffraction theory; the droplet size spectra are gamma distributions with the mean radii $\bar{r}_d = 5, 10, 20$, and 50 μm and index $p = 6$.

crystals with a radius of 5–10 μm), the expansion of (Eqns. 2.5.19a and 2.5.20a) by this parameter up to second order yields

$$\alpha_{Li} = \frac{4\pi\kappa_{\lambda i}}{\lambda\rho_i} - \frac{2\bar{r}_i}{3\rho_i}\frac{(p_i+4)}{(p_i+1)}\left(\frac{4\pi\kappa_{\lambda i}}{\lambda}\right)^2, \quad (2.5.21)$$

$$\sigma_{\lambda i}^{sc} = \frac{3}{2\rho_i\bar{r}_i}\frac{p_i+1}{p_i+3}\left[1 + \frac{1}{2}\frac{p_i+1}{p_i+2}\left(\frac{2\pi\bar{r}_i(m_{\lambda i}-1)}{\lambda}\right)^{-2}\right]$$
$$- \frac{4\pi\kappa_{\lambda i}}{\lambda\rho_i} + \frac{2\bar{r}_i}{3\rho_i}\frac{(p_i+4)}{(p_i+1)}\left(\frac{4\pi\kappa_{\lambda i}}{\lambda}\right)^2. \quad (2.5.22)$$

In bands with weak absorption and with small particles, when $X_\lambda = (16\pi/3)\bar{r}_i\kappa_{\lambda i}/\lambda \ll 1$, the major contribution to the mass absorption coefficient α_{Li} comes from the first term in (Eqn. 2.5.21), which does not depend on the mean radius. For example, in the infrared window at $\lambda = 10$ μm with $k_{\lambda i} \approx 5.1 \times 10^{-2}$ (Warren, 1984; Liou, 1992), and $\bar{r}_i = 10\,\mu$m, $p_i = 6$, using (Eqn. 2.5.20a) for water drops or small crystals with $\rho_w = 1$ g cm^{-3} gives $\alpha_{Li} \sim 380$ cm^{2}g^{-1}, and an estimate with the approximate (Eqn. 2.5.21) gives $\alpha_{Li} \approx 430$ cm^{2}g^{-1}—i.e., the error of (Eqn. 2.5.21) is ~12%. In the bands with strong absorption, $(16\pi/3)\bar{r}_i\kappa_{\lambda i}/\lambda \geq 1$, it follows from (Eqns. 2.5.19a and 2.5.20a) that

$$\alpha_{Li} \approx \sigma_{\lambda i}^{sc} \approx \frac{3}{4\rho_i\bar{r}_i}\frac{p_i+1}{p_i+3} = \frac{3}{4\rho_i r_e} = \frac{3}{2\rho_i D_e}, \qquad \omega_\lambda \approx 0.5. \quad (2.5.23)$$

This simple analytical limit corresponds to (Eqn. 2.5.2) or to a parameterization $\alpha_{Li}(r_e) = \alpha_{Li}(r_e = 11\ \mu\text{m}) \times (11\ \mu\text{m}/r_e)$ suggested in Stephens et al. (1990) for crystals in cirrus clouds, and its radius dependence is similar to the parameterizations (Eqn. 2.5.2) by Slingo (1989) and Ebert and Curry (1992).

These two limiting cases show that there are two different dependencies of the absorption coefficient on the mean or effective radius in the bands with weak absorption or with sufficiently small particles, $X_\lambda \ll 1$, and in strong absorption bands or with large particles, $X_\lambda \geq 1$. In the first case, α_{Li} can be parameterized as a polynomial by $\bar{r}_i$ or r_e, such as the linear function (Eqn. 2.5.21). In the second case, α_{Li} becomes a function of the inverse radius, $1/\bar{r}_i$ or $1/r_e$, as shown in (Eqn. 2.5.23) and used in the parameterizations similar to (Eqn. 2.5.2). A rapid transition from one regime to another can occur even at small changes of wavelengths due to increase in $k_{\lambda i}$. In the example considered earlier at $\lambda = 10$ μm, we had $X_\lambda = 0.85 < 1$, while a small shift to $\lambda = 11$ μm with $k_{\lambda i} = 0.248$ for ice (Warren, 1984) yields $X_\lambda = 3.77 \gg 1$ and (Eqn. 2.5.23) shows that $\alpha_{Li} \sim 1/\bar{r}_i$ in agreement with (Eqn. 2.5.2). An estimate with (Eqn. 2.5.23) gives $\alpha_{Li} \sim 540$ cm^2 g^{-1} for small crystals with $\bar{r}_i = 10\,\mu$m.

It is seen that (Eqns. 2.5.19a,b)–(2.5.23) predict dependence on the mean or effective radius similar to the broad-band parameterizations (Eqns. 2.5.1 and 2.5.2) by Slingo (1989), Stephens et al. (1990), Ebert and Curry (1992), Fu and Liou (1993), and others. Expressions (Eqns. 2.5.19a,b)–(2.5.21) in various spectral regions or their expansions of by $\bar{r}_i$ or r_e of the form (Eqn. 2.5.21)–(2.5.23) for various wavelengths and models of the size spectra provide the functional form and coefficients for such parameterizations. Note that due to the different functional dependencies of α_{Li} on r_e or on $1/r_e$, α_{Li} may change even at small changes of λ or r_{ef}.

Cloud emissivity ε_i can be written as

$$\varepsilon_i = 1 - \exp\left[-\int_{z_b}^{z_t} \alpha_{Li}[\overline{r}_i(z)] q_{li}(z)\, dz \right] \tag{2.5.23a}$$

$$\approx 1 - \exp[-\alpha_{Li}(\overline{r}_i)\,(Lwp)_i], \tag{2.5.23b}$$

where z is the height in the cloud, z_b and z_t are the cloud boundaries (and the second equation is valid if α_{Li} weakly depends on height), $(Lwp)_i$ is the liquid or ice water path, and α_{Li} is described by (Eqn. 2.5.20a,b). The formulae for emissivities are used in cloud and climate models and in remote sensing techniques for parameterization of longwave radiation. This equation with (Eqn. 2.5.20a,b) allows determination of the effects on ε_i of the cloud microstructure and phase state. It also enables determination of the different emissivities in the different channels of remote sensing devices. The different dependence of α_{Li} on r_e at various wavelengths or bands may substantially influence the values of ε_i. As seen from (Eqn. 2.5.23b), emissivity decreases with increasing $\overline{r}_i$ and constant water path.

Another form of the volume absorption coefficient $\tilde{\alpha}_{abs,i}$ is obtained by integration of the van de Hulst absorption factor (Eqn. 2.5.8b) with the gamma distribution (Eqn. 2.5.11), which yields

$$\tilde{\alpha}_{abs,i} = \int_0^\infty \pi r^2 K_{abs}^H \left(\frac{2\pi r}{\lambda} \right) f_i(r_i) dr$$

$$= \pi N_i \overline{r}_i^2 \frac{p_i + 2}{p_i + 1} \left\{ 1 + \frac{p_i + 1}{p_i + 2} \left(\frac{2}{\overline{\psi}} \right) \left(1 + \frac{\overline{\psi}}{p_i + 1} \right)^{-(p_i+2)} + \frac{p_i + 1}{p_i + 2} \left(\frac{2}{\overline{\psi}^2} \right) \left[\left(1 + \frac{\overline{\psi}}{p_i + 1} \right)^{-(p_i+1)} - 1 \right] \right\}, \tag{2.5.24}$$

where $\overline{\psi} = 8\pi\kappa_{\lambda i}\overline{r}/[(p_i + 1)\lambda]$. The corresponding mass absorption coefficient is

$$\alpha_{Li}^{(H)} = \frac{\tilde{\alpha}_{abs,i}}{q_{Li}} = \frac{3}{4\rho_w \overline{r}} \frac{p_i + 1}{p_i + 3} \left\{ 1 + \frac{p_i + 1}{p_i + 2} \left(\frac{2}{\overline{\psi}} \right) \left(1 + \frac{\overline{\psi}}{p_i + 1} \right)^{-(p_i+2)} \right.$$

$$\left. + \frac{p_i + 1}{p_i + 2} \left(\frac{2}{\overline{\psi}^2} \right) \left[\left(1 + \frac{\overline{\psi}}{p_i + 1} \right)^{-(p_i+1)} - 1 \right] \right\}. \tag{2.5.25}$$

Its value at small k_λ or small r—i.e., at $\overline{\psi} << 1$—is obtained by expanding $\overline{\psi}$. Keeping the terms up to the 3rd order by $\overline{\psi}$, we obtain

$$\alpha_{Li}^{(H)} \approx \frac{4\pi\kappa_{\lambda i}}{\lambda \rho_i}, \tag{2.5.26}$$

which coincides with the major first term in the expansion (Eqn. 2.5.21) of the absorption coefficient (Eqn. 2.5.20a) obtained with the interpolation absorption efficiency (Eqn. 2.5.4). The analogous analytical expressions for the extinction and absorption coefficients of polydisperse ensembles of the volume equivalent spheres were obtained by Ackerman and Stephens (1987) and Stephens et al. (1990) in a somewhat different form, as a real part of the two complex-conjugated expressions.

There have been many refinements to anomalous diffraction theory and its combinations with the other methods for calculation of the optical coefficients. Stephens (1984) applied ADT for calculation of the optical properties of circular cylinders that may approximate columnar crystals. Ackerman and

Stephens (1987) incorporated into ADT a correction due to refraction and found that this modified anomalous diffraction theory (MADT) does not appreciably increase the accuracy of the extinction factor K_{ext} over ADT but improves the accuracy of the absorption factor K_{abs}. Van de Hulst's (1957) and Bryant and Latimer (1969) generalized the anomalous diffraction theory for a variety of nonspherical particles. The extinction and absorption cross-sections for this case can be written as integrals over the projected area, A, of a particle on the plane perpendicular to the direction of the incoming radiation,

$$\sigma_{ext} = 2\int_A \left\{1 - \exp\left(-\frac{2\pi\kappa_{\lambda i} d_i}{\lambda}\right)\cos\left[\frac{2\pi}{\lambda} d_i (m_{\lambda i} - 1)\right]\right\} dA\,, \qquad (2.5.27)$$

$$\sigma_{abs} = \int_A [1 - \exp(-4\pi\kappa_{\lambda i} d_i/\lambda)]\, dA\,. \qquad (2.5.28)$$

Here d_i is the geometrical path of an individual ray through the particle. Chylek and Klett (1991a,b) applied this version of ADT to hexagonal columns and plates that can approximate ice crystals in clouds, integrated over the angles that determine d_i, and derived the extinction and absorption cross-sections for these crystal types. Chylek and Videen (1994) integrated these cross-sections with the gamma distributions and derived expressions for the extinction and absorption coefficients for the polydisperse ensembles of crystals. These solutions considered only horizontal orientation of the columns and plates and with the beam of incoming radiation perpendicular to the long axes of a column or hexagonal side of a plate.

Mitchell and Arnott (1994, hereafter MA94) have chosen a somewhat different method following Bryant and Latimer (1969). They suggested that for randomly oriented crystals tumbling through space, the dependence on the angles between incident radiation and crystal axes is smoothed, and the absorption efficiency is simplified and is reduced to the integrand in (Eqn. 2.5.28), then

$$K_{abs} = (1/A)\sigma_{abs} = 1 - \exp(-4\pi\kappa_{\lambda i} d_i/\lambda),$$
$$= 1 - \exp[-4\pi\kappa_{\lambda i}(V_i/A_i)/\lambda], \qquad (2.5.29)$$

where the geometrical path d_i is expressed via the crystal volume V_i and the projected area A_i as $d_i = V_i/A_i$. The corresponding absorption cross-section is

$$\sigma_{abs} = A_i\{1 - \exp[-4\pi\kappa_{\lambda i}(V_i/A_i)/\lambda]\}. \qquad (2.5.30)$$

It is seen that (Eqn. 2.5.29 and Eqn. 2.5.30) generalize (Eqn. 2.5.4 and Eqn. 2.5.6) for non-spherical particles and revert to the spherical particular case when $d_i = V_i/A_i = (4/3)r_i$. The two limits of (Eqn. 2.5.30) for weak and strong absorption or for small and large particles ($4pk_{\lambda i}d_i/\lambda << 1$ and $>> 1$, respectively) reduce to the same two expressions in (Eqn. 2.5.7) as in the spherical case that determine the volume-dominated and surface-dominated absorption. For evaluation of the absorption coefficient, MA94 parameterized the volume and projected area for various crystal habits and various size regions based on the experimental data, and using some equations from Takano and Liou (1989), approximated crystal size spectra with gamma distributions and derived analytical expressions for the crystal absorption coefficient in terms of the incomplete gamma functions. For the extinction coefficient, MA94 assumed the volume equivalent spheres, integrated with gamma distributions, and obtained expressions similar to those in Stephens et al. (1990). An assumption in Mitchell and Arnott (1994) on the chaotic crystal orientation is not always justified. It is known that ice crystals are generally not randomly oriented in space. Large ice crystals fall preferentially with their major axis oriented horizontal, whereas smaller crystals may fall randomly oriented, or also exhibit preferred orientations depending on their Reynolds number (Landau and Lifshitz, v. 6, 1959; Sassen, 1980; Pruppacher and Klett, 1997; Cotton et al., 2011).

Several methods have been developed for calculation of the optical properties of nonspherical crystals. Mischenko et al. (1991, 1996) performed detailed calculations based on the *T*-matrix of light scattering and the exact solutions of electrodynamics equations. Fu (1996) calculated the SW single scattering properties of hexagonal ice crystals using an improved geometric ray-tracing program that can produce accurate results for size parameters larger than 15. Then, the crystal extinction and absorption coefficients $\bar{\sigma}_{ext}$, $\bar{\sigma}_{abs}$, and the asymmetry factor $\bar{g}$ were averaged over the selected 28 crystal size spectra observed in tropical and midlatitude cirrus clouds and were parameterized as the linear and cubic polynomials of the generalized effective size D_e:

$$\bar{\sigma}_{ext} = IWC(a_0 + a_1/D_e), \quad (2.5.31a)$$

$$1 - \tilde{\omega} = (\bar{\sigma}_{abs}/\bar{\sigma}_{ext}) = b_0 + b_1 D_e + b_2 D_e^2 + b_3 D_e^3, \quad (2.5.31b)$$

$$\bar{g} = c_0 + c_1 D_e + c_2 D_e^2 + c_3 D_e^3, \quad (2.5.31c)$$

where the overbar denotes averaging over the size spectra. The coefficients of this parameterization were given in Fu (1996) for 25 and 6 bands. Fu et al. (1998) developed a similar parameterization for the infrared radiative properties of cirrus clouds, using for randomly oriented hexagonal ice crystals a composite scheme. This scheme employed a linear combination of single-scattering properties from four methods: the Mie theory, the anomalous diffraction theory, the geometric optics method (GOM), and the finite-difference time domain technique, which is accurate for a wide range of size parameters. Similar refinements in calculations of the optical parameters of nonspherical crystals were done by Yang et al. (2001) using the improved geometric optics approximation, and in a number of subsequent works that are continuing at present.

Mitchell et al. (2006, hereafter M06) calculated extinction efficiencies K_{ext} for nonspherical crystals using the anomalous diffraction approximation (ADA) corrected for tunneling effects. Photon tunneling accounts for radiation beyond the physical cross-section of a particle that is either absorbed or scattered outside the forward diffraction peak. This correction results in multiplication of the efficiencies or optical coefficients by the tunneling efficiency factor T_e. Tunneling factors T_e can range from 0 to 1.0, with zero indicating no tunneling and 1.0 corresponding to spheres and the maximum tunneling contribution. This new analytical form of ADA was referred to as modified ADA or MADA. M06 compared MADA for a laboratory-grown ice cloud with measurements of K_{ext} and K_{abs} over the wavelength range 2–14 μm, and compared MADA with three popular schemes used for predicting the radiative properties of cirrus clouds, Ebert and Curry (1992), Fu et al. (1998), and Yang et al. (2001).

Fig. 2.7 shows a comparison of the MADA predictions from M06 (the data taken from M06) with the scheme by Ebert and Curry (1992, EC92) and some its modification for the atmospheric window region 8.0–12.5 μm. The whole curve from M06 (spheres) lies lower than the EC92 curve (triangles). It is not surprising because a tunneling efficiency $T_e = 0.6$ was used in M06 corresponding to hexagonal columns, while parameterization in Ebert and Curry was fitted for volume-equivalent spheres with $T_e = 1$. The curve "EC92, hex. col." was calculated by multiplication of the equation from Ebert and Curry ([Eqn. 2.5.2] here) by $T_e = 0.6$, which corresponds to the same crystal shape of hexagonal

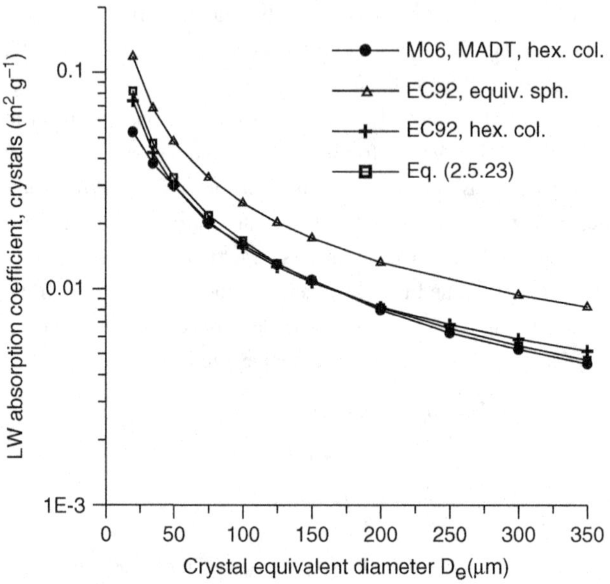

Figure 2.7. Longwave absorption coefficient α_{Li} for crystals in the wavelength band 8–12 μm calculated with various methods. Modified anomalous diffraction theory (MADT), Mitchell et al. (2006), hexagonal columns (M06); Ebert and Curry (1992), equivalent spheres (triangles, EC92); Ebert and Curry (1992) corrected for hexagonal columns, multiplied by $T_e = 0.6$ (EC92, hex. col., crosses); calculation with Equation (2.5.23) of this section, $\alpha_{Li} = (3/2\rho_i D_e)$ with $\rho_i = 0.916$ g cm^{-3}.

columns as in M06. It is seen that EC92, hex. col., and M06 almost coincide, except for the region of the smallest crystals. Also plotted in Fig. 2.7 is the absorption coefficient calculated with the simple asymptotic limit (Eqn. 2.5.23) of the usual ADT described earlier. This parameterization also almost coincides with both M06 and EC92, hex. col., because it accounts for the sufficiently strong absorption averaged over the window, being closer to EC92, hex. col., for small sizes and to M06 for larger sizes. This example illustrates that various parameterizations may provide comparable accuracy.

These fine features of the optical coefficients can be masked in natural clouds by the tremendous spatial and temporal variability of cloud microphysical properties. The absorption and scattering coefficients measured in cirrus clouds varied by more than one order of magnitude and the reason for this was not quite clear until recently (e.g., Stephens et al., 1990; CIRRUS-2002). Fig. 2.8 depicts vertical profiles of the mass and volume scattering coefficient at $\lambda = 0.5$ μm and LW absorption coefficient calculated with ADT using equations of the type (Eqns. 2.5.17)–(2.5.20a) for a typical mid-latitude cirrus cloud in the layer 6–9.5 km simulated in Khvorostyanov and Sassen (1998a,b, 2002) with a 2D cloud model with spectral bin microphysics.

The properties of this cloud were similar to those observed in cirrus clouds described in Sassen et al. (1989). Both mass coefficients decrease from cloud top to bottom by an order of magnitude due to the inverse size dependence and increasing downward crystal size. Further, they exhibit a significant horizontal variability due to variations in crystal sizes caused by the cellular convection and horizontal variability of vertical velocity as simulated in Khvorostyanov and Sassen (1998a,b, 2002).

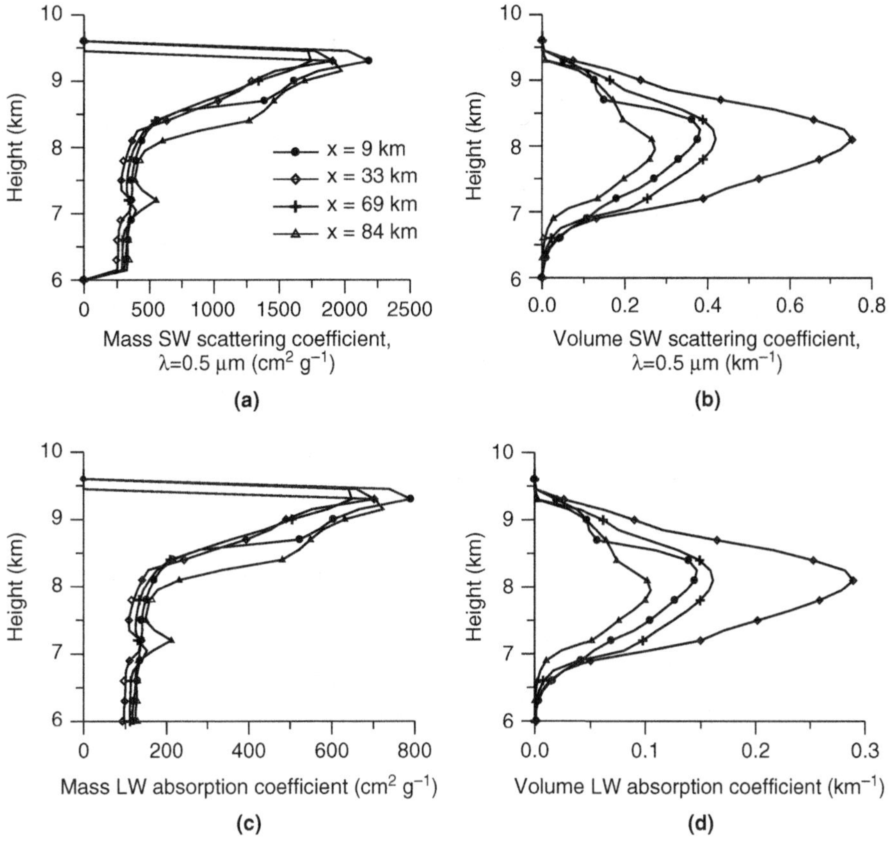

Figure 2.8. Vertical profiles of the mass and volume scattering coefficients in the SW region ($\lambda = 0.5$ μm) and LW absorption coefficients ($\lambda = 10.75$ μm) in a midlatitude cirrus cloud with boundaries of 6 and 9.5 km at 4 horizontal locations, $x = 9$, 33, 64, and 84 km. Modeling results from Khvorostyanov and Sassen (1998b, 2002), with changes.

The maxima of volume coefficients are located closer to the middle cloud layer with a maximum of IWC, and exhibit even greater horizontal variability. This example explains the wide variability of the measured optical coefficients and shows a necessity of fine vertical and horizontal resolution in the modeling of microphysical and radiative cloud properties.

The equations for the cloud extinction, scattering, and absorption coefficients described in this section can be used in the radiative codes of cloud and climate models of various complexity, both with bulk parameterization of cloud microphysics (e.g., Fowler et al., 1996; Lohmann et al., 1999; Mokhov et al., 2002; Morrison et al., 2005a,b; Sud et al., 2007, 2009; Gettelman et al., 2008; Morrison and Gettelman, 2008; Klein et al., 2009; Yun and Penner, 2012) and with spectral bin microphysics (e.g., Kondratyev et al., 1990a,b; Khvorostyanov, 1982, 1995; Jensen et al., 1994, 2005; Feingold et al., 1994; Khvorostyanov and Sassen, 1998b, 2002; Khvorostyanov et al., 2003, 2006; Fridlind et al., 2004; Krakovskaia and Pirnach, 2004; Monier et al., 2006). Such an approach allows us to account for the interactions of evolving cloud microstructure and cloud radiative properties.

Appendix A.2 for Chapter 2. Evaluation of the Integrals with Lognormal Distribution

The lognormal size spectrum of aerosol $f_a(r_a)$ by the radii r_a can be presented as

$$f_a(r_a) = \frac{N_a}{\sqrt{2\pi}(\ln\sigma_a) r_a} \exp\left[-\frac{\ln^2(r_a/r_{a0})}{2\ln^2\sigma_a}\right], \tag{A.2.1}$$

where N_a is the aerosol number concentration, r_{a0} is the mean geometric radius and σ_a is the dispersion of the spectrum. The moment $M^{(n)}$ is

$$M^{(n)} = \int_0^\infty r^n f(r)\, dr = \frac{N_a}{\sqrt{2\pi}(\ln\sigma_a)} \int_0^\infty r^{n-1} \exp\left[-\frac{\ln^2(r_a/r_{a0})}{2\ln^2\sigma_a}\right] dr. \tag{A.2.2}$$

Introducing a new variable $x = r/r_0$, (Eqn. A.2.2) can be rewritten as

$$M^{(n)} = \frac{N_a r_{a0}^n}{\sqrt{2\pi}(\ln\sigma_a)} \int_0^\infty x^{n-1} \exp\left[-\frac{\ln^2 x}{2\ln^2\sigma_a}\right] dx. \tag{A.2.3}$$

After introducing a new variable $z = (\ln x)/(\sqrt{2}\ln\sigma_a)$ and transformations, this becomes

$$M^{(n)} = \frac{N_a r_{a0}^n}{\sqrt{\pi}} \int_{-\infty}^{\infty} \exp(n\sqrt{2}(\ln\sigma_a)z - z^2)\, dz. \tag{A.2.4}$$

This integral is simplified with the new variable $t = n\ln\sigma_a/\sqrt{2}$, and then

$$M^{(n)} = \frac{N_a r_{a0}^n}{\sqrt{\pi}} \exp\left(\frac{n^2\ln^2\sigma_a}{2}\right) \int_{-\infty}^{\infty} \exp[-(z-t)^2\, dz. \tag{A.2.5}$$

The last integral after substitution $z - t = z'$ is reduced to the Gaussian integral

$$\int_{-\infty}^{\infty} \exp(-z'^2)\, dz' = \sqrt{\pi}. \tag{A.2.6}$$

(Gradshteyn and Ryzhik, 1994) and we obtain the n-th moment of the lognormal distribution

$$M^{(n)} = N_a r_{a0}^n \exp\left(\frac{n^2\ln^2\sigma_a}{2}\right). \tag{A.2.7}$$

All the particular moments are obtained by prescribing n the corresponding values, e.g., $n = 1, 2, 3$.

3

Thermodynamic Relations

This chapter describes the major thermodynamic functions and relations that provide the foundation for the following chapters.

3.1. Thermodynamic Potentials

The state of a thermodynamic system is characterized by the three thermodynamic variables, pressure p, temperature T, and volume V. For a homogeneous system of constant composition, these three variables are related by an equation of state, so that only two of variables are independent. Entropy S_η, internal energy E, enthalpy H, and the Gibbs function G are also state variables, and depend only on the current state of the system and not on the process by which the system acquired that state. Thermodynamic variables and functions are called extensive if they depend on mass (e.g., volume), and intensive if they do not depend on mass and can be defined for every point of the system (e.g., temperature, density). The intensive variables are related to a unit mass or to a mole. The thermodynamic system is further characterized by its chemical composition and phase state. If the composition and phase state of a system is variable, there is an additional characteristic, the chemical potential μ.

The processes occurring in a thermodynamic system can be described using scalar functions of the thermodynamic variables called the thermodynamic potentials. We will use the following extensive potentials, denoted by capital letters: internal energy E, enthalpy H, Helmholtz free energy F, the Gibbs free energy G, and the Landau potential Ω. The corresponding intensive variables are obtained by reducing to the corresponding mass. The corresponding intensive quantities are denoted here by lowercase letters; the molar quantities are obtained by dividing extensive quantities by the number of moles n (e.g., $\breve{e} = E/n$, $h = H/n$, $f = F/n$, $g = G/n$, $\varpi = \Omega/n$), and specific quantities are related to a unit mass. For the molar potentials, it is convenient to use the molar entropy $s_\eta = S_\eta/n$ and molar volume $v = V/n$.

The internal energy E is a measure of energy stored in a system and characterizes its internal state—e.g., speed of the molecules, internal motion in the molecules (rotation and vibration), etc. The first law of thermodynamics can be written for a reversible process as

$$dE = dQ + dW + \sum_{k=1}^{c} \mu_k dn_k, \tag{3.1.1}$$

where Q is the heat and W is the expansion work done on the system that can be expressed for quasi-static changes via the temperature T, entropy S_η, pressure p, and volume V as:

$$dQ = TdS_\eta, \qquad dW = -pdV. \tag{3.1.2}$$

μ_k and n_k are the chemical potential and the number of moles of the k-th chemical component. Equation (3.1.1) shows that the chemical potential μ_k characterizes a change dE when the number of moles changes by dn_k. Now E can be specified as

$$dE = TdS_\eta - pdV + \sum_{k=1}^{c} \mu_k dn_k . \tag{3.1.3}$$

Integration at the constant intensive variables T, p, and μ_k yields

$$E = TS_\eta - pV + \sum_{k=1}^{c} \mu_k n_k . \tag{3.1.3a}$$

The other thermodynamics potentials to be used—the heat function or enthalpy H, the Helmholtz free energy F, the Gibbs free energy G, and the Landau potential Ω—are related to E, p, V, T, and S_η:

$$H = E + pV, \tag{3.1.4}$$

$$F = E - TS_\eta, \tag{3.1.5}$$

$$G = E + pV - TS_\eta = H - TS_\eta, \tag{3.1.6}$$

$$\Omega = E - TS_\eta - \mu N = F - \mu N, \tag{3.1.7}$$

where N is the total number of particles in the system. The differential of the enthalpy H in a system with a variable number of moles is:

$$dH = TdS_\eta + Vdp + \sum_{k=1}^{c} \mu_k dn_k . \tag{3.1.8}$$

This equation allows calculation of the natural variables of E: T, V, and μ_k

$$T = \left(\frac{\partial H}{\partial S_\eta}\right)_{p,n_k}, \quad V = \left(\frac{\partial H}{\partial p}\right)_{S_\eta,n_k}, \quad \mu_k = \left(\frac{\partial H}{\partial n_k}\right)_{S_h,p,\mu_i}, \tag{3.1.9}$$

where the subscripts denote constant corresponding values and $\mu_i \neq \mu_k$. The differential of the Helmholtz free energy is

$$dF = -S_\eta dT - pdV + \sum_{k=1}^{c} \mu_k dn_k . \tag{3.1.10}$$

This allows calculations of S_η, p, and μ_k using F:

$$S_\eta = -\left(\frac{\partial F}{\partial T}\right)_{V,n_k}, \quad p = -\left(\frac{\partial F}{\partial V}\right)_{T,n_k}, \quad \mu_k = \left(\frac{\partial F}{\partial n_k}\right)_{T,V,n_i}. \tag{3.1.11}$$

The relation for p is used in Section 3.4 for derivation of the equation of state, and the equations for μ_k are used in Section 3.10 and in subsequent chapters for evaluation of the chemical potentials of the systems with variable numbers of particles.

The differential of the Gibbs free energy is

$$dG = -S_\eta dT + Vdp + \sum_{k=1}^{c} \mu_k dn_k . \tag{3.1.12}$$

This gives the relations for S_η, V, and μ_k

$$S_\eta = -\left(\frac{\partial G}{\partial T}\right)_{p,n_k}, \qquad V = \left(\frac{\partial G}{\partial p}\right)_{T,n_k}, \qquad \mu_k = \left(\frac{\partial G}{\partial n_k}\right)_{T,p,n_i}. \tag{3.1.13}$$

The differential of Ω includes μ_k as a variable and is

$$d\Omega = -S_\eta dT - pdV - \sum_{k=1}^{c} n_k d\mu_k. \tag{3.1.14}$$

This gives the relations

$$S_\eta = -\left(\frac{\partial \Omega}{\partial T}\right)_{V,n_k}, \qquad p = -\left(\frac{\partial \Omega}{\partial V}\right)_{T,n_k}, \qquad n_k = -\left(\frac{\partial \Omega}{\partial \mu_k}\right)_{T,V}. \tag{3.1.15}$$

This allows a particular calculation of n_k, therefore Ω is especially useful for calculations of the statistical distributions in systems with a variable number of particles, as shown in Section 3.2.

The molar Gibbs free energy is related to the chemical potential μ_k

$$g_k = \left(\frac{\partial G}{\partial n_k}\right)_{T,p,n} = \mu_k. \tag{3.1.16}$$

That is, the chemical potential is equal to the molar Gibbs free energy. Similar relations can be written for the other potentials. Using (Eqn. 3.1.16), the Gibbs energy G can be rewritten as the sum over the components

$$G = \sum_{k=1}^{c} n_k \mu_k = \sum_{k=1}^{c} n_k g_k. \tag{3.1.17}$$

Now, using these definitions and relations, we can introduce several useful quantities and equations. Equation (3.1.8) shows that H characterizes the heat content in the system if p = const, and all n_k = const, and h characterizes the heat content per one mole at p = const. At constant pressure, $dp = 0$, and a constant number of moles, $dn_k = 0$, it follows from (Eqn. 3.1.8) that

$$dH = dQ = T\left(\frac{\partial S_\eta}{\partial T}\right)_{p,n_k} dT = C_p dT, \qquad C_p = T\left(\frac{\partial S_\eta}{\partial T}\right)_{v,n_k}, \tag{3.1.18a}$$

where C_p is *the heat capacity at constant pressure*. It characterizes an increase in enthalpy or the heat of a system with an increase of the temperature by one degree at constant pressure. A similar relation is valid also for the molar or specific quantities at p = const,

$$dh = dq = T\left(\frac{\partial s_\eta}{\partial T}\right)_p dT = c_p dT, \qquad c_p = T\left(\frac{\partial s_\eta}{\partial T}\right)_p, \tag{3.1.18b}$$

For the process with constant volume, $dV = 0$, and n_k = const, $dn_k = 0$, (Eqn. 3.1.3) for dE reduces to

$$dE = dQ = T\left(\frac{\partial S_\eta}{\partial T}\right)_{v,n_k} dT = C_v dT, \qquad C_v = T\left(\frac{\partial S_\eta}{\partial T}\right)_{v,n_k}. \tag{3.1.19a}$$

where C_V is *the heat capacity at constant volume.* Equation (3.1.19a) shows that it characterizes increase in the internal energy or heat of the body if the temperature increases by one degree at constant volume. For a system of one mole or unit mass with just one substance, (Eqn. 3.1.3) becomes

$$d\tilde{e} = dq + dw = Tds_\eta - pdv, \tag{3.1.19b}$$

where s_η and v are the entropy and volume per unit mole or mass. A relation similar to (Eqn. 3.1.19a) is valid also for the molar or specific quantities with $v = \text{const}$,

$$d\tilde{e} = dq = T\left(\frac{\partial s_\eta}{\partial T}\right)_v dT = c_v dT, \qquad c_v = T\left(\frac{\partial s_\eta}{\partial T}\right)_v, \tag{3.1.19c}$$

where c_v is the heat capacity at constant volume per unit mole or mass. It was shown experimentally that for an ideal gas, $(\partial\tilde{e}/\partial v)_T = 0$ (i.e., $\tilde{e} = \tilde{e}(T)$) and $(\partial h/\partial p)_T = 0$ (i.e., $h = h(T)$). These equations are used in Section 3.11 for consideration of adiabatic processes.

Two additional useful relations can be obtained for the molar entropy s_η and molar volume v_k. It follows from (Eqns. 3.1.13 and 3.1.12) that

$$\begin{aligned}\left(\frac{\partial \mu_k}{\partial T}\right)_{p,n_{k\neq i}} &= \frac{\partial}{\partial T}\left(\frac{\partial G}{\partial n_k}\right)_{p,n_{k\neq i}} \\ &= \frac{\partial}{\partial n_k}\left(\frac{\partial G}{\partial T}\right)_{p,n_{k\neq i}} = -\left(\frac{\partial S_\eta}{\partial n_k}\right)_{T,p,n_{k\neq i}} = -s_{\eta k}.\end{aligned} \tag{3.1.20}$$

Thus, the derivative $\partial\mu_k/\partial T$ gives the molar entropy $s_{\eta k}$. Similarly, the derivative $\partial\mu_k/\partial p$ gives the molar volume v_k:

$$\begin{aligned}\left(\frac{\partial \mu_k}{\partial p}\right)_{T,n_{k\neq i}} &= \frac{\partial}{\partial p}\left(\frac{\partial G}{\partial n_k}\right)_{T,n_{k\neq i}} \\ &= \frac{\partial}{\partial n_k}\left(\frac{\partial G}{\partial p}\right)_{T,n_{k\neq i}} = \left(\frac{\partial V}{\partial n_k}\right)_{T,n_{k\neq i}} = v_k.\end{aligned} \tag{3.1.21}$$

The molar Gibbs free energy g_k, chemical potential μ_k, and enthalpy h_k are related similarly to (Eqn. 3.1.6) for G:

$$\mu_k = g_k = h_k - Ts_{\eta k}, \qquad h_k = \mu_k + Ts_{\eta k}. \tag{3.1.22}$$

This expression allows us to obtain a relation between μ_k and h_k using (Eqn. 3.1.20), $\partial\mu_k/\partial T = -s_{\eta k}$,

$$\begin{aligned}\frac{\partial}{\partial T}\left(\frac{\mu_k}{T}\right)_{p,n_{i\neq k}} &= -\frac{1}{T}\left[\frac{\mu_k}{T} - \left(\frac{\partial \mu_k}{\partial T}\right)\right]_{p,n_{i\neq k}} \\ &= -\frac{1}{T}\left(\frac{\mu_k}{T} + s_{\eta k}\right) = -\frac{1}{T}\left(\frac{\mu_k + Ts_{\eta k}}{T}\right) = -\frac{h_k}{T^2}.\end{aligned} \tag{3.1.23}$$

A similar relation follows from (Eqn. 3.1.21) for μ_k/T and v_k

$$\left(\frac{\partial(\mu_k/T)}{\partial p}\right)_{T,n_{i\neq k}} = \frac{1}{T}\left(\frac{\partial \mu_k}{\partial p}\right)_{p,n_{i\neq k}} = \frac{v_k}{T}. \tag{3.1.24}$$

Equations (3.1.23 and 3.1.24) will be used in derivations of the conditions of equilibrium and several fundamental laws in the following sections.

A useful relation can be derived from the expressions (Eqns. 3.1.3 and 3.1.3a) for the energy E. Taking the total differential of (Eqn. 3.1.3a)

$$dE = TdS_\eta + S_\eta dT - pdV - Vdp + \sum_{k=1}^{c} \mu_k dn_k + \sum_{k=1}^{c} n_k d\mu_k , \tag{3.1.25}$$

and subtracting (Eqn. 3.1.3) from this expression yields

$$\sum_{k=1}^{c} n_k d\mu_k = -S_\eta dT + Vdp. \tag{3.1.26}$$

This is the *Gibbs–Duhem relation*. It can be written in another form introducing the mole fractions for the k-th substance

$$x_k = n_k / \sum_{k=1}^{c} n_k . \tag{3.1.27}$$

Dividing (Eqn. 3.1.26) by Σn_k, we obtain this relation in terms of mole fractions

$$\sum_{k=1}^{c} x_k d\mu_k = -s_\eta dT + vdp, \tag{3.1.28}$$

where s_η and v are the molar entropy and volume. This relation is simplified for an isothermic (T = const) and isobaric (p = const) process,

$$\sum_{k=1}^{c} x_k d\mu_k = 0. \tag{3.1.29}$$

The Gibbs–Duhem relation can be a particularly useful tool in calculations and analyses of the laboratory experiments when it is necessary to relate the properties of various substances—e.g., solute and solvent as will be illustrated in Section 3.6 and in Chapter 11 on aerosol phase transitions.

3.2. Statistical Energy Distributions

There are five basic statistical distributions that play fundamental roles in statistical physics. The Gibbs distribution was derived by Gibbs in 1901 for classical statistics, which assumes a continuous energy spectrum. The Gibbs distribution was subsequently generalized for the quantum case with a discrete energy spectrum and for the combined mixed continuous and discrete energy. The Gibbs distribution is the most general of the statistical distributions; the other four distributions can be derived from the Gibbs distribution, as shown next.

3.2.1. The Gibbs Distribution

The Gibbs energy distribution for a thermodynamic system can be obtained from the micro-canonic distribution. For a system having a constant number of particles at a quantum energy level n with energy E_n, the micro-canonic distribution is

$$w_{nN} = \exp\left(\frac{F - E_n}{kT} \right), \tag{3.2.1}$$

where k is the Boltzmann's constant, F is the Helmholtz free energy, and w_{nN} is the probability of this state. The sum of all these probabilities should equal unity, thus the normalization condition is

$$\sum_n w_{nN} = \exp\left(\frac{F}{kT}\right)\sum_n \exp\left(\frac{-E_n}{kT}\right) = 1. \tag{3.2.2}$$

This yields

$$Z = \sum_n \exp\left(\frac{-E_n}{kT}\right) = \exp\left(\frac{-F}{kT}\right). \tag{3.2.3}$$

The ensemble described by the distribution (Eqn. 3.2.1) is called the *canonical ensemble*, and the quantity Z is called the *statistical sum* or the *canonical partition function*. The free energy F now can be expressed via the statistical sum Z

$$F = -kT \ln Z. \tag{3.2.4}$$

This form for F is used in thermodynamical applications of the Gibbs formula (Eqn. 3.2.1), in particular, for derivation of the equation of state, as will be illustrated further on.

In classical physics with a continuous energy spectrum, the Gibbs distribution is

$$d\rho(p,q) = \rho(p,q)dp\,dq = \exp\left(\frac{F - E(p,q)}{kT}\right)d\Gamma, \tag{3.2.5}$$

where p and q are generalized coordinates, their number is s, $dpdq$ is the differential element of the phase space—i.e., $dpdq = (dp_1dp_2\ldots dp_s)(dq_1dq_2\ldots dq_s)$, $d\Gamma = dpdq\,/(h)^s$ is the number of states in this differential element of the phase space, h is the Planck constant, and it plays the role of an elementary cell in the phase space.

The differential form of the Gibbs distribution in the quasi-classical case is

$$dw_n(p,q) = \exp\left(\frac{F - E_n(p,q)}{kT}\right)d\Gamma. \tag{3.2.6}$$

The form (Eqn. 3.2.6) is written for the quasi-classical case and implies that the energy $E_n(p, q)$ has two components: the continuous quasi-classical (e.g., the kinetic and potential energy of the motion of gas molecules) and the internal energy that may have quantum levels (e.g., vibrational and rotational) indicated by the subscript "n".

From the normalization condition

$$\sum_n \int_\Gamma dw_n(p,q) = 1, \tag{3.2.7}$$

we obtain F in the quasi-classical case

$$F = -kT \ln \int \exp\left(\frac{-E_n(p,q)}{kT}\right)d\Gamma. \tag{3.2.8}$$

The system with a variable number of particles N is called the *grand canonical ensemble*. The Gibbs distribution for the grand canonical ensemble is

$$w_{nN} = \exp\left(\frac{\Omega + \mu N - E_{nN}}{kT}\right) \tag{3.2.9}$$

where w_{nN} is the differential probability of the energy E_{nN} at quantum level n with the number of particles N, and Ω is the Landau thermodynamic potential (Eqn. 3.1.7). The normalization condition is

$$\sum_N \sum_n w_{nN} = \sum_N \sum_n \exp\left(\frac{\Omega + \mu N - E_{nN}}{kT}\right) = 1. \tag{3.2.10}$$

Equation (3.2.10) gives for Ω

$$\Omega = -kT \ln \sum_N \left[\exp\left(\frac{\mu N}{kT}\right) \sum_n \exp\left(-\frac{E_{nN}}{kT}\right)\right]. \tag{3.2.11}$$

This equation defines Ω as a function of T, μ, N. The number of particles N in each state n can be calculated as $\partial\Omega/\partial\mu$.

3.2.2. The Maxwell Distribution

In classical statistical mechanics, the energy $E(p, q)$ can be presented as a sum of two terms, kinetic energy $E_{kin}(p_m)$ that depends only on momentum p_m, and potential energy $U_{pot}(q)$ depending only on coordinate q. We can choose p_m and q as generalized coordinates and write

$$E(p_m, q) = K_{kin}(p_m) + U_{pot}(q). \tag{3.2.12}$$

Then, the energy distribution is factorized

$$dw(p_m, q) = A_G e^{-\frac{E_{kin}(p_m)}{kT}} e^{-\frac{U_{pot}(q)}{kT}} = dw_p(p_m) dw_q(q). \tag{3.2.13}$$

where A_G is the normalization constant, and dw_p and dw_q are the corresponding energy distributions by momentum and coordinates

$$dw_p(p_m) = a_p e^{-\frac{E_{kin}(p_m)}{kT}} dp_m, \tag{3.2.14}$$

$$dw_q(q) = a_q e^{-\frac{U_{pot}(q)}{KT}} dq. \tag{3.2.15}$$

The constants a_p and a_q are determined from the conditions that each of these probabilities should be normalized to unity.

The kinetic energy of an atom or molecule with mass m is $E_{kin} = (p_x^2 + p_y^2 + p_z^2)/2m$, where p_x, p_y, and p_z, are the Cartesian components of the momentum, then

$$dw_p(p_m) = a_p \exp\left(-\frac{p_x^2 + p_y^2 + p_z^2}{2mkT}\right) dp_x dp_y dp_z. \tag{3.2.16}$$

The normalizing constant is calculated in Appendix A.3. It is $a_p = (2\pi mkT)^{-3/2}$, and the momentum distribution is

$$dw_p(p_m) = (2\pi mkT)^{-3/2} \exp\left(-\frac{p_x^2 + p_y^2 + p_z^2}{2mkT}\right) dp_x dp_y dp_z. \tag{3.2.17}$$

It is convenient sometimes to use instead of (Eqn. 3.2.16) the momentum distribution $d\tilde{w}_p(p_m)$ with another normalization, $\tilde{a}_p = 1$:

$$d\tilde{w}_p(p_m) = \exp\left(-\frac{p_x^2 + p_y^2 + p_z^2}{2mkT}\right) dp_x\, dp_y\, dp_z. \tag{3.2.17a}$$

It is clear that the integral of this distribution is equal to $a_p^{-1} = (2\pi mkT)^{3/2}$.

Transformation to the velocity distribution can be done using the relations $dp_{x,y,z} = mdv_{x,y,z}$. Then, we obtain from (Eqn. 3.2.17) in the Cartesian coordinate system in the velocities space

$$dw_v(v_x, v_y v_z) = \left(\frac{m}{2\pi kT}\right)^{3/2} \exp\left[-\frac{m(v_x^2 + v_y^2 + v_z^2)}{2kT}\right] dv_x\, dv_y\, dv_z, \tag{3.2.18}$$

This is the Maxwell distribution derived by Maxwell in 1860. In the polar coordinate system of the velocities space with the polar angles θ, φ, and velocity v it converts into

$$dw_v(v, \theta, \varphi) = \left(\frac{m}{2\pi kT}\right)^{3/2} \exp\left(-\frac{mv^2}{2kT}\right) v^2 \sin\theta\, d\theta\, d\varphi\, dv. \tag{3.2.19}$$

The normalization constant $a_v = (m/2\pi kT)^{3/2}$ for (Eqns. 3.2.18 and 3.2.19) is calculated in Appendix A.3. The Maxwell distribution for water vapor normalized to the molecular concentration $c_v = \rho_v/m_w$ in this polar coordinate system can be derived in a similar way or is obtained just by multiplication by c_v:

$$f_v(v)dv = c_v \left(\frac{m_w}{2\pi kT}\right)^{3/2} \exp\left(-\frac{m_w v^2}{2kT}\right) v^2 \sin\theta\, d\theta\, d\varphi, \tag{3.2.20}$$

where m_w is the mass of a water molecule and ρ_v is the vapor density. The Maxwell distribution is used in Chapter 5 to evaluate the kinetic molecular flux to the drops and crystals.

3.2.3. The Boltzmann Distribution

The Boltzmann distribution is applied to the ideal gas, whereby interaction among molecules can be neglected. This implies that the interaction is really small at any distances, or the density of the gas is small. A further requirement is that the quantum mechanical interactions caused by the particles' identity can be also neglected. All of these requirements are fulfilled if the average number of particles n_k in each energy state ε_k is small, $n_k << 1$. If particle interactions can be neglected, then the Gibbs distribution can be applied to individual molecules:

$$n_k = a_B \exp(-\varepsilon_k/kT). \tag{3.2.21}$$

The normalizing coefficient a_B can be expressed through the thermodynamic parameters of the gas. The condition $n_k << 1$ is equivalent to $\varepsilon_k >> kT$ (i.e., this distribution becomes invalid at sufficiently low temperatures).

This equation can be also directly derived from the Gibbs distribution (Eqn. 3.2.9) for a given quantum state k. Since interaction among the molecules can be neglected, their total energy is a sum of the molecular energies, $E = n_k \varepsilon_k$. Assuming also $N = n_k$, and $\Omega = \Omega_k$ for the given k-state, we can write from (Eqn. 3.2.9)

$$w_{n_k} = \exp\left[\frac{\Omega_k + n_k(\mu - \varepsilon_{n_k})}{kT}\right]. \tag{3.2.22}$$

When particles are absent at any state k, then $n_k = 0$, and the condition $w_{n_k} = \exp(\Omega_k/kT) = 1$ is the probability of such a state. Since $n_k << 1$, this means that $\exp(\Omega_k/kT) \approx 1$, that is, $\Omega_k \approx 0$. Then, the Boltzmann distribution to find one particle in k-th state ($n_k = 1$) can be simplified as

$$w_1 = \exp\left(\frac{\mu - \varepsilon_k}{kT}\right). \tag{3.2.23}$$

The mean number of particles in this state is

$$\bar{n}_k = \sum_{n_k} w_{n_k} n_k = w_1 = \exp\left(\frac{\mu - \varepsilon_k}{kT}\right), \tag{3.2.24}$$

since the probability for one particle coincides with w_1. This expression is the Boltzmann distribution (derived by Boltzmann in 1877).

For the classical case, the Boltzmann distribution can be written for the phase space of the generalized coordinates p, q as

$$dn(p,q) = \exp\left[\frac{\mu - \varepsilon(p,q)}{kT}\right] dp\, dq. \tag{3.2.25}$$

If the energy of a molecule can be presented as a sum of the potential and kinetic energies as in (Eqn. 3.2.12), then the probability is factorized into two dependencies by momentum p_m or velocity and by coordinates q as in (Eqn. 3.2.13):

$$dN(p_m,q) = A_G e^{-\frac{E_{kin}(p_m)}{kT}} e^{-\frac{U_{pot}(q)}{kT}} = dN_p(p_m)\, dN_q(q). \tag{3.2.26}$$

If the gas is not located in an external field or if this field depends only on coordinates, then the kinetic part $dN_p(p_m)$ is described by the Maxwell distribution discussed earlier. If the generalized coordinates q are Cartesian (x, y, z), the distribution $dN_q(q) = dN_q(x,y,z)$ is

$$dN_q(x,y,z) = n_0 e^{-\frac{U_{pot}(x,y,z)}{kT}} dV, \tag{3.2.27}$$

where n_0 is a normalization parameter. Since $dN_q(x,y,z) = n(x,y,z)dV$ with $n(x,y,z)$ being the density of the molecules, this gives for the spatial distribution $n(x,y,z)$

$$n(x,y,z) = n_0 e^{-\frac{U_{pot}(x,y,z)}{kT}}. \tag{3.2.28}$$

This equation determines the concentration of the molecules in the point (x,y,z). Equations (3.2.27) and (3.2.28) are called the Boltzmann distribution.

In particular, in the gravitational field directed along the z-axis, the potential energy is $U = mgz$, and the Boltzmann distribution describes the known barometric formula for the density of the molecules

$$n(z) = n_0 e^{-\frac{mgz}{kT}}, \tag{3.2.29}$$

where n_0 is the density at the surface level. This equation can be rewritten for the gas density $\rho = mn$. By multiplying the numerator and denominator in the exponent by Avogadro's number, N_{Av}, and using the relations $mN_{Av} = M$ and $kN_{Av} = R$, where M is the gas molecular weight and R is the universal gas constant, this becomes

$$\rho(z) = \rho_0 e^{-\frac{Mgz}{RT}}. \tag{3.2.30}$$

This equation shows that the density of a gas with greater molecular weight decreases with height faster than the density of a lighter gas. Note that convection and turbulence in the atmosphere may redistribute and mix the gases, so that their vertical distribution may not obey this relation.

3.2.4. Bose–Einstein Statistics

If the temperature of an ideal gas becomes sufficiently low or the density is high, then the criterion for Boltzmann statistics $n_k << 1$ is not satisfied. Bose–Einstein statistics apply when quantum effects are important, the particles are indistinguishable, and they have an integer spin (internal momentum of rotation); that is, counted in integer numbers of Planck constant h. Bose–Einstein statistics was introduced for photons and then later generalized to atoms. Important applications of Bose–Einstein statistics include the following: black-body radiation, which can be considered as an ideal gas of photons; evaluation of the heat capacities of the solids and gases; and ice nucleation at low temperatures.

Bose–Einstein statistics can be derived from the thermodynamic potential Ω in (Eqn. 3.2.11) for an ideal gas. The energy E_{nN} of gas particles in the k-th quantum state can be presented as a sum of the energies ε_k of individual particles, the number of particles is $N = n_k$, that is $E_{nN} = n_k\varepsilon_k$. Substitution of these relations into (Eqn. 3.2.11) yields for the thermodynamics potential Ω_k of this state

$$\Omega_k = -kT \ln Z, \qquad Z = \sum_{n_k} \left[\exp\left(\frac{\mu - \varepsilon_k}{kT} \right) \right]^{n_k}. \tag{3.2.31}$$

The statistical sum Z here represents a geometric progression by the powers n_k and can be different depending on the particles' rotation moment, or spin. The spins of the particles can be either "integer" (equal to Planck's constant h—an example are photons—then the particles wave functions are symmetric) or "half-integer" (equal to $h/2$, then their wave functions are anti-symmetric—examples are electrons, and the constituents of atomic nuclei, protons and neutrons). In this section, we consider a system that has a symmetric function and the integer spin, as is the case with photons. Then, the number of particles in each state can be unlimited. The geometric progression in (Eqn. 3.2.31) has an infinite number of terms from $n_k = 0$ to ∞ and converges for all terms if $(\mu - \varepsilon_k) < 0$. Summing up, we obtain $Z_k = \left\{1 - \exp\left[(\mu - \varepsilon_k)/kT\right]\right\}^{-1}$, and

$$\Omega_k = kT \ln \left[1 - \exp\left(\frac{\mu - \varepsilon_k}{kT} \right) \right]. \tag{3.2.32}$$

The number of particles in each k-state is

$$\bar{n}_k = -\frac{\partial \Omega_k}{\partial \mu} = \frac{1}{\exp[(\varepsilon_k - \mu)/kT - 1]}. \tag{3.2.33}$$

This distribution is called Bose–Einstein statistics. It was derived by Bose in 1924 for photons and generalized later by Einstein for atoms.

The particles that obey this statistics are called *bosons* (e.g., photons are bosons, and the recently discovered Higgs particle that gives inertia to all the bodies is also a boson). If $(\varepsilon_k - \mu) >> kT$, then 1 in the denominator of (Eqn. 3.2.33) can be neglected, and we obtain

$$\bar{n}_k = \exp[-(\varepsilon_k - \mu)/kT], \tag{3.2.34}$$

which is the Boltzmann distribution. Therefore, Bose–Einstein statistics is a generalization of the Boltzmann statistics for arbitrary relations of $(\varepsilon_k - \mu)$ and kT, which is in agreement with the correspondence principle. Boltzmann statistics has been used in classical derivations of the ice nucleation rates (see Chapters 8 and 9). At very low temperatures or other conditions, when a relation $(\varepsilon_k - \mu) \leq kT$ is satisfied, the Boltzmann distribution can become inapplicable and the Bose–Einstein distribution should be used.

A well-known application of the Bose–Einstein distribution (Eqn. 3.2.33) is black-body radiation, which represents the ideal gas of photons with a variable number of particles. The condition of energy minimum gives $\mu = 0$ for this system, the energy of a photon is $\varepsilon = h\nu$, where ν is the frequency of radiation, and the density of oscillators with the frequency ν is $dN_\nu = (8\pi/c^3)\nu^2 d\nu$ (Landau and Lifshitz, 1958b, v.5; Born, 1963). Multiplying this dN_ν by $\bar{n}_\nu$ from (Eqn. 3.2.33), we obtain the energy density of radiation:

$$d\rho_\varepsilon(\nu,T) = \rho_\varepsilon(\nu,T)d\nu = \frac{8\pi h}{c^3}\frac{\nu^3 d\nu}{\exp(h\nu/kT) - 1}. \tag{3.2.35}$$

This is the Planck function for the black-body radiation, which follows from the Bose–Einstein distribution of the photon gas. Another application of the Bose–Einstein distribution for ice nucleation will be considered in Chapters 8 and 9.

3.2.5. Fermi–Dirac Statistics

For completeness, we briefly outline the last of the major statistics. If the spin of the particles is half-integer, $h/2$, like the electron spin, the wave function of the system is antisymmetric, changing its sign at the interchange of any 2 particles. Then, the *Pauli's exclusion principle* acts, which states that not more than one particle can be in any quantum state. Thus, the number of particles n_k in each quantum energy level can only be 0 and 1. The statistical sum Z in (Eqn. 3.2.31) has only two terms and this yields the Ω-potential

$$\Omega_k = -kT\ln\left[1 + \exp\left(\frac{\mu - \varepsilon_k}{kT}\right)\right]. \tag{3.2.36}$$

Thus, the average number of particles in this k-state is

$$\bar{n}_k = -\frac{\partial\Omega_k}{\partial\mu} = \frac{1}{\exp[(\varepsilon_k - \mu)/kT + 1]}. \tag{3.2.37}$$

This is called *Fermi–Dirac statistics*. The particles obeying Fermi–Dirac statistics are called *fermions*. A gas or any system with this distribution is called a Fermi-gas or Fermi-system. An important example of such Fermi-gas is the electron cloud in an atom or free electrons in metals; the protons and neutrons in atomic nuclei represent Fermi-liquid. The atoms with electrons fulfilling one atomic electron shell constitute one period in the periodic table of the elements discovered by Dmitry Mendeleev in 1869 (Landau and Lifshitz, v. 3, 1958a; Born, 1963). Due to Fermi–Dirac statistics, the number of electrons in each quantum state is limited (0 or 1), quantum mechanics predicts that the number of electrons in each atomic shell is limited and equal to $n_e = 2N_p^2$, where N_p is the principal quantum number of the shell that determines the period in the periodic table. For example, for the

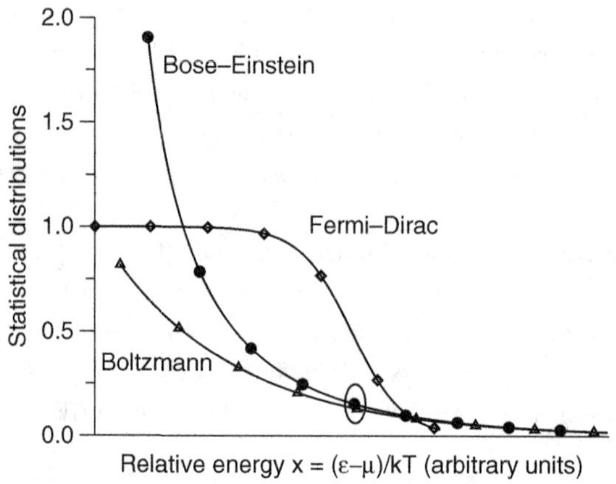

Figure 3.1. Three major statistical distributions by Bose–Einstein, Fermi–Dirac, and Boltzmann. At sufficiently high energies $x = (\varepsilon - \mu)/kT$, all three statistics become close—that is, the Bose–Einstein and Fermi–Dirac statistics convert into the Boltzmann distribution at the high energy limit $(\varepsilon - \mu) >> kT$.

periods with $N_p = 1$, 2, and 3 in the table, these are: $n_e = 2$ (hydrogen and helium in the 1st period), 8 and 18 in the 2nd and 3rd periods respectively, which explains the periodic table (when the number of electrons in a shell becomes large, their mutual interactions, spin-orbital and other interactions make the picture more complicated and influence the filling of the shell, the number of electrons in each shell and the number of the elements in one period of the periodic table). The number of electrons in each shell determines the chemical properties of the elements. Thus, all chemical properties of all chemical elements follow from the Fermi–Dirac statistics and Pauli's principle. Additional important applications of this statistics are the theories of superconductivity and superfluidity. In cloud physics, application of this statistics may be desirable—e.g., for consideration of cloud electrification with strong ionization of ice molecules at low temperatures, when a large number of free electrons may form, then their behavior may diverge from the Boltzmann distribution and obey the Fermi–Dirac distribution.

As in the case with Bose–Einstein statistics, if $(\varepsilon_k - \mu) >> kT$, then 1 in the denominator of (Eqn. 3.2.37) can be neglected, and we obtain again the Boltzmann distribution. Thus, Fermi–Dirac statistics is another generalization of the Boltzmann distribution in agreement with the correspondence principle, and the predictions of all three statistics coincide for the case $(\varepsilon_k - \mu) >> kT$. These three distributions are shown in Fig. 3.1. It shows that at sufficiently low T or small ε, Boltzmann statistics becomes invalid and Bose–Einstein or Fermi–Dirac statistics should be used.

3.3. Phase Rules

The state of a thermodynamic system can be described using several intensive variables like temperature T, pressure p, chemical potentials μ_k, etc. Not all of these variables are independent. The number of independent variables is called the *number of thermodynamic degrees of freedom* and obeys *phase rules*. These rules are different for systems consisting of bulk phases with and without boundary surfaces.

3.3.1. Bulk Phases

Consider a system of c components and φ bulk phases that do not have the boundary surfaces among the phases or components. The system can be described with the following variables:

$$T;\, p_1,\ldots,p_\varphi;\, x_1^1,\ldots,x_c^\varphi, \tag{3.3.1}$$

where T is the temperature, common for the entire system, p_α is the pressure inside the α-th phase (where $\alpha = 1,\ldots, \varphi$—i.e., total φ variables), and x_γ^α is the weight fraction of component γ in phase α ($\alpha = 1,\ldots, \varphi$, and $\gamma = 1,\ldots, c$, which gives the total $c\varphi$ variables). Thus, (Eqn. 3.3.1) contains $1 + \varphi + c\varphi$ variables, but not all of them are independent.

The weight fractions are constrained by φ relations of mass conservation

$$\sum_\gamma x_\gamma^\alpha = 1, \tag{3.3.2}$$

where $\alpha = 1,\ldots, \varphi$, that is, (Eqn. 3.3.2) forms φ relations. Thus, (Eqns. 3.3.1 and 3.3.2) show that the system is defined by

$$N_v = 1 + \varphi + c\varphi - \varphi = 1 + c\varphi \tag{3.3.3}$$

variables. There are two additional constraints: equality of pressures among the phases

$$p_1 = .. = p_\alpha = \ldots = p_\varphi, \tag{3.3.4}$$

which gives $n_p = \varphi - 1$ relations, and equality of the φ chemical potentials for c components,

$$\mu_\gamma^1 = .. = \mu_\gamma^\alpha = \ldots = \mu_\gamma^\varphi, \tag{3.3.5}$$

which gives $n_\mu = c(\varphi - 1)$ relations. So, we obtain the number of independent variables, or thermodynamic degrees of freedoms or variance

$$N_w = N_v - n_p - n_\mu = 1 + c\varphi - (\varphi - 1) - c(\varphi - 1) = c + 2 - \varphi. \tag{3.3.6}$$

This is the Gibbs' phase rule for bulk phases without curved interfaces.

If we have a homogeneous liquid or gas (e.g., air or water vapor), there is one component, $c = 1$, and one phase, $\varphi = 1$. Thus, $N_w = 2$, and there are two independent variables. This agrees with the fact that the three variables describing the liquid, the density ρ, pressure p, and temperature T, are related by one equation of state; that is, only two variables are independent, so the system is *bivariant*.

Consider liquid water in equilibrium with its vapor. Then, $c = 1$, $\varphi = 2$, thus, $N_w = 1$ and there is one independent variable, the system is *monovariant*. We can therefore consider vapor pressure as a function of temperature. The same is valid for ice in equilibrium with vapor, $c = 1$, $\varphi = 2$, thus, $N_w = 1$. If liquid water is in equilibrium with its vapor and ice, $c = 1$, $\varphi = 3$, thus, $N_w = 0$. In this case, there are no independent variables, and equilibrium is possible at only one pressure and temperature, which defines the *triple point temperature*.

3.3.2. Systems with Curved Interfaces

A more complicated case is a system of c components and φ bulk phases separated by ψ curved interfaces, each comprising only a single surface phase, and having the radius of mean curvature r_β. There are no chemical reactions in the system. The system is characterized by the temperature T, pressures

p_α in the bulk phase α ($\alpha = 1, \ldots, \varphi$), and the mass fractions dissolved in the bulk and adsorbed on the surfaces. The components in the bulk phase are characterized by x_γ^α, the weight fractions of component γ in phase α, which gives a total of $c\varphi$ variables. The components adsorbed on the interfaces can be characterized by the adsorbed mole fraction n_γ^β ($\beta = 1, \ldots, \psi$), or by their density, $\Gamma_\gamma^\beta = n_\gamma^\beta / A_\beta$, called *adsorption*, where A_β is the surface area of the β-th surface. So, the variables describing the system include the following:

$$T; p_1, \ldots, p_\varphi; x_1^1, \ldots, x_c^\varphi; \Gamma_1^1, \ldots, \Gamma_c^\psi; r_1, \ldots, r_\psi, \tag{3.3.7}$$

that is, φ pressures, $c\varphi$ volume fractions x_γ^α, $c\psi$ adsorptions, and ψ radii of curvature. The total number in this list is $1 + \varphi + c\varphi + c\psi + \psi$. The same mass conservation (Eqn. 3.3.2) for the weight fractions yields φ constraints. Thus, the number of variables is

$$N_v = 1 + \varphi + c\varphi + c\psi + \psi - \varphi = 1 + c\varphi + \psi(c+1). \tag{3.3.8}$$

There are also another two types of constraints. The Laplace's relations at each surface are

$$p_\alpha - p'_\alpha = \frac{2\sigma_\beta}{r_\beta}, \tag{3.3.9}$$

where p_α and p'_α are pressures on the two sides of the surface, and σ_β is the surface tension on the surface. This yields $n_p = \psi$ relations. The chemical potentials are related now as

$$\mu_\gamma^1 = \ldots = \mu_\gamma^\alpha = \ldots = \mu_\gamma^\varphi = \mu_\gamma^{\alpha 1} = \ldots = \mu_\gamma^{\alpha\beta} = \ldots = \mu_\gamma^{\alpha\psi}, \tag{3.3.10}$$

where μ_γ^α is the chemical potential of component γ in the bulk phase α ($\gamma = 1, \ldots, c$; and $\alpha = 1, \ldots, \varphi$) and $\mu_\gamma^{\alpha\beta}$ is the chemical potential of component γ in the phase α adsorbed in the surface phase β ($\beta = 1, \ldots, \psi$).

Equations (3.3.9) and (3.3.10) express the conditions of equilibrium for this system. The conditions for μ_γ^α give $c(\varphi - 1)$ relations, and the conditions for $\mu_\gamma^{\alpha\beta}$ give $c\psi$ relations. Thus, (Eqn. 3.3.10) gives total $n_\mu = c(\varphi + \psi - 1)$ relations. So, we obtain the number of degrees of freedom to be

$$N_w = N_v - n_p - n_\mu = 1 + c\varphi + \psi(c+1) - \psi - c(\varphi + \psi - 1). \tag{3.3.11}$$

The summation on the right-hand side yields a simple result:

$$N_w = 1 + c. \tag{3.3.12}$$

This is the phase rule for a system with the interfaces. It expresses an important property of the system: the number of degrees of freedom is independent of the number of both bulk and surface phases.

When a component is absent in some phase (e.g., is insoluble), then one of the relations (Eqn. 3.3.10) vanishes. However, since a condition of insolubility appears, the fraction $x_{\gamma\alpha} = 0$; thus the number of constraints remains the same and the phase rule (Eqn. 3.3.12) holds.

Consider a few practically important applications of the phase rule (Eqn. 3.3.12).

1. A system comprised by a pure water drop of radius r surrounded by pure water vapor of pressure e_r at a temperature T. The number of components is $c = 1$ (water), and $N_w = 2$. We may choose $T = \text{const}$ and study the dependence of e_r on r. This will give us Kelvin's law for the dependence e_r on r for the drop (Section 3.8). The same consideration of a crystal in water vapor will bring us the dependence e_r on r for the crystal.

2. A system consisting of a pure water drop of radius r surrounded by humid air at total pressure p. Now, $c = 2$ (water and air), and $N_w = 3$ from (Eqn. 3.3.12). To study the dependence e_r on r for a drop or crystal, we have to take both T = const and p = const. This will also give us Kelvin's law.
3. A system of an aqueous solution drop in humid air at pressure p and temperature T. The number of components $c = 3$ (salt, water, air), thus $N_w = 4$. However, the mass of salt in the drop is constant despite possible changes of its radius due to growth or evaporation. This adds one more constraint, and the actual number of thermodynamic degrees of freedom is $N_w = 3$ as in the previous case.
4. A system of pure water drop with radius r_w in humid air at pressure p and temperature T with a spherical crystal of radius r_c embedded in the drop. Then, $c = 2$ (water and air) and $N_w = 3$. This system is considered in Chapters 8, 9, and 10 in connection with immersion freezing, along with the other examples, and the phase rule helps to determine the degrees of freedom in such systems.

3.4. Free Energy and Equations of State

3.4.1. An Ideal Gas

The ideal gas equation of state is usually given in books on atmospheric physics either without derivation, or sometimes with simple physical derivations from molecular-kinetic theory. A more general derivation can be done from the thermodynamic potentials (Landau and Lifshitz, 1958b), which allows further generalizations for non-ideal gases and for the liquids.

The Helmholtz free energy F of the gas with the Boltzmann distribution by the energies ε_k is given by Equations (3.2.3) and (3.2.4). The energy E_n in (Eqn. 3.2.3) can be written as a sum of the energies ε_k of all N particles in the gas, and summing up independently over all the combinations of quantum energies and over all particles, we obtain from the statistical sum Z in (Eqn. 3.2.3)

$$Z = \frac{1}{N!}\prod_N \sum_n \exp\left(\frac{-E_n}{kT}\right) = \frac{1}{N!}\left[\sum_n \exp\left(\frac{-\varepsilon_k}{kT}\right)\right]^N, \tag{3.4.1}$$

where the sign "N!" means the factorial, product of the natural numbers from 1 to N. The factor $1/N!$ arises here because in this summation, each quantum state with the same energy distributions over the N particles is counted N! times. This arises because the particles are identical; an interchange of any two particles does not produce a new quantum state, but all such identical states are accounted for in the sum independently. The number of such interchanges in a system of N particles is N!. Thus, each state is counted in the sum N! times but should be counted just one time. Therefore, a correct statistical sum is obtained by dividing by N!

Substituting this Z into (Eqn. 3.2.4), we obtain the free energy for a Boltzmann gas:

$$F = -kTN \ln \sum_i \exp(-\varepsilon_k / kT) + kT \ln N!\,. \tag{3.4.2}$$

Since N is a very large number, it is possible to use an approximate Stirling formula

$$\ln N! \approx N \ln(N/e)\,, \tag{3.4.3}$$

where $e \approx 2.718$ is the base of the natural logarithm. Then, (Eqn. 3.4.2) can be rewritten as

$$F = -kTN \ln\left[\frac{e}{N}\sum_i \exp(-\varepsilon_i/kT)\right]. \tag{3.4.4}$$

In classical statistics, this equation becomes

$$F = -kTN \ln\left[\frac{e}{N}\int e^{-\frac{\varepsilon(p_m,q)}{kT}}\frac{dp_m dq}{h^3}\right], \tag{3.4.5}$$

where p_m and q are the momentum and coordinate that form the phase space of the generalized coordinates, $dp_m dq$ is the differential volume of the phase space, h is the Planck constant, and $dp_m dq/(h)^3$ is the number of states in the phase space.

The energy of the gas molecules can be written as

$$\varepsilon(p_x, p_y, p_z) = \frac{p_x^2 + p_y^2 + p_z^2}{2m} + \varepsilon_k', \tag{3.4.6}$$

where p_x, p_y, p_z are the 3 components of momentum, and ε_k' is the internal energy independent on the speed of the molecules. In this case, the phase space $[p_m - q]$ consists of the three-dimensional momentum space and three-dimensional Cartesian space with an element $[dp_m dq] = dp_x dp_y dp_z dV$. Accounting for this fact and substituting (Eqn. 3.4.6) into the integral (Eqn. 3.4.5) gives

$$F = -kTN \ln\left[\frac{e}{N}Z_1\right], \tag{3.4.7}$$

where the statistical sum Z_1 is

$$Z_1 = \sum_i e^{-\frac{\varepsilon_i'}{kT}}\int dV \int_{-\infty}^{\infty}\int_{-\infty}^{\infty}\int_{-\infty}^{\infty}\exp\left(-\frac{p_x^2 + p_y^2 + p_z^2}{2mkT}\right)\frac{dp_x dp_y dp_z}{h^3}, \tag{3.4.8}$$

and the expression in exp represents the kinetic energy of the molecules.

The internal integral over the momentum is calculated in Appendix A.3, and gives

$$\int_{-\infty}^{\infty}\int_{-\infty}^{\infty}\int_{-\infty}^{\infty}\exp\left(-\frac{p_x^2 + p_y^2 + p_z^2}{2mkT}\right)\frac{dp_x dp_y dp_z}{h^3} = \left(\frac{2\pi mkT}{h^2}\right)^{3/2}. \tag{3.4.9}$$

Substitution of this into (Eqn. 3.4.8) and integration also over dV yields

$$Z_1 = V\left(\frac{2\pi mkT}{h^2}\right)^{3/2}\sum_i e^{-\frac{\varepsilon_i'}{kT}}. \tag{3.4.10}$$

The factor V arises from the integration because molecular energy ε_i' and dw_p are independent on coordinates. The integral over momentum space is calculated in Appendix A.3. Then, we obtain for the free energy F:

$$F = -kTN \ln\left[\frac{eV}{N}\left(\frac{2\pi mkT}{h^2}\right)^{3/2}\sum_i e^{-\frac{\varepsilon_i'}{kT}}\right]. \tag{3.4.11}$$

Its general evaluation can be rather complicated, as it requires knowledge of the internal energies ε_i'.

However, calculation of the gas pressure is simple. Separating now the term with V, the free energy of an ideal gas can be rewritten

$$F_{id} = -kTN \ln\left[\left(\frac{eV}{N}\right)\left(\frac{2\pi mkT}{h^2}\right)^{3/2}\right] + Nf(T), \tag{3.4.12}$$

where $f(T)$ is some function of T—its exact form is unimportant because it is independent on V. (If the internal energy of molecules is much smaller than the kinetic energy and can be neglected, $\varepsilon_i' \approx 0$, then the sum in (Eqn. 3.4.11) is equal to 1, and $f(T)$ is simply evaluated). The pressure is now obtained from the relation (Eqn. 3.1.6):

$$p = -\frac{\partial F_{id}}{\partial V} = \frac{kTN}{V}, \tag{3.4.13}$$

or in another form

$$pV = kNT. \tag{3.4.14}$$

This is the equation of state for an ideal gas.

Several other forms of this equation are useful in applications. For air, the mass is $m_a = \nu_a M_a$, where M_a is the molecular weight and ν_a is the number of moles, and the density $\rho_a = m_a/V$. If the mass is one gram-molecule, $\nu_a = 1$ mole, then the number of molecules N is equal to Avogadro's number $N_{Av} = 6.023 \times 10^{23}$ (molecules mole^{-1}), the molar volume v_{am} is related to the molecular mass M_a as $v_{am} = M_a/\rho_a$, and the equation of state for one gram-molecule of the air at pressure p_a can be written as

$$p_a v_a = kN_{Av}T = RT, \qquad \text{or} \qquad p_a = \rho_a R_a T, \tag{3.4.15}$$

where $R = kN_{Av} = 8.3144 \times 10^7$ erg mole^{-1} K^{-1} = 1.9858 cal mole^{-1} K^{-1} is the universal gas constant, $R_a = R/M_a = 2.8706 \times 10^6$ erg g^{-1} K^{-1} = 287 J K^{-1} kg^{-1} is the gas constant for the air.

If a gas consists of a mixture of gases that occupy a common volume V and have partial pressures p_k with a number of moles n_k, then this equation becomes

$$p_k V = n_k RT. \tag{3.4.15a}$$

For water vapor, the density is $\rho_v = m_v/V$, and the pressure is e_v. For the mass of one gram-molecule, the molar volume is $v_m = M_w/\rho_v$, and the equation of state for water vapor considered as an ideal gas can be written as

$$e_v v_m = kN_{Av}T = RT, \qquad e_v = \rho_v R_v T \tag{3.4.16}$$

where $R_v = R/M_w = 4.6150 \times 10^6$ erg g^{-1} K^{-1} is the gas constant for the vapor. The same equations of state are valid for the saturated over water and ice vapor pressures e_{ws}, e_{is}, and the densities ρ_{ws}, ρ_{is}:

$$e_{ws} = \rho_{ws} R_v T, \tag{3.4.17}$$

$$e_{is} = \rho_{is} R_v T. \tag{3.4.18}$$

The *Clausius–Clapeyron equation* for e_{ws}, e_{is} describing their temperature dependence, and other atmospheric humidity variables will be described in Section 3.7.

3.4.2. Free Energy and the van der Waals Equation of State for a Non-ideal Gas

The atmospheric gases at normal conditions in Earth's atmosphere are close to ideal. The error of the ideal gas equation for atmospheric vapor is about 1% under atmospheric conditions; therefore the equations of state for the non-ideal gases usually are not used in most applications to Earth atmospheric physics under moderate conditions. However, non-ideality of the gases increases as the pressure increases or temperature decreases, and as a gas approaches the phase transition point into a liquid (e.g., Curry and Webster, 1999). The non-ideal gases and liquids can be described with the *van der Waals equation of state* and its modifications.

Since the 1990s, various modifications of the van der Waals equation of state have begun to play an increasingly important role in studies of liquid water, its equilibrium with the vapor especially at low temperatures, and the processes of ice nucleation (see Chapter 4). Therefore, a derivation is provided in this subsection of the free energy F and of the van der Waals equation of state that describes the non-ideal gases and liquids. This derivation provides a platform for understanding the modifications of the van der Waals equation described in Chapter 4 and their applications for supercooled liquids.

The van der Waals equation can also be used for studies of the atmospheres and clouds of other planets where gases behave as non-ideal because of the great variety of pressures and temperatures. For example, the temperature and pressure in the atmosphere of Venus vary from ~740 K and 88 atm at the surface to less than 200 K and very low pressure at altitudes more that 100 km. The major constituent of the Venusian atmosphere, CO_2, has a critical point at $T_{cr} = 304.2$ K and $p_{cr} = 72.9$ atm. The critical point of CO_2 is within the parameter range of the Venusian atmosphere, where its behavior is different from the ideal gas—in particular, the lapse rate may substantially exceed the value for the ideal gas (e.g., Callen, 1960; Curry and Webster, 1999, Chapter 14 therein). The atmospheres of the Jovian planets (Jupiter, Saturn, Uranus, and Neptune) also have high pressures, and the gases may behave as non-ideal. This may influence cloud formation on these planets.

For derivation of the van der Waals equation, consider a monatomic gas—a generalization for the polyatomic gases can be derived in a similar way (Landau and Lifshitz, v. 5, 1958b). The motion of its molecules can be considered as classical and its energy can be written as a sum of the kinetic energies $p_m^2/2m$ and the potential energy U of the molecular interactions:

$$E(p,q) = \sum_{i=1}^{N} \frac{p_{mi}^2}{2m} + U, \tag{3.4.19}$$

where p_m and q denote the momentum and coordinates of a molecule, and N is the number of molecules. For a monatomic gas, U is a function of only the distances between the atoms.

The statistical integral $\int e^{-E(p_M,q)/kT}\, dp_m dq$ in the Equation (3.4.5) for the free energy F becomes a product of the integral over the atomic momentum p_m and the integral over the coordinates, which has the form

$$\int \ldots \int e^{-U/kT}\, dv_1 \ldots dv_N, \tag{3.4.20}$$

where integration over each volume $dV = dx_a dy_a dz_a$ is performed over all the volume occupied by the gas. For an ideal gas, $U = 0$, and this integral would be equal to V^N. It is clear then that evaluation of the free energy F from Equation (3.2.8) would give an expression

$$F = F_{id} - kT \ln\left(\frac{1}{V^N} \int \ldots \int e^{-U/kT}\, dV_1 \ldots dV_N\right). \tag{3.4.21}$$

Adding and subtracting 1 in the integrand and using the relation $\int ... \int dV_1 ... dV_N = V^N$, we can rewrite (Eqn. 3.4.21) as

$$F = F_{id} - kT \ln \left[\frac{1}{V^N} \int ... \int (e^{-U/kT} - 1)\, dV_1 ... dV_N + 1 \right]. \quad (3.4.22)$$

We assume further that the density of the gas is sufficiently small, so that only the collisions between the two molecules are possible, while the collisions among the three and greater number of molecules can be neglected. Interactions between the atoms become substantially strong and the integrand in (Eqn. 3.4.22) is nonzero only when the two atoms are sufficiently close to each other. The number of such couples of the atoms is $C_N^2 = N(N - 1)/2$. They give the non-zero contribution into the integral in (Eqn. 3.4.22), which therefore can be written as

$$\int ... \int (e^{-U_{12}/kT} - 1)\, dV_1 ... dV_N, \quad (3.4.23)$$

where U_{12} is the energy of interaction of the two atoms. U_{12} depends on the coordinates of any two atoms, and we can integrate over the coordinates of all the other $(N - 2)$ atoms, which yields $V^{(N-2)}$. Since N is very large, we can assume that $N(N - 1) \approx N^2$.

Then, we can write (Eqn. 3.4.22) as

$$F = F_{id} - kT \ln \left[\frac{N^2}{2V^2} \int ... \int (e^{-U_{12}/kT} - 1)\, dV_1\, dV_2 + 1 \right], \quad (3.4.23a)$$

where $dV_1 dV_2$ is a product of volume differentials of the first two atoms. We assumed that the gas density is small, $N/V \ll 1$, and the first term under the logarithm sign is much smaller than 1 due to the factor $(N/V)^2$. Using the relation $\ln(1 + x) \approx x$ at $x \ll 1$, we can write (Eqn. 3.4.23a) as

$$F = F_{id} - kT \frac{N^2}{2V^2} \int ... \int \left(e^{-U_{12}/kT} - 1\right) dV_1 dV_2. \quad (3.4.24)$$

The potential energy U_{12} is a function only of the distance between the atoms—i.e., of the difference of their coordinates. We can perform a transformation of the coordinates, and instead of the coordinates of each atom we introduce the coordinate of their center of inertia R_{12} and mutual distance r_{12}. Thus, U_{12} will depend only on r_{12}, and we denote its differential as dV. We can integrate over R_{12}, which again gives V, and we finally obtain from (Eqn. 3.4.24)

$$F = F_{id} + \frac{N^2 kT}{V} B_W(T), \quad (3.4.25)$$

$$B_W(T) = \frac{1}{2} \int (1 - e^{-U_{12}/kT})\, dV. \quad (3.4.26)$$

The second term in (Eqn. 3.4.25) is a correction for non-ideal gas behavior, and these expressions represent a platform for derivation of the van der Waals equation.

A schematic picture of the potential energy $U_{12}(r)$ versus distance r between the atoms for a monatomic gas is given in Fig. 3.2. It shows that $U_{12}(r)$ is positive and very large at $r < 2r_0$ (where the repulsive forces dominate), passes zero at $r = 2r_0$, and becomes negative at $r > 2r_0$ (where attractive forces dominate), and so reaches a minimum $-U_0$ at a larger r and tends to zero at a very large r. $U_{12}(r)$ increases rapidly, with decreasing r at $r < 2r_0$ and forms almost a potential wall, thereby prohibiting mutual penetration of the atoms. Therefore, r_0 can be interpreted as the atomic radius. Equations (3.4.25) and (3.4.26) make sense if the integral B_W converges—that is, if $U_{12}(r)$ decreases

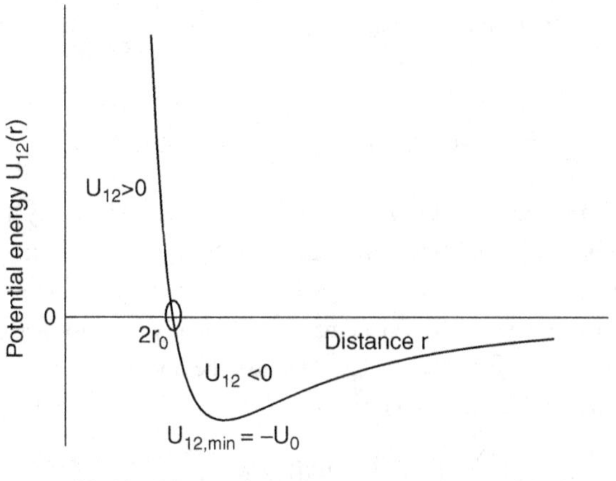

Figure 3.2. Potential energy $U_{12}(r)$ of interaction of two atoms of a non-ideal gas as a function of the distance r between the atoms. $U_{12}(r)$ is positive and large at $r < 2r_0$, $U_{12}(r) = 0$ at the point $r = 2r_0$ (denoted by the ellipse) becomes negative and reaches a minimum U_0 at a larger r and tends to zero at a very large r.

fast enough with r. If $U_{12}(r)$ decreases asymptotically as $U_{12}(r) \sim 1/r^n$, then B_W converges for $n > 3$. This occurs since the electrical forces between the electrically neutral atoms and the potential energy of their interaction are often approximated as $U_{12}(r) = a_u/r^n - b_u/r^6$, where the first term describes repulsion due to interaction of the two atomic electronic shells at small distances that grows with decreasing radius as $1/r^n$, and the second term describes attraction at large distances, which arises due to interaction of the electrical dipole moments and decrease at large r as $U_{12}(r) \sim 1/r^6$ (Landau and Lifshitz, v. 3, 1958a; Born, 1963).

Further derivation is based on the transformation of the integral $B_W(T)$. We assume that $U_0/kT << 1$, and since $U_{12}(r)$ depends only on r and does not depend on the angles, we can integrate over the angles, write in spherical coordinates $dV = 4\pi r^2 dr$ and present $B_W(T)$ as a sum of the two integrals:

$$B_W(T) = \frac{1}{2}\int (1 - e^{-U_{12}(r)/kT})\,dV = I_{B1} + I_{B2} \tag{3.4.27}$$

$$I_{B1} = 2\pi \int_0^{2r_0} (1 - e^{-U_{12}(r)/kT})r^2 dr, \quad I_{B2} = 2\pi \int_{2r_0}^{\infty} (1 - e^{-U_{12}(r)/kT})r^2 dr \tag{3.4.28}$$

In the integral I_{B1}, $U_{12}(r)$ is very large (see Fig. 3.2), thus we can assume $U_{12}(r)/kT >> 1$ and neglect $\exp(-U_{12}(r)/kT)$. Then, I_{B1} is simply evaluated as

$$I_{B1} = \frac{16\pi}{3} r_0^3 = 4v_0 = b. \tag{3.4.29}$$

Thus, $I_{B1} = b$ is four times greater than the volume of an atom $v_0 = (4/3)\pi r_0^3$. In the second integral, I_{B2}, the value $U_{12}(r)$ does not exceed U_0, and, as we assumed $U_0/kT << 1$, we have a condition $|U_{12}(r)|/kT << 1$. Then, we can expand into the power series $\exp(-U_{12}(r)/kT)$, and keeping only the first two terms, we have $\exp(-U_{12}(r)/kT) \approx 1 - U_{12}(r)/kT$, so substitution into I_{B2} yields

$$I_{B2} \approx \frac{2\pi}{kT} \int_{2r_0}^{\infty} U_{12}(r) r^2 dr = -\frac{a}{kT}, \qquad a = -2\pi \int_{2r_0}^{\infty} U_{12}(r) r^2 dr. \tag{3.4.30}$$

The minus sign is chosen because $U_{12}(r) < 0$ in the integrand and the integral I_{B2} is negative. Substituting Equations 3.4.27–3.4.30 for $B_W(T)$ into (Eqn. 3.4.25) for F, we obtain

$$F = F_{id} + \frac{N^2}{V}(kTb - a). \tag{3.4.31}$$

Substitution of Equation (3.4.12) for the free energy of the ideal gas F_{id} yields

$$F = Nf(T) - kTN\ln\frac{e}{N} - kTN\left\{\ln\left[V\left(\frac{2\pi mkT}{h^2}\right)^{3/2}\right] - \frac{Nb}{V}\right\} - \frac{N^2 a}{V}. \tag{3.4.32}$$

This equation still does not account for the limited compressibility of the gas; it allows $V \to 0$ instead of some minimum V_{min} determined by the volume of all molecules. This equation can be modified as follows. In the preceding derivation, we assumed that although the gas is non-ideal, its density is so small that the triple, quadruple, and higher order collisions can be neglected. This is equivalent to the statement that the total volume occupied by the molecules Nv_0 is much smaller than the gas volume V (i.e., $Nv_0 << V$). Since $b = 4v_0$, this condition is equivalent to the condition $x = Nb/V << 1$. For small x, an approximate relation can be used $x \approx \ln(1 + x)$, and we can transform the expression in (Eqn. 3.4.32) as

$$\begin{aligned}\ln\left[V\left(\frac{2\pi mkT}{h^2}\right)^{3/2}\right] - \frac{Nb}{V} &\approx \ln\left[V\left(\frac{2\pi mkT}{h^2}\right)^{3/2}\right] + \ln\left(1 - \frac{Nb}{V}\right)\\ &= \ln\left[\left(\frac{2\pi mkT}{h^2}\right)^{3/2}(V - Nb)\right].\end{aligned} \tag{3.4.33}$$

Substituting this relation into (Eqn. 3.4.32), we obtain F as

$$\begin{aligned}F &= Nf(T) - kTN\ln\left[\frac{eV}{N}\left(\frac{2\pi mkT}{h^2}\right)^{3/2}\left(1 - \frac{Nb}{V}\right)\right] - \frac{N^2 a}{V}\\ &= \left\{-kTN\ln\left[\frac{eV}{N}\left(\frac{2\pi mkT}{h^2}\right)^{3/2}\right] + Nf(T)\right\} - kTN\ln\left(1 - \frac{Nb}{V}\right) - \frac{N^2 a}{V}.\end{aligned} \tag{3.4.34}$$

The first two terms here in the figure parentheses represent F_{id} according to (Eqn. 3.4.12), and finally

$$F = F_{id} - kTN\ln\left(1 - \frac{Nb}{V}\right) - \frac{N^2 a}{V}. \tag{3.4.35}$$

This equation already has the property of final compressibility, the term under the logarithm sign should be positive, thus $V > Nb$. Equation (3.4.35) is also used in terms of the number of moles n. Using the relations $n = N/N_{Av}$, $k = R/N_{av}$, and the notations $a_m = aN_{Av}^2$, $b_m = bN_{Av}$, we obtain

$$F = F_{id} - RTn\ln\left(1 - \frac{nb_m}{V}\right) - \frac{n^2 a_m}{V}. \tag{3.4.36}$$

This gives for one mole

$$F_m = F_{id} - RT\ln\left(1 - \frac{b_m}{V}\right) - \frac{a_m}{V}. \tag{3.4.37}$$

Equations like (Eqns. 3.4.35) and (3.4.37) are used in modifications of the free energy and van der Waals equation for liquids (see Chapter 4). The gas pressure can now be evaluated from (Eqn. 3.4.35):

$$p = -\frac{\partial F}{\partial V} = -\frac{\partial F_{id}}{\partial V} + \frac{kTN}{V}\frac{Nb}{V - Nb} - \frac{N^2 a}{V^2}. \tag{3.4.38}$$

Substituting here Equation (3.4.13) for the ideal gas, $p_{id} = -\partial F_{id}/\partial V = kTN/V$, and rearranging the terms, we obtain

$$p = \frac{kTN}{V}\left(1 + \frac{Nb}{V - Nb}\right) - \frac{N^2 a}{V^2} = \frac{kTN}{V - Nb} - \frac{N^2 a}{V^2}. \tag{3.4.39}$$

Or, finally,

$$\left(p + \frac{N^2 a}{V^2}\right)(V - Nb) = kTN. \tag{3.4.40}$$

This is the *van der Waals equation of state* for the real gas. This equation is also used in terms of the number of moles n,

$$\left(p + \frac{n^2 a_m}{V^2}\right)(V - nb_m) = RTn, \tag{3.4.41a}$$

where a_m and b_m are the same as defined after (Eqn. 3.4.35). For one mole, we have

$$\left(p + \frac{a_m}{V^2}\right)(V - b_m) = RT. \tag{3.4.41b}$$

The van der Waals equation of state generalizes the equation of state for an ideal gas (Eqn. 3.4.15) and converts into it for $a_m = 0$, $b_m = 0$. These are two corrections to the equation of state for the ideal gas. The pressure correction accounts for the pressure reduction $\delta p = a_m/V^2$ due to interaction of the gas molecules (the van der Waals forces); the volume correction accounts for the available volume decrease $\delta V = -b_m$ due to the finite volume occupied by all gas molecules, called *the excluded volume*. Both pressure and volume corrections were expressed in (Eqns. 3.4.29) and (3.4.30) via the parameters of potential energy of intermolecular interactions.

In practice, the parameters a_m and b_m are determined from measurements in order to reach the best agreement of predictions of the van der Waals equation with experimental quantities. For water, $a_m = 5.26$ cm^3 atm mole^{-2}, and $b_m = 30.5$ cm^3 mole^{-1}. The corresponding values of b_m for the major air components, oxygen and nitrogen, are comparable, $\approx$32.6 and $\approx$38.5 cm^3 mole^{-1}, but the values of a_m are almost 4 times smaller: 1.36 and 1.34 cm^3 atm mole^{-2}. This indicates that the excluded volume only slightly varies among different gases. Calculations show that the molecular volumes of many gases vary between $v_0 \sim 1.7$ and 4.9 (Å)^3. Variation of the intermolecular forces characterized by a_m is much greater.

Many modifications and generalizations of the van der Waals equation are used for studies of liquids and solids, in particular for liquid water and ice. Some applications for water will be considered in Chapter 4.

3.5. Thermodynamics of Solutions

All these processes of drop and crystal nucleation substantially depend on the particles' chemical composition and concentrations of the solutions. The concentrations are characterized by the *mole fractions and molalities*. In general, there can be several solutes in a solution, and the mole fraction x_{kl} of the substance k in solution is

$$x_{kl} = \frac{n_{kl}}{n_{solv} + n_{kl}}, \qquad (3.5.1)$$

where n_{solv} and n_{kl} are the number of moles of solvent and the k-th solute in solution. The mole fractions of solute x_s and water x_w in aqueous solution are defined via the corresponding number of moles n_s and n_w in solution:

$$x_w = \frac{n_w}{n_w + n_s}, \qquad x_s = \frac{n_s}{n_w + n_s}. \qquad (3.5.2)$$

The molality $\hat{M}$ is defined as the number of moles of salt dissolved in 1000 g of water. Then, $n_w = 1000/M_w$, with M_w being the molecular weight of water, $n_s = \hat{M}$, and

$$x_s = \frac{n_s}{n_s + n_w} = \frac{\hat{M}}{\hat{M} + 1000/M_w}. \qquad (3.5.3)$$

The mole fraction x_k of the k-th gas in the mixture of gases with the total pressure p is determined from *Dalton's law* as

$$p_k = p x_k. \qquad (3.5.4)$$

In an "ideal" solution without interaction between the molecules of solvent and solute, *Raoult's law* acts, which states that the equilibrium vapor pressure p_k is proportional to the mole fraction x_{kl} of this substance in solution (its derivation is given in Section 3.10),

$$p_k = x_{kl} p_{k0}, \qquad (3.5.4a)$$

where pk_0 is the partial vapor pressure of component k over a pure liquid of k-th component. Obviously, if only this pure component exists in the system, $x_{kl} = 1$ and $p_k = pk_0$.

Now we can proceed to the derivation of the chemical potentials in solutions. From Equation (3.1.21) of Section 3.1, we have for the molar chemical potential μ_{vk} and volume v_k of an ideal gas k (vapor phase, subscript "v"):

$$\left(\frac{\partial \mu_{kv}}{\partial p}\right)_{T, n_{k \neq i}} = v_k. \qquad (3.5.5)$$

Substituting here $v_k = RT/p_k$ from the equation of state (Eqn. 3.4.15a), we obtain

$$\left(\frac{\partial \mu_{kv}}{\partial p}\right)_{T, n_{k \neq i}} = \frac{RT}{p_k}. \qquad (3.5.6)$$

Integration of this equation for μ_{kv} yields

$$\mu_{kv}(p_k, T) = \mu_{kv0}(T) + RT \ln p_k. \qquad (3.5.7)$$

Substituting here Raoult's law (Eqn. 3.5.4a) for p_k, we obtain for the vapor phase

$$\mu_{kv}(p_k,T) = \mu_{kv0}(T) + RT\ln p_{k0} + RT\ln x_k\,. \tag{3.5.8}$$

When a gas and liquid are in equilibrium, the chemical potentials of the k-th component should be equal in both phases, otherwise this component will be redistributed between the phases until equilibrium is reached. Therefore, the chemical potential of the k-th substance in the liquid (subscript "l") with the partial pressure p_k will be

$$\mu_{kl}(p_k,T) = \mu_{kl0}(T) + RT\ln p_k\,. \tag{3.5.9}$$

The value of p_k in a solution is determined by Raoult's law. Substituting (Eqn. 3.5.4a) for Raoult's law into (Eqn. 3.5.9), we obtain

$$\mu_{kl}(p_k,T,x_{kl}) = \mu_{k0}(p,T) + RT\ln x_{kl}\,, \tag{3.5.10}$$

where $\mu_{k0}(p, T)$ corresponds to zero x_{kl}—that is, it depends on p and T, but not on x_{kl}. The solutions described by (Eqn. 3.5.10) are called *ideal solutions*. The chemical potential of water in an aqueous solution can be written according to (Eqn. 3.5.10) as

$$\mu_w(p,T,x_w) = \mu_{w0}(p,T) + RT\ln x_w\,. \tag{3.5.11}$$

Most dilute non-electrolyte solutions are satisfactorily described by Equation (3.5.11).

However, solute present in deliquescent aerosol or in the drops may be salts or acids that dissociate into several ions. There are several methods to account for the degree of ionic dissociation of the solutions. One way is introduction of the van't Hoff factor i as

$$x_{kl} = \frac{n_{kl}}{n_{solv} + i n_{kl}}\,. \tag{3.5.12}$$

The van't Hoff factor i should account for the effective increase or decrease of the number of ions in a dissociated solution and was applied in earlier studies on cloud physics. This factor was introduced in van't Hoff's equation for osmotic pressure. The disadvantages of this factor are that the values of i are known for very few substances, and its effective values increase or vary non-monotonically with decreasing solution molality (see Table 1 in Low, 1969). For example, for NaCl, a component of the marine aerosol and cloud condensation nuclei, the number of ions is $\nu = 2$, while $i \neq 2$, but $i = 1.867$ at molality $\hat{M} = 5$, and $i = 2.659$ at molality $\hat{M} = 0.1$.

A more accurate modern approach is based on application of the concept of the *activity of water* a_w and the *molal or practical osmotic potential* Φ_s, which are related as

$$a_w = \exp\left(-\nu\Phi_s \frac{m_s M_w}{m_w M_s}\right) = \exp\left(-\nu\Phi_s \frac{n_s}{n_w}\right), \tag{3.5.13}$$

where ν is the number of potential ions in a salt or acid, m_s and M_s are the mass and molecular weight of the soluble fraction, and m_w is the mass of water in the solution. The water activity a_w is also expressed as $a_w = k_w x_w$, where k_w is the *rational activity coefficient*. The values of a_w, Φ_s, and k_w can be determined from the measurements and the tabulated data are available (e.g., Low, 1969; Robinson and Stokes, 1970; Pruppacher and Klett, 1997). In general, calculation of a_w in an aqueous solution drop is not a simple process and is often accompanied by calculations of physico-chemical properties of solutions and particles growth.

After deliquescence and hygroscopic or diffusion growth, the aerosol particle becomes a solution drop with the mass m_{aw}, volume V_{aw}, and radius r_{aw}. The drop gains a mass of water m_w with the volume V_w. Evaluation of water activity and many other related aerosol properties (e.g., hygroscopic growth, activation into cloud drops) is based on (Eqn. 3.5.13) and is often performed using two approximations.

One approximation is based on the simplifying assumptions on the constancy of the water molar volume in solution and additivity of the volume V_w of solvent (water) and the original volume V_d of the dry particle, including solute (e.g., salt or acid) and insoluble core. These are good approximations for many common ionic solutes (Dufour and Defay, 1963; Pruppacher and Klett, 1997; Brechtel and Kreidenweis, 2000a,b). Using the additivity property, the volume of the solution droplet V_{aw} can be written as $V_{aw} = V_w + V_d$. Then, the volume V_w and mass m_w of water are

$$V_w = V_{aw} - V_d, \tag{3.5.14}$$

$$m_w = \rho_w (V_{aw} - V_d) = (4/3)\pi\rho_w (r_{aw}^3 - r_d^3). \tag{3.5.15}$$

Substituting this m_w into (Eqn. 3.5.13), we obtain

$$a_w = \exp\left(-\frac{3\nu\Phi_s M_w}{4\pi M_s \rho_w} \frac{m_s}{(r_{aw}^3 - r_d^3)} \right) = \exp\left(-\frac{B}{r_{aw}^3 - r_d^3} \right), \tag{3.5.16}$$

where B is a parameter that describes the physico-chemical properties of the aerosol

$$B = \frac{3\nu\Phi_s M_w m_s}{4\pi M_s \rho_w}. \tag{3.5.17}$$

The parameter B is called the activity of a nucleus. It describes the effects of the soluble fraction of an aerosol particle or cloud condensation nucleus (CCN) and determines the activity of this CCN for activation of a cloud droplet (see Chapters 6 and 7).

In the other approximation, used for fully soluble particles with dry radius r_s, the mass of water m_w is calculated as the difference between the total mass of solution $m_{aw} = (4/3)\pi\rho_s'' r_{aw}^3$, where ρ_s'' is the solution density, and the mass of the soluble fraction $m_s = (4/3)\pi r_s r_s^3$—i.e., $m_w = (4/3)\pi(\rho_s'' r_{aw}^3 - \rho_s r_s^3)$—and water activity is

$$a_w = \exp\left(-\frac{\nu\Phi_s M_w m_s}{M_s[(4/3)\pi\rho_s'' r_{aw}^3 - m_s]} \right). \tag{3.5.18}$$

For sufficiently diluted solutions, $m_w >> m_s$, and $\rho_s'' \approx \rho_w$, this expression is simplified as

$$a_w = \exp\left(-\frac{3\nu\Phi_s M_w m_s}{4\pi M_s \rho_w r_{aw}^3} \right) = \exp\left(-\frac{B}{r_{aw}^3} \right), \tag{3.5.19}$$

which is a particular case of (Eqn. 3.5.16) with $r_d = 0$.

After introducing a_k, the dependence of μ_k on the chemical composition can be rewritten with a_k instead of x_{kl}

$$\mu_k(p,T,a_k) = \mu_{k0}(p,T) + RT\ln a_k. \tag{3.5.20}$$

In particular, for water,

$$\mu_w(p,T,a_k) = \mu_{w0}(p,T) + RT\ln a_w. \tag{3.5.21}$$

Another characteristic of solutions used in calculations of their chemical and thermodynamic properties is the weight concentration or weight percent w of solute defined as

$$w = \frac{m_s}{m_s + m_w} \cdot 100\%, \tag{3.5.22}$$

where m_s and m_w are the masses of solute and water in solution. Molality $\hat{M}$ can be related to the weight concentration w, the water saturation ratio S_w, molecular weights of solute M_s and water M_w, and the number of ions ν_s and osmotic potential Φ_s of solvent as described earlier, and can be expressed in the forms (e.g., Atkins, 1982; Martin, 2000)

$$\hat{M} = \frac{10^3 m_s}{M_s m_w}, \quad \hat{M} = \frac{10w}{M_s(1 - w/100)}, \quad \hat{M} = \frac{10^3(1 - S_w)}{\nu\Phi_s M_w}. \tag{3.5.23}$$

Using these relations, w can be expressed in terms of S_w:

$$w = \frac{\hat{M}M_s}{\hat{M}M_s + 10^3} 100\ \% = \frac{M_s(1 - S_w)}{M_s(1 - S_w) + \nu_s\Phi_s M_w} 100\ \%. \tag{3.5.24}$$

Another useful representation of w via mole fraction x_s is

$$w = \frac{M_s x_s}{M_w + x_s(M_s - M_w)} 100\%. \tag{3.5.25}$$

3.6. General Phase Equilibrium Equation for Solutions

3.6.1. General Equilibrium Equation

It is convenient to study phase equilibrium in a thermodynamic system using the quantity μ_k/T. An equation for phase equilibrium can be obtained from Equation (3.5.20) for chemical potential dividing by T and taking the differential

$$\begin{aligned} d\left(\frac{\mu_k(p,T,a_k)}{T}\right) &= d\left(\frac{\mu_{k0}(p,T)}{T}\right) + R d\ln a_k \\ &= \frac{\partial(\mu_{k0}/T)}{\partial T} dT + \frac{\partial(\mu_{k0}/T)}{\partial p} dp + R d\ln a_k. \end{aligned} \tag{3.6.1}$$

It was shown in Section 3.1 that the first derivative here is $\partial/\partial T(\mu_{k0}/T) = -h_{k0}/T^2$, and the second derivative is $\partial/\partial p(\mu_{k0}/T) = v_{k0}/T$, where h_{k0} and v_{k0} are the enthalpy and molar volume of the pure component k. Using these relations, we obtain

$$d\left(\frac{\mu_k}{T}\right) = -\frac{h_{k0}}{T^2} dT + \frac{v_{k0}}{T} dp + R d\ln a_k. \tag{3.6.2}$$

Note that the subscript "0" here relates to the pure component k.

Now we consider an equilibrium thermodynamic system consisting of a spherical germ of the bulk phase (2) of k-th component with the chemical potential $\mu_{k,2}$ and radius r_2 embedded within a k-th component of spherical bulk phase (1) with chemical potential $\mu_{k,1}$ and radius r_1, with phase (1) itself being in equilibrium with the environment. An example is formation of an ice germ with radius r_2 (phase 2) within a supercooled liquid drop (phase 1) with radius r_1, which is in equilibrium with the environmental air.

If phase 2 is a solid (e.g., ice or salt crystal) that nucleates heterogeneously on the surface of a foreign substrate, an additional term should be added to $\mu_{k,2}$, which arises from the elastic strain ε caused by the misfit between ice and substrate lattices (Turnbull and Vonnegut, 1952; Young, 1993). This strain increases the density of activation energy by $C_\varepsilon \varepsilon^2$ [erg cm^{-3}] where $C_\varepsilon = 1.7 \times 10^{11}$ erg cm^{-3} is the Turnbull-Vonnegut constant (Turnbull and Vonnegut, 1952), and ε is measured in percent. The term in $\mu_{k,2}/T$ associated with the elastic strain can be formulated as $M_k C_\varepsilon \varepsilon^2/(T\rho_{k2})$ (Khvorostyanov and Curry, 2004a,b) with M_k and ρ_{k2} being the molecular weight and the density of phase 2.

Both phases are assumed to be in equilibrium, thus their temperatures and chemical potentials are equal, $T_1 = T_2$, $\mu_{k,1} = \mu_{k,2}$, and

$$d\left(\frac{\mu_{k,1}}{T_1}\right) = d\left(\frac{\mu_{k,2}}{T_2}\right). \tag{3.6.3}$$

Substituting here (Eqn. 3.6.2) for both phases, we can write

$$-\frac{(h_{k0,1} - h_{k0,2})}{T^2} dT + \frac{v_{k0,1}}{T} dp_1 - \frac{v_{k0,2}}{T} dp_2 - \frac{M_k}{T} d\left(\frac{C_\varepsilon \varepsilon^2}{\rho_{k,2}}\right) + Rd\ln\left(\frac{a_{k0,1}}{a_{k0,2}}\right) = 0 . \tag{3.6.4}$$

This is the general equilibrium or entropy equation that determines the equilibrium in the system. The pressures p_1 and p_2 in these bulk phases are not equal, but obey Laplace's equation or the conditions of mechanical equilibrium:

$$p_1 = p + \frac{2\sigma_{1e}}{r_1}, \qquad p_2 = p_1 + \frac{2\sigma_{12}}{r_2}, \tag{3.6.5}$$

where σ_{1e} is the surface tension at the boundary between phase 1 and the environment (e.g., between the drop and the air) and σ_{12} is the surface tension at the boundary between phases 1 and 2 (e.g., the liquid-ice interface), and p is the pressure in the environment (e.g., air).

The difference of molar enthalpies can be expressed via the latent heat of phase transition between phases 1 and 2 as

$$h_{k0,1} - h_{k0,2} = \hat{L}_{12} = L_{12} M_k, \tag{3.6.6}$$

where $\hat{L}_{12}$ and L_{12} are the molar and specific latent heats and M_k is the molecular weight of component k. Substituting this relation and the Laplace's relations (Eqn. 3.6.5) for pressure into (Eqn. 3.6.4), we obtain

$$-\frac{\hat{L}_{12}}{T^2} dT + \frac{(v_{k0,1} - v_{k0,2})}{T} d\left(p + \frac{2\sigma_{1e}}{r_1}\right) - \frac{v_{k0,2}}{T} d\left(\frac{2\sigma_{12}}{r_2}\right)$$
$$- \frac{M_k}{T} d\left(\frac{C_\varepsilon \varepsilon^2}{\rho_{k,2}}\right) + Rd\ln\left(\frac{a_{k,1}}{a_{k,2}}\right) = 0. \tag{3.6.7}$$

This equation of the general equilibrium slightly generalizes the previous equation given in Dufour and Defay (1963) and Pruppacher and Klett (1997), and can be further generalized by taking into account additional affects.

It can be convenient to use this equation in a slightly different form, with the densities ρ_{k0} instead of the molar volumes v_{k0}. Substituting into (Eqn. 3.6.7) the relation $v_{k0} = M_k/\rho_{k0}$, with M_k being the molecular weight of the k-th component, dividing by M_k and multiplying by T, we obtain

$$-\frac{L_{12}(T)}{T}dT + \left(\frac{2}{\rho_{k0,1}} - \frac{2}{\rho_{k0,2}}\right)dp + \left(\frac{2}{\rho_{k0,1}} - \frac{2}{\rho_{k0,2}}\right)d\left(\frac{\sigma_{1e}}{r_1}\right)$$
$$-\frac{2}{\rho_{k0,2}}d\left(\frac{\sigma_{is}}{r_2}\right) - d\left(\frac{C_\varepsilon \varepsilon^2}{\rho_{k0,2}}\right) + \frac{RT}{M_k}d\ln\left(\frac{a_{k,1}}{a_{k,2}}\right) = 0. \tag{3.6.8}$$

It is important to emphasize that, although we began with the chemical potentials for solutions, the final quantities (the latent heats, molar volumes, and densities) relate to the pure component k in the mixture, which is indicated by the subscript "0". These quantities for pure substances are independent on solution concentrations, which makes calculations much easier. Equations (3.6.7), (3.6.8) describe a large variety of situations pertinent to cloud processes. Many of them are considered in Dufour and Defay (1963), Defay, Prigogine, and Bellemans (1966), and Pruppacher and Klett (1997). In this chapter, we will consider the most important processes and laws: the Clausius–Clapeyron equation, the Kelvin equation, and Köhler's equation. These three equations are required for derivation of the supersaturation equations that describes cloud kinetics. Equations (3.6.7) and (3.6.8) will be used in Chapters 8, 9, 10 and 11 for derivations of the critical radii, energies, and nucleation rates of homogeneous and heterogeneous nucleation processes.

3.6.2. The Gibbs–Duhem Relation

Equations (3.6.7) and (3.6.8) describe the relations between the properties of the two phases of the same substance. However, in many cases it is desirable to use relations between the properties of two different substances—e.g., between the solvent and solute. This can be done using the Gibbs–Duhem relation described in Section 3.1, which is written for an isothermal ($T = \text{const}$) and isobaric ($p = \text{const}$) process as

$$\sum_{k=1}^{c} x_k d\mu_k = 0. \tag{3.6.9}$$

The differential $d\mu_k$ follows from Equation (3.6.2), where the first two terms on the right-hand side vanish for such a process, and the rest is

$$d\mu_k = RTd\ln a_k. \tag{3.6.10}$$

Substitution of (Eqn. 3.6.9) into (3.6.10) yields the Gibbs–Duhem relation in the form

$$\sum_{k=1}^{c} x_k d\ln a_k = 0. \tag{3.6.11}$$

In particular, for an aqueous solution that contains only two components ($c = 2$), water and solute, this relation is

$$x_s d\ln a_s = -x_w d\ln a_w. \tag{3.6.12}$$

This equation allows determination of solute activities and concentrations with measurements of the water activities or relative humidities over solutions of various concentrations. It has been used in studies of aerosol liquid-solid phase transitions, as will be described in Chapter 11.

3.7. The Clausius–Clapeyron Equation

An important application of the general equilibrium Equation (3.6.7) is the temperature-pressure relations for the three systems: liquid-ice, vapor-liquid, and vapor-ice.

3.7.1. Equilibrium between Liquid and Ice Bulk Phases

Consider the equilibrium of the water and ice bulk phases, where phase 1 is water (subscript "w"), and phase 2 is ice ("i"). The radii $r_1 = r_2 = \infty$ for the bulk substances, the phase transition is melting, and the difference of the molar enthalpies is the molar melting latent heat, $h_w - h_i = \hat{L}_m = M_w L_m$, with L_m being the specific melting heat [cal g^{-1}], the molar volumes are those of liquid, $v_{k0,1} = v_w = M_w/\rho_w$, and of ice, $v_{k0,2} = v_i = M_w/\rho_i$. No foreign substance is present, and the misfit strain is absent, $\varepsilon = 0$. No salt is present and both activities $a_{k,1} = a_{k,2} = 1$. Under these conditions, (Eqn. 3.6.7) is

$$-\frac{\hat{L}_m}{T^2}dT + \frac{(v_w - v_i)}{T}dp = 0\,. \tag{3.7.1}$$

The number of components in this system is $c = 1$, the number of phases $\varphi = 2$, and, according to the phase rule (Eqn. 3.3.6), the variance $N_w = 2 + 1 - 2 = 1$. We have one independent variable and can study dependence of the melting pressure p_m on the temperature T. Thus, the general equilibrium Equation (3.7.1) reduces to the *Clapeyron equation* used in the following forms

$$\frac{dp_m}{dT} = \frac{h_w - h_i}{T(v_w - v_i)} = \frac{\Delta S_{\eta wi}}{\Delta v_{wi}} = \frac{M_w L_m}{T(v_w - v_i)} = \frac{L_m}{T(1/\rho_w - 1/\rho_i)}, \tag{3.7.2}$$

where the second equality makes use of the relation for the enthalpies h and entropies S_η, $h_w - h_i = T(S_{\eta w} - S_{\eta i})$ that follows from (Eqn. 3.1.22). This equilibrium between liquid and ice can exist at the melting point with p_m, T_m, and this Clapeyron equation describes the variation of the melting pressure p_m with temperature, $p_m(T_m)$.

Rewritten as

$$\frac{dT_m}{dp_m} = \frac{T(v_w - v_i)}{M_w L_m} = \frac{T(1/\rho_w - 1/\rho_i)}{L_m}, \tag{3.7.3}$$

it describes variations of the melting temperature with pressure. The sign of this dependence depends on the difference in the molar volumes of liquid and solid. Most known substances contract at freezing, the relation of their liquid and solid volumes is $v_w > v_i$. Equation (3.7.3) shows that $dT_m/dp_m > 0$ in this case (i.e., the melting temperature increases with increasing pressure, and the slopes of the curves are positive on the p-T diagram). However, water is an anomalous substance. When liquid water freezes, thus forming the ordinary known hexagonal ice I_h, it expands, the ice I_h has a greater molar volume and smaller density than liquid water, and there is the reverse relation $v_i > v_w$, or $\rho_i < \rho_w$. Therefore, $dT_m/dp_m < 0$; thus, the melting temperature decreases with increasing pressure, and the slopes of the curves are negative on the p-T diagram. However, there are many other forms of different ices with various molar volumes, and various signs of dT_m/dp_m. These will be considered in Chapter 4.

3.7.2. Equilibrium of a Pure Water Drop with Saturated Vapor

We can apply again the general equilibrium Equation (3.6.7) to this system. In this case, phase 1 is water vapor; phase 2 is liquid water; the equilibrium exists with respect to evaporation or condensation; radius $r_1 = \infty$ for the vapor; the drop radius $r_2 = r$; pressure p is equal to the saturated vapor pressure $e_{ws}(r)$ over the drop (which depends on the drop radius); the difference in enthalpies Δh_{12} is the molar latent heat of evaporation $\hat{L}_{12}$ and is related to the specific evaporation heat L_e [cal g^{-1}] as $\hat{L}_{12} = h_v - h_w = M_w L_e$; the molar volumes are those of vapor, $v_{k0,1} = v_v$, and of liquid water, $v_{k0,2} = v_w = M_w/\rho_w$; and both activities $a_{k,1} = a_{k,2} = 1$. Thus, the general equilibrium Equation (3.6.7) is simplified as

$$-\frac{M_w L_e}{T} dT + (v_v - v_w) de_{ws}(r) - v_w d\left(\frac{2\sigma_{wv}}{r}\right) = 0, \tag{3.7.4}$$

where σ_{wv} is the surface tension at the water-vapor interface. The number of components in this system is $c = 1$, and according to the phase rule for the curved interfaces (Eqn. 3.3.12), the variance $N_w = 2$. So we have two independent variables. We will vary T and r and study dependence of vapor pressure on the temperature and the drop radius r. Equation (3.7.4) can be rewritten as

$$\frac{de_{ws}(r)}{dT} = \frac{M_w L_e}{T(v_v - v_w)} + \frac{v_w}{v_v - v_w} \frac{d}{dT}\left(\frac{2\sigma_{wv}}{r}\right), \tag{3.7.5}$$

where σ_{wv} is the surface tension at the water-vapor interface. The molar volume of vapor is much greater than the volume of condensed water, $v_v >> v_w$, and v_w can be neglected in the denominators. For v_v, we can write from the equation of state (Eqn. 3.4.16) $v_v = RT/e_{ws}$, and this equation becomes

$$\frac{de_{ws}(r)}{dT} = \frac{e_{ws} M_w L_e}{RT^2} + \frac{e_{ws} v_w}{RT} \frac{d}{dT}\left(\frac{2\sigma_{wv}}{r}\right). \tag{3.7.6}$$

This is the *generalized Clausius–Clapeyron equation*. It describes the temperature dependence of the saturated vapor pressure. The second term on the right-hand side contains r and represents a correction to de_{ws}/dT due to the curvature of the drop surface. It was estimated by Dufour and Defay (1963) to be $\sim 10^{-8}/r$ with r in cm. Thus, its contribution is less than 10% for $r > 10^{-7}$ cm, and less than 1% for $r > 10^{-6}$ cm. For larger drops or for the bulk water, the r-dependence can be neglected with the corresponding accuracy, so the rest of this equation is

$$\frac{1}{e_{ws}} \frac{de_{ws}}{dT} = \frac{d \ln e_{ws}}{dT} = \frac{M_w L_e}{RT^2} = \frac{L_e}{R_v T^2}, \tag{3.7.7}$$

where $R_v = R/M_w$ is the specific gas constant for the water vapor. This is the *Clausius–Clapeyron equation* for water vapor pressure saturated over a plane water surface that plays an important role in cloud physics. The application of the generalized Clausius–Clapeyron Equation (3.7.6) may be desirable for very small droplets—that is, nanoparticles.

The Clausius–Clapeyron Equation (3.7.7) is written for the saturated vapor pressure e_{ws}. In some applications, a modification of the Clausius–Clapeyron equation for the vapor density ρ_{ws} saturated over water is desired. It can be obtained expressing ρ_{ws} from the equation of state (Eqn. 3.4.17) as $\rho_{ws} = e_{ws}/(R_v T)$ and using the Clausius–Clapeyron equation. Thus,

$$\frac{d\rho_{ws}}{dT} = \frac{1}{R_v} \frac{d}{dT}\left(\frac{e_{ws}}{T}\right) = \frac{1}{R_v}\left(\frac{1}{T}\frac{de_{ws}}{dT} - \frac{e_{ws}}{T^2}\right)$$

$$= \frac{1}{R_v T}\left(\frac{e_{ws} L_e}{R_v T^2} - \frac{e_{ws}}{T}\right) = \frac{\rho_{ws}}{T}\left(\frac{L_e}{R_v T} - 1\right), \tag{3.7.7a}$$

where we again used the equation of state in the last equality.

3.7.3. Equilibrium of an Ice Crystal with Saturated Vapor

For this system, phase 1 is vapor (subscript "v"), phase 2 is ice ("i"), $r_1 = \infty$, $r_2 = r_c$ is the radius of an ice crystal (assuming that it has a spherical shape), the environmental pressure is the pressure e_{is} of vapor saturated over ice, and the enthalpies difference is $h_v - h_i = \hat{L}_s = M_w L_s$, where L_s is the specific latent heat of sublimation or deposition. We can again apply the general equilibrium Equation (3.6.7). As in the case with the drop earlier, the number of components in this system is $c = 1$, and according to the phase rule for the curved interfaces (Eqn. 3.3.12), the variance $N_w = 2$. So we have two independent variables. We will vary again T and r and study dependence of vapor pressure on the temperature and the crystal radius r. Just as with the vapor and droplet system earlier, starting from the general equilibrium Equation (3.6.7) we arrive at a similar generalized Clausius–Clapeyron equation for the vapor pressure over ice e_{is}

$$\frac{de_{is}(r_c)}{dT} = \frac{e_{is} M_w L_s}{RT^2} + \frac{e_{is} v_i}{RT} \frac{d}{dT}\left(\frac{2\sigma_{iv}}{r_c}\right), \tag{3.7.8}$$

where σ_{iv} is the surface tension at the ice-vapor interface. The second term on the right-hand side is also typically much smaller than the first term, and can be on the order of 10% for crystals of size ~10^{-7} cm. Its contribution can be noticeable—e.g., for small ice germs nucleating in water at very low temperatures. For larger crystals or for bulk ice, the second term can be ignored, and so we arrive at the Clausius–Clapeyron equation for the vapor pressure $e_{is}(T)$ saturated over a plane ice surface:

$$\frac{1}{e_{is}} \frac{de_{is}}{dT} = \frac{d \ln e_{is}}{dT} = \frac{M_w L_s}{RT^2} = \frac{L_s}{R_v T^2}. \tag{3.7.9}$$

A modification of the Clausius–Clapeyron equation for the vapor density ρ_{is} saturated over ice can be derived similarly to the derivation for ρ_{ws} just discussed. Expressing ρ_{is} from the equation of state (Eqn. 3.4.18) as $\rho_{is} = e_{is}/(R_v T)$ and using the Clausius–Clapeyron equation, we obtain

$$\frac{d\rho_{is}}{dT} = \frac{1}{R_v} \frac{d}{dT}\left(\frac{e_{is}}{T}\right) = \frac{1}{R_v}\left(\frac{1}{T}\frac{de_{is}}{dT} - \frac{e_{is}}{T^2}\right)$$
$$= \frac{1}{R_v T}\left(\frac{e_{is} L_s}{R_v T^2} - \frac{e_{is}}{T}\right) = \frac{\rho_{is}}{T}\left(\frac{L_s}{R_v T} - 1\right). \tag{3.7.9a}$$

Equations (3.7.7a) and (3.7.9a) are used in subsequent chapters for derivation of the particle growth rates and supersaturation equations.

The transition vapor-ice can be viewed as a sequence of transitions between vapor-liquid and liquid-ice. Then, from the energy or enthalpy conservation,

$$\hat{L}_s = M_w L_s = h_v - h_s = (h_v - h_w) + (h_w - h_s) = M_w (L_e + L_m), \tag{3.7.10}$$

we obtain a relation among the three melting heats of deposition, condensation, and melting:

$$\hat{L}_s = \hat{L}_e + \hat{L}_m, \qquad L_s = L_e + L_m. \tag{3.7.11}$$

This relation is a constraint that allows determination of any of the three quantities, L_s, L_e and L_m, if the other two are known.

The right-hand sides of Equations (3.7.7) and (3.7.9) are positive; therefore the saturated pressures e_{ws} and e_{is} increase with increasing temperatures. If it is assumed that L_e, L_s do not depend on temperature, Equations (3.7.7) and (3.7.9) can be easily integrated in the following manner:

$$e_{ws}(T) = e_{ws}(T_0)\exp\left[-\frac{L_e}{R_v}\left(\frac{1}{T}-\frac{1}{T_0}\right)\right] = e_{ws}(T_0)\exp\left(-\frac{L_e\Delta T}{R_v TT_0}\right), \tag{3.7.12}$$

$$e_{is}(T) = e_{is}(T_0)\exp\left[-\frac{L_s}{R_v}\left(\frac{1}{T}-\frac{1}{T_0}\right)\right] = e_{is}(T_0)\exp\left(-\frac{L_s\Delta T}{R_v TT_0}\right), \tag{3.7.13}$$

where T_0 is some reference point, which is usually the triple point 273.16 K, and $\Delta T = T_0 - T$. These equations can be used with sufficient accuracy in many cloud physics applications where the temperature change is sufficiently small.

However, in many problems, especially at low temperatures, variations of L_e, L_s with temperature become substantial and should be accounted for. To calculate the derivative dL_{12}/dt, we use the relations (Eqn. 3.1.22) from Section 3.1, which with the condition of equilibrium $\mu_1 = \mu_2$ give $\hat{L}_{12} = h_1 - h_2 = T(s_{\eta 1} - s_{\eta 2})$, and (Eqn. 3.1.20), $s_\eta = -(\partial\mu/\partial T)_p$, and a definition (Eqn. 3.1.18) of the molar heat capacity, $C_p = (\partial h/\partial T)_p = -T(\partial s_\eta/\partial T)_p$. Now we take the differential

$$\frac{d\hat{L}_{12}}{dT} = \frac{d(h_1 - h_2)}{dT} = \frac{\partial(h_1 - h_2)}{\partial T} + \frac{\partial(h_1 - h_2)}{\partial p}\frac{dp}{dT}. \tag{3.7.14}$$

The first term on the right-hand side is the difference of the heat capacities between the phases 1 and 2—i.e., $\partial\Delta h/\partial T = \Delta C_p = C_{p1} - C_{p2}$. The derivative $\partial h_k/\partial p$ in the second term can be transformed using (Eqn. 3.1.22) from Section 3.1, $h_k = \mu_k + Ts_{\eta k}$, giving us

$$\frac{\partial h_k}{\partial p} = \frac{\partial(\mu_k + TS_{\eta k})}{\partial p} = \frac{\partial\mu_k}{\partial p} + T\frac{\partial s_{\eta k}}{\partial p}. \tag{3.7.15}$$

Using the relation (Eqn. 3.1.21), we see that the first term on the right-hand side is v_k, and the second term can be expressed via μ_k, so

$$\frac{\partial h_k}{\partial p} = v_k + T\left(\frac{\partial^2\mu_k}{\partial T\partial p}\right)_p = v_k - T\frac{\partial}{\partial T}\left(\frac{\partial\mu_k}{\partial p}\right)_p = v_k - T\left(\frac{\partial v_k}{\partial T}\right)_p. \tag{3.7.16}$$

Collecting all the terms, we obtain

$$\frac{d\hat{L}_{12}}{dT} = (C_{p1} - C_{p2}) + [(v_1 - v_2) - T(v_1\kappa_{v1} - v_2\kappa_{v2})]\frac{dp}{dT}, \tag{3.7.17}$$

where k_{vk} is the isobaric compressibility defined as

$$\kappa_{vk} = \frac{1}{v_k}\left(\frac{\partial v_k}{\partial T}\right)_p. \tag{3.7.18}$$

The second term on the right-hand side of (Eqn. 3.7.17), proportional to dp/dT, is usually much smaller than the first term under typical cloud conditions. However, at low temperatures or at high pressures this term can give a noticeable correction to dL/dT and to the saturated vapor pressures. If the second term is neglected, we obtain

$$\frac{d\hat{L}_{12}}{dT} = C_{p1} - C_{p2}. \tag{3.7.19}$$

Dividing both sides by M_w yields the same relation for the specific latent heat L and capacity c_p. For equilibria among the three pairs of water phases, vapor, liquid and solid, this gives the relations for the corresponding specific latent heats and isobaric heat capacities known as *Kirchoff's equations*:

$$\frac{dL_e}{dT} = c_{pv} - c_{pw}, \qquad \frac{dL_s}{dT} = c_{pv} - c_{pi}, \qquad \frac{dL_m}{dT} = c_{pw} - c_{pi}, \tag{3.7.20}$$

where subscripts "v", "w", and "i" mean vapor, water, and ice.

Using Kirchoff's equations, $L_e(T)$, $L_s(T)$, and $L_m(T)$ can be obtained by integration of (Eqn. 3.7.20), then they can be substituted into the Clausius–Clapeyron equations to give

$$e_{ws}(T) = e_{ws}(T_0)\exp\left[-\int_T^{T_0} \frac{L_e(T)dT}{R_v T^2}\right], \tag{3.7.21}$$

$$e_{is}(T) = e_{is}(T_0)\exp\left[-\int_T^{T_0} \frac{L_s(T)dT}{R_v T^2}\right]. \tag{3.7.22}$$

The vapor pressures of water and ice coincide at the triple point $T_0 = 273.16$ K, where $e_{ws} = e_{is} =$ 6.11657 hPa $\approx$ 6.11 mb.

Variations of c_p with temperature are much smaller than that of L_{12} at moderate temperatures. However, variations of c_p may become important at low temperatures, when c_p decreases and tends to 0 as T approaches 0 Kelvin (Landau and Lifshitz, v.5, 1958b; Hobbs, 1974). This may influence the saturated vapor pressures. Reviews and reconstructions of the enthalpies, entropies, saturation pressures, heat capacities, and other quantities will be considered in Chapter 4. For most derivations in the next chapters, it is sufficient to use the Clausius–Clapeyron Equations (3.7.7) and (3.7.9) for bulk phases. In practical calculations used in numerical models or when analyzing measurement data, various empirical fits are usually used that account for these additional temperature dependencies. We will return to a discussion of the liquid-ice transitions in Chapter 4 devoted to the properties of water, and in Chapters 8 and 9 devoted to ice nucleation in clouds.

Consider now a situation when all three water phases are in equilibrium. The number of components $c = 1$, the number of phases $\varphi = 3$, and according to the phase rule (Eqn. 3.3.6) for the bulk phases, this system has $N_w = c + 2 - \varphi = 0$, zero of thermodynamics degrees of freedom. The condition of equilibrium is

$$\mu_v(T,p) = \mu_w(T,p) = \mu_i(T,p). \tag{3.7.23}$$

These are two equations with the two unknowns, T and p, which are uniquely defined by these equations. Thus, such an equilibrium is possible in only one point T_0, p_0 on the $T - p$ diagram, called the triple point. The solution of these equations gives $T_0 = 273.16$, $p_0 = 6.11$ mb.

A schematic representation of the p-T phase diagram for water around the triple point T_{tr} is shown in Fig. 3.3. The diagram is divided into three areas: solid ice, liquid, and vapor. Within the single-phase areas, the system is *bivariant* ($N_w = 2$), meaning pressure and temperature can be independently varied. The solid lines separating the areas with $N_w = 2$ connect the points of equilibrium coexistence of the two phases, representing the *univariant* systems with $N_w = 1$. All the lines join at the triple point T_{tr} ($T = 273.16$ K, $p = 6.11$ mb), where all three phases coexist and the system is *invariant*, $N_w = 0$. The line "liquid-vapor" is the vapor pressure curve of liquid water at $T > T_{tr}$. This line below T_{tr} denotes an extension of the vapor pressure curve to temperatures below 273.16 K. Where water still does not freeze and can be in a liquid metastable state (as in this case), it is called *supercooled water*. A large fraction of atmospheric clouds and fogs exist at temperatures below 0 °C. Some may stay liquid down to −25 to −30 °C in this liquid supercooled metastable state without

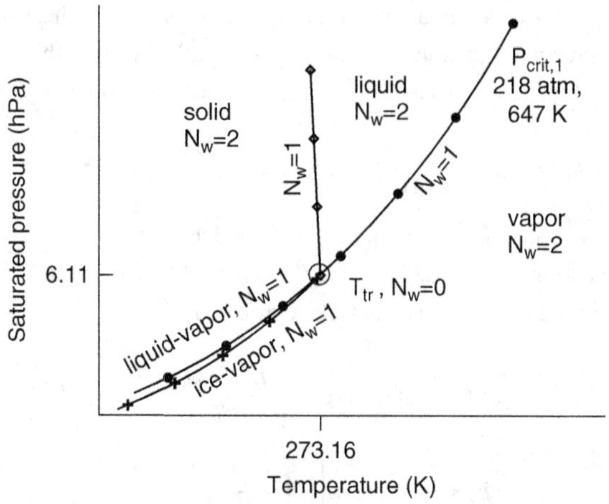

Figure 3.3. *T-p* phase diagram of water around the triple point T_{tr}. The three curves indicate the lines of coexistence of the two phases of water in equilibrium, and N_w indicates the number of thermodynamic degrees of freedom in the corresponding areas. At the triple point T_{tr} ($T = 273.16$ K, $p = 6.11$ hPa), all three phases coexist and $N_w = 0$. The line "liquid-vapor" below T_{tr} denotes extension of the vapor pressure curve to temperatures below 273.16 K. The line below, marked "ice-vapor," denotes the line of equilibrium coexistence of vapor and ice. $P_{\text{crit},1}$ denotes the first critical point at $p_{cr1} = 218$ atm, $T_{cr1} = 647$ K, and the liquid and vapor phases become indistinguishable at $p \geq p_{cr1}$ and $T \geq T_{cr1}$.

natural freezing. This makes them suitable for *cloud seeding* with *crystallizing agents*, as will be described in subsequent chapters.

The line below, marked "ice-vapor," denotes the line of equilibrium coexistence of vapor and ice. $P_{\text{crit},1}$ denotes the first critical point at $p_{cr1} = 218$ atm, $T_{cr1} = 647$ K. The liquid and vapor phases become indistinguishable at $p \geq p_{cr1}$ and $T \geq T_{cr1}$. Here, this critical point of water is called "first," because there can be another hypothesized "second" critical point at low temperatures where the phase states of water, and the transitions between them, can be substantially different, as will be discussed in Chapter 4.

3.7.4. Humidity Variables

Cloud formation depends on the moisture content of the air that is characterized by several humidity variables. The absolute humidity is equal to the vapor density ρ_v, which is the vapor mass in a unit volume of the air. The other variables are the water vapor mixing ratio w_v, its saturated values over water and ice, w_{ws}, w_{is}, and the mole fraction of vapor in the air, x_v, which are defined via vapor pressure e_v or density ρ_v, their saturated values, air density ρ_a, and pressure p:

$$w_v = \frac{\rho_v}{\rho_a} = \varepsilon_w \frac{e_v}{p - e_v}, \qquad w_{ws} = \frac{\rho_{ws}}{\rho_a} = \varepsilon_w \frac{e_{ws}}{p - e_{ws}},$$

$$w_{is} = \frac{\rho_{is}}{\rho_a} = \varepsilon_w \frac{e_{is}}{p - e_{is}}, \qquad x_v = \frac{n_v}{n_a + n_v} = \frac{e_v}{p}, \tag{3.7.24}$$

where $\varepsilon_w = M_w/M_d = 0.622$, n_v and n_a are the number of moles of the vapor and air in a volume and we used the equation of state for water vapor and air from Section 3.4. The air density at normal conditions is on the order $\rho_a \sim 10^{-3}$ cm^{-3}, and the vapor density ρ_v is 3 to 4 orders smaller. Since $p >> e$, $p >> e_{ws}$, and $p >> e_{is}$, the approximate relations are

$$w_v \approx \varepsilon_w \frac{e_v}{p}, \qquad w_{ws} \approx \varepsilon_w \frac{e_{ws}}{p}, \qquad w_{is} \approx \varepsilon_w \frac{e_{is}}{p}. \tag{3.7.24a}$$

The specific humidity q_v and its saturated values over water and ice q_{ws}, q_{is} are defined via the masses m_v of vapor and m_d of dry air or their densities:

$$q_v = \frac{m_v}{m_a + m_v} = \frac{\rho_v}{\rho_a + \rho_v} = \varepsilon_w \frac{e_v}{p - (1 - \varepsilon_w)e_v} \approx \varepsilon_w \frac{e_v}{p} \approx w_v \approx \frac{\rho_v}{\rho_a}, \tag{3.7.25a}$$

$$q_{ws} = \frac{m_{ws}}{m_a + m_{ws}} = \frac{\rho_{ws}}{\rho_a + \rho_{ws}} \approx \varepsilon_w \frac{e_{ws}}{p} \approx w_{ws} \approx \frac{\rho_{ws}}{\rho_a}, \tag{3.7.25b}$$

$$q_{is} = \frac{m_{is}}{m_a + m_{is}} = \frac{\rho_{is}}{\rho_a + \rho_{is}} \approx \varepsilon_w \frac{e_{is}}{p} \approx w_{is} \approx \frac{\rho_{is}}{\rho_a}. \tag{3.7.25c}$$

Since $\rho_v << \rho_a$, the specific humidities are very close to the mixing ratios and are often used as $q_v \approx w_v = \rho_v/\rho_a$. Thus, specific humidity and its values saturated over water and ice are approximately equal to the corresponding mixing ratios.

The water and ice saturation ratios, S_w, S_i, are defined as

$$S_w = \frac{\rho_v}{\rho_{ws}} = \frac{e_v}{e_{ws}}, \qquad S_i = \frac{\rho_v}{\rho_{is}} = \frac{e_v}{e_{is}}. \tag{3.7.26}$$

The fractional relative humidities over water and ice are equal to the corresponding saturation ratios. The relative humidities in percent are

$$RHW = S_w \times 100\%, \qquad RHI = S_i \times 100\%. \tag{3.7.27}$$

The water and ice supersaturations, s_w, s_i, are used in cloud physics in various forms, as fractional, absolute, and specific, depending on the problem considered or the model used. The fractional supersaturations are

$$s_w = \frac{\rho_v - \rho_{ws}}{\rho_{ws}} = \frac{e_v - e_{ws}}{e_{ws}} = S_w - 1, \qquad s_i = \frac{\rho_v - \rho_{is}}{\rho_{is}} = \frac{e_v - e_{is}}{e_{is}} = S_i - 1. \tag{3.7.28}$$

The supersaturations in percent are obtained by multiplication of these quantities by 100%. The specific supersaturations over water Δ_w and ice Δ_i are defined via specific humidity q_v and its saturated values over water and ice q_{ws}, q_{is}

$$\Delta_w = q_v - q_{ws}, \qquad \Delta_i = q_v - q_{is}. \tag{3.7.29}$$

The absolute supersaturations over water and ice are related to the vapor densities,

$$\delta_w = \rho_v - \rho_{ws} = \Delta_w \rho_a, \qquad \delta_i = \rho_v - \rho_{is} = \Delta_i \rho_a. \tag{3.7.30}$$

The supersaturations are the small differences of the two much larger quantities, humidity and saturated humidity. Therefore, solving equations directly for supersaturations yields more precise results than solving equations for humidities. The equations for supersaturations in various forms suitable for various models will be derived and used in the subsequent sections.

3.8. Phase Equilibrium for a Curved Interface—The Kelvin Equation

Consider a system consisting of a pure water drop in equilibrium with an environment of humid air and again apply the general equilibrium Equation (3.6.7). For this system, phase 1 is water vapor, phase 2 is liquid water. Then, the latent heat of phase transition is condensation or evaporation heat L_e, the molar volumes $v_{k0,1}$ and $v_{k0,w}$ are the molar volumes of vapor v_v and water v_w, $r_1 = \infty$ (bulk phase), $r_2 = r$ (the radius of the drop), $\sigma_{12} = \sigma_{wa}$ (the surface tension at the water-air interface), and the misfit strain is absent, $\varepsilon = 0$. The activity of water vapor a_v, assuming that it behaves as an ideal gas, is just its mole fraction, $a_v = x_v = e_{wr}(r)/p$, where e_{wr} is the vapor pressure over the drop surface, and p is the air pressure. Thus, (Eqn. 3.5.7) is reduced to the form:

$$-\frac{\hat{L}_e}{T^2}dT + \frac{(v_v - v_w)}{T}dp - \frac{v_w}{T}d\left(\frac{2\sigma_{wa}}{r}\right) + Rd\ln(e_{wr}/p) = 0. \qquad (3.8.1)$$

The number of components in this system is $c = 2$ (water and air), and according to the phase rule (Eqn. 3.3.12) for the curved substances, the variance of the system is $N_w = 3$. There are 4 differentials in (Eqn. 3.8.1) of 4 variables (T, p, r, e_{wr}), but only 3 of them are independent, and for the integration of (Eqn. 3.8.1), we have to choose any 3 of 4 variables, and the number of combinations is $C_4^3 = 4 \cdot 3 \cdot 2/1 \cdot 2 \cdot 3 = 4$. Integration can yield the dependencies among these 4 possible combinations—that is, 4 possible equations.

If we consider an isobaric process and keep $p = \text{const}$, $dp = 0$, then (Eqn. 3.8.1) is simplified as

$$-\frac{\hat{L}_e}{T^2}dT - \frac{v_w}{T}d\left(\frac{2\sigma_{wa}}{r}\right) + Rd\ln(e_{wr}) = 0. \qquad (3.8.2)$$

If we divide it by dT and multiply by R, we obtain again the generalized Clausius–Clapeyron equation (3.7.6). Then, neglecting the curvature term for large r, we would obtain the Clausius–Clapeyron equation (3.7.7) describing the dependence of e_{ws} on T and ignoring its dependence on r. If, vice versa, we assume $T = \text{const}$, $dT = 0$, and study the dependence of e_{wr} on the curvature r, we obtain another equation. For this case, using the relation $v_w = M_w/\rho_w$, (Eqn. 3.8.2) becomes

$$d\ln[e_{wr}(r)] = \frac{v_w}{RT}d\left(\frac{2\sigma_{wa}}{r}\right) = \frac{M_w 2\sigma_{wa}}{\rho_w RT}d\left(\frac{1}{r}\right). \qquad (3.8.3)$$

Integration from ($r = \infty$, e_{ws})—i.e., from the plane water surface—to (r, $e_{wr}(r)$) yields

$$\ln\frac{e_{wr}(r)}{e_{ws}} = \frac{2M_w\sigma_{wa}}{RT\rho_w r} = \frac{A_K}{r}, \qquad A_K = \frac{2M_w\sigma_{wa}}{RT\rho_w} = \frac{2\sigma_{wa}}{R_v T\rho_w}, \qquad (3.8.4)$$

or in exponential form

$$\frac{e_{wr}(r)}{e_{ws}} = \exp\left(\frac{2M_w\sigma_{wa}}{RT\rho_w r}\right) = \exp\left(\frac{A_K}{r}\right). \qquad (3.8.5)$$

This equation describes the dependence of the vapor pressure around the drop on its radius. It was derived by William Thomson (Lord Kelvin) and is called the *Kelvin equation*; the parameter A_K is called the Kelvin curvature parameter. Since $A_K > 0$, the Kelvin equation describes an increase of the vapor pressure over the curved surface relative to the plane surface. An estimate of $\sigma_{wa} \sim 72$ dyn cm^{-2} gives for the Kelvin's curvature parameter $A_K \sim 1.2 \times 10^{-7}$ cm at $T \sim 20\,°$C. Therefore, an increase of vapor pressure over a drop with $r = 10^{-6}$ cm is ~exp(0.1) ~ 10%, and less that 1% for the drops with $r \geq 10^{-5}$ cm.

The ratio $e_{wr}(r)/e_{ws}$ is the water saturation ratio S_w, and the Kelvin equation can also be written as

$$\ln S_w = \frac{2M_w \sigma_{wv}}{RT\rho_w r} = \frac{A_K}{r}. \tag{3.8.6}$$

Kelvin equations in this forms are used in many applications, as will be illustrated later in equations for the drop activation (Chapters 6 and 7) and crystal nucleation (Chapters 8 and 9).

3.9. Solution Effects and the Köhler Equation

Consider a system consisting of a solution water drop in equilibrium with an environment of humid air and again apply the general equilibrium Equation (3.6.7). For this system, phase 1 is water vapor, phase 2 is aqueous solution. The latent heat of phase transition is condensation or evaporation heat $h_v - h_w = \hat{L}_e = M_w L_e$, with L_e being the specific evaporation heat, the molar volumes $v_{k0,1}$ and $v_{k0,w}$ are the molar volumes of vapor v_v and water v_w, $r_1 = \infty$ (bulk phase), $r_2 = r$ (the radius of the drop), $\sigma_{12} = \sigma_{sa}$ (the surface tension at the solution-air interface), and the misfit strain is absent, $\varepsilon = 0$. The activity of water vapor a_v, assuming its behavior as an ideal gas, is its mole fraction, $a_v = x_v = e_{wr}(r)/p$, where $e_{wr}(r)$ is the vapor pressure over the drop surface, and p is the air pressure. The activity of water $a_w \neq 1$, in contrast to the case of pure water, and is determined by Equations (3.5.16) or (3.5.18) and (3.5.19). Thus, (Eqn. 3.6.7) is reduced to the form:

$$-\frac{\hat{L}_e}{T^2}dT + \frac{(v_v - v_w)}{T}dp - \frac{v_w}{T}d\left(\frac{2\sigma_{sa}}{r}\right) + Rd\ln[e_{wr}(r)/p] - Rd\ln a_w = 0, \tag{3.9.1}$$

which is a generalization of (Eqn. 3.8.1). The number of components in this system is $c = 3$ (water, salt, and air), and according to the phase rule (Eqn. 3.3.12) for the curved substances, the variance of the system is N_w, which would be 4. However, if we assume that mass of salt m_s = const (neglecting scavenging at the earlier stages of drop growth that could change the salt mass), this imposes one more constraint, and so the variance $N_w = 3$, as in the case of pure water drop. Thus, of 5 differentials in (Eqn. 3.9.1) we have only 3 independent variables. We keep T = const, and p = const, and so this equation is reduced to:

$$-\frac{v_w}{T}d\left(\frac{2\sigma_{sa}}{r}\right) + Rd\ln[e_{wr}(r)] - Rd\ln a_w = 0. \tag{3.9.2}$$

Substitution $v_w = M_w/\rho_w$ and integration from ($r = \infty$, e_{ws}, $a_w = 1$)—i.e., from the plane surface of pure water, to (r, $e_{wr}(r)$, a_w) yields

$$\ln\frac{e_{wr}(r)}{e_{ws}} = \frac{2M_w\sigma_{sa}}{RT\rho_w r} + \ln a_w = \frac{A_K}{r} + \ln a_w. \tag{3.9.3}$$

Using the definition of the water saturation ratio $S_w = e_{rs}/e_{ws}$ and water activity (Eqn. 3.5.18) for fully soluble CCN, this equation can also be written in the form

$$\ln S_w = \frac{A_K}{r} + \ln a_w = \frac{2M_w\sigma_{sa}}{RT\rho_w r} - \frac{\nu\Phi_s M_w m_s}{M_s[(4/3)\pi\rho_s'' r_{aw}^3 - m_s]}, \tag{3.9.4}$$

or

$$S_w = a_w \exp\left(\frac{A_K}{r}\right) = \exp\left(\frac{2M_w\sigma_{sa}}{RT\rho_w r} - \frac{\nu\Phi_s M_w m_s}{M_s[(4/3)\pi\rho_s'' r_{aw}^3 - m_s]}\right). \tag{3.9.5}$$

For sufficiently diluted solutions, using approximation (Eqn. 3.5.19) for a_w, we obtain

$$\ln S_w = \frac{2M_w\sigma_{sa}}{RT\rho_w r} - \frac{3\nu\Phi_s M_w m_s}{4\pi M_s \rho_w r^3} = \frac{A_K}{r} - \frac{B}{r^3}, \tag{3.9.6}$$

where A_K is the Kelvin's or curvature parameter defined in (Eqn. 3.8.4) and B is the nucleus activity defined in (Eqn. 3.5.17),

$$A_K = \frac{M_w 2\sigma_{sa}}{RT\rho_w}, \qquad B = \frac{3\nu\Phi_s M_w m_s}{4\pi M_s \rho_w}. \tag{3.9.7}$$

For the aerosol particles containing insoluble fractions, substituting water activity (Eqn. 3.5.16) for mixed CCN into (Eqn. 3.9.3), we obtain

$$\ln S_w = \frac{2M_w\sigma_{sa}}{RT\rho_w r} - \frac{3\nu\Phi_s M_w}{4\pi M_s \rho_w}\frac{m_s}{(r^3 - r_d^3)} = \frac{A_K}{r} - \frac{B}{r^3 - r_d^3}, \tag{3.9.8}$$

and the soluble mass m_s in B is expressed via the mass soluble fraction ε_m, and the mass of a dry nucleus m_d as $m_s = \varepsilon_m m_d$.

Equations (3.9.3) to (3.9.8) are various forms of the Köhler equation, derived by the Swedish scientist Köhler (1921, 1936). They represent an invaluable tool in atmospheric physics for calculations of the aerosol hygroscopic growth, drop activation on CCN, and in many other applications considered in the subsequent chapters.

3.10. Thermodynamic Properties of Gas Mixtures and Solutions

3.10.1. Partial Gas Pressures in a Mixture of Gases

The partial pressure p_k of a k-th component in a mixture of gases (Eqn. 3.5.4) and Raoult's law (Eqn. 3.5.4a) were given in Section 3.5, as in most textbooks, as empirical laws, without derivation. The mathematical derivation can be useful since it is based on the equation of equilibrium for bulk phases, and shows the approximations and applicability of these laws and their relation to the other properties of gas mixtures and solutions.

Consider first a mixture of ideal gases in the volume V with partial pressures p_i, number of moles n_i, and a common temperature T. We can write the equation of state (Eqn. 3.4.15a) for each gas and then sum the equations over all N_i gases,

$$p_k V = n_k RT, \qquad \sum_{i=1}^{N_i} p_i V = \sum_{i=1}^{N_i} n_i RT. \tag{3.10.1}$$

Dalton's law states that the total pressure p in a mixture of gases is equal to the sum of the partial pressures p_i of all gases, $\Sigma p_i = p$. Dividing the lhs and rhs of the first equation in (Eqn. 3.10.1) by the lhs and rhs of the second equation and using Dalton's law for the partial gas pressures p_i, we obtain

$$\frac{p_k}{\sum_{i=1}^{N_i} p_i} = \frac{p_k}{p} = \frac{n_k}{\sum_{i=1}^{N_i} n_i} = x_k . \tag{3.10.2}$$

The expression on the right-hand side is the mole fraction x_k of the k-th gas, thus

$$p_k = p x_k . \tag{3.10.3}$$

We have therefore derived Equation (3.5.4.a), whereby the partial pressure of a gas in the gas mixture is proportional to the total pressure and the gas mole fraction.

3.10.2. Equilibrium of Two Bulk Phases around a Phase Transition Point

Several useful equations for bulk solutions can be derived from the general equilibrium Equation (3.6.7). For bulk solutions, $r_1 = r_2 = \infty$, so the terms with surface tension vanish. We do not consider effects of the misfit strain and suppose $\varepsilon = 0$; thus the equation becomes

$$-\frac{\hat{L}_{12}}{T^2} dT + \frac{(v_{k0,1} - v_{k0,2})}{T} dp + R d \ln\left(\frac{a_{k,1}}{a_{k,2}}\right) = 0. \tag{3.10.4}$$

Consider the equilibrium between two bulk phases around the phase transition point T_0, p_0 in a pure solvent. Addition of solute will cause a shift of the equilibrium point towards other values T and p. If a solution is sufficiently weak, the changes $\Delta T = T - T_0$ and $\Delta p = p_0 - p$ are small compared to T and p. Integration of (Eqn. 3.10.4) within these small intervals ΔT and Δp and multiplication by T yields

$$-\frac{M_w L_{12}}{T} \Delta T + (v_{k0,1} - v_{k0,2}) \Delta p + RT \ln\left(\frac{a_{k,1}}{a_{k,2}}\right) = 0, \tag{3.10.5}$$

where the molar latent heat $\hat{L}_{12}$ is expressed via the specific latent heat L_{12}. This equation describes the equilibrium in bulk solvent around the phase transition point.

If we consider ideal solutions, then approximately $a_{k1} \approx 1 - x_{s1}$, and $a_{k2} \approx 1 - x_{s2}$, where x_{s1} and x_{s2} are the mole fractions of solute in phases 1 and 2. For sufficiently weak solutions, $\ln a_{k1} \approx \ln(1 - x_{s1}) \approx -x_{s1}$, $\ln a_{k2} \approx -x_{s2}$, and $\ln(a_{k1}/a_{k2}) \approx -(x_{s1} - x_{s2})$. Substituting into (Eqn. 3.10.5) yields

$$-\frac{M_w L_{12}}{T} \Delta T + (v_{k0,1} - v_{k0,2}) \Delta p = RT(x_{s1} - x_{s2}). \tag{3.10.6}$$

Equations (3.10.5) and (3.10.6) are in agreement with the Clapeyron equation for pure solvent since with $a_{k1} = a_{k2} = 1$, or $x_{s1} = x_{s2} = 0$, they convert into $\Delta p/\Delta T = (M_w L_{12}/[(v_{k0,1} - v_{k0,2})T]$ (Section 3.7). Consider three particular cases with the presence of solute.

3.10.3. Raoult's Law for Solutions

Case (a). Suppose that phase 1 is liquid, phase 2 is vapor, the phases are at the same temperature in equilibrium, and $\Delta T = 0$. If the liquid is pure, then the vapor pressure is equal to its saturated value. We denote the vapor pressure over pure solvent with $e_{ws,0}$. If solvent is present, the vapor pressure in solution will change by an amount $\Delta e_{ws} \equiv \Delta p$ that can be expressed with (Eqn. 3.10.6). Neglecting the volume v_{w1} of the liquid that is much smaller than the vapor volume $v_{wv} \equiv v_{w2}$, (Eqn. 3.10.6) with $\Delta T = 0$ yields

$$\Delta e_{ws} v_{wv} = -RT(x_{s1} - x_{s2}). \tag{3.10.7}$$

That is, pressure change in the vapor phase is determined by the difference of solute mole fractions in both phases.

If the solute is nonvolatile and is absent in vapor, $x_{s2} = 0$, then

$$\Delta e_{ws} = -\frac{RT}{v_{wv}} x_{s1}. \tag{3.10.8}$$

According to the equation of state in Section 3.4, for one mole of vapor, $RT/v_{wv} = e_{ws,0}$ and we obtain

$$\Delta e_{ws} = -e_{ws,0} x_{s1}. \tag{3.10.9}$$

This is *Raoult's law*, which states that the vapor pressure over a solution is depressed compared to the pressure over pure solvent; a depression is proportional to the saturated pressure over pure solvent and to the molar fraction of solute in solution.

Raoult's law can be written for a full vapor pressure:

$$p_w = e_{ws,0} + \Delta e_{ws} = e_{ws,0} - e_{ws,0} x_{s1} = e_{ws,0}(1 - x_{s1}) = e_{ws,0} x_w. \tag{3.10.10}$$

where $e_{ws,0}$ is the vapor pressure over pure solvent. That is, the pressure of a solvent (e.g., water) is proportional to its molar fraction. The same is valid for any substance in solution, as was formulated by (Eqn. 3.5.4a), $p_k = x_k p_{k0}$, which is another formulation of Raoult's law. Raoult's law is a key starting point in derivation of Köhler's equation, considered in Section 3.9, which governs aerosol hygroscopic growth and drop activation in clouds (Chapters 6 and 7). This equation is also used in derivation of the critical radii and energies of ice nucleation in Chapters 8 and 9.

3.10.4. Freezing Point Depression and Boiling Point Elevation

Consider now another particular ***case (b)*** of Equation (3.10.6), the process of freezing when $\Delta p = 0$,

$$-M_w L_{12} \Delta T/T = RT(x_{s1} - x_{s2}). \tag{3.10.11}$$

For freezing, phase 1 is water with the molar fraction of solute $x_s = x_{s1}$, phase 2 is ice, and assume that the solute is rejected from ice upon freezing, $x_{s2} = 0$, since usually the salt concentration in ice is very small. The latent heat is released upon freezing, and the enthalpy of ice is lower than that of water; thus $M_w L_{12} = h_w - h_i = M_w L_m > 0$. Equation (3.10.11) predicts negative ΔT, while *the freezing point depression* is defined as a positive $\delta T_f = T_0 - T_f$. Thus, the last equation yields

$$\delta T_f = T_0 - T_f = -\Delta T = \frac{RT^2}{M_w L_m} x_s, \tag{3.10.12}$$

where T_0 is the freezing point of a pure solvent. This equation shows that the freezing point in the presence of solution shifts to lower temperatures, and the depression δT_f is proportional to the molar fraction of solute in solution.

This equation is also used with the solute molality $\hat{M}_s$ instead of the mole fraction x_s, which can be written for dilute solutions as $\hat{M}_s \approx x_s/M_w$. If we replace x_s with $\hat{M}_s$, this equation becomes

$$\delta T_f = \frac{RT^2}{L_m}\hat{M}_s = K_f\hat{M}_s, \qquad K_f = \frac{RT^2}{L_m}, \tag{3.10.13}$$

where K_f is the *cryoscopic constant* (recall that L_m is here the specific heat per unit mass, not per 1 mole). The cryoscopic constants were measured and tabulated for various substances (e.g., Atkins, 1982). For water, $K_f = 1.86$ K kg mol^{-1}. It is easy to show that the same is valid for a reverse process of melting, and then the melting point depression $\delta T_m \sim \delta T_f$.

This effect of freezing (melting) point depression provides the basis for the preparation of "*cooling mixtures*" consisting of milled ice or snow and *cooling agents* that freeze at temperatures much lower 0 °C. For example, a mixture of snow with salt NaCl (in the ratio 2:1) freezes at −21 °C, and a mixture of snow with $CaCl_2$ (in the ratio 7:10) freezes at −50 °C. Cooling agents have numerous applications. For example, when a cooling agent is dispersed over the roads covered with snow, a cooling mixture is formed, and the snow begins to melt at temperatures much lower than 0 °C, thus cleaning the road of snow. Natural aerosols usually contain many various salts and acids; therefore deliquescent aerosol particles may remain in liquid state at very cold temperatures. Their freezing point depressions and thresholds of ice nucleation in clouds can be calculated in a first approximation with the equations given earlier.

As follows from the preceding derivation, Equations (3.10.12) and (3.10.13) are an approximation for sufficiently weak solutions without strong non-ideality. For concentrated and non-ideal solutions, these equations may lead to errors; more detailed equations will be considered in Chapters 8, 9, and 10.

Consider another particular ***case (c)*** of Equation (3.10.6), when the phase transition is evaporation or boiling and $\Delta p = 0$. In this case, phase 1 is liquid and phase 2 is vapor. The enthalpy of vapor is higher than that of liquid water, $h_v > h_w$; therefore the molar latent heat is negative, $M_w L_{12} = h_w - h_v = -M_w L_e$. Suppose that the solute is nonvolatile and is absent in the vapor phase, so that $x_{s2} = 0$, its mole fraction in solute is x_s, and (Eqn. 3.10.6) yields the boiling temperature increase due to solute

$$\delta T_b = T_b - T_0 = \Delta T = \frac{RT^2}{M_w L_e} x_s, \tag{3.10.14}$$

where T_0 is the boiling point of pure solvent (e.g., water). This is *the boiling point elevation*, it shows that the boiling point of a liquid shifts to higher temperatures in the presence of solute.

This shift is proportional to the mole fraction of solute. With molality $\hat{M}_s \approx x_s/M_w$, this gives

$$\delta T_b = \frac{RT^2}{L_e}\hat{M}_s = K_b\hat{M}_s, \qquad K_b = \frac{RT^2}{L_e}, \tag{3.10.15}$$

where K_b is *the ebullioscopic constant*. The ebullioscopic constants have been measured and tabulated for various substances (Atkins, 1982). For water, $K_b = 0.51$ K kg mol^{-1}. Boiling point elevation can be used to measure the relative molecular mass of soluble non-volatile materials in atmospheric aerosols that participate in cloud formation (Atkins, 1982).

A similar consideration shows that the condensation point shifts to higher temperatures by δT_c in the presence of solutes. The effects of elevated boiling and condensation points may play a role in the formation of clouds from superheated and polluted sources. When the vapor jet is ejected from the source, it cools, eventually condenses at some distance from the source, and forms a cloud at temperature T_c. This temperature of condensation is higher by δT_c in the presence of solute, and condensation may occur closer to the hot source.

Examples are formations of clouds from volcanic fumaroles or eruptions, from massive natural forest and prairie fires, and from special devices such as *meteotrons*, which are designed for artificial rain production. *Meteotrons* are constructions consisting of several vertically pointed jet engines that can produce a powerful convective plume. Under appropriate conditions, this plume may create a convective cloud and rain. Some native tribes in Africa and South America create such artificial fires to make clouds and rain, and thus fight long droughts. These attempts are often successful. These clouds may spread over long distances (up to regional and global scales) and can significantly affect the atmosphere and surface since they contain various chemicals. Due to boiling (condensation) point elevation, the condensation level of these clouds may be lower above the hot surface, when the pollutants' concentration is higher. This may influence the properties of these clouds, the intensity of precipitation formation, and their subsequent effects. Another example is the formation of *condensation trails* or *contrails* from jet aircraft or *shiptracks* from ship exhausts, which may cause noticeable effects on atmospheric radiation on a regional and perhaps global scale. Due to boiling or condensation point elevation, contrails and shiptracks may form closer to the hot source, which may influence their sizes, shapes, and effects.

3.10.5. Relation of Water Activity and Freezing Point Depression

Previous consideration was valid for ideal solutions, while real liquids exhibit deviation from ideal behavior. There are several ways to account for the non-ideality of solutions. One is to introduce van't Hoff's factor i on the right-hand side of equations for freezing point depression and boiling point elevation for ionic solutions:

$$\delta T_f = i\frac{RT^2}{M_w L_m}x_s, \qquad \delta T_b = i\frac{RT^2}{M_w L_e}x_s, \tag{3.10.16}$$

Another way is to use water activity instead of the molar fraction, as described in the following.

If a pure ice (free of solute) and an aqueous solution are in equilibrium, the chemical potentials of ice, $\mu_i(p, T)$, and water, $\mu_w(p, T)$, should be equal, $\mu_i(p, T) = \mu_w(p, T)$. Using (Eqn. 3.5.21) for μ_w, we obtain

$$\mu_i(p,T,a_w) = \mu_{w0}(p,T) + RT\ln a_w. \tag{3.10.17}$$

Equation (3.10.17) allows derivation of a useful relation for the water activity. Dividing each term by T and taking the derivative by T at p = const, we obtain

$$\frac{\partial}{\partial T}\left(\frac{\mu_i}{T}\right)_p = \frac{\partial}{\partial T}\left(\frac{\mu_{w0}}{T}\right)_p + R\frac{\partial \ln a_w}{\partial T}. \tag{3.10.18}$$

According to (Eqn. 3.1.23), $\partial(\mu_k/T)/\partial T = -h_k/T^2$. Substituting this relation into (Eqn. 3.10.18), we obtain

$$\frac{\partial \ln a_w}{\partial T} = \frac{1}{R}\frac{\partial}{\partial T}\left[\left(\frac{\mu_i}{T}\right)_p - \left(\frac{\mu_{w0}}{T}\right)_p\right] = -\frac{h_i - h_w}{RT^2} = \frac{M_w L_m}{RT^2} = \frac{L_m}{R_v T^2}, \tag{3.10.19}$$

where $R_v = R/M_w$ is the gas constant for water vapor. Integration of this equation by T from the triple point T_0 to T with the boundary condition $a_w(T_0) = 1$ yields

$$\ln a_w(T) = -\frac{L_m}{R_v}\left(\frac{1}{T} - \frac{1}{T_0}\right) = -\frac{L_m \delta T}{R_v T_0 T}, \tag{3.10.20}$$

where $\delta T = T_0 - T$. This gives two relations, for δT_f and a_w. The first one is

$$\delta T_f = -\frac{RT_0 T}{M_w L_m}\ln a_w(T). \tag{3.10.21}$$

This equation describes the freezing point depression, taking into account non-ideality effects. Using (Eqn. 3.5.13) for a_w, we obtain

$$\delta T_f = \frac{RT_0 T}{M_w L_m}\left(\nu\Phi_s \frac{m_s M_w}{m_w M_s}\right) = \frac{RT_0 T}{M_w L_w}\nu\Phi_s \frac{n_s}{n_w} \approx \frac{RT^2}{M_w L_m}\nu\Phi_s x_s, \tag{3.10.22}$$

which differs from (Eqn. 3.10.12) by the factor $\nu\Phi_s$ and from (Eqn. 3.10.16) by a more rigorous account for non-ideality and allows more accurate calculations of the freezing point depression, taking into account the non-ideality of solutions. Equation (3.10.22) is used in many applications. However, this is still a linear approximation to δT_f by x_s or a_w for sufficiently dilute solutions. We will return to an evaluation of δT_f in Chapters 8 and 9, where a non-linear approximation will be derived for a wider range of solution concentrations.

The other equation follows from (Eqn. 3.10.20) for $a_w(T)$,

$$a_w(T) = \exp\left[-\frac{M_w L_m \delta T}{RT_0 T}\right]. \tag{3.10.23}$$

This equation shows that water activity a_w in a bulk aqueous solution at a fixed pressure is a function of the temperature—that is, the solution concentration in equilibrium with pure ice adjusts to the temperature. The argument in exponent in (3.10.23) is negative and $a_w(T)$ decreases with decreasing temperature (increasing supercooling δT). Since $L_m = L_s - L_e$, using the Clausius–Clapeyron Equations (3.7.12) and (3.7.13) for saturated water vapor pressures over water e_{ws} and ice e_{is}, Equation (3.10.23) can be rewritten as

$$a_w^i(T) = \exp\left(-\frac{M_w L_s \delta T}{RT_0 T}\right)\left[\exp\left(-\frac{M_w L_e \delta T}{RT_0 T}\right)\right]^{-1} = \frac{e_{is}}{e_{ws}}. \tag{3.10.24}$$

That is, the water activity in equilibrium with ice $a_w^i(T)$ is expressed via the ratio of saturated vapor pressures over water and ice, $a_w^i(T) = e_{is}/e_{ws}$, this allows much easier calculations and parameterizations of a_w in solutions that are in equilibrium with ice, which can be required for calculations of ice nucleation and freezing and melting temperatures (Chapters 8 and 9).

3.11. Adiabatic Processes

Thermodynamic processes are called *adiabatic* if they occur in an air parcel without exchange of heat with the environment—that is, $dQ = dq = 0$ in equations of Section 3.1. A process is called a *dry adiabatic process* if it proceeds without phase transitions of water. If such transitions take place in the air with the presence of saturated water vapor, the process is called a *wet adiabatic process.* Dry and wet adiabatic processes are considered in this section.

3.11.1. Dry Adiabatic Processes

The first law of thermodynamics (Eqn. 3.1.19b) for air with intensive variables is

$$dq = d\tilde{e} - dw, \tag{3.11.1}$$

where $d\tilde{e} = c_v dT$ is the internal energy, c_v is the heat capacity at constant volume, and $dw = -pdv$ is the work done on gas expnasion. Thus,

$$dq = c_v dT + pdv. \tag{3.11.2}$$

Considering the air as an ideal gas and using the equation of state for an ideal gas, $pv = R_a T$, the last term can be transformed, $pdv = R_a dT - vdp$, and substitution into (Eqn. 3.11.2) yields

$$dq = (c_v + R_a)dT - vdp. \tag{3.11.3}$$

If we consider an isobaric process, $dp = 0$, the second term on the right-hand side vanishes and Equation (3.11.3) shows that the first term is $dq = c_p dT$. Thus, we obtained the relations

$$dq = c_p dT = (c_v + R_a)dT, \tag{3.11.4a}$$

$$c_p = (c_v + R_a)\,. \tag{3.11.4b}$$

For dry air, $c_v = 717\ \mathrm{J\ kg^{-1}\ K^{-1}} = 0.17\ \mathrm{cal\ g^{-1}\ K^{-1}}$, $c_p = 1004\ \mathrm{J\ kg^{-1}\ K^{-1}} = 0.24\ \mathrm{cal\ g^{-1}\ K^{-1}}$, the difference $c_p - c_v = 288\ \mathrm{J\ kg^{-1}\ K^{-1}}$, and the ratio $k = c_p/c_v = (c_v + R_a)/c_v = 1.4$.

For adiabatic process with $dq = 0$, it follows from (Eqn. 3.11.3) using the equation of state that

$$c_p dT = vdp = R_a T(dp/p). \tag{3.11.5}$$

If an air parcel rises from a level with initial T_1, p_1 to a level with T_2, p_2, this equation allows us to find a relation between the initial and final states. Dividing (Eqn. 3.11.5) by T and integrating yields

$$c_p \int_{T_1}^{T_2} \frac{dT}{T} = R_a \int_{p_1}^{p_2} \frac{dp}{p}, \tag{3.11.6}$$

or

$$\frac{T_2}{T_1} = \left(\frac{p_2}{p_1}\right)^{R_a/c_p} = \left(\frac{p_2}{p_1}\right)^{(c_p - c_v)/c_p} = \left(\frac{p_2}{p_1}\right)^{(\kappa-1)/\kappa}, \tag{3.11.7}$$

where $R_a/c_p = (k-1)/k = 0.286$. This equation is called the *equation of the dry adiabat* or *Poisson's equation.* If we begin with Equation (3.11.2) for adiabatic processes with $dq = 0$,

$$c_v dT = -p dv = -\frac{R_a T}{v} dv, \tag{3.11.8}$$

where we use the equation of state for p. Dividing by T, and integration from (T_1, v_1) to (T_2, v_2) gives

$$c_v \int_{T_1}^{T_2} \frac{dT}{T} = -R_a \int_{v_1}^{v_2} \frac{dv}{v}, \tag{3.11.9}$$

$$\frac{T_2}{T_1} = \left(\frac{v_1}{v_2}\right)^{R_a/c_v} = \left(\frac{v_1}{v_2}\right)^{\kappa-1}. \tag{3.11.10}$$

Combining (Eqns. 3.11.7 and 3.11.10) gives

$$\frac{p_2}{p_1} = \left(\frac{v_1}{v_2}\right)^{c_p/c_v}, \qquad \frac{T_2}{T_1} = \left(\frac{p_2}{p_1}\right)^{R/c_p}. \tag{3.11.11}$$

Equations (3.11.10) and (3.11.11) are other forms of the Poisson equation for dry adiabatic processes. They show that when the temperature decreases in an air parcel with increasing height, pressure also decreases and volume increases—that is, the parcel expands with height.

These equations allow us to determine the temperature lapse rate that would be in the atmosphere without phase transitions and the absence of turbulent mixing and non-adiabatic heating. It follows from (Eqn. 3.1.3) for the dry adiabatic process with $dq = 0$ that

$$dq = c_p dT - v_a dp = c_p dT - R_a T \frac{dp}{p} = 0. \tag{3.11.12}$$

Using the equation of state, $p = \rho_a R_a T_e$, where T_e is the external temperature, and the hydrostatic equation, $dp = -\rho_a g dz$, the second term can be transformed as $-(R_a T)(dp/p) = (T/T_e) g dz$, and we obtain

$$c_p dT = -g \frac{T}{T_e} dz. \tag{3.11.13}$$

Dividing by $c_p dz$, assuming that the temperature inside the parcel T is equal to the external temperature T_e, and denoting the gradient $-dT/dz = \gamma_a$, we obtain

$$\gamma_a = -\frac{dT}{dz} = \frac{g}{c_p}. \tag{3.11.14}$$

The gradient γ_a is called the *dry adiabatic lapse rate*. Its value is $\approx 9.8\,°\mathrm{C\ km^{-1}}$. Thus, an air parcel cools by about 1 °C for every 100 m of lift. The values of γ_a depend on g and c_p and are different for various planets. For example, estimates and measurements give $\gamma_a \sim 3.5$–$4\,°\mathrm{C\ km^{-1}}$ for Mercury, $\gamma_a \sim 7.6$–$11\,°\mathrm{C\ km^{-1}}$ for Venus, $\sim 4.5\,°\mathrm{C\ km^{-1}}$ for Mars, ~ 2–$2.5\,°\mathrm{C\ km^{-1}}$ for Jupiter, and $\sim 13.4\,°\mathrm{C\ km^{-1}}$ for the Sun.

The temperature of an air parcel with initial temperature T and pressure p after adiabatic ascent or descent to the level with pressure p = 1000 hPa is called the *potential temperature* and is usually denoted by θ. According to (Eqn. 3.11.7),

$$\frac{T}{\theta} = \left(\frac{p}{1000}\right)^{R_a/c_p}, \qquad \text{or} \qquad \theta = T\left(\frac{1000}{p}\right)^{R_a/c_p}. \tag{3.11.15}$$

Taking the logarithm and the differential of (Eqn. 3.11.15), we obtain

$$\frac{d\theta}{\theta} = \frac{dT}{T} - \frac{R_a}{c_p}\frac{dp}{p}. \tag{3.11.16}$$

The right-hand side according to (Eqn. 3.11.12) is zero, and we have

$$d\theta = 0, \qquad \text{or} \qquad \theta = \text{const}, \tag{3.11.17}$$

that is, the potential temperature is conserved in adiabatic processes. This determines its usefulness for atmospheric thermodynamics, whereby many of these equations are formulated using θ.

3.11.2. Wet Adiabatic Processes

The presence of phase transitions of moisture influences the thermodynamic processes and changes the vertical lapse rate. Consider an isolated parcel with volume v at pressure p containing one gram of dry air ($v = 1/\rho_a$), saturated water vapor with a mixing ratio q_{ws}, and some amount of liquid water mixing ratio q_l that can evaporate. If an amount of heat dq is added to the parcel, it will cause an increase in the temperature dT with the corresponding increase in enthalpy $dh = c_p dT$. The term $q_{ws}c_{pv}$, where c_{pv} is the heat capacity of the vapor, should be added to c_p, but it is typically less than 10^{-2} of c_p and can be neglected. The work of expansion is $dw = -vdp$. The vapor will become subsaturated and evaporation of liquid occurs, so that its amount will decrease by $-dq_l$ until saturation of vapor is reached again and the saturated vapor mixing ratio q_{ws} is increased by $dq_{ws} = -dq_l$ with the energy $L_e dq_{ws}$. For an adiabatic process, when $dq = 0$, it follows from the first law of thermodynamics (Eqn. 3.11.3) that

$$c_p dT + L_e dq_{ws} - vdp = 0. \tag{3.11.18}$$

This equation can be rewritten in terms of the rates dividing by dt and using the relations $v = 1/\rho_a$, $dq_{ws} = -dq_l$, multiplying by ρ_a and introducing the condensation rate $I_{con} = \rho_a dq_l/dt$, thus

$$\rho_a c_p \frac{dT}{dt} = \frac{dp}{dt} + L_e I_{con}. \tag{3.11.19}$$

This equation of heat balance will be used in subsequent chapters for derivations of supersaturation equations in liquid clouds.

Using the equation of state for the air, $pv = R_a T$, and the hydrostatic equation, $dp = -\rho_a g dz$, the last term on the left-hand side of (Eqn. 3.11.18) can be transformed as $-vdp = -(R_a T)(dp/p) = gdz$, and we obtain

$$c_p dT + L_e dq_{ws} + gdz = 0. \tag{3.11.20}$$

Dividing by $c_p dz$ and denoting the *wet adiabatic gradient* $\gamma_w = -dT/dz$, we obtain

$$\gamma_w = -\frac{dT}{dz} = \frac{g}{c_p} + \frac{L_e}{c_p}\frac{dq_{ws}}{dz}. \tag{3.11.21}$$

Since $g/c_p = \gamma_a$, this yields

$$\gamma_w = \gamma_a + \frac{L_e}{c_{pa}}\frac{dq_{ws}}{dz}. \tag{3.11.22}$$

To derive a more detailed formula for γ_w, we take a logarithmic derivative of $q_{ws} = 0.622(e_{ws}/p)$:

$$\frac{1}{q_{ws}}\frac{dq_{ws}}{dz} = \frac{1}{e_{ws}}\frac{de_{ws}}{dz} - \frac{1}{p}\frac{dp}{dz}. \qquad (3.11.23)$$

Using the chain rule $de_{ws}/dz = (de_{ws}/dT)(dT/dz)$, the Clausius–Clapeyron equation and hydrostatic equation yield

$$\frac{dq_{ws}}{dz} = q_{ws}\left(\frac{1}{e_{ws}}\frac{de_{ws}}{dz} - \frac{1}{p}\frac{dp}{dz}\right) = q_{ws}\left(-\frac{L_e}{R_v T^2}\gamma_w + \frac{g}{R_a T}\right). \qquad (3.11.24)$$

Substituting this dq_{ws}/dz into (Eqn. 3.11.22) gives us

$$\gamma_w = \gamma_a + \frac{L_e}{c_p}q_{ws}\left(-\frac{L_e}{R_v T^2}\gamma_w + \frac{g}{R_a T}\right). \qquad (3.11.25)$$

Solving relative to γ_w, we obtain the wet adiabatic lapse rate:

$$\gamma_w = \gamma_a \frac{1 + \dfrac{L_e}{R_a T}q_{ws}}{1 + \dfrac{L_e^2}{c_p R_v T^2}q_{ws}}. \qquad (3.11.26)$$

At $T \approx 275$ K (2 °C) and $p_a \sim 1000$ mb, the second term in the numerator, $L_e/(R_a T)q_{ws} \sim 0.046$ (i.e., the numerator is close to 1). The second term in the denominator, $L_e^2/(c_p R_v T^2)q_{ws} \sim 0.64$, thus $\gamma_w/\gamma_a = 1.046/1.64 = 0.64$ and $\gamma_w \approx 10$ K km^{-1} × 0.64 = 6.4 K km^{-1}. Since the numerator in (Eqn. 3.11.26) for γ_w is close to 1, it can be approximately written as

$$\gamma_w = \gamma_a \frac{1}{1 + \dfrac{L_e}{c_p}\dfrac{dq_{ws}}{dT}} = \frac{\gamma_a}{\Gamma_1}, \qquad (3.11.27)$$

where

$$\Gamma_1 = 1 + \frac{L_e}{c_p}\frac{dq_{ws}}{dT} = 1 + \frac{L_e^2}{c_p R_v T^2}q_{ws}. \qquad (3.11.28)$$

This quantity Γ_1 approximately coincides with the psychrometric correction to the drop diffusion growth rate (see Section 5.1) and will participate in many equations in the next chapters. Equation (3.11.27) shows that the value of γ_w depends both on temperature and pressure via $q_{ws}(T, p)$. The value of $\gamma_w \approx 6.14$ °K km^{-1} at $T = 0$ °C, $p = 800$ hPa; γ_w increases with decreasing temperature and increasing pressure. The reason is that $q_{sw} = 0.622e_s(T)/p$, and q_{sw} decreases toward low T or high p, the numerator and denominator of γ_w tend to 1, $\Gamma_1 \to 1$, and γ_w approaches γ_a at low T below −40 °C. For example, $\gamma_w \approx 9.44$ K km^{-1} at $T = -40$ °C, $p = 800$ hPa, and $\gamma_w \approx 9.59$ K km^{-1} at $T = -60$ °C, $p = 200$ hPa, at cirrus altitudes near the tropopause.

It follows from the mass conservation law in the condensation process that the increment dq_l of the adiabatic liquid water mixing ratio is related to the dq_{ws} as $dq_l = -dq_{ws}$. Taking this into account, (Eqn. 3.11.20) can be rewritten as

$$c_p dT - L_e dq_l + g dz = 0. \qquad (3.11.29)$$

This gives for the adiabatic gradient $(dq_l/dz)_{ad}$

$$\left(\frac{dq_l}{dz}\right)_{ad} = \frac{c_p}{L_e}\left(\frac{dT}{dz} + \frac{g}{c_p}\right). \tag{3.11.30}$$

Introducing the dry and wet adiabatic temperature gradients, we obtain for the adiabatic gradient of the liquid water mixing ratio

$$\left(\frac{dq_l}{dz}\right)_{ad} = \frac{c_p}{L_e}(\gamma_a - \gamma_w). \tag{3.11.31}$$

Integration of this equation over height from the cloud base gives the adiabatic liquid water mixing ratio at a height z. The adiabatic profile of LWC is observed sometimes in clouds with suppressed mixing with the environment (e.g., Verlinde et al., 2007; Klein et al., 2009) or above the bases of convective clouds (Pruppacher and Klett, 1997), but LWC is usually smaller than the adiabatic value. The reason for this is that clouds form via non-adiabatic processes, and mixing with the drier environment "dilutes" the cloud parcels and leads to a decrease in liquid water content relative to the adiabatic value. These processes and profiles of LWC will be considered in subsequent chapters.

In pure crystalline (ice) clouds, vapor excess is deposited on the cloud crystals. If all the vapor was deposited on the crystals, adiabatic ascent would be characterized by the wet adiabatic lapse rate in ice clouds, $-dT/dz = \gamma_{is}$, which can be evaluated similar to water clouds. We can again use the first thermodynamic law, which can be written for ice clouds with the heat of sublimation L_s instead of the heat of condensation L_e as

$$c_p dT + L_s dq_{is} - vdp = 0, \tag{3.11.32}$$

where $q_{is} = 0.622e_{is}/p$ is the vapor mixing ratio saturated over ice, and e_{is} is the vapor pressure saturated over ice. This equation can also be rewritten in terms of the rates dividing by dt and using the relations $v = 1/\rho_a$, $dq_{is} = -dq_i$, where q_i is the ice water content. Multiplying by ρ_a and introducing the deposition rate $I_{dep} = r_a dq_i/dt$,

$$\rho_a c_p \frac{dT}{dt} = \frac{dp}{dt} + L_s I_{dep}. \tag{3.11.33}$$

This equation of heat balance will be used in subsequent chapters for derivation of the supersaturation equations in crystalline clouds. Using the equation of state for air, $pv = R_a T$, and the hydrostatic equation, $dp = -\rho_a g dz$, the last term on the left-hand side of (Eqn. 3.11.32) can be transformed into $vdp = -gdz$, and so we obtain

$$c_p dT + L_s dq_{is} + gdz = 0. \tag{3.11.34}$$

Dividing by dz, we obtain the wet adiabatic lapse rate in ice clouds, $-dT/dz = \gamma_{is}$, similar to (Eqn. 3.11.21)

$$\gamma_{is} = \frac{g}{c_p} + \frac{L_s}{c_p}\frac{dq_{is}}{dz} = \gamma_a + \frac{L_s}{c_p}\frac{dq_{is}}{dz}. \tag{3.11.35}$$

Taking again a logarithmic derivative of $q_{is} = 0.622e_{is}/p$ yields

$$\frac{dq_{is}}{dz} = q_{is}\left(-\frac{L_s}{c_p}\gamma_{is} + \frac{g}{R_a T}\right). \tag{3.11.36}$$

Using the relation for dq_{is}/dz similar to (Eqn. 3.11.24) and substituting into (Eqn. 3.11.35) and solving for γ_{is}, we obtain

$$\gamma_{is} = \gamma_a \frac{1 + \dfrac{L_s}{R_a T} q_{is}}{1 + \dfrac{L_s^2}{c_p R_v T^2} q_{is}}. \tag{3.11.37}$$

The values of q_{is} at low T in ice clouds are substantially smaller than the values of q_{ws} in water clouds, and the numerator of γ_{is} is closer to 1 than in the equation for γ_w. Thus, γ_{is} is approximately

$$\gamma_{is} = \gamma_a \frac{1}{1 + \dfrac{L_s}{c_p}\dfrac{dq_{is}}{dT}} = \frac{\gamma_a}{\Gamma_2}, \tag{3.11.38}$$

where

$$\Gamma_2 = 1 + \frac{L_s}{c_p}\frac{dq_{is}}{dT} = 1 + \frac{L_s^2}{c_p R_v T^2} q_{is}. \tag{3.11.39}$$

This quantity Γ_2 coincides with the psychrometric correction to the crystal growth rate (Section 5.2) and will be used many times in upcoming chapters.

The mass conservation law in the deposition process in ice clouds states that the increment dq_i of the adiabatic ice water mixing ratio is related to dq_{is} as $dq_i = -dq_{is}$. Taking this into account, (Eqn. 3.11.34) can be rewritten as

$$c_p dT - L_s dq_i + g dz = 0. \tag{3.11.40}$$

This gives for the adiabatic gradient $(dq_i/dz)_{ad}$ of the ice water mixing ratio

$$\left(\frac{dq_i}{dz}\right)_{ad} = \frac{c_p}{L_s}\left(\frac{dT}{dz} + \frac{g}{c_p}\right) = \frac{c_p}{L_s}(\gamma_a - \gamma_{is}). \tag{3.11.41}$$

Integration of this equation over the height from the cloud base gives the adiabatic ice water mixing ratio at a height z. This is an upper limit of the ice water content (IWC), but IWC is usually smaller than the adiabatic value due to mixing with the environment. These processes and profiles of IWC will be considered in subsequent chapters.

There is a fundamental difference between water and ice clouds in values of supersaturation and saturated humidities. In water clouds, the supersaturation s_w is small and humidity is close to saturated over water. In ice clouds, supersaturation over ice is typically above 0.2–0.3 and can reach 0.7–0.8 or more in cirrus clouds; thus saturation ratios are 1.3 to 1.7, and humidity is substantially higher than saturated over ice (e.g., Khvorostyanov and Sassen, 1998b, 2002; Khvorostyanov et al.,

2001, 2006; Jensen et al., 2001, 2005; Gierens et al., 2003; Comstock et al., 2004; Krämer et al., 2009; Spichtinger and Krämer, 2012). In mixed-phase clouds, supersaturation can be intermediate between saturated over water and ice (e.g., Khvorostyanov et al., 2001, 2003). This is caused by the slow deposition process and will be discussed in more detail in upcoming chapters. Therefore, the deposition heat in ice clouds is smaller than it would be in adiabatic ascent, the lapse rate is higher than the wet adiabatic, and closer to γ_a. Besides, it depends on time and is determined by the *supersaturation relaxation time* in ice clouds (Section 5.3). More precise evaluation of γ_{is} requires use of the non-stationary equations and will be discussed in subsequent chapters.

Consider now the adiabatic process in mixed-phase clouds. In this case, both condensation and deposition latent heats are present and the first law of thermodynamics is

$$c_p dT + L_e dq_{ws} + L_s dq_{is} - v dp = 0. \tag{3.11.42}$$

This equation again can be rewritten in terms of the rates dividing by dt and using the relations $v = 1/\rho_a$, $dq_{ws} = -dq_l$, $dq_{is} = -dq_i$, multiplying by ρ_a, thus giving

$$\rho_a c_p \frac{dT}{dt} = \frac{dp}{dt} + L_e I_{con} + L_s I_{dep}. \tag{3.11.43}$$

This equation of heat balance will be used later for derivation of the supersaturation equations in the mixed-phase clouds.

Appendix A.3 for Chapter 3. Calculation of Integrals with Maxwell Distribution

The n-th moment integrals of the type

$$I_n = \int_0^\infty x^n e^{-\alpha x^2} \, dx \tag{A.3.1}$$

can be calculated using a substitution $\alpha x^2 = z$. Thus

$$I_n = \frac{1}{2}\alpha^{-(n+1)/2} \int_0^\infty z^{(n-1)/2} e^{-z} \, dz = \frac{1}{2}\alpha^{-(n+1)/2} \Gamma\left(\frac{n+1}{2}\right), \tag{A.3.2}$$

which follows from the definition of Euler's gamma function $\Gamma(x)$ (Gradshtein and Ryzhik, 1994). The first 3 moments are especially important in applications, since they follow from (Eqn. A.3.2) for a particular n. For $n = 0$ (normalization integrals)

$$I_0 = \frac{1}{2}\alpha^{-1/2}\Gamma\left(\frac{1}{2}\right) = \frac{1}{2}\sqrt{\frac{\pi}{\alpha}}. \tag{A.3.3}$$

Here the relation $\Gamma(1/2) = \sqrt{\pi}$ was used. This gives us another integral:

$$J_0 = \int_{-\infty}^\infty e^{-\alpha x^2} \, dx = 2\int_0^\infty e^{-\alpha x^2} \, dx = 2I_0 = \sqrt{\frac{\pi}{\alpha}}. \tag{A.3.4}$$

For $n = 1$ (mean velocity components),

$$I_1 = \frac{1}{2\alpha}\Gamma(1) = \frac{1}{2\alpha}. \tag{A.3.5}$$

For $n = 2$ (mean components of kinetic energy),

$$I_2 = \frac{1}{2}\alpha^{-3/2}\Gamma\left(\frac{3}{2}\right) = \frac{\sqrt{\pi}}{4}\alpha^{-3/2}. \tag{A.3.6}$$

For $n = 3$ (mean velocity),

$$I_3 = \frac{1}{2}\alpha^{-2}\Gamma(2) = \frac{1}{2}\alpha^{-2}. \tag{A.3.7}$$

The 0-th moments (Eqns. A.3.3 and A.3.4). These relations are used for evaluation of normalization constants of the Maxwell distribution in their various forms. For the momentum distribution, it is

$$dw_p(p) = a_p \exp\left(-\frac{p_x^2 + p_y^2 + p_z^2}{2mkT}\right) dp_x \, dp_y \, dp_z. \tag{A.3.8}$$

The constant a_p is determined from the normalization condition

$$\begin{aligned} \int_{-\infty}^\infty dw_p(p) &= a_p \int_{-\infty}^\infty \int_{-\infty}^\infty \int_{-\infty}^\infty \exp\left(-\frac{p_x^2 + p_y^2 + p_z^2}{2mkT}\right) dp_x \, dp_y \, dp_z \\ &= a_p \left[\int_{-\infty}^\infty \exp\left(\frac{-p_x^2}{2mkT}\right) dp_x\right]^3 = a_p \left[\int_{-\infty}^\infty \exp(-\alpha p_x^2) \, dp_x\right]^3 = a_p [J_0(\alpha)]^3 = 1. \end{aligned} \tag{A.3.9}$$

We can use (Eqn. A.3.4) for $J_0(\alpha)$ with $p_x = x$, $\alpha = (2mkT)^{-1}$. Thus, it follows from (Eqns. A.3.4 and A.3.9) that

$$J_0 = (2\pi mkT)^{1/2}, \qquad a_p = J_0^{-3} = (2\pi mkT)^{-3/2}. \tag{A.3.10}$$

This yields the normalized momentum distribution

$$dw_p(p) = (2\pi mkT)^{-3/2} \exp\left(-\frac{p_x^2 + p_y^2 + p_z^2}{2mkT}\right) dp_x\, dp_y\, dp_z. \tag{A.3.10a}$$

The Maxwell distribution by the velocity components is

$$dw_v(v_x, v_y v_z) = a_v \exp\left[-\frac{m(v_x^2 + v_y^2 + v_z^2)}{2kT}\right] dv_x\, dv_y\, dv_z. \tag{A.3.11}$$

The constant a_v is determined from the normalization condition

$$\int_{-\infty}^{\infty} dw_v(v) = a_v \int_{-\infty}^{\infty}\int_{-\infty}^{\infty}\int_{-\infty}^{\infty} \exp\left(-\frac{v_x^2 + v_y^2 + v_z^2}{2mkT}\right) dv_x\, dv_y\, dv_z$$

$$= a_v \left[\int_{-\infty}^{\infty} \exp\left(\frac{-mv_x^2}{2kT}\right) dp_x\right]^3 = a_v \left[\int_{-\infty}^{\infty} \exp(-\alpha v_x^2)\, dv_x\right]^3 = a_v [J_0(\alpha)]^3 = 1. \tag{A.3.11a}$$

The constant a_v is again determined from (Eqn. A.3.4) with $\alpha = (m/2kT)$. Then, from (Eqns. A.3.4, A.3.11a),

$$J_0(\alpha) = (2\pi kT/m)^{1/2}, \qquad a_v = [J_0(\alpha)]^{-3} = (m/2\pi kT)^{3/2}. \tag{A.3.12}$$

The 1-st moment (Eqn. A.3.5). The relation (Eqn. A.3.5) could be used for evaluation of the mean components of velocity $\bar{v}_x$, $\bar{v}_y$, $\bar{v}_z$; however, the integrals of the odd power of velocity from $-\infty$ to $+\infty$ are equal to zero—that is,

$$\bar{v}_x = \int_{-\infty}^{\infty} v_x\, dw_v(v_x, v_y v_z)\, dv_x = \bar{v}_y = \bar{v}_z = 0. \tag{A.3.13}$$

The first moment can be used for evaluation of the 1st moments in the limited bounds $(0, \infty)$.

The 2-nd moment (Eqn. A.3.6). It is used for calculation of the normalizing constant for distribution in the polar system and for calculation of the mean kinetic energy. The Maxwell distribution in the polar system in the velocities space converts into

$$dw_v(v, \theta, \varphi) = a_{vp} \exp\left(-\frac{mv^2}{2kT}\right) v^2 \sin\theta\, d\theta\, d\varphi\, dv. \tag{A.3.14}$$

Its integral over the angles is 4π, and this integral over v yields

$$\int d\Omega \int_0^{\infty} dw_v(v, \theta, \varphi)\, dv = 4\pi a_{vp} I_2(\alpha), \tag{A.3.15}$$

with $\alpha = (m/2kT)$ again. From (Eqn. A.3.6), $I_2 = \dfrac{\sqrt{\pi}}{4}\left(\dfrac{m}{2\pi kT}\right)^{-3/2}$. Thus, from (Eqn. A.3.15) $a_{vp} = \left(\dfrac{m}{2\pi kT}\right)^{3/2}$, which coincides with a_v earlier in (Eqn. A.3.12)—that is, normalizations in the components space and polar system are the same, $a_{vp} = a_v$.

The mean components of the kinetic energy along the x-axis can be calculated as

$$\varepsilon_x = < \frac{mv_x^2}{2} > = \frac{m}{2}\int_{-\infty}^{\infty} v_x^2\, dw_v(v_x, v_y v_z)\, dv_x$$
$$= a_v \frac{m}{2}\int_{-\infty}^{\infty} dv_x v_x^2 e^{-\frac{mv_x^2}{2kT}} \int_{-\infty}^{\infty} dv_y e^{-\frac{mv_y^2}{2kT}} \int_{-\infty}^{\infty} dv_z e^{-\frac{mv_z^2}{2kT}}. \qquad \text{(A.3.16)}$$

This is expressed via $I_2(\alpha)$ and $I_0(\alpha)$ with $\alpha = (m/2kT)$, as shown earlier, and

$$\varepsilon_x = a_v \frac{m}{2}[2I_2(\alpha)][2I_0(\alpha)]^2 = \left(\frac{m}{2\pi kT}\right)^{3/2} m\frac{\sqrt{\pi}}{4}\alpha^{-3/2}\frac{\pi}{\alpha} = \frac{kT}{2}. \qquad \text{(A.3.17)}$$

This is a known result for the Maxwell distribution: the mean energy of each component of velocity is $(1/2)kT$, and the full kinetic energy per three degrees of freedom is $(3/2)kT$.

The 3-rd moment (Eqn. A.3.7). This is used for evaluation of the mean velocity. As we saw earlier, the velocity components have zero mean values. But the velocity module is not zero, it is calculated using the distribution in the polar coordinate system (Eqn. A.3.14):

$$\bar{V}_v = \int d\Omega \int_0^{\infty} v\, dw_v(v,\theta,\varphi) dv,$$
$$= a_v \int \sin\theta\, d\theta\, d\varphi \int_0^{\infty} v^3 \exp\left(-\frac{mv^2}{2kT}\right) dv$$
$$= a_v 4\pi I_3(\alpha) = 2(\pi\alpha)^{-1/2} = \left(\frac{8kT}{\pi m}\right)^{1/2} = \left(\frac{8RT}{\pi M}\right)^{1/2}, \qquad \text{(A.3.18)}$$

where we used Equation (A.3.7) for I_3 with $\alpha = (m/2kT)$ and (Eqn. A.3.12) for a_v. Here, R is the universal gas constant, and M is the gas molecular weight. The last equation is obtained from the previous one by multiplication of the numerator and denominator by Avogadro's number N_{Av}. This equation is used in many applications of cloud physics since it determines the molecular fluxes of cloud particles.

4

Properties of Water and Aqueous Solutions

4.1. Properties of Water at Low Temperatures and High Pressures

4.1.1. Forms of Water at Low Temperatures

A general schematic of various states and transitions of water at atmospheric pressure (1 bar) and lower pressures is shown in Fig. 4.1. The stable state of liquid water is located in the region from 0 °C to the boiling point, T_{boil} = 100 °C, and a superheated metastable liquid is found at temperatures higher than the boiling point.

A region of metastable supercooled water begins below 0 °C and extends to about –40 °C. Heterogeneous freezing of water drops begins mostly below –5 to –12 °C on small foreign particles called ice nuclei (IN), although liquid drops can exist down to about –38 to –42 °C, with smaller droplets freezing at colder temperatures (Hagen et al., 1981; Pruppacher and Klett, 1997). At these temperatures, rapid homogeneous freezing begins, and the probability of existence of pure water in the liquid phase sharply decreases at $T < 235$ K.

Two new forms of ice, not crystalline but amorphous, were discovered at very low temperatures: low-density amorphous ice (LDA) and high-density amorphous ice (HDA). LDA was discovered by Barton and Oliver (1936) who studied water vapor deposits on a cold plate at very low T. They observed the diffraction patterns of X-rays of these deposits and found that the distinct pattern at –80 °C characteristic of crystalline hexagonal ice I_h gradually became less perfect with lowering temperature and converted into two diffuse rings at –110 °C, resembling the pattern for liquid water. They concluded that it was an *amorphous* of *vitreous* solid. The term *amorphous* means a substance with non-crystalline structure. The term *vitreous* means a solid that has amorphous structure and in addition may undergo a *glass transition* with changing temperature, which is a transition from the crystal-like to liquid-like structure with a corresponding change in the values of heat capacity or expansion coefficients. The amorphous ice is also called *glassy water*, which is a solid that has a liquid-like disordered arrangement. It was found that another way to obtain LDA is via rapid cooling at a rate > 10^5 K s^{-1} to temperatures below about –170 °C (100 K), which produces not a crystalline but a glassy water or LDA (Brüggeller and Mayer, 1980). Subsequent warming causes a glass transition at about T_g = –140 °C (130 K), and ultraviscous metastable water exists between T_g and $T_x \approx$ –120 °C (150 K), where again rapid crystallization begins. A region extending more than 80 K between T_x and the spinodal limit T_s is called "*No man's land*" because experiments on liquid water cannot be performed. Particles of LDA ice may form in polar mesospheric (noctilucent) clouds (Murray and Jensen, 2010).

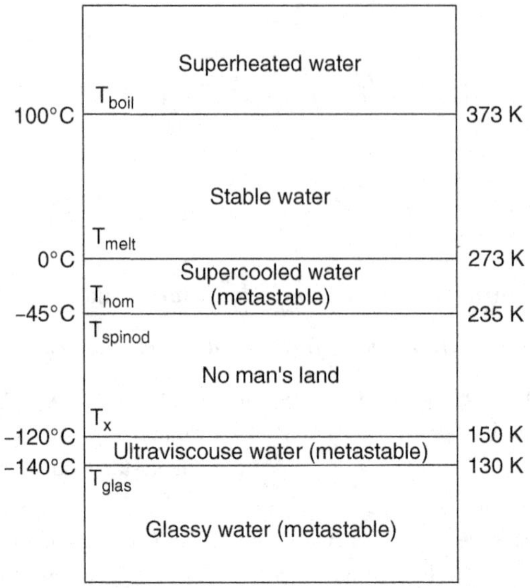

Figure 4.1. Schematic diagram of different water states at atmospheric pressure 1 atm. The points mean: $T_{boil} = 100\,°C$ is the boiling point, $T_{melt} = 0\,°C$ is the melting point, $T_{hom} \approx -38\,°C$ is the temperature of homogeneous freezing, $T_{spinod} \approx -45\,°C$ is the spinodal temperature (the temperature that separates metastable and unstable states), $T_x \approx -120\,°C$ is the temperature of crystallization of amorphous water at reheating, and T_{glas} is the temperature of the glass transition.

Using an electron diffraction camera, König (1942) found in the range −140 to −80 °C the diffraction patterns of sharp rings corresponding to the cubic structure of the diamond type. Thus, these experiments indicated existence of the new, cubic form of ice, called ice I_c. Upon further heating above −80 °C, this ice converted into ordinary hexagonal ice I_h. Subsequent experiments showed that ice I_c may form simultaneously with ice I_h or amorphous LDA (Vertsner and Zhdanov, 1966), or after cooling the other forms of high-pressure ice (ices II, III, and V, described later) and reducing the pressure to atmospheric (Hobbs, 1974). Studies of the crystalline structure of I_c showed that its structure is similar to that of I_h (described, for example, in Hobbs, 1974; Young, 1993; Curry and Webster, 1999), being tetrahedral around each oxygen atom. Each layer is identical to that of the hexagonal phase, but each layer is shifted by one half of the diameter of the hexagonal ring; thus, the oxygen atoms in I_c have the same configuration as the carbon atoms in the diamond cubic structure. The measured unit cell parameter a_0 was near 6.35 Å and density was 0.9343 g cm^{-3} at −130 °C, close to ice I_h. At temperatures above −80 °C, ice I_c was unstable, having a saturated vapor pressure higher than ice I_h, and converted into ice I_h upon heating during the times ~ 10^3–10^2 min at T = 170–190 K to a few minutes at T > 210 K. However, the occurrence and time of transition may depend on the environmental conditions and heating rates (Murray and Bertram, 2006; Murray et al., 2005, 2010). The metastable state of cubic ice I_c and conversion into I_h may cause an increase in dehydration of the upper troposphere where this temperature range is observed (Murphy, 2003).

4.1.2. Forms of Water at High Pressures

A great variance of forms of various solid phases of water (ices) can occur at high pressures. Beyond I_h, ice possesses at least 12 other equilibrium (crystalline) phases and two amorphous states (Whalley, 1969; Hobbs, 1974; Johari, 1998; Tse et al., 1999; Klug, 2002; Rosenberg, 2005), all discovered in the laboratory. A transient formation of cubic ice I_c was observed in cirrus clouds and could be confirmed experimentally (Mayer and Hallbrucker, 1987; Murray and Bertram, 2006; Malkin et al., 2012).

HDA was discovered almost 50 years after LDA (Mishima et al., 1984, 1985; Heide, 1984). It was obtained by compressing hexagonal ice I_h below 150 K or by bombarding an ice I_h layer with an electron beam. It was found that a "*polyamorphic transition*" LDA–HDA is possible between these two forms of ice when the thermodynamic parameters (temperature and pressure) change infinitesimally, but this is accompanied by sharp and large (20%) volume change and hence a first-order phase transition. Existence of a pure substance in more than one amorphous form is called *polyamorphism*.

A general schematic of these ice forms is given in Fig. 4.2 following Whalley (1969), Hobbs (1974), and Rosenberg (2005). Besides the ordinary hexagonal ice I_h, there are another eight ices that exist at pressures higher than 2×10^3 bar. These are labeled in Fig. 4.2 with the Roman numerals from II to IX. The stable and metastable boundaries among these forms are indicated with the various line types. Tammann (1900) was the first to discover this pressure-induced ice transformation. He observed I_h in the temperature range −30 to −50 °C at increasing pressure and found that in the pressure range 2.0–2.2 kbar there was a sharp decrease in volume, which corresponded to formation of a new ice form called ice III. At lower temperatures of −70 to −80 °C and p just below 2 kbar, Tammann observed the transformation of ice I_h into a solid ice II slightly denser than ice III. This series of experiments was continued later by Bridgman (1912, 1937), who extended applied pressure to 20.5 and then to 45 kbar, and discovered several other forms of ice: ices IV, V, VI, and VII. Fig. 4.2 shows that the slopes on the p-T diagram between various pairs of ices can be either positive or negative.

The Clapeyron Equation (3.7.3) is written for the boundary of ices "a" and "b"

$$\frac{dT_m}{dp_m} = \frac{\Delta V}{\Delta S_\eta} = \frac{T(v_b - v_a)}{M_w L_m} = \frac{T(1/\rho_b - 1/\rho_a)}{L_m}, \tag{4.1.1}$$

where L_m is the specific heat of transition, and ρ_a and ρ_b are the densities, all of which describe the boundary slopes between the ices and explain the different signs of the slopes by the different relations of the densities of the ices. Whalley and Davidson (1965) and Brown and Whalley (1966) discovered ice VIII, and found that ice VII transforms into ice VIII below about 0 °C. Whalley et al. (1968) discovered a new solid that they called ice IX. In contrast to ice I_h, these forms of ice have densities greater than liquid water. The ices X and XI were discovered later and their mutual transformations with other forms of ice were studied (e.g., Matsuo et al., 1986; Johari, 1998). Thus, ice exhibits a rich variety of crystalline and glassy structures in at least 12 distinct phases or polymorphs at different pressures and temperatures (Rosenberg, 2005), the largest number among all known substances.

An interesting feature of ices VI and VII is that they can exist at positive Celsius temperatures, approaching +80 to +100 °C, but this requires very high applied pressures Δp = 6–25 kbar (Fig. 4.2).

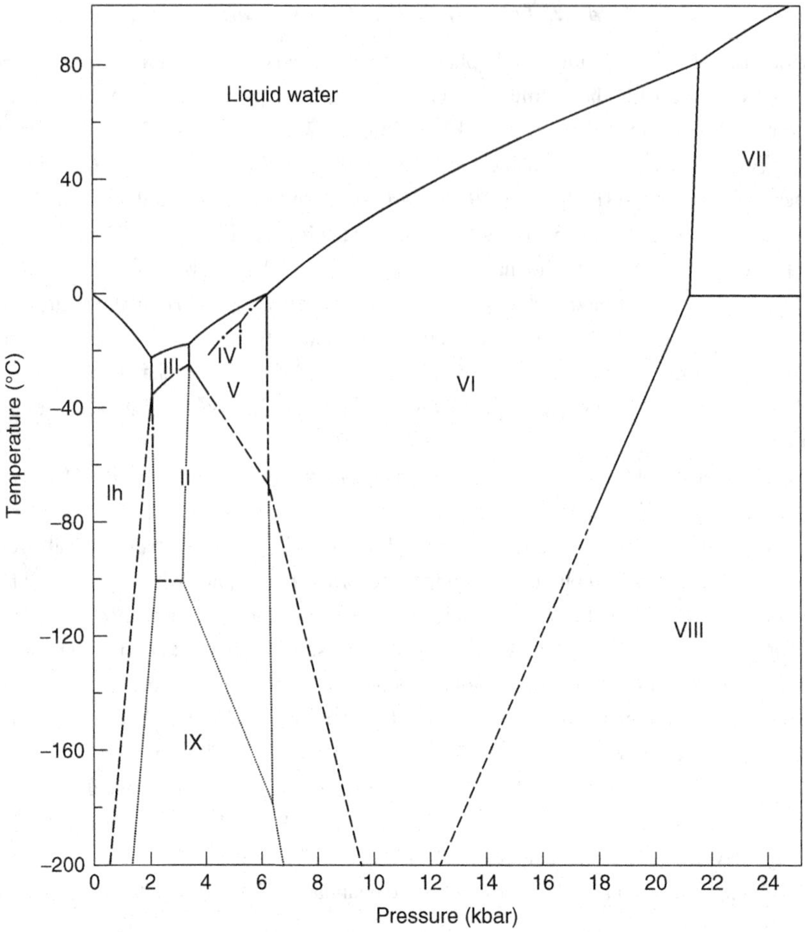

Figure 4.2. Phase diagrams of the solid states of water. Solid lines are measured stable lines; dash-dotted lines are measured metastable lines; dashed lines are extrapolated or estimated stable lines; dotted lines are extrapolated or estimated metastable lines. Ice IV is metastable in the region of stability of ice V; the indicated field for ice IV is inferred from the D_2O system. Ice I_c and vitreous ice are not indicated. Adapted from Whalley (1969), Hobbs (1974), and Rosenberg (2005) with changes.

In contrast to ordinary ice I_h, all high-pressure forms of ice, including the ices VI and VII, have greater density than liquid water at the same conditions. The slopes dT_m/dp_m at the boundaries of the ices VI and VII with liquid water are positive, and they can transform into liquid water (melt) at increasing temperature or decreasing pressure. Note that if creation of the ices VI and VII and other high-temperature ices were technologically possible on aircraft, they could be considered potential candidates for cloud seeding programs.

Many experimental studies of selected thermodynamic properties of the ambient hexagonal ice I_h have been reviewed in several books and articles (e.g., Dorsey, 1968; Fletcher, 1970a; Hobbs, 1974; Petrenko and Whitworth, 1999; Curry and Webster, 1999; Sastry, 2002). Reviews of the properties of cubic ice I_c and its transformations into I_h are given in Murray et al. (2005, 2010), Murray and Bertram, (2006), and Malkin et al. (2012).

4.2. Theories of Water

Various models and hypotheses have been developed to explain water structure, anomalous behavior, and numerous forms in the supercooled region. Three of the most often discussed hypotheses are (1) stability limit conjecture, (2) the singularity free hypothesis, and (3) the liquid–liquid critical point hypothesis. These hypotheses are illustrated in the *p-T* diagrams in Figs. 4.3a, b, and c. The first hypothesis is based on the measured sharp increase near 235 K (−38 °C) in many thermodynamic properties of water: isobaric heat capacity c_p, isothermal compressibility k_T, and the coefficient of thermal expansion α_p:

$$c_p = T\left(\frac{\partial s_\eta}{\partial T}\right)_p, \qquad \kappa_T = \frac{1}{\rho}\left(\frac{\partial \rho}{\partial T}\right)_p, \qquad \alpha_p = \frac{1}{\rho}\left(\frac{\partial \rho}{\partial T}\right)_p. \tag{4.2.1}$$

The increase of these properties toward colder temperatures is similar to the power law singularity $(T - T_s)^{-\lambda}$; that is, the characteristics of water tend to infinity as T approaches T_s (Angell, 1982), analogous to the second- or higher-order phase transitions (Landau and Lifshitz, v.5, 1958; Lifshitz and Pitaevskii, 1997). The *stability limit hypothesis* (Speedy and Angell, 1976; Angell, 1982; Speedy, 1982) suggests that these thermodynamic properties of supercooled water (c_p, k_T, and α_p) tend to

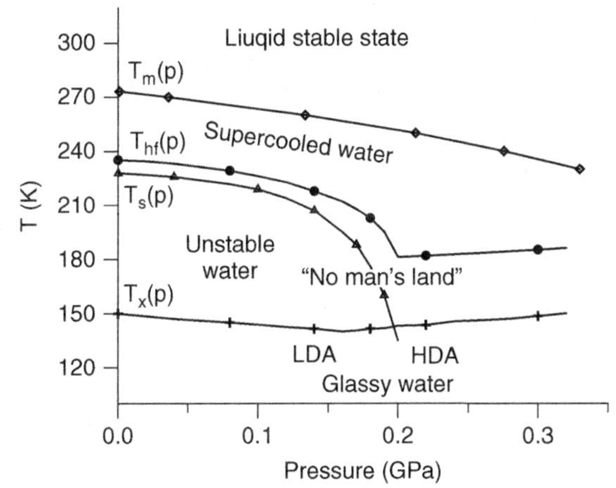

(a)

Figure 4.3. (a) The *p-T* diagram illustrating the thermodynamic behavior of water predicted by the stability limit conjecture. $T_m(p)$ is the melting temperature, $T_{hf}(p)$ is the temperature of homogeneous freezing (235 K (−38 °C) at p = 1 bar). The line $T_s(p)$ is the spinodal temperature, $T_s \approx 228$ K (−45 °C) at p = 1 bar, $T_x(p)$ is the temperature of crystallization of ultraviscous water at warming; the glass temperature T_g is located below $T_x(p)$ (not shown). LDA and HDA indicate the domains where the low-density and high-density amorphous solids may exist. Both $T_m(p)$ and $T_{hf}(p)$ decrease with increasing pressure at $p < 0.2$ GPa as predicted by the Clapeyron equation because the density of I_h is smaller than the density of liquid water. At $p > 0.2$ GPa, dT_{hf}/dp becomes positive and grows with increasing pressure because water freezes into the other forms of ice denser than liquid water (see Fig. 4.2). The sign or value of $dT_m(p)/dp$ on the melting curve $T_m(p)$ change at higher $p \sim 0.6 = 0.7$ GPa and $T \sim 150$ K (not shown here) is where liquid water can melt into denser solids (see Chapter 8). Adapted from Mishima and Stanley (1998), Koop (2004) and Stanley et al. (2007), with changes.

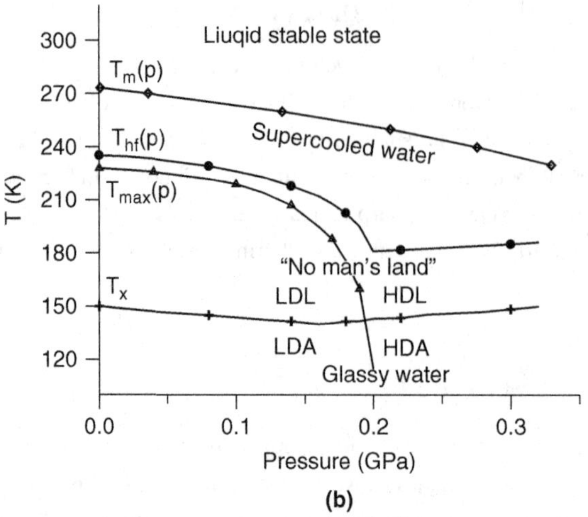

Figure 4.3. (b) The *p-T* diagram illustrating the thermodynamic behavior of water predicted by the singularity-free hypotheses. $T_m(p)$ is the melting temperature, $T_{hf}(p)$ is the temperature of homogeneous freezing (235 K or −38 °C at p = 1 bar), and $T_{max}(p)$ is the temperature of the maximum values of the thermodynamic properties (c_p, k_T, α_v). $T_x(p)$ is the temperature of crystallization of ultraviscous water at warming; the glass temperature T_g is located below $T_x(p)$ (not shown). LDL and HDL indicate the domains where the low-density and high-density liquid waters may exist. LDA and HDA indicate the domains where the low-density and high-density amorphous solids may exist. The $T_m(p)$ and $T_{hf}(p)$ decrease or increase with varying pressure as described in Fig. 4.3a. Adapted from Mishima and Stanley (1998, Koop (2004) and Stanley et al. (2007).

infinity at some limiting temperature. At a normal pressure of 1 atm, this limit was hypothesized to be near $T_s \approx -45 \pm 3$ °C (228 ± 3 K), called the *spinodal temperature*, which is located a few degrees below the homogeneous freezing temperature T_{hf} (Fig. 4.3a). This hypothesis assumes that the spinodal temperature connects at negative pressures to the locus of the liquid-to-gas spinodal (the limit of metastability) for supercooled water. However, the limit T_s cannot be reached in experiments because liquid water rapidly crystallizes earlier, at −38 °C (235 K) or 2–4 degrees above the spinodal temperature.

The temperature of singularities is a function of pressure, $T_s(p)$, and decreases with increasing pressure to about −110 °C (160 K) at $p = 0.2$ GPa (Fig. 4.3a), which is a result of the increase in the *freezing point depression* $\Delta T_f(p)$ with increasing pressure. The liquid water becomes unstable and therefore cannot exist below this temperature $T_s(p)$. The line $T_s(p)$ that separates the metastable and unstable water is called the *spinodal line* (the line with triangles in Fig. 4.3a). Existence of the spinodal temperature was hypothesized also for aqueous solutions by Rasmussen (1982), and Rasmussen and Liang (1993).

The second hypothesis is called the "*singularity free*" or *percolation hypothesis*. Since measurements of liquid water are possible only a few degrees above the hypothesized spinodal but not at the spinodal temperature itself, this hypothesis assumes that thermodynamic properties of water have a

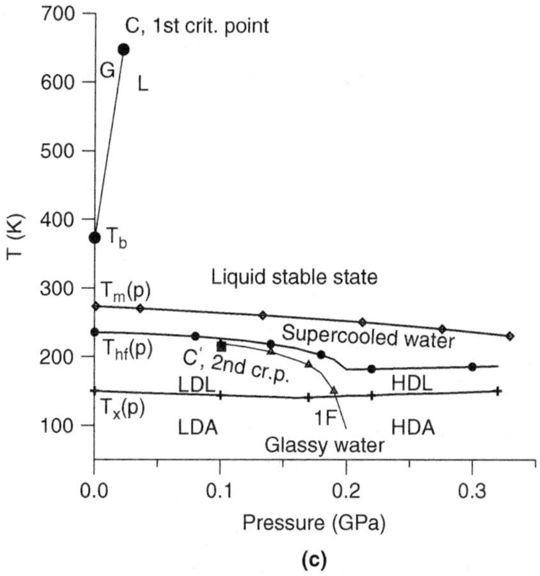

Figure 4.3. (c) The *p*-*T* diagram illustrating the thermodynamic behavior of water predicted by the liquid–liquid transition or the second critical point hypotheses. *C* is the first critical point of water at $T = 647$ K, $p = 220$ atm (0.022 GPa), *G* and *L* mean the gaseous and liquid phases ("low-density and high-density liquids") around *C*. $T_b = 373$ K is the boiling temperature, $T_m(p)$ is the melting temperature, $T_{hf}(p)$ is the temperature of homogeneous freezing (235 K (−38 °C) at p = 1 bar), $T_x(p)$ is the temperature of crystallization of ultraviscous water at warming; the glass temperature T_g is located below $T_x(p)$ (not shown). The black square with label *C′* near $T = 220$ K and $p = 0.1$ GPa (1000 bar) denotes the hypothesized second critical point. LDL and HDL indicate the domains where the low- and high-density liquid waters may exist. LDA and HDA indicate the domains where the low- and high-density amorphous solids may exist. The label 1F denotes the line of first-order phase transition that emanates from *C′* and separates the high-density and low-density phases that occur at temperatures below $T_{hf}(p)$. (After Mishima and Stanley (1998), Poole et al. (1992, 1994), with changes.)

sharp increase near T_s, and have local maxima near T_s at some temperature $T_{max}(p)$ (Fig. 4.3b), but are finite without a singularity, and decrease below T_{max}. This second conjecture assumes that the polyamorphic changes between LDA and HDA are relaxation phenomena, although they resemble genuine phase transitions.

The third hypothesis is based on an analogy with the well-known critical point *C* of water located at $T_{cr1} = 647$ K (374 °C), and $p_{cr1} = 218$ atm (22 MPa), and have a critical density $\rho_{cr1} = 0.328$ g cm^{-3} (Fig. 4.3c). Above this point, there is no difference between liquid water and vapor. It can be said that below the point *C*, water is "separated" into two "liquids": one is the "high-density liquid" (i.e., true liquid water), and the other is "low-density liquid"—that is, water vapor, meaning both "liquids" can coexist. By analogy, this theory hypothesizes the existence of *a second critical point C′* at low temperature and high pressure, near $T_{cr2} \approx 220$ K (−53 °C) and $p_{cr2} \approx 0.1$ GPa (1000 atm) with the density $\rho_{cr2} \approx 1$ g cm^{-3} (Fig. 4.3c). This hypothesis is based on molecular dynamics simulations and the equation of the state of liquid water (Poole et al., 1992, 1994) and was supported by the other approaches (Ponyatovsky et al., 1994; Mishima, 1996; Borick et al., 1995; Moynihan, 1997; Tejero and Baus, 1998).

By analogy with the water structure around the first critical point, this third hypothesis states that two liquid waters exist below the second critical temperature T_{cr2}: a low-density liquid (LDL) at low pressure, and a high-density liquid (HDL) at a higher pressure, with possible transitions between them similar to transitions between liquid water and vapor below the first critical point C (Fig. 4.3c). Therefore, this theory is also called *"the liquid–liquid phase transition"* hypothesis. Near the first known critical point, water is a fluctuating mixture of the local molecule structures that resemble structures of liquid water and vapor. The analogous structure is hypothesized to exist near the second critical point C', consisting of a fluctuating mixture of LDL and HDL structures.

The thermodynamic response functions (e.g., isobaric heat capacity c_p) still have maxima in the domains of the phase diagram with one phase and far enough from the critical points. These maxima are located along the lines $T_{max}(p)$ that continue the spinodal or coexistence lines into the domains with one phase. These asymptotic lines continuing the spinodal line are sometimes called the *Widom lines*, and are often regarded as an extension of the coexistence line into the "one-phase regime." The lines of the maxima for different response functions asymptotically approach one another as the critical point is approached, since all response functions become expressible in terms of the correlation length (Stanley et al., 2007).

One possible explanation of existence and coexistence of the two phases of water is the following (Stanley et al., 2007). Because water is both tetrahedral and charged, it means that the simple potentials of intermolecular interactions with one minimum as in the van der Waals theory considered in Section 3.4 (e.g., Lennard-Jones potential) are not sufficient to describe the complexity of water. One way to modify the Lennard-Jones potential and to provide at least a simplified description is to bifurcate the single minimum (as shown in Fig. 3.2) into two minima. The first minimum, at a closer distance, corresponds to two pentamers (a water molecule and its four neighbors) of water interacting with each other in a rotated configuration. The second minimum, at a greater distance, occurs in the unrotated position. This second position is a deeper minimum because although the pentamers are farther apart there is the potential for hydrogen bonding between the molecules and we can see the beginnings of an ice-like hexagonal structure. The existence of the two minima in the potential instead of one minimum in Fig. 3.2 may cause the formation of two polymorphs or polyamorphs of water when temperature and pressure reach some critical values, with a low-density form corresponding to a second minimum at a greater distance, and a high-density form corresponding to a first minimum at a closer distance between the molecules.

Another analogy for the coexistence of LDL and HDL can be seen in the theory of superfluidity at very low temperatures, where two liquids also exist: a normal viscous component and a superfluid component without viscosity. The two liquids coexist with mutual percolation without friction, owing to the specific energy spectrum with existence of a finite threshold for an elementary energy excitation (Landau and Lifshitz, v. 5, 1958; Lifshitz and Pitaevskii, 1997).

The properties and theory of the vitreous state of water (thermodynamics, structure, and crystallization) are described by Gutzow and Schmelzer (1995, 2011) and Debenedetti (2003).

4.3. Temperature Ranges in Clouds and Equivalence of Pressure and Solution Effects

Some fraction of clouds and fogs form and evolve at "warm" temperatures above 0 °C, without participation of ice crystals. Parameterization of microphysical processes in this temperature domain is called "*warm cloud microphysics*." However, the vast majority of clouds or cloud layers forms entirely or partially at temperatures below 0 °C, where liquid water is metastable and is called "*supercooled water*." Clouds with liquid drops and ice crystals coexisting below 0 °C are called *mixed-phase clouds* and can exist in such state for a long time.

Pure supercooled water may exist down to about –40 °C, but droplets of aqueous supercooled solutions may exist without freezing at much lower temperatures. Interest in the processes and forms of water at low temperatures was stimulated in the 1980s up to now by the intensive studies of clouds that form at temperatures around and below –40 °C and their importance in the climate system. These clouds include the anvils of deep convective clouds and various cirriform clouds at altitudes from 6 to 12 km in the Arctic and midlatitudes to 15–20 km in the tropics with the corresponding temperature range from –30 to –80 °C (CIRRUS-2002), and "pearl" clouds formed over mountains in orographic waves at altitudes of 15–25 km at similar temperatures.

Polar stratospheric clouds may occur at temperatures of –80 to –100 °C (193 to 173 K), especially in the Antarctic winter, and may play an important role in ozone depletion and formation of the ozone hole. Even colder cloud types are noctilucent or polar mesospheric clouds that form in the upper mesosphere at altitudes of 80–90 km at temperatures that can be near a climatological mean of –150 °C (120 K) for this layer or a few tens of degrees lower (Lübken et al., 2009; Murray and Jensen, 2010). Recent studies suggest that ice I_h is metastable at temperatures below 72 K and can be transformed into higher-ordered, ferroelectric ice XI (Petrenko and Whitworth, 1999; Kuo et al., 2001, 2004; Singer et al., 2005). The lowest temperatures observed in terrestrial clouds are on the order of 80–90 K (Lübken et al., 2009), still above the supposed transition temperature to ice XI. Clouds on other planets may form at even colder temperatures (e.g., Lewis and Prinn, 1984; Lewis, 1995; Ibragimov, 1990; Curry and Webster, 1999).

Another important area of low-temperature studies related to cloud physics is weather modification by cloud seeding. Some earlier studies used various artificial ice nuclei like AgI, PbI, or similar substances with a crystalline structure close to natural ice, which were active at environmental temperatures (e.g., Schaeffer, 1949; Vonnegut, 1947; Hobbs, 1974; Kachurin, 1978; Dennis, 1980; Cotton and Pielke, 2007). Subsequently, many different crystallizing or cooling agents were tested, including solid granulated carbon dioxide (CO_2), also called "dry ice" and used to keep ice cream in summer (sublimation temperature T_{subl} = –78.9 °C or 194.25 K), liquefied propane (boiling temperature T_b = –40.2 °C or 233 K), liquefied nitrogen (boiling temperature T_b = –200 °C or 73.15 K), and similar substances. These cooling agents, being relatively cheap and convenient to use, have very low boiling or sublimation points and create huge amounts of tiny ice crystals when evaporating, with efficiency up to 10^{11}–10^{14} per 1 gram of evaporated substance.

These crystals cause rapid glaciation of a cloud or fog under appropriate conditions, which can lead to precipitation enhancement (applied for agriculture or fighting wildfires) or decrease (applied for decrease of snowfall in the cities in winter) or to fog dispersal in airports, seaports and along the roads. These cooling agents and techniques were tested in laboratories and have been widely applied in field experiments (e.g., Gaivoronsky and Seregin, 1962; Serpolay, 1969; Bigg et al., 1969;

Krasnovskaya et al., 1964, 1987; Hicks and Vali, 1973; Silverman and Weinstein, 1973; Dennis, 1980; Berjulev et al., 1989; Vlasiuk et al., 1994; Cotton and Pielke, 2007).

Field experiments on cloud seeding can be expensive, time consuming, and require continuation for many years to obtain and estimate a statistically significant effect. Therefore, cloud seeding programs stimulated development of numerical models with detailed accounts of the microphysical processes of cloud glaciation (e.g., Orville and Chen, 1982; Cotton et al., 1986; Khvorostyanov, 1984, 1987; Bakhanov and Buikov, 1985; Khvorostyanov et al., 1989; Kondratyev et al., 1990; Khvorostyanov and Khairoutdinov, 1990; Khairoutdinov and Khvorostyanov, 1991; Krakovskaia and Pirnach, 2004; Cotton and Pielke, 2007). However, both field experiments and the numerical modeling of cloud seeding are hampered by the lack of understanding of the mechanisms of action of the seeding agents, their efficiency under various conditions, and even lack of knowledge about the forms of ice that are nucleated by these agents. This has stimulated research on cloud microphysical processes at low temperatures, which may help to improve cloud seeding methods.

A closely related problem is the high-pressure effects on water forms and transitions. A similarity was found in experiments and calculations in the freezing ΔT_f and melting ΔT_m point depressions caused by the addition of solutes to pure water and by applied high pressure. The solution concentration of ~2.7 mole of NaCl caused effects similar to an applied pressure of ~10^3 bar; and molality of 4 moles NaCl caused $\Delta T_m \approx 15$ K and $\Delta T_f \approx 27$ K, equivalent to a pressure of 1,400 bar (Kanno and Angell, 1977; Leberman and Soper, 1995). Parameterizations of this equivalence were developed by Koop et al. (2000) and Baker and Baker (2004), and a simple quantitative equivalence between these two effects was theoretically derived and discussed by Khvorostyanov and Curry (2004a) and Curry and Khvorostyanov (2012), as described in Chapter 8. Therefore, studies of the forms and transitions of water under high pressure can elucidate the freezing of solutions droplets and mechanisms of ice formation in clouds.

4.4. Parameterizations of Water and Ice Thermodynamic Properties

In this section, we describe empirical parameterizations of water and ice thermodynamical properties used in numerical modeling and analyses of experiments on ice nucleation and drop activation.

4.4.1. Saturated Vapor Pressures

The saturated vapor pressures are used in various units. The relations among these units are given in the Appendix for Chapter 4 for convenience. Since the melting heats and heat capacities depend on the temperature and pressure, the saturated humidities over water and ice obtained by integration of the Clausius–Clapeyron equations described in Chapter 3 may require corrections. The more precise parameterizations of the saturated pressures of water, e_{ws}, and ice, e_{is}, account for such empirical or semi-empirical corrections or are based on various fits used in practical calculations. These include several versions of the Magnus equation and parameterizations by Gratch and Goff, Wexler and Hyland, and others where e_{ws} and e_{is} were approximated with exponential functions with arguments as the polynomial and logarithmic functions of temperature. A detailed analysis and comparison of many of these parameterizations and their accuracy were performed by Murphy and Koop (2005).

Wexler (1976) fitted the temperature dependence of $e_{ws}(T)$ as

$$e_{ws} = \exp(a_0 T^{-2} + a_1 T^{-1} + a_2 + a_3 T + a_4 T^2 + a_5 T^3 + a_6 T^4 + a_7 \log T), \tag{4.4.1}$$

with e_{ws} in Pa, $a_0 = -2.9912729 \times 10^3$, $a_1 = -6.0170128 \times 10^3$, $a_2 = 1.887643854 \times 10^1$, $a_3 = -2.8354721 \times 10^{-2}$, $a_4 = 1.7838301 \times 10^{-5}$, $a_5 = -8.4150417 \times 10^{-10}$, $a_6 = 4.4412543 \times 10^{-13}$, and $a_7 = 2.858487$. The values of e_{ws} in dyn cm^{-2} (= 10^{-3} mb) are obtained using the relation 1 dyn cm^{-2} = 10 Pa—i.e., multiplying e_{ws} in (Eqn. 4.4.1) and subsequent similar equations by 10. Murphy and Koop (2005) fitted $e_{ws}(T)$ for the range $123 < T < 332$ K as:

$$\ln e_{ws}(T) = a_0 - a_1/T - a_2 \ln T + a_3 T$$
$$+ \tanh[a_4(T - a_5)](a_6 - a_7/T - a_8 \ln T + a_9 T), \qquad (4.4.2)$$

with e_{ws} in Pa, $a_0 = 54.842763$, $a_1 = 6763.22$, $a_2 = 4.210$, $a_3 = 3.67 \times 10^{-4}$, $a_4 = 0.0415$, $a_5 = 218.8$, $a_6 = 53.878$, $a_7 = 1331.22$, $a_8 = 9.44523$, and $a_9 = 0.014025$.

The saturated over ice vapor pressure was approximated by Hyland and Wexler (1983) in the range $173.16 < T < 273.16$ as

$$e_{is} = \exp(a_0/T + a_1 + a_2 T + a_3 T^2 + a_4 T^3 + a_5 T^4 + a_6 \log T), \qquad (4.4.3)$$

with e_{is} in Pa, $a_0 = -5.6745359 \times 10^3$, $a_1 = 6.3925247$, $a_2 = -9.6778430 \times 10^{-3}$, $a_3 = 6.2215701 \times 10^{-7}$, $a_4 = 2.0747825 \times 10^{-9}$, $a_5 = -9.4840240 \times 10^{-13}$, and $a_6 = 4.1635019$. Murphy and Koop (2005) suggested the parameterization for e_{is} for $T > 110$ K

$$e_{is} = \exp(a_0 + a_1/T + a_2 \ln T + a_3 T), \qquad (4.4.4)$$

with e_{is} in Pa, $a_0 = 9.550426$, $a_1 = -5723.265$, $a_2 = 3.53068$, and $a_3 = -0.00728332$. A simpler exponential fit is also used for e_{is}

$$e_{is}(T) = \exp(a_1 + b_1/T). \qquad (4.4.5)$$

The values of the parameters vary slightly in the fits by various authors; Murphy and Koop (2005) give $a_1 = 28.9074$ and $b_1 = -6143.7$ to fit the recent measurements in the range $164 < T < 273.16$ K. The values of e_{is} in dyn cm^{-2} are obtained by multiplying by 10. For reference, the values of e_{is} in hPa are 6.106 (0 °C), 2.597 (−10 °C), 1.032 (−20 °C), 0.3797 (−30 °C), 0.1283 (−40 °C), and 0.0393 (−50 °C).

4.4.2. Heat Capacity of Water and Ice

Pruppacher and Klett (1997) provide the following empirical fit for the heat capacity of water c_w in cal g^{-1},

$$c_w = c_{w0} + a_1(T_c - T_{c1})^2 + a_2(T_c - T_{c1})^4, \qquad 0 \le T_c \le 35\ ^\circ\text{C}$$

with T_c in Celsius, $c_{w0} = 0.9979$, $a_1 = 3.1 \times 10^{-6}$, $a_2 = 3.8 \times 10^{-9}$, $T_{c1} = 35$, and

$$c_w = \sum_0^4 a_n T_c^n, \qquad -37 \le T_c \le 0\ ^\circ\text{C}, \qquad (4.4.6a)$$

with $a_0 = 1.000938$, $a_1 = -2.7052 \times 10^{-2}$, $a_2 = -2.3235 \times 10^{-5}$, $a_3 = 4.3778 \times 10^{-6}$, and $a_4 = 2.7136 \times 10^{-7}$. The isobaric molar heat capacity of ice $c_{pi,m}$ was approximated by Murphy and Koop (2005) as

$$c_{pi,m} = a_1 + a_2 T + a_3 T \exp[-(T/a_4)^2], \qquad T > 20\ \text{K}, \qquad (4.4.6b)$$

with $c_{pi,m}$ in J mol^{-1} K^{-1}, T in Kelvins, $a_1 = -2.0572$, $a_2 = 0.14644$, $a_3 = 0.06163$, and $a_4 = 125.1$. The values of specific heat capacity c_{pi} in cal g^{-1} K^{-1} are obtained as $c_{pi} = c_{pi,m}/(4.18 M_w)$—i.e., dividing

a_1, a_2, and a_3 by $4.18M_w \approx 75.24$, yields $a_1 = -0.02734$, $a_2 = 194.6305$, $a_3 = 8.1911 \times 10^{-4}$ for c_{pi}. Pruppacher and Klett (1997) using the data by Flubacher et al. (1960) approximated specific c_{pi} (cal g^{-1} K^{-1}) in the range 0 to −40 °C as

$$c_{pi} = 0.503 + 0.00175T_c. \tag{4.4.6c}$$

Equations (4.4.6a) and (4.4.6c) show that the heat capacity of ice near 0 °C is about half that of liquid water, which is caused by the difference in their internal structure.

The difference in molar heat capacity in J mol^{-1} K^{-1} between vapor and ice was approximated by Murphy and Koop (2005), similar to (Eqn. 4.4.6b), as

$$\Delta c_{p,m} = a_1 + a_2T + a_3T\exp[-(T/a_4)^2], \tag{4.4.7}$$

with $a_1 = -35.319$, $a_2 = 0.14457$, $a_3 = 0.06155$, and $a_4 = 129.85$. The difference in specific heat capacities is obtained by dividing coefficients a_1, a_2, a_3 by $4.18M_w \approx 75.24$, with $a_1 = -0.46941$, $a_2 = 1.92145 \times 10^{-3}$, $a_3 = 8.18049 \times 10^{-4}$, and a_4 is the same.

4.4.3. *Latent Heats of Phase Transitions*

The latent heat of evaporation L_v was approximated in Pruppacher and Klett (1997) based on measurements of the energies of the hydrogen bonds, fractions of broken bonds, and enthalpies of the vacancies formation as

$$L_v = 597.3 - 0.561T_c, \qquad -20 < T_c < 40\,°\text{C}, \tag{4.4.8a}$$

$$L_v = \sum_0^5 a_n T_c^n, \qquad -44 < T_c < -20\,°\text{C}, \tag{4.4.8b}$$

with L_v in cal g^{-1}, and $a_0 = -1412.3$, $a_1 = -338.82$, $a_2 = -122.347$, $a_3 = -0.7526$, $a_4 = -1.1595 \times 10^{-2}$, and $a_5 = -7.313 \times 10^{-5}$. These expressions predict an increase in L_v toward low T and give $L_v = 597.2$ cal g^{-1} at $T_c = 0$ °C and $L_v = 625$ cal g^{-1} at $T_c = -40$ °C. The limit of −44 °C in (Eqn. 4.4.8b) is caused by the assumption in Pruppacher and Klett (1997) that −45 °C is a singularity point. As discussed in Section 4.2, more recent experimental data and the current theories of water show the absence of a singularity at this temperature.

Murphy and Koop (2005) suggested another approximation for supercooled water

$$L_v(T) = b_0 + b_1T + \exp[b_2(b_3 - T)], \qquad 236\text{ K} < T < 273.16\text{ K}, \tag{4.4.9}$$

with L_v in J mol^{-1}, $b_0 = 56579$, $b_1 = -42.212$, $b_2 = 0.1149$, and $b_3 = 281.6$.

The molar latent heat of sublimation L_s can be obtained from (Eqn. 3.7.17) by integration of dL_s/dT over T; that is, using Kirchoff's relations and $\Delta c_{p,m}$ in (Eqn. 4.4.7):

$$L_s = L_{s0} + \int_{T_0}^{T} [c_{p1,m}(T') - c_{p2,m}(T')]dT'$$
$$+ \int_{T_0}^{T} \frac{dp}{dT'}\left[(v_1 - v_2) - T'\left(\frac{\partial(v_v - v_i)}{\partial T'}\right)_p\right]dT' \tag{4.4.10}$$

Murphy and Koop (2005) found the following fit to this integral

$$L_s(T) = b_0 + b_1 T + b_2 T^2 + b_3 \exp[-(T/b_4)^2], \qquad T > 30 \text{ K}. \tag{4.4.11}$$

with L_s in J mol^{-1}, $b_0 = 46782.5$, $b_1 = 35.8925$, $b_2 = -0.07414$, $b_3 = 541.5$, and $b_4 = 123.75$. L_s in cal g^{-1} is obtained dividing b_0, b_1, b_2, b_3 by $\approx 4.18 \times 18 = 75.24$, and b_4 remains the same. For reference, the value of L_s is 678.5 cal g^{-1} near 0 °C.

Several parameterizations have been developed for the specific melting heat $L_m(T)$. Pruppacher and Klett (1997) approximated it as

$$L_m(T) = \sum_0^4 a_n T_c^n, \qquad T > 229 \text{ K}, \tag{4.4.12}$$

with L_m in cal g^{-1}, $a_0 = 79.7$, $a_1 = -0.12$, $a_2 = -8.0481 \times 10^{-2}$, $a_3 = -3.2376 \times 10^{-3}$, and $a_4 = -4.2553 \times 10^{-5}$. This function describes a decrease in L_m with decreasing temperature; L_m = 79.7 cal g^{-1} ($\approx$1.43 kcal mole^{-1}) at $T_c = 0$ °C and $L_m = 53.9$ cal g^{-1} ($\approx$0.97 kcal mole^{-1}) at $T_c = -40$ °C.

Khvorostyanov and Curry (2004a) proposed an expression for the molar and specific melting heat, which is based on a compilation of experimental data from Johari et al. (1994) extrapolated to very low temperatures of amorphous ice:

$$L_m(T) = C_L T\{\tanh[(T - T_{L1})/T_{L2}] + c_{L0}\}, \tag{4.4.13}$$

with T in Kelvin, $T_{L1} = 215$ K, $T_{L2} = 40$ K, and $c_{L0} = 1.6$; the value $C_L = 8.82$ for L_m in J mole^{-1}, and 0.1172 for L_m in cal g^{-1}. This yields $L_m \approx 79.8$ cal g^{-1} at 0 °C, which well approximates the data from Johari et al. (1994) down to 160 K and is suitable for calculations of ice nucleation in very cold clouds.

Hellmuth et al. (2013, Fig. 37 therein) compared four parameterizations of L_m in the range $253 < T < 273$ K: (4.4.12) by Pruppacher and Klett (1997, PK97), (Eqn. 4.4.13) by Khvorostyanov and Curry (2004a, KC04a), L_m calculated as $L_m(T) = L_s(T) - L_v(T)$ using (Eqns. 4.4.9) and (4.4.11) for L_v and L_s from Murphy and Koop (2005, MK05), and L_m calculated with the Thermodynamic Equation of State TEOS-10 (IAPWS, 2009a,b). The differences between these different parameterizations were found to be not very large at $T > 253$ K, which indicate that each of them can be used in calculations in this temperature range. However, as will be shown in Chapters 8 and 9, the differences dramatically increase at lower temperatures. This is very important for correct calculations of parameters of relevance to homogeneous and heterogeneous ice nucleation, and special attention will be paid to the correct choice of $L_m(T)$ and its averaging in those chapters.

4.4.4. Surface Tension between Water and Air or Vapor

Pruppacher and Klett (1997) suggested the following expression for the surface tension between water and air, σ_{wa}, at $T_c > -40$ °C:

$$\sigma_{wa}(T) = \sum_{n=0}^{6} a_n T_c^n, \tag{4.4.14}$$

with σ_{wa} in erg cm^{-2}, a_0 = 75.93, a_1 = 0.115, $a_2 = 6.818 \times 10^{-2}$, $a_3 = 6.511 \times 10^{-3}$, $a_4 = 2.933 \times 10^{-4}$, $a_5 = 6.283 \times 10^{-6}$, and $a_6 = 5.285 \times 10^{-8}$. For reference, $\sigma_{wa} \approx \sigma_{wv}$ = 75.93 erg cm^{-2} (0 °C) and $\approx$88 erg cm^{-2} (−40 °C).

4.4.5. Surface Tension between Ice and Water or Solutions

Pruppacher and Klett (1997) recommend the following expressions for the surface tension between ice and water, σ_{iw}, in erg cm^{-2},

$$\sigma_{iw} = 28 + 0.25T_c, \qquad -36 < T_c < 0\ °\mathrm{C}, \tag{4.4.15a}$$

$$\sigma_{iw} = a_0 + a_1T_c + a_2T_c^2 + a_3T_c^3 \qquad -44 < T_c < -36\ °\mathrm{C}, \tag{4.4.15b}$$

with T_c in Celsius, $a_0 = 189.081$, $a_1 = 13.1625$, $a_2 = 0.3469$, and $a_3 = 3.125 \times 10^{-3}$.

Dufour and Defay (1963) compared calculations of the nucleation rates and freezing temperatures with observations and suggested a linear parameterization with a smaller slope

$$\sigma_{iw} = 23.8 - |\sigma'_{iw}|\, T_c, \tag{4.4.15c}$$

with the average value $|\sigma'_{iw}| = 0.102\,\mathrm{erg\,cm^{-2}K^{-1}}$ and the limits $0.06 < |\sigma'_{iw}| < 0.13\,\mathrm{erg\,cm^{-2}K^{-1}}$. This yields $\sigma_{iw} = 23.8$ at 0 °C and 19.7 erg cm^{-2} at −40 °C. Dufour and Defay did not limit this parameterization at −40 °C. Zobrist (2007) and Hoose and Möhler (2012) found the following fit in the T- range of 229 K to 238 K, which can be extended over the entire supercooled region up to the melting point $T_0 = 273.15$ K

$$\sigma_{iw} = a_0 + a_1T_r + a_2T_r^2, \qquad T_r = (T - T_0)/T_0, \tag{4.4.16}$$

where T_r is the reduced temperature, σ_{iw} is in erg cm^{-2}, T is in Kelvins, $a_0 = 32.98$, $a_1 = 12.048$, and $a_2 = -4.6705 \times 10^2$. This choice gives $\sigma_{iw} = 32.98$ erg cm^{-2} (0 °C) in agreement with the measurements described by Hobbs (1974) and $\sigma_{iw} \approx 18$ erg cm^{-2} at $T = -44$ °C; the last value is close to that from Dufour and Defay (1963).

The concept of water-ice surface tension becomes uncertain below the temperature of homogeneous freezing T_{fhom}, about −38 to −42 °C, because pure water freezes below these temperatures, as described in Section 4.2. Liquid aqueous solutions still can exist at lower temperatures. Several parameterizations have been suggested for the surface tension at the solution–air interfaces σ_{sa}. Tabazadeh et al. (1997) found the following fit to the experimental data by Sabinina and Terpugov (1935)

$$\sigma_{sa} = a_0 + a_1w - T(a_2 + a_3w), \tag{4.4.17}$$

where σ_{sa} is in erg cm^{-2}, T is in K, w is the percentage concentration of solute by weight, $a_0 = 142.35$, $a_1 = -0.96525$, $a_2 = 0.22954$, and $a_3 = -0.0033948$. This gives $\sigma_{sa} = 75.06$ erg cm^{-2} at $w = 0$, $T =$ 293.15 K, which is close to the prediction of (Eqn. 4.4.14) for the interface of pure water and vapor.

Myhre et al. (1998) performed detailed measurements of the surface tension σ_{sa} and density ρ_{sa} of aqueous solutions of sulfuric acid, approximated as the polynomials

$$\sigma_{sa}(w,T) = \sum_{i=0}^{7}\sum_{j=0}^{3} \sigma_{ij} w^i T_c^j, \tag{4.4.18}$$

$$\rho_{sa}(w,T) = \sum_{i=0}^{10}\sum_{j=0}^{4} \rho_{ij} w^i T_c^j. \tag{4.4.19}$$

The coefficients σ_{ij} and ρ_{ij} of these polynomials were given in the Tables in Myhre et al. (1998) for the domain $-53 < T_c < 27$ °C. A comparison with the earlier data by Sabinina and Terpugow (1935) at $T_c = 20$ °C showed a good agreement. In particular, the behavior of $\sigma_{sa}(w, T_c = 20\,°\mathrm{C})$ was characterized by an increase from ≈72.5 erg cm^{-2} at $w = 0$ (pure water) to the maximum of ≈77 erg cm^{-2} at $w = 0.4$ to 0.6, and then by monotonous decrease with increasing w to ~52.5 erg cm^{-2} at $w = 1$.

4.4.6. Surface Tension between Ice and Air or Vapor

Surface tension between ice and air, σ_{ia}, can be determined from Antonoff's rule, which can be considered as a limiting case of Young's relation (see Section 9.2) and states that $\sigma_{iw} = \sigma_{iv} - \sigma_{wv}$. Using (Eqn. 4.4.14) for σ_{wv} and (Eqn. 4.4.15a) for σ_{iw}, we can write for $\sigma_{iv} = \sigma_{iw} + \sigma_{wv}$,

$$\sigma_{iv}(T) = \sigma_{wv} + \sigma_{iw} = \sum_{n=0}^{6} a_n T_c^n + (28 - 0.25T_c), \qquad (4.4.20a)$$

where the coefficients a_0 to a_6 are given after (Eqn. 4.4.14). Hoose and Möhler (2012) combined the equations from Pruppacher and Klett (1997) and found the following fit for $\sigma_{iv} = \sigma_{iw} + \sigma_{wv}$,

$$\sigma_{iv} = (a_0 + a_1 T_c) + (a_2 + a_3 T_c), \qquad (4.4.20b)$$

with σ_{iv} in erg cm^{-2}, T_c in Celsius, $a_0 = 76.1$, $a_1 = -0.155$, $a_2 = 28.5$, and $a_3 = 0.25$. More detailed thermodynamic models (e.g., Clegg and Brimblecombe, 1995; Clegg et al., 1998) allow more precise calculations of various properties of multicomponent aqueous solutions over the wide regions of compositions and temperatures.

4.4.7. Density of Water

For liquid water above 0 °C, Kell (1975) suggests the following polynomial expression at $p = 1$ atm:

$$\rho_w(T_c) = (1 + B_{pw} T_c)^{-1} \sum_{n=0}^{5} A_n T_c^n, \qquad 0 \le T_c \le 100\ °C, \qquad (4.4.21a)$$

with ρ_w in g cm^{-3}, T_c in degrees C, $A_0 = 0.9998396$, $A_1 = 1.8224944 \times 10^{-2}$, $A_2 = -7.922210 \times 10^{-6}$, $A_3 = -5.544846 \times 10^{-8}$, $A_4 = 1.497562 \times 10^{-10}$, $A_5 = -3.932952 \times 10^{-13}$, and $B_{pw} = 1.8159725 \times 10^{-2}$. Hare and Sörensen (1987) found a polynomial for supercooled water to −33 °C,

$$\rho_w(T_c) = \sum_{n=0}^{6} a_n T_c^n, \qquad -33 \le T_c \le 0\ °C, \qquad (4.4.21b)$$

with $a_0 = 0.99986$, $a_1 = 6.690 \times 10^{-5}$, $a_2 = -8.486 \times 10^{-6}$, $a_3 = 1.518 \times 10^{-7}$, $a_4 = -6.9984 \times 10^{-9}$, $a_5 = -3.6449 \times 10^{-10}$, and $a_6 = -7.497 \times 10^{-12}$. This yields $\rho_w = 0.99986$ g cm^{-3} at 0 °C and ≈0.978 g cm^{-3} at −33 °C. More detailed data on water density at various temperatures and pressures can be obtained by solving the thermodynamic equation of state (TEOS) considered in the next section.

4.4.8. Density of Ice

The density of ice can be determined either from measurements of the volume V_{uc} of the unit cell of ice using the relation $\rho_i = 4M_w/(N_{Av}V_{uc})$, or from measurements of a volume v_i' of a single molecule of ice, or from direct measurements of the density. The ice density can also be calculated from the equation of state for water in the ice phase or from the direct numerical simulations (DNS) of water structure using various intermolecular potentials. Pruppacher and Klett (1997) obtained an empirical fit for ρ_i for hexagonal ice I_h

$$\rho_i(T) = \sum_{0}^{2} a_n T_c^n, \qquad (4.4.22)$$

with ρ_i in g cm^{-3}, $a_0 = 0.9167$, $a_1 = -1.75 \times 10^{-4}$, $a_2 = -5.0 \times 10^{-7}$, for $-180 \le T_c \le 0$ °C.

Zobrist (2007) and Hoose and Möhler (2012), based on data from Lide (2001), parameterized v_i' using the relations

$$v_i'(T) = \frac{m_w}{\rho_i} = \frac{M_w}{N_{Av}\rho_{i0}}\left[\sum_0^3 a_{vn}T_r^n\right]^{-1}, \tag{4.4.23}$$

where $m_w = M_w/N_{Av}$ is the mass of a water molecule, v_i' in cm^3, $T_r = (T - T_0)/T_0$ is the reduced temperature, $T_0 = 273.15$ K, $a_{v0} = 1$, $a_{v1} = -0.05294$, $a_{v2} = -0.05637$, $a_{v3} = -0.002913$. The ice density follows from the relation

$$\rho_i(T) = \frac{m_w}{v_i'(T)} = \frac{M_w}{N_{Av}v_{ice}(T)} = \rho_{i0}\left[\sum_0^3 a_{vn}T_r^n\right], \tag{4.4.24}$$

with $\rho_{i0} = 0.9167$ g cm^{-3} and $v_{ice} = v_i'$.

A few other parameters used in calculations or estimates are (approximate values at normal conditions): the molar volume of water $v_w = M_w/\rho_w \approx 18$ cm^3 g^{-1}; the molar volume of ice $v_i = M_w/\rho_w \approx 19.63$ cm^3 g^{-1}; the concentration of molecules in liquid $N_l = N_{Av}/v_w \approx 3.34 \times 10^{22}$ cm^{-3}; the concentration of molecules in ice $N_I = N_{Av}/v_i \approx 3.06 \times 10^{22}$ cm^{-3}; the molecular volume in liquid water $v_{w1} = v_w/N_{Av} \approx 3 \times 10^{-23}$ cm^3; the molecular volume in ice $v_{i1} = v_i/N_{Av} \approx 3.27 \times 10^{-23}$ cm^3.

4.5. Heat Capacity and Einstein–Debye Thermodynamic Equations of State for Ice

The energy and heat capacity of a solid in classical physics is evaluated by assuming that every atom has vibrational degrees of freedom that can be described as a linear oscillator. This model is realistic in many cases because the displacement of an atom from the equilibrium state is accompanied by an elastic force that tends to return it to the original state as in a real oscillator. It is assumed in classical physics that each atom has the mean energy $\varepsilon = kT$ for each vibrational degree of freedom. If a solid consists of N molecules with l atoms in each, and each atom has three degrees of freedom, then the total number of degrees of freedom is $3Nl$. Three degrees of freedom correspond to the translation, and three correspond to the rotation of the body as a whole. Thus, the number of vibrational degrees of freedom is $3Nl - 6 \approx 3Nl$ since Nl is very large. The vibrational energy is $E_{vib} = 3NlkT$, and the heat capacity is therefore

$$C_v = \frac{dE_{vib}}{dT} = \frac{d}{dT}(3NlkT) = 3Nkl, \tag{4.5.1}$$

or the heat capacity per one molecule is $c_v' = 3kl$. If the amount of solid is one mole, $N = N_{Av}$, then using the relation $N_{Av}k = R$, (Eqn. 4.5.1) can be written as

$$C_v = 3Rl \tag{4.5.2}$$

Equations (4.5.1) and (4.5.2) express Dulong and Petit's law, which predicts that the heat capacity is independent of temperature. This law satisfactorily describes the experimental data on C_v for many simple atomic or ionic crystals at sufficiently high temperatures and is also applicable for some gases.

However, the heat capacities of solids decrease with decreasing temperature, and tend to zero at zero temperatures, in conflict with Dulong and Petit's law. Einstein (1906) suggested not using the equipartition energy distribution, kT for each oscillation, but the quantum Planck's expression considered in Section 3.2

$$E_{vib} = 3Nl\bar{\varepsilon}, \quad \bar{\varepsilon} = \frac{h\nu}{\exp(h\nu/kT) - 1} \tag{4.5.3}$$

where ν is the frequency of oscillations, and $\bar{\varepsilon}$ is the mean energy per vibration. Einstein for simplicity assumed the same frequency for all oscillations in a solid. Then, the heat capacity becomes

$$C_V = \frac{dE_{vib}}{dT} = 3Nkl\frac{(h\nu/kT)^2 \exp(h\nu/kT)}{[\exp(h\nu/kT) - 1]^2}. \tag{4.5.4}$$

This equation correctly predicts decrease C_v with decreasing T. At high temperatures, $kT >> h\nu$, and expanding the denominator by the small quantity $x = h\nu/kT << 1$ also yields $C_v = 3Nkl$; that is, Einstein's formula contains Dulong and Petit's law as its high temperature limit in agreement with the correspondence principle. At low T, when $kT << h\nu$, Einstein's formula predicts that C_v tends to zero as $\exp(-h\nu/kT)$, while experimental data show that $C_v \sim T^3$ at low T, called *Debye's cubic law for heat capacity*. This disagreement is caused by the assumptions that all vibrations have the same frequency and all vibrations are independent.

A more advanced thermodynamic theory of the heat properties of the solids has been developed by Debye (1912). Debye again assumed that the major contribution to the energy of a solid comes from the vibrations of the atoms around their equilibrium positions, and each vibration can be considered an oscillator. However, the vibrational energy spectrum in Debye's theory was characterized not by a single frequency but by a spectrum of frequencies. If the number of molecules in a solid is N and the number of atoms in each molecule is l, then the number of vibrational degrees is $3Nl - 6 \approx 3Nl$, as discussed earlier.

The derivation of Debye's expressions for heat capacity can be based on the Bose–Einstein energy distribution (see Section 3.2.4) and using the concepts of quantum mechanics (Landau and Lifshitz, v. 5, 1958). If the frequency of an oscillator is ν, the quantum of energy of a vibration is $h\nu$, and the number of an energy level is n, then the energy of each oscillator is $\varepsilon_n = h\nu(n + 1/2)$. Summation over the frequencies leads to rather complicated expressions for the statistical sum Z and Helmholtz free energy F. Debye (1912) derived an interpolation formula between the low and high temperature limits based on two assumptions: introducing a continuous density of oscillators, and assuming some maximum frequency of oscillators ν_m. With these assumptions, Debye obtained an expression for F

$$F = N\varepsilon_0 + NlkT\left\{3\ln\left[1 - \exp\left(-\frac{\theta_D}{T}\right)\right] - D_D\left(\frac{\theta_D}{T}\right)\right\}, \tag{4.5.5}$$

where θ_D is Debye's temperature defined as $\theta_D = h\nu_m/k$, and $D_D(z)$ is Debye's function,

$$D_D(z) = \frac{3}{z^3}\int_0^z \frac{x^3 dx}{e^x - 1}. \tag{4.5.6}$$

The integrand here represents integration over the frequencies with the Bose–Einstein distribution and is similar to those arising in Plank's theory of black-body radiation. As described in Section 3.1, the entropy $S_\eta = -(\partial F/\partial T)_V$ and the energy $E = F + TS_\eta$ are both calculated with (Eqns. 4.5.5) and (4.5.6), and finally, the heat capacity C_V in Debye's theory is

$$C_V = \left(\frac{\partial E}{\partial T}\right)_V = 3Nlk\left[4\left(\frac{\theta_D}{T}\right) - 3\frac{\theta_D}{T}\frac{1}{\exp(\theta_D/T)-1}\right]. \tag{4.5.7}$$

The limits of C_V can be obtained using the limits of $D_D(\theta_D/T)$. In the high temperature limit, $x = \theta_D/T << 1$, the function D_D tends to 1. For one mole, $N = N_{av}$, this gives the molar heat capacity $C_V^h = 3Rl$, or for ice with $l = 3$, $C_V^h = 9R$. That is, the limit of Debye's equation gives at high T the Dulong and Petit's law (Eqn. 4.5.2). At low T, so that $z = \theta_D/T >> 1$, the limit of Debye's function (Eqn. 4.5.6) is expressed via Riemann's zeta function, and yields the low temperature limit:

$$C_V^l = \frac{12Nlk\pi^4}{5}\left(\frac{T}{\theta_D}\right)^3. \tag{4.5.8}$$

Thus, the heat capacity gives the correct cubic Debye's law $C_v \sim T^3$ in the low temperature limit, in agreement with experimental data (Landau and Lifshitz, v.5, 1958; Hobbs, 1974).

Debye's formula appeared to be applicable to many solids in predicting the specific heat capacities of many simple atomic and ionic crystals. However, as noted by Hobbs (1974), it is less successful in the case of molecular crystals such as ice, and the entire T-curve of the heat capacity calculated with Debye's (Eqn. 4.5.7) cannot be fitted with a single value of θ_D. The values $\theta_D = 192$ K or 318 K better reproduce $C_V(T)$ either at lower or higher temperatures but cannot reproduce in the entire T-region and the experimentally measured curves such as those approximated by the fit (Eqn. 4.4.6b). It was clarified later that several rotational and translational vibrations can contribute to C_p of ice (Hobbs, 1974). Then, it may be necessary to introduce several of Debye's temperatures, or θ_D as the function of T for a description of ice properties in various T-regions, and the theory becomes too complicated. A detailed analysis performed in Hobbs' (1974) fundamental monograph showed that a unified theory of ice thermodynamic properties for the entire T-range was still missing at the beginning of the 1970s. This situation existed until a substantial breakthrough occurred only in the last two decades, as described in the next sections.

4.6. Equations of State for Ice in Terms of Gibbs Free Energy

The development of the new theory of ice was stimulated by the International Association for Properties of Water and Steam (IAPWS) and was based on construction of the thermodynamic potential for ice. The specific Gibbs energy for hexagonal ice I_h (or the Gibbs function of ice) expressed as a function of temperature and pressure, $G_{Ih}(T, p)$, represents the fundamental equation of state for ice. This definition of the equation of state differs from the equations of state for gases and liquids described in Chapter 3, which are usually formulated as relations between the density ρ or volume V with temperature and pressure, $\rho = \rho(p, T)$, or $V = V(p, T)$. The Gibbs function $G(T, p)$ also represents an equation of state of the liquid or solid and allows evaluation of various properties of the substance using the thermodynamic derivatives described in Section 3.1.

The following brief review of the history of development and of the essence of the new ice theory follows Feistel and Wagner (2005), IAPWS (2009a) and Hellmuth et al. (2013). The first Gibbs functions for ice I_h were published by Feistel and Hagen (1995, 1998, 1999) and Tillner-Roth (1998) and were valid only in the vicinity of the melting curve of ice. A new comprehensive Gibbs function that for the first time covered the entire range of existence of ice I_h was developed consistently with the IAPWS-95 formulation of fluid water (Wagner and Pruß, 2002; IAPWS 2009b), a detailed description of the construction of the Gibbs function and of the experimental data used was provided by Feistel and Wagner (2005). A second, slightly improved version was published by Feistel and Wagner (2006) and adopted as an international standard formulation by the International Association for the Properties of Water and Steam (IAPWS, 2009a) as part of the Thermodynamic Equation of Seawater–2010 (TEOS-10) (IOC, SCOR, and IAPSO, 2010). A review of the scientific development of this standard including a description of the literature involved in its development was given by Pawlowicz et al. (2012) in the preface of a special issue of Ocean Science. The perfect consistency of the TEOS-10 ice formulation with those for liquid water, water vapor, humid air, and seawater (Feistel et al., 2008) permits very accurate calculation of phase equilibria between ice and those substances, such as with humid air (Feistel et al. 2010b) and with liquid and gaseous water (IAPWS, 2011; Wagner et al., 2011) even at extremely low temperatures (Feistel and Wagner, 2007; Feistel, 2012; IAPWS, 2012).

In contrast to Debye's (1912) theory that calculates first the free energy and then the entropy and heat capacity as its derivatives (see Section 4.5), Feistel and Wagner (2005) constructed the theory in the reverse order. First, Feistel and Wagner (2005) found a very precise approximation for the specific heat capacity of ice c_{pi} by interpolation between the exponents of Debye's cubic law at low temperatures and the linear behavior at high temperatures using the rational Pade' approximation structure

$$\frac{c_{pi}}{c_{piu}} = T_r^3 \frac{c_{1FW} + c_{2FW} T_r^2 + c_{3FW} T_r^4}{1 + c_{4FW} T_r^2 + c_{5FW} T_r^4 + c_{6FW} T_r^6}, \tag{4.6.1}$$

where $T_r = T/T_t$ is the reduced temperature, $T_t = 273.16$ K, $c_{piu} = 1$ J kg^{-1} K^{-1}, $c_{1FW} = 185656.573673404$, $c_{2FW} = 159038807.871411$, $c_{3FW} = 2321798783.50605$, $c_{4FW} = 154.188350854909$, $c_{5FW} = 62945.966529576$, and $c_{6FW} = 1120863.23495484$. The resulting r.m.s. (root mean square) of the fit corresponding to (Eqn. 4.6.1) is 0.8% for the data of Giauque and Stout (1936), and 3.0% for those of Flubacher et al. (1960). So (Eqn. 4.6.1) by Feistel and Wagner (2005) approximates c_{pi} with very good accuracy over the entire temperature range from 0 °C to almost 0 K and can be used for high-accuracy calculations of ice I_h properties. Feistel and Wagner's (Eqn. 4.6.1) is in good agreement with the empirical approximations for c_{pi} described in Section 4.4 and extends them down to 0 K.

The heat capacity function (Eqn. 4.6.1) can be integrated analytically to obtain expressions for entropy s_η and free enthalpy (Gibbs energy) G_{Ih} for ice I_h. For this purpose, Feistel and Wagner (2005) represented the rational function (Eqn. 4.6.1) in a mathematically equivalent way by the real part of two pairs of complex partial fractions

$$\frac{c_{pi}}{c_{piu}} = T_r \,\mathrm{Re} \sum_{k=1}^{2} r_k \left[\frac{1}{T_r - T_{rk}} - \frac{1}{T_r + T_{rk}} + \frac{2}{T_{rk}} \right] \tag{4.6.2}$$

This expression takes the form of (Eqn. 4.6.1) if the real part of the complex sum is outlined explicitly. The two complex weight factors, r_1, r_2, and the corresponding root pairs, T_{r1}, T_{r2}, are responsible for bell-shaped low-temperature corrections, r_1 and T_{r1} at about 0 K, and r_2 and T_{r2} centered at about 100 K, modifying the otherwise asymptotically linear behavior for high temperatures at normal pressure, lim (c_{pi}/T) = 7.583 J kg^{-1} K^{-2} at $T \to \infty$. The low-temperature expansion of (Eqn. 4.6.2) is $\lim(c_{pi}/c_{piu}) = -2\,\mathrm{Re}\sum_{k=1}^{2} r_k (T_r/T_{rk})^3$, yielding Debye's cubic temperature dependence.

The complex numbers r_1, T_{r1}, and T_{r2} were found by fitting to the cited experimental data. The fourth one, r_2, was obtained by Feistel and Wagner by expanding into powers of pressure leading to the full Gibbs function, thus providing pressure-dependent heat capacities derived from the temperature dependence of compressibility. Using the relation from Section 3.1, $(\partial S_\eta/\partial T)_p = c_{pi}/T$, and integrating once over T, Feistel and Wagner obtained the entropy S_η; using the relation $S_\eta = -(\partial G_{Ih}/\partial T)_p$, the second integration yields the free enthalpy or specific Gibbs energy for hexagonal ice I_h in the following form

$$G_{Ih}(T,p) = G_{0Ih}(p) - S_{\eta 0} T_r T_t + T_t\,\mathrm{Re}\left\{\sum_{k=1}^{2} r_k[(T_{rk} - T_r)\ln(T_{rk} - T_r) + (T_{rk} + T_r)\ln(T_{rk} + T_r) - 2T_{rk}\ln T_{rk} - T_r^2/T_{rk}]\right\}, \tag{4.6.3}$$

$$G_{0Ih}(p) = \sum_{k=0}^{4} G_{0k}(p_r - p_{0r})^k, \tag{4.6.4}$$

$$r_2(p) = \sum_{k=0}^{2} r_{2k}(p_r - p_{0r})^k, \tag{4.6.5}$$

where $p_r = p/p_t$, $p_{0r} = p_0/p_t$, and p_t = 611.657 Pa. The complex numbers r_1, r_{20} to r_{22}, and T_{r1}, T_{r2}, are given in Table 4.1 (Feistel and Wagner, 2005b,c, 2006; IAPWS, 2009a; IOC et al., 2010). Here, $T_r = T/T_t$ denotes the reduced temperature, $p_r = p/p_t$ the reduced pressure, $p_{r0} = p_0/p_t$, the reduced normal pressure with T_t = 273.16 K, p_t = 611.657 Pa, and p_0 = 101 325 Pa. The complex logarithm ln(z) is meant as the principal value—i.e., it evaluates to imaginary parts in the interval $-\pi < \mathrm{Im}\,[\ln(z)] \leq +\pi$.

Table 4.1. *Coefficients of the Gibbs function of ice I_h in Feistel–Wagner (Eqns. 4.6.3) to (4.6.5).*

Coefficient	Real part	Imaginary part	Unit
G_{00}	$-0.632\,020233\,335\,886 \times 10^{6}$		J kg^{-1}
G_{01}	0.655 022 213 658 955		J kg^{-1}
G_{02}	$-0.189\,369\,929\,326\,131 \times 10^{-7}$		J kg^{-1}
G_{03}	$0.339\,746\,123\,271\,053 \times 10^{-14}$		J kg^{-1}
G_{04}	$-0.556\,464\,869\,058\,991 \times 10^{-21}$		J kg^{-1}
$S_{\eta 0}$ (absolute)	$0.189\,13 \times 10^{3}$		J kg^{-1} K^{-1}
$S_{\eta 0}$ (IAPWS-95)	$-0.332\,733\,756\,492\,168 \times 10^{4}$		J kg^{-1} K^{-1}
T_{r1}	$0.368\,017\,112\,855\,051 \times 10^{-1}$	$0.510\,878\,114\,959\,572 \times 10^{-1}$	
r_1	$0.447\,050\,716\,285\,388 \times 10^{2}$	$0.656\,876\,847\,463\,481 \times 10^{2}$	J kg^{-1} K^{-1}
T_{r2}	0.337 315 741 065 416	0.335 449 415 919 309	
r_{20}	$0.725\,974\,574\,329\,220 \times 10^{2}$	$-0.781\,008\,427\,112\,870 \times 10^{2}$	J kg^{-1} K^{-1}
r_{21}	$-0.557\,107\,698\,030\,123 \times 10^{-4}$	$0.464\,578\,634\,580\,806 \times 10^{-4}$	J kg^{-1} K^{-1}
r_{22}	$0.234\,801\,409\,215\,913 \times 10^{-10}$	$-0.285\,651\,142\,904\,972 \times 10^{-10}$	J kg^{-1} K^{-1}

The list of 18 adjusted parameters in Table 4.1 contains two unnecessary ones that formally appeared during the transformation of six real parameters describing heat capacity into four complex numbers (see Feistel and Wagner (2005) for more details). Two of the parameters reported in Table 4.1, $S_{\eta 0}$ and G_{00}, are adjusted to arbitrary reference-state conditions (Feistel et al., 2008). The IAPWS-95 reference state specifies entropy and internal energy of liquid water to vanish at the triple point (Wagner and Pruß, 2002; IAPWS- 2009b). The related values of the coefficients must be used if phase equilibria are calculated from equating the chemical potential of ice with that of liquid water or of water vapor provided by IAPWS-95 and TEOS-10. In contrast, the "absolute" reference state defines the residual entropy of ice I_h at 0 K to coincide with the value calculated statistically (Pauling, 1935; Nagle 1966; see also Gutzow and Schmelzer, 2011, for a review of the third law). The remaining 14 adjustable coefficients were fitted to 522 measured values of 32 different groups of measurements, 5 of those groups at elevated pressures (Feistel and Wagner, 2006).

The Feistel–Wagner equation of state (Eqn. 4.6.3) is valid over the entire range of naturally abundant, hexagonal ice I_h of H_2O, covering the temperature and pressure range $0\ \text{K} \le T \le 273.16\ \text{K}$ and $0 \le p \le 210$ MPa. The function also behaves reasonably at high pressures and low temperatures where no experimental data are available. The thermodynamic properties of ice I_h are computed from the Gibbs function by general thermodynamic relations given in Section 3.1. Additional relations, in particular those required for the calculation of phase equilibria with liquid water, seawater, water vapor, or humid air are available from the articles of Feistel et al. (2010a,b).

The Gibbs function of ice, together with the thermodynamic potentials of liquid water, water vapor, seawater, and humid air, as well as numerous properties derived thereof, are implemented equivalently in Fortran and in Visual Basic (or VBA for Excel) in the open-source Sea-Ice-Air (SIA) Library (Feistel et al. 2010a; Wright et al. 2010), whose latest update is freely available from the TEOS-10 website (www.teos-10.org). For convenience and higher computation speed, it is often useful to calculate a set of values from the full TEOS-10 equations over the particular range of interest, and then fit a separate correlation to those points as a "secondary" equation. For the frequently used phase-equilibrium curves of ice with liquid water and with water vapor, such simple and compact equations are already available from an additional IAPWS formulation (IAPWS 2011; Wagner et al. 2011).

The "secondary" equation for the sublimation pressure for H_2O ice I_h is (Wagner et al. 2011):

$$\ln p_r = (T_r)^{-1}\sum_{i=1}^{3} a_i (T_r)^{b_i}, \qquad p_r = \frac{p_{subl}}{p_t}, \tag{4.6.6}$$

where $T_r = T/T_t$, $T_t = 273.16$ K, and $p_t = 611.657$ Pa. The coefficients a_i and exponents b_i are: $a_1 = -0.212\,144\,006 \times 10^2$, $a_2 = 0.273\,203\,819 \times 10^2$, $a_3 = -0.610\,598\,130 \times 10^1$, $b_1 = 0.333\,333\,333 \times 10^{-2}$, $b_2 = 0.120\,666\,667 \times 10^1$, and $b_3 = 0.170\,333\,333 \times 10^1$. Equation (4.6.6) is valid down to 50 K based on a low-temperature extension for the water-vapour heat capacity (Feistel and Wagner, 2007; Feistel et al., 2010b; Wagner et al., 2011). The "secondary" equation for the melting pressure for H_2O ice I_h is (Wagner et al., 2011):

$$p_r = 1 + \sum_{i=1}^{3} a_i (1 - T_r^{b_i}), \qquad p_r = \frac{p_{melt}}{p_t}. \tag{4.6.7}$$

The coefficients a_i and exponents b_i are: $a_1 = 0.119\,539\,337 \times 10^7$, $a_2 = 0.808\,183\,159 \times 10^5$, $a_3 = 0.333\,826\,860 \times 10^4$, $b_1 = 0.300\,000 \times 10^1$, $b_2 = 0.257\,500 \times 10^2$, and $b_3 = 0.103\,750 \times 10^3$.

The intersection point of the two curves—(Eqns. 4.6.6) and (4.6.7)—is the experimental triple point at 273.16 K and 6.11657 hPa, which deviates slightly by 0.02 Pa from the pressure of the numerical triple point calculated directly from the thermodynamic potentials (Feistel et al., 2008; Wagner et al., 2011). Equations (4.6.6) and (4.6.7) describe the corresponding saturated vapor pressures with high accuracy and can be used along with the parameterizations described in Section 4.4.

The validity of the liquid–water formulation IAPWS-95 of TEOS-10 is restricted to temperatures above the triple point (Wagner and Pruß, 2002; IAPWS-2009b). But, because the function extrapolates well to lower temperatures, its use down to −20 °C is considered as safe, at least at low pressures.

4.7. Generalized Equations of State for Fluid Water

4.7.1. Equations of the van der Waals Type and in Terms of Helmholtz Free Energy

The equation of state (EOS) is one of the tools used for description of the properties of liquid water, water vapor, and their equilibrium in the wide regions of the temperature and pressure variations. More than thirty various versions of EOS for liquid water have been developed since the 1980s for applications in studies of water in the atmosphere, seawater, and in the fundamental properties of water. Reviews of EOS are provided, e.g., in IAPWS (1995), Jeffery and Austin (1997, 1999), Curry and Webster (1999), Kiselev (2000), Kiselev and Ely (2001), Wagner and Pruß (2002), IAPWS (2008), Holten et al. (2011, 2012), and Shi and Mao (2012). The analytical formulations of EOS have different forms; some represent generalizations of the van der Waals equation, while others are formulated in terms of the free Helmholtz energy, or as the thermodynamic models for supercooled water by adopting suitable physical scaling fields relative to the location of a liquid–liquid critical point in metastable water (Section 4.2).

For example, Song and Mason (1989, 1990a, b) and Ihm et al. (1991) generalized the van der Waals equation considered in Section 3.4. Their derivations were analogous to those described in Section 3.4 but with a more detailed consideration of the intermolecular interaction and expanding the free energy into a power series by the density (called *virial expansion*) including terms of higher order, which resulted in additional terms in the equation. The empirically determined equation for seawater in the form of the density as a function of temperature, pressure, and salinity adopted in UNESCO (1981) was described in Curry and Webster (1999). These forms of EOS have been further modified to describe the deeply supercooled water as discussed further on in this chapter.

Another approach is similar to that described in Section 4.6 for ice and is based on construction of thermodynamic potentials for liquid water. As described in Wagner and Pruß (2002), the Helmholtz free energy equation developed by Haar et al. (1982, 1984) has been adopted as a new international standard by the International Association for the Properties of Water and Steam (IAPWS). The fundamental equation of state for water developed by Saul and Wagner (1989) was optimized regarding its functional structure and that was fitted simultaneously to the experimental data of many different properties. In 1995, IAPWS adopted the equation of state developed by Pruß and Wagner as the new scientific standard under the name "The IAPWS Formulation 1995 for the Thermodynamic Properties of Ordinary Water Substance for General and Scientific Use."

The IAPWS-95 formulation EOS with its subsequent modifications (Wagner and Pruß, 2002; IAPWS- 2008, IAPWS-2009b; Shi and Mao, 2012) is the most elaborate EOS in terms of the

Helmholtz function. A compact description of the IAPWS-1995 formulation according to these sources is as follows. EOS is expressed in the form of dimensionless Helmholtz free energy $\phi(\delta_r, \tau_r)$ that is separated into two parts, an ideal gas part $\phi^0(\delta_r,\tau_r)$ and a residual part $\phi^r(\delta_r,\tau_r)$, so that

$$\phi(\delta_r,\tau_r) = \frac{f(\rho,T)}{R_{sp}T} = \phi^0(\delta_r,\tau_r) + \phi^r(\delta_r,\tau_r), \tag{4.7.1}$$

where T is temperature, ρ is the density, $f_h(\rho, T)$ is the specific Helmholtz free energy, $\delta_r = \rho/\rho_{cr}$ is the reduced dimensionless density, $\tau_r = T_{cr}/T$ is the inverse reduced temperature, $\rho_{cr} = 322$ kg m^{-3} is the critical density, $T_{cr} = 647.096$ K is the critical temperature, and $R_{sp} = 0.46151805$ kJ kg^{-1} K^{-1} is the specific gas constant. The ideal gas part $\phi^0(\delta_r,\tau_r)$ is written as (Eqn. 6.5) in Wagner and Pruß, 2002)

$$\phi^0(\delta_r,\tau_r) = \ln\delta_r + n_1^0 + n_2^0\tau_r + n_3^0\ln\tau_r + \sum_{i=4}^{8} n_i^0 \ln[1-\exp(-\gamma_i^0\tau_r)], \tag{4.7.2}$$

where the eight parameters n_i^0 and five parameters γ_i^0 are given in Wagner and Pruß (2002, Table 6.1), and IAPWS-2009b (Table 1). The residual part $\phi^r(\delta_r,\tau_r)$ can be written following (Eqn. 6.6) in Wagner and Pruß (2002):

$$\phi^r(\delta_r,\tau_r) = \sum_{i=1}^{7} n_i(\delta_r)^{d_i}(\tau_r)^{t_i} + \sum_{i=8}^{51} n_i(\delta_r)^{d_i}(\tau_r)^{t_i}\exp[(\delta_r)^{c_i}]$$
$$+\sum_{i=52}^{54} n_i(\delta_r)^{d_i}(\tau_r)^{t_i}\exp[-\alpha_i(\delta_r-\varepsilon_i)^2 - \beta_i(\tau_r-\gamma_i)^2] + \sum_{55}^{56} n_i\Delta^{b_i}\delta_r\psi_r, \tag{4.7.3}$$

where

$$\Delta = \theta_r^2 + B_i[(\delta_r-1)^2]^{a_i}, \theta_r = (1-\tau_r) + A_i[(\delta_r-1)^2]^{1/(2\beta_i)},$$
$$\psi_r = \exp[-C_i(\delta_r-1)^2 - D_i(\tau_r-1)^2]. \tag{4.7.4}$$

About 250 parameters—n_i, d_i, t_i, c_i, α_i, β_i, γ_i, ε_i, C_i, D_i, and A_i—are given in Wagner and Pruß (2002, Table 6.2), and IAPWS (2009b) (Table 2). The optimal values of the parameters were found by the multiparametric fits to the various properties of water over wide domains of temperature and pressure.

Relations of thermodynamic properties (entropy, enthalpy, pressure, isohoric and isobaric heat capacities, second and third virial coefficients, and others) to the ideal-gas part, $\phi^0(\delta_r,\tau_r)$, and the residual part $\phi^r(\delta_r,\tau_r)$, of the dimensionless Helmholtz free energy are given in Table 6.3 of Wagner and Pruß (2002). The analytical derivatives of $\phi^0(\delta_r,\tau_r)$ and $\phi^r(\delta_r,\tau_r)$ with respect to δ_r and τ_r are given in Tables 6.3 and 6.4, respectively, of Wagner and Pruß (2002).

As a result of comprehensive tests carried out by IAPWS, the following statements on the range of validity of IAPWS-95 were made (Wagner and Pruß, 2002; IAPWS-2009b). The formulation is valid in the entire stable fluid region of H_2O from the melting curve to 1273 K at pressures up to 1,000 MPa; the lowest temperature on the melting curve is 251.165 K at ~209.9 MPa. In the stable fluid region, the formulation can also be extrapolated beyond these limits. Tests show that (Eqn. 4.7.1) behaves reasonably when extrapolated to pressures up to about 100 GPa and temperatures up to about 5,000 K. This holds at least for the density and enthalpy of undissociated H_2O. In the gas region at pressures below the triple-point pressure, (Eqn. 4.7.1) behaves reasonably when extrapolated to the sublimation-pressure curve for temperatures down to 130 K (IAPWS-2011). As far as can be tested with experimental data, the formulation behaves reasonably when extrapolated into the metastable regions. Equation (4.7.1) represents the currently available experimental data of the supercooled liquid (solid–liquid metastable region) and of the superheated liquid (liquid–gas metastable region) to within the experimental uncertainty. In the case of the supercooled gas (gas–liquid metastable region), no experimental data are available. In this region, for pressures below 10 MPa, (Eqn. 4.7.1) produces reasonable values close to the saturation line. For calculations further away from the saturation line, an alternative equation (the so-called gas equation) is given in Wagner and Pruß (2002). The modification described in the Guideline (IAPWS-2012) extends the validity of the heat-capacity equation down to 50 K and permits the computation of the gas phase properties also in this low-temperature range.

Holten et al. (2011, 2012) reviewed the available data on various water properties and compared them with the calculations using the IAPWS-95 EOS. They noted that while the IAPWS-95 formulation does indeed represent the experimental density data at ambient pressure, the deviations from the formulation become larger with increasing pressure. Especially at higher pressures, there is a sizable discrepancy between the IAPWS-95 formulation and the experimental data; the slope (or the expansivity) has a different sign. As described earlier, the major domain of applicability of IAPWS-95 and its predecessors for liquid water was mostly limited at $T < 251$ K, and the accuracy in predictions of the density with IAPWS-95 decreased in the supercooled region. Therefore, several formulations of EOS for the supercooled region were suggested, many of them based on the concept of the second critical point for liquid water (Section 4.2).

4.7.2. Equations of State Based on the Concept of the Second Critical Point

Poole et al. (1992, 1994) proposed another extension of the van der Waals equation by incorporating the effects of the network of hydrogen bonds that exist in liquid water. They developed a simple model for the free energy of the hydrogen bonds F_{hb} that was superimposed on the free energy F_{EOS} of a simple liquid described by the equation of state from van der Waals theory. The total free energy was $F = F_{EOS} + 2F_{hb}$, where the factor of 2 accounts for the fact that there are two moles of hydrogen bonds for every mole of molecules. Based on this additivity and the relation (Eqn. 3.1.11), $p = -(\partial F/\partial V)_T$, Poole et al. presented the equation of state in a simple form:

$$p = p_{EOS} + 2p_{hb}, \tag{4.7.5}$$

where p is the is the total pressure in the liquid, p_{EOS} is the pressure determined from the van der Waals equation of state, p_{hb} is the pressure exerted by the hydrogen bonds and again the factor of 2 reflects 2 moles of bonds for each mole of water. Equation (4.7.5) is a "mixture model" where the hydrogen bonds act as a separate species of water exerting their own pressure p_{hb}. The hydrogen bond fraction is temperature dependent. Above freezing point T_f, the bond fraction falls to zero and the only contribution to the pressure is the background pressure p_{EOS}. Poole et al. assumed that the population of the hydrogen bonds consists of two species, strong (open) bonds with energy ε_{hb} and entropy $S_{\eta,hb}$ and weak bonds with energy $\varepsilon_{hb} = 0$ and entropy S_0, the number of configurations in each species are Z_{hb} and Z_0, respectively. They constructed the Helmholtz free energy F_{hb} for a combination of these two species as a simple partition function. As described in Chapter 3, the number of configurations (statistical sum) in a statistical ensemble of each bond type is

$$Z_0 = \exp(-S_{\eta,0}/R), \qquad Z_{hb} = \exp(-S_{\eta,hb}/R). \tag{4.7.6}$$

The total hydrogen bond free energy was assumed to be a mixture of the two bond types

$$F_{hb} = -f_{hb}RT\ln[Z_0 + \exp(-\varepsilon_{hb}/RT)] - (1 - f_{hb})RT\ln(Z_0 + 1), \tag{4.7.7}$$

where the mixture function f_{hb} is a fraction of the hydrogen bonds that are capable of forming strong (open) bonds. Poole et al. argued that strong bonds are most likely to occur when the bulk molar volume V is equal to the specific volume V_{hb} of ice I_h, and approximated $f_{hb}(V)$ as a Gaussian function of V with some dispersion σ_{hb}.

Poole et al. (1994) found that this model was able to: 1) qualitatively reproduce the known thermodynamic behavior of water; 2) consolidate proposals for the behavior of supercooled water by showing that they may be different realizations of a single behavior; 3) improve understanding liquid and amorphous solid water.

Jeffery and Austin (1997, 1999) combined both these approaches in one model. They also assumed that the equation of state could be written as the sum of the two terms as (Eqn. 4.7.5). For calculation of p_{EOS}, Jeffery and Austin (1999) modified the equations by Song and Mason (1989, 1990a, b) of the van der Waals type and presented it in the form

$$\frac{p_{EOS}}{\rho RT} = 1 + \left(\alpha_{EOS} - b^* - \frac{a_{EOS}}{RT}\right)\rho + \alpha_{EOS}\rho\left(\frac{1}{1-\lambda_{EOS}b_{EOS}\rho} - 1\right). \tag{4.7.8}$$

The three parameters here, b_{EOS}, b^*, and α_{EOS} were expressed via the Boyle volume v_B and temperature T_B as

$$b_{EOS}/v_B = 0.2\exp[-21.4(T/T_B + 0.445)^3] - b_1\exp(1.016T/T_B) + b_2, \tag{4.7.9}$$

$$\alpha_{EOS}/v_B = 2.145, \qquad b^*/v_B = 1.0823 \tag{4.7.10}$$

The values of the coefficients were determined from fitting them to experimental data as described next. For calculation of p_{hb}, Jeffery and Austin modified the free energy from Poole et al. as

$$F_{hb} = -f_{hb}RT\ln[Z_0 + Z_{hb}\exp(-\varepsilon_{hb}/RT)] - (1 - f_{hb})RT\ln(Z_0 + Z_{hb}), \tag{4.7.11}$$

which includes the possibility that there are Z_{hb} configurations of strong bonds. It was suggested that the bond fraction is a function of the temperature and density. These dependencies were separated by Jeffery and Austin as $f_{hb}(T, \rho_w) = f^*(\rho_w)f^{**}(T)$:

$$f^*(\rho_w) = \frac{1 + C_{hb1}}{\exp[(\rho_w - \rho_{hb})/\sigma_{hb}]^2 + C_{hb1}} \tag{4.7.12a}$$

$$f^{**}(T) = \exp[-0.18(T/T_f)^8], \tag{4.7.12b}$$

where $T_f = 273.15$. That is, the density dependence was approximated by a modified Gaussian function centered around density ρ_{hb}, and the temperature function accounted for the fact that the maximum specific volume of water is reached near T_f. Using a sophisticated fit to the experimental data on the temperature, pressure, density, and entropy of liquid water in the extended range −34 °C < 1,200 °C, and 1 bar < 3,000 bars, Jeffery and Austin (1997, 1999) found the best set of the parameters: $\varepsilon_{hb} = -11.490$ kJ mole^{-1}, $S_0 = -61.468$ J mole^{-1} K^{-1}, $S_{hb} = -5.128$ J mole^{-1} K^{-1}, $\rho_{hb} = 0.8447$ g cm^{-3}, $C_{hb1} = 0.7140$, $\sigma_{hb} = 0.1425$ g cm^{-3} = $0.1687 \times \rho_{hb}$, $b_1 = 0.2581$, $b_2 = 0.99859$, and $\lambda_{EOS} = 0.3241$, $T_B = 1408.4$ K, $v_B = 4.1782 \times 10^{-5}$ m^{-3} mole^{-1}.

Having calculated p_{EOS} from EOS (Eqn. 4.7.8) and p_{hb} from (Eqn. 4.7.11) with the use of Equation (3.1.11), $p = -(\partial F/\partial V)_T$, the total pressure p can be calculated as a function of the temperature and density or volume, which allows a plot of the phase diagrams of water in various states. The results of calculations of water properties in the supercooled region presented in Jeffery and Austin (1997, 1999) were in reasonable agreement with the available data.

An analogous approach for supercooled region was developed by Kiselev (2000) and Kiselev and Ely (2002). Holten et al. (2011, 2012) developed a different model based on introducing the scaling fields for supercooled water and the hypothesis of the second critical point and the liquid–liquid transition. Their model makes use of the fact that fluids belong to the universality class of Ising-like systems (Landau and Lifshitz, v.5, 1958; Lifshitz and Pitaevskii, 1997) whose critical behavior is characterized by two independent scaling fields: a "strong" scaling field h_1 (ordering field) and a "weak" scaling field h_2, and by a dependent scaling field, which asymptotically close to the critical point becomes a generalized homogeneous function of h_1 and h_2. By allowing the slope of the liquid–liquid transition line and the critical pressure to be freely adjustable parameters, their model was able to represent almost all available thermodynamic property data for supercooled water. A compact review of the theory by Holten et al. (2011, 2012) was given by Hellmuth et al. (2013). The theories of water in the supercooled region and at high pressures are being extensively developed, and a substantial breakthrough in this area can be expected in the near future.

Appendix A.4 for Chapter 4. Relations among Various Pressure Units

Saturated vapor pressures are used in various units. The relations among these units are given here for convenience. In SI units, the unit of pressure is the pascal (Pa), 1 Pa = 1 N m^{-2}, where N is the newton, and pressure is often measured in hectopascals (hPa), 1 hPa = 10^2 Pa. In high-pressure experiments, the following units are used: megapascal (MPa), 1 MPa = 10^6 Pa, and gigapascal (GPa), 1 GPa = 10^9 Pa. In CGS units, the pressure is often measured in dyn cm^{-2}. The ratio between these units is Pa = 10 dyn cm^{-2}, and 1 hPa = 10^3 dyn cm^{-2}. Alternative units of pressure include: 1 bar = 10^5 Pa = 10^6 dyn cm^{-2}; 1 mb (millibar) = 10^2 Pa = 1 hPa = 10^3 dyn cm^{-2}; 1 atm (atmosphere) = 1.01325 bar = 1.013×10^6 dyn cm^{-2} = 1.013×10^5 Pa = 0.1013 MPa; 1 MPa = 10 bar = 10^7 dyn cm^{-2} $\approx$ 10 atm; 1 GPa = 10^4 bar = 10^{10} dyn cm^{-2} $\approx$ 10^4 atm.

5

Diffusion and Coagulation Growth of Drops and Crystals

In this chapter, we consider the condensation growth and evaporation of drops and the depositional growth and sublimation of crystals by diffusion of water vapor from or to environmental moist air. These processes require knowledge of the water and ice supersaturation, the equations for which are derived and studied for various cases. Cloud glaciation due to crystal growth and evaporation of drops is considered. The kinetic equations of regular condensation and deposition are derived. The coagulation/accretion growth of drops and crystals due to the particles' collisions with subsequent coalescence is considered.

5.1. Diffusional Growth of Individual Drops

The theory of drop diffusional growth and evaporation in a field of water vapor originated from Maxwell (1890), who solved the problem for the steady-state condensational growth of a spherical drop assuming saturated vapor pressure at the drop surface. This theory was subsequently developed to account for various additional effects. The equations for drop growth by diffusion are derived in the following.

5.1.1. Diffusional Growth Regime

Consider the simplest case of a water drop in a field of water vapor, with the following two assumptions: 1) the vapor field around a drop can be described in the diffusion approximation; 2) neglecting the latent heat released at condensation. Then, the vapor diffusion toward (from) the drop can be described with the convective-diffusion equation for the vapor density ρ_v:

$$\frac{\partial \rho_v}{\partial t} + \vec{u}\nabla \rho_v = D_v \nabla^2 \rho_v, \qquad (5.1.1)$$

where t is time, u is velocity relative to the drops, D_v is the vapor diffusion coefficient, and ∇ is the differential divergence operator. For a spherical drop in a still air and neglecting drop fall speed, (Eqn. 5.1.1) can be written in spherical coordinates as

$$\frac{\partial \rho_v}{\partial t} = D_v \left(\frac{\partial^2 \rho_v}{\partial r^2} + \frac{2}{r}\frac{\partial \rho_v}{\partial r} \right) \rho_v, \qquad (5.1.2)$$

where r is the distance from the drop center. The initial and boundary conditions for this problem are

$$\rho_v(r,0)\,|_{r\to\infty} = \rho_{v\infty}, \qquad \rho_v(r_d,t) = \rho_{ws}, \qquad (5.1.3)$$

where r_d is the drop radius, ρ_{ws} is the saturated vapor density at the drop surface, and $\rho_{v\infty}$ is the vapor density at infinite distance from the drop. A solution to (Eqn. 5.1.2) with the conditions (Eqn. 5.1.3) is

$$\rho_v(r,t) = \rho_{v\infty} + (\rho_{ws} - \rho_{v\infty})\frac{r_d}{r}\left[1 - erf\left(\frac{r - r_d}{2\sqrt{D_v t}}\right)\right], \tag{5.1.4}$$

where $erf(z)$ is the error function. The second term in the brackets accounts for the time dependence and is much less than 1 for $t = t_{eq} >> r_d^2/\pi D_v$. For standard atmospheric conditions $t_{eq} \sim 10^{-6}$ s for $r_d \sim 100$ μm, and $t_{eq} \sim 10^{-4}$ s for $r_d \sim 1000$ μm. These time scales are much shorter than the characteristic time scales of vapor fluctuations; therefore, the second term with time can be neglected, so that for $t >> t_{eq}$, the solution is close to the steady state

$$\rho_v(r) = \rho_{v\infty} + (\rho_{ws} - \rho_{v\infty})\frac{r_d}{r}. \tag{5.1.5}$$

The solution of this type found by Maxwell resembles the expression for the electrostatic potential outside of a conducting sphere of radius r_d with electrical charge, and is therefore referred to as an *electrostatic analogy* of the water vapor field around a droplet. As discussed in Section 5.2, this analogy helps to formulate an appropriate description of the vapor field around the crystals.

The diffusional vapor flux to or from the drop surface can be evaluated as a surface integral of the vapor density gradient over the drop surface S_d

$$J_{c,dif}|_{r_d} = -\int_{S_d} D_v \frac{\partial \rho_v(r)}{\partial r} dS_d$$
$$= -4\pi r_d^2 D_v \frac{\partial \rho_v(r)}{\partial r}|_{r_d} = 4\pi r_d D_v(\rho_{ws} - \rho_{v\infty}). \tag{5.1.6}$$

Note that this flux is also similar to the flux of the electrostatic field around a conducting sphere with capacitance C_e, which is equal to its radius, $C_e = r_d$. The flux (Eqn. 5.1.6) determines the drop mass growth rate

$$\dot{m}_{dif} \equiv \left(\frac{dm_d}{dt}\right)_{dif} = -J_{c,dif} = 4\pi D_v r_d(\rho_{v\infty} - \rho_{ws}) = 4\pi D_v r_d \delta_w = 4\pi D_v r_d \rho_{ws} s_w, \tag{5.1.7}$$

where $\delta_w = (\rho_{v\infty} - \rho_{ws})$ and $s_w = (\rho_{v\infty} - \rho_{ws})/\rho_{ws}$ are the absolute and fractional water supersaturations introduced in Chapter 3. Using the equation of state for vapor from Chapter 3, $e_v = R_v\rho_v T$, this can be rewritten

$$\left(\frac{dm_d}{dt}\right)_{dif} = 4\pi D_v r_d \frac{M_w}{R}\left(\frac{e_{v\infty}}{T_\infty} - \frac{e_{ws}}{T_0}\right). \tag{5.1.8}$$

Equation (5.1.7) yields for the radius growth rate

$$\dot{r}_{d,dif} \equiv \left(\frac{dr_d}{dt}\right)_{dif} = \frac{D_v \delta_w}{r_d \rho_w} = \frac{D_v s_w \rho_{ws}}{r_d \rho_w}. \tag{5.1.9}$$

When $\rho_{v\infty} > \rho_{ws}$, and the supersaturation $s_w > 0$, the drop grows, and when $\rho_{v\infty} < \rho_{ws}$, and $s_w < 0$, the drop evaporates. The value of D_v depends on the temperature and pressure. Hall and Pruppacher (1976) approximated D_v as

$$D_v = 0.211\left(\frac{T}{T_0}\right)^{1.94}\left(\frac{p_0}{p}\right), \tag{5.1.10}$$

where $T_0 = 273.15$ K, $p_0 = 1013.25$ mb, and D_v is in cm^2 s^{-1}. Equation (5.1.9) is the simplest form of the equation for dr_d/dt. Subsequent derivations have accounted for various corrections, including kinetic and psychrometric corrections, which are considered in the next sections.

5.1.2. The Kinetic Regime and Kinetic Corrections to the Growth Rate

We assumed earlier in this chapter that the diffusion approximation is valid for both the vapor and temperature fields, which is characterized by abundant collisions between the vapor molecules and the diffusion coefficient. However, this regime becomes invalid at small distances from the drop surface on the order of the molecular mean free path length, $\lambda_f \sim 0.1$ μm at typical conditions. The kinetic regime acts at distances $\sim\lambda_f$ from the drop surface, which is characterized by the absence of collisions between the molecules. Several methods have been developed to overcome this problem, with the best known of these being the boundary sphere method developed by Fuchs (1959, 1964), which is illustrated in Fig. 5.1. This method assumes that the space outside the drop can be separated into two parts by a "*boundary sphere*" with the radius $r_b = r_d + \Delta_v$, where Δ_v is some length comparable to λ_f. The kinetic regime acts at $r_d < r < r_d + \Delta_v$, and the diffusion regime acts at $r > r_d + \Delta_v$. The main premise of the method is that the kinetic and diffusion vapor fluxes are matched at the surface of the boundary sphere at $r = r_b$.

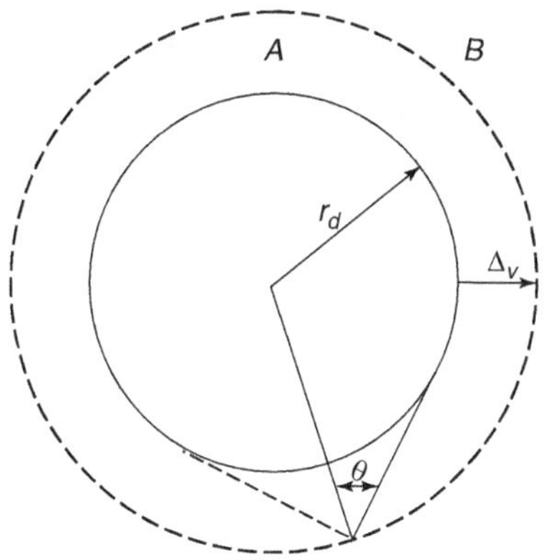

Figure 5.1. Illustration of the method of the boundary sphere for calculation of the drop growth. The domain with kinetic vapor fluxes is denoted with A, the domain with diffusion fluxes is B, the droplet radius is r_d, and the thickness of the boundary sphere is Δ_v. The radius of the boundary sphere is $r_b = r_d + \Delta_v$. θ is the maximum angle at which water molecules flying to the sphere still may collide with it.

We can now repeat the solution of the diffusion (Eqn. 5.1.2) but with the boundary conditions at the boundary sphere, and instead of (Eqn. 5.1.6) we obtain a solution for the diffusion flux:

$$J_{c,dif}|_{r_d+\Delta_v} = 4\pi D_v (r_d + \Delta_v)[\rho_v(r_d + \Delta_v) - \rho_{v\infty}]. \quad (5.1.11)$$

The kinetic flux is a sum of the condensation flux J^+_{kin} toward the drop and evaporation flux J^-_{kin} from the drop. Fig. 5.1 illustrates the geometry of the fluxes. The molecular flux J^+_{kin} is proportional to the vapor density at $r < r_d + \Delta_v$, and to the thermal velocity V_w of vapor molecules (see Chapter 3), and can be evaluated using the velocity distribution function $f_v(v)$ that is assumed to be a Maxwell distribution. We consider the polar coordinate system and assume that the polar axis is directed toward the drop center.

The Maxwell distribution normalized to the molecular concentration $c_v = \rho_v/m_w$ in this coordinate system can be written as (see Appendix A to Chapter 3)

$$f_v(v)dv = c_v \left(\frac{m_w}{2\pi kT}\right)^{3/2} \exp\left(-\frac{m_w v^2}{2kT}\right) \sin\theta \, d\theta \, d\varphi, \quad (5.1.12)$$

where m_w is the mass of a water molecule. Consider a unit area S_1 on the boundary sphere $r_b = r_d + \Delta_v$. The molecular density flux toward the drop through this S_1 is proportional to the velocity component v_N normal to this area. We denote θ as the angle between the axis normal to the surface and a direction of a vapor molecule velocity. Then, the normal projection of the velocity is $v_N = v\cos\theta$, and the concentration flux density to the unit area is

$$j_c = -c_v \int_0^\infty dv f_v(v) \int_0^{\theta_0} \sin\theta \cos\theta d\theta \int_0^{2\pi} d\varphi, \quad (5.1.13)$$

where θ_0 is the maximum angle at which molecules flying from the boundary sphere still can collide with the drop (Fig. 5.1). The minus sign occurs because the flux is directed inside and the normal to the drop surface is directed outward. Evaluation of these integrals given in Appendix A of Chapter 3 yields

$$j_c = -c_v \left[\left(\frac{m_w}{2\pi kT}\right)^{1/2} \frac{1}{2}\left(\frac{m_w}{2kT}\right)^{-2}\right]\left[\frac{1}{2}\sin^2\theta_0\right][2\pi], \quad (5.1.14)$$

where each square bracket denotes each of the three integrals in (Eqn. 5.1.13). The resulting flux density is

$$j_c = -\frac{1}{4} c_v V_w \sin^2\theta_0, \qquad V_w = \left(\frac{8kT}{\pi m_w}\right)^{1/2} = \left(\frac{8RT}{\pi M_w}\right)^{1/2}. \quad (5.1.15)$$

Here, V_w is the mean thermal velocity of vapor molecules. It is seen from Fig. 5.1 that

$$\sin^2\theta_0 = \frac{r_d^2}{(r_d + \Delta_v)^2}. \quad (5.1.16)$$

The vapor density ρ_v and concentration c_v are related as $\rho_v = c_v m_w$; thus, the vapor mass flux is $j_\rho = m_w j_c$, which results in replacement of c_v with ρ_v in the mass flux density:

$$j_\rho = -\frac{1}{4}\rho_v V_w \sin^2\theta_0. \quad (5.1.17)$$

Not all of the molecules colliding with the drop will be absorbed on its surface, some will bounce back. The *condensation coefficient*, α_c, is the ratio of the molecular flux j_{cond} condensed on the water drop to the incident molecular flux j_p; that is, $\alpha_c = |j_{cond}/j_p|$, where | | means the absolute value. Numerous measurements since the 1930s have shown significant variations of α_c from 1 to 0.01 for relatively pure drops, with much smaller values for polluted drops. Values $\alpha_c = 0.036$–1 are most frequently used in drop growth calculations. The values of α_c can decrease to 10^{-5}–10^{-3} for drops covered by films of surface-active substances like cetyl alcohol and similar agents used for artificial fogs suppression (e.g., Deryagin and Kurgin, 1969; Bigg et al., 1969; Buikov and Khvorostyanov, 1979). The kinetic condensation flux J^+_{kin} is obtained from the product of j_p, α_c and the boundary sphere surface $S_b = 4\pi(r_d + \Delta_v)^2$, yielding the fraction of colliding molecules absorbed by the drop.

$$J^+_{kin} = -\alpha_c S_b j_p = -\alpha_c S_b \frac{1}{4} V_w \rho_v (r_d + \Delta_v)\sin^2\theta_0$$
$$= -\pi r_d^2 \alpha_c V_w \rho_v (r_d + \Delta_v), \tag{5.1.18}$$

where we used (Eqn. 5.1.16) for $\sin\theta_0$. The evaporation flux from the drop surface is evaluated similarly

$$J^-_{kin} = \pi r_d^2 \alpha_e V_w \rho_{ws}(T_0), \tag{5.1.19a}$$

where α_e is the evaporation coefficient that is the fraction of the molecules emitted from the surface.

Assuming $\alpha_e = \alpha_c$, we can write the total kinetic flux

$$J_{kin} = J^+_{kin} + J^-_{kin} = \pi r_d^2 \alpha_c V_w [\rho_{ws}(T_0) - \rho_v(r_d + \Delta_v)] \tag{5.1.19b}$$

This kinetic flux should be equal to the diffusion flux (Eqn. 5.1.11) at the boundary surface. Equating them, we obtain

$$\rho_v(r_d + \Delta_v) = \frac{4D_v(r_d + \Delta_v)\rho_{v\infty} + \alpha_c r_d^2 V_w \rho_{ws}(T_0)}{4D_v(r_d + \Delta_v) + \alpha_c r_d^2 V_w}. \tag{5.1.20}$$

Substituting this expression into (Eqn. 5.1.11), we obtain $J_{c,dif}$ and dm_d/dt, taking into account the kinetic correction

$$\frac{dm_d}{dt} = -J_{c,dif} = 4\pi r_d D_v^*(\rho_{v\infty} - \rho_{ws}) = 4\pi r_d D_v^* \rho_{ws} s_w, \tag{5.1.21}$$

$$D_v^* = \frac{D_v}{\dfrac{r_d}{r_d + \Delta_v} + \dfrac{\xi_{con}}{r_d}} = D_v k_{kin}, \quad k_{kin} = \left(\frac{r_d}{r_d + \Delta_v} + \frac{\xi_{con}}{r_d}\right)^{-1}, \quad \xi_{con} = \frac{4D_v}{\alpha_c V_w}, \tag{5.1.22a}$$

where s_w is the water supersaturation, and we introduced an effective diffusion coefficient D_v^* and the kinetic correction ξ_{con} to the drop condensation growth rate. As mentioned before, $\Delta_v \sim \lambda_f \sim 0.1$ μm, and for sufficiently large drops, $r_d > 0.5$–1 μm, this expression is simplified

$$D_v^* = \frac{D_v}{1 + \xi_{con}/r_d} = D_v k_{kin}. \qquad k_{kin} = \frac{r_d}{r_d + \xi_{con}}, \tag{5.1.22b}$$

where k_{kin} is the simplified kinetic correction. Using (Eqn. 5.1.22b), the mass growth rate (Eqn. 5.1.21) can be written as

$$\frac{dm_d}{dt} = 4\pi D_v \frac{r_d^2}{r_d + \xi_{con}} \rho_{ws} s_w = 4\pi D_v r_d \rho_{ws} s_w k_{kin}. \tag{5.1.23}$$

5.1.3. Psychrometric Correction Due to Latent Heat Release

The saturated vapor density in (Eqn. 5.1.21) for dm_d/dt depends on the drop temperature, which is influenced by the latent heat release during condensation or evaporation. This results in occurrence of the psychrometric correction to the drop diffusion growth rate. To account for this effect, we have to consider the heat transfer in the vicinity of the drop. The heat flux density is given by *Fourier's law*, $j_h = -k_a \nabla T$, where k_a is the *thermal conductivity coefficient* of humid air parameterized as $k_a = (5.69 + 0.017T) \times 10^{-5}$, k_a is in cal/(cm sec °C), and T is in °C (Pruppacher and Klett, 1997). With this flux, the temperature field around a drop is described by a heat balance equation for the temperature similar to the Equation (5.1.1) for ρ_v

$$\frac{\partial T}{\partial t} + \vec{u}\nabla T = -\frac{1}{\rho_a c_p}\nabla j_h = \kappa_T \nabla^2 T, \tag{5.1.24}$$

where κ_T is the *thermal diffusivity coefficient* [cm^2 s^{-1}] related to k_a as $\kappa_T = k_a/(\rho_a c_p)$.

The heat fluxes and their relations to the mass fluxes are considered in cloud physics (as in many other aerosol studies), introducing the dimensionless numbers referred to as *Peclet* (Pe), *Prandtl* (Pr), and *Schmidt* (Sc) numbers,

$$\mathrm{Pe} = \frac{Du}{\kappa_T}, \qquad \mathrm{Pr} = \frac{\eta_a}{\rho_a \kappa_T} = \frac{\nu_a}{\kappa_T}, \qquad \mathrm{Sc} = \frac{\eta_a}{\rho_a D_v} = \frac{\nu_a}{D_v}, \tag{5.1.24a}$$

where $\eta_a = \nu_a \rho_a$ is the dynamic viscosity of the air, ν_a is the air kinematic viscosity, and D is the particle diameter, which serves here as a length scale. The Peclet number characterizes the ratio of the convective and diffusive heat transfer processes, and the Prandtl number characterizes the ratio of the diffusive fluxes of the mass and heat. For air, Pr ≈ 0.8 and Sc ≈ 0.71, and $D_v/\kappa_T =$ Pr/Sc ≈ 1.12. At $T = 273.15$ K and $p = 1$ atm, as follows from (Eqn. 5.1.10), $D_v \approx 0.21$ cm^2 s^{-1}; thus $\kappa_T \approx 1.12 \times D_v = 0.19$ cm^2 s^{-1}. That is, the coefficients D_v and κ_T are close, and it is possible in many cases to assume that $\kappa_T \approx D_v$, which simplifies the growth rate equations.

Neglecting again the air and drop velocities, and using the initial and boundary conditions for the steady state problem,

$$T(r,0)|_{r=r_d} = T_0, \qquad T|_{r\to\infty} = T_\infty, \tag{5.1.25}$$

where T_0 and T_∞ are the temperatures at the drop surface and at infinity. Thus, we obtain a solution similar to (Eqn. 5.1.5) for ρ_v:

$$T(r) = T_\infty + (T_0 - T_\infty)\frac{r_d}{r}, \tag{5.1.26}$$

where again we neglect the time dependence and assume steady state. The rate of heat transfer dQ/dt is equal to the heat flux from the drop surface $J_{heat}(r_d)$,

$$\frac{dQ}{dt} = -J_{heat}(r_d) = -\int_{S_d} k_a \frac{\partial T(r)}{\partial r} dS_d$$

$$= -4\pi r_d^2 k_a \left.\frac{\partial T(r)}{\partial r}\right|_{r_d} = 4\pi k_a r_d (T_0 - T_\infty). \tag{5.1.27}$$

Under quasi-steady conditions, this heat flux from the drop should be equal to the condensation heat flux $L_e J_{c,dif}$ with $J_{c,dif}$ given by (Eqn. 5.1.6). Equating them, we obtain

$$\Delta T = T_0 - T_\infty = (D_v L_e / k_a)(\rho_{v\infty} - \rho_{ws}). \tag{5.1.28}$$

This is the psychrometric temperature difference ("thermal jump") between the drop surface and environment, which shows that a growing droplet at $\rho_{v\infty} > \rho_{ws}$ is warmer than the environment, and an evaporating drop at $\rho_{v\infty} < \rho_{ws}$ is cooler than the environment. This expression was obtained by considering only the diffusion process.

Accounting for kinetic effects in the temperature field near the drop surface can be done analogously to that for the vapor field, by considering the boundary sphere that separates the domains with diffusion and kinetic fluxes. The radius of the boundary sphere is $r_b = r_d + \Delta_T$, where Δ_T is the thickness of the boundary sphere with the kinetic molecular heat flux (Fig. 5.1). Evaluating the kinetic heat flux near the drop at the distance $r_b = r_d + \Delta_T$, and matching the diffusion and kinetic heat fluxes leads to an equation similar to (Eqn. 5.1.22a)

$$k_a^* = \frac{k_a}{\dfrac{r_d}{r_d + \Delta_T} + \dfrac{\beta_T}{r_d}}, \qquad \beta_T = \frac{4k_a}{\alpha_T c_p \rho_a V_w} = \frac{4\kappa_T}{\alpha_T V_w}, \tag{5.1.29}$$

where we used the relation $\kappa_T = k_a/(\rho_a c_p)$, β_T is the kinetic correction to the growth rate in thermal conductivity and α_T is *the thermal accommodation coefficient.* The thermal accommodation coefficient is the ratio of the number of water vapor molecules that achieve thermal equilibrium upon collision with a drop, to a total number of the molecules colliding with the surface. The value of Δ_T is also ~0.1 μm, and for sufficiently large drops, $r_d >> \Delta_T$, this expression is simplified as

$$k_a^* = \frac{k_a}{1 + \beta_T / r_d} = k_a \frac{r_d}{r_d + \beta_T}. \tag{5.1.30}$$

To account for the psychrometric correction for latent heat release to the mass or radius growth rate, the system of (Eqns. 5.1.21) and (5.1.27) should be solved together. To solve this system and obtain dm_d/dt with a psychrometric correction, it is convenient to express first ΔT via dm_d/dt using the energy balance between the heat transfer dQ/dt and mass transfer dm_d/dt

$$\frac{dQ}{dt} = 4\pi r_d k_a^* (T_0 - T_\infty) = L_e \frac{dm_d}{dt}, \tag{5.1.31}$$

which gives

$$\Delta T = (T_0 - T_\infty) = \frac{L_e}{4\pi r_d k_a^*} \frac{dm_d}{dt}. \tag{5.1.32}$$

The correction $\Delta T << T_\infty$, and the saturated density $\rho_{ws}(T_\infty + \Delta T)$ can be expanded into a power series by $\Delta T / T$. Using the Clausius–Clapeyron (Eqn. 3.7.7), and omitting further for brevity the subscript "∞", we obtain for $e_{ws}(T + \Delta T)$:

$$e_{ws}(T + \Delta T) = e_{ws}(T) \exp\left(\frac{L_e}{R_v T} \frac{\Delta T}{T} \right) \approx e_{ws}(T) \left(1 + \frac{L_e}{R_v T} \frac{\Delta T}{T} \right). \tag{5.1.33}$$

And for ρ_v using the equation of state $e_{ws} = R_v \rho_{ws} T$:

$$\rho_{ws}(T+\Delta T) = \frac{e_{ws}(T+\Delta T)}{R_v(T+\Delta T)} \approx \frac{e_{ws}(T)}{R_v T}\left(1+\frac{L_e}{R_v T}\frac{\Delta T}{T}\right)\left(1-\frac{\Delta T}{T}\right) \tag{5.1.34}$$

$$\approx \rho_{ws}(T)\left[1+\left(\frac{L_e}{R_v T}-1\right)\frac{\Delta T}{T}\right]. \tag{5.1.35}$$

Substitution of (Eqn. 5.1.32) for ΔT into (Eqn. 5.1.35) yields

$$\rho_{ws}(T+\Delta T) \approx \rho_{ws}(T)\left[1+\left(\frac{L_e}{R_v T}-1\right)\frac{L_e}{4\pi r_d k_a^* T}\frac{dm_d}{dt}\right]. \tag{5.1.36}$$

Taking into account that ρ_{ws} in (Eqn. 5.1.21) depends on the temperature at the drop surface, $(T+\Delta T)$, and substituting (Eqn. 5.1.36) into (Eqn. 5.1.21) for dm/dt, we obtain

$$\frac{dm_d}{dt} = 4\pi r_d D_v^*[\rho_{v\infty} - \rho_{ws}(T+\Delta T)]$$

$$= 4\pi r_d D_v^*[\rho_{v\infty} - \rho_{ws}(T)] - \frac{D_v^* L_e \rho_{ws}(T)}{k_T^* T}\left(\frac{L_e}{R_v T}-1\right)\frac{dm_d}{dt}. \tag{5.1.37}$$

Solving for dm_d/dt and introducing the fractional supersaturation s_w, we get finally

$$\frac{dm_d}{dt} = \frac{4\pi r_d D_v^* \rho_{ws} s_w}{\Gamma_w}, \tag{5.1.38}$$

where Γ_w is the psychrometric *correction* that occurs due to the latent heat release at condensation,

$$\Gamma_w = 1 + \frac{D_v^* L_e \rho_{ws}}{k_a^* T}\left(\frac{L_e}{R_v T}-1\right) \approx 1 + \frac{L_e}{c_p T}\frac{\rho_{ws}}{\rho_a}\left(\frac{L_e}{R_v T}-1\right). \tag{5.1.39a}$$

The second equality uses the relation $k_a^* = c_p \rho_a \kappa_T^*$ and the approximate relation $\kappa_T^* \approx D_v^*$ discussed earlier. The first term in the parentheses is ~20 >> 1 and the second term, −1, can be neglected with the error of ~5%. Then, Γ_w can be written in the form that coincides with Γ_1 introduced in Section 3.11 for the wet adiabat

$$\Gamma_w \approx 1 + \frac{L_e^2}{c_p R_v T^2} q_{ws} = 1 + \frac{L_e}{c_p}\frac{dq_{ws}}{dT} = \Gamma_1, \tag{5.1.39b}$$

where q_{ws} is the specific humidity saturated over water. The phase heat release resulted in the second term; it would be $\Gamma_w = 1$ without account for this effect.

When considering growth of the small solution drops, it is necessary to account for the solution and curvature effects. This can be done simply by adding the corresponding terms to supersaturation as described in Chapter 3; then (Eqn. 5.1.38) becomes

$$\frac{dm_d}{dt} = \frac{4\pi r_d D_v^* \rho_{ws}(s_w - s_{salt})}{\Gamma_w}, \tag{5.1.40a}$$

$$s_{salt} = \left(\frac{A_K}{r_d} - \frac{B r_d^3}{m_w^3 - m_{dry}^3}\right), \tag{5.1.40b}$$

where A_K and B are the Kelvin curvature parameter and the nucleus activity defined in Section 3.9, and m_{dry} is the mass of a dry aerosol particle. For drops with r_d of a few microns, the solution and curvature effects s_{salt} are usually ignored in simulations although more precise calculations may require accounting for these terms.

In many papers and textbooks (e.g., Mason, 1971; Young, 1993; Pruppacher and Klett, 1997), the equation for the mass growth rate is given in the form

$$\frac{dm_d}{dt} = \frac{4\pi r_d(s_w - s_{salt})}{\dfrac{RT}{D_v^* e_{ws} M_w} + \dfrac{L_e}{k_a^* T}\left(\dfrac{L_e M_w}{RT} - 1\right)}, \tag{5.1.41a}$$

or in simplified forms with $s_{salt} = 0$ and omitting the term -1 in the parentheses. It can be shown that this form (Eqn. 5.1.41a) is equivalent to the form (Eqn. 5.1.40a). Substituting here the equation of state, $e_{ws} = \rho_{ws} R_v T$, using the relation $R_v = R/M_w$ and multiplying by $D_v^* \rho_{ws}$, we obtain

$$\begin{aligned}\frac{dm_d}{dt} &= \frac{4\pi r_d(s_w - s_{salt})}{\dfrac{RT}{D_v^* \rho_{ws} M_w R_v T} + \dfrac{L_e}{k_a^* T}\left(\dfrac{L_e}{R_v T} - 1\right)} \\ &= \frac{4\pi r_d D_v^*(s_w - s_{salt})\rho_{ws}}{1 + \dfrac{D_v^* L_e \rho_{ws}}{k_a^* T}\left(\dfrac{L_e}{R_v T} - 1\right)} = \frac{4\pi r_d D_v^*(s_w - s_{salt})\rho_{ws}}{\Gamma_w},\end{aligned} \tag{5.1.41b}$$

where we used (Eqn. 5.1.39a) for Γ_w. Using the approximate relation $k_a^* = \kappa_T^* \rho_a c_p \approx D_v^* \rho_a c_p$, the factor Γ_w can be replaced with Γ_1 as in (Eqn. 5.1.39b). Thus, both forms of the growth rate equation, derived here in a simpler form (Eqn. 5.1.40a) and commonly used (Eqn. 5.1.41a) are equivalent. The forms (Eqns. 5.1.38) and (5.1.40a) for dm_d/dt are more convenient in many cases because they are compact and show that the growth rate is proportional to the major driving force: saturated vapor density and supersaturation. Further, the growth rate equation is written in a factorized form, the variables (supersaturation, kinetic, and psychrometric factors) are separated and are written as three multipliers, which simplifies various analytical solutions as is illustrated in the following chapters.

5.1.4. Radius Growth Rate

Equation (5.1.40a) yields for the droplet radius growth rate

$$\dot{r}_d \equiv \frac{dr_d}{dt} = \frac{D_v^* \rho_{ws}(s_w - s_{salt})}{r_d \rho_w \Gamma_w} \approx \frac{D_v^* \rho_{ws}(s_w - s_{salt})}{r_d \rho_w \Gamma_1}, \tag{5.1.42}$$

or without solute and curvature effects

$$\dot{r}_d \equiv \frac{dr_d}{dt} = \frac{D_v^* \rho_{ws} s_w}{r_d \rho_w \Gamma_w} \approx \frac{D_v^* \rho_{ws} s_w}{r_d \rho_w \Gamma_1}. \tag{5.1.43}$$

These simplified forms are equivalent to the other forms commonly used for dr/dt that follow from (Eqn. 5.1.41a) for dm_d/dt (e.g., Young, 1993; Pruppacher and Klett, 1997; Cotton et al., 2011)

$$r_d \frac{dr_d}{dt} = \frac{s_w - s_{salt}}{\dfrac{\rho_w RT}{D_v^* e_{ws} M_w} + \dfrac{\rho_w L_e}{k_a^* T}\left(\dfrac{L_e M_w}{RT} - 1\right)}. \tag{5.1.44}$$

The equivalence of (Eqn. 5.1.44) with (Eqns. 5.1.42) and (5.1.43) for dr_d/dt can be shown again substituting into (Eqn. 5.1.44) the equation of state $e_{ws} = \rho_{ws} R_v T$, the relation $R_v = R/M_w$, multiplying by $D_v^* \rho_{ws}$, and dividing by r_d,

$$\frac{dr_d}{dt} = \frac{s_w - s_{salt}}{\dfrac{\rho_w R T r_d}{D_v^* \rho_{ws} R_v T M_w} + \dfrac{\rho_w L_e r_d}{k_a^* T}\left(\dfrac{L_e M_w}{RT} - 1\right)}$$
$$= \frac{(s_w - s_{salt}) D_v^* \rho_{ws}}{r_d \rho_w \left[1 + \dfrac{D_v^* L_e \rho_{ws}}{k_a^* T}\left(\dfrac{L_e M_w}{RT} - 1\right)\right]}$$
$$= \frac{(s_w - s_{salt}) D_v^* \rho_{ws}}{r_d \rho_w \Gamma_w} \approx \frac{(s_w - s_{salt}) D_v^* \rho_{ws}}{r_d \rho_w \Gamma_1}, \tag{5.1.44a}$$

where we used (Eqn. 5.1.39a) for Γ_w, and the last equation is obtained using again the relations $k_a^* = \kappa_T^* \rho_a c_p \approx D_v^* \rho_a c_p$. Equations (5.1.42) and (5.1.43) are written in the factorized form so that the factors describing the supersaturation, kinetic, and psychrometric corrections are separated. This factorized form of (Eqns. 5.1.42) and (5.1.43) and separation of the dependencies is especially convenient in integration of the equations when developing analytical parameterizations of drop and crystal nucleation, as will be shown in the next chapters. Equation (5.1.42) can be rewritten with explicit dependence on the kinetic correction as

$$\frac{dr_d}{dt} = \frac{D_v \rho_{ws} s_w}{\rho_w \Gamma_w (r_d + \xi_{con})}. \tag{5.1.45}$$

It is also used in a slightly different form:

$$\frac{dr_d}{dt} \approx \frac{c_{3w} s_w}{r_d + \xi_{con}}, \quad c_{3w} = \frac{D_v \rho_{ws}}{\rho_w \Gamma_1}. \tag{5.1.46}$$

Equation (5.1.40a) for dm_d/dt and (Eqn. 5.1.46) for dr_d/dt shows that the droplet growth rate may substantially depend on ξ_{con}, which is determined by α_c. At $T = 0\,°C$, $p = 1013$ mb, an estimate with (Eqn. 5.1.15) gives $V_w \sim 570$ m s^{-1}, then (Eqn. 5.1.22a) with $D_v = 0.21$ cm s^{-2} yields the values of the kinetic correction of $\xi_{con} \approx 0.15$ μm for $\alpha_c = 1$ and $\xi_{con} \approx 3.7$ μm for $\alpha_c = 0.04$. Thus, the kinetic correction has little influence on drop growth rate for $r_d > 1$–2 μm if $\alpha_c = 1$, but can substantially suppress the drop growth rates with $r_d \sim 4$ μm if $\alpha_c = 0.04$. Therefore, further measurements and refinements of the values of the condensation coefficient under various conditions are needed. If the drops are polluted with some substances with low condensation coefficients, the kinetic correction can be very large. For example, some surface-active agents like cetyl alcohol have $\alpha_c \sim 10^{-5}$ and $\xi_{con} \sim 10^4$ cm. When $\xi_{con} >> r_d$, the droplets grow in the kinetic regime

$$\frac{dr_d}{dt} = \frac{D_v \rho_{vs} s_w}{\rho_w \xi_{con} \Gamma_w}. \tag{5.1.47}$$

The drop radius growth is very slow in this regime. In the Maxwellian diffusion regime, $dr_d/dt \sim 1/r_d$, the smaller droplets grow faster than the larger droplets, and the size spectrum narrows with time. In contrast, the radius growth rate in kinetic regime (Eqn. 5.1.47) does not depend on the radius, and the size spectrum can broaden. This can be one reason for spectral broadening in polluted clouds.

Extensive theoretical, laboratory, and field studies of α_c were performed during the 1960s and 1970s in the context of weather modification projects on warm fog suppression, and it was found that α_c can decrease by 2–5 orders of magnitude (down to 3.5×10^{-5}) in the presence of surfactants (e.g., Juisto, 1964; Deryagin et al., 1966; Bigg et al., 1969; Deryagin and Kurgin, 1972; Silverman and Weinstein, 1974; Buikov and Khvorostyanov, 1979). The slow growth of the droplets covered with the surface-active agents was used for artificial fog suppression. However, if supersaturation generation continues (e.g., due to radiative or advective cooling or moistening), this leads to a substantial increase in supersaturation that compensates for the slow drop growth, thus hampering fog development only for a short period (Buikov and Khvorostyanov, 1979).

5.1.5. Ventilation Corrections

In the preceding derivations, the effects of drop motion and convective velocities u on the drop growth and evaporation were neglected. Large drops fall at sufficiently high velocities so that the effects of convective velocities become significant, especially on the drop evaporation in the subsaturated air. These effects are accounted for by introducing the *ventilation correction* to the drop growth rate or *the ventilation coefficient* f_v. These corrections in principle can be derived analytically by solving the convective-diffusion equation for the vapor field. However, such derivations require detailed calculations of the airflow around a drop, are rather complicated, and may need additional fitting to the measured evaporation rates.

Hence, it has been customary in cloud physics to use empirical expressions obtained from measurements and to express them using dimensionless numbers: the Schmidt number Sc introduced in (Eqn. 5.1.24a), the Reynolds number Re, the mean transfer coefficient k_v, and the Sherwood number Sh,

$$\mathrm{Re} = \frac{uD}{\nu_a}, \quad f_v = \frac{(dm_d/dt)}{(dm_d/dt)_0}, \quad k_v = \frac{(dm_d/dt)}{4\pi r_d^2(\rho_{v\infty} - \rho_{ws})},$$

$$\mathrm{Sh} = \frac{2k_v r_d}{D_v} = \frac{2r_d}{D_v}\frac{(dm_d/dt)}{4\pi r_d^2(\rho_{v\infty} - \rho_{ws})} = 2f_v, \tag{5.1.48}$$

where u is the convective velocity around the drop, $D = 2r_d$ is the drop diameter, and $(dm_d/dt)_0$ is the mass growth rate without account for the ventilation correction.

The experimental data on the ventilation coefficient f_v were approximated in Beard and Pruppacher (1971) as

$$f_v = 1.0 + 0.108(\mathrm{Sc}^{1/3}\,\mathrm{Re}^{1/2})^2 \approx 1.00 + 0.86\,\mathrm{Re}, \tag{5.1.49a}$$

for Re < 2.46, that is, $r_d < 60$ μm and

$$f_v = 0.78 + 0.308(\mathrm{Sc}^{1/3}\,\mathrm{Re}^{1/2}) \approx 0.78 + 0.275\,\mathrm{Re}^{1/2}, \tag{5.1.49b}$$

for $2.46 < \mathrm{Re} < 3.34 \times 10^2$, that is, 60 μm $< r_d <$ 1500 μm. The second equalities here and the limits by Re and r_d are given for $\mathrm{Sc} \approx 0.71$ for the air—i.e., $\mathrm{Sc}^{1/3} = 0.89$.

The mass growth rate with account for the ventilation correction is

$$\frac{dm_d}{dt} = \left(\frac{dm_d}{dt}\right)_0 f_v, \tag{5.1.50}$$

where $(dm_d/dt)_0$ is described by (Eqn. 5.1.38). The factor $f_v \sim 5$ at $r_d = 0.5$ mm and $f_v \sim 10$ at $r_d = 1$ mm; hence, the ventilation correction substantially increases the growth-evaporation rates of sufficiently large drops. Equation (5.1.50) with (Eqns. 5.1.38), (5.1.48), and (5.1.49a,b) describe droplet growth or evaporation rates in the field of the supersaturation with account for the kinetic, psychrometric, and ventilation corrections.

5.2. Diffusional Growth of Crystals

5.2.1. Mass Growth Rates

The growth rates of crystals are derived from the *electrostatic analogy* following Jeffreys (1918). This analogy was mentioned already in Section 5.1 with respect to droplet growth rates, and becomes especially useful for consideration of the problem for nonspherical crystals. The analogy is based on the similarity of the fields of water vapor density ρ_v and temperature T around a crystal to the fields and fluxes of the electrostatic potential Φ_e around a charged conductor with electrical charge Q_e and capacitance C_e. The relations of electrical field $\vec{E}$ to Φ_e and to the density of electrical charge ρ_e are described by the Maxwell equations (e.g., Landau and Lifshitz, v. 2, 2005)

$$\vec{E} = -\nabla\Phi_e, \qquad \mathrm{div}\vec{E} = 4\pi\rho_e. \tag{5.2.1}$$

Substitution of the second equation into the first one yields

$$\mathrm{div}\vec{E} = \mathrm{div}(-\nabla\Phi_e) = -\Delta\Phi_e = 4\pi\rho_e, \qquad \text{or} \qquad \Delta\Phi_e = -4\pi\rho_e, \tag{5.2.2}$$

which is the *Poisson* equation that relates the electrical potential and the charge. Integration of the first equation of (Eqn. 5.2.1) over the surface surrounding the charge can be done using (Eqn. 5.2.2) and the Gauss–Ostrogradsky equation, which expresses the integral over the surface S via the integral over the volume V,

$$\oint_S \vec{E}d\vec{S} = \int_V \mathrm{div}\vec{E}dV. \tag{5.2.3}$$

This yields

$$\oint_S \nabla\Phi_e d\vec{S} = \oint_S (-\vec{E})d\vec{S} = -\int_V \mathrm{div}\vec{E}dV = -4\pi\int_V \rho_e dV = -4\pi Q_e. \tag{5.2.4}$$

Here, Q_e is the total integral charge that is the integral of the charge density over the volume and can be expressed via the capacity C_e and the difference of the potentials at the surface Φ_{es} and at the infinite distance from the conductor $\Phi_{e\infty}$:

$$Q_e = C_e(\Phi_{es} - \Phi_{e\infty}). \tag{5.2.5}$$

Combining (Eqns. 5.2.4) and (5.2.5) yields

$$\oint_S \vec{\nabla}\Phi_e d\vec{S} = -4\pi Q_e = -4\pi C_e(\Phi_{es} - \Phi_{e\infty}). \tag{5.2.6}$$

According to the electrostatic analogy, the water vapor density ρ_v is analogous to Φ_s, the vapor flux $\vec{j}_v = D_v\vec{\nabla}\rho_v$ is analogous to the flux $\vec{\nabla}\Phi_e$, and we can write by analogy with (Eqn. 5.2.6)

$$\oint_S \vec{j}_v d\vec{S} = \oint_S D_v\vec{\nabla}\rho_v d\vec{S} = -4\pi D_v C_e(\rho_{is} - \rho_{v\infty}), \tag{5.2.7}$$

where ρ_{is} is the saturated over ice vapor pressure at the crystal surface, $\rho_{v\infty}$ is the vapor pressure at infinity. The surface integral over the vapor flux j_v is the crystal mass growth rate dm_c/dt, and we obtain

$$\frac{dm_c}{dt} = \oint_S \vec{j}_v d\vec{S} = 4\pi D_v C_e (\rho_{v\infty} - \rho_{is}). \qquad (5.2.8)$$

Using the equation of state for the vapor from Chapter 3, $e_v = R_v \rho_v T$, this can be rewritten

$$\left(\frac{dm_c}{dt}\right)_{dif} = 4\pi D_v C_e \frac{M_w}{R}\left(\frac{e_{v\infty}}{T_\infty} - \frac{e_{is}}{T_0}\right). \qquad (5.2.9)$$

To account for the temperature difference between the crystal and environment, we need to consider the heat flux due to the latent heat release (or consumption) at deposition (or sublimation). The psychrometric correction can be derived in the same way as in Section 5.1 for the drops, but the particle radius should be replaced again with the capacity C_e, and this leads to a generalization of (Eqn. 5.1.27) for the rate of heat flux transfer

$$\frac{dQ}{dt} = -J_{heat} = -\int_{S_a} k_a \frac{\partial T(r)}{\partial r} dS_a = 4\pi k_a C_e (T_0 - T_\infty), \qquad (5.2.10)$$

where the notation is the same as in Section 5.1 but r_d is replaced with C_e. This equation also shows that a growing crystal is warmer than environment ($T_0 > T_\infty$) and a sublimating crystal is colder ($T_0 < T_\infty$). Equating again the heat flux dQ/dt to the latent heat flux $L_s(dm_c/dt)$, expressing as in (Eqn. 5.1.28) $\Delta T = (T_0 - T_\infty)$ via $\Delta\rho_v = (\rho_{v\infty} - \rho_{is})$, expanding the saturated humidity $\rho_{is}(T + \Delta T)$ into the power series and rearranging the terms we obtain similar to (Eqn. 5.1.38) for the mass growth rate

$$\frac{dm_c}{dt} = \frac{4\pi C_e D_v \rho_{is} s_i}{\Gamma_i}, \qquad (5.2.11)$$

where s_i is the fractional ice supersaturation, and Γ_i is the psychrometric correction that occurs due to the latent heat release at condensation:

$$\Gamma_i = 1 + \frac{D_v L_s \rho_{is}}{k_a T}\left(\frac{L_s}{R_v T} - 1\right) \approx 1 + \frac{L_s}{c_p T}\left(\frac{L_s}{R_v T} - 1\right)\frac{\rho_{is}}{\rho_a}. \qquad (5.2.12)$$

The second equality uses the relation $k_a = c_p \rho_a \kappa_T$ and the approximate equality $\kappa_T \approx D_v$ discussed in Section 5.1. The second term in parentheses, -1, is approximately 4–5% of magnitude of the first term and can be neglected in many cases.

The generalization of the spherical vapor and heat fluxes for the drops to the case of nonspherical crystals was relatively simple. However, a generalization of the kinetic corrections is more difficult, because the method of the boundary sphere cannot be used since spherical symmetry cannot be applied for the crystals. For spherical drops in Section 5.1, the kinetic correction was introduced by replacement of the diffusion D_v and heat conductivity k_a coefficients with the "effective coefficients" D_v^* and k_a^* that depend on the drop radius r_d as $D_v^*/D_v = r_d/(r_d + \xi_{con})$ with $\xi_{con} = 4D_v/\alpha_c V_w$. This approach does not work for crystals since we cannot characterize a nonspherical crystal with any fixed radius. The kinetic corrections to the crystal mass growth rate are also described in many books and papers by introducing the "effective coefficients" D_v^* and k_a^*, but it is unclear what "radius" or size is meant. Thus, we need some scaling length l_c for D_v^* and k_a^*. The simplest option is to assume that the boundary sphere is inscribed around a crystal, and then the scaling length is the crystal major semi-axis r_c, and the effective D_v^* and k_a^* can be defined similarly to Section 5.1 as

$$D_v^* = D_v k_{kin}, \qquad k_{kin} = \frac{r_c}{r_c + \xi_{dep}} \qquad \xi_{dep} = \frac{4D_v}{\alpha_d V_w}, \tag{5.2.13a}$$

$$k_a^* = k_a \frac{r_c}{r_c + \beta_T}, \qquad \beta_T = \frac{4k_a}{\alpha_T c_p \rho_a V_w} = \frac{4\kappa_T}{\alpha_T V_w}. \tag{5.2.13b}$$

Here ξ_{dep} is the kinetic correction to the crystal deposition growth rate, k_{kin} is the reducing factor of the growth rate due to kinetic correction, β_T is the kinetic correction to the growth rate in thermal conductivity, and α_d is the deposition coefficient, which is the ratio of the vapor molecular flux deposited on a crystal to the incident molecular flux to the crystal; α_d is analogous to the condensation coefficient α_c in Section 5.1.

Another approximate evaluation of the kinetic corrections can be based on using the same electrostatic analogy. We have seen that this results in the mass growth rate by replacement of the radius by the electrical capacity. We can assume that the boundary sphere is also generalized as a surface similar to the surface of constant electrostatic potential and $r_c \to C_e$, then

$$D_v^* = D_v \frac{C_e}{C_e + \xi_{dep}}, \qquad k_a^* = k_a \frac{C_e}{C_e + \beta_T}. \tag{5.2.13c}$$

Yet another evaluation of the kinetic corrections can be based on the choice for l_c as the spherically equivalent radius of a crystal $r_{c,equ} = (3m_c/4\pi\rho_i)^{1/3}$ or as the minor crystal semi-axis b_c. Replacement in (Eqn. 5.2.11) of D_v with D_v^* and substitution of (Eqn. 5.2.13a) for D_v^* yields

$$\frac{dm_c}{dt} = \frac{4\pi C_e D_v^* \rho_{is} s_i}{\Gamma_i} = \frac{4\pi C_e D_v \rho_{is} s_i}{\Gamma_i} k_{kin}. \tag{5.2.14}$$

This equation accounts for the three major factors that govern crystal growth: ice supersaturation s_i, psychrometric correction Γ_i and the kinetic correction ξ_{dep}. The dependencies on these three factors are separated, which can be convenient in many cases. It is easy to show that (Eqn. 5.2.14) derived here is equivalent to the form of the mass growth rate equation that is commonly used (e.g., Young, 1993; Pruppacher and Klett, 1997),

$$\frac{dm_c}{dt} = \frac{4\pi C_e s_i}{\dfrac{RT}{D_v^* e_{is} M_w} + \dfrac{L_s}{k_a^* T}\left(\dfrac{L_s M_w}{RT} - 1\right)}. \tag{5.2.15a}$$

Substituting here the equation of state, $e_{is} = \rho_{is} R_v T$, using the relation $R_v = R/M_w$, multiplying by $D_v^* \rho_{is}$, and using (Eqn. 5.2.13a) for D_v^*, we obtain

$$\frac{dm_c}{dt} = \frac{4\pi C_e s_i}{\dfrac{RT}{D_v^* \rho_{is} M_w R_v T} + \dfrac{L_s}{k_a^* T}\left(\dfrac{L_s}{R_v T} - 1\right)}$$

$$= \frac{4\pi C_e D_v^* s_i \rho_{is}}{1 + \dfrac{D_v^* L_s \rho_{is}}{\kappa_T^* \rho_a c_p T}\left(\dfrac{L_s}{R_v T} - 1\right)} \approx \frac{4\pi C_e D_v s_i \rho_{is}}{\Gamma_i} k_{kin}. \tag{5.2.15b}$$

where we used the definition (Eqn. 5.2.12) for Γ_i. This equation coincides with (Eqn. 5.2.14). Thus, both forms of the growth rate equation, (Eqn. 5.2.14) and (Eqn. 5.2.15a) are equivalent. Using the approximate relation $k_a^* = \kappa_T^* \rho_a c_p \approx D_v^* \rho_a c_p$, the factor Γ_i can be approximately replaced with Γ_2 introduced in Section 3.11. The advantage of (Eqn. 5.2.14) is that the dependencies on the three major factors, supersaturation s_i, psychrometric correction Γ_i and the kinetic correction ξ_{dep}, are separated and factorized, which makes this equation convenient for deriving analytical parameterizations.

5.2.2. Axial Growth Rates

Axial growth rates of a crystal may be required for some applications (e.g., for evaluation of the crystal optical properties, radar reflectivities, and development of analytical parameterizations of ice nucleation). Consideration of the crystal axes growth rates requires specification of the crystal habit or shape. Detailed classifications of the crystal habits such as the scheme of Magono and Lee (1966) include up to 80–100 different crystal shapes. However, analytical expressions for the electrical capacities C_e are known only for a few shapes that can approximate crystal shapes and are considered in the following text.

Simple ice plates can be approximated as oblate spheroids. We denote the major and minor axes as r_c and b_c, and the axis ratio as $\xi_f = b_c/r_c < 1$. The volume and mass of the oblate spheroid are $V_c = (4/3)\pi r_c^2 b_c = (4/3)\pi r_c^3 \xi_f$ and $m_c = (4/3)\pi r_c^3 \xi_f \rho_i$. The capacitance of this spheroid is written as

$$C_e = \frac{r_c \varepsilon_c}{\arcsin(\varepsilon_c)}, \qquad \varepsilon_c = (1-\xi_f^2)^{1/2}. \tag{5.2.16}$$

The mass and major semi-axis growth rates are related as

$$\frac{dm_c}{dt} = 4\pi r_c^2 \dot{r}_c \xi_f \rho_i, \qquad \dot{r}_c = \frac{1}{4\pi r_c^2 \xi_f \rho_i} \frac{dm_c}{dt}. \tag{5.2.17}$$

Substituting here (Eqn. 5.2.14) for dm_c/dt and (Eqn. 5.2.16) for C_e, we obtain the major axis growth rate dr_c/dt for a plate-like crystal:

$$\dot{r}_c = \frac{k_f}{\xi_f} \frac{D_v \rho_{is} s_i}{\rho_i \Gamma_i (r_c + \xi_{dep})}, \qquad k_f = \frac{C_e}{r_c} = \frac{\varepsilon_c}{\arcsin(\varepsilon_c)}, \tag{5.2.18}$$

where we introduced the dimensionless *shape factor* k_f.

Columnar crystal shapes can be approximated by prolate spheroids, with the major and minor axes being r_c and b_c, respectively, and the axes ratio is $\xi_f = b_c/r_c < 1$. Then, $V_c = (4/3)\pi r_c b_c^2 = (4/3)\pi r_c^3 \xi_f^2$ and $m_c = (4/3)\pi r_c^3 \xi_f^2 \rho_i$. The capacitance of the prolate spheroid is written as

$$C_e = \frac{2 r_c \varepsilon_c}{\ln\frac{1+\varepsilon_c}{1-\varepsilon_c}} = \frac{r_c \varepsilon_c}{\ln\frac{1+\varepsilon_c}{\xi_f}}, \qquad \varepsilon_c = (1-\xi_f^2)^{1/2}. \tag{5.2.19}$$

The mass and major axis growth rates are related as

$$\frac{dm_c}{dt} = 4\pi r_c^2 \dot{r}_c \xi_f^2 \rho_i, \qquad \dot{r}_c = \frac{1}{4\pi r_c^2 \xi_f^2 \rho_i} \frac{dm_c}{dt}. \tag{5.2.20}$$

Substituting here (Eqn. 5.2.14) for dm_c/dt and (Eqn. 5.2.19) for C_e, we obtain dr_c/dt for a columnar crystal

$$\dot{r}_c = \frac{k_f}{\xi_f^2} \frac{D_v^* \rho_{is} s_i}{\rho_i \Gamma_i (r_c + \xi_{dep})}, \quad k_f = \frac{C_e}{r_c} = \frac{2\varepsilon_c}{\ln[(1+\varepsilon_c)/(1-\varepsilon_c)]}. \tag{5.2.21}$$

It is easy to show that for spherical particles, as the axis ratio $\xi_f \to 1$, the shape factor $k_f \to 1$ and (Eqns. 5.2.18), (5.2.21) transform to the equations for the spherical drops derived in Section 5.1. Thus, these equations for oblate and prolate spheroids generalize the equations from Section 5.1 for the non-spherical cases.

A thin hexagonal plate can be approximated by a disc with radius r_c and height h_c, so the axis ratio $\xi_f = h_c/r_c < 1$. The volume and mass of the disc are $V_c = \pi r_c^2 h_c = (4/3)\pi r_c^3 \xi_f$ and $m_c = \pi r_c^3 \xi_f \rho_i$. The electrical capacitance of the disc is $C_e = (2/\pi) r_c$. The mass and major axis growth rates are related as

$$\frac{dm_c}{dt} = 3\pi r_c^2 \dot{r}_c \xi_f \rho_i, \quad \dot{r}_c = \frac{1}{3\pi r_c^2 \xi_f \rho_i} \frac{dm_c}{dt}. \tag{5.2.22}$$

Substituting here (Eqn. 5.2.14) for dm_c/dt and C_e, we obtain dr_c/dt for the disc like crystal

$$\dot{r}_c = \frac{4k_f}{3\xi_f} \frac{D_v^* \rho_{is} s_i}{\rho_i \Gamma_i (r_c + \xi_{dep})}, \quad k_f = \frac{C_e}{r_c} = \frac{2}{\pi}. \tag{5.2.23}$$

The equations for the axial growth rates for these three crystal shapes can be written in a generalized form similar to (Eqn. 5.1.46) for the drops as

$$\frac{dr_c}{dt} \approx \frac{c_{3i} s_i}{r_c + \xi_{dep}}, \quad c_{3i} = \zeta_i \frac{D_v \rho_{is}}{\rho_i \Gamma_i}, \quad \zeta_i = k_{0i} \frac{k_{fi}}{\xi_f^n}, \tag{5.2.24}$$

where ζ_i describes the effect of the shape, $n = 1$ for the oblate spheroids and thin plates (disks), $n = 2$ for prolate spheroids, $k_{0i} = 1$ for the oblate and prolate spheroids, and $k_{0i} = 4/3$ for the thin plates. For equivalent spheres, $\zeta_i = 1$.

These equations show that the crystal mass and axial growth rates depend on the kinetic correction ξ_{dep} that is inversely proportional to the deposition coefficient α_d. The measured values of α_d vary over a wide range (Hobbs, 1974; Stephens, 1983; Pruppacher and Klett, 1997), from $\alpha_d = 0.014$ at $T = -2$ to $-13\,°C$, yielding $\xi_{dep} \approx 11\ \mu m$, to 0.04 at -6 to $-7\,°C$ ($\xi_{dep} \approx 3.9\ \mu m$), to 0.7 at $T \approx -10$ to $-11\,°C$ ($\xi_{dep} \approx 0.22\ \mu m$), to values near 1 at temperatures below $-40\,°C$ ($\xi_{dep} \sim 0.15\ \mu m$). Experimental data on α_d are contradictory, often indicating substantially different values for the same temperature and pressure. Measured values of α_d are higher for pure ice and decrease substantially in the presence of impurities, which are always abundant in haze particles. Another reason for lower α_d in low-temperature cirrus clouds can be adsorption of nitric acid or other foreign gases on the crystal surface (e.g., Gao et al., 2004). Yet another reason for low α_d can be the formation of liquid films on the crystal surface. As the solution droplets are cooled and ice nucleation begins inside haze particles, a phase separation occurs, solute and concentrated solutions are rejected from the ice, which leads to the formation of a residual solution coating around the ice crystals formed. The coating can serve as a shield, slowing the rate of ice growth up to ~10^3 in comparison with uncoated ice, and this can be one of the reasons for the persistence of the large in-cloud RHI (Räisänen et al., 2006; Bogdan et al., 2009, 2013).

These adsorbed gases or liquid films of impurities act as surfactants and may cause a decrease of α_d. Note that there is a close analogy with the action of surfactants to decrease the condensation coefficient of water drops mentioned in Section 5.1. A similar effect may be caused by the surfactants on the surface of the cirrus crystals.

Values of α_d used in the cloud models may substantially vary. In the Cirrus Parcel Models Comparison Project (CPMCP, see review and comparison in Lin et al., 2002), α_d varied in different models from 1 (AMES model, Jensen et al.), to 0.1 (GSFC model, Lin, and U. Michigan model, Liu), to 0.04 (CSU model, DeMott et al.), and down to 10^{-3} (Gierens et al., 2003). The values of the deposition coefficient can substantially influence the growth rates of the crystals, especially of the small-size fraction; these variations indicate the necessity of further experiments for quantitative assessment of α_d.

5.2.3. Ventilation Corrections

Parameterizing the ventilation corrections to crystal diffusional growth can be accomplished similarly to the method applied in Section 5.1 for drops. A characteristic length scale D_{sc} is introduced based on the fact that ventilation depends on the particle projected area normal to the flow, which is determined by the crystal total area A_t and the perimeter p_c of the projected area normal to the flow. Therefore, a natural scaling length is determined from their ratio—i.e., $D_{sc} = A_t/p_c$.For spherical particles with radius r, we have $A_t = 4\pi r^2$, $p_c = 2\pi r$, and $D_{sc} = 2r$, the diameter. Thus, this definition of D_{sc} is a generalization for nonspherical particles. This allows generalizations of the mass transfer coefficient k_v, ventilation coefficient f_v, and the Sherwood number Sh to be formulated similarly to Section 5.1,

$$k_v = \frac{(dm/dt)}{A_t(\rho_{v\infty} - \rho_{ws})}, \quad f_v = \frac{(dm/dt)}{(dm/dt)_0} = \frac{p_c}{4\pi C_e}\mathrm{Sh},$$

$$\mathrm{Sh} = \frac{k_v D_{sc}}{D_v} = \frac{(dm/dt)}{A_t(\rho_{v\infty} - \rho_{ws})}\frac{D_{sc}}{D_v}, \tag{5.2.25}$$

where $(dm/dt)_0$ is the mass growth rate without ventilation correction defined by (Eqn. 5.2.11) or (Eqn. 5.2.14).

Using these definitions, the crystal mass growth rate with ventilation correction can be written as

$$\frac{dm}{dt} = \left(\frac{dm}{dt}\right)_0 f_v = \left(\frac{dm}{dt}\right)_0 \frac{p_c}{4\pi C_e}\mathrm{Sh}. \tag{5.2.26}$$

Calculations using this equation require knowledge of the perimeter p_c. For a plate or other crystal habit approximated by an oblate spheroid with the major axis r_c, a suitable expression is that for a sphere, $p_c = 2\pi r_c$. For prolate spheroids or other more complicated crystal shapes, A_t and p_c can be evaluated using the mass-dimension and area-dimension relations (e.g., Auer and Veal, 1970; Locatelli and Hobbs, 1974)

$$m_c = \alpha D^{\beta}, \qquad A_t = \gamma D^{\sigma}, \tag{5.2.27}$$

where D is the crystal characteristic dimension. These relations and their parameters are considered in more detail in Chapter 12 in application to ice crystal fall velocities.

Measurements of the ventilation corrections are scarce and are still missing for most crystal habits. The coefficients f_v were calculated theoretically solving the convective-diffusion equation for some idealized shapes. Hall and Pruppacher (1976) approximated these calculated coefficients similar to those for the drops described in Section 5.1 as

$$f_v = 1.0 + 0.14(\mathrm{Sc}^{1/3}\,\mathrm{Re}^{1/2})^2, \qquad \mathrm{Sc}^{1/3}\,\mathrm{Re}^{1/2} < 1, \tag{5.2.28a}$$

$$f_v = 0.86 + 0.28(\mathrm{Sc}^{1/3}\,\mathrm{Re}^{1/2}), \qquad \mathrm{Sc}^{1/3}\,\mathrm{Re}^{1/2} \geq 1. \tag{5.2.28b}$$

With $\mathrm{Sc} \approx 0.71$ for the air, these equations are simplified,

$$f_v \approx 1.00 + 0.11\mathrm{Re}, \qquad \mathrm{Re}^{1/2} < 1.12, \tag{5.2.29a}$$

$$f_v \approx 0.86 + 0.25\mathrm{Re}^{1/2}, \qquad \mathrm{Re}^{1/2} \geq 1.12. \tag{5.2.29b}$$

Subsequently, Wang and Ji (1992) parameterized the ventilation coefficient f_v based on more detailed calculations of the air flow around the crystals of various shapes. They found for the columnar crystals at $0.2 \leq \mathrm{Re} \leq 20$

$$f_v \approx 1.0 - 0.00668X_1 + 2.394X_1^2 + 0.7341X_1^3 - 0.7391X_1^4, \tag{5.2.30a}$$

for simple hexagonal plates P1a at $1 \leq \mathrm{Re} \leq 120$

$$f_v \approx 1.0 - 0.6042X_2 + 2.7982X_2^2 - 0.3193X_2^3 - 0.06247X_2^4, \tag{5.2.30b}$$

and for broad branched crystals P1c at $1 \leq \mathrm{Re} \leq 120$

$$f_v \approx 1.0 + 0.3546X_2 + 3.5533X_2^2, \tag{5.2.30c}$$

where $X_1 = \mathrm{Sc}^{1/3}\mathrm{Re}^{1/2}/4$ and $X_2 = \mathrm{Sc}^{1/3}\mathrm{Re}^{1/2}/10$. The mass growth rates can be calculated with (Eqn. 5.2.26) using these ventilation coefficients and the expressions for $(dm/dt)_0$ derived earlier, where a similar ventilation correction should be introduced for the temperature change.

5.3. Equations for Water and Ice Supersaturations

As we have seen in Sections 5.1 and 5.2, calculation of the droplets and crystal growth requires knowledge of the water and ice supersaturations. Here we derive equations for the fractional water supersaturation $s_w = (\rho_v - \rho_{ws})/\rho_{ws}$ and fractional ice supersaturation $s_i = (\rho_v - \rho_{is})/\rho_{is}$, where ρ_v is the vapor density, and ρ_{ws} and ρ_{is} are saturated vapor densities over water and ice.

5.3.1. General Form of Equations for Fractional Water Supersaturation

Consider a mixed-phase cloud, consisting of droplets, crystals, water vapour, and air. The equation for s_w can be derived from the relation

$$\frac{ds_w}{dt} = \frac{1}{\rho_{ws}}\frac{d\rho_v}{dt} - \frac{\rho_v}{\rho_{ws}^2}\frac{d\rho_{ws}}{dt}. \tag{5.3.1}$$

The continuity equation for the density $\rho_m = \rho_a + \rho_v$ of the mixture of the air and water vapor is written as

$$\frac{\partial \rho_m}{\partial t} + \mathrm{div}(\rho_m \vec{V}) = -I_{con} - I_{dep}, \tag{5.3.2}$$

where I_{con} and I_{dep} are the vapor fluxes onto the drops and crystals due to condensation and deposition, div means the divergence, and $\vec{V}$ is the air velocity vector. The air density continuity equation is

$$\frac{d\rho_a}{dt} + \rho_a \text{div}(\vec{V}) = 0. \tag{5.3.3}$$

The continuity equation for the density ρ_v of water vapor follows from these two equations

$$\frac{d\rho_v}{dt} + \rho_v \text{div}(\vec{V}) = -I_{con} - I_{dep}. \tag{5.3.4}$$

Substituting the expression for $\text{div}(\vec{V})$ from (Eqn. 5.3.3) into (Eqn. 5.3.4), we obtain

$$\frac{d\rho_v}{dt} - \frac{\rho_v}{\rho_a}\frac{d\rho_a}{dt} = -I_{con} - I_{dep}. \tag{5.3.5}$$

The term $d\rho_a/dt$ can be transformed using the equation of state for air, $p = \rho_a R_a T$ (see Section 3.4.1), where p is the air pressure, and R_a is the gas constant for the air, which gives us

$$\frac{d\rho_a}{dt} = \frac{1}{R_a T}\frac{dp}{dt} - \frac{\rho_a}{T}\frac{dT}{dt} = \rho_a\left(\frac{1}{p}\frac{dp}{dt} - \frac{1}{T}\frac{T}{dt}\right). \tag{5.3.6}$$

Substituting (Eqn. 5.3.6) into (Eqn. 5.3.5), we obtain

$$\frac{d\rho_v}{dt} = \rho_v\left(\frac{1}{p}\frac{dp}{dt} - \frac{1}{T}\frac{dT}{dt}\right) - I_{con} - I_{dep}. \tag{5.3.7}$$

We can transform (Eqn. 5.3.1) to express the terms via environmental temperature and pressure. The term $d\rho_{ws}/dt$ in (Eqn. 5.3.1) can be expressed via the cooling rate dT/dt using (Eqn. 3.7.7a) for the water vapor density ρ_{ws} that follows from the Clausius–Clapeyron equation

$$\frac{d\rho_{ws}}{dt} = \frac{d\rho_{ws}}{dT}\frac{dT}{dt} = \frac{\rho_{ws}}{T}\left(\frac{L_e}{R_v T} - 1\right)\frac{dT}{dt}, \tag{5.3.8}$$

where R_v is the specific gas constant for water vapor, and L_e is the specific latent heat of evaporation. Substituting (Eqns. 5.3.7) and (5.3.8) into (Eqn. 5.3.1), we obtain

$$\frac{ds_w}{dt} = \frac{\rho_v}{\rho_{ws}}\left(\frac{1}{p}\frac{dp}{dt} - \frac{1}{T}\frac{dT}{dt}\right) - \frac{I_{dep} + I_{con}}{\rho_{ws}} - \frac{\rho_v}{\rho_{ws}}\frac{1}{T}\left(\frac{L_e}{R_v T} - 1\right)\frac{dT}{dt}. \tag{5.3.9}$$

Using the relation from the definition of fractional supersaturation, $\rho_v/\rho_{ws} = (s_w + 1)$, and collecting the terms at dT/dt, this equation can be rewritten as

$$\frac{ds_w}{dt} = (s_w + 1)\left(\frac{1}{p}\frac{dp}{dt} - \frac{L_e}{R_v T^2}\frac{dT}{dt}\right) - \frac{I_{dep} + I_{con}}{\rho_{ws}}. \tag{5.3.10}$$

Dividing by $(s_w + 1)$ and using again the relation $\rho_{ws}(s_w + 1) = \rho_v$, we obtain

$$\frac{1}{(s_w + 1)}\frac{ds_w}{dt} = \frac{1}{p}\frac{dp}{dt} - \frac{L_e}{R_v T^2}\frac{dT}{dt} - \frac{I_{con} + I_{dep}}{\rho_v}. \tag{5.3.11}$$

The term dT/dt can be excluded using the heat balance equation from Section 3.11

$$\rho_a c_p \frac{dT}{dt} = \frac{dp}{dt} + L_e I_{con} + L_s I_{dep}, \tag{5.3.12}$$

where L_s is the specific sublimation (deposition) latent heat rate. Substituting dT/dt from (Eqn. 5.3.12) into (Eqn. 5.3.11) and collecting the terms at I_{con} and I_{dep}, we obtain

$$\frac{1}{(s_w+1)}\frac{ds_w}{dt} = \frac{1}{p}\frac{dp}{dt} - \frac{L_e}{R_v T^2}\frac{1}{c_p \rho_a}\frac{dp}{dt}$$

$$-\frac{I_{con}}{\rho_v}\left(1+\frac{L_e^2}{R_v c_p T^2}\frac{\rho_v}{\rho_a}\right) + \frac{I_{dep}}{\rho_v}\left(1+\frac{L_e L_s}{R_v c_p T^2}\frac{\rho_v}{\rho_a}\right). \tag{5.3.13}$$

The second term on the right-hand side can be transformed using the equation of state for air, $1/\rho_a = R_a T/p$, and definitions $R_a = R/M_a$, $R_v = R/M_w$, and then the second term on the RHS becomes

$$-\frac{L_e}{c_p T}\frac{M_w}{M_a}\frac{1}{p}\frac{dp}{dt}. \tag{5.3.14}$$

It is convenient to use the notations Γ_1 and Γ_2 for the psychrometric corrections to the drop and crystal growth rates from Section 5.1 and Section 5.2, and a similar notation Γ_{12} in the fourth term on the RHS in (Eqn. 5.3.13)

$$\Gamma_1 = 1+\frac{L_e^2}{R_v c_p T^2}\frac{\rho_v}{\rho_a}, = 1+\frac{L_e}{c_p}\frac{\partial q_{ws}}{\partial T}, \quad \Gamma_{12} = 1+\frac{L_e L_s}{R_v c_p T^2}\frac{\rho_v}{\rho_a} = 1+\frac{L_s}{c_p}\frac{\partial q_{ws}}{\partial T},$$

$$\Gamma_2 = 1+\frac{L_s^2}{R_v c_p T^2}\frac{\rho_v}{\rho_a} = 1+\frac{L_s}{c_p}\frac{\partial q_{is}}{\partial T}. \tag{5.3.15}$$

Then (Eqn. 5.3.13) can be rewritten using (Eqns. 5.3.14) and (5.3.15),

$$\frac{1}{(1+s_w)}\frac{ds_w}{dt} = \frac{1}{p}\frac{dp}{dt}\left(1-\frac{L_e}{c_p T}\frac{M_w}{M_a}\right) - \frac{\Gamma_1 I_{con}}{\rho_v} - \frac{\Gamma_{12} I_{dep}}{\rho_v}. \tag{5.3.16}$$

The vapor fluxes I_{con} and I_{dep} to the droplets and crystals (the sources or sinks in the supersaturation equation) are the integrals of the mass growth rates. The integral mass growth rates can be expressed using equations for the growth rates dr_d/dt and dr_c/dt of the droplet radii and crystal equivalent radii over the corresponding size spectra of the drops $f_d(r_d, t)$ and crystals $f_c(r_c, t)$, where r_d and r_c are corresponding radii:

$$I_{con} = \int_0^\infty \frac{dm_d}{dt} f(r_d,t)\,dr_d = 4\pi\rho_w \int_0^\infty r_d^2 \frac{dr_d}{dt} f(r_d,t)\,dr_d, \tag{5.3.17a}$$

$$I_{dep} = \int_0^\infty \frac{dm_c}{dt} f(r_c,t)\,dr_c = 4\pi\rho_i \int_0^\infty r_c^2 \frac{dr_c}{dt} f(r_c,t)\,dr_c, \tag{5.3.17b}$$

where the second equalities assume approximate spherical symmetry of the particle. We use dr_d/dt and dr_c/dt in the form (Eqns. 5.1.46) and (5.2.24) derived in Sections 5.1 and 5.2,

$$\frac{dr_d}{dt} = \frac{c_{3w} s_w}{r_d + \xi_{con}}, \qquad c_{3w} = \frac{D_v \rho_{ws}}{\rho_w \Gamma_1}, \qquad \xi_{con} = \frac{4D_v}{\alpha_c V_w}, \tag{5.3.18a}$$

$$\frac{dr_c}{dt} = \frac{c_{3i}s_i}{r_c + \xi_{dep}}, \qquad c_{3i} = k_{0i}\frac{k_f}{\xi_f^n}\frac{D_v \rho_{is}}{\rho_i \Gamma_2}, \qquad \xi_{dep} = \frac{4D_v}{\alpha_d V_w}, \tag{5.3.18b}$$

where all notation is the same as in Sections 5.1 and 5.2.

5.3.2. Supersaturation Relaxation Times and Their Limits

Substituting (Eqns. 5.3.18a,b) into (Eqns. 5.3.17a,b), we obtain

$$I_{con} = \frac{\rho_{ws}s_w}{\Gamma_1}\tau_{fd}^{-1} = \frac{\Delta_w \rho_a}{\Gamma_1}\tau_{fd}^{-1}, \qquad I_{dep} = \frac{\rho_{is}s_i}{\Gamma_2}\tau_{fc}^{-1} = \frac{\Delta_i \rho_a}{\Gamma_2}\tau_{fd}^{-1}, \tag{5.3.19}$$

where $\Delta_w = q_v - q_{ws}$ and $\Delta_i = q_v - q_{is}$ are the specific supersaturations over water and ice, and we have introduced the *supersaturation absorption or relaxation times* for the droplets τ_{fd} and crystals τ_{fc} (also called in the literature the *phase relaxation times*):

$$\tau_{fd}^{-1} = 4\pi D_v \int_0^\infty \frac{r_d^2}{r_d + \xi_{con}} f_d(r_d, t)\, dr_d, \tag{5.3.20a}$$

$$\tau_{fc}^{-1} = 4\pi D_v k_{0i} \frac{k_f}{\xi_f^n} \int_0^\infty \frac{r_c^2}{r_c + \xi_{dep}} f_c(r_c, t)\, dr_c. \tag{5.3.20b}$$

This form of τ_{fd}, τ_{fc} account for the kinetic corrections and crystal shapes. The more general expressions are different for cases with and without nucleation of the new droplets and crystals. If droplet or crystal nucleation takes place, then the droplet or crystal size spectra can be expressed using the concentration conservation laws in differential form as

$$f_d(r_d, t)dr_d = \varphi_s(s_w)ds_w, \qquad f_c(r_c, t)dr_c = \psi_s(s_w)ds_w, \tag{5.3.21}$$

where $\varphi_s(s_w)$ is the supersaturation activity spectrum of CCN (Chapter 6), $\psi_s(s_w)$ is the supersaturation activity spectrum of IN (Chapters 8 and 9). Substitution of these expressions into (Eqns. 5.3.17a,b) gives the expressions for the relaxation times

$$\tau_{fd}^{-1} = 4\pi D_v \int_0^\infty \frac{r_d^2}{r_d + \xi_{con}} \varphi_s(s_w)\, ds_w, \tag{5.3.22a}$$

$$\tau_{fc}^{-1} = 4\pi D_v k_{0i} \frac{k_f}{\xi_f^n} \int_0^\infty \frac{r_c^2}{r_c + \xi_{dep}} \psi_s(s_w)\, ds_w. \tag{5.3.22b}$$

Substitution of these expressions into (Eqn. 5.3.16) gives the equations for supersaturation that contain the terms with s_w in the integrand—i.e. the integro-differential equations. The form of the equations including activation is more complicated but allows evaluation of the nucleation laws for the droplets and crystals. These equations with various activity spectra φ_s and ψ_s and their solutions will be considered in detail in Chapters 6, 7, 8, and 9.

Here we consider the simpler situation when nucleation is absent and the vapor fluxes I_{con} and I_{dep} are determined by preexisting droplets with $f_d(r_d)$ and crystals with $f_c(r_c)$, which determine the relaxation times τ_{fd} and τ_{fc} in (Eqns. 5.3.20a,b). The expressions and values of τ_{fd} and τ_{fc} are different in the diffusion and kinetic regimes of particle growth. If the growth of droplets and crystals

Table 5.1. *Supersaturation relaxation times in liquid clouds, seconds (upper line) and equilibrium water supersaturations, % (lower line) for various drop concentrations* N_d *and mean radius* $\bar{r}_d$.

	Droplet concentration N_d (cm^{-3})			
Mean radius $\bar{r}_c$ (μm)	50	100	200	500
5	14.4 0.072	7.2 0.036	3.6 0.018	1.44 0.72×10^{-2}
10	7.2 0.036	3.6 0.018	1.8 0.009	0.72 0.36×10^{-2}

proceeds in the diffusion regime, then the expressions (Eqns. 5.3.20a,b) for the relaxation times become

$$\tau_{fd} \approx \left(4\pi D_v \int_0^\infty r_d f_d(r_d,t)\,dr_d\right)^{-1} = (4\pi D_v N_d \bar{r}_d)^{-1}, \tag{5.3.23a}$$

$$\tau_{fc} \approx \left(4\pi D_v \int_0^\infty r_c f_c(r_c,t)\,dr_c\right)^{-1} = (4\pi D_v k_{0i} \frac{k_f}{\xi_f^n} N_c \bar{r}_c)^{-1}, \tag{5.3.23b}$$

where N_d, N_c are droplet and crystal concentrations, and $\bar{r}_d$ and $\bar{r}_c$ are their mean radii. Typical values of τ_{fd} in liquid or mixed clouds vary from 1–2 seconds with large droplet concentrations to 10–15 seconds with small concentrations (Table 5.1). Typical values of τ_{fc} in ice clouds vary from 5–10 minutes to 5–10 hours (Table 5.2). The supersaturation relaxation times in a cloud represent in general a three-dimensional field and may vary from a few seconds to a few hours in various parts of the cloud (e.g., Khvorostyanov and Sassen, 1998a,b, 2002; Khvorostyanov, Curry et al., 2001, 2003, 2006; Krämer et al., 2009).

In the kinetic growth regime, the expressions for τ_{fd} and τ_{fc} become

$$\tau_{fd} \approx \left(\frac{4\pi D_v}{\xi_{con}} \int_0^\infty r_d^2 f_d(r_d,t)\,dr_d\right)^{-1} = \left(\frac{4\pi D_v}{\xi_{con}} N_d < r_d^2 >\right)^{-1} \sim L_{vis,d}, \tag{5.3.24a}$$

$$\tau_{fc} \approx \left(\frac{4\pi D_v}{\xi_{dep}} k_{0i} \frac{k_f}{\xi_f^n} \int_0^\infty r_c^2 f_c(r_c,t)\,dr_d\right)^{-1}$$

$$= \left(\frac{4\pi D_v}{\xi_{dep}} k_{0i} \frac{k_f}{\xi_f^n} N_c < r_c^2 >\right)^{-1} \sim L_{vis,c}, \tag{5.3.24b}$$

where $< r_d^2 >, < r_c^2 >$ are the mean squared radii of the size spectra, and $L_{vis,d}$ and $L_{vis,c}$ are the corresponding visibility ranges that are inversely proportional to the second moments of the size spectra (see Chapter 2).

Table 5.2. *Supersaturation relaxation times in crystalline clouds for spherical crystals with various concentrations* N_c *and mean radii* $\bar{r}_c$. *The units are hours (h), minutes (min), and seconds (s)*

Mean radius	Crystal concentration N_c (L^{-1})							
$\bar{r}_c$ (μm)	1	5	10	100	200	500	1000	3000
2	500 h	100 h	50 h	5 h	2.5 h	1 h	30 min	10 min
5	200 h	40 h	20 h	2 h	1 h	24 min	12 min	4 min
10	100 h	20 h	10 h	1 h	30 min	12 min	6 min	3 min
20	50 h	10 h	5 h	30 min	15 min	6 min	3 min	1 min
50	20 h	4 h	2 h	12 min	6 min	144 s	72 s	24 s
100	10 h	2 h	1 h	6 min	3 min	72 s	36 s	12 s
500	2 h	24 min	12 min	72 s	36 s	14 s	3.6 s	2.5 s
1000	1 h	12 min	6 min	36 s	18 s	7.2 s	3.6 s	1.1 s

5.3.3. Equation for Water Supersaturation in Terms of Relaxation Times

Substituting I_{con} and I_{dep} from (Eqn. 5.3.19 into 5.3.16), and using the relations $\rho_v/\rho_{ws} = 1/(1+s_w)$, $\rho_v/\rho_{is} = 1/(1+s_i)$, we obtain the equation for s_w:

$$\frac{1}{(1+s_w)}\frac{ds_w}{dt} = \frac{1}{p}\frac{dp}{dt}\left(1 - \frac{L_e}{c_p T}\frac{M_w}{M_a}\right) - \frac{1}{\tau_d}\frac{s_w}{(1+s_w)} - \frac{1}{\tau_c}\frac{\Gamma_{12}}{\Gamma_2}\frac{s_i}{(1+s_i)}. \tag{5.3.25}$$

If the cooling rate is governed in a rising parcel by the vertical velocity w like in a Lagrangian parcel model or at the adjustment stage of the process- and coordinate-splitting method in a multidimensional Eulerian model, then the problem is reduced to a zero-dimensional Lagrangian task, and the term with pressure can be expressed using the hydrostatic equation, $dp = -\rho_a g dz$, and the relation $dp/dt = (dp/dz)(dz/dt) = -\rho_a g w$, where g is acceleration of gravity. Using equation of state for the air, the term with pressure can be written as $(1/p)(dp/dt) = -(g/R_a T)w$. Substitution of this into (Eqn. 5.3.25) yields:

$$\frac{1}{(1+s_w)}\frac{ds_w}{dt} = c_{1w}w - \frac{1}{\tau_{fd}}\frac{s_w}{(1+s_w)} - \frac{1}{\tau_{fc}}\frac{\Gamma_{12}}{\Gamma_2}\frac{s_i}{(1+s_i)}, \tag{5.3.26a}$$

where

$$c_{1w}(T) = \left(\frac{L_e}{c_p T}\frac{M_w}{M_a} - 1\right)\frac{g}{R_a T}. \tag{5.3.26b}$$

Using the definitions of the dry γ_a and wet γ_w adiabatic lapse rates from Section 3.11, a simple transformation yields

$$c_{1w}(T) = \frac{c_p}{L_e}(\gamma_a - \gamma_w)\Gamma_1\frac{\rho_a}{\rho_{ws}} = \left(\frac{dq_l}{dz}\right)_{ad}\Gamma_1\frac{\rho_a}{\rho_{ws}}, \tag{5.3.26c}$$

where $(dq_l/dz)_{ad}$ is the adiabatic gradient of the liquid water content introduced in Section 3.11. Equation (5.3.26a) can be written in a slightly different form by multiplying by $(1 + s_w)$, and using the relation

$$s_i\frac{1+s_w}{1+s_i} = s_w + \frac{\rho_{ws} - \rho_{is}}{\rho_{ws}}. \tag{5.3.27}$$

Thus, we obtain

$$\frac{ds_w}{dt} = (1+s_w)c_{1w}w - s_w\left(\frac{1}{\tau_{fd}} + \frac{\Gamma_{12}}{\Gamma_2}\frac{1}{\tau_{fc}}\right) - \frac{\Gamma_{12}}{\Gamma_2}\frac{1}{\tau_{fc}}\frac{\rho_{ws}-\rho_{is}}{\rho_{ws}}. \quad (5.3.28a)$$

This supersaturation equation has a clear physical meaning. The first term on the right-hand side of (Eqn. 5.3.26a) or (Eqn. 5.3.28a) describes the supersaturation generation rate in the parcel in adiabatic ascent. This rate is proportional to the vertical velocity w and to c_{1w}, and as (Eqn. 5.3.26c) shows, to the adiabatic gradient of the liquid water content. The second term represents the rate of supersaturation absorption or relaxation by diffusion growth of the droplets and crystals. The last term describes the flux from the droplets to the crystals, called the *Wegener–Bergeron–Findeisen process* (*WBF*), therefore it is proportional to the difference $\rho_{ws} - \rho_{is}$ in (Eqn. 5.3.28a).

We can introduce the "effective" supersaturation relaxation time in a mixed cloud as

$$\tau_{f,mix} = \left(\frac{1}{\tau_{fd}} + \frac{\Gamma_{12}}{\Gamma_2}\frac{1}{\tau_{fc}}\right)^{-1}. \quad (5.3.28b)$$

Then, (Eqn. 5.3.28a) can be written as

$$\frac{ds_w}{dt} = (1+s_w)c_{1w}w - \frac{s_w}{\tau_{f,mix}} - \frac{\Gamma_{12}}{\Gamma_2}\frac{1}{\tau_{fc}}\frac{\rho_{ws}-\rho_{is}}{\rho_{ws}}. \quad (5.3.29)$$

The second term on the right-hand side in (Eqn. 5.3.29), $s_w/\tau_{f,mix}$, shows that the rate of vapor absorption (relaxation) is determined by $\tau_{f,mix}$, therefore it is termed the supersaturation "absorption" or "relaxation" time. Equation (5.3.26a) can be written in a slightly different form that is more convenient for analytical solutions:

$$\frac{ds_w}{dt} = (1+s_w)c_{1w} - s_w\tau_{fd}^{-1} - c_{2w}s_i\tau_{fc}^{-1}, \quad (5.3.30a)$$

$$c_{2w}(T) = \frac{\Gamma_{12}}{\Gamma_2}\frac{1+s_w}{1+s_i}. \quad (5.3.30b)$$

In pure liquid clouds, $f_c(r_c) = 0$ and $\tau_{fc}^{-1} = 0$, and so the terms with τ_{fc}^{-1} on the right-hand side in (Eqns. 5.3.26a) and (5.3.30a) vanish. Since $s_w << 1$ and $(1 + s_w) \approx 1$ in such clouds, the supersaturation (Eqn. 5.3.30a) is simplified to be

$$\frac{ds_w}{dt} = c_{1w}w - \frac{s_w}{\tau_{fd}}, \quad (5.3.31a)$$

that is, the supersaturation balance is determined by its generation $c_{1w}w$ and absorption s_w/τ_{fd}. At small times, or if vapor absorption is weak and the second term here is small, the solution is

$$s_w(t) = c_{1w}wt. \quad (5.3.31b)$$

The solution of this type is used as a first iteration for parameterization of the initial stage of drop activation (see Chapter 7) and of ice nucleation in homogeneous and heterogeneous freezing (Chapters 8–10). In the absence of supersaturation generation, if vertical velocity $w = 0$, and radiative and other sources of cooling are absent, then the solution is

$$s_w(t) = s_w(t_0)\exp[-(t-t_0)/\tau_{fd}]. \quad (5.3.31c)$$

That is, in the absence of supersaturation generation, the supersaturation e-fold decrease (relaxation) occurs during τ_{fd}. This originated the term "relaxation time." The same is valid for ice supersaturation in crystalline clouds.

The equation for the water saturation ratio S_w is used in some cases instead of the equation for s_w. It is easily obtained from the previous equations and the relation $s_w = S_w - 1$. Substitution of this relation into (Eqn. 5.3.28a) gives

$$\frac{dS_w}{dt} = S_w c_{1w} w - (S_w - 1)\left(\frac{1}{\tau_{fd}} + \frac{\Gamma_{12}}{\Gamma_2}\frac{1}{\tau_{fc}}\right) - \frac{\Gamma_{12}}{\Gamma_2}\frac{1}{\tau_{fc}}\frac{\rho_{ws} - \rho_{is}}{\rho_{ws}}. \tag{5.3.32a}$$

In pure liquid clouds with $1/\tau_{fc} = 0$, the equation is simplified to

$$\frac{dS_w}{dt} = S_w c_{1w} w - \frac{S_w - 1}{\tau_{fd}}. \tag{5.3.32b}$$

5.3.4. Equivalence of Various Forms of Supersaturation Equations

In most textbooks and papers, the supersaturation equation is derived and used in the following form

$$\frac{ds_w}{dt} = A_1 w - A_2 \frac{dq_l}{dt}. \tag{5.3.33}$$

Here, q_l is the liquid water content (LWC), the second term on the RHS (sink or source of supersaturation) is written proportional to the LWC change rate dq_l/dt, and the coefficients are (Pruppacher and Klett, 1997, eq. (13-31), page 513)

$$A_1 = \frac{L_e g}{R_a T^2 c_p}\frac{M_w}{M_a} - \frac{g}{R_a T}, \tag{5.3.34}$$

$$A_2 = \frac{R_a T}{(M_w/M_a)e_{ws}} + \frac{(M_w/M_a)L_e^2}{pTc_p}, \tag{5.3.35}$$

where e_{ws} is the vapor pressure saturated over water. The coefficients A_1 and A_2 describe supersaturation generation and absorption, respectively. It can be shown that this form of the supersaturation equation for the liquid cloud is equivalent to (Eqn. 5.3.31a) derived earlier. The coefficient A_1 can be rewritten as

$$A_1 = \left(\frac{L_e}{c_p T}\frac{M_w}{M_a} - 1\right)\frac{g}{R_a T} = c_{1w}(T). \tag{5.3.36}$$

In other words, A_1 is equal to c_{1w} defined in (Eqn. 5.3.26b). The coefficient A_2 can be transformed and simplified using the equations of state for the water vapor, $e_{ws} = \rho_{ws} R_v T$, and for the air, $p = \rho_a R_a T$, and the relations $R_a = R/M_a$ and $R_v = R/M_w$,. Thus, A_2 becomes

$$\begin{aligned} A_2 &= \frac{(R_a M_a)T}{\rho_{ws}(M_w R_v)T} + \frac{L_e^2}{\rho_a (R_a M_a/M_w)T^2 c_p} \\ &= \frac{1}{\rho_{ws}} + \frac{L_e^2}{\rho_a R_v T^2 c_p} = \frac{1}{\rho_{ws}}\left(1 + \frac{L_e^2}{R_v T^2 c_p}\frac{\rho_{ws}}{\rho_a}\right) = \frac{\Gamma_1}{\rho_{ws}}, \end{aligned} \tag{5.3.37}$$

where Γ_1 is the psychrometric correction defined in (Eqn. 5.3.15). The rate of liquid water content change, dq_l/dt, by definition is equal to the vapor flux of the droplets I_{con}, so it can be transformed using the relations (Eqns. 5.3.17a), (5.3.19), (5.3.20a)

$$\frac{dq_l}{dt} = \int_0^\infty \frac{dm_d}{dt} f_d(r_d) dr_d = I_{con} = \frac{\rho_{ws} s_w}{\Gamma_1} \tau_{fd}^{-1}. \tag{5.3.38}$$

This expression shows that the term dq_l/dt is proportional to supersaturation. Thus, the second term in (Eqn. 5.3.33) is

$$A_2 \frac{dq_l}{dt} = A_2 I_{con} = \frac{\Gamma_1}{\rho_{ws}} \frac{\rho_{ws} s_w}{\Gamma_1 \tau_{fd}} = \frac{s_w}{\tau_{fd}}. \tag{5.3.39}$$

Substituting these expressions for A_1 and $A_2 dq_l/dt$ into (Eqn. 5.3.33), we obtain

$$\frac{ds_w}{dt} = A_1 w - A_2 \frac{d\rho_L}{dt} = c_{1w} w - \frac{s_w}{\tau_{fd}}, \tag{5.3.40}$$

which coincides with (Eqn. 5.3.31a). Thus, (Eqn. 5.3.33) is equivalent to (Eqn. 5.3.31a) for liquid clouds. The convenience of (Eqn. 5.3.31a) or the second form of (Eqn. 5.3.40) is that the sink term dq_l/dt is expressed via s_w, thus this equation contains only one variable and does not require evaluation of dq_l/dt. Equations (5.3.25) and (5.3.28a) generalize these equations for mixed-phase clouds containing drops and crystals.

Water supersaturation governs crystal nucleation during the freezing of haze particles even at cold temperatures (below −40 °C) in cirrus in the upper troposphere, when submicron solution droplets may exist but liquid water drops are absent. Under these conditions, s_w is needed for calculations of ice nucleation rates (see Chapters 8–10). In pure crystalline clouds, $f_d(r_d) = 0$, $\tau_{fd}^{-1} = 0$, and the terms with τ_{fd}^{-1} on the right-hand side in (Eqns. 5.3.28a and 5.3.30a) vanish. Thus, (Eqn. 5.3.28a) becomes

$$\frac{ds_w}{dt} = (1 + s_w) c_{1w} w - \frac{\Gamma_{12}}{\Gamma_2} \frac{s_w}{\tau_{fc}} - \frac{\Gamma_{12}}{\Gamma_2} \frac{1}{\tau_{fc}} \frac{\rho_{ws} - \rho_{is}}{\rho_{ws}}. \tag{5.3.41}$$

5.3.5. Equation for Fractional Ice Supersaturation

In simulations of mixed clouds or crystalline clouds like cirrus, the supersaturation over ice $s_i = (\rho_v - \rho_{is})/\rho_{is}$ is required for calculation of the crystal growth (Section 5.2). It can be determined in two ways. One way to determine s_i is by solving the equation for s_w given earlier (if s_w is needed for the other purposes—e.g., drop activation and crystal nucleation) and using the relation that follows from the definitions of s_w and s_i and the Clausius–Clapeyron equation:

$$s_i = \frac{\rho_{ws}}{\rho_{is}}(s_w + 1) - 1 = c_{wi}(s_w + 1) - 1, \qquad c_{wi} = \exp\left[\frac{L_m (T_0 - T)}{R_v T_0 T}\right], \tag{5.3.42a}$$

where $T_0 = 273.15$ K is the triple point temperature. If s_w and the equation for s_w are not needed, another direct way is to use the equation for s_i that can be derived by differentiating s_i:

$$\frac{ds_i}{dt} = \frac{1}{\rho_{is}} \frac{d\rho_v}{dt} - \frac{\rho_v}{\rho_{is}^2} \frac{d\rho_{is}}{dt}. \tag{5.3.42b}$$

Analogously to the derivation of the equation for s_w earlier, using the equations of continuity and state for the air, water vapor, and their mixture (the Clausius–Clapeyron equation), we arrive at an equation

$$\frac{1}{(1+s_i)}\frac{ds_i}{dt} = \frac{1}{p}\frac{dp}{dt}\left(1-\frac{L_s}{c_p T}\right) - \frac{I_{dep}}{\rho_v}\Gamma_2 - \frac{I_{con}}{\rho_v}\Gamma_{12}. \quad (5.3.43a)$$

Substituting here I_{con} and I_{dep} from (Eqn. 5.3.19), and using the relation $\rho_v = (1 + s_i)\rho_{is}$ and then multiplying by $(1 + s_i)$ yields

$$\frac{ds_i}{dt} = (1+s_i)\frac{1}{p}\frac{dp}{dt}\left(1-\frac{L_s}{c_p T}\right) - \frac{s_i}{\tau_{fc}} - \frac{\Gamma_{12}}{\Gamma_2}\frac{(1+s_i)}{(1+s_w)}\frac{s_w}{\tau_{fd}}. \quad (5.3.43b)$$

Now s_w can be excluded using the relation

$$\frac{1+s_i}{1+s_w}s_w = s_i + \frac{\rho_{is}-\rho_{ws}}{\rho_{is}}, \quad (5.3.43c)$$

and we obtain an equation for s_i:

$$\frac{ds_i}{dt} = (1+s_i)\frac{1}{p}\frac{dp}{dt}\left(1-\frac{L_s}{c_p T}\right) - s_i\left(\frac{1}{\tau_{fc}} + \frac{\Gamma_{12}}{\Gamma_2}\frac{1}{\tau_{fd}}\right) + \frac{\Gamma_{12}}{\Gamma_2}\frac{1}{\tau_{fd}}\frac{\rho_{ws}-\rho_{is}}{\rho_{is}}. \quad (5.3.44)$$

As in the case with s_w earlier, (Eqn. 5.3.44) can be further transformed using the hydrostatic approximation for the case with adiabatic parcel ascent in a Lagrangian model or for the step of adjustment in the splitting procedure in a multidimensional Eulerian model. Then, $(1/p)(dp/dt) = -(g/R_a T)w$, and we obtain

$$\frac{ds_i}{dt} = (1+s_i)c_{1i}w - s_i\left(\frac{1}{\tau_{fc}} + \frac{\Gamma_{12}}{\Gamma_2}\frac{1}{\tau_{fd}}\right) + \frac{\Gamma_{12}}{\Gamma_2}\frac{1}{\tau_{fd}}\frac{\rho_{ws}-\rho_{is}}{\rho_{is}}, \quad (5.3.45)$$

where

$$c_{1i}(T) = \left(\frac{L_s}{c_p T}\frac{M_w}{M_a} - 1\right)\frac{g}{R_a T}. \quad (5.3.46a)$$

Using the equations for the dry γ_a and wet (in ice clouds) γ_{is} adiabatic lapse rates and adiabatic gradient of the ice water content $(dq_i/dz)_{ad}$ in ice clouds introduced in Section 3.11, this coefficient also can be expressed via the difference $(\gamma_a - \gamma_{is})$ similar to c_{1w} in (Eqn. 5.3.26c) for the case of liquid clouds:

$$c_{1i}(T) = \frac{c_p}{L_s}(\gamma_a - \gamma_{is})\Gamma_2\frac{\rho_a}{\rho_{is}} = \left(\frac{dq_i}{dz}\right)_{ad}\Gamma_2\frac{\rho_a}{\rho_{is}}. \quad (5.3.46b)$$

This equation shows that the ice supersaturation generation rate in ice clouds is proportional to the difference $(\gamma_a - \gamma_{is})$, or to the adiabatic gradient of the ice water content. For pure crystalline clouds, or when the vapor flux to the liquid substance can be neglected, then $\tau_{fd} = \infty$, $1/\tau_{fd} = 0$, and (Eqn. 5.3.45) is simplified as

$$\frac{ds_i}{dt} = (1+s_i)c_{1i}w - \frac{s_i}{\tau_{fc}}, \quad (5.3.47)$$

which is similar to the water supersaturation (Eqn. 5.3.31a) in liquid clouds. In contrast to liquid clouds where s_w is close to zero, s_i in crystalline clouds is not close to 0, so it can be 0.3–0.7 (30–70%) or more in cold cirrus clouds, and the term $(1 + s_i)$ cannot be replaced with 1.

The equation for the ice saturation ratio S_i is easily obtained from the previous equations and the relation $s_i = S_i - 1$. Substitution of this relation into (Eqn. 5.3.45) gives

$$\frac{dS_i}{dt} = S_i c_{1i} w - (S_i - 1)\left(\frac{1}{\tau_{fc}} + \frac{\Gamma_{12}}{\Gamma_2}\frac{1}{\tau_{fd}}\right) + \frac{\Gamma_{12}}{\Gamma_2}\frac{1}{\tau_{fd}}\frac{\rho_{ws} - \rho_{is}}{\rho_{is}}. \tag{5.3.47a}$$

For a pure ice cloud, $1/\tau_{fd} = 0$ and it is simplified as

$$\frac{dS_i}{dt} = c_{1i} w S_i - \frac{S_i - 1}{\tau_{fc}}. \tag{5.3.47b}$$

5.3.6. Equilibrium Supersaturations over Water and Ice

Liquid clouds. Cloud layers may exist in quasi-steady states that can be characterized by equilibrium supersaturations. The equilibrium supersaturation s_w^{eq} in a liquid cloud is determined from (Eqn. 5.3.31a) with the steady state condition $ds_w/dt = 0$. Using also (Eqns. 5.3.26b,c) for c_{1w}, we obtain

$$s_{w,liq}^{eq} = c_{1w} w \tau_{fd} = \frac{c_p}{L_e}\frac{(\gamma_a - \gamma_w)}{4\pi D_v N_d \bar{r}_d}\frac{\Gamma_1 \rho_a}{\rho_{ws}} w = \left(\frac{dq_l}{dz}\right)_{ad}\frac{\Gamma_1}{q_{ws}}\frac{w}{4\pi D_v N_d \bar{r}_d}. \tag{5.3.48a}$$

This simple equation shows that the equilibrium supersaturation is proportional to the vertical velocity, the difference of the dry and wet adiabatic lapse rates, or to the adiabatic gradient of LWC, and to the supersaturation relaxation time (which is inversely proportional to the droplet concentration and mean radius). Hence, typical values of equilibrium water supersaturation in clouds can be estimated from (Eqn. 5.3.48a) using a value $c_{1w} \sim 10^{-5}$ cm^{-1}. The equilibrium supersaturation relaxation times for various drop concentrations and two values of the mean radius are given in Table 5.1, which shows that s_w^{eq} with $\bar{r}_d = 5$ μm decreases from 0.072% at $N_d = 50$ cm^{-3} ($\tau_{fd} = 14.4$ s) to 0.72×10^{-2} % at $N_d = 500$ cm^{-3} ($\tau_{fd} = 1.44$ s), and s_w^{eq} is a factor of two smaller for $\bar{r}_d = 10\,\mu$m.

The equilibrium condensation rate can be calculated substituting (Eqn. 5.3.48a) into (Eqn. 5.3.19):

$$I_{con}^{eq} = \frac{\rho_{ws}}{\Gamma_1 \tau_{fd}} s_{w,liq}^{eq} = \frac{\rho_{ws}}{\Gamma_1 \tau_{fd}} c_{1w} w \tau_{fd} = \frac{\rho_{ws}}{\Gamma_1 \tau_{fd}}\frac{c_p}{L_e}(\gamma_a - \gamma_w)\tau_{fd}\frac{\Gamma_1 \rho_a}{\rho_{ws}} w$$

$$= \frac{c_p}{L_e}(\gamma_a - \gamma_w)\rho_a w = \left(\frac{dq_l}{dz}\right)_{ad}\rho_a w. \tag{5.3.48b}$$

Ice clouds. In contrast to water supersaturation, the equilibrium ice supersaturation in crystalline clouds like cirrus, contrails, and some As-Ac or diamond dust is not small. It can be determined from (Eqn. 5.3.47) under the condition $ds_i/dt = 0$,

$$s_{i,ice}^{eq} = \frac{c_{1i} w \tau_{fc}}{1 - c_{1i} w \tau_{fc}}$$

$$= \frac{c_p(\gamma_a - \gamma_{is})\Gamma_2}{L_s}\frac{\rho_a}{\rho_{is}} w \tau_{fc}\left[1 - \frac{c_p(\gamma_a - \gamma_{is})\Gamma_2}{L_s}\frac{\rho_a}{\rho_{is}} w \tau_{fc}\right]^{-1}. \tag{5.3.49a}$$

The crystal relaxation times τ_{fc} calculated for spherical crystals are given in Table 5.2. The values of τ_{fc} vary over a wide range from 1–10 s to 1–10 min for high crystal concentrations ~10^3–3×10^3 L^{-1} (e.g., in cloud-seeding experiments) to several hundred hours for small crystal concentrations and radii. The corresponding equilibrium ice supersaturations in crystalline clouds calculated with (Eqn. 5.3.49a) are given in Table 5.3. For $\tau_{fc} \leq$ 10–600 s and $w \leq$ 10–20 cm s^{-1}, supersaturation values are smaller than one percent. For larger τ_{fc} and w, the supersaturation values increase to several percent or tens of percent (right lower corner in Table 5.3). These values may exceed the critical (or threshold) values of ice supersaturations $s_{i,cr}$, ~ 25–70%, required for homogeneous or heterogeneous ice nucleation (see Chapters 8 and 9). This indicates that such states are unstable and new crystals should nucleate until equilibrium can be established.

At $w >$ 1–2 m s^{-1} and $\tau_{fc} >$ 1–3 hours, when $c_{i1} w \tau_{fc} > 1$, the denominator in (Eqn. 5.3.49a) becomes negative and the ice supersaturations determined from the condition $ds_i/dt = 0$ become negative. This means that even if equilibrium is reached at some moment in a cloud, a cloud would evaporate via the sublimation process, and the equilibrium would be violated, because equilibrium supersaturation cannot exist for a long time with such w and τ_{fc}. Thus, the states with positive equilibrium ice supersaturations can be reached under two conditions: a) $c_{i1} w \tau_{fc} < 1$ and b) $s_i < s_{i,cr}$.

The equilibrium deposition rate I_{dep} can be calculated by substituting the equilibrium ice supersaturation into (Eqn. 5.3.19) for I_{dep}. This is illustrated here for the case without high crystal concentrations and fast glaciation when $c_{1i} w \tau_{fc} << 1$. Then, the denominator in (Eqn. 5.3.49a) is unity and the expression for equilibrium supersaturation is simplified as

$$s_{i,ice}^{eq} \approx c_{1i} w \tau_{fc} = \frac{c_p(\gamma_a - \gamma_{is})\Gamma_2}{L_s} \frac{\rho_a}{\rho_{is}} w \tau_{fc}. \tag{5.3.49b}$$

Substituting this expression into (Eqn. 5.3.19) for I_{dep}, we obtain the equilibrium deposition rate

$$\begin{aligned} I_{dep}^{eq} &= \frac{\rho_{is}}{\Gamma_2 \tau_{fc}} s_{i,ice}^{eq} = \frac{\rho_{is}}{\Gamma_2 \tau_{fc}} \frac{c_p}{L_s} (\gamma_a - \gamma_{is}) \tau_{fc} \frac{\Gamma_2 \rho_a}{\rho_{is}} w \\ &= \frac{c_p}{L_s} (\gamma_a - \gamma_{is}) \rho_a w = \left(\frac{dq_i}{dz} \right)_{ad} \rho_a w, \end{aligned} \tag{5.3.49c}$$

where $(dq_i/dz)_{ad}$ is the adiabatic gradient of ice water content.

Table 5.3. *Approximate equilibrium ice supersaturations (%) in crystalline clouds with various crystal supersaturation relaxation times* τ_c *(s, min) and vertical velocities* w *(cm s^{-1}) for spherical ice crystals*

Supersaturation relaxation time τ_c (sec, min)	Vertical velocity w (cm s^{-1})						
	1	5	10	20	50	100	250
10 s	0.12E-02	0.60E-02	0.12E-01	0.024	0.060	0.12	0.30
60 s	0.72E-02	0.36E-01	0.72E-01	0.145	0.36	0.73	1.84
100 s	0.012	0.060	0.12	0.24	0.61	1.22	3.11
600 s	0.072	0.36	0.73	1.47	3.75	7.81	22.11
1000 s	0.12	0.607	1.22	2.47	6.42	13.73	43.22
30 min	0.22	1.098	2.22	4.54	12.19	27.76	118.9
60 min	0.44	2.22	4.54	9.517	27.76	76.84	−1259
120 min	0.88	4.54	9.517	21.04	76.84	663.6	−185.3
180 min	1.32	6.97	14.99	35.27	187.2	−429.4	−144.3

Mixed phase clouds. Equilibrium supersaturation in a mixed phase cloud can be obtained from (Eqn. 5.3.28a) with $ds_w/dt = 0$, which yields

$$s_{w,mix}^{eq} = \frac{c_{1w}w - (\Gamma_{12}/\Gamma_2)\tau_{fc}^{-1}(\rho_{ws} - \rho_{is})/\rho_{ws}}{\tau_{fd}^{-1} + (\Gamma_{12}/\Gamma_2)\tau_{fc}^{-1} - c_{1w}w}. \quad (5.3.50)$$

Since $c_{1w} \sim 10^{-5}$ cm^{-1} and $\tau_{fd}^{-1} \sim 0.1$–1 s^{-1} (Table 5.1), then for $w \le 10$ m s^{-1}, the term $c_{1w}w$ in the denominator can be usually neglected. A comparison of Tables 5.1 and 5.2 shows that τ_{fc} usually is much greater than τ_{fd} (except for the cases with very high crystal concentrations and rapid cloud glaciation), thus $\tau_{fd}^{-1} >> \tau_{fc}^{-1}$. The factor $\Gamma_{12}/\Gamma_2 \sim 1$, and the term $(\Gamma_{12}/\Gamma_2)\tau_{fc}^{-1}$ can be also neglected in the denominator. Then, the equilibrium supersaturation in a mixed cloud becomes

$$s_{w,mix}^{eq} = c_{1w}w\tau_{fd} - \frac{\tau_{fd}}{\tau_{fc}}\frac{\rho_{ws} - \rho_{is}}{\rho_{ws}}. \quad (5.3.51)$$

The first term coincides with the equilibrium supersaturation in a liquid cloud (Eqn. 5.3.48a), while the second term describes the decrease of equilibrium supersaturation due to the presence of crystals. This decrease is proportional to the difference $\Delta\rho_s = \rho_{ws} - \rho_{is}$ and to the ratio of the drop and crystal relaxation times. Since usually $\tau_{fd} << \tau_{fc}$, the equilibrium supersaturation in a mixed cloud can only be slightly lower than the supersaturation in a liquid cloud (Eqn. 5.3.48a). For example, if $\tau_{fd} \approx 2$ s ($N_d = 200$ cm^{-3}, $\bar{r}_d = 10\,\mu$m) and if $\tau_{fc} \approx 1$ h ($N_c = 10$ L^{-1}, $\bar{r}_c = 100\,\mu$m, Tables 5.1. and 5.2) and $\Delta\rho_s/\rho_{ws} \approx 0.1$ at $T = -10\,°$C, and $c_{1w} \sim 10^{-5}$ cm^{-1}, then according to (Eqn. 5.3.51), $s_{w,mix}^{eq} \approx 10^{-4} - 0.5\times10^{-4} = 0.5\times10^{-4} = 0.5\times10^{-2}$ %—i.e., the second negative term compensates for the first term with supersaturation generation by about one half, but a mixed cloud exists at positive water supersaturation. Note that some models calculated the supersaturation in mixed clouds as weighted by the water and ice masses. Equations (5.3.50) and (5.3.51) show that this approach may overestimate glaciation rates, and more accurate calculations should be based on the use of supersaturation relaxation times—i.e., drop and crystal concentrations and mean sizes.

The condition when supersaturation in a mixed cloud is positive follows from (Eqn. 5.3.51),

$$c_{1w}w - \tau_{fc}^{-1}(\rho_{ws} - \rho_{is})/\rho_{ws} \ge 0. \qquad s_{w,mix}^{eq} \ge 0, \quad (5.3.52a)$$

which gives two criteria for w or τ_{fc}:

$$w \ge (\Delta\rho_{is}/\rho_{ws})/(c_{1w}\tau_{fc}), \qquad \tau_{fc} \ge (\Delta\rho_s/\rho_{ws})/(c_{1w}w). \quad (5.3.52b)$$

For example, if $N_c \approx 70$ L^{-1} and $\bar{r}_c = 50\,\mu$m, then $\tau_{fc} = 1{,}000$ s (Table 5.2), and with $\Delta\rho_s/\rho_{ws} \approx 0.1$ at $T = -10\,°$C and $c_{1w} \sim 10^{-5}$ cm^{-1}, the first of these equations gives $w \ge 10$ cm s^{-1} as a condition for positive water supersaturation. At a smaller w with this microstructure, the supersaturation will be negative and gradual cloud glaciation will occur. At a smaller $N_c \sim 5$–10 L^{-1}, an estimate from (Eqns. 5.3.52a,b) shows that s_w can be positive even with a w of a few cm s^{-1}, and a cloud may exist in a mixed state without full glaciation for a long time, which may explain long lifetimes of polar clouds in the mixed phase. Such supercooled clouds and fogs in liquid state can be appropriate objects for cloud seeding with cooling agents, which cause cloud crystallization, and precipitation enhancement or dispersal of clouds and fogs.

5.3.7. Adiabatic Lapse Rates with Non zero Supersaturations

We derived in Chapter 3 the wet adiabatic gradients in liquid and ice clouds under the assumption of instantaneous condensation and zero supersaturations. A question arises if finite condensation or deposition rates and the nonzero supersaturations in clouds can change the values of wet adiabatic lapse rates. Consider again (Eqn. 3.11.22) for γ_w and rewrite it for a process with finite rate—i.e., with time dependence:

$$\gamma_w = \gamma_a + \frac{L_e}{c_p}\frac{dq_{ws}}{dz} = \gamma_a + \frac{L_e}{c_p}\frac{dq_{ws}}{dt}\frac{dt}{dz} \tag{5.3.53}$$

In general, the value of γ_w can be different from the wet adiabatic value. The situation is simplified for the case with equilibrium supersaturation. The term dq_{ws}/dt can be expressed via the condensation rate using the relations for adiabatic process and expression (Eqn. 5.3.48a) for I_{con}:

$$\frac{dq_{ws}}{dt} = -\frac{dq_l}{dt} \approx -\frac{I_{con}^{eq}}{\rho_a} = -\frac{c_p}{L_e}(\gamma_a - \gamma_w)w. \tag{5.3.54}$$

Substituting this equation and the relation $dz/dt = 1/w$ into (Eqn. 5.3.53), we obtain for the wet adiabat in equilibrium

$$\gamma_w^{eq} = \gamma_a + \frac{L_e}{c_p}\left[-\frac{c_p}{L_e}(\gamma_a - \gamma_w)w\right]\frac{1}{w} = \gamma_a - (\gamma_a - \gamma_w) = \gamma_w. \tag{5.3.55}$$

That is, for condensation in liquid clouds with finite condensation rates and equilibrium supersaturation, the wet adiabatic gradient is equal to its value in the wet adiabatic process with instantaneous condensation.

The wet adiabatic gradient in ice clouds with equilibrium ice supersaturation can be derived in a similar way. Consider again (Eqn. 3.11.35) for γ_{is} and rewrite it for a process with finite deposition rate—i.e., with time dependence:

$$\gamma_{is} = \gamma_a + \frac{L_s}{c_p}\frac{dq_{is}}{dz} = \gamma_a + \frac{L_s}{c_p}\frac{dq_{is}}{dt}\frac{dt}{dz}. \tag{5.3.56}$$

In general, the value of γ_{is} can be different from the wet adiabatic value in ice clouds but is simplified for the case with equilibrium ice supersaturation (Eqn. 5.3.49a) and deposition rate (Eqn. 5.3.49c) valid for the case without high crystal concentrations and rapid glaciation, when $c_{1i} w \tau_{fc} << 1$. The term dq_{is}/dt can be transformed for an adiabatic process as

$$\frac{dq_{is}}{dt} = -\frac{dq_i}{dt} \approx -\frac{I_{dep}^{eq}}{\rho_a} = -\frac{c_p}{L_e}(\gamma_a - \gamma_{is})w. \tag{5.3.57}$$

Substituting this expression and the relation $dz/dt = 1/w$ into (Eqn. 5.3.57), we obtain

$$\gamma_{is}^{eq} = \gamma_a + \frac{L_e}{c_p}\left[-\frac{c_p}{L_e}(\gamma_a - \gamma_{is})w\right]\frac{1}{w} = \gamma_a - (\gamma_a - \gamma_{is}) = \gamma_{is}. \tag{5.3.58}$$

That is, for deposition with a finite rate in ice clouds and equilibrium ice supersaturation, and without high crystal concentrations, the wet adiabatic temperature gradient is equal to its value in the wet adiabatic process with instantaneous deposition in ice clouds. The values of γ_{is}^{eq} with high crystal concentrations and negative ice supersaturations should be evaluated using time-dependent models.

These equalities, $\gamma_w^{eq} = \gamma_w$ and $\gamma_{is}^{eq} = \gamma_{is}$, mean that under the quasi-equilibrium conditions, the latent heats and ice nucleation rates can be calculated in models using the equilibrium ice supersaturations.

5.4. The Wegener–Bergeron–Findeisen Process and Cloud Crystallization

After the occurrence of sufficient crystal concentrations in a mixed phase cloud, the process of cloud crystallization (or glaciation) begins. During the first stage, a mixed cloud exists and the net vapor flux is directed from the drops to the crystals because the saturated vapor pressure over water is higher than over ice. This is the *Wegener–Bergeron–Findeisen* or *WBF process* (Wegener, 1911; Bergeron, 1935; Findeisen, 1938), when the drops evaporate and crystals grow at their expense. If the crystal concentration is sufficiently high and the time for this process is sufficiently long, the second stage occurs after the drops have been completely evaporated, and then a pure crystalline cloud exists and supersaturation relaxation continues. Detailed consideration of the evolution of a mixed phase cloud parcel with a WBF process followed by a crystalline cloud are given in Chapters 8, 9, and 14. Here, a simplified description is provided following Shifrin and Perelman (1960), with modifications that help illustrate the physics of the glaciation process.

First stage. Suppose for simplicity that an adiabatic cloud parcel contains initially a monodisperse ensemble of drops with radius $r_d(0) = r_{d0}$ and concentration N_d, which exist at a temperature T, and small water supersaturation s_w. Then, a number of crystals with concentration N_c and negligibly small radius $r_{c0} << r_{d0}$ are nucleated at the time $t = 0$ and N_c is subsequently constant with time, and so the processes of coagulation and accretion are insignificant and can be neglected. Such a situation may occur after natural ice nucleation as described in Chapters 8 and 9, or after cloud seeding with artificial crystallizing agents. The crystallization (WBF) process can be characterized by the specific time that can be called "*crystallization time*" τ_{cr}. All drops will evaporate by the time τ_{cr}, and the parcel will contain only crystals with the same initial concentration and the radius r_{cf1}. We assume that the vertical velocities or other sources of supersaturation generation are absent, so that during drops evaporation until $t = \tau_{cr}$, the water supersaturation is small, $|s_w| << 1$, and can be neglected, and vapor density is equal to the vapor saturated over water, $\rho_v = \rho_{ws}$. It is also assumed that the temperature change is small during evaporation, $\rho_{ws} = \text{const}$, and the crystals have a spherical shape.

The time τ_{cr} and the radius r_{cf1} can be evaluated using the water balance equation and the equations for growth rates of drops and crystals from Sections 5.1 and 5.2. At the beginning of the first stage, the cloud parcel contains liquid drops with the liquid water content $q_{L1}(0) = (4\pi/3)\rho_w N_d r_{d0}^3$ and vapor saturated over water with the density ρ_{ws}. At the end of the first stage, when all the drops have been evaporated, $q_{L1}(\tau_{cr}) = 0$, the parcel contains ice crystals with the ice water content $q_{L2}(\tau_{cr}) = (4\pi/3)\rho_i N_i r_{cf1}^3$ and vapor saturated over water with the same density ρ_{ws}. The water balance equation for the first stage can be written as:

$$\rho_{ws} + \frac{4\pi}{3}\rho_w N_d r_{d0}^3 = \rho_{ws} + \frac{4\pi}{3}\rho_i N_i r_{cf1}^3. \tag{5.4.1}$$

The equations for the drops evaporation and crystal growth rates can be written using equations from Sections 5.1 and 5.2, neglecting kinetic and psychrometric corrections:

$$\rho_w r_d \frac{dr_d}{dt} = D_v(\rho_v(t) - \rho_{ws}), \tag{5.4.2}$$

$$\rho_i r_c \frac{dr_c}{dt} = D_v(\rho_v(t) - \rho_{is}). \tag{5.4.3}$$

The initial and final conditions for this system of equations are:

$$\begin{aligned} &r_d|_{t=0} = r_{d0}, \quad && r_c|_{t=0} = 0, \quad && \rho_v|_{t=0} = \rho_{ws}, \\ &r_d|_{t=\tau_{cr}} = 0, \quad && r_c|_{t=\tau_{cr}} = r_{cf1}, \quad && \rho_v|_{t=\tau_{cr}} = \rho_{ws}, \end{aligned} \tag{5.4.4}$$

The system (Eqns. 5.4.1–5.4.4) allows determination of two unknown quantities: r_{cf1} and τ_{cr}. The final crystal radius after the first stage can be evaluated from (Eqn. 5.4.1):

$$r_{cf1} = r_{d0}\left(\frac{\rho_w N_d}{\rho_i N_c}\right)^{1/3}. \tag{5.4.5}$$

Subtracting (Eqn. 5.4.3) from (Eqn. 5.4.2) and integration by time yields

$$\rho_w[r_d^2(\tau_{cr}) - r_d^2(0)] - \rho_i[r_c^2(\tau_{cr}) - r_c^2(0)] = -2D_v(\rho_{ws} - \rho_{is})\tau_{cr}. \tag{5.4.6}$$

Using the conditions (Eqn. 5.4.4), (Eqn. 5.4.5) for r_{cf1} and solving for τ_{cr}, we obtain the time of cloud glaciation:

$$\tau_{cr} = \tau_{cr,0}\left[1 + \left(\frac{\rho_i}{\rho_w}\right)^{1/3}\left(\frac{N_d}{N_c}\right)^{2/3}\right] \approx \tau_{cr,0}\left(\frac{\rho_i}{\rho_w}\right)^{1/3}\left(\frac{N_d}{N_c}\right)^{2/3}, \tag{5.4.7a}$$

where

$$\tau_{cr,0} = \frac{\rho_w r_{d0}^2}{2D_v(\rho_{ws} - \rho_{is})}. \tag{5.4.7b}$$

The second approximate equality in (Eqn. 5.4.7a) makes use of the condition $(N_d/N_c)^{2/3} >> 1$, equivalent to the condition $r_{cf1}/r_{d0} >> 1$, which is usually valid for natural clouds and even in cloud-seeding experiments. The physical meaning of $\tau_{cr,0}$ is the time during which a drop with initial radius r_{d0} would evaporate if the humidity is equal to the saturation over ice. Equations (5.4.5) and (5.4.7a,b) allow approximate estimates of the glaciation time and the crystal size after glaciation.

The second stage begins at the end of the first stage. Thus, the total water mass content at the beginning is the sum of the ice water content $q_{L2}(\tau_{cr})$ and vapor saturated over water with ρ_{ws}, so that the ice supersaturation $s_i = (\rho_{ws} - \rho_{is})/\rho_{is}$. It is assumed that s_i relaxes to zero by the end of the second stage due to vapor absorption by the crystals and $\rho_v = \rho_{is}$, the crystal concentration remains N_c, and their radius r_{cf2} after the second stage should be determined. The water balance for the second stage is

$$\rho_{ws} + \frac{4\pi}{3}\rho_i N_c r_{cf1}^3 = \rho_{is} + \frac{4\pi}{3}\rho_i N_i r_{cf2}^3. \tag{5.4.8}$$

Solving for r_{cf2} and using (Eqn. 5.4.5) for r_{cf1}, we obtain the crystal radius at the end of the second stage:

$$r_{cf2} = \left[r_{cf1}^3 + \frac{(\rho_{ws} - \rho_{is})}{(4\pi/3)\rho_i N_c} \right]^{1/3} = r_{d0} \left(\frac{\rho_w N_d}{\rho_i N_c} \right)^{1/3} \left(1 + \frac{\rho_{ws} - \rho_{is}}{q_{L1}} \right)^{1/3}. \quad (5.4.9)$$

Theoretically, the duration of the second stage is infinite since ρ_v exponentially approaches ρ_{is} as described in Section 5.3, but the initial ice supersaturation ($\rho_{ws} - \rho_{is}$) decreases e times during the crystal phase relaxation time $\tau_{f,cr} = (4\pi D_v N_c r_{cf1})^{-1}$ and ≈ 20 times during $3\tau_{f,cr}$. Thus, $\tau_{f,cr}$ can serve as a characteristic time of the second stage.

Example calculations are given in Table 5.4 for two droplet concentrations and two temperatures. With $r_{d0} = 5$ μm, the ratio r_{cf1}/r_{d0} varies from 35.5 to 7.7 at $N_d = 500$ cm^{-3} and from 20.8 to 4.5 at $N_d = 100$ cm^{-3} when N_c increases from 10 L^{-1} to 1,000 L^{-1}. The crystallization time τ_{cr} varies from 30–60 min to more than 7 hours at small crystal concentrations of 10–20 L^{-1} and decreases to 100–150 s at very high $N_c \sim 1{,}000$ L^{-1}. Equations (5.4.7a and b) show that $\tau_{cr} \sim r_{d0}^2$; therefore, the glaciation time rapidly increases for larger drops. Since $\tau_{cr} \sim [D_v(\rho_{ws} - \rho_{is})]^{-1}$, the glaciation time is minimum at the temperature with a maximum difference of vapor densities, $T \approx -12$ °C, and decreases at lower temperatures. Although the ice nucleation rates increase at low temperatures (see Chapters 8, 9, and 10), the crystallization time τ_{cr} with the same crystal concentrations at −30 °C is a factor of two larger than at −10 °C, and glaciation proceeds slower. This difference can explain the relative stability and long persistence of the mixed phase arctic clouds at low temperatures and not very high crystal concentrations.

Table 5.4. *The ratio* r_{cf1}/r_{d0} *and the crystallization time* τ_{cr} *(seconds) for various* N_c *and liquid cloud properties:* $N_d = 500$ *cm*$^{-3}$*, two initial drop radii,* $r_{d0} = 5$ *μm, (*$q_{L1} = 0.26$ *g m*$^{-3}$*), and* $r_{d0} = 10$ *μm (*$q_{L1} = 2.09$ *g m*$^{-3}$*);* $N_d = 100$ *cm*$^{-3}$*, the drop radius* $r_{d0} = 5$ *μm, (*$q_{L1} = 0.052$ *g m*$^{-3}$*), at two temperatures, −10 °C and −30 °C, and pressure* p = *800 mb.*

	N_c, L^{-1}					
	10	20	50	100	500	1000
Quantity	$r_{d0} = 5$ μm, $N_d = 500$ cm^{-3}, $T = -10$ °C					
r_{cf1}/r_{d0}	35.5	28.2	20.8	16.5	9.6	7.7
τ_{cr}, sec	3120	1970	1070	670	230	150
	$r_{d0} = 10$ μm, $N_d = 500$ cm^{-3}, $T = -10$ °C					
τ_{cr}, sec	12480	7880	4280	2680	920	600
	$r_{d0} = 5$ μm, $N_d = 500$ cm^{-3}, $T = -30$ °C					
τ_{cr}, sec	6614	4176	2268	1420	487	318
	$r_{d0} = 10$ μm, $N_d = 500$ cm^{-3}, $T = -30$ °C					
τ_{cr}, sec	26457	16705	9073	5681	1950	1272
	$r_{d0} = 5$ μm, $N_d = 100$ cm^{-3}, $T = -30$ °C					
r_{cf1}/r_{d0}	20.8	16.5	12.1	9.6	5.6	4.5
τ_{cr}, sec	2260	1430	775	485	166	110

5.5. Kinetic Equations of Condensation and Deposition in the Adiabatic Process

5.5.1. Derivation of the Kinetic Equations

Temporal evolution of the droplet and crystal size spectra in clouds can be described with the dynamic equations termed the *kinetic equations for the size distributions functions*. Several processes govern transformation of the cloud particle size spectra: nucleation of water and ice, diffusion growth or evaporation, collisions and coalescence, breakup. The general kinetic equation includes the terms that describe all these processes, but each of these processes can be considered in isolation—that is, using a specific kinetic equation. In this section, we introduce the kinetic equation that describes the diffusional growth or evaporation of the drops and crystals due to their interaction with the vapor field, which is termed the *kinetic equation of condensation or deposition*. We consider in this chapter the kinetic equations in a laminar flow in an adiabatic process without turbulence or mixing, which are also termed the kinetic equations of regular condensation.

In general, the distribution function of an ensemble of the cloud particles depends on three spatial coordinates, with a fourth coordinate being the particle radius r (or mass m). The distribution function may include several different parameters, such as soluble fraction, electrical charge, shape, and velocity. We assume that there are n additional variables and denote them as ξ_i. These $3 + 1 + n$ variables form a $(n + 4)$-dimensional phase space. (If distributions are over only three coordinates and the radius or mass are considered, we deal with the 4-dimensional space.) An ensemble of cloud particles represents a thermodynamic system that can be characterized by a distribution function of all these variables $f(\vec{x},r,\xi_i)$ in this phase space, where $\vec{x}$ denotes the vector of the three spatial coordinates. Each possible state of the system corresponds to a uniquely determined point in this $(n + 4)$-dimensional space, and the temporal evolution of $f(\vec{x},r,\xi_i)$ represents a trajectory in this phase space. If this system can be considered closed, then the conservation law for the function $f(\vec{x},r,\xi_i)$ along the trajectory in this phase space is described by the *Liouville theorem*, which is equivalent to the continuity equation in the phase space

$$\frac{df(\vec{x},r,\xi_i)}{dt} = 0. \qquad (5.5.1)$$

Expanding the full derivative, we obtain

$$\frac{\partial f(\vec{x},r,\xi_i,t)}{\partial t} + \frac{\partial}{\partial \vec{x}}\left[\dot{x} f(\vec{x},r,\xi_i)\right] + \frac{\partial}{\partial r}\left[\dot{r} f(\vec{x},r,\xi_i)\right] + \sum_i \frac{\partial}{\partial \xi_i}\left[\dot{\xi} f(\vec{x},r,\xi_i)\right] = 0, \qquad (5.5.2)$$

where $\dot{x}$ denote the particles' velocities, and $\dot{r} = dr/dt$ and $\dot{\xi}_i = d\xi_i/dt$ are the corresponding "velocities" in the space of radii and the variable ξ_i. The term $\partial(\dot{r}f)/\partial r$ represents the divergence of $f(r)$ due to condensation or deposition growth. If there are sources and sinks in the system, then an additional term J_s should be added to the right-hand side that describes the source or sink of the particles.

$$\frac{\partial f(\vec{x},r,\xi_i,t)}{\partial t} + \frac{\partial}{\partial \vec{x}}\left[\dot{x} f(\vec{x},r,\xi_i)\right] + \frac{\partial}{\partial r}\left[\dot{r} f(\vec{x},r,\xi_i)\right] + \sum_i \frac{\partial}{\partial \xi_i}\left[\dot{\xi} f(\vec{x},r,\xi_i)\right] = J_s. \qquad (5.5.3)$$

If we consider only the size (or mass) distribution function and neglect the other variables ξ_i, then the phase space is 4-dimensional and this equation is simplified as

$$\frac{\partial f(\vec{x},r,t)}{\partial t} + \frac{\partial}{\partial \vec{x}}\left[\dot{x} f(\vec{x},r,t)\right] + \frac{\partial}{\partial r}\left[\dot{r} f(\vec{x},r,t)\right] = J_s. \qquad (5.5.4)$$

If a particle falls at a fall speed $V_t(r)$ along the 3rd (z) coordinate, the i-th component of its full velocity is $\dot{x} = u_i - V_t(r)\delta_{i3}$, and the kinetic equation can be written in the form

$$\frac{\partial f}{\partial t} + \frac{\partial}{\partial x_i}[(u_i - V_t(r)\delta_{i3})f] + \frac{\partial f}{\partial r}(\dot{r}f) = J_s, \tag{5.5.4a}$$

where the usual convention of summation over double subscript $i = 1, 2, 3$ is assumed. If we consider a spatially homogeneous system without dependence on the space variables, $\partial/\partial x_i = 0$, then the phase space contains only one variable, r, and this equation is further simplified as

$$\frac{\partial f(r,t)}{\partial t} + \frac{\partial}{\partial r}\left[\dot{r}f(r,t)\right] = J_s. \tag{5.5.5}$$

If we consider the size distribution by masses rather than by radii, then we should replace r with m, $\dot{r}$ with $\dot{m} = dm/dt$, and then the term $\partial[\dot{r}f(r)]/\partial r$ becomes $\partial[\dot{m}f_m(m)]/\partial m$ and (Eqn. 5.5.5) becomes

$$\frac{\partial f_m(m)}{\partial t} + \frac{\partial}{\partial m}\left[\dot{m}f_m(m)\right] = J_s. \tag{5.5.6}$$

Equations (5.5.2)–(5.5.6) represent various forms of the *kinetic equation of regular condensation or deposition* in a laminar flow. The functions introduced earlier are related as

$$f(\vec{x},r) = \int f(\vec{x},r,\xi_i)d\xi_1 d\xi_2 ... d\xi_n, \quad f(r) = \int f(\vec{x},r)dV,$$
$$f_m(m) = \int f(\vec{x},m)dV, \qquad f_m(m)dm = f(r)dr, \tag{5.5.7}$$

where $dV = dx_1dx_2dx_3$ is the differential of the volume. The corresponding growth rates, described in the preceding Sections 5.1 and 5.2, are proportional to the supersaturation. Thus, the system of equations for calculations of the growth or evaporation by the condensation and deposition of an ensemble of the cloud particles includes kinetic equations, the supersaturation equation, and the equation of the growth of an individual particle. This system of equations will be used in the subsequent chapters for studies of the drop and crystal nucleation. The term *regular condensation* here means that these equations are written for the case of a flow without turbulence.

The kinetic equations of regular condensation (Eqns. 5.5.4–5.5.6) were introduced in cloud physics at the beginning of 1960s by Buikov (1961, 1963, 1966a,b) by analogy with the kinetic equation formulated by Lifshitz and Slezov (1958, 1961) for the description of efflorescence in the polydisperse ensemble of the salt crystals in supersaturated solutions (see Chapter 11). The corresponding kinetic equations for the turbulent atmosphere are termed the *kinetic equations of stochastic condensation* and will be considered in Chapters 13 and 14.

The method of kinetic equations is a powerful tool for calculation of the condensation growth of polydisperse ensembles of the drops and crystals and in simulations of cloud evolution (e.g., Buikov and Pirnach, 1973, 1975; Buikov and Khvorostyanov, 1976, 1977; Tzivion et al., 1987; Feingold et al., 1994; Fridlind et al., 2004; Jensen et al., 1994; Khvorostyanov et al., 2001, 2003, 2006; Xue et al., 2010; Curry and Khvorostyanov, 2012). Both numerical and analytical solutions of the kinetic equations can be used for these calculations. The numerical solutions to the kinetic equations serve as the basis for the spectral bin calculations of the evolution of the particles' size spectra. The method of moments widely used in cloud modeling (Chapter 14) can be rigorously derived from the kinetic equations, multiplying the kinetic equation by various powers of radius or mass, integrating over the size spectrum and splitting the higher moments with various approximations. The kinetic equation itself is equivalent to the method with an infinite number of moments. Further, the kinetic equations can be solved analytically to provide insights into the condensation process.

5.5.2. Some Properties of Regular Condensation

Two important features of regular condensation are 1) conservation of the particle concentration in the absence of sources and sinks, and 2) narrowing of the size spectra with time (Howell, 1949; Mordy, 1959; Neiburger and Chien, 1960). These features can be illustrated with the kinetic (Eqn. 5.5.5).

Integrating (Eqn. 5.5.5) by radius without the source term, $J_s = 0$, and using the condition of normalization of $f(r)$ to the particle concentration N, we obtain

$$\frac{\partial}{\partial t}\int_0^\infty f(r)dr = \frac{\partial N}{\partial t} = -\left[\dot{r}f(r)\right]_0^\infty = \dot{r}(0)f(0) - \dot{r}(\infty)f(\infty). \tag{5.5.8}$$

As seen from (Eqn. 5.1.46), the growth rate $\dot{r}$ is finite at $r = 0$ and $r = \infty$, and $f(r)$ is zero at zero and infinite radii. The right-hand side of (Eqn. 5.5.8) is zero and this yields $\partial N/\partial t = 0$. That is, the kinetic equation of condensation without sources describes the temporal evolution of the size spectrum with conservation of the particle concentration.

The second feature of the spectra, narrowing with time, is easily seen. In the diffusion regime of the particles' growth, when $r >> \xi_{con}$, which is valid in most situations, dr/dt is inversely proportional to the particle radius (see Eqn. 5.1.46), the larger particles grow more slowly than the smaller particles, and thus the size spectrum should become narrower with time. The quantitative evaluation of the spectral narrowing has usually been done using numerical techniques (e.g., Howell, 1949; Mordy, 1959; Neiburger and Chien, 1960).

The kinetic Equation (5.5.5) allows a simple analytical interpretation of the spectral narrowing. Consider the spectral dispersion σ_r^2

$$\sigma_r^2 = \frac{1}{N}\int_0^\infty (r-\bar{r})^2 f(r)dr. \tag{5.5.9}$$

Differentiating both sides by time yields

$$\frac{\partial \sigma_r^2}{\partial t} = I_1 + I_2, \tag{5.5.10a}$$

$$I_1 = \frac{1}{N}\int_0^\infty (r-\bar{r})^2 \frac{\partial f(r)}{\partial t} dr, \qquad I_2 = \frac{1}{N}\int_0^\infty f(r)\frac{\partial}{\partial t}(r-\bar{r})^2\, dr. \tag{5.5.10b}$$

It will be shown in the following text that the major contribution here comes from I_1, while I_2 is much smaller or vanishes.

Substituting $\partial f/\partial t$ from the kinetic (Eqn. 5.5.5) into I_1, integrating by parts and using the condition $\dot{r}(r)f(r)|_0^\infty = 0$ yields

$$I_1 = -\frac{1}{N}\int_0^\infty (r-\bar{r})^2 \frac{\partial}{\partial r}[\dot{r}f(r)]dr = \frac{2}{N}\int_0^\infty \dot{r}f(r)(r-\bar{r})dr. \tag{5.5.11}$$

We use hereafter (Eqn. 5.1.46), assuming the diffusion growth regime, and so (Eqn. 5.1.46) can be written as

$$\frac{dr_d}{dt} = \frac{b_d}{r_d}, \qquad b_d = \frac{D_v \rho_{ws} s_w}{\rho_w \Gamma_1}, \qquad \xi_{con} \approx 0. \tag{5.5.11a}$$

Using this growth rate in the simplified form, we obtain

$$I_1 = \frac{2b_d}{N} \int_0^\infty f(r)\left(1 - \frac{\bar{r}}{r}\right) dr = \frac{2b_d}{N}\left(N - \bar{r}M^{(-1)}\right). \tag{5.5.12}$$

We accounted for the fact that the integral of the first term in the integrand gives N, and $M^{(-1)}$ is the minus first moment of the spectrum. Further quantification requires some choice of the spectrum. We choose the gamma distribution $f(r) = cr^p \exp(-\beta r)$; the moment $M^{(-1)}$ for this spectrum was calculated in Section 2.4 as

$$M^{(-1)} = \int_0^\infty \frac{f(r)}{r} dr = N\frac{\beta}{p} = \frac{N}{r_m}, \tag{5.5.13}$$

where $r_m = p/\beta$ is the modal radius (see Section 2.4). Substituting this into (Eqn. 5.5.12) and using an expression for the mean radius $\bar{r} = (p+1)/\beta$, we obtain

$$I_1 = 2b_d\left(1 - \frac{\bar{r}}{r_m}\right) = 2b_d\left(1 - \frac{p+1}{p}\right) = -\frac{2b_d}{p}. \tag{5.5.14}$$

Thus, I_1 is negative, and, as seen from (Eqn. 5.5.10a), I_1 causes narrowing of the spectrum.

For calculation of I_2, we first need to know $\partial\bar{r}/\partial t$. Differentiating by time $\bar{r}(t)$ defined as the first moment of the spectrum $f(r, t)$, and using (Eqn. 5.5.5) with $J_s = 0$, integration by parts yields

$$\begin{aligned}\frac{\partial \bar{r}(t)}{\partial t} &= \frac{1}{N}\int_0^\infty r\frac{\partial f(r,t)}{\partial t} dr = -\frac{1}{N}\int_0^\infty r\frac{\partial(\dot{r}f)}{\partial r} dr \\ &= -\dot{r}f\,|_0^\infty + \frac{1}{N}\int_0^\infty \dot{r}f\,dr = \frac{b_d}{N}M^{(-1)} = \frac{b_d}{r_m},\end{aligned} \tag{5.5.15}$$

where we again use the condition $\dot{r}(r)f(r)|_0^\infty = 0$ and (Eqn. 5.5.13) for $M^{(-1)}$. This result shows that the growth rate of the mean radius of the gamma spectrum at regular condensation is equal to that in a monodisperse ensemble with the modal radius r_m. Now we can evaluate I_2, differentiating the integrand by time and using (Eqn. 5.5.15):

$$\begin{aligned}I_2 &= \frac{2}{N}\int_0^\infty f(r)\left(-r\frac{\partial\bar{r}}{\partial t} + \bar{r}\frac{\partial\bar{r}}{\partial t}\right) dr \\ &= \frac{2}{N}\left(-M^{(1)}\frac{b_d}{r_m} + \bar{r}\frac{b_d}{r_m}\right) = -\frac{2b_d\bar{r}}{r_m} + \frac{2b_d\bar{r}}{r_m} = 0.\end{aligned} \tag{5.5.16}$$

where we use the expression from Section 2.4 for the first moment, $M^{(1)} = N\bar{r}$. Thus, the term I_2 vanishes and we have

$$\frac{\partial\sigma_r^2}{\partial t} = I_1 = -\frac{2b_d}{p}. \tag{5.5.17}$$

So, using the kinetic equation, we have demonstrated that the dispersion is decreasing and the size spectrum is narrowing with time during regular condensation. This expression can be further clarified. As was shown in Section 2.4, the spectral dispersion of the gamma distribution can be expressed as $\sigma_r^2 = \bar{r}^2/(p+1)$. We can express p as $p = \bar{r}^2/\sigma_r^2 - 1$. If the spectrum is not very wide, then $\bar{r}^2/\sigma_r^2 >> 1$, and approximately $p \approx \bar{r}^2/\sigma_r^2$. Substitution of this expression into (Eqn. 5.5.17) yields an equation for σ_r^2,

$$\frac{\partial \sigma_r^2}{\partial t} \approx -\frac{2b_d}{\bar{r}^2}\sigma_r^2 = -\frac{\sigma_r^2}{\tau_\sigma}. \tag{5.5.18a}$$

Its solution is

$$\sigma_r^2(t) \approx \sigma_r^2(t_0)\exp(-(t-t_0)/\tau_\sigma), \tag{5.5.18b}$$

where we introduced the dispersion relaxation or narrowing time τ_σ

$$\tau_\sigma = \frac{\bar{r}^2}{2b_d} = \frac{\bar{r}^2}{(2D_v\rho_{ws}/\rho_w\Gamma_1)s_w}. \tag{5.5.19}$$

Here we used (Eqn. 5.5.11a) for b_d and neglected the kinetic correction ξ_{con}. Equations (5.5.18b) and (5.5.19) show that the dispersion σ_r^2 of the initial gamma distribution decreases with time exponentially, which determines the rate of the spectral narrowing.

Thus, using the kinetic equation of regular condensation, we determined that particle growth is accompanied by the narrowing of the size spectrum. If the spectrum can be approximated by a gamma distribution, the characteristic narrowing time τ_σ of the dispersion is $\tau_\sigma \sim \bar{r}^2/s_w$ —i.e., the time is proportional to the square of the mean radius and inversely proportional to the supersaturation s_w. Therefore, the rate of spectral narrowing increases for smaller mean radii and for faster growth rates with larger supersaturations.

The theoretical prediction of spectral narrowing in regular condensation in an isolated cloud parcel predicted by the calculations of Howell (1949), Mordy (1959), and others are in conflict with measurements in clouds. Numerous experiments *in situ* have shown that the size spectra are sufficiently broad at all stages of cloud developments. Therefore, it became clear that regular condensation is an idealized model of condensation that may occur in an isolated cloud parcel or in cloud chambers. Various hypotheses have been suggested to explain the broad size spectra in clouds. These hypotheses are discussed in Chapters 13 and 14.

5.5.3. Analytical Solution of the Kinetic Equations of Regular Condensation

The analytical solution of (Eqn. 5.5.5) for a homogeneous cloud without the sources, $J_s = 0$, can be obtained using the general method of characteristics for the differential equations, $dt/1 = df/\dot{f} = dr/\dot{r}$ (Buikov, 1961, 1963, 1966a,b; Sedunov, 1967, 1974; Kovetz and Olund, 1969). A simpler solution is based on the differential conservation law, which expresses the conservation of the particles $dN(r, t) = dN(r_0, t_0)$, or

$$f(r,t)dr = f_0(r_0,t_0)dr_0. \tag{5.5.20}$$

Here $r_0 = r(t_0)$, $f_0(r_0, t_0)$ are the radius and size spectrum at the initial time t_0, and $r(t)$, $f(r, t)$ are the radius and size spectrum at a subsequent time t. Integrating (Eqn. 5.1.46) for the radius growth rate by time from t_0 to t, we obtain

$$(r_0 + \xi_{con})^2 = (r + \xi_{con})^2 - Q_c(t), \qquad Q_c(t) = 2c_{3w}y_w(t), \tag{5.5.21}$$

where $Q_c(t)$ describes an increase in the squared radius, c_{3w} and ξ_{con} were defined in Section 5.1, and $y_w(t)$ is the integral water supersaturation

$$c_{3w} = \frac{D_v \rho_{ws}}{\rho_w \Gamma_1}, \quad \xi_{con} = \frac{4D_v}{\alpha_c V_v}, \qquad y_w(t) = \int_{t_0}^{t} s_w(t')\,dt'. \tag{5.5.22}$$

Equation (5.5.21) expresses r_0 as a function of r and we can calculate $f(r, t)$ using (Eqn. 5.5.20):

$$f(r,t) = f_0(r_0, t_0)\frac{dr_0}{dr}. \tag{5.5.23}$$

Finally, expressing $r(t_0)$ via $r(t)$ from (Eqn. 5.5.21), we obtain

$$f(r,t) = \frac{r + \xi_{con}}{[(r + \xi_{con})^2 - Q_c(t)]^{1/2}} f_0([(r + \xi_{con})^2 - Q_c(t)]^{1/2} - \xi_{con}). \tag{5.5.24}$$

This is the analytical solution of the kinetic equation of the regular condensation for a spatially homogeneous case. It expresses the size spectrum $f(r, t)$ at any time t via the known function $f_0(r_0, t_0)$ at the time t_0. Note that the solution $f(r, t)$ exists only when the denominator and argument of f_0 are positive—that is, at $r > r_{min} = Q_c^{1/2}(t) - \xi_{con}$. This means that the smaller droplets have grown above this lower limit r_{min} by time t, and there are no small drops in the absence of new drop nucleation. Thus, the solution (Eqn. 5.5.24) also predicts narrowing of the spectrum.

This solution (Eqn. 5.5.24) can be conveniently used both for analytical studies and in numerical simulations of evolution of the size spectrum at the condensation stage. If $t = t_0 + \Delta t$, where Δt is the time step in a numerical model, (Eqn. 5.5.24) allows us to calculate $f(r, t + \Delta t)$ using the function $f_0(r_0, t_0)$ from the previous time step. This requires a choice of the finite-difference grid by radius r_0, a finite-difference representation of $f_0(r_0, t_0)$ on this grid, and a recalculation from every grid point by r_0 to the grid points by the argument $[(r + \xi_{con})^2 - Q_c(t)]^{1/2}$. This can be easily done using numerical recalculation between the two grids (e.g., linear, quadratic, spline interpolation, or any other that ensures sufficient accuracy). This procedure can be repeated iteratively for every time step in the model, and allows simulation for a long time. Examples of such numerical simulation will be given in the subsequent chapters.

As discussed in Section 5.1 for pure drops in the absence of the surface-active substances, the condensation coefficient $\alpha_c \sim 1$, the kinetic correction $\xi_{con} \sim 0.2$ μm, and hence these terms can be neglected for the drops with $r_d > 1$–2 μm that grow in the diffusion regime. Thus, (Eqn. 5.5.24) is simplified as

$$f(r,t) = \frac{r}{\sqrt{[r^2 - Q_c(t)]^{1/2}}} f_0(\sqrt{[r^2 - Q_c(t)]^{1/2}}). \tag{5.5.25}$$

As discussed in Sections 5.1 and 5.2, when the drops or crystal are covered with a surface-active substance or film, the condensation and deposition coefficients can be as low as $\alpha_c \sim 10^{-3}$ to 10^{-5}, and so $\xi_{con} \sim 1/\alpha_c$ can be large. Then, the particles grow in the kinetic regime, and (Eqn. 5.5.24) becomes in this limit

$$f(r,t) = \frac{\xi_{con}}{\sqrt{\xi_{con}^2 - Q_c(t)}} f_0(\sqrt{\xi_{con}^2 - Q_c(t)} - \xi_{con}). \tag{5.5.26}$$

Equations (5.5.25) and (5.5.26) show that the solutions for $f(r, t)$ exist only at $r > r_{min} = Q_c^{1/2}(t)$ in the diffusion limit and at $Q_c^{1/2}(t) < \xi_{con}$ in the kinetic limit. It is not a limitation of the solution itself, but it does impose a limit for the integration time steps.

5.5.4. Equation for the Integral Supersaturation

The solution (Eqn. 5.5.24) to the kinetic (Eqn. 5.5.5) or its particular cases (Eqns. 5.5.25) and (5.5.26) require the integral supersaturation $y_w(t)$. This equation can be obtained from (Eqn. 5.3.28a) for the differential supersaturation s_w in the liquid cloud (without the crystal phase, $\tau_{fc} = \infty$) using the analytical solution of the kinetic equation. Replacing in (Eqn. 5.3.28a) $s_w = y_w'(t)$, and using (Eqn. 5.3.20a) for the droplet supersaturation relaxation time τ_f (we omit hereafter the subscript d for brevity), we obtain

$$\frac{dy_w'(t)}{dt} = [1 + y_w'(t)]c_{1w}w - y_w'(t)4\pi D_v \int_0^\infty \frac{r^2}{r + \xi_{con}} f(r,t)\,dr. \tag{5.5.27}$$

This equation expresses $y_w'(t)$ via $f(r, t)$ at the same time t; this $f(r, t)$ enters on the right-hand side in the integral I_{con} defined in (Eqn. 5.3.17a). In numerical solution, it is convenient to express $f(r, t)$ in I_{con} via $f(r, t_0)$ at the previous time step t_0. Using the relation

$$\frac{r^2}{r + \xi_{con}} = \frac{\{[(r_0 + \xi_{con})^2 - Q_c]^{1/2} - \xi_{con}\}^2}{[(r_0 + \xi_{con})^2 - Q_c]^{1/2}}, \tag{5.5.28}$$

which follows from (Eqn. 5.5.21), and using the relation $f(r)dr = f_0(r_0)dr_0$ and substituting the solution (Eqn. 5.5.24) into (Eqn. 5.5.27) yields the integral I_{con}:

$$I_{con} = \int_0^\infty \frac{\{[(r_0 + \xi_{con})^2 - Q_c]^{1/2} - \xi_{con}\}^2}{[(r_0 + \xi_{con})^2 - Q_c]^{1/2}} f_0(r_0)\,dr_0. \tag{5.5.29}$$

Substituting I_{con} into (Eqn. 5.5.27), we obtain

$$\frac{dy'_w(t)}{dt} = [1 + y_w'(t)]c_{1w}w - y_w'(t)4\pi D_v \int_0^\infty \frac{\{[(r_0 + \xi_{con})^2 + Q_c(t)]^{1/2} - \xi_{con}\}^2}{[(r_0 + \xi_{con})^2 + Q_c(t)]^{1/2}} f_0(r_0)\,dr_0. \tag{5.5.30}$$

This is the integro-differential equation for the integral supersaturation $y_w(t)$, which includes the size spectrum $f_0(r_0)$ from the previous time step. For the drop growth in the diffusion regime, $\xi_{con} << r$, the equation is simplified as

$$\frac{dy_w'(t)}{dt} = [1 + y_w'(t)]c_{1w}w - y_w'(t)4\pi D_v \int_0^\infty \sqrt{r_0^2 + Q_c(t)}\, f_0(r_0)\,dr_0. \tag{5.5.31}$$

These two equations, the kinetic (Eqn. 5.5.24) and the integral supersaturation (Eqn. 5.5.30), form a closed system of equations where the droplet growth rate equation has been incorporated. This system of equations allows simulation of the cloud droplet size spectra in a Lagrangian parcel model in a spatially homogeneous cloud during characteristic times of condensation before the particles' collisions and coagulation become important. The kinetic equation combined with the supersaturation equation constitute a platform for cloud numerical modeling with detailed accounting for the microphysical processes. This method with direct calculation of the size spectra and their evolution is termed the *spectral microphysics* method in order to distinguish it from *the parameterized or bulk microphysics* that is based on calculation of the bulk cloud properties, usually of the liquid or ice water content and particle concentration, and using some a priori prescribed size spectra and their parameterizations considered in Sections 2.3 and 2.4.

Further generalizations of this basic system of equations are straightforward. The coagulation processes are considered in Section 5.6 of this chapter. The source terms J_s in Equations (5.5.3)–(5.5.6) describes activation of the new droplets considered in Chapters 6 and 7, and nucleation of the crystals is considered in Chapters 8–10, where these equations are generalized for a mixed-phase cloud. A generalization for spatially inhomogeneous clouds is done based on (Eqn. 5.5.4) in Chapters 13 and 14, where effects of turbulent mixing are also included. These equations can be used in the Eulerian one-, two-, and three-dimensional models. For this purpose, these equations should be presented using finite-difference approximations and various numerical techniques can be used. The finite-difference approximation of the size spectrum consists in dividing the entire spectrum into many small domains by radius or mass, called bins. Therefore, this method is also called the *bin microphysics method*, or sometimes these two terms are combined in the name "*spectral bin microphysics method.*"

5.6. Kinetic Equations of Coagulation

5.6.1. Various Forms of the Coagulation Equation

Another important process that governs evolution of the size spectra of cloud particles are the processes of particles' interactions: *coagulation, riming,* and *accretion*. The term *coagulation* denotes the process of collision, with subsequent merging of the two liquid drops; the term *riming* means the collision and merging of a drop and a crystal; and the term *accretion* means the collision and merging of the two crystals. These three processes are sometimes collectively referred to as coagulation.

The coagulation of drops can be caused by various processes. *Brownian coagulation* occurs due to random thermal motions of the molecules that strike large particles, so that the resulting momentum forces these particles to move randomly, collide with each other, and coagulate. The kinetic equation for Brownian coagulation was derived by Smoluchowski (1916) in the following way. Consider an ensemble of identical particles with equal volumes v_1 (monomers), and assume that the collision of any two particles results in their coagulation and the adding of their volumes. If particle collisions and coagulation continue for some time, a number n_2 of the particles will occur with the volume $v_2 = 2v_1$ (dimers), a number n_3 of the particles with the volume $v_3 = 3v_1$ (trimers), and so on, and a number n_k of the particles with $v_k = kv_1$ (k-mers). There are two types of collisions that govern the concentration of the k-mers: 1) their collision with the i-mers leads to formation of the particles with the volume $v_{ik} = (v_i + v_k)$, that is, to the loss of the k-mers; and 2) the collision of the i-mers (volume v_i) with the $(k - i)$-mers (volume $v_k - v_i$), by adding their volumes, produce k-mers.

Obviously, the rate of loss $(dn_k/dt)_{\text{loss}}$ of the k-mers due to their collisions with all the i-mers is proportional to their concentration n_k and can be written as

$$\left(\frac{dn_k}{dt}\right)_{loss} = -n_k \sum_{i=1}^{\infty} K_{ik} n_i, \tag{5.6.1}$$

where K_{ik} is some coefficient, called *the collection kernel*, that depends on the physical properties of the particles and their motion. The rate of gain of the k-mers is proportional to the concentration of the i-mers and $(k-i)$-mers, thus

$$\left(\frac{dn_k}{dt}\right)_{gain} = \frac{1}{2}\sum_{i=1}^{k-1} K_{i,k-i} n_i n_{k-i}. \tag{5.6.2}$$

The factor 1/2 arises here in order to avoid counting twice the same collisions $(i, k\text{-}i)$ and $(k\text{-}i, i)$. Now the rate of change of the k-mers can be written by summing up these two rates:

$$\left(\frac{dn_k}{dt}\right) = \left(\frac{dn_k}{dt}\right)_{gain} + \left(\frac{dn_k}{dt}\right)_{loss} = \frac{1}{2}\sum_{i=1}^{\infty} K_{i,k-i} n_i n_{k-i} - n_k \sum_{i=1}^{\infty} K_{ik} n_i. \tag{5.6.3}$$

Equation (5.6.3) is the Smoluchowski (1916) coagulation equation that has been used in numerous studies of the coagulation process. Smoluchowski derived this equation in its discrete form, and its continuous analog was obtained by Müller (1928) for the distribution function by particles' masses. Based on the conservation of particle masses during collisions, Müller replaced the sums over i-mers with the integrals over their masses dm, and the discrete variables n_i with continuous function $f(m)$. this transformed the equation into

$$\frac{\partial f(m,t)}{\partial t} = \frac{1}{2}\int_0^m K(m-m',m) f(m-m') f(m')\,dm' - f(m\int_0^{\infty} K(m, m') f(m')\,dm', \tag{5.6.4}$$

where $K(m, m')$ is the collection kernel in the mass space. The first term on the right-hand side describes the gain of the particles with the mass m due to the collision and merging of the particles with $(m\text{-}m')$ and m', while the second term describes the particles' loss due to collisions of m and m'. Assuming that the particles are identical and have the same density, this equation can be rewritten in terms of particle volumes:

$$\frac{\partial f_v(v,t)}{\partial t} = \frac{1}{2}\int_0^v K(u, v-u) f_v(u,t) f_v(v-u)\,du - f_v(v,t)\int_0^{\infty} K(u,v) f_v(u,t)\,du, \tag{5.6.5}$$

where again the first and second terms describe the gain and loss of the particles at collisions.

The coagulation equation is sometimes used in terms of the particles' radii (e.g., Junge, 1963). The transformation to radii can be done using again the differential conservation law, $f_v(v)dv = f_r(r)dr$, where the subscripts indicate the corresponding distribution functions. Let r, r', and r'' denote the radii corresponding in (Eqn. 5.6.5) to the volumes v, u, and $(v-u)$, so that $v = (4/3)\pi r^3$, $u = (4/3)\pi r'^3$, and $(u-v) = (4/3)\pi r''^3$. Then, the relations among the functions are

$$f_v(u)du = f_r(r')dr', \qquad f_v(v) = f_r(r')\frac{dr'}{du} = \frac{1}{4\pi r'^2} f_r(r'),$$

$$f_v(v-u) = f_r(r'')\frac{dr''}{d(v-u)} = f_r(r'')\frac{1}{4\pi r''^2} = f_r(r'')\frac{1}{4\pi(r^3 - r'^3)^{2/3}}. \tag{5.6.6}$$

Substituting these relations into (Eqn. 5.6.5) and multiplying by $4\pi r^2$, we obtain a coagulation equation in terms of radii,

$$\frac{\partial f_r(r,t)}{\partial t} = \frac{1}{2}\int_0^r K[(r^3 - r'^3)^{1/3}, r']\left(1 - \frac{r'^3}{r^3}\right)^{-2/3} f_r[(r^3 - r'^3)^{1/3}]f_r(r')dr'$$

$$- f_r(r,t)\int_0^\infty K(r,r')f_r(r')dr'. \quad (5.6.7)$$

5.6.2. Collection Kernels for Various Coagulation Processes

There are several types of processes that cause particle collisions with subsequent coagulation. The processes relevant to cloud physics include Brownian coagulation, gravitational coagulation, laminar and turbulent shear coagulation, and electrical coagulation. They differ by the collection kernels that describe the physics of the corresponding processes. Note that the dimension of the coagulation kernel $K(r, r')$ is [$\text{cm}^3\ \text{s}^{-1}$], which characterizes it as the volume per unit time available for two particles to collide and merge. Thus, the goal in determination of $K(r, r')$ is reduced to calculation of this volume in various processes. A brief description of some processes and kernels used in cloud models is given next.

Brownian coagulation is one of the most important for cloud physics. It governs evolution of the aerosol size spectra and aerosol fluxes to the drops, thereby influencing ice nucleation in the contact mode (see Chapter 10). The collection kernel $K(r, r')$ for Brownian coagulation represents the diffusion flux of the aerosol particles. The aerosol flux onto a particle with radius r can be derived from the diffusion equation, as was done for the vapor flux in Section 5.1 and, similar to the molecular flux in (Eqn. 5.1.6), is proportional to the Brownian diffusion coefficient D_r and particle radius. The diffusion coefficient D_r for Brownian diffusion is described by Einstein's formula

$$D_r = \frac{kT}{6\pi\eta_a r}, \quad (5.6.8)$$

where k is the Boltzmann constant and η_a is the dynamic viscosity of the air. Experimental studies showed that the diffusion to the small particles is better described using Cunningham's slip-flow correction $\alpha_{sl}(r)$ to Einstein's diffusion coefficient,

$$D_r = \frac{kT}{6\pi\eta_a r}\alpha_{sl}(r), \quad \alpha_{sl}(r) = A_C + B_C \exp(-C_C/N_{Kn}), \quad (5.6.9)$$

where A_C, B_C, and C_C, are the empirical constants, Kn = λ_f/r is the Knudsen number, and λ_f is the mean free molecular path defined in Section 5.1. If we consider mutual diffusion of the aerosol particles with radii r and r', their diffusion coefficients, fluxes, and the collection kernels are added,

$$D_{rr'} = D_r + D_{r'} = \frac{kT}{6\pi\eta_a}\left(\frac{\alpha_{sl}(r)}{r} + \frac{\alpha_{sl}(r')}{r'}\right). \quad (5.6.10)$$

$$K(r,r') = 4\pi D_{rr'}(r + r') = \frac{2kT}{3\eta_a}(r + r')\left(\frac{\alpha_{sl}(r)}{r} + \frac{\alpha_{sl}(r')}{r'}\right). \quad (5.6.11)$$

These equations are used in Chapter 10 to describe contact ice nucleation.

Gravitational coagulation. This type of coagulation process is caused by the different fall velocities of a larger drop or crystal with radius R and fall velocity $V_t(R)$ relative to a smaller particle with radius r and fall velocity $V_t(r)$. The simplest model of this process is the following. The large particle in its free fall passes in one second a path equal to $V_t(R)$ and sweeps out a volume of air in a vertical cylinder with the height $z = V_t(R)$. It can touch and collect all the smaller drops that are located at a distance from the axis of the cylinder $x \le (R + r)$. The relative velocity of the two particles with R and r is $\Delta V_t = V_t(R) - V_t(r)$. It is clear that the larger particle can collide in unit time with all of the smaller particles in the volume δV of the cylinder with the height $\Delta V_t = V_t(R) - V_t(r)$ and a cross-section $\pi(R + r)^2$, so that $\delta V = \pi(R + r)^2[V_t(R) - V_t(r)]$.

However, not all of these collisions will lead to coagulation, and some smaller particles may bounce off. The probability of coagulation after collision is characterized by the collision efficiency $E(R, r) \le 1$, the ratio of the number of coagulation events to the number of collisions, and the kernel K_G of gravitational coagulation, which is

$$K_G(r,r') = E(R,r)\delta V = \pi(R+r)^2 E(R,r)[V_t(R) - V_t(r)]. \tag{5.6.12}$$

Refinements of this model include accounting for the more complicated flow of the smaller particle along the stream functions around the larger particle and the parameterizations of $E(R, r)$. Fuchs (1959, 1964) and some other authors considered this flow in more detail by accounting for the angular dependence, and so developed a parameterization of $E(R, r)$ in terms of the ratio $p_{rR} = r/R$ as

$$E(R,r) = \frac{p_{rR}^2}{2(1+p_{rR}^2)}. \tag{5.6.13}$$

A comparison of (Eqn. 5.6.13) with these detailed calculations showed a sufficient accuracy of (Eqn. 5.6.13) for $p_{rR} \le 0.5$.

The methods of calculation and reviews of various approximations for $E(R, r)$ for gravitational Brownian coagulation, laminar and turbulent shear coagulation, and electrical coagulation are given by Jonas (1972), Long (1974), Voloshchuk and Sedunov (1975), Hall (1980), Low and List (1982a,b), Voloshchuk (1984), Pruppacher and Klett (1997), Bott (1998, 2000, 2001), MacFarquhar (2004), Straka (2009), and Cotton et al. (2011). These methods and calculations can be used for evaluation of the coagulation kernels and rates.

5.7. Thermodynamic and Kinetic Equations for Multidimensional Models

In multidimensional models, as in many cloud, mesoscale, and large-scale models, the heat and moisture transfer are often calculated using the equations for the potential temperature θ and the specific humidity q and related specific variables (supersaturation, condensation rates, etc.). The equations for θ and q in the turbulent atmosphere can be written in the curvilinear coordinate system over a complex terrain with accounting for orographic effects as

$$\frac{\partial \theta}{\partial t} + \frac{\partial u\theta}{\partial x} + \frac{\partial v\theta}{\partial y} + \frac{\partial w\theta}{\partial z} - (uh_x + vh_y)\frac{\partial \theta}{\partial z} = \frac{\partial}{\partial z} k_z \frac{\partial \theta}{\partial z} + \hat{\Delta}\theta$$

$$+ \frac{L_e}{c_p}\varepsilon_{cw} + \frac{L_s}{c_p}\varepsilon_{ci} + \frac{L_m}{c_p}\varepsilon_f + \frac{L_m}{c_p}\varepsilon_m + R_{rad}, \tag{5.7.1}$$

$$\frac{\partial q}{\partial t} + \frac{\partial uq}{\partial x} + \frac{\partial vq}{\partial y} + \frac{\partial wq}{\partial z} - (uh_x + vh_y)\frac{\partial q}{\partial z} = \frac{\partial}{\partial z} k_z \frac{\partial q}{\partial z} + \hat{\Delta} q - \varepsilon_{cw} - \varepsilon_{ci}. \tag{5.7.2}$$

Here x, y, and z are the coordinates; t is time; u, v, and w are three components of wind speed; k_z is the coefficient of vertical turbulent diffusion, and

$$\hat{\Delta} = \frac{\partial}{\partial x} k_x \frac{\partial}{\partial x} + \frac{\partial}{\partial y} k_y \frac{\partial}{\partial y} \tag{5.7.3}$$

is the operator of horizontal eddy diffusion. k_x, k_y are the horizontal components of the turbulent exchange coefficient, $h(x, y)$ is the height of surface relief, and $h_x = \partial h/\partial x$ and $h_y = \partial h/\partial y$ are the slopes of the relief height. ε_{cw} and ε_{ci} are the specific condensation and deposition growth rates [g g^{-1} s^{-1}] related to the condensation and deposition rates I_{con} and I_{dep} considered in Section 5.3 as $\varepsilon_{cw} = I_{con}/\rho_a$ and $\varepsilon_{ci} = I_{dep}/\rho_a$; ε_f and ε_m are the specific freezing and melting rates; $R_{rad} = (\partial T/\partial t)_{rad}$ is the radiative temperature change rate.

The microphysics of a mixed-phase cloud in the models using spectral bin microphysics method is calculated using the two kinetic equations for the size distribution functions of the droplets $f_d(x, y, z, r_d, t)$ and crystals $f_c(x, y, z, r_c, t)$ that are solved at each time step. As discussed in Section 5.5 and later in Chapter 13, these equations are written here in the simple form that account for the turbulent mixing in clouds, as practically in all cloud models based on the spectral bin microphysics. The kinetic equation for the drop size spectrum $f_d(r_d)$ is

$$\begin{aligned} &\frac{\partial f_d}{\partial t} + \frac{\partial u f_d}{\partial x} + \frac{\partial v f_d}{\partial y} + \frac{\partial [w - v_d(r_d)] f_d}{\partial z} - (u h_x + v h_y) \frac{\partial f_d}{\partial z} \\ &= \frac{\partial}{\partial z} k_z \frac{\partial f_d}{\partial z} + \hat{\Delta} f_d + \left(\frac{\partial f_d}{\partial t}\right)_{act} + \left(\frac{\partial f_d}{\partial t}\right)_{cond/evap} \\ &+ \left(\frac{\partial f_i}{\partial t}\right)_{aggr} + \left(\frac{\partial f_i}{\partial t}\right)_{frz} + \left(\frac{\partial f_i}{\partial t}\right)_{break/mult}. \end{aligned} \tag{5.7.4}$$

The kinetic equation for the crystals size spectrum $f_c(r_c)$ is

$$\begin{aligned} &\frac{\partial f_c}{\partial t} + \frac{\partial u f_c}{\partial x} + \frac{\partial v f_c}{\partial y} + \frac{\partial [w - v_c(r_c)] f_c}{\partial z} - (u h_x + v h_y) \frac{\partial f_c}{\partial z} \\ &= \frac{\partial}{\partial z} k_z \frac{\partial f_c}{\partial z} + \hat{\Delta} f_c + \left(\frac{\partial f_c}{\partial t}\right)_{nucl} + \left(\frac{\partial f_c}{\partial t}\right)_{cond/evap} \\ &+ \left(\frac{\partial f_c}{\partial t}\right)_{aggr} + \left(\frac{\partial f_c}{\partial t}\right)_{melt} + \left(\frac{\partial f_c}{\partial t}\right)_{break/mult}. \end{aligned} \tag{5.7.5}$$

Here, the indices d and c mean drops and crystals, r_d is the droplet radius and r_c is the crystal equivalent radius (or semi-axis), $\dot{r}_d$ and $\dot{r}_c$ are their growth rates, and $v_d(r_d)$ and $v_c(r_c)$ are the terminal velocities evaluated as described in Chapter 12. The last five terms on the right-hand side of (Eqns. 5.7.4 and 5.7.5) describe various microphysical processes: drop activation $(\partial f_d/\partial t)_{act}$ is considered in Chapters 6 and 7; heterogeneous and homogeneous nucleation of crystals, $(\partial f_c/\partial t)_{nucl}$ and drop freezing $(\partial f_d/\partial t)_{frz}$ (Chapters 8–10); drop and crystal diffusion growth and evaporation $(\partial f_d/\partial t)_{cond/evap}$ and $(\partial f_c/\partial t)_{cond/evap}$ (Sections 5.1 and 5.2); aggregation (coagulation and accretion), $(\partial f_d/\partial t)_{aggr}$ and $(\partial f_c/\partial t)_{aggr}$ (Section 5.6); and crystal melting $(\partial f_c/\partial t)_{melt}$; and drop breakup and ice multiplication, $(\partial f_d/\partial t)_{break/mult}$ and $(\partial f_c/\partial t)_{break/mult}$.

After calculating the distribution functions f_d and f_c, many characteristics can be calculated as the integrals (moments) over the size spectra at each grid point and time step, such as liquid water content (LWC), ice water content (IWC), concentrations N_d and N_c, mean radii, radar and lidar reflectivities and visibility (as described in Section 2.3), absorption and extinction coefficients, optical thickness and emissivity, and integral terminal velocities. Calculation of the diffusion growth of drops and crystal requires supersaturation that is calculated using the splitting method described in the next section (Section 5.8). Once the size spectra and supersaturation have been calculated, a detailed evaluation can be done of the supersaturation budget (generation and absorption rates) along with the crystal mass budget: the gravitational flux of crystals (precipitation rate), the regular flux due to vertical velocities, the turbulent flux, and the total budget of crystal mass, which is the sum of four gradients (influxes) of the corresponding fluxes defined previously (e.g., Khvorostyanov and Sassen, 1998c, 2002; Khvorostyanov, Curry, et al., 2001, 2003, 2006).

This relatively simple method for accounting for cloud microphysics described here is based on solving the two kinetic equations for drop and crystal size distribution functions, each function, besides the spatial variables, depends on one size parameter, radius or mass. This method can be easily generalized for several types of crystals, but a separate size distribution function should be added for each crystal type taking into account its shape. Another possible generalization would be to introduce additional variables for f_d and f_c like the soluble and insoluble fractions, electrical charge, surface-active substances, etc.

5.8. Fast Algorithms for Microphysics Modules in Multidimensional Models

The system of thermodynamic and kinetic equations for a multidimensional model described in Section 5.7 can be solved by using a splitting method according to physical processes and components as described, e.g., in Buikov and Pirnach (1973, 1975), Buikov and Khvorostyanov (1977), Khvorostyanov (1982, 1995), Kondratyev et al. (1990a,b), Khvorostyanov and Sassen (1998a), Khvorostyanov, Curry et al. (2001, 2003, 2006), Curry and Khvorostyanov (2012). The splitting method provides fast algorithms that allow us to combine large and small time steps and significantly accelerate simulations.

The equations for temperature, humidity, and for the spectra of droplets and crystals are solved by splitting the major time step Δt into six substeps, denoted hereafter as $k/6$, so that the results of calculations from the previous of these six substeps serve as initial conditions for the next substep. The first three substeps account for the evolution of the variables due to turbulent and wind transport along the x, y, and z directions:

$$\frac{\partial \varphi^{1/6}}{\partial t} + \frac{\partial (u\varphi^{1/6})}{\partial x} = \frac{\partial}{\partial x} k_x \frac{\partial \varphi^{1/6}}{\partial x}, \qquad \frac{\partial \varphi^{2/6}}{\partial t} + \frac{\partial (v\varphi^{2/6})}{\partial y} = \frac{\partial}{\partial y} k_y \frac{\partial \varphi^{2/6}}{\partial y},$$

$$\frac{\partial \varphi^{3/6}}{\partial t} + \frac{\partial (\tilde{w}\varphi^{3/6})}{\partial z} = \frac{\partial}{\partial z} k_z \frac{\partial \varphi^{3/6}}{\partial z}. \tag{5.8.1}$$

Here φ denotes any of the four parameters θ, q, f_d, or f_c, the superscripts 1/6, 2/6, and 3/6 denote the substeps of splitting: $\tilde{w} = w - uh_x - vh_y$ for θ and q, and $\tilde{w} = w - v_{d,c} - uh_x - vh_y$ for f_d and f_c.

During the fourth substep, the growth of droplets and crystals by condensation, deposition, or evaporation is calculated after a transition to the equations for specific supersaturations with respect to water $\Delta_w = q - q_{ws}(T)$ and with respect to ice $\Delta_i = q - q_{is}(T)$. These equations are being obtained in the way similar to that in Section 5.3 for fractional supersaturations. During the fourth step of splitting, after removal of the space variables, a Lagrangian zero-dimensional problem is actually being solved. Equations for the temperature and specific humidity can be written after removal of the space variables in (Eqns. 5.7.1) and (5.7.2) as

$$\frac{dT}{dt} = \frac{L_e}{c_p}\varepsilon_{cw} + \frac{L_s}{c_p}\varepsilon_{ci} - w\,\gamma_a + R_{\mathrm{rad}}, \qquad \frac{dq}{dt} = -\varepsilon_{cw} - \varepsilon_{ci}. \tag{5.8.2}$$

The condensation ε_{cw} and deposition ε_{ci} growth rates are obtained from (Eqn. 5.3.19):

$$\varepsilon_{cw} = \frac{I_{con}}{\rho_a} = \frac{\Delta_w}{\Gamma_1 \tau_{fd}}, \qquad \varepsilon_{ci} = \frac{I_{dep}}{\rho_a} = \frac{\Delta_i}{\Gamma_2 \tau_{fc}}, \tag{5.8.3}$$

where $\Gamma_1 \approx \Gamma_w$ and $\Gamma_2 \approx \Gamma_i$ are the psychrometric corrections (Eqn. 5.3.15), τ_{fd} and τ_{fc} are the relaxation times of supersaturation by drop and crystals. Differentiating specific supersaturation $\Delta_w = q_v - q_{ws}(T)$ by time and substituting (Eqn. 5.8.2), we obtain:

$$\begin{aligned}
\frac{d\Delta_w}{dt} &= \frac{dq_v}{dt} - \frac{\partial q_{ws}}{\partial T}\frac{dT}{dt} \\
&= -(\varepsilon_{cw} + \varepsilon_{ci}) - \frac{\partial q_{ws}}{\partial T}\left(\frac{L_e}{c_p}\varepsilon_{cw} + \frac{L_s}{c_p}\varepsilon_{ci} - \gamma_a w + R_{rad}\right) \\
&= -\varepsilon_{cw}\left(1 + \frac{L_e}{c_p}\frac{\partial q_{ws}}{\partial T}\right) - \varepsilon_{ci}\left(1 + \frac{L_s}{c_p}\frac{\partial q_{ws}}{\partial T}\right) + \frac{\partial q_{ws}}{\partial T}(\gamma_a w - R_{rad}) \\
&= -\varepsilon_{cw}\Gamma_1 - \varepsilon_{ci}\Gamma_2 + \frac{\partial q_{ws}}{\partial T}(\gamma_a w - R_{rad}),
\end{aligned} \tag{5.8.4}$$

where we used definitions (Eqn. 5.3.15) for psychrometric corrections Γ_1 and Γ_2. Substituting here (Eqn. 5.8.3) for ε_{cw} and ε_{ci}, we obtain an equation for the specific supersaturation Δ_w for a mixed-phase cloud, containing water vapor, droplets, and crystals:

$$\frac{d\Delta_w}{dt} = -\frac{\Delta_w}{\tau_d} - \frac{\Gamma_{12}}{\Gamma_2}\frac{\Delta_i}{\tau_c} + \frac{\partial q_{ws}}{\partial T}(\gamma_a w - R_{rad}), \tag{5.8.5}$$

where $\Gamma_{12} = 1 + (L_s/c_p)(\partial q_{sw}/\partial T)$. Taking into account the relation $\Delta_i = \Delta_w + (q_{ws} - q_{is})$, (Eqn. 5.8.5) can be rewritten for the 4/6-th substep of splitting as

$$\frac{d\Delta_w^{4/6}}{dt} = -\left(\tau_{fd}^{-1} + \frac{\Gamma_{12}}{\Gamma_2}\tau_{fc}^{-1}\right)\Delta_w^{4/6} - \frac{\Gamma_{12}}{\Gamma_2}\tau_{fc}^{-1}(q_{ws} - q_{is}) + \frac{\partial q_{ws}}{\partial T}(\gamma_a w - R_{rad}). \tag{5.8.6}$$

As discussed in Section 5.3, this equation has clear physical meaning. The first term on the right-hand-side of (Eqn. 5.8.6) describes supersaturation relaxation due to absorption of the vapor by droplets and crystals, the second term describes the vapor flux from drops to crystals (the Wegener–Bergeron–Findeisen process), and the third term describes supersaturation generation due to upward vertical velocities w and radiative cooling R_{rad}, or the occurrence of negative supersaturation and evaporation of the cloud, if $(\gamma_a w - R_{rad}) < 0$. Equations for the ice supersaturation Δ_i and for the temperature T at the substep 4/6 are obtained in the same way:

$$\frac{d\Delta_i^{4/6}}{dt} = -\left(\frac{\Gamma_{21}}{\Gamma_1}\tau_{fd}^{-1} + \tau_{fc}^{-1}\right)\Delta_i^{4/6} + \frac{\Gamma_{21}}{\Gamma_1}\frac{(q_{ws} - q_{is})}{\tau_{fc}} + \frac{\partial q_{is}}{\partial T}(\gamma_a w - R_{rad}), \tag{5.8.7}$$

$$\frac{dT^{4/6}}{dt} = \left(\frac{L_e}{c_p}\frac{1}{\Gamma_1 \tau_{fd}} + \frac{L_s}{c_p}\frac{1}{\Gamma_2 \tau_{fc}}\right)\Delta_w^{4/6} + \frac{L_s}{c_p}\frac{(q_{ws} - q_{is})}{\Gamma_2 \tau_{fc}} - (\gamma_a w - R_{rad}), \tag{5.8.8}$$

where $\Gamma_{21} = 1 + (L_e/c_p)(\partial q_{si}/\partial T)$. In (Eqn. 5.8.8), the first term is responsible for the heating due to the condensation and deposition, the second term describes heating caused by the WBF process, and the third term means cooling due to updraft velocities and radiative exchange or heating in the downdrafts.

The preceding equations describe the thermodynamics and kinetics of the mixed-phase cloud. Reduction to the liquid cloud is with $\tau_{fc} = \infty$ $(N_c = 0)$, and reduction to the pure crystalline cloud is with $\tau_{fd} = \infty$ $(N_d = 0)$. For the calculation of temporal evolution of droplet and crystal size spectra, it is convenient to introduce the normalized integral specific supersaturations Q_w and Q_i that describe an increase in the squared radius (see Section 5.5) and are related to the specific supersaturations as follows:

$$Q_w(t) = \frac{2D_v \rho_a}{\rho_w \Gamma_w}\int_{t_j}^{t} \Delta_w(t')dt', \qquad Q_i(t) = \frac{2D_v \rho_a \zeta_i}{\rho_i \Gamma_i}\int_{t_j}^{t} \Delta_i(t')dt', \tag{5.8.9}$$

$$\frac{dQ_w}{dt} = \frac{2D_v \rho_a}{\rho_w \Gamma_w}\Delta_w, \qquad \frac{dQ_i}{dt} = \frac{2D_v \rho_a \zeta_i}{\rho_i \Gamma_i}\Delta_i, \tag{5.8.10}$$

where ζ_i describes the effect of the crystal shape defined in (Eqn. 5.2.24). As discussed in Section 5.5, using Maxwell's equations from Sections 5.1, 5.2 for the growth rates, and the conservation law in the differential form $f[(r(t)]dr(t) = f[(r(t_0)]dr(t_0)$, we can express the function $f(t)$ at a time t via the function at an initial time t_0 as described in Section 5.5 by (Eqns. 5.5.20)–(5.5.24). Applying these relations to the substep 4/6 with diffusion and deposition growth, we can simply express the function $f^{4/6}(r)$ after the end of the substep 4/6, via the function $f^{3/6}(r)$ at the beginning of this substep—i.e., after the three substeps of spatial transport. Using (Eqn. 5.5.24), we obtain for the droplet spectrum

$$f_d^{4/6}(r_d, t) = f_d^{3/6}(r_d^{3/6})\frac{dr_d^{3/6}}{dr_d^{4/6}}$$
$$= \frac{r_d + \xi_{con}}{[(r_d + \xi_{con})^2 - Q_w(t)]^{1/2}} f_d^{3/6}([(r_d + \xi_{con})^2 - Q_w(t)]^{1/2} - \xi_{con}), \tag{5.8.11}$$

where $f_d^{3/6}(r_d)$ is the "old" distribution function after the 3/6 substeps of transport, and $f_d^{4/6}(r_d)$ is the "new" distribution function after the 4/6-th substep of condensation or evaporation. The crystal size spectrum is calculated in the same way,

$$f_c^{4/6}(r_c, t) = f_c^{3/6}(r_c^{3/6})\frac{dr_c^{3/6}}{dr_c^{4/6}}$$
$$= \frac{r_c + \xi_{dep}}{[(r_c + \xi_{dep})^2 - Q_i(t)]^{1/2}} f_c^{3/6}([(r_c + \xi_{dep})^2 - Q_i(t)]^{1/2} - \xi_{dep}), \tag{5.8.12}$$

where $f_c^{3/6}(r_c)$ is the "old" distribution function after the three substeps of transport, and $f_c^{4/6}(r_c)$ is the "new" distribution function after the 4/6-th step of deposition or evaporation.

In cloud modeling, it is often more convenient to use equations for the integral supersaturations Q_w and Q_i instead of equations for Δ_w, Δ_i and to solve directly equations for Q_w and Q_i. The equations for Q_w and Q_i can be derived by the substitution of relations (Eqns. 5.8.9) and (5.8.10) into the (Eqns. 5.8.6) and (5.8.7) for Δ_w and Δ_i. To derive the equation for Q_w, we first transform the expressions for the relaxations times. Substitution of (Eqn. 5.8.11) into (Eqn. 5.3.20a) for τ_{fd} and accounting also for the ventilation corrections gives (we omit here the superscripts of the substeps for brevity)

$$\tau_{fd}^{-1} = 4\pi D_v \int_{r_{d\min}}^{\infty} \frac{r_d^2 f_d\{[(r_d+\xi_{con})^2 - Q_w(t)]^{1/2} - \xi_{con}\}}{[(r_d+\xi_{con})^2 - Q_w(t)]^{1/2}} F_{ven}(r_d)\, dr_d. \tag{5.8.13}$$

The lower limit here is determined from the condition that the argument of f_d and the denominator should not be negative, which yields

$$[(r_d+\xi_{con})^2 - Q_w(t)]^{1/2} \geq 0, \qquad \text{or} \qquad r_{d\min} = Q_w^{1/2}(t) - \xi_{con}. \tag{5.8.14}$$

It is convenient now to introduce a new variable r_d' so that

$$r_d' = [(r_d+\xi_{con})^2 - Q_w(t)]^{1/2} - \xi_{con}. \tag{5.8.15}$$

Substituting (Eqn. 5.8.15) into (Eqn. 5.8.13), and taking into account that, as follows from (Eqn. 5.8.14), the limit r_{dmin} in the integral over r_d corresponds to the limit $r_d' = 0$ in integration over r_d', and omitting the prime at r_d' in the final equation, we obtain

$$\tau_{fd}^{-1} = 4\pi D_v \int_0^{\infty} R_d(r_d, \xi_{con}) f_d^{3/6}(r_d, t) F_{ven}(r_d)\, dr_d,$$
$$R_d(r_d, \xi_{con}) = \frac{\{[(r_d+\xi_{con})^2 + Q_w(t)]^{1/2} - \xi_{con}]\}^2}{[(r_d+\xi_{con})^2 + Q_w(t)]^{1/2}} \tag{5.8.16}$$

where the superscript 3/6 indicates that the droplet size spectrum is taken from this substep. A similar expression can be derived for τ_{fc}. Substitution of (Eqn. 5.8.12) into (Eqn. 5.3.20a) for τ_{fc} gives

$$\tau_{fc}^{-1} = 4\pi D_v \zeta_i \int_{r_{c\min}}^{\infty} \frac{r_c^2 f_c\{[(r_c+\xi_{dep})^2 - Q_i(t)]^{1/2}, t\}}{[(r_d+\xi_{dep})^2 - Q_i(t)]^{1/2}} F_{ven}(r_c)\, dr_c, \tag{5.8.17}$$

where $r_{c,min}$ is again defined so that the expressions under square root are positive. Introducing a new variable r_c',

$$r_c' = [(r_c+\xi_{dep})^2 - Q_i(t)]^{1/2} - \xi_{dep}. \tag{5.8.18}$$

and substituting this into (Eqn. 5.8.17) and omitting the prime at r_c' in the final equation, we obtain

$$\tau_{fc}^{-1} = 4\pi D_v \zeta_i \int_0^{\infty} R_c[(r_c, \xi_{dep}) f_d^{3/6}(r_c, t) F_{ven}(r_c)\, dr_c,$$
$$R_c(r_c, \xi_{dep}) = \frac{\{[(r_c+\xi_{dep})^2 + Q_i(t)]^{1/2} - \xi_{dep}]\}^2}{[(r_c+\xi_{dep})^2 + Q_i(t)]^{1/2}} \tag{5.8.19}$$

Thus, the quantities τ_{fd}^{-1} and τ_{fc}^{-1} are expressed now via the size spectra $f_d(r_d)$ and $f_c(r_c)$ after the substep 3/6 of the splitting—i.e., after spatial transport and before the condensation substep.

Differentiating (Eqn. 5.8.10) for Q_w by time and substituting into (Eqn. 5.8.6) for Δ_w, we obtain an equation for Q_w of the second order by time,

$$\frac{d^2Q_w}{dt^2} = -\left(\tau_{fd}^{-1} + \frac{\Gamma_{12}}{\Gamma_2}\tau_{fc}^{-1}\right)\frac{dQ_w}{dt} - \frac{2D_v\rho_a}{\rho_w}\frac{\Gamma_{12}}{\Gamma_1\Gamma_2}\tau_{fc}^{-1}(q_{ws} - q_{is})$$

$$+\frac{2D_v\rho_a}{\rho_w\Gamma_w}\frac{\partial q_{ws}}{\partial T}(\gamma_a w - R_{rad}), \qquad (5.8.20a)$$

where τ_{fd}^{-1} and τ_{fc}^{-1} are defined in (Eqns. 5.8.16) and (5.8.19). Substituting these expressions for τ_{fd}^{-1} and τ_{fc}^{-1}, we obtain the equation for the integral normalized supersaturation Q_w in a closed form

$$\frac{d^2Q_w}{dt^2} = -\frac{dQ_w}{dt}4\pi D_v\left\{\int_0^\infty R_d(r_{d,}\xi_{con})f_d(r_d,t)F_{ven}(r_d)dr_d\right.$$

$$\left.+\zeta_i\frac{\Gamma_{wi}}{\Gamma_i}\int_0^\infty R_c(r_{c,}\xi_{dep})f_c(r_c,t)F_{ven}(r_c)dr_c\right\}$$

$$-\frac{8\pi D_v^2\rho_a(q_{sw} - q_{si})}{\rho_w}\frac{\Gamma_{wi}\zeta_i}{\Gamma_w\Gamma_i}\int_0^\infty R_c(r_{c,}\xi_{dep})f_c(r_c,t)F_{ven}(r_c)dr_c$$

$$+\frac{2D_v\rho_a}{\rho_w\Gamma_w}\frac{\partial q_{sw}}{\partial T}(\gamma_a w - R_{rad}), \qquad (5.8.20b)$$

This is the integro-differential equation for Q_w of the second order by time analogous to the equations for the integral supersaturations y_w, y_i derived in Section 5.5. This equation is also used for studies of drop activation in Chapters 6 and 7 and ice nucleation and the parcel modeling in Chapters 8–10.

Despite a slightly complicated form, the solution of this equation is sufficiently simple and consists of several steps.

1. The supersaturation relaxation times τ_{fd}^{-1} and τ_{fc}^{-1} are calculated using (Eqn. 5.8.16), (Eqn. 5.8.19) and the size spectra after the splitting substep 3/6—that is, after the spatial transport.
2. Then a simpler (Eqn. 5.8.20a) is solved instead of (Eqn. 5.8.20b). The solution is performed within a special microphysical module, which solves together a system of equations including (Eqn. 5.8.10) for dQ_w/dt, (Eqn. 5.8.20a) for dQ_w^2/dt^2, and (Eqn. 5.8.8) for dT/dt. This system is solved using the Runge–Kutta method. The major advantage of this scheme developed by M. V. Buikov with pupils (Buikov and Pirnach, 1973, 1975; Buikov and Khvorostyanov, 1976, 1977) is that the algorithm is very economic and fast. The main (dynamical) time step Δt in the model can be rather large, from 10 s to 5–10 min, comparable to those used in mesoscale or even climate models, and (Eqns. 5.8.1) for spatial transport of temperature, humidity, and size spectra are solved with this large time step at the first three of six substeps of splitting. Then, the system of (Eqns. 5.8.10) for dQ_w/dt, (Eqn. 5.8.20a) for dQ_w^2/dt^2, and (Eqn. 5.8.8) for dT/dt are solved in a separate subroutine with a smaller (microphysical)

time step Δt_{mic}. This small step Δt_{mic} is chosen in this subroutine by comparison with relaxation times τ_{fd}, τ_{fc}. It is obtained by dividing the main time step Δt by some number N_t (e.g., 2 to 5) and is compared with τ_{fd} and τ_{fc}. If $\Delta t_{mic} > \tau_{fd}$, or $\Delta t_{mic} > \tau_{fc}$, Δt is divided again by N_t, and this procedure is repeated until the conditions are satisfied $\Delta t_{mic} < \tau_{fd}$, and $\Delta t_{mic} < \tau_{fc}$. This ensures computational stability in calculations of diffusion growth as was mathematically proven in cited works by Buikov et al. After evaluation of τ_{fd} and τ_{fc}, the time step Δt_{mic} is chosen automatically in this microphysical subroutine to satisfy the described stability conditions, the number of these small substeps is calculated as $n_t = \Delta t/\Delta t_{mic}$, and this system of equations is integrated n_t times with Δt_{mic} over the entire Δt. Generally, τ_{fd} and τ_{fc} are different at every grid point, therefore Δt_{mic} and n_t should be chosen separately at each grid point. If the liquid phase is present, then $\Delta t_{mic} < \tau_{fd} \sim 1$–$10$ s, and if a cloud is pure crystalline like cirrus, then $\Delta t_{mic} < \tau_{fc}$ and can be much greater, a few tens or hundreds of seconds or a few minutes (see Section 5.3, Tables 5.1 and 5.2). After integration, $Q_w(\Delta t)$, $Q_i(\Delta t)$, $dQ_w(\Delta t)/dt$, and $dQ_i(\Delta t)/dt$ are known, and then Δ_w, s_w and Δ_i, s_i are calculated with (Eqn. 5.8.10).

3. Finally, the "new" size spectra $f_d^{4/6}$ and $f_c^{4/6}$ after the condensation and deposition substep 4/6 are calculated with known $Q_w(\Delta t)$, $Q_i(\Delta t)$ using (Eqns. 5.8.11) and (5.8.12). The advantage of this method is that the use of the integral supersaturations $Q_w(t)$, $Q_i(t)$ along with (Eqns. 5.8.11) and (5.8.12) allows us to recalculate the distribution functions only once during the complete and relatively large time step Δt and enables us to substantially decrease computational diffusivity in the space of radii, which may occur in the other numerical techniques with multiple recalculations at each small time step. Note that the same method for microphysics is used in the parcel model simulations described in Chapters 8 and 9, where the substeps from 1/6 to 3/6 with spatial transport are absent.

During the fifth substep 5/6, coalescence and accretion of droplets and crystals are calculated using the kinetic equations of coagulation described in Section 5.6 using size spectra $f_d^{4/6}$ and $f_c^{4/6}$ after the condensation substep as the initial data. Various algorithms were developed for solution of the coagulation equation, one of the best is the fast and precise effective flux method developed by Bott (1998, 2000, 2001) that can be used for a numerical solution of the collection equations. A simpler but less precise approach for calculations of the coagulation and accretion rates, the approximation of continuous growth, is still used in some cloud models. It is described in Chapter 14.

At the end of the time step, at substep 6/6, activation of droplets is calculated as described in Chapters 6 and 7, and nucleation of crystals is calculated as described in Chapters 8, 9, and 10.

The final values of T, q, f_d, and f_c obtained after all six substeps are used as initial data at the beginning of the next time step.

In the microphysical units of the model, a finite-difference grid is used, which includes N_x, N_y, and N_z gridpoints in the horizontal and vertical directions with spacing of Δx, Δy, and Δz. These parameters are varied as described in the numerical experiments. For example, in the 3D version of the model described in Khvorostyanov, Curry et al. (2003) with simulations of fog and Sc clouds observed over Arctic polynya in the Beaufort Sea in the SHEBA-FIRE field experiment, there were 21×21 horizontal grid points with the steps $\Delta x = \Delta y = 3$ km, and 31 vertical levels with resolution

33.3 m. The droplet and crystal size ranges were divided into 30 intervals from 0.1 μm to 3.5 mm with logarithmically increasing spacing. The initial T and q fields are specified for each simulated case using the data of sounding, aircraft, and surface observations. With the configuration of this spectral 3D model, due to fast microphysics algorithms with combinations of large and small time steps, it was possible to perform simulations on a PC Pentium. The model satisfactorily reproduced the characteristics of the cloudy boundary layer, evolving size spectra of drops and crystals and their moments, supersaturation, phase state, and surface heat fluxes, as was shown in this work by comparison with observations for that case, and was confirmed later by Raddatz et al. (2011, 2012) in observations of similar cloudy boundary layers and heat fluxes over Arctic polynyas.

6

Wet Aerosol Processes

6.1. Introduction

Interactions of atmospheric aerosol particles with humidity result in the transformation of aerosol properties and size spectra. At low humidities, a thin water film can be *adsorbed* on the surface of an aerosol particle. The result of this adsorption can be a monolayer of water molecules or a water film of several water molecules thick on the surface of the aerosol particle. The particles of atmospheric aerosol usually consist of soluble and insoluble fractions and are termed "*mixed*" aerosol particles. The aerosol population of one sort with the same properties and chemical composition of each particle is called "*internally mixed*," and aerosol consisting of several sorts of particles with different composition and properties is called "*externally mixed*." Detailed reviews of various types of atmospheric aerosols, their properties, and the mechanisms and models of their formation are given, for example, in Fuchs (1959, 1964), Levin (1954), Junge (1963), Pruppacher and Klett (1997), Sedunov (1974), and Seinfeld and Pandis (1998).

Aerosol particles may have an adsorbed liquid film on the surface but are solid inside if they are insoluble or if environmental water saturation ratio S_w (or relative humidity $H = 100\ S_w$) is below some critical value called the *threshold of deliquescence saturation ratio* $S_{w,del}$ (or *threshold deliquescence humidity*). If the saturation ratio increases and exceeds the threshold of deliquescence $S_{w,del}$, the soluble fraction deliquesces (see Chapter 11), and the *hygroscopic growth* of the particles begins as they accumulate more water. These mixed aerosol particles convert into *solution* or *haze particles* comprised of a liquid solution and an *insoluble core*. The solution contains water and an aerosol soluble fraction, while the insoluble fraction or insoluble core still remains in the solid state. Haze droplets are generally of submicron sizes.

The equilibrium size at increasing humidity is reached rapidly. The time of relaxation to equilibrium size for a particle with wet radius r_w at saturation ratio S_w can be estimated as $\tau_{eq} \approx 0.4 r_w^2 / [D_v(\rho_v/\rho_w)(1-S_w)]$ (Sedunov, 1974), and are ~10^{-4}, 10^{-2}, and 1–2 s, for the submicron particles with the radii ~0.01, 0.1, and 1 μm (Sedunov, 1974; Pruppacher and Klett, 1997). If the relative humidity increases further and reaches values slightly above 100%, or water supersaturation reaches some *critical* value s_{cr}, the aerosol particles are said to be *activated* and become capable of unlimited growth. Aerosol particles that have appropriate properties to form drops by activation are termed *cloud condensation nuclei* (CCN).

Hygroscopic growth of aerosol particles influences aerosol radiative properties and the *direct aerosol effect* on climate. Activation of cloud drops influences cloud microphysical and radiative properties and influences the *indirect aerosol effects on climate*: an increase in aerosol concentrations (e.g., anthropogenic pollutions, volcanic eruptions) causes *the first indirect* or *Twomey effect* (*albedo* increase, Twomey, 1977). This increase in aerosol concentrations may result in the formation of higher concentrations of smaller droplets, which hamper coagulation and cause precipitation decrease and longer cloud life times. This is termed *the second indirect or Albrecht effect* (Albrecht, 1989). Both of these indirect effects are important for the studies of climate change (Intergovernmental Panel on Climate Change, IPCC-2007; Lohmann and Feichter, 2005; McFiggans et al., 2006).

Both hygroscopic growth and CCN activation are described by the *Köhler* equation. The *Köhler* equation for the saturation ratio expresses the equilibrium size of the solution droplet as the balance between the curvature (Kelvin) effect and the solute (Raoult) effect as described in Chapter 3, Section 3.9. Some features and previous parameterizations of aerosol hygroscopic growth and CCN activation are briefly reviewed in this section, while theoretical descriptions of these processes based on the Köhler theory and its generalizations are presented in other sections of this chapter.

6.1.1. Empirical Parameterizations of Hygroscopic Growth

Numerous measurements of aerosol properties at varying humidity have shown that the radii of aerosol particles, in general, increase with increasing humidity. This feature, called *hygroscopic growth*, is usually parameterized either with equations for the dependence of the radius r_w of wet aerosol on the water saturation ratio S_w (or on the relative humidity $H = 100S_w$) as

$$r_w(S_w) = r_d[A_w + B_w(1 - S_w)^{-\gamma_1}], \tag{6.1.1a}$$

$$r_w(S_w) = \alpha_g(S_w) r_d^{\gamma_1}, \tag{6.1.1b}$$

where $\alpha_g(S_w)$ is some function of the water saturation ratio, or by the growth factor $GF(S_w) = r_w(S_w)/r_d$, where r_w and r_d are the wet and dry radii, and S_w is the water saturation ratio.

The function $GF(S_w)$ and the humidity impact on aerosol extinction coefficient σ_{ext} are often parameterized with empirical relations of the type

$$GF(S_w) = A_w + B_w(1 - S_w)^{-\gamma_1}, \tag{6.1.2}$$

$$\sigma_{ext}(S_w) \sim (1 - S_w)^{-\gamma_2}, \tag{6.1.3}$$

or nonlinear expressions similar to (Eqn. 6.1.1b). The quantities A_w, B_w, γ_1, and γ_2 are empirical parameters determined by fitting experimental data (e.g., Kasten, 1969; Hegg et al., 1996; Kotchenruther et al., 1999; Swietlicki et al., 1999; Zhou et al., 2001) or by fitting functional dependencies for $GF(S_w)$ found from approximate analytical solutions of the Köhler equation combined with analysis of the experimental data (e.g., Fitzgerald, 1975; Khvorostyanov and Curry 1999a,b, 2006; Swietlicki et al., 1999; Cohard et al., 2000; Kreidenweis et al., 2005; Rissler et al., 2006). Some parameterizations assumed $A_w = 0$ (e.g., Kasten, 1969), and some tested more complicated functions of S_w and other aerosol properties (e.g., Hänel, 1976; Kotchenruther et al., 1999).

A more detailed evaluation of hygroscopic growth and $GF(S_w)$ was based on numerical solutions of the Köhler equation with comprehensive parameterizations of its parameters as functions of the solution concentration. This approach is also used to constrain aerosol physico-chemical properties from the hygroscopic data for subsequent evaluation of the *critical radii* r_{cr} and *critical supersaturations* s_{cr} that determine activation of an aerosol particle into a cloud drop (e.g., Fitzgerald, 1975; Hänel, 1976; Fitzgerald et al., 1982; Chen, 1994; Brechtel and Kreidenweis, 2000a,b; Hämeri et al.,

2000, 2001; Snider et al., 2003). Numerical solutions may be more precise and complete, but are more time-consuming. The analytical solutions are desirable since they provide a basis for development of simple and fast parameterizations for cloud and climate models. Besides evaluation of the sizes of individual wet particles, one of the major tasks in descriptions of hygroscopic growth is the evaluation of transformation of the initially dry spectra $f_d(r_d)$ by the dry radius r_d into the wet spectra $f_w(r_w)$ by the wet radius r_w and analytical representation of the size spectra $f_w(r_w)$ of wet aerosol.

6.1.2. Empirical Parameterizations of Droplet Activation

One of the major results of the Köhler theory was the derivation of analytical expressions for the CCN critical radii r_{cr} and supersaturations s_{cr}. These quantities play a fundamental role in aerosol-cloud interactions since they determine the concentration of cloud drops, and have been widely used in cloud and climate studies for parameterizations of cloud drop formation (for overviews of earlier works, see, for example, Mason, 1961; Dufour and Defay, 1963; Sedunov, 1974; Matveev, 1984; Cotton et al., 2011; Rogers and Yau, 1989; Houze, 1993; Young, 1993; Pruppacher and Klett, 1997; Seinfeld and Pandis, 1998; Curry and Webster, 1999).

The two key functions of drop activation are the *differential*, $\varphi_s(s_{cr})$, and *integral or cumulative*, $N_{CCN}(s_{cr})$, *CCN activity spectra*. The differential spectrum $\varphi_s(s_{cr})$ determines the increase in activated CCN per a small increase of supersaturation and is a starting point for evaluation of the cumulative spectrum $N_{CCN}(s_{cr})$ that determines the drop concentration formed at a given s_{cr}. Knowledge of analytical expressions for r_{cr} and s_{cr} allow analytical evaluation of $\varphi_s(s_{cr})$ and $N_{CCN}(s_{cr})$ from the dry aerosol size spectrum $f_d(r_d)$.

The first CCN activity spectrum was suggested by Squires (1958) and Twomey (1959) in the form of a power law by the water supersaturation s_w reached in a cloud parcel:

$$N_{CCN}(s) = Cs_w^k. \tag{6.1.4}$$

This parameterization has been the most commonly used parameterization of the concentration of cloud drops or CCN activity spectrum for more than five decades. Numerous studies have provided a wealth of data on the parameters k and C for various geographical regions (see, e.g., Hegg and Hobbs, 1992, and Table 9.1 in Pruppacher and Klett, 1997). Parameterizations of the type (Eqn. 6.1.4) were derived by Squires (1958) and Twomey (1959) using several assumptions and similar power laws for the differential CCN activity spectrum, $\varphi_s(s_w)$:

$$\varphi_s(s_w) = dN_{CCN}/ds_w = Cks_w^{k-1}. \tag{6.1.5}$$

Equations (6.1.4) and (6.1.5) have been used in many cloud models with empirical values of k and C that are assumed to be constant for a given air mass and are usually constant during model runs.

The empirically derived power law supersaturation activity spectra (Eqns. 6.1.4) and (6.1.5) explained some features of drop nucleation, but did not have a theoretical foundation. To interpret these empirical dependencies, models were developed of partially soluble CCN with the size spectra of Junge-type power law $f(r) = C_N r^{-\mu}$ (total aerosol concentration $N_a \sim r_{\min}^{-(\mu-1)}$) and the index k was expressed as a function of μ. Jiusto and Lala (1981) found a linear relation

$$k = 2(\mu - 1)/3, \tag{6.1.6}$$

which implies $k = 2$ for a typical *Junge* index $\mu = 4$, while the experimental values of k compiled in that work varied over the range 0.2–4. Levin and Sedunov (1966a), Sedunov (1974), Smirnov (1978), Cohard et al. (1998, 2000), and Shipway and Abel (2010) derived more general power laws or algebraic equations for $\varphi_s(s_w)$ and expressed k as a function of μ and CCN soluble fraction. Khvorostyanov and Curry (1999a,b) derived power laws for the size spectra of the wet and interstitial aerosol, for the activity spectra $\varphi_s(s_w)$, $\varphi_{CCN}(s)$ and for the Ångström (1929, 1964) wavelength λ index k_A of the aerosol extinction coefficient σ_{ext} and optical thickness τ_a

$$\sigma_{ext} \sim \lambda^{-k_A}, \qquad \tau_a \sim \lambda^{-k_A} \tag{6.1.6a}$$

and found linear relations among these indices expressed in terms of the index μ and the aerosol soluble fraction.

A deficiency of (Eqn. 6.1.4) is that it overestimates the droplet concentration at large s_w and predicts unlimited values of droplet concentration that exceed total aerosol concentration. This occurs because of the functional form of the power law and use of a single value of k for the entire s_w range. Many field and laboratory measurements have shown that a more realistic $N_{CCN}(s_w)$ spectrum in log-log coordinates is not linear as it would be with k = const, but has a concave curvature—i.e., the index k decreases with increasing s_w (e.g., Jiusto and Lala, 1981; Hudson, 1984; Yum and Hudson, 2001). This deficiency in (Eqn. 6.1.4) has been corrected in various ways.

Braham (1976) introduced the *C-k* space with wide ranges for both parameters and constructed nomograms in this space. Cohard et al. (1998, 2000) and Shipway and Abel (2010) introduced an empirical correction to Twomey's spectrum $\varphi_s(s_w)$ that rapidly decreases at high s_w:

$$\varphi_s(s_w) = kCs_w^{k-1}(1+\eta s_w^2)^{-\lambda}, \tag{6.1.7}$$

where the first term is Twomey's differential power law (Eqn. 6.1.5), and the term in parentheses was added by the authors as an empirical correction with the two additional empirical fitting parameters η, λ in order to cut $\varphi_s(s_w)$ at high s_w. Ji and Shaw (1998) suggested another analogous integral empirical spectrum $N_{CCN}(s_w)$ with two empirical parameters B_{JS} and k:

$$N_{CCN}(s) = N_a[1-\exp(-B_{JS}s_w^k)], \tag{6.1.8a}$$

which tends to a correct finite limit N_a at high s_w and corresponds to the differential activity spectrum

$$\varphi_s(s_w) = N_a B_{JS} k s_w^{k-1} \exp(-B_{JS}s_w^k) \tag{6.1.8b}$$

Khvorostyanov and Curry (1999a) considered the change in the indices μ, k in different CCN size ranges. Several authors proposed using a lognormal CCN size spectrum instead of the power law, which yields lognormal concave activity spectra and finite N_{dr} at high s_w (von der Emde and Wacker, 1993; Ghan et al., 1993, 1995, 1997; Feingold et al., 1994; Abdul-Razzak et al., 1998; Abdul-Razzak and Ghan, 2000, 2004; Nenes and Seinfeld, 2003; Rissman et al., 2004; Fountoukis and Nenes, 2005). Based on these studies, the prognostic equations for the drop concentration were incorporated into climate models (Ghan et al., 1997; Lohmann et al., 1999).

Analytical expressions for r_{cr} and s_{cr} have usually been determined from the *Köhler* equation with the following approximations: 1) high dilution in a haze drop (i.e., neglecting the insoluble CCN fraction in the denominator of Raoult's term); 2) soluble fraction proportional to the drop volume; and 3) small supersaturations. Concerns about each of these three assumptions are described next.

1. Many recent field experiments have found aerosols with very small aerosol soluble fractions or aerosol particles that are nearly hydrophobic, often constituting a significant fraction of the total aerosol load. These field experiments have been conducted over many different regions: in desert areas (Levin et al., 1996; Rosenfeld et al., 2001), the Arctic (Curry et al., 2000; Pinto et al., 2001; Leck et al., 2001; Bigg and Leck, 2001a,b), and the Aerosol Characterization Experiments 1 and 2, ACE-1 and ACE-2, (Swietlicki et al., 2000; Snider et al., 2003). Earlier detailed numerical calculations show that even at the time of activation, the degree of dilution (ratio of gained water mass to the dry mass) for the aerosol particles with the small insoluble fraction ~0.01 can be as low as 0.8 and ~2 for the dry radii $r_d = 0.02$ and 0.1 μm—i.e., near the modal radii for many aerosols (e.g., Hänel, 1976; Pruppacher and Klett, 1997, Table 6.3). Therefore, the high dilution approximation may not be satisfied, and therefore analytical solutions to the Köhler equation without high dilution approximation are desirable.
2. The model of internally mixed aerosol particles with soluble fraction proportional to the particle volume has been challenged in several papers. A surface-proportional soluble shell was found in CCN measurements in several regions of Eastern Europe (e.g., Laktionov, 1972; Sedunov 1974). Levin et al. (1996), Falkovich et al. (2001), and Rosenfeld et al. (2001) found that mineral dust particles coated with thin sulfate shells are typical in the eastern Mediterranean. It was hypothesized that several mechanisms may be responsible for such soluble shells—e.g., coagulation of the mineral dust with sulfate particles, deposition of sulfate on desert particles with oxidation of SO_2 or SO_4 on the particle surface, and nucleation of cloud drops on sulfate CCN with subsequent accumulation of mineral dust and evaporation of drops leaving sulfate-coated dust. Simulations of heterogeneous chemical reactions and soluble shells on the dust surface showed their significant impact on the aerosol composition (Wurzler et al., 2000) and on radiative forcing in the GISS GCM (Bauer and Koch, 2005).
3. In some aerosol chambers, very high supersaturations are reached, $s_w \sim 15$–25%, violating the assumption of small s_w. Since the lower size boundary of activated CCN is inversely proportional to s_w (Sedunov 1974; Ghan et al., 1993; Khvorostyanov and Curry, 1999a), this may lead to activation of very small aerosol particles with radii $r_d << 0.01$ μm, whose r_{cr} and s_{cr} can be different than those in the accumulation mode. Thus, the approximation of a small s_w should be eliminated for accurate interpretation of aerosol/cloud chamber experiments.

The Köhler theory and its generalizations without these limitations are described in this chapter. The older power law and newer lognormal parameterizations of aerosol size spectra and supersaturation activity spectra are widely used in studies of aerosol optical and radiative properties, cloud physics, and climate research. These two kinds of parameterizations existed in parallel, and implicitly compete, but their relation was unclear. A bridge between these two types of the size spectra and activity spectra was established by Khvorostyanov and Curry (2006, 2007), where a new class of size spectra and activity spectra was found: the algebraic distributions. These new distributions allowed: a) to show the equivalence and applicability of the power laws and lognormal spectra; and b) to develop a generalization of the power law (Eqn. 6.1.4), and to find the k-index and C-coefficient as continuous analytical functions of supersaturation and of the aerosol microphysics and chemistry.

The approach developed by Khvorostyanov and Curry (2006, 2007) appeared to be convenient for use in both bin and bulk cloud models (e.g., Morrison and Grabowski, 2007, 2008; Curry and Khvorostyanov, 2012). In Sections 6.2–6.7, the Köhler theory and its modification are described. A model of mixed (partially soluble) dry CCN is developed with parameterization of the soluble fraction as a function of the dry nucleus radius; analytical solutions are found for the equilibrium wet radius in the hygroscopic growth and of unactivated particles in the cloud (Section 6.2). The Köhler theory for the critical radius and supersaturation is described and generalized, with account taken for the insoluble fraction and various geometries of soluble fractions (Section 6.3). The lognormal size spectrum of the dry aerosol is described and its two equivalents are derived as a power law Junge-type spectrum and a new algebraic spectrum (Section 6.4). The analytical transformations of the aerosol size spectra at varying humidity are derived for the general shape of the dry spectrum, lognormal, and algebraic spectrum (Section 6.5). The differential and integral CCN activity spectra are derived in the lognormal, equivalent algebraic, and modified power law forms (Section 6.6 and Section 6.7). Some recent modifications of the Köhler theory are described in the next section.

6.2. Equilibrium Radii

We considered in Chapter 3 the Kelvin effect of vapor pressure increase over a droplet due to its curvature, and the pressure decrease due to Raoult's solution effect. Köhler (1921, 1936) wrote an equation for water saturation ratio S_w, which accounted for the curvature and solution effects, and considered their joint effect on a solution drop. This equation was called the *Köhler equation*, which determines the critical conditions when deliquesced aerosol particles (CCN) in their hygroscopic growth reach critical radii and supersaturations and thus "activate" into cloud droplets.

Subsequent to Köhler's pioneering work, there have been numerous derivations of the Köhler equation from basic thermodynamic principles with a variety of refinements (e.g., Dufour and Defay, 1963; Defay et al., 1966; Low, 1969; Sedunov, 1974; Young and Warren, 1992; Pruppacher and Klett, 1997; Chylek and Wong, 1998; Curry and Webster, 1999). Various modifications to the Köhler theory accounted for the solubility limitation of the soluble CCN fraction that allowed description of deliquescence and humidity hysteresis (Chen, 1994); absorption of the soluble gases by the haze drops that led to additional terms and yielded multimodal Köhler-type curves for CCN activation (Kulmala et al., 1993; Schulman et al., 1996; Laaksonen et al., 1998); effects of the insoluble fraction of the equilibrium CCN radii and critical saturations (Khvorostyanov and Curry, 1999a,b; 2006, 2007, 2008a, 2009a); and adsorption of the water film on the surface of insoluble aerosol particles (e.g., Henson, 2007; Sorjamaa and Laaksonen, 2007). Prior to 2000, a detailed analysis of several versions of the Köhler equation and of approximations in evaluation of its several basic parameters, solvent and solute volume additivity, surface tension, osmotic potential, and other facets are given in Brechtel and Kreidenweis (2000a), Charlson et al. (2001), and Kreidenweis et al. (2005).

In this section, we consider the version of the Köhler equation that takes into account the insoluble fraction, which describes the equilibrium water vapor pressure over a solution drop that consists of soluble and insoluble components and is in equilibrium with ambient humid air. Then, following mostly the approach developed in Khvorostyanov and Curry (2006, 2007) with convenient parameterization of the soluble fraction, we find the analytical solutions for the equilibrium radii of deliquescent aerosol at subsaturation, and for the equilibrium and critical radii and supersaturations of interstitial aerosol at water saturation in a cloud. These solutions are used for derivation of the equilibrium size spectra of the dry and wet aerosol and of the CCN activity spectra. The modifications of the Köhler equation and their solutions, as developed in the works of authors cited earlier, are described in subsequent sections.

6.2.1. Equilibrium Radii at Subsaturation

Below a certain threshold saturation ratio $S_{w,th}$, the aerosol particles are solid. They can be covered by a thin adsorbed water film but they do not deliquesce. The thresholds of deliquescence for the most common substances met in the atmosphere are, for example, $S_w \sim 0.75$ for sodium chloride, NaCl, and ~0.8 for ammonium sulfate, $(NH_4)_2SO_4$, at room temperatures of 15 to 25 °C, which slightly grow with decreasing temperature. The process of deliquescence and a reverse process of efflorescence are considered in Chapter 11. As the water saturation ratio S_w (equal to fractional relative humidity, $S_w = H$) exceeds the threshold of deliquescence $S_{w,th}$ of the soluble fraction, the hygroscopic growth of the particles begins, and an initially dry CCN with radius r_d converts into a wet aerosol particle with radius r_w called a "haze particle."

The equilibrium radius of the wet aerosol $r_w(S_w)$ as a function of the ambient saturation ratio S_w (or of the relative humidity H) and of the dry radius r_d can be obtained using the Köhler equation for S_w or supersaturation $s_w = (\rho_v - \rho_{ws})/\rho_{ws} = S_w - 1$ (see Section 3.9),

$$\ln S_w = \frac{2M_w \sigma_{sa}}{RT\rho_w r} - \frac{\nu \Phi_s M_w}{M_s} \frac{m_s}{m_w}, \qquad m_s = \varepsilon_m m_d = \varepsilon_v \rho_s V_d, \tag{6.2.1}$$

where m_s is the mass of the soluble fraction, ε_m and ε_v are the mass and volume fractions of the soluble fraction, ρ_s is the density of the soluble fraction, and all the notations here are the same as in Section 3.9. The first term here describes the Kelvin curvature effect, and the second term describes Raoult's solution effect. The volume fraction ε_v is related to the mass fraction ε_m as $\varepsilon_v = \varepsilon_m(\rho_d/\rho_s)$, where ρ_d is the effective density of a dry aerosol particle, weighted by the densities of its soluble ρ_s and insoluble ρ_u fractions, $\rho_d = \varepsilon_v\rho_s + (1 - \varepsilon_v)\rho_u$.

Using the simplifying assumptions on the constancy of the water molar volume in solution and the volume additivity of solvent (water) and solute (salt), $V = V_w + V_d$, which are good approximations for many common ionic solutes (Dufour and Defay, 1963; Pruppacher and Klett, 1997; Brechtel and Kreidenweis, 2000a; Kreidenweis et al., 2005), the ratio m_d/m_w can be expressed as $m_d/m_w = (\rho_d/\rho_w)[(r_w/r_d)^3 - 1]^{-1}$. Substitution of these relations into (Eqn. 6.2.1) yields:

$$\ln S_w = \frac{A_k}{r_w} - \frac{B}{r_w^3 - r_d^3}. \tag{6.2.2a}$$

$$A_k = \frac{2M_w \sigma_{sa}}{RT\rho_w}, \qquad B = \frac{3\nu\Phi_s \varepsilon_m m_d M_w}{4\pi M_s \rho_w} = \frac{3\nu\Phi_s m_s M_w}{4\pi M_s \rho_w}. \tag{6.2.2b}$$

Here A_k is the Kelvin curvature parameter, and the parameter B, called the activity of a nucleus, describes effects of the soluble fraction. Note that here the dimension of parameter B is [cm^3] and this definition of B is different from the dimensionless B in Pruppacher and Klett (1997, p. 178).

It is written here in a form that allows generalization for various soluble fractions, proportional to the volume or surface of CCN as clarified in the following text.

The solute effect in the second term in (Eqn. 6.2.1) is often parameterized using the hygroscopicity parameter k_h. It is introduced with an assumption that the soluble fraction is proportional to the volume of an aerosol particle, thus the mass of the soluble fraction $m_s = \varepsilon_m m_d = \varepsilon_v \rho_s V_d$. Then, the second term in (Eqn. 6.2.1), equal to $\ln(a_w)$, can be written as

$$\ln a_w = -\frac{\nu\Phi_s M_w \varepsilon_m}{M_s}\frac{m_d}{m_w} = -\frac{\nu\Phi_s M_w \varepsilon_m \rho_d}{M_s \rho_w}\frac{V_d}{V_w}, \quad (6.2.3a)$$

where V_d and V_w are the volumes of the dry particulate mass and water in a solution droplet. The hygroscopicity parameter k_h is introduced by a relation (e.g., Petters and Kreidenweis, 2007)

$$a_w^{-1} = 1 + \kappa_h \frac{V_d}{V_w}. \quad (6.2.3b)$$

Combining (Eqns. 6.2.3a) and (6.2.3b) yields

$$a_w^{-1} = \exp\left(\frac{\nu\Phi_s \varepsilon_m M_w \rho_d}{M_s \rho_w}\frac{V_d}{V_w}\right) \approx 1 + \frac{\nu\Phi_s \varepsilon_m M_w \rho_d}{M_s \rho_w}\frac{V_d}{V_w} = 1 + \kappa_h \frac{V_d}{V_w}. \quad (6.2.3c)$$

This gives the expression for k_h:

$$\kappa_h = \frac{\nu\Phi_s \varepsilon_m M_w \rho_d}{M_s \rho_w} = \frac{\nu\Phi_s \varepsilon_v M_w \rho_s}{M_s \rho_w}. \quad (6.2.3d)$$

A generalization for several substances is done by summation over n soluble components in an aerosol particle (e.g., Ghan et al., 2011):

$$\kappa_h = \frac{\sum_i q_i \kappa_{hi} / \rho_{di}}{q_i / \rho_{di}}, \quad (6.2.3e)$$

where k_{hi} is the hygroscopicity parameter for the i-th soluble species determined by (Eqn. 6.2.3d), q_i and ρ_{di} are the mass fractions, and the density of the component i.

The deficiencies of the approach based on the hygroscopicity parameter are: expansion of a_w in (Eqn. 6.2.3c) is valid for sufficiently diluted solutions, and the explicit assumption of proportionality of soluble fraction to the particle volume. Both assumptions have limitations. 1) The expansion (Eqn. 6.2.3c) for a_w becomes invalid for sufficiently concentrated solutions in the drop and introduces noticeable errors, which is especially pronounced for smaller soluble fractions. 2) The assumption of the soluble fraction proportional to the volume is invalid for the particles whose soluble fraction represents a shell on the surface of an insoluble core.

We employ a more general and convenient parameterization of the soluble fraction and nucleus activity (Levin and Sedunov, 1966a; Sedunov, 1974; Smirnov, 1978; Khvorostyanov and Curry, 1999a,b, 2006, 2007, 2008a, 2009a) that easily allows incorporation of alternative assumptions regarding the soluble fraction of the aerosol particle:

$$B = b r_d^{2(1+\beta)}, \quad (6.2.4)$$

where the parameters b and β depend on the chemical composition and physical properties of the soluble part of an aerosol particle. According to (Eqn. 6.2.2b), $B \sim m_s$, the mass of the soluble fraction $m_s \sim r_d^{2(1+\beta)}$ and the parameter β describes the soluble fraction in a particle. Generally, the solubility decreases with increasing particle size (e.g., Laktionov, 1972; Sedunov, 1974; Pruppacher and Klett, 1997), so that β decreases with increasing r_d and may vary from 0.5, when the soluble fraction is proportional to the volume, to 0, when the soluble fraction is proportional to the surface, and to −1, when the soluble fraction is independent of the radius. This variation describes the change of the dominant mechanisms of accumulation of the soluble fraction with growing particle size. A detailed analysis of these mechanisms is given, e.g., in Junge (1963), Pruppacher and Klett (1997), and Seinfeld and Pandis (1998), and an interested reader can find the details in these books. We consider in more detail two particular and most important cases.

1. $\beta = 0.5$. The value $\beta = 0.5$ corresponds to the case $m_s \sim B = br_d^3$. When the soluble fraction is distributed within the particle volume, the soluble mass is proportional to the volume, and this is usually implicitly assumed in most parameterizations of CCN deliquescence and drop activation (e.g., von der Emde and Wacker, 1993; Ghan et al., 1993, 1995; Abdul-Razzak et al., 1998; Abdul-Razzak and Ghan, 2000; Ghan et al., 2011). The parameter b can be found by equating the second equation of (Eqns. 6.2.2b) and (6.2.4) (Khvorostyanov and Curry, 1999a, 2006, 2007).

$$B = \frac{3v\Phi_s m_s M_w}{4\pi M_s \rho_w} = br_d^{2(1+\beta)}, \tag{6.2.5}$$

and writing the mass of the soluble fraction as $m_s = (4/3)\pi\varepsilon_v\rho_s r_d^3$, where ρ_s is the density of the soluble fraction, and ε_v is its volume fraction. This gives with $\beta = 0.5$ that b is a dimensionless parameter:

$$b = (v\Phi_s)\varepsilon_v \frac{\rho_s}{\rho_w}\frac{M_w}{M_s} = (v\Phi_s)\varepsilon_m \frac{\rho_d}{\rho_w}\frac{M_w}{M_s}, \tag{6.2.6}$$

where ε_v and ε_m are the volume and mass fractions of a CCN. For $\beta = 0.5$, ε_v and ε_m do not depend on the dry radius r_d. The value of Φ_s, in general, also depends on r_w, however this effect on r_{cr} and s_{cr} is weaker than those of the other factors (Brechtel and Kreidenweis, 2000a). When evaluating b, Φ_s can be assigned some appropriate mean constant value, e.g., $v\Phi_s \approx 2.1$ for ammonium sulfate, rather than 3, which yields a good approximation (Snider et al., 2003), or the available parameterizations for Φ_s (e.g., Brechtel and Kreidenweis, 2000a) can be substituted into the final equations for r_w, r_{cr}, and s_{cr}. For example, using the typical parameters in (Eqn. 6.2.6) for the case $\beta = 0.5$ yields for ammonium sulfate: $b \approx 0.5$ for fully soluble nuclei ($\varepsilon_v = 1$), and $b \approx 0.25$ with $\varepsilon_v = 0.5$. For NaCl, the values are: $b \approx 1.33$ with $\varepsilon_v = 1$, and $b \approx 0.67$ with $\varepsilon_v = 0.5$. This case with $m_s \sim r_d^3$ is analogous to that usually considered in the literature, and b is analogous to the hygroscopicity parameter k_h.

2. $\beta = 0$. Then $B = br_d^2$—i.e., mass of soluble fraction is proportional to the surface area where it is accumulated as a soluble film or shell at the surface of an aerosol particle. The particle volume fraction ε_v and b can be parameterized as

$$\varepsilon_v = \varepsilon_{v0}\frac{r_{d,sc}}{r_d}, \qquad b = r_{d,sc}\varepsilon_{v0}(v\Phi_s)\frac{\rho_s}{\rho_w}\frac{M_w}{M_s}, \tag{6.2.7}$$

where $r_{d,sc}$ is some scaling radius and ε_{v0} is the reference soluble fraction (dimensionless). For this case, b has the dimension of length and is proportional to the scaling radius $r_{d,sc}$ that characterizes the thickness l_0 of the soluble film (but is not equal to it). Such a model is based on the experimental data by Laktionov (1972), Sedunov (1974), Levin et al. (1996), Falkovich et al. (2001), Rosenfeld et al. (2001) and some theoretical models (Wurzler et al., 2000; Bauer and Koch, 2005) discussed earlier. A detailed chemical analysis in Levin et al. (1996) showed that the surface density $P_s = m_s/S_p$ ($S_p = 4\pi r_d^2$ is the particle surface area) of the sulfates was fairly constant with particle size, $P_s \sim (2-6)\times 10^{-6}$ g cm^{-2}, in the range $r_d = 0.15$ to 10 μm. This indicates that $m_s \sim S_p \sim r_d^2$ and supports parameterization of soluble fraction mass m_s proportional to the surface area with $\beta = 0$. Assuming a thin shell with the thickness $l_0 << r_d$ and volume $V_s = 4\pi r_d^2 l_0$, we can estimate l_0 from the relation $P_s = m_s/S_p \approx 4\pi\rho_s r_d^2 l_0/4\pi r_d^2 = l_0\rho_s$, or $l_0 = P_s/\rho_s$. With $\rho_s \sim 2$ g cm^{-2}. This yields an estimate $l_0 \sim 0.01$–0.03 μm, where $l_0 << r_d$ in this radii range.

For this model of aerosol particles with a thin soluble shell, the soluble mass fraction is inversely proportional to the dry radius

$$\varepsilon_m = \frac{m_s}{m_d} \approx \frac{4\pi r_d^2 l_0 \rho_s}{(4/3)\pi r_d^3 \rho_d} = \frac{3l_0}{r_d}\frac{\rho_s}{\rho_d}, \tag{6.2.8}$$

Substituting this relation into (Eqn. 6.2.7), we obtain the parameter b that has the dimension of length

$$b = 3l_0(\nu\Phi_s)\frac{\rho_d}{\rho_w}\frac{M_w}{M_s}. \tag{6.2.9}$$

This parameterization requires knowledge of the soluble film thickness l_0 that can be obtained from experimental data (e.g., Levin et al., 1996; Falkovich et al., 2001), or simulations (e.g., Wurzler et al., 2000; Bauer and Koch, 2005). If the soluble and insoluble masses are known, as in many current climate and pollution transport models, then the hygroscopicity can be expressed using parameterization (Eqns. 6.2.4), (6.2.6), and (6.2.9) assuming some insoluble core geometry and proportionality of the soluble mass to the volume ($\beta = 0.5$) or to the surface, ($\beta = 0$). Then, b can be determined and the equations given next applied.

A generalization of the parameter b for several soluble substances comprising a particle is easily done similar to this procedure for the hygroscopicity parameter in (Eqn. 6.2.3e). In the following text, for simplicity, we consider one soluble fraction.

Using (Eqn. 6.2.4), Köhler's equation (Eqn. 6.2.2a) can be rewritten as

$$\ln S_w = \ln(1+s_w) = \frac{A_k}{r_w} - \frac{b r_d^{2(1+\beta)}}{r_w^3 - r_d^3}. \tag{6.2.10}$$

For dilute mixed haze particles, when $\ln S_w = \ln(1+s_w) \approx s_w = S_w - 1$ and $r_w^3 >> r_d^3$, this equation is reduced to the commonly used dilute approximation

$$s_w = S_w - 1 = \frac{A_k}{r_w} - \frac{b r_d^{2(1+\beta)}}{r_w^3}, \tag{6.2.11}$$

where the second term is generalized for various soluble fractions. Various solutions for the humidity dependence of the wet particle radius $r_w(S_w)$ at subsaturation have been obtained in algebraic and trigonometric forms using Kohler's equation (Eqn. 6.2.11) for dilute solutions or fully soluble particles (e.g., Levin and Sedunov, 1966a; Sedunov, 1974; Fitzgerald et al., 1975, 1982; Hänel, 1976; Smirnov, 1978; Khvorostyanov and Curry, 1999a, 2006, 2007; Swietlicki et al., 1999; Cohard et al., 2000; Kreidenweis et al., 2005; Shipway and Abel, 2010). Here, we generalize the previous expressions and following Khvorostyanov and Curry (2007), find analytical expressions for $r_w(S_w)$ without assuming a dilute solution and accounting for the insoluble fraction in the Raoult term—i.e., from (Eqn. 6.2.10).

Due to the small value of A_k, the solution for $r_w(S_w)$ can be obtained by finding a positive real root of the cubic (Eqn. 6.2.10) with the simplifying assumption $A_k = 0$, and then a correction due to A_k can be obtained by expansion into the power series:

$$r_w(S_w) = r_d\left\{1+\frac{br_d^{2(1+\beta)-3}}{(-\ln S_w)}\left[1+C_w\left(-\ln S_w\right)^{-2/3}\right]^{-3}\right\}^{1/3}, \tag{6.2.12}$$

$$C_w = \frac{A_k}{3b^{1/3} r_d^{2(1+\beta)/3}}. \tag{6.2.13}$$

This expression can be simplified for a dilute wet particle, when $r_w >> r_d$ and (Eqn. 6.2.11) is applicable, allowing us to neglect term 1 in the figure bracket of (Eqn. 6.2.12), and with $-\ln S_w \approx 1 - S_w$,

$$r_{w1}(S_w) = \frac{b^{1/3} r_d^{2(1+\beta)/3}}{\left(1-S_w\right)^{1/3}}\left[1+C_w\left(1-S_w\right)^{-2/3}\right]^{-1}. \tag{6.2.14}$$

This solution is valid under the condition of sufficiently large subsaturation $s_w < s_{lim}$, when the term with A_k is smaller than 1. An estimate from (Eqn. 6.2.14) with $b = 0.5$, $\beta = 0.5$, and $A_k = 10^{-7}$ cm yields $s_{lim} = -0.8 \times 10^{-3}$ (−0.08%) for $r_d = 0.1$ μm and $s_{lim} = -2.4 \times 10^{-2}$ (−2.4%) for $r_d = 0.01$ μm. This is the upper limit for application of (Eqns. 6.2.12)–(6.2.14), hence these equations can be used up to $S_w \le 0.95$–0.97 ($H \le 95$–97%) for the aerosol spectrum with $r_d \ge 0.01$ μm.

The expression (Eqn. 6.2.14) was derived in Khvorostyanov and Curry (1999a) assuming a dilute solution. It predicts humidity dependence for the growth factor $GF(S_w) \sim (1 - S_w)^{-1/3}$ for S_w lower than s_{lim}. The more general expression (Eqn. 6.2.12) predicts the same humidity dependence in the intermediate region of S_w, but weaker dependencies at lower values of $S_w < 0.7$–0.8 (in supersaturated solutions below deliquescence points in humidity hysteresis) and at higher values of $S_w > 0.9$. In contrast to expressions with empirical parameters, (Eqns. 6.2.12) and (6.2.14) express $GF(S_w)$ directly in terms of the primary variables of the Köhler equation. The additional corrections due to dependencies on the solution concentration of the surface tension σ_{sa} (in A_k) and of the osmotic potential Φ_s (in b) can be introduced in (Eqns. 6.2.12)–(6.2.14) using the known parameterizations for these quantities (e.g., Chen, 1994; Tang and Munkelwitz, 1994; Brechtel and Kreidenweis, 2000a; Hämeri et al., 2000, 2001). Thus, these analytical solutions can be used directly for evaluation of the equilibrium aerosol radii or for constraints of the empirical parameterizations.

For particles with $r_d > 0.01$ μm and saturation ratios lower than $S_w \sim 0.95$–0.97, the last term in the brackets in (Eqn. 6.2.12) can be neglected and the wet radius can be approximated as

$$r_{w2}(S_w) = r_d[1+br_d^{2(1+\beta)-3}(-\ln S_w)^{-1}]^{1/3}. \tag{6.2.15}$$

For dilute particles, we can neglect the first term in the bracket, thus

$$r_{w3} = b^{1/3} r_d^{2(1+\beta)/3} \left(1 - S_w\right)^{-1/3}. \tag{6.2.16}$$

For $\beta = 0.5$, this equation reduces to $r_w = r_d \cdot b^{1/3}(1 - S_w)^{-1/3}$. This equation is similar to the empirical parameterization formulated by Kasten (1969) and used for calculation of the aerosol extinction coefficient $\sigma_{ext}(S_w)$ and visibility, $r_w(S_w) = r_w(0.1)[0.9/(1 - S_w)]^{1/\mu}$, with $1/\mu = 0.25$–0.3 for an average aerosol, and is similar to Fitzgerald's (1975) $r_w = \alpha r_d^{\gamma}$, found as a parameterization of the numerical calculations with Köhler's theory. Thus, (Eqn. 6.2.16) provides a theoretical basis for the empirical dependencies for r_w, σ_{ext}, and visibility. The empirical coefficients before the factor $(1 - S_w)^{-1/3}$ found in these works by fitting the experimental data are expressed now via the primary aerosol parameters, and (Eqns. 6.2.12)–(6.2.15) generalize these expressions and take into account the insoluble fraction in the wet aerosol and the different geometry of the soluble fraction-volume and surface proportional.

A comparison of (Eqns. 6.2.12)–(6.2.15) with the empirical parameterizations and calculations of $r_w(S_w)$ and $\sigma_{ext}(S_w)$ show that these equations still can be used as a reasonable approximation down to the lower humidities (30–40%) as can be reached by the deliquescent aerosol in the humidity hysteresis before spontaneous salt crystallization, efflorescence (see Chapter 11).

Fig. 6.1a,b shows the humidity dependence of the wet radii $r_w(S_w)$ and the growth factor $r_w(S_w)/r_d$ calculated with the complete Equation (6.2.12) for $\beta = 0.5$ and r_{w1}, r_{w2} calculated with approximate (Eqns. 6.2.14) and (6.2.15). A comparison of the calculated growth factor with precise numerical calculations from Kreidenweis et al. (2005) for pure ammonium sulfate particles with $\varepsilon_v = 1$ (Fig. 6.1b, solid and open circles) shows very good agreement and indicates a sufficiently high accuracy of (Eqn. 6.2.12). If the soluble fraction ε_v of ammonium sulfate is 0.3–1, r_{w1}, r_{w2}, and r_w/r_d increase by a factor of 2 to 3 in the region $0.3 < S_w < 1$, and increase especially rapidly at $S_w > 0.9$, in agreement with numerous experimental data and previous parameterizations cited earlier. The aerosol extinction coefficient $\sigma_{ext}(S_w) \sim r_w^2(S_w)$ and optical thickness $\tau \sim \sigma_{ext}(S_w) \sim r_w^2(S_w)$. This implies that an increase in r_w of 2 to 3 times causes an increase of aerosol optical thickness by a factor of 4 to 9 and illustrates the strong humidity effect on aerosol radiative properties and the direct aerosol effect on climate. The growth is much smaller for $\varepsilon_v = 0.1$, and is negligible for $\varepsilon_v = 0.01$. The error $\delta_1 = (r_w - r_{w1})/r_w \times 100$ in r_{w1} using the approximate (Eqn. 6.2.14) for dilute solutions increases with decreasing soluble fraction and reaches 60–80% for $\varepsilon_v = 0.1$ to 0.01 (Fig. 6.1c). The error of (Eqn. 6.2.15), $\delta_2 = (r_w - r_{w2})/r_w \times 100$, does not exceed 2–6% when $S_w < 0.97$ (Fig. 6.1d). Thus (Eqn. 6.2.15), without the assumption of a dilute solution, appears to be a much better approximation than (Eqn. 6.2.14), especially for evaluation of *GF* for less hygroscopic or nearly hydrophobic aerosol particles.

This shows that these analytical solutions for the equilibrium radii $r_w(S_w)$ can be used with good accuracy for calculations of visibility, aerosol extinction, and optical thickness, and for evaluation of the direct aerosol effects on climate with account for aerosol humidity transformation in the wide regions of humidities and aerosol soluble fractions if an appropriate model of aerosol is chosen with the parameters b and β.

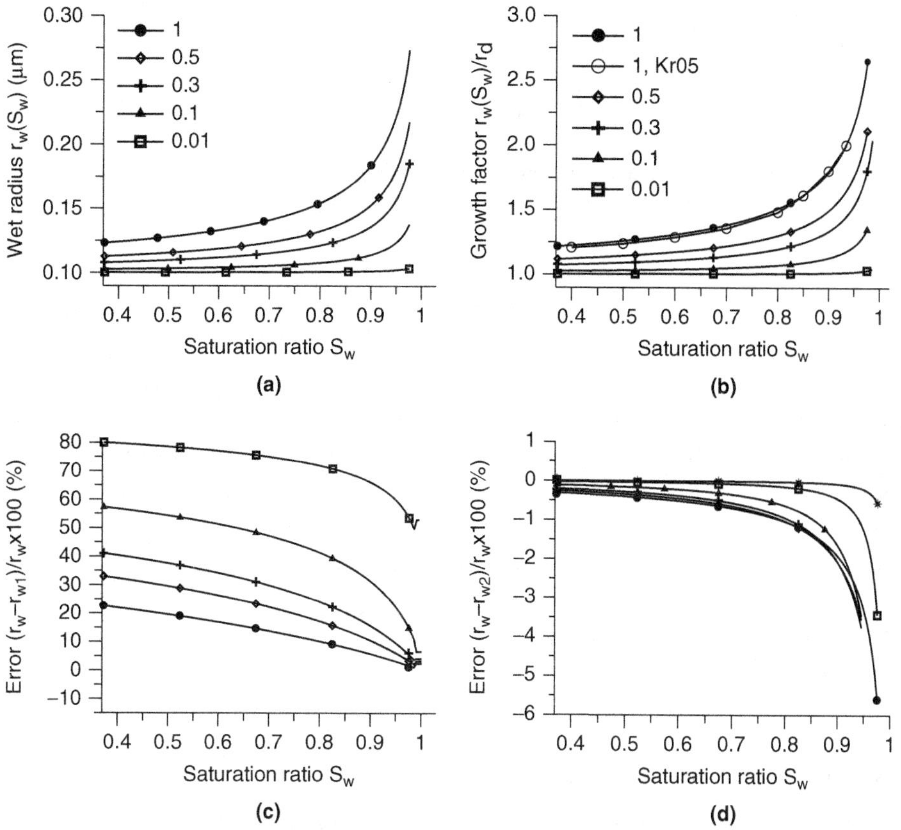

Figure 6.1. (a) Humidity dependence of the wet radii $r_w(S_w)$ calculated with (Eqn. 6.2.12), solid circles; (b) growth factor r_w/r_d calculated with (Eqn. 6.2.12), solid circles, in comparison with numerical calculations from Kreidenweis et al. (2005), open circle; (c) the errors $\delta_1 = (r_w - r_{w1})/r_w \times 100$ with r_{w1} calculated from (Eqn. 6.2.14) and (d) $\delta_2 = (r_w - r_{w2})/r_w \times 100$ with r_{w2} calculated from (Eqn. 6.2.15). The values r_w, r_{w1}, and r_{w2} are calculated with $r_d = 0.1$ μm, $\beta = 0.5$, for various volume soluble fractions ε_v of ammonium sulfate indicated in the legend and b evaluated with (Eqn. 6.2.6). From Khvorostyanov and Curry (2007), *J. Geophys. Res.*, 112(D5), D05206, reproduced with permission of John Wiley & Sons, Inc.

6.2.2. Equilibrium Radii of Interstitial Aerosol in a Cloud

Equations (6.2.12)–(6.2.16) describe humidity transformations of the majority of wet CCN except for the smallest particles at very high humidities close to water saturation—i.e., at values of S_w approaching 1. At higher humidity within a cloud, $S_w \to 1$ ($s_w \to 0$), an alternative approximation should be used. Consider first the case of nearly soluble aerosol, when the insoluble fraction can be neglected.

At high humidity, water supersaturation s_w tends to zero, and the left-hand side of (Eqn. 6.2.11) is zero. Thus, (Eqn. 6.2.11) becomes

$$s_w = \frac{A_k}{r_w} - \frac{B}{r_w^3} = 0, \tag{6.2.17}$$

and the radius r_{wi} of a wet interstitial aerosol particle (subscripts w and i) can be found from the quadratic equation

$$r_{wi0}(S_w \cong 1) = \left(\frac{B}{A_k}\right)^{1/2} = r_d^{1+\beta}\left(\frac{b}{A_k}\right)^{1/2}, \tag{6.2.18}$$

where the subscript "0" means zero insoluble fraction. The second equality here is written with use of (Eqn. 6.2.5) for B and generalizes the known expression, $r_{wi0} \sim r_d^{3/2}$ with $\beta = 0.5$, for the other values of β; in particular, $r_{wi0} \sim r_d$ with $\beta = 0$ (surface-proportional soluble fraction). A generalization of this solution for r_{wi}, taking into account the insoluble fraction, can be done with the use of Köhler's equation (Eqn. 6.2.10) at $S_w = 1$,

$$\ln S_w = \frac{A_k}{r_w} - \frac{b r_d^{2(1+\beta)}}{r_w^3 - r_d^3} = 0, \qquad \text{or} \qquad \frac{r_w}{A_k} = \frac{r_w^3 - r_d^3}{b r_d^{2(1+\beta)}}. \tag{6.2.19}$$

This gives a cubic equation for $x = r_w/r_d$,

$$x^3 - 3Z_d^2 x - 1 = 0, \qquad \text{with} \qquad Z_d = \left(\frac{b r_d^{2\beta}}{3A_k}\right)^{1/2} = r_d^{\beta}\left(\frac{b}{3A_k}\right)^{1/2}. \tag{6.2.20}$$

This is an incomplete cubic equation for x of the form

$$y^3 + py + q = 0, \qquad p = -3Z_d^2, \qquad q = -1. \tag{6.2.21}$$

Its solution is described in Appendix A.6 and can be chosen in trigonometric or algebraic form, depending on the sign of Q,

$$Q = \left(\frac{p}{3}\right)^3 + \left(\frac{q}{2}\right)^2 = -Z_d^6 + \frac{1}{4}. \tag{6.2.22}$$

A trigonometric solution is convenient if $Q < 0$, or $Z_d^6 > 1/4$—i.e., with a sufficiently large soluble fraction. Then, there is one physical real root; the other two roots are complex conjugated and should be rejected. The real root is

$$y_1 = 2\left(-\frac{p}{3}\right)^{1/2} \cos\frac{\alpha}{3}, \qquad \text{with} \qquad \cos\alpha = -\frac{q}{2}\left(-\frac{p}{3}\right)^{-3/2}. \tag{6.2.23}$$

Substituting here p and q from (Eqns. 6.2.21) and (6.2.20), we obtain $x = y_1$:

$$x = y_1 = 2Z_d \cos\left[\frac{1}{3}\arccos\left(\frac{1}{2Z_d^3}\right)\right]$$

$$= 2r_d^{\beta}\left(\frac{b}{3A_k}\right)^{1/2} \cos\left\{\frac{1}{3}\arccos\left[\frac{1}{2}\left(\frac{3A_k}{b r_d^{2\beta}}\right)^{3/2}\right]\right\}. \tag{6.2.24}$$

Finally, the equilibrium interstitial aerosol radius $r_{wi} = xr_d$ is

$$r_{wi} = 2r_d^{1+\beta}\left(\frac{b}{3A_k}\right)^{1/2} \cos\left\{\frac{1}{3}\arccos\left[\frac{1}{2}\left(\frac{3A_k}{br_d^{2\beta}}\right)^{3/2}\right]\right\}. \quad (6.2.25)$$

For an aerosol particle with $\beta = 0.5$ and $\varepsilon_m = 0.5$ of ammonium sulfate ($b = 0.25$), the condition $Q < 0$ is equivalent to $r_d > 0.01$ μm, which is a typical case for many mixed CCN. For an aerosol particle with a very small soluble fraction $\varepsilon_m = 10^{-2}$ and an insoluble fraction of SiO_2 ($\rho_u = 2.65$ g cm^{-3}, $b = 8 \times 10^{-2}$), the condition $Q < 0$ is equivalent to $r_d > 0.3$ μm, which can be a typical case for dust particles with a thin soluble shell. Note that the interstitial radius $r_{wi} \sim r_d^{3/2}$ for the volume-proportional soluble fraction, $\beta = 0.5$, and $r_{wi} \sim r_d$ for the surface-proportional soluble fraction, $\beta = 0$.

The algebraic Cardano's solution of (Eqn. 6.2.20) is more convenient if $Q > 0$, or $Z_d^6 < 1/4$—i.e., with the small soluble fraction. Then, the solution of incomplete cubic equation is written as described in Appendix A.6,

$$y_1 = \tilde{A} + \tilde{B}, \quad \tilde{A} = \left(-\frac{q}{2} + Q^{1/2}\right)^{1/3} = \left[\frac{1}{2} + \left(\frac{1}{4} - Z_d^6\right)^{1/2}\right]^{1/3} \quad (6.2.26)$$

$$\tilde{B} = \left(-\frac{q}{2} - Q^{1/2}\right)^{1/3} = \left[\frac{1}{2} - \left(\frac{1}{4} - Z_d^6\right)^{1/2}\right]^{1/3}, \quad (6.2.27)$$

$$r_{wi} = r_d y_1 = \frac{r_d}{2^{1/3}}\{[1 + (1 - 4Z_d^6)^{1/2}]^{1/3} + [1 - (1 - 4Z_d^6)^{1/2}]^{1/3}\}, \quad (6.2.28)$$

with Z_d defined in (Eqn. 6.2.20). Note that (Eqn. 6.2.28) can also be used for $Q < 0$, and then $Q^{1/2}$ is imaginary and the expressions $(-q/2 \pm Q^{1/2})$ become complex. However, the expressions $(-q/2 \pm Q^{1/2})^{1/3}$ in the brackets $\tilde{A}$ and $\tilde{B}$ are complex-conjugate, the entire expression is real, and can be easily calculated.

Consider the two limits for the particular cases with high and low soluble fractions.

1. The first limit is $Z_d^2 = br_d^{2\beta}/3A_k >> 1/4^{1/3}$, or $Z_d >> 1$, high soluble fraction. Then, the trigonometric solution (Eqn. 6.2.25) can be simplified. For $b = 0.25$, $\beta = 0.5$ (volume soluble fraction), $A_k \approx 10^{-7}$ cm, and $r_d = 0.1$ μm, the argument of arccos in (Eqn. 6.2.25) is ~ 0.017 << 1, and arccos ≈ π/2, thus the last term in (Eqn. 6.2.25) is $\approx \cos(\pi/6) = \sqrt{3}/2$. Substitution of this value into (Eqn. 6.2.25) yields

$$r_{wi} \approx 2r_d^{1+\beta}\left(\frac{b}{3A_k}\right)^{1/2} \cos\left(\frac{\pi}{6}\right) = 2r_d^{1+\beta}\left(\frac{b}{3A_k}\right)^{1/2}\frac{\sqrt{3}}{2} = r_d^{1+\beta}\left(\frac{b}{A_k}\right)^{1/2}. \quad (6.2.29)$$

Thus, our general solution (Eqn. 6.2.25) reduces to the classical case (Eqn. 6.2.18) derived without account for an insoluble fraction, and (Eqn. 6.2.25) generalizes it. A similar estimate is obtained from the algebraic solution (Eqn. 6.2.28). Since Z_d is large, expanding it into a power series by $1/Z_d$, we obtain

$$r_{wi} \approx r_d Z_d[i^{1/3} + (-i^{1/3})], \quad (6.2.30)$$

where $i = \sqrt{-1}$. Using the trigonometric Moivre's formula $i^n = \cos\left(\frac{\pi}{2}n\right) + i\sin\left(\frac{\pi}{2}n\right)$, gives $\pm i^{1/3} = \cos(\pi/6) \pm i\sin(\pi/6) = \sqrt{3}/2 \pm i/2$, and $i^{1/3} + (-i^{1/3}) = \sqrt{3}$. Substituting this into (Eqn. 6.2.30) yields

$$r_{wi} \approx r_d Z_d \sqrt{3} = r_d^{1+\beta} \left(\frac{b}{A_k} \right)^{1/2}, \tag{6.2.31}$$

that is, we arrive again at the classical case (Eqn. 6.2.18) for pure soluble CCN. Thus, (Eqns. 6.2.25 and 6.2.28) generalize this case for the arbitrary amount of insoluble fractions.

2. The second limit is $Z_d^2 = br_d^{2\beta}/3A_k << 1/4^{1/3}$, or $Z_d << 1$, low soluble fraction b or low radius r_d, or low both. Then, expansion of (Eqn. 6.2.28) by Z_d gives

$$r_{wi} \approx r_d \left[1 + \frac{br_d^{2\beta}}{3A_k} - \frac{1}{3}\left(\frac{br_d^{2\beta}}{3A_k} \right)^3 \right]. \tag{6.2.32}$$

Thus, the wet interstitial radius r_{wi} is only slightly larger than the initial dry radius r_d. This case is relevant for particles of small sizes and/or small soluble fractions.

Table 6.1 along with Fig. 6.1 illustrate rapid increase of r_w in the region $0.98 < S_w < 1$. Calculations are performed with (Eqns. 6.2.12) and (6.2.28) for the same parameters as in Fig. 6.1: $\beta = 05$, $r_d = 0.1$ µm, and various values of ε_v. For $\varepsilon_v = 1$ (fully soluble CCN), r_w increases by a factor of 3 over the range $S_w = 0.3$ to 1, and increases by a factor of 2.2 over the range $S_w = 0.98$ to 1. The growth factors decrease with a decreasing soluble fraction: for $\varepsilon_v = 0.01$, r_w increases by 0.4% only in the range $S_w = 0.3$ to 1 (Fig. 6.1), and increases by 38% from $S_w = 0.98$ to 1 (Table 6.1). This rapid growth just below $S_w \to 1$ explains the sharp decrease in visibility and growth of aerosol optical thickness observed when aerosol transforms into "a milky haze" prior to activation (condensation) (e.g., Kasten, 1969; Fitzgerald, 1975) that can be seen just below the low boundaries of non-precipitating clouds.

Table 6.1 shows calculations of the relative error of the classical equation (Eqn. 6.2.18), $\delta_3 = (r_{wi} - r_{wi0})/r_{wi} \times 100$, where r_{wi0} is calculated with (Eqn. 6.2.18), and r_{wi} with the new equation (Eqn. 6.2.28). The errors at $\varepsilon_v = 1$ are small (1–2%), and grow to 7–11% at $\varepsilon_v = 0.1$. For these values of ε_v, the accuracy of the classical equation (Eqn. 6.2.18) is acceptable. The errors grow above 10% at $\varepsilon_v < 0.1$ and to ~50% at $\varepsilon_v = 0.01$; thus, for very small ε_v, it is better to use the new equations (Eqns. 6.2.28) or (6.2.25). It is interesting to note that the relative error δ_2 defined in previous sections and δ_3 given in Table 6.1 are quite comparable, although they are calculated with completely different equations. This indicates that the equations derived in Section 6.2.1 describe $r_w(S_w)$ as a smooth function at transition from subsaturation to interstitial conditions despite its rapid growth as $S_w \to 1$.

Table 6.1. *Transformation of wet radii* r_w *(µm) at high saturation ratios* S_w *for* $\beta = 0.5$, $r_d = 0.1$ *µm, and various soluble fractions* ε_v *and the errors in using approximate equations for* r_w.

ε_v	1	0.5	0.3	0.1	0.01
$r_w(S_w = 0.98)$	0.32	0.25	0.21	0.15	0.10
$r_w(S_w = 1.0)$	0.69	0.49	0.39	0.23	0.14
$GF = r_w(1.0)/r_w(0.98)$	2.19	1.99	1.84	1.53	1.38
$\delta_2 = (r_w - r_{w2})/r_w \times 100$, (%) $S_w = 0.98$	1.05	2.18	3.72	10.9	49.6
$\delta_3 = (r_{wi} - r_{wi0})/r_{wi} \times 100$, (%) $S_w = 1.0$	2.31	3.20	4.04	6.55	52.9

The expressions (Eqns. 6.2.25) and (6.2.28) and the particular cases (Eqns. 6.2.29) and (6.2.31) are valid for S_w close to 1, and can be used at the stage of CCN activation at slight subsaturation, $S_w < 1$, or for interstitial but unactivated cloud aerosol. In the latter case, $S_w > 1$ ($s_w > 0$), the spectra are limited by the boundary radius $r_b = 2A_k/3s_w$.

6.3. Critical Radius and Supersaturation

A typical behavior of water supersaturation $s_w(r)$ over a solution droplet calculated with (Eqn. 6.2.10) is shown in Fig. 6.2. Calculations were performed for $b = 0.25$ (CCN with 50% soluble fraction of ammonium sulfate), $\beta = 0.5$ (volume soluble fraction) and five masses of dry particles: $m_d = 10^{-16}$ g, 10^{-15} g, 10^{-14} g, 10^{-13} g, 10^{-12} g ($r_d = 0.0238$ μm, 0.0512 μm, 0.11 μm, 0.24 μm, and 0.51 μm, respectively). This figure shows that the curve $s_w(r)$ consists of the two branches, ascending and descending, which is caused by the presence of two terms in Köhler's equation: the positive curvature term and negative solution term. The ascending branch at smallest radii approach the asymptotic formed by Raoult's solution term in (6.2.10), $br_d^{2(1+\beta)}/(r_w^3 - r_d^3)$, which is more singular at small radii than the Kelvin's curvature term, A_k/r_w. At large radii, vice versa, Raoult's term is small, and the asymptotic is determined by the curvature term. Hence, the supersaturation curves have maxima at some intermediate radii, about 0.04 to 10 μm in the considered examples. The maximum supersaturation

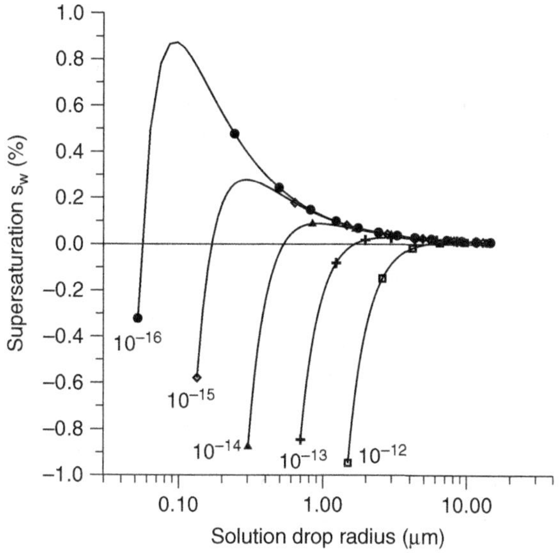

Figure 6.2. Equilibrium supersaturation over aqueous solution drops as a function of droplet radius. Calculations are performed with equation (Eqn. 6.2.10) for aerosol particles with dry masses $m_d = 10^{-16}$, 10^{-15}, 10^{-14}, 10^{-13}, and 10^{-12} g indicated near the curves, containing 50% of ammonium sulfate as a soluble fraction ($r_d = 0.0238$ μm, 0.0512 μm, 0.11 μm, 0.24 μm, and 0.51 μm, respectively).

on Köhler's curve $s_w(r)$ is called the *critical supersaturation* s_{cr}, and the corresponding radius on the x-axis is called the *critical radius* r_{cr}. The solution droplets with radii $r < r_{cr}$, on the ascending left branch, are in stable equilibrium, and those at $r > r_{cr}$, on the descending right branch, are in metastable equilibrium because a small perturbation in s_w causes further growth or evaporation until the critical values are reached. This figure shows that as the dry radius r_d increases, the critical s_{cr} decreases and shifts to greater r_{cr}.

The expressions for the critical radius r_{cr} and supersaturation s_{cr} for CCN activation can be found using Köhler's equations from the condition of the maximum, $ds_w/dr_w = 0$. For dilute solutions when $r_{cr} >> r_d$, or neglecting the insoluble fraction, (Eqn. 6.2.11) yields a quadratic equation relative to r_{cr}:

$$\left.\frac{ds_w}{dr_w}\right|_{r_{cr}} = -\frac{A_k}{r_w^2} + \frac{3B}{r_w^4} = 0. \tag{6.3.1}$$

Solving for $r_w = r_{cr}$, we obtain the critical radius

$$r_{cr} = \left(\frac{3B}{A_k}\right)^{1/2} = r_d^{1+\beta}\left(\frac{3b}{A_k}\right)^{1/2}. \tag{6.3.2}$$

Substitution into (Eqn. 6.2.11) gives

$$s_{cr} = \frac{A_k}{r_{cr}} - \frac{B}{r_{cr}^3} = \left(\frac{4A_k^3}{27B}\right)^{1/2} = \frac{2}{3}\frac{A_k}{r_{cr}} = r_d^{-(1+\beta)}\left(\frac{4A_k^3}{27b}\right)^{1/2}. \tag{6.3.3}$$

The expressions with B on the right-hand side of (Eqns. 6.3.2) and (6.3.3) are the classical Köhler's solutions for r_{cr} and s_{cr} and are valid for dilute solutions or for CCN without an insoluble fraction. The expressions with b and β on the right-hand sides of (Eqns. 6.3.2) and (6.3.3) are modifications of the Köhler expressions obtained using the preceding parameterization of B via b and β in (Eqns. 6.2.4)–(6.2.7).

The information on the soluble fraction chemistry is in the parameter b, and information on its geometry is in β. The latter representation shows, in particular, that r_{cr} is proportional to $r_d^{(1+\beta)}$, and s_{cr} is inversely proportional to $r_d^{(1+\beta)}$. This explains why r_{cr} shifts to larger radii and s_{cr} decreases with an increase of r_d in Fig. 6.2. These changes are different for different β, and are stronger for the volume-proportional soluble fraction ($r_d^{(1+\beta)} \sim r_d^{3/2}$) than for the surface-proportional soluble fraction ($r_d^{(1+\beta)} \sim r_d$). The explicit analytical dependence of these effects is seen with the parameterization (Eqns. 6.2.4)–(6.2.7) of B via b and β, but is not seen without this parameterization, just with B. This, in particular, illustrates the convenience of using the parameterization of B via b and β.

If the soluble fraction on CCN is small (e.g., a small volume fraction or a thin soluble shell on the surface of an insoluble dust particle), the assumption of high dilution and the condition $r_{cr} >> r_d$ can be invalid even at the time of activation, then the more complete (Eqn. 6.2.10) for S_w should be used instead of (Eqn. 6.2.11). We will obtain now an analytical solution for s_{cr} and r_{cr} without approximation of high dilution at activation, $r_{cr} >> r_d$. This will allow estimation of the accuracy and applicability of the classical solutions (Eqns. 6.3.2) and (6.3.3). The equation $ds(r_w)/dr_w = 0$ with $s_w(r_w)$ defined by (Eqn. 6.2.10) yields a condition at $r_w = r_{cr}$ of

$$\left.\frac{d\ln S_w}{dr_w}\right|_{r_{cr}} = -\frac{A_k}{r_w^2} + \left.\frac{3r_w^2 b r_d^{2(1+\beta)}}{(r_w^3 - r_d^3)^2}\right|_{r_{cr}} = 0, \tag{6.3.4}$$

which gives a sixth-order equation for r_{cr}:

$$\frac{A_k}{r_{cr}^2} = \frac{3br_d^{2(1+\beta)} r_{cr}^2}{(r_{cr}^3 - r_d^2)^2}. \tag{6.3.5}$$

Taking the square root, this can be reduced to the cubic equation by r_{cr},

$$r_{cr}^3 - (3r_d Z_d) r_{cr}^2 - r_d^3 = 0, \qquad Z_d = \left(\frac{br_d^{2\beta}}{3A_k}\right)^{1/2} = r_d^{\beta}\left(\frac{b}{3A_k}\right)^{1/2}. \tag{6.3.6}$$

It is convenient to divide all terms by r_d^3 and solve an equation for a dimensionless variable $x = r_{cr}/r_d$:

$$x^3 - 3Z_d x^2 - 1 = 0. \tag{6.3.6a}$$

This is an incomplete cubic equation without a linear power of x. The solution to (Eqns. 6.3.6) and (6.3.6a) is obtained in Appendix A.6 in the algebraic Cardano's form. It is expressed in terms of Z_d and gives a solution for r_{cr},

$$r_{cr} = r_d \chi(Z_d), \qquad x = \chi(Z_d) = Z_d + P_+(Z_d) + P_-(Z_d), \tag{6.3.7}$$

$$P_{\pm}(Z_d) = \left(Z_d^3 \pm \left(Z_d^3 + \frac{1}{4}\right)^{1/2} + \frac{1}{2}\right)^{1/3}. \tag{6.3.8}$$

Equation (6.3.7) is a generalization of (Eqn. 6.3.2) for r_{cr} without an assumption of high dilution. It is valid for an arbitrary soluble fraction and reduces to the classical expression for the particular case $Z_d \gg 1$ (high b or high soluble fraction). Then, expanding (Eqns. 6.3.7) and (6.3.8) into the power series yields

$$\begin{aligned} r_{cr} &\approx r_d[Z_d + Z_d(1 + Z_d^{-3/2}/3) + Z_d(1 - Z_d^{-3/2}/3)] \\ &= 3r_d Z_d = \left(\frac{3br_d^{2(1+\beta)}}{A_k}\right)^{1/2} = \left(\frac{3B}{A_k}\right)^{1/2}, \end{aligned} \tag{6.3.9}$$

which is the classical expression (Eqn. 6.3.2) for r_{cr}, confirming the validity of (Eqn. 6.3.7). In the opposite limit, $Z_d \ll 1$, which may happen at very small soluble fractions or radii, another expansion of (Eqn. 6.3.7) by Z_d yields:

$$r_{cr} \approx r_d\left(1 + Z_d + \frac{2}{3}Z_d^3\right) = r_d\left[1 + \left(\frac{br_d^{2\beta}}{3A_k}\right)^{1/2} + \frac{2}{3}\left(\frac{br_d^{2\beta}}{3A_k}\right)^{3/2}\right]. \tag{6.3.10}$$

Using r_{cr} from (Eqn. 6.3.7), the critical supersaturation is then calculated from the equation

$$s_{cr} = \exp\left(\frac{A_k}{r_{cr}} - \frac{br_d^{2(1+\beta)}}{r_{cr}^3 - r_d^3}\right) - 1. \tag{6.3.11}$$

This equation can be simplified using the previous relations between r_d and r_{cr} as

$$s_{cr} = \exp\left(\frac{2A_k}{3r_{cr}}\right) - 1. \tag{6.3.11a}$$

Fig. 6.3a, b depicts the critical supersaturations, s_{cr}, and critical radii, r_{cr}, calculated with the new (Eqns. 6.3.11) and (6.3.7) as functions of the dry radius r_d and volume soluble fraction ε_v from 1 to 10^{-3} for ammonium sulfate. Shown in Fig. 6.3c,d are the relative errors in calculations of s_{cr} and r_{cr} defined as $\delta s_{cr} = (s_{cr,1} - s_{cr,2})/s_{cr,2}$, and $\delta r_{cr} = (r_{cr,2} - r_{cr,1})/r_{cr,2}$, where the subscript "1" denotes the classical Köhler expressions (Eqns. 6.3.2) and (6.3.4) that do not account for the insoluble fraction at the time of activation, and subscript "2" denotes the new (Eqns. 6.3.11) and (6.3.7) that account for the insoluble fraction. The accuracy of the classical equations is reasonably good for the soluble

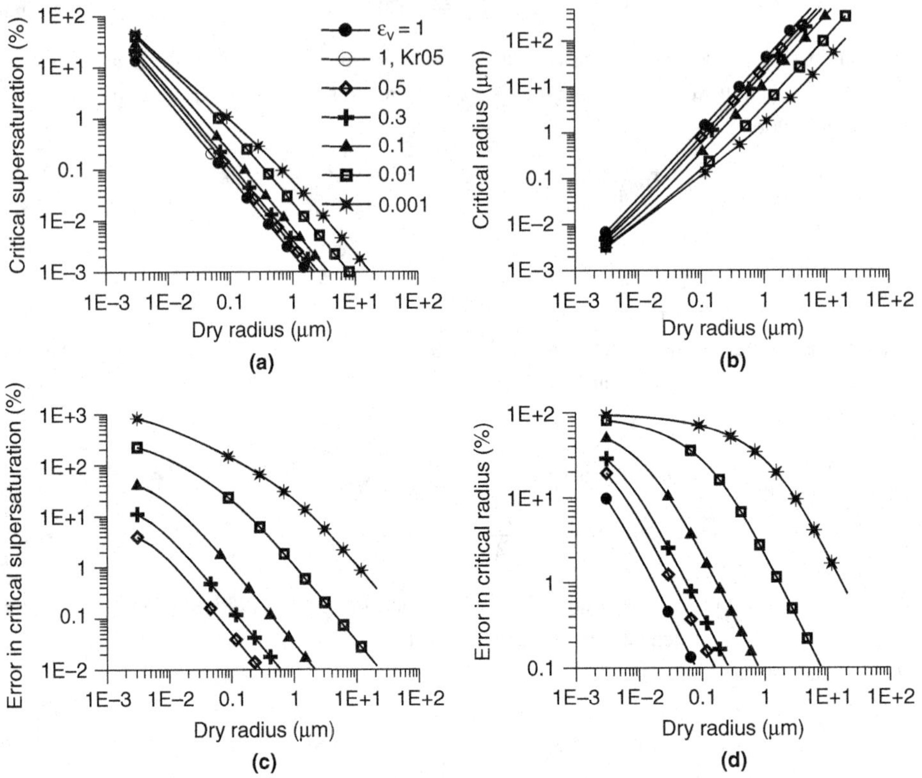

Figure 6.3. (a) Critical supersaturation s_{cr} calculated with the new (Eqn. 6.3.11). The open circle is a precise numerical calculation from Kreidenweis et al. (2005, Kr05) for pure ammonium sulfate ($\varepsilon_v = 1$). (b) Radius r_{cr} (μm) calculated with the new (Eqn. 6.3.7). (c) and (d): Relative errors (%) of calculation with the old expressions (Eqns. 6.3.3) and (6.3.2) for s_{cr} and r_{cr} as functions of the dry radius and volume soluble fraction ε_v from 1 to 10^{-3} indicated in the legend. From Khvorostyanov and Curry (2007), *J. Geophys. Res.*, 112(D5), D05206, reproduced with permission of John Wiley & Sons, Inc.

fractions $\varepsilon_v \geq 0.1$ in accumulation (0.1–1 μm) and coarse (>1 μm) modes. The errors in s_{cr} and r_{cr} are smaller than 1–2% (indicating also the correct limits of the new equations). For a particle of pure ammonium sulfate ($\varepsilon_v = 1$), the solution (Eqn. 6.3.11) exactly coincides with the precise numerical calculation from Kreidenweis et al. (2005, open circle in Fig. 6.3a). The errors of the classical diluted approximation grow with decreasing r_d and ε_v. For $\varepsilon_v = 0.01$, the errors reach 15–25% in the accumulation mode. At $r_d = 0.01$ μm, they grow to 20–30% with $\varepsilon_v = 0.1$, and to 80–200% at $\varepsilon_v = 0.01$. For the very small solubility of 10^{-3}, the errors are substantially greater, 100–500% for the submicron fraction 0.01–0.1 μm, and decrease to 1–20% for the coarse fraction.

The accuracy and applicability of the classical expressions is further illustrated in Fig. 6.4a,b,c. The ratio r_{cr}/r_d decreases with decreasing solubility. For small $\varepsilon_v \sim 10^{-3}$ to 10^{-2}, the values of the critical and dry radii are comparable (lower curves in Fig. 6.4a). The ratios r_{cr}/r_d evaluated with the new (Eqn. 6.3.7) for these small solubilities lie in accumulation mode slightly above the curve $r_{cr} = r_d$. So, $r_{cr}/r_d \sim 1$ and the relation $r_{cr} >> r_d$, usually used in the derivation of r_{cr} from (Eqn. 6.2.11) for dilute solutions, is not satisfied, hence (Eqns. 6.2.10) and (6.3.5) and the solution (Eqn. 6.3.7) should be used instead under these conditions. Note that values $\varepsilon_v \sim 10^{-3}$ to 10^{-2} (the lower curves in Fig. 6.4a) represent the case of mineral dust with thin soluble coating. The ratio r_{cr}/r_d calculated for these cases of small r_d and ε_v with the classical (Eqn. 6.3.2) lie below the curve $r_{cr} = r_d$ (Fig. 6.4b)—i.e., $r_{cr} < r_d$, and the classical expression fails as it predicts critical radii smaller than the dry ones. According to Rosenfeld et al. (2001), the fine-dispersed fractions of dust with small solubility may have a significant climatic effect, since they may suppress precipitation in semi-arid and arid areas and cause a desertification feedback loop. Hence, its correct account in the models is important, and drop activation should be calculated with (Eqns. 6.3.7) and (6.3.11) rather than with (Eqns. 6.3.2) and (6.3.3).

For $\varepsilon_v \geq 0.1$ and $r_d > 0.1$ μm, the ratio $r_{cr}/r_d \geq 4$, and the approximation $r_{cr}^3 >> r_d^3$ with classical equations for r_{cr}, s_{cr} given by (Eqns. 6.3.2) and (6.3.3) become valid. Note that the 1000-fold decrease in the soluble fraction from 1 to 10^{-3} for $r_d = 1$ μm causes r_{cr} to decrease by only ~22 times from 37 μm to 1.6 μm—i.e., a mineral dust particle with thin soluble shell requires much less time for growth to activation size than a fully soluble CCN of the same size. Fig. 6.4c shows that $s_{cr,new}$ predicted by (Eqn. 6.3.11) for small $\varepsilon_v = 10^{-2}$ to 10^{-3} and small r_d is 3–9 times less than $s_{cr,old}$ predicted by the classical (Eqn. 6.3.3). Thus, calculations with the new (Eqn. 6.3.11) may substantially accelerate the activation process of such CCN than by the old (Eqn. 6.3.3).

These estimates show that the classical expressions (Eqns. 6.3.2) and (6.3.3) should be used with caution for particles with small solubilities and small sizes. A more precise approach for determining r_{cr} and s_{cr} is based on (Eqns. 6.3.7) and (6.3.11).

Fig. 6.3a shows that the values of s_{cr} required for activation of a given CCN increase with decreasing r_d from 0.1–0.3% at $r_d = 0.1$ μm and $\varepsilon_v = 1$ to 10^{-2} (a typical cloud case) to 2–12% at $r_d = 0.01$ μm and up to 10–50% at $r_d = 0.003$ μm. Thus, supersaturations $s_w = 10$–25%, as can be reached in a cloud chamber, may cause activation of accumulation and Aitken modes of CCN with $r_d = 0.1$ to 0.003 μm even with a very small soluble fraction of 10^{-2}–10^{-3}. Hence, analysis or numerical simulations of such cloud chamber experiments may require the more precise (Eqns. 6.3.7) and (6.3.11).

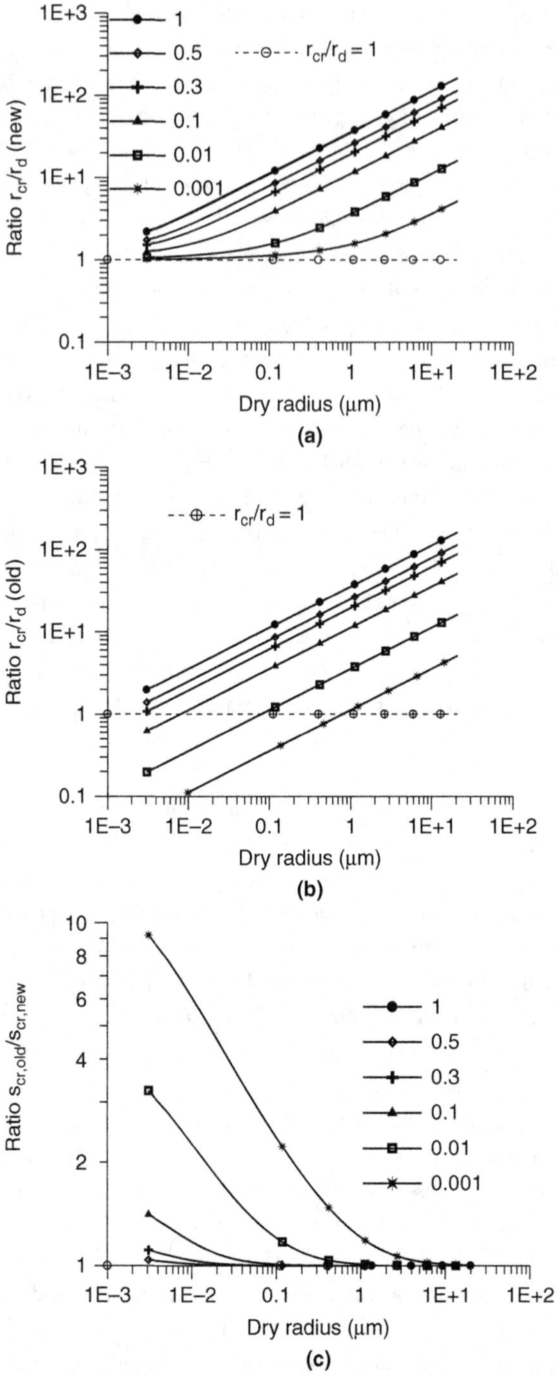

Figure 6.4. (a) The ratio of r_{cr}/r_d calculated with the new (Eqn. 6.3.7) (denoted "new"). (b) Same ratio calculated with the old (Eqn. 6.3.2) (denoted "old"); the dashed line with open circles is $r_{cr} = r_d$. (c) The ratio $s_{cr,old}/s_{cr,new}$ with $s_{cr,old}$ and $s_{cr,new}$ is calculated with (Eqns. 6.3.3) and (6.3.11), respectively. The various soluble fractions ε_v from 1 to 0.001 are indicated in the legend. From Khvorostyanov and Curry (2007), *J. Geophys. Res.,* 112(D5), D05206, reproduced with permission of John Wiley & Sons, Inc.

6.4. Aerosol Size Spectra

6.4.1. Lognormal and Inverse Power Law Size Spectra

The atmospheric aerosol generally represents a *polydisperse ensemble* of internally and externally mixed aerosol particles, each consisting of soluble and insoluble fractions. Historically, one of the first analytical functions suggested by Junge (1952, 1963) based on measurements of aerosol size spectra and widely used in aerosol sciences for several decades was the inverse power law:

$$f_a(r_a) = c_f r_a^{-\mu}, \tag{6.4.1}$$

where c_f is the normalization constant, and μ is the index usually determined from the fitting of measured size spectra. The values $4 < \mu < 5$ were found in most of the measurements, although this index could vary in a wider range (Junge, 1963, Pruppacher and Klett, 1997; Seinfeld and Pandis, 1998). These inverse power law size spectra allow us to describe a number of aerosol properties, including Ångström's (1929, 1964) inverse power law dependence of the optical extinction coefficients (Eqn. 6.1.6a) (e.g., van de Hulst, 1957; Junge, 1963; Tomasi et al., 1983, 2007; Khvorostyanov and Curry, 1999b). The deficiencies of the power law (Eqn. 6.4.1) is that, as measurements and theory showed, the index μ depends itself on the size range and moments (integrals) of this distribution and may diverge at small and large sizes.

At present, the most common function used to describe a polydisperse aerosol is the lognormal distribution, which allows us to account for variations of μ with size and gives us finite integrals. The lognormal size spectrum of dry aerosol $f_d(r_d)$ by the dry radii r_d or spectrum $f_w(r_w)$ of wet aerosol by wet radii r_w can be presented in the form:

$$f_{d,w}(r_{d,w}) = \frac{N_a}{\sqrt{2\pi}(\ln\sigma_{d,w})r_{d,w}} \exp\left[-\frac{\ln^2(r_{d,w}/r_{d,w0})}{2\ln^2\sigma_{d,w}}\right], \tag{6.4.2}$$

where the subscripts "*d, w*" mean dry or wet aerosol, N_a is the aerosol number concentration, σ_{d0} and σ_{w0} are the dispersions of the dry or wet spectrum, and r_{d0} and r_{w0} are the mean geometric radii. The properties of the lognormal distributions were considered in Chapter 2.

The measured aerosol distributions usually are multimodal—that is, have several local maxima at several radii. To describe this feature, several versions of multimodal distributions by aerosol radius r_a were suggested of the type

$$f_a(r_a) = \sum_{i=1}^{n} \frac{N_{ai}}{\sqrt{2\pi}(\ln\sigma_{ai})r_{ai}} \exp\left[-\frac{\ln^2(r_a/r_{ai0})}{2\ln^2\sigma_{ai}}\right], \tag{6.4.3}$$

where n is the number of modes, N_{ai}, r_{ai0}, and σ_{ai} are respectively the number concentration, mean geometric radius, and dispersion of the i-th mode, and summation is taken over all n modes. For example, Jaenicke (1988) and Whitby (1978) suggested the trimodal distributions, and Vignati et al. (2004) suggested a model with seven modes.

6.4.2. Approximation of the Lognormal Size Spectra by the Inverse Power Law

In various applications, either lognormal distributions or inverse power laws are used and it is desirable to establish a quantitative correspondence between them. A correspondence between them can be found using the same method as for a continuous power law representation of the fall velocities (see Chapter 12 here). Suppose $f(r)$ is smooth and has a smooth derivative $f'(r)$. Then, they can be presented at a point r as the power law functions

$$f(r) = c_f r^{-\mu}, \quad f'(r) = -\mu f(r)/r. \tag{6.4.4}$$

Solving (Eqn. 6.4.3) for μ and c_f, we obtain:

$$\mu(r) = -\frac{rf'(r)}{f(r)}, \qquad c_f(r) = f(r) r^{\mu(r)}. \tag{6.4.5}$$

The power index μ and c_f are here the functions of radius. If aerosol size spectrum $f(r)$ is described by the dry $f_d(r_d)$ or wet $f_w(r_w)$ lognormal distribution of the form (Eqn. 6.4.2), substitution of $f_{d,w}$ and $f'_{d,w}$ into (Eqn. 6.4.5) yields the power index for the dry and wet aerosol spectra

$$\mu_{d,w}(r_{d,w}) = 1 + \frac{\ln(r_{d,w}/r_{d,w0})}{\ln^2 \sigma_{d,w}}, \tag{6.4.6}$$

where the indices "*d, w*" denote dry or wet aerosol, and r_{w0}, σ_{w0} are the corresponding mean geometric radii and dispersions. Thus, the lognormal dry and wet spectra of the form (Eqn. 6.4.2) can be presented at every point r_d, r_w, in the power law form (Eqn. 6.4.1) with the index $\mu_{d,w}$ (Eqn. 6.4.6). This will be illustrated in the next section for dry and wet spectra.

6.4.3. Examples of the Lognormal Size Spectra, Inverse Power Law, and Power Indices

Fig. 6.5a shows four lognormal size spectra of the dry aerosol with various modal radii and dispersions along with the corresponding critical radii and supersaturations calculated with $\beta = 0.5$, $b = 0.55$ (~50% soluble fraction of ammonium sulfate). All spectra have a modal shape, which is asymmetric with respect to small and large sizes with a long tail in the region of larger radii. This modal shape of lognormal spectra along with the inverse dependence of the critical supersaturation on dry radius described in Section 6.3 by (Eqns. 6.3.3) or (6.3.11a) are important for understanding any quantitative description of CCN activation; this also determines the separation between activated and unactivated CCN fractions. One can see in Fig. 6.5a that if the maximum supersaturation does not exceed ~0.1%, the right branch of the dry spectrum with $r_m = 0.1$ μm can be activated down to the mode, and the left branch ($r_d < r_m$) remains unactivated. Such situations and s_w may occur in stratiform clouds. For the spectrum with $r_m = 0.03$ μm, activation of the right branch down to r_m requires a supersaturation of 0.6%. Such a supersaturation may occur in convective clouds. Thus, we can expect in these examples of the size spectra that the right branches of CCN will be activated in natural clouds. Activation of the left branches with $r_d < 0.01$–0.03 μm requires noticeably higher supersaturations and may be typical of the CCN measurements in cloud chambers where $s_w \sim 5$–20% can be reached (e.g., Jiusto and Lala 1981; Hudson 1984, Yum and Hudson 2001). A more detailed study of contributions of various ranges of the size spectra will be done in Section 6.6 in terms of the supersaturation activity spectra.

Fig. 6.5b shows a lognormal dry aerosol size spectrum $f_d(r_d)$ with modal radius $r_m = 0.03$ μm and dispersion $\sigma_d = 2$, close to Whitby (1978) for the accumulation mode, and concentration

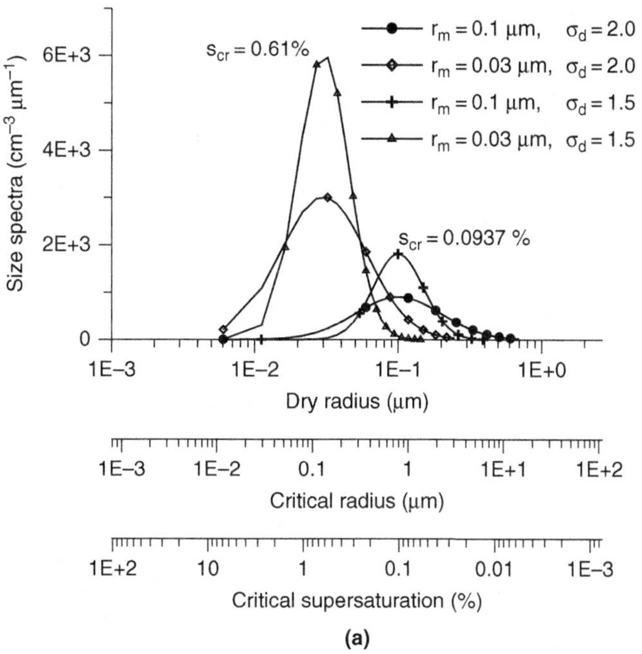

Figure 6.5. (a) Lognormal aerosol size spectra for $r_m = 0.1$ µm, $\sigma_d = 2$ (solid circles), $r_m = 0.1$ µm, $\sigma_d = 1.5$ (crosses), $r_m = 0.03$ µm, $\sigma_d = 2$ (diamonds), and $r_m = 0.03$ µm, $\sigma_d = 1.5$ (triangles), plotted as the functions of dry radii and of corresponding critical radii and critical supersaturations. Critical supersaturations s_{cr} corresponding to the maxima at 0.1 µm and 0.03 µm are shown near the curves. From Khvorostyanov and Curry (2006), *J. Geophys. Res.*, 111, D12202, reproduced with permission of John Wiley & Sons, Inc.

$N_a = 200$ cm^{-3}. Its Junge-type inverse power law approximations calculated with (Eqn. 6.4.5) and (Eqn. 6.4.6) shows the radius dependence of the power index μ: It changes from $\mu = -1.27$ at $r_1 = 0.016$ µm $< r_m$, to $\mu = 0$ at the modal radius $r_2 = r_m$, and μ is positive at all $r > r_m$; in particular, $\mu = 2.5$ at $r_3 = 0.1$ µm, and $\mu = 5.0$ at $r_4 = 0.33$ µm. These results well agree with the dimensional analysis of the coagulation equation and its asymptotic solutions that yield the inverse power laws with the index $\mu = 2.5$ near $r_d \sim 0.1$ µm due to the dominant effects of Brownian coagulation, and $\mu = 4.5$ for $r_d > 1$ µm due to prevailing sedimentation (e.g., Pruppacher and Klett, 1997, Chapter 11).

The lognormal spectrum with the described parameters is in general agreement with these power law solutions. Thus, the lognormal size spectrum can be considered as the superposition of the power laws with the indices from (Eqn. 6.4.6). The transition of the dry lognormal size spectrum to the limit $S_w = 1$ (curve with open circles) is fairly smooth, although there is a distinct decrease of the slopes. This resembles a corresponding effect for the power law spectra, when the transition to $S_w = 1$ is accompanied by a decrease of the Junge power index, e.g., $\mu = 4$ at $S_w < 1$ converts into $\mu = 3$ at $S_w \to 1$ over a very narrow humidity range. The humidity dependence of the inverse power law size spectra will be considered in more detail in Section 6.5.

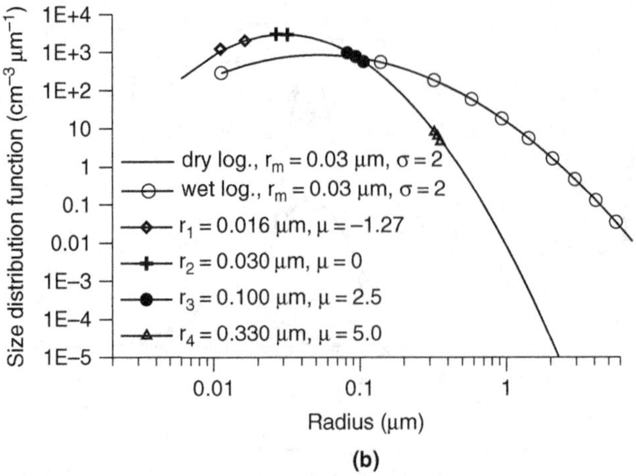

Figure 6.5. (b) Lognormal dry (solid curve) and interstitial (open circles) aerosol size spectra with the parameters: modal radius $r_m = 0.03$ μm, $\sigma_{d,w} = 2$, $N_a = 200$ cm^{-3}, and the Junge-type power law approximations to the dry spectrum calculated from Equation (6.4.6) at 4 points: before modal radius, $r_1 = 0.016$ μm, negative index $\mu = -1.27$; modal radius $r_2 = 0.03$ μm, $\mu = 0$; $r_3 = 0.1$ μm, $\mu = 2.5$; and $r_4 = 0.33$ μm, $\mu = 5.0$). From Khvorostyanov and Curry (2006), *J. Geophys. Res.*, 111, D12202, reproduced with permission of John Wiley & Sons, Inc.

The effective power indices calculated for lognormal distributions with various modal radii and dispersions are shown for the dry aerosol in Fig. 6.6, which illustrates the following general features of (Eqn. 6.4.6). The indices (the slopes of the spectra) increase with radii for all r. The indices μ increase with decreasing dispersions and with increasing r_{d0} of the dry lognormal spectrum. The calculated μ intersect the average Junge's curve $\mu = 4$ in the range of radii 0.06 μm to 0.7 μm—i.e., mostly in the accumulation mode. Since the values chosen here for r_{d0} and σ_d are realistic (close to Whitby's multimodal spectra), this may explain how a superposition of several lognormal spectra may lead to an "effective average" $\mu = 4$, observed by Junge. The growth of the indices above 4 for larger r is in agreement with the coagulation theory that predicts steeper slopes $\mu = 19/4$ of the power laws for greater radii $r \geq 5$ μm (Pruppacher and Klett, 1997). The described relation between the lognormal and power law size spectra might be useful for analysis and parameterizations of the measured spectra and of solutions to the coagulation equation, for sectional representations of the size spectra as well as for comparison of parameterizations based on lognormal and inverse power law distributions.

6.4.4. Algebraic Approximation of the Lognormal Distribution

A disadvantage of the lognormal size spectra is the difficulty of evaluating analytical asymptotics and moments of this distribution. A convenient algebraic equivalent of the lognormal distribution was obtained by Khvorostyanov and Curry (2006, 2007) using two representations of the smoothed (bell-shaped) Dirac's delta-function $\delta(x)$. The first one was taken as (Korn and Korn, 1968):

$$\delta_1(x,\alpha) = \frac{\alpha}{\sqrt{\pi}} \exp(-\alpha^2 x^2), \qquad (6.4.7)$$

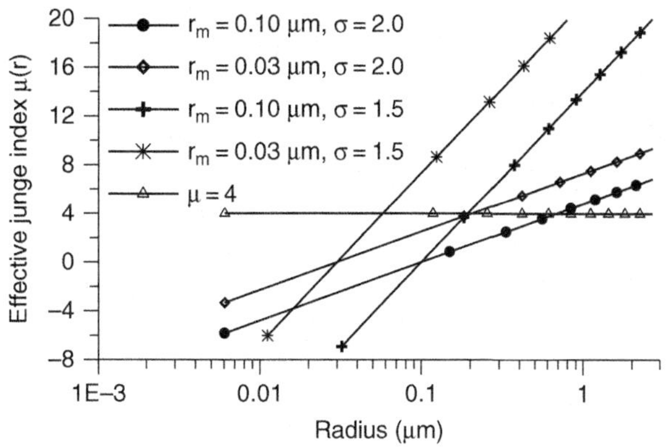

Figure 6.6. Effective power indices calculated from (Eqn. 6.4.5) for lognormal distributions with modal radii and dispersions indicated in the legend; the constant value $\mu = 4$ is given for comparison. From Khvorostyanov and Curry (2006), *J. Geophys. Res.*, 111, D12202, reproduced with permission of John Wiley & Sons, Inc.

where α determines the width of the function. Another representation as an algebraic function of $\exp(x/\alpha)$ is given in Levich (1969). We transformed it to the following form:

$$\delta_2(x,\alpha) = \frac{\alpha}{\sqrt{\pi}} \frac{1}{\mathrm{ch}^2(2\alpha x/\sqrt{\pi})}, \tag{6.4.8}$$

It can be shown that both functions $\delta_1(x,\alpha)$ and $\delta_2(x,\alpha)$ are normalized to unity, tend to the Dirac's delta-function when $\alpha \to \infty$ and exactly coincide in this limit:

$$\lim_{\alpha\to\infty} \delta_1(x,\alpha) = \lim_{\alpha\to\infty} \delta_2(x,\alpha) = \delta(x). \tag{6.4.9}$$

Thus, we obtain the following approximate equality:

$$\frac{\alpha}{\sqrt{\pi}} \exp(-\alpha^2 x^2) \approx \frac{\alpha}{\sqrt{\pi}} \frac{1}{\mathrm{ch}^2(2\alpha x/\sqrt{\pi})}. \tag{6.4.10}$$

A comparison shows that both functions are sufficiently close already at values $\alpha \sim 1$, which is illustrated in Fig. 6.7a. Where these functions divided by $\alpha/\sqrt{\pi}$ are plotted, the difference generally does not exceed 3–5%. The approximate equality of these smoothed delta-functions is used further for an algebraic representation of the aerosol size and CCN activity spectra.

Integration of (Eqn. 6.4.10) from 0 to x yields another useful relation:

$$erf(\alpha x) \approx \tanh(2\alpha x/\sqrt{\pi}), \tag{6.4.11}$$

where $erf(z)$ is the Gaussian function of errors:

$$erf(z) = \frac{2}{\sqrt{\pi}} \int_0^z e^{-x^2}\, dx. \tag{6.4.12}$$

Equation (6.4.11) with $\alpha = 1$ was found by Ghan et al. (1993), who showed its accuracy, which is also illustrated in Fig. 6.7b, for two values of α. It is known that integration of Dirac's delta-function

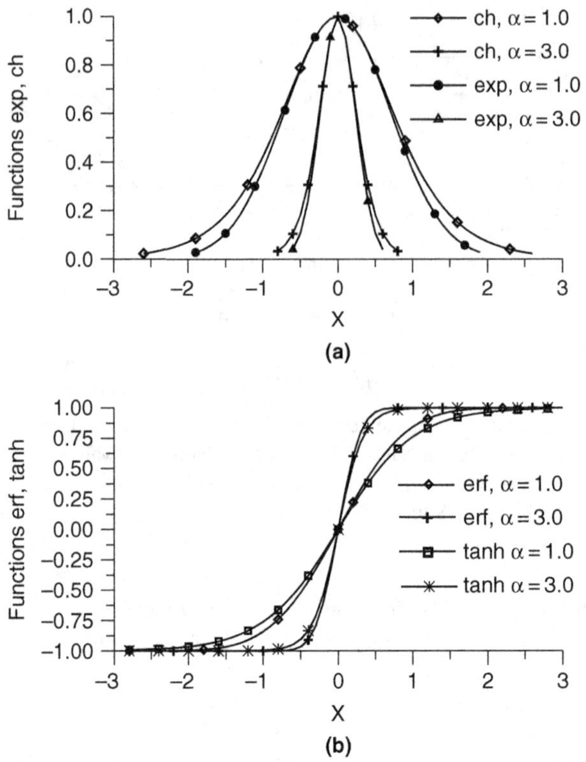

Figure 6.7. (a) Comparison of the smoothed Dirac delta-functions: $\exp(-\alpha^2 x^2)$ (denoted exp), and its approximation $1/\mathrm{ch}^2(2\alpha x/\sqrt{\pi})$ (denoted ch), for two values $\alpha = 1$ and 3. The difference generally does not exceed 3–5%. (b) Comparison of the smoothed Heaviside step-functions, $\mathrm{erf}(\alpha x)$, and $\tanh(2\alpha x/\sqrt{\pi})$. From Khvorostyanov and Curry (2006), *J. Geophys. Res.*, 111, D12202, reproduced with permission of John Wiley & Sons, Inc.

yields the Heaviside step-function. Comparison of Figs. 6.7a and 6.7b shows that the same is valid for the correspondent smoothed functions; therefore, the earlier derivation shows that the relation (Eqn. 6.4.11) is an approximate equality of the two representations of the smoothed Heaviside step-functions. Equation (6.4.11) is used in the following for the proof of equivalence of the lognormal and algebraic representations of the drop concentration and *k*-index.

The lognormal distribution can be approximated by an algebraic form introducing in (Eqn. 6.4.10) the new variables $x = \ln(r/r_0)$ and $\alpha = (\sqrt{2}\ln\sigma)^{-1}$. Thus,

$$\frac{2\alpha x}{\sqrt{\pi}} = \frac{1}{2}\frac{4}{\sqrt{2\pi}\ln\sigma}\ln\frac{r}{r_0} = \frac{k_0}{2}\ln\left(\frac{r}{r_0}\right), \tag{6.4.13}$$

where k_0 is a parameter introduced by Ghan et al. (1993) as

$$k_0 = \frac{4}{\sqrt{2\pi}\ln\sigma}. \tag{6.4.14}$$

Then, the left-hand side of (Eqn. 6.4.10) can be rewritten as

$$Z_1 \equiv \frac{\alpha}{\sqrt{\pi}} \exp(-\alpha^2 x^2) = \frac{1}{\sqrt{2\pi}\ln\sigma} \exp\left[-\frac{\ln^2(r/r_0)}{2\ln^2\sigma}\right]. \tag{6.4.15}$$

The right-hand side of (Eqn. 6.4.10) can be rewritten using (Eqn. 6.4.13):

$$\begin{aligned} Z_2 &\equiv \frac{\alpha}{\sqrt{\pi}} \frac{1}{\mathrm{ch}^2(2\alpha x/\sqrt{\pi})} \\ &= \frac{k_0}{4}\left\{\frac{1}{2}\left[\exp\left(\frac{2\alpha x}{\sqrt{\pi}}\right) + \exp\left(-\frac{2\alpha x}{\sqrt{\pi}}\right)\right]\right\}^{-2} \\ &= k_0\left[\exp\left(\frac{k_0}{2}\ln\frac{r}{r_0}\right) + \exp\left(-\frac{k_0}{2}\ln\frac{r}{r_0}\right)\right]^{-2} \\ &= k_0\left\{\exp\left[\left(\ln\frac{r}{r_0}\right)^{k_0/2}\right] + \exp\left[\left(\ln\frac{r}{r_0}\right)\right]^{-k_0/2}\right\}^{-2}. \end{aligned} \tag{6.4.16}$$

Using the equality $\exp[a\ln x] = [\exp(\ln x)]^a = x^a$, the expression (Eqn. 6.4.16) can be transformed as

$$Z_2 = k_0\left[\left(\frac{r}{r_0}\right)^{k_0/2} + \left(\frac{r}{r_0}\right)^{-k_0/2}\right]^{-2} = \frac{k_0(r/r_0)^{k_0}}{[1+(r/r_0)^{k_0}]^2}. \tag{6.4.16a}$$

Equating now the left- and right-hand sides of (Eqn. 6.4.10)—that is, Z_1 from (Eqn. 6.4.15) and Z_2 from (Eqn. 6.4.16a), we obtain

$$\frac{1}{\sqrt{2\pi}\ln\sigma}\exp\left[-\frac{\ln^2(r/r_0)}{2\ln^2\sigma}\right] \approx \frac{k_0(r/r_0)^{k_0}}{[(r/r_0)^{k_0}+1]^2}. \tag{6.4.17}$$

Now, the lognormal spectrum (Eqn. 6.4.2) can be transformed using (Eqn. 6.4.17) as

$$\begin{aligned} f(r) &= \frac{N_a}{r}\left\{\frac{1}{\sqrt{2\pi}(\ln\sigma)}\exp\left[-\frac{\ln^2(r/r_0)}{2\ln^2\sigma}\right]\right\} \\ &\approx \frac{N_a}{r}\left\{\frac{k_0(r/r_0)^{k_0}}{[(r/r_0)^{k_0}+1]^2}\right\} = \frac{N_a k_0}{r_0}\frac{(r/r_0)^{k_0-1}}{[(r/r_0)^{k_0}+1]^2}. \end{aligned} \tag{6.4.18}$$

The left-hand side of (Eqn. 6.4.18) is a lognormal spectrum with the concentration N_a, mean geometric radius r_0, and dispersion σ, and the right-hand side is its algebraic equivalent, which depends on the same parameters. These two distributions are different but are equivalent representations of the smoothed delta-function. That is, we found a new class of the size distribution functions that approximate the lognormal spectra in algebraic form. When the dispersion tends to its lower limit $\sigma \to 1$, then in (Eqn. 6.4.13) $\alpha \to \infty$, and according to (Eqn. 6.4.9), both distributions tend to the delta-function—i.e., become infinitely narrow (monodisperse) at $\sigma = 1$. For $\sigma > 1$, they represent distributions of the finite width. Using Equation (6.4.18), the dry lognormal aerosol size spectra (Eqn. 6.4.2) with $\sigma = \sigma_d$ and $r = r_d$ can be rewritten in algebraic form as

$$f_d(r_d) = \frac{k_{d0}N_a}{r_{d0}} \frac{(r_d/r_{d0})^{k_{d0}-1}}{[1+(r_d/r_{d0})^{k_{d0}}]^2}, \tag{6.4.19}$$

where r_{d0} and σ_d are the geometric radius and dispersion of the dry spectrum and the index k_{0d} is

$$k_{d0} = \frac{4}{\sqrt{2\pi}\ \ln\sigma_d}. \tag{6.4.20}$$

The size spectra (Eqns. 6.4.18) and (6.4.19) were derived in Khvorostyanov and Curry (2006), where they were called "*algebraic distributions*," and they will be called so hereafter. A comparison of the algebraic (Eqn. 6.4.19) and lognormal (Eqn. 6.4.2) aerosol size spectra calculated for the two pairs of spectra with the same parameters is shown in Fig. 6.8. One can see that algebraic and lognormal size spectra are in good agreement over the sufficiently wide region from the small radii to the maxima $r_m = 0.043$–0.1 μm and to the larger radii. Some discrepancy is seen at larger radii in the tails, but this region provides a small contribution into the number density and activity spectrum because $f_d(r_d)$ are here by two to three orders of magnitude smaller. An important feature of these distributions is that they have finite asymptotics at both small and large radii, and tend to zero as $(r/r_0)^{k_0-1}$ at small r, and $(r/r_0)^{-(k_0+1)}$ at large r. Thus, at large radii, they resemble the inverse power law Junge's spectra with the index $\mu = (k_0 + 1)$, but in contrast to these, the algebraic spectra are finite at small r. In summary, the new algebraic distributions are equivalent to both lognormal and inverse power laws, serve for their "reconciliation," and establish another link between them.

These algebraic representations of the size spectra are useful in various applications. Evaluation of their analytical and asymptotic properties is simpler than those of the lognormal distributions, and various expressions with these size spectra can be expressed via elementary functions rather than via the Gaussian integral *erf* as in the case with the lognormal distributions. Being a good approximation to the lognormal distribution but simpler in use, the algebraic functions (Eqns. 6.4.18)–(6.4.20) can be

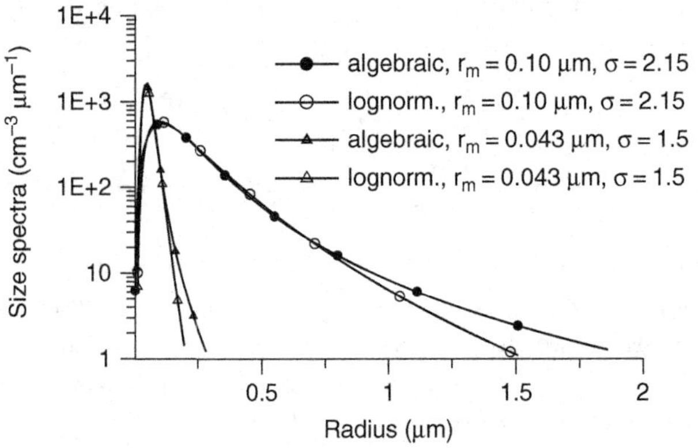

Figure 6.8. Comparison of the algebraic and lognormal aerosol size spectra for the two pairs of parameters: $r_m = 0.1$ μm, $\sigma_d = 2.15$ (solid and open circles for the algebraic and lognormal, respectively), and $r_m = 0.0.43$ μm, $\sigma_d = 1.5$ (solid and open triangles). From Khvorostyanov and Curry (2006), *J. Geophys. Res.*, 111, D12202, reproduced with permission of John Wiley & Sons, Inc.

used for approximation of the aerosol, droplet, and crystal size spectra as a supplement, and an alternative to the traditionally used power law, lognormal, and gamma distributions. Some applications for calculations of the CCN activity spectra and drop activation are illustrated in the next sections.

6.5. Transformation of the Size Spectra of Wet Aerosol at Varying Humidity

6.5.1. Arbitrary Initial Spectrum of Dry Aerosol

We consider the humidity transformation of the size spectra of polydisperse ensemble of mixed aerosol particles consisting of soluble and insoluble fractions. As the saturation ratio S_w increases and exceeds the threshold of deliquescence $S_{w,del}$ of the soluble fraction, the hygroscopic growth of the particles begins, and initially dry aerosol particles convert into wet "haze particles." The initial size spectrum $f_d(r_d)$ of dry mixed aerosol particles transforms into the size spectrum $f_w(r_w)$ of the wet haze particles. The wet spectrum $f_w(r_w)$ can be found from the concentration conservation equation in differential form

$$f_w(r_w)dr_w = f_d(r_d)dr_d, \qquad (6.5.1)$$

which can be rewritten in the form that allows evaluation of $f_w(r_w)$:

$$f_w(r_w) = f_d[r_d(r_w)](dr_d/dr_w). \qquad (6.5.2)$$

This equation requires knowledge of r_d as a function of r_w, which is the reverse problem relative to that considered in Section 6.2, and which again can be obtained from the cubic (Eqn. 6.2.10). For $\beta = 0.5$ (volume-proportional soluble fraction), we obtain from (Eqn. 6.2.10)

$$\frac{br_d^3}{r_w^3 - r_d^3} = \frac{A_k}{r_w} - \ln S_w. \qquad (6.5.3)$$

This equation contains only r_d^3 and no other powers of r_d, and its solution is simple. Solving for r_d we obtain

$$r_d(r_w) = r_w\varphi(r_w), \qquad \varphi(r_w) = \left[\frac{(-\ln S_w)r_w + A_k}{(-\ln S_w)r_w + A_k + br_w}\right]^{1/3}, \qquad (6.5.4)$$

$$\frac{dr_d}{dr_w} = \varphi(r_w) - \frac{r_w\varphi^{-2}(r_w)}{3}\frac{A_k b}{[(-\ln S_w)r_w + A_k + br_w]^2}. \qquad (6.5.5)$$

Substitution of (Eqns. 6.5.4) and (6.5.5) into (Eqn. 6.5.2) allows calculation of the wet spectrum $f_w(r_w)$ with any arbitrary analytical expression for the dry spectrum $f_d(r_d)$.

For the case $\beta = 0$ (surface-proportional soluble fraction), we obtain from (Eqn. 6.2.10) a cubic equation for $r_d(r_w)$ similar to (Eqn. 6.3.6) for r_{cr} in Section 6.3,

$$r_d^3 - \frac{br_w}{r_w \ln S_w - A_k} r_d^2 - r_w^3 = 0. \qquad (6.5.6)$$

Dividing by r_d^3 (Eqn. 6.5.6) can be rewritten for a dimensionless variable $x = r_d/r_w$

$$x^3 - 3Z_w x^2 - 1 = 0, \qquad Z_w = \frac{b}{3[r_w \ln S_w - A_k]}. \qquad (6.5.7)$$

This incomplete cubic equation has the same form as (Eqn. 6.3.6a). Its parameter $Q = Z_w^3 + 1/4$. Since $Z_w < 0$ at $S_w < 1$ and $\ln S_w < 0$, the solution can be chosen in algebraic or trigonometric form. If $Q > 0$—i.e., $-Z_w^3 < 1/4$ or $|Z_w| < (1/4)^{1/3}$—then it is convenient to choose the same algebraic Cardano's solution for x as described in Section 6.3 and Appendix A.6 with the replacement $Z_d \to Z_w$. Using these solutions for x, we obtain for $r_d = xr_w$:

$$r_d = r_w \chi(r_w), \qquad \chi(Z_w) = Z_w + P_+(Z_w) + P_-(Z_w), \tag{6.5.8}$$

$$\frac{dr_d}{dr_w} = \chi(Z_w) + r_w \left\{ \frac{dZ_w}{dr_w} + Z_w^2 [P_+^{-2}(Z_w) Q_+(Z_w) + P_-^{-2}(Z_w) Q_-(Z_w)] \right\}, \tag{6.5.9}$$

where the functions $P_\pm(x)$ are defined in (Eqn. 6.3.8), and

$$Q_\pm(Z_w) = \left(1 \pm \frac{1}{2} \left(Z_w^3 + \frac{1}{4} \right)^{-1/2} \right), \tag{6.5.10}$$

$$\frac{dZ_w}{dr_w} = -\frac{b(-\ln S_w)}{3(-\ln S_w)[(-\ln S_w) r_w + A_k]^2}. \tag{6.5.11}$$

As noted earlier, the functions $P_\pm$, $r_d(r_w)$ and $Q_\pm$, are real under the condition $|Z_w| < (1/4)^{1/3}$, or

$$|Z_w^3| < 1/4, \qquad \text{or} \qquad b < (3/2^{2/3})(A_k - r_w \ln S_w). \tag{6.5.11a}$$

This ensures the physical condition $r_d < r_w$. If $Q < 0$—i.e., $-Z_w^3 > 1/4$ or $|Z_w| > (1/4)^{1/3}$—then it is convenient to choose the trigonometric solution described in Appendix A.6, which gives

$$x = y_1 + Z_w = Z_w + 2Z_w \cos\left[\frac{1}{3} \arccos\left(1 + \frac{1}{2Z_w^2} \right) \right], \tag{6.5.11b}$$

and $r_d = r_w x(Z_w)$. Substitution of (Eqns. 6.5.8)–(6.5.11b) into (Eqn. 6.5.2) yields $f_w(r_w)$. The function $r_d(r_w)$ for the interstitial aerosol is obtained as the limit $-\ln S_w \approx 0$ in the earlier equation, which gives the smooth transition to size spectra of the wet interstitial aerosol.

It should be emphasized that this method for recalculation from the dry to the wet spectrum is not tied to any specific shape of the spectrum, but is suitable for any dry aerosol spectrum. These equations provide a solution and easy algorithm for analytical and numerical calculations of $f_w(r_w)$ with any initial $f_d(r_d)$. This method can accommodate any analytical parameterization (e.g., Junge power law, lognormal, algebraic, etc.) or a measured spectrum of any shape, e.g., a sectional representation. Despite some algebraic complexity, these analytical equations are easy for coding and can be used for numerical calculation of the wet spectrum from the dry spectrum. Particular cases of this approach are the calculations of hygroscopic growth of the dry power law spectra (e.g., Levin and Sedunov, 1966a; Sedunov, 1974; Fitzgerald, 1975; Khvorostyanov and Curry, 1999a; Cohard et al., 2000), or lognormal spectra (e.g., von der Emde and Wacker, 1993; Ghan et al., 1993; Khvorostyanov and Curry, 2006; Shipway and Abel, 2010) or sectional representation of the measured spectra (e.g., Nenes, and Seinfeld, 2003).

6.5.2. Lognormal Initial Spectrum of Dry Aerosol

As discussed in Section 6.2, for $\varepsilon_m \geq 0.1$–0.2, at sufficient dilution and not very high humidity, $S_w < 0.95$–0.97, the terms with A_k can be neglected, and the relation between r_w and r_d can be described by (Eqns. 6.2.15) or (6.2.16), which can be written in the form

$$r_w = \alpha r_d^\gamma, \tag{6.5.12}$$

where α and γ are determined in (Eqns. 6.2.15) and (6.2.16). The transformation to the wet spectra becomes especially simple if the size spectrum of dry aerosol $f_d(r_d)$ by the dry radii r_d can be represented by the lognormal distribution (Eqn. 6.4.2). We can derive the wet lognormal spectrum substituting (Eqn. 6.5.12) into the relation (Eqn. 6.5.2) in the inverse form—that is, $r_d = (r_w/\alpha)^{1/\gamma}$ and $dr_d/dr_w = (1/\gamma r_w)(r_d/\alpha)^{1/\gamma}$. Thus, we can express $f_w(r_w)$ via $f_d[r_d(r_w)]$ with r_d as a function of r_w:

$$f_w(r_w) = f_d[r_d(r_w)](dr_d/dr_w)$$
$$= \frac{N_a}{\sqrt{2\pi}(\ln\sigma_d)(r_w/\alpha)^{1/\gamma}} \exp\left\{-\frac{\{\ln[(r_w/\alpha)^{1/\gamma}/r_{d0})]\}^2}{2\ln^2\sigma_d}\right\} \frac{1}{\gamma r_w}\left(\frac{r_w}{\alpha}\right)^{1/\gamma}. \quad (6.5.13)$$

The logarithm in the numerator of the argument in exponent here can be transformed as

$$\{\ln[(r_w/\alpha)^{1/\gamma}/r_{d0})]\}^2 = \left(\frac{1}{\gamma}\ln r_w - \frac{1}{\gamma}\ln\alpha - \frac{1}{\gamma}\ln r_{d0}^{\gamma}\right)^2$$
$$= \frac{1}{\gamma^2}\left(\ln\frac{r_w}{\alpha r_{d0}^{\gamma}}\right)^2 = \frac{\ln^2(r_w/r_{w0})}{\gamma^2}, \quad (6.5.14)$$

where we introduced the "wet mean geometric radius" related to the dry mean geometric radius r_{d0} according to the same general humidity transformation (Eqn. 6.5.12):

$$r_{w0} = \alpha r_{d0}^{\gamma}. \quad (6.5.15)$$

Now the argument of the exponent in (Eqn. 6.5.13) can be rewritten as

$$-\frac{\{\ln[(r_w/\alpha)^{1/\gamma}/r_{d0})]\}^2}{2\ln^2\sigma_d} = -\frac{\ln^2(r_w/r_{w0})}{2\gamma^2\ln^2\sigma_d} = -\frac{\ln^2(r_w/r_{w0})}{2\ln^2\sigma_w}, \quad (6.5.16)$$

where we introduced the "wet dispersion"

$$\sigma_w = \sigma_d^{\gamma}. \quad (6.5.17)$$

The terms in front and after the exponent in (Eqn. 6.5.13) can be rewritten as

$$\frac{N_a}{\sqrt{2\pi}(\ln\sigma_d)(r_w/\alpha)^{1/\gamma}} \frac{1}{\gamma r_w}\left(\frac{r_w}{\alpha}\right)^{1/\gamma} = \frac{N_a}{\sqrt{2\pi}(\gamma\ln\sigma_d)}\frac{1}{r_w} = \frac{N_a}{\sqrt{2\pi}(\ln\sigma_w)r_w}, \quad (6.5.17a)$$

where we used again the relation for the wet dispersion $\gamma\ln(\sigma_d) = \ln(\sigma_d^{\gamma}) = \ln(\sigma_w)$. Substituting now (Eqn. 6.5.16) into the exponent in (Eqn. 6.5.13) and using (Eqn. 6.5.17a) as a preexponential factor, we obtain

$$f_w(r_w) = \frac{N_a}{\sqrt{2\pi}(\ln\sigma_w)r_w}\exp\left[-\frac{\ln^2(r_w/r_{w0})}{2\ln^2\sigma_w}\right]. \quad (6.5.18)$$

So, we arrived again at the lognormal distribution, but with the new mean geometric radius (Eqn. 6.5.15) and dispersion (Eqn. 6.5.17) that are expressed via the corresponding properties of the initial dry spectrum.

Thus, we derived a general and useful feature of the lognormal distributions: the invariance relative to the nonlinear transformation of the argument (radius or diameter) of the type (Eqn. 6.5.12). This feature is hereafter referred to as *"the 1st invariance property of the lognormal distribution"* and allows great simplification of calculations of the wet lognormal spectra. Using this invariance, the transition to the wet spectra for a lognormal distribution can be based on the simple rule. If we have a lognormal spectrum of radii r_d with the mean geometric radius r_{d0} and dispersion σ_d (or corresponding terms for diameters), then a nonlinear transformation of the variable of the form (Eqn. 6.5.12) that satisfies the conservation law (Eqn. 6.5.2) leads again to the lognormal distribution with the new mean geometric radius r_{w0} and the wet dispersion σ_w defined by (Eqns. 6.5.15) and (6.5.17). This is valid not only for humidity transformation but for any other transformation of this form.

We can use this invariance to relate $f_d(r_d)$ and $f_w(r_w)$. For example, if the relation $r_w(r_d)$ for the wet aerosol with $\beta = 0.5$ at subsaturation is described by the hygroscopic growth law (Eqn. 6.2.15), it is a transformation (Eqn. 6.5.12) with

$$\alpha = [1 + b(-\ln S_w)^{-1}]^{1/3}, \qquad \gamma = 1, \quad . \tag{6.5.19}$$

In many cases for sufficiently dilute solutions, (Eqn. 6.2.16) is a good approximation as discussed in Section 6.2, and thus

$$\alpha = b^{1/3}(1 - S_w)^{-1/3}, \qquad \gamma = 2(1+\beta)/3. \tag{6.5.20}$$

For the interstitial aerosol at $S_w \sim 1$, we find from (Eqn. 6.2.18) that

$$\alpha = (b/A_k)^{1/2}, \qquad \gamma = 1 + \beta, \tag{6.5.21}$$

Then, from the invariance property of the lognormal distribution, we obtain that the dry lognormal spectrum (Eqn. 6.4.2) transforms into the wet lognormal spectrum (Eqn. 6.5.18), and its mean geometric radius r_{w0} and dispersion σ_w are related to the corresponding quantities r_{d0} and σ_d of the dry aerosol as:

$$r_{w0} = r_{d0}[1 + b(-\ln S_w)^{-1}]^{1/3}, \qquad \sigma_w = \sigma_d, \qquad S_w < 0.97, \tag{6.5.22}$$

which corresponds to the relations (Eqn. 6.5.19) for $r_w(r_d)$. In approximations (Eqns. 6.2.16 and 6.2.18) with α, γ from (Eqns. 6.5.20) and (6.5.21), this yields another form: outside of a cloud,

$$r_{w0} = r_{d0}^{2(1+\beta)/3} b^{1/3}(1 - S_w)^{-1/3}, \qquad \sigma_w = \sigma_d^{2(1+\beta)/3}, \qquad S_w < 0.97, \tag{6.5.23}$$

and for the interstitial aerosol,

$$r_{w0} = r_{d0}^{(1+\beta)}(b/A_k)^{1/2}, \qquad \sigma_w = \sigma_d^{(1+\beta)}, \qquad S_w \to 1. \tag{6.5.24}$$

The analytic lognormal form of the size spectrum (Eqn. 6.5.18) in these cases is the same at subsaturation and for the interstitial aerosol, but the values of the mean geometric radius r_{w0} and dispersion σ_w are different, as shown by (Eqns. 6.5.22)–(6.5.24). However, the transition from the subsaturation $S_w < 1$ to $S_w \sim 1$ is sufficiently smooth. The same relations allow easy recalculation between the two saturation ratios S_{w1} and S_{w2}, and show that a lognormal spectrum at S_{w1} transforms to a lognormal spectrum at S_{w2}.

The relations for hygroscopic growth $r_w(r_d)$ of the form (Eqn. 6.5.12) can be used for recalculation from the dry to the wet lognormal aerosol size spectrum or between the two wet spectra using the invariance property described earlier.

The effect of humidity increase on the lognormal wet aerosol size spectra is shown in Fig. 6.9. As predicted by (Eqns. 6.2.12) and (6.5.22), it results in a gradual shift of the modal radius to larger values, with a simultaneous decrease of the maxima. The transition to the limit $S_w = 1$ is fairly smooth,

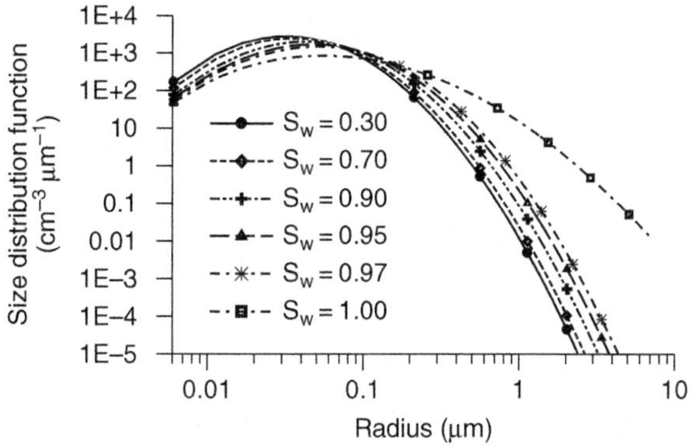

(a)

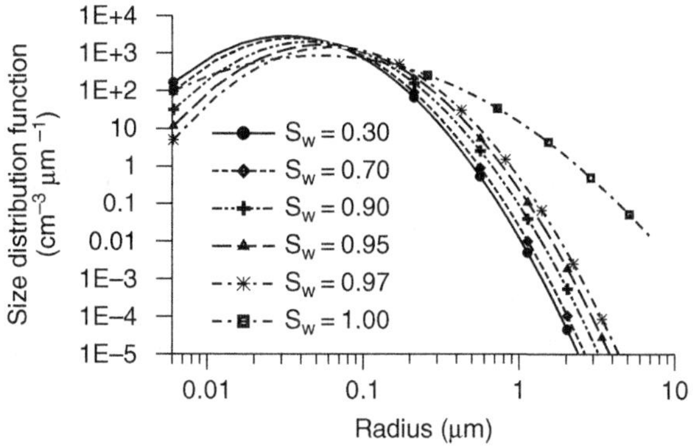

(b)

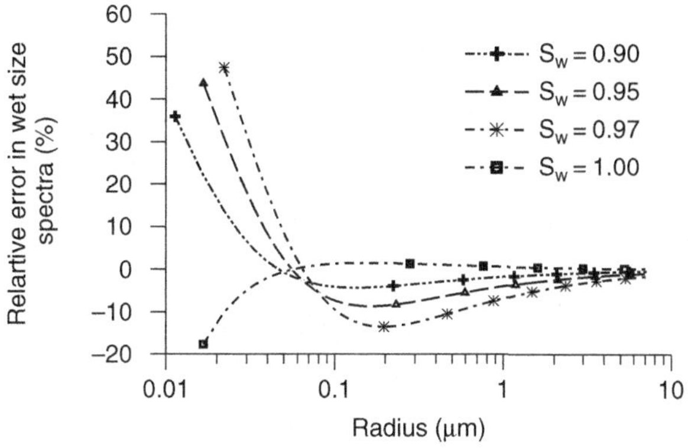

(c)

Figure 6.9. Humidity dependence of the wet aerosol size spectra. (a) full spectra $f_{w1}(r_w)$ calculated with (Eqns. 6.5.2)–(6.5.4) and complete equation (Eqn. 6.2.12) for $r_w(S_w)$; (b) approximate spectra $f_{w2}(r_w)$ calculated with (Eqn. 6.5.18) and r_{w0}, σ_w from (Eqns. 6.5.15) and (6.5.17) using shortened (Eqn. 6.2.15) for r_w; (c) relative error $\delta_{fw} = (f_{w1} - f_{w2})/f_{w1} \times 100$ (%). Calculations are made for the initial lognormal dry spectrum ($r_{d0} = 0.03$ μm, $\sigma_d = 2.0$), $\beta = 0.5$, $b = 0.25$ (50% of ammonium sulfate), and several S_w indicated in the legend. In case $S_w > 1$ ($s_w > 0$), the spectra should be limited by the boundary radius $r_b = 2A/3s_w$. From Khvorostyanov and Curry (2007), *J. Geophys. Res.*, 112(D5), D05206, reproduced with permission of John Wiley & Sons, Inc.

although there is a distinct decrease of the slopes. This resembles a corresponding effect for the power law spectra, when the transition to $S_w = 1$ is accompanied by a decrease of the Junge power index—e.g., $\mu = 4$ at $S_w < 1$ converts into $\mu = 3$ at $S_w \to 1$ over a very narrow humidity range (see previous Section 6.4 and Fig. 6.5).

The simplified (Eqns. 6.5.12) to (6.5.24) can be used in many cases for not very low dilutions and soluble fractions, while the more complete (Eqns. 6.5.2)–(6.5.11) can be used otherwise. The application of any of these methods depends on the specific aerosol spectra, and the choice of the method can be justified for a given situation under consideration.

6.5.3. Inverse Power Law Spectrum

As discussed in Section 6.4, the aerosol size spectra are often represented by the inverse power laws:

$$f_{d,w}(r_{d,w}) = c_{d,w} r^{-\mu_{d,w}}, \tag{6.5.25a}$$

$$c_{d,w} = N_a(\mu - 1) r_{\min}^{\mu-1}[1 - (r_{\min}/r_{\max})^{\mu-1}], \tag{6.5.25b}$$

where the subscripts "*d*" and "*w*" relate to the dry and wet aerosol, $\mu_{d,w}$ is the power index, and $c_{d,w}$ is the normalizing factor. The spectrum is normalized to the aerosol concentration N_a by integration from r_{min} to r_{max}. The second term in the brackets in (Eqn. 6.5.25b) can be usually neglected since $r_{min} << r_{max}$. Consider the case with $\beta = 0.5$. If $S_w < 0.97$ and $r_d \geq 0.01$ μm, then (Eqns. 6.5.2)–(6.5.4) can be simplified by neglecting A_k in φ and in the denominator of dr_d/dr_w. Substituting (Eqns. 6.5.25a) and (6.5.25b) for the dry aerosol into (Eqn. 6.5.2), and using (Eqns. 6.5.4) and (6.5.5) with these simplifications, we obtain after some transformations:

$$f_w(r_w) = c_d \varphi^{-(\mu-1)}\left[1 - \frac{bA_k}{3r_w(-\ln S_w)^2}\varphi^3\right]$$

$$\approx N_a(\mu-1) r_{\min}^{\mu-1} r_w^{-\mu}\left(1 + \frac{b}{-\ln S_w}\right)^{(\mu-1)/3} \times \left[1 - \frac{bA_k}{3r_w(-\ln S_w)^2}\left(1 + \frac{b}{-\ln S_w}\right)\right], \tag{6.5.26}$$

where $\varphi(r_w)$ is defined in (Eqn. 6.5.4). Hence, the dry power law spectrum (Eqn. 6.5.25a) with the index $-\mu$ transforms at subsaturation into the wet spectrum, which is a superposition of the two power laws with the indices $-\mu$ and $-(\mu + 1)$; the second of these becomes significant at $S_w \geq 0.95$. The wet inverse power law spectra were derived in Levin and Sedunov (1966a), Sedunov (1974), Fitzgerald (1972), Smirnov (1978), and Khvorostyanov and Curry (1999a) for sufficiently diluted solutions or when neglecting the insoluble fraction. Equation (6.5.26) is a generalization of these works that allows using this spectrum down to substantially smaller humidities. The impact of the insoluble fraction is described by the first term "1" in the rounded parentheses, and lower humidities are more accurately represented using $-\ln S_w$ instead of $(1 - S_w)$ as in the previous works.

If $b/(-\ln S_w) >> 1$, then 1 can be neglected in the rounded parentheses; if also $-\ln S_w \approx (1 - S_w)$, then using (Eqn. 6.2.14) for $r_w(S_w)$, making the corresponding simplifications in φ and dr_d/dr_w in (Eqns. 6.5.4) and (6.5.5), and substituting them into the general conservation (Eqn. 6.5.2), we obtain after some algebraic transformations a wet spectrum valid for any β:

$$f_w(r_w) = 3N_a r_{\min}^{\mu-1}\left(\frac{b}{1-S_w}\right)^{R_s} R_s\left[r_w^{-\mu_{wet}} - \left(R_s + \frac{1}{3}\right)\frac{A_k}{(1-S_w)} r_w^{-(\mu_{wet}+1)}\right], \tag{6.5.27a}$$

$$\mu_{wet} = \frac{3\mu + 2\beta - 1}{2(1+\beta)}, \qquad R_s = \frac{\mu - 1}{2(1+\beta)}, \tag{6.5.27b}$$

where μ_{wet} is the modified inverse power law index of the wet aerosol at subsaturation, and R_s is the parameter that determines the major humidity dependence of the size spectrum (it could be called the *humidity index*). Equation (6.5.27a) shows that the wet spectrum again consists of the two inverse power laws with the indices μ_{wet} and (μ_{wet} + 1), but the estimates show that the second term gives a noticeable contribution at high humidities, $S_w \geq 0.95$–0.97. At lower humidities, it can be simplified as

$$f_w(r_w) = 3N_a r_{\min}^{\mu-1} \left(\frac{b}{1 - S_w} \right)^{R_s} R_s r_w^{-\mu_{wet}}. \tag{6.5.27c}$$

The Equations (6.5.27a) and (6.5.27c) become invalid in the cloud, at $S_w \to 1$, but substitution of the corresponding approximations in (Eqns. 6.5.4) and (6.5.5) for φ and dr_d/dr_w into (Eqn. 6.5.2) gives the wet size spectrum of the interstitial wet aerosol in the cloud:

$$f_w(r_w) = N_a r_{\min}^{\mu-1} (2R_s) \left(\frac{b}{A_k} \right)^{R_s} r_w^{-\mu_{\text{int}}}, \qquad \mu_{\text{int}} = \frac{\mu + \beta}{1 + \beta}, \tag{6.5.28}$$

where μ_{int} is the index of the interstitial wet aerosol. The spectra of deliquescent but not activated interstitial aerosol at $S_w > 1$ are limited by the boundary radius $r_b = (2A_k/3)/(S_w - 1)$. These equations also can be used for evaluation of the aerosol optical properties with account for the simultaneous humidity and wavelength dependencies of aerosol extinction coefficients (e.g., Khvorostyanov and Curry, 2007).

The Junge-type inverse power laws were and are used for evaluation of the dependence of the wavelength λ of aerosol extinction coefficients in Ångström's law $\sigma_{ext} \sim \lambda^{-k_A}$ and optical depth $\tau_a \sim \lambda^{-k_A}$ (e.g., van de Hulst, 1957; Junge, 1963; Tomasi et al., 1983; Khvorostyanov and Curry, 1999a,b). Substitution of the Junge inverse power law spectra $f(r) \sim r^{-\mu}$ into the expression (Eqn. 2.3.8a) (see Chapter 2), for σ_{ext} yields the relation between the Junge index μ and Ångström's k_A coefficient, $k_A = \mu - 3$ (van de Hulst, 1957; Junge, 1963). The usual methods give the Ångström coefficients but do not allow evaluation of the humidity dependence and a relation between the wavelength and humidity dependencies. Equations (6.5.27a), (6.5.27b), and (6.5.27c) are convenient for evaluation of the simultaneous dependencies of σ_{ext} and τ_a on wavelength and humidity, and allow us to establish a relation between them. Substitution of (Eqn. 6.5.27a) into (Eqn. 2.3.8a) for σ_{ext} described in Sections 2.3 and 2.5 gives the expressions (Khvorostyanov and Curry, 1999b)

$$\begin{aligned} \sigma_{ext}(S_w, \lambda) = {} & D_{ext,1}^a N_a \lambda^{-k_A} (1 - S_w)^{-R_s} [\psi_{ref}(S_w)]^{k_A} \\ & - D_{ext,2}^a N_a \lambda^{-(k_A+1)} (1 - S_w)^{-(R_s+1)} [\psi_{ref}(S_w)]^{k_A}, \end{aligned} \tag{6.5.28a}$$

where $D_{ext,1}^a$, $D_{ext,2}^a$ are the coefficients dependent on aerosol microstructure,

$$\begin{aligned} D_{ext,1}^a &= 3\pi \cdot (2\pi)^{k_A} b^{R_s} r_{\min}^{\mu-1} \left[1 - (r_{\min}/r_{\max})^{\mu-1} \right] R_s I_1^{ext} \\ D_{ext,2}^a &= 3\pi \cdot (2\pi)^{k_A+1} b^{R_s} r_{\min}^{\mu-1} \left[1 - (r_{\min}/r_{\max})^{\mu-1} \right] A_k R_s \left(R_s + \frac{1}{3} \right) I_2^{ext}, \end{aligned} \tag{6.5.28b}$$

I_1^{ext} and I_2^{ext} are dimensionless integrals

$$I_1^{ext} = \int_{x_{\min}}^{x_{\max}} x^{-(k_A+1)} K_{ext}(x, n_d)\, dx, \qquad I_2^{ext} = \int_{x_{\min}}^{x_{\max}} x^{-(k_A+2)} K_{ext}(x, n_d)\, dx, \tag{6.5.28c}$$

and $\psi_{ref}(S_w)$ are the corrections due to variations of the refractive index with humidity. The modified Ångström's coefficient k_A, the power index μ_{wet}, and the humidity index R_s are linearly related

$$k_A = \mu_{wet} - 3 = \frac{3\mu - 4\beta - 7}{2(1+\beta)}, \qquad k_A = 3R_s - 2, \qquad \mu_{wet} = 3R_s + 1. \tag{6.5.28d}$$

These expressions establish relations between the size spectra, humidity, and wavelength dependencies. For the volume-proportional soluble fraction with $\beta = 0.5$, the first of Equations (6.5.28d) for k_A reduces to the previous relation $k_A = \mu - 3$. For the surface-proportional soluble fraction with $\beta = 0$, (Eqn. 6.5.28d) gives a different relation, $k_A = (3\mu - 7)/2$. For example, with $\mu = 4$, the coefficient $k_A = 1$ for $\beta = 0.5$ ($\sigma_{ext} \sim \lambda^{-1}$), but $k_A = 2.5$ for $\beta = 0$ ($\sigma_{ext} \sim \lambda^{-2.5}$). That is, the wavelength dependence is much stronger for $\beta = 0$—e.g., for the dust particles covered with a soluble shell. These relations can be useful for estimates of the aerosol direct and indirect effects on climate. The expression for σ_{ext} (Eqn. 6.5.28a) depends on a single value of μ. However, if the aerosol size spectrum is more complicated, it can be approximated by a superposition of several power laws in various size ranges with different μ as described in the previous sections, and the equations for σ_{ext} can be applied for each part of the spectrum, the indices k_A should be different in each size interval in the general case. Thus, σ_{ext} or τ_a in this case can be presented as a sum of several Ångström's laws with different coefficients.

Both lognormal and inverse power laws have their own advantages and deficiencies and were used for analysis or simulation of various aerosol properties, usually as independent tools, although in many cases the correspondence between them is desirable. A link between the Junge and lognormal spectra was found in Khvorostyanov and Curry (2006), as described by (Eqns. 6.4.4)–(6.4.6). The properties of these effective power law indices for the wet aerosol are illustrated in Fig. 6.10. The indices are negative at smaller r (corresponding to the growing branch on the left from the modal radius of lognormal spectra) and become positive at larger r (to the right of the mode of lognormal spectrum). An increase in S_w at $S_w < 1$ results in the parallel shift of the curves, since the increasing humidity causes the growth of the mean radius but the dispersion does not depend on S_w and does not change at $S_w < 1$ [see (Eqns. 6.5.22) and (6.5.23)]. When $S_w \approx 1$, the μ-indices decrease similarly to the power laws for dry aerosol.

6.5.4. Algebraic Size Spectra

If the dry aerosol spectrum is described by the lognormal distribution (Eqn. 6.4.2), then its algebraic analog for dry aerosol is (Eqn. 6.4.19). If the wet aerosol is described by the corresponding wet spectrum (Eqn. 6.5.18), its algebraic analog is obtained in the same way as the dry spectrum (Eqn. 6.4.19) and is

$$f_w(r_w) = \frac{k_{w0} N_a}{r_{w0}} \frac{(r_w/r_{w0})^{k_{w0}-1}}{[1 + (r_w/r_{w0})^{k_{w0}}]^2}, \tag{6.5.29}$$

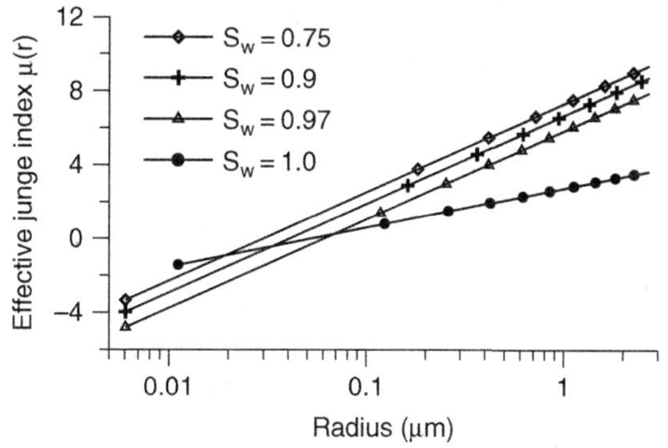

Figure 6.10. Dependence on the saturation ratio S_w of the power indices $\mu(S_w)$ of the wet aerosol approximating the lognormal distributions calculated from (Eqn. 6.4.6) and r_{w0}, σ_w from (Eqns. 6.5.22) and (6.5.24); the parameters of the dry aerosol are modal radius $r_m = 0.03$ μm, dispersion $\sigma_d = 2$, $\beta = 0.5$, and $b = 0.25$. From Khvorostyanov and Curry (2007), *J. Geophys. Res.*, 112(D5), D05206, reproduced with permission of John Wiley & Sons, Inc.

$$k_{w0} = \frac{4}{\sqrt{2\pi}\ln\sigma_w}, \tag{6.5.30}$$

where r_{w0} and $\sigma_w = \sigma_d^\gamma$ are the geometric radius and dispersion of the wet spectrum defined in (Eqns. 6.5.15) and (6.5.17) or in (Eqns. 6.5.19)–(6.5.21), and k_{0w} is the corresponding index of the wet spectrum. In particular, if the hygroscopic growth law (Eqn. 6.5.12) is valid and $\gamma = 1$, then the dispersion of the spectrum does not change, while the radius r_{w0} increases according to (Eqn. 6.5.19). The advantage of the spectrum (Eqn. 6.5.29) is that it tends to zero at small r_w as $r_w^{k_{w0}-1}$ in contrast to the inverse power law spectrum that tends to infinity at small r_w. At large r_w, the spectrum behaves as $r_w^{-(k_{w0}+1)}$, similar to Junge's inverse power law. Small values at small and large radii ensure better convergence of the integrals with the algebraic spectra at small radii than that with the inverse power laws; no limitations are needed at small radii as with the Junge's spectra.

6.6. CCN Differential Supersaturation Activity Spectrum

6.6.1. Arbitrary Dry Aerosol Size Spectrum

The CCN differential supersaturation activity spectrum $\varphi_s(s_{cr})$ can be obtained from the size spectrum of the dry CCN similar to the wet spectra in Section 6.5 using the conservation law in differential form:

$$\varphi_s(s_{cr})ds_{cr} = -f_d(r_d)dr_d. \tag{6.6.1}$$

Here s_{cr} is the critical supersaturation required to activate a dry particle with radius r_d. The minus sign occurs since an increase in r_d ($dr_d > 0$) corresponds to a decrease in s_{cr} ($ds_{cr} < 0$). This is seen from (Eqns. 6.3.3) or (6.3.11a) and was illustrated in Fig. 6.3a. The activity spectrum $\varphi_s(s_{cr})$ can be calculated from this equation if to rewrite it as

$$\varphi_s(s_{cr}) = -f_d[(r_d(s_{cr}))](dr_d/ds_{cr}). \tag{6.6.2}$$

In order to use (Eqn. 6.6.2), we need to know r_d as a function of s_{cr} or of r_{cr}, since the relation of s_{cr} and r_{cr} is given by (Eqn. 6.3.11). This is a reverse problem relative to that considered in Section 6.3, where we searched $s_{cr}(r_d)$. Taking the logarithm of (Eqn. 6.3.11), we obtain the equation

$$\ln(1+s_{cr}) = \frac{A_k}{r_{cr}} - \frac{b r_d^{2(1+\beta)}}{r_{cr}^3 - r_d^3}. \tag{6.6.3}$$

Solving it relative to r_d, we obtain

$$r_d^3 + a_{cr} r_d^{2(1+\beta)} - r_{cr}^3 = 0, \qquad a_{cr} = \frac{b r_{cr}}{A_k - r_{cr}\ln(1+s_{cr})}. \tag{6.6.4}$$

Using the relation that follows from (Eqn. 6.3.11a),

$$r_{cr} = \frac{2}{3}\frac{A_k}{\ln(1+s_{cr})}, \tag{6.6.5}$$

we can exclude r_{cr} and obtain the coefficient a_{cr} as

$$a_{cr} = \frac{2b}{\ln(1+s_{cr})}. \tag{6.6.6}$$

For $\beta = 0.5$, we obtain from (Eqn. 6.6.4) an equation that contains only r_d^3. Thus, the solution for $r_d(s_{cr})$ is

$$r_d(s_{cr}) = r_{cr}\xi(s_{cr}) = \frac{2A_k}{3\ln(1+s_{cr})}\xi(s_{cr}), \tag{6.6.7}$$

$$\xi(r_{cr}) = \left[1 + \frac{2b}{(1+\ln s_{cr})}\right]^{-1/3}. \tag{6.6.7a}$$

The function dr_d/ds_{cr} required for (Eqn. 6.6.2) can be calculated from (Eqn. 6.6.7) as

$$\frac{dr_d}{ds_{cr}} = -\frac{2A_k\xi(s_{cr})}{3(1+s_{cr})\ln^2(1+s_{cr})}\left[1 - \frac{2b\xi^3(s_{cr})}{3\ln(1+s_{cr})}\right]. \tag{6.6.8}$$

For a CCN with $r_d >> 0.01$ μm and a sufficiently large soluble fraction, these equations reduce to the classical case. At these conditions, $s_{cr} << b$, $1 + \ln(s_{cr}) \approx s_{cr}$, and approximately

$$\xi(s_{cr}) \approx (1 + 2b/s_{cr})^{-1/3} \approx (s_{cr}/2b)^{1/3}. \tag{6.6.9}$$

Substituting this into (Eqn. 6.6.7), we obtain

$$r_d(s_{cr}) = (4A_k^3/27b)^{1/3} s_{cr}^{-2/3}. \tag{6.6.10}$$

Inverting (Eqn. 6.6.10) for $s_{cr}(r_d)$, we obtain

$$s_{cr} = r_d^{-(1+\beta)}\left(\frac{4A_k^3}{27b}\right)^{1/2} = \left(\frac{4A_k^3}{27B}\right)^{1/2}, \tag{6.6.11}$$

which coincides with the classical Köhler's expression (Eqn. 6.3.3). Hence, (Eqn. 6.6.7) is an inversion of the generalized $s_{cr} - r_d$ relation (Eqn. 6.3.11a) with r_{cr} from (Eqn. 6.3.7). For this limiting case, we obtain from (Eqn. 6.6.8):

$$dr_d/ds_{cr} = (32A_k^3/729b)^{1/3}\, s_{cr}^{-5/3}, \tag{6.6.12}$$

which can be obtained directly from (Eqn. 6.6.10) and verifies the validity of (Eqn. 6.6.8).

Now, the differential activity spectrum $\varphi_s(s_{cr})$ (Eqn. 6.6.2) can be calculated for $\beta = 0.5$ using (Eqns. 6.6.7) and (6.6.8) for any shape of the initial dry size spectrum $f_d(r_d)$.

For the case $\beta = 0$, (Eqn. 6.6.4) becomes a cubic equation similar to (Eqn. 6.3.6) but with unknown r_d,

$$r_d^3 + a_{cr} r_d^2 - r_{cr}^3 = 0, \tag{6.6.13}$$

with a_{cr} defined in (Eqn. 6.6.6). It can be reduced to the dimensionless form for $x = r_d/r_{cr}$,

$$x^3 + \tilde{a}x^2 - 1 = 0, \qquad \tilde{a} = (a_{cr}/r_{cr}) = 3b/A_k. \tag{6.6.13a}$$

This is a cubic equation of the type (Eqn. A.6.1) from Appendix A.6 with $\tilde{a} = 3b/A_k$, $\tilde{b} = 0$, and $\tilde{c} = -1$. It can be reduced to the incomplete cubic equation considered in Appendix A.6 with defined there parameters $p = -3(b/A_k)^2$, $q = 2(b/A_k)^3 - 1$, and $Q = 1/4 - (b/A_k)^3$. As discussed in Section 6.3 and Appendix A.6, it is convenient to choose a solution depending on the sign of $Q = 1/4 - (b/A_k)^3$. If $Q < 0$, or $b/A_k > (1/4)^{1/3} \approx 0.63$, then it is suitable to choose a trigonometric solution

$$r_d(s_{cr}) = \frac{2b}{3\ln(1+s_{cr})}\psi(U), \qquad U = \frac{b}{A_k}, \tag{6.6.14}$$

$$\psi(U) = 2\cos\left[\frac{1}{3}\arccos\left(\frac{1}{2U^3} - 1\right)\right] - 1, \tag{6.6.15}$$

$$\frac{dr_d}{ds_{cr}} = -\frac{r_d(s_{cr})}{(1+s_{cr})\ln(1+s_{cr})}. \tag{6.6.16}$$

If $Q > 0$, or $b/A_k < (1/4)^{1/3} \approx 0.63$, then a convenient solution is Cardano's algebraic form

$$r_d = r_{cr}\chi(Z_{cr}), \qquad \chi(Z_{cr}) = \left[Z_{cr} + P_+(Z_{cr}) + P_-(Z_{cr})\right], \qquad Z_{cr} = -U = -b/A_k, \tag{6.6.17}$$

and the functions $P_\pm(z)$ are defined in (Eqn. 6.3.8), where Z_d should be replaced with Z_{cr}. The definitions of Z_{cr} and (Eqn. 6.3.8) for $P_\pm(z)$ show that $P_\pm(z)$ becomes complex for $Q < 0$. Although $P_+(z)$ and $P_-(z)$ are complex-conjugate and the solution is real, it can be more convenient to use the trigonometric solution (Eqns. 6.6.14)–(6.6.16) for this case. The functions $r_d(ds_{cr})$, dr_d/ds_{cr} are expressed for this case by (Eqns. 6.6.14) and (6.6.16), and the activity spectrum $\varphi(s_{cr})$ is given by (Eqn. 6.6.2) for arbitrary $f_d(r_d)$.

6.6.2. Lognormal Activity Spectrum

The calculation of the CCN activity spectrum is simpler for a dry lognormal size spectrum $f_d(r_d)$ with a mean geometric radius r_{d0} and dispersion σ_d if the critical supersaturation s_{cr} is related to the dry radius r_d by a nonlinear transformation of the variable of the form

$$s_{cr} = \alpha r_d^{-\gamma}. \tag{6.6.18}$$

Then, the reverse transformation, $r_d(s_{cr})$, is

$$r_d = \left(\frac{s_{cr}}{\alpha}\right)^{-1/\gamma}, \qquad \frac{dr_d}{ds_{cr}} = -\frac{1}{\gamma s_{cr}}\left(\frac{s_{cr}}{\alpha}\right)^{-1/\gamma}. \tag{6.6.19}$$

The derivation of the activity spectrum is similar to the derivation of the lognormal wet spectrum in Section 6.5. Substitution of (Eqn. 6.6.19) into (Eqn. 6.6.2) gives

$$\varphi_s(s_{cr}) = -f_d[(r_d(s_{cr})](dr_d/ds_{cr})$$

$$= -\frac{N_a}{\sqrt{2\pi}(\ln\sigma_d)(s_{cr}/\alpha)^{-1/\gamma}}\exp\left\{-\frac{\{\ln[(s_{cr}/\alpha)^{-1/\gamma}r_{d0}^{-1}]\}^2}{2\ln^2\sigma_d}\right\}\times\left(-\frac{1}{\gamma s_{cr}}\right)\left(\frac{s_{cr}}{\alpha}\right)^{-1/\gamma}. \tag{6.6.20}$$

The logarithm in the numerator of the argument in exponent here can be transformed as

$$\{\ln[(s_{cr}/\alpha)^{-1/\gamma}r_{d0}^{-1}]\}^2 = \left(-\frac{1}{\gamma}\ln s_{cr} + \frac{1}{\gamma}\ln\alpha + \frac{1}{\gamma}\ln r_{d0}^{-\gamma}\right)^2$$

$$= \frac{1}{\gamma^2}\left(\ln\frac{s_{cr}}{\alpha r_{d0}^{-\gamma}}\right)^2 = \frac{\ln^2(s_{cr}/s_0)}{\gamma^2}, \tag{6.6.21}$$

where we introduced the "mean geometric supersaturation" s_0 related to the dry mean geometric radius r_{d0} according to the same relation (Eqn. 6.6.18) as in the general variable transformation

$$s_0 = \alpha r_{d0}^{-\gamma}. \tag{6.6.22}$$

Now the argument of the exponent in (Eqn. 6.6.20) can be rewritten as

$$-\frac{\{\ln[(s_{cr}/\alpha)^{-1/\gamma}r_{d0}^{-1}]\}^2}{2\ln^2\sigma_d} = -\frac{\ln^2(s_{cr}/s_0)}{2\gamma^2\ln^2\sigma_d} = -\frac{\ln^2(s_{cr}/s_0)}{2\ln^2\sigma_s}, \tag{6.6.23}$$

where we introduced the "supersaturation dispersion" σ_s

$$\sigma_s = \sigma_d^{\gamma}. \tag{6.6.24}$$

The terms before and after the exponent in (Eqn. 6.6.20) can be combined as

$$-\frac{N_a}{\sqrt{2\pi}(\ln\sigma_d)(s_{cr}/\alpha)^{-1/\gamma}}\left(-\frac{1}{\gamma s_{cr}}\right)\left(\frac{s_{cr}}{\alpha}\right)^{-1/\gamma} = \frac{N_a}{\sqrt{2\pi}(\gamma\ln\sigma_d)}\frac{1}{s_{cr}} = \frac{N_a}{\sqrt{2\pi}(\ln\sigma_s)s_{cr}}, \tag{6.6.25}$$

where we used again the relation for the supersaturation dispersion $\gamma\ln(\sigma_d) = \ln(\sigma_d^{\gamma}) = \ln(\sigma_s)$. Substituting now (Eqn. 6.6.23) into the exponent in (Eqns. 6.6.20) and (6.6.25) as a preexponential factor, we obtain

$$\varphi_s(s_{cr}) = \frac{N_a}{\sqrt{2\pi}(\ln\sigma_s)s_{cr}}\exp\left[-\frac{\ln^2(s_{cr}/s_0)}{2\ln^2\sigma_s}\right]. \tag{6.6.26}$$

So, we arrive again at the lognormal distribution, but with the new mean geometric supersaturation s_0 (Eqn. 6.6.22) and supersaturation dispersion σ_s (Eqn. 6.6.24) that are expressed via the corresponding properties of the initial dry spectrum.

Thus, if the initial CCN size spectrum is lognormal and satisfies the conservation law (Eqn. 6.6.1), and the critical supersaturation is related to the dry radius by a nonlinear transformation (Eqn. 6.6.18), then the differential activity spectrum $\varphi_s(s_{cr})$ is also lognormal distribution by the variable s_{cr} with the new mean geometric supersaturation s_0, and supersaturation dispersion σ_s expressed via the dry radius and dispersion as in (Eqns. 6.6.22) and (6.6.24). This feature is referred hereafter to as *"the 2nd invariance property of lognormal distributions"* and is similar to the 1st invariance property in Section 6.5, except for the different power of s_0.

In the classical Köhler case, for dilute solutions when $r_{cr} >> r_d$, it follows from (Eqn. 6.3.3) that the parameters α, γ of this transformation (Eqn. 6.6.18) are

$$\alpha = \left(\frac{4A_K^3}{27b}\right)^{1/2}, \qquad \gamma = 1 + \beta. \tag{6.6.27}$$

Using the 2nd invariance property, we find that the dry lognormal CCN size spectrum with r_{d0} and σ_d corresponds to the lognormal CCN activity spectrum (Eqn. 6.6.26) with the mean geometric supersaturation s_0 and dispersion σ_s

$$s_0 = \alpha r_{d0}^{-\gamma} = \left(\frac{4A_k^3}{27b}\right)^{1/2} r_{d0}^{-(1+\beta)}, \qquad \sigma_s = \sigma_d^{\gamma} = \sigma_d^{(1+\beta)}. \tag{6.6.28}$$

Using the 2nd invariance property makes the transition from the dry size spectra to the activity spectra automatic and very simple once we define the $s_{cr} - r_d$ relation—e.g. (Eqns. 6.6.18) and (6.6.27).

Fig. 6.11 shows a comparison of the differential activity spectra $\varphi_s(s_{cr})$, integral (or cumulative) activity spectra

$$N_{CCN}(s_{cr}) = \int_0^{s_{cr}} \varphi_s(s)\, ds, \tag{6.6.29}$$

and relative error $\delta_{NCCN} = (N_{CCN}(f) - N_{CCN}(K))/N_{CCN}(f) \times 100$, calculated with the 2 different methods: 1) using the dry lognormal size spectrum (Eqn. 6.4.2) and the full (Eqns. 6.6.2)–(6.6.8) (symbol *f*), and 2) as the lognormal distribution for $\varphi_s(s_{cr})$ with s_0 and σ_s based on the approximation of the classical Köhler theory (Eqn. 6.6.28) (symbol *K*). Calculations are performed with β = 0.5, N_d =150 cm^{-3}, r_m = 0.03 μm, σ_d = 2.15, and various volume soluble fractions of ammonium sulfate in % from 50% to 0.1%. The approximate classical method overestimates $N_{CCN}(s)$ by 10–40% at $s < 0.02$, but the general agreement of both methods is good enough for $\varepsilon_v > 10^{-2}$. For smaller ε_v, the classical approximation underestimates $N_{CCN}(s)$ by 40–55% in the region s = 0.1–0.6%, where the major activation process occurs. Thus, the new more complete (Eqns. 6.6.2)–(6.6.8) predict much more rapid activation at small soluble fractions than Köhler's method.

Shown in Fig. 6.12 are the differential and cumulative CCN activity spectra calculated for β = 0 with (Eqns. 6.6.2) and (6.6.14)–(6.6.17) for CCN with the soluble film of thickness l_0 on the surface of an insoluble core. The dry size spectra are chosen as a model of the coarse mode, lognormal with r_m = 0.3 μm and σ_d = 2.15. The values of b are calculated with (Eqn. 6.2.9) and l_0 estimated from the data in Levin et al. (1996) as described in Section 6.2.1. The results for β= 0 are compared with the case β= 0.5, ε_v = 0.1, and the same dry spectrum. Fig. 6.12 shows that N_{CCN} with

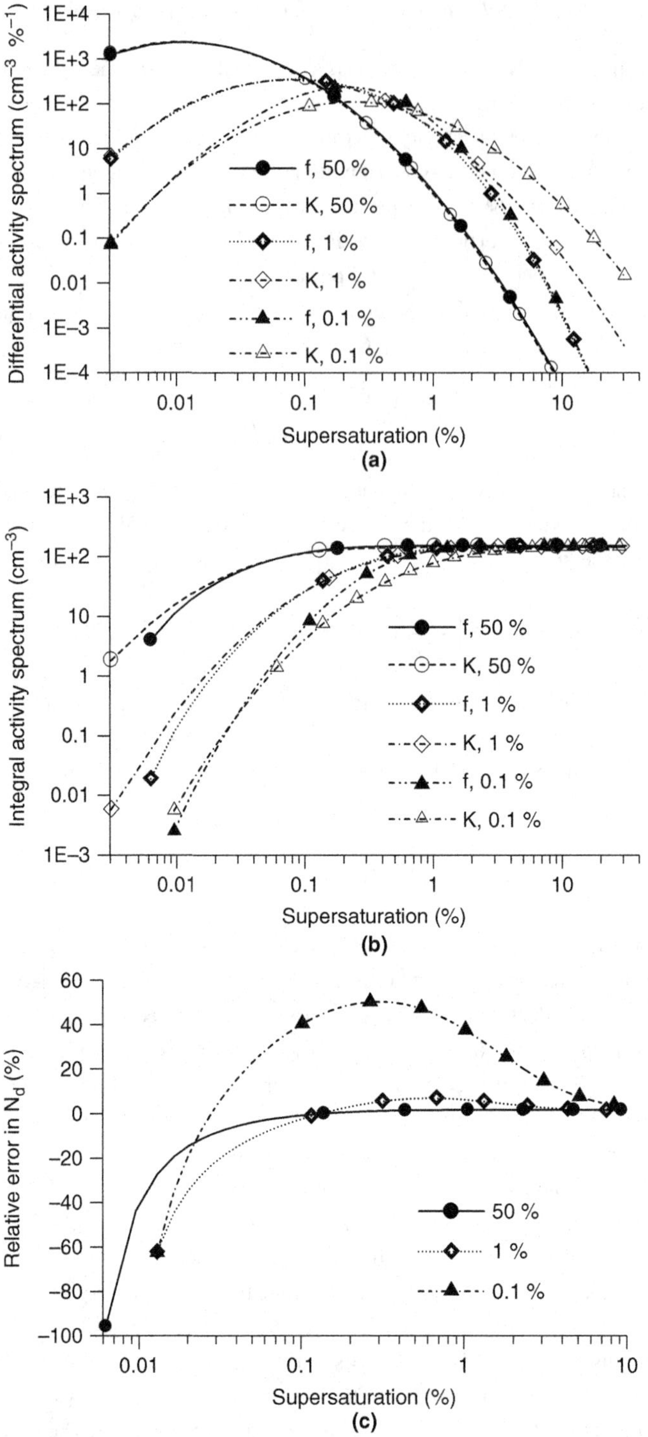

Figure 6.11. (a) Differential activity spectra $\varphi_s(s_{cr})$; (b) integral activity spectra $N_{CCN}(s_{cr})$; and (c) relative error defined in the text, calculated with the 2 methods: 1) for the dry lognormal size spectrum (Eqn. 6.4.2) with the full equations (Eqns. 6.6.2)–(6.6.8) (symbol *f*, solid symbols), and 2) as the lognormal distribution for $\varphi_s(s_{cr})$ with s_0 and σ_s based on the approximation of the classical Köhler theory (Eqn. 6.6.28) (symbol *K*, open symbols). Calculations are performed with $\beta = 0.5$, $N_d = 150\ \text{cm}^{-3}$, $r_m = 0.03\ \mu\text{m}$, and $\sigma_d = 2.15$. The numbers in the legends indicate volume soluble fraction of ammonium sulfate in %. From Khvorostyanov and Curry (2007), *J. Geophys. Res.*, 112(D5), D05206, reproduced with permission of John Wiley & Sons, Inc.

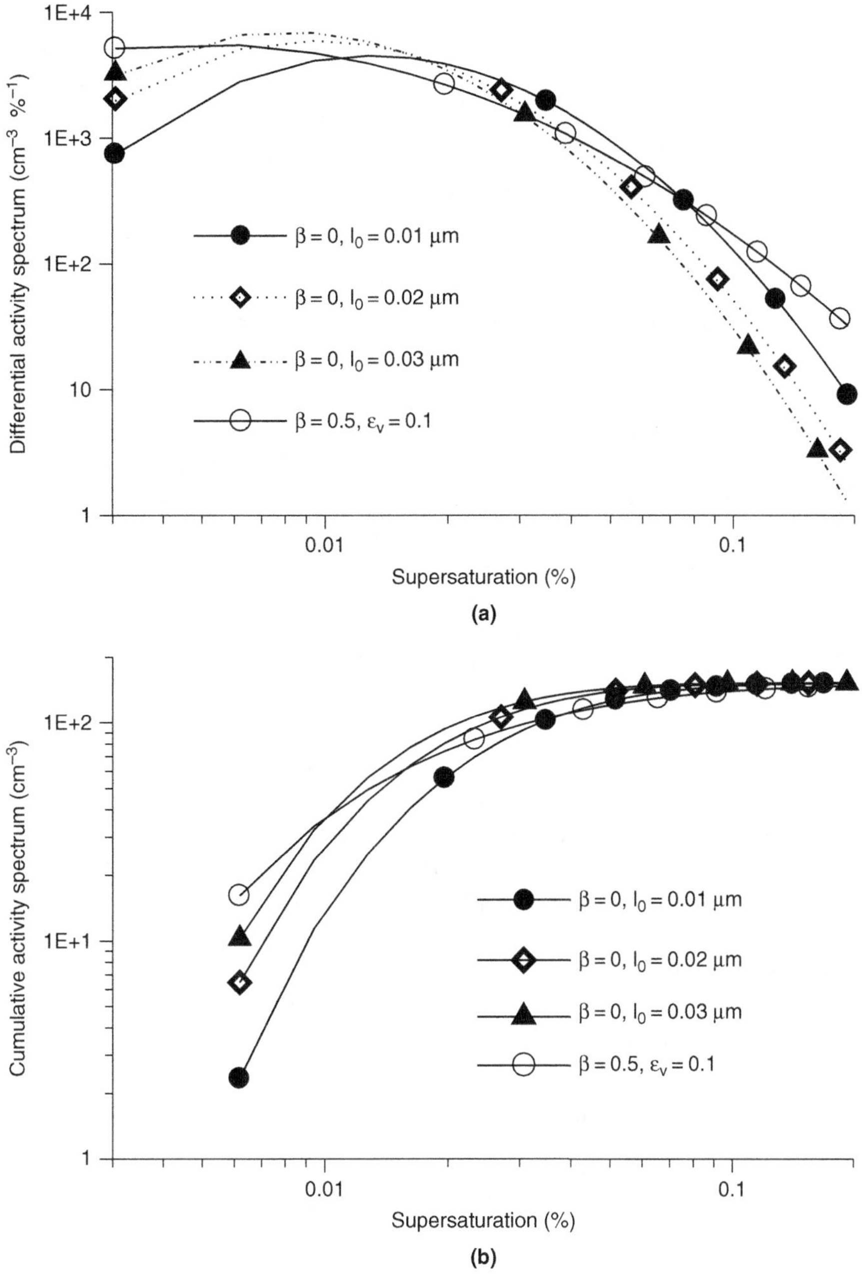

Figure 6.12. Differential (a) and cumulative (b) CCN activity spectra calculated with $\beta = 0$, (Eqns. 6.6.2) and (6.6.14)–(6.6.17) and various thicknesses l_0 of the soluble film on the surface of an insoluble core, and compared with the case $\beta = 0.5$ with soluble fraction $\varepsilon_v = 0.1$. The dry size spectra are lognormal (Eqn. 6.4.2) with the modal radius $r_m = 0.3$ μm and $\sigma_d = 2.15$. From Khvorostyanov and Curry (2007), *J. Geophys. Res.*, 112(D5), D05206, reproduced with permission of John Wiley & Sons, Inc.

$\beta = 0$ (surface-proportional soluble fraction) and thicknesses l_0 = 0.03, 0.02, and 0.01 μm become greater than that with $\beta = 0.5$ (volume-proportional soluble fraction) at $s_w > 0.01$, 0.015 and 0.04%, respectively, although in the latter two cases the soluble fractions in CCN with $\beta = 0$ are substantially smaller. This example shows that CCN such as dust particles coated with thin soluble films can be more effective than CCN with soluble fractions homogeneously distributed in volume, suggesting an important role of mineral dust in cloud nucleation.

6.6.3. Algebraic Activity Spectrum

If the dry aerosol spectrum is described by the lognormal distribution (Eqn. 6.4.2), then its algebraic analog for dry aerosol is (Eqn. 6.4.19) and the algebraic wet spectrum is described by (Eqn. 6.5.29). The corresponding differential activity spectrum can be obtained in various ways. The first way is the same as was used for derivation of the lognormal activation spectrum (Eqn. 6.6.26) from the lognormal size spectrum (Eqn. 6.4.2) of the dry aerosol—i.e., using the 2nd invariance property. As we could see, the analytical form of the activation spectrum is the same as that of the size spectrum, with the replacements $r_{d0} \to s_0$, $\sigma_d \to \sigma_s$. Applying this simple procedure to the dry algebraic spectrum (Eqn. 6.4.19) with an additional replacement

$$k_{d0} = \frac{4}{\sqrt{2\pi}\,\ln\sigma_d} \to k_{s0} = \frac{4}{\sqrt{2\pi}\,\ln\sigma_s}, \qquad (6.6.30a)$$

we obtain the algebraic activity spectrum

$$\varphi_s(s_{cr}) = \frac{N_a k_{s0}}{s_0} \frac{(s_{cr}/s_0)^{k_{s0}-1}}{[1+(s_{cr}/s_0)^{k_{s0}}]^2}. \qquad (6.6.30b)$$

The other way is to use a general transformation (Eqn. 6.6.18) in the conservation (Eqn. 6.4.2). The advantage of this way is that it allows us to check if this transformation of the variable in the algebraic size spectrum leads again to the algebraic activity spectrum. This is illustrated in the following text. Substituting $r_d(s_{cr}) = (s_{cr}/\alpha)^{-1}/\gamma$ and dr_d/ds_{cr} from (Eqn. 6.6.19) in the conservation Equation (6.6.2) with the dry algebraic size spectrum (Eqn. 6.4.19), we obtain

$$\varphi_s(s_{cr}) = -\frac{k_{d0}N_a}{r_{d0}} \frac{[(s_{cr}/\alpha)^{-1/\gamma}/r_{d0}]^{k_{d0}-1}}{[1+[(s_{cr}/\alpha)^{-1/\gamma}/r_{d0}]^{k_{d0}}]^2} \left(-\frac{1}{\gamma s_{cr}}\right)\left(\frac{s_{cr}}{\alpha}\right)^{-1/\gamma}. \qquad (6.6.31)$$

The following relation allows us to simplify this expression:

$$(s_{cr}/\alpha)^{-1/\gamma}/r_{d0} = (s_{cr}/\alpha r_{d0}^{-\gamma})^{-1/\gamma} = (s_{cr}/s_0)^{-1/\gamma}, \qquad (6.6.32)$$

where

$$s_0 = \alpha r_{d0}^{-\gamma} \qquad (6.6.33)$$

is the same mean geometric supersaturation s_0, and dispersion σ_s as introduced for lognormal distribution. The factor k_{d0}/γ in (Eqn. 6.6.31) is transformed as

$$\frac{k_{d0}}{\gamma} = \frac{4}{\sqrt{2\pi}(\gamma\ln\sigma_d)} = \frac{4}{\sqrt{2\pi}\,\ln\sigma_s} = k_{s0}, \qquad \sigma_s = \sigma_d^{\gamma}, \qquad (6.6.33a)$$

where k_{s0} is the analog of the k_{d0}-index and σ_s is the supersaturation dispersion. Substituting these relations into (Eqn. 6.6.31), we obtain

$$\varphi_s(s_{cr}) = \frac{N_a k_{s0}}{s_{cr}(s_{cr}/s_0)^{1/\gamma})} \frac{[(s_{cr}/s_0)^{-1/\gamma}]^{\gamma k_{s0}-1}}{\{1+[s_{cr}/s_0)^{-1/\gamma}]^{\gamma k_{s0}}\}^2} \tag{6.6.34}$$

Reducing the powers and canceling $(s_{cr}/s_0)^{1/\gamma}$ in the numerator and denominator, we have

$$\varphi_s(s_{cr}) = \frac{N_a k_{s0}}{s_{cr}} \frac{(s_{cr}/s_0)^{-k_{s0}}}{[1+(s_{cr}/s_0)^{-k_{s0}}]^2} = \frac{N_a k_{s0}}{s_0} \frac{(s_{cr}/s_0)^{k_{s0}-1}}{[1+(s_{cr}/s_0)^{k_{s0}}]^2}. \tag{6.6.35a}$$

So, we arrive again at the algebraic distribution (Eqn. 6.6.30b). That is, we proved that the general transformation of the variable (Eqn. 6.6.18), $s_{cr} = \alpha r_d^{-\gamma}$, in the algebraic size spectrum leads to the algebraic activity spectrum. If this transformation is based on Köhler's relations (Eqn. 6.3.3) for the critical supersaturation, then the parameters α and γ are expressed by (Eqn. 6.6.27), s_0 and σ_s are defined by (Eqn. 6.6.28) as in the case with lognormal spectrum, and the k_{s0} index is

$$s_0 = \left(\frac{4A_k^3}{27b}\right)^{1/2} r_{d0}^{-(1+\beta)}, \qquad \sigma_s = \sigma_d^{(1+\beta)}, \tag{6.6.35b}$$

$$k_{s0} = \frac{4}{\sqrt{2\pi}\ln\sigma_s} = \frac{4}{\sqrt{2\pi}(1+\beta)\ln\sigma_d}. \tag{6.6.35c}$$

The algebraic activity spectrum (Eqn. 6.6.35a) is a generalization of Twomey's power law activity spectrum $\varphi_s(s) = Cks^{k-1}$. It is shown in the following that k_{s0} is the asymptotic value of the k-index in the CCN activity power law at small s, while at large supersaturations, $\varphi_s(s)$ decreases with s and ensures finite drop concentrations.

A slightly different form of the algebraic activity spectrum (Eqns. 6.6.30b) and (6.6.35a) can be more convenient for comparison with the other models (the subscript "cr" at s is omitted in the following text for brevity and convenience of comparison). Equation (6.6.35a) can be rewritten as

$$\varphi_s(s) = k_{s0} C_0 s^{k_{s0}-1}(1+\eta_0 s^{k_{s0}})^{-2}. \tag{6.6.36}$$

Here, taking into account (Eqns. 6.6.35b) and (6.6.35c),

$$C_0 = N_a s_0^{-k_{s0}} = N_a \left(\frac{27b}{4A_k^2}\right)^{k_{s0}/2} (r_{d0})^{k_{s0}(1+\beta)}, \tag{6.6.37a}$$

$$\eta_0 = s_0^{-k_{s0}} = \left(\frac{27b}{4A_k^2}\right)^{k_{s0}/2} (r_{d0})^{k_{s0}(1+\beta)}. \tag{6.6.37b}$$

The first term on the right-hand side of (Eqn. 6.6.36), $k_{s0}C_0 s^{k_{s0}-1}$, represents Twomey's (1959) power law (Eqn. 6.1.5) for the differential activity spectrum. In contrast to the C_0 and k in previous empirical parameterizations based on fits to experimental data, here the analytical dependence is derived from the modified Köhler theory and the algebraic approximation of the lognormal functions, and parameters C_0, k_{s0} are expressed directly via aerosol parameters. This first term describes drop activation at small supersaturations $s << s_0$ (as in the clouds with weak updrafts and in the "haze chambers") and grows with s as $s^{k_{s0}-1}$. The second term in parentheses in (Eqn. 6.6.36) decreases at large $s >> s_0$ as

$s^{-2k_{s0}}$, overwhelms the growth of the first term, and ensures the asymptotic decrease $\varphi_s(s) \sim s^{-(k_{s0}+1)}$. Thus, the term in parentheses serves effectively as a correction to Twomey's law and prevents an unlimited growth of concentration of activated drops at large supersaturations. The asymptotic behavior can be summarized as

$$\varphi_s(s << s_0) \sim s^{k_{s0}-1}, \qquad \varphi_s(s >> s_0) \sim s^{-(k_{s0}+1)}. \tag{6.6.37c}$$

The entire CCN spectrum (Eqn. 6.6.36) resembles the corresponding equation from Cohard et al. (1998, 2000) or Shipway and Abel (2010):

$$\varphi_s(s) = kCs^{k-1}(1+\eta s^2)^{-\lambda}, \tag{6.6.38}$$

where the first term is also Twomey's differential power law, and the term in parentheses was added by the cited authors as an empirical correction with the two parameters η, λ, which were varied and chosen by fitting the experimental data. Despite the general functional similarity, the powers in the parentheses are different in (Eqns. 6.6.36) and (6.6.38). Therefore, integration of (Eqn. 6.6.36) yields an expression as algebraic fraction for the drop concentration $N_d(s)$ (Ghan et al., 1993, 1995; Khvorostyanov and Curry, 2006, 2008a), and integration of (Eqn. 6.6.38) yields an expression that contains the hypergeometric function (Cohard et al., 1998, 2000; Shipway and Abel, 2010). The algebraic representations (Eqns. 6.6.36) or (6.6.35a) of $\varphi_s(s)$ found in Khvorostyanov and Curry (2006) allow us to establish a bridge between Twomey's power law, the lognormal parameterizations of drop activation (Ghan et al., 1993, 1995; von der Emde and Wacker, 1993; Abdul-Razzak et al., 1998, 2000; Fountoukis and Nenes, 2005), and the modifications of the power law in Cohard et al. (1998, 2000) and Shipway and Abel (2010).

Shown in Fig. 6.13 is an example of the CCN differential algebraic activity spectrum $\varphi_s(s_{cr})$ (Eqns. 6.6.35a) or (6.6.36) calculated with $b = 0.25$ (CCN with ~50% of ammonium sulfate or ~20% of NaCl) and four combinations of modal radii and dispersions. The area beneath each curve represents the drop concentration N_{dr} that could be activated at any given supersaturation s, so that the

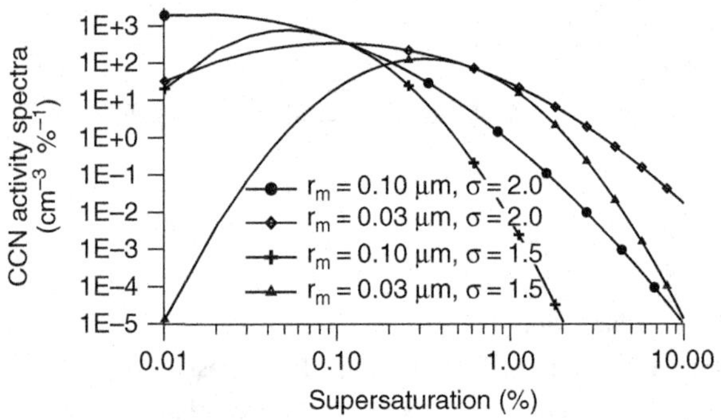

Figure 6.13. CCN algebraic differential activity spectrum $\varphi_s(s_{cr})$ calculated with (Eqn. 6.6.36), $b = 0.25$, and four combinations of modal radii r_m and dispersions σ_d of the dry aerosol spectra indicated in the legend. From Khvorostyanov and Curry (2006), *J. Geophys. Res.,* 111, D12202, reproduced with permission of John Wiley & Sons, Inc.

maximum of $\varphi_s(s_{cr})$ indicates the region of s_{cr} where its increase leads to the most effective activation. A remarkable feature of this figure is a substantial sensitivity to the variations of r_m, σ_d. The maximum for the maritime-type spectrum with $r_m = 0.1$ μm, $\sigma_d = 2$, lies at very small $s_{cr} \sim 0.01\%$. When σ_d decreases to 1.5, or r_m decreases to 0.03 (closer to the continental spectrum), the maximum of $\varphi_s(s_{cr})$ shifts to $s \sim 0.04$–0.1%. With $r_m = 0.03$ μm, $\sigma_d = 1.5$, the region $s_{cr} \sim 0.01$–0.1% becomes relatively inactive and the maximum shifts to 0.4%. Thus, even moderate narrowing of the dry spectra or a decrease in the modal radius may require much greater vertical velocities for activation of the drops with the same concentrations.

A comparison of the algebraic (Eqns. 6.6.35a) and (6.6.36) and lognormal (Eqn. 6.6.26) differential activity spectra in Fig. 6.14 shows their general good agreement. The discrepancy increases in both tails, but their contributions to the number density is small, as will be illustrated in the following. Thus, the algebraic functions (Eqns. 6.6.35a) and (6.6.36) approximate the lognormal spectrum (Eqn. 6.6.26) with good accuracy.

The differential activity spectrum for a multimodal aerosol consisting of I fractions is a simple generalization of (Eqn. 6.6.35a) by summation over all modes

$$\varphi_s(s) = \sum_{i=1}^{I} \frac{N_{ai} k_{s0i}}{s_{0i}} \frac{(s/s_{0i})^{k_{s0i}-1}}{[1+(s/s_{0i})^{k_{s0i}}]^2} \tag{6.6.39}$$

where N_{ai} is the CCN concentration in the i-th mode, s_{0i} and k_{s0i} are defined as in (Eqns. 6.6.35b) and (6.6.35c), but with A_{ki}, β_i, and b_i specific for each i-th mode

$$s_{0i} = \left(\frac{4A_{ki}^3}{27b_i}\right)^{1/2} r_{d0i}^{-(1+\beta_i)}, \tag{6.6.40a}$$

$$k_{s0i} = \frac{4}{\sqrt{2\pi}\ln\sigma_{si}} = \frac{4}{\sqrt{2\pi}(1+\beta_i)\ln\sigma_{di}}. \tag{6.6.40b}$$

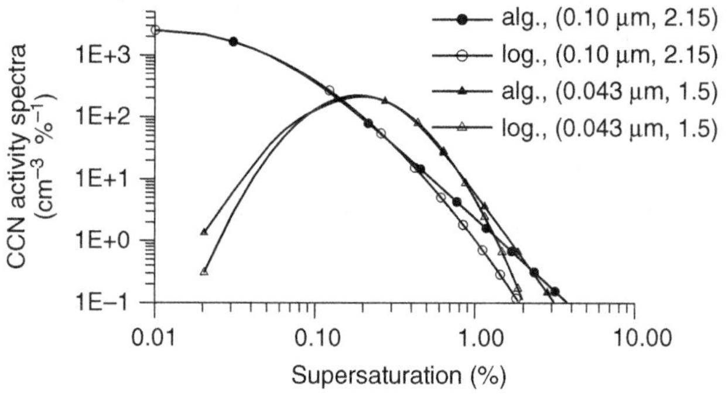

Figure 6.14. Comparison of the algebraic (alg.) (Eqn. 6.6.36) and lognormal (log.) (Eqn. 6.6.26) differential CCN activity spectra $\varphi_s(s_{cr})$. A pair of digits in the parentheses in the legend denotes modal radius r_m in μm and dispersion σ_d of the corresponding dry aerosol spectrum. From Khvorostyanov and Curry (2006), *J. Geophys. Res.*, 111, D12202, reproduced with permission of John Wiley & Sons, Inc.

6.7. Droplet Concentration and the Modified Power Law for Drops Activation

6.7.1. Lognormal and Algebraic CCN Spectra

The concentration of CCN or activated drops based on the lognormal size spectrum can be obtained by integration of the lognormal differential activity spectrum $\varphi_s(s)$ (Eqn. 6.6.26) over the supersaturations to the maximum value s reached in a cloudy parcel (we omit here also the subscript "*cr*" for brevity) (Ghan et al., 1993, 1995; von der Emde and Wacker, 1993; Fountoukis and Nenes, 2005; Khvorostyanov and Curry, 2006, 2008a),

$$N_d(s) = N_{CCN}(s) = \int_0^s \varphi_s(s')ds' = \frac{N_a}{2}\left\{1 + erf\left(\frac{1}{\sqrt{2}}\frac{\ln(s/s_0)}{\ln\sigma_s}\right)\right\}. \tag{6.7.1}$$

Droplet or CCN concentration in algebraic form can be obtained from integration of the algebraic CCN spectrum $\varphi_s(s)$ (Eqn. 6.6.35a) or (6.6.36) by s (Khvorostyanov and Curry, 2006, 2008a),

$$N_d(s) = N_a \frac{(s/s_0)^{k_{s0}}}{[1+(s/s_0)^{k_{s0}}]} = C_0 s^{k_{s0}}(1+\eta_0 s^{k_{s0}})^{-1}. \tag{6.7.2}$$

The parameters s_0 and σ_s in (Eqns. 6.7.1) and (6.7.2) are defined by (Eqn. 6.6.28). C_0 and η_0 are defined in (Eqns. 6.6.37a) and (6.6.37b) and expressed via aerosol microphysical properties.

The algebraic form allows a simple but clear physical interpretation. According to (Eqn. 6.7.2), the asymptotic of $N_{CCN}(s)$ at $s << s_0$ has the form of Twomey's power law $N_{CCN}(s) = C_0 s^{k_{s0}}$. The deficiency of the previous parameterization was an unlimited growth of $N_{CCN}(s)$ at high s, which stimulated several attempts to bound $N_{CCN}(s)$ at high s as in the just cited works. The new algebraic parameterization (Eqn. 6.7.2) removes this deficiency. The term in parentheses is a correction to the power law: Its asymptotic at $s >> s_0$ is $s^{-k_{s0}}$, it compensates the growth of the first term, and it prevents unlimited drop concentration. The asymptotic limit of drop concentration is $N_{CCN}(s) \to N_a$ at large s, and hence the number of activated drops is limited by the total aerosol concentration.

If the dry aerosol size spectrum is multimodal with I modes (e.g., $I = 3$ in three-modal distributions in Whitby, 1978, or seven modes as in Vignati et al., 2004), then it can be written as a superposition of I lognormal fractions (Eqn. 6.4.2) with N_{ai}, σ_{di}, and r_{d0i} being the number concentration, dispersion, and mean geometric radius of the i-th fraction as in (Eqn. 6.4.3). The mean geometric supersaturations s_{0i}, dispersions σ_{si}, and other parameters for each fraction are then defined by the same equations as described earlier for a single fraction. Then, the algebraic dry and wet size spectra, and the differential and integral activity spectra, are obtained for each fraction as described earlier, and the corresponding multimodal spectra are superpositions of the I fractions. The multimodal droplet concentration in algebraic form is derived from (Eqns. 6.6.39) or (6.7.2):

$$N_d(s) = \sum_{i=1}^{I} N_{ai} \frac{(s/s_{0i})^{k_{s0i}}}{[1+(s/s_{0i})^{k_{s0i}}]} = \sum_{i=1}^{I} C_{0i} s^{k_{s0i}}[1+\eta_{0i} s^{k_{s0i}}]^{-1}. \tag{6.7.3}$$

The parameters N_{ai}, s_{0i}, b_i, and β_i can be different for each fraction—e.g., $\beta_i = 0.5$ may correspond to the accumulation mode, and $\beta_i = 0$ may correspond to the coarse mode consisting of the dust particles coated with the soluble film.

6.7.2. Modified Power Law for the Drop Concentration

The power law $N_{CCN} = Cs^k$ is often used for parameterization of the results of CCN measurements and of drop activation in many cloud models. An analytical function for the k-index has been found (similar to the μ-index (Eqn. 6.4.5)) by Khvorostyanov and Curry (2006) using the method for the power law representation of the continuous functions. The power law for CCN and its derivative are written as

$$N_{CCN}(s) = C_s s^k, \qquad dN_{CCN}(s)/ds = kC_s s^{k-1}. \tag{6.7.4}$$

Solving these two equations for $k(s)$, we obtain

$$k(s) = s\frac{dN_{CCN}(s)/ds}{N_{CCN}(s)} = s\frac{\varphi_s(s)}{N_{CCN}(s)}. \tag{6.7.5}$$

Equation (6.7.5) allows evaluation of $k(s)$ with known differential $\varphi_s(s)$ and cumulative $N_{CCN}(s)$ CCN spectra. Substituting here (Eqn. 6.6.26) for $\varphi_s(s)$ and (Eqn. 6.7.1) for $N_{CCN}(s)$, we obtain $k(s)$ in terms of lognormal functions:

$$k(s) = \sqrt{\frac{2}{\pi}}\frac{1}{\ln\sigma_s}\exp\left[-\frac{\ln^2(s/s_0)}{2\ln^2\sigma_s}\right]\left[1+erf\left(\frac{\ln(s/s_0)}{\sqrt{2}\ln\sigma_s}\right)\right]^{-1}. \tag{6.7.6}$$

Here s_0 and σ_s are defined by (Eqn. 6.6.28). The algebraic form of the index $k(s)$ is obtained by substitution of (Eqn. 6.6.30b) for $\varphi_s(s)$ and (Eqn. 6.7.2) for $N_{CCN}(s)$ into (Eqn. 6.7.5):

$$k(s) = \frac{k_{s0}}{1+(s/s_0)^{k_{s0}}} = k_{s0}\left[1+s^{k_{s0}} r_{d0}^{k_{s0}(1+\beta)}\left(\frac{27b}{4A_k^3}\right)^{k_{s0}/2}\right]^{-1}. \tag{6.7.7}$$

This equation shows that the effective index $k(s)$ decreases with increasing s. If the aerosol size spectrum is multimodal and is given by a superposition of I algebraic distributions, then the k-index is derived from (Eqns. 6.7.5) and (6.7.3) by summation over I modes:

$$k(s) = \sum_{i=1}^{I}\frac{k_{s0i}N_{ai}\left(s/s_{0i}\right)^{k_{s0i}}}{[1+(s/s_{0i})^{k_{s0i}}]^2}\left\{\sum_{i=1}^{I}\frac{N_{ai}\left(s/s_{0i}\right)^{k_{s0i}}}{[1+(s/s_{0i})^{k_{s0i}}]}\right\}^{-1}, \tag{6.7.8}$$

where the parameters k_{s0i}, s_{0i}, and N_{ai} are defined as before in (Eqn. 6.6.35b), but for each i-th mode.

Equations (6.7.5)–(6.7.8) provide a continuous representation of k over the entire s-range and allow evaluation of $k(s)$ for any s with given σ_s and s_0 that are expressed via r_d, σ_d, b, and β—i.e., directly via aerosol physico-chemical properties. That is, knowledge of the size spectra and composition of the dry or wet CCN allows direct evaluation of $k(s)$ over the entire s-range. This solves the problem outlined in Jiusto and Lala (1981), Yum and Hudson (2001), and other studies: various values of the k-index at various s found experimentally show a general tendency of a decrease of k with increasing s. Note that the k-index is not constant for a given air mass or aerosol type as often assumed, but depends on a given s at which it is measured. Equation (6.7.7) shows that for small $s << s_0$, the asymptotic value $k(s) = k_{s0}$. For large $s >> s_0$, the index decreases with s as $k(s) = k_{s0}(s/s_0)^{-k_{s0}}$ and tends to zero. This behavior is in agreement with the experimental data from Jiusto and Lala (1981), Yum and Hudson (2001), and others. That is, with the same CCN properties, the values of $k(s)$ are different at different locations and vertical velocities in a cloud.

The coefficient $C_s(s)$ can now be calculated using (Eqn. 6.7.2) for $N_{CCN}(s)$ and (Eqn. 6.7.7) for $k(s)$ as

$$C_s(s) = N_{CCN}(s)s^{-k(s)} = C_0 s^{\chi(s)}[1+(s/s_0)^{k_{s0}}]^{-1}, \tag{6.7.9}$$

where $C_0 = N_a s_0^{-k_{s0}}$ is expressed in (Eqn. 6.6.37a) via aerosol microstructure. The power index $\chi(s)$ of s in (Eqn. 6.7.9) is

$$\chi(s) = k_{s0} - k(s) = k_{s0}\frac{(s/s_0)^{k_{s0}}}{1+(s/s_0)^{k_{s0}}}. \tag{6.7.10}$$

Now we can write the CCN (or droplet) concentration in the form of a modified power law as

$$N_{CCN}(s) = C_s(s)s^{k(s)}, \qquad s \text{ in share of unit.} \tag{6.7.11}$$

This equation is usually used with supersaturation s in percent, related to s_{su} in share of the unit as $s = 10^2\, s_{su}$. Substitution of this relation into (Eqn. 6.7.11) yields a more conventional form:

$$N_{CCN}(s) = C_{pc}(s)s^{k(s)}, \qquad s \text{ in percent.} \tag{6.7.12}$$

In (Eqn. 6.7.11), both $k(s)$ and $C_s(s)$ depend on the ratio s/s_0 and thus do not depend on units; $k(s)$ is calculated again from (Eqn. 6.7.7) but with s and s_0 in percent, and the coefficient

$$C_{pc}(s_{pc}) = 10^{-2k(s)}C(s). \tag{6.7.13}$$

Equation (6.7.12) is similar to power law (Eqn. 6.1.4); however, now with the parameters depending on s as just described. Hence, k(s) and $C(s)$ are now expressed directly via the parameters of the lognormal spectrum (N_a, r_{d0}, σ_d) or its algebraic equivalent and its physico-chemical properties via simple parameterization of b in (Eqns. 6.2.5)–(6.2.7).

Finally, the modified power law for the multimodal aerosol types is described again by (Eqns. 6.7.11) and (6.7.12) with the k(s)-index evaluated from (Eqn. 6.7.8) and the coefficient $C_s(s)$

$$C(s) = \sum_{i=1}^{I} C_{0i}s^{\chi_i(s)}[1+(s/s_{0i})^{k_{s0i}}]^{-1}, \tag{6.7.14}$$

where $\chi_i = k_{s0i} - k(s)$, and k_{s0i}, C_{0i}, s_{0i}, and N_{ai} are defined for each i-th mode. The coefficient $C_{pc}(s)$ with supersaturation in percent is defined by (Eqn. 6.7.13). The number of modes I is determined by the nature of aerosol and is the same as used for multimodal size spectra with the lognormal approach.

This analytical s-dependence of $C_s(s)$ and $k(s)$ in (Eqns. 6.7.11) and (6.7.12) provides a theoretical basis for many previous findings and improvements of the drop activation power law. In particular, the C-k space constructed by Braham (1976) based on experimental data follows now from the equations for $k(s)$ and $C_s(s)$ with use of any relation of maximum s to vertical velocity. The decrease of the k-slopes with increasing s analyzed by Jiusto and Lala (1981), Yum and Hudson (2001), Cohard et al. (1998, 2000), Shipway and Abel (2010), and others is predicted by (Eqns. 6.7.5)–(6.7.8), as well as its dependence on the nature of aerosol. The proportionality of coefficient C_s to the aerosol concentration explains substantially greater values of C_s observed in continental rather than maritime air masses as compiled by Twomey and Wojciechowski (1969) and Hegg and Hobbs (1992), and used by cloud modelers.

6.7.3. Supersaturation Dependence of Power Law Parameters

The equivalent power index $k(s)$, calculated with (Eqn. 6.7.7) and a monomodal algebraic (equivalent to lognormal) size spectrum, is shown in Fig. 6.15a and b and is compared with the experimental data from Jiusto and Lala (1981) who collected the data from several chambers. Given here is also the field data collected by Yum and Hudson (2001) with a continuous flow diffusion chamber (CFDC) at two altitudes in the Arctic in May 1998 during the First International Regional Experiment-Arctic Cloud Experiment (FIRE-ACE). It is seen that the k-index is not constant as assumed in many cloud models. The calculated k-indices reach the largest values of 1.5–4.2 at $s \le 0.01\%$, where they tend to the asymptotic k_{s0} (Eqn. 6.6.35c) and increase with decreasing dispersion of the dry CCN spectrum since $k_{s0} \sim (\ln\sigma_d)^{-1}$. As s increases, $k(s)$ is initially almost constant, and then decreases slowly up to $s \sim 0.03$–0.05%, and then decreases more rapidly in the region $s \sim 0.05$–0.12% to values less than 0.1–0.2 at $s > 1\%$. The k-indices increase with decreasing modal radii r_m and dispersions σ_d indicated in the legends. Agreement of calculated $k(s)$ with haze chamber data is reached at $\sigma_d = 1.3$, $r_m = 0.085$ μm, and with CFDC at $\sigma_d = 1.5$ and $r_m = 0.043$ μm. Agreement with the Arctic field data by Yum and Hudson (2001) is reached with $r_m = 0.045$ μm, $\sigma_d = 2.15$ in the layer 960–1010 mb and $r_m = 0.043$ μm, and $\sigma_d = 1.9$ in the layer 560–660 mb.

Thus, this model explains and describes quantitatively the observed decrease in the k-index with increasing supersaturation reported previously by many investigators (e.g., Twomey and Wojciechowski, 1969; Braham, 1976; Jiusto and Lala, 1981) and relates it to the aerosol size spectra and physico-chemical properties. A more detailed analysis of experimental data with these equations should be based on simultaneous use of the measured multimodal aerosol size spectra and CCN supersaturation activity spectra.

In contrast to most previous models where k and C are constant, the values of "effective" $C(s)$, $C_{pc}(s)$ in this model also depend on supersaturation. The spectral behavior of $C_{pc}(s)$ calculated from (Eqn. 6.7.9) with different r_m, σ_d and the same $N_a = 100$ cm^{-3} is shown in Fig. 6.16. For $s \ll s_0$, the coefficients tend to the asymptotic limit $C_{pc}(0) = 10^{-2k_{s0}} C$ and reach values of 10^3–10^5 cm^{-3} caused by the dependence $C_0 \sim A_k^{-2k_{s0}} r_{d0}^{k_{s0}(1+\beta)}$ with $A_k \sim 10^{-7}$ cm and $r_{d0} \sim 3 \times 10^{-6}$ to 10^{-5} cm. This dependence causes an increase in C_{pc} with decreasing σ_d (increasing k_{s0}) and with increasing r_{d0} seen at small s in Fig. 6.16; i.e., the narrower spectra and the larger modal radius, the faster the drop activation at small s. The values of $C_{pc}(s)$ decrease with increasing s by 1–3 orders of magnitude and tend to N_a at $s_{pc} > 0.1$–0.7%, since the asymptotics at large s is $C_{pc}(s_{pc}) = 10^{-2k(s)} N_a s_{pc}^{-k(s)}$, and $k(s) \to 0$ at large s. Hence, all the coefficients C_{pc} tend to N_a at $s_{pc} \ge 0.5\%$ as seen in Fig. 6.16. Thus, $N_{CCN}(s) = C_{pc} s_{pc}^{k(s)} \sim 10^{-2k(s)} N_a$ and tends to the limiting value N_a, which ensures finite values of N_a in contrast to the models with constant k, C.

The data on k and C measured at $s = 1\%$ in various air masses and geographical regions were compiled by Hegg and Hobbs (1992) and Pruppacher and Klett (1997). The values mostly varied in the wide ranges: in a maritime air mass, $k = 0.3$–1.3, $C = 100$–400; and in a continental air mass, $k = 0.4$–0.9, $C = 600$–5000. Such values are often used in cloud models at much smaller than 1% maximum supersaturations. The preceding consideration shows that these values of k and C may be substantially higher at smaller s (Figs. 6.15 and 6.16). Equations (6.7.6)–(6.7.10) allow recalculations between any values of supersaturation and direct calculations using the measured physico-chemical properties of aerosol.

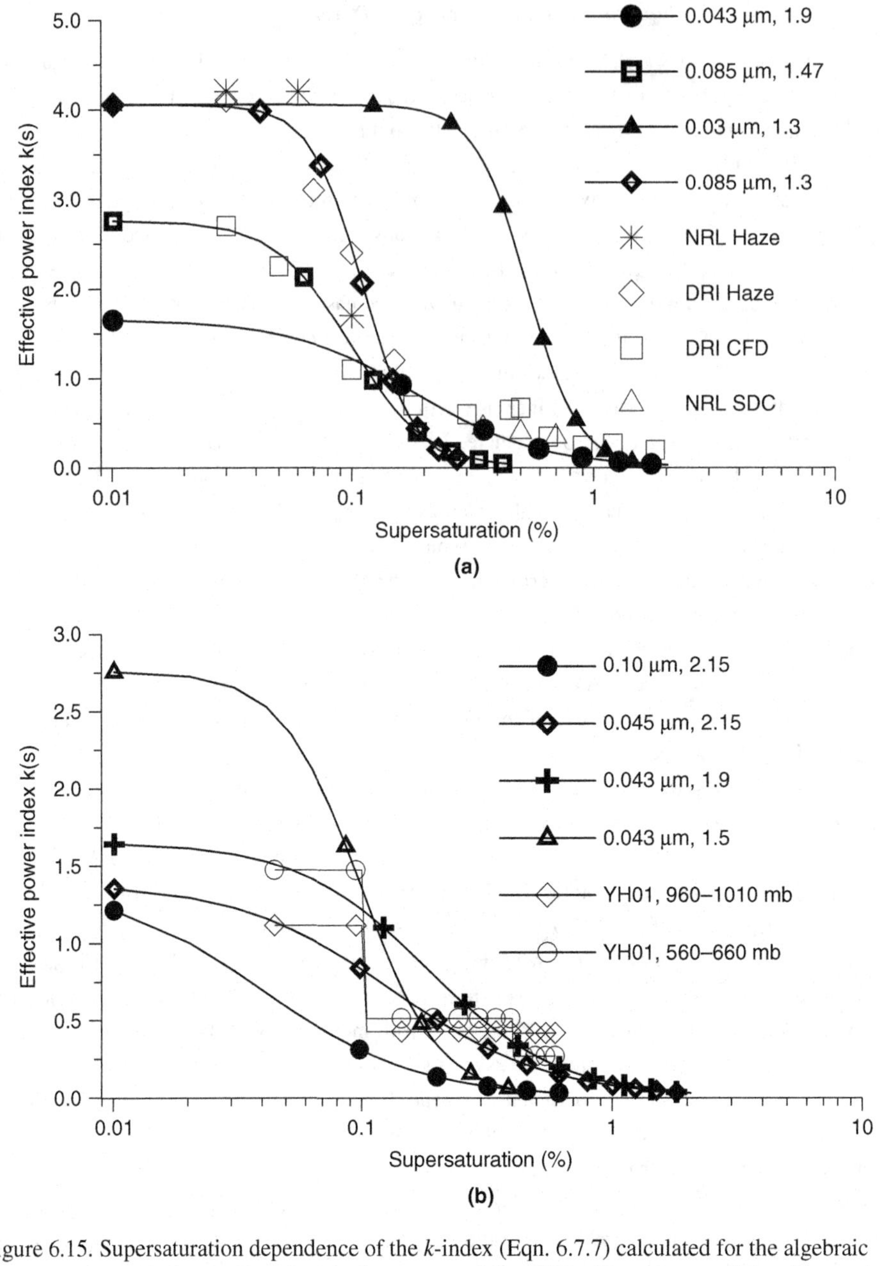

Figure 6.15. Supersaturation dependence of the *k*-index (Eqn. 6.7.7) calculated for the algebraic size spectrum equivalent to the lognormal spectrum of dry CCN with various modal radii r_m and dispersions σ_d indicated in the legend and comparison: (a) with the experimental data from Jiusto and Lala (1981) from 4 chambers: haze chambers (NRL Haze and DRI Haze), continuous flow diffusion chamber (DRI CFD), and static diffusion chamber (NRL SDC), and (b) with the data collected in FIRE-ACE in 1998 (Yum and Hudson, 2001) (YH01) at the heights 960–1010 mb and 560–660 mb. From Khvorostyanov and Curry (2006), *J. Geophys. Res.*, 111, D12202, reproduced with permission of John Wiley & Sons, Inc.

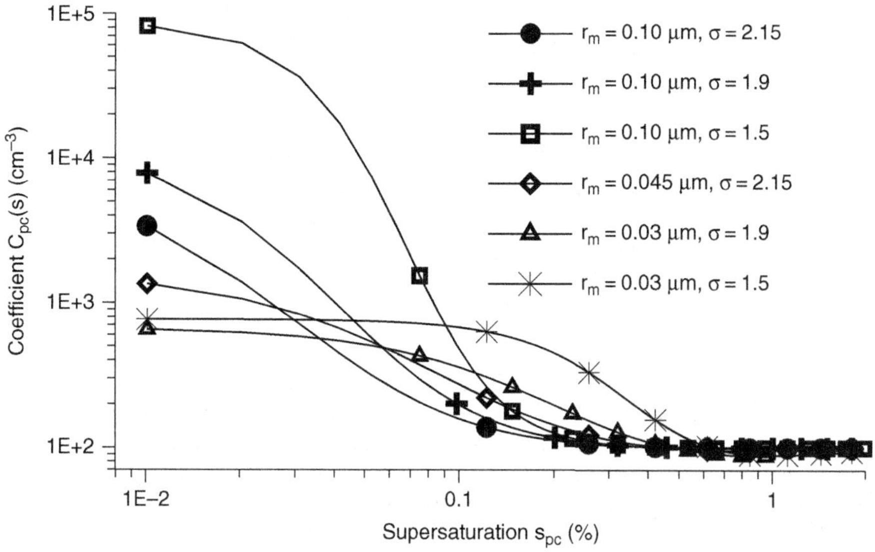

Figure 6.16. Equivalent coefficient $C_{pc}(s)$ (Eqns. 6.7.9) and (6.7.13) in the modified power law for N_{CCN} (Eqns. 6.7.11) and (6.7.12) calculated for the algebraic size spectrum equivalent to the lognormal size spectrum of dry CCN with modal radii r_m and dispersions σ_d indicated in the legend and other parameters described in the text. For the convenience of comparison, all curves are calculated with $N_a = 10^2$ cm^{-3}. From Khvorostyanov and Curry (2006), *J. Geophys. Res.*, 111, D12202, reproduced with permission of John Wiley & Sons, Inc.

A comparison of the concentrations $N_{CCN}(s)$ calculated using the generalized power law (Eqn. 6.7.11) and *erf* function based on the lognormal distribution (Eqn. 6.7.1) shows in Fig. 6.17 their good agreement and illustrates good accuracy of the algebraic distributions and renewed generalized power law as compared to the direct integration of the lognormal distributions. The curves $N_{CCN}(s)$ calculated with the newer power law are not linear in log-log coordinates as they are with the old power law (Eqn. 6.1.4) and constant C_0, k, but are concave in agreement with the measurements (e.g., Jiusto and Lala, 1981; Hudson, 1984; Yum and Hudson, 2001) and other models (e.g., Ghan et al., 1993; 1995; Feingold et al., 1994; Cohard et al., 1998, 2000; Shipway and Abel, 2010). The reason for this concavity is quite clear from Figs. 6.5a,b and 6.11–6.14 for the size and activity spectra: the rate of increase in the concentration of activated drops $dN_d(s)/ds$ is greatest at small supersaturations in the left branch of $\varphi_s(s)$ (Figs. 6.11–6.14), which correspond to the right branch of $f_d(r_d)$ at $r_d > r_m$ (Fig. 6.5a,b). Thus, $dN_d(s)/ds$ decreases with growing s, especially at $s > 1\%$—i.e., in the left branch of the small dry CCN with $r_d < 0.03$ μm (Fig. 6.5a). Effectively, some "saturation" with respect to supersaturation occurs, causing $k(s)$, $C(s)$, and $dN_d(s)/ds$ to decrease with s and leading to finite $N_d(s)$ at large s.

Finally, Fig. 6.18 shows $N_{CCN}(s)$ calculated with the lognormal size spectra and generalized power law (Eqn. 6.11) and compared to the corresponding field data obtained by Yum and Hudson (2001) in the FIRE-ACE campaign in May 1998 in the Arctic. The parameters r_m, σ_d, and N_a were varied

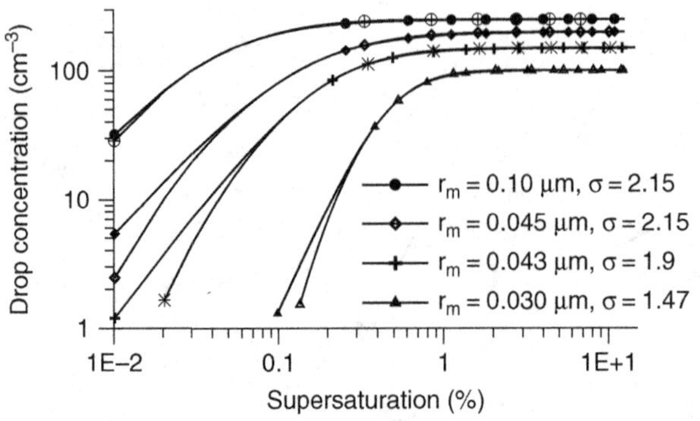

Figure 6.17. Comparison of the accumulated CCN spectra $N_{CCN}(s)$ calculated using the modified power law (Eqn. 6.7.11) (solid symbols) and using (Eqn. 6.7.1) with the *erf* function (open symbols). The values of the corresponding modal radii r_m and dispersions σ_d of the dry aerosol spectra are indicated in the legend. For convenience of comparison, N_{CCN} is calculated with the dry aerosol concentrations of 250, 200, 150, and 100 cm^{-3} from the upper to lower curves. From Khvorostyanov and Curry (2006), *J. Geophys. Res.*, 111, D12202, reproduced with permission of John Wiley & Sons, Inc.

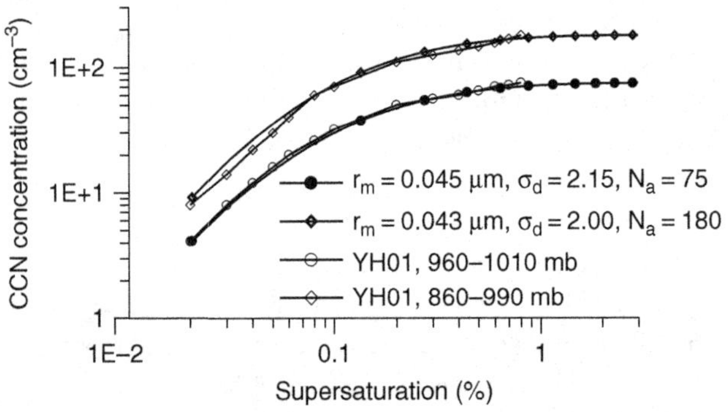

Figure 6.18. Supersaturation dependence of the CCN integral activity spectrum $N_{CCN}(s)$ calculated with the modified power law (Eqn. 6.7.11) for the lognormal size spectra with r_m, σ_d, and N_a (in cm^{-3}) indicated in the legend (solid diamonds and circles), compared to the corresponding field data obtained by Yum and Hudson (2001, YH01) in the FIRE-ACE campaign in May 1998 in the Arctic in two flights at altitudes 960–1010 mb and 860–990 mb (open diamonds and circles). From Khvorostyanov and Curry (2006), *J. Geophys. Res.*, 111, D12202, reproduced with permission of John Wiley & Sons, Inc.

slightly in the calculations, although the calculations were constrained to the measured *k*-indices (Fig. 6.15b), so that the values of the parameters had to provide an agreement for both $k(s)$ and $N_{CCN}(s)$. Good agreement over the entire range of supersaturations was found with the values indicated in the legend. The values of modal radii and dispersions are typical for the accumulation mode. This comparison shows that this modified power law (Eqns. 6.7.11)–(6.7.12) can reasonably reproduce the experimental data on CCN activity spectra and concentrations of activated drops and can be used for parameterization of drop activation in cloud models.

Appendix A.6 for Chapter 6. Solutions of Cubic Equations for Equilibrium and Critical Radii

A cubic equation of the form

$$x^3 + \tilde{a}x^2 + \tilde{b}x + \tilde{c} = 0. \tag{A.6.1}$$

has algebraic and trigonometric solutions.

The algebraic Cardano's solution is sought by substitution:

$$x = y - \tilde{a}/3. \tag{A.6.2}$$

(e.g., Korn and Korn, 1966). This gives an incomplete cubic equation for y

$$y^3 + py + q = 0, \tag{A.6.3}$$

where p and q are expressed via the primary parameters

$$p = -\frac{\tilde{a}^2}{3} + \tilde{b}, \qquad q = 2\left(\frac{\tilde{a}}{3}\right)^3 - \frac{\tilde{a}\tilde{b}}{3} + \tilde{c}. \tag{A.6.4}$$

The three roots y_1, y_2, and y_3 of the incomplete cubic (Eqn. A.6.3) are

$$y_1 = \tilde{A} + \tilde{B}, \quad y_{2,3} = -\frac{\tilde{A} + \tilde{B}}{2} \pm i\frac{\tilde{A} - \tilde{B}}{2}\sqrt{3}, \tag{A.6.5}$$

with

$$\tilde{A} = \left(-\frac{q}{2} + Q^{1/2}\right)^{1/3}, \qquad \tilde{B} = \left(-\frac{q}{2} - Q^{1/2}\right)^{1/3}, \tag{A.6.6}$$

$$Q = \left(\frac{p}{3}\right)^3 + \left(\frac{q}{2}\right)^2, \tag{A.6.7}$$

where $i = \sqrt{-1}$. The choice of the cubic roots in $\tilde{A}$ and $\tilde{B}$ is determined by the relation $\tilde{A}\tilde{B} = -p/3$. If the coefficients of (Eqn. A.6.1) are real, then the equation has: a) one real root and two complex-conjugated roots if $Q > 0$; or b) three real roots, at least two of which are equal, if $Q = 0$; or c) three real different roots if $Q < 0$.

The trigonometric solution of (Eqns. A.6.3) or (6.2.21) with real coefficients, $Q < 0$ and $p < 0$ is

$$y_1 = 2\left(-\frac{p}{3}\right)^{1/2} \cos\frac{\alpha}{3}, \qquad y_{2,3} = -2\left(-\frac{p}{3}\right)^{1/2} \cos\left(\frac{\alpha}{3} \pm \frac{\pi}{3}\right), \tag{A.6.8}$$

with

$$\alpha = \arccos\left[-\frac{q}{2}\left(-\frac{p}{3}\right)^{-3/2}\right]. \tag{A.6.9}$$

Only the real root y_1 is physical and should be used, and the roots $y_{2,3}$ are complex and should be rejected.

These two types of solutions, algebraic (Eqns. A.6.5)–(A.6.7) and trigonometric (Eqn. A.6.8), are used in Sections 6.2 and 6.3.

Solution (Eqn. 6.2.21) for the Equilibrium Radius r_{wi} of Interstitial Aerosol

Equation (6.2.19) for the equilibrium radius r_{wi} of interstitial aerosol is reduced to the cubic (Eqn. 6.2.20) for the dimensionless variable $x = r_w/r_d$:

$$x^3 - 3Z_d^2 x - 1 = 0, \quad \text{with} \quad Z_d = \left(\frac{br_d^{2\beta}}{3A_k}\right)^{1/2} = r_d^{\beta}\left(\frac{b}{3A_k}\right)^{1/2}. \tag{A.6.10}$$

This is an incomplete cubic equation for x of the form (Eqn. A.6.3) with $\tilde{a} = 0$ and the coefficients

$$p = -3Z_d^2, \quad q = -1. \tag{A.6.11}$$

A solution can be chosen in trigonometric or algebraic form, depending on the sign of Q, defined in (Eqn. A.6.7) as

$$Q = \left(\frac{p}{3}\right)^3 + \left(\frac{q}{2}\right)^2 = -Z_d^6 + \frac{1}{4}. \tag{A.6.12}$$

The trigonometric solution is convenient if $Q < 0$, or $Z_d^6 > 1/4$ —i.e., with a sufficiently large soluble fraction. Thus, there is one physical real root; the other two roots are complex-conjugated and should be rejected. The real root is

$$y_1 = 2\left(-\frac{p}{3}\right)^{1/2} \cos\frac{\alpha}{3}, \quad \text{with} \quad \cos\alpha = -\frac{q}{2}\left(-\frac{p}{3}\right)^{-3/2}. \tag{A.6.13}$$

Substituting here p and q from (Eqn. A.6.11), we obtain $x = y_1$, since $\tilde{a} = 0$:

$$\begin{aligned} x = y_1 &= 2Z_d \cos\left[\frac{1}{3}\arccos\left(\frac{Z_d^{-3}}{2}\right)\right] \\ &= 2r_d^{\beta}\left(\frac{b}{3A_k}\right)^{1/2} \cos\left\{\frac{1}{3}\arccos\left[\frac{1}{2}\left(\frac{3A_k}{br_d^{2\beta}}\right)^{3/2}\right]\right\}. \end{aligned} \tag{A.6.14}$$

Finally, the equilibrium interstitial aerosol radius $r_{wi} = xr_d$ is

$$r_{wi} = r_d^{1+\beta}\left(\frac{b}{3A_k}\right)^{1/2} \cos\left\{\frac{1}{3}\arccos\left[\frac{1}{2}\left(\frac{3A_k}{br_d^{2\beta}}\right)^{3/2}\right]\right\}. \tag{A.6.15}$$

The algebraic Cardano's solution is more convenient if $Q > 0$, or $Z_d^6 < 1/4$ —i.e., for a small soluble fraction. Then, the solution of an incomplete cubic equation is described by (Eqns. A.6.5) and (A.6.6) with Q from (Eqn. A.6.12). Thus

$$y_1 = \tilde{A} + \tilde{B}, \qquad \tilde{A} = \left(-\frac{q}{2} + Q^{1/2}\right)^{1/3} = \left[\frac{1}{2} + \left(\frac{1}{4} - Z_d^6\right)^{1/2}\right]^{1/3}, \tag{A.6.16}$$

$$\tilde{B} = \left(-\frac{q}{2} - Q^{1/2}\right)^{1/3} = \left[\frac{1}{2} - \left(\frac{1}{4} - Z_d^6\right)^{1/2}\right]^{1/3}, \tag{A.6.17}$$

$$r_{wi} = r_d y_1 = \frac{r_d}{2^{1/3}}\{[1+(1-4Z_d^6)^{1/2}]^{1/3} + [1-(1-4Z_d^6)^{1/2}]^{1/3}\}, \tag{A.6.18}$$

with Z_d from (Eqn. A.6.10).

Solution (Eqns. 6.3.7)–(6.3.8) for the Critical Radius r_{cr}

Equation (6.3.6) for the critical radius

$$r_{cr}^3 - (3r_d Z_d)r_{cr}^2 - r_d^3 = 0, \qquad Z_d = \left(\frac{br_d^{2\beta}}{3A_k}\right)^{1/2}, \tag{A.6.19}$$

can be rewritten as an equation for a dimensionless variable $x = r_{cr}/r_d$ as

$$x^3 + \tilde{a}x^2 - 1 = 0, \qquad \tilde{a} = -3Z_d. \tag{A.6.20}$$

It is a particular case of the complete cubic (Eqn. A.6.1) for x with coefficients $\tilde{a} = -3Z_d$, $\tilde{b} = 0$, and $\tilde{c} = -1$. Then, the coefficients p, q, and Q of the general algebraic Cardano's solution (Eqns. A.6.4)–(A.6.7) can be expressed via the primary parameters

$$p = -\frac{\tilde{a}^2}{3} + \tilde{b} = -3Z_d^2, \tag{A.6.21}$$

$$q = 2\left(\frac{\tilde{a}}{3}\right)^3 - \frac{\tilde{a}\tilde{b}}{3} + \tilde{c} = -2Z_d^3 - 1, \tag{A.6.22}$$

$$Q = \left(\frac{p}{3}\right)^3 + \left(\frac{q}{2}\right)^2 = -Z_d^6 + \left(Z_d^3 + \frac{1}{2}\right)^2 = Z_d^3 + \frac{1}{4}. \tag{A.6.23}$$

Since the coefficients of the original equation are real and $Q > 0$ in (Eqn. A.6.23), the case a) of the preceding three cases is valid and the equation has one real root. Substitution of these p, q, and Q into (Eqn. A.6.6) gives $\tilde{A}$ and $\tilde{B}$ after some algebraic transformations

$$\tilde{A} = \left(-\frac{q}{2} + Q^{1/2}\right)^{1/3} = P_+(Z_d), \tag{A.6.24}$$

$$P_+(Z_d) = \left(Z_d^3 + \left(Z_d^3 + \frac{1}{4}\right)^{1/2} + \frac{1}{2}\right)^{1/3}, \tag{A.6.25}$$

$$\tilde{B} = \left(-\frac{q}{2} - Q^{1/2}\right)^{1/3} = P_-(Z_d), \tag{A.6.26}$$

$$P_-(Z_d) = \left(Z_d^3 - \left(Z_d^3 + \frac{1}{4}\right)^{1/2} + \frac{1}{2}\right)^{1/3}, \tag{A.6.27}$$

Now, the single real solution of the incomplete cubic (Eqn. A.6.3) is

$$y_1 = \tilde{A} + \tilde{B} = P_+(Z_d) + P_-(Z_d). \tag{A.6.28}$$

The solution of the original equation of the cubic (Eqn. A.6.19) for r_{cr} is obtained after returning to the variable x according to (Eqn. A.6.2):

$$x = y_1 - \frac{\tilde{a}}{3} = y_1 + Z_d. \qquad (A.6.29)$$

Substituting here y_1 from (Eqn. A.6.28), and the relation $r_{cr} = r_d x$, we finally obtain

$$r_{cr}(r_d) = r_d \chi(Z_d), \qquad \chi(Z_d) = \left[Z_d + P_+(Z_d) + P_-(Z_d) \right]. \qquad (A.6.30)$$

This is the final solution (Eqn. 6.3.7) for $r_{cr}(r_d)$ given in Section 6.3.

7
Activation of Cloud Condensation Nuclei into Cloud Drops

7.1. Introduction

As discussed in Chapter 6, the number of cloud drops is an important determinant of cloud optical properties and the formation of precipitation. In the initial stages of cloud formation, the number of cloud drops is directly related to the number of cloud condensation nuclei (CCN) activated. Drop activation schemes used prior to the 1990s were based mostly on Twomey's (1959) power law for the concentration of drops N_{dr} activated on cloud condensation nuclei (CCN). This scheme is still applied in some atmospheric models. The drawback of Twomey's parameterization is that N_{dr} is unlimited at high supersaturations and can exceed CCN concentration N_a.

A substantial revision and intensive development of new cloud drop activation parameterizations and their effects on climate modeling began in the 1990s (see reviews in Charlson et al., 2001; Lohmann and Feichter, 2005; McFiggans et al., 2006; Svenningsson et al., 2006; IPCC, 2007; and in Chapter 6 of this book). Along with the analytical modifications of Twomey's power law described in Chapter 6, parameterizations of drop nucleation have been developed using results from parcel models that are fit to various empirical functions of a few nondimensional parameters (Abdul-Razzak et al., 1998; Abdul-Razzak and Ghan, 2000, 2004; Fountoukis and Nenes, 2005; Ming et al., 2006). Saleeby and Cotton (2004) developed another approach consisting of lookup tables compiled from several thousand runs of a parcel model. A comparison of various recent parameterizations of drop activation and their performance was conducted by Ghan et al. (2011).

The necessity to run parcel models many times for compiling lookup tables or tuning the parameterizations may make the procedure time-consuming, and does not easily reveal the relevant physical relationships. Two alternative approaches were suggested by Khvorostyanov and Curry (2008a, 2009a), who derived the integro-differential equation for supersaturation using the algebraic size and activity spectra from Khvorostyanov and Curry (2006, 2007) described in Chapter 6. In the first approach, Khvorostyanov and Curry (2008a) found a fast numerical solution of this equation for the time of CCN activation t_m, maximum supersaturation s_m, and concentration of activated droplets N_{dr} as functions of the physico-chemical properties and size spectrum of the CCN and the vertical velocity w, and studied the kinetics of drop activation using this numerical solution. This method can be much faster than methods based on running numerous parcel models and compiling the lookup tables.

A deficiency of all numerical solutions is that they do not demonstrate the general analytical dependencies of the drop activation on various parameters. The search for analytical solutions for t_m, s_m, and N_{dr} is desirable because they allow the following applications: a) control for the accuracy of various numerical solutions, especially in cases where general dependencies are unclear and the spread of the parcel model runs can be significant (e.g., at strong variations of the condensation coefficient or surface tension); b) development of newer parameterizations based on the characteristic parameters occurring in these solutions; c) construction of the general solutions by interpolation among the particular analytical solutions; and d) the design of laboratory and field experiments. A method based on analytical solutions of the supersaturation equation was developed in Khvorostyanov and Curry (2009a).

It has been known since the pioneering work by Twomey (1959) that the supersaturation equation does not allow general analytical solutions, even with the simplest power law activity spectrum, unless certain simplifications are made. The situation becomes even more complicated when more complex activity spectra are used (e.g., lognormal or algebraic) because of the impossibility of exact analytical evaluation of the characteristic integrals of this problem. Twomey (1959) considered droplet growth with the condensation coefficient $\alpha_c \sim 1$—i.e., in the diffusion regime of droplet growth. Subsequent laboratory and field experiments and theoretical works showed that α_c can be as low as 10^{-5}—i.e., droplets grow in the kinetic regime (Deryagin et al., 1966; Deryagin and Kurgin, 1972; Pruppacher and Klett, 1997; Seinfeld and Pandis, 1998; Buikov and Khvorostyanov, 1979; Feingold and Chuang, 2002; Chuang, 2003). However, parameterizations of drop activation considered mostly the cases with $\alpha_c \sim 1$, with some numerical estimates of the effects of α_c (Nenes et al., 2002; Ming et al., 2006), but the general analytical dependencies of t_m, s_m, and N_{dr} on α_c have not been derived.

The CCN differential activity spectrum is written as described in Chapter 6:

$$\varphi_s(s) = dN_{dr}(s)/ds, \tag{7.1.1}$$

where N_{dr} is the number of cloud drops nucleated, which corresponds directly to the number of CCN activated at supersaturation s. The number of CCN actually activated is determined by the maximum supersaturation s_m reached in a cooling cloud parcel, which is determined by two competing processes: supersaturation generation due to cooling, and supersaturation absorption due to condensation growth of the newly formed drops. The final N_{dr} formed after the supersaturation has reached s_m is given by the integral CCN activity spectrum:

$$N_{dr}(s_m) = \int_0^{s_m} \varphi_s(s')\,ds'. \tag{7.1.2}$$

Thus, parameterization of N_{dr} according to (Eqn. 7.1.2) requires knowledge of both $\varphi_s(s)$ and s_m, and therefore a solution of the supersaturation equation in order to find s_m.

Among the various methods for parameterizing the differential CCN spectrum $\varphi_s(s)$, we consider three different mathematical approaches: power law, lognormal, and algebraic. The most commonly used power law approach was introduced by Twomey (1959) (see Chapter 6):

$$\varphi_s(s) = C_T k s^{k-1}, \tag{7.1.3}$$

where the constant C_T and the index k are typically empirically determined. Substitution of (Eqn. 7.1.3) into (Eqn. 7.1.2) yields the corresponding power law

$$N_{dr}(s_m) = C_T s_m^k, \tag{7.1.4}$$

which has been widely used in cloud physics. As mentioned before, the drawbacks of this parameterization are the following: the value of N_{dr} is not limited at high s_m and can exceed total CCN concentration N_a; and the parameters C_T and k are usually determined empirically from measurements of N_{dr}

at varying values of s_m and are not related directly to the aerosol microphysical properties. Besides, as shown in Chapter 6 , cloud chamber experiments have found a decrease of the index k with increasing s_m (e.g., Jiusto and Lala, 1981; Ji and Shaw, 1998; Yum and Hudson, 2001; Pruppacher and Klett, 1997), and hence there is no unique value of k even for the same aerosol type.

We considered in Chapter 6 the activity spectra $\varphi_s(s)$, and accumulated spectra $N_{dr}(s)$ and treated supersaturation as a continuous variable. Hence, the method developed in Chapter 6 can be used as a refinement to the power law used in cloud models with prognostic supersaturation, or in the analysis of chamber experiments, where the maximum supersaturation s_m can be calculated or measured and then $N_{dr}(s_m)$ determined. However, most cloud models, numerical weather prediction models, and climate models do not have prognostic supersaturation equations and do not evaluate supersaturation explicitly. Hence, they use parameterizations for s_m and N_{dr} in terms of the predicted variables, most often in terms of the vertical velocity w similar to Twomey's (1959) original parameterization (e.g., Ghan et al., 1997; Lohmann et al., 1999; Saleeby and Cotton, 2004; Morrison et al., 2005a,b).

When s is not an independent continuous variable, but is constrained by s_m due to interaction of the dynamical and microphysical factors, the applicability of the power law (Eqn. 7.1.4) becomes unclear. To ensure finite $N_{dr} \le N_a$ at high values of s, the indices k and coefficients C_T should be different at low and high values of s_m. However, it is uncertain in general if s_m and the function $N_{dr}(s_m)$ can be described by the power law or quasi-power law with variable k and C_T. If so, analytical expressions are needed for s_m and N_{dr} that relate the values of k and C_T to aerosol microphysics and cloud dynamics.

For models with longer time steps that cannot accommodate a prognostic equation for supersaturation, a parameterization is needed that allows us to calculate s_m, along with the time t_m during which maximum supersaturation is reached and CCN activation proceeds, and the final parameterization of $N_{dr}(s_m)$. This requires analysis of the dependencies of t_m, s_m, N_{dr}, and other characteristics on the parameters of the CCN size spectra and chemical composition. Toward improving parameterization of cloud droplet activation in cloud and climate models, Chapter 7 extends the quasi-power law for drop activation obtained in Chapter 6. The integro-differential equation for integral water supersaturation is derived and solved together with the other equations of drop activation kinetics, which yields relatively simple equations for t_m, s_m, and N_{dr} in the form of the quasi-power laws that are functions of the vertical velocity and parameters of aerosol size spectra and chemical composition.

7.2. Integral Supersaturation in Liquid Clouds with Drop Activation

When an air parcel rises adiabatically at a vertical velocity w, the air cools, and the relative humidity increases and reaches saturation at some level. Then, the parcel becomes slightly supersaturated and CCN activation begins. The parcel supersaturation is governed by two competing processes: supersaturation generation by the rising motion and supersaturation absorption by the drops in the condensation process.

This process can be described by the equation for water supersaturation, which we use in the form from Section 5.3. Using (Eqn. 5.3.16) for a liquid cloud ($I_{dep} = 0$), and the relation $(1/p)(dp/dt) = -(g/R_aT)w$ based on the equation of state and hydrostatic approximation, we can write

$$\frac{1}{(1+s_w)}\frac{ds_w}{dt} = c_{1w}w - \frac{\Gamma_1}{\rho_v}I_{con}, \qquad (7.2.1)$$

where

$$c_{1w}(T) = \left(\frac{L_e}{c_p T} \frac{M_w}{M_a} - 1 \right) \frac{g}{R_a T}. \tag{7.2.2}$$

Γ_1 is the psychrometric correction (Eqn. 5.3.15), and all notation is the same as in Section 5.3. The condensation vapor flux I_{con} to the droplets is the integral of the mass growth rate over the size spectrum that is expressed via growth rate (dr_d/dt) of the droplet radius r_d:

$$I_{con} = 4\pi\rho_w \int_0^\infty r_d^2 \frac{dr_d}{dt} f(r_d, t)\, dr_d, \tag{7.2.3}$$

where $f_d(r_d)$ is the size distribution function of the newly formed drops. We use (dr_d/dt) in the form (Eqn. 5.1.46) of Section 5.1

$$\frac{dr_d}{dt} = \frac{c_{3w} s_w}{r_d + \xi_{con}}, \qquad c_{3w} = \frac{D_v \rho_{ws}}{\rho_w \Gamma_1}, \tag{7.2.4}$$

$$\xi_{con} = \frac{4D_v}{\alpha_c V_w}, \qquad V_w = \left(\frac{8RT}{\pi M_w} \right)^{1/2}, \tag{7.2.5}$$

where the notation is the same as in Section 5.1.

Substitution of dr_d/dt from (Eqn. 7.2.4) into (Eqn. 7.2.3) yields

$$I_{con}(t) = s_w(t) \frac{4\pi D_v \rho_{vs}}{\Gamma_1} \int_0^\infty \frac{r_d^2(t, t_0)}{r_d(t, t_0) + \xi_{con}} f_d(r_d, t)\, dr_d. \tag{7.2.6}$$

Here $r_d(t,t_0)$ denotes the radius at time t of a drop activated at time t_0. The radius $r_d(t,t_0)$ is evaluated by integration of (Eqn. 7.2.4) by the time from t_0 to t

$$r_d(t, t_0) = \{[r_d(t_0) + \xi_{con}]^2 + 2c_{3w}[y_w(t) - y_w(t_0)]\}^{1/2} - \xi_{con}, \tag{7.2.7}$$

where $r_d(t_0)$ is the initial drop radius at the activation time t_0, and $y_w(t)$ is the integral supersaturation:

$$y_w(t) = \int_0^t s_w(t')\, dt'. \tag{7.2.8}$$

To express the size spectrum $f(r_d)$ via the activity spectrum, we apply a kinetic equation for the diffusion growth of the droplet size spectrum:

$$\frac{\partial f_d(r_d)}{\partial t} + \frac{\partial}{\partial r} \left(\frac{dr_d}{dt} f_d \right) = \varphi_s(s_w) \frac{ds_w}{dt} \delta(r_d - r_d(t_0)), \tag{7.2.9}$$

where $\varphi_s(s_w)$ is the differential activity spectrum, and $\delta(x)$ is the Dirac delta-function. A solution to this equation describes the conservation law for the concentration of the newly formed drops and CCN:

$$f_d(r_d) = \varphi_s(s_w) \frac{ds_w}{dt_0} \frac{dt_0}{dr_d}. \tag{7.2.10}$$

Substitution of $r_d^2(t, t_0)$ and $[r_d(t, t_0) + \xi_{con}]$ from (Eqn. 7.2.7) and $f_d(r_d)$ from (Eqn. 7.2.10) into (Eqn. 7.2.6) yields

$$I_{con}(t) = s_w(t)\frac{4\pi D_v \rho_{vs}}{\Gamma_1}\int_0^t r_{d,ef}(t,t_0)\varphi_s(s_w)\frac{ds_w(t_0)}{dt_0}dt_0, \quad (7.2.11a)$$

where the integration is performed over the initial times of activation t_0 and

$$r_{d,ef}(t,t_0) = \frac{\{[(r_d(t_0)+\xi_{con})^2 + 2c_{3w}(y_w(t) - y_w(t_0))]^{1/2} - \xi_{con}\}^2}{[(r_d(t_0)+\xi_{con})^2 + 2c_{3w}(y_w(t) - y_w(t_0))]^{1/2}}. \quad (7.2.11b)$$

Note that the integrand in (Eqn. 7.2.11a) represents a product of the "source function" $\varphi(s_w)ds_w/dt_0$, which describes the "birth of particles" at a time t_0, and of $r_{d,ef}(t, t_0)$, which plays here the role of the Green function and describes the transformation of the particles formed at time t_0 to the time t.

As shown in the following, variations of α_c (or ξ_{con}) can have substantial effects on the activation process, especially for small α_c. The method developed in Khvorostyanov and Curry (2008a, 2009a) and described here provides analytical solutions for any given value of α_c as described in Sections 7.6–7.8 and allows examination of variable α_c and the kinetic correction ξ_{con} without running parcel models.

Substituting the algebraic activity spectrum $\varphi_s(s_w)$ from (Eqn. 6.6.36) into (Eqn. 7.2.11a) and using the relations $s_w = y_w'$, $\rho_v = \rho_{ws}(1 + s_w) = \rho_{ws}(1 + y_w')$, or $\rho_{ws}/\rho_v = (1 + s_w)^{-1} = (1 + y_w')^{-1}$, yields the 2nd (relaxation) term on the right-hand side of the supersaturation (Eqn. 7.2.1) in the form

$$\frac{\Gamma_1}{\rho_v}I_{con}(t) = \frac{y_w'(t)}{[1+y_w'(t)]}[4\pi D_v N_a J_0(t)] = \frac{y_w'(t)[1+y_w'(t)]^{-1}}{\tau_{f,ac}(t)}, \quad (7.2.12)$$

$$\tau_{f,ac}(t) = [4\pi D_v N_a r_{act}(t)]^{-1}, \quad (7.2.13)$$

where we introduced the effective supersaturation relaxation time $\tau_{f,ac}$ during this stage of drop activation, an activation radius $r_{act}(t)$, and an integral $J_0(t)$ that has the dimension of length

$$J_0(t) \equiv r_{act}(t) = \frac{k_{s0}}{s_0^{k_{s0}}}\int_0^t r_{d,ef}(t,t_0)\frac{[y_w'(t_0)]^{k_{s0}-1}y_w''(t_0)}{\{1+\eta_0[y_w'(t_0)]^{k_{s0}}\}^{\mu}}dt_0. \quad (7.2.14)$$

Here $\eta_0 = s_0^{-k_{s0}}$ is expressed via CCN microphysical properties in (Eqn. 6.6.37b), k_{s0} is defined in (Eqn. 6.6.30a), $k_{s0} = 4/(\sqrt{2\pi}\ln\sigma_s)$, and s_0 and $\sigma_s = \sigma_d^{(1+\beta)}$ are the mean geometric supersaturation and its dispersion defined in (Eqn. 6.6.28). The single and double primes at $y_w(t_0)$ hereafter denote the first and second time derivatives. Note that $J_0(t)$ is written in somewhat more general form with the parameter μ in the denominator of the integrand. With the Khvorostyanov and Curry (2006) algebraic activity spectrum (Eqns. 6.6.36 or 6.6.30b), we have $\mu = 2$ in the denominator of integrand, but μ can have different values in other models (e.g., $\mu = 0$ in the power law models by Twomey (1959), Buikov (1966a,b), Sedunov (1967, 1974), and subsequent similar models).

Using (Eqns. 7.2.12)–(7.2.14), the integral water supersaturation equation (Eqn. 7.2.1) becomes

$$y_w''(t) = c_{1w}w[1+y_w'(t)]$$
$$-y_w'(t)\left(4\pi D_v N_a \frac{k_{s0}}{s_0^{k_{s0}}}\int_0^t r_{d,ef}(t,t_0)\frac{[y_w'(t_0)]^{k_{s0}-1}y_w''(t_0)}{\{1+\eta_0[y_w'(t_0)]^{k_{s0}}\}^{\mu}}dt_0\right). \quad (7.2.15)$$

This equation can be written in a more compact form via J_0 or $\tau_{f,ac}$ as

$$y''_w(t) = c_{1w}w[1 + y'_w(t)] - y'_w(t)[4\pi D_v N_a J_0(t)]. \tag{7.2.16}$$

$$y''_w(t) = c_{1w}w[1 + y'_w(t)] - y'_w(t)[\tau_{p,ac}(t)]^{-1}, \tag{7.2.17}$$

This is a nonlinear integro-differential equation for the integral supersaturation $y_w(t)$ that governs the kinetics of the drop formation. With $\mu = 0$, neglecting the kinetic correction, $\xi_{con} = 0$, and the initial radius, $r_d(t_0) = 0$, the integral $J_0(t)$ reduces to the form similar to Twomey's formulation

$$J_0(t) = \frac{k_{s0}}{s_0^{k_{s0}}}(2c_{w3})^{1/2}\int_0^t [y_w(t) - y_w(t_0)]^{1/2}[y'_w(t_0)]^{k_{s0}-1} y''_w(t_0)dt_0. \tag{7.2.18}$$

Using the definition $C_0 = N_a s_0^{-k_{s0}}$ from (Eqn. 6.6.37a), the supersaturation equation can be written as

$$y''_w(t) = c_{1w}w[1 + y'_w(t)]$$
$$-y'_w(t)\left(4\pi D_v C_0 k_{s0}\int_0^t [y_w(t) - y_w(t_0)]^{1/2}[y'_w(t_0)]^{k_{s0}-1} y''_w(t_0)\, dt_0\right), \tag{7.2.19}$$

which is close to the form used by Twomey (1959) (see also Sedunov, 1967, 1974, and Pruppacher and Klett, 1997, Chapter 13 therein), which leads to a solution $N_{dr}(s_m) \sim C_0 s_m^{k_{s0}}$. Equations (7.2.15)–(7.2.17) generalize the previous (Eqn. 7.2.19) by the presence of the new term $\{1 + \eta_0[y'_w(t_0)]^{k_{s0}}\}^{\mu}$ with $\mu = 2$ in the denominator of the integrand. This term ensures correct asymptotic behavior of the activation spectrum and gives finite droplet concentrations at large supersaturations s_m or large vertical velocities. Besides, (Eqns. 7.2.15)–(7.2.17) account for the initial radius $r_d(t_0)$, and for the kinetic correction ξ_{con}, which allows us to study the effect of the condensation coefficient α_c on the kinetics of drop activation as will be shown next.

7.3. Analytical Solutions to the Supersaturation Equation

Solutions to the supersaturation (Eqn. 7.2.15) should yield the maximum value s_m of supersaturation in a rising parcel, the time t_m when s_m is reached, and the concentration N_{dr} of the activated drops. This solution requires evaluation of the integral J_0 in (Eqn. 7.2.14). It is convenient to introduce a new integral J_1 and rewrite (Eqn. 7.2.15) as

$$[1 + y'_w(t)]^{-1} y''(t) = c_{1w}w - c_4 y'_w(t) J_1(t). \tag{7.3.1}$$

Here

$$J_1(t) = \frac{s_0^{k_{s0}}}{(2c_{3w})^{1/2} k_{s0}} J_0(t)$$
$$= \int_0^t \frac{r_{d,ef}(t,t_0)}{(2c_{3w})^{1/2}} \frac{[y'_w(t_0)]^{k_{s0}-1} y''_w(t_0)}{\{1 + \eta_0[y'_w(t_0)]^{k_{s0}}\}^{\mu}} dt_0, \tag{7.3.2}$$

with $r_{ef}(t,t_0)$ defined in (Eqn. 7.2.11b), and

$$c_4 = 4\pi D_v N_a (2c_{3w})^{1/2} k_{s0} s_0^{-k_{s0}}. \tag{7.3.3}$$

c_{3w} is defined in (Eqn. 7.2.4) and we retain the general index μ in the denominator. Recall that $\mu = 2$ in KC algebraic spectra, and $\mu = 0$ (the denominator vanishes) in Twomey's and similar activity spectra.

Note that the form of (Eqn. 7.3.1) differs from the usual supersaturation equations (e.g., Pruppacher and Klett, 1997, Seinfeld and Pandis, 1998): a) It deals with the integral y_w and not with the differential s_w supersaturation; b) it is of 2nd order with respect to the integral supersaturation y_w but operates only with y_w and its derivatives; c) the last term, which describes the sink of vapor onto the droplets, is expressed via y_w' (i.e., directly supersaturation) in contrast to typically used derivative $d(LWC)/dt$ with implicit dependence on s_w, as was discussed in Chapter 5, Section 5.3.4. This form of the equation is convenient both for analytical solutions and numerical models of various complexity.

A solution to (Eqns. 7.2.15) or (7.3.1) can yield the values of time t_m, s_m, and $N_{dr}(s_m)$. As noted by Twomey (1959), Sedunov (1967, 1974), and others, an analytical solution of the complicated nonlinear integral-differential (Eqns. 7.3.1) or (7.2.15) is a formidable task. Numerical solutions of the supersaturation equation in parcel models may lead to substantial differences among various models due to different numerical approximations (e.g., Lin et al., 2002).

Two approaches to the solution of (Eqn. 7.3.1) are described here: a) a direct numerical solution with some simplifications obtained in Khvorostyanov and Curry (2008a); and b) analytical solutions based on Khvorostyanov and Curry (2009a) similar to the approach developed by Twomey (1959) and Sedunov (1967, 1974). Both methods of solution follow the original idea by Twomey (1959) and are based on approximations of the right-hand side of (Eqn. 7.3.1), which allow us to obtain analytically the lower and upper bounds for solutions to (Eqn. 7.3.1). The advantage of the analytical approach is that it provides the general analytical dependences of solutions on the input parameters that are often hidden in numerical solutions.

The simplest solution to (Eqn. 7.3.1) can be obtained by substituting the 1st term $c_{1w}w$ on the RHS into the 2nd term—i.e., to search iteratively for a solution of the form $y_w'(t) = s_w(t) = c_{1w}wt$. Linear functions of this form were used by Twomey (1959) and Sedunov (1967, 1974) to obtain analytical bounds. We transform the 2nd term on the RHS of (Eqn. 7.3.1) using the linear approximation for $y_w'(t) = s_w(t)$; then

$$y_w'(t) = s_w(t) = a_m t, \quad y_w'' = a_m, \quad y_w(t) = (a_m / 2)t^2. \tag{7.3.4}$$

This method with a linear approximation for $s_w(t)$ developed by Twomey (1959) and promoted later by many authors (e.g., Sedunov, 1967, 1974; Ghan et al., 1993; Abdul-Razzak et al., 1998; Abdul-Razzak and Ghan, 2000; Khvorostyanov and Curry, 2008a, 2009a; Shipway and Abel, 2010; Ghan et al., 2011) gives an upper limit for $J_1(t)$, and a lower limit for $s_w(t)$, and approximates a solution to good accuracy. The form of a_m will be specified in the following text; the choice (Eqn. 7.3.4) is sufficiently general and allows determination of the upper and lower bounds of t_m and s_m. Substitution of (Eqn. 7.3.4) into (Eqn. 7.3.1) yields after simple algebraic transformations:

$$[1 + y_w'(t)]^{-1} y_w''(t) = c_{1w}w - \Lambda(t)t^{k_{s0}+2}, \tag{7.3.5}$$

where

$$\Lambda(t) = c_5 N_a s_0^{-k_{s0}} a_m^{k_{s0}+3/2} J_3(t) \tag{7.3.6a}$$

$$= c_5 N_a s_0^{-k_{s0}} a_m^{k_{s0}+3/2} [2J_2(t)], \tag{7.3.6b}$$

the new coefficient c_5 is

$$c_5 = (c_4 / 2\sqrt{2}) N_a^{-1} s_0^{k_{s0}} = 2\pi D_v c_{3w}^{1/2} k_{s0}. \tag{7.3.7}$$

The integral J_1 in (Eqn. 7.3.2) is related to J_2 and J_3 as

$$J_1 = \frac{1}{\sqrt{2}}(a_m)^{k_{s0}+1/2} t^{k_{s0}+1} J_2, \tag{7.3.7a}$$

J_2 and J_3 are the dimensionless integrals, obtained from J_1 by substitution $r_{d,ef}(t, t_0)$ into J_1, and introducing the variables $x = t/t_0$ in J_2 and $x^2 = z$ or $z = x^{1/2}$ in J_3, which yields

$$J_2(t) = \int_0^1 \frac{\{[U_s + (1-x^2)]^{1/2} - V_s\}^2}{[U_s + (1-x^2)]^{1/2}} \frac{x^{k_{s0}-1}}{(1+\lambda_s x^{k_{s0}})^2} dx, \tag{7.3.8}$$

$$J_3(t) = 2J_2(t) = \int_0^1 \frac{\{[U_s + (1-z)]^{1/2} - V_s\}^2}{[U_s + (1-z)]^{1/2}} \frac{z^{k_{s0}/2-1}}{(1+\lambda_s z^{k_{s0}/2})^2} dz, \tag{7.3.9}$$

and λ_s, U_s, and V_s are the nondimensional parameters

$$\lambda_s = \left(\frac{a_m t}{s_0}\right)^{k_{s0}} = \left(\frac{3\sqrt{3}}{2} \frac{a_m t b^{1/2} r_{d0}^{1+\beta}}{A_k^{3/2}}\right)^{k_{s0}}, \tag{7.3.10}$$

$$U_s = \frac{[r_d(t_0) + \xi_{con}]^2}{c_{3w} a_m t^2}, \qquad V_s = \frac{\xi_{con}}{(c_{3w} a_m)^{1/2} t}, \tag{7.3.11}$$

where we used (Eqn. 6.6.28) for s_0. The time t_m when supersaturation in the parcel reaches a maximum s_m is determined from the condition $s_w'(t_m) = 0$, or $y_w''(t_m) = 0$. Then, the left-hand side of (Eqn. 7.3.5) is zero and we obtain an algebraic expression

$$\Lambda(t) t^{k_{s0}+2} = c_{1w} w, \qquad \text{or} \qquad t_m = \left[\frac{c_{1w} w}{\Lambda(t_m)}\right]^{1/(k_{s0}+2)}. \tag{7.3.12}$$

After t_m is determined, it can be substituted into $\Lambda(t_m)$ in (Eqn. 7.3.5), which can be integrated neglecting $s_w = y' << 1$ on the left-hand side because of small water supersaturation in clouds, and we obtain a time-dependent solution for $s_w(t)$

$$y_w'(t) \equiv s_w(t) = c_{1w} w t \left[1 - \frac{1}{k_{s0}+3}\left(\frac{t}{t_m}\right)^{k_{s0}+2}\right], \tag{7.3.13}$$

where we used (Eqn. 7.3.12) to express $\Lambda(t_m)$ via t_m. Equation (7.3.13) is a parabola by t with the two terms of the powers t and $t^{k_{s0}+3}$; it describes with good accuracy the time evolution of supersaturation to the time t_m and slightly beyond, and gives s_m at the time $t = t_m$:

$$y_w'(t_m) \equiv s_m(t_m) = c_{1w} w t_m \left(1 - \frac{1}{k_{s0}+3}\right) = \alpha_{k1} c_{1w} w t_m, \tag{7.3.14}$$

where

$$\alpha_{k1} = \frac{k_{s0}+2}{k_{s0}+3}. \tag{7.3.15}$$

Equation (7.3.15) shows that $\alpha_{k1} < 1$, and hence (Eqn. 7.3.14) shows that the maximum supersaturation $s_m(t_m)$ in the lower limit is smaller than what would be obtained without this correction.

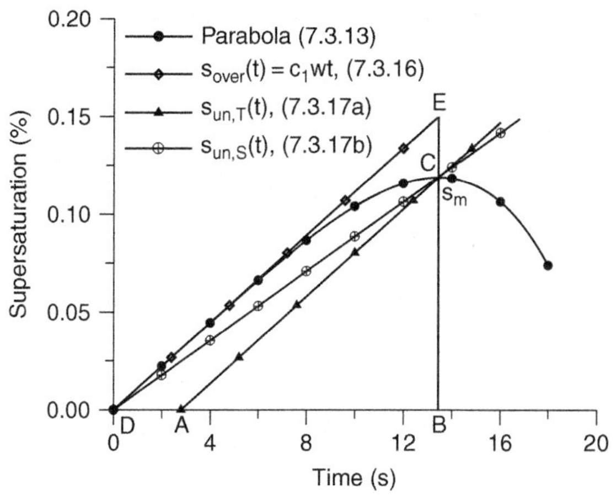

Figure 7.1. Illustration of the upper and lower bounds for the solution of supersaturation equation. The parabolic analytical solution (Eqn. 7.3.13) (solid circles); linear approximation $s_{over}(t) = c_1 wt$ (Eqn. 7.3.16) yielding upper bound for the integrals J_2, J_3 and lower bound for s_m (diamonds); linear approximations by Twomey (1959), $s_{un,T}(t) = c_1 w(t - t_m) + s_m$ (Eqn. 7.3.17a) (triangles), and Sedunov (1967, 1974) $s_{un,S}(t) = s_m(t/t_m)$ (Eqn. 7.3.17b) (open circles) yielding lower bound for the integrals and upper bound for s_m, t_m. The last two straight lines intersect parabola (Eqn. 7.3.13) at s_m. Calculations are performed with the input parameters: $N_a = 500$ cm^{-3}; $r_{d0} = 0.05$ μm; $\sigma_d = 1.8$; $\beta = 0.5$ (soluble fraction is proportional to the volume); $b = 0.25$ (50% of soluble fraction of ammonium sulfate); $T = 10\,°$C; $p = 800$ hPa. From Khvorostyanov and Curry (2009a), *J. Atmos. Sci.*, 66, © American Meteorological Society. Used with permission.

Equations (7.3.12) and (7.3.14) form a system of two equations for the two unknown quantities, t_m and s_m, that allow their evaluation in various particular cases when the integrals J_2 and J_3 are specified. The asymptotic expressions for the integrals J_2 and J_3 will be obtained later in Sections 7.7–7.10 that allow determination of 4 limits of the activation process, depending on λ_s and α_c.

The lower and upper bounds for t_m, s_m, and N_{dr} can be evaluated using (Eqn. 7.3.4) $y'_w(t) = s_w(t) = a_m t$, and (Eqn. 7.3.14) to relate s_m and t_m. This is illustrated in Fig. 7.1, showing upper and lower bounds for $s_m(t)$. It is usually presented as a schematic diagram (e.g., Fig. 1 in Twomey, 1959; or Fig. 13-3 in Pruppacher and Klett, 1997), but this Fig. 7.1 is plotted here based on concrete calculations of all curves using (Eqns. 7.3.13), (7.3.16), (7.3.17a), and (7.3.17b). It was shown by Twomey (1959) and Sedunov (1967, 1974) that a choice

$$a_m = c_{1w} w, \qquad \text{i.e.,} \qquad s_{over}(t) = c_{1w} wt \tag{7.3.16}$$

overestimates the integral J_0 of the type (Eqn. 7.2.18). Although Twomey and Sedunov considered the integral (Eqn. 7.2.18) and supersaturation equation (Eqn. 7.2.19) that are much simpler than (Eqns. 7.2.14) and (7.2.15), with $\mu = 0$, $\xi_{con} = 0$, $r_d(t_0) = 0$, this statement is valid also for a more complicated J_0 in (Eqn. 7.2.14)—i.e., (Eqn. 7.3.16) overestimates J_0 in (Eqn. 7.2.14) and therefore underestimates s_m.

The proof follows from Fig. 7.1, because the straight line $s_{over}(t)$ (DE in Fig. 7.1) lies above the exact curve $s_w(t)$ (the curve DC) and the area of the triangle DEB in Fig. 7.1 is greater than the area under the curve DC, equal to J_0. Therefore, (Eqn. 7.3.16) underestimates s_m and yields a lower bound for s_m. Twomey (1959) showed that another approximation,

$$s_{un,T}(t) = c_{1w}w(t - t_m) + s_m, \tag{7.3.17a}$$

underestimates the J_0, because the straight line $s_{un,T}(t)$ (AC in Fig. 7.1) lies below $s(t)$ and the area of the triangle ABC is less than the area below $s(t)$ equal to J_0. Therefore, using $s_{un,T}(t)$ overestimates $s_m(t_m)$ and gives the upper bound for s_m. Sedunov (1967, 1974) proved that another choice shown in Fig. 7.1,

$$s_{un,S}(t) = s_m(t / t_m), \qquad a_m = s_m / t_m, \tag{7.3.17b}$$

also underestimates the RHS of (Eqns. 7.2.15) or (7.3.1), and gives another upper bound for t_m and s_m. Fig. 7.1 shows that this bound is better than $s_{un,T}(t)$ because it is closer to the exact solution, the strait line $s_{un,S}(t)$ (DC) lies above $s_{un,T}(t)$ (AC) and better approximates J_0. Owing to the coefficient $\alpha_{k1} < 1$, calculated with (Eqn. 7.3.14) s_m, is lower than it would be in the point E and is much closer to the point C. Thus, (Eqn. 7.3.14) would be used for evaluation of the bounds. Although the integral J_0 in (Eqn. 7.2.14) and the integrals J_2, J_3 in (Eqns. 7.3.8) and (7.3.9) are much more complicated than J_0 (Eqn. 7.2.18) considered by Twomey and Sedunov (which corresponds to the 1st analytical limit considered in Section 7.7), the same proof is valid for J_0 in (Eqn. 7.2.14) in the other limits: (Eqn. 7.3.16) yields an upper bound for J_0 and a lower bound for s_m, while (Eqns. 7.3.17a) and (7.3.17b) vice versa, yield a lower bound for J_0 and an upper bound for s_m. This is seen in Fig. 7.1 and will be proven for all four analytical limits in the subsequent Sections 7.6–7.10.

As will be shown in the following pages, the low bound may underestimate the exact solution by 10–20%, and the upper bound overestimate the solution by the same amount (see Table. 7.1 later in the text). The average error would be 5–10%, which is lower than in the other methods.

7.4. Analytical Solutions for the Activation Time, Maximum Supersaturation, and Drop Concentration

As discussed in Section 7.3, the form of (Eqn. 7.3.1) suggests that the first iteration to the solution can be sought by substituting the first term on the right-hand side into the 2nd term. Such a numerical solution gives a lower bound for s_m and N_{dr}. Approximation $a_m = c_{1w}w$ in (Eqn. 7.3.4) yields

$$y_w'' = c_{1w}w, \qquad y_w'(t) = s_w(t) = c_1 wt, \qquad y_w(t) = (c_{1w}w/2)t^2. \tag{7.4.1}$$

The numerical solution of the supersaturation equation for CCN activation is based on (Eqn. 7.3.12) for the time t_m when supersaturation in the parcel reaches a maximum s_m, which is determined from (Eqn. 7.3.5) with the condition $s_w'(t_m) = 0$, or $y_w''(t_m) = 0$. The first of algebraic expressions in (Eqn. 7.3.12) for t_m can be written in more detail as

$$t_m^{k_{s0}+2} = c_5^{-1} N_a^{-1} s_0^{k_{s0}} (c_{1w}w)^{-(k_{s0}+1/2)} [J_3(t_m)]^{-1}, \tag{7.4.2}$$

Solving for t_m, after some transformations we obtain

$$t_m = K_{tms} N_a^{\frac{-1}{(k_{s0}+2)}} s_0^{\frac{k_{s0}}{(k_{s0}+2)}} w^{\frac{-(k_{s0}+1/2)}{(k_{s0}+2)}}, \tag{7.4.3}$$

$$K_{tms} = [c_{1w}^{(k_{s0}+1/2)} c_5 J_3(t_m)]^{\frac{-1}{(k_{s0}+2)}}. \tag{7.4.4}$$

The maximum supersaturation can now be evaluated as $s_m = c_{1w} t_m w$, using (Eqn. 7.4.3) for t_m, which yields

$$s_m = K_{sms} N_a^{\frac{-1}{(k_{s0}+2)}} s_0^{\frac{k_{s0}}{(k_{s0}+2)}} w^{\frac{3}{2(k_{s0}+2)}} \tag{7.4.5}$$

$$K_{sms} = [c_{1w}^{-3/2} c_5 J_3(t_m)]^{\frac{-1}{(k_{s0}+2)}}. \tag{7.4.6}$$

The drop concentration can be calculated using (Eqn. 6.7.2) based on the algebraic activity spectrum (Eqns. 6.6.30b) or (6.6.36) with $s = s_m$, and can be written as a function of maximum supersaturation s_m reached in the parcel:

$$N_{CCN}(s_m) = N_{dr}(s_m) = \int_0^{s_m} \varphi_s(s')\,ds'$$

$$= N_a \frac{(s/s_0)^{k_{s0}}}{[1+(s/s_0)^{k_{s0}}]} = \frac{C_0 s_m^{k_{s0}}}{(1+\eta_0 s_m^{k_{s0}})}, \tag{7.4.7}$$

$$N_{CCN}(s_m) = C(s_m) s_m^{k(s_m)}. \tag{7.4.8}$$

The first form (Eqn. 7.4.7) is similar to Twomey's power law with correction at large s_m as derived by Ghan et al. (1993, 1995), and modified in Khvorostyanov and Curry (2006, 2008a) with account for the drop growth after activation and for various soluble fractions. The second form (Eqn. 7.4.8) is the generalized power law, with the coefficient $C(s_m)$ and index $k(s_m)$ being continuous functions of the maximum supersaturation or of the vertical velocity. The equations for $C(s_m)$, $k(s_m)$ are given in Chapter 6. Now, in terms of s_m, they become:

$$C(s_m) = C_0 s^{\chi(s_m)} [1 + (s_m/s_0)^{k_{s0}}]^{-1}, \tag{7.4.9}$$

$$k(s_m) = k_{s0} [1 + (s_m/s_0)^{k_{s0}}]^{-1}, \tag{7.4.10}$$

$$\chi(s_m) = k_{s0} (s_m/s_0)^{k_{s0}} [1 + (s_m/s_0)^{k_{s0}}]^{-1}. \tag{7.4.11}$$

Equations (7.4.5)–(7.4.8) for s_m, $N_{dr}(s_m)$ represent a generalization of the corresponding equations from Twomey (1959) and Sedunov (1967, 1974) that were based on the power law (Eqns. 6.1.4) and (6.1.5) for the activity spectrum and convert into them if $\lambda_s = 0$ or $\mu = 0$ instead of $\mu = 2$ in denominators of the integrals (Eqns. 7.3.8) and (7.3.9), or if it is assumed that $U_s = 0$, and $V_s = 0$ (i.e., initial radius $r_d(t_0) = 0$, and kinetic correction $\xi_{con} = 0$) in the integrals (Eqns. 7.3.8) and (7.3.9). Without these simplifications, (Eqn. 7.4.3) for t_m depends on the coefficient K_{tms} (Eqn. 7.4.4) that contains the integrals $J_2(t_m)$ or $J_3(t_m)$ (Eqns. 7.3.8) or (7.3.9), which themselves depend on t_m and in general cannot be evaluated analytically without some simplifications. However, these equations are easily solved numerically and lookup tables are easily calculated.

The primary equation is (Eqn. 7.4.3) for t_m, since once we know t_m, we can calculate s_m and $N_{dr}(s_m)$. The method of calculating t_m is illustrated in Fig. 7.2. Since the terms on the RHS are functions of t_m via $J_3(t_m)$ in (Eqn. 7.4.4), the algorithm of the solution is the following. First, the cycle over time t is organized from $t = 0$ with some step $\Delta t \sim 0.1$–0.2 s. For each t, $J_3(t)$ is calculated using (Eqns. 7.3.9)–(7.3.11), then K_{tms} is calculated with (Eqn. 7.4.4). Finally, the RHS of (Eqn. 7.4.3) is calculated and compared with the LHS of (Eqn. 7.4.3). If their difference is greater than some value—e.g., $\varepsilon_t = 0.1 - 0.2$ s—the RHS of (Eqn. 7.4.3) is calculated with another $t + \Delta t$ and this procedure is continued with increasing

t until the difference between the left- and right-hand sides of (Eqn. 7.4.3) becomes smaller than ε_t. When the difference is smaller than ε_t, the solution for t_m is found, and $t_m = t$ at which the RHS is evaluated. The numerical tests showed that this algorithm is very fast since it includes a minimum of numerical operations and allows us to find a solution for t_m rapidly.

This algorithm is illustrated in Fig. 7.2, where the left-hand side of (Eqn. 7.4.3)—that is, t_m—is plotted along with the terms on the right-hand side of (Eqn. 7.4.3), which are calculated as functions of t and extended to larger t to illustrate their behavior. Calculations are performed with $\beta = 0.5$, $r_m = 0.03$ μm, $\sigma_d = 2.1$, $k_{s0} = 1.44$ according to (Eqn. 6.6.30a), $b = 0.25$ (ammonium sulfate and solubility 50%), and with three values of $w = 3.5$, 21, and 101 cm s^{-1}. As seen in Fig. 7.2, t_m begins at zero and linearly increases with time, while the right-hand side of (Eqn. 7.4.3) begins at nonzero values and grows almost linearly but much more slowly. The intersection of these two curves is a solution of (Eqn. 7.4.3) for t_m, which are 38.7, 14.9, and 6.9 seconds, respectively, for these values of w. Then, with known t_m, we can calculate s_m using (Eqns. 7.4.5) and (7.4.6) and then calculate N_{dr} using the algebraic law (Eqn. 7.4.7) or the generalized power law (Eqn. 7.4.8).

Calculations using this algorithm with the generalized power law (Eqns. 7.4.8–7.4.11) were made using the following parameters: $T = 10\,°\text{C}$, $p = 800$ mbar, and updraft $w = 5$ m s^{-1}. Aerosol consisted of fully soluble ammonium sulfate particles, the size spectrum is lognormal with r_{d0} varied in the range indicated on the x-axis, dispersion $\sigma_d = 2.5$, and total concentration $N_a = 200$ cm^{-3}. The results of calculations from Khvorostyanov and Curry (2008a) using this method are presented in Figs. 7.3a,b (labeled KC) and are compared with two methods of calculations from Abdul-Razzak et al. (1998; hereafter AGR), using detailed parcel model simulations and based on them parameterization. The fraction activated, N_{dr}/N_a, is shown in Fig. 7.3a as a function of the mean geometric radius r_{d0} varied in

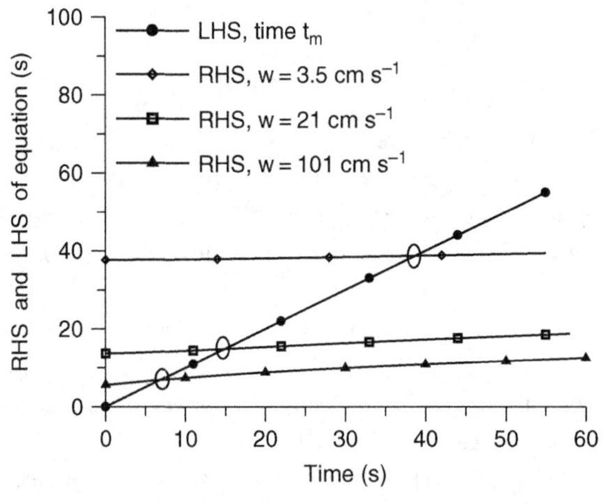

Figure 7.2. Illustration of the method of the solution of the (Eqn. 7.4.3) for t_m. The left-hand side (LHS), time, is denoted by the solid circles. The right-hand side (RHS) is calculated with $r_m = 0.03$ μm, $\sigma_d = 2.1$, $\beta = 0.5$, $b = 0.25$, and 3 values of $w = 3.5$, 21, and 101 cm s^{-1}, as indicated in the legend. The intersections of the RHS and LHS, marked with the ellipses, are the solutions for t_m, here, respectively, 38.7, 14.9, and 6.9 seconds. From Khvorostyanov and Curry (2008a), *J. Atmos. Sci.*, 65, © American Meteorological Society. Used with permission.

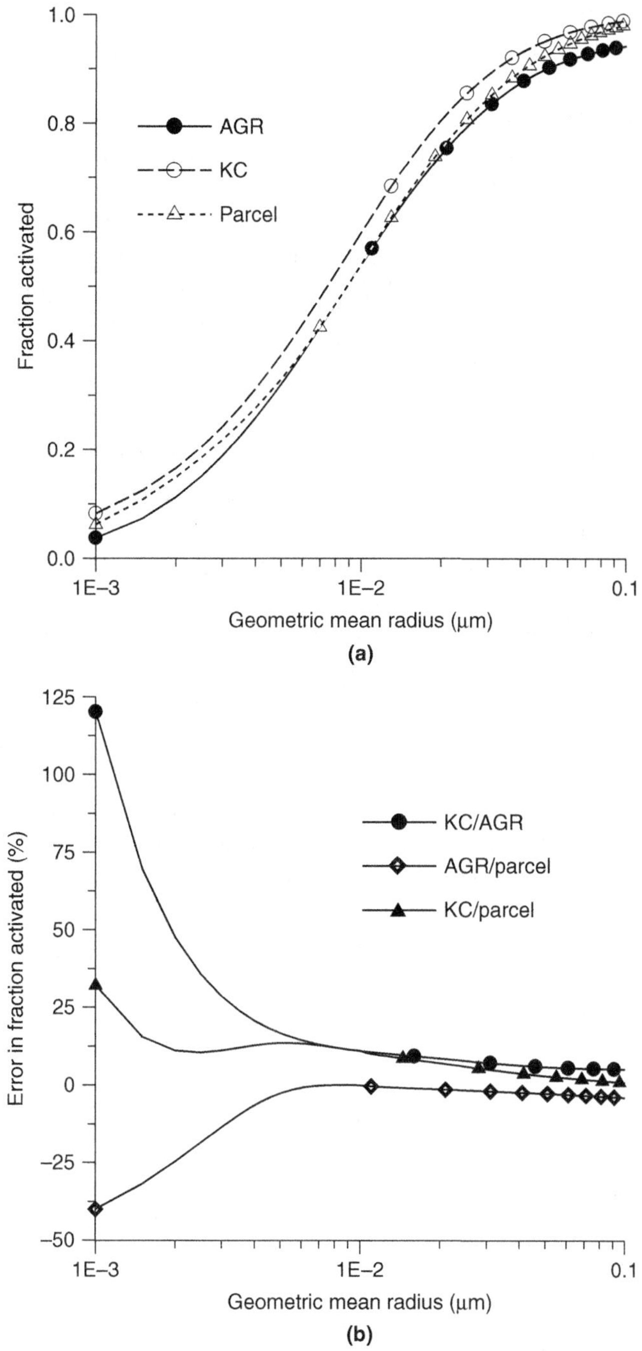

Figure 7.3. Comparison of the CCN fraction activated as a function of the mean geometric radius r_{d0} calculated using three methods: the modified power law (Eqn. 7.4.8) of Khvorostyanov and Curry (2008a) (KC), the parameterization from Abdul-Razzak et al. (1998) (AGR) and the results of detailed parcel simulations from AGR (parcel). (a) fraction activated, N_{dr}/N_a; and (b) relative errors in percent between each two methods as indicated in the legend. Calculations are made at $T = 10\,°C$, $p = 800$ mbar, and updraft of 5 m s^{-1}. Aerosol consists of fully soluble ammonium sulfate particles, the size spectrum is lognormal with r_{d0} varied in the range indicated on the x-axis, dispersion $\sigma_d = 2.5$, total concentration $N_a = 200$ cm^{-3}. The error between the methods does not exceed 10% for realistic spectra at $r_{d0} \geq 0.01$ μm and 5% at $r_{d0} \geq 0.03$ μm. From Khvorostyanov and Curry (2008a), *J. Atmos. Sci.*, 65, © American Meteorological Society. Used with permission.

the range 0.001–0.1 μm, calculated with the modified power law (Eqn. 7.4.8) here, and with the two methods from AGR. Fig. 7.3a shows that the curve obtained with (Eqn. 7.4.8), labeled KC, is close to those from AGR. The precise AGR parcel results lie between the parameterizations AGR and KC.

Fig. 7.3b shows the differences between the methods calculated as $(x_1 - x_2)/x_2 \times 100$, where x_1 and x_2 represent each compared pair of results. One can see that the difference between the methods does not exceed 10% for realistic spectra at $r_{d0} \geq 0.01$ μm and 5% at $r_{d0} \geq 0.03$ μm. The errors increase toward smaller values of r_{d0}, mostly because the fraction N_{dr}/N_a (denominator x_2) becomes small, but these results are shown rather for illustration, and size spectra with small $r_{d0} < 0.01$ μm play little role in drop activation. Thus, the method KC based on the algebraic size and activity spectra is in good agreement with both methods from AGR, despite their completely different mathematical formulations.

7.5. Calculations of CCN Activation Kinetics

Using the described numerical algorithm, a series of calculations is performed using (Eqns. 7.4.3)–(7.4.11) to study various characteristics of CCN activation kinetics with varying vertical velocity. The effects of variations of any other single parameter of the set (r_m, σ_d, N_a, b, α_c) are studied by varying this parameter while the others were fixed as described next. The results shown in Figs. 7.4–7.8 are obtained with the condensation coefficient $\alpha_c = 1$, and the effect of variations of α_c is shown in Figs. 7.9 and 7.10.

Fig. 7.4 demonstrates the effect of variations of the modal radius r_m on the kinetics of drop activation. The maximum time t_m (Fig. 7.4a) decreases with increasing w and with increasing r_m from 50–150 s and to 5–7 seconds at $w = 5$ m s^{-1}. The maximum supersaturation s_m (Fig. 7.4b) increases from 0.02–0.08% at $w = 1$ cm s^{-1} to 1–1.5% s at $w = 5$ m s^{-1}, and decreases with increasing r_m. Both t_m and s_m when plotted versus w in double-log coordinates show nearly straight lines with slightly changing slopes. The droplet concentrations and fraction activated (Fig. 7.4c,d) increase with increasing w and r_m, and roughly resemble smoothed Heaviside step-functions as was described in Chapter 6. Note that N_{dr} is limited by N_a, in contrast to the usual power law expression for the CCN activity spectra.

The indices $k(w)$ (or $k(s_m)$) (Eqn. 7.4.10) of the generalized power law (Eqn. 7.4.8) substantially decrease with increasing w (Fig. 7.4e), from 1.5–1.8 at small values of $w \sim 1$ cm s^{-1} to much smaller values at $w = 5$ m s^{-1}: $k(s_m) \sim 0.4$ at $r_m = 0.02$ μm and $k(s_m)$ is close to zero at $r_m = 0.1$ μm. The coefficients $C(s_m)$ (Eqn. 7.4.9) also decrease with w from 10^3–2×10^4 cm^{-3} at small w to values $\sim N_a = 500$ cm^{-3} at $w > 1$–2 m s^{-1} (Fig. 7.4f); variations of $C(s_m)$ are much smaller at smaller values of r_m. These two Figures (7.4e,f) for $k(w)$ and $C(w)$ may provide a physical basis for the C-k space of drop activation suggested by Braham (1976) based on the generalization of experimental data. However, the consideration here shows that this space is actually multidimensional since $N_{dr}(w)$ depends also on several other parameters: r_{d0}, σ_d, N_a, ε_v, Φ_s, α_c, etc. The method developed here allows fast calculation of $C(s_m)$ and $k(s_m)$ for any given w and any fixed set of aerosol microphysical parameters (r_m, σ_d, N_a, b, α_c) that together with w form a 6-dimensional C-k space.

The values of the integral $J_0(w) = r_{\text{act}}(w)$ defined in (Eqn. 7.2.14) and shown in Fig. 7.5a increase from 0.04–0.2 μm at small values of w to 1–2 μm at large values of w and increase with increasing

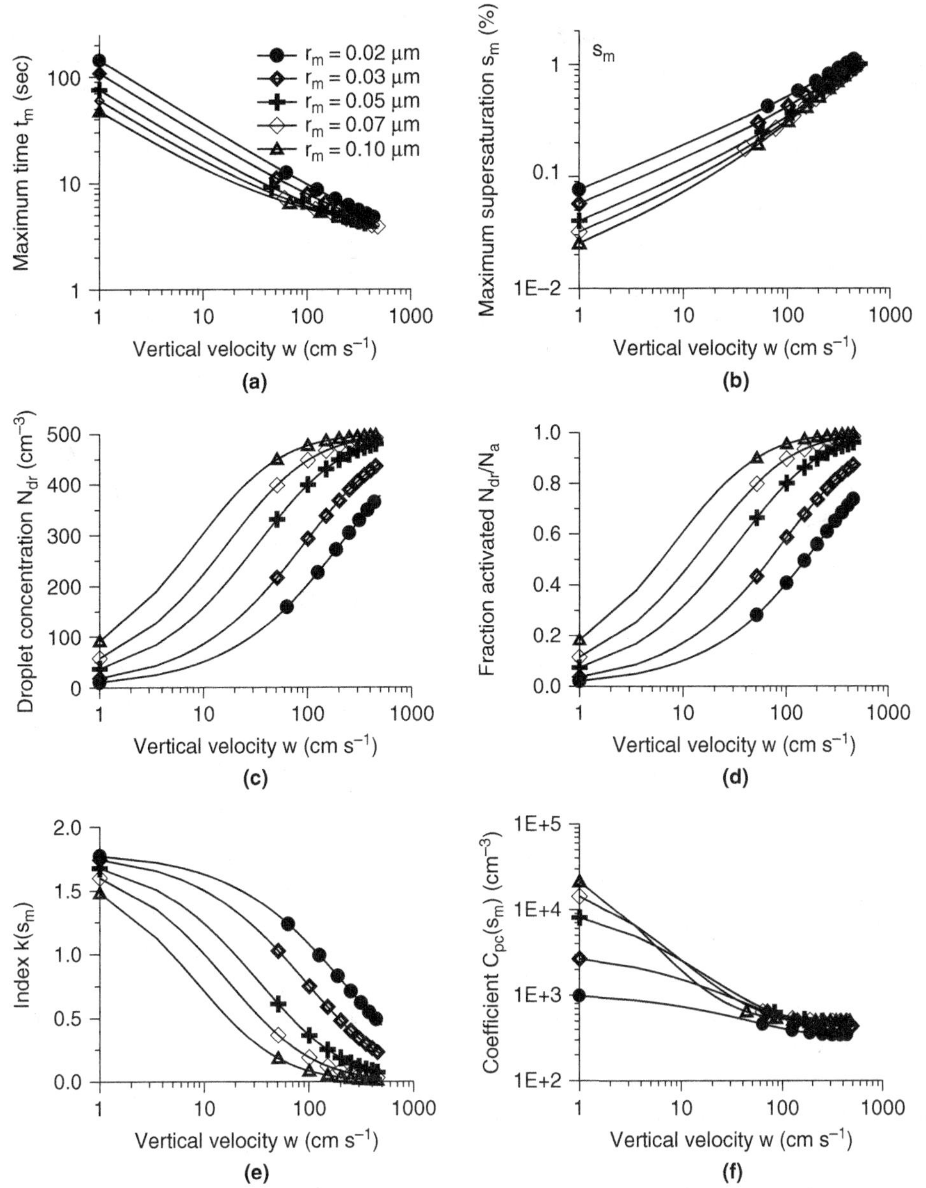

Figure 7.4. Effect on the drop activation kinetics of variations of the modal radius r_m indicated in the legend. (a) maximum time t_m, sec; (b) supersaturation s_m, %; (c) droplet concentration N_{dr}, cm^{-3}; (d) fraction activated; (e), index $k(s_m)$; (f) coefficient $C_{pc}(s_m)$, cm^{-3} as functions of vertical velocity w calculated with the KC06 algebraic CCN activity spectrum. The parameters: size dispersion $\sigma_d = 1.8$; $\beta = 0.5$ (volume soluble fraction); $b = 0.25$ (50% of ammonium sulfate); CCN concentration is $N_a = 500$ cm^{-3}. From Khvorostyanov and Curry (2008a), *J. Atmos. Sci.*, 65, © American Meteorological Society. Used with permission.

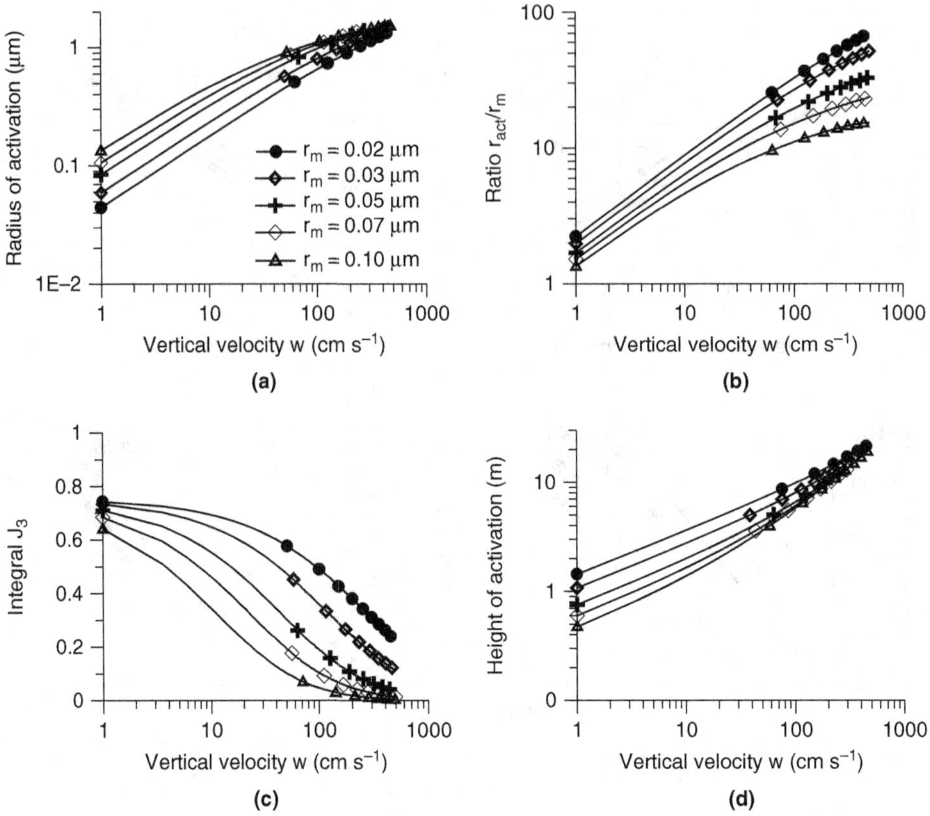

Figure 7.5. Additional characteristics of activation. (a) Effective radius of activation r_{act} (integral J_0); (b) ratio r_{act}/r_m; (c) Integral J_3; (d) height of activation H_{lift}; input parameters are the same as in Fig. 7.4. From Khvorostyanov and Curry (2008a), *J. Atmos. Sci.*, 65, © American Meteorological Society. Used with permission.

values of r_m. The relaxation time $\tau_{f,ac}$ during activation calculated with (Eqn. 7.2.13) decreases from ~70 seconds at $w = 1$ cm s^{-1} to 4–10 seconds at $w \geq 0.3$ m s^{-1}. The latter value of $\tau_{f,ac}$ is quite comparable to the drop relaxation time in a developed cloud—i.e., the rate of vapor absorption by the newly formed drops is comparable to that of the "old" drops.

An important characteristic of the increase in CCN radii during activation is the ratio $r_{act}(w)/r_m$. Fig. 7.5b shows that $r_{act}(w)/r_m$ increases from 1–4 at $w = 1$ cm s^{-1} to the values ≥10 at $w \geq 0.2$–0.4 m s^{-1}—i.e., this ratio is >>1 at the typical turbulent updrafts in Sc clouds and vertical velocities in cumulus clouds. Fig. 7.5c shows that the defined in (Eqn. 7.3.9) integral $J_3 < 1$ and decreases with increasing w and r_m. These dependencies are useful in analysis of the analytical limits of solutions. The depth of the activation layer calculated as $H_{act} = wt_m$ (Fig. 7.5d) increases from less than 1 m at small values of w to 20–25 m at large values of w. Recall, these results were obtained with $\alpha_c = 1$, and H_{act} is greater for smaller values of α_c as described in the following.

The effect of varying size dispersion σ_d is shown in Fig. 7.6a–f. The dependencies on w for all values of σ_d are the same as described earlier. An increase in σ_d (broadening of the CCN size spectra) is

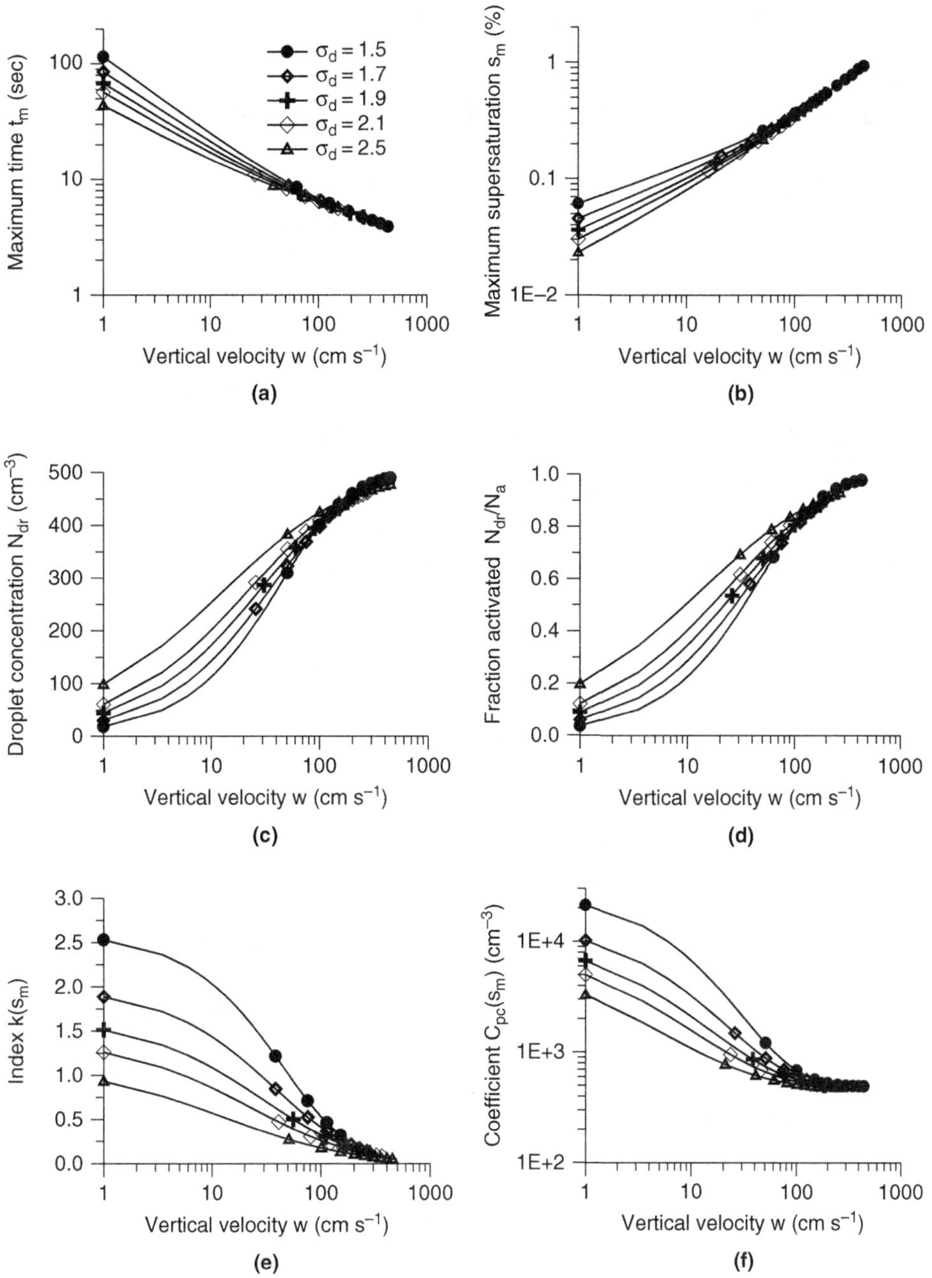

Figure 7.6. Effect on the drop activation kinetics of variations of size dispersion σ_d indicated in the legend. (a) maximum time t_m, sec; (b) supersaturation s_m, %; (c) droplet concentration N_{dr}, cm^{-3}; (d) fraction activated; (e), index $k(s_m)$; (f) coefficient $C_{pc}(s_m)$, cm^{-3} as functions of vertical velocity w calculated with the KC06 CCN activity spectrum. The mean geometric radius $r_{d0} = 0.05$ μm, $\beta = 0.5$ (volume soluble fraction), $b = 0.25$ (50% of ammonium sulfate), and CCN concentration is $N_a = 500$ cm^{-3}. From Khvorostyanov and Curry (2008a), *J. Atmos. Sci.*, 65, © American Meteorological Society. Used with permission.

most pronounced at small values of w, where it shortens activation time t_m, reduces maximum supersaturation, index $k(s_m)$ and coefficient $C(s_m)$, and causes an increase in droplet concentration and the fraction activated. The effect of σ_d decreases at $w > 1$ m s^{-1} and almost vanishes at $w > 2$ m s^{-1}—i.e., the width of the size spectra is important in stratiform clouds and becomes much less significant in the large updrafts in convective clouds.

Fig. 7.7 depicts the effect of variations of CCN concentration N_a in the range 100 cm^{-3} (e.g., maritime or arctic clouds) to 2000 cm^{-3} (continental or moderately polluted atmosphere). A 20-fold increase in N_a causes a decrease in t_m and s_m by 2.5 times at small values of w and 4.5 times at large values of w (Fig. 7.7a)—i.e., the effect is smaller than linear, and increases with w. The drop concentration increases with N_a but more slowly than linearly, so that the fraction activated decreases with increasing N_a. This decrease is greater at moderate values of w on the order of turbulent fluctuations ~0.3–0.5 m s^{-1}, and smaller at small and large values of w (Fig. 7.7c,d). Both the index $k(w)$ and coefficient $C(w)$ substantially increase at larger values of N_a, and their variations are also largest at moderate values of w.

The effect of variations of the soluble fraction ε_v of ammonium sulfate from fully soluble nuclei with $\varepsilon_v = 100\%$ varying to 6% is shown in Fig. 7.8. The values of b calculated from (Eqn. 6.2.6) for the soluble fraction proportional to the volume ($\beta = 0.5$) vary from ≈ 0.5 to 0.03, respectively. The effect of solubility is nonlinear in w, and decreases with increasing w. A more than 16-fold decrease in the soluble fraction leads to a decrease in t_m and s_m by only a factor of two at small w. This difference further decreases with increasing w, and becomes almost negligible at $w = 5$ m s^{-1}. With this 16-fold decrease in ε_v, the drop concentration and fraction activated decrease by a factor of 3.5 at small w, a factor of 2–2.5 at $w = 0.2$–1 m s^{-1} (turbulent fluctuations), and this difference is only 15–20% at $w = 4$–5 m s^{-1} (convective clouds). The value for convective clouds with large updrafts is close to that in Abdul-Razzak et al. (1998). Fig. 7.8 shows the effect of solubility over a wide range of w and its increase at smaller w indicates that the solubility of CCN may play a much more important effect in stratiform than in convective cloud types. The index $k(w)$ increases and the coefficient $C(w)$ decreases for lower solubility. An interesting feature of activation is that even CCN with very small solubility can serve as effective nuclei for drop activation under certain conditions.

The results presented in Figs. 7.4–7.8 were obtained with the condensation coefficient $\alpha_c = 1$ that corresponds to drops of pure water. The values of α_c in the range 1 to 0.001 and smaller have been measured in various experiments (e.g., Pruppacher and Klett, 1997, Table 5.4); α_c may become smaller than 10^{-3} to 10^{-5} in the presence of surfactants and organic substances at the CCN surface (Deryagin et al., 1966; Deryagin and Kurgin, 1972; Bigg et al., 1969; Silverman and Weinstein, 1974; Buikov and Khvorostyanov, 1979; Charlson et al., 2001; Feingold and Chuang, 2002; Chuang, 2003). To study the effects of α_c, we tested five values: $\alpha_c = 10^{10}$, 1, 0.1, 0.04, 0.01. The corresponding kinetic corrections ξ_{con} in dr/dt in (Eqns. 7.2.4) and (7.2.5) can be written for the standard conditions as $\xi \approx 0.15/\alpha_c$ μm, and for these 5 cases are respectively 0, 0.15, 1.5, 3.8, and 15 μm. (The real value of α_c is of course always ≤ 1, but the choice $\alpha_c = 10^{10}$ allows us to study the case with the absence of kinetic corrections, $\xi_{con} \approx 0$—i.e., the diffusion approximation used in some works).

The impacts of varying the condensation coefficient α_c are shown in Figs. 7.9 and 7.10. The difference in t_m and s_m and in all the other characteristics shown in Figs. 7.9 and 7.10 for the case

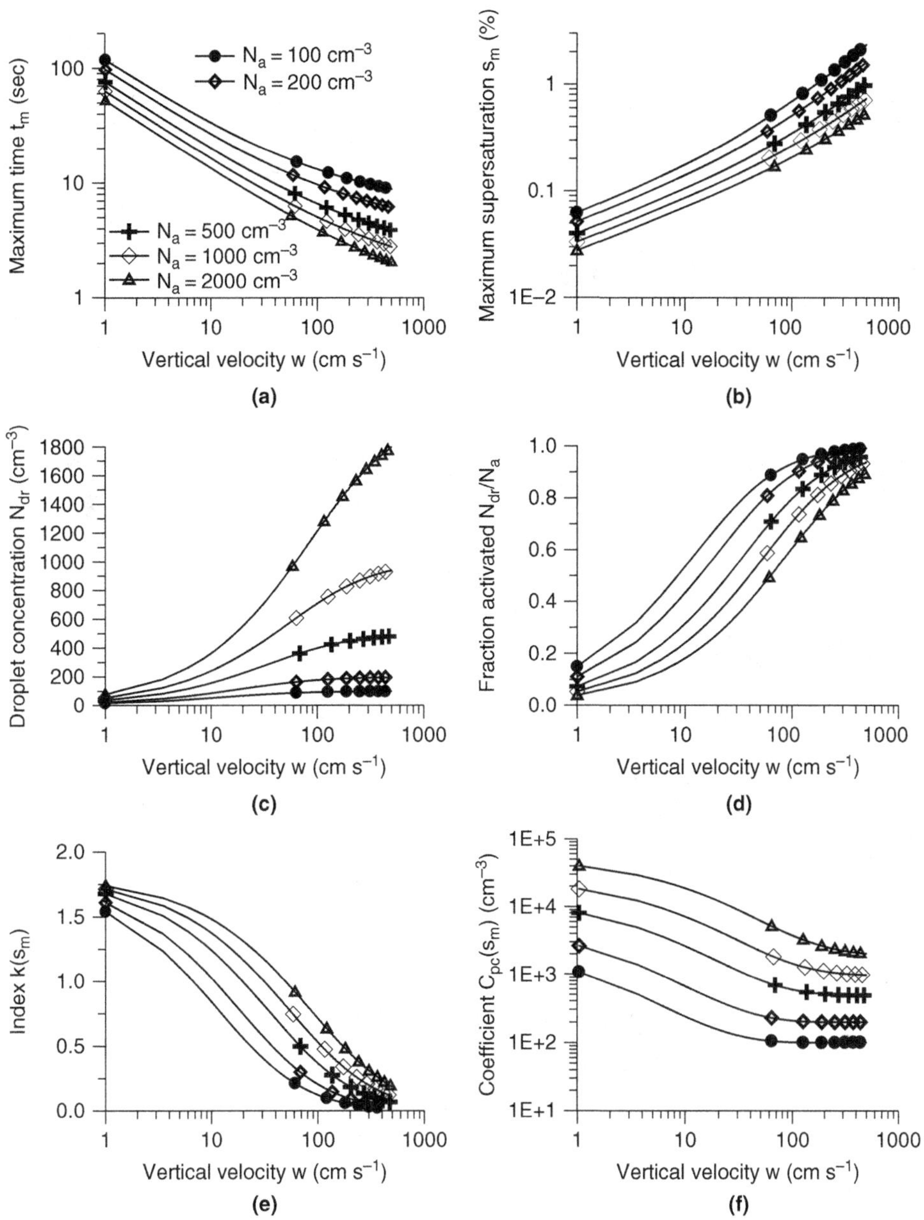

Figure 7.7. Effect on the drop activation of CCN concentration N_a indicated in the legend. (a) maximum time t_m, sec; (b) supersaturation s_m, %; (c) droplet concentration N_{dr}, cm^{-3}; (d) fraction activated; (e), index $k(s_m)$; (f) coefficient $C_{pc}(s_m)$, cm^{-3} as functions of vertical velocity w calculated with the KC06 CCN activity spectrum. The mean geometric radius $r_{d0} = 0.05$ μm, $\sigma_d = 1.8$, $\beta = 0.5$ (volume soluble fraction), $b = 0.25$ (50% of ammonium sulfate). From Khvorostyanov and Curry (2008a), *J. Atmos. Sci.*, 65, © American Meteorological Society. Used with permission.

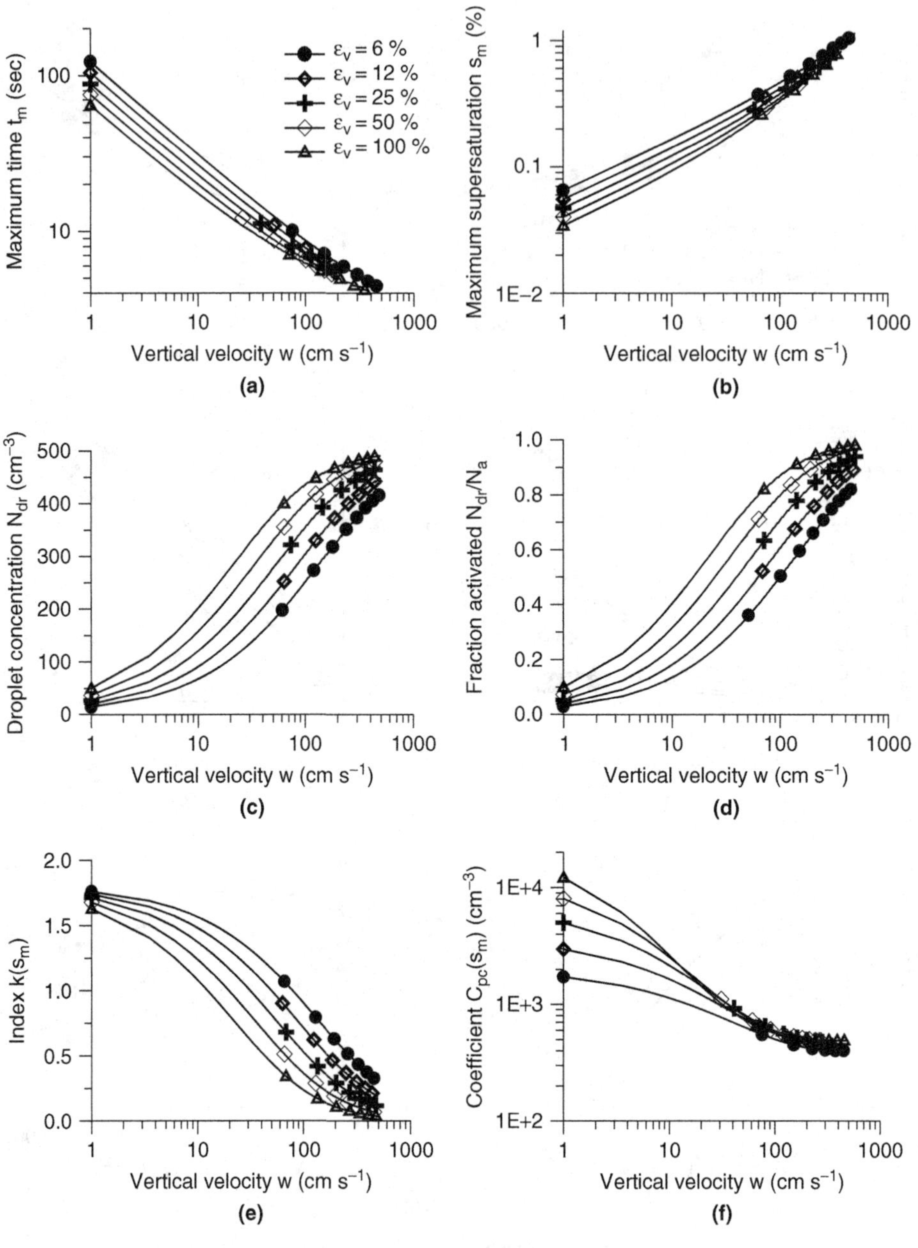

Figure 7.8. Effect on the drop activation of variations of the soluble fraction ε_v of ammonium sulfate from 6% to fully soluble nuclei (100%) indicated in the legend. (a) maximum time t_m, sec; (b) supersaturation s_m, %; (c) droplet concentration N_{dr}, cm^{-3}; (d) fraction activated; (e), index $k(s_m)$; (f) coefficient $C_{pc}(s_m)$, cm^{-3} as functions of vertical velocity w calculated with KC06 CCN activity spectrum. The mean geometric radius $r_{d0} = 0.05$ μm, $\sigma_d = 1.8$, $\beta = 0.5$ (volume soluble fraction), $N_a = 500$ cm^{-3}. From Khvorostyanov and Curry (2008a), *J. Atmos. Sci.*, 65, © American Meteorological Society. Used with permission.

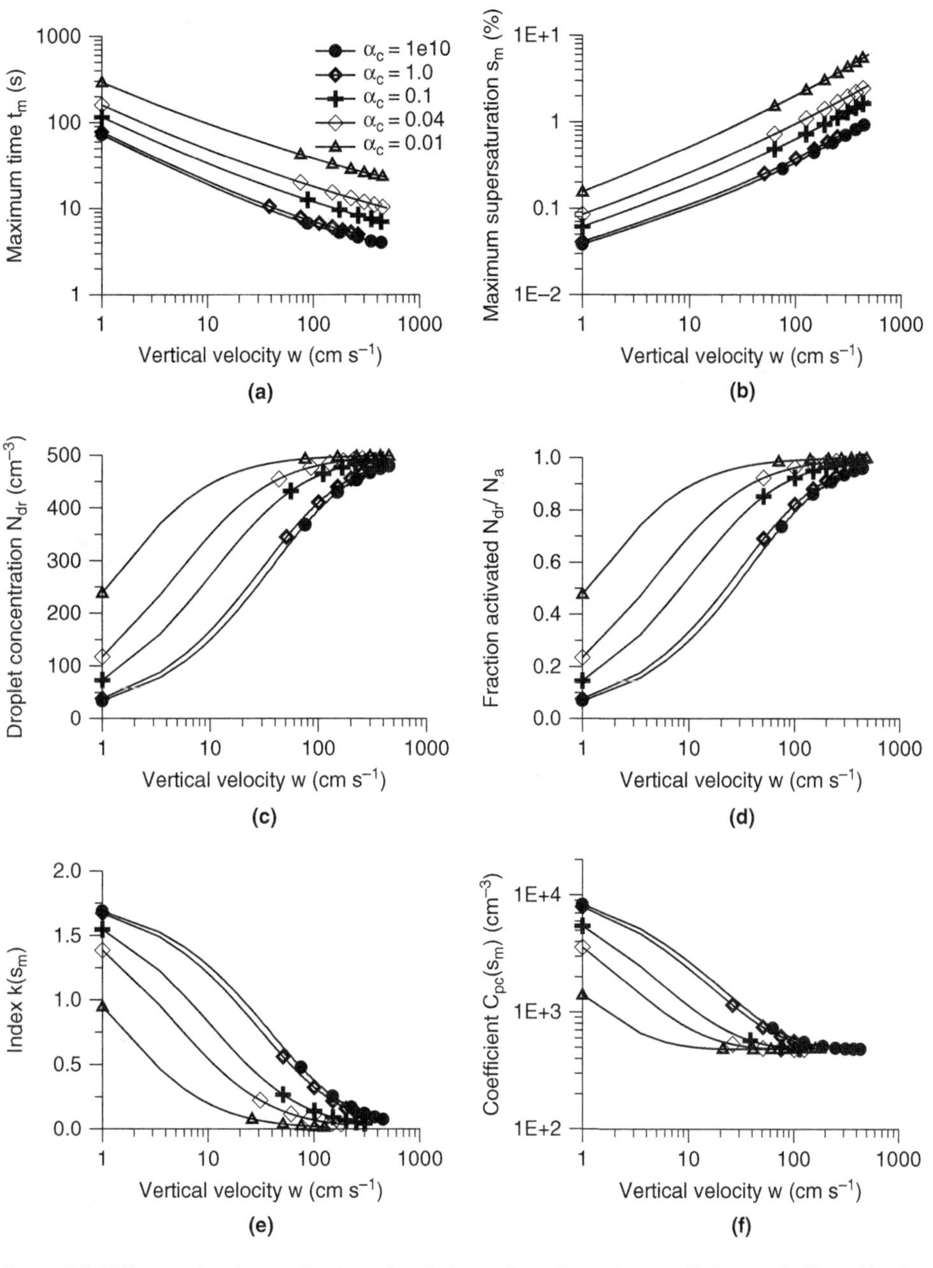

Figure 7.9. Effect on the drop activation of variations of condensation coefficient α_c indicated in the legend ($\alpha_c = 10^{10}$ corresponds to $\xi = 0$). (a) maximum time t_m, sec; (b) supersaturation s_m, %; (c) droplet concentration N_{dr}, cm^{-3}; (d) fraction activated; (e), index $k(s_m)$; (f) coefficient $C_{pc}(s_m)$, cm^{-3} as functions of vertical velocity w calculated with the KC06 algebraic CCN activity spectrum. The mean geometric radius $r_{d0} = 0.05$ µm, $\sigma_d = 1.8$, $\beta = 0.5$ (volume soluble fraction), $b = 0.25$ (50% of ammonium sulfate), CCN concentration is $N_a = 500$ cm^{-3}. From Khvorostyanov and Curry (2008a), *J. Atmos. Sci.*, 65, © American Meteorological Society. Used with permission.

$\alpha_c = 10^{10}$ (absence of kinetic correction) and $\alpha_c = 1$ generally does not exceed 4–7%. That is, the case $\alpha_c = 1$ is close to the absence of kinetic correction as in Twomey (1959) and subsequent similar parameterizations. A decrease in α_c from 1 to 0.1 causes a significant increase in t_m and s_m of 50–70%; the droplet concentration and fraction activated are almost twice as large at small values of w with a gradual decrease of the difference toward larger values of w. Both the index $k(w)$ and coefficient $C(w)$ decrease with increasing w but in such a way that N_{dr} increases. A decrease in α_c from 1 to 0.04 (the typical value used in many cloud models) results in 2–2.5-fold increase in t_m, s_m; the corresponding increase in N_{dr} is 1.5–2 times at w = 10–50 cm s^{-1} (comparable or greater than the indirect aerosol effect), and gradually vanishes at large w. When α_c decreases from 1 to 0.01, which may correspond to substantially polluted clouds, there is a 5- to 6-fold increase in t_m, s_m. This leads to an increase in N_{dr} by 50–100% at w = 10–50 cm s^{-1}, which tends to zero at large w. Any decrease in α_c is accompanied by a significant decrease in $k(w)$ and $C(w)$ similar to those just described (Fig. 7.9a–f).

The values of $r_{act}(w) = J_0(w)$ and r_{act}/r_m become much smaller with decreasing α_c (Fig. 7.10a), and the assumption $r_{act}/r_m >> 1$ with $\alpha_c \le 0.04$ made in some previous works is not satisfied even

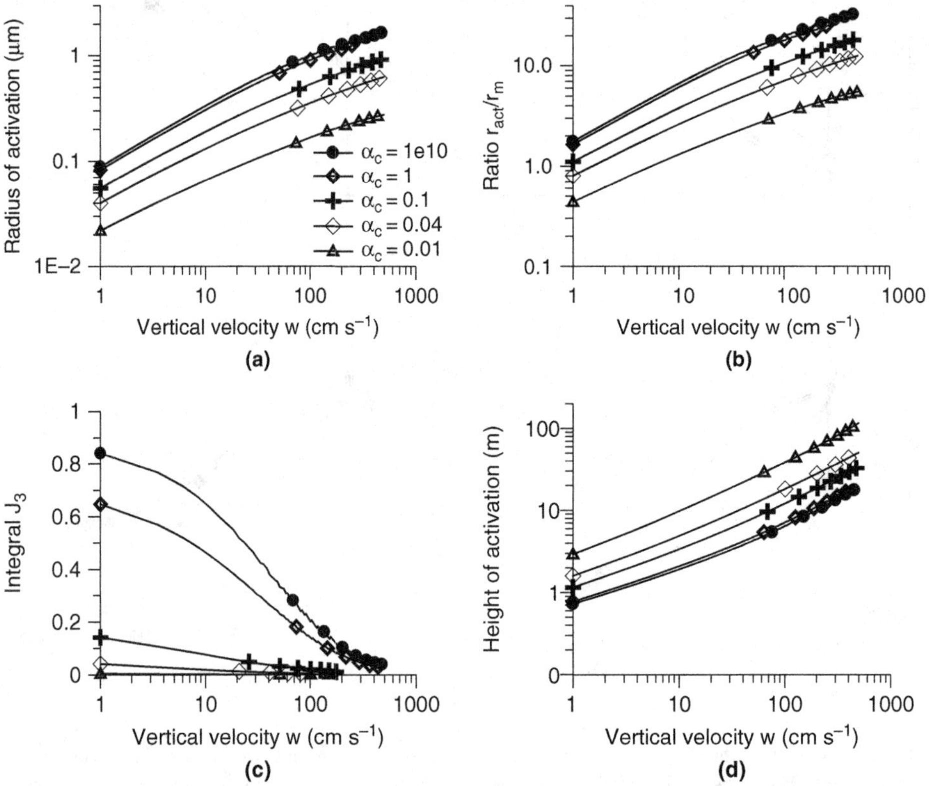

Figure 7.10. Effect of variations of condensation coefficient α_c indicated in the legend ($\alpha_c = 10^{10}$ corresponds to $\xi = 0$). (a) effective radius of activation r_{act} (integral J_0); (b) ratio r_{act}/r_m; (c) Integral J_3; (d) height of activation H_{lift}; input parameters are the same as in Fig.7.8. From Khvorostyanov and Curry (2008a), *J. Atmos. Sci.*, 65, © American Meteorological Society. Used with permission.

for sufficiently high values of w. The height of activation substantially increases for small $\alpha_c \leq 0.1$ (Fig. 7.10d), and may exceed 30–100 m, so that the process of activation should be distributed among several levels in models with fine vertical resolution for polluted clouds with low α_c.

Thus, a decrease in the condensation coefficient leads to slower CCN activation, and higher maximum supersaturation, droplet concentration, and CCN fraction activated. Therefore, correct measurements of the condensation coefficient and use of its proper values for evaluation of the concentration of activated drops is very important. The method developed here allows direct fast calculations of CCN activation in a model or compilation of lookup tables for cloud and climate models for any values of condensation coefficient.

7.6. Four Analytical Limits of Solution

Another, analytical, way of studying the kinetics of drop nucleation is based on simplifying J_2 or J_3 in (Eqns. 7.3.8 and 7.3.9) and finding the asymptotic analytical limits of the solutions, and then constructing interpolations among them. Analysis shows that there are four limiting cases for which approximate analytical solutions for t_m, s_m, and N_{dr} can be found (Khvorostyanov and Curry, 2009a). These cases are naturally separated by the values of the vertical velocities (or maximum supersaturations s_m) that are determined by the values of λ_s in (Eqn. 7.3.10), and of the condensation coefficients α_c that determine diffusion growth of the activated droplets (sufficiently pure water drops) or kinetic growth regime (polluted water drops). The expression for $\lambda_s(s_m)$ for the lower bound with $a_m = c_{1w}w$, or $s_w(t_m) = c_{1w}wt_m$, follows from (Eqn. 7.3.10):

$$\lambda_s(s_m) = \left(\frac{s_m}{s_0}\right)^{k_{s0}} = \left(\frac{c_{1w}wt_m}{s_0}\right)^{k_{s0}} = \left[(c_{1w}wt_m)\frac{3\sqrt{3}}{2}\frac{b^{1/2}r_{d0}^{1+\beta}}{A_k^{3/2}}\right]^{k_{s0}}, \tag{7.6.1}$$

where (Eqn. 6.6.28) was used for s_0. Thus, these four limits are:

1. $\lambda_s << 1$ (small vertical velocities w), $\alpha_c \sim 1$ (diffusion droplet growth limit);
2. $\lambda_s << 1$ (small vertical velocities w), $\alpha_c << 1$ (kinetic droplet growth limit);
3. $\lambda_s >> 1$ (large vertical velocities w), $\alpha_c \sim 1$ (diffusion droplet growth limit);
4. $\lambda_s >> 1$ (large vertical velocities w), $\alpha_c << 1$ (kinetic droplet growth limit).

Equations (7.4.7) and (7.4.8) for $N_{dr}(s_m)$, and solutions for s_m, represent a generalization of the corresponding expressions from Twomey (1959) and Sedunov (1967, 1974) that were based on the power law for $\varphi_s(s)$ and correspond to the 1st limit here as described in Section 7.1. These 4 limits, along with the situations intermediate among them, cover the vast majority of the situations that can be met in atmospheric clouds. Thus, if we find these limits and construct interpolations among them, the problem of parameterization can be solved. However, these limits have physical meaning without interpolation since each of them corresponds to some particular type of cloud with a certain range of vertical velocities. For example, limits 1 and 2 may correspond to stratiform clouds or fogs with low vertical velocities w, and limits 3 and 4 correspond to convective clouds with higher w.

The applicability of these limits is determined by the scaling parameters, s_m/s_0, and $\lambda_s = (s_m/s_0)^{k_{s0}}$. Fig. 7.11 shows that the values $\alpha_c = 10^{10}$ (the physical value of $\alpha_c \leq 1$, but $\alpha_c = 10^{10}$ means absence

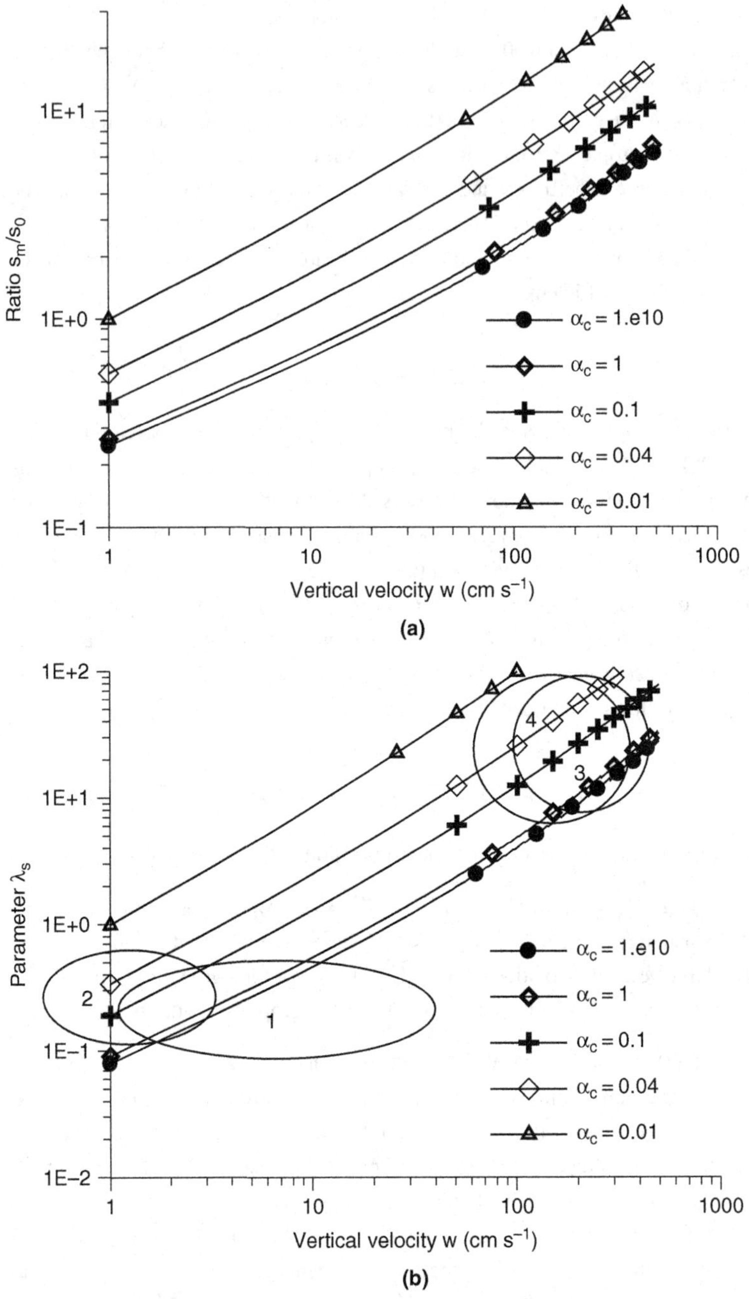

Figure 7.11. Scaling parameters: (a) s_m/s_0, and (b) $\lambda_s(s_m) = (s_m/s_0)^{k_{s0}}$, calculated with the input parameters: $N_a = 500$ cm^{-3}; $r_{d0} = 0.05$ μm; $\sigma_d = 1.8$; $\beta = 0.5$ (soluble fraction is proportional to the volume); $b = 0.25$ (50% of soluble fraction of ammonium sulfate); $T = 10$°C; $p = 800$ hPa, and five values of the condensation coefficient indicated in the legend. The regions of the four limits are approximately denoted with the ellipses in Fig. 7.11b on the $\lambda_s - w$ diagram, the digits are the numbers of the limits. From Khvorostyanov and Curry (2009a), *J. Atmos. Sci.*, 66, © American Meteorological Society. Used with permission.

of kinetic correction, $\xi_{con} = 0$) and $\alpha_c = 1$ yield close results—i.e., the value $\alpha_c = 1$ that may be characteristic of pure water drops is close to Twomey's (1959) case with $\xi_{con} = 0$. The regions of the four limits are depicted in Fig. 7.11b on the $\lambda_s - w$ diagram. One can see that the 1st limit, $\lambda_s << 1$, is valid with $\alpha_c \sim 1$ and small w. The 2nd limit with $\alpha_c << 1$ is marked in Fig. 7.11 at $w \le 1$ cm s^{-1} around the ordinate axis. The third and fourth limits are reached at $w >$ 1–2 m s^{-1}. This representation is somewhat schematic and these regions may extend to larger areas with varying parameters. The applicability of these limits will be illustrated in more detail in the next sections by comparison of these limits with the exact solutions.

7.7. Limit #1: Small Vertical Velocity, Diffusional Growth Regime

This limit corresponds to $\lambda_s << 1$ (low w or low s) and $\alpha_c \sim 1$ (the diffusion growth regime of activated drops), so that the denominator of the 2nd fraction in the integrand of J_3 (Eqn. 7.3.9) tends to unity. This is equivalent to the absence of the denominator in the differential CCN activity spectrum (Eqn. 6.6.36), which is then similar to Twomey's (1959) power law

$$\varphi_s(s) = k_{s0} C_0 s^{k_{s0}-1}. \tag{7.7.1}$$

When $\alpha_c \sim 1$, we can neglect small ξ_{con}, and we also assume small $r_d(t_0)$. With these assumptions and (Eqn. 7.7.1), this limit #1 is equivalent to the model of activation considered by Twomey, Sedunov, and all subsequent similar models, but their empirical coefficient C_0 and index k_{s0} were expressed in (Eqns. 7.4.9)–(7.4.11) via aerosol parameters. As shown in Appendix A.7, J_3 in this case is reduced to Twomey's integral with Euler's beta function $\mathrm{B}(x,z)$:

$$J_3^{(1)} = \mathrm{B}\left(\frac{k_{s0}}{2}, \frac{3}{2}\right). \tag{7.7.2}$$

7.7.1. Lower Bound

Substitution of (Eqn. 7.7.2) into (Eqns. 7.3.12) or (7.4.3) for t_m, using Λ from (Eqn. 7.3.6a) with $a_m = c_{1w} w$ from (Eqn. 7.3.16), solving for t_m, and using c_5 from (Eqn. 7.3.7) yields the lower bound $t_{ml}^{(1)}$ in the 1st limit:

$$t_{ml}^{(1)} = K_{tm}^{(1)} N_a^{\frac{-1}{k_{s0}+2}} s_0^{\frac{k_{s0}}{k_{s0}+2}} w^{\frac{-(k_{s0}+1/2)}{k_{s0}+2}}, \tag{7.7.3}$$

$$K_{tm}^{(1)}(T,p) = \left[c_{1w}^{k_{s0}+1/2} 2\pi D_v c_{3w}^{1/2} k_{s0} \mathrm{B}\left(\frac{k_{s0}}{2}, \frac{3}{2}\right)\right]^{-1/(k_{s0}+2)}. \tag{7.7.4}$$

(The superscripts "(i)" hereafter denote the i-th limit with i = 1–4, and the subscript "l" means lower bound, while the subscript "m" means maximum value). The maximum supersaturation is from (Eqn. 7.3.14) in this limit:

$$s_{ml}^{(1)} = \alpha_{k1} K_{sm}^{(1)} N_a^{\frac{-1}{k_{s0}+2}} s_0^{\frac{k_{s0}}{k_{s0}+2}} w^{\frac{3}{2(k_{s0}+2)}}, \tag{7.7.5}$$

$$K_{sm}^{(1)}(T,p) = \left[2\pi D_v c_{1w}^{-3/2} c_{3w}^{1/2} k_{s0} \mathrm{B}\left(\frac{k_{s0}}{2}, \frac{3}{2}\right)\right]^{\frac{-1}{k_{s0}+2}}, \tag{7.7.6}$$

and α_{k1} was defined in (Eqn. 7.3.15). Note that maxima $s_{ml}^{(1)}$ and $t_{ml}^{(1)}$ are related as $s_{ml}^{(1)} = \alpha_{k1} c_1 w t_{ml}^{(1)}$—i.e., almost as in Twomey's upper approximation (Eqn. 7.3.16) for $s_{over}(t)$ but with the coefficient α_{k1}.

Table 7.1 shows that $s_{ml}^{(1)}$ is only by 11–24% lower than $s_{over}(t)$. The droplet concentration from (Eqn. 7.4.7) is

$$N_{dr,l}^{(1)}(s_m) = N_a (s_m^{(1)} / s_0)^{k_{s0}} = \alpha_{k1}^{k_{s0}} [K_{sm}^{(1)}]^{k_{s0}} N_a^{\frac{2}{k_{s0}+2}} s_0^{\frac{-2k_{s0}}{k_{s0}+2}} w^{\frac{3k_{s0}}{2(k_{s0}+2)}}. \tag{7.7.7}$$

The powers of w in the expressions for s_m and N_{dr} are the same as derived by Twomey (1959), while his empirical indices and coefficients are expressed now via aerosol microphysical parameters. The coefficient α_{k1} is similar to the 1st major term in Twomey's power series expansion for the lower bound of s_m (Twomey, 1959, Eq. (7)). More detailed expressions can be obtained if the mean geometric supersaturation s_0 is expressed via the mean geometric radius r_{d0} from (Eqn. 6.6.35b), the parameter of chemical composition b from (Eqn. 6.2.6), and the curvature parameter A_k is expressed via the surface tension σ_{sa} and temperature T using (Eqn. 6.2.2b):

$$t_{ml}^{(1)} = K_{tm}^{(1)} Q_a^{\frac{k_{s0}}{2(k_{s0}+2)}} N_a^{\frac{-1}{k_{s0}+2}} r_{d0}^{\frac{-k_{s0}(1+\beta)}{k_{s0}+2}}$$
$$\times b^{\frac{-k_{s0}}{2(k_{s0}+2)}} \sigma_{sa}^{\frac{3k_{s0}}{2(k_{s0}+2)}} T^{\frac{-3k_{s0}}{2(k_{s0}+2)}} w^{\frac{-(k_{s0}+1/2)}{k_{s0}+2}}, \tag{7.7.8}$$

and $Q_a = (2^{5/3} M_w / 3R\rho_w)^3$. Equation (7.7.8) shows that t_m decreases (i.e., activation ceases more rapidly) with the increase of N_a, r_{d0}, solubility b (i.e., ε_v), updraft w, and with the decrease of the surface tension σ_{sa} in the presence of surfactants. In particular, with $k_{s0} \sim 2$ ($\sigma_d = 1.7$, $\beta = 0.5$), (Eqn. 7.7.8) yields

$$t_{ml}^{(1)} \sim K_{tm}^{(1)}(T) N_a^{-1/4} r_{d0}^{-3/4} b^{-1/4} \sigma_{sa}^{3/4} T^{-3/4} w^{-5/8}. \tag{7.7.9}$$

Table 7.1. *Index k_{s0} and coefficients α_{k1}, $\alpha_{up}^{(1)}$, $\alpha_{low,up}^{(1)}$, $\alpha_{up}^{(2)}$, $\alpha_{low,up}^{(2)}$, $\alpha_{up}^{(3)}$, $\alpha_{low,up}^{(3)}$, $\alpha_{up}^{(4)}$, $\alpha_{low,up}^{(4)}$ at various size dispersions σ_d and $\beta = 0.5$ (soluble fraction proportional to volume).*

σ_d	1.2	1.5	1.7	1.8	2.0	2.2	2.5
k_{s0}	5.84	2.63	2.00	1.81	1.54	1.35	1.16
α_{k1}	0.89	0.82	0.80	0.79	0.78	0.77	0.76
$\alpha_{up}^{(1)}$	0.993	0.979	0.972	0.969	0.966	0.961	0.957
$\alpha_{low,up}^{(1)}$	1.116	1.193	1.215	1.227	1.238	1.248	1.260
$\alpha_{up}^{(2)}$	0.979	0.949	0.944	0.933	0.927	0.921	0.915
$\alpha_{low,up}^{(2)}$	1.099	1.158	1.180	1.181	1.188	1.196	1.203
$\alpha_{up}^{(3)}$	0.971	0.952	0.946	0.943	0.940	0.937	0.934
$\alpha_{low,up}^{(3)}$	1,091	1.160	1.182	1.194	1.205	1.216	1.228
$\alpha_{up}^{(4)}$	0.954	0.924	0.915	0.910	0.905	0.901	0.896
$\alpha_{low,up}^{(4)}$	1.072	1.126	1.143	1.152	1.161	1.170	1.179

The analytical dependencies for another σ_d can be found using the $k_{s0} - \sigma_d$ relation from Table 7.1. For s_m, we obtain from (Eqn. 7.7.5), expressing again s_0 via r_{d0} and using (Eqn. 6.2.6) for b:

$$s_{ml}^{(1)} = \alpha_{k1} K_{sm}^{(1)}(T) Q_a^{\frac{k_{s0}}{2(k_{s0}+2)}} N_a^{\frac{-1}{k_{s0}+2}} r_{d0}^{\frac{-k_{s0}(1+\beta)}{k_{s0}+2}}$$
$$\times b^{\frac{-k_{s0}}{2(k_{s0}+2)}} \sigma_{sa}^{\frac{3k_{s0}}{2(k_{s0}+2)}} T^{\frac{-3k_{s0}}{2(k_{s0}+2)}} w^{\frac{3}{2(k_{s0}+2)}}. \quad (7.7.10)$$

Thus, s_m increases with increasing w and σ_{sa}, and decreases with increasing supersaturation absorption rate—i.e., increasing N_a, r_{d0}, and b (or ε_v). In particular, with $k_{s0} \sim 2$, we have

$$s_{ml}^{(1)} \sim \alpha_{k1} K_{sm}^{(1)}(T) N_a^{-1/4} r_{d0}^{-3/4} b^{-1/4} \sigma_{sa}^{3/4} T^{-3/4} w^{3/8}. \quad (7.7.10a)$$

For N_{dr}, we have from (Eqn. 7.7.7) expressing s_0 via r_{d0} and b, and using (Eqn. 6.2.6) for b and (Eqn. 7.7.10) for s_m

$$N_{dr,l}^{(1)} = \alpha_{k1}^{k_{s0}} [K_{sm}^{(1)}(T)]^{k_{s0}} Q_a^{\frac{-k_{s0}}{k_{s0}+2}} N_a^{\frac{2}{k_{s0}+2}} r_{d0}^{\frac{2k_{s0}(1+\beta)}{k_{s0}+2}}$$
$$\times b^{\frac{k_{s0}}{k_{s0}+2}} \sigma_{sa}^{\frac{-3k_{s0}}{2(k_{s0}+2)}} T^{\frac{3k_{s0}}{2(k_{s0}+2)}} w^{\frac{3k_{s0}}{2(k_{s0}+2)}}. \quad (7.7.11)$$

The index of σ_{sa} is negative—i.e., the maximum drop concentration grows with decreasing surface tension (as with increasing concentration of organics or surface-active substances). The other 5 power indices are positive, thus N_{dr} increases with an increase of any of the 4 factors: N_a, r_{d0}, b (or ε_v), or w. The temperature increase due to the positive power of T is overwhelmed by decreasing $K_{sm}^{(1)}(T)$, and N_{dr} decreases with increasing T. A similar negative temperature dependence was obtained in Saleeby and Cotton (2004) with parcel model simulations. In particular, with $k_{s0} \sim 2$, we have from (Eqn. 7.7.11)

$$N_{dr,l} \sim [K_{sm}^{(1)}(T)]^{k_{s0}} N_a^{1/2} r_{d0}^{3/2} b^{1/2} \sigma_{sa}^{-3/4} T^{3/4} w^{3/4}. \quad (7.7.11a)$$

The variations δN_{dr} predicted by (Eqn. 7.7.11) at small variations of the surface tension $\delta\sigma_{sa}$ are

$$\frac{\delta N_{dr,l}^{(l1)}}{N_{dr,l}^{(l1)}} \approx -C_{N\sigma} \frac{\delta\sigma_{sa}}{\sigma_{sa}}, \qquad C_{N\sigma} = \frac{3k_{s0}}{2(k_{s0}+2)}. \quad (7.7.11b)$$

In particular, with $k_{s0} \sim 2$, then $C_{N\sigma} = 3/4$, and $\delta N_{dr,l}/N_{dr,l} \sim (-3/4)\delta\sigma_{sa}/\sigma_{sa}$, which exactly coincides with the expression derived in Facchini et al. (1999). For a broader size spectrum with $\sigma_d = 2.5$ and $\beta = 0.5$, we have $k_{s0} \sim 1.16$ (Table 7.1), and $C_{N\sigma} \approx 0.55$—i.e., the dependence of $N_{dr,l}$ on surface tension becomes weaker for broader CCN spectra, but the general qualitative dependence of $N_{dr,l}$ on σ_{sa} in the 1st limit is similar to that in Facchini et al. (1999), and Abdul-Razzak and Ghan (2004). Note that these conclusions are drawn since we consider here σ_{sa} as a variable, and do not consider effects of its variations due to redistribution of organics between the surface and volume in CCN as in Li et al. (1998) and Abdul-Razzak and Ghan (2004). According to Facchini et al. (1999), a 30% decrease in σ_{sa} should cause a ~20% increase in N_{dr}, which should lead to an increase in top-of-the-atmosphere albedo locally by ~1% and could cause a global mean forcing with an upper limit of −1 W m^{-2}. The opposite effects of decreasing N_{dr} with an increasing organic coating of CCN and a decrease of σ_{sa} were predicted by Feingold and Chuang (2002); various studies of coating effects on drop activation were reviewed in IPCC (2007). Equations here allow estimations of the effects of surface tension with various other CCN properties. These effects can be noticeable for small w, but it will be shown in Sections 7.9 and 7.10 that N_{dr} is highly insensitive to surface tension at high $w > 1$–2 m s^{-1}.

7.7.2. Upper Bound

Evaluation of the upper bound in this and other limits is a little more complicated and is done using the system of two (Eqns. 7.3.12) and (7.3.14) to determine the two unknowns, t_m and s_m. Substitution of $J_3^{(1)}$ from (Eqn. 7.7.2) and $a_m=s_m/t_m$ from (Eqn. 7.3.17b) into (Eqn. 7.3.6a) yields $\Lambda(t_m)$. Substituting it into (Eqns. 7.3.12) and (7.3.14) and solving for t_m, we obtain for $t_{mu}^{(1)}$, $s_{mu}^{(1)}$, and $N_{dr,u}^{(1)}$ in the upper bound (subscript "u") of the first limit almost the same expressions as (Eqns. 7.7.8), (7.7.10), and (7.7.11) but multiplied by the coefficient $\alpha_{low,up}^{(1)}$ for $t_{mu}^{(1)}$ in (Eqn. 7.7.8), by $\alpha_{up}^{(1)}$ instead of α_{k1} for $s_{ml}^{(1)}$ in (Eqn. 7.7.10), and $(\alpha_{up}^{(1)})^{k_{s0}}$ instead of $\alpha_{k1}^{k_{s0}}$ for $N_{dr,u}^{(1)}$ in (Eqn. 7.7.11). The relation $s_{mu}^{(1)} = \alpha_{up}^{(1)} c_{1w} w t_{mu}^{(1)}$ similar to Twomey's upper approximation (Eqn. 7.3.16) arises again, but with coefficient $\alpha_{up}^{(1)} \sim 1$. The relation of the lower and upper bounds in the 1st limit is determined by the coefficient $\alpha_{low,up}^{(1)}$:

$$t_{mu}^{(1)} = \alpha_{low,up}^{(1)} t_{ml}^{(1)}, \qquad s_{mu}^{(1)} = \alpha_{low,up}^{(1)} s_{ml}^{(1)},$$

$$N_{dr,u}^{(1)} = (\alpha_{low,up}^{(1)})^{k_{s0}} N_{dr,l}^{(1)}, \tag{7.7.12}$$

where

$$\alpha_{up}^{(1)} = \alpha_{k1}^{1/[2(k_{s0}+2)]} = [(k_{s0}+2)/(k_{s0}+3)]^{1/[2(k_{s0}+2)]}, \tag{7.7.13a}$$

$$\alpha_{low,up}^{(1)} = \alpha_{up}^{(1)} / \alpha_{k1} = [(k_{s0}+3)/(k_{s0}+2)]^{(2k_{s0}+3)/[2(k_{s0}+2)]}. \tag{7.7.13b}$$

The coefficients α_{k1}, $\alpha_{up}^{(1)}$, and $\alpha_{low,up}^{(1)}$ along with the index k_{s0} are given in Table 7.1 for various size dispersions σ_d. One can see that k_{s0} decreases from 5.84 at $\sigma_d = 1.2$ (narrow spectra) to 1.16 at $\sigma_d = 2.5$ (wide spectra). All the coefficients are close to 1, and their variations are much smaller: α_{k1} decreases from 0.89 to 0.76, $\alpha_{up}^{(1)}$ decreases from 0.993 to 0.957.

Thus, the lower bound for s_m obtained with (Eqn. 7.3.14) is 11–24% lower than would be obtained with the linear relation (Eqn. 7.3.16) $s_m = c_{1w} w t_m$, and the upper bound is only 1–4% lower. Calculated variations of the coefficients with $\beta = 0$ (not shown here) are even smaller. For all σ_d, $\alpha_{up}^{(1)} > a_{k1}$—i.e., the condition is satisfied that the upper bound is higher than the lower bound; however, the difference between the bounds is rather small. The coefficient $\alpha_{low,up}^{(1)}$, which characterizes the ratio of the upper to the lower bounds, increases from 1.116 to 1.260 in this range—that is, the difference does not exceed 12% and 26% for the narrow and wide spectra, respectively. Assuming that the true value is close to the mean of the upper and lower bounds, the error would be a factor of two smaller: ~6 and 13%.

A comparison was performed of these approximate analytical limits for t_m, s_m, and N_{dr} with more precise numerical solution of the complete supersaturation equation (Eqn. 7.2.15) with the modified power law (Eqn. 7.4.8) and coefficients from (Eqns. 7.4.9)–(7.4.11), hereafter referred to as "exact solution." Fig. 7.12 shows that the agreement between the 1st limit and exact solution for t_m, s_m is good for both bounds at $w \le 30$–50 cm s^{-1}. A comparison of s_m with parameterization from Abdul-Razzak et al. (1998, hereafter AGR98) also shows a satisfactory agreement; the curve calculated using AGR98 algorithm is closer to our lower bound for s_m at small w, and coincides with the exact solution at $w > 0.5$ m s^{-1}.

Fig. 7.12c shows a comparison of N_{dr} calculated with three methods. The 1st limit for N_{dr} (Eqn. 7.7.11)—i.e., the traditional power law (solid squares)—is close to the exact solution only at $w = 1$ to 4 cm s^{-1} (triangles). At greater w, using the power law may significantly overestimate droplet concentrations.

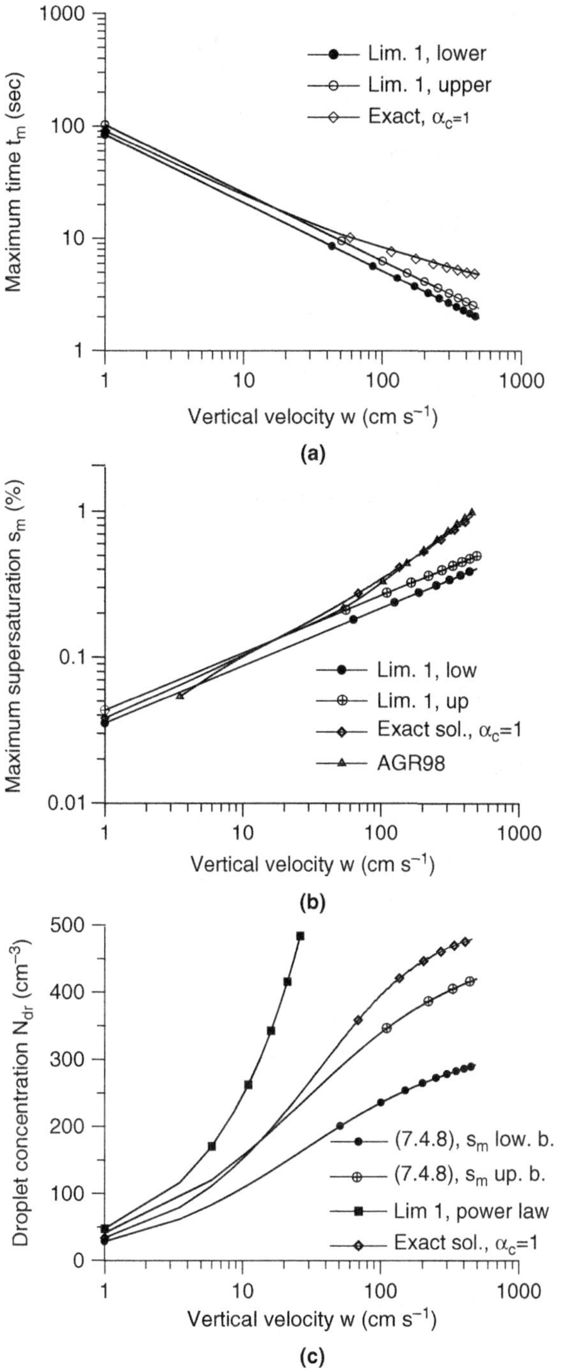

Figure 7.12. Comparison of the lower and upper bounds of the 1st limit (solid and open circles) at $\alpha_c = 1$ with the exact solutions (diamonds). (a) maximum time, t_m; (b) maximum supersaturation, s_m, compared with parameterization from Abdul-Razzak et al. (1998, AGR98); (c) concentration of activated droplets, N_{dr}, calculated as the power law in the 1st limit, lower bound (Eqn. 7.7.11) (solid squares), modified power law (Eqn. 7.4.8) with the lower (Eqn. 7.7.10) and upper (Eqn. 7.7.12) bounds for s_m from Fig. 7.12b, and exact solution as modified power law (Eqn. 7.4.8) with the exact s_m and $\alpha_c = 1$. Calculations are performed with the same parameters as in Fig. 7.11. From Khvorostyanov and Curry (2009a), *J. Atmos. Sci.*, 66, © American Meteorological Society. Used with permission.

This curve and conclusion are in close agreement with those from Ghan et al. (1993, Fig. 7), and this was the major reason of revisions of the Twomey's theory. The modified power law (Eqn. 7.4.8) for N_{dr} with the lower bound for s_m (Eqn. 7.7.10) (solid circles) gives the results that are lower than both the old power law and the exact solution. The modified power law (Eqn. 7.4.8) with the upper bound for s_m (Eqn. 7.7.12) is close to the exact solution to $w \sim 0.5$ m s^{-1}, the error does not exceed 10–15%, but is still lower than the exact solution. The exact solution based on the modified power law (Eqn. 7.4.8) with "exact" s_m is the most precise.

7.7.3. Comparison with Twomey's Power Law

A more detailed comparison with Twomey's power law allows expression of Twomey's empirical coefficients C_T and k via aerosol microphysical parameters. Twomey (1959) provided a solution for s_m and N_{dr} with numerical coefficients for some specified conditions, p and T. A more general solution with analytical coefficients was given in Pruppacher and Klett (1997, PK97), and we will compare this with the PK97 derivation and solution. The supersaturation equation in PK97 was used in the form [PK97, eq. (13-30)],

$$\frac{ds_w}{dt} = A_1 w - A_2 \frac{d\rho_L}{dt}, \tag{7.7.14}$$

where ρ_L is the liquid water content, and the coefficients A_1 and A_2 (eq. (13-31) in PK97) describe supersaturation generation and absorption in the supersaturation equation. The expression for maximum supersaturation, derived in more general form in Pruppacher and Klett (1997, eq. (13-40)), can be rewritten with two coefficients as

$$s_m = C_T^{-1/(k_{s0}+2)} C_{ST} w^{3/[2(k+2)]}, \tag{7.7.15a}$$

$$C_{ST} = \left[\frac{A_1^{3/2}}{2\pi \rho_w A_2 A_3^{3/2} k \mathrm{B}(k/2, 3/2)} \right]^{1/(k+2)}, \tag{7.7.15b}$$

where C_T is Twomey's empirical constant of the activation spectrum (see Eqns. 7.1.3 and 7.1.4) and we introduced a new coefficient C_{ST} for convenience of comparison (a misprint in PK97 in the sign of the power index of C_T is corrected), and A_3 in (Eqn. 7.7.15b) is the coefficient in the drop radius growth rate (PK97, eq. (13-32)). The equation (Eqn. 7.7.14) was compared in Chapter 5 with the supersaturation equation (Eqn. 7.2.1) derived in Chapter 5 and used here. According to (Eqns. 5.3.36) and (5.3.37):

$$A_1 = c_{1w}, \quad A_2 = \Gamma_1/\rho_{ws}, \quad A_3 = c_{3w}, \tag{7.7.15c}$$

where Γ_1 is the psychrometric correction. Substitution of A_1, A_2, and A_3 into (Eqn. 7.7.15b) yields

$$C_{ST} = \left[\frac{c_{1w}^{3/2} \rho_{ws}}{2\pi \rho_w \Gamma_1 c_{3w}^{3/2} k \mathrm{B}(k/2, 3/2)} \right]^{1/(k+2)}. \tag{7.7.16a}$$

It follows from (Eqns. 7.7.1) or (7.4.10) that $k = k_{s0}$ in this limit $(s_m/s_0)^{k_{s0}} \ll 1$. Then, writing $c_{3w}^{3/2} = c_{3w}^{1/2} c_{3w}$ and using (Eqn. 7.2.4) for $c_{3w} = (D_v \rho_{ws}/\Gamma_1 \rho_w)$, we see that

$$C_{ST} = \frac{(c_{1w}^{3/2}\rho_{ws})^{1/(k+2)}}{\left[2\pi\rho_w\Gamma_1 c_{3w}^{1/2}\left(\frac{D_v\rho_{ws}}{\Gamma_1\rho_w}\right)k\mathrm{B}\left(\frac{k}{2},\frac{3}{2}\right)\right]^{1/(k+2)}}$$

$$= \left[2\pi D_v c_{1w}^{-3/2} c_{3w}^{1/2} k_{s0}\mathrm{B}\left(\frac{k}{2},\frac{3}{2}\right)\right]^{\frac{-1}{k+2}} = K_{sm}^{(1)}(T,p). \qquad (7.7.16b)$$

That is, C_{ST} exactly coincides with our $K_{sm}^{(1)}$ in (Eqn. 7.7.6). Equating our expression for the lower bound of s_m (Eqn. 7.7.10) with the upper correction (Eqn. 7.7.13b), and Twomey's s_m in (Eqn. 7.7.15a), dividing by $C_{ST} = K_{sm}^{(1)}$, and using the definitions of b (Eqns. 6.2.6) and (6.2.7) and Q_a (defined after Eqn. 7.7.8) we obtain for Twomey's empirical coefficient:

$$C_T = 10^{-2k_{s0}}\alpha_{k1}^{-1/2}\left(\frac{2^{5/3}M_w}{3R\rho_w}\right)^{-3k_{s0}/2} N_a$$

$$\times\, \sigma_{sa}^{\frac{-3k_{s0}}{2}} T^{\frac{3k_{s0}}{2}} r_{d0}^{k_{s0}(1+\beta)}\left(\nu\Phi_s\varepsilon_v\frac{\rho_s}{\rho_w}\frac{M_w}{M_s}\right)^{k_{s0}/2}. \qquad (7.7.17)$$

Equation (7.7.17) is valid for both $\beta = 1/2$ and $\beta = 0$ and s_m measured in percent. Thus, Twomey's coefficient C_T, used for more than five decades as an empirical parameter, is now expressed via CCN microphysical parameters: N_a, r_{d0}, σ_d (in k_{s0}), T, σ_{sa}, β and b (or ε_v), ρ_s, and other properties of CCN. In addition, these dependencies are nonlinear, governed by the powers of k_{s0}. It is not surprising that measurements made over a period of decades have found such different values for C_T, varying by several orders of magnitude in various air masses and clouds (Hegg and Hobbs, 1992; Pruppacher and Klett, 1997, Table 9.1 therein; Seinfeld and Pandis, 1998). Twomey's empirical index k, equal to k_{s0} at $s << 1\%$, is also expressed with (Eqns. 6.6.35c) or (7.4.10) via σ_d and β, $k_{s0} = \frac{4}{\sqrt{2\pi}\ln\sigma_s} = \frac{4}{\sqrt{2\pi}(1+\beta)\ln\sigma_d}$. Equations (7.7.17) and (7.4.10) explain many features of the observed values of C_T and k. In particular, C_T is proportional to N_a, therefore C_T has higher values in continental than in maritime air masses, and greater values in polluted areas. C_T also increases with increasing soluble fraction ε_v as $\varepsilon_v^{k_{s0}/2}$, with the mean geometric radius of CCN r_{d0} as $r_{d0}^{k_{s0}(1+\beta)}$, and with decreasing surface tension in the presence of surfactants as $\sigma_{sa}^{-3k_{s0}/2}$.

Drop concentration is evaluated according to Twomey as $N_{dr} = C_T s_m^{\,k}$, which can be written using (Eqn. 7.7.15a,b) as

$$N_{dr} = C_T^{2/(2+k)} C_{ST}^{k} w^{3k/2(k+2)}. \qquad (7.7.18)$$

This expression is equivalent to that derived in Twomey (1959) and in Pruppacher and Klett (1997), (Eqn. 13-40 therein), the power indices of C_T and w are the same as in cited works, and additionally, all the empirical parameters are expressed via CCN microphysical properties. It should be noted that the typical calculated values of C_T and k in the limit $s << 1\%$ are usually much greater (k = 1.5–2, and up to 4) than often measured k = 0.2–0.6 at s = 0.2–1% and then used in parameterizations (Pruppacher and Klett, 1997). This reflects a substantial difference of this model with variable $k(s)$ relative to Twomey's model using typical values $k \sim 0.2$–0.6.

Fig. 7.13 depicts $C(s_m)$ and $k(s_m)$ calculated with (Eqns. 7.4.9)–(7.4.11), analogous to Fig. 6.15 with $k(s)$, but presented now as functions of s_m and w simultaneously. $C(s_m) \approx 8000$ cm^{-3} at

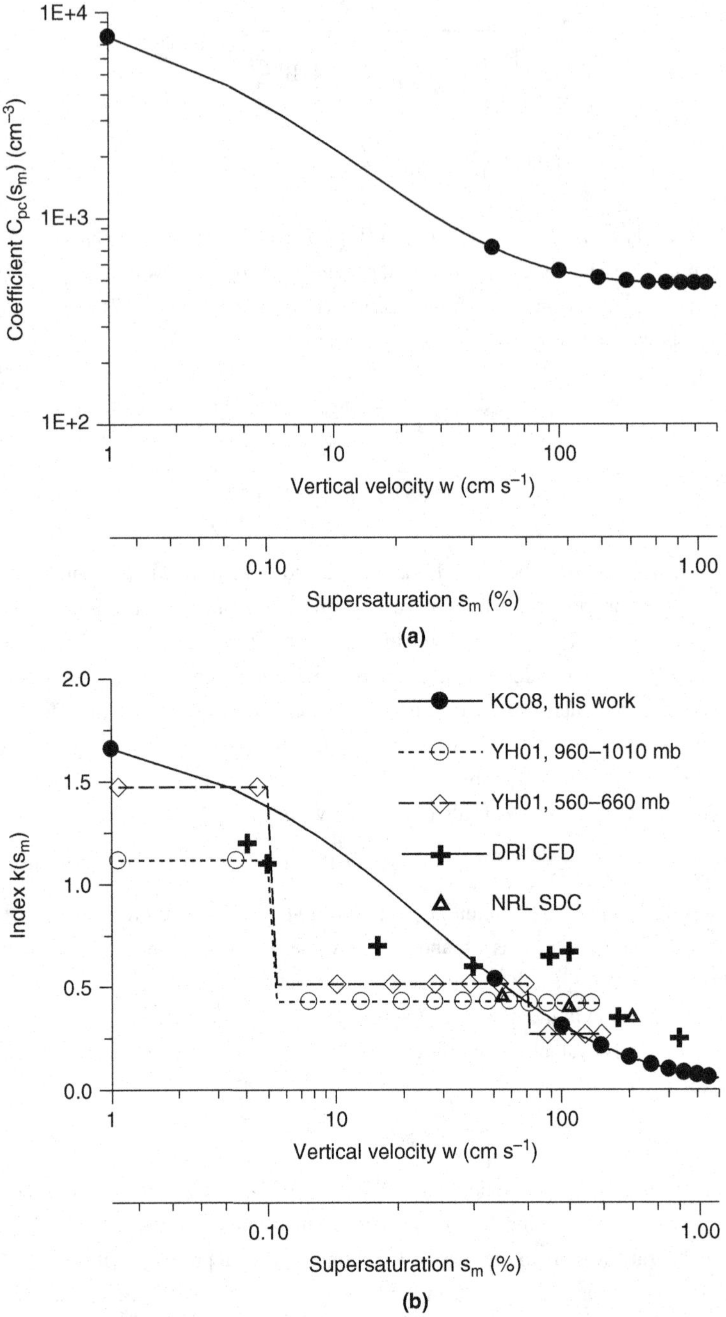

Figure 7.13. (a) Coefficient $C_{pc}(s_m)$ and $C_{pc}(w)$, and (b) index $k(s_m)$ and $k(w)$ as functions simultaneously of the vertical velocities w and maximum supersaturations s_m. Calculations are conducted with the same parameters as in Fig. 7.12 where $s_m(w)$ is also given. Examples of measured $k(s)$ are the data collected in FIRE-ACE in 1998 (Yum and Hudson, 2001, YH01) at the heights 960–1010 mb and 560–660 mb, and the data from Jiusto and Lala (1981) from 2 chambers (continuous flow diffusion chamber, CFD; and static diffusion chamber, SDC). From Khvorostyanov and Curry (2009a), *J. Atmos. Sci.*, 66, © American Meteorological Society. Used with permission.

$w = 1$ cm s^{-1} ($s_m = 0.043\%$) and decreases to the limit $N_a \approx 500$ cm^{-3} at $w \approx 1$–2 m s^{-1} ($s_m = 0.5$–1%). Experimental data on $C(s_m)$ are not readily available. However, the calculated index $k(s_m)$, which also decreases from 1.67 to below 0.2 in this range of s_m or w (Fig. 7.13b), is in good agreement with data both from cloud chamber and field experiments (e.g., Jiusto and Lala, 1981; Yum and Hudson, 2001). This indicates that this model generalizes Twomey's model and correctly predicts decreasing $C(s_m)$, $k(s_m)$ with increasing s_m or increasing w. If these parameters are measured at some s_m, they can be easily recalculated for any value of s_m using (Eqns. 7.4.9)–(7.4.11), but this requires additional measurements or hypotheses on CCN size spectrum and chemical composition. If measurements of all the variables are available, then (Eqns. 7.4.9)–(7.4.11) for $C(s_m)$, $k(s_m)$ can be verified.

7.8. Limit #2: Small Vertical Velocity, Kinetic Growth Regime

For small w, the activation spectrum is also approximated by the power law (Eqn. 7.7.1), but the first fraction and integrand in J_3 in (Eqn. 7.3.9) are substantially different than that in Twomey's model due to the kinetic drop growth rate. The integral J_3 for this case is evaluated in Appendix A.7:

$$J_3^{(2)}(t) = 4^{-4} D_v^{-3} V_w^3 \alpha_c^3 (c_{3w} a_m)^{3/2} t^3 \mathrm{B}\left(\frac{k_{s0}}{2},\ 3\right). \tag{7.8.1}$$

In contrast to the 1st limit (Eqn. 7.7.2) for $J_3^{(1)}$ in the diffusion regime that contained only the beta function, now $J_3^{(2)}$ also contains several coefficients in dimensionless combination.

7.8.1. Lower Bound

Substituting (Eqn. 7.8.1) into (7.3.12) for t_m with $a_m = c_{1w} w$ from (Eqn. 7.3.16), we obtain the lower bound (superscript "2" means the 2nd limit, subscript "l" means lower bound)

$$t_{ml}^{(2)} = K_{tm}^{(2)} N_a^{\frac{-1}{k_{s0}+5}} s_0^{\frac{k_{s0}}{k_{s0}+5}} w^{\frac{-(k_{s0}+2)}{k_{s0}+5}} \alpha_c^{\frac{-3}{k_{s0}+5}}, \tag{7.8.2}$$

$$K_{tm}^{(2)}(T,p) = \left[\frac{\pi k_{s0}}{128} c_{1w}^{k_{s0}+2} c_{3w}^2 D_v^{-2} V_w^3 \mathrm{B}\left(\frac{k_{s0}}{2},\ 3\right)\right]^{\frac{-1}{k_{s0}+5}}. \tag{7.8.3}$$

The lower bound for s_m is calculated using (Eqn. 7.3.14):

$$s_{ml}^{(2)} = \alpha_{k1} K_{sm}^{(2)} N_a^{\frac{-1}{k_{s0}+5}} s_0^{\frac{k_{s0}}{k_{s0}+5}} w^{\frac{3}{k_{s0}+5}} \alpha_c^{\frac{-3}{k_{s0}+5}}, \tag{7.8.4}$$

$$K_{sm}^{(2)}(T,p) = \left[\frac{\pi k_{s0}}{128} c_{1w}^{-3} c_{3w}^2 D_v^{-2} V_w^3 \mathrm{B}\left(\frac{k_{s0}}{2},\ 3\right)\right]^{\frac{-1}{k_{s0}+5}}. \tag{7.8.5}$$

The lower bound for N_{dr} is calculated from (Eqn. 7.4.7) taking into account that $s_m << s_0$ (low s limit):

$$N_{dr,l}^{(2)}(s_m) = (\alpha_{k1})^{k_{s0}} K_{Nd}^{(2)} N_a^{\frac{5}{k_{s0}+5}} s_0^{\frac{-5k_{s0}}{k_{s0}+5}} w^{\frac{3k_{s0}}{k_{s0}+5}} \alpha_c^{\frac{-3k_{s0}}{k_{s0}+5}}, \tag{7.8.6}$$

$$K_{Nd}^{(2)}(T,p) = \left[\frac{\pi k_{s0}}{128} c_{1w}^{-3} c_{3w}^2 D_v^{-2} V_w^3 \mathrm{B}\left(\frac{k_{s0}}{2},\ 3\right)\right]^{\frac{-k_{s0}}{k_{s0}+5}}. \tag{7.8.7}$$

As in the previous limit, these equations can be expressed via r_{d0} and b instead of s_0 using (Eqn. 6.6.35b) for s_0, and via σ_{sa} and T instead of A_k using (Eqn. 6.2.2b). Thus, we obtain for t_m

$$t_{ml}^{(2)} = K_{tm}^{(2)} Q_a^{\frac{k_{s0}}{k_{s0}+5}} N_a^{\frac{-1}{k_{s0}+5}} r_{d0}^{\frac{-(1+\beta)k_{s0}}{k_{s0}+5}} b^{\frac{-k_{s0}}{2(k_{s0}+5)}}$$
$$\times \sigma_{sa}^{\frac{3k_{s0}}{2(k_{s0}+5)}} T^{\frac{-3k_{s0}}{2(k_{s0}+5)}} w^{\frac{-(k_{s0}+2)}{k_{s0}+5}} \alpha_c^{\frac{-3}{k_{s0}+5}}. \qquad (7.8.8)$$

For $k_{s0} = 2$, it follows from (Eqn. 7.8.8) in the kinetic regime:

$$t_{ml}^{(2)} \sim K_{tm}^{(2)}(T) N_a^{-1/7} r_{d0}^{-3/7} b^{-1/7} \sigma_{sa}^{3/7} T^{-3/7} w^{-4/7} \alpha_c^{-3/7}. \qquad (7.8.8a)$$

Comparing with the diffusion limit (Eqn. 7.7.9), we see that all six dependencies (except for α_c) are weaker in the kinetic regime. The maximum supersaturation from (Eqns. 7.8.4), (7.8.5), and (6.3.35b) becomes

$$s_{ml}^{(2)} = \alpha_{k1} K_{sm}^{(2)} Q_a^{\frac{k_{s0}}{k_{s0}+5}} N_a^{\frac{-1}{k_{s0}+5}} r_{d0}^{\frac{-(1+\beta)k_{s0}}{k_{s0}+5}}$$
$$\times b^{\frac{-k_{s0}}{2(k_{s0}+5)}} \sigma_{sa}^{\frac{3k_{s0}}{2(k_{s0}+5)}} T^{\frac{-3k_{s0}}{2(k_{s0}+5)}} w^{\frac{3}{k_{s0}+5}} \alpha_c^{\frac{-3}{k_{s0}+5}}. \qquad (7.8.9)$$

For $k_{s0} = 2$, (Eqn. 7.8.9) gives in this kinetic regime

$$s_{ml}^{(2)} \sim \alpha_{k1} K_{sm}^{(2)}(T) N_a^{-1/7} r_{d0}^{-3/7} b^{-1/7} \sigma_{sa}^{3/7} T^{-3/7} w^{3/7} \alpha_c^{-3/7}. \qquad (7.8.9a)$$

Comparing with the diffusion limit (Eqn. 7.7.10a), we see that the dependencies on N_a, r_{d0}, σ_{sa}, T, and b are substantially weaker in the kinetic regime, and the dependence on w is slightly stronger. The droplet concentration from (Eqns. 7.8.6) and (7.8.7) converts into

$$N_{dr,l}^{(2)}(s_m) = (\alpha_{k1})^{k_{s0}} K_{Nd}^{(2)} Q_a^{\frac{-5k_{s0}}{k_{s0}+5}} N_a^{\frac{5}{k_{s0}+5}} r_{d0}^{\frac{5(1+\beta)k_{s0}}{k_{s0}+5}}$$
$$\times b^{\frac{5k_{s0}}{2(k_{s0}+5)}} \sigma_{sa}^{\frac{-15k_{s0}}{2(k_{s0}+5)}} T^{\frac{15k_{s0}}{2(k_{s0}+5)}} w^{\frac{3k_{s0}}{k_{s0}+5}} \alpha_c^{\frac{-3k_{s0}}{k_{s0}+5}}. \qquad (7.8.10)$$

In particular, for $k_{s0} = 2$, (Eqn. 7.8.10) gives in the kinetic regime:

$$N_{dr}^{(2)} \sim K_{nd}^{(2)}(T) N_a^{5/7} r_{d0}^{15/7} b^{5/7} \sigma_{sa}^{15/7} T^{-15/7} w^{6/7} \alpha_c^{-6/7}. \qquad (7.8.10a)$$

Comparing this with the diffusional growth limit (Eqn. 7.7.11a), we see that the dependencies on the 6 parameters (N_a, r_{d0}, b, σ_{sa}, T, w) become somewhat stronger in the kinetic limit. In addition, this limit includes explicit dependence on the condensation coefficient α_c. An advantage of this method of asymptotic limits is that it allows for simple estimates of the effects of variations of the condensation coefficients on t_m, s_m, and N_{dr}. It follows from (Eqns. 7.8.8)–(7.8.10) that

$$t_m^{(2)} \sim s_m^{(2)} \sim \alpha_c^{-3/(k_{s0}+5)}, \qquad (7.8.11)$$
$$N_{dr}^{(2)} \sim \xi_{con}^{3k_{s0}/(k_{s0}+5)} \sim \alpha_c^{-3k_{s0}/(k_{s0}+5)}. \qquad (7.8.12)$$

This inverse dependence of the concentration of activated drops on the condensation coefficient is in agreement with the results of the numerical simulation of radiative fog suppression by injection of surfactants performed in Buikov and Khvorostyanov (1979) with a 1D spectral bin model. It was assumed that surfactants cause a decrease in the condensation coefficient by 1–3 orders of magnitude as indicated by laboratory experiments (Juisto, 1964; Deryagin et al., 1966; Deryagin and Kurgin, 1972; Silverman and Weinstein, 1974). The simulation showed that this decrease in α_c resulted in two different stages of fog development. In the first stage, continuing for about half an hour to an hour, the droplet growth was suppressed and visibility in the fog increased. In the second stage, due to a continuing radiative cooling equivalent to vertical velocities of a few cm s^{-1} and suppressed drop growth, supersaturation increased by 1–3 orders of magnitude (Eqn. 7.8.9), the droplet concentration increased up to 2–3 times, as described by (Eqn. 7.8.10), and visibility decreased and became lower than in the natural fog without artificial modification.

This inverse dependence of the drop concentration on α_c provides the basis for quantitative estimates of the indirect aerosol effects (Twomey's, Albrecht's, and others) that are caused by an increase in drop concentration in polluted clouds due to anthropogenic, volcanic, and other emissions into the atmosphere, which cause a decrease in the condensation coefficient.

7.8.2. Upper Bound

Derivation of the upper bound, as for the 1st limit, is based on the system of (Eqns. 7.3.12) and (7.3.14). Substituting (Eqn. 7.8.1) for $J_3^{(2)}$ with $a_m = s_m/t_m$ from (Eqn. 7.3.17b) into this system and solving for t_m and s_m, we obtain for $t_{mu}^{(2)}$, $s_{mu}^{(2)}$, and $N_{dr,u}^{(2)}$ in the upper bound (subscript "u") of the 2nd limit (superscript "2") almost the same expressions as (Eqns. 7.8.8)–(7.8.10) but multiplied by the coefficient $\alpha_{low,up}^{(2)}$ for $t_{mu}^{(2)}$ in (Eqn. 7.8.8), by $\alpha_{up}^{(2)}$ instead of α_{k1} for $s_{ml}^{(2)}$ in (Eqn. 7.8.9), and $(\alpha_{up}^{(2)})^{k_{s0}}$ instead of $\alpha_{k1}^{k_{s0}}$ for $N_{dr,u}^{(1)}$ in (Eqn. 7.8.10). That is, the relation of the limits is

$$t_{mu}^{(2)} = \alpha_{low,up}^{(2)} t_{ml}^{(2)}, \qquad s_{mu}^{(2)} = \alpha_{low,up}^{(2)} s_{ml}^{(2)}, \qquad N_{dr,u}^{(2)} = (\alpha_{low,up}^{(2)})^{k_{s0}} N_{dr,l}^{(2)}, \tag{7.8.13}$$

where

$$\alpha_{up}^{(2)} = \alpha_{k1}^{2/(k_{s0}+5)} = [(k_{s0}+2)/(k_{s0}+3)]^{2/(k_{s0}+5)}, \tag{7.8.14}$$

$$\alpha_{low,up}^{(2)} = \alpha_{up}^{(2)}/\alpha_{k1} = [(k_{s0}+3)/(k_{s0}+2)]^{(k_{s0}+3)/(k_{s0}+5)}. \tag{7.8.15}$$

The coefficients α_{k1}, $\alpha_{up}^{(2)}$, and $\alpha_{low,up}^{(2)}$ are given in Table 7.1. As in the 1st limit, all the coefficients are close to 1, and $\alpha_{up}^{(2)}$ is 2–8% lower than 1. In this range of σ_d, $\alpha_{low,up}^{(2)}$ increases from 1.099 to 1.20, and characterizes the ratio of t_m and s_m in upper to lower bounds. Shown in Fig. 7.14 is a comparison of the upper bound of the 2nd limit with the exact solution to the supersaturation equation (Eqn. 7.2.15a). The agreement becomes sufficiently good at $\alpha_c \le 0.04$, and small $w \le 2$–3 cm s^{-1}. A surprising finding is that the value $\alpha_c = 0.04$, used in many cloud models, is close to the boundary of the kinetic regime, and the values of t_m, s_m, and N_{dr} are substantially greater in this regime than in the 1st limit with $\alpha_c = 1$ and the diffusion regime (compare Figs. 7.12 and 7.14).

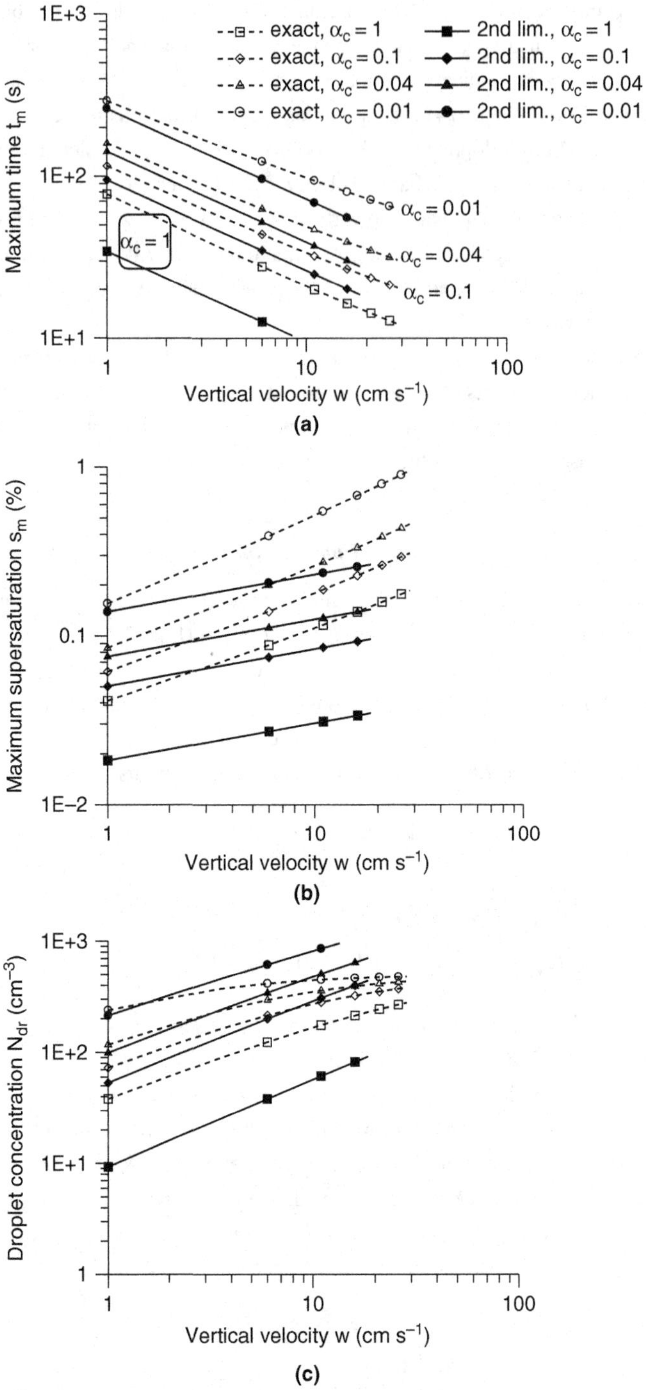

Figure 7.14. Comparison of the upper bound of the 2nd limit (solid symbols and lines) with the corresponding exact solutions to the supersaturation equation (corresponding open symbols, dashed lines). (a) Maximum time, t_{max}; (b) maximum supersaturation, s_{max}; (c) concentration of activated droplets, N_{dr}. Calculations are performed with the parameters: $N_a = 500$ cm^{-3}; $r_{d0} = 0.05$ μm; $\sigma_d = 1.8$; $\beta = 0.5$; $b = 0.25$; $T = 10$ °C; $p = 800$ hPa. Agreement (accuracy of the 2nd limit) becomes better at $\alpha_c \leq 0.04$—i.e., $\alpha_c = 0.04$ is close to the boundary of the kinetic regime. From Khvorostyanov and Curry (2009a), with changes, *J. Atmos. Sci.*, 66, © American Meteorological Society. Used with permission.

7.9. Limit #3: Large Vertical Velocity, Diffusional Growth Regime

At large w—i.e., in the limit $s_m >> s_0$ ($\lambda_s >> 1$)—the integral J_3 in (Eqn. 7.3.9) in general cannot be reduced to the beta function, and it is more convenient to use (Eqn. 7.3.12) for evaluation of t_m with (Eqn. 7.3.6b) for Λ and (Eqn. 7.3.8) for J_2. The integral $J_2^{(3)}(t)$ in this limit is evaluated in Appendix A.7 as

$$J_2^{(3)}(t) \approx \frac{1}{k_{s0}\lambda_s} = \frac{1}{k_{s0}}\left(\frac{s_0}{a_m t}\right)^{k_{s0}}. \tag{7.9.1}$$

The error of this analytical $J_2^{(3)}(t)$ relative to the exact numerical calculation is less than 4% for $\lambda_s \approx 30$. Substituting (Eqn. 7.9.1) into (Eqn. 7.3.6b) for Λ and then in to (Eqn. 7.3.12) yields an expression for t_m

$$t_m^{(3)} = (4\pi D_v c_{1w}^{-1} c_{3w}^{1/2} a_m^{3/2})^{-1/2} N_a^{-1/2} w^{1/2}. \tag{7.9.2}$$

7.9.1. Lower Bound

Substituting $a_m = c_{1w}w$ from (Eqn. 7.3.16), we obtain a simple asymptotic for the lower bound (subscript "l") of the 3rd limit (superscript "3") at large w:

$$t_{ml}^{(3)} = K_{tm}^{(3)} N_a^{-1/2} w^{-1/4}, \tag{7.9.3}$$

$$K_{tm}^{(3)} = [4\pi D_v c_{1w}^{1/2} c_{3w}^{1/2}]^{-1/2}. \tag{7.9.4}$$

This gives for s_m from (Eqn. 7.3.14)

$$s_{ml}^{(3)} = \alpha_{k1} K_{sm}^{(3)} N_a^{-1/2} w^{3/4}, \tag{7.9.5}$$

$$K_{sm}^{(3)} = [4\pi D_v c_{1w}^{-3/2} c_{3w}^{1/2}]^{-1/2}. \tag{7.9.6}$$

7.9.2. Upper Bound

As in the previous sections, the upper bound is obtained by substituting $a_m = s_m/t_m$ from (Eqn. 7.3.17b) into (Eqn. 7.9.2), and then substituting the resulting relation $t_m(s_m)$ into the system of (Eqns. 7.3.12)–(7.3.14) and solving for t_m, s_m. Then, we obtain t_{mu} similar to (Eqn. 7.9.3), but multiplied by the coefficient $\alpha_{low,up}^{(3)}$ and s_{mu} similar to (Eqn. 7.9.5), but with the coefficient $\alpha_{up}^{(3)}$ instead of α_{k1}, so that

$$t_{mu}^{(3)} = \alpha_{low,up}^{(3)} t_{ml}^{(3)}, \qquad s_{mu}^{(3)} = \alpha_{low,up}^{(3)} s_{ml}^{(3)}, \tag{7.9.7}$$

where

$$\alpha_{up}^{(3)} = \alpha_{k1}^{1/4} = [(k_{s0}+2)/(k_{s0}+3)]^{1/4}, \tag{7.9.8}$$

$$\alpha_{low,up}^{(3)} = \alpha_{up}^{(3)}/\alpha_{k1} = \alpha_{k1}^{-3/4} = [(k_{s0}+3)/(k_{s0}+2)]^{3/4}. \tag{7.9.9}$$

The coefficients α_{k1}, $\alpha_{up}^{(3)}$, and $\alpha_{low,up}^{(3)}$ are given in Table 7.1. They are also close to 1. The coefficient α_{k1} for the lower bound is the same for all 4 limits, for the upper bound is lower than 1 by 3–7%, and the ratio of the upper to lower limits, $\alpha_{low,up}^{(3)}$, is 1.09–1.23. That is, the error of any solution average between these bounds does not exceed 5–11%.

Equations (7.9.3)–(7.9.9) show that t_m and s_m in the limit of high w are also related by the linear relation similar to (Eqn. 7.3.14), and also have the form of power laws but only by N_a and w. The dependencies of t_m and s_m on r_{d0}, σ_d, b, β, and on curvature parameter A_k or surface tension σ_{sa}, which were very substantial in the 1st and 2nd limits at small w, disappear from the coefficients and power indices in the limit of high w. Thus, for the 3rd limit with large w and $\alpha_c \sim 1$ (sufficiently pure drops), we come to an unexpected result: the maximum supersaturation s_m, and activation time t_m, are independent of the CCN size spectrum and chemical composition. Thus, t_m and s_m depend only on CCN concentration N_a and vertical velocities w, and the indices of these dependencies are constant and universal for various CCN—i.e., are the same for different values of k_{s0}. This somewhat paradoxical result can be explained by the fact that the 3rd limit deals with the region $\lambda_s >> 1$, where s_m is much greater than the mean geometric s_0, or the minimum radius activated $r_{min} << r_m$—i.e., lies well below the modal radius. Thus, the vast majority of CCN will be activated, no matter what their individual properties are. This independence on the physical and chemical properties of CCN indicates that the effects of the surface tension and chemical composition on the global albedo and radiative balance should be evident mostly in stratiform clouds and become weaker in convective clouds.

Fig. 7.15 shows that the t_m and s_m in the upper bound of the 3rd limit approach the corresponding exact solutions at $w > 1$–2 m s^{-1}, but this conclusion is valid only for $\alpha_c \sim 1$, and becomes invalid at $\alpha_c = 0.1$. At $w < 0.7$–1 m s^{-1}, the 3rd limit is invalid. This tendency to independence of t_m and s_m in this analytical limit on r_m, σ_d, and b (or ε_v) at high w was illustrated with numerical calculations in Figs. 7.4–7.8, where $t_m(w)$, $s_m(w)$ in double-log coordinates represented almost straight lines with slopes slightly increasing toward large w. The corresponding curves $t_m(w)$ and $s_m(w)$ converged at $w \geq 0.5$–2 m s^{-1} toward the limits with fixed slopes. The simple equations of the 3rd limit here provide an explanation for this and exact values of the slopes. The expression (Eqn. 7.4.7) for N_{dr} shows that N_{dr} tends to N_a at $\lambda_s >> 1$ ($s_m >> s_0$). We can approximate it by expanding the denominator in (Eqn. 7.4.7) into a power series by $s_0/s_m \sim 1/\lambda_s$ and retaining two terms—i.e., N_{dr} or fraction activated N_{dr}/N_a is a parabolic function of $(s_m/s_0)^{-k_{s0}}$:

$$N_{dr} = N_a[1 - (s_m/s_0)^{-k_{s0}} + (1/2)(s_m/s_0)^{-2k_{s0}}], \qquad (7.9.10)$$

and s_m is defined in (Eqns. 7.9.5) and (7.9.7). Fig. 7.15 shows that N_{dr} in the 3rd limit at $\alpha_c \sim 1$ also approaches the exact solution at $w > 1$–2 m s^{-1}, but at $w \leq 0.5$ m s^{-1}, N_{dr} increases with decreasing w—i.e., the number of terms in expansion (Eqn. 7.9.10) becomes insufficient and (Eqn. 7.9.10) becomes invalid. In contrast to t_m and s_m, the value of N_{dr} depends on the CCN size spectrum and chemical composition via s_0, but this dependence is substantially weaker than in the 1st and 2nd limits.

7.10. Limit #4: Large Vertical Velocity, Kinetic Growth Regime

When $\alpha_c << 1$, then $\xi_{con} >> r_d(t_0)$, and $r_d(t_0)$ can be neglected in U_s in (Eqn. 7.3.11). Then, $U_s = V_s^2$. Further, $(1 - x^2) << U_s^{1/2}$ in J_2 (Eqn. 7.3.8) and $(1 - x^2) << V_s$. The integral J_2 in the 4th limit is evaluated in Appendix A.7:

$$2J_2^{(4)}(t_m) \approx \frac{1}{128} \frac{a_m^{3/2-k_{s0}} c_{3w}^{3/2} t_m^{3-k_{s0}} \alpha_c^3 V_w^3 s_0^{k_{s0}}}{k_{s0} D_v^3}. \qquad (7.10.1)$$

7.10.1. *Lower Bound*

Substituting (Eqn. 7.10.1) into (Eqns. 7.3.6b) and (7.3.12) for Λ and using (Eqn. 7.3.16) for a_m, we obtain an algebraic equation for t_m, which gives the lower bound

$$t_{ml}^{(4)} = K_{tm}^{(4)} N_a^{-1/5} w^{-2/5} \alpha_c^{-3/5}, \qquad (7.10.2)$$

$$K_{tm}^{(4)} = [(\pi/64) c_{1w}^2 c_{3w}^2 V_w^3 D_v^{-2}]^{-1/5}. \qquad (7.10.3)$$

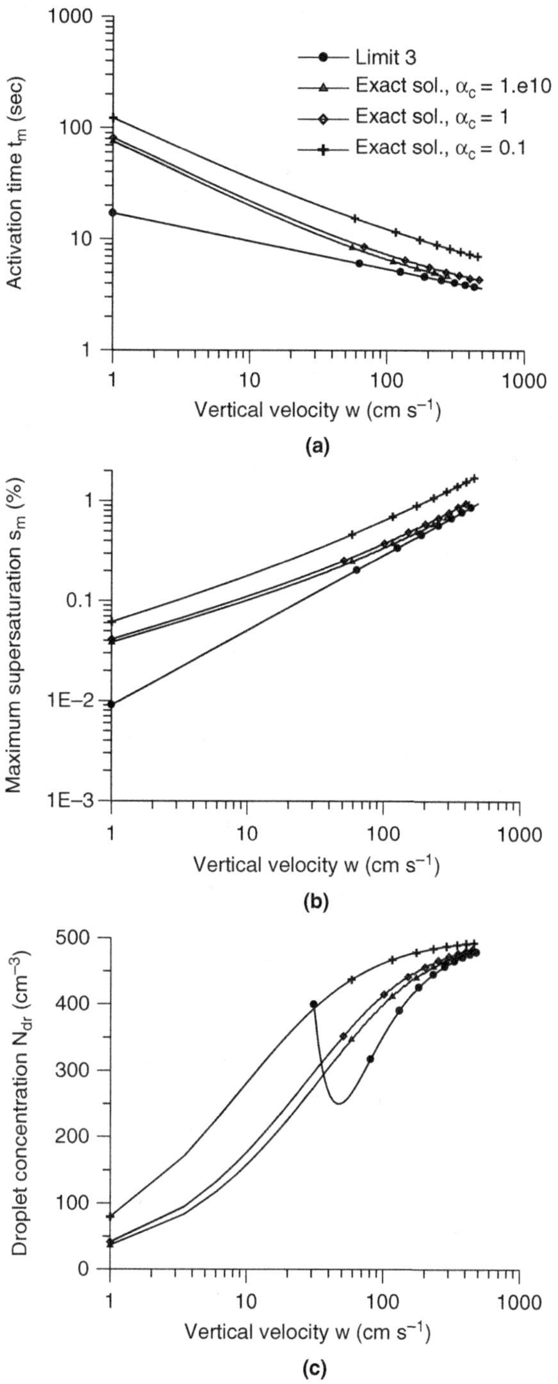

Figure 7.15. Comparison of the upper bound of the 3rd limit (solid circles) with the corresponding exact solutions to the supersaturation equation at various values of condensation coefficient indicated in the legend. (a) Maximum time, t_m; (b) maximum supersaturation, s_m; (c) concentration of activated droplets, N_{dr}. Calculations are performed with the same parameters as in Fig. 7.2. The 3rd limit approaches the exact solution at $w > 1$–2 m s^{-1} and is invalid at $w < 0.7$–1.0 m s^{-1}. From Khvorostyanov and Curry (2009a), *J. Atmos. Sci.*, 66, © American Meteorological Society. Used with permission.

The lower bound for maximum supersaturation s_m is calculated from (Eqns. 7.10.2) and (7.3.14):

$$s_{ml}^{(4)} = \alpha_{k1} K_{sm}^{(4)} N_a^{-1/5} w^{3/5} \alpha_c^{-3/5}, \tag{7.10.4a}$$

$$K_{sm}^{(4)} = [(\pi/64) c_{1w}^{-3} c_{3w}^{2} V_w^3 D_v^{-2}]^{-1/5}. \tag{7.10.4b}$$

The drop concentration can be calculated similar to (Eqn. 7.9.10),

$$N_{dr} = N_a[1 - (s_{ml}^{(4)}/s_0)^{-k_{s0}} + (1/2)(s_{ml}^{(4)}/s_0)^{-2k_{s0}}], \tag{7.10.5}$$

7.10.2. Upper Bound

The upper bound for t_{mu}, s_{mu} is obtained as in the previous sections using (Eqn. 7.3.17b) for a_m, substituting J_2 from (Eqn. 7.10.1) into (Eqns. 7.3.12) and (7.3.14), and solving the system of the two equations. Then, we obtain $t_{mu}^{(4)}$ similar to (Eqn. 7.10.2), but multiplied by the coefficient $\alpha_{low,up}^{(4)}$ and s_{mu} similar to (Eqn. 7.10.4), but with coefficient $\alpha_{up}^{(4)}$ instead of α_{k1}, so that

$$t_{mu}^{(4)} = \alpha_{low,up}^{(4)} t_{ml}^{(4)}, \qquad s_{mu}^{(4)} = \alpha_{low,up}^{(4)} s_{ml}^{(4)}, \tag{7.10.6}$$

where

$$\alpha_{up}^{(4)} = \alpha_{k1}^{2/5} = [(k_{s0}+2)/(k_{s0}+3)]^{2/5}, \tag{7.10.7}$$

$$\alpha_{low,up}^{(4)} = \alpha_{up}^{(4)}/\alpha_{k1} = \alpha_{k1}^{-3/5} = [(k_{s0}+3)/(k_{s0}+2)]^{3/5}. \tag{7.10.8}$$

The coefficients α_{k1}, $\alpha_{up}^{(4)}$, and $\alpha_{low,up}^{(4)}$ are given in Table 7.1. They also are close to 1. The coefficient α_{k1} for the lower limit is the same, $\alpha_{up}^{(4)}$ for the upper limit is lower than 1 by 5–10%, and the ratio of the upper to lower limits, $\alpha_{low,up}^{(4)}$, is 1.07–1.18. That is, the error of any solution mean between these limits does not exceed 3.5–9%.

Thus, the activation time t_m and maximum supersaturation s_m in both bounds of the 4th limit do not depend on the CCN size spectrum (r_d, σ_d), surface tension, soluble fraction β, b or ε_v, or chemical composition. They depend only on CCN concentration N_a, vertical velocities w, and condensation coefficient α_c. These dependencies have the form of the power laws, their indices are constant and universal for various CCN similar to the 3rd limit, and in contrast to the 1st and 2nd limits at small w, where the indices depend on the size dispersions. The drop concentration N_{dr} can be calculated from (Eqn. 7.10.5) as a parabolic function of λ_s below N_a. It depends on physico-chemical properties of CCN as in the 3rd limit, but the dependence is much weaker than in the 1st and 2nd limits.

The characteristics of the 4th limit are compared with the exact solution in Fig. 7.16 for various values of α_c. The 4th limit calculated with $\alpha_c = 1$ is far from the exact solution at all w, which is quite natural as $\alpha_c = 1$ is characteristic of the diffusion regime. With $\alpha_c = 0.04$, the t_m, s_m in the 4th limit are already close to the exact solution. This means that $\alpha_c = 0.04$, often used in the models, is closer to the kinetic than to the diffusion regime. For $\alpha_c = 0.01$, that can be characteristic of polluted clouds: the parameters t_m, s_m, and N_{dr} almost merge with exact solutions at $w \geq 10$ cm s^{-1}.

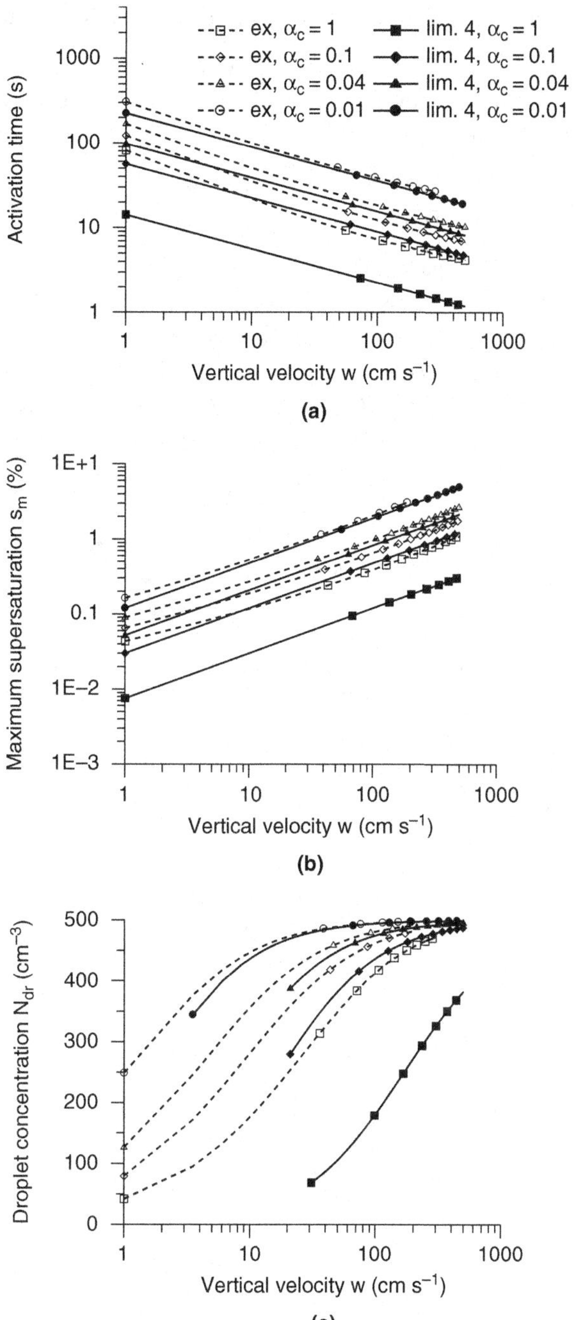

Figure 7.16. Comparison of the upper bound of the 4th limit (solid symbols) with the corresponding exact solutions at various values of the condensation coefficient indicated in the legend (open symbols). (a) activation time, t_m; (b) maximum supersaturation, s_m; (c) concentration of activated droplets, N_{dr}. Calculations are performed with the same parameters as in Fig. 7.11. From Khvorostyanov and Curry (2009a), with changes, *J. Atmos. Sci.*, 66, © American Meteorological Society. Used with permission.

7.11. Interpolation Equations and Comparison with Exact Solutions

General solutions for t_m and s_m can be constructed by interpolating between the limiting solutions with some interpolation function $\Omega_w(w, w_{sc})$ or $\Omega_\alpha(\alpha, \alpha_{sc})$. For example, interpolations between the 1st and 3rd limits can be constructed as

$$s_m^{(13)}(w) = s_m^{(1)}\Omega_w(w, w_{sc}) + s_m^{(3)}[1 - \Omega_w(w, w_{sc})]$$
$$+\Delta s_{m13}\Omega_w(w, w_{sc})[1 - \Omega_w(w, w_{sc})], \qquad (7.11.1)$$

where w_{sc} and α_{sc} are the scaling parameters. The function $\Omega_w(w, w_{sc})$ should satisfy the conditions $\Omega_w(w = 0) \to 1$, $\Omega_w(w >> w_{sc}) \to 0$. This ensures corresponding limits at small and large w. The 3rd term on the RHS of each equation is added to make these fitting functions closer to the exact solutions at the intermediate w, and the product $\Omega_w(1 - \Omega_w)$ vanishes at $w \to 0$ and $w >> w_{sc}$.

Fig. 7.17a shows s_m calculated with (Eqn. 7.11.1), and $\Omega_w(w, w_{sc}) = \exp(-w/w_{sc})$, with $w_{sc} = 1.8$ m s^{-1}, and $\Delta s_{m13} = s_m^{(3)}(5\,m\,s^{-1}) - s_m^{(3)}(1\,cm\,s^{-1})$. The fitted interpolated curve (solid circles) tends

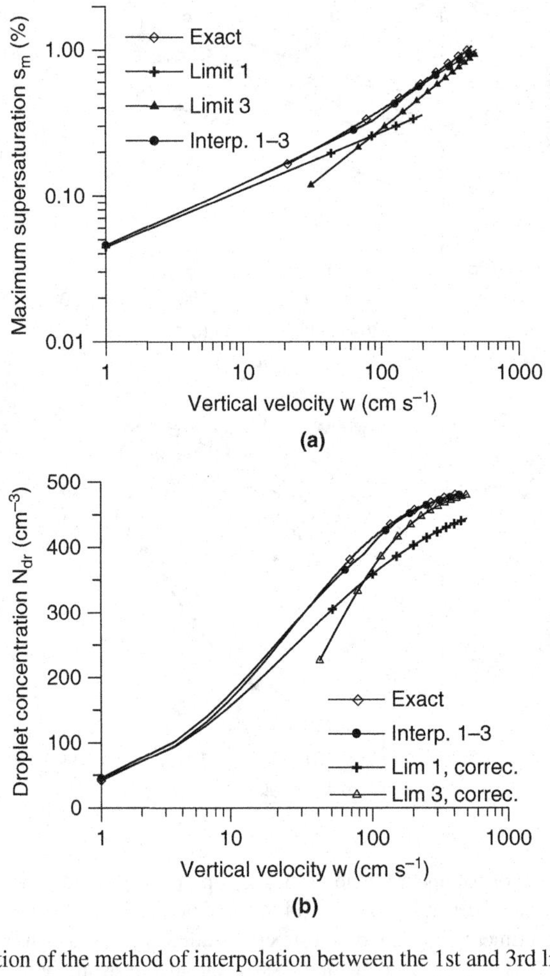

Figure 7.17. Illustration of the method of interpolation between the 1st and 3rd limits and comparison with exact solutions. (a) maximum supersaturation s_m; (b) droplet concentration N_{dr} calculated as described in the text. The parameters are the same as in Fig. 7.11. From Khvorostyanov and Curry (2009a), *J. Atmos. Sci.*, 66, © American Meteorological Society. Used with permission.

to the corresponding limits at $w \rightarrow 0$ and $w >> w_{sc}$, lies above both limits owing to addition of the term with Δs_{m13}, and is very close to the exact curve. The $N_{dr}(w)$ calculated with this interpolated $s_m(w)$ and the modified power laws (Eqns. 7.4.7) or (7.4.8) is also close to the exact curve. This example indicates that more precise interpolation expressions for N_{dr} can be obtained using (Eqns. 7.4.7) or (7.4.8) and interpolations among various analytical limits for s_m. A detailed elaboration of this interpolation procedure among all 4 limits requires a special careful choice of several functions Ω_w, and can be done for every particular case as in the described example.

Appendix A.7 for Chapter 7. Evaluation of the Integrals J_2 and J_3 for Four Limiting Cases

The J_2 and J_3 are the dimensionless integrals introduced in Section 7.3,

$$J_2(t) = \int_0^1 \frac{\{[U_s + (1-x^2)]^{1/2} - V_s\}^2}{[U_s + (1-x^2)]^{1/2}} \frac{x^{k_{s0}-1}}{(1+\lambda_s x^{k_{s0}})^2} dx, \tag{A.7.1}$$

$$J_3(t) = 2J_2(t) = \int_0^1 \frac{\{[U_s + (1-z)]^{1/2} - V_s\}^2}{[U_s + (1-z)]^{1/2}} \frac{z^{k_{s0}/2-1}}{(1+\lambda_s z^{k_{s0}/2})^2} dz. \tag{A.7.2}$$

Here $x = t/t_0$, $z = x^{1/2}$, and the 3 nondimensional parameters λ_s, U_s, and V_s are

$$\lambda_s = \left(\frac{a_m t}{s_0}\right)^{k_{s0}} = \left(\frac{3\sqrt{3}}{2} \frac{a_m t b^{1/2} r_{d0}^{1+\beta}}{A_k^{3/2}}\right)^{k_{s0}}, \tag{A.7.3}$$

$$U_s = \frac{[r_d(t_0) + \xi_{con}]^2}{c_{3w} a_m t^2}, \quad V_s = \frac{\xi_{con}}{(c_{3w} a_m)^{1/2} t}. \tag{A.7.4}$$

The limiting values of J_2 and J_3 are evaluated in the following text for the four limits described in Sections 7.6–7.10.

Limit # 1. Small *w*, diffusional growth regime. In this case, $\lambda_s << 1$, the denominator of the second fraction in the integrand of J_3 tends to unity, which corresponds to the power law activity spectrum (Eqn. 7.7.1). We make additional approximations in (Eqn. A.7.2), consider the case $\alpha_c \sim 1$, neglect small ξ_{con}, and assume small $r_d(t_0)$. Then, U_s and V_s vanish, and the integral J_3 (Eqn. A.7.2) is simplified and reduced to Euler's beta function $\mathrm{B}(x,z)$ as in the Twomey model:

$$J_3^{(1)} = \int_0^1 z^{k_{s0}/2-1}(1-z)^{1/2} dz = \mathrm{B}\left(\frac{k_{s0}}{2}, \frac{3}{2}\right). \tag{A.7.5}$$

Limit #2: Small *w*, $\alpha_c << 1$. When $\alpha_c << 1$, then $\xi_{con} >> r_d(t_0)$ and $r_d(t_0)$ can be neglected in (Eqns. 7.3.11) or (A.7.4) for U_s, and then $U_s \approx \xi_{con}^2/(a_m c_{3w} t^2) = V_s^2$. We also have in (Eqn. A.7.2) for J_3 the inequalities $(1-z) << U_s$ and $(1-z) << V_s^2$. Expansion of the first fraction in the integrand of J_3 in (Eqn. A.7.2) by the small parameter $(1-z)/V_s^2$ yields

$$\frac{(1-z)^2}{4V_s^3} = \frac{(1-z)^2 (c_{3w} a_m)^{3/2} t^3}{4\xi_{con}^3}. \tag{A.7.6}$$

Substituting this expression for the first fraction in the integrand in J_3, we obtain

$$J_3^{(2)}(t) = \frac{(c_{3w} a_m)^{3/2} t^3}{4\xi_{con}^3} \int_0^1 (1-z)^2 \frac{z^{k_{s0}/2-1}}{(1+\lambda_s z^{k_{s0}/2})^{\mu}} dz. \tag{A.7.7}$$

For $\lambda_s << 1$, we can neglect in the denominator of the integrand in (Eqn. A.7.7) the second term—that is, the activation spectrum is also approximated by the power law (Eqn. 7.7.1). Then, using (Eqn. 7.2.5) for ξ_{con}

$$J_3^{(2)}(t) = \frac{(c_{3w} a_m)^{3/2} t^3}{4\xi_{con}^3} \int_0^1 z^{k_{s0}/2-1}(1-z)^2 dz$$

$$= 4^{-4} D_v^{-3} V_w^3 \alpha_c^3 (c_{3w} a_m)^{3/2} t^3 \mathrm{B}\left(\frac{k_{s0}}{2}, 3\right). \tag{A.7.8}$$

Limit #3: Large *w*, diffusional growth regime. At large *w*—i.e., in the limit $s_m >> s_0$ ($\lambda_s >> 1$)—the integral J_3 in (Eqns. 7.3.9) or (A.7.2) in general cannot be reduced to the beta function, and it is more convenient to use $J_2(t)$. Let us consider two functions in the subintegral expression for J_2 in (Eqn. A.7.1):

$$\varphi_{nd}(x) = \frac{x^{k_{s0}-1}}{(1+\lambda_s x^{k_{s0}})^2}, \qquad \psi(x) = \frac{\{[U_s + (1-x^2)]^{1/2} - V_s\}^2}{[U_s + (1-x^2)]^{1/2}}. \tag{A.7.9}$$

Then, J_2 can be written as

$$J_2(t) = \int_0^1 \psi(x)\varphi_{nd}(x)dx. \tag{A.7.10}$$

The function $\varphi_{nd}(x)$ represents the algebraic CCN activity spectrum in nondimensional form. Its maximum evaluated from the condition $d[\varphi_{nd}(x)]/dx = 0$ is located at

$$x_m = \left(\frac{1}{\lambda_s}\frac{k_{s0}-1}{k_{s0}+1}\right)^{1/k_{s0}}. \tag{A.7.11}$$

For example, if $k_{s0} = 2$, then $x_m \approx 0.1$ for $\lambda_s \sim 30$ and $x_m \approx 0.18$ for $\lambda_s \sim 10$—i.e., $x_m << 1$ at $\lambda_s >> 1$. It was shown in Khvorostyanov and Curry (2006) and in Chapter 6 that this differential CCN activity spectrum has the shape of a "smoothed", bell-shaped Dirac delta-function, and the nondimensional function $\varphi_{nd}(x)$ also has this property. This is illustrated in Fig. 7.18, which shows that $\varphi_{nd}(x)$ has a sharp maximum at $x = x_m$, while $\psi(x)$ is a smooth function. Thus, the major contribution into J_2 comes

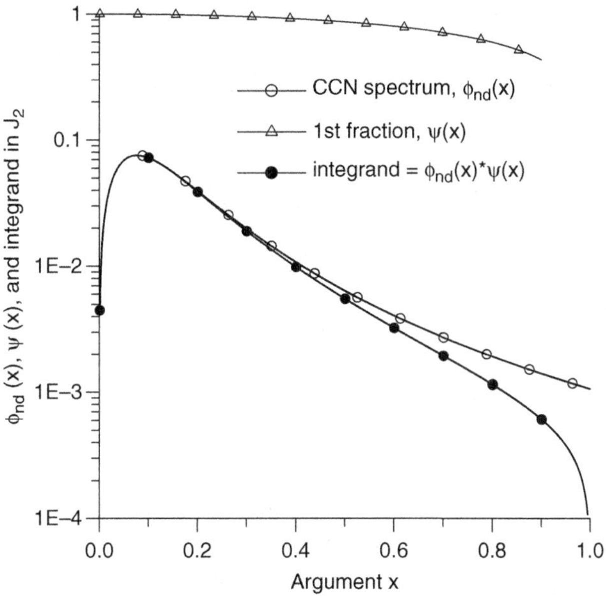

Figure 7.18. Functions $\varphi_{nd}(x)$ (dimensionless CCN spectrum, open circles), $\psi(x)$ defined in (Eqn. A.7.9), triangles, and the integrand $\varphi_{nd}(x)\ \psi(x)$ (solid circles) of the integral J_2 in (Eqn. A.7.10). The values of the parameters are similar to those used earlier: $r_{d0} = 0.05$ μm, $\sigma_d = 1.8$, $r_d(t_0) = 0$, $b = 0.25$, $\beta = 0.5$, $t = 10$ s, but with greater $w = 200$ cm s^{-1}. Then, $\lambda_s \sim 29.7 >> 1$. From Khvorostyanov and Curry (2009a), *J. Atmos. Sci.*, 66, © American Meteorological Society. Used with permission.

from the region around x_m, and we can use for evaluation of J_2 the "quasi δ-function" property of $\varphi_{nd}(x)$, or "the mean value theorem" from the mathematical analysis. Thus, this limit for J_2 is:

$$J_2^{(3)}(t) = \int_0^1 \psi(x)\varphi_{nd}(x)dx \approx \psi(x_m)\int_0^1 \varphi_{nd}(x)\,dx. \qquad \text{(A.7.12)}$$

The last integral is not equal to 1 and still should be evaluated since $\varphi_{nd}(x)$ is proportional to the smoothed delta-function but not equal. To find this integral, we use the property found in KC06 and described in Chapter 6 that the CCN differential spectrum is a smoothed δ-function and is a derivative of the droplet concentration defined in (Eqn. 7.4.7), which is a smoothed Heaviside step-function, found in a similar form by Ghan et al. (1993, 1995). The corresponding fraction activated, $F_{nd}(x) = N_{dr}(s_m)/N_a$, can be written from (Eqn. 7.4.7) in nondimensional form as

$$F_{nd}(x) = \frac{\lambda_s x^{k_{s0}}}{1+\lambda_s x^{k_{s0}}}. \qquad \text{(A.7.13)}$$

One can see that

$$\varphi_{nd}(x) = \frac{1}{k_{s0}\lambda_s}\frac{dF_{nd}(x)}{dx}. \qquad \text{(A.7.14)}$$

Thus,

$$\int_0^1 \varphi_{nd}(x)dx = \frac{1}{k_{s0}\lambda_s}\int_0^1 \frac{dF_{nd}(x)}{dx}dx$$
$$= \frac{1}{k_{s0}\lambda_s}\left[F_{nd}(1) - F_{nd}(0)\right] = \frac{1}{k_{s0}(1+\lambda_s)} \approx \frac{1}{k_{s0}\lambda_s}. \qquad \text{(A.7.15)}$$

The last approximate equation uses the fact that $\lambda_s >> 1$. With $\alpha_c \sim 1$ and $r_d(t_0) = 0$, the values $U_s << 1$, and $V_s << 1$, and $\psi(x_m) \approx (1 - x_m^2)$. Fig. 7.18 shows that at $\lambda_s >> 1$ and $x_m << 1$, the value of $\psi(x_m)$ is very close to 1—e.g., $\psi(x_m) = 0.997$ in this example. Thus, it can be assumed $\psi(x_m) = 1$. Substituting (Eqn. A.7.15) into (Eqn. A.7.12) with $\psi(x_m) = 1$, and using (Eqn. 7.3.10) for λ_s yields

$$J_2^{(3)}(t) \approx \frac{1}{k_{s0}\lambda_s} = \frac{1}{k_{s0}}\left(\frac{s_0}{a_m t}\right)^{k_{s0}}. \qquad \text{(A.7.16)}$$

The exact numerical calculation of J_2 with the functions $\varphi_{nd}(x)$ and $\psi(x)$ shown in Fig. 7.18 yields $J_2^{(3)} = 1.726\times10^{-2}$, and calculation with approximation (Eqn. A.7.16) and $\lambda_s = 29.7$, corresponding to Fig. 7.18, yields $J_2^{(3)} = 1.795\times10^{-2}$—i.e., the error is less than 4%, so the limit (Eqn. A.7.16) for $J_2^{(3)}$ can be used with good accuracy in this limit.

Limit #4: Large w ($\lambda_s >> 1$), kinetic growth regime ($\alpha_c << 1$). Now we consider the analytical evaluation of the integral J_2 in (Eqns. 7.3.8) or (A.7.1) in the 4th limit, $s_m >> s_0$ ($\lambda_s >> 1$), and $\alpha_c << 1$. When $\alpha_c << 1$, then $\xi_{con} >> r_d(t_0)$, and $r_d(t_0)$ can be neglected in U_s, then $U_s = V_s^2$; besides, $(1 - x^2) << U_s$, and $(1 - x^2) << V_s^2$. Similar to the 3rd limit in the previous section, we present the integrand of J_2 (Eqn. A.7.1) as $\varphi_{nd}(x)\psi(x)$ with functions defined in (Eqn. A.7.9). Similar to the 2nd limit, expansion of $\psi(x)$ by the small parameter $(1 - x^2)/V_s^2$ yields

$$\psi(x) \approx \frac{(1-x^2)^2}{4V_s^3} = \frac{(a_m c_{3w})^{3/2} t^3}{4\xi_{con}^3}(1-x^2)^2. \qquad \text{(A.7.17)}$$

As for the 3rd limit, the function $\varphi_{nd}(x)$ has a sharp maximum at $x = x_m << 1$ when $\lambda_s >> 1$, and $\psi(x)$ is a smooth function. Using again the "quasi δ-function" property of $\varphi_{nd}(x)$ and the mean value theorem, we can present $J_2^{(4)}$ as

$$J_2^{(4)}(t) = \int_0^1 \psi(x)\varphi_{nd}(x)dx \approx \psi(x_m)\int_0^1 \varphi_{nd}(x)\,dx. \tag{A.7.17a}$$

Substituting (Eqn. A.7.17) into (Eqn. A.7.17a) with account for (Eqn. A.7.15), yields

$$J_2^{(4)}(t) \approx \frac{(a_m c_{3w})^{3/2} t^3}{4\xi_{con}^3}(1 - x_m^2)^2 \frac{1}{k_{s0}\lambda_s}. \tag{A.7.18}$$

Using (Eqn. 7.3.10) for λ_s, (Eqn. 7.2.5) for ξ_{con}, and using again the relation $(1 - x_m^2) \approx 1$, $J_2^{(4)}$ becomes finally

$$2J_2^{(4)}(t) \approx \frac{1}{128}\frac{a_m^{3/2-k_{s0}} c_{3w}^{3/2} t^{3-k_{s0}} \alpha_c^3 V_w^3 s_0^{k_{s0}}}{k_{s0} D_v^3}. \tag{A.7.19}$$

8

Homogeneous Nucleation

Phase transitions in a substance may proceed with or without the presence of foreign substances. If formation (nucleation) of a new phase in the original phase is catalyzed by a foreign substance, this process is called *heterogeneous nucleation.* IIn this chapter, we consider phase transitions without participation of a foreign substance. Formation of a new phase within the original phase under such conditions is termed *homogeneous nucleation.* The most important examples of homogeneous nucleation for cloud physics are: formation of ice crystals (*homogeneous ice nucleation*) or liquid drops from supersaturated water vapor (*homogeneous drop nucleation*); formation of ice crystals by the freezing of liquid water or solution drops (*homogeneous freezing*); and formation of drops by the melting of ice crystals.

In Chapters 3 and 4, we discussed phase transitions and conditions of equilibrium between various phases. However, a phase transition from a metastable phase A into a stable phase B does not occur instantaneously or exactly at equilibrium. A bulk phase A of some substance may exist during a sufficient time in a metastable state; that is, not in thermodynamic equilibrium, but still without transition into a more stable phase B. Examples of relevance to water are: a) supersaturated vapor (phase A) at a relative humidity greater than 100% but still without condensation into the bulk liquid water (phase B); b) superheated liquid water (phase A) above the boiling point without full transition into the vapor (phase B); c) supercooled liquid water at temperatures below the freezing point (phase A) without freezing into ice (phase B); and d) a salt solution (phase A) supersaturated with respect to dissolved salt but without crystallizing salt into a solid (phase B).

In this chapter, the concepts of metastable states and nucleation of a new phase are introduced based on classical nucleation theory. Equations are derived for nucleation rates of condensation, deposition, and homogeneous ice nucleation. Empirical and semi-empirical parameterizations of homogeneous ice nucleation are reviewed. Equations for water and ice supersaturation with homogeneous ice nucleation are derived to study ice nucleation kinetics. The homogeneous freezing and melting of water and aqueous solutions are examined in the framework of extended classical nucleation theory. Based on thermodynamic principles, general equations for the critical germ radius and free energy are derived that express these properties as functions of temperature, solution molality, pressure, and the finite size of freezing and melting particles. This theory is applied to the study of liquid–solid phase transitions: the homogeneous freezing of aqueous solutions and the surface melting of ice. Simple analytical expressions are derived for the corresponding freezing and melting critical temperatures $T_{f,hom}$, T_m, freezing and melting point depressions, $\Delta T_{f,hom}$, ΔT_m, for their relation, and for the critical water and ice saturation ratios $S_{w,hom}$, $S_{i,hom}$. The kinetics of homogeneous ice nucleation and microphysical processes of formation of a cirrus-like cloud are studied with use of a cloud parcel model with spectral bin microphysics and a detailed supersaturation equation.

Based on the general analytical equations of the classical nucleation theory and some simplifications prompted by the parcel model simulations, an analytical parameterization of homogeneous ice nucleation is developed.

8.1. Metastable States and Nucleation of a New Phase

In a metastable phase A (called the *mother* or *parent* phase), fluctuations exist that lead to formation of small clusters or *nuclei* consisting of several molecules or atoms of the stable phase B, termed *embryos*. Phase transition begins with formation of the small embryos of phase B inside phase A, which have the form of small drops, bubbles, or crystals (i.e., small embryos of a new phase B)—in other words, generally small volumes surrounded by a newly formed surface at the interface of the phases A and B. Before the embryos reach some critical size r_{cr}, the energy ΔF of their formation increases with size. The embryos are also metastable and disrupt in further fluctuations. But after they reach the critical size, ΔF begins to decrease with the size of the embryo, and the embryos can grow spontaneously. Such supercritical embryos are called *germs*, although the subcritical embryos are sometimes also called germs.

Suppose that the molecular chemical potentials in phases A and B are μ_{A1} and μ_{B1}, and phase A is comprised of $(N_A + N_B)$ molecules, of which N_B molecules can form an embryo in phase B. Then, the thermodynamic potentials before and after nucleation, F_1 and F_2, are $F_1 = \mu_{A1}(N_A + N_B)$, and $F_2 = \mu_{A1}N_A + \mu_{B1}N_B$. Formation of an embryo of radius r causes two kinds of energy changes: a decrease in the volume energy $\Delta F_V = F_2 - F_1 = -N_B(\mu_{A1} - \mu_{B1})$, and an increase in the surface energy $\Delta F_s = 4\pi r^2\sigma_{AB}$ due to formation of the new surface surrounding the embryo of phase B, with σ_{AB} being the surface tension between the two phases. The total energy change ΔF is

$$\Delta F = \Delta F_V + \Delta F_s = F_2 - F_1 + 4\pi r^2\sigma_{AB} = -N_B(\mu_{A1} - \mu_{B1}) + 4\pi r^2\sigma_{AB}. \qquad (8.1.1)$$

This expression can also be written for the molar quantities using the relation $N_B = N_{Av}n_B$, where N_{Av} is Avogadro's number, and n_B is the number of moles of phase B. Thus,

$$\Delta F = \Delta F_V + \Delta F_s = -n_B(\mu_A - \mu_B) + 4\pi r^2\sigma_{AB}, \qquad (8.1.1a)$$

where $\mu_A = N_{Av}\mu_{A1}$ and $\mu_B = N_{Av}\mu_{B1}$ are the molar chemical potentials of phases A and B. If phase A is metastable, and B is stable, then $\mu_{A1} > \mu_{B1}$, and the first term on the RHS of these equations is negative, it causes a decrease in energy. The second term with σ_{AB} is positive and causes an energy increase.

Now, the embryo contains N_B molecules. If the volume of a germ is V_B, and the volume of a molecule in phase B is $v_{B1} = M_B/(\rho_B N_{Av})$, where M_B and ρ_B are the molecular weight and the density of phase B, then N_B can be expressed via r as $N_B = V_B/v_{B1} = (4/3)\pi r^3/v_{B1}$, and ΔF can be rewritten as

$$\Delta F = \Delta F_V + \Delta F_s = -\frac{(\mu_{A1} - \mu_{B1})}{v_{B1}}V_B + 4\pi r^2\sigma_{AB}$$

$$= -(\mu_{A1} - \mu_{B1})\frac{4\pi}{3}\frac{r^3}{v_{B1}} + 4\pi r^2\sigma_{AB}. \qquad (8.1.2)$$

The energy change therefore consists of two components: latent heat release and surface tension work. At very small r, the second term proportional to r^2 is greater than the first term, $\Delta F > 0$, and will increase with increasing r. However, the absolute value of the first term on the RHS, proportional to r^3, will increase faster, eventually exceeding the magnitude of the first term, and ΔF will reach a maximum at some critical radius r_{cr} and then will begin to decrease at $r > r_{cr}$. The maximum or critical value of ΔF_{cr} is determined from the condition $d(\Delta F_{cr})/dr = 0$, which yields from (Eqn. 8.1.2) an expression for r_{cr} of

$$r_{cr} = \frac{2\sigma_{AB} v_{B1}}{\mu_{A1} - \mu_{B1}}. \tag{8.1.3}$$

This can be rewritten in molar quantities by multiplying the numerator and denominator in (Eqn. 8.1.3) by the Avogadro number N_{Av} to obtain

$$r_{cr} = \frac{2\sigma_{AB} v_{B\mu}}{\mu_A - \mu_B}, \tag{8.1.4}$$

where $v_{B\mu} = N_{Av} v_{B1} = M_B/\rho_B$ is the molar volume of phase B. The difference of the molar chemical potentials $A_\infty = \Delta\mu_{AB} = \mu_A - \mu_B$ is called *the ordinary affinity of a phase transition*, and the previous equations show that the critical radius can be evaluated using known values of affinity A_∞, surface tension, and the molar volume of the more stable phase.

When the affinity $\Delta\mu_{AB} = \mu_A - \mu_B$ is known, then r_{cr} can be evaluated along with the corresponding critical energy ΔF and the nucleation rate J. If r_{cr} is known, then the affinity $\Delta\mu_{AB}$ can be evaluated, and then ΔF_{cr} and J. In many cases, r_{cr} can be evaluated without considering separately μ_A and μ_B, or the difference $\Delta\mu_{AB}$; the values of r_{cr}, ΔF, and J can be calculated directly from the entropy equation or other equations. This is illustrated in the following text considering the homogeneous nucleation of the drops from a supersaturated vapor.

In Chapter 3, we obtained Kelvin's equation (Eqn. 3.8.4) for the equilibrium saturation ratio S_w over a droplet of radius r in the vapor or in a humid air:

$$\ln S_w = \frac{2M_w \sigma_{vw}}{RT\rho_w r}. \tag{8.1.5}$$

This equation can be solved for the equilibrium radius $r_{eq,w}$ of a drop at a saturation ratio S_w

$$r_{eq,w} = \frac{2M_w \sigma_{vw}}{RT\rho_w \ln S_w} = \frac{2\sigma_{vw} v_{w\mu}}{RT \ln S_w} = \frac{2\sigma_{vw} v_{w1}}{kT \ln S_w}, \tag{8.1.6}$$

where $v_{w\mu} = M_w/\rho_w$ and $v_{w1} = M_w/(N_{av}\rho_w)$ are the molar and molecular volumes of water, and k is the Boltzmann's constant. Comparing (Eqns. 8.1.4) and (8.1.6), we see that the difference of molecular chemical potentials of vapor μ_{v1} and water μ_{w1} (the affinity) is

$$\Delta\mu_{vw,1} = \mu_{v1} - \mu_{w1} = kT \ln S_w. \tag{8.1.7}$$

In terms of molar chemical potentials, this relation is

$$\Delta\mu_{vw} = \mu_v - \mu_w = RT \ln S_w. \tag{8.1.7a}$$

Substituting (Eqn. 8.1.7a) into (Eqn. 8.1.2) for ΔF with superheated vapor as phase A and condensed water as phase B, we obtain the free energy ΔF of an embryo formation

$$\Delta F = -\frac{kT \ln S_w}{v_{w1}} \frac{4\pi r^3}{3} + 4\pi r^2 \sigma_{vw}. \tag{8.1.8}$$

This expression has a maximum value at r_{cr} that is determined from the condition $d(\Delta F_{cr})/dr = 0$, which yields

$$r_{cr,w} = \frac{2\sigma_{vw} v_{w1}}{kT \ln S_w} = \frac{2M_w \sigma_{vw}}{RT\rho_w \ln S_w}. \tag{8.1.9}$$

Thus, the critical radius r_{cr} coincides in this case with the equilibrium Kelvin's radius (Eqn. 8.1.6). Substituting this r_{cr} into (Eqn. 8.1.8), we obtain the corresponding critical energy

$$\Delta F_{cr} = \frac{4\pi\sigma_{vw} r_{cr}^2}{3} = \frac{\sigma_{vw} A_{cr}}{3} = \frac{16\pi M_w^2 \sigma_{vw}^3}{3[RT\rho_w (\ln S_w)]^2}, \tag{8.1.10}$$

where A_{cr} is the surface area of the sphere with radius r_{cr}, and we used the first of (Eqn. 8.1.6) for r_{cr}. Thus, the second equality in (Eqn. 8.1.10) shows that the critical energy of a germ formation is equal to one third of the product of the surface tension by the surface area of the germ.

Fig. 8.1a,b illustrate the radius dependence of a liquid embryo formation from supersaturated vapor: an increase and subsequent decrease of the embryo energies with increasing radius r (Fig. 8.1a) and a corresponding increase of the number of molecules g (Fig. 8.1b) inside the embryo for various saturation ratios S_w from 0.9 to 6. It is seen that the energy maxima shift to smaller r and g with increasing S_w—i.e., the germs' sizes decrease with increasing S_w. These maxima correspond to the critical energies, and are given in more detail in Table 8.1 along with the critical radii r_{cr} and the number of molecules g_{cr}. Table 8.1 shows that as S_w increases from 0.9 to 6, the critical radius r_{cr} decreases from 40.8Å (40.8 × 10^{-8} cm) to 6.2Å, and the corresponding number of molecules in an embryo decreases from 12496 to 43. The energy ΔF_{cr} decreases almost 170 times, from 1.89 × 10^{-10} erg to 1.15 × 10^{-12} erg as S_w increases from 0.9 to 6. The smallest radius (6.2Å) at S_w = 6 is already comparable to the size of the molecules (~3.5Å). However, even a small germ still contains 43 molecules, which can make justifiable to some extent application of the bulk concepts of surface tension, bulk density, germ surface, and volume.

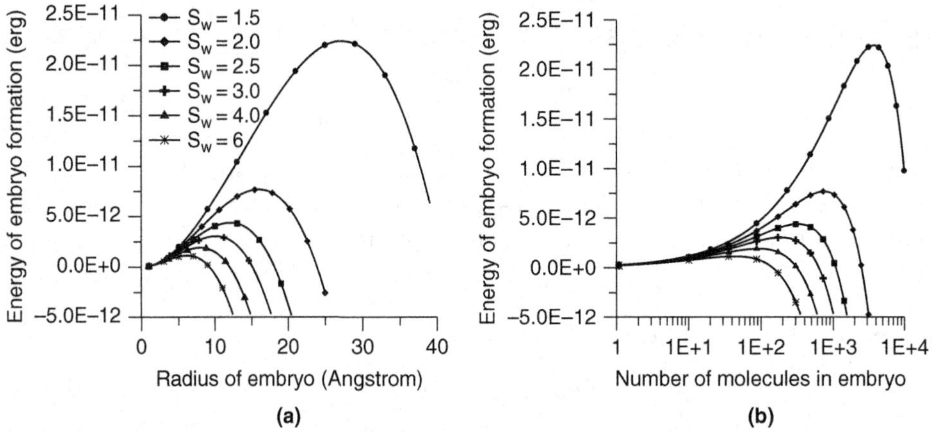

Figure 8.1. Dependence of the energy of a liquid embryo formation from supersaturated vapor on the embryo radius r (a), and on the number of molecules g in the embryo (b) at various water saturation ratios S_w indicated in the legend.

Table 8.1. *Critical radii* r_{cr} *(Å), critical energies* ΔF_{cr} *(erg), and the number of molecules* g_{cr} *in the critical germs of water droplet in supersaturated vapor at various saturation ratios* S_w. *The critical values correspond to the maxima of the curves in Fig. 8.1.*

S_w	0.9	1.5	2.0	2.5	3.0	4.0	6.0
r_{cr} (Å)	40.8	27	16	12	10	8	6.2
ΔF_{cr} (10^{-12} erg)	189	22.4	7.66	4.38	3.05	1.92	1.15
g_{cr} (number of molecules)	12496	3702	753	317	184	94	43

The critical radius $r_{cr,i}$ and energy of a critical germ for ice formation from supersaturated water vapor are derived analogously and differ from the properties of the liquid droplet germs by replacing all liquid water parameters with ice parameters:

$$r_{cr,i} = \frac{2M_w \sigma_{iv}}{RT\rho_i \ln S_i} = \frac{2\sigma_{iv} v_{i1}}{kT \ln S_i}, \tag{8.1.11}$$

where the subscript "*i*" denotes the quantities relevant to ice. The critical energy of an ice germ formation instead of (Eqn. 8.1.10) is

$$\Delta F_{cr,i} = \frac{4\pi\sigma_{iv} r_{cr,i}^2}{3} = \frac{\sigma_{iv} A_{cr,i}}{3} = \frac{16\pi M_w^2 \sigma_{iv}^3}{3[RT\rho_i (\ln S_i)]^2}. \tag{8.1.12}$$

8.2. Nucleation Rates for Condensation and Deposition

8.2.1. Application of Boltzmann Statistics

The rate of formation of the new embryos of the more stable daughter phase within the unstable parent phase (per unit volume per unit time) J_{hom} is termed *the nucleation rate*. The classical derivation of the nucleation rate was given by Volmer and Weber (1926) based on thermodynamic and kinetic arguments. It was subsequently modified by several researchers (e.g., Farkas, 1927; Becker and Döring, 1935; Volmer, 1939; Zeldovich, 1942; Frenkel, 1946), who refined calculations of the reverse process: evaporation of molecules from the germs. These derivations were based on two major assumptions: 1) the germs before nucleation have a spherical form with the usual macroscopic densities and surface tensions; and 2) the germs are distributed according to the Boltzmann statistics. The first assumption was later revised and some correction factors were introduced to account for the nonsphericity of ice germs (e.g., Fletcher 1962). The second assumption can also be revised, and more general statistical distributions considered, as will be discussed later in this chapter.

Application of Boltzmann's statistics for the number N_g of g-mers (embryos containing g molecules or atoms with g-th energy state ΔF_g) in a volume V is

$$N_g = N_{sat} \exp(-\Delta F_g / kT), \tag{8.2.1}$$

or of the concentrations $c_g = N_g/V$ of the g-mers

$$c_g = c_{sat,w} \exp(-\Delta F_g / kT), \tag{8.2.2}$$

where N_{sat} and $c_{sat,w}$ are the number and concentration of the monomers (water molecules) in a water-saturated environment. The nucleation rate J_g of transition from the metastable parent phase A into the stable daughter phase B can be evaluated as the number of embryos (cm^{-3}s^{-1}) per unit of time entering the category containing g molecules or atoms of the parent phase. Then, J_g can be calculated from the kinetic equation

$$J_g = f_{g-1} j_{g-1}^{(+)} A_{g-1} - f_g j_g^{(-)} A_g, \tag{8.2.3}$$

where g is the number of molecules or atoms of the parent phase in an embryo of the new phase, $A_g = 4\pi r_g^2$ is the surface area of an embryo with radius r_g containing g particles of the parent phase, f_g is the time-dependent concentration of such embryos (different from c_g), $j_{g-1}^{(+)}$ is the molecular flux from the $(g-1)$-mer embryo into g-mer, and $j_g^{(-)}$ is the flux from a g-mer into a $(g+1)$-mer embryo. The first and second terms in (Eqn. 8.2.3) describe respectively the influx and depletion of the germs of class g. In equilibrium with $J_g = 0$, these two fluxes are equal and the condition of equilibrium is

$$c_{g-1} j_{g-1}^{(+)} A_{g-1} - c_g j_g^{(-)} A_g = 0. \tag{8.2.4}$$

Equation (8.2.4) allows exclusion of the depletion flux,

$$j_g^{(-)} A_g = (c_{g-1}/c_g) j_{g-1}^{(+)} A_{g-1}. \tag{8.2.5}$$

Substituting this expression into (Eqn. 8.2.3) yields

$$J_g = f_{g-1} A_{g-1} j_{g-1}^{(+)} - f_g \frac{c_{g-1}}{c_g} A_{g-1} j_{g-1}^{(+)} = c_{g-1} j_{g-1}^{(+)} A_{g-1} \left(\frac{f_{g-1}}{c_{g-1}} - \frac{f_g}{c_g} \right). \tag{8.2.6}$$

The time variation of the concentration f_g in the g-state is determined by the equation of balance

$$\frac{\partial f_g}{\partial t} = J_g - J_{g+1}, \tag{8.2.7}$$

where the first term on the RHS describes the arrival into the embryo class g, and the second term describes outflow from class g—that is, arrival into embryo class $(g+1)$.

When the number g of molecules in a germ is sufficiently large (this requirement usually is fulfilled), we can replace the discrete number g by a continuous variable g and write it as an argument of the function instead of the subscript. The finite differences in (Eqns. 8.2.6) and (8.2.7) are replaced by the derivatives

$$\frac{\partial f(g)}{\partial t} = J(g) - J(g+1) = -\frac{\partial J(g)}{\partial g}, \tag{8.2.8}$$

$$J_g = c_{g-1} j_{g-1}^{(+)} A_{g-1} \left(\frac{f_{g-1}}{c_{g-1}} - \frac{f_g}{c_g} \right) = -c_{g-1} j_{g-1}^{(+)} A_{g-1} \frac{\partial}{\partial g} \left(\frac{f(g)}{c(g)} \right). \tag{8.2.9}$$

We now introduce a diffusion coefficient in the g-space:

$$D(g) = j_{g-1}^{(+)} A_{g-1} \approx j_g^{(+)} A_g. \tag{8.2.10}$$

The last approximate equality holds if the number g is sufficiently high. Then, J in (Eqn. 8.2.9) can be rewritten as

$$J(g) = -c(g) D(g) \frac{\partial}{\partial g} \left(\frac{f(g)}{c(g)} \right) \tag{8.2.11a}$$

$$= -D(g) \frac{\partial f(g)}{\partial g} + D(g) f(g) \frac{\partial [\ln c(g)]}{\partial g}. \tag{8.2.11b}$$

It follows from (Eqn. 8.2.2) that

$$\frac{\partial[\ln c(g)]}{\partial g} = -\frac{\partial}{\partial g}\left(\frac{\Delta F(g)}{kT}\right). \tag{8.2.12}$$

Substituting this equation into (Eqn. 8.2.11b), we obtain

$$J(g) = -D(g)\frac{\partial f(g)}{\partial g} - \left\{\frac{D(g)}{kT}\frac{\partial[\Delta F(g)]}{\partial g}\right\} f(g). \tag{8.2.13}$$

Thus, we arrive at an interesting result: the nucleation rate can be presented in the form of the flux in the g-space. This flux can also be written in the form

$$J(g) = -D(g)\frac{\partial f(g)}{\partial g} - f(g)V_g(g), \tag{8.2.14}$$

which includes drift and diffusion. The diffusion part (the first term on the RHS) is proportional to $D(g)$ and describes fluxes in both directions by g, toward larger g and back to smaller g, broadening the function $f(g)$. Drift (the second term on the RHS) is proportional to $V_g(g)$, which is an effective velocity in g-space:

$$V_g(g) = \frac{D(g)}{kT}\frac{\partial[\Delta F(g)]}{\partial g}. \tag{8.2.15}$$

8.2.2. The Fokker–Planck, and the Frenkel–Zeldovich Kinetic Equations and the Zeldovich Factor

The expressions (Eqns. 8.2.13) and (8.2.14) are typical for physical kinetics, and are analogous to the usual formulae in the *Fokker–Planck kinetic equation* for the flux of the particles. In the Fokker–Planck equation, the particles are distributed along the g-axis with density $f(g)$ and move along the axis g by the joint action of diffusion $D(g)$ and drift $V_g(g)$ induced by an external potential force $\Delta F(g)$, while the mobility coefficient q of the particles is determined by Einstein's relation $q = D(g)/kT$. The diffusion coefficient $D(g)$ defined by (Eqn. 8.2.10) represents a function of g that is related to the germ radius as

$$g v_B = (4\pi/3) r_g^3, \qquad \text{or} \qquad r_g = g^{1/3} v_B^{1/3} (3/4\pi)^{1/3}, \tag{8.2.16}$$

where v_B is the volume of a molecule in phase B. According to (Eqn. 8.2.10), $D(g)$ is proportional to the surface area A_g. Substituting this r_g from (Eqn. 8.2.16) into A_g in (Eqn. 8.2.10), $D(g)$ can be written via g as

$$D(g) = (4\pi)^{1/3} 3^{2/3} v_B^{2/3} g^{2/3} j_g^{(+)}. \tag{8.2.17}$$

Thus, $D \sim g^{2/3}$ (or $D \sim r_g^2$ since $g \sim r^3$) and resembles a diffusion coefficient in an inhomogeneous medium with coordinate dependence—e.g., the turbulent diffusion coefficient in the atmospheric surface layer $k_z \sim z^{\varepsilon}$, or the Richardson turbulent diffusion coefficient with the dependence $k_x \sim x^{4/3}$ (e.g., Monin and Yaglom, 2007a, b).

Substituting (Eqn. 8.2.13) for J into the kinetic equation (Eqn. 8.2.8) for f, we obtain a kinetic equation for the distribution $f(g)$ known as *the Frenkel–Zeldovich kinetic equation*:

$$\frac{\partial f}{\partial t} = \frac{\partial}{\partial g}\left(D \frac{\partial f_g}{\partial g} \right) + \frac{1}{kT}\frac{\partial}{\partial g}\left(Df \frac{\partial[\Delta F(g)]}{\partial g} \right). \tag{8.2.18}$$

This is a kinetic equation of the Fokker–Planck type with drift and diffusion, and the second derivative by g, and can be written in the form of a continuity equation:

$$\frac{\partial f}{\partial t} = -\frac{\partial J(g)}{\partial g}, \tag{8.2.19}$$

where the flux $J(g)$ is defined in (Eqns. 8.2.13) or (8.2.14).

The steady state distribution, $\partial f/\partial t = 0$, corresponds to $J(g) = \text{const}$. Then, we can rewrite (Eqn. 8.2.11a) as

$$\frac{\partial}{\partial g}\left(\frac{f(g)}{c(g)} \right) = -\frac{J}{c(g)D(g)}. \tag{8.2.20}$$

The integration of this equation yields a solution to the Frenkel–Zeldovich equation:

$$f(g) = Jc(g)\int_g^G \frac{dg'}{D(g')c(g')}, \tag{8.2.21}$$

where the upper limit of integration G is chosen such that $f(G) = 0$. Using the dependence (Eqn. 8.2.2) $c(g) = c_{sat}\exp[-\Delta F(g)/kT]$ in front of the integral and in the integrand, this expression can be transformed as

$$f(g) = J\exp\left(-\frac{\Delta F(g)}{kT} \right)\int_g^G \frac{1}{D(g')}\exp\left(\frac{\Delta F(g')}{kT} \right) dg'. \tag{8.2.22}$$

The function $\exp[\Delta F(g)/kT]$ in the integrand has a very sharp maximum near $g = g^*$, which corresponds to a critical radius. Therefore, the value $1/D(g')$ with rather weak dependence on g can be removed out of the integral and can be taken as $1/D(g^*)$. Then, ΔF can be expanded into the power series near $g = g^*$:

$$\Delta F(g) \approx \Delta F_{\max} + \left.\frac{d\Delta F(g)}{dg}\right|_{g=g^*}(g-g^*) + \frac{1}{2}\left.\frac{d^2\Delta F}{dg^2}\right|_{g=g^*}(g-g^*)^2. \tag{8.2.23}$$

Since $d\Delta F/dg = 0$ near the maximum point at $g = g^*$, this expansion reduces to

$$\Delta F(g) \approx \Delta F_{\max}(g^*) - \gamma\frac{1}{2}(g-g^*)^2, \qquad \gamma = -\left(\frac{d^2\Delta F}{dg^2} \right)_{g=g^*}. \tag{8.2.24}$$

The minus sign here is chosen because $d^2\Delta F/dg^2 < 0$ near the maximum point $g = g^*$, so $\gamma > 0$. Substituting (Eqn. 8.2.24) into (Eqn. 8.2.22), replacing $1/D(g')$ in the integrand with $1/D(g^*)$ before the integral, and introducing a new variable $x = g' - g^*$, we obtain

$$f(g) = \frac{J}{D(g^*)}\exp\left(\frac{\Delta F(g^*)}{kT} - \frac{\Delta F(g)}{kT} \right)\int_{-(g^*-g)}^{G-g^*} \exp\left(-\frac{x^2}{(2kT/\gamma)} \right) dx. \tag{8.2.25}$$

The quantity $(kT/\gamma)^{1/2}$ represents the width of the maximum of the integrand near $g = g^*$.

If $(G - g^*)$ and $(g^* - g)$ are large compared to $(kT/\gamma)^{1/2}$, then the limits of the integral can be extended to plus and minus infinities. Then, the integral can be evaluated using the Gaussian formula as

$$\int_{-(g^*-g)}^{G-g^*} \exp\left(-\frac{\gamma x^2}{2kT}\right) dx \approx \left(\frac{2kT}{\gamma}\right)^{1/2} \int_{-\infty}^{\infty} e^{-x^2}\, dx = \left(\frac{2kT}{\gamma}\right)^{1/2} \pi^{1/2} = \left(\frac{2\pi kT}{\gamma}\right)^{1/2}. \quad (8.2.26)$$

Equation (8.2.25) can be rewritten using (Eqn. 8.2.2) for $c(g)$ and (Eqn. 8.2.26) for the integral as

$$\frac{f(g)}{c(g)} = \frac{J}{c_{sat,w} D(g^*)} \exp\left(\frac{\Delta F(g^*)}{kT}\right)\left(\frac{2\pi kT}{\gamma}\right)^{1/2} = const. \quad (8.2.27)$$

Since $f(g) = c(g)$ at $g << g^*$, the value of the constant is 1. Then, equating the middle part of this equation to 1, we obtain for J

$$J = c_{sat,w} D(g^*) \exp\left(-\frac{\Delta F(g^*)}{kT}\right)\left(\frac{\gamma}{2\pi kT}\right)^{1/2}. \quad (8.2.28)$$

Or, using again (Eqn. 8.2.2)

$$J = c(g^*) D(g^*) \left(\frac{\gamma}{2\pi kT}\right)^{1/2} = c(g^*) D(g^*) Z, \quad (8.2.29)$$

where we introduced a multiplier called the "*Zeldovich factor*" (Zeldovich, 1942):

$$Z = \left(\frac{\gamma}{2\pi kT}\right)^{1/2} = \left[\frac{1}{2\pi kT}\left(\frac{-d^2 \Delta F(g^*)}{dg^2}\right)\right]^{1/2}. \quad (8.2.30)$$

Equation (8.2.29) allows calculation of the nucleation rates for various situations and plays an important role in cloud kinetics. Numerical calculations show that $Z \sim O(10^{-1})$—that is, taking into account germ evaporation decreases the nucleation rate.

For calculations of the Zeldovich factor and nucleation rates, we need to calculate $d^2\Delta F/dg^2$, which requires some auxiliary relations. If the masses of a germ and of a single molecule are m_g and m_{w1}, then the following relations are valid

$$g = \frac{v_g}{v_{w1}} = \frac{4\pi}{3v_{w1}} r_g^3, \qquad \text{or} \qquad r_g = g^{1/3}\left(\frac{3v_{w1}}{4\pi}\right)^{1/3}. \quad (8.2.31)$$

These relations also yield the following equations for the surface area A_g:

$$A_g = 4\pi r_g^2 = g^{2/3}\left(\frac{3v_{w1}}{4\pi}\right)^{1/2}, \qquad \frac{dA_g}{dg} = \frac{2}{3}\frac{A_g}{g}. \quad (8.2.32)$$

Now using the expression (Eqn. 8.1.1) for ΔF written for $N_B = g$

$$\Delta F = -(\mu_{A1} - \mu_{B1}) g + A_g \sigma_{vw}, \quad (8.2.33)$$

we can calculate the derivatives for vapor (v) as metastable phase A and water (w) as stable phase B

$$\frac{d\Delta F}{dg} = (\mu_{w1} - \mu_{v1}) + \frac{2A_g}{3g}\sigma_{vw}, \quad (8.2.34)$$

$$\frac{d^2 \Delta F}{dg^2} = -\frac{2A_g \sigma_{vw}}{3g^2} + \frac{4A_g \sigma_{vw}}{9g^2} = -\frac{2A_g \sigma_{vw}}{9g^2}. \quad (8.2.35)$$

Substituting (Eqn. 8.2.35) into (Eqn. 8.2.30) for Z and using the relation $A_g\sigma_{vw} = 3\Delta F$ from (Eqn. 8.1.10) yields

$$Z = \left(\frac{A_g\sigma_{vw}}{9\pi kTg^2}\right)^{1/2} = \left(\frac{\Delta F_g}{3\pi kTg^2}\right)^{1/2}. \quad (8.2.36)$$

Substituting into (Eqn. 8.2.36) the first equality from (Eqn. 8.1.10) for ΔF_g and (Eqn. 8.2.31) for g, we obtain the Zeldovich factor:

$$Z = \frac{2v_{w1}}{A_g}\left(\frac{\sigma_{vw}}{kT}\right)^{1/2} = \frac{2M_w}{A_g\rho_w N_{av}}\left(\frac{\sigma_{vw}}{kT}\right)^{1/2}, \quad (8.2.37)$$

where we used the relation for the molecular volume $v_{w1} = M_w/(\rho_w N_{av})$.

The nucleation rate for the homogeneous condensation of the drops from supersaturated vapor is obtained from (Eqn. 8.2.29) using the Boltzmann distribution for g-mers with the critical energy ΔF_{cr}:

$$J = c(g^*)D(g^*)Z_z = c_{sat,w}j_g^{(+)}A_g Z\exp(-\Delta F_{cr}/kT). \quad (8.2.38)$$

Using the first form of Z in (Eqn. 8.2.37), this can be written as

$$J = 2c_{sat,w}j_g^{(+)}v_{w1}\left(\frac{\sigma_{vw}}{kT}\right)^{1/2}\exp(-\Delta F_{cr}/kT). \quad (8.2.38a)$$

The flux of the monomers $j_g^{(+)}$ to the germ in this case is just the molecular vapor flux, the same as the molecular flux of condensation onto a droplet that was calculated in Chapter 5. Using also the equation of state for water vapor, $c_w = e_w/kT$, we obtain

$$j_g^{(+)} = \frac{\alpha_c c_w V_w}{4} = \frac{\alpha_c e_w}{(2\pi m_{w1}kT)^{1/2}} = \frac{\alpha_c e_{ws} S_w}{(2\pi m_{w1}kT)^{1/2}}, \quad (8.2.39)$$

where $V_w = (8RT/\pi M_w)^{1/2} = (8kT/\pi m_{w1})^{1/2}$ is the mean thermal molecular speed (see Chapter 5), m_{w1} is the mass of a water molecule, e_w is the vapor pressure, and α_c is the condensation coefficient.

Substituting (Eqn. 8.2.39) for $j_g^{(+)}$ into (Eqn. 8.2.38a), we obtain for the homogeneous nucleation rate $J_{cond,hom}$ of the drops from supersaturated vapor

$$J_{cond,hom} = 2c_{sat,w}\frac{\alpha_c e_w S_w v_{w1}}{(2\pi m_{w1}kT)^{1/2}}\left(\frac{\sigma_{vw}}{kT}\right)^{1/2}\exp(-\Delta F_{cr}/kT). \quad (8.2.40)$$

Using the equation of state for the saturated vapor, $c_{sat,w} = e_{ws}/kT$, the relation for the volume of a water molecule $v_{w1} = M_w/(\rho_w N_{av})$, and the energy ΔF_{cr} defined in (Eqn. 8.1.10), it can be rewritten as

$$J_{cond,hom} = \frac{\alpha_c e_{ws}^2 M_w S_w}{(kT)^2\rho_w N_{av}}\left(\frac{2\sigma_{vw}}{\pi m_{w1}}\right)^{1/2}\exp\left(-\frac{16\pi M_w^2\sigma_{vw}^3}{3kT[RT\rho_w(\ln S_w)]^2}\right), \quad (8.2.41)$$

This relation is also used in several other forms. Substituting here the relations $m_{w1} = M_w/N_{Av}$, $k = R/N_{Av}$, and rearranging the terms, we obtain

$$J_{cond,hom} = \frac{\alpha_c S_w N_{Av}}{\rho_w}\left(\frac{e_{ws}}{RT}\right)^2\left(\frac{2N_{Av}M_w\sigma_{vw}}{\pi}\right)^{1/2}\exp\left(-\frac{16\pi M_w^2\sigma_{vw}^3}{3kT[RT\rho_w(\ln S_w)]^2}\right). \quad (8.2.41a)$$

The nucleation rate of the spherical ice germs from the vapor by deposition $J_{dep,hom}$ can be calculated in the same way, or, just with replacement in (Eqn. 8.2.41) of the water properties with the ice properties:

$$J_{dep,\mathrm{hom}} = \frac{\alpha_d e_{is}^2 M_w S_i}{(kT)^2 \rho_i N_{av}} \left(\frac{2\sigma_{iv}}{\pi m_{w1}} \right)^{1/2} \exp\left(-\frac{16\pi M_w^2 \sigma_{iv}^3}{3kT[RT\rho_i (\ln S_i)]^2} \right), \tag{8.2.42}$$

where α_d is the deposition coefficient (Section 5.2), σ_{iv} is the surface tension at the ice-vapor interface, and the energy of an ice germ formation ΔF_{cr} is described by (Eqn. 8.1.12) instead of (Eqn. 8.1.10). Using the same relations as for (Eqn. 8.2.41a), $m_{w1} = M_w/N_{Av}$, $k = R/N_{Av}$, this equation can be written as

$$J_{dep,\mathrm{hom}} = \frac{\alpha_d S_i N_{Av}}{\rho_i} \left(\frac{e_{is}}{RT} \right)^2 \left(\frac{2N_{Av} M_w \sigma_{iv}}{\pi} \right)^{1/2} \exp\left(-\frac{16\pi M_w^2 \sigma_{iv}^3}{3kT[RT\rho_i (\ln S_i)]^2} \right). \tag{8.2.42a}$$

Numerical calculations with (Eqns. 8.2.41) and (8.2.41a) show that the nucleation rates of drops from supersaturated vapor are very small at $S_w < 5$–6; $J_{con.hom}$ is ~ $7 \times 10^{-2}\,\mathrm{cm}^{-3}\,\mathrm{s}^{-1}$ at $S_w = 5$ and $J_{con.hom}$ is ~ $6 \times 10^3\,\mathrm{cm}^{-3}\,\mathrm{s}^{-1}$ at $S_w = 6$. The nucleation rates of crystals from the vapor are even smaller, ~ $3 \times 10^{-52}\,\mathrm{cm}^{-3}\,\mathrm{s}^{-1}$ at $S_i = 6$, and noticeable nucleation begins at $S_i > 12$–15. Such high saturation ratios normally are not reached in natural clouds, thus drops and crystals do not form in clouds via homogeneous nucleation from the vapor. However, this and higher saturation ratios can be created in the cloud chambers, where various properties of homogeneous nucleation from the vapor are studied.

8.2.3. Application of Bose–Einstein Statistics for Condensation and Deposition

The previous derivation of the nucleation rate was based on the Boltzmann distribution of the g-mers. However, as we have seen in Chapter 3, the Boltzmann distribution is a particular case of the more general Bose–Einstein distribution:

$$c(g) = c_{sat,w} \frac{1}{\exp(\Delta F_g / kT) - 1}. \tag{8.2.43}$$

With the Bose–Einstein statistics, the homogeneous nucleation rate of drops nucleation in supersaturated vapor (Eqn. 8.2.38) can be rewritten as

$$J_{\mathrm{hom}} = c_{sat,w} j_g^{(+)} A_g Z \frac{1}{\exp(\Delta F_{cr} / kT) - 1}. \tag{8.2.44}$$

For $\Delta F_{cr} >> kT$, we have an inequality $\exp(\Delta F_{cr}/kT) >> 1$, the Bose–Einstein distribution converts into the Boltzmann distribution, $c(g) = c_{sat,w}\exp(-\Delta F_g/kT)$, and (Eqn. 8.2.44) converts into (Eqn. 8.2.38). The relation $\Delta F_{cr} >> kT$ is valid in most cases. However, in some cases, when σ_{vw} is small (e.g., at very low temperatures or in the presence of strong surfactants or other pollutants that decrease surface tension), then $\Delta F \sim \sigma_{vw}^3$ may become comparable to or smaller than kT. Then, all of the previous derivations should be repeated using the Bose–Einstein statistics for particles with integer spin as the water molecules instead of the Boltzmann statistics.

For the other opposite case, $\Delta F_g << kT$, the Bose–Einstein statistics has another limit obtained by expanding the exponent into the power series, $\exp(\Delta F_g/kT) - 1 \approx 1 + \Delta F_g/kT - 1 = \Delta F_g/kT$, then

$$c(g) = c_{sat,w} \frac{kT}{\Delta F_g}. \tag{8.2.45}$$

The derivation of the nucleation rates should be performed for this distribution and will have the form

$$J_{\text{hom}} = c_{sat,w} j_g^{(+)} A_g Z \frac{kT}{\Delta F_{cr}}. \tag{8.2.46}$$

8.3. Nucleation Rates for Homogeneous Ice Nucleation

8.3.1. Nucleation Rates with the Boltzmann Distribution

The process of homogeneous ice nucleation in supercooled water has similarities and differences from the process of droplet nucleation in supersaturated vapor. The similarity is that nucleation also proceeds via the formation and decay of the embryos of the new stable phase (ice) within the metastable phase (liquid water). The difference is in the process of growth of these *g-mers*. In the case of vapor–liquid transition, growth of the embryos proceeds via the molecular flux of monomers from the vapor. In the case of the liquid–ice transition, the water molecules have to pass through three stages. First, they have to escape from the attraction of their neighbors in the liquid; that is, to break the water-to-water bonds. Second, they have to pass through a position of higher energy, ΔF_{act}, to overcome some potential barrier, which is called the *energy of activation*. And finally, the water molecules join the crystalline structure, and align in a position of minimum energy in the solid ice lattice.

Glasstone, Laidler, and Eyring (1941) developed the theory of viscosity, where it was shown using the methods of statistical mechanics that the shift of a molecule from its equilibrium state in viscosity also requires that it overcomes some potential barrier of activation. The frequency with which a molecule crosses this potential barrier is

$$\nu_m = \frac{kT}{h} \frac{\zeta_{act}}{\zeta_{init}} \exp(-\varepsilon_{act}/kT), \tag{8.3.1}$$

where ε_{act} is the activation energy for viscosity, ζ_{init} is the partition function of molecules in the initial state, ζ_{act} is the partition function in the activated state, and h is Planck's constant.

Turnbull and Fischer (1949) used this analogy with the theory of viscosity to show that the diffusive flux of the molecules across the liquid–ice boundary can be written as

$$j_{g,dif}^{(+)} = N_{cont} \nu_{dif}, \qquad \nu_{dif} = \nu_0 \exp\left(-\frac{\Delta F_{act}}{kT}\right), \qquad \nu_0 = \frac{kT}{h}. \tag{8.3.2}$$

where ΔF_{act} is the energy of activation of water molecules when crossing the liquid–ice boundary, and N_{cont} is the number of monomers of water in contact with the unit area of the ice's surface, ν_{dif} is the jump rate, and ν_0 is the frequency of vibration of an absorbed molecule at the surface. Now, the nucleation rate of the crystals in supercooled water can be written by analogy with (Eqn. 8.2.38a), replacing corresponding quantities with those for liquid and ice:

$$J_{f,\text{hom}} = 2c_w j_g^{(+)} v_{i1} \left(\frac{\sigma_{iv}}{kT}\right)^{1/2} \exp\left(-\frac{\Delta F_{cr}}{kT}\right), \tag{8.3.3}$$

where c_w is the concentration of water molecules in the liquid phase. Substituting here the flux (Eqn. 8.3.2) and using the relations for c_w and volume v_{i1} of a water molecule in ice,

$$c_w = \frac{N_{Av}}{v_w} = \frac{N_{Av}\rho_w}{M_w}, \qquad v_{i1} = \frac{m_{i1}}{\rho_i} = \frac{M_w}{N_{Av}\rho_i}, \tag{8.3.4}$$

we obtain the homogeneous freezing nucleation rate $J_{f,hom}$—i.e., the rate of formation of ice crystals in supercooled water:

$$J_{f,\mathrm{hom}} = 2\frac{N_{cont}kT}{h}\left(\frac{\rho_w}{\rho_i}\right)\left(\frac{\sigma_{iw}}{kT}\right)^{1/2}\exp\left(-\frac{\Delta F_{act}}{kT} - \frac{\Delta F_{cr}}{kT}\right)$$
$$= C_{\mathrm{hom}}\exp\left(-\frac{\Delta F_{act}}{kT} - \frac{\Delta F_{cr}}{kT}\right), \tag{8.3.5}$$

where

$$C_{\mathrm{hom}} = 2\frac{N_{cont}kT}{h}\left(\frac{\rho_w}{\rho_i}\right)\left(\frac{\sigma_{iw}}{kT}\right)^{1/2}. \tag{8.3.5a}$$

The value of $N_{cont} \sim 5.85 \times 10^{12}\,\mathrm{cm}^{-2}$ was estimated by Dufour and Defay (1963) from classical nucleation theory, and a close value $N_{cont} \sim 5.30 \times 10^{12}\,\mathrm{cm}^{-2}$ was obtained by Eady (1971) using a molecular model.

Fig. 8.2 shows homogeneous freezing nucleation rates $J_{f,hom}$ measured or estimated for pure water with various methods by several authors: Bigg (1953), Langham and Mason (1958), Butorin and Skripov (1972), Sassen and Dodd (1988, 1989), Krämer et al. (1996), and Stöckel et al. (2005).

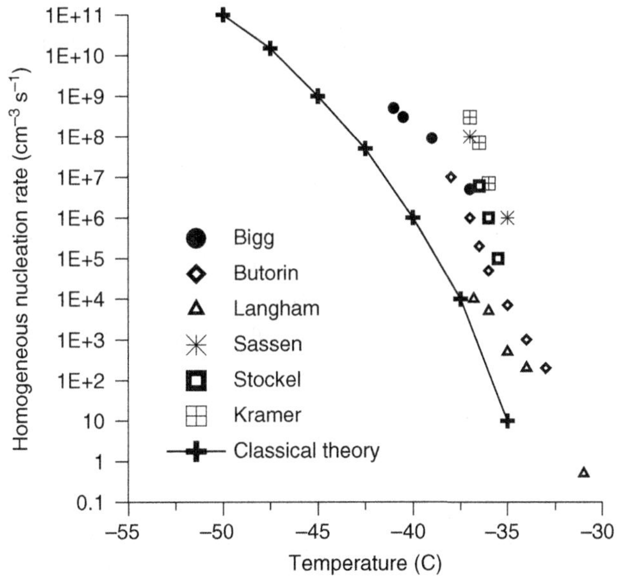

Figure 8.2. Homogeneous freezing nucleation rates $J_{f,hom}$ for pure water: old classical theory before 1995 (crosses) versus various experimental data: Bigg (1953), Langham and Mason (1958), Butorin and Skripov (1972), Sassen and Dodd (1988, 1989), Krämer et al. (1996), Stöckel et al. (2005).

A comparison of these data with the predictions from the old classical theory formulated before the 1990s (crosses) shows that the old theory predicts systematically lower nucleation rates than those measured. It was shown in the 1990s that this discrepancy can be avoided with the appropriate choice of parameters for the classical theory, as described next.

The probability of freezing of a haze particle or a drop with radius r_a and volume $v(r_a)$ during the time interval from t_0 to t is

$$P_{f,\mathrm{hom}}(r_a,t) = 1-\exp\left(-\int_{t_0}^{t} J_{f,\mathrm{hom}}(t')v(r_a)\,dt'\right). \tag{8.3.6}$$

If the dependence of $J_{f,hom}$ and $v(r_a)$ on time is weak, this expression is simplified as

$$P_{f,\mathrm{hom}}(r_a,t) \approx 1-\exp[J_{f,\mathrm{hom}}v(r_a)(t-t_0)]. \tag{8.3.7}$$

If the product $J_{f,hom}v(r_a)t$ is small, as for slow freezing or small particles or small time intervals, the argument in the exponent can be expanded into the power series, and so we have

$$P_{f,\mathrm{hom}}(r_a,t) \approx J_{f,\mathrm{hom}}v(r_a)(t-t_0). \tag{8.3.8}$$

The crystal concentration N_c in a polydisperse aerosol with uniform size and surface properties can be calculated by integrating the probability of freezing $P_{f,hom}$ of an individual haze or cloud droplet over the size spectrum $f(r_a)$ of aerosol or droplets normalized to the aerosol or drop concentration N_a:

$$\begin{aligned} N_{c,\mathrm{hom}}(t) &= \int_{r_{\min}}^{r_{\max}} P_{f,\mathrm{hom}}(r_a,t)f_a(r_a)\,dr_a, \\ &= \int_{r_{\min}}^{r_{\max}} \left[1-\exp\left(-\int_0^t J_{f,\mathrm{hom}}(t')v(r_a)\,dt'\right)\right] f_a(r_a)\,dr_a. \end{aligned} \tag{8.3.9}$$

The polydisperse freezing nucleation rate $R_{f,hom} = dN_c(t)/dt$ (cm^{-3} s^{-1}) can be calculated as

$$\begin{aligned} R_{f,\mathrm{hom}}(t_0) = \frac{dN_c}{dt} &= \int_{r_{\min}}^{r_{\max}} dr_a f_a(r_a)v(r_a) \\ &\times J_{f,\mathrm{hom}}(t_0)\exp\left(-\int_0^t J_{f,\mathrm{hom}}(t')v(r_a)\,dt'\right). \end{aligned} \tag{8.3.10}$$

The expressions (Eqns. 8.3.9) and (8.3.10) can be simplified for the processes of slow freezing using (Eqns. 8.3.7) and (8.3.8) as discussed in the following sections. If the aerosol size spectrum is partitioned into multiple size fractions, which in addition may have different chemical compositions, then these equations are modified by summation over all of these size fractions as

$$N_{c,\mathrm{hom}}(t) = \sum_i \int_{r_{\min,i}}^{r_{\max,i}} \left[1-\exp\left(-\int_0^t J_{f,\mathrm{hom}}(t')v(r_{ai})\,dt'\right)\right] f_{ai}(r_{ai})\,dr_{ai}, \tag{8.3.10a}$$

$$\begin{aligned} R_{f,\mathrm{hom}}(t_0) = \sum_i \frac{dN_{ci}}{dt} &= \sum_i \int_{r_{\min,i}}^{r_{\max,i}} dr_{ai} f_{ai}(r_{ai})v(r_{ai}) \\ &\times J_{f,\mathrm{hom}}(t_0)\exp\left(-\int_0^t J_{f,\mathrm{hom}}(t')v(r_{ai})\,dt'\right) \end{aligned} \tag{8.3.10b}$$

8.3.2. Application of Bose–Einstein Statistics for Freezing

If $\Delta F_{act} \le kT$, then the more general Bose–Einstein distribution should be used to describe the effect of activation energy on the crystal nucleation in supercooled water. The new diffusive flux with Bose–Einstein statistics is

$$j_g^{(+)} = \frac{N_{cont} kT}{h} \frac{1}{\exp(\Delta F_{act} / kT) - 1}. \quad (8.3.11)$$

If $\Delta F_{cr} \le kT$, then the Bose–Einstein distribution should be used also for the distribution of the germs:

$$c_g = c_w \frac{1}{\exp(\Delta F_g / kT) - 1}. \quad (8.3.12)$$

Then, the freezing nucleation rate (Eqn. 8.3.5) can be rewritten as

$$J_{f,\mathrm{hom}} = C_{\mathrm{hom}} \frac{1}{[\exp(\Delta F_{act} / kT) - 1]} \frac{1}{[\exp(\Delta F_{cr} / kT) - 1]}. \quad (8.3.13)$$

The normalizing factor C_{hom} and the Zeldovich factor, derived earlier with the Boltzmann distribution, should be refined for this case, taking into account the new distribution. The relations $\Delta F_{cr} >> kT$ and $\Delta F_{act} >> kT$ are valid in most cases. However, as discussed earlier, when σ_{iw} is small—e.g., at low temperatures or in the presence of strong surfactants—then $\Delta F_{cr} \sim \sigma_{vw}^3$ may become comparable to or smaller than kT. Then, the Bose–Einstein distribution may give higher accuracy than the Boltzmann distribution. For $\Delta F_{cr} >> kT$, and $\Delta F_{act} >> kT$, we have the inequalities $\exp(\Delta F_{cr}/kT) >> 1$, $\exp(\Delta F_{act}/kT) >> 1$, the Bose–Einstein distribution converts into the Boltzmann distribution used in the previous derivations, and (Eqn. 8.3.13) converts into (Eqn. 8.3.5).

For the opposite case, $\Delta F_g << kT$, the Bose–Einstein statistics has another limit, obtained by expanding the exponent into the power series, $\exp(\Delta F_g/kT) - 1 \approx 1 + \Delta F_g/kT - 1 = \Delta F_g/kT$. Then, $c_g = c_w(kT/\Delta F_g)$ and the nucleation rate has the form

$$J_{f,\mathrm{hom}} = C_{\mathrm{hom}} \frac{kT}{\Delta F_{act}} \frac{kT}{\Delta F_{cr}}. \quad (8.3.14)$$

8.3.3. Parameterizations of Activation Energy

The diffusion activation energy $\Delta F_{act}(T)$ of a water molecule to cross the water–ice embryo interface plays a fundamental role in studies of water properties and ice nucleation. $\Delta F_{act}(T)$ is parameterized most often by two methods. The first method is based on an analogy with the viscous flow of water and its self-diffusion because both phenomena are caused by the similar structure breaking mechanisms and presumably may have the same energy of activation. Using this analogy and the theory of water viscosity, the molar energy $\Delta F_{act}(T)$ is related to the self-diffusion coefficient of water D_w or to the viscosity of water η_w:

$$\Delta F_{act}(T) = R \frac{d \ln(\eta_w / T)}{d(1/T)} = -R \frac{d \ln D_w}{d(1/T)} = N_{Av} kT^2 \frac{d \ln D_w}{dT}, \quad (8.3.15)$$

where $R = kN_{Av}$ is the universal gas constant, k is the Boltzmann constant, and N_{Av} is the Avogadro number.

The values of D_w were measured and parameterized at various temperatures (Pruppacher and Klett, 1997) as

$$D_w = \sum_0^3 a_{Dwn} T_c^n, \qquad 0\,°\mathrm{C} < T_c < 50\,°\mathrm{C}, \tag{8.3.16}$$

with D_w in cm^2 s^{-1}, T_c in degrees C, $a_{Dw0} = 1.076 \times 10^{-5}$, $a_{Dw1} = 4.260 \times 10^{-7}$, $a_{Dw2} = 2.667 \times 10^{-9}$, $a_{Dw3} = -2.667 \times 10^{-11}$, and

$$D_w = a_{Dw0} \exp\left(\sum_1^3 a_{Dwn} T_c^n\right), \qquad -40\,°\mathrm{C} < T_c < 0\,°\mathrm{C}, \tag{8.3.17}$$

with D_w in cm^2 s^{-1} and $a_{Dw0} = 1.076 \times 10^{-5}$, $a_{Dw1} = 4.14 \times 10^{-2}$, $a_{Dw2} = 2.048 \times 10^{-4}$, and $a_{Dw3} = 2.713 \times 10^{-5}$. This led to a parameterization (Pruppacher and Klett, 1997)

$$\Delta F_{act}(T) = a_{Fa0} \exp\left(\sum_1^3 a_{Fan} T_c^n\right), \tag{8.3.18}$$

with $\Delta F_{act}(T)$ in kcal mole^{-1}, $a_{Fa0} = 5.55$, $a_{Fa1} = -8.423 \times 10^{-3}$, $a_{Fa2} = 6.384 \times 10^{-4}$, and $a_{Fa3} = 7.891 \times 10^{-6}$ in the range at $-40\,°\mathrm{C} < T_c < 40\,°\mathrm{C}$—i.e., $\Delta F_{act} \approx 5.55$ kcal mole^{-1} at 0 °C. For $\Delta F_{act}(T)$ in erg, using the relation 1 kcal $\approx 4.18 \times 10^{10}$ erg, and dividing by $N_{Av} = 6.023 \times 10^{23}$, we obtain $a_{Fa0} \approx 3.85 \times 10^{-13}$ erg, which is ΔF_{act} at 0 °C. The empirical Vogel–Fulcher–Tammann relation is also used for D_w,

$$D_w = D_{w0} \exp\left(-\frac{C_Z}{T - T_Z}\right). \tag{8.3.19}$$

Zobrist et al. (2007) and Hoose and Möhler (2012) give $C_Z = 892$ K and $T_Z = 118$ K.

Dividing (Eqn. 8.3.15) by N_{Av} gives us a ΔF_{act} in ergs:

$$\Delta F_{act}(T) = \frac{kT^2 C_Z}{(T - T_Z)^2}. \tag{8.3.20}$$

This yields $\approx 0.4 \times 10^{-12}$ erg at 0 °C, close to the value estimated earlier, and increases to $\approx 0.6 \times 10^{-12}$ erg at -60 °C. The other parameterization of ΔF_{act} is based on fitting calculated to the measured ice nucleation rates described in Chapters 8–10. Jensen et al. (1994) suggested an exponential fit for the activation energy in ergs

$$\Delta F_{act}(T) = 1.29 \times 10^{-12} \exp[0.05(T_c + 36)]. \tag{8.3.21}$$

Tabazadeh et al. (1997) suggested the following polynomial fit for $F_{act}(T)$ in ergs based on the ice nucleation rates data from Bertram et al. (1996):

$$\begin{aligned}\Delta F_{act}(T) = (&1.1992475202 \times 10^4 - 5.2298196745 \times 10^2 T + 8.232842046 T^2 \\ &-6.4173902638 \times 10^{-2} T^3 + 2.6889968134 \times 10^{-4} T^4 - 5.8279763451 \times 10^{-7} T^5 \\ &+5.1479983159 \times 10^{-10} T^6) \times 10^{-13}.\end{aligned} \tag{8.3.22}$$

Khvorostyanov and Sassen (1998c) and Khvorostyanov and Curry (2000) fitted the experimental nucleation rates to those calculated with the extended classical nucleation theory (Chapters 8–10) with ΔF_{act} in ergs as

$$\Delta F_{act}(T) = 0.694\times 10^{-12}[1+0.027(T_c+30)], \qquad T_c > -30\,^\circ\mathrm{C}, \tag{8.3.23a}$$

$$\Delta F_{act}(T) = 0.694\times 10^{-12}[1+0.027(T_c+30)] \times \exp[0.01(T_c+30)], \qquad T_c < -30\,^\circ\mathrm{C}. \tag{8.3.23b}$$

Chen et al. (2008) fitted various experimental data with simpler fits with ΔF_{act} independent on T, assuming $\Delta F_{act}(T)$ = const = 10^{-12} erg or $\Delta F_{act}(T)$ = const = 1.57×10^{-12} erg in the temperature range from 0 to −60 °C. Fornea et al. (2009) suggested another approximation:

$$\Delta F_{act}(T) = a_0 \exp(a_1 T_c + a_2 T_c^2 + a_2 T_c^2), \tag{8.3.24}$$

with $a_0 = 0.3856 \times 10^{-12}$ erg, $a_1 = -8.423 \times 10^{-3}$, $a_2 = 6.384 \times 10^{-4}$, and $a_3 = 7.891 \times 10^{-6}$, which yields $\Delta F_{act}(T) = 0.3856 \times 10^{-12}$ erg at 0 °C, and increases to $\approx 1.15 \times 10^{-12}$ erg at −60 °C. Hoose and Möhler (2012) compared various parameterizations of ΔF_{act} and noted that its choice as a fitting parameter may depend on the other parameters used for calculation of the nucleation rates. This dependence of fitted ΔF_{act} on the other fitted parameters was also emphasized by Lin et al. (2002) in comparison of various ice nucleation schemes. Therefore, measurements of ΔF_{act} or its evaluation from the measurements of nucleation rates are desirable for various conditions and processes.

8.4. Semi-empirical Parameterizations of Homogeneous Ice Nucleation

The homogeneous freezing of haze particles and cloud droplets plays an important role in crystal formation in cirrus, orographic, deep convective clouds, and other clouds under low temperatures. Development of parameterizations of various features of homogeneous ice nucleation suitable for cloud and climate models has been underway for the past several decades. These parameterizations have been mostly semi-empirical, based on heuristic relations for various properties of ice nucleation: nucleation rates, critical humidities, nucleated crystal concentrations, etc. These parameterizations were developed using parcel model simulations and either experimental data or some relations of the classical nucleation theory or alternative nucleation theories. A brief review of several such parameterizations is given in this section.

These parameterizations are separated into two types. The first type provides equations for the characteristics of the nucleation process at any given intermediate time of nucleation—e.g., nucleation rates and concentrations can be calculated over the course of simulation. A few parameterizations of the second type considered the entire nucleation process as a sub-step process (taking less than one time step in a model) and found equations for the final characteristics of the nucleation process after the nucleation has ceased: crystal concentrations, radii, masses.

Parameterizations of the first type. One of the most important characteristics of freezing is the nucleation rate J_{hom}. Heymsfield and Miloshevich (1993) used the results from the statistical molecular model of Eadie (1971) and fitted $J_{f,hom0}$ for *pure water* with an expression

$$J_{f,\mathrm{hom}0}(T) = 10^{-X(T)}, \qquad X(T) = \sum_{i=0}^{4} A_{i,HM} T_c^i, \tag{8.4.1}$$

where $J_{f,hom0}$ is in $cm^{-3}s^{-1}$, T_c the temperature in degrees Celsius, and $A_{0,HM} = 606.3952$, $A_{1,HM} = 52.6611$, $A_{2,HM} = 1.7439$, $A_{3,HM} = 0.0265$, and $A_{4,HM} = 1.536 \times 10^{-4}$. Experimental data show that the freezing rates of the haze particles are smaller than given by this equation, since they are depressed by the presence of solute.

Sassen and Dodd (1988, 1989) describe this depression of the nucleation rate by introducing an effective freezing temperature

$$T^* = T + \lambda_{SD}\Delta T_m, \quad \text{or} \quad \Delta T_f = T_f^* - T = \lambda_{SD}\Delta T_m, \tag{8.4.2}$$

where ΔT_m and ΔT_f are the depressions of the melting and freezing temperatures respectively, and $\lambda_{SD} = 1.7$ was chosen as an average from experimental data obtained by Rasmussen (1982) on the relationship between depressions of the nucleation and melting temperatures for a number of salts. It was clarified later that the coefficient 1.7 is not a universal, and can vary in the range 1.4–2.4 depending on the chemical composition of solute in the haze particle (Martin, 2000; Chen et al., 2000; Lin et al., 2002; DeMott, 2002), and may reach 3 to 5 for some organic substances with high molecular weight, as will be discussed in Section 8.7. Thus, the nucleation rate for haze particles can be calculated using (Eqn. 8.4.1) but with the effective freezing temperature T_f* instead of T

DeMott et al. (1994, hereafter DMC94) suggested a parameterization of ΔT_m for ammonium sulfate as a function of molality $\hat{M}$:

$$\Delta T_m = 0.102453 + 3.48484\hat{M}. \tag{8.4.3}$$

Molality in this work was evaluated as described in Section 3.5, in terms of the equilibrium wet particle diameters based on Köhler's equation, and the freezing point depression was calculated with (Eqn. 8.4.2). DMC94 used (Eqns. 8.4.1) to (8.4.3) to calculate the frozen fraction F_{hf} of the haze particles at various T and S_w

$$F_{hf} = 1 - \exp(-aD_s^b), \qquad a = [(\pi\rho_s)^2/6]10^{c_{D1}+c_{D2}(1-S_w)} \tag{8.4.4a}$$

$$c_{D1} = -14.65 - 1.045(T - T_0), \quad b = 6 \tag{8.4.4b}$$

$$c_{D2} = -492.35 - 8.34(T - T_0) - 0.061(T - T_0)^2, \tag{8.4.4c}$$

with the density of dry ammonium sulfate $\rho_s = 1.769\,g\,cm^{-3}$, and $T_0 = 273.16\,K$. Having calculated F_{hf} at various T and S_w and assuming an exponential size spectrum of the haze particles by the diameters D with the maximum $D_m = 1.02 \times 10^{-4}\,cm$, DMC94 suggested a fit for the concentration of nucleated crystals in the form

$$N_f = N_a \int_0^{y_{max}} \exp(-y)\left\{1 - \exp\left[-y^b\left(\frac{D_n}{D_m}\right)^b\right]\right\}, \tag{8.4.5}$$

where $y = D/D_n$, and $D_n = 7.5 \times 10^{-6}\,cm$ is the scaling parameter, and D_m is defined such that $a = D_m^{-b}$; N_a is the total aerosol concentration, which is taken to be $200\,cm^{-3}$. This scheme reproduced the experimental data on ice nucleation in haze particles and could be used in cloud models.

An important characteristic of homogeneous ice nucleation is the critical humidity or the critical water saturation ratio $S_{w,cr}^{hom}$. Sassen and Dodd (1988, 1989), based on a generalization of their parcel model simulations, and Heymsfield and Miloshevich (1995), based on the field measurements in orographic and cirrus clouds at low temperatures, parameterized $S_{w,cr}^{\text{hom}}$ as a polynomial fit by the temperature

$$S_{w,cr}^{\text{hom}} = A_{0T} + A_{1T}T + A_{2T}T^2, \tag{8.4.6}$$

with $A_{0T} = -27.6$, $A_{1T} = 5.36 \times 10^{-1}$, $A_{2T} = 0$ and T in degrees Kelvin in Sassen and Dodd (1989), and $A_{0T} = 1.8892$, $A_{1T} = 2.81$, $A_{2T} = 0.013336$, and T in degrees Celsius in Heymsfield and Miloshevich (1995). Sassen and Benson (2000), based on parcel model simulations, generalized this equation with account for the wind shear.

Based on the results of laboratory experiments on solutions freezing, Koop et al. (2000) suggested a parameterization of the homogeneous nucleation rate $J_{f,hom}$ similar to Heymsfield and Miloshevich (1993) for pure water but accounted for the solute effects parameterized in terms of water activity

$$J_{f,\text{hom}}(T) = 10^{-X_K(T)}, \quad X_K(T) = \sum_{i=0}^{3} A_{i,K}(\Delta a_w)^i, \tag{8.4.7}$$

with $A_{0,K} = 906.7$, $A_{1,K} = -8502$, $A_{2,K} = 26624$, $A_{3,K} = -29180$, and

$$\Delta a_w = a_w - a_w^i, \tag{8.4.8}$$

where a_w is the water activity in the liquid solution and a_w^i is the activity of water in solution in equilibrium with ice. Koop et al. (2000) assumed that in equilibrium a_w is equal to the environmental water saturation ratio S_w and the Kelvin curvature effect can be neglected for sufficiently large drops, and a_w^i was parameterized as

$$a_w^i = \exp\{[\mu_w^i(T, p) - \mu_w^0(T, p)]/RT\}, \tag{8.4.9}$$

where μ_w^i and μ_w^0 are the chemical potentials of water in pure ice and pure liquid water, respectively. In calculation for atmospheric conditions, $p \approx 1$ atm. The difference of these chemical potentials was parameterized as a function of temperature following Johari et al. (1994):

$$\Delta\mu(T) = \mu_w^i(T, 0) - \mu_w^0(T, 0) = B_{K0} + B_{K1}T + B_{K2}T^{-1} + B_{K3}\ln(T), \tag{8.4.10}$$

with $B_{K0} = 210368$, $B_{K1} = 131.438$, $B_{K2} = -3.32373 \times 10^6$, $B_{K3} = -41729.1$, T in degrees Kelvin, and μ_w^i and μ_w^0 in J mol^{-1}. This parameterization is assumed to be valid in the range $150 < T < 273$, but exhibits a somewhat unphysical maximum around 265–270 K.

Parameterizations of homogeneous freezing of the second type as a sub-step process include more intermediate steps and assumptions. Such parameterizations are also semi-empirical, and as examples we describe the parameterizations developed by Kärcher and Lohmann (2002a,b) and Ren and MacKenzie (2005).

The basis of these parameterizations is the equation for the ice saturation ratio, similar to (Eqn. 5.3.47a) described in Section 5.3,

$$\frac{dS_i}{dt} = c_{1i}S_i - (a_2 + a_3S_i)R_i, \tag{8.4.11}$$

where c_{1i} was defined by (Eqn. 5.3.46a) in Section 5.3, and

$$a_2 = n_{sat}^{-1} = \left(\frac{\rho_{is}}{m_{w1}}\right)^{-1}, \qquad a_3 = \frac{L_s^2 M_w m_{1w}}{c_p p T M_a}, \tag{8.4.12}$$

n_{sat} is the number density of water molecules at ice saturation, and all the notations were introduced earlier. The freezing/growth term R_i is defined similar to the integrals I_{dep} and I_{con} in Chapters 5 and 7:

$$R_i = \frac{4\pi}{v_{i1}} \int_{r_s}^{\infty} dr_0 \int_{-\infty}^{t} dt_0 r_i^2(t_0, t) \frac{dn_i(r_0, t_0)}{dt_0} \frac{dr_i}{dt}(t_0, t), \tag{8.4.13}$$

where v_{i1} is a specific volume of a water molecule in ice. The term $dn_i(t_0)/dt_0$ is an analog of the polydisperse nucleation rate $R_{f,hom}$ in Section 8.3, and $[dn_i(t_0)/dt_0]dt_0$ is the number of particles freezing during the time interval $(t_0, t_0 + dt_0)$. It was written in the form

$$\frac{dn_i(t_0)}{dt_0} = \int_{r_s}^{\infty} dr_0 \frac{4\pi}{3} r_0^3 J_{f,\text{hom}} \frac{dn_a}{dr_0} = J_{f,\text{hom}}(t) V_a(t), \tag{8.4.14}$$

where $dn_a/dr_0 = f_a(r_0)$ is the aerosol size distribution function, V_a is the aerosol volume in a unit volume of air.

To solve this non-linear system of equations, the authors introduce several additional hypotheses.

1. Following Ford (1998a,b), a hypothesis on the exponential time behavior of the nucleation rate dn_i/dt_0 was introduced

$$\frac{dn_i(t_0)}{dt_0} = \frac{dn_i(t)}{dt} \exp\left(-\frac{t - t_0}{\tau_{nuc}} \right), \tag{8.4.15}$$

where τ_{nuc} is some characteristic time scale of the ice nucleation event, unknown for now, which has to be determined. Integration of (Eqn. 8.4.15) by t yields

$$N_i = \int_{-\infty}^{t} dt_0 \frac{dn_i}{dt}(t_0) = \frac{dn_i(t)}{dt} \tau_{nuc}. \tag{8.4.16}$$

2. An additional heuristic relation was introduced for τ_{nuc} by Kärcher and Lohmann (2002a,b):

$$\tau_{nuc}^{-1} = c_\tau \left(\left| \frac{\partial \ln J_{\text{hom}}}{\partial T} \right| \right)_{S_i = S_{i,cr}} \frac{dT}{dt}. \tag{8.4.17}$$

The unknown parameter c_τ was parameterized in Kärcher and Lohmann (2002a) as a function of temperature, and was replaced with a constant value $c_\tau = 50$ in Kärcher and Lohmann (2002b). Ren and MacKenzie (2005), taking a logarithm of (Eqn. 8.4.15) and differentiating, arrived at a simpler expression:

$$\tau_{nuc}^{-1} = \frac{d \ln J_{\text{hom}}(t)}{dt} + \frac{d \ln V_w(t)}{dt} \approx c_\tau(T) \frac{dT}{dt}, \tag{8.4.18}$$

$$c_\tau(T) \approx -0.004T^2 + 2T - 304.4, \tag{8.4.19}$$

where c_τ is the approximate value of $d\ln(J_{f,hom})/dt$ at the difference $\Delta a_w \approx 0.306$.

3. The third hypothesis was that the ice saturation ratio S_i changes only slightly around its critical value $S_{i,cr}$ during the nucleation event. Its variations can be neglected and it can be assumed that $S_i(t) \approx S_{i,cr}(T)$.

4. The diffusional growth of the nucleated crystals is described by equations similar to (Eqn. 5.2.18) for the diffusional growth regime with kinetic correction but for equivalent spherical crystals:

$$\frac{dr_i}{dt} = \frac{b_1}{1+b_2 r_i},$$
$$b_1 = \frac{\alpha_d V_w n_{sat}(S_i - 1)}{4}, \qquad b_2 = \frac{\alpha_d V_w}{4D_v} = \frac{1}{\xi_{dep}}, \tag{8.4.20}$$

where V_w is the thermal speed of the vapor molecules, and the other notations are the same as in (Eqn. 5.2.18). The assumption that $S_i \approx \text{const} \approx S_{i,cr}$ relates to b_1 in (Eqn. 8.4.20).

5. The next hypothesis was that, similar to Twomey's theory for drop nucleation (see Chapter 7), homogeneous ice nucleation stops when S_i reaches a maximum:

$$dS_i/dt\big|_{S_i \approx S_{i,cr}} = 0. \tag{8.4.21}$$

(This assumption needs further tuning because ice nucleation does not cease at this point, as will be discussed in Section 8.10). From (Eqn. 8.4.11) and the condition (Eqn. 8.4.21), we obtain

$$\frac{c_{1w} S_{i,cr}}{a_2 + a_3 S_{i,cr}} w = R_i. \tag{8.4.22}$$

After substituting (Eqn. 8.4.15) into (Eqn. 8.4.13) for R_i and some transformations, Ren and MacKenzie (2005) obtained from (Eqn. 8.4.22) an equation

$$\frac{c_{1w} S_{i,cr}}{a_2 + a_3 S_{i,cr}} w = \int_{r_s}^{\infty} R_{im}(r_0) \frac{dn_a}{dr_0}(r_0, t_0 = -\infty)\, dr_0, \tag{8.4.23}$$

where the monodisperse (at radius r_0) freezing/growth term is defined by

$$R_{im} = \frac{4\pi}{v_{i1}} \int_{-\infty}^{t} dt_0 \frac{1}{\tau} \exp\left(-\frac{t - t_0}{\tau}\right) r_i^2(r_0, t_0, t) \frac{dr_i}{dt}(r_0, t_0, t). \tag{8.4.24}$$

Solving (Eqn. 8.4.23) produces the number of nucleated crystals:

$$N_i = \int_{r_s}^{\infty} n_a(r_0, t_0 = -\infty)\, dr_0 \le N_a. \tag{8.4.25}$$

Integration in (Eqn. 8.4.24) by t_0 yields the monodisperse R_{im}:

$$R_{im}(r_0) = \frac{4\pi}{v_{i1}} \frac{b_1}{b_2^2} \left\{ \left[\frac{1+\delta}{2} \sqrt{\kappa} + \frac{1}{1+\delta} \frac{1}{\sqrt{\kappa}} \right] \exp\left(\frac{1}{\kappa}\right) \sqrt{\pi}\, erfc\left(\frac{1}{\sqrt{\kappa}}\right) + \delta - 1 \right\}. \tag{8.4.26}$$

where

$$\delta = b_2 r_0 = \frac{r_0}{\xi_{dep}}, \qquad \kappa = \frac{2 b_1 b_2 \tau}{(1+\delta)^2}, \tag{8.4.27}$$

and $erfc(x) = 1 - erf(x)$ is the complimentary error function. Kärcher and Lohmann (2002b) derived a similar expression. For studies of R_{im}, Ren and MacKenzie (2005, 2007) found a representation of $erfc(x)$ via elementary functions with error less than 0.7%:

$$\exp\left(\frac{1}{\kappa}\right) \sqrt{\pi}\, erfc\left(\frac{1}{\sqrt{\kappa}}\right) \approx \frac{3}{2/\sqrt{\kappa} + \sqrt{1/\kappa + 9/\pi}}. \tag{8.4.28}$$

Substitution of (Eqn. 8.4.28) into (Eqn. 8.4.26) yields for R_{im} an expression in elementary functions

$$R_{im}(r_0) = \frac{4\pi}{v_{i1}} \frac{b_1}{b_2^2} \left[\frac{1+\delta}{2} \phi(\kappa) + \frac{1}{1+\delta} \phi(\kappa) + \delta - 1 \right], \quad (8.4.29a)$$

$$\phi(\kappa) = \frac{3\kappa}{2 + (1 + 9\kappa/\pi)^{1/2}}. \quad (8.4.29b)$$

Substitution of (Eqn. 8.4.29a,b) into (Eqn. 8.4.23), using (Eqn. 8.4.25) and integration over the aerosol size spectrum yields the number of nucleated crystals N_i. The analysis becomes simpler in the case of a monodisperse freezing aerosol with radius r_0. The expression for N_i then becomes

$$N_i = \frac{S_{i,cr}}{(S_{i,cr} - 1)} \frac{c_{1i} w}{[R_{im}(r_0)(b_2^2/b_1)(v_{i1}/4\pi)](4\pi D_v/b_2)}. \quad (8.4.30)$$

Kärcher and Lohmann (2002b) and Ren and MacKenzie (2005) found for $R_{im}(r_0)$ four various limiting cases for $\kappa \to 0$, $\kappa \to \infty$, $\delta \to 0$ (which correspond to the diffuse regime of crystal growth) and $\delta \to \infty$ (kinetic regime of crystal growth). Substitution of these limits into (Eqn. 8.4.30) shows dependencies of N_i on w, and T. In particular, for $\kappa >> 1$, approximately corresponding to the diffusion growth regime,

$$N_i \sim w^{3/2}, \qquad N_i \sim n_{sat}^{-1/2} \sim \rho_{is}^{-1/2}. \quad (8.4.31)$$

For $\kappa << 1$, approximately corresponding to the kinetic crystal growth regime, Ren and MacKenzie (2005) found that $N_i \sim w$ for large particles ($r_0 >> \xi_{dep}$), and $N_i \sim w^2$, $N_i \sim \rho_{is}^{-2}$ for small particles ($r_0 << \xi_{dep}$).

Barahona and Nenes (2008) developed a similar substep parameterization of homogeneous ice nucleation, using Twomey's (1959) upper limit approximation for ice supersaturation, and a representation for the nucleation rate similar to Khvorostyanov and Curry (2004b):

$$\ln \frac{J_{hom}(S_i)}{J_{hom}(S_{i,cr})} = b_\tau(T)(S_i - S_{i,cr}). \quad (8.4.32)$$

Barahona and Nenes derived an empirical temperature dependence for $b_\tau(T)$ from Koop et al. (2000), made several auxiliary simplifications, and arrived at a parameterization that required an iterative numerical solution. All of the parameterizations described earlier used parcel models for tuning the parameters of the final parameterization equations.

In the studies reviewed earlier, it was assumed that stable hexagonal ice I_h nucleates in supercooled water or solution droplets (see Section 4.1). The thermodynamic parameters associated with ice I_h were therefore used. Evidence was provided recently that metastable cubic ice I_c may form first in some cases at low temperatures, especially at $T < 200$ K, with subsequent relaxation to the stable ice I_h (e.g., Murray et al., 2005, 2010; Murray and Bertram, 2006; Malkin et al., 2012). However, uncertainty remains in the general conceptual picture of this sequence of the processes, and is widespread in the current data on the thermodynamic parameters for I_c, so that the nucleation rates for I_c estimated with classical nucleation theory may significantly vary (e.g., Murray et al., 2010). We therefore assume here that hexagonal ice I_h nucleates in droplets and use the corresponding parameters for I_h. Calculations for I_c or any other type of ice described in Chapter 4 can be done using the same equations derived here with corresponding changes to the thermodynamic parameters: the surface tension, melting heat, saturated vapor pressure, etc.

8.5. Equations for Water and Ice Supersaturations with Homogeneous Ice Nucleation

Both water and ice supersaturations govern ice nucleation kinetics: water supersaturation determines the nucleation process as specified in Section 8.6, and the growth of ice particles is determined by ice supersaturation. We will consider the equations for fractional water and ice supersaturations, $s_w = (\rho_v - \rho_{ws})/\rho_{ws}$, and $s_i = (\rho_v - \rho_{is})/\rho_{is}$, where ρ_v is the environmental water vapor pressure, ρ_{ws} and ρ_{is} are the saturated over water and ice vapor pressures, respectively. In a rising air parcel, supersaturation is governed by two competing processes: supersaturation generation by the cooling in an updraft and supersaturation absorption by the crystals in the deposition process.

This process can be described by the supersaturation equations, which we use in the form (Eqns. 5.3.16), (5.3.43a), and (5.3.47) derived and discussed in Chapter 5 with modifications that account for homogeneous ice nucleation:

$$\frac{1}{(1+s_w)}\frac{ds_w}{dt} = c_{1w}w - \frac{\Gamma_{12}}{\rho_v}I_{dep}, \tag{8.5.1}$$

$$\frac{1}{(1+s_i)}\frac{ds_i}{dt} = c_{1i}w - \frac{\Gamma_2}{\rho_v}I_{dep} = c_{1i}w - \frac{s_i}{\tau_{fc}}, \tag{8.5.2}$$

where τ_{fc} is the crystal supersaturation relaxation time defined in Section 5.3 and

$$c_{1w}(T) = \left(\frac{L_e}{c_p T}\frac{M_w}{M_a} - 1\right)\frac{g}{R_a T}, \tag{8.5.3}$$

$$c_{1i}(T) = \left(\frac{L_s}{c_p T}\frac{M_w}{M_a} - 1\right)\frac{g}{R_a T}. \tag{8.5.4}$$

Here Γ_{12} and Γ_2 are the psychrometric corrections associated with the latent heat release at condensation defined by (Eqn. 5.3.15) in Chapter 5.

The deposition vapor flux I_{dep} to the crystals is the integral of the mass growth rate over the crystal size spectrum. We assume that crystal size can be characterized by an equivalent radius r_c, and then I_{dep} is expressed via crystal growth rate (dr_c/dt) as

$$I_{dep}(t) = 4\pi\rho_i \int_0^{\infty} \frac{dr_c(t,t_0)}{dt} r_c^2(t,t_0) f(r_c,t_0)\, dr_c, \tag{8.5.5}$$

where $f_c(r_c, t_0)$ is the size distribution function of the crystals nucleated at a time t_0, and $r_c(t,t_0)$ denotes the radius at time t of a crystal nucleated at time t_0. We use (dr_c/dt) in the form of (Eqn. 5.2.24):

$$\frac{dr_c}{dt} = \frac{c_{3i}s_i}{r_c + \xi_{dep}}, \qquad c_{3i} = \frac{D_v \rho_{is}}{\rho_i \Gamma_2}, \tag{8.5.6}$$

$$\xi_{dep} = \frac{4D_v}{\alpha_d V_w}, \qquad V_w = \left(\frac{8RT}{\pi M_w}\right)^{1/2}. \tag{8.5.7}$$

where D_v is the water vapor diffusion coefficient, ξ_{dep} is the kinetic correction to the radius growth rate, V_w is the thermal speed of water vapor molecules, and α_d is the deposition coefficient. Substitution of (Eqn. 8.5.6) into (Eqn. 8.5.5) yields

$$I_{dep}(t) = s_i(t)\frac{4\pi D_v \rho_{is}}{\Gamma_2}\int_0^{\infty} \frac{r_c^2(t,t_0)}{r_c(t,t_0)+\xi_{dep}} f_c(r_c,t)\,dr_c. \tag{8.5.8}$$

The radius $r_c(t, t_0)$ at time t of a crystal nucleated at time t_0 is evaluated by integrating (Eqn. 8.5.6)

$$r_c(t,t_0) = \{(r_{c0}+\xi_{dep})^2 + 2c_{3i}[y_i(t)-y_i(t_0)]\}^{1/2} - \xi_{dep}, \tag{8.5.9}$$

where $r_{c0} = r_i(t_0)$ is the initial crystal radius at the activation time t_0, and $y_i(t)$ is the integral ice supersaturation defined as

$$y_i(t) = \int_{t_0}^{t} s_i(t')\,dt'. \tag{8.5.10}$$

Ice nucleation via haze freezing depends simultaneously on T and s_w, and we can generalize this relation using a kinetic equation for the crystal size spectrum and introducing two activity spectra, by supersaturation $\varphi_s(s_w)$ and by temperature $\varphi_T(T)$:

$$\frac{\partial f_c(r_c)}{\partial t} + \frac{\partial}{\partial r}\left(\frac{dr_c}{dt} f_c\right)$$
$$= [\varphi_s(T,s_w)\frac{ds_w}{dt} + \varphi_T(T,s_w)\frac{dT}{dt}]\delta(r_c - r_c(t_0)), \tag{8.5.11}$$

where the Dirac delta-function $\delta(r_c - r_c(t_0))$ describes nucleation of a crystal with radius $r_c(t_0)$. Equation (8.5.11) can be viewed as a generalization of the known relation for the drop activation on the CCN, where usually only the supersaturation activity spectrum is accounted for (e.g., Twomey, 1959; Sedunov, 1974; Khvorostyanov and Curry, 2008a, 2009b; Ghan et al., 2011; Tao et al., 2012; see Chapters 6 and 7 here). Both s_w– and T– spectra for homogeneous ice nucleation will be derived and considered simultaneously in Section 8.10, but for now a simpler and faster way is to use an equivalent equation for concentration conservation:

$$dN_{fr}(t_0) = f_c(r_c)dr_c = \varphi_s(T,s_w)ds_w + \varphi_T(T,s_w)dT = R_{f,\text{hom}}(t_0)dt_0, \tag{8.5.12}$$

where $R_{f,hom}$ is the homogeneous freezing rate defined in (Eqn. 8.3.10) of Section 8.3 describing the effects of both T and s_w on freezing.

Substituting the equality $f_c(r_c)dr_c = R_{f,hom}(t_0)dt_0$ from (Eqn. 8.5.12) into (Eqn. 8.5.8) for I_{dep} and using (Eqn. 8.5.9), we obtain

$$I_{dep} = s_i\frac{4\pi D_v \rho_{is}}{\Gamma_2}\int_0^t r_{c,ef}(t,t_0)R_{f,\text{hom}}(t_0)\,dt_0, \tag{8.5.13}$$

where we introduced the effective radius $r_{c,ef}(t, t_0)$:

$$r_{c,ef}(t,t_0) = \frac{\{[(r_{c0}+\xi_{dep})^2 + 2c_{3i}(y_i(t)-y_i(t_0))]^{1/2} - \xi_{dep}\}^2}{[(r_{c0}+\xi_{dep})^2 + 2c_{3i}(y_i(t)-y_i(t_0))]^{1/2}}. \tag{8.5.14}$$

Substituting (Eqn. 8.5.13) into (Eqn. 8.5.2) and using the definition of (Eqn. 8.5.10) of $y_i(t)$, we obtain an equation for integral ice supersaturation that participates in Equation (Eqn. 8.5.9)

$$\frac{1}{(1+y_i')}\frac{dy_i'}{dt} = c_{1i}w - \frac{\Gamma_2}{\rho_v}I_{dep} = c_{1i}w - \frac{y_i'}{\tau_{fc}}, \tag{8.5.15}$$

where

$$I_{dep} = y_i' \frac{4\pi D_v \rho_{is}}{\Gamma_2} \int_0^t r_{c,ef}(t,t_0) R_{f,\mathrm{hom}}(t_0)\, dt_0, \tag{8.5.16}$$

Specific application of (Eqn. 8.5.16) for homogeneous ice nucleation is considered in the next sections. Substitution of $R_{f,hom}$ from (Eqn. 8.3.10) into (Eqn. 8.5.16) yields

$$I_{dep} = y_i' \frac{4\pi D_v \rho_{is}}{\Gamma_2} \left[\int_0^t r_{c,ef}(t,t_0) \int_{r_{\min}}^{r_{\max}} f_a(r_a) v(r_a) \right.$$
$$\left. \times J_{f,\mathrm{hom}}(t_0) \exp\left(-\int_0^t J_{f,\mathrm{hom}}(t') v(r_a) dt' \right) dr_a dt_0 \right]. \tag{8.5.17}$$

Substitution of this expression into the supersaturation equation (Eqn. 8.5.15) and using the relation $\rho_v = (1 + y_i')\rho_{is}$ yields

$$\frac{1}{(1+y_i')} \frac{dy_i'}{dt} = c_{1i} w - \frac{y_i'}{1+y_i'} (4\pi D_v) \left[\int_0^t r_{c,ef}(t,t_0) \int_{r_{\min}}^{r_{\max}} f_a(r_a) v(r_a) \right.$$
$$\left. \times J_{f,\mathrm{hom}}(t_0) \exp\left(-\int_0^t J_{f,\mathrm{hom}}(t') v(r_a) dt' \right) dr_a dt_0 \right]. \tag{8.5.18}$$

This equation describes the evolution of integral ice supersaturation. It is analogous to the integral supersaturation equation (Eqn. 7.2.15) for the drop activation in Chapter 7, but includes a more complicated integrand with a description of crystal nucleation. The first term on the RHS describes supersaturation generation by the cooling action of updrafts, and the second term accounts for its depletion by the newly nucleated and growing crystals.

8.6. Critical Germ Size, Energy, and Homogeneous Freezing Rate

Adequate description and parameterization of homogeneous ice nucleation was hampered for many years by the absence of the equations for the critical radius r_{cr}, energy ΔF_{cr}, and nucleation rates $J_{f,hom}$ with simultaneous analytical dependence on the temperature T and the relative humidity or the saturation ratio S_w. Expressions for $r_{cr}(T, S_w)$, $\Delta F_{cr}(T, S_w)$, and $J_{f,hom}(T, S_w)$ can be derived by developing the classical nucleation theory and using specific features of the deliquescent haze particles considered in Chapters 6 and 7. In this section, we derive analytical expressions for the critical radii, energies, and homogeneous nucleation (freezing) rates taking into account solution and curvature effects, and then compare this method with previous parameterizations.

8.6.1. Derivation of the Critical Germ Size, Energy, and Nucleation Rate

The nucleation rate and number of crystals formed by homogeneous nucleation due to the freezing of supercooled droplets of pure water or deliquescent haze particles is determined by the equation for $J_{f,hom}$ derived in Section 8.3. It was shown that $J_{f,hom}$ is determined by the activation energy ΔF_{act} for diffusion across the liquid–ice boundary, and the critical energy ΔF_{cr} of nucleation which depends on the critical radius of an ice germ formed in supercooled liquid.

As discussed in Section 8.3, the expression for ΔF_{cr} in classical theory is related to the critical radius of an ice germ r_{cr}, which can be obtained from the equilibrium condition at the ice–water interface. The theory for pure water derived by J. J. Thomson (1888) yields

$$r_{cr}(T) = \frac{2\sigma_{iw}}{\bar{L}_m \rho_i \ln(T_0 / T)}, \tag{8.6.1}$$

$$\Delta F_{cr}(T) = \frac{4}{3}\pi\sigma_{iw} r_{cr}^2 = \frac{16\pi\sigma_{iw}^3}{3[\bar{L}_m \rho_i \ln(T_0/T)]^2}. \tag{8.6.2}$$

Here, $\bar{L}_m$ is the melting heat averaged over the temperature interval (T_0, T), T_0 is the triple-point temperature, T_0 and T are in degrees Kelvin, σ_{iw} is the surface tension at the ice–water interface, and ρ_i is the bulk ice density.

These equations do not account for solution effects, and therefore in Thomson's formulation the critical radius and energy depend only on temperature T, and not on the humidity or saturation ratio S_w. Dufour and Defay (1963) accounted for the effects of solution in terms of the osmotic coefficient Φ_s, calculations of which are rather complicated, but still not in terms of environmental humidity or S_w. A more general derivation of the critical radius, energy, and nucleation rate for the homogeneous freezing of aqueous solutions was performed by Khvorostyanov and Sassen (1998c) with some simplifications that allowed derivation of expressions for $r_{cr}(T, S_w)$, energy $\Delta F_{cr}(T, S_w)$, and $J_{f,hom}(T, S_w)$. In Khvorostyanov and Curry (2000, 2004a,b, 2005a, 2009b, 2012), these expressions were generalized to account for dependencies on the finite size of the freezing particles and on external pressure, a simple quantitative equivalence between the pressure and solution effects was found, and these equations were applied for evaluation of critical temperatures and humidities for ice nucleation.

Here, extending further the cited works, we consider in more detail the processes of freezing taking into account the difference between ice nucleation in the small solution drops (haze particles) or unactivated deliquescent interstitial CCN in a cloud, and ice nucleation in the much larger activated cloud drops. The difference between the freezing of the small solution drops and larger activated cloud drops is determined by the difference in their equilibrium states. It was shown in Chapter 3, that the equilibrium of unactivated small solution droplets is determined by the Köhler equation, while the equilibrium of the larger cloud drops is determined by the Kelvin equation. The freezing of unactivated solution drops around Köhler's equilibrium is considered in Sections 8.6.1 to 8.6.3, while the critical radii and energies of cloud drop freezing around Kelvin's equilibrium are considered in Section 8.6.4.

To derive the size of a critical germ that takes into account the solution effect, we use the general equilibrium or entropy equation (Eqn. 3.6.4) for the condition of equilibrium for an aqueous solution drop with radius r_{as} containing an ice germ of radius r_{cr}. Here, phase 1 is liquid water and phase 2 is ice. Then, the molar volumes v_1 and v_2 in phases 1 and 2 are those of water and ice, v_w and v_i, the molar enthalpies h_1 and h_2 are those of water and ice, h_w and h_i, the molar melting latent heat $L_{12,mol} = h_w - h_i = M_w L_m$, where L_m is the specific melting latent heat and M_w the molecular weight of water, and the misfit strain $\varepsilon = 0$ for homogeneous nucleation without a foreign substance. The pressures inside the solution drop and ice crystals are p_s and p_i, and the activities of water and ice are $a_{k,1} = a_w$, $a_{k,2} = a_i$.

The fraction of salt left in ice after freezing is characterized by the *segregation* or *partition* or *retention coefficient*. Observations show that upon freezing, salt is mostly rejected from ice into surrounding brine and *the retention coefficient* is generally small, ~10^{-2} to 10^{-4}, if freezing is not rapid and the pockets of solution are not trapped in ice. We assume that the solute is rejected from the ice crystal upon freezing, so that $a_i = 1$ and $\ln(a_i) = 0$, while $a_w \neq 1$. We obtain from (Eqn. 3.6.4) for a solution drop with embedded ice crystal

$$-\frac{M_w L_m}{T^2} dT + \frac{v_w}{T} dp_s - \frac{v_i}{T} dp_i + R d \ln a_w = 0. \qquad (8.6.3)$$

If an ice germ is approximated by a spherical shape, the internal pressures inside a crystal, p_i, and inside a liquid solution drop, p_s, can be expressed in terms of external pressure p with use of Laplace's conditions of mechanical equilibrium:

$$dp_i = dp_s + d\left(\frac{2\sigma_{is}}{r_{cr}}\right), \qquad dp_s = dp + d\left(\frac{2\sigma_{sa}}{r_{as}}\right), \qquad (8.6.4)$$

subscripts i and s refer to ice and solution, respectively, σ_{is} and σ_{sa} are the surface tensions at the ice–solution and solution–air interfaces, r_{as} is the radius of an aqueous solution droplet, and r_{cr} is the radius of a crystal formed inside it. Substituting Laplace's relations (Eqn. 8.6.4) and the relations for the molar volumes, $v_w = M_w/\rho_w$, $v_i = M_w/\rho_i$, into (Eqn. 8.6.3), and then multiplying by T and dividing by M_w, we obtain

$$-\frac{L_m(T)}{T} dT + \left[\frac{1}{\rho_w(T)} - \frac{1}{\rho_i(T)}\right]\left[dp + d\left(\frac{2\sigma_{sa}(T)}{r_{as}}\right)\right]$$
$$-\frac{1}{\rho_i(T)} d\left(\frac{2\sigma_{is}(T)}{r_{cr}}\right) + \frac{RT}{M_w} d\ln a_w = 0. \qquad (8.6.5)$$

The activity of water in solution a_w was defined in (Eqn. 3.5.18) and in this case

$$\ln a_w = -\frac{v\Phi_s m_s M_w}{M_s[(4\pi/3)\rho_s''(r_{as}^3 - r_{cr}^3) - m_s]}, \qquad (8.6.6)$$

where Φ_s is the osmotic coefficient, v is the number of ions in solution, m_s and M_s are the mass and molecular weight of solute in a droplet, and ρ_s'' is the solution density.

The number of components in this thermodynamic system is $c = 3$ (water, solute, air); thus, the number of thermodynamic degrees of freedom should be $N_w = c + 1 = 4$ according to the phase rule (Eqn. 3.3.12) for a system with curved substances. If we consider the case when the mass of solute m_s is constant, this imposes an additional constraint and $N_w = 3$; if m_s is variable due to transfer processes, then $N_w = 4$. The number of variables in (Eqn. 8.6.5) is 5 (T, r_{cr}, r_{as}, p, a_w), and only 3 or 4 of them are independent according to the phase rule. We can consider the critical radius r_{cr} of a crystal as a function of 3 or 4 variables and choose several possible combinations of variables and integrate (Eqn. 8.6.5), yielding several analytical dependencies of the critical radius r_{cr} on various variables. To shorten and generalize calculations, we can integrate (Eqn. 8.6.5), considering all variables as independent; a selection of the possible combinations of independent variables can be done in the final result.

Integration of (Eqn. 8.6.5) from the initial conditions ($T = T_0$, $r_{as} = \infty$, $r_{cr} = \infty$, $a_w = 1$, $p = p_0$) to (T, r_{as}, r_{cr}, a_w, p) yields

$$L_m^{ef}(T)\ln\frac{T_0}{T} - \frac{\Delta\rho}{\rho_w\rho_i}\Delta p + \frac{2\sigma_{sa}}{\rho_w r_{as}} - \frac{2\sigma_{sa}}{\rho_i r_{as}} - \frac{2\sigma_{is}}{\rho_i r_{cr}} + \frac{R\bar{T}}{M_w}\ln a_w = 0, \tag{8.6.7}$$

where $\Delta p = p - p_0$, $\Delta\rho = \rho_w - \rho_i$, $\bar{T}$ is the mean temperature in the range (T_0, T), and we introduce an "effective" melting heat, averaged over the range (T_0, T):

$$L_m^{ef}(T) = \left(\int_T^{T_0}\frac{L_m(T)}{T}dT\right)\times\left(\int_T^{T_0}\frac{dT}{T}\right)^{-1} = \left(\int_T^{T_0}\frac{L_m(T)}{T}dT\right)\times\left[\ln\left(\frac{T_0}{T}\right)\right]^{-1}. \tag{8.6.8}$$

Using data on the temperature behavior of $L_m(T)$, the function $L_m^{ef}(T)$ can be parameterized as $L_m^{ef}(T) = L_{m0}(T_0)\varphi(T)$, which depends on temperature and is related to $L_{m0}(T_0)$ through some algebraic function $\varphi(T)$.

If we assume that $p = \text{const}$ and $dp = 0$ in (Eqn. 8.6.5), pressure is excluded from the final equation. Substituting (Eqn. 8.6.6) for a_w into (Eqn. 8.6.7), and writing $m_s = (4/3)\pi\rho_s r_s^3$, we obtain

$$L_m^{ef}(T)\ln\frac{T_0}{T} = -\frac{2\sigma_{sa}}{\rho_w r_{as}} + \frac{2\sigma_{sa}}{\rho_i r_{as}},$$
$$-\frac{2\sigma_{is}}{\rho_i r_{cr}} - \frac{R\bar{T}}{M_w}\frac{\nu\Phi_s m_s M_w/M_s}{(4\pi/3)[\rho_s''(r_{as}^3 - r_{cr}^3) - \rho_s r_s^3]} = 0. \tag{8.6.9}$$

This equation determines r_{cr} as a function of T, r_{as}, and a_w, although this dependence is implicit because r_{cr} is also contained in the denominator of the last term. If we consider a large drop or a bulk solution with $r_{as} \to \infty$, then the terms with $1/r_{as}$ vanish and we obtain

$$r_{cr} = \frac{2\sigma_{is}}{\rho_i\left[L_m^{ef}\ln\frac{T_0}{T} - \frac{RT}{M_w}\frac{\nu\Phi_s m_s M_w/M_s}{(4\pi/3)\rho_s''(r_{as}^3 - r_{cr}^3) - m_s}\right]}. \tag{8.6.10}$$

The calculation of r_{cr} with (Eqns. 8.6.9) and (8.6.10) requires knowledge of Φ_s, ν, and ρ_s'' as functions of the solution molality or the droplet radius and may be a rather complicated task.

A simple algebraic transformation allows us to significantly simplify the equations for r_{cr} and ΔF_{cr} and to avoid using chemical properties of the solutions. Equation (8.6.7) can be rewritten after regrouping the terms as

$$\frac{2\sigma_{is}}{\rho_i r_{cr}} = L_m^{ef}(T)\ln\frac{T_0}{T} + \frac{R\bar{T}}{M_w}\left(\frac{2M_w\sigma_{sa}}{\rho_w R\bar{T} r_{as}} + \ln a_w\right) - \frac{\Delta\rho\Delta p}{\rho_w\rho_i} - \frac{2\sigma_{sa}}{\rho_i r_{as}}. \tag{8.6.11}$$

The expression in parentheses can be expressed via the water saturation ratio S_w using the Köhler equation (Eqn. 3.9.4):

$$\frac{2M_w\sigma_{sa}}{\rho_w R\bar{T} r_{as}} + \ln a_w = \frac{A_K}{r_{as}} + \ln a_w = \ln S_w, \tag{8.6.12}$$

where $A_K = 2M_w\sigma_{sa}/\rho_w R\bar{T}$ is the Kelvin's curvature parameter. The applicability of the Köhler equation here is justified by the assumption made earlier that the solution drop is in equilibrium with environmental air. A wet haze particle just before formation of an ice germ will be in Köhler's equilibrium, whereby the curvature and solution terms should be balanced and determined by S_w. Nucleation of an ice germ cannot substantially violate this equilibrium because the size of the germ ($\sim 10^{-7}$ cm) is much smaller than the size of even a submicron haze particle ($\sim 10^{-6}$–10^{-4} cm), so that the volume of the germ is 10^3–10^9 smaller than the drop volume and cannot seriously violate the thermodynamic Köhler equilibrium. For cloud drops with radii of a few microns, the curvature and solution terms are usually sufficiently small, and saturation ratio S_w is close to 1 in the cloud environment with small water supersaturation.

Using (Eqn. 8.6.12) allows significant simplifications of the expression for the ice germ critical radius and energy. Equation (8.6.11) is simplified as

$$\frac{2\sigma_{is}}{\rho_i r_{cr}} = L_m^{ef}(T)\ln\frac{T_0}{T} + \frac{R\bar{T}}{M_w}\ln S_w - \frac{\Delta\rho\Delta p}{\rho_w\rho_i} - \frac{2\sigma_{sa}}{\rho_i r_{as}}. \tag{8.6.13}$$

This gives the critical radius of an ice germ

$$r_{cr} = \frac{2\sigma_{is}}{\rho_i L_m^{ef}(T)\ln\frac{T_0}{T} + \rho_i\frac{R\bar{T}}{M_w}\ln S_w - \frac{\Delta\rho\Delta p}{\rho_w} - \frac{2\sigma_{sa}}{r_{as}}} \tag{8.6.14a}$$

$$= \frac{2\sigma_{is}}{\rho_i L_m^{ef}\left[\ln\left(\frac{T_0}{T}S_w^{G_n}\right) - A_p\Delta p - \frac{A_f}{r_{as}}\right]}, \tag{8.6.14b}$$

where we introduce a new dimensionless parameter G_n, which characterizes the dependence on S_w, the coefficient A_p that characterizes pressure effect, and a new curvature parameter for freezing $A_f = r_{fr}$ (scaling freezing radius) that characterizes the effect of the drop radius on freezing. A_f is analogous and proportional to the Kelvin parameter

$$G_n = \frac{R\bar{T}}{M_w L_m^{ef}}, \quad A_p = \frac{\Delta\rho}{\rho_w\rho_i L_m^{ef}}, \quad A_f = r_{fr} = \frac{2\sigma_{sa}}{\rho_i L_m^{ef}} = G_n\frac{\rho_w}{\rho_i}A_K. \tag{8.6.15}$$

An estimate at $T = -40\,^\circ$C yields $G_n \sim 0.4$–0.5, $A_p \sim 0.44 \times 10^{-10}\,\text{cm}^3\,\text{erg}^{-1} = 0.44 \times 10^{-4}\,\text{atm}^{-1}$, and $r_{fr} \sim 0.6A_K \sim 0.8 \times 10^{-7}$ cm. It is also convenient to introduce an auxiliary function $H_{v,fr}$ of Δp and r_{as}:

$$H_{v,fr} = A_p\Delta p + \frac{A_f}{r_{as}} = \frac{\Delta\rho\Delta p}{\rho_w\rho_i L_m^{ef}} + \frac{2\sigma_{sa}}{\rho_i L_m^{ef} r_{as}}, \tag{8.6.16}$$

and to write r_{cr} in the forms

$$r_{cr} = \frac{2\sigma_{is}}{\rho_i L_m^{ef}\left[\ln\left(\frac{T_0}{T}S_w^{G_n}\right) - H_{v,fr}\right]} \tag{8.6.17a}$$

$$= \frac{2\sigma_{is}}{\rho_i L_m^{ef}\left[\ln\left(\frac{T_0}{T}S_w^{G_n}\exp(-H_{v,fr})\right)\right]}. \tag{8.6.17b}$$

Equations (8.6.14a, b) and (8.6.17a, b) express the critical radius as a function of 4 variables: T, S_w, p, r_{as}. Recall that due to the limited number of independent variables, only 3 (with m_s = const) or 4 ($m_s \neq$ const) are independent. In cases with 3 variables, the most informative are T and S_w, and we can choose either pressure p or drop radius r_{as}. It will be shown in the following text that the use of T, S_w, and p allows us to establish equivalence between the solution and pressure effects, or describe theoretically the decrease of the freezing and melting temperatures on the external pressure.

The critical energy ΔF_{cr} can now be evaluated analogously to the cases considered in Section 8.1—i.e., writing the volume and surface energies of an embryo, and then minimizing their sum as $d[\Delta F_{cr}(r)]/dr = 0$. After this procedure, similar to (Eqn. 8.1.12), ΔF_{cr} is equal to one-third of the product of the surface tension σ_{is} by the crystal surface area $A_{cr,i} = 4\pi r_{cr}^2$:

$$\Delta F_{cr} = \frac{4}{3}\pi\sigma_{is} r_{cr}^2 = \frac{\sigma_{is} A_{cr,i}}{3} = \frac{16\pi}{3} \frac{\sigma_{is}^3}{\left\{\rho_i L_m^{ef} \ln\left[\frac{T_0}{T} S_w^{G_n} \exp(-H_{v,fr})\right]\right\}^2}. \tag{8.6.18}$$

For the particular case when we keep the drop radius r_{as} = const ($dr_{as} = 0$) or for a bulk solution ($r_{as} \to \infty$) and isobaric processes with external pressure p = const ($dp = 0$), these expressions are simplified, and r_{cr} and ΔF_{cr} can be expressed as the functions of the temperature T and saturation ratio S_w:

$$r_{cr} = \frac{2\sigma_{is}}{\rho_i [L_m^{ef}(\ln T_0/T + \ln S_w^{G_n})]} = \frac{2\sigma_{is}}{\rho_i L_m^{ef}(T) \ln[(T_0/T) S_w^{G_n}]}, \tag{8.6.19}$$

$$\Delta F_{cr} = \frac{4}{3}\pi\sigma_{is} r_{cr}^2 = \frac{16\pi}{3} \frac{\sigma_{is}^3}{[\rho_i L_m^{ef}(T) \ln[(T_0/T) S_w^{G_n}]]^2}. \tag{8.6.20}$$

The nucleation freezing rate $J_{f,hom}$ follows from (Eqn. 8.3.5) with this ΔF_{cr} from (Eqn. 8.6.18):

$$J_{f,\text{hom}} = 2\frac{N_{cont} kT}{h}\left(\frac{\rho_w}{\rho_i}\right)\left(\frac{\sigma_{is}}{kT}\right)^{1/2} \exp\left(-\frac{\Delta F_{act}}{kT}\right)$$
$$\times \exp\left(-\frac{1}{kT} \frac{(16\pi/3)\sigma_{is}^3}{\{\rho_i L_m^{ef} \ln[(T_0/T) S_w^{G_n} \exp(-H_{v,fr})]\}^2}\right). \tag{8.6.21}$$

The equation for r_{cr} yields an expression for the ordinary affinity A_∞, equal to the difference $\Delta\mu_{si}$ of the chemical potentials of the phases. Applying again (Eqn. 8.1.4), here for the ice and solution (subscripts "i" and "s"),

$$r_{cr} = \frac{2\sigma_{is} v_{i1}}{\mu_{s1} - \mu_{i1}} = \frac{2\sigma_{is} v_{i\mu}}{\mu_s - \mu_i}, \tag{8.6.22}$$

where again the subscript "1" denotes molecular potentials, and the quantities without "1" denote the molar properties, and v_{i1} and $v_{i\mu}$ denote the molecular and molar volumes. Comparing (Eqns. 8.6.17b) and (8.6.22), we obtain $\Delta\mu_{si}$

$$A_{\infty,1} = \Delta\mu_{si,1} = \mu_{s1} - \mu_{i1} = \frac{M_w L_m^{ef}(T)}{N_{Av}} \ln\left[\left(\frac{T_0}{T} S_w^{G_n}\right) \exp(-H_{v,fr})\right], \tag{8.6.23}$$

$$A_\infty = \Delta\mu_{si} = \mu_s - \mu_i = M_w L_m^{ef}(T) \ln\left[\left(\frac{T_0}{T} S_w^{G_n}\right) \exp(-H_{v,fr})\right], \tag{8.6.24}$$

where N_{Av} is Avogadro's number. The corresponding differences of the chemical potentials (affinities) for the case of pure water are obtained from (Eqns. 8.6.23) and (8.6.24) at $S_w = 1$.

8.6.2. Analysis and Properties of the Solution

Equations (8.6.14a, b) or (8.6.17a, b) for r_{cr} and (Eqn. 8.6.18) for ΔF_{cr} include several particular cases. In the first particular case for pure water droplets ($S_w = 1$) at atmospheric pressure ($p = p_0$ or $\Delta p = 0$), and sufficiently large drops or bulk solution ($r_{as} \to \infty$), (Eqns. 8.6.17a, b) and (8.6.18) transform into the classical Thomson's expressions (Eqns. 8.6.1) and (8.6.2) for homogeneous ice nucleation in pure water. Compared to previous parameterizations reviewed in Section 8.4, (Eqns. 8.6.17a), (8.6.17b), and (8.6.18) do not contain such quantities as water activity, the osmotic coefficient, and droplet mass or molality. So, this method is explicit and simpler to use. A comparison with the previous parameterizations may help to find the best fits for the parameters of classical nucleation theory: L_m, σ_{is}, ΔF_{act}.

In the other particular case, at $T \sim T_0$ (slightly below 0 °C, which can be around melting equilibrium), after substituting G_n from (Eqn. 8.6.15) into (Eqn. 8.6.19), it is simplified to $r_{cr} = 2\sigma_{is}/(R_v T \rho_i \ln S_w)$. Thus, it resembles the usual expression for the germ radius in the homogeneous nucleation of a crystal (or droplet) directly from the supersaturated vapor in the two-phase system—i.e., a transition from the vapor to water or ice phase, considered in Section 8.1. However, in the three-phase system (vapor, solution, and ice germ), the surface tension at the solution–ice interface is $\sigma_{is} \sim 18$–$25\,\mathrm{erg\,cm^{-2}}$, while the surface tension in the two-phase system at the ice–vapor interface is $\sigma_{iv} \sim 100$–$104\,\mathrm{erg\ cm^{-2}}$ [see Section 4.4, (Eqns. 4.4.20a) and (4.4.20b)]. So, the critical germ size r_{cr} is 4 times greater, ΔF_{cr} is 16 times greater, and nucleation rate $J_{f,hom} \sim \exp(-\Delta F_{cr}/kT)$ is many orders of magnitude smaller for the direct transitions into ice from the vapor than from the liquid. Thus, homogeneous nucleation via the three-phase system or via the two-step process (vapor–liquid and liquid–ice) is energetically easier than via the one-step vapor–ice transition; both processes require high S_w, but homogeneous freezing requires lower S_w than homogeneous nucleation via deposition from the vapor. This property of this equation for r_{cr} is in agreement with *Ostvald's rule*, which predicts that the phase transformation may occur as a two-step process: the first transition into a less stable phase (liquid), and then into a more stable phase (ice).

In summary, the equations of this theory, (Eqns. 8.6.14a), (8.6.14b), (8.6.17a), (8.6.17b), and (8.6.18) along with (Eqn. 8.6.21) for $J_{f,hom}$, provide the quantitative tools for description of homogeneous nucleation both for haze particles and pure water, and convert into the classical Thomson equations for freezing (Eqns. 8.6.1) and (8.6.2) at $S_w \to 1$, $\Delta p \to 0$, $r_{as} \to \infty$. At $T \to T_0$, these equations become similar to those for the homogeneous condensation or deposition nucleation from the vapor. Thus, this generalization of the theory of homogeneous freezing satisfies the generalized Bohr's correspondence principle described in Chapter 1 of this book (Section 1.2). These newer equations: a) do not reject but generalize the classical equations and contain them as the particular cases; b) convert into the classical equations when the parameters of the new theory tend to the certain limits: $S_w \to 1$, or $T \to T_0$, or $\Delta p \to 0$, or $r_{sa} \to \infty$.

Invariance to chemical composition. The final equations for homogeneous freezing show an important and somewhat surprising feature. The specific chemical properties (density of dry particle ρ_s, dry mass m_s, osmotic potential Φ_s, number of ions ν, and surface tension σ_{sa}) of an aerosol (haze) particle were present in the initial equations but vanished from the final (Eqns. 8.6.14a,b)–(8.6.21) for r_{cr}, ΔF_{cr}, and $J_{f,hom}$. The major quantity that accounts for all specific chemicals and describes unified chemical action is the environmental water saturation ratio S_w. The surface tension σ_{is} is less variable than S_w and its value also adjusts to the environmental S_w.

This dependence on the water saturation ratio and independence of the chemical composition is caused by the basic behavior of Kohler curves (Eqn. 8.6.12), which show that the radius of a deliquescent haze particle in equilibrium assumes a Kohler's radius through the combined solution and curvature effects in such a way that it can be expressed through relative humidity or the water saturation ratio alone. Otherwise, supersaturation or subsaturation occurs around a droplet and it will grow or evaporate until its radius reaches such a value that Köhler's equation (Eqn. 8.6.12) is satisfied.

The relaxation time τ_{eq} during which equilibrium radius is reached can be estimated by:

$$\tau_{eq} \sim \frac{0.4 r_{as}^2}{D_v(\rho_v/\rho_w)(1-S_w)}, \tag{8.6.25}$$

where D_v is the vapor diffusion coefficient and ρ_v the vapor density. This gives an estimate for the relaxation time of homogeneous ice nucleation at cirrus altitudes ($p \sim 200$ mb, $T \sim -40\,°C$), of $\tau_{eq} \sim 10^{-3}$ to ~ 2 s in the range of haze particles radii $0.01\,\mu m < r_{as} < 0.5\,\mu m$, and $0.9 < S_w < 0.98$. Thus, haze particles adjust sufficiently rapidly to equilibrium conditions. At lower altitudes and higher vapor pressures, τ_{eq} is smaller because $\tau_{eq} \sim \rho_v^{-1}$. Thus, Köhler's equilibrium is reached very rapidly for submicron liquid particles, therefore its application along with the equilibrium assumption is justified. Note that these assumptions can be valid even at sufficiently low temperatures and high vertical velocities as in the upper layers of convective clouds and anvils.

The "Saturation ratio criterion" for freezing. Since the environmental saturation ratio S_w is the major factor (along with temperature) that governs homogeneous freezing rates for solutions, this prediction can be called *the water saturation ratio criterion.* The physical reason for this is the rapid relaxation of the haze particles to Köhler's equilibrium, so that the combined action of chemical properties (the Raoult effect) and curvature (Kelvin) effect results in a single major parameter S_w. The value of S_w for the haze droplets is rather close to water activity a_w. Thus, this theory predicts that the homogeneous freezing behavior of the haze droplets could also be approximately described in terms of a_w. This prediction is confirmed by Koop et al. (2000), who analyzed a good deal of experimental data on homogeneous freezing temperatures $T_{f,hom}$ of solution droplets with radii of a few microns for 18 chemical substances and showed that their behavior was close to invariant relative to a_w. That is, $T_{f,hom}$ plotted as a function of the molality (solution concentration) exhibited significant differences for various substances, but when plotted versus a_w, all curves $T_{f,hom}$ in Koop et al. (2000) were close to each other and to one universal curve. Note, however, that for submicron haze particles, S_w and a_w may differ due to the curvature effects and predictions according to the water activity criterion may be different from predictions made by the water saturation ratio.

Effects of solution, pressure, and finite drop size on homogeneous nucleation. The "saturation ratio" or "water activity" criterion is an important criterion that regulates the freezing temperature $T_{f,hom}$, but a few other factors may noticeably change $T_{f,hom}$. Estimates of the effects of solution, pressure, and finite drop size on homogeneous nucleation can be done using (Eqn. 8.6.14b) for r_{cr}. It is convenient to transform the terms in the denominator of (Eqn. 8.6.14b) or $\Delta\mu_{si}$ in (Eqn. 8.6.24). We can write $T = T_0 - \Delta T$; at supercooling ~ 40 K, when homogeneous nucleation begins, $\Delta T << T_0$, and $\ln(T_0/T) = \ln[T_0/(T_0 - \Delta T)] \approx \Delta T/T_0$. The saturation ratio is related to water supersaturation as $S_w = 1 + s_w$; since $|s_w| << 1$, an expansion into the power series gives $S_w^{G_n} = (1+s_w)^{G_n} \approx 1 + G_n s_w$. With these simplifications, we can rewrite the denominator in (Eqn. 8.6.14b) as

$$\Delta\mu_{si} = \mu_s - \mu_i = M_w L_m^{ef}(T)\left[\frac{\Delta T}{T_0} + G_n s_w - A_p \Delta p - \frac{A_f}{r_{as}}\right]. \tag{8.6.26}$$

At $T \approx -40\,°C$, the contribution of the first term in parentheses in (Eqn. 8.6.26) is $\approx \Delta T/T_0 \approx 0.14$ and the value of $G_n \sim 0.4$. Haze particles usually freeze homogeneously in cirrus at subsaturation, and the second term in (Eqn. 8.6.26) is always negative at $S_w < 1$. For example, at $s_w = -0.1$ (RHW = −10%), the contribution of the term with s_w is $G_n s_w \sim -0.04$—i.e., about 25% of the temperature term. The signs of the ΔT– and s_w– terms are different; thus, the same values of r_{cr} (therefore, of ΔF_{cr} and $J_{f,hom}$) are reached at greater ΔT, which is equivalent to the depression of the freezing point—i.e., the solution effect (subsaturation) causes a shift of the freezing process to the lower temperatures. The contribution of $G_n s_w$ decreases as humidity increases and tends to 100% (s_w tends to 0), which is reasonable as the concentration of the solution decreases.

An estimate for the coefficient A_p in (Eqn. 8.6.14b) at $T \approx -40\,°C$ gives $A_p \sim 0.44 \times 10^{-4}\,atm^{-1}$. Therefore, the contribution of the pressure term is negligible at atmospheric pressure, $\Delta p = 1$ atm, and becomes significant only at very high pressures—e.g., $A_p \Delta p \approx 0.13$ at $\Delta p = 3000$ atm, which is comparable with the temperature term. Since the contributions of the terms with ΔT and Δp have different signs, this high pressure is equivalent in r_{cr} to the temperature increase, or freezing at high pressures should occur at lower temperatures. The effects of solutions and high applied pressure are similar.

An estimate for A_f with this L_m^{ef} yields $A_f = r_{fr} \sim 0.8 \times 10^{-7}$ cm. The freezing curvature term with A_f causes the freezing process to shift to lower temperatures. The value of the parameter A_f is slightly smaller than the Kelvin parameter A_K, namely, $A_f \sim 0.8 \times 10^{-7}$ cm $\sim (0.6–0.7) \times A_K$; thus, the effect of A_f is comparable with the effect of A_K. For the drops with radius $r_{sa} = 10^{-6}$ cm, the term $A_f/r_{sa} \sim 0.08$, or about half of the temperature term. In summary, the effects of solution, high pressure, and finite drop size result in the freezing point depression.

The general behavior of the germ critical radius r_{cr} and energy ΔF_{cr} computed from (Eqns. 8.6.19) and (8.6.20) for pure water at $S_w = 1$ and the solution (haze) particles at environmental $S_w = 0.9$ is presented in Figs. 8.3a and b. Differences between the pure water and solution particles are shown, whereby solution effects increase both quantities by ~ 25–30%. These differences slightly decrease with lowering temperature. Because this difference in ΔF_{cr} is in the exponent of the nucleation rate

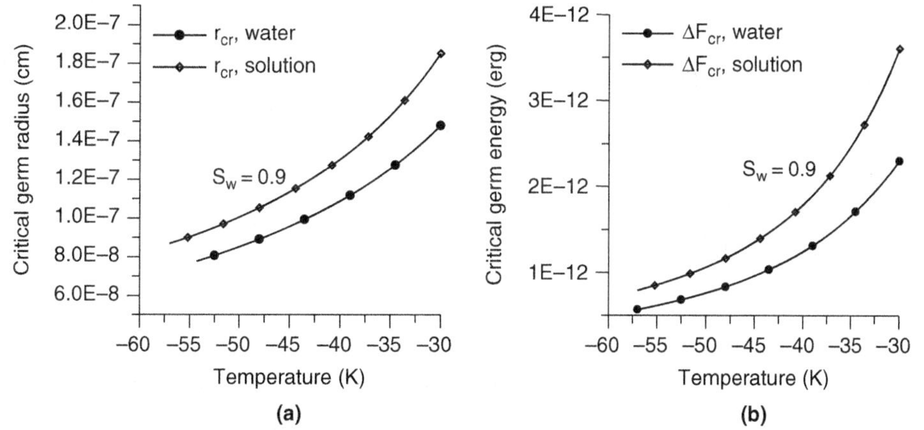

Figure 8.3. Effect of solution on the critical radius (a) and energy (b) of an ice germ at various temperatures at environmental saturation ratio $S_w = 0.9$ (upper curves) with pure water at $S_w = 1$ (lower curves).

$J_{f,hom}$ in (Eqn. 8.6.21), the solution effect causes a strong suppression of the nucleation rate by several orders of magnitude occurs (see the following).

8.6.3. Comparison with Other Models and Observations

As described in Section 8.4, Sassen and Dodd (1988, SD88 model) parameterized the effective freezing temperature of the haze particles as $T_f^* = T + \Delta T_f$, relating the freezing and melting point depressions $\Delta T_f = \lambda_{SD}\Delta T_m$ and $\lambda_{SD} = 1.7$. DeMott et al. (1994) related the melting point temperature ΔT_m to the molality μ_s of deliquesced haze particles.

It can be shown that the equations for $r_{cr}(T, S_w)$ derived by Khvorostyanov and Sassen (1998c) and Khvorostyanov and Curry (2004a, b) (KS98-KC04 model) extending the classical theory yield relations equivalent to the parameterization in the SD88 model and allow ΔT_f to be expressed via the primary variables. The critical radius r_{cr} in the SD88 model can be written in the form of Thomson's equation (Eqn. 8.6.1), but with T_f^* instead of T,

$$r_{cr}(T) = \frac{2\sigma_{iw}}{L_m^{ef}\rho_i \ln(T_0/T_f^*)} = \frac{2\sigma_{iw}}{L_m^{ef}\rho_i \ln[T_0/(T+\Delta T_f)]}. \tag{8.6.27}$$

The term with $\ln(T_0/T_f^*)$ can be expanded by the small ratio $\Delta T_f/T << 1$,

$$\ln\frac{T_0}{T+\Delta T_f} = \ln\frac{T_0}{T(1+\Delta T_f/T)} \approx \ln\frac{T_0}{T} + \ln\left(1-\frac{\Delta T_f}{T}\right) \approx \ln\frac{T_0}{T} - \frac{\Delta T_f}{T}. \tag{8.6.28}$$

In the KS98 model, the term with $\ln(T_0/T)$ in (Eqn. 8.6.19) for $r_{cr}(T, S_w)$ at $(1 - S_w) << 1$ (or $S_w \geq 0.85$, when homogeneous freeing mostly occurs in cirrus) can be approximated as

$$\ln(T_0/T) - G_n(1-S_w). \tag{8.6.29}$$

Equating the last two expressions, we obtain

$$\Delta T_f \approx G_n T(1-S_w) = \frac{RT^2}{M_w L_m^{ef}}(1-S_w). \tag{8.6.30}$$

Thus ΔT_f is expressed through fundamental constants and the saturation ratio. An estimate around −40 to −50 °C gives $G_n \sim 0.4$–0.6, which yields from (Eqn. 8.6.30) for the haze solution particles freezing at $S_w = 0.95$ a freezing point depression $\Delta T_f \sim 5$–6 °C. It will be shown further on that this estimate is close to more precise calculations.

Freezing and melting point depressions are often parameterized through molality (e.g., Sassen and Dodd, 1988; DeMott et al., 1994). It can be shown that molality can be expressed via the saturation ratio. If the solution is sufficiently dilute, the solution density $\rho_s'' \approx \rho_w$ and molality $\hat{M}_s$ can be written as:

$$\hat{M}_s = 10^3(\varepsilon_m \rho_s / M_s \rho_w)[(r_w/r_d)^3 - 1]^{-1}, \tag{8.6.31}$$

where r_d and r_w are the radii of the dry and wet aerosol particles. We can use the results of Chapter 6 to calculate r_w/r_d. Using (Eqn. 6.2.15) with $\beta = 0.5$ (the soluble fraction proportional to the volume), we obtain for r_w/r_d:

$$r_w(S_w)/r_d = (1 - b/\ln S_w)^{1/3}\theta(S_{th}) \approx [1 + b/(1-S_w)]^{1/3}\theta(S_{th}). \tag{8.6.32}$$

where the parameter $b = \nu\Phi_s\varepsilon_m(\rho_d M_w/\rho_w M_s)$ was introduced in (Eqn. 6.2.6). Typical values of b for substances that can serve as the ice germs in cirrus clouds (e.g., ammonium sulfate and sulfuric acid) were estimated from (Eqn. 6.2.6). For $\beta = 0.5$ and ammonium sulfate, $b \approx 0.5$ for fully soluble nuclei. The Heaviside step function $\theta(x)$ accounts for the lower humidity threshold (subscript "*th*") of deliquescence ($S_{w,th} \approx 0.8$ for ammonium sulfate, see Chapter 11), and the second equation in (Eqn. 8.6.32) is valid at $S_w > 0.85$–0.9.

This permits evaluation of the equilibrium radius of a haze particle—e.g., for ammonium sulfate, the ratio of the wet r_w to dry radii r_d calculated from (Eqn. 8.6.32) at $S_w = 0.95$ and $b = 0.5$ is $r_w/r_d \approx (1 + 0.5/0.05)^{1/3} \approx 2.2$, or the volume grows $(r_w/r_d)^3 \sim 10$ times.

Now we can relate molality to S_w. Substituting r_w/r_d from (Eqn. 8.6.32) into (Eqn. 8.6.31) yields the molality

$$\hat{M}_s \approx 10^3(\nu\Phi_s M_w)^{-1}(1 - S_w) \approx 30(1 - S_w) \tag{8.6.33}$$

in the units (mol g^{-1}). The second relation here is based on the fact that the osmotic coefficient for ammonium sulfate varies slightly with S_w and $\Phi_s \sim 0.63$–0.73 for $S_w > 0.9$ to 0.95. Such relations can be used to verify or simplify various numerical models. This equation can be inverted for S_w as

$$S_w = 1 - 10^{-3}(\nu\Phi_s M_w)\hat{M}_s. \tag{8.6.34}$$

Substituting it into (Eqn. 8.6.30), we obtain

$$\Delta T_f \approx 10^{-3} G_n T(\nu\Phi_s M_w)\hat{M}_s = 10^{-3}(\nu\Phi_s M_w)\frac{RT^2}{M_w L_m^{ef}}\hat{M}_s, \tag{8.6.35}$$

$$\Delta T_f \approx 5.43\hat{M}_s, \qquad \Delta T_m \approx \Delta T_f/1.7 = 3.2\hat{M}_s. \tag{8.6.36}$$

The estimate (Eqn. 8.6.36) was calculated for ammonium sulfate at $T \approx -40\,°\mathrm{C}$ for a dilute solution. The agreement between the empirical and semi-empirical parameterizations described in Section 8.4 and the extension of the classical nucleation theory described in Section 8.6.1 can be illustrated in calculations of $J_{f,hom}(T)$ by the two methods as illustrated in Fig. 8.4. Method 1 uses an empirical fit for $J_{f,hom}(T)$ by temperature (Eqn. 8.4.1) from Heymsfield and Miloshevich (1993) for pure water with correction δT_f for solutions from Sassen and Dodd (1988)—i.e., $J_{f,hom0}(T_f)$ and $J_{f,hom}(T_f^*)$ were calculated. The value of δT_f was in this case calculated from (Eqn. 8.6.30) for a given S_w. Method 2 uses the theory KS98-KC04 with (Eqns. 8.6.19)–(8.6.21) and the temperature-dependent parameters described in Chapter 4, Section 4.4. The theoretical and parameterization results presented in Fig. 8.4 are in good agreement both for pure water and haze particles with solution. The upper curves for pure water indicate that an appropriate choice of the temperature dependence of the parameters allows us from the theory to extend the empirical parameterizations to lower temperatures. The agreement of the lower curves for haze particles indicates that this newer theory described in Section 8.6.1 correctly accounts for the effects of solution in a wide temperature range. The general agreement supports the validity of this extension of the classical theory for the case of solutions.

As described in Section 8.4, DeMott et al. (1994, DMC94) parameterized the frozen fraction F_{hf} and the freezing rate of haze particles in the form $J_{f,hom} \sim (1 - F_{hf}) \sim \exp[2.3\cdot\{c_{D1} + c_{D2}(1 - S_w)\}]$ as indicated in (Eqn. 8.4.4a) with c_{D1} and c_{D2} being some empirical functions of T given by (Eqns. 8.4.4b) and (8.4.4c). Comparing this $J_{f,hom}$ with Equation (8.6.21) derived from the theory, we

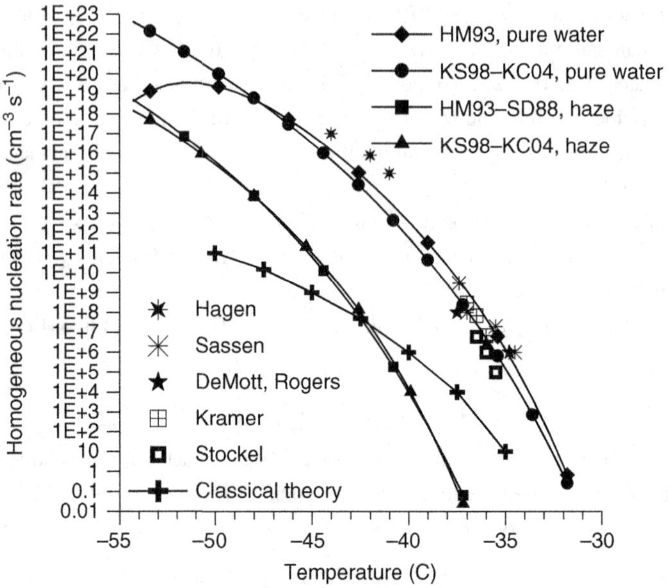

Figure 8.4. Comparison of the homogeneous nucleation rate $J_{f,hom}(T)$ calculated with the two methods. Method 1 uses an empirical fit for $J_{f,hom}(T)$ by temperature (Eqn. 8.4.1) from Heymsfield and Miloshevich (1993) for pure water (HM93, pure water), and with correction δT_f for solutions from Sassen and Dodd (1988)—i.e., $J_{f,hom}(T_f^*)$ (HM93-SD88). The value of δT_f in this case was not taken from tables but was calculated from (Eqn. 8.6.30) for a given S_w. Method 2 uses the theory of Khvorostyanov and Sassen (1998c), and Khvorostyanov and Curry (2000, 2004a,b), with (Eqns. 8.6.19)–(8.6.21) and the temperature-dependent parameters described in Chapter 4, also for pure water (KS98–KC04, pure water) and haze particles (KS98-KC04, haze). Also shown are the experimental points by various authors: Hagen et al. (1981), Sassen and Dodd (1988, 1989), DeMott and Rogers (1990), Krämer et al. (1996), and Stöckel et al. (2001, 2005). The old classical theory (crosses) is given for comparison.

see again the same functional dependence on S_w in the exponent, while the constants c_{D1} and c_{D2} can be found from (Eqn. 8.6.21). Further, DMC94 suggested a parameterization of the melting point depression ΔT_m via the molality $\hat{M}_s$ as $\Delta T_m = 0.1 + 3.485\hat{M}_s$ based on experimental data. Comparing this expression with (Eqns. 8.6.35) and (8.6.36), and the expression obtained from the classical theory $\Delta T_m = 3.2\hat{M}_s$, we see again a reasonable agreement.

Finally, Fig. 8.5 shows a comparison of the crystals concentrations nucleated per 1 s (comparable to a cloud model time step) from the polydisperse haze of ammonium sulfate with exponential size spectrum calculated using two methods: $J_{f,hom}$ taken from DeMott et al. (1994); and $J_{f,hom}$ from the theory described here with equations for $r_{cr}(T, S_w)$, $\Delta F_{cr}(T, S_w)$, and $J_{f,hom}(T, S_w)$ in Section 8.6.1. The comparison in Fig. 8.5 shows good agreement. The curves for haze particles are surprisingly close, even though they were computed with completely different methods.

This good agreement of the various semi-empirical parameterizations and this approach based on an extension of the classical nucleation theory indicates again their equivalence.

A good agreement of various empirical parameterizations with a newer theory of homogeneous ice nucleation described in Section 8.6.1, illustrated in this Section 8.6.3 with several examples, confirms

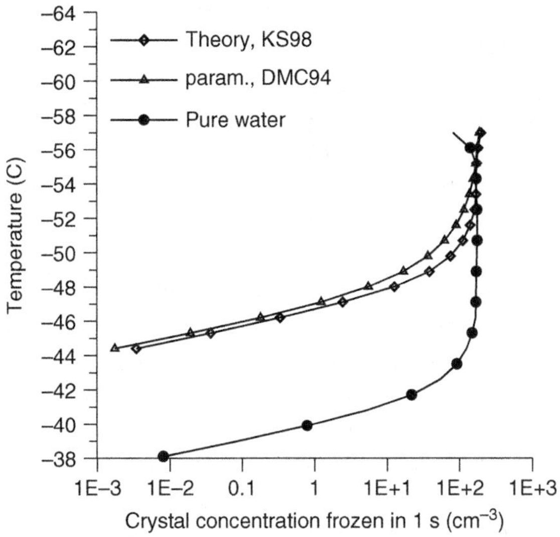

Figure 8.5. Crystal concentrations nucleated in 1 s from droplets of pure water (lower curve) and from a polydisperse haze of ammonium sulfate particles with an exponential size spectrum from DeMott et al. (1994, DMC94). Comparison of calculations based on the homogeneous freezing theory by Khvorostyanov and Sassen (1998, KS98) using (Eqns. 8.6.17a), (8.6.18), and (8.6.21) here with parameterization from DMC94, and (Eqns. 8.4.4a,b,c) and (8.4.5) here. From Khvorostyanov and Sassen (1998), *Geophys. Res. Lett.*, 25, (16), reproduced with permission of John Wiley & Sons, Inc.

again the general conclusion following from the correspondence principle (Section 1.2) that many empirical and seemingly independent parameterizations can be derived from a corresponding theory, which may help unify and clarify different parameterizations and create a joint basis for them.

The method described here for parameterizing the homogeneous freezing process is relatively simple. It has been incorporated into several bin and bulk cloud models and been used in numerical simulations of cirrus clouds and multi-layer systems, where homogeneous freezing may play an important role (e.g., Khvorostyanov and Sassen, 1998a,b, 2002; Khvorostyanov et al., 2001, 2006; Sassen et al., 2002; Morrison et al., 2005a,b; Comstock et al., 2008). It can also be used in parcel models (e.g., Khvorostyanov and Curry, 2005; see Section 8.9 here), or for analysis of droplet freezing experiments. Note that the effects of solution on the surface tension, melting heat, the chemical composition of the haze particles, and other thermodynamic parameters may be important, but there are still many uncertainties in these parameters, as is seen in numerous different parameterizations described in Section 4.4. Therefore, further studies are desirable, especially at very low temperatures, and more experimental data at low T are required.

8.6.4. The Freezing of Cloud Drops

Thus far, we have considered the nucleation of ice crystals inside submicron solution droplets, haze particles. The process of the nucleation of ice crystals inside cloud drops differs in several aspects: cloud drops have much larger radii; the solution inside the drops is very dilute; and Köhler's equilibrium is not applicable to such drops. Derivation of the critical radii and energies of ice crystals nucleating inside cloud drops can be considered similar to the derivation of the Kelvin equation in Section 3.8, with some generalization.

Consider again the general equilibrium or entropy equation (Eqn. 3.6.4) for the condition of equilibrium for a cloud drop with radius r_d containing an ice germ of radius r_{cr}. As in Section 3.8, we assume that the drop is in equilibrium with the environment's humid air, and assume additionally that the nucleating ice crystal inside the drop is in equilibrium with the drop. In (Eqn. 3.6.4), the substances are, in this case: phase 1 is liquid water; phase 2 is ice. The molar volumes v_1 and v_2 in phases 1 and 2 are those of water and ice, v_w and v_i; the molar enthalpies h_1 and h_2 are those of water and ice, h_w and h_i; the molar melting latent heat $L_{12,mol} = h_w - h_i = M_w L_m$, where L_m is the specific melting latent heat, and M_w is the molecular weight of water; and the misfit strain is $\varepsilon = 0$ for homogeneous nucleation without a foreign substance. The pressures inside the solution drop and ice crystals are p_s and p_i. Since the solution in the drop is very dilute, its activity $a_w \approx 1$. Since solute is rejected from ice upon freezing, we also have the condition $a_i \approx 1$. Thus, we can neglect the term $d\ln(a_w/a_i)$ in (Eqn. 3.6.4). However, since we assumed that the cloud drop is in equilibrium with the environment's humid air, as in the derivation of the Kelvin equation in Section 3.8, this term should be replaced with the term describing the contribution of the water vapor activity, $d\ln(e_v/p)$, where e_v is the vapor pressure, p is the atmospheric pressure, and (e_v/p) is the mole fraction of water vapor equal to its activity a_v, assuming an ideal gas behavior.

Then, we obtain for a cloud drop with embedded ice crystal from (Eqn. 3.6.4) multiplying it by T,

$$-\frac{M_w L_m}{T} dT + v_w dp_s - v_i dp_i + RTd\ln(e_v/p) = 0. \tag{8.6.37}$$

Although the solution in the cloud drop is very dilute, it is called hereafter a "solution drop" for purposes of uniformity with the previous consideration of haze particles. If an ice germ is approximated by a spherical shape, the internal pressures inside a crystal, p_i, and inside a liquid solution drop, p_s, can be expressed as earlier in terms of external pressure p by using Laplace's conditions of mechanical equilibrium:

$$dp_s = dp + d\left(\frac{2\sigma_{sa}}{r_d}\right), \qquad dp_i = dp_s + d\left(\frac{2\sigma_{is}}{r_{cr}}\right), \tag{8.6.38}$$

where subscripts i and s refer to ice and solution, respectively; σ_{is} and σ_{sa} are the surface tensions at the ice–solution and solution–air interfaces; r_d is the radius of a droplet; and r_{cr} is the radius of a crystal formed inside it. Substituting Laplace's relations (Eqn. 8.6.38) and the relations for the molar volumes, $v_w = M_w/\rho_w$, $v_i = M_w/\rho_i$, into (Eqn. 8.6.37), and dividing by M_w, we obtain

$$-L_m(T)\frac{dT}{T} + \frac{\Delta\rho}{\rho_w \rho_i} dp - \frac{\Delta\rho}{\rho_w \rho_i} d\left(\frac{2\sigma_{sa}}{r_d}\right)$$

$$-\frac{1}{\rho_i} d\left(\frac{2\sigma_{is}}{r_{cr}}\right) + \frac{RT}{M_w} d\ln\left(\frac{e_v}{p}\right) = 0, \tag{8.6.39}$$

where $\Delta\rho = \rho_w - \rho_i$. The number of components in this thermodynamic system is $c = 3$ (water, solute, air); thus, the number of thermodynamic degrees of freedom is $N_w = 4$ according to the phase rule for a system with curved substances. The variables in (Eqn. 8.6.39) are T, p, e_v, and one of the radii

(r_{cr} or r_d), the other radius is a function of the other 4 variables. If we choose r_d as a variable, we can search for the critical crystal radius $r_{cr}(T, p, e_v, r_d)$.

Suppose that the initial conditions correspond to the state $T = T_0$, $r_{d0} = \infty$ (plane surface), $r_{cr} = \infty$, $e_v = e_{ws}$ (saturated vapor pressure over a plane water surface), and $p = p_0$. Integration of (Eqn. 8.6.39) from the initial conditions to (T, r_d, r_{cr}, e_v, p) yields

$$L_m^{ef}(T)\ln\frac{T_0}{T} - \frac{\Delta\rho\Delta p}{\rho_w\rho_i} - \frac{2\Delta\rho\sigma_{sa}}{\rho_w\rho_i}\frac{1}{r_d} - \frac{2\sigma_{is}}{\rho_i r_{cr}} + \frac{R\bar{T}}{M_w}\ln\left(S_w\frac{p_0}{p}\right) = 0, \tag{8.6.40}$$

where $\Delta p = p - p_0$, $\bar{T}$ is the mean temperature in the range (T_0, T), $L_m^{ef}(T)$ is the effective melting heat introduced earlier, and $S_w = e_v/e_{ws}$ is the water saturation ratio. If, as in Section 3.8, we assume T = const, p = const, the terms with T and p vanish in (Eqn. 8.6.40). If we do not consider nucleation of an ice crystal, then the term $dp_i = 0$ in (Eqn. 8.6.38) and all terms related to ice (v_i, ρ_i, σ_{si}, r_{cr}) also vanish in (Eqn. 8.6.40). Then, the rest of (Eqn. 8.6.40) is

$$\ln S_w = \frac{2\sigma_{sa}M_w}{r_d R\bar{T}\rho_w}. \tag{8.6.41}$$

That is, with these simplifying assumptions, we arrived again at the Kelvin equation, and (Eqn. 8.6.40) can be viewed as its modification, taking into account ice nucleation inside the drop. Solving (Eqn. 8.6.40) relative to r_{cr}, we obtain

$$r_{cr} = \frac{2\sigma_{is}}{\rho_i L_m^{ef}\left[\ln\left(\frac{T_0}{T}S_w^{G_n}\frac{p_0}{p}\right) - \frac{\Delta\rho}{\rho_w\rho_i}\frac{2\sigma_{sa}}{L_m^{ef}}\frac{1}{r_d}\right]}, \tag{8.6.42}$$

where $G_n = R\bar{T}/M_w L_m^{ef}$ is defined as in (Eqn. 8.6.15). An estimate of the terms in (Eqn. 8.6.42) at $r_d \sim 10-30$ μm, $T \sim 238$ K using L_m^{ef} from Fig. 8.14a, and σ_{sa} from Section 4.4 shows that the second term in the denominator of (Eqn. 8.6.42) ~ –(1 to 3) × 10^{-6}. It is much smaller than the first term with ln, and can be neglected. Thus, (Eqn. 8.6.42) is simplified as

$$r_{cr} = \frac{2\sigma_{is}}{\rho_i L_m^{ef}\left[\ln\left(\frac{T_0}{T}S_w^{G_n}\frac{p_0}{p}\right)\right]}. \tag{8.6.43}$$

If droplet freezing occurs with a substantial uplift, the factor $p_0/p > 1$ in (Eqn. 8.6.43). This leads to a smaller critical radius and energy and accelerates droplet freezing. If we consider the processes without strong changes of pressure, then $p \approx p_0$, and this equation is further simplified as

$$r_{cr} = \frac{2\sigma_{is}}{\rho_i L_m^{ef}\left[\ln\left(\frac{T_0}{T}S_w^{G_n}\right)\right]}. \tag{8.6.44}$$

This equation coincides with the simplified versions of r_{cr} in (Eqns. 8.16.14b), (8.6.17a), and (8.6.17b) neglecting there the effects of pressure and droplet curvature ($H_{v,fr} = 0$). Thus, the effects of the temperature and water saturation ratio are similar for both haze particles and activated cloud drops.

The effects of T and S_w can be compared if they are presented as $T = T_0 - \Delta T$, and $S_w = 1 + s_w$, where ΔT is supercooling and s_w is the water supersaturation. The terms under the logarithm sign in (Eqn. 8.6.44) can be presented as

$$\frac{T_0}{T} S_w^{G_n} \approx \left(1 + \frac{\Delta T}{T_0}\right)(1 + G_n s_w) \approx 1 + \frac{\Delta T}{T_0} + G_n s_w = 1 + \frac{\Delta T_{ef}}{T_0}, \quad (8.6.45)$$

where we introduced the "effective supercooling"

$$\Delta T_{ef} = \Delta T + \delta_s \Delta T, \qquad \delta_s \Delta T = T_0 G_n s_w. \quad (8.6.46)$$

Thus, positive supersaturation increases effective supercooling by $\delta_s \Delta T = T_0 G_n s_w$. Fig. 8.14c shows that $G_n \sim 0.45$ at $T \approx -40$ °C, and so $T_0 G_n \sim 120$. Therefore, if $s_w > 0$, this leads to an increase of the effective supercooling by $\delta_s \Delta T \approx 1.2 \times s_w$ (%), where the supersaturation is measured in percent. In general, the supersaturation is small in a developed cloud with activated drops, $s_w \sim 10^{-3}$ to 10^{-1}%, and its effect on the drop freezing is insignificant. However, there can be at least two situations when the effect of s_w can be noticeable or large. It was hypothesized by Hobbs and Rangno (1990) and Rangno and Hobbs (1991, 2001) that s_w in a cloud may reach 5% or more after intensive coagulation and precipitation. Then, $\delta_s \Delta T \sim 6$ °C, which can cause a noticeable acceleration of freezing. In cloud chamber experiments, s_w can reach 10–25%; thus, $\delta_s \Delta T \sim 12$–30 K and may cause a significant acceleration of freezing. If the temperature in the chamber is sufficiently low (–20 to –30 °C), ΔT_{ef} may become lower than –40 °C and this may result in homogeneous drop freezing. This should be taken into account in analyses of drop freezing experiments at high humidity.

The critical energy ΔF_{cr} of cloud drop freezing and the nucleation rate $J_{f,hom}$ can be calculated using (Eqns. 8.6.20) and (8.6.21), respectively, with $H_{v,fr} = 0$ and $\sigma_{is} \approx \sigma_{iw}$. In subsequent sections, the discussion will be concentrated mostly on the submicron haze particles since the processes of their homogeneous freezing are especially important in the formation of upper tropospheric clouds at low temperatures. Applications for cloud drops are also discussed, taking into account their larger sizes, as well as the applications for bulk solutions, which require an appropriate modification of the equations.

8.7. Critical Freezing and Melting Temperatures of Homogeneous Freezing

8.7.1. General Expressions Based on Classical Theory

The critical or threshold temperatures and saturation ratios of homogeneous freezing depending on the nucleation rates can be derived from the two equations for the critical energy ΔF_{cr}. One expression follows from (Eqn. 8.3.5) for the freezing nucleation rate $J_{f,hom}$:

$$\Delta F_{cr} = -kT \ln \frac{J_{f,\mathrm{hom}}}{C_{\mathrm{hom}}} - \Delta F_{act}, \quad (8.7.1)$$

with C_{hom} defined in (Eqn. 8.3.5a). We use the other expression for ΔF_{cr} in the form (Eqn. 8.6.18):

$$\Delta F_{cr} = \frac{(16\pi/3)\sigma_{is}^3}{\left\{\rho_i L_m^{ef}\left[\ln\left(\frac{T_0}{T} S_w^{G_n}\right) - H_{v,fr}\right]\right\}^2}, \quad (8.7.2)$$

with $H_{v,fr}$ defined in (Eqn. 8.6.16). Solving (Eqn. 8.7.2) relative to the logarithmic term in the denominator, we obtain

$$\ln\left(\frac{T_0}{T}S_w^{G_n}\right) = H_{v,fr} + \frac{1}{\rho_i L_m^{ef}}\left[\frac{(16\pi/3)\sigma_{is}^3}{\Delta F_{cr}}\right]^{1/2}. \tag{8.7.3}$$

Substituting ΔF_{cr} here from (Eqn. 8.7.1) yields

$$\ln\left(\frac{T_0}{T}S_w^{G_n}\right) = H_{v,fr} + H_{f,\mathrm{hom}}, \tag{8.7.4}$$

where

$$H_{f,\mathrm{hom}} = \frac{1}{\rho_i L_m^{ef}}\left[\frac{-(16\pi/3)\sigma_{is}^3}{kT\ln(J_{f,\mathrm{hom}}/C_{\mathrm{hom}}) + \Delta F_{act}}\right]^{1/2}. \tag{8.7.5}$$

Equation (8.7.4) is a key relation for the threshold T and S_w, which allows expression of the critical or threshold temperature of homogeneous freezing $T_{f,hom}$ as a function of the saturation ratio, or lets us alternatively express the critical saturation ratio $S_{w,cr}$ as a function of the temperature.

Solving (Eqn. 8.7.4) for T, we obtain an equation for the critical temperature $T_{f,hom}$ of homogeneous freezing:

$$T_{f,\mathrm{hom}}(\Delta p) = T_0 S_w^{G_n}\exp\left(-H_{v,fr} - H_{f,\mathrm{hom}}\right). \tag{8.7.6}$$

In a more expanded form, substituting $H_{v,fr}$, $H_{f,hom}$, from (Eqns. 8.6.16) and (8.7.5) yields

$$T_{f,\mathrm{hom}}(J_{\mathrm{hom}}, S_w, r_{as}, \Delta p) = T_0 S_w^{G_n}\exp\left(-\frac{\Delta\rho\Delta p}{\rho_w\rho_i L_m^{ef}} - \frac{A_f}{r_{as}}\right)$$
$$\times\exp\left\{-\frac{1}{L_m^{ef}\rho_i}\left[-\frac{16\pi}{3}\frac{\sigma_{is}^3}{kT\ln(J_{f,\mathrm{hom}}/C_{\mathrm{hom}}) + \Delta F_{act}}\right]^{1/2}\right\}, \tag{8.7.7}$$

where A_f was defined in (Eqn. 8.6.15), $G_n = RT/(M_w L_m^{ef})$, and L_m^{ef}, ρ_i, C_{hom}, σ_{is}, and ΔF_{act} are functions of temperature and should be evaluated at the same values of $T_{f,hom}$ as on the left-hand side of (Eqn. 8.7.7). The solution to (Eqn. 8.7.7) requires an iteration procedure, but iterations converge rapidly as illustrated in the following text.

Note that the expression in square brackets contains a minus sign under the square root. Since the numerator σ_{is}^3 is positive, this means that the denominator should be negative, or $\ln(J_{f,hom}/C_{hom})$ should be negative and its absolute value should be sufficiently large to satisfy the condition $|\ln(J_{f,hom}/C_{hom})| > \Delta F_{act}$. This implies that the threshold or critical temperature exists if the freezing rate is large enough. Equation (8.7.7) shows that critical temperature depends on the freezing rate; thus, experiments with different cooling rates should measure different freezing temperatures, which has been found in various experiments (e.g., Cziczo and Abbatt, 1999; Martin, 2000; Kashchiev, 2000; Kashchiev et al., 2010).

A differential equation follows from (Eqn. 8.7.7):

$$\frac{dT_{f,\mathrm{hom}}}{dp} = -\frac{\Delta\rho T_{f,\mathrm{hom}}}{\rho_w\rho_i L_m^{ef}}, \tag{8.7.8}$$

which formally resembles the Clapeyron equation that is valid for the melting transition in a one-component system, but is derived here for freezing with a finite nucleation rate and a multi-component system. For the particular case of constant external pressure ($\Delta p = 0$) and large drops ($r_{as} \to \infty$), the first exponent on the right-hand side of (Eqn. 8.7.7) vanishes. For pure water drops, $S_w = 1$, and only the nucleation rate in the last exponent regulates $T_{f,hom}$. For solutions, $S_w < 1$, which leads to the lowering of $T_{f,hom}$ as shown in the next section.

For very slow freezing, (Eqn. 8.7.7) can be simplified. When $J_{f,hom} \to 0$, and $\ln(J_{f,hom}) \to -\infty$, the square root in the second exponent tends to zero, the second exponent tends to 1, and we obtain

$$T_{f,\mathrm{hom}}(\Delta p) = T_0 S_w^{G_n} \exp\left(-\frac{\Delta\rho\Delta p}{\rho_w \rho_i L_m^{ef}} - \frac{A_f}{r_{as}} \right). \tag{8.7.9}$$

Melting can be considered using a similar approach, whereby it is a reverse process with respect to freezing and can be described as a formation of a melting liquid germ on the surface of an ice crystal with radius r_{cr}. Thus, an approximate equation for the melting temperature is similar to (Eqn. 8.7.9) with replacement $r_{sa} \to r_{cr}$:

$$T_m(\Delta p) = T_0 S_w^{G_n} \exp\left(-\frac{\Delta\rho\Delta p}{\rho_w \rho_i L_m^{ef}} - \frac{A_f}{r_{cr}} \right). \tag{8.7.10}$$

The expressions (Eqns. 8.7.9) and (8.7.10) do not depend on J_{hom}, C_{hom}, owing to the very small nucleation rate. Equation (8.7.10) allows for a simple calculation of the melting point depression with increasing pressure. Differentiating (Eqn. 8.7.10) by p, we obtain

$$\frac{dT_m}{dp} = -\frac{\Delta\rho T_0}{\rho_i \rho_w L_m^{ef}} S_w^{G_n(T)} \exp\left[-\frac{1}{\rho_i L_m^{ef}} \left(\frac{\Delta\rho\Delta p}{\rho_w} + \frac{2\sigma_{ia}}{r_{cr}} \right) \right]. \tag{8.7.11}$$

For a particular case of bulk pure ice ($r_{cr} \to \infty$), (Eqn. 8.7.11) is simplified and converts into the generalization of the Clapeyron equation that takes into account solution effects:

$$\frac{dT_m}{dp} = -\frac{\Delta\rho T_0}{\rho_i \rho_w L_m^{ef}} S_w^{G(T)}. \tag{8.7.11a}$$

For pure water ($S_w = 1$), it converts into the Clapeyron equation:

$$\frac{dT_m}{dp} = -\frac{\Delta\rho T_0}{\rho_i \rho_w L_m^{ef}}, \tag{8.7.11b}$$

which is used to explain the dependence $T_m(p)$ near the triple point $T_0 = 273.15\,\mathrm{K}$ (see Chapter 4). Substituting here the values $L_m^{ef} = 79.7\,\mathrm{cal\,g^{-1}}$, $\rho_i = 0.92\,\mathrm{g\,cm^{-3}}$, and $\Delta\rho = 0.08\,\mathrm{g\,cm^{-3}}$ at the triple point $T_0 = 273.15\,\mathrm{K}$, we have $dT_m/dp \approx -1/138.3\,\mathrm{K\,atm^{-1}}$, close to the experimental values of $1/146.7\,\mathrm{K\,atm^{-1}}$ or $1/136\,\mathrm{K\,bar^{-1}}$ (Hobbs, 1974). The slopes may be different for solutions when the relative humidity in experiments is less than 100%. It is seen from (Eqn. 8.7.11a) that the slope dp/dT_m increases with decreasing S_w—i.e., the melting point shifts to lower temperatures at the same pressure but at a lower humidity or with increasing solution concentration.

8.7.2. Liquidus Curves

Equations (8.7.9) and (8.7.10) for $T_{f,hom}$, T_m at $\Delta p = 0$, large r_{as}, and r_{cr} or bulk solutions and very low nucleation rates J_{hom} are reduced to the following:

$$T_{f,\mathrm{hom}}(\Delta p = 0) = T_0 S_w^{G_n(T)} = T_0 S_w^{RT/M_w L_m^{ef}} \approx T_0 a_w^{RT/M_w L_m^{ef}} \tag{8.7.12a}$$

$$T_m(\Delta p = 0) = T_0 S_w^{G_n(T)} = T_0 S_w^{RT/M_w L_m^{ef}} \approx T_0 a_w^{RT/M_w L_m^{ef}}, \tag{8.7.12b}$$

which indicates the reversibility of slow freezing and melting. Equations (8.7.12a) and (8.7.12b) are nonlinear in the water saturation ratio or in the solution concentration, and are valid not only for dilute solutions but also for concentrated solutions. The curves $T_m(x_s)$ or $T_m(S_w)$ on the phase diagrams $T - x_s$ or $T - S_w$ are called *liquidus curves*, which separate the domain on the phase diagrams with a liquid solution located above the liquidus (also called *brine* in the case of seawater) from the domain with a mixture of liquid solution and ice that is located beneath the liquidus curve. (See also Chapter 11, Figs. 11.4 and 11.6).

At sufficiently small supersaturation or subsaturation, $|s_w| < 0.2$–0.3, we can expand (Eqn. 8.7.12b) into a Taylor series by $s_w = S_w - 1$. Keeping only the first three terms in s_w and using (Eqn. 8.6.15) for G_n, we have a cubic polynomial by s_w:

$$T_m \approx T_0\left[1 + G_n s_w + \frac{G_n(G_n - 1)}{2} s_w^2 + \frac{G_n(G_n - 1)(G_n - 2)}{6} s_w^3\right]$$

$$= T_0 + \Delta T_{m1} + \Delta T_{m2} + \Delta T_{m3}, \tag{8.7.12c}$$

$$\Delta T_{m1} = \frac{R_v T_0^2}{L_m} s_w, \quad \Delta T_{m2} = \frac{R_v T_0^2}{2L_m}\left(\frac{R_v T_0}{L_m} - 1\right) s_w^2, \tag{8.7.12d}$$

$$\Delta T_{m3} = \frac{R_v T_0^2}{6L_m}\left(\frac{R_v T_0}{L_m} - 1\right)\left(\frac{R_v T_0}{L_m} - 2\right) s_w^3. \tag{8.7.12e}$$

where R_v is the specific gas constant for water vapor.

Equations (8.7.12b)–(8.7.12e) describe the shift of the freezing or melting point temperature in a bulk solution due to the solute effects; ΔT_{m1}, ΔT_{m2}, and ΔT_{m3} are corrections of the first, second, and third order by supersaturation or solution concentration, respectively. Since $G_n \sim 0.4 < 1$ and $s_w < 0$ for solutions, (Eqn. 8.7.12c) shows that all these three corrections are negative. It is easy to see that ΔT_{m1} is equivalent to the traditional formulation for the freezing point depression ΔT_f in the bulk water solution that is derived usually by the equating chemical potentials of ice and solution (e.g., Pruppacher and Klett, 1997; Curry and Webster 1999). Using the value $G_n \approx 0.38$ near $T_0 = 273.15$ (see Fig. 8.14c), (Eqn. 8.7.12c) can be rewritten as

$$T_m \approx T_0 + 103.8 s_w - 32.18 s_w^2 + 17.37 s_w^3. \tag{8.7.12f}$$

For small s_w, using an approximation $s_w \approx \ln(1 + s_w) \approx \ln S_w \approx \ln a_w$, (Eqn. 8.7.12f) can be rewritten as

$$T_m \approx T_0 + 103.8 \ln a_w - 32.18(\ln a_w)^2 + 17.37(\ln a_w)^3. \tag{8.7.12g}$$

This functional form is similar to the empirical relation found by Chukin et al. (2010, 2012) by fitting the experimental data

$$T_m \approx T_0 + 103.6 \ln a_w + 15.613(\ln a_w)^2 + 54.118(\ln a_w)^3, \tag{8.7.12h}$$

that well approximates the function $T_m(a_w)$ (see Fig. 8.14c).

The liquidus curve determines the freezing or melting point depressions $\Delta T_f(x_s)$ or $\Delta T_m(x_s)$ as a function of solution concentration or equivalent S_w. The freezing point depressions are usually described by a linear function of solution molality, which is valid for dilute solutions. At higher concentrations, the $\Delta T_f(x_s)$ or $\Delta T_m(x_s)$ grow much faster than the solution concentrations, and a nonlinear description is required. Equations (8.7.12a–h) give such a nonlinear approximation, and (Eqn. 8.7.9)–(Eqn. 8.7.10) generalize it and take into account the pressure and droplet size effects. These equations describe liquidus curves or the freezing and melting point depressions with high accuracy over a wide range of solution concentration, as will be illustrated in Section 8.7.4.

For a particular case of a dilute solution, the saturation ratio S_w is related to the mole fraction x_s as $S_w \approx 1 - x_s$, and expanding (Eqn. 8.7.12a) into the power series, $T_{f,hom}$ can be presented in a linear approximation as

$$T_{f,\text{hom}} = T_0(1-x_s)^{G_n} T_0 - G_n T_0 x_s T_0 + m_l x_s,$$

$$\Delta T_f = T_0 - T_{f,\text{hom}} = -m_l x_s, \quad m_l = -R_v T_0^2 / L_m^{ef} = -103.1\,\text{K}, \tag{8.7.13}$$

where (Eqn. 8.6.15) is used for G_n, and m_l is a slope of the liquidus curve. Equation (8.7.13) is a known linear approximation for the liquidus or for the ideal freezing point depression (e.g., Pruppacher and Klett, 1997; Curry and Webster, 1999). Note that the difference between $T_{f,hom}$ and T_m for ice may be caused by the small values of the segregation coefficient k_s (the ratio of the amount of salt trapped in ice during freezing to the amount in the solution), which is typically ~10^{-6} to 10^{-4}. If a concentrated solution exhibits strong non-ideality, a correction may be required for (Eqn. 8.7.13).

8.7.3. Relation of the Freezing and Melting Point Depressions

The freezing and melting point depressions, $\Delta T_{f,hom}$ and ΔT_m, arise due to the presence of solutes. The relations between $\Delta T_{f,hom}$ and ΔT_m are often given as:

$$\Delta T_{f,\text{hom}} = \lambda \Delta T_m, \tag{8.7.14}$$

$$\Delta T_{f,\text{hom}} = T_0 - T_{f0} - T_{f,\text{hom}}, \quad \Delta T_m = T_0 - T_m, \tag{8.7.15}$$

where $T_0 = 273.15\,\text{K}$ and $T_{f0} = 38\,\text{K}$. Spontaneous freezing is usually assumed to start around −38 °C, or $T_0 - T_{f0} = 235\,\text{K}$, which is therefore a reference point for the freezing depression, and $T_{f,hom}$ and T_m are evaluated from experimental data on freezing and melting (e.g., Rasmussen and MacKenzie, 1972; Rasmussen, 1982; Bertram et al., 1996; Koop et al., 1998, 2000; Martin, 2000; DeMott, 2002; Kimizuka and Suzuki, 2007; Koop and Zobrist, 2009). The definition of λ from (Eqn. 8.7.14) is an empirical relation, and the value of $\Delta T_{f,hom}$ required for numerical cloud models or analysis of the laboratory experiments is taken usually from experimental data on ΔT_m. The mean value $\lambda = 1.7$ is often used (e.g., Sassen and Dodd, 1988, 1989; DeMott et al., 1994; Martin, 2000; Chen et al., 2000; Lin et al., 2002) as an average over the experimental data by Rasmussen (1982) on the relationship between depressions of the freezing and melting temperatures for a number of salts. As discussed in Section 8.4, it was clarified later that the coefficient $\lambda = 1.7$ is not universal, and can vary over the range 1.4–2.4 and may reach 3–5 for some organic substances, depending on the chemical composition and increasing with molecular weight and solute concentration (e.g., Kimizuka and Suzuki, 2007; Koop and Zobrist, 2009).

We can now calculate λ from the theory as:

$$\lambda(T) = \frac{T_0 - T_{f0} - T_{f,\text{hom}}(S_w, r)}{T_0 - T_m(S_w)}, \tag{8.7.16}$$

where $T_{f,hom}$ and T_m are evaluated from (Eqns. 8.7.7) and (8.7.12b). In a more extended form, it becomes

$$\lambda(T, S_w, r_{sa}, \Delta p, J_{f,\text{hom}})$$
$$= \left\{ T_0 - T_{f0} - T_0 S_w^{G_n} \exp(-A_p \Delta p - A_f / r_{sa}) \times \exp\left[-\frac{1}{L_m^{ef} \rho_i} \left(\frac{-(16\pi/3)\sigma_{is}^3}{kT \ln(J_{f,\text{hom}}/C_{\text{hom}}) + \Delta F_{act}} \right)^{1/2} \right] \right\}$$
$$\times [T_0 (1 - S_w^{RT/M_w L_m^{ef}})]^{-1}. \tag{8.7.16a}$$

This expression is a generalization of λ that accounts for the effects of temperature, solution concentration (S_w), pressure, finite droplet size, and finite nucleation rate. In particular, λ as a function of T can be different at different pressures, cooling rates, particles sizes, or solution concentrations. Equations (8.7.16) and (8.7.16a) allow determination of these dependencies and may explain the scatter of the experimentally derived values of λ.

The equations derived here based on classical nucleation theory suggest a direct relation between $T_{f,hom}$ and T_m using (Eqns. 8.7.7) and (8.7.10):

$$\frac{T_{f,\text{hom}}(J_{\text{hom}}, S_w, r_{as}, \Delta p)}{T_m(\Delta p)} = \exp\left(-\frac{A_f}{r_{as}} + \frac{A_f}{r_{cr}} \right)$$
$$\times \exp\left\{ -\frac{1}{L_m^{ef} \rho_i} \left[-\frac{16\pi}{3} \frac{\sigma_{is}^3}{kT \ln(J_{\text{hom}}/C_{\text{hom}}) + \Delta F_{act}} \right]^{1/2} \right\}. \tag{8.7.17}$$

This relation allows evaluation of $T_{f,hom}$ using measured or evaluated T_m. For the bulk solutions and large crystals, $r_{cr} \to \infty$, $r_{as} \to \infty$, the first parenthesis on the right-hand side vanishes.

8.7.4. Comparison with Observations

The equations derived in this section represent $T_{f,hom}$ and T_m as functions of S_w and the nucleation rates $J_{f,hom}$. These equations contain several functions of $T_{f,hom}$ or T_m on the right-hand side. They are transcendental algebraic equations and can be solved using an iterative procedure. Numerical tests show that two to three iterations provide a convergent solution. Calculation of $T_{f,hom}$ and T_m requires data on the temperature dependence of the melting heat $L_m(T)$ or $L_m^{ef}(T)$, ΔF_{act}, water and ice densities, and data on the temperature and composition dependence of the surface tensions.

Several sets of the thermodynamic parameters lead to calculated properties that agree with those observed, but these parameters should be mutually consistent. In the calculations described in this section, we use analytical functions $L_m^{ef}(T)$ and σ_{is} parameterized following Khvorostyanov and Curry (2004a) and based on extrapolation from Johari et al. (1994) as described in Section 4.4 and their appropriate parameterization and averaging. The values of σ_{is} were approximated with σ_{iw} and calculated from *Antonoff's rule*. Numerical tests show that the values of $T_{f,hom}$ are less sensitive to the ΔF_{act}. The choice of ΔF_{act} from Jeffery and Austin (1997) was satisfactory. In the calculations described here, the parameters are tuned to the data that are in the middle of that diverging range.

If the initial concentration of drops before freezing is N_0, then the concentrations of frozen N_f and unfrozen N_u drops are determined by the relations:

$$N_f = N_0 - N_u, \quad dN_f = -dN_u$$
$$dN_f = -N_u V J_{\mathrm{hom}} dt = (N_0 - N_f) V J_{\mathrm{hom}} dt. \tag{8.7.18a}$$

Integration with the initial condition $N_u(t = 0) = N_0$ yields $N_u = N_0 \exp\left(-VJ_{f,\mathrm{hom}} \tau_{fr}\right)$,

$$N_f = N_0[1 - \exp(-VJ_{f,\mathrm{hom}} \tau_{fr})] \tag{8.7.18b}$$

where $V = (4/3)\pi r^3$ is the drop volume, r is the droplet radius, and τ_{fr} is the characteristic "freezing time." We assumed in most calculations that $J_{f,hom}\, V\tau_{fr} = 1$—i.e., the freezing temperature $T_{f,hom}$ is defined by an e-folding decrease of N_0 during τ_{fr}. This assumption implies that the nucleation rate is determined by $J_{f,hom} = (V\tau_{fr})^{-1}$ or $\tau_{fr} = (J_{f,hom}V)^{-1}$. If we suppose that 99% of drops freeze during time τ, and 1% of N_0 are unfrozen, then we have a relation $N_u = 10^{-2} N_0 = N_0\exp(-VJ_{f,hom}\tau)$, or $\tau = -\ln(10^{-2})(1/VJ_{f,hom}) = 4.6(1/VJ_{f,hom})$. Substituting here $(J_{f,hom}V)^{-1} = \tau_{fr}$, we obtain $\tau = 4.6\tau_{fr}$; that is, (Eqn. 8.7.18b) along with the condition $J_{f,hom}\, V\tau_{fr} = 1$ means that 99% of drops freeze during $4.6\tau_{fr}$. In most calculations described in the following text, $\tau_{fr} = 1$ s, r was varied, and then $J_{f,hom}$ was evaluated.

For ease of comparing with observations, the results of the calculations are presented as functions of the water saturation ratio S_w and mass weighted concentration w. The relations between S_w, molality $\hat{M}$, and w were given in Section 3.5. The parameterizations of water activity a_w via $\hat{M}$ or w for ammonium sulfate were taken from Tang and Munkelwitz (1993), for sulfuric acid from Chen (1994), DeMott et al. (1997), and for NaCl from (Curry and Webster, 1999, and Millero, 1978).

The homogeneous freezing temperature $T_{f,hom}$ for $p = 1$ atm is presented in Fig. 8.6a,b. Calculations were performed using (Eqn. 8.7.7) for sulfuric acid ($M_s = 98$) and ammonium sulfate ($M_s = 132$), $\tau_{fr} = 1$ s, and for the solution drop radii $r = 5\,\mu$m and $0.2\,\mu$m. Fig. 8.6a,b shows that the values of $T_{f,hom}$ for pure water ($w = 0$, $S_w = 1$) are 231 K (−42 °C) for $r = 0.2\,\mu$m and 235.2 K (−38 °C) for $r = 5\,\mu$m—i.e., (Eqn. 8.7.7) predicts that the freezing point depression increases with a decreasing drop radius. Thus, (Eqn. 8.7.7) along with the fits for the parameters described earlier yields the well-known temperatures near −40 °C, referred to as the temperature of spontaneous freezing of pure water drops. Fig. 8.6a shows that calculated variations of $T_{f,hom}$ with solute concentration w are in good agreement with parameterizations of laboratory data from Koop et al. (1998) and Bertram et al. (2000) and are different for ammonium sulfate and sulfuric acid, depending on the chemical nature of the solute. However, Fig. 8.6b shows that the curves $T_{f,hom}(S_w)$ for $r = 5\,\mu$m plotted as functions of saturation ratio S_w almost coincide and merge with the experimental data and parameterizations from Koop et al. (1998), Bertram et al. (2000), and Larson and Swanson (2006) presented as functions of S_w (or a_w). That is, $T_{f,hom}(S_w)$ exhibits *colligative* properties and does not depend on the nature of the solute. This is a consequence of the dependence of r_{cr} and ΔF_{cr} on S_w, and not on w. Fig 8.6b shows that the decrease in r from $5\,\mu$m to $0.2\,\mu$m results in a decrease of $T_{f,hom}$ by 4 K, which is also in agreement with experimental data (Hagen et al., 1981; Young, 1993; Pruppacher and Klett, 1997). The curve $T_{f,hom}$ calculated for $r = 0.2\,\mu$m lies lower as a whole at each S_w, which can contribute to the residual difference for polydisperse drop spectra (seemingly non-colligative—i.e., depending not only on S_w). This is in agreement with some residual non-colligative features of freezing that were noticed in the experiments by Miyata et al. (2001).

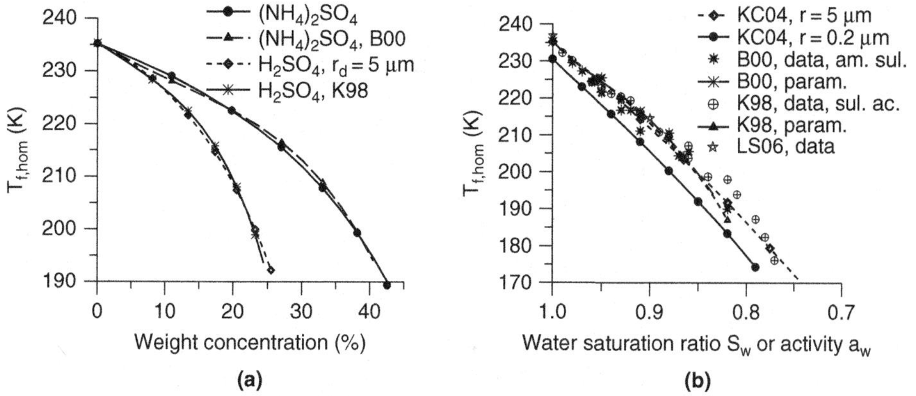

Figure 8.6. (a) Homogeneous freezing temperature $T_{f,hom}(w)$ as a function of the weight concentration w. Calculations using the theory from Khvorostyanov and Curry (2004a) using (Eqn. 8.7.7) for solution drops of 5 μm for ammonium sulfate (solid circles) compared to the parameterization from Bertram et al. (2000, B00) and calculations for sulfuric acid droplets of 5 μm (diamonds), compared to the parameterization from Koop et al. (1998, K98). (b). Homogeneous freezing temperatures $T_{f,hom}$ as functions of the water saturation ratio: calculated as in Khvorostyanov and Curry (2004a, KC04) from (Eqn. 8.7.7) as described in the text after a second T-iteration for the haze drop radius $r_d = 5\,\mu m$ (diamonds), and for $r_d = 0.2\,\mu m$ (solid circles), compared to the experimental data and parameterizations of $T_{f,hom}$ for ammonium sulfate of Bertram et al. (2000) (B00), of Larson and Swanson (2006) (LS06), and for sulfuric acid of Koop et al. (1998) (K98) plotted as a function of $S_w \approx a_w$.

This approach allows estimation of the dependence of freezing temperature on the drop radius, $T_{f,hom}(r)$, along with estimation of the lower temperature limit of freezing of water and determination of the factors that influence it. The upper solid line in Fig. 8.7 is calculated using (Eqn. 8.7.7) with a constant freezing time $\tau_{fr} = 1$ s, $p = 1$ atm, and $S_w = 1$ (pure water) for the drops with radii from 10Å to 5 mm (the drops with greater r are hydrodynamically unstable). The calculations are in good agreement with the experimental data by Mossop et al. (1955), Hagen et al. (1981), and DeMott and Rogers (1990), denoted by the symbols near the curve. The figure shows that the upper limit of freezing temperature does not rise above −30 °C even for the largest drops with $r \sim 5$ mm; that is, homogeneous freezing begins only at low temperatures.

At $r = 3.1 \times 10^{-6}$ cm (310Å), the curve reaches the limit $T_{f,th} = -45\,°C$ that was hypothesized to be the "stability limit," where the thermodynamic parameters tend to infinity, and below which liquid water does not exist (Speedy and Angell, 1976; Angell, 1982; Speedy, 1982; Pruppacher and Klett, 1997; see Chapter 4 here). At $r \approx 10$Å, close to the minimum possible drop size, $T_{f,hom} \approx -54\,°C$, well below the hypothesized stability limit. Recent measurements by Huang and Bartell (1995) found liquid drops with $r_d = 30$Å near −70 °C, and molecular dynamics simulations by Tanaka (1996) indicated the possible existence of liquid water down to $T = 193$ K (−80 °C), but the freezing time of such drops was very small. To further explore the lower limit of $T_{f,hom}$, we performed another series of calculations with a constant radius of 30Å (same as in Huang and Bartell) and a freezing time τ_{fr} decreasing from 1 s to 10^{-7} s (an almost vertical line with triangles in Fig. 8.7). The results show

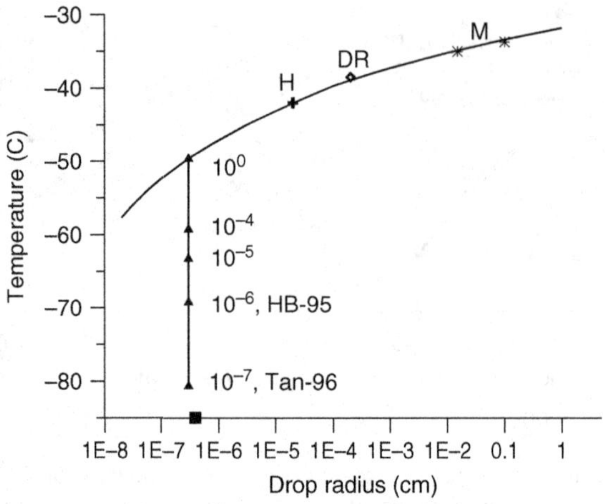

Figure 8.7. Homogeneous freezing temperature $T_{f,hom}$ of pure water drops ($S_w = 1$) as a function of drop radius calculated with (Eqn. 8.7.7) with the freezing time $\tau_{fr} = 1$ sec, $p = 1$ atm for the drop radii from 10Å to 5 cm (upper solid line) compared to the experimental data by Hagen et al. (1981) (H, cross), DeMott and Rogers (1990) (D, diamond), and Mossop (1955) (M, asterisks). Dotted vertical line with triangles shows decrease of T_{fhom} for a drop of radius 30Å (solid square) with decreasing freezing time τ_{fr}, indicated by the numbers near the curve ($\tau_{fr} = 10^0$ sec, 10^{-4} sec, etc.). $\tau_{fr} = 10^{-6}$ sec with $T_f \approx -70\,°C$ corresponds to the experiment by Huang and Bartell (1995, HB-95), and $\tau_{fr} = 10^{-7}$ sec with $T_f \approx -81\,°C$ corresponds to the molecular dynamics simulation of Tanaka (1996, Tan-96). These cases are indicated at corresponding τ_{fr}.

a decrease of $T_{f,hom}$ from −50 °C at $\tau_{fr} = 1$ s to $T_f = -81\,°C$ at $\tau_{fr} = 10^{-7}$ s. Thus, (Eqn. 8.7.7) shows that liquid drops could exist at very low temperatures but their lifetime τ_{fr} is very short, which is in agreement with the observations by Huang and Bartell (1995) and molecular dynamics simulations by Tanaka (1996). The calculations shown in Fig. 8.7 can explain both the older experimental data (for larger τ_{fr}) with the new data and concepts based on computer simulations by Tanaka (1996).

Note that the increase of the nucleation rate $J_{f,hom}$ or the decrease of the time τ_{fr} are not unlimited. Since the denominator in brackets of the second exponent in (Eqn. 8.7.7) should be negative (to yield a positive value under the square root), we have the following conditions for J_{hom} and τ_{fr}:

$$J_{\mathrm{hom}} \le C_{\mathrm{hom}} \exp(-\Delta F_{act}/kT), \tag{8.7.19}$$

which, along with assuming $J_{f,hom} V \tau_{fr} = 1$, yields

$$\tau_{fr} \ge (C_{\mathrm{hom}} V)^{-1} \exp(\Delta F_{act}/kT). \tag{8.7.20}$$

These conditions may determine the minimum (spinodal) temperature of homogeneous freezing.

Similar calculations with (Eqn. 8.7.12b) of the melting temperature T_m (liquidus curves) at $p = 1$ atm for ammonium sulfate, sulfuric acid, and NaCl are shown in Fig. 8.8a,b, again exhibiting a good agreement with experimental data. Note that these results do not depend on surface tension or droplet radius. Again, T_m as a function of mass weighted concentration w differs for the three indicated chemical species (Fig. 8.8a), but merges as a function of S_w, exhibiting a colligative property

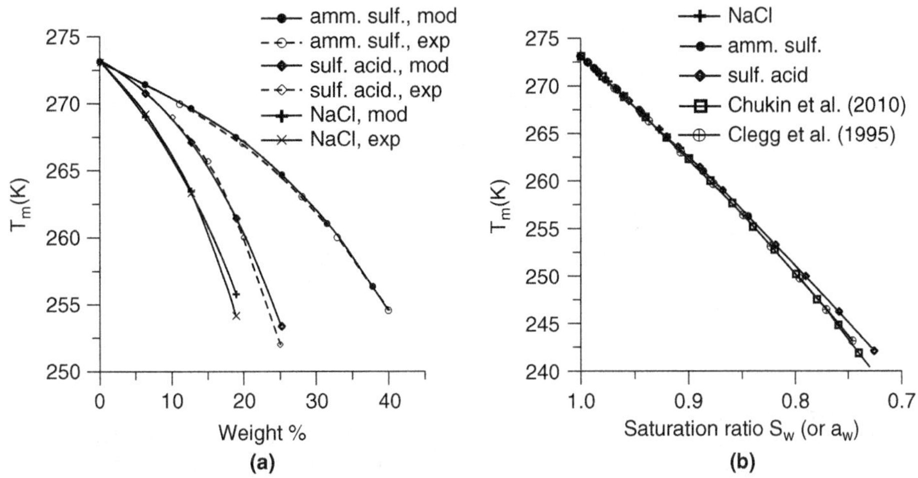

Figure 8.8. Melting temperatures T_m calculated with the KC04a model from (Eqn. 8.7.12b) for ammonium sulfate (closed circles), sulfuric acid (closed diamonds), and NaCl (crosses). (a) T_m calculated as functions of weight percent w (labeled "mod") compared to the corresponding experimental data (labeled "exp"); (b) the same $T_m(S_w)$ calculated with (Eqn. 8.7.12b) as in Fig 8.8a, but with weight percent recalculated to the saturation ratio S_w or water activity a_w and compared to the parameterizations from Clegg et all. (1995) and Chukin et al. (2010). Saturation ratio S_w is assumed to be equal to water activity, which is calculated for ammonium sulfate with Eqns. from Tang and Munkelvitz (1993), for sulfuric acid from Chen (1994) and DeMott et al. (1997). Experimental data on T_m for ammonium sulfate and sulfuric acid are from DeMott (2002). The experimental parameterization of $T_m(s)$ for NaCl as a function of salt concentration s (psu) is taken from Millero (1978) and Curry and Webster (1999) and recalculated from s to water activity. $T_m(S_w)$ in Fig. 8.8b is in good agreement with the experimental parameterization from Chukin et al. (2010) and with the thermodynamic model of composite solutions from Clegg et al. (1995).

of T_m (Fig. 8.8b). Thus, Fig. 8.8 shows that (Eqn. 8.7.12b) can be used with sufficient accuracy for calculations of T_m for standard pressure.

One of the most interesting and important applications of this theory is calculation of the freezing and melting point depressions, $\Delta T_{f,hom}$ and ΔT_m, and their ratio, λ, which is used in cloud models. Rasmussen (1982) found that measured values of $\Delta T_{f,hom}$ and ΔT_m for micron-sized drops for several substances exhibit a linear relation. Rasmussen hypothesized that this is in conflict with classical nucleation theory, and it could be explained by the relation of thermodynamic and kinetic processes and by the spinodal decomposition in ice formation of aqueous electrolytes. This conjecture has been challenged by several researchers. Here, we use (Eqns. 8.7.7) and (8.7.12b) to calculate $\Delta T_{f,hom}$ and ΔT_m for $r_d = 5\,\mu$m with the same parameters as used for Figs. 8.6 and 8.8, and then calculate λ with (Eqns. 8.7.16) and (8.7.16a) and compare with laboratory data from Rasmussen (1982) and DeMott (2002) and with the experimental parameterization for sulfuric acid from Koop et al. (1998) (Fig. 8.9).

To also illustrate the effect of drop radii, we have chosen the freezing threshold $T_{f0} = 38\,$K. As seen in Fig. 8.9a, the calculated freezing point depression exhibits a relation that looks quasi-linear on the $\Delta T_{f,hom} - \Delta T_m$ diagram. The experimental data for $T_{f,hom}$ lie mostly between the regression lines

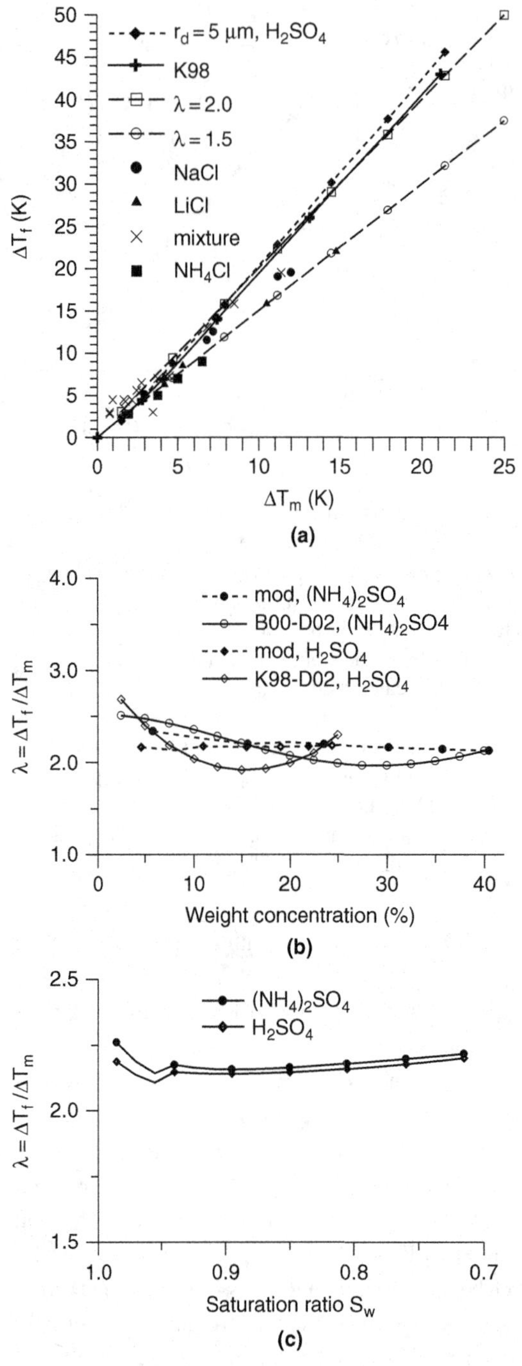

Figure 8.9. (a) Calculated relation $\Delta T_f = \lambda \Delta T_m$ for the 5 μm solution droplets of sulfuric acid (diamonds) compared to the experimental data by Koop et al. (1998) for sulfuric acid and to the experimental data for indicated substances compiled by DeMott (2002); correlation curves $\lambda = \Delta T_f / \Delta T_m = 1.5$ and $\lambda = 2$ are given for comparison.
(b) Parameter $\lambda(w)$ calculated with (Eqn. 8.7.16) for the drop radii of 5 μm as a function of weight concentration w for ammonium sulfate and sulfuric acid (denoted "mod") compared to λ calculated with the experimental $T_{f,hom}$ from Koop et al. (1998) and DeMott (2002) for sulfuric acid (K98-D02) and from Bertram et al. (2000) and DeMott (2002) (B00-D02) for ammonium sulfate, and with T_m shown in Fig. 8.8.
(c) $\lambda(S_w)$ calculated with (Eqn. 8.7.16) for the drop radii of 5 μm for ammonium sulfate and sulfuric acid, the same as in Fig. 8.9b, but as a function of the water saturation ratio S_w. From Khvorostyanov and Curry (2004a), Thermodynamic theory of freezing and melting of water and aqueous solutions. *J. Phys. Chem. A*, 108 (50), 11073–11085. Copyright 2004, American Chemical Society, reprinted with permission of the American Chemical Society.

$\lambda = 1.5$ and 2 up to $\Delta T_f \sim 10\,\text{K}$, and the curve calculated here for $r = 5\,\mu\text{m}$ matches the observations. For higher values of ΔT_f up to $\sim 45\,\text{K}$, experimental data from Koop et al. (1998) lie closer to $\lambda = 2$, and so does the calculated curve. We can draw the following conclusions from Fig. 8.9a: a) classical nucleation theory with the appropriate extension described here is capable of describing the quasi-linear relation between the freezing and melting point depressions and can be used in cloud models for evaluation of λ or directly for $T_{f,hom}$; b) a more detailed theoretical analysis should include kinetic simulation of the freezing of a polydisperse drop ensemble as it is done in some cloud models, since freezing proceeds from larger to smaller drop sizes and the values of $T_{f,hom}$ and λ determined in experiments may depend on the fraction and size of frozen drops.

Direct calculation of the parameter $\lambda = \Delta T_f / \Delta T_m$ with (Eqn. 8.7.16) (Fig. 8.9b,c) shows that it is not a constant since it depends on the chemical composition and S_w. The calculated $\lambda(w)$ for ammonium sulfate and sulfuric acid (Fig. 8.9b) depends on the chemical composition, but the curves are relatively close to each other and to the curves derived from laboratory data (Koop et al., 1998; Bertram, et al., 2000). $\lambda(w)$ varies mostly between 1.9–2.2 in agreement with the previous analysis of experimental data for these substances. When plotted as $\lambda(S_w)$ (Fig. 8.9c), the curves almost merge, exhibiting again colligative properties as both $\Delta T_{f,hom}$ and ΔT_m (Figs. 8.6b and 8.8b). The values $\lambda(S_w)$ are not constant but exhibit a monotonous growth with decreasing S_w, and use of constant λ in cloud models may lead to errors in $\Delta T_{f,hom}$. The difference slightly increases at small w or $S_w \to 1$, but both calculations and measurements become less reliable at small solute concentrations (or $S_w \to 1$) since both $\Delta T_{f,hom}$ and ΔT_m decrease, and even a small error in each of these quantities may lead to an increasing error in their ratio λ.

Kimizuka and Suzuki (2007) tested many various chemical species and found that a good linear relationship $\Delta T_{f,hom}(S_w) = \lambda \Delta T_m(S_w)$ was observed for all samples. They noted that this relationship and the slopes λ were in good agreement with predictions from the Khvorostyanov and Curry (2004a) theory for the substances with sufficiently low molar mass. Recent studies showed that λ can generally increase with increasing molar mass and may reach values of 3.5–5.1. Kimizuka and Suzuki (2007) proposed that this large variation in λ is due to the dependence of λ on the solute's diffusion coefficient, which in turn depends on the solute's molar mass. One possible way of describing this effect within classical nucleation theory could be based on an appropriate account for the surface tension and other thermodynamic parameters on the solute chemical composition.

Calculations with varying r (not shown here) demonstrate that λ also may depend on drop size, since T_{f0} increases with decreasing drop size r (recall, the threshold temperature $T_{f0} \approx 38\,\text{K}$ for $r = 5\,\mu\text{m}$ and 42 K for $0.2\,\mu\text{m}$, see Fig. 8.6), and no single value is representative for the polydisperse ensemble of drops. So, different thresholds T_{f0} should be chosen for various drop radii, which illustrates a problem in the analysis of experimental data obtained with polydisperse drops and in the evaluation of λ using the theory as it is used here.

8.7.5. Equivalence of Solution and Pressure Effects

Calculations of freezing and melting point temperatures under conditions of variable pressure are sensitive to variations in the density of liquid and solid water, $\rho_w(T, p)$ and $\rho_i(T, p)$. Here we use a parameterization from Pruppacher and Klett (1997) and results from Jeffery and Austin (1997, 1999) and Holten et al. (2011, 2012) for $\rho_i(T)$ of hexagonal ice I_h (see Chapter 4). The data on $\rho_w(T, p)$ below $T = -40\,°C$ are scarce and $\rho_w(T, p)$ can be taken or extrapolated from the equation of state for liquid water (e.g., Poole et al., 1994; Jeffery and Austin, 1997, 1999; Kiselev and Ely, 2002), molecular dynamics simulations (Tse, 1992, 1999; Poole et al., 1994), or from experimental data where possible (Mishima and Stanley, 1998) (see Chapter 4). We have used available data and (Eqn. 8.7.8) to estimate variations of $\rho_w(T, p)$ from the slopes $dT_{f,hom}/dp$ in the experimental data of Kanno and Angell (1977) for $T_{f,hom}$. We have from (Eqn. 8.7.8) for $\Delta\rho = \rho_w - \rho_i$

$$\Delta\rho \approx -\frac{\rho_w \rho_i L_m}{T_{f,\text{hom}}}\left(\frac{dT_{f,\text{hom}}}{dp}\right), \quad (8.7.21)$$

The results described in the following were obtained using the values of $\Delta\rho$ given in these works and values estimated from (Eqn. 8.7.21) using the corresponding data from the cited works.

Equation (8.7.7) includes the simultaneous effects of pressure and chemical composition (solute concentration described by $S_w < 1$) on $T_{f,hom}(\Delta p)$. The effects of composition at p = const were described in the previous section, and now Fig. 8.10 presents freezing temperatures $T_{f,hom}(p)$ calculated with (Eqn. 8.7.7) as a function of pressure for pure water ($S_w = 1$). A comparison of theoretical and experimental results in Fig. 8.10 exhibits good agreement over the entire temperature and pressure ranges, indicating the validity of (Eqn. 8.7.7). Note that the experimental data show the change of the sign of the slope at $p \sim 2$ kb. This is caused by the nucleation of ice III, which begins

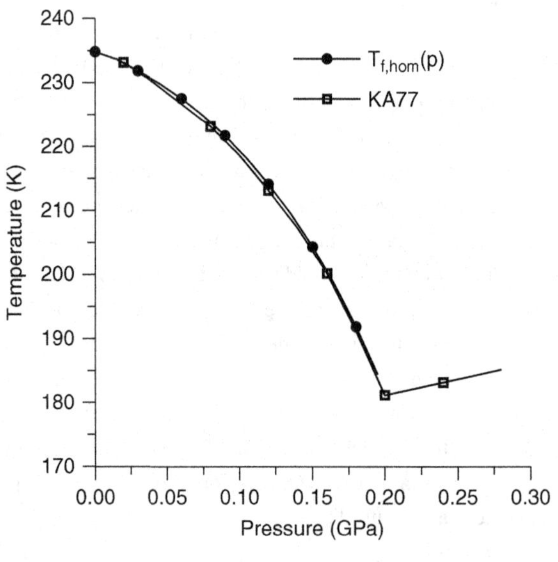

Figure 8.10. Homogeneous freezing nucleation temperatures $T_{f,hom}$ calculated with (Eqn. 8.7.7) for pure water ($S_w = 1$) as a function of pressure (solid circles) and compared to the experimental data from Kanno, and Angell (1977) (KA77). From Khvorostyanov and Curry (2004a). Thermodynamic theory of freezing and melting of water and aqueous solutions. *J. Phys. Chem. A*, 108 (50), 11073–11085. Copyright 2004, American Chemical Society. Reprinted with permission of the American Chemical Society.

at p ~ 1.8 kb and prevails at pressures higher than 2 kb (Kanno and Angell, 1977; see Section 4.1 here); a possible relatively wide temperature range of coexistence of ice I and ice III at p = 1 atm was explained by Johari (1998) by the polydisperse structure of finely dispersed ices. This transition in Fig. 8.10 at p = 2 kb is explained simply by (Eqns. 8.7.7), (8.7.8), and (8.7.21): Since the pressure dependence of freezing temperature is determined by the difference $(\rho_w - \rho_i)$ and the density of ice III is greater than that of ice I_h (Hobbs, 1974), the slopes $dT_{f,hom}/dp$ may vary and change sign at abrupt changes of the ice densities. A more detailed analysis of this effect requires more precise data on ρ_i, ρ_w in this pressure range.

Similar calculations were performed for pressure-induced melting ice temperature T_m using (Eqn. 8.7.10) for bulk ice ($r_{cr} = \infty$) and $S_w = 1$ (pure water). The results are compared in Fig. 8.11 with experimental data (Wagner et al., 1994; Mishima, 1996; Mishima and Stanley, 1998). One can see that the experimental curve from (Mishima, 1996; Mishima and Stanley, 1998) has two distinct branches and the slopes change at p ~ 0.5 GPa, which also is explained in these works by the change near this point of the ice type from hexagonal I_h to the other types of ice with higher densities (ice III or ice V). Fig. 8.11 shows that the values of T_m calculated from (Eqn. 8.7.10) are very close to both experimental curves up to p = 0.5 GPa, but the difference between the calculated and observed values increases at higher pressures if we are to keep $\rho_i(T)$ = const. Agreement becomes a little better with

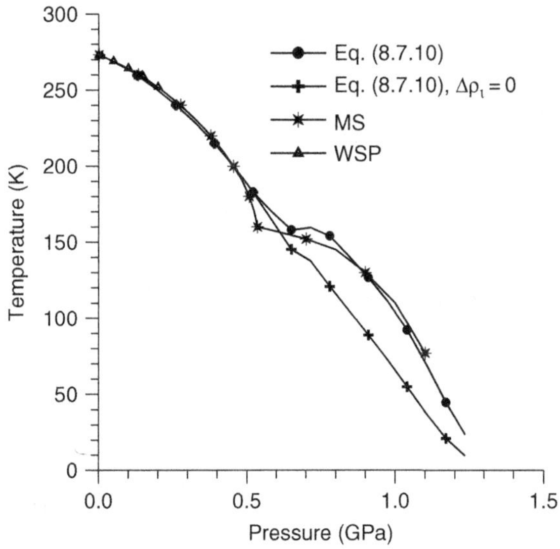

Figure 8.11. The melting temperature for the pressure-induced melting of ice calculated using (Eqn. 8.7.10) with $\rho_i(T)$ = const ($\Delta\rho_i = 0$, crosses) and with a linear increase of ice density $\rho_i(T)$ by 0.04 g cm^{-3} in the range p = 0.6–0.8 GPa (solid circles) compared to the experimental data of Mishima (1996) and Mishima and Stanley (1998) down to ~80 K (MS) and to Wagner, Saul, and Pruss (1994) down to 251 K (triangles, WSP). Reprinted with permission of the American Chemical Society from Khvorostyanov and Curry (2004a). Thermodynamic theory of freezing and melting of water and aqueous solutions. *J. Phys. Chem. A*, **108** (50), 11073–11085. Copyright 2004, American Chemical Society.

a linear increase in $\rho_i(T)$ by 0.04 g cm^{-3} over the range $p = 0.6 - 0.8$ GPa (this could imitate admixture of the other denser ice III or ice V); however, some discrepancy still remains. Computer lattice dynamics simulations (Tse, 1994; Tse et al., 1999) determined that melting is caused by thermodynamic instability up to $p \sim 0.5$ GPa and by mechanical instability at higher pressures. The results shown in Fig. 8.11 are consistent with this conclusion, satisfactorily describing the upper branch of the T_m curve to $p \sim 0.5$ GPa but worsening for higher p values. Thus, (Eqn. 8.7.10) can serve for simple calculations of $T_m(p)$ for pure water up to $p = 0.5$ GPa; its predictions for pressure-induced melting taking into account the solutions ($S_w < 1$) could be verified experimentally.

As discussed in Chapter 4, an interesting question on the equivalence of the pressure and solution effects on freezing and melting temperatures was explained by the similar effect of solution and applied pressure on the hydrogen bonding network (Kanno and Angell, 1977; Leberman and Soper, 1995), and by showing that the plots of $T_f(a_w)$ and $T_f(p)$ are similar. This effect was described by introducing the "effective" solution concentration (Koop et al., 2000). A simple but sufficiently general quantitative expression for this equivalence was found by Khvorostyanov and Curry (2004a) from (Eqns. 8.6.14a), (8.6.14b) or (8.6.17b) for $r_{cr}(T, S_w, p)$, and is described in the following text. It shows that variations in the solution concentration (or in S_w) may be viewed as equivalent to pressure variations. Recalculations from the solution molality to S_w and back can be done with (Eqns. 8.6.33) and (8.6.34). It is seen from (Eqn. 8.6.17b) that this pressure–solution equivalence can be expressed as $S_w^{G_n} = \exp(-H_{v,fr})$, which, using (Eqn. 8.6.16) for $H_{v,fr}$, can be written as

$$S_w^{G_n} = \exp\left(-\frac{\Delta\rho\Delta p}{\rho_i \rho_w L_m^{ef}} - \frac{2\sigma_{sa}}{\rho_i L_m^{ef} r_d} \right). \tag{8.7.22}$$

Equation (8.7.22) expresses S_w (or a corresponding molality) as a function of equivalent pressure Δp or vice versa. Solving relative to Δp, we obtain

$$\Delta p = -\frac{RT\rho_i\rho_w}{M_w(\rho_w - \rho_i)} \ln S_w - \frac{2\sigma_{sa}}{r_d}\frac{\rho_w}{\rho_w - \rho_i} = -Q_{\rho T} \ln S_w + \Delta p_{cur}, \tag{8.7.23}$$

where $Q_{\rho T}$ is the function of T, ρ_w, ρ_i, and Δp_{cur} is the correction to pressure due to curvature of a freezing drop with radius r_d:

$$Q_{\rho T}(T, \rho_w, \rho_i) = \frac{RT\rho_i\rho_w}{M_w(\rho_w - \rho_i)}, \qquad \Delta p_{cur} = -\frac{2\sigma_{sa}}{r_d}\frac{\rho_w}{\rho_w - \rho_i}. \tag{8.7.23a}$$

Equations (8.7.23) and (8.7.23a) were derived in Khvorostyanov and Curry (2004a) and are generalized here by adding the term Δp_{cur}. Equations (8.7.23)–(8.7.23a) show that a decrease in S_w or a_w (an increase in solution molality) is equivalent to an increase in Δp, with proportionality determined by the function $Q_{\rho T}$ that depends on the densities and temperature. The proportionality is $\Delta p \sim -T \ln S_w$ with the constant densities, although they in turn depend on p, T. The value of $Q_{\rho T}$ is very large, an estimate at $T \sim -60\,°\mathrm{C}$ gives $Q_{\rho T} \sim 0.9$ GPa $= 0.9 \times 10^4$ atm, and $Q_{\rho T}$ further increases with decreasing T. Equation (8.7.23) shows that a solution with saturation ratio $S_w = 0.9$ ($\ln S_w \approx -0.1$) is equivalent to an applied external pressure $\Delta p \approx Q_{\rho T} \ln S_w \approx -0.9 \times 10^4 \times (-0.1)$ atm $= 0.9 \times 10^3$ atm ≈ 0.9 kbar at $T \sim -60\,°\mathrm{C}$. The curvature term Δp_{cur} is generally smaller than the first term with $Q_{\rho T}$ or osmotic pressure Π; an estimate yields $\Delta p_{cur} \sim -10$ atm for a drop with $r = 1.5\,\mu$m and -100 atm for $r = 0.15\,\mu$m.

Its sign is opposite the sign of the term with $-\ln(S_w)$, and partially compensates for the solution effect. For the smaller drops, the derivation should be refined accounting for the dependence of the surface tension on the drop radius. The term Δp_{cur} vanishes for the bulk solutions ($r = \infty$).

Thus, (Eqns. 8.7.22)–(8.7.23a) establish the equivalence of the deformation of the hydrogen bonds network by the chemical forces due to solutes and mechanical pressure. This is illustrated in Fig. 8.12, which presents the $\Delta p - S_w$ relation calculated with (Eqn. 8.7.23) in the same way as the previous two figures. For comparison, we also plotted the two experimental points from Kanno and Angel (1977, KA77) who presented the data in units $R = 55.5/\hat{M}$, with $\hat{M}$ being the molality. We recalculated from R to the molalities, and then to S_w for NaCl, and the two experimental points from KA77 are shown in Fig. 8.12, which shows the equivalence of $p = 1000\,\text{bar}$ (0.1 GPa) to the molality of NaCl $\hat{M} = 2.75$ $(R = 20)$ and $p = 1500\,\text{bar}$ (0.15 GPa) to the molality $\hat{M} = 4.65$ $(R = 12)$ for both freezing and melting temperatures. Fig. 8.12 shows good agreement of the curve calculated with (Eqn. 8.7.23) with the experimental data and confirms the validity of these equations, which therefore can be used for the prediction of the pressure effects on $T_{f,hom}$, T_m given the solution effects, or vice versa.

This agreement and thermodynamic equivalence of solution (or S_w) and pressure effects leads to an important conclusion. Kanno and Angell (1977), Leberman and Soper (1995), Koop et al. (2000),

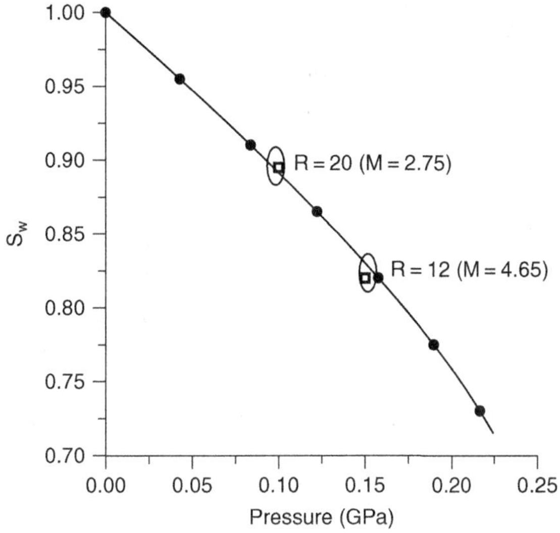

Figure 8.12. Equivalence of pressure and saturation ratio or molality expressed with (Eqns. 8.7.23) and (8.7.23a). The curve with solid circles is calculated with (Eqn. 8.7.23) here, and two points (squares) are experimental data for NaCl from Kanno and Angell (1977) expressed in that work in terms of the ratio $R = 55.5/\hat{M}$, with $\hat{M}$ being the molality. These two points from Kanno and Angell (1977) establish the equivalence of $p = 1.0\,\text{kbar}$ (0.1 GPa) to $R \approx 20$ ($\hat{M} = 55.5/R = 2.75$) and $p = 1.5\,\text{kbar}$ (0.15 GPa) to $R \approx 12$ ($\hat{M} = 4.65$). The experimental points lie almost on the theoretical curve (denoted by ellipses) that confirms the validity of (Eqn. 8.7.23). Reprinted with permission of the American Chemical Society from Khvorostyanov and Curry (2004a). Thermodynamic theory of freezing and melting of water and aqueous solutions. *J. Phys. Chem. A*, **108** (50), 11073–11085. Copyright 2004, American Chemical Society.

and others interpreted this equivalence at the molecular level in terms of the distortion of the hydrogen bonds and quantitative relationships between the solution and pressure effects for required calculations like molecular dynamics simulations by specifying the intermolecular potentials, or similar methods. The consideration in Khvorostyanov and Curry (2004a) and this book shows that this equivalence can be interpreted and evaluated more simply, without considering the molecular effects, from the general bulk thermodynamic relations based on the new expressions for the critical radius in Section 8.6.

Equation (8.7.7) includes the simultaneous effects of pressure and chemical composition on $T_{f,hom}(\Delta p)$. Fig. 8.13 presents freezing temperatures $T_{f,hom}(\Delta p)$ as a function of pressure for pure water and three solutions of NaCl and LiCl calculated from (Eqn. 8.7.7) for varying concentrations in terms of R units as in Kanno and Angell (1977), for R = 55.5 (molality $\hat{M} = 1$), R = 20 ($\hat{M} = 2.775$), and R = 12 ($\hat{M} = 4.625$). A comparison of theoretical and experimental results in Fig. 8.13 exhibits generally good agreement with observations for the corresponding molalities over the entire temperature and pressure ranges, and supports the validity of (Eqn. 8.7.7); that is, both dependencies on pressure and solution concentration correctly describe the observations.

The experimental data from Kanno and Angell (1977) in Fig. 8.13 show an abrupt change of slopes, labeled as the points T_g, which was interpreted in Kanno and Angell (1977) as the temperature of glassy transition points or regions at pressures $p_g = 0.14 - 0.2\,\mathrm{GPa}$ and $T_g = -95$ to $-115\,°\mathrm{C}$. For $p > p_g$, the slope changes the sign, $dT_{f,hom}/dp$ becomes positive, and we see a second branch with a linear increase of $T_{f,hom}$ with increasing p at a constant gradient. The term in (Eqn. 8.7.7) that determines the change in slope arises from the $\exp(-X_p)$, where X_p is

$$X_p = -\frac{A_p \Delta p}{\rho_i L_m^{ef}} = -\frac{\Delta\rho \cdot \Delta p}{\rho_w \rho_i L_m^{ef}}. \tag{8.7.24}$$

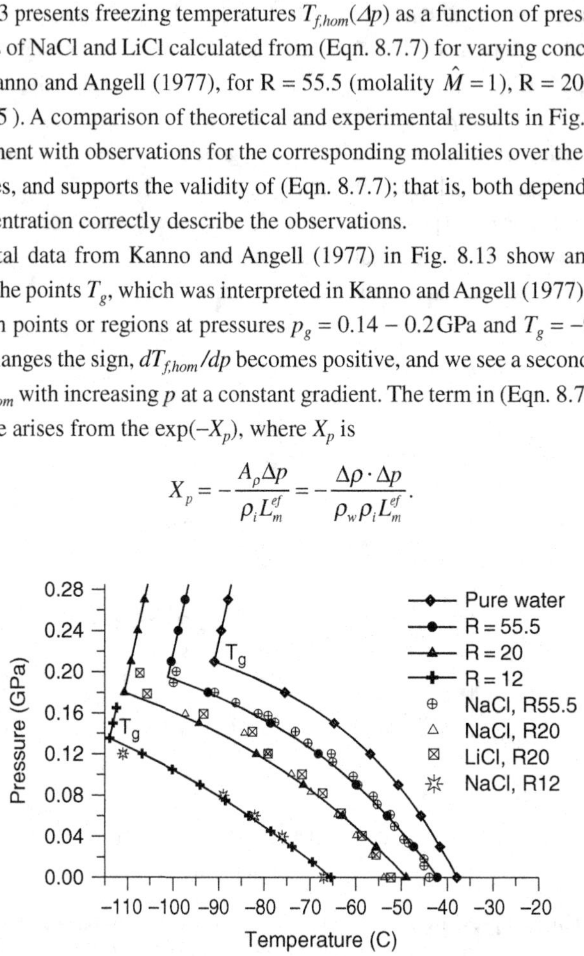

Figure 8.13. Homogeneous freezing temperatures $T_{f,hom}(p, \hat{M})$ calculated with (Eqn. 8.7.7) for pure water (diamonds) and solution drops as the functions of pressure p and various molalities $\hat{M}$ of NaCl. The results are presented in the form similar to Fig. 3 in Kanno and Angell (1977) in terms of R units as in that work (R = moles water/moles salt = $55.5/\hat{M}$): for R = 55.5 ($\hat{M} = 1$, solid circles), R = 20 ($\hat{M} = 2.775$, triangles), and R = 12 ($\hat{M} = 4.625$, crosses). These calculations are compared to the four sets of experimental data for the same R and $\hat{M}$ from Kanno and Angell (1977) kindly provided by Prof. Hitoshi Kanno and shown with open symbols for NaCl with R = 55.5 (open circles), R = 20 (triangles), and R = 12 (asterisks), and LiCl, R = 20 (squares). T_g are the temperatures of glassy transition, as described in the text.

Equation (8.7.21) for $\Delta\rho$ shows that the decrease of $T_{f,hom}$ or the negative gradient $dT_f/dp < 0$ at $p < p_g$ means that $\Delta\rho = \rho_w - \rho_i > 0$—i.e., the density of ice is less than the water density. However, (Eqns. 8.7.21) and (8.7.24) imply that $\Delta\rho$ changes its sign at $p > p_g$, along with the sign of the gradient $dT_{f,hom}/dp$. Since the solution or water density is the same at $T = T_g$, slightly below and above p_g, we see that the ice density increases in a step-like manner in the vicinity of T_g, and the sign change of $\Delta\rho$ arises solely due to the change in ice density. This behavior can be explained by the formation of the "glassy" ice germs with increased density in water or solution drops at $p > p_g$. At these high pressures, critical radii and free energies of the germs are described by the same (Eqns. 8.6.17a), (8.6.17b), and (8.6.18) and the freezing temperature by (Eqn. 8.7.7); however, the increase of $T_{f,hom}$ with p requires that X in (Eqn. 8.7.24) becomes positive at $p > p_g$, $T > T_g$.

Denoting the density difference as $\Delta\rho_c = \rho_w - \rho_{ic}$ at $T < T_g$ (before the glass transition) and $\Delta\rho_g = \rho_w - \rho_{hda}$ at $T > T_g$ (after transition) with ρ_{ic} and ρ_{hda} being ice densities in crystalline I_h and in high-density amorphous (HDA) states (see Chapter 4), and assuming that $\rho_w = \text{const}$ at the glassy transition, we obtain from (Eqn. 8.7.21) the ice density increase $\Delta\rho_{ice}$ as a difference of $\Delta\rho_c$ and $\Delta\rho_g$ below and above p_g:

$$\Delta\rho_{ice} = \Delta\rho_c - \Delta\rho_g \approx \frac{\rho_i \rho_w L_m}{T_g}\left[\left.\frac{dT_f}{dp}\right|_{Tg+} - \left.\frac{dT_f}{dp}\right|_{Tg-}\right], \tag{8.7.25}$$

where the subscripts T_{g-} and T_{g+} denote derivatives just before and just after T_g, p_g.

Hence, analysis of the experimental data from Kanno and Angell (1977) in the context of (Eqns. 8.7.21) and (8.7.25) implies that: 1) the sign of the ice density jump is positive—i.e., the density of ice increases abruptly at the point or region of glassy transition; 2) since the experimental gradients $dT_{f,hom}/dp \approx \text{const}$ at $p > p_g$, $dT_{f,hom} > T_g$, the value $\Delta\rho_{ice}$ does not depend here on pressure, and the density of amorphous ice is nearly constant in some region of p, T after the glassy transition; and 3) the increase in ice density is 20–25% at the transition point and yields $\rho_{hda} \sim 1.17$–$1.22\,\text{g cm}^{-3}$. These consequences are in agreement with current views on this type of glassy transition (Mishima and Stanley, 1998; Stanley et al., 2007). The pressure Δp equivalent to the solution according to (Eqns. 8.7.23)–(8.7.23a) is proportional to $\Delta\rho = \rho_w - \rho_i$. This quantity changes signs and becomes negative after the glassy transition. Therefore, the pressure equivalent to the solution becomes negative.

The theory described here has certain limitations. It is based on general thermodynamics, does not consider the properties of crystalline lattices and cannot explain why formation of amorphous ice is preferred rather than more conventional forms of ice. This theory also does not predict the transition point T_g, p_g itself, since the symmetry of a phase is not accounted for in this approach. However, this theory based on thermodynamic expressions for the critical radii and energies explains and describes many features of homogeneous freezing and melting without invoking more complicated methods or empirical parameterizations with many unknown parameters that require fitting, or time consuming molecular dynamics simulations with various potentials.

8.8. Threshold or Critical Saturation Ratios for Homogeneous Freezing

Knowledge of the critical (or threshold) humidities or saturation ratios for the onset of ice nucleation is crucial for parameterization in cloud and climate models, where simulations of mixed phase and crystalline clouds depend critically on the choice of the threshold relative humidity. Critical humidities over water and ice, or the corresponding saturation ratios, for homogeneous ice nucleation at cold temperatures below –40 °C in cirrus clouds have been studied and parameterized using results of parcel model simulations (Sassen and Dodd, 1988, 1989; Sassen and Benson, 2000; Lin et al., 2002; Barahona and Nenes, 2008), data collected in aircraft measurements of cirrus or wave clouds (Heymsfield and Miloshevich, 1995; Heymsfield et al., 1998), extensions of classical homogeneous ice nucleation theory (Khvorostyanov and Sassen, 1998, 2002; Khvorostyanov and Curry, 2000, 2004a,b, 2005a, 2009b, 2012; Curry and Khvorostyanov, 2012), and data from laboratory measurements (e.g., DeMott et al., 1998, 2002; Koop et al., 1998, 1999, 2000; Bertram et al., 2000; Martin, 2000; Möhler et al., 2006; Hoose and Möhler, 2012).

Until recently, classical ice nucleation theory has not yielded simple equations for the critical humidity because the equations for the critical radius and energy of freezing in both modes written in the form derived by J. J. Thomson (1888) depend on temperature only and do not include humidity dependence. Although the more detailed differential entropy equations outlined in classical theory in principle can account for the humidity or activity dependence (Dufour and Defay, 1963, Defay et al., 1966; Pruppacher and Klett, 1997), these equations can only be solved numerically. A generalization of classical theory for both homogeneous and heterogeneous ice nucleation presented here provides analytical expressions for the critical radius and energy of an ice embryo and nucleation rates as functions of temperature, the water saturation ratio, pressure, and the finite size of freezing particles simultaneously. This allows to invert these equations and express the critical water and ice saturation ratios or humidities as the functions of the corresponding variables.

In this section, we consider the critical humidities and saturation ratios for homogeneous ice nucleation. Generalized equations are derived that express the critical saturation ratios or relative humidities as functions of temperature, nucleation rate (or cooling rate), the size of the freezing particles, and environmental pressure. Variations of the critical humidities with varying input parameters are illustrated with an extensive series of calculations.

8.8.1. General Equations

The critical saturation ratios and humidities for homogeneous freezing can be derived from (Eqn. 8.7.3), which accounts for the dependence on the nucleation rates $J_{f,hom}$ given by (Eqn. 8.3.5). Transforming (Eqn. 8.7.4),

$$\frac{T_0}{T} S_w^{G_n} = \exp(H_{v,fr} + H_{f,\mathrm{hom}}), \tag{8.8.1}$$

and solving it for S_w, we obtain the critical or threshold water saturation ratio of homogeneous freezing:

$$S_{w,cr}^{\mathrm{hom}} = [(T/T_0)\exp(H_{v,fr} + H_{f,\mathrm{hom}})]^{1/G_n}, \tag{8.8.2}$$

where $H_{v,fr}$ is defined in (Eqn. 8.6.16) and $H_{f,hom}$ is defined in (Eqn. 8.7.5). In a more complete form,

$$S_{w,cr}^{\mathrm{hom}} = \left\{ \left(\frac{T}{T_0}\right) \exp\left(\frac{\Delta\rho\Delta p}{\rho_w \rho_i L_m^{ef}} + \frac{r_{fr}}{r_{as}}\right) \right.$$
$$\left. \times \exp\left[\frac{1}{L_m^{ef}\rho_i}\left(\frac{(-16\pi/3)\sigma_{is}^3}{kT\ln(J_{f,\mathrm{hom}}/C_{f,\mathrm{hom}}) + \Delta F_{act}}\right)^{1/2}\right]\right\}^{M_w L_m^{ef}/RT}, \tag{8.8.3}$$

where $\Delta\rho = \rho_w - \rho_i$, r_{as} is the radius of a solution drop, and r_{fr} was defined in (Eqn. 8.6.15). Equation (8.8.3) expresses $S_{w,cr}^{\text{hom}}$ as a function of T, r_{as}, r_{fr}, Δp, and nucleation rate $J_{f,hom}$. We can prescribe $J_{f,hom}$, and then calculate $S_{w,cr}^{\text{hom}}$ as a function of T and other parameters. In practice, it is more convenient to prescribe $J_{f,hom}V$, where $V = (3\pi/4)r_d^3$ is the volume of an aerosol particle.

8.8.2. Parameterization of Effective Melting Heat

As discussed in Chapter 4, there are still substantial uncertainties in the literature on the values of melting heat $L_m(T)$, especially in solutions for $T < -45\,°\text{C}$ because direct measurements are difficult at these temperatures. The functions $L_m(T)$ may substantially differ in various sources (e.g., Dufour and Defay, 1963; Defay et al., 1966; Hobbs, 1974; Johari et al., 1994; Pruppacher and Klett, 1997; Jeffery and Austin, 1997, 1999; Murphy and Koop, 2005; Feistel and Wagner, 2005a,b; Wagner et al., 2011; see Sections 4.4 and 4.6). Evaluation of the critical radii, energies, and nucleation rates requires effective (properly averaged) $L_m^{ef}(T)$; however, there were different definitions of $L_m^{ef}(T)$ and only a few attempts to evaluate it. Dufour and Defay (1963, Chapter XI therein) defined it as a simple arithmetic average, noting that it is not a true value, but rather a simple mean for estimates of r_{cr}. Khvorostyanov and Sassen (1998c) defined it similarly to (Eqn. 8.6.8) as a fit to a function of T using measured $J_{f,hom}$ for pure water and solution droplets.

The equations for the critical saturation ratios derived in this section from the classical nucleation theory, applied to conditions of a slow process for a bulk substance, can be used to derive an analytical parameterization for $L_m^{ef}(T)$ and compare it with other parameterizations. The two similar simple methods of calculation and parameterization of $L_m^{ef}(T)$ are described in the following text.

The first method follows Khvorostyanov and Curry (2009b). As discussed earlier, for very slow processes, $J_{f,hom} \to 0$ and $\ln(J_{f,om}) \to -\infty$, it follows from (Eqn. 8.7.5) for $H_{f,hom}$ that the denominator of $H_{f,hom}$ tends to $-\infty$ and $H_{f,hom} \to 0$. Then, for a bulk substance ($r_d = \infty$) and atmospheric pressure ($\Delta p = 0$), (Eqn. 8.8.3) is simplified as

$$S_{w,cr}^{\text{hom}}(T) = (T/T_0)^{M_w L_m^{ef}/RT}. \tag{8.8.4}$$

Note that the process with $J_{f,hom} \to 0$ can be identified with the equilibrium state of bulk water with ice—i.e., melting—and (Eqn. 8.8.4) represents the equation for the melting curve $S_w(T_m)$ or $T_m(S_w)$ of the aqueous solution in the bulk state, where T_m is the melting temperature (see Section 8.7). The environmental S_w then is equal to the water activity $a_w^i(T)$ of bulk solution in equilibrium with ice, thus we obtain:

$$S_{w,cr}^{\text{hom}}(T) = a_w^i(T) = a_w(T_m) = (T/T_0)^{M_w L_m^{ef}/RT}. \tag{8.8.5}$$

This simple equation describes the temperature dependence of $a_w^i(T_m)$ along the melting curve. Equations (8.8.4) and (8.8.5) can also be rewritten in an explicit form relative to T_m, as in (Eqn. 8.7.12b):

$$T_m(S_w) = T_0 S_w^{RT/M_w L_m^{ef}}. \tag{8.8.6}$$

Thus, (Eqns. 8.8.4)–(8.8.6) also describe the melting point temperature in the presence of solute. Parameterizations for $a_w^i(T_m)$ along with (Eqns. 8.8.4) and (8.8.5) can be used to determine $L_m^{ef}(T)$. By definition,

$$\mu_w^i(T) = \mu_w^0(T) + RT \ln a_w^i(T), \tag{8.8.6a}$$

where $\mu_w^i(T)$ and $\mu_w^0(T)$ are the chemical potentials of water in pure ice and of liquid water (see Section 3.5). Thus, $a_w^i(T)$ can be expressed from (Eqn. 8.8.6a) via the difference of the chemical potentials (ordinary affinity), $\Delta\mu = \mu_w^i(T) - \mu_w^0(T)$:

$$a_w^i(T) = \exp\left(\frac{\Delta\mu}{RT}\right). \quad (8.8.7)$$

Taking the logarithm of the relation $S_{w,cr}^{\text{hom}}(T) = a_w^i(T)$, and using (Eqns. 8.8.5) and (8.8.7), we obtain

$$L_m^{ef}(T) = \frac{\Delta\mu(T)}{M_w \ln(T/T_0)}. \quad (8.8.8)$$

This equation expresses effective L_m^{ef} as a function of T. Khvorostyanov and Curry (2009b) used for the evaluation of $L_m^{ef}(T)$ an expression for $\Delta\mu(T)$ from Koop et al. (2000), who utilized $\Delta\mu(T)$ from Johari et al. (1994) as was described by (Eqn. 8.4.10).

The second, newer method for evaluation of $L_m^{ef}(T)$ is even simpler and is based on the use of the expression (Eqn. 3.10.24) for water activity in a solution in equilibrium with ice—i.e., the water activity along the melting curve $a_w^i(T_m)$—as

$$a_w^i(T_m) = \frac{e_{is}(T)}{e_{ws}(T)}. \quad (8.8.9)$$

This expression is simpler than (Eqn. 8.8.7) based on $\Delta\mu(T)$ and requires only saturated pressures that are known well enough (see Chapter 4, Section 4.4). Moreover, equating (Eqns. 8.8.7) and (8.8.9), yields a useful expression for $\Delta\mu(T)$:

$$\mu_w^i(T) - \mu_w^0(T) = \Delta\mu(T) = RT \ln\left(\frac{e_{is}(T)}{e_{ws}(T)}\right), \quad (8.8.10a)$$

which allows refinement of $\Delta\mu(T)$ and its calculation down to very low temperatures using parameterizations of $e_{is}(T)$ and $e_{ws}(T)$ (see Sections 4.4 and 4.6). Now, we have an equation from (Eqns. 8.8.9) and (8.8.5) for the melting curve $T = T_m$:

$$S_{w,cr}^{\text{hom}}(T) = a_w^i(T) = \frac{e_{is}(T)}{e_{ws}(T)} = \left(\frac{T}{T_0}\right)^{M_w L_m^{ef}/RT}. \quad (8.8.10b)$$

Taking a logarithm, we obtain a simple relation:

$$L_m^{ef}(T) = \frac{RT}{M_w} \frac{\ln[e_{is}(T)/e_{ws}(T)]}{\ln(T/T_0)}. \quad (8.8.11)$$

The advantage of (Eqn. 8.8.10b) for $S_{w,cr}^{\text{hom}}(T_m)$ and the water activity $a_w^i(T)$ at melting compared to (Eqn. 8.8.7) and the advantage of (Eqn. 8.8.11) for $L_m^{ef}(T)$ compared to (Eqn. 8.8.8) is that they do not require knowledge of $\Delta\mu(T)$, nucleation rates, or surface tensions, and are based only on the saturated pressures for water and ice that are known and parameterized better than the difference of the chemical potentials $\Delta\mu(T)$. Various precise parameterizations for $e_{is}(T)/e_{ws}(T)$ were described in Chapter 4.

Figure 8.14a shows a comparison of various $L_m(T)$ and $L_m^{ef}(T)$. The three functions $L_m(T)$ are parameterizations from Dufour and Defay (1963, DD63) down to 220 K, from Johari et al. (1994,

Joh94), which is very close to DD63 but is extended down to 170 K, and from Pruppacher and Klett (1997). The $L_m(T)$ chosen by Pruppacher and Klett (1997) from several available data sets are lower than in DD63 and Joh94 and are given to 228 K because PK97 assumed a stability limit conjecture at ≈ 228 K. The four effective (averaged) $L_m^{ef}(T)$ include: the arithmetic mean from Dufour and Defay (1963, DD63, L_m-aver); similar to the DD63 arithmetic mean of Johari's (1994) parameterization (Joh94, aver); and two newer calculations based on the definition (Eqn. 8.6.8), calculations with (Eqn. 8.8.11) using (Eqns. 4.4.2) and (4.4.4) for $e_{is}(T)$ and $e_{ws}(T)$ by Murphy and Koop (2005), and calculations with (Eqn. 8.8.8) from Khvorostyanov and Curry (2009b) with (Eqn. 8.4.10) for $\Delta\mu$ from Koop et al. (2000) (marked KC09–K00). Fig. 8.14a shows that the last two methods, (Eqns. 8.8.8) and (8.8.11) give very close $L_m^{ef}(T)$, which is close to a linear function, while both simple arithmetic means overestimate $L_m^{ef}(T)$.

The $L_m^{ef}(T)$ calculated with (Eqn. 8.8.8) and $\Delta\mu$ (Eqn. 8.4.10) from Koop et al. (2000) was approximated in Khvorostyanov and Curry (2009b) by the polynomial of the fourth order:

$$L_m^{ef}(T) = c_{L0} + c_{L1}T_c + c_{L2}T_c^2 + c_{L4}T_c^4. \qquad (8.8.12a)$$

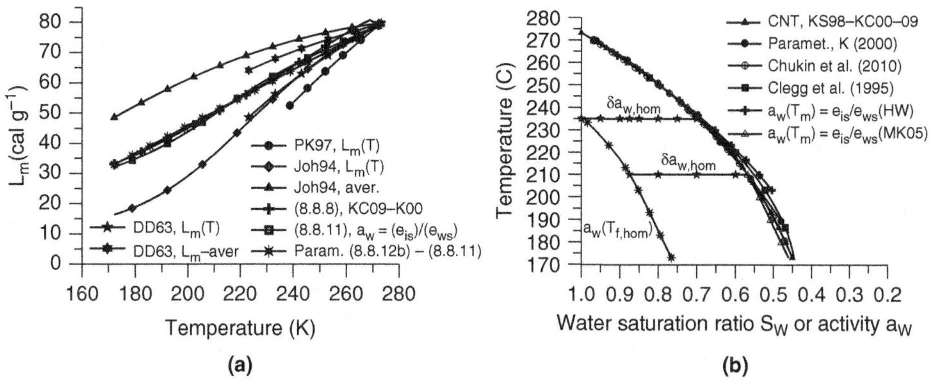

Figure 8.14. (a) Various $L_m(T)$ and $L_m^{ef}(T)$. The three functions $L_m(T)$ are parameterizations from Dufour and Defay (1963, DD63) down to 220 K, from Johari et al. (1994) down to 170 K, and from Pruppacher and Klett (1997) down to 228 K. The four effective (averaged) $L_m^{ef}(T)$ are calculated with new (Eqn. 8.8.11) with $a_w^i(T) = e_{is}(T)/e_{ws}(T)$ using (Eqns. 3.10.24), (4.4.2), and (4.4.4) for $e_{is}(T)$ and $e_{ws}(T)$ by Murphy and Koop (2005); calculated with (Eqn. 8.8.8) as in Khvorostyanov and Curry (2009b) with (Eqn. 8.8.8) for $\Delta\mu$ from Koop et al. (2000) (KC09-K00); Johari's (1994) parameterization averaged analytically as described in KC09b, and $L_m^{ef}(T)$ averaged as described in Dufour and Defay (1963, DD63). Parameterization (Eqn. 8.8.12b) is marked with asterisks. (b) melting temperatures $T_m(S_w)$ in equilibrium with ice calculated with various methods: parameterization from Koop et al. (2000, param., K00); calculation with CNT from Khvorostyanov and Sassen (1998) and Khvorostyanov and Curry (2000) with (Eqn. 8.8.10b), $S_{w,cr}^{hom}(T) = (T/T_0)^{M_w L_m^{ef}/RT}$, and (Eqn. 8.8.8) for L_m^{ef} from KC09b shown in Fig. 8.14b (CNT, KS98-KC00-09); parameterization from Clegg et al. (1995); parameterization from Chukin et al. (2010); and water activity $a_w(T_m)$ in equilibrium with ice calculated with (Eqn. 3.10.24) of Section 3.10 as $a_w(T_m) = e_{is}(T)/e_{ws}(T)$ calculated with $e_{is}(T)$ and $e_{ws}(T)$ from Wexler (1976) and Hyland and Wexler (1983) (HW-76-83) and with $e_{is}(T)$, $e_{ws}(T)$ from Murphy and Koop (2005, MK05) described in Section 4.4.

The newer refinement showed that (Eqn. 8.8.11) can be well approximated by a simpler linear parameterization

$$L_m^{ef}(T) = c_{L0} + c_{L1}T_c, \tag{8.8.12b}$$

with $L_m^{ef}(T)$ in cal g^{-1}, T_c in °C, $c_{L0} = 79.7$, and $c_{L1} = 0.463$. Fig. 8.14a shows that this parameterization practically coincides with more exact calculations using (Eqns. 8.8.8) and (8.8.11).

A verification of (Eqns. 8.8.8) and (8.8.11) for $L_m^{ef}(T)$ and its parameterizations (Eqn. 8.8.12a) and (8.8.12b) can be done by substituting these $L_m^{ef}(T)$ into the equations for the melting curve $S_w(T_m)$ or $a_w(T_m)$ and comparing them with the other curves $a_w(T_m)$ or $T_m(a_w)$ evaluated with various methods. This is illustrated in Fig. 8.14b, which presents water activities or saturation ratios along the melting curve obtained with 6 different methods: a) calculation with (Eqn. 8.8.5) derived by the authors from this modified CNT, as described by $S_{w,cr}^{\text{hom}}(T_m) = (T/T_0)^{M_w L_m^{ef}/RT}$, and $L_m^{ef}(T)$ from (Eqn. 8.8.11) shown in Fig. 8.14b (marked as CNT, KS98-KC00-09); b) parameterization from Koop et al. (2000); c) parameterization from Clegg et al. (1995) adapted from the textbook on chemical thermodynamics (Klotz and Rosenberg, 1972), and based on the measurements:

$$\log(a_w) = a_{c1}\Delta T + a_{c2}\Delta T^2 + a_{c3}\Delta T^3 + a_{c4}\Delta T^4 + a_{c5}\Delta T^5 + a_{c6}\Delta T^6, \tag{8.8.13}$$

where $\Delta T = T_0 - T_m$ is the melting point depression, $a_{c1} = -4.2091 \times 10^{-3}$, $a_{c2} = -0.2152 \times 10^{-5}$, $a_{c3} = 0.32233 \times 10^{-7}$, $a_{c4} = 0.3446 \times 10^{-9}$, $a_{c5} = 0.1758 \times 10^{-11}$, and $a_{c6} = 0.765 \times 10^{-14}$; and d) parameterization from Chukin et al. (2010, 2012) in (Eqn. 8.7.12h) based on experimental data. The other two curves for $a_w(T_m)$ are calculated much more simply using (Eqn. 8.8.9) and two different parameterizations for $e_{is}(T)$ and $e_{ws}(T)$: from Wexler (1976) and Hyland and Wexler (1983) (label HW) and from Murphy and Koop (2005, MK05) described in Section 4.4. The homogeneous freezing curve $a_w(T_{f,hom})$ is given for comparison and illustration of the difference between the freezing and melting curves and of the water activity shift discussed later.

Fig. 8.14b plotted as $T_m(S_w)$ shows that all of these 6 curves $a_w(T_m)$ are sufficiently close to each other down to about 220 K or $a_w = 0.60$, which supports the validity of (Eqn. 8.8.8) and (Eqn. 8.8.11) for $L_m^{ef}(T)$ based on classical nucleation theory and the accuracy of $L_m^{ef}(T)$ parameterizations (Eqn. 8.8.12a) and (8.8.12b). The discrepancies among several curves increase a little below 220 K and $a_w < 0.60$. This may indicate decreasing accuracy under such extreme conditions (low T or concentrated solutions with small a_w) of the parameterizations of $\Delta\mu(T)$ (Eqn. 8.4.10), some earlier parameterizations of pressures e_{is}, e_{ws}, and the accuracy of the data given in the books on physical chemistry and cited by Clegg et al. (1995). However, the close agreement between $S_w(T)$ from (Eqn. 8.8.4) with $L_m^{ef}(T)$ from (Eqn. 8.8.11) and e_{is}, e_{ws} from MK05, and $a_w^i(T) = e_{is}(T)/e_{ws}(T)$ from (Eqn. 8.8.9) is seen down to 170 K and $a_w = 0.45$. Thus, $L_m^{ef}(T)$ determined from this extended classical nucleation theory can be used for nucleation calculations down to these T and a_w with sufficient accuracy.

8.8.3. Derivation from Classical Theory of the Water Activity Shift Method

The major factors of (Eqn. 8.8.2) for $S_{w,cr}^{\text{hom}}$ can be parameterized as functions of temperature. The temperature dependence of various factors in (Eqn. 8.8.2) is shown in Fig. 8.14c, whereby calculations are performed with $r_d = 0.05\,\mu$m and $J_{hom}V = 1$ s^{-1}. In the range 273.15 to 170 K, variations of all parameters are not very large, $a_w(T_m) = (T/T_0)^{1/G_n}$ decreases from 1 at $T_0 = 273.15$ to 0.45 at 170 K.

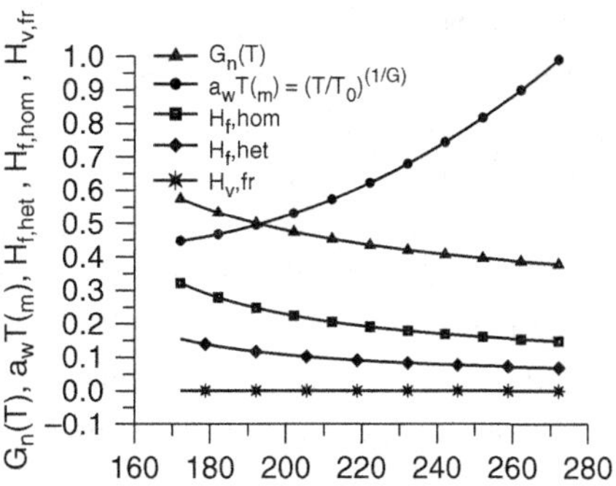

(c)

Figure 8.14. (c) Temperature dependence of various factors in equations for critical temperatures and saturation ratios (or water activities) of homogeneous ice nucleation, $T_{f,hom}$, $S_{w,cr}^{\text{hom}}(T)$ (Sections 8.7 and 8.8) and heterogeneous ice nucleation $T_{f,het}$, $S_{w,cr}^{het}(T)$ (Sections 9.11 and 9.12) calculated with the effective $L_m^{ef}(T)$ defined by (Eqn. 8.8.11): $G_n(T)$ in (Eqn. 8.6.15), $H_{v,fr}$, in (Eqn. 8.6.16), $H_{f,hom}$, in (Eqn. 8.7.5), $H_{f,het}$ in (Eqn. 9.11.4), and $a_w(T_m) = (T/T_0)^{1/G_n}$. Calculations are performed with the drop radius $r = 0.05\,\mu\text{m}$, $\Delta p = 0$, and for $H_{f,het}$: $\sigma_{sa} = 76\,\text{dyn}\,\text{cm}^{-2}$, active sites area $\alpha_a = 0$, misfit strain $\varepsilon = 0$ (so that $H_{v,fr}$ is the same for heterogeneous and homogeneous nucleation), $J_{f,het} = 1\,\text{s}^{-1}$, $J_{f,hom}V = 1\,\text{s}^{-1}$. The values of $H_{f,hom}$, $H_{f,het}$, and $H_{v,fr}$ are sufficiently smaller than 1, so that expansions by $H_{f,hom}$ in Chapter 8 and by $H_{f,het}$ in Chapter 9 are justified.

The other parameters increase with decreasing T: G_n increases from 0.38 to 0.58, $H_{f,hom}$ increases from 0.148 to 0.322, and $H_{v,fr} < 0.05$ over the entire temperature range. Therefore, (Eqn. 8.8.2) can be simplified by expansion of the exponents into the power series

$$S_{w,cr}^{\text{hom}} \approx \left(\frac{T}{T_0}\right)^{1/G_n}\left(1+\frac{H_{v,fr}}{G_n}+\frac{H_{f,\text{hom}}}{G_n}\right). \tag{8.8.14a}$$

Only the first-order terms are kept in the expansion, since an estimate shows that the next terms give a contribution of less than 3–6%. Thus, $S_{w,cr}^{\text{hom}}$ can be presented in the form

$$S_{w,cr}^{\text{hom}}(T_{f,\text{hom}}) \approx S_{w,cr}^{m}(T_m)(1+\Delta S_{w,cr}^{\text{hom}}) = S_{w,cr}^{m}(T_m) + \delta S_{w,cr}^{\text{hom}}. \tag{8.8.14b}$$

Here, $S_{w,cr}^{m}(T_m) = (T/T_0)^{M_w L_m^{ef}/RT}$ corresponds to the melting curve described earlier, and

$$\delta S_{w,cr}^{\text{hom}} = S_{w,cr}^{m}(T_m)\Delta S_{w,cr}^{\text{hom}}, \tag{8.8.15a}$$

$$\Delta S_{w,cr}^{\text{hom}} = (\Delta S_{w,J} + \Delta S_{w,rd} + \Delta S_{w,p}). \tag{8.8.15b}$$

(The components $\Delta S_{w,J}$, $\Delta S_{w,rd}$, and $\Delta S_{w,p}$, are defined in (Eqn. 8.8.17) and analyzed later). Thus, $\delta S_{w,cr}^{\text{hom}} = \delta a_w = S_{w,cr}^{m}\Delta S_{w,cr}^{\text{hom}}$ is the activity difference or *the activity shift* between the critical saturation ratios or activities of melting and homogeneous freezing,

$$\delta S_{w,cr}^{\text{hom}} = \delta a_{w,\text{hom}} = S_{w,cr}^{\text{hom}}(T_{f,\text{hom}}) - S_{w,cr}^{m}(T_m), \tag{8.8.16a}$$

or, in terms of activities,

$$\delta a_{w,\text{hom}} = a_w(T_{f,\text{hom}}) - a_w(T_m) = a_{w,\text{hom}} - a_w(T_m). \tag{8.8.16b}$$

The physical meaning of these quantities is simple: They describe the freezing point depression due to the presence of solute at each temperature expressed in terms of $\delta a_{w,hom}$. Fig. 8.14b shows that the curves $a_w(T_{f,hom})$ and $a_w(T_m)$ are "quasi-parallel" and the freezing curve can be obtained by shifting the melting curve as a whole toward greater a_w by a shift $\delta a_{w,hom}$, which is illustrated at $T = 235$ K and 210 K in Fig. 8.14b. Koop et al. (2000), based on analysis of several empirical data sets, suggested that an average shift $\delta a_{w,hom} \approx 0.305$ can be representative for various substances and conditions.

Equations (8.8.2), (8.8.3), (8.8.14a), (8.8.14b), (8.8.15a), and (8.8.15b) show that the critical humidity $S_{w,cr}^{\text{hom}}(T_{f,\text{hom}})$ and the shift $\delta a_{w,hom}$ in reality are more complicated quantities, depending on several factors, and extended classical theory provides more precise descriptions. Fig. 8.14d shows the shifts of water activity $\delta a_{w,hom} = a_{w,hom}(T_{f,hom}) - a_w(T_m)$ calculated with (Eqns. 8.8.15a), (8.8.15b), (8.8.16a), and (8.8.16b) for two nucleation rates, $J_{f,hom}V = 10^{-6}\,\text{s}^{-1}$ and $J_{f,hom}V = 1\,\text{s}^{-1}$, with V for the drops with two radii, $r = 0.02\,\mu\text{m}$ and $5\,\mu\text{m}$. Calculations are compared with the three experimental data sets that were retrieved from the data on $T_{f,hom}(a_{w,hom})$ by Larson and Swanson (2006) and Bertram et al. (2000) for ammonium sulfate and Koop et al. (2000) for sulfuric acid. The experimental shifts $\delta a_{w,hom}$ were calculated as the difference in activities between each individual experimental point $T_{f,hom}(a_{w,hom})$ from the freezing data sets and the same temperature at the melting curve $T_m(a_{w,m})$. Given for comparison are also two average constant shifts, $\delta a_{w,hom} = 0.305$ recommended in Koop et al. (2000), and $\delta a_w = 0.246$ determined by Zuberi et al. (2002) for heterogeneous nucleation.

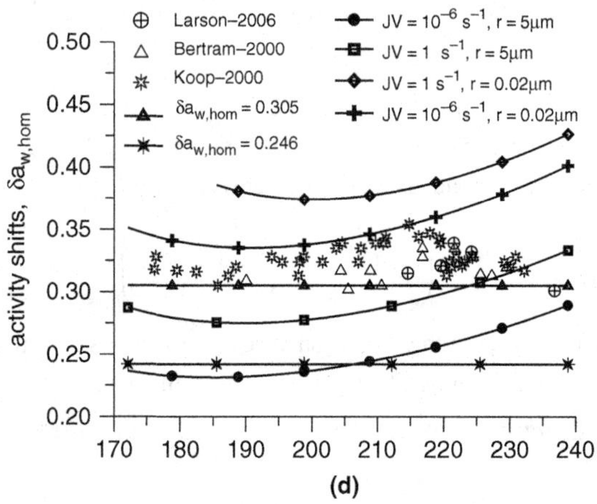

Figure 8.14. (d) The shifts of water activity $\delta a_{w,hom} = a_{w,hom}(T_{f,hom}) - a_w(T_m)$ calculated with equations of extended CNT (Eqns. 8.8.15a), (8.8.15b), (8.8.16a), and (8.8.16b) described in Section 8.8 for two nucleation rates, $J_{f,hom}V = 10^{-6}\,\text{s}^{-1}$ and $J_{f,hom}V = 1\,\text{s}^{-1}$, with the volume V for the drops with two radii, $r = 0.02\,\mu\text{m}$ and $5\,\mu\text{m}$. The three experimental data sets were retrieved from the data on $T_{f,hom}(a_{w,hom})$ by Bertram et al. (2000) and Larson and Swanson (2006) for ammonium sulfate and Koop et al. (2000) for sulfuric acid. The experimental shifts $\delta a_{w,hom}$ were calculated using the freezing $T_{f,hom}(a_{w,hom})$ and melting $T_m(a_{w,m})$ points from these data sets as the difference in activities for each individual experimental point $T_{f,hom}(a_{w,hom})$ on the freezing curve and the corresponding temperature at the melting curve $T_m(a_{w,m})$. Given for comparison are also two average constant shifts, $\delta a_{w,hom} = 0.305$ recommended in Koop et al. (2000), and $\delta a_w = 0.246$ determined by Zuberi et al. (2002) for heterogeneous nucleation.

Fig. 8.14d shows a noticeable scatter of the experimental points of δa_w, most of them are located in the range $0.3 < \delta a_{w,hom} < 0.36$. This picture agrees with Swanson (2009) who compared many experimental data sets on $T_{f,hom}$ obtained with different devices, also found a noticeable scatter in the shifts of activities, and noticed that a better average shift could be 0.32, which is closer to Fig. 8.14d. As could be expected, the shifts calculated from classical nucleation theory substantially depend on both the nucleation rate $J_{f,hom}$ and the radius of freezing particles and vary from a minimum of 0.27 at $J_{f,hom}V = 10^{-6}\,\mathrm{s}^{-1}$, $r = 5\,\mu\mathrm{m}$ to 0.42 at $J_{f,hom}V = 1\,\mathrm{s}^{-1}$, $r = 0.02\,\mu\mathrm{m}$. The calculated from CNT shift $\delta a_{w,hom}$ decreases by 0.05–0.1 when $J_{f,hom}$ decreases by 6 orders of magnitude, and by 0.1–0.12 when the radius increases from 0.02 μm to 5 μm. This indicates that more precise calculations of the nucleation thresholds for polydisperse aerosols and various cooling rates should be based on classical nucleation theory rather than on some constant average shift $\delta a_{w,hom}$. This is especially true for small particles with radii ~ 0.01–0.05 μm that nucleate ice in cold clouds. The submicron haze particles in cirrus, polar stratospheric clouds or diamond dust may freeze at substantially greater $\delta a_{w,hom}$—i.e., at lower temperatures than those determined in the laboratory conditions for the larger particles with $r = 5$–20 μm. The lower freezing temperature for submicron particles was illustrated in Fig. 8.6b. The lowering of the freezing temperature with increasing cooling rate was also determined experimentally (e.g., Cziczo and Abbatt, 1999; Martin, 2000; Kashchiev, 2000).

8.8.4. Effects of Various Factors on the Critical Humidity $S_{w,cr}^{\mathrm{hom}}$

The equations derived earlier allow estimation of the contribution of various terms to the critical humidity and water activity shift. The correction to, or the shift in, the critical saturation ratio $\delta S_{w,cr}^{\mathrm{hom}}$ in (Eqns. 8.8.15a) and (8.8.15b) includes three terms, $\Delta S_{w,J} = \dfrac{H_{f,\mathrm{hom}}}{G_n}$,

$$\Delta S_{w,rd} = \frac{\rho_w}{\rho_i}\frac{A_K}{r_d} \sim \frac{3.6\times 10^{-5}}{Tr_d}, \qquad \Delta S_{w,p} = \frac{R_v}{T}\frac{\Delta\rho\Delta p}{\rho_w \rho_i}. \tag{8.8.17}$$

The terms $\Delta S_{w,J}$, $\Delta S_{w,rd}$, and $\Delta S_{w,p}$ describe respectively the effects of the finite cooling rate $J_{f,hom}$, the finite radius of the freezing particle, and the effect of external pressure. If we express the temperature via supercooling ΔT, so that $T = T_0 - \Delta T$, and $\Delta T << T_0$, then the factor T/T_0 in (Eqn. 8.8.14a) can be also expanded in a power series:

$$S_{w,cr}^{\mathrm{hom}} \approx 1 + \Delta S_{w,T} + \Delta S_{w,J} + \Delta S_{w,rd} + \Delta S_{w,p}, \tag{8.8.18}$$

and $\Delta S_{w,T}$ is

$$\Delta S_{w,T} = -\frac{1}{G_n}\frac{\Delta T}{T} = -\frac{L_m^{ef}}{R_v}\frac{\Delta T}{T^2}. \tag{8.8.19}$$

For very slow processes, when $J_{f,hom} \to 0$ and $\ln(J_{f,hom}) \to -\infty$, it is seen from (Eqn. 8.7.5) that $H_{f,hom} \to 0$, and at atmospheric pressure, $\Delta p = 0$, it follows from (Eqn. 8.8.3) that

$$S_{w,cr}^{\mathrm{hom}} = \left[\left(\frac{T}{T_0}\right)\exp\left(\frac{r_{fr}}{r_d}\right)\right]^{M_w L_m^{ef}/RT}. \tag{8.8.20}$$

As discussed in Section 8.7, this equation for the bulk substance ($r_d = \infty$ and $\exp(r_{fr}/r_d) = 1$) describes the melting curve $S_w(T_m)$ or $T_m(S_w)$ and the term with r_{fr}/r_d describes the effect of the drop size. An estimate shows that $S_{w,cr}^{\text{hom}}$ increases by 0.075–0.15 for $r_d = 0.02$–$0.01\ \mu\text{m}$—i.e., the curvature correction $\Delta S_{w,rd} = (\rho_w/\rho_i)A_K/r_d \approx 7.5 - 15\%$ gives a noticeable effect in the case with homogeneous nucleation, and gradually vanishes at $r_d > 0.5$–$1\ \mu\text{m}$.

Each of the terms in (Eqns. 8.8.17) and (8.8.18) is a function of temperature, and these expressions resemble previous empirical parameterizations of $S_{w,cr}^{\text{hom}}$ for homogeneous ice nucleation in the form of polynomials of temperature described in Section 8.4 (e.g., Sassen and Dodd, 1988, 1989; Heymsfield and Miloshevich, 1995):

$$S_{w,cr}^{\text{hom}} = \sum_{k=0}^{3} A_{kT} T^k. \tag{8.8.21a}$$

Koop et al. (1998, 1999) and Bertram et al. (2000) parameterized the critical saturation ratio over ice $S_{i,cr}^{\text{hom}}$ as polynomials of the saturated vapor pressure e_{ws}, and Koop et al. (2000) parameterized it via water activity. Kärcher and Lohmann (2002a,b) suggested an empirical linear parameterization for $S_{i,cr}^{\text{hom}}(T)$ that is very close to Sassen and Dodd (1989). Thus, (Eqns. 8.8.17) and (8.8.18) serve as a physical justification for previous empirical parameterizations and as a basis for their improvement. Based on (Eqns. 8.8.17) and (8.8.18), the terms $(\Delta S_{w,rd} + \Delta S_{w,p})$ can be added to the T-polynomial (Eqn. 8.8.21a) to account for the curvature and pressure effects in the simple parameterizations, so that it can be presented in the form

$$S_{w,cr}^{\text{hom}} = A_{0T} + A_{1T}T + A_{2T}T^2 + \Delta S_{w,rd} + \Delta S_{w,p}. \tag{8.8.21b}$$

However, (Eqns. 8.8.2) and (8.8.3) are sufficiently simple for direct use in cloud and climate models and analysis of the chamber experiments and show that there are several other dependencies, indicating that the critical humidities are not functions only of temperature. The dependencies on r_d and Δp show that $S_{w,cr}^{\text{hom}}$ depends on the size of a freezing particle and on the external pressure, in agreement with experimental data (e.g., Kanno and Angell, 1977; Miyata et al., 2001, 2002). In addition, (Eqns. 8.8.2 and 8.8.3) show that the measured threshold saturation ratio depends on the nucleation rate $J_{f,hom}$—i.e., on the cooling rate. All these factors may at least partially explain the differences in $S_{i,cr}^{\text{hom}}$ determined by various experiments. The critical values of the ice saturation ratio can be obtained from the relation

$$S_{i,cr}^{\text{hom}} = S_{w,cr}^{\text{hom}}(e_{ws}/e_{is}). \tag{8.8.21c}$$

8.8.5. Calculations of Critical Relative Humidities over Water and Ice

Calculations of the critical values of RHW and RHI for homogeneous nucleation were performed with (Eqn. 8.8.3) and the following set of parameters in the baseline case: $\sigma_{sa} = 76\,\text{dyn}\,\text{cm}^{-2}$, $\Delta p = 0$, and two prescribed values of the homogeneous nucleation rate, so that $J_{f,hom}V = 10^{-6}\,\text{s}^{-1}$ and $1\,\text{s}^{-1}$, which characterize slow and fast freezing rates respectively, and V was calculated for $r_d = 5\,\mu\text{m}$. The averaged $L_m^{ef}(T)$ was calculated as described in the previous section.

The RHW and RHI for the baseline case are shown in Fig. 8.15. The general feature of RHW is a monotonic decrease with decreasing temperature from ~100% at $T \sim -40\,°\text{C}$ to ~ 72–82% at −75 °C. The corresponding RHI increases from 145–155% at −40 °C to 165–175% at −75 °C, which arises from a combination of the temperature dependence of RHW and $e_{ws}(T)/e_{is}(T)$. Fig. 8.15a shows that the expansion (Eqn. 8.8.18) is a reasonable approximation at these two nucleation rates. The values of

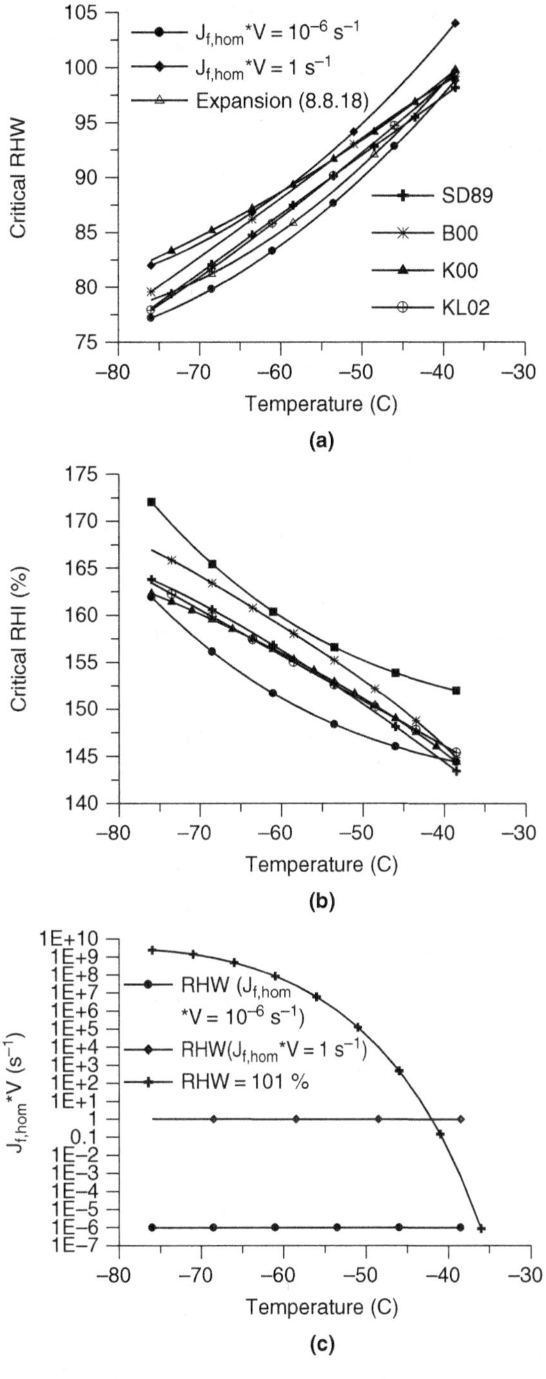

Figure 8.15. Critical RHW (a) and RHI (b) for homogeneous ice nucleation calculated with (Eqn. 8.8.3) for $r = 5\,\mu m$ and the other parameters as in Fig. 8.14, compared with expansion (Eqn. 8.8.18) and with parameterizations by Sassen and Dodd (1989, SD89), Koop et al. (1999, 2000, K00), Bertram et al. (2000, B00), and Kärcher and Lohmann (2002a, KL02) with their RHI or RHW recalculated to the corresponding RHW or RHI as described in the text. (c) Verification of the solutions: homogeneous nucleation rates $J_{f,hom}V$ calculated using (Eqn. 8.6.21) with these RHW(T) in Fig. 8.15a corresponding to $J_{f,hom}V = 10^{-6}\,s^{-1}$ and $1\,s^{-1}$, and with RHW = const = 101%. From Khvorostyanov and Curry (2009b), *J. Geophys. Res.*, 114, D04207, reproduced with permission of John Wiley & Sons, Inc.

RHW and RHI calculated here are compared with 4 parameterizations from (Sassen and Dodd, 1989, SD89), (Koop et al., 1999, 2000, K00), (Bertram et al., 2000, B00), and (Kärcher and Lohmann, 2002a, KL02), whereby the RHW and RHI given in these works are recalculated to the corresponding RHI and RHW as $RHI_{cr} = RHW_{cr}(e_{ws}/e_{is})$. Agreement of the present calculations with the previous parameterizations is sufficiently good, and the difference generally does not exceed 3–5%. The curves calculated here with $J_{hom} = 10^{-6}$ and 1 s^{-1} and $r_d = 5$ μm differ by not more than 5% and bracket these parameterizations—in particular, because the cited chamber experiments were performed with particles of comparable $r_d \sim 2$–6 μm.

However, even this small difference in humidity causes great differences in the nucleation rates. To illustrate the important role of humidity in the homogeneous freezing process and to verify the calculated RHW_{cr}, the reverse calculation was performed as shown in Fig. 8.15c; that is, the product of the nucleation rate and the volume, $J_{f,hom}(T, RHW_{cr})V$, for $r_d = 5$ μm was calculated using the same equations for $J_{f,hom}$ and ΔF_{cr} as in Fig. 8.15a, but with the two critical values of $RHW_{cr}(T)$ from Fig. 8.15a as input. This calculation yields $J_{f,hom}V$ = const = 10^{-6} s^{-1}, and 1 s^{-1} (Fig. 8.15c) for the two different critical values. This result was compared with $J_{f,hom}(T, RHW_{cr})V$ calculated with RHW_{cr} = const = 101% and also presented in Fig. 8.15c. The nucleation rate with $RHW = 101\%$ is very high at $T < -75$ to -50 °C but falls below 10^{-3} s^{-1} at $T > -40$ °C, which is similar to several experimental data sets compiled in (Pruppacher and Klett, 1997, Chapter 7). This behavior is typical at constant relative humidity; however, if humidity is close to the critical value or slightly above it, the T-dependence of the nucleation rate is smooth, in contrast to the explosive increase predicted by the classical theory based on r_{cr} with T-dependence but without S_w-dependence.

The critical humidities also depend on the particle radius, which is illustrated in Fig. 8.16, where the results of calculations with 0.05 μm and 5 μm for two values of $J_{hom}V = 10^{-6}$ and 1 s^{-1} are

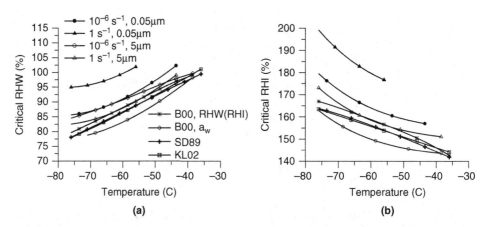

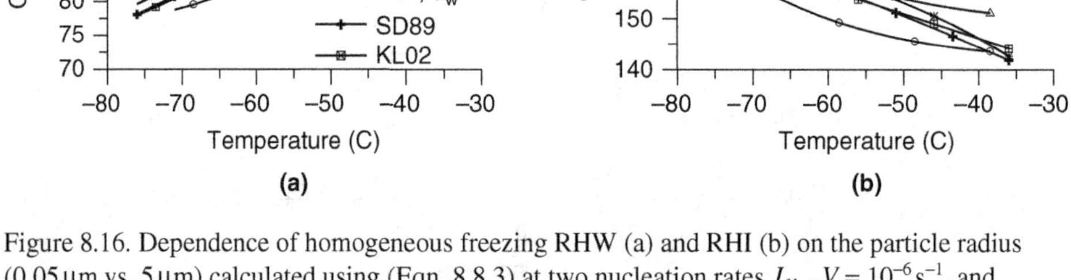

Figure 8.16. Dependence of homogeneous freezing RHW (a) and RHI (b) on the particle radius (0.05 μm vs. 5 μm) calculated using (Eqn. 8.8.3) at two nucleation rates $J_{f,hom}V = 10^{-6}$ s^{-1}, and $J_{f,hom}V = 10^{-6}$ s^{-1}, compared to parameterizations by Sassen and Dodd (1989, SD89), Kärcher and Lohmann (2002, KL02), and RHI and a_w from Bertram et al. (2000, B00). From Khvorostyanov and Curry (2009b), *J. Geophys. Res.*, 114, D04207, reproduced with permission of John Wiley & Sons, Inc.

presented and again compared to parameterizations (SD89), (B00), and (KL02). Fig. 8.16 shows that when the particle radius increases by 2 orders of magnitude, *RHW* decreases by 7–15%; a corresponding decrease is 17–27% for *RHI*. The dependence of the homogeneous freezing rate on the particle radius is in agreement with experimental data (e.g., Pruppacher and Klett, 1997, Miyata et al., 2001, 2002). It is seen in Figs. 8.15 and 8.16 that both humidities calculated with $r_d = 5$ μm are close to the parameterizations (SD89), (B00), (KL02), and (K00). This agreement provides empirical support for the validity of this theoretical method for evaluation of the critical humidities. However, the critical humidities strongly depend on the prescribed nucleation rate and on the radius of freezing particles, which is usually not accounted for in the empirical parameterizations.

8.9. Parcel Model Simulations of the Kinetics of Homogeneous Ice Nucleation

The kinetics of homogeneous ice nucleation can be studied numerically or analytically or by combining both methods. For numerical analysis, the cloud models can be used, and the major tool in these studies is usually *a cloud parcel model* that is used for simulation of a cloud process involving ice nucleation.

8.9.1. Parcel Model Description

The simplest model used in cloud physics for simulation of various cloud properties is a parcel model. This is a model of a rising air parcel where droplets or crystals may form. A parcel model is a zero-dimensional or Lagrangian model, where all variables depend only on time t and the dependence on the coordinates is excluded. There are two major types of parcel models: with and without entrainment. Models with entrainment account for the partial exchange of the rising or descending parcel with environment by the air, heat, moisture, aerosol, and condensed matter, droplets, and crystals. The rates of this exchange are parameterized with various mixing lengths. Examples of parcel models of these types are described in Flossmann, Hall, and Pruppacher (1986), Young (1993), and Pruppacher and Klett (1997). Models without entrainment (e.g., Lin et al., 2002 and references herein) consider an isolated adiabatic air parcel that rises and cools, which causes nucleation and growth of the drops and crystals.

We describe here an example of such a parcel model with spectral bin microphysics. The microphysical formulation used here is similar to the Eulerian numerical 1D, 2D, and 3D models with spectral bin microphysics developed and used previously for the simulation of mixed-phase clouds of various types (e.g., Buikov and Pirnach, 1973, 1975; Young, 1974a; Buikov and Khvorostyanov, 1976, 1977; Hall, 1980; Khvorostyanov, 1982, 1987, 1995; Kondratyev and Khvorostyanov, 1989; Kondratyev, Ovtchinnikov and Khvorostyanov, 1990a,b; Khvorostyanov and Khairoutdinov, 1990; Khairoutdinov and Khvorostyanov, 1991; Feingold et al., 1994; Pirnach and Krakovskaya, 1994, 1998; Jensen et al., 1994, 1998, 2005; Ovtchinnikov et al., 2000; Khvorostyanov et al., 2001, 2003, 2006; Fridlind et al., 2004; Khain et al., 2004; Flossmann and Wobrock, 2010; Curry and Khvorostyanov, 2012), with some differences owing to the Lagrangian approach in the parcel model. The dynamics in parcel models is usually parameterized by prescribing vertical velocity w, which is usually constant in time. The main thermodynamic equations are the prognostic equations for supersaturation and temperature. This system of equations includes terms that describe the phase transitions and is closed using the kinetic equations of condensation and deposition for the drop and ice crystal size distribution functions that account for nucleation, and equations for the droplet and crystal growth rates, as described in Chapter 5. Here we exclude from consideration coagulation among droplets and aggregation between the droplets and crystals, sedimentation, entrainment, turbulent exchange, etc. to isolate the effects directly related to nucleation processes. We describe in the following text the parcel model based on spectral bin microphysics, developed by the authors of this book and used for studies of the kinetics of ice nucleation (Khvorostyanov and Curry, 2005a, hereafter KC05a).

The heat balance is calculated using the equation for the temperature T in a wet adiabatic process considered in Chapters 3 and 5, and has the form similar to (Eqn. 3.11.43):

$$\frac{dT}{dt} = -\gamma_a w + \frac{L_e}{c_p \rho_a} I_{con} + \frac{L_s}{c_p \rho_a} I_{dep}. \tag{8.9.1}$$

This equation is solved along with the integral ice supersaturation (Eqn. 8.5.18) that is required to describe crystal growth:

$$\frac{1}{(1+y_i')}\frac{dy_i'}{dt} = c_{1i} w - \frac{y_i'}{1+y_i'}(4\pi D_v)\left[\int_0^t r_{c,ef}(t,t_0) \int_{r_{\min}}^{r_{\max}} f_a(r_a) v(r_a) \right.$$

$$\left. \times J_{f,\mathrm{hom}}(t_0) \exp\left(-\int_0^t J_{f,\mathrm{hom}}(t') v(r_a) dt'\right) dr_a dt_0 \right]. \tag{8.9.2}$$

Evolution of the crystal $f_c(r_c)$ and droplet $f_d(r_d)$ size spectra is described using two kinetic equations for these spectra in the form (Eqn. 5.5.5) as in KC05a:

$$\frac{\partial f_c}{\partial t} + \frac{\partial}{\partial r_c}\left(\frac{dr_c}{dt} f_c\right) = \psi_{fc}(r_c, t) \tag{8.9.3a}$$

$$\frac{\partial f_d}{\partial t} + \frac{\partial}{\partial r_d}\left(\frac{dr_d}{dt} f_d\right) = \psi_{fd}(r_d, t). \tag{8.9.3b}$$

Here we consider homogeneous ice nucleation at cold temperatures and not very vigorous updrafts when the haze solution particles freeze at water subsaturation, so that the drops do not form and (Eqn. 8.9.3b) for drops is not used. In deep convective clouds with the stronger updrafts of several meters per second, water saturation may be reached, homogeneous ice nucleation may occur simultaneously with drop activation, and (Eqn. 8.9.3b) may be needed. The terms of growth rates in (Eqn. 8.9.3a,b) are calculated using (Eqn. 5.1.46) for dr_d/dt and (Eqn. 5.2.24) for dr_c/dt:

$$\frac{dr_d}{dt} = \frac{c_{3w} s_w}{r_d + \xi_{con}}, \qquad c_{3w} = \frac{D_v \rho_{ws}}{\rho_w \Gamma_1}, \qquad \xi_{con} = \frac{4D_v}{\alpha_c V_w} \tag{8.9.4a}$$

$$\frac{dr_c}{dt} = \frac{c_{3i} s_i}{r_c + \xi_{dep}}, \qquad c_{3i} = \frac{D_v \rho_{is}}{\rho_i \Gamma_2}, \qquad \xi_{dep} = \frac{4D_v}{\alpha_d V_w} \tag{8.9.4b}$$

The crystal nucleation term can be calculated as

$$\psi_{fc} = \Delta N_{c,fr}(\Delta t)/\Delta r_c/\Delta t. \tag{8.9.5}$$

where $\Delta N_{c,fr}$ is the number concentration of the crystals nucleated in a time step Δt and calculated using (Eqn. 8.3.9). In the finite difference scheme, the crystal source term is calculated for the homogeneous freezing mode as $\psi_{fc} = \Delta N_{c,fr}/\Delta r_c/\Delta t$, where Δr_c denotes the smallest size range of crystal radii (0.1–1 μm). The crystal size spectrum includes 30 radius intervals with 10 increments of 0.1–1 μm, and the next 20 increments increasing logarithmically to 100–350 μm. Numerical experiments show that this size division allows coverage of both small and large size ranges without losing accuracy.

8.9.2. Simulation Results

To simulate the ice crystal nucleation process, the parcel model was run for 1 hour with most initial conditions specified following the Cirrus Parcel Model Comparison Project (CPMCP) (Lin, Starr, et al., 2002), and varying some parameters to estimate the sensitivity of the results. We describe and compare the results for three values of vertical velocity, w = 4, 20, and 100 cm s^{-1}, two values of the initial temperature, T_0 = −40 °C and −60 °C (CPMCP), and two values of the aerosol concentration, N_a = 200 cm^{-3} (CPMCP) and N_a = 500 cm^{-3}. For convenience of comparison at different temperatures, the initial RHW_0 was chosen in such a way that at both temperatures homogeneous ice nucleation processes occurred at comparable times. This required RHW_0 = 90% for T_0 = −40 °C and RHW_0 = 78% for T_0 = −60 °C. The initial pressure p_0 was specified to be 340 hPa. The parcel model includes the option of isolating specific ice crystal nucleation modes, by turning off the other modes. In this chapter, we deal with the homogeneous freezing of deliquescent haze particles. The haze size spectrum was assumed to be a lognormal size spectrum of soluble particles with a mean geometric radius of 0.02 μm and dispersion σ_s = 2.5. The time steps were 0.01–0.2 s in the main program, but the time step can be divided further in the nucleation or condensation subroutines to meet the stability conditions. The accuracy of the calculations was controlled by comparing the total number of crystals nucleated with those obtained by integration over the size spectrum of the grown crystals at the end of a parcel run. If the error exceeded 5% (especially at low temperatures), the time and radius steps were varied and several additional runs were performed until the error became less than 5%.

Figs. 8.17 and 8.18 illustrate the effect of the vertical velocity (w = 4 and 20 cm s^{-1}) on the kinetics of homogeneous freezing at T_0 = −40 °C and N_a = 200 cm^{-3}. It is seen that the nucleation process has two branches. At the ascending branch, the first term on the right-hand side of (Eqn. 8.9.2) with supersaturation generation dominates. Therefore, the relative humidity and both supersaturations, s_w and s_i, increase from the initial values to the maximum values reached at the time t_m. At the descending branch at $t > t_m$, the RHW, s_w and s_i decrease due to domination of the second term on the RHS of (Eqn. 8.9.2) with supersaturation depletion. This feature is similar to the drop activation process (Chapter 7). As a result of adiabatic cooling, RHW increases in the ascending branch and reaches at w = 4 cm s^{-1} a maximum of 97.7% at a time of about 35 min and then begins to decrease (Fig. 8.17a). The water and ice supersaturation pass the first critical values in the ascending branch of $s_{w,cr1}$ = −4.2% and $s_{i,cr1}$ = 42% at about $t \approx$ 22 min, reach maxima of −2.45% and 46% respectively at t = 33.67 min, and then decrease in the descending branch to the second critical values reached at about t = 40 min (Fig. 8.18a). Note that the change in ice supersaturation $\Delta s_i = s_{i,max} - s_{i,cr1} \approx 4\%$, or $\Delta s_i/s_{i,max}$ is less than 10%. Thus, it can be assumed that nucleation occurs at almost constant ice supersaturation.

Noticeable ice nucleation with w = 4 cm s^{-1} begins after the first critical point $s_{w,cr1}$ at $t \approx$ 22 min (Fig. 8.17d,e,f). At the time of maximum RHW and s_w, the crystal critical radius and energy reach a minima of 1.36×10^{-7} cm and 1.38×10^{-12} erg, respectively (Fig. 8.17b,c), while the nucleation rate

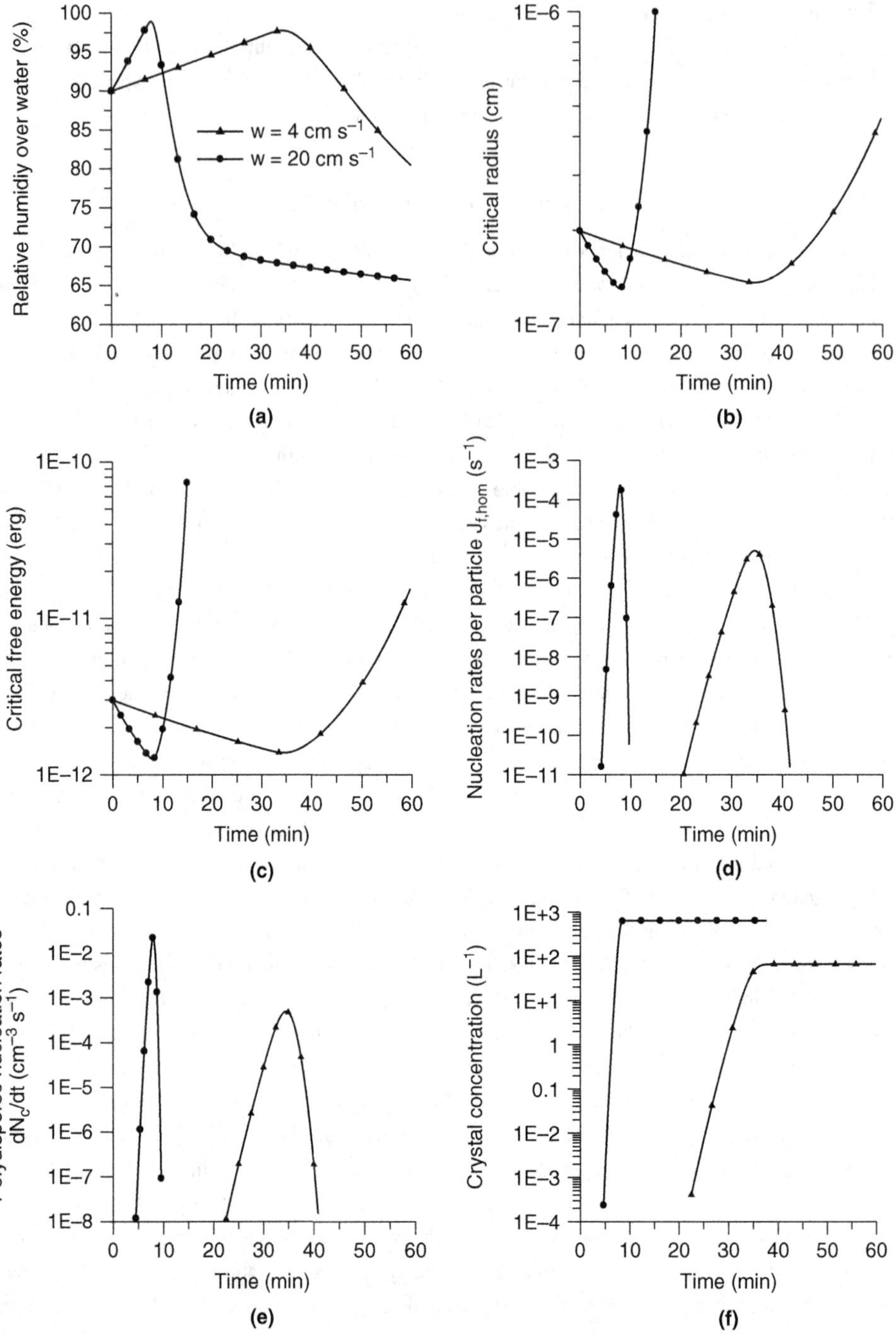

Figure 8.17. Kinetics of homogeneous nucleation at $T_{0c} = -40\,°C$, $RHW_0 = 90\%$, $p_0 = 340\,\text{hPa}$, $N_a = 200\,\text{cm}^{-3}$ and two vertical velocities, $w = 4\,\text{cm s}^{-1}$, and $w = 20\,\text{cm s}^{-1}$. (a) Relative humidity over water *RHW*; (b) critical radius r_{cr}; (c) critical free energy ΔF_{cr}; (d) homogeneous $J_{f,hom}r_a^3$ nucleation rates for a particle with radius of 0.11 μm; (e) polydisperse nucleation rates, $R_{f,hom} = dN_{fr}/dt$, defined by (Eqn. 8.3.10); (f) crystal concentration. From Khvorostyanov and Curry, Atmos. Chem. Phys. (2012), reproduced with permission.

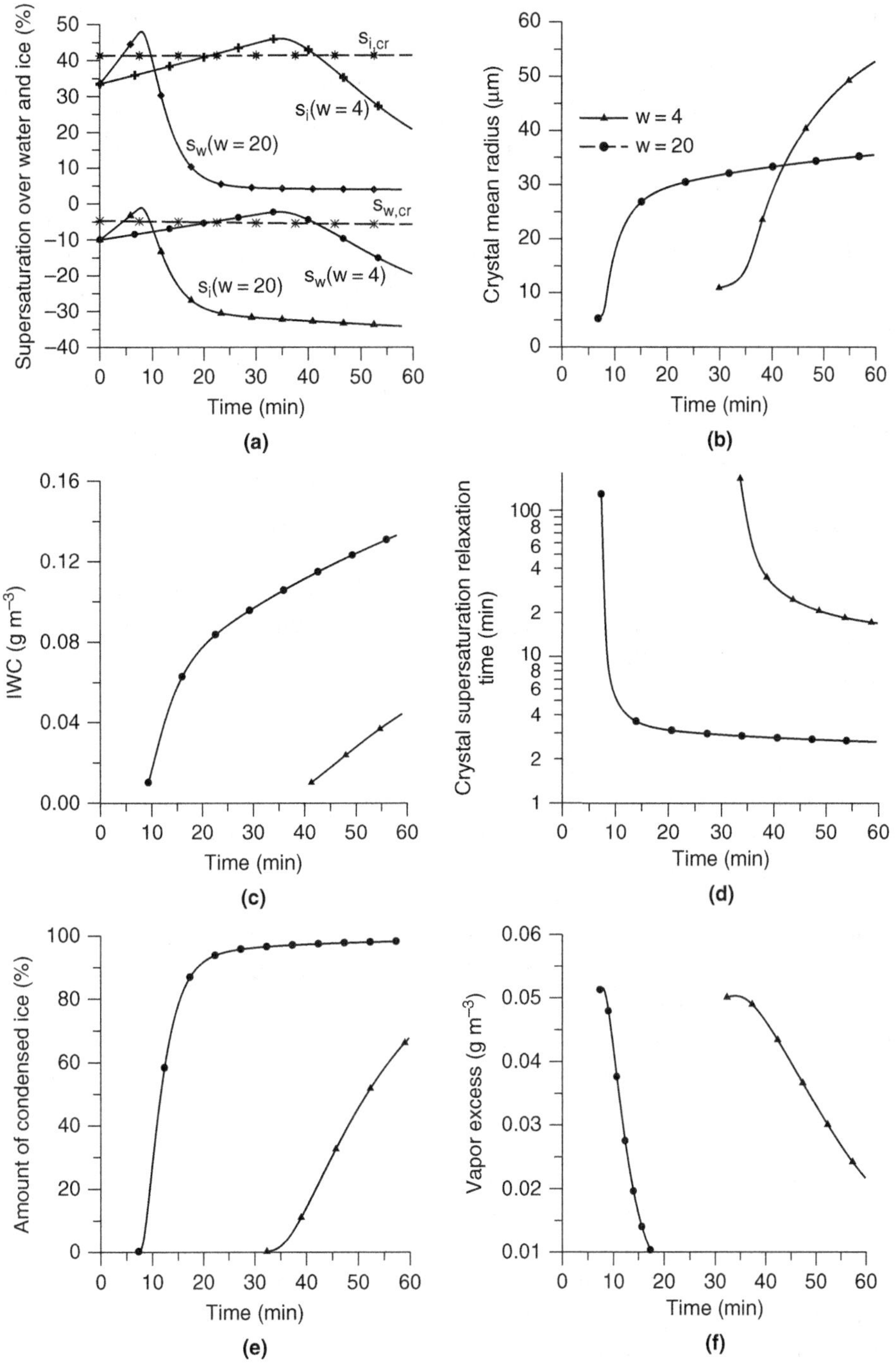

Figure 8.18. Comparison (continuation) of kinetics of homogeneous ice nucleation at $w = 20\,\mathrm{cm\,s^{-1}}$ (solid circles) and $w = 4\,\mathrm{cm\,s^{-1}}$ (triangles) at −40 °C and the other parameters as in Fig. 8.17.
(a) Supersaturations over water, s_w, and ice, s_i, % and corresponding critical supersaturations; (b) crystal mean radius, μm; (c) ice water content, g m^{-3}; (d) crystal supersaturation relaxation time, min; (e) relative amount of condensed ice, %; (f) vapor excess, mg m^{-3}. From Khvorostyanov and Curry, Atmos. Chem. Phys. (2012), reproduced with permission.

per particle ($J_{f,hom}r_d^3$, with $r_d = 0.11\,\mu m$) and the polydisperse nucleation rate $R_{f,hom}$ reach a maxima of $4.90 \times 10^{-6}\,s^{-1}$ and $4.93 \times 10^{-4}\,cm^{-3}\,s^{-1}$ (Fig. 8.17d,e). The values of r_{cr} and ΔF_{cr} are substantially greater, while $J_{f,hom}r_d^3$ and $R_{f,hom}$ are smaller at the later times, although the temperature continues to decrease. This illustrates an important key role of humidity in ice nucleation.

In contrast to the drop activation, the nucleation process continues also after t_m along the descending branch till the point when the second critical values $s_{w,cr}$ and $s_{i,cr}$ are reached, and ceases after that. The entire nucleation process takes 15–20 min with $w = 4\,cm\,s^{-1}$, and the final crystal concentration is $66\,L^{-1}$ (Fig. 8.17f). The crystal mean radius grows by $t = 1$ hour to 53 microns, the ice water content (IWC) grows to $0.044\,g\,m^{-3}$ and the supersaturation relaxation time τ_{fc} decreases from more than 3 hours at the beginning of nucleation to 17 min by the end of simulation. This indicates that deposition of the vapor is not instantaneous and a significant amount of vapor is not deposited for a long time. For quantitative illustration, it is convenient to introduce the two quantities, vapor excess, M_v, and the relative amount, or percentage of uncondensed ice, Fr_{con}, as

$$M_v = \rho_v s_i, \qquad Fr_{con} = IWC/(IWC + M_v) \times 100. \tag{8.9.6}$$

These quantities characterize the mass of uncondensed ice (mass of ice supersaturation) and the fraction of condensed ice. In a bulk model with instantaneous condensation and deposition, $M_v = 0$, and $Fr_{con} = 100\%$, but it is not so in this microphysical model with explicit calculation of supersaturation. Fig. 8.18f shows that the vapor excess is greater or comparable to IWC and the fraction of condensed ice is less than 50% during 30 minutes. This means that optical thickness and emissivity of cirrus clouds at the initial stages of formation are significantly smaller than would be in a bulk model, the radiation of the cloud is not radiation of the black body, and the optical properties should be accordingly corrected (decreased) in climate and weather forecast models.

The corresponding curves for the case with $w = 20\,cm\,s^{-1}$ (solid circles in Figs. 8.17 and 8.18) show that increased w causes much faster nucleation, which takes now about 5 min. The other features of the nucleation process are qualitatively similar, with some quantitative differences. The minimum critical radius and energy become smaller, the nucleation rates increase by almost two orders of magnitude, and the final crystal concentration increases by almost an order of magnitude to $649\,L^{-1}$, ~10 times greater than with $w = 4\,cm\,s^{-1}$; that is, N_c increases with w approximately as $N_c \sim w^{3/2}$ in this case. Because of more numerous crystals and their competition for vapor, the mean crystal radius is smaller than with $w = 4\,cm\,s^{-1}$, but the relaxation time τ_{fc} is also smaller with a minimum of 2.6 min. The deposition is faster in this case, but the vapor excess and percentage of condensed ice are still smaller for 15–20 min that would be in a bulk model (Fig. 8.18e,f).

A comparison of the results with $N_a = 200\,cm^{-3}$ (described earlier) and $500\,cm^{-3}$ at $T_0 = -40\,°C$, $w = 4\,cm\,s^{-1}$ is shown in Figs. 8.19 and 8.20; all the other parameters are as before. This comparison shows that a significant increase in N_a causes a very weak effect on nucleation kinetics and all the quantities. Nucleation with higher N_a begins and ceases a little earlier, and the resulting crystal concentration is $68.6\,L^{-1}$ vs. $66\,L^{-1}$ with $N_a = 200\,cm^{-3}$; that is, an increase of 2.5 times in N_a causes an increase of only 4% in N_c. This remarkable insensitivity to the initial concentration of deliquescent freezing aerosol indicates that a "saturation" with respect to N_a occurs at much smaller values of N_a than is typical for the upper troposphere.

The fraction of nucleated haze particles (the ratio N_c/N_a), is really tiny $(66\,L^{-1})/(200{,}000\,L^{-1}) = 3.3 \times 10^{-4}$, which is much smaller than the typical fraction of CCN activated into drops, ~0.3–0.7 (see Chapter 7).

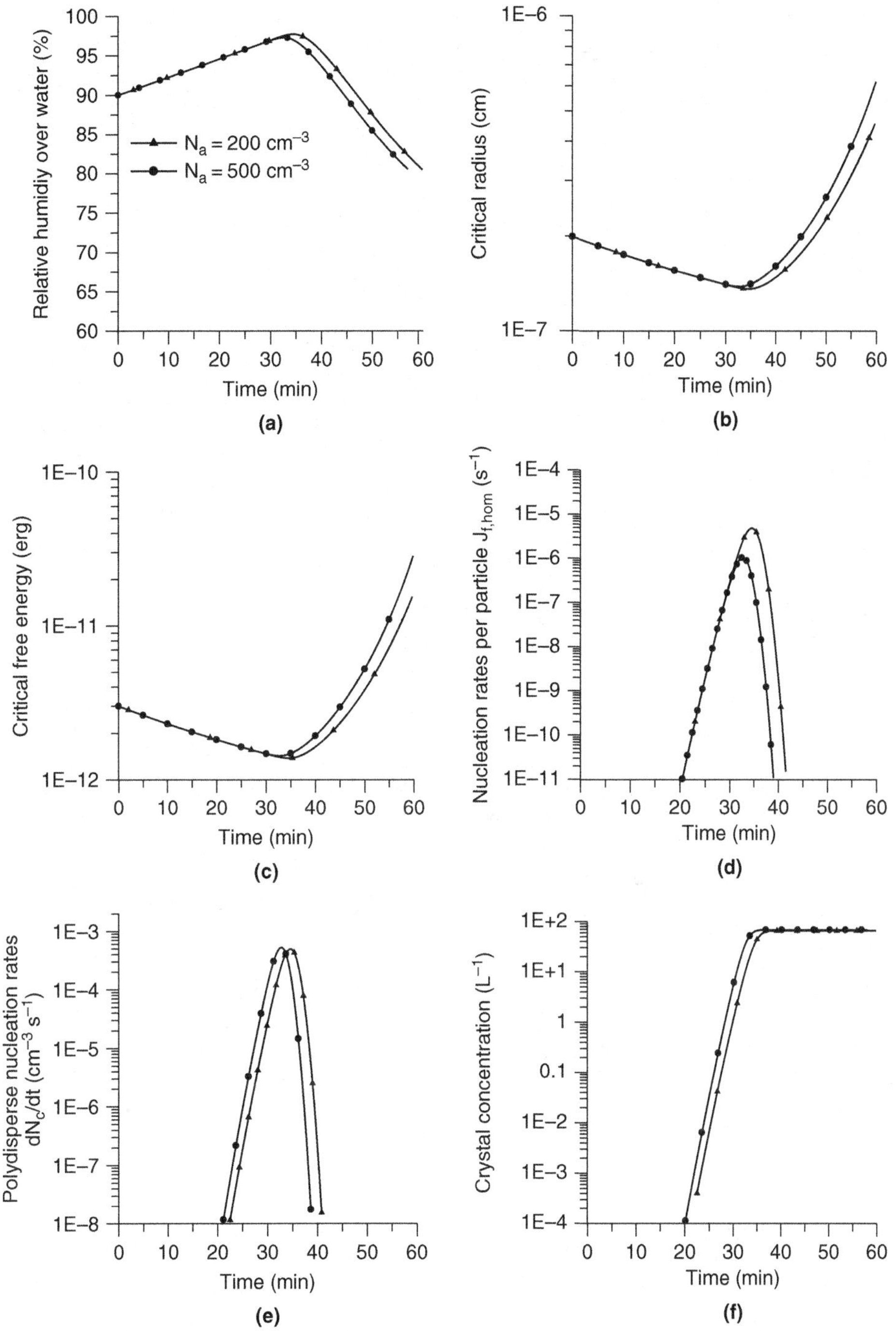

Figure 8.19. Comparison of homogeneous nucleation kinetics at $N_a = 200\,\mathrm{cm}^{-3}$ and $500\,\mathrm{cm}^{-3}$. The other parameters are $T_{0c} = -40\,°\mathrm{C}$, $RHW_0 = 90\%$, $p_0 = 340\,\mathrm{hPa}$, and $w = 4\,\mathrm{cm\,s}^{-1}$. (a) Relative humidity over water RHW; (b) critical radius r_{cr}; (c) critical free energy ΔF_{cr}; (d) homogeneous nucleation rates $J_{f,hom} r_a^3$ for a particle with radius $r_a = 0.11\,\mu\mathrm{m}$; (e) polydisperse nucleation rates, $R_{f,hom} = dN_{fr}/dt$, defined by (Eqn. 8.3.10); (f) crystal concentration. From Khvorostyanov and Curry, Atmos. Chem. Phys. (2012), reproduced with permission.

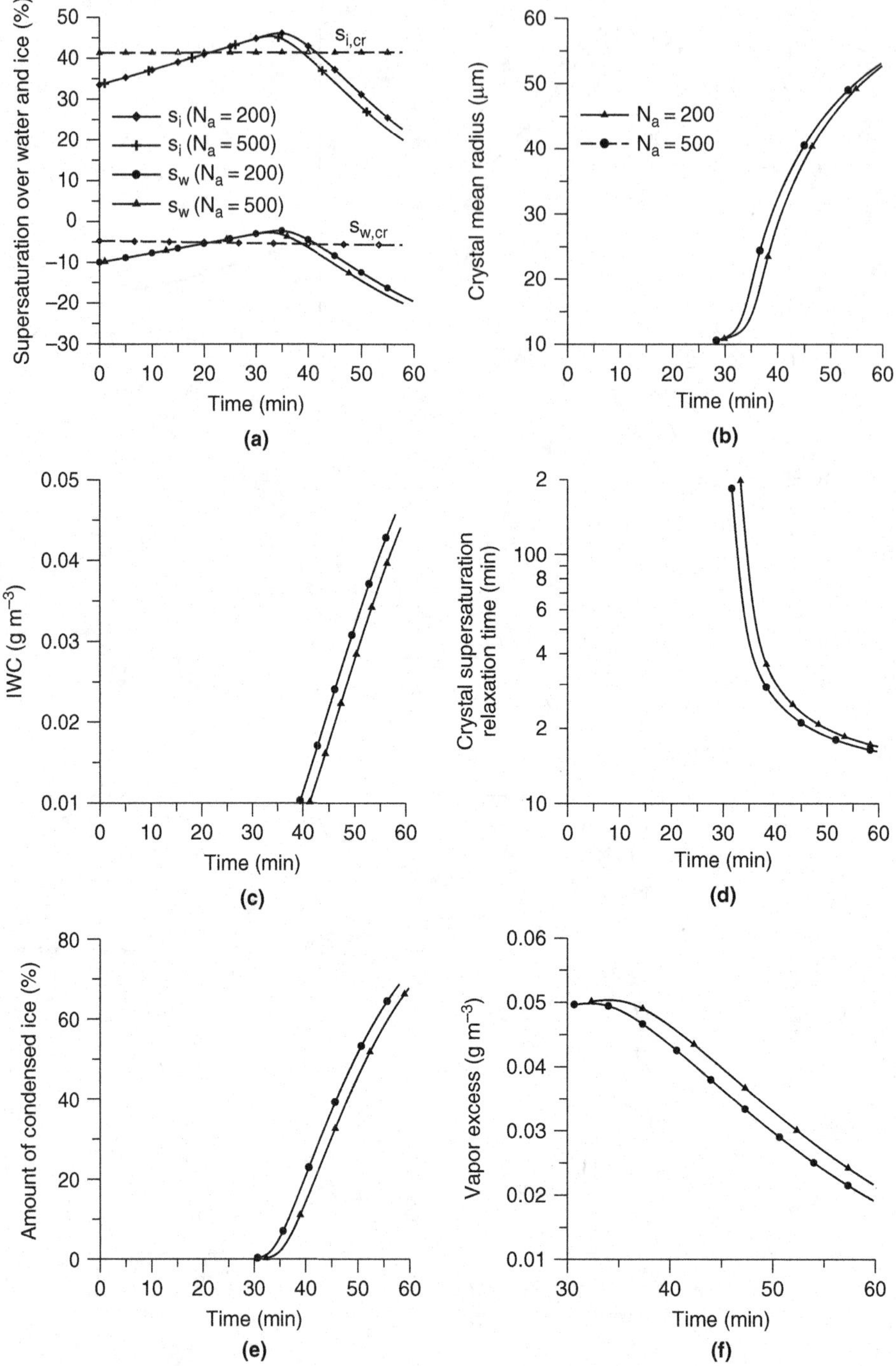

Figure 8.20. Comparison (continuation) of homogeneous ice nucleation kinetics at $N_a = 200\,\text{cm}^{-3}$ (solid circles) and $500\,\text{cm}^{-3}$ (triangles), at $-40\,°\text{C}$ and the other parameters as in Fig. 8.19. (a) Supersaturations over water, s_w, and ice, s_i, and critical supersaturations, $s_{w,cr}$, and ice, $s_{i,cr}$, %; (b) crystal mean radius, μm; (c) ice water content, $\text{g}\,\text{m}^{-3}$; (d) crystal supersaturation relaxation time, min; (e) relative amount of condensed ice, %; (f) vapor excess, $\text{mg}\,\text{m}^{-3}$. From Khvorostyanov and Curry, Atmos. Chem. Phys. (2012), reproduced with permission.

This very small fraction of freezing solution particles is explained by the following factors: a) very strong negative feedback by the water supersaturation: even a small decrease in s_w causes a significant decrease in the nucleation rate $J_{f,hom}$, where the dependence on s_w is exponential with a large coefficient, as will be demonstrated in Section 8.10; b) much faster crystal growth at high ice supersaturation than drop growth at small water supersaturation, so that the second term in (Eqn. 8.9.2) with supersaturation absorption becomes greater than the first term with production much faster than in the case with drops activation and their much slower growth at small water supersaturation. This explanation, based on numerical simulations, is confirmed by analytical parameterizations described in the next section.

The effect of temperature is illustrated in Figs. 8.21 and 8.22, where a comparison is made for the cases –40 °C and –60 °C, at $w = 4\,\mathrm{cm\,s^{-1}}$, and all the other parameters as before. The initial $RHW_0 =$ 78% at –60 °C, and maximum RHW = 89%. The critical $s_{w,cr} = -17\%$ and $s_{i,cr} = 49\%$ vs. –4% and 42%, respectively, at $T_0 = -40$ °C; maximum supersaturations are $s_{w,max} = -11\%$ vs. –2.4% and $s_{i,max} = 62.1\%$ vs. $s_{i,max} = 46.0\%$ at $T = -40$ °C. Thus, the critical and maximum water supersaturations (negative) decrease and ice supersaturations increase with decreasing temperature. Minimum critical radius and energy are comparable at both temperatures, while the nucleation rates increase 4–7 times at lower T. The crystal concentration grows almost 4 times to $242\,\mathrm{L^{-1}}$ at lower T (Fig. 8.21f), but crystal growth is slower. Therefore, the mean radius is about 4 times smaller and the fraction of condensed ice is lower by the end of the simulation at $t = 1$ hour, and the supersaturation relaxation times are close, ~15–17 min, since the increase in crystal concentration is balanced by a decrease in the mean radius (Fig. 8.22). Thus, the amount of condensed ice is again smaller than would be the case in a bulk model, the cloud radiates less than a black body, and the error of the black-body approximation increases with lowering temperature.

Some properties of the nucleation rates allow simplifications of the nucleation equations. The nucleation rates are very small at all stages of the process ($J_{f,hom} r_d^3 < 10^{-5} - 10^{-4}\,\mathrm{s^{-1}}$ with the radius of a deliquescent haze particle $r_d = 0.11$ μm, and $R_{f,hom} < 10^{-3} - 10^{-1}\,\mathrm{cm^{-3}\,s^{-1}}$ even at their maxima; see Figs. 8.17, 8.19, 8.21, units (d) and (e)). Therefore, the expressions introduced in Section 8.3 can be substantially simplified since

$$\exp\left(-\int_0^t J_{f,\mathrm{hom}}(t')v(r_d)dt'\right) \approx 1 - \int_0^t J_{f,\mathrm{hom}}(t')v(r_d)dt'. \tag{8.9.7}$$

The probability $P_{f,hom}(r_d, t)$ (Eqn. 8.3.6) of homogeneous freezing of a haze particle or a drop with radius r_d and volume $v(r_d)$ during the time interval from t_0 to t can be simplified as

$$P_{f,\mathrm{hom}}(r_d,t) = 1 - \exp\left(-\int_{t_0}^t J_{f,\mathrm{hom}}(t')v(r_d)dt'\right) \approx \int_{t_0}^t J_{f,\mathrm{hom}}(t')v(r_d)dt'. \tag{8.9.8}$$

Equation (8.3.9) for crystal concentration $N_{c,hom}$ in a polydisperse aerosol (deliquescent haze particles or drops) can be simplified as

$$\begin{aligned} N_{c,\mathrm{hom}}(t) &= \int_{r_{\min}}^{r_{\max}} P_{f,\mathrm{hom}}(r_d,t) f_d(r_d)\,dr_d, \\ &\approx \int_{r_{\min}}^{r_{\max}} \int_{t_0}^t J_{f,\mathrm{hom}}(t')v(r_d) f_d(r_d)\,dt'dr_d. \end{aligned} \tag{8.9.9}$$

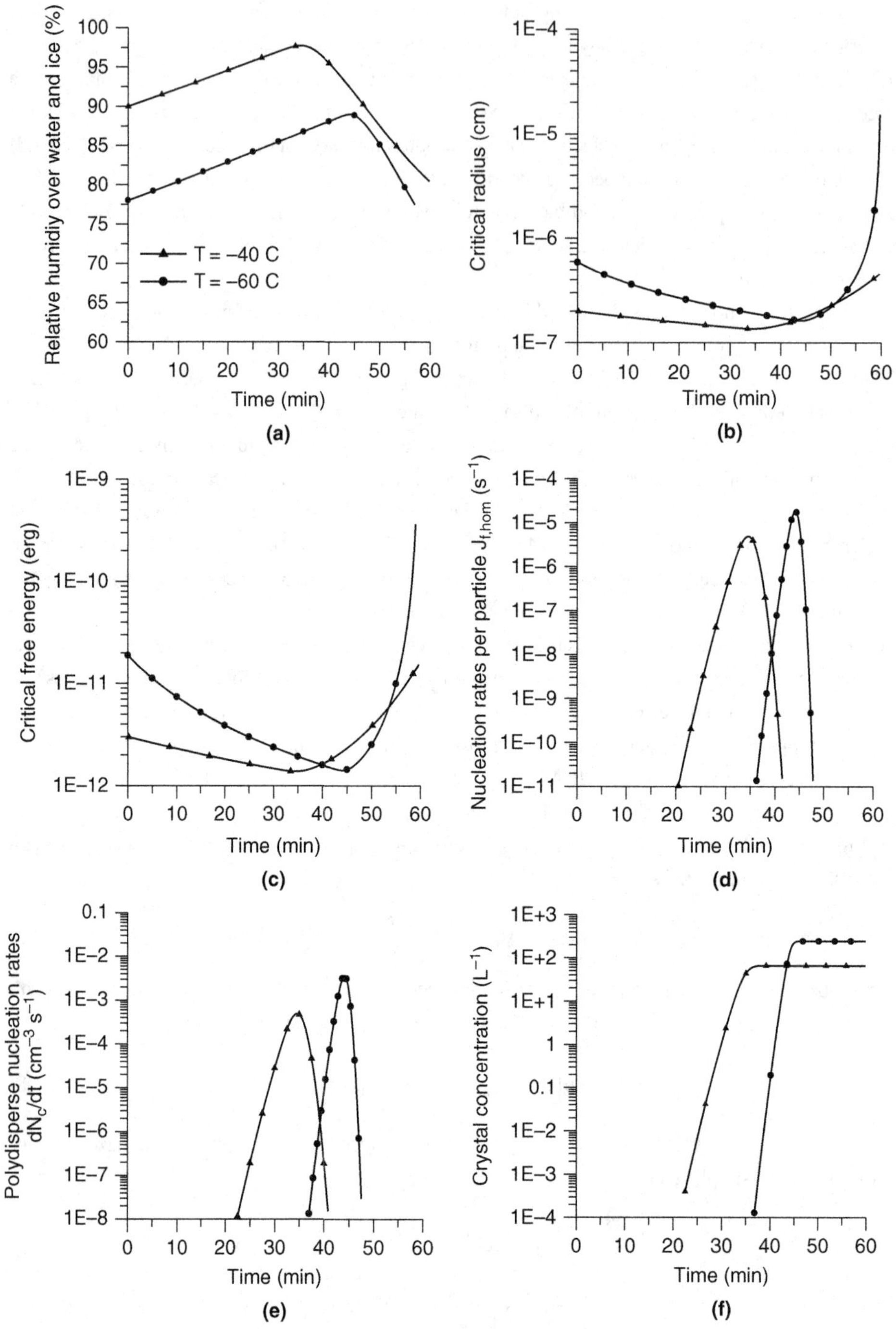

Figure 8.21. Comparison of homogeneous nucleation kinetics at $T = -40\,°C$ and $-60\,°C$. The other parameters are $RHW_0(-40\,°C) = 90\%$ and $RHW_0(-60\,°C) = 78\%$, $p_0 = 340$ hPa, and $w = 4\,cm\,s^{-1}$. (a) Relative humidity over water RHW; (b) critical radius r_{cr}; (c) critical free energy ΔF_{cr}; (d) homogeneous $J_{f,hom}r_a^3$ nucleation rates for a particle with radius of 0.11 μm; (e) polydisperse nucleation rates, $R_{f,hom} = dN_{fr}/dt$, defined by (Eqn. 8.3.10); (f) crystal concentration. From Khvorostyanov and Curry, Atmos. Chem. Phys. (2012), reproduced with permission.

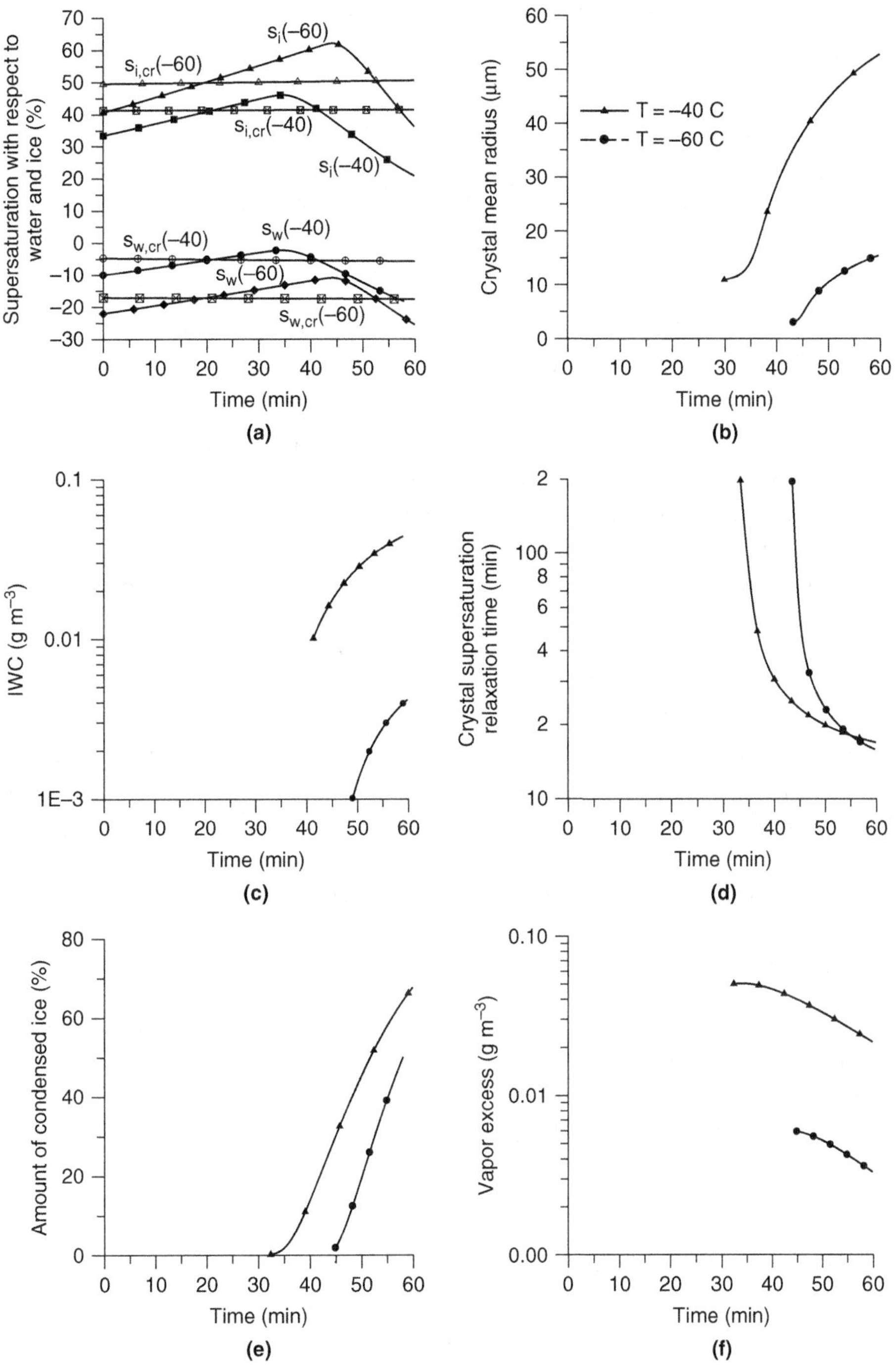

Figure 8.22. Comparison of homogeneous nucleation kinetics at $T = -40\,°C$ and $-60\,°C$ (continuation). The other parameters are $RHW_0(-40\,°C) = 90\%$ and $RHW_0(-60\,°C) = 78\%$, $p_0 = 340\,hPa$, $w = 4\,cm\,s^{-1}$. (a) Supersaturations over water, s_w, and ice, s_i, and critical supersaturations, $s_{w,cr}$, $s_{i,cr}$, %; (b) crystal mean radius, μm; (c) ice water content, $g\,m^{-3}$; (d) supersaturation relaxation time, min; (e) relative amount of condensed ice, %; (f) vapor excess, $mg\,m^{-3}$. From Khvorostyanov and Curry, Atmos. Chem. Phys. (2012), reproduced with permission.

The crystal nucleation rate (Eqn. 8.3.10) in a polydisperse aerosol can be simplified as:

$$R_{f,\mathrm{hom}}(t) = \frac{dN_{f,\mathrm{hom}}}{dt} = \int_{r_{\min}}^{r_{\max}} f_d(r_d) v(r_d) J_{f,\mathrm{hom}}(t) dr_d. \quad (8.9.10)$$

These simplifications can be very useful since they significantly shorten the number of arithmetic operations in numerical modeling and allow analytical calculations. These features of the homogeneous ice nucleation kinetics are used for development of the analytical ice nucleation parameterization derived next in Section 8.10.

8.10. Analytical Parameterization of Homogeneous Ice Nucleation Kinetics Based on Classical Nucleation Theory

Previous parameterizations of homogeneous ice nucleation suitable for the cloud and climate models described in Section 8.4 have mostly been semi-empirical, whereby they used either experimental data or elements of the classical nucleation theory along with the supersaturation equation and a few additional heuristic relations tuned by fitting either experimental data or data from parcel model simulations. In this section, a new parameterization of homogeneous nucleation kinetics is derived, which is based directly on the extended classical nucleation theory and on the analytical solutions of the supersaturation equation.

8.10.1. General Features of Homogeneous Ice Nucleation Kinetics

The general features of homogeneous ice nucleation kinetics described in Section 8.9 are illustrated with better resolution in Fig. 8.23, which shows parcel model simulations of a typical general evolution of water and ice supersaturations and crystal concentration. The symbols $t_{cr,1}$ and $t_{cr,2}$ denote the 1st and 2nd times when the critical (threshold) ice supersaturations $s_{i,cr1}$ and $s_{i,cr2}$ are reached (i.e., the start and end of nucleation), and t_{max} is the time when maximum ice and water supersaturations, $s_{i,max}$ and $s_{w,max}$, are reached. A comparison of Fig. 8.23 with the drop nucleation kinetics described in Chapters 6 and 7 shows that homogeneous ice nucleation has features that are both similar to and different from drop nucleation. In both cases, supersaturation increases due to adiabatic cooling updraft, but in contrast to drop activation, ice nucleation begins at water subsaturations of a few percent (about –3 to –4%), at the time $t_{cr,1}$, when some critical ice supersaturation $s_{i,cr1} \sim 0.42$ (42%) is reached.

As in the case with drop activation, the $s_i(t)$ and $s_w(t)$ curves consist of two branches with increasing and decreasing supersaturations. However, in contrast to drop activation, nucleation does not cease at t_{max}, when maximum $s_{i,max}$ and $s_{w,max}$ are reached. Only about 50–60% of the final crystal concentration has been nucleated by this time (the ellipse in Fig. 8.23b), and nucleation continues along the branch with decreasing supersaturation to the point $t_{cr,2}$, $s_{i,cr2}$ when $s_i(t)$ again intersects the line $s_{i,cr}(t)$. It is seen that an increase in both s_w and s_i is linear almost to the point of maximum.

As discussed in Section 8.7, an important feature of ice nucleation is that an increase in s_i from minimum $s_{i,cr1} \sim 0.42$ to the maximum $s_{max} = 0.46$ is only 0.04, less than 10% of the initial critical value $s_{i,cr1}$; that is, crystal growth proceeds at almost constant supersaturation. With varying

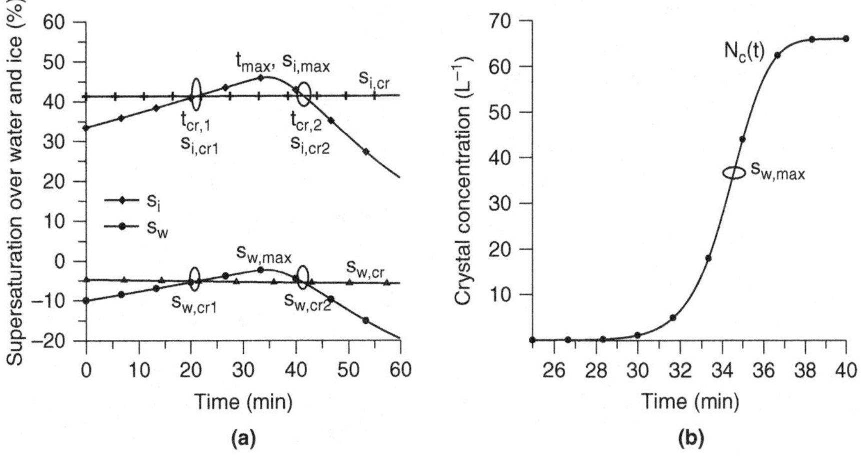

Figure 8.23. General features of homogeneous ice nucleation kinetics (evolution of water and ice supersaturations and crystal concentration) illustrated with a parcel model run with the parameters: initial temperature $T_c = -40\,°\mathrm{C}$, $s_w(t = 0) = -0.1$ (−10%), lognormal size spectrum of haze particles with mean geometric radius of 0.02 μm and concentration $N_a = 200\,\mathrm{cm}^{-3}$. The symbols $t_{cr,1}$ and $t_{cr,2}$ (marked with ellipses) denote the first and second times when critical (threshold) ice supersaturations $s_{i,cr1}$ and $s_{i,cr2}$ are reached—that is, the start and end of nucleation; t_{max} is the time when maximum ice and water supersaturations $s_{i,max}$, $s_{w,max}$ are reached; $s_{w,cr}$, $s_{i,cr}$, denote the curves of critical (threshold) water and ice supersaturations. From Khvorostyanov and Curry, Atmos. Chem. Phys. (2012), reproduced with permission.

temperature, pressure, vertical velocity, and aerosol properties, this picture varies quantitatively, but the general qualitative features remain similar. Accounting for these features allows significant simplifications of calculations of kinetics of the homogeneous ice nucleation.

The basic equations describing the kinetics of homogeneous ice nucleation include the integro-differential equations for water and ice supersaturations derived in Section 8.5, and the equation for crystal radius growth rate taking into account the kinetic effects, also given in Section 8.5. Additionally, we need the equation for the homogeneous nucleation rate of haze particles taking into account solution effects, the equation for the critical supersaturation $s_{w,cr}$, and the equations for the critical radius and energy of homogeneous nucleation with some simplifications that are described in the following text.

8.10.2. The Freezing Rate and Its Simplification

The equation for the critical water supersaturation $s_w = S_w - 1$ follows from (8.8.2):

$$s_{w,cr}^{\mathrm{hom}} = S_{w,cr}^{\mathrm{hom}} - 1 = [(T/T_0)\exp(H_{v,fr} + H_{f,\mathrm{hom}})]^{1/G_n} - 1, \tag{8.10.1}$$

where $H_{v,fr}$ and $H_{f,hom}$ are defined in (Eqns. 8.6.16) and (8.7.5). The corresponding ice saturation ratio S_i and supersaturation s_i can be obtained using (Eqn. 8.8.21c). In the following text, we denote the critical supersaturation of homogeneous freezing with the superscript "hom" to distinguish from heterogeneous freezing considered in Section 10.1. Hereafter, we use a common notation r_a for the radius of the aerosol particles, and $f_a(r_a)$ for their size spectra.

The polydisperse freezing rate $R_{f,hom} = dN_c(t_0)/dt_0$ can be calculated using the classical nucleation theory as described by (Eqn. 8.3.10) in Section 8.3. It was illustrated in Section 8.9 (Figs. 8.17, 8.19, and 8.21) that at typical cooling rates (w), the homogeneous nucleation rates per particle, $J_{f,hom}r_a^3 < 10^{-5} - 10^{-4}\,\mathrm{s}^{-1}$ and $R_{f,hom} < 10^{-3} - 10^{-1}\,\mathrm{cm}^{-3}\,\mathrm{s}^{-1}$, and the inner integral in the exponent in (Eqn. 8.3.10) is close to 1. Therefore, the approximation (Eqn. 8.9.10) can be used as a good approximation for $R_{f,hom}$:

$$R_{f,\mathrm{hom}}(t_0) = \frac{dN_{fr}}{dt} = \int_{r_{\min}}^{r_{\max}} dr_a f_a(r_a) v(r_a) J_{f,\mathrm{hom}}(t_0). \tag{8.10.2}$$

This expression can be further simplified if the depletion of $v(r_a)$ and $f_a(r_a)$ are small during freezing, which is usually a good approximation with abundant concentrations of the freezing particles

$$R_{f,\mathrm{hom}}(t_0) \approx J_{f,\mathrm{hom}}(t_0) \int_{r_{\min}}^{r_{\max}} dr_a f_a(r_a) v(r_a) = N_a \overline{v}_a J_{f,\mathrm{hom}}(t_0). \tag{8.10.3}$$

where $\overline{v}_a$ is the mean aerosol volume averaged over the aerosol size spectrum:

$$\overline{v}_a = \frac{4}{3}\pi \frac{1}{N_a} \int_{r_{\min}}^{r_{\max}} r_a^3 f_a(r_a) dr_a. \tag{8.10.4}$$

In general, N_a and $\overline{v}_a$ depend on time; however, the fraction of haze particles nucleated into crystals is very small compared to the initial haze population. Therefore, I_{dep} from (Eqn. 8.5.16) can be further simplified assuming $N_a \approx$ const, $\overline{v}_a \approx$ const:

$$I_{dep} \approx y_i' \frac{4\pi D_v \rho_{is}}{\Gamma_2} N_a \overline{v}_a \int_0^t r_{c,ef}(t, t_0) J_{f,\mathrm{hom}}(t_0)\, dt_0, \tag{8.10.5}$$

where $r_{c,ef}(t, t_0)$ is defined in (Eqn. 8.5.14).

8.10.3. Separation of Temperature and Supersaturation Dependencies

The nucleation rate $J_{f,hom}(T, s_w)$ defined by (Eqn. 8.6.21) depends on temperature and water saturation ratio or supersaturation. Analytical solution of the supersaturation equation requires some simplifications; in particular, it is desirable to find a representation of $J_{f,hom}$ with separated T- and s_w-dependencies.

The quantity $J_{f,hom}$ can be calculated using (Eqn. 8.3.5) from the classical nucleation theory as derived in Section 8.3:

$$J_{f,\mathrm{hom}} = C_{\mathrm{hom}} \exp\left[-\frac{\Delta F_{act}(T) + \Delta F_{cr}(T, S_w, \Delta p, r_a)}{kT} \right], \tag{8.10.6}$$

where C_{hom} is defined in (Eqn. 8.3.5a), ΔF_{act} and ΔF_{cr} are the activation and critical energies of an ice germ freezing, and we wrote explicitly about their arguments according to Section 8.6. The energy ΔF_{act} is expressed as a function of temperature. The critical free energy ΔF_{cr} of homogeneous freezing can be expressed via the critical radius r_{cr} of a nucleated ice germ; it is also a function of T. The analytical expressions for r_{cr} and ΔF_{cr} were generalized in Section 8.6, taking into account the effect of the solution in haze particles, so that they can be expressed as the functions of T and the water saturation ratio S_w simultaneously. The effects of pressure in (Eqn. 8.6.18) for ΔF_{cr} become substantial only at very high pressures that can be achieved in laboratory experiments and so can be neglected under atmospheric conditions. Neglecting also the effects of curvature for submicron particles—i.e., assuming $H_{v,fr} = 0$ in (Eqn. 8.6.18), we can use (Eqn. 8.6.20) for ΔF_{cr}. Here, we express ΔF_{cr} via water supersaturation s_w using (Eqn. 8.6.20) and the relation $S_w = 1 + s_w$:

$$\Delta F_{cr} = \frac{(16\pi/3)\sigma_{is}^3}{\left\{\rho_i L_m^{ef}(T)\ln\left[\frac{T_0}{T}(1+s_w)^{G_n}\right]\right\}^2}, \tag{8.10.7}$$

where $G_n(T) = RT/(M_w L_m^{ef})$ is a dimensionless parameter defined in (Eqn. 8.6.15); the other quantities were defined in Section 8.6.

The equation for ΔF_{cr} can be transformed so that the dependencies of T and s_w will be separated. It was found in observations and model simulations (see Section 8.9) that the homogeneous freezing of haze particles in cirrus clouds usually occurs at small water subsaturations of −2% to −10%—i.e., $s_w = -2 \times 10^{-2}$ to -10×10^{-2}, so that $|s_w| << 1$. Since $|s_w| << 1$, we can expand the denominator in (Eqn. 8.10.7) into a power series in s_w. The logarithmic term can thus be transformed as

$$\ln\left[\frac{T_0}{T}(1+s_w)^{G_n}\right] \approx \ln\left[\frac{T_0}{T}(1+G_n s_w)\right]$$
$$\approx \ln\left(\frac{T_0}{T}\right) + G_n s_w = \ln\left(\frac{T_0}{T}\right)\left(1+\frac{G_n s_w}{\ln(T_0/T)}\right), \tag{8.10.8}$$

where we use a relation $\ln(1 + G_n s_w) \approx G_n s_w$ for $|s_w| << 1$ and $G_n \sim 0.4 - 0.6$. Substituting this expansion into (Eqn. 8.10.7), we obtain

$$\Delta F_{cr}(T, s_w) \approx \Delta F_{cr,0}^{\text{hom}}(T)[1 - \kappa_s s_w], \tag{8.10.9}$$

where

$$\kappa_s = \frac{2G_n}{\ln(T_0/T)} = \frac{2RT}{M_w L_m^{ef}\ln(T_0/T)}, \tag{8.10.9a}$$

$$\Delta F_{cr,0}^{\text{hom}} = \frac{(16\pi/3)\sigma_{is}^3}{[\rho_i L_m^{ef}(T)\ln(T_0/T)]^2}. \tag{8.10.10}$$

That is, $\Delta F_{cr,0}^{\text{hom}}$ is the critical energy for pure water defined by (Eqn. 8.10.7) but at $S_w = 1$ or $s_w = 0$—i.e., it depends only on temperature but does not depend on supersaturation. For $T \sim -50\,°\text{C}$, an estimate gives $G_n \sim 0.5$, and $\kappa_s \sim 5$, that is, with $s_w = -3 \times 10^{-2}$ (−3%), this yields $\kappa_s s_w \sim -0.15 << 1$. An estimate shows that the second-order term in expansion by $\kappa_s s_w$ in (Eqn. 8.10.9) contributes ~ 2–3%; therefore, retaining only the first term in expansion (Eqn. 8.10.9) is justified. The substitution of (Eqn. 8.10.9) into (Eqn. 8.10.6) yields

$$J_{f,\text{hom}}(T, s_w) = J_{f,\text{hom}}^{(0)}(T)\exp[u_{s\,\text{hom}}(T)s_w(t)], \tag{8.10.11a}$$

$$J_{f,\text{hom}}(T, s_w) = J_{f,\text{hom}}^{(0)}(T)[b_{\text{hom}}(T)]^{s_w(t)}, \tag{8.10.11b}$$

so that $J_{f,hom}$ can be written such that the s_w-dependence is presented in the exponential or power law forms via the parameters b_{hom} and u_{shom}:

$$u_{s\,\mathrm{hom}}(T) = \frac{\Delta F^{\mathrm{hom}}_{cr,0}}{kT}\frac{2G_n}{\ln(T_0/T)}$$
$$= \frac{2R}{kM_w L^{ef}_m}\frac{\Delta F^{\mathrm{hom}}_{cr,0}}{\ln(T_0/T)} = \frac{2N_{Av}}{M_w L^{ef}_m}\frac{\Delta F^{\mathrm{hom}}_{cr,0}}{\ln(T_0/T)}, \tag{8.10.12}$$

$$b_{\mathrm{hom}}(T) = \exp(u_{s\,\mathrm{hom}}), \tag{8.10.12a}$$

where $J^{(0)}_{f,\mathrm{hom}}(T)$ is the nucleation rate for pure water—i.e., is defined by (Eqn. 8.10.6) with $S_w = 1$ or $s_w = 0$, and N_{Av} is the Avogadro number. Thus, $J_{f,hom}(T, s_w)$ is presented in a separable form as a product of the two factors: $J^{(0)}_{f,\mathrm{hom}}$ depends on T but does not depend on s_w, and the dependence on s_w is separated into the exponent in (Eqn. 8.10.11a). An estimate shows that at cirrus conditions $u_{shom} \sim (2-4)\times 10^2 >> 1$. Since $s_w < 0$ in the nucleation process, the value of $u_{shom}s_w$ is negative. If $s_w \sim -(4 \text{ to } 10)\times 10^{-2}$, as estimated earlier under typical nucleation conditions, the value of $|u_{shom}s_w| > 10$, and we have an inequality $\exp(u_{shom}s_w) << 1$.

Numerical simulation with the parcel model shows that changes in $J^{(0)}_{f,\mathrm{hom}}(T)$ in (Eqns. 8.10.11a) and (8.10.11b) are several orders of magnitude smaller than variations in $\exp(u_{shom}s_w)$. This is illustrated in Fig. 8.24, which shows that $J^{(0)}_{f,\mathrm{hom}}(T) \sim (4-5)\times 10^5\ \mathrm{cm}^{-3}\ \mathrm{s}^{-1}$ and only slightly varies during the nucleation event, while $J_{f,\mathrm{hom}}(T, s_w) \sim 10$ in maximum and varies (decreases from maximum) by 10 orders of magnitude at the beginning and at the end of nucleation. This is caused by the effect of $\exp(u_{shom}s_w)$, which reaches in maximum $\sim 10^{-5}$ at $t = 34.5$ min, the time of maximum of s_w, and decreases by 10 orders of magnitude otherwise. Fig. 8.24 shows that: a) the ratio $J_{f,\mathrm{hom}}(T, s_w)/J^{(0)}_{f,\mathrm{hom}}(T)$ is very close to $\exp(u_{shom}s_w)$, supporting the validity of the analytical representation of the nucleation rate

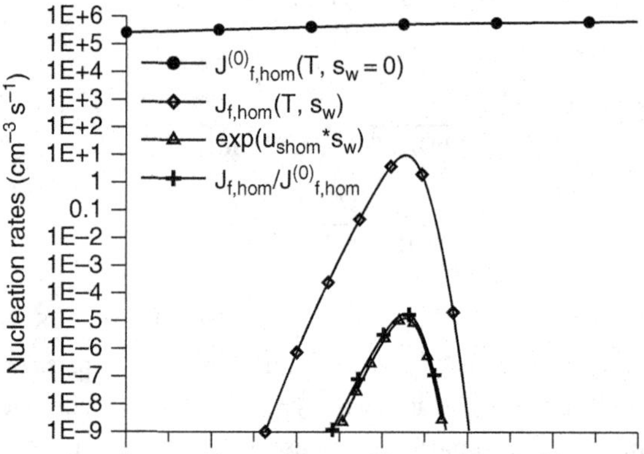

Figure 8.24. Homogeneous nucleation rates $J_{f,\mathrm{hom}}(T, s_w)$, $J^{(0)}_{f,\mathrm{hom}}(T, s_w = 0)$, their ratio $J_{f,\mathrm{hom}}(T, s_w)/J^{(0)}_{f,\mathrm{hom}}(T, s_w = 0)$, and $\exp(u_{shom}s_w)$ that determines this ratio. Calculations for the same conditions as in Fig. 8.23. It is seen that $J_{f,hom}/J^{(0)}f,hom$ is very close to $\exp(u_{shom}s_w)$, which is a good approximation to this ratio. From Khvorostyanov and Curry, Atmos. Chem. Phys. (2012), reproduced with permission.

in (Eqn. 8.10.11a) in the separable form by T and s_w; b) the major variations in $J_{f,\mathrm{hom}}(T,s_w)$ occur due to variations in humidity via s_w, while changes due to the temperature are several orders smaller. Therefore, the deposition integral I_{dep} in (Eqn. 8.10.5) can be presented in the form that substantially simplifies calculations:

$$I_{dep} = y_i' \frac{4\pi D_v \rho_{is}}{\Gamma_2} N_a \bar{v}_a J_{f,\mathrm{hom}}^{(0)} \int_0^t r_{c,ef}(t,t_0) \exp[u_{s\mathrm{hom}} s_w(t_0)] dt_0, \tag{8.10.13}$$

or in the other form, introducing the integral J_{0i} as

$$I_{dep} = y_i' \frac{4\pi D_v}{\Gamma_2} N_a \bar{v}_a J_{f,\mathrm{hom}}^{(0)} J_{0i}, \tag{8.10.13a}$$

$$J_{0i} = \int_0^t r_{c,ef}(t,t_0) \exp[u_{s\mathrm{hom}} s_w(t_0)] dt_0. \tag{8.10.13b}$$

8.10.4. The Evolution of the Nucleation Rate and Crystal Concentration

The integro-differential supersaturation (Eqn. 8.5.18) for the case of homogeneous ice nucleation is similar to that used for drop activation in Chapters 6 and 7. In the theory of drop activation, the time dependence of supersaturation from the beginning of activation to the maximum at $t = t_{max}$ is often approximated by a liner function, $s_w = c_{1w}w$, which is determined by the first term on the RHS of (Eqn. 8.5.18). The initial value is assumed to be zero since drop activation begins at a very small positive s_w. In contrast, as Fig. 8.23 shows, homogeneous ice nucleation begins at slightly negative water supersaturation s_w but high ice supersaturation s_i and proceeds at almost constant s_i.

We seek a solution to the supersaturation equation similar to that used for the drops as a linear by t approximation for the differential supersaturations s_w, s_i and a quadratic by t approximation for the integral supersaturations y_w, y_i, but with the initial critical (threshold) values, which account for the specifics of ice nucleation:

$$s_i(t) = y_i'(t) = s_{i,cr}^{\mathrm{hom}} + a_{1i}t, \qquad y_i(t) = s_{i,cr}^{\mathrm{hom}}t + (a_{1i}/2)t^2. \tag{8.10.14}$$

$$s_w(t) = y_w'(t) = s_{w,cr}^{\mathrm{hom}} + a_{1w}t, \qquad y_w(t) = s_{w,cr}^{\mathrm{hom}}t + (a_{1w}/2)t^2. \tag{8.10.15}$$

As described in Chapter 7, the parameters a_{1w} and a_{1i} can be specified in various ways, which yield the lower and upper limits of the solution. An approximation that gives a lower bound of the solution can be obtained with $a_{1w} = c_{1w}w$. The difference between the limits is on the order of 10–20% or smaller, but for simplicity we will consider the approximations $a_{1w} = c_{1w}w$ and $a_{1i} = c_{1i}w$ as prompted by the supersaturation equations (Eqns. 8.5.1), (8.5.2), and (8.5.18), and described by (Eqns. 8.10.14) and (8.10.15). As discussed earlier, Fig. 8.23 shows that the increase $\Delta s_i = c_{1i}w(t_{max} - t_0) \sim 0.04$ (4%) during ice nucleation from t_0 to t_{max} is much smaller than the initial critical $s_{i,cr} \sim 0.042$ (42%) or maximum $s_{i,max} \sim 0.046$ (46%). Since $\Delta s_i << s_{i,cr}$, we can neglect Δs_i in y_i in (Eqn. 8.10.14), which was also neglected in the previous parameterizations of homogeneous ice nucleation (Kärcher and Lohmann, 2002a,b; Ren and MacKenzie, 2005). We also assume that $s_i(t) \approx \mathrm{const} \approx s_{i,cr}$ and neglect increases in s_i during the nucleation event. In contrast, we cannot neglect increases in s_w—i.e., the term

$\Delta s_w = c_{1w}w(t_{max} - t_0)$—because water supersaturation varies substantially and determines variations in $J_{f,hom}$ (see Fig. 8.24). Thus, we use s_i, y_i, s_w, and y_w as

$$s_i(t) = y_i'(t) = s_{i,cr}^{\text{hom}}, \qquad y_i(t) = s_{i,cr}^{\text{hom}} t, \tag{8.10.16a}$$

$$s_w(t) = y_w'(t) = s_{w,cr}^{\text{hom}} + c_{1w}wt, \qquad y_w(t) = s_{w,cr}^{\text{hom}} t + (c_{1w}w/2)t^2. \tag{8.10.16b}$$

Substitution of $s_w(t)$ into the separable nucleation rate in (Eqn. 8.10.11a) yields $J_{f,hom}(T, s_w)$ as a function of time in the form

$$J_{f,\text{hom}}[T, s_w(t)] = J_{f,\text{hom}}^{(0)}(T_{cr})\exp(u_{s\,\text{hom}} s_{w,cr}^{\text{hom}})\exp(u_{s\,\text{hom}} c_{1w} wt), \tag{8.10.17a}$$

where u_{shom} is defined in (Eqn. 8.10.12). We assumed here, based on Fig. 8.24, that the major time dependence is determined by s_w, and the temperature dependence is determined near T_{cr}. Dividing $J_{f,hom}(t)$ by $J_{f,hom}(t_0)$ at some initial t_0, we obtain the time dependence $J_{f,hom}(t)$ of the form

$$J_{f,\text{hom}}(t) = J_{f,\text{hom}}(t_0)\exp[u_{s\,\text{hom}} c_{1w} w(t - t_0)]. \tag{8.10.17b}$$

For $t_0 = t_{cr}$, (Eqn. 8.10.17b) can be rewritten with (Eqn. 8.10.17a) as

$$\ln\frac{J_{f,\text{hom}}[s_w(t)]}{J_{f,\text{hom}}[s_{w,cr}^{\text{hom}}(t_{cr})]} = u_{s\,\text{hom}} c_{1w} w(t - t_{cr}) = u_{s\,\text{hom}}(T)[s_w(t) - s_{w,cr}^{\text{hom}}]. \tag{8.10.17c}$$

Using the relation following from the Clausius–Clapeyron equation of Chapter 3,

$$s_w + 1 = c_{iw}(s_i + 1), \qquad c_{iw} = \exp[-L_m(T_0 - T)/R_v T_0 T], \tag{8.10.17d}$$

where R_v is the vapor gas constant and $T_0 = 273.15$, we express s_w in (Eqn. 8.10.17c) via the ice saturation ratio $S_i = s_i + 1$ and obtain

$$\ln\frac{J_{f,\text{hom}}[s_w(t)]}{J_{f,\text{hom}}[s_{w,cr}^{\text{hom}}(t_{cr})]} = u_{s\,\text{hom}}(T)c_{iw}(T)[S_i(t) - S_{i,cr}^{\text{hom}}]. \tag{8.10.17e}$$

This expression has the same form as (Eqn. 8.4.32) hypothesized by Barahona and Nenes (2008, BN08) by analogy with the separable representation of the heterogeneous nucleation rate in Khvorostyanov and Curry (2004b). Now, this functional form is derived from the theory of homogeneous freezing and the empirical coefficient b_τ introduced in BN08 and fitted with experimental data is expressed from the extended classical nucleation theory as $b_\tau(T) = u_{shom}(T)c_{iw}(T)$.

Equation (8.10.17b) can also be written as

$$J_{f,\text{hom}}(t) = J_{f,\text{hom}}(t_0)\exp[(t - t_0)/\tau_{nuc,\text{hom}}], \tag{8.10.18}$$

where we introduce the characteristic "homogeneous nucleation time" $\tau_{nuc,hom}$

$$\tau_{nuc,\text{hom}} = (c_{1w} w u_{s\,\text{hom}})^{-1} = \left[c_{1w} w \frac{\Delta F_{cr,0}^{\text{hom}}}{kT}\frac{2G_n}{\ln(T_0/T)}\right]^{-1}$$
$$= (c_{1w}w)^{-1}\frac{kL_m^{ef}\ln(T_0/T)}{2R_v \Delta F_{cr,0}^{\text{hom}}} = (c_{1w}w)^{-1}\frac{M_w L_m^{ef}\ln(T_0/T)}{2N_{Av}\Delta F_{cr,0}^{\text{hom}}}. \tag{8.10.19}$$

The temporal dependence of $J_{f,hom}(t)$ in (Eqn. 8.10.18) was hypothesized by Ford (1998a,b) and utilized by Kärcher and Lohmann (2002a,b) and Ren and MacKenzie (2005); the time $\tau_{nuc,hom}$ was introduced as some empirical parameter and was found by fitting to some auxiliary relations (Eqns. 8.4.17)–(8.4.19). Here, the time dependence of $J_{f,hom}(t)$ and the nucleation time τ_{nuc} are derived in terms of the extended classical nucleation theory with a dependence on S_w. Equation (8.10.19) shows that $\tau^{-1}_{nuc,\mathrm{hom}} \sim c_{1w}w$; that is, according to (Eqns. 8.4.17) and (8.4.18), $\tau^{-1}_{nuc\ hom}$ is proportional to (dT/dt), in agreement with the temperature equation, the other factors in (Eqn. 8.10.19) determine $\partial \ln J_{hom}/\partial T$ and the empirical coefficient c_τ in (Eqn. 8.4.18). Thus, the approach based on extended classical nucleation theory provides a theoretical basis for the functional forms hypothesized in the previous parameterizations by Ford (1998a,b), Kärcher and Lohmann (2002a,b), Ren and MacKenzie (2005), and Barahona and Nenes (2008) and allows their expression in terms of fundamental thermodynamic parameters. This is again in agreement with the correspondence principle (Section 1.2), which states that the particular theories and empirical parameterizations can be derived from a more general theory, which does not reject the previous theories and parameterizations but unifies and clarifies them and allows us to express their empirical parameters via fundamental constants and functions.

The linear approximation (Eqn. 8.10.16b) for $s_w(t)$ allows description of the time evolution of the polydisperse nucleation rate $R_{f,hom}(t)$ and crystal concentration $N_c(t)$. Substitution of (Eqn. 8.10.17a) into (Eqn. 8.10.3) yields

$$\begin{aligned} R_{f,\mathrm{hom}}(t_0) &\approx N_a \bar{v}_a J^{(0)}_{f,\mathrm{hom}}(T_{cr}) \exp(u_{s\,\mathrm{hom}} s^{\mathrm{hom}}_{w,cr}) \exp(\beta_{\mathrm{hom}} t) \\ &\approx N_a \bar{v}_a J^{(0)}_{f,\mathrm{hom}}(T_{cr}) \exp[u_{s\,\mathrm{hom}} s_w(t)], \end{aligned} \tag{8.10.20}$$

$$\beta_{\mathrm{hom}} = u_{s\,\mathrm{hom}} c_{1w} w = \tau^{-1}_{nuc,\mathrm{hom}}. \tag{8.10.21}$$

Integration over time gives $N_c(t)$:

$$\begin{aligned} N_c(t) &= \int_0^t R_{f,\mathrm{hom}}(t)\,dt \\ &\approx N_a \bar{v}_a J^{(0)}_{f,\mathrm{hom}}(T_{cr}) \beta^{-1}_{\mathrm{hom}} \exp(u_{s\,\mathrm{hom}} s^{\mathrm{hom}}_{w,cr})[\exp(\beta_{\mathrm{hom}} t) - 1] \\ &= N_a \bar{v}_a J^{(0)}_{f,\mathrm{hom}}(T_{cr}) \beta^{-1}_{\mathrm{hom}} \{\exp[u_{s\,\mathrm{hom}} s_w(t)] - \exp(u_{s\,\mathrm{hom}} s^{\mathrm{hom}}_{w,cr})\}, \end{aligned} \tag{8.10.22a}$$

which is the parameterization for $N_c(t)$ that was sought.

The separation of the dependencies of N_c on s_w and T allows introduction of the supersaturation and temperature activity spectra of ice nuclei $\phi_s(T, s_w)$ and $\phi_T(T, s_w)$ by s_w and T defined in (Eqns. 8.5.11) and (8.5.12). Differentiation of (Eqn. 8.10.22a) by s_w and T yields

$$\phi_s(T, s_w) = \frac{\partial N_c}{\partial s_w} \approx N_a \bar{v}_a J^{(0)}_{f,\mathrm{hom}}(T_{cr})(c_{1w} w)^{-1} \exp(u_{s\,\mathrm{hom}} s^{\mathrm{hom}}_{w,cr}) \exp(\beta_{\mathrm{hom}} t), \tag{8.10.22b}$$

$$\begin{aligned} \phi_T(T, s_w) &= \frac{\partial N_c}{\partial T} \approx N_a \bar{v}_a J^{(0)}_{f,\mathrm{hom}}(T_{cr}) \\ &\times \frac{\partial}{\partial T} \{\beta^{-1}_{\mathrm{hom}} \exp(u_{s\,\mathrm{hom}} s^{\mathrm{hom}}_{w,cr})[\exp(\beta_{\mathrm{hom}} t) - 1]\}. \end{aligned} \tag{8.10.22c}$$

The activity spectrum $\phi_s(T, s_w)$ characterizes the rate of ice nucleation with increasing humidity and constant temperature, (similar to that considered for drop activation in Chapters 6 and 7), and the spectrum $\phi_T(T, s_w)$, vice versa, characterizes the rate of ice nucleation with decreasing temperature and constant humidity. Such processes may occur under natural conditions of cirrus or mid-level Ac-As clouds formation with advection of humid air and weak variations of T, as was observed, for example, with lidar near Fairbanks (Sassen et al., 2006; Sassen and Khvorostyanov, 2008), or with the advection of cold air and weak changes of humidity. Using (Eqns. 8.10.22b) and (8.10.22c), the relative role of variations of the temperature and humidity can be estimated, or these processes can be studied in isolation in a cloud chamber.

The relation between β_{hom} and t determines the regime of growth of N_c with time. For example, at $T = -40\,°\mathrm{C}$ with $u_s \sim 250$, $c_{1w} \sim 10^{-5}\,\mathrm{cm}^{-1}$, and $w \sim 10\,\mathrm{cm\,s}^{-1}$, an estimate gives $\beta_{hom} \sim 2.5 \times 10^{-2}\,\mathrm{s}^{-1}$ and $\tau_{nuc,\mathrm{hom}} = \beta_{\mathrm{hom}}^{-1} \sim 40$ s. Thus, for small times, $t << \beta_{\mathrm{hom}}^{-1} \sim 40$ s, (Eqn. 8.10.22a) gives a linear growth of $N_c(t)$ with time:

$$N_c(t) = N_a \bar{v}_a J_{f,\mathrm{hom}}^{(0)}(T_{cr})\, t \exp(u_{s\,\mathrm{hom}} s_{w,cr}^{\mathrm{hom}}). \tag{8.10.23}$$

For large times, $t >> \beta_{\mathrm{hom}}^{-1} = 40$ s, (Eqn. 8.10.22a) produces an exponential time dependence

$$N_c(t) = N_a \bar{v}_a J_{f,\mathrm{hom}}^{(0)}(T_{cr})\tau_{nuc,\mathrm{hom}} \exp(u_{s\,\mathrm{hom}} s_{w,cr}^{\mathrm{hom}} + \beta_{\mathrm{hom}} t)$$
$$\sim \exp[u_{s\,\mathrm{hom}} s_w(t)]. \tag{8.10.24}$$

In this regime, $\ln[N_c(t)] \sim t$, which may explain the linear dependence of $\ln[N_c(t)]$ with the time seen in the Figures of Section 8.9 with parcel model simulations.

It is interesting to note that (Eqn. 8.10.24) for homogeneous nucleation can be presented in the form similar to some empirical parameterizations. We present $N_c(t)$ in (Eqn. 8.10.24) as

$$N_c(s_w) = \exp(\ln A_M + u_{s\,\mathrm{hom}} s_w), \qquad A_M = N_a \bar{v}_a J_{f,\mathrm{hom}}^{(0)}(T_{cr})\tau_{nuc,\mathrm{hom}}. \tag{8.10.25}$$

Using the relation (Eqn. 8.10.17d) as $s_w = c_{iw} s_i + (c_{iw} - 1)$, we replace s_w with s_i and obtain

$$N_c(s_i) = \exp(a_M + b_M s_i), \qquad b_M = u_{s\,\mathrm{hom}} c_{iw}, \tag{8.10.26a}$$

$$a_M = \ln A_M + u_{s\,\mathrm{hom}}(c_{iw} - 1) = \ln[N_a \bar{v}_a J_{f,\mathrm{hom}}^{(0)}(T_{cr})\tau_{nuc,\mathrm{hom}}] + u_{s\,\mathrm{hom}}(c_{iw} - 1) \tag{8.10.26b}$$

The aerosol concentration N_a is included in A_M but can be placed also in front of the exponent as

$$N_c(s_i) = N_a \exp(\tilde{a}_M + b_M s_i), \tag{8.10.26c}$$

$$\tilde{a}_M = \ln[\bar{v}_a J_{f,\mathrm{hom}}^{(0)}(T_{cr})\tau_{nuc,\mathrm{hom}}] + u_{s\,\mathrm{hom}}(c_{iw} - 1). \tag{8.10.26d}$$

Thus, these theoretical parameterizations (Eqns. 8.10.26a–d) are written in the form similar to Meyers et al. (1992) for heterogeneous nucleation, or in a modified form, proportional to N_a, but all the parameters are expressed now via the primary atmospheric and aerosol quantities. They are not constant and substantially vary with temperature, aerosol volume, and cooling rate via w in u_{shom}. Similar formulations will be given in Chapters 9 and 10 for heterogeneous nucleation.

The consideration stated earlier is valid when the effects of the vapor depletion are not strong and are neglected. However, if we consider the nucleation process at longer times and near the point of the maximum supersaturations in the Figures of Section 8.9, the vapor depletion becomes substantial and finally exceeds supersaturation production. Then, a more accurate consideration should include evaluation of the deposition integral I_{dep} and the supersaturation equation, as we did in Chapters 6 and 7 for drop activation, which is done in the following sections.

8.10.5. Evaluation of the Deposition Integral I_{dep}

Evaluation of I_{dep} is analogous to that developed in Khvorostyanov and Curry (2008a, 2009a) for the drop nucleation, described in Chapter 7; however, integration for ice nucleation is more complicated due to the exponential activity spectrum and can be done following Khvorostyanov and Curry (2012). Substitution of $y_i(t)$ from (Eqn. 8.10.16a) into (Eqn. 8.5.14) for $r_{c,ef}(t, t_0)$ yields

$$r_{c,ef}(t,t_0) = \frac{\{[(r_0 + \xi_{dep})^2 + B_{i\,\mathrm{hom}}(t-t_0)]^{1/2} - \xi_{dep}\}^2}{[(r_0 + \xi_{dep})^2 + B_{i\,\mathrm{hom}}(t-t_0)]^{1/2}}, \tag{8.10.27}$$

where

$$B_{i\,\mathrm{hom}} = 2c_{i3}s_{i,cr}^{\mathrm{hom}}. \tag{8.10.28}$$

To evaluate the integral in I_{dep} in (Eqn. 8.10.13), we present $r_{c,ef}(t, t_0)$ in (Eqn. 8.10.27) in the integrand of (Eqn. 8.10.13) as a sum of three terms:

$$r_{c,ef}(t,t_0) = \frac{r^2(t,t_0)}{r(t,t_0)+\xi_{dep}} = r_{c,ef}^{(1)} + r_{c,ef}^{(2)} + r_{c,ef}^{(3)}, \tag{8.10.29}$$

where

$$r_{c,ef}^{(1)}(t,t_0) = [(r_0 + \xi_{dep})^2 + B_{i\,\mathrm{hom}}(t-t_0)]^{1/2}, \tag{8.10.30a}$$

$$r_{c,ef}^{(2)}(t,t_0) = -2\xi_{dep}, \tag{8.10.30b}$$

$$r_{c,ef}^{(3)}(t,t_0) = \xi_{dep}^2[(r_0 + \xi_{dep})^2 + B_{i\,\mathrm{hom}}(t-t_0)]^{-1/2}. \tag{8.10.30c}$$

Substitution of (Eqns. 8.10.29) and (8.10.30a), (8.10.30b), and (8.10.30c) into (Eqn. 8.10.13) for I_{dep} yields

$$I_{dep}(t) = y_i{}' \frac{4\pi D_v \rho_{is}}{\Gamma_2} N_a \bar{v}_a J_{f,\mathrm{hom}}^{(0)} \exp(u_{s\,\mathrm{hom}} s_{w,cr}^{\mathrm{hom}}) J_{0i,\mathrm{hom}}(t), \tag{8.10.31}$$

where $J_{0i,hom}(t)$ represents a sum of the three terms:

$$J_{0i,\mathrm{hom}}(t) = \sum_{k=1}^{3} J_{0i,\mathrm{hom}}^{(k)} = \sum_{k=1}^{3} \int_0^t r_{c,ef}^{(k)}(t,t_0) \exp(\beta_{\mathrm{hom}} t_0)\, dt_0. \tag{8.10.32}$$

Substitution of (Eqn. 8.10.29) with (Eqns. 8.10.30a), (8.10.30b), and (8.10.30c) into (Eqn. 8.10.32) and evaluation of the integral $J_{0i,\mathrm{hom}}(t)$ given by (Eqn. A.8.21) in Appendix A.8 yields

$$J_{0i,\mathrm{hom}}(t) = \exp(\beta_{\mathrm{hom}} t)\Psi_{\mathrm{hom}}, \tag{8.10.33}$$

where Ψ_{hom} is defined by the (Eqns. A.8.22)–(A.8.26),

$$\Psi_{\mathrm{hom}} = \Psi_{1\mathrm{hom}} + \Psi_{2\mathrm{hom}} + \Psi_{3\mathrm{hom}}, \tag{8.10.34}$$

$$\Psi_{1\mathrm{hom}} = e^{\lambda_{\mathrm{hom}}} B_{i\,\mathrm{hom}}^{1/2} \beta_{\mathrm{hom}}^{-3/2} \left[\Gamma\left(\frac{3}{2}, \lambda_{\mathrm{hom}}\right) - \Gamma\left(\frac{3}{2}, \lambda_{\mathrm{hom}} + \beta_{\mathrm{hom}} t\right) \right], \tag{8.10.35}$$

$$\Psi_{2\mathrm{hom}} = 2\xi_{dep} \beta_{\mathrm{hom}}^{-1} (e^{-\beta_{\mathrm{hom}} t} - 1), \tag{8.10.36}$$

$$\Psi_{3\mathrm{hom}} = e^{\lambda_{\mathrm{hom}}} \xi_{dep}^2 (\beta_{\mathrm{hom}} B_{i\,\mathrm{hom}})^{-1/2} \left[\Gamma\left(\frac{1}{2}, \lambda_{\mathrm{hom}}\right) - \Gamma\left(\frac{1}{2}, \lambda_{\mathrm{hom}} + \beta_{\mathrm{hom}} t\right) \right], \tag{8.10.37}$$

$$\lambda_{\mathrm{hom}} = \frac{\beta_{\mathrm{hom}} (r_0 + \xi_{dep})^2}{B_{i\,\mathrm{hom}}} = \frac{(u_{s\,\mathrm{hom}} c_{1w} w)(r_0 + \xi_{dep})^2}{2c_{i3} s_{i,cr}^{\mathrm{hom}}}. \tag{8.10.38}$$

Here, $\Gamma(\alpha, x)$ is the incomplete gamma function—its properties and asymptotics are defined in Appendix A.8. Using (Eqn. 8.10.33) for $J_{0i}(t)$, and the relation $u_{shom}s_{w,cr} + \beta_{hom}t = u_{shom}s_w(t)$, the deposition term I_{dep} can be written as

$$I_{dep}(t) = y_i' \frac{4\pi D_v \rho_{is}}{\Gamma_2}(N_a \bar{v}_a J_{\rm hom}^{(0)}) \exp[u_{s\rm hom} s_w(t)] \Psi_{\rm hom}. \tag{8.10.39}$$

Note that the water supersaturation $s_w(t)$ at a time t is present in the exponent.

The function Ψ_{hom} defined by (Eqns. 8.10.34)–(8.10.37) can be transformed and reduced to the functions more convenient for calculations. Using the recurrent relation for $\Gamma(3/2, x)$ (the next and subsequent relations for the gamma functions $\Gamma(x, \alpha)$ and the error function $\Phi(x) = erf(x)$ are from Gradshteyn and Ryzhik, 1994, see Appendix A.8)

$$\Gamma(\alpha+1, \lambda) = \alpha\Gamma(\alpha, \lambda) + \lambda^{\alpha} e^{-\lambda}, \tag{8.10.40a}$$

and the relation between the gamma function and error function $erf(x)$,

$$\Gamma(1/2, \lambda) = \sqrt{\pi}[1 - erf(\sqrt{\lambda})], \tag{8.10.40b}$$

we can transform the gamma function in Ψ_{1hom} into

$$\Gamma\left(\frac{3}{2}, \lambda\right) = \frac{1}{2}\Gamma\left(\frac{1}{2}, \lambda\right) + \lambda^{1/2} e^{-\lambda} = \frac{\sqrt{\pi}}{2}[1 - erf(\sqrt{\lambda})] + \lambda^{1/2} e^{-\lambda}. \tag{8.10.41}$$

Substituting this relation into (Eqns. 8.10.35)–(8.10.37), we can rewrite Ψ_{1hom} and Ψ_{3hom} with use of only $erf(x) \equiv \Phi(x)$ and without the gamma function, which is more convenient for applications:

$$\begin{aligned}\Psi_{1\rm hom} = {} & e^{\lambda_{\rm hom}} B_{i\rm hom}^{1/2} \beta_{\rm hom}^{-3/2} \{(\sqrt{\pi}/2)[\Phi(\sqrt{\lambda_{\rm hom} + \beta_{\rm hom} t}) - \Phi(\sqrt{\lambda_{\rm hom}})] \\ & + e^{-\lambda_{\rm hom}} [\lambda_{\rm hom}^{1/2} - (\lambda_{\rm hom} + \beta_{\rm hom} t)^{1/2} e^{-\beta_{\rm hom} t)}]\},\end{aligned} \tag{8.10.42}$$

$$\Psi_{3\rm hom} = e^{\lambda_{\rm hom}} \xi_{dep}^2 (\beta_{\rm hom} B_{i\rm hom})^{-1/2} \sqrt{\pi}[\Phi(\sqrt{\lambda_{\rm hom} + \beta_{\rm hom} t}) - \Phi(\sqrt{\lambda_{\rm hom}})]. \tag{8.10.43}$$

Then, the function Ψ_{hom} is expressed with use of only $\Phi(x) = erf(x)$:

$$\begin{aligned}\Psi_{\rm hom} = {} & e^{\lambda_{\rm hom}} \beta_{\rm hom}^{-1/2} \sqrt{\pi}[\Phi(\sqrt{\lambda_{\rm hom} + \beta_{\rm hom} t}) - \Phi(\sqrt{\lambda_{\rm hom}})] \\ & \times [(1/2) B_{i\rm hom}^{1/2} \beta_{\rm hom}^{-1} + \xi_{dep}^2 B_{i\rm hom}^{-1/2}] \\ & + B_{i\rm hom}^{1/2} \beta_{\rm hom}^{-3/2} [\lambda^{1/2} - (\lambda_{\rm hom} + \beta_{\rm hom} t)^{1/2} e^{-\beta_{\rm hom} t}] \\ & + 2\xi_{dep} \beta_{\rm hom}^{-1} (e^{-\beta_{\rm hom} t} - 1).\end{aligned} \tag{8.10.44}$$

This expression can be further simplified by expressing the transcendent function $erf(x)$ via the elementary function $\tanh(x)$ as suggested in Ghan et al. (1993) and used also in Chapter 6 for derivation of the algebraic size and activity spectra for drop activation:

$$erf(x) \approx \tanh[(2/\sqrt{\pi})x]. \tag{8.10.45}$$

Then, Ψ_{hom} becomes

$$\begin{aligned}\Psi_{\rm hom} = {} & e^{\lambda_{\rm hom}} \beta_{\rm hom}^{-1/2} \sqrt{\pi}[\tanh(2\sqrt{(\lambda_{\rm hom} + \beta_{\rm hom} t)/\pi}) - \tanh(2\sqrt{\lambda_{\rm hom}/\pi})] \\ & \times [(1/2) B_{i\rm hom}^{1/2} \beta_{\rm hom}^{-1} + \xi_{dep}^2 B_i^{-1/2}] \\ & + B_{i\rm hom}^{1/2} \beta_{\rm hom}^{-3/2} [\lambda_{\rm hom}^{1/2} - (\lambda_{\rm hom} + \beta_{\rm hom} t)^{1/2} e^{-\beta_{\rm hom} t}] + 2\xi_{dep} \beta_{\rm hom}^{-1} (e^{-\beta_{\rm hom} t} - 1).\end{aligned} \tag{8.10.46}$$

Now, with this Ψ_{hom}, the deposition integral I_{dep} (Eqn. 8.10.39) is expressed only via the elementary functions. Another transition to the elementary functions can be done using equations for $erf(x)$ and $erfc(x)$ given in Ren and MacKenzie (2005, 2007) described by (Eqn. 8.4.28). In the next sections, the solutions of equations for supersaturation and crystal concentration will be expressed via Ψ_{hom}. Although these expressions may look a little complicated, the advantages of analytical representation (Eqns. 8.10.44) and (8.10.46) are that they allow us to reduce the unavoidable errors caused by the finite difference representations and numerical calculations, and enable the derivation of solutions for the asymptotic limits of I_{dep} and N_c for diffusion and kinetics regimes of crystal growth.

8.10.6. Solution of Equations for the Supersaturation and for Crystal Concentration

Substituting the expression for I_{dep} (Eqn. 8.10.39) into the integral supersaturation (Eqn. 8.5.15), multiplying it by $(1 + y_i')$, and using the relation $\rho_v = \rho_{is}(1 + y_i')$ yields

$$\frac{dy_i'}{dt} = c_{1i}w(1+y_i') - \frac{\Gamma_2}{\rho_{is}} I_{dep}$$
$$= c_{1i}w(1+y_i') - (4\pi D_v)y_i'(N_a\bar{v}_a J^{(0)}_{f,\text{hom}})\exp[u_{s\,\text{hom}}s_w(t)]\Psi_{\text{hom}}. \quad (8.10.47)$$

At time t_{max} with $s_{i,max}$ and $s_{w,max}$, there is a condition $ds_i/dt = dy_i'/dt = 0$; thus, the LHS of (Eqn. 8.10.47) is zero, which yields

$$\exp[u_{s\,\text{hom}}s_{w,\max}(t_{\max})] = c_{1i}w(1+s_{i,\max})s_{i,\max}^{-1}(4\pi D_v)^{-1}(N_a\bar{v}_a J^{(0)}_{f,\text{hom}})^{-1}\Psi^{-1}_{\text{hom}}. \quad (8.10.48)$$

Now we can rewrite (Eqn. 8.10.3) for $R_{f,hom}(t)$ using (Eqn. 8.10.17a) for $J_{f,hom}$ as

$$R_{f,\text{hom}}(t_0) \approx N_a\bar{v}_a J_{\text{hom}}(t_0) = N_a\bar{v}_a J^{(0)}_{f,\text{hom}}\exp[u_{s\,\text{hom}}s_w(t_0)]$$
$$= N_a\bar{v}_a J^{(0)}_{f,\text{hom}}\exp[u_{s\,\text{hom}}s^{\text{hom}}_{w,cr} + \beta_{\text{hom}}t_0]. \quad (8.10.49)$$

The crystal concentration at the time t_{max} is obtained by integrating over t_0

$$N_c(t_{\max}) = \int_{t_{cr,1}}^{t_{\max}} R_{f,\text{hom}}(t_0)\,dt_0,$$
$$\approx N_a\bar{v}_a J^{(0)}_{\text{hom}}(T)\beta^{-1}\exp[u_{s\,\text{hom}}s_{w,\max}(t_{\max})]$$
$$\times\{1-\exp[-\beta_{\text{hom}}(t_{\max}-t_{cr,1})]\}$$
$$\approx N_a\bar{v}_a J^{(0)}_{\text{hom}}(T)\beta^{-1}_{\text{hom}}\exp[u_{s\,\text{hom}}s_{w,\max}(t_{\max})]. \quad (8.10.50)$$

The last equation accounts for the fact that $(t_{max} - t_{cr,1}) >> \beta^{-1}_{\text{hom}}$ or $\beta_{\text{hom}}(t_{max} - t_{cr,1}) >> 1$ and N_c can be written as in (Eqn. 8.10.24). Substituting $\exp[u_{shom}s_w(t_{max})]$ from (Eqn. 8.10.48) into (Eqn. 8.10.50) and using the approximate equality $s_{i,\max} \approx s^{\text{hom}}_{i,cr}$ due to small variations of s_i during nucleation as discussed earlier, we obtain a parameterization of the crystal concentration in homogeneous freezing nucleation:

$$N_c(t_{\max}) = K_{gen}(1+s^{\text{hom}}_{i,cr})(s^{\text{hom}}_{i,cr})^{-1}\Psi^{-1}_{\text{hom}}, \quad (8.10.51)$$

$$K_{gen} = (4\pi D_v)^{-1}u^{-1}_{s\,\text{hom}}(c_{1i}/c_{1w}). \quad (8.10.52)$$

Equation (8.10.51) gives N_c at the time t_{max} with maximum supersaturation—i.e., at the end of the 1st stage with increasing s_i. Some previous parameterizations assumed that $N_c(t_{max})$ at the time t_{max} of maximum supersaturations is the final crystal concentration. However, as we have seen in Fig. 8.23, during the descending branch at $t_{max} < t < t_{cr,2}$, $s_w(t)$ decreases but still exceeds $s_{w,cr}$; therefore, nucleation continues after t_{max} until $t_{cr,2}$, and $N_c(t_{max})$ is approximately half the magnitude of the total $N_{c,tot}(t_{cr,2})$ after nucleation ceases at $t > t_{cr,2}$. Evaluation of the 2nd stage at $t > t_{max}$ with decreasing supersaturation in principle can be done in a similar way as for $t < t_{max}$, although it is somewhat more complicated. To simplify the solution, we can use the solutions for $t = t_{max}$ and slightly tune them using results of the parcel model runs. Their detailed analysis shows that the total $N_{c,tot}(t_{cr,2})$ at $t > t_{cr2}$, when nucleation has ceased, is proportional to $N_c(t_{max})$:

$$N_{c,tot} \approx K_{cor} N_c(t_{\max}) \tag{8.10.53a}$$

Numerical experiments with the parcel model show that $K_{cor} \sim 1.8$ to 2.2 (Fig. 8.23). A more precise fit shows that this coefficient can be chosen as a function of the vertical velocity w as

$$K_{cor}(w) = 1.85 + (2 - 1.85 w/w_{sc}) \quad \text{at} \quad w < 2 \text{ m s}^{-1}, \tag{8.10.53b}$$

$$K_{cor}(w) = 2.0, \quad \text{at} \quad w \geq 2 \text{ m s}^{-1}, \tag{8.10.53c}$$

and $w_{sc} = 2\,\text{m s}^{-1}$. An even simpler choice of the average $K_{cor} \sim 2$, which takes into account that about half of the crystals nucleate at decreasing supersaturation at $t_{max} < t < t_{cr,2}$, gives satisfactory results.

The method developed here enables estimation of the supersaturation relaxation time τ_{fc} and the mean crystal radius $\bar{r}_c$ at the time t_{max} and then after the end of nucleation. This can be done by equating in the supersaturation (Eqn. 8.5.2) the two expressions for the second term I_{dep} that describes supersaturation relaxation, which yields $(\Gamma_2/\rho_v)I_{dep} = -s_i/\tau_{fc}$. Substituting here (Eqn. 8.10.39) for I_{dep} and (Eqn. 8.10.48) for $\exp[u_{shom}s_{w,max}(t_{max})]$, and assuming $s_{i,max} \approx s_{w,cr}^{\text{hom}}$, we obtain I_{dep} and τ_{fc} at t_{max} as $I_{dep} = c_{1i}w(1 + s_{w,cr}^{\text{hom}})\rho_{is}/\Gamma_2$ and

$$\tau_{fc}(t_{\max}) = [c_{1i}w(1 + s_{w,cr}^{\text{hom}})]^{-1} s_{w,cr}^{\text{hom}}. \tag{8.10.54}$$

It was shown in Section 5.3, that in the diffusion approximation of crystal growth, $\tau_{fc} \approx [4\pi D_v N_c \bar{r}_c]^{-1}$. Applying this equation at t_{max} and using (Eqn. 8.10.54), we obtain the mean crystal radius at $t = t_{max}$,

$$\begin{aligned} \bar{r}_c(t_{\max}) &= [4\pi D_v N_{c\text{hom}}(t_{\max})]^{-1} \tau_{fc}^{-1} \\ &= [4\pi D_v N_{c\text{hom}}(t_{\max})]^{-1} c_{1i} w(1 + s_{i,cr}^{\text{hom}})(s_{i,cr}^{\text{hom}})^{-1}. \end{aligned} \tag{8.10.55}$$

Using the values from Figs. 8.17, 8.18, and 8.23, $N_{cm} \approx 35\text{L}^{-1}$, at $t = t_{max} \approx 34.5\,\text{min}$, and $c_{1i} \sim 10^{-5}\text{cm}^{-1}$, $s_{i,cr}^{\text{hom}} \approx 0.42, w = 4$ cm s^{-1}, an estimate from (Eqn. 8.10.55) gives $\bar{r}_c(t_{\max}) \sim 11.8$ µm, which is in good agreement with $\bar{r}_c \approx 12.5$ µm in Fig. 8.18b at 34.5 min. The final value of $\bar{r}_c$ after the end of nucleation can be evaluated as $\bar{r}_c(t_{cr2}) \approx K_{cor}(w)\bar{r}_c(t_{\max})$ or simply as $\bar{r}_c(t_{cr2}) \approx 2\bar{r}_c(t_{\max})$. These values of N_{chom} and $\bar{r}_c$ can be used in the models for calculation of the IWC at the end of nucleation.

8.10.7. Particular Limiting Cases

Diffusion growth limit. The important asymptotic limits can be obtained by analysis of the characteristic parameters of the solution (Eqns. 8.10.51) and (8.10.52) with Ψ_{hom} from (Eqn. 8.10.44). The parameter λ_{hom} in (Eqn. 8.10.38) can be rewritten in the form

$$\lambda_{\mathrm{hom}} = \frac{\beta_{\mathrm{hom}}(r_0+\xi_{dep})^2}{B_{i\,\mathrm{hom}}} = \left(\frac{r_0+\xi_{dep}}{\Lambda_{\mathrm{hom}}}\right)^2,$$

$$\Lambda_{\mathrm{hom}} = \left(\frac{B_{i\,\mathrm{hom}}}{\beta_{\mathrm{hom}}}\right)^{1/2} = \left(\frac{2c_{i3}s_{i,cr}^{\mathrm{hom}}}{u_{s\,\mathrm{hom}}c_{1w}w}\right)^{1/2}. \tag{8.10.56}$$

Here Λ_{hom} is a scaling length that characterizes the ratio of the crystal growth rate (Eqn. 8.5.6) to the supersaturation generation rate (the first term on the RHS of (Eqn. 8.10.47)). Now we present asymptotics of the solution (Eqn. 8.10.51) at $\lambda_{\mathrm{hom}} << 1$ and $\lambda_{\mathrm{hom}} >> 1$. The values of λ_{hom} and Λ_{hom} and physical meaning of the asymptotics are analyzed in the following text.

The values $\lambda_{\mathrm{hom}} << 1$ mean small ξ_{dep} and r_0, and are typical of the diffusion regime of crystal growth with deposition coefficient $\alpha_d \sim 1$ or $\alpha_d > 0.1$ at not very large w and not very low T. In this case, we can neglect in (Eqn. 8.10.44) for Ψ_{hom} all terms with ξ_{dep} and r_0 and note also that $erf(\sqrt{\lambda_{\mathrm{hom}}}) \to 0$ at $\lambda_{\mathrm{hom}} << 1 \to 0$. Based on the estimates earlier, we can also assume that $\beta_{\mathrm{hom}}t_{max} >> 1$ and the terms with $\exp(-\beta_{\mathrm{hom}}t_{max})$ can be neglected. Thus, Ψ_{hom} is simplified in this diffusion regime

$$\Psi_{\mathrm{hom},dif} \approx (\sqrt{\pi}/2)\, e^{\lambda_{\mathrm{hom}}} B_{i\mathrm{hom}}^{1/2}\beta_{\mathrm{hom}}^{-3/2} \approx (\pi/2)^{1/2}(c_{i3}s_{i.cr}^{\mathrm{hom}})^{1/2}u_{s\mathrm{hom}}^{-3/2}(c_{1w}w)^{-3/2}. \tag{8.10.57}$$

Substitution of this expression into (Eqns. 8.10.51) and (8.10.52) yields for the crystal concentration $N_{cm,dif}^{\mathrm{hom}}$ in the diffusion regime of crystal growth at t_{max}

$$N_{cm,dif}^{\mathrm{hom}}(t_{\max}) = K_{i,dif}^{\mathrm{hom}}(1+s_{i.cr}^{\mathrm{hom}})(s_{i.cr}^{\mathrm{hom}})^{-3/2}(c_{1i}w)^{3/2}, \tag{8.10.58}$$

$$K_{i,dif}^{\mathrm{hom}} = (2\pi D_v)^{-3/2}\left(\frac{\rho_i\Gamma_2}{\rho_{is}}\right)^{1/2} u_{s\mathrm{hom}}^{1/2}\left(\frac{c_{1w}}{c_{1i}}\right)^{1/2}$$

$$= (2\pi D_v)^{-3/2}\left(\frac{\rho_i\Gamma_2}{\rho_{is}}\right)^{1/2}\left[\frac{2R}{kM_wL_m^{ef}}\frac{\Delta F_{cr,0}^{\mathrm{hom}}(T)}{\ln(T_0/T)}\right]^{1/2}\left(\frac{c_{1w}}{c_{1i}}\right)^{1/2}. \tag{8.10.59}$$

The properties of this solution are discussed in the following text and compared with the other limits. The total crystal concentration in the diffusion regime after the end of nucleation at $t > t_{cr,2}$ is $N_{tot,dif}^{\mathrm{hom}} \approx 2N_{cm,dif}^{\mathrm{hom}}$.

Kinetic growth and large particle limits. The diffusion limit was at $\lambda_{\mathrm{hom}} << 1$, and the opposite limit is lhom >> 1. It is seen from the definition of λ_{hom} (Eqn. 8.10.56) that this case can be typical of a kinetic regime with large ξ_{dep} (small α_d) or with a large initial particle radius r_0 of freezing particles. It can be studied using the asymptotic property of $erf(x)$ at $x >> 1$ (see Appendix A.8):

$$erf(\sqrt{\lambda}) = 1 - \frac{1}{\sqrt{\pi}}\lambda^{-1/2}e^{-\lambda}\left(1-\frac{1}{2\lambda}\right). \tag{8.10.60}$$

Expanding in (Eqn. 8.10.44) for Ψ_{hom} the functions $erf(\sqrt{\lambda_{\text{hom}}})$ and $erf(\sqrt{\lambda_{\text{hom}} + \beta_{\text{hom}} t_{\max}})$ with (Eqn. 8.10.60), neglecting the terms with $\exp(-\beta_{\text{hom}} t_{max})$ and the terms $(\lambda_{hom})^{-3/2}$ compared to $(\lambda_{hom})^{-1/2}$, and collecting the terms of the same order, Ψ_{hom} can be written as

$$\begin{aligned}\Psi_{\text{hom},kin} &= \beta_{\text{hom}}^{-1/2}\lambda_{\text{hom}}^{-1/2}[(1/2)B_{i\,\text{hom}}^{1/2}\beta_{\text{hom}}^{-1} + \xi_{dep}^{2}B_{i}^{-1/2}] + B_{i\,\text{hom}}^{1/2}\beta_{\text{hom}}^{-3/2}\lambda_{\text{hom}}^{1/2} - 2\xi_{dep}\beta_{\text{hom}}^{-1} \\ &= (r_0 + \xi_{dep})^{-1}[(1/2)B_{i\,\text{hom}}\beta_{\text{hom}}^{-2} + \xi_{dep}^{2}\beta_{\text{hom}}^{-1}] + (r_0 - \xi_{dep})\beta_{\text{hom}}^{-1}. \qquad (8.10.61)\end{aligned}$$

This case is divided into two subcases: a) ξ_{dep} is large (small deposition coefficient α_d) but r_0 is small (small particles limit)—that is, $\xi_{dep} >> r_0$; and b) r_0 is large (large particles limit)—that is, $\xi_{dep} << r_0$, which may correspond to diffusion or kinetic regimes. These limits are considered in the following text.

a. $\lambda_{hom} >> 1$, $\xi_{dep} >> r_0$, kinetic regime, small particle limit
With these conditions, r_0 can be neglected compared to ξ_{dep} and Ψ_{hom} in (Eqn. 8.10.61) is simplified as

$$\Psi_{\text{hom},kin,s} = (1/2)B_{i\text{hom}}\,\beta_{\text{hom}}^{-2}\xi_{dep}^{-1} = c_{i3}s_{i,cr}^{\text{hom}}u_{s\text{hom}}^{-2}(c_{1w}w)^{-2}(\alpha_d V_w/4D_v). \qquad (8.10.62)$$

Substitution into the general (Eqn. 8.10.51) yields the crystal concentration $N_{cm,kin,s}^{\text{hom}}(t_{\max})$ at maximum s_{wmax}

$$N_{cm,kin,s}^{\text{hom}}(t_{\max}) = K_{i,kin,s}^{\text{hom}}(1 + s_{i.cr}^{\text{hom}})(s_{i.cr}^{\text{hom}})^{-2}(c_{1w}w)^{2}, \qquad (8.10.63)$$

$$K_{i,kin,s}^{\text{hom}} = \frac{1}{(\pi D_v)}\frac{u_{s\,\text{hom}}}{\alpha_d V_w}\left(\frac{\rho_i \Gamma_2}{\rho_{is}}\right)\left(\frac{c_{1i}}{c_{1w}}\right). \qquad (8.10.64)$$

The total crystal concentration after the end of nucleation at $t \geq t_{cr,2}$ is $N_{c,kin,s}^{\text{hom}} \approx 2N_{cm,kin,s}^{\text{hom}}(t_{\max})$. Thus, in this limit $N_{c,kin,s}^{\text{hom}} \sim w^{2}$, which is in agreement with Ren and McKenzie (2005), but all coefficients are expressed now via the primary fundamental parameters, without empirical constants. Note also that the crystal concentration is inversely proportional to the saturated vapor pressure, $N_{c,kin,s}^{\text{hom}} \sim \rho_{is}^{-1}(T)$ and to the deposition coefficient, $N_{c,kin,s}^{\text{hom}} \sim \alpha_d^{-1}$; that is, the smaller α_d or the more polluted the clouds, the greater the nucleated crystal concentration. The wide scatter of measured α_d was described in Hobbs (1974) and Pruppacher and Klett (1997). Gierens et al. (2003) discussed possible reasons for α_d as small as 10^{-3}; in these cases, the dependence $N_c \sim 1/\alpha_d$ can be significant. This is in agreement with data from the Interhemispheric Differences in Cirrus Properties from Anthropogenic Emissions (INCA) field experiment (Ovarlez et al., 2002; Ström et al., 2003; Haag et al., 2003; Gayet et al., 2004; Monier et al., 2006) that found greater ice crystal concentrations in cirrus in the more polluted northern hemisphere than in the cleaner southern hemisphere. This could be caused not only by the heterogeneous ice nucleation mode, but also by a small deposition coefficient in homogeneous nucleation in polluted areas.

b. Initial r_0 is large and $r_0 >> \xi_{dep}$; large particles limit
Neglecting ξ_{dep} compared to r_0, (Eqn. 8.10.61) can be further transformed

$$\begin{aligned}\Psi_{\text{hom},kin,l} &= \frac{1}{2r_0}B_{i\,\text{hom}}\beta_{\text{hom}}^{-2} + r_0\beta_{\text{hom}}^{-1} = r_0\beta_{\text{hom}}^{-1}\left(\frac{B_{i\,\text{hom}}}{\beta_{\text{hom}} r_0^{2}} + 1\right) \\ &= r_0\beta_{\text{hom}}^{-1}(\lambda_{\text{hom}}^{-1} + 1) \approx r_0\beta_{\text{hom}}^{-1}. \qquad (8.10.65)\end{aligned}$$

The last equality takes into account that $\lambda_{hom} >> 1$, the first term in the parentheses is much smaller than the second and can be neglected. Substituting $\Psi_{hom.kin,l}$ into the general solution (Eqn. 8.10.51), we obtain

$$N_{cm,l}^{\text{hom}}(t_{\max}) = (4\pi D_v r_0)^{-1}(1+s_{i.cr}^{\text{hom}})(s_{i.cr}^{\text{hom}})^{-1}(c_{1i}w). \quad (8.10.66)$$

The total crystal concentration after the end of nucleation at $t \geq t_{cr,2}$ is $N_{c,l}^{\text{hom}} \approx 2N_{cm,l}^{\text{hom}}(t_{\max})$. That is, the dependence on w is linear, $N_{c,l}^{\text{hom}} \sim w$, in agreement with predictions in Kärcher and Lohmann (2002a,b) and in Ren and MacKenzie (2005). The term $\rho_{is}(T)$ is absent now, thus the temperature dependence is much weaker than in the previous cases, and is caused now by the T-dependence on D_v, c_{1i}, and $s_{i.cr}^{\text{hom}}$.

8.10.8. Physical Interpretation

Two example calculations using this new parameterization are shown in Figs. 8.25 and 8.26. The crystal concentrations $N_c(w)$ calculated using the diffusion approximation with (Eqns. 8.10.57)–(8.10.59) and $\alpha_d = 1$ (denoted KC2012) for an ascending air parcel is shown in Fig. 8.25. The applicability of the diffusion approximation is justified by the small $\lambda_{hom} \sim 10^{-3}$ to 0.03 with $\alpha_d = 1$ for all w. It is compared with the parameterizations by Sassen and Benson (2000, SB2000, to $w = 1\ \text{m s}^{-1}$), Liu and Penner (2005, LP2005), Kärcher and Lohmann (2002a, KL2002). Also shown here are the results of several parcel model simulations from Lin et al. (2002) according to the protocols of CPMCP for the three values of $w = 4$, 20 and $100\ \text{cm s}^{-1}$. Simulations were performed by Cotton, DeMott, Jensen, Kärcher, Lin, Sassen, and Liu as indicated in Fig. 8.25 (the models are described in Sassen and Dodd, 1988; DeMott et al., 1994; Jensen et al., 1994; Lin, 1997; Khvorostyanov and Sassen 1998a,b; Spice et al., 1999; Kärcher and Lohmann, 2002a,b; Liu and Penner, 2005); the results of parcel simulations from Khvorostyanov and Curry (2005a; KC2005) are added. This figure shows that the new parameterization KC2012 lies within the spread of the parcel models results, being closer to the lower limit, and to the parcel simulations by Jensen, who also used the model with spectral microphysics and supersaturation (Jensen et al., 1994). The KC2012 is in a qualitative agreement with Sassen and Benson (2000) at small w and is especially close to the parameterization by Kärcher and Lohmann (2002a,b) at all w and at both temperatures, although it was based on a substantially different approach. This agreement supports the validity of the new parameterization (Eqns. 8.10.51)–(8.10.53c), based on an extension of the classical nucleation theory and shows that the semi-empirical approaches lead to results that can be derived from classical nucleation theory.

Fig. 8.26 shows a comparison of the full solution (Eqns. 8.10.51)–(8.10.53c) with diffusion limit (Eqns. 8.10.57)–(8.10.59) at $\alpha_d = 1$ and a kinetic limit (Eqns. 8.10.62)–(8.10.64) at $\alpha_d = 0.04$, 0.01, and 0.001. The diffusion approximation (solid circles) is valid at $\lambda_{hom} << 1$, and limited at $w \leq 170\ \text{cm s}^{-1}$; the kinetic limit is valid at $\lambda_{hom} >> 1$ and with $\alpha_d = 0.04$ is limited at $w > 30\ \text{cm s}^{-1}$; both limits are denoted with ellipses. This figure illustrates good accuracy of the two approximations for corresponding values λ_{hom} and underscores the important role of the deposition coefficient. With small α_d, such as in polluted clouds, the crystal concentrations are substantially higher (2 times at $\alpha_d = 0.04$, and 8 times at $\alpha_d = 0.01$) than with $\alpha_d = 1$ for relatively clean clouds. So, polluted

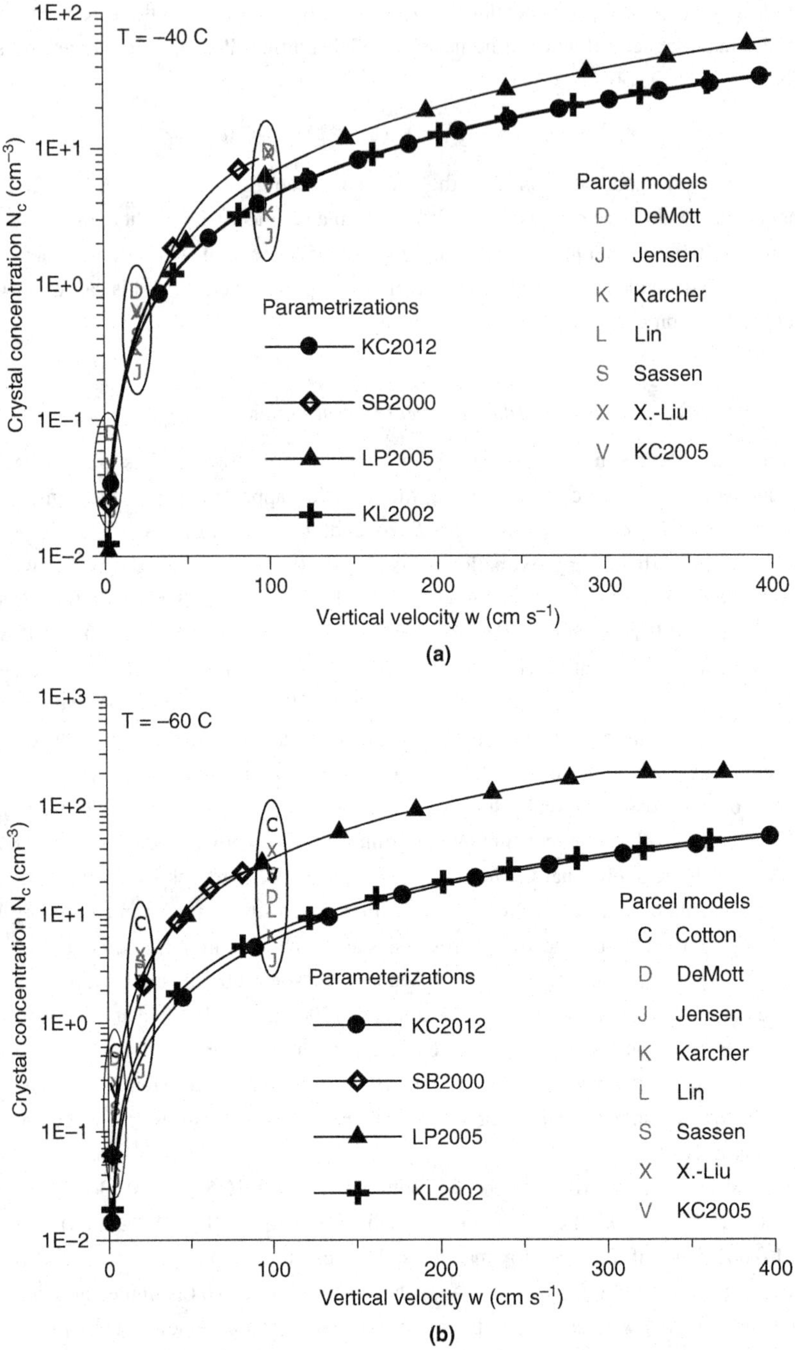

Figure 8.25. Comparison of the diffusion approximation (Eqns. 8.10.57)–(8.10.59) of the crystal concentration $N_c(w)$ as a function of w (KC2012) with the parameterizations by Sassen and Benson (2000, SB2000, limited to $w = 100\,\mathrm{cm\,s^{-1}}$), Kärcher and Lohmann (2002, KL02), Liu and Penner (2005, LP2005), and with parcel model simulations from Lin et al. (2002) (Cotton, DeMott, Jensen, Kärcher, Lin, Sassen, X.-Liu, as indicated in the legend) and from Khvorostyanov and Curry, 2005 (KC2005) for the three values of $w = 4$, 20, and $100\,\mathrm{cm\,s^{-1}}$. From Khvorostyanov and Curry, Atmos. Chem. Phys. (2012), reproduced with permission.

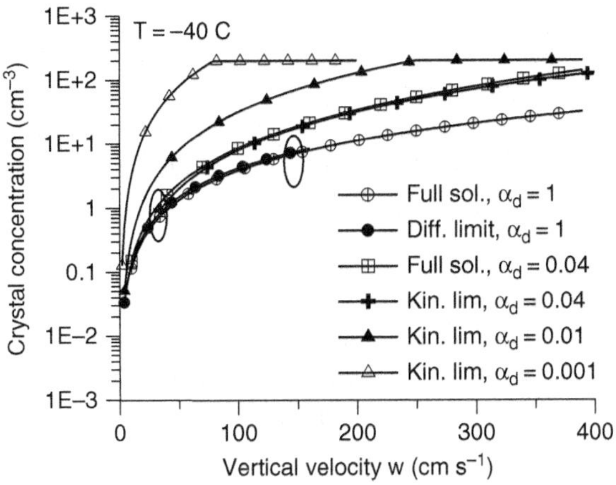

Figure 8.26. Comparison of the full solution (Eqns. 8.10.51)–(8.10.53) with Ψ defined in (Eqn. 8.10.46) at two values of $\alpha_d = 1$ and 0.04 (Full sol.) with diffusion limit (Eqns. 8.10.57)–(8.10.59) and kinetic limit (Eqns. 8.10.62)–(8.10.64). The diffusion approximation (solid circles) is valid at $\lambda_{hom} << 1$, and limited here at $w \le 170\,\mathrm{cm\,s^{-1}}$; the kinetic limit is valid at $\lambda_{hom} >> 1$ and with $\alpha_d = 0.04$ is limited here at $w > 30\,\mathrm{cm\,s^{-1}}$. Both limits are denoted with ellipses. From Khvorostyanov and Curry, Atmos. Chem. Phys. (2012), reproduced with permission.

crystalline clouds may have substantially greater crystal concentrations and albedo effects, which is similar to the Twomey (1997a,b) effect for liquid clouds. The possibility of this effect for crystalline and mixed clouds was discussed—e.g., by Lohmann and Feichter (2005). The parameterization described here provides a quantitative tool for estimation of these effects for crystalline clouds. Fig. 8.26 shows that these effects can be significant for small α_d—e.g., if the nucleated crystals are covered by the layers of adsorbed gases or the liquid films, which may act as the surface active substances (e.g., Gao et al., 2004; Räisänen et al., 2006; Bogdan and Molina, 2009; Bogdan et al., 2013).

The accuracy of the parameterization equations for N_c derived in this section was estimated by comparison with the data on N_c from the International CPMCP Project (Lin et al., 2002) and parcel simulation results described in Section 8.9. The average error of this parameterization relative to the parcel runs described in Section 8.9 is about ±5–15%. This is a satisfactory accuracy that takes into account the fact that the difference in N_c among various models in CPMCP was much greater.

The new analytical parameterization of homogeneous freezing derived from extended classical nucleation theory and analytical solutions to the supersaturation equation includes the time dependence, and can therefore be used for calculations of the crystal concentrations in cloud models with small time steps (e.g., Jensen et al., 1994; Khvorostyanov and Sassen, 1998b, 2002; Spice et al., 1999; Khvorostyanov et al., 2001, 2003, 2006; Lin et al. 2002; Sassen et al., 2002; Randall et al., 2003; Khairoutdinov and Randall, 2003; Fridlind et al., 2004; Krakovskaia and Pirnach, 2004; Khain et al., 2004; Morrison et al., 2005a,b; Seifert and Beheng, 2006; Monier et al., 2006; Grabowski and Morrison, 2008; Klein et al., 2009; Spichtinger and Gierens, 2009; Xue et al., 2010; Fan et al., 2011; Curry and Khvorostyanov, 2012; Tao et al., 2012; Spichtinger and Krämer, 2012), for substep parameterizations in the mesoscale models like the Weather Research and Forecast Model (WRF) (e.g., Zhang et al., 2011; Veltishchev et al., 2011) or the Consortium for Small-Scale Modeling (COSMO Model) (Zubler et al., 2011), and in large-scale climate models and GCMs with time steps that can be greater than the nucleation time τ_{nuc} (e.g., Lohmann and Kärcher, 2002; Morrison and Gettelman, 2008; Gettelman et al., 2008; Sud et al., 2009; Hoose et al., 2010).

The analytical parameterization of homogeneous freezing derived in this section has the following properties.

1. We can delineate three different regimes of crystal homogeneous nucleation in cold clouds, depending on the cooling time of an air parcel in ascending motion. At small times, $t << \tau_{nuc,hom} = (u_{shom}c_{1w}w)^{-1}$ (~40 s in the example considered earlier), the crystal concentrations increase linearly with time. At larger times, $t >> \tau_{nuc,hom}$, but smaller then the time t_{max} of maximum supersaturation in the parcel, $t < t_{max}$, the number N_c exponentially grows with time. Crystal concentrations in these two regimes are proportional to the homogeneous nucleation rate and concentration of the aerosol (haze) particles as described by (Eqns. 8.10.23) and (8.10.24). These regimes could be also created in some diffusion chambers at slow cooling rates; the updraft $w = 10\,\text{cm s}^{-1}$ corresponds to the cooling rate $dT/dt \sim -\gamma_a w \sim -0.06\,\text{K min}^{-1}$. Note that the cooling rates in the diffusion chambers are usually greater, and the particle residence times in the chambers are smaller.
   If a parcel rises and cools without mixing, so that $t > t_{max}$ and $t > t_{cr,2}$, the supersaturation reaches and passes a maximum and falls below the threshold value, then a third limiting regime occurs where the maximum possible crystal concentrations are nucleated in the parcel at the thermodynamic conditions. This regime is described by (Eqns. 8.10.51), (8.10.52), and (8.10.53a), (8.10.53b), and (8.10.53c). The dependence on the nucleation rate and haze concentration vanishes in this regime, although concentration of nucleated crystals is much smaller than the concentration of haze particles. As parcel model simulation showed (Section 8.9), the times t_{max} at $T = -40\,°\text{C}$ and $w = 4$, 20 and $100\,\text{cm s}^{-1}$ are about 530, 120, and 40 seconds, and the total times of nucleation required for this regime are $(t_{cr,2} - t_{cr,1}) \sim 800$, 190, and 60 seconds, respectively. At $T = -60\,°\text{C}$, these times are about half the magnitude.
2. Expressions for the crystal concentration in the third limiting regime are very simple, and somewhat surprising. They do not include most of the basic factors present in the original supersaturation equation: neither the nucleation rate $J_{hom}(T, s_w)$, nor the concentration N_a of the haze particles, nor any info on their volume, size spectra, or chemical composition. The reason why N_c does not depend on N_a is because N_c is usually on the order of a few or a few tens per liter (rarely, a few hundred, and very rarely, a few thousand), while N_a is typically on the order of a few hundred per cubic centimeter. That is, only a very small fraction of haze particles freezes, and the dependence of N_c on N_a vanishes at values of N_a much smaller than those in the upper troposphere studied here. However, if N_a is small, then N_c is limited by N_a.
3. The major three factors that govern homogeneous ice nucleation in the third limiting regime are the vertical velocity, w, the temperature, T, and the critical (threshold) saturation ratio $s_{i.cr}^{\text{hom}}$. The equations for N_c derived in this section show that in a first approximation in diffusion limit, $N_c \sim w^{3/2}$, and $N_c \sim \rho_{si}^{-1/2}(T)$, both dependencies are the same as in Kärcher and Lohmann (2002a,b) and in Ren and MacKenzie (2005) in the diffusion growth limit. However, the actual dependence of N_c on w and T is more complicated and somewhat different since $(s_{i.cr}^{\text{hom}})^{-3/2}$ also

includes dependence on w and T, the critical supersaturation $s_{i,cr}^{\text{hom}}$ also depends on T, and the coefficient $Ki_{,\text{dif}}$ also depends on T via factors D_v, c_{1w}, c_{1i}, and u_{shom}. In the kinetic limit or in the large particle limit, N_c can be proportional to w^2 or to w, depending on the initial particle radius, in agreement with previous semi-empirical parameterizations. An additional explicit dependence on the deposition coefficient, derived here, allows estimates of the effects of atmospheric pollutions on the properties of the crystalline and mixed phase clouds.

4. The nucleation rate $J_{f,hom}(T, s_w)$, derived here, varies exponentially with time, and this dependence is characterized by some scaling nucleation time $\tau_{nuc,hom}$ that is expressed here directly via the parameters of classical nucleation theory and fundamental physical constants.

8.11. Temperature Effects and the Homogeneous Freezing of Cloud Drops

The effects of temperature on the homogeneous freezing of drops were not accounted for in Section 8.10 since they were small compared to the supersaturation effects in the freezing of the deliquescent haze particles. These effects may become dominant if supersaturation variations are small, as in an ensemble of water drops, or if the temperature changes are large enough. Evaluation of the temperature effects on the nucleation rates and crystal concentrations can be done in a similar manner to the derivation for the supersaturation effects. Equation (8.10.7) shows that small variations in the critical energy of nucleation $\Delta F_{cr}(T)$ due to the temperature change ΔT are caused primarily by variations in four parameters: surface tension σ_{is}, ice density ρ_i, effective melting heat, and $\ln(T_0/T)$.

Parcel model simulations in Section 8.9 and analysis of experimental data in the preceding sections show that temperature changes ΔT during nucleation are usually much smaller than T. If $\Delta T \ll T$, the perturbations due to temperature decrease ΔT can be approximated as linear:

$$\sigma_{is}^3(T-\Delta T) \approx \sigma_{is}^3(T)[1 - 3\sigma_{is}'(T)/\sigma_{is}(T)\Delta T], \tag{8.11.1}$$

$$L_m^{ef}(T-\Delta T) \approx L_m^{ef}(T)[1 - (L_m^{ef})'/L_m^{ef}(T)\Delta T], \tag{8.11.2}$$

$$\rho_i(T-\Delta T) \approx \rho_i(T)[1 - \rho_i'/\rho_i(T)\Delta T], \tag{8.11.3}$$

$$\ln\frac{T_0}{T-\Delta T} \approx \ln\frac{T_0}{T}\left[1 + \frac{\Delta T}{T\ln(T_0/T)}\right]. \tag{8.11.4}$$

Here the primes indicate absolute values of the derivatives by T (i.e., the derivatives and ΔT are assumed to be positive for the quantities decreasing with T), and are themselves functions of T. Substitution of these expressions in (Eqn. 8.10.10) for $\Delta F_{cr,0}^{\text{hom}}$ with $s_w \approx 0$, and expanding the denominator into the power series to the first-order terms by $\Delta T/T$ yields

$$\Delta F_{cr,0}^{\text{hom}}(T-\Delta T) \approx \Delta F_{cr,0}^{\text{hom}}(T)[1 - \kappa_{Tn}(T)\Delta T(t)], \tag{8.11.5a}$$

$$\kappa_{Tn}(T) = 3\frac{\sigma_{is}'}{\sigma_{is}} - 2\left(\frac{\rho_i'}{\rho_i} + \frac{(L_m^{ef})'}{L_m^{ef}}\right) + \frac{2}{T\ln(T_0/T)}. \tag{8.11.5b}$$

The temperature variations in the activation energy can be presented in a similar form

$$\Delta F_{act}(T-\Delta T)\approx \Delta F_{act}(T)[1-\kappa_{Ta}(T)\Delta T(t)], \tag{8.11.6a}$$

$$\kappa_{Ta}(T)=\Delta F'_{act}(T)/\Delta F_{act}(T). \tag{8.11.6b}$$

Substituting these expressions into (Eqn. 8.10.6) for $J_{f,hom}$, and using the approximation for the denominator $1/[k(T-\Delta T)]\approx 1/(kT)(1+\Delta T/T)$ yields

$$J_{f,\hom}(T-\Delta T)=J_{f,\hom}(T)\exp[B_{T\hom}\Delta T(t)], \tag{8.11.7}$$

where we denoted

$$B_{T\hom}=u_{Tact}(T)+u_{T\hom}(T), \tag{8.11.8a}$$

$$u_{Tact}(T)=\left(\kappa_{Ta}(T)-\frac{1}{T}\right)\frac{\Delta F_{act}(T)}{kT}, \tag{8.11.8b}$$

$$u_{T\hom}(T)=\left(\kappa_{Tn}(T)-\frac{1}{T}\right)\frac{\Delta F_{cr,0}^{\hom}(T)}{kT}. \tag{8.11.8c}$$

The parameters B_{Thom}, κ_{Ta}, κ_{Tn}, u_{Tact}, and u_{Thom} have the dimension K^{-1}, describe a decrease in ΔF_{cr} and increase in $J_{f,hom}$ with temperature decrease ΔT. These parameters are the analogs of κ_s and u_{shom} introduced in Section 8.10 that describe variations in ΔF_{cr} and $J_{f,hom}$ due to supersaturation variations. B_{Thom} describes the major exponential increase in $J_{f,hom}$ with temperature decrease ΔT similar to several empirical parameterizations of ice nucleation. The values of these parameters are estimated in the following.

The temperature decrease ΔT can be related to the time dependence via the temperature lapse rate γ and vertical velocity w using the temperature equations from Section 3.11, and integrating with respect to time

$$dT/dt=\dot{T}=-\gamma w,\quad \Delta T(t)=-\dot{T}t=\gamma wt. \tag{8.11.9}$$

The minus sign in the second equation arises because we defined the temperature decrease as positive, $\Delta T>0$, when $dT/dt<0$. Substituting (Eqn. 8.11.9) into (Eqn. 8.11.7), we obtain the time dependent $J_{f,hom}$:

$$J_{f,\hom}(t)=J_{f,\hom}^{(0)}(t_0)\exp(\beta_{T\hom}t)=J_{f,\hom}^{(0)}(t_0)\exp(t/\tau_{T\hom}), \tag{8.11.10}$$

where we introduce again the characteristic freezing frequency β_{Thom} and time τ_{Thom}:

$$\beta_{T\hom}(T)=(u_{Tact}+u_{T\hom})\gamma w=(u_{Tact}+u_{T\hom})(-dT/dt), \tag{8.11.11a}$$

$$\tau_{T\hom}=[(u_{Tact}+u_{T\hom})\gamma w]^{-1}=[(u_{Tact}+u_{T\hom})(-dT/dt)]^{-1}. \tag{8.11.11b}$$

The first form of these equations is convenient for analysis in numerical models where γ and w are known, and the second form can be more convenient for analysis of laboratory experiments where dT/dt is known. As in Section 8.10, the polydisperse nucleation rate for freezing due to the temperature change can be written as

$$R_{f,\hom}=\frac{dN_c}{dt}\approx N_a\bar{v}_aJ_{f,\hom}(t)=N_a\bar{v}_aJ_{f,\hom}^{(0)}(t_0)\exp(\beta_{T\hom}t). \tag{8.11.12}$$

Integration over time yields

$$N_c(t) = N_a \bar{v}_a J^{(0)}_{f,\text{hom}}(t_0)\tau_{T\text{hom}}[\exp(\beta_{T\text{hom}}t) - 1]. \tag{8.11.13}$$

So, the time dependence of the crystal concentration due to temperature decrease is similar to that due to the supersaturation variations described in Section 8.10. If we consider the freezing of deliquescent haze particles, the joint effect of the supersaturation and temperature variations can be simply accounted for by replacing of the frequency β_{shom} with ($\beta_{shom} + \beta_{Thom}$) in the equations for $J_{f,hom}$ and $N_c(t)$ in Section 8.10. If we consider homogeneous freezing of water drops near the water saturation with $s_w \approx 0$, then only the terms with β_{Thom} are essential and the equations of this section can be used.

For small times or temperature changes, if $\beta_{Thom}t << 1$, the exponent in (Eqn. 8.11.13) can be expanded and (Eqn. 8.11.13) is reduced to

$$N_c(t) = N_a \bar{v}_a t J^{(0)}_{f,\text{hom}}(t_0), \tag{8.11.14}$$

that is, the time dependence of N_c is linear in this limit. For large times or temperature changes, if $\beta_{Thom}t >> 1$, the time dependence becomes exponential,

$$N_c(t) = N_a \bar{v}_a J^{(0)}_{f,\text{hom}}(t_0)\tau_{T\text{hom}} \exp(\beta_{T\text{hom}}t). \tag{8.11.15}$$

The last two equations may explain the difference in the nucleation rates at small and large ΔT. If a permanent and significant cooling rate is present, the rate of crystal formation is exponential according to (Eqn. 8.11.15). If such a source of cooling is absent, and there are only small random temperature fluctuations, the rate of crystal production is close to linear with time as in (Eqn. 8.11.14). These features are similar to those described in experiments on the heterogeneous freezing of drops with an alternating cooling rate by Vali (1994, 2008), who found different growth regimes of $N_c(t)$ at high and low cooling rates. The nature of the homogeneous and heterogeneous freezing is similar and the last two equations provide some explanation for these observations, which will also be discussed in Section 10.2.

It follows from (Eqn. 8.11.11a) that

$$\beta_{T\text{hom}}\Delta t = (u_{T\text{hom}} + u_{Tact})(-dT/dt)\Delta t = (u_{T\text{hom}} + u_{Tact})\Delta T = B_{T\text{hom}}\Delta T, \tag{8.11.16}$$

or $B_{Thom} = u_{Thom} + u_{Tact}$. Using this relation, (Eqn. 8.11.15) can be rewritten for ice crystals nucleated during interval Δt as

$$N_c(\Delta t) = N_a \bar{v}_a J^{(0)}_{f,\text{hom}}(t_0)\tau_{T\text{hom}} \exp(B_{T\text{hom}}\Delta T). \tag{8.11.17}$$

The exponential temperature dependence of N_c derived here is theoretically similar to the parameterization $N_c(T) \sim \exp(\beta_F \Delta T)$ with $\Delta T = T_0 - T$ for heterogeneous ice nucleation derived empirically by Fletcher (1962) and several other researchers (see Section 9.4). Equation (8.11.17) shows that there is similarity in the temperature dependence of the homogeneous and heterogeneous ice nucleation. The parameter B_{Thom} [K^{-1}] is a theoretically derived analogue of the empirical parameter β_F introduced by Fletcher (1962) and others in similar parameterizations. The average value of the exponent in Fletcher (1962) was $\beta_F \approx 0.6\,\text{K}^{-1}$, although wide variations were reported (Pruppacher and Klett, 1997).

We can compare now the T-dependencies of homogeneous and heterogeneous nucleation. The values of the coefficients κ_{Tn}, κ_{Ta}, and B_{Thom} in (Eqns. 8.11.8a)–(8.11.8c) can be estimated using the parameterizations of $\sigma_{is}(T)$, $\rho_i(T)$ from Section 4.4 and of $L_m^{ef}(T)$ from Section 8.8 and $\Delta F_{act}(T)$ from Section 8.3. These estimates may vary depending on the parameterizations for the key parameters of classical nucleation theory. An estimate at T = 233 K using (Eqn. 4.4.15a) for $\sigma_{is}(T)$, (Eqn. 4.4.22) for $\rho_i(T)$ and (Eqn. 8.8.12a) for $L_m^{ef}(T)$ gives $\kappa_{Tn} - 1/T \approx 7 \times 10^{-2}\,\mathrm{K}^{-1}$. Fig. 8.3 shows that $\Delta F_{cr,0}^{hom} \approx 1.2 \times 10^{-12}$ erg for pure water drops at 233 K; thus, $\Delta F_{cr,0}^{hom}/kT \approx 37.3$. Substituting this value into (Eqn. 8.11.8c), along with the estimated κ_{Tn} yields $u_{Thom} \approx 2.1$.

An estimate for κ_{Ta} is more uncertain because several different parameterizations suggested for $\Delta F_{act}(T)$ differ even by the sign of $d(\Delta F_{act})/dT$ (see Section 4.4). An estimate using (Eqn. 8.3.20) for $\Delta F_{act}(T)$ from Zobrist et al. (2007) at $T = -40\,°\mathrm{C}$ yields $\kappa_{Ta} - 1/T \approx -1.3 \times 10^{-2}\,\mathrm{K}^{-1}$, $\Delta F_{act}/(kT) \approx 15.6$, and $u_{Tact} \approx -0.195\,\mathrm{K}^{-1}$, so that $B_{Thom} = u_{Thom} + u_{Tact} \approx 1.92\,\mathrm{K}^{-1}$, with the small contribution of u_{Tact} being approximately 7% of u_{Thom}. Chen et al. (2008) suggested $\Delta F_{act}(T) = \mathrm{const}$; thus, $\kappa_{Ta} = 0$, and $u_{Tact} = -0.06$. Using $\Delta F_{act}(T)$ as (Eqn. 8.3.23b) from Khvorostyanov and Sassen (1998c) or Khvorostyanov and Curry (2000) also yields a small value, $\kappa_{Ta} \approx 5 \times 10^{-5}$. Thus, in these cases, the contribution of u_{Tact} is small and the sum $B_{Thom} \approx 2\,\mathrm{K}^{-1}$, which is about a factor 3 higher than Fletcher's mean $0.6\,\mathrm{K}^{-1}$. Using Equation (8.3.24) for $\Delta F_{act}(T)$ from Fornea et al. (2009), faster increasing to low T yields an estimate $(\kappa_{Ta} - 1/T) \approx -2.65 \times 10^{-2}\,\mathrm{K}^{-1}$, $u_{Tact} \approx -0.74\,\mathrm{K}^{-1}$, and $B_{Thom} \approx 1.36\,\mathrm{K}^{-1}$.

An estimate using (Eqn. 4.4.15c) from Dufour and Defay (1963) for $\sigma_{iw}(T)$ and (Eqn. 8.8.12b) for $L_m^{ef}(T)$ gives smaller $\sigma'_{is}(T)/\sigma_{is}(T)$ and $[L_m^{ef}(T)]'/L_m^{ef}(T)$ and with the same $\rho'_i(T)/\rho_i(T)$ this yields $\kappa_{Tn} \approx 4.35 \times 10^{-2}\,\mathrm{K}^{-1}$ and $u_{Thom} \approx 1.21\,\mathrm{K}^{-1}$. Using $u_{Tact} \approx -0.195\,\mathrm{K}^{-1}$, based on $\Delta F_{act}(T)$ from Zobrist et al. (2007), we obtain $B_{Thom} \approx 1.0\,\mathrm{K}^{-1}$, and using $u_{Tact} \approx -0.74\,\mathrm{K}^{-1}$, based on $\Delta F_{act}(T)$ from Fornea et al. (2009), we obtain $B_{Thom} \approx 0.47\,\mathrm{K}^{-1}$, which is closer to Fletcher's mean $\beta_F = 0.6\,\mathrm{K}^{-1}$ for heterogeneous nucleation.

Thus, this extended classical nucleation theory allows derivation of an exponential temperature dependence for crystal concentrations and makes clear its physical reason: this is the exponential time dependence of N_c in (Eqn. 8.11.15) and the temperature dependence of $\sigma_{iw}(T)$, $L_m^{ef}(T)$, $\rho_i(T)$, and $\Delta F_{act}(T)$. However, the parameter B_{Thom} in the exponent in (Eqn. 8.11.17) is sensitive to the parameters of classical theory, which emphasizes the necessity of further refining these parameters based on newer experimental data on ice nucleation. It will be shown in Chapters 9 and 10 that a similar analysis for heterogeneous freezing allows an explanation of the exponential temperature dependence of $N_c(t)$ and also gives the values of the coefficients closer to those experimentally derived.

Appendix A.8 for Chapter 8. Evaluation of the Integrals

$$J_{0i}^{(k)} = \int_0^t r_{c,ef}^{(k)}(t,t_0)\exp(\beta t_0)\,dt_0$$

These integrals are defined in (Eqn. 8.10.32) with β defined in (Eqn. 8.10.21), and $r_{c,ef}^{(k)}(t,t_0)$ with $(k) = 1, 2, 3$ defined in (Eqns. 8.10.30a), (8.10.32b), and (8.10.32c) as

$$\beta = u_s c_{1w} w, \qquad B_i = 2c_{i3}s_{i,cr}, \tag{A.8.1}$$

$$r_{c,ef}^{(1)}(t,t_0) = [(r_0 + \xi_{dep})^2 + B_i(t-t_0)]^{1/2}, \tag{A.8.2}$$

$$r_{c,ef}^{(2)}(t,t_0) = -2\xi_{dep}, \tag{A.8.3}$$

$$r_{c,ef}^{(3)}(t,t_0) = \xi_{dep}^2[(r_0 + \xi_{dep})^2 + B_i(t-t_0)]^{-1/2}. \tag{A.8.4}$$

We omit here the subscript "*hom*" at β and B_i, λ, J_{0i}, Ψ, $\Psi_{1,2,3}$, and other quantities for brevity. Besides, the same integrals will occur in Chapters 9 and 10, with all these quantities specific for heterogeneous nucleation, and the same solutions will be valid there, with the superscript "*het*." The first of these integrals is

$$J_{0i}^{(1)} = \int_0^t r_{c,ef}^{(1)}(t,t_0)\exp(\beta t_0)\,dt_0$$
$$= \int_0^t [(r_0 + \xi_{dep})^2 + B_i(t-t_0)]^{1/2}\exp(\beta t_0)\,dt_0, \tag{A.8.5}$$

Introducing a new variable $x = t_0/t$, it is transformed into

$$J_{0i}^{(1)} = B_i^{1/2}t^{3/2}\int_0^1 (1-x+a)^{1/2}\exp(\beta t x)\,dx, \tag{A.8.6}$$

$$a = \frac{(\xi_{dep} + r_0)^2}{B_i t}. \tag{A.8.6a}$$

Introducing now a new variable, $z = 1 - x$, this integral transforms into

$$J_{0i}^{(1)} = B_i^{1/2}t^{3/2}e^{\beta t}J_{1i}^{(1)} \qquad J_{1i}^{(1)} = \int_0^1 (z+a)^{1/2}\exp(-\beta t z)\,dz. \tag{A.8.7}$$

The next change of the variable, $z' = z + a$, yields

$$J_{1i}^{(1)} = e^{\lambda}\int_a^{1+a} z'^{1/2}\exp[-\beta t z']\,dz', \tag{A.8.8}$$

and λ does not depend on t

$$\lambda = a\beta t = \frac{(\xi_{dep} + r_0)^2\beta}{B_i} = \frac{(u_s c_{1w} w)(\xi_{dep} + r_0)^2}{2c_{1i}s_{i,cr}}. \tag{A.8.9}$$

We introduce a new variable $x' = \beta t z'$. The limits $z' = a$ and $z' = (1 + a)$ transform into $x' = a\beta t = \lambda$ and $x' = \beta t(1 + a) = \lambda + \beta t$. Thus, we have

$$J_{1i}^{(1)} = \frac{\exp(\lambda)}{(\beta t)^{3/2}}\left[\int_{\lambda}^{\infty} x'^{1/2}\exp(-x')\,dx' + \int_{\infty}^{\lambda+\beta t} x'^{1/2}\exp(-x')\,dx'\right],$$

$$= \frac{\exp(\lambda)}{(\beta t)^{3/2}}\left[\Gamma\left(\frac{3}{2},\lambda\right) - \Gamma\left(\frac{3}{2},\lambda+\beta t\right)\right]. \quad \text{(A.8.10)}$$

Here, $\Gamma(\mu, \lambda)$ is the incomplete Euler's gamma function (Gradshteyn and Ryzhik, 1994):

$$\Gamma(\mu,\lambda) = \int_{\lambda}^{\infty} x^{\mu-1}\exp(-x)\,dx. \quad \text{(A.8.10a)}$$

Substitution of (Eqn. A.8.10) into (Eqn. A.8.7) yields

$$J_{0i}^{(1)} = e^{\beta t} B_i^{1/2} \beta^{-3/2} e^{\lambda}\left[\Gamma\left(\frac{3}{2},\lambda\right) - \Gamma\left(\frac{3}{2},\lambda+\beta t\right)\right]. \quad \text{(A.8.11)}$$

Calculation of the second integral $J_{0i}^{(2)}$ is much easier:

$$J_{0i}^{(2)} = \int_0^t r_{c,ef}^{(2)}(t,t_0)\exp(\beta t_0)\,dt_0 = -2\xi_{dep}\int_0^t \exp(u_s c_{1w} w t_0)\,dt_0$$

$$= -2\xi_{dep}\frac{\exp(\beta t)-1}{\beta} = -2\xi_{dep}\frac{\exp(u_s c_{1w} w t)-1}{u_s c_{1w} w} \approx -2\xi_{dep}\frac{\exp(\beta t)}{\beta}. \quad \text{(A.8.12)}$$

The last approximation here is valid only at $\beta t \gg 1$. The third integral is

$$J_{0i}^{(3)} = \int_0^t r_{c,ef}^{(3)}(t,t_0)\exp(\beta t_0)\,dt_0$$

$$= \xi_{dep}^2 \int_0^t \frac{\exp(\beta t_0)}{[B_i(t-t_0)+\xi_{dep}^2]^{1/2}}\,dt_0. \quad \text{(A.8.13)}$$

Similar to evaluation of the first integral, introducing a new variable $x = t_0/t$, and then $z = 1 - x$, this integral is reduced to

$$J_{0i}^{(3)} = \frac{\xi_{dep}^2 t^{1/2} e^{\beta t}}{B_i^{1/2}} J_{1i}^{(3)}, \qquad J_{1i}^{(3)} = \int_0^1 \frac{\exp(-\beta t z)}{(z+a)^{1/2}}\,dz, \quad \text{(A.8.14)}$$

where a is the same as in (Eqn. A.8.6a). Introducing now a new variable $z' = \beta t z$, we obtain

$$J_{1i}^{(3)} = (\beta t)^{-1/2} J_{2i}^{(3)}, \qquad J_{2i}^{(3)} = \int_0^{\beta t} \frac{\exp(-z')}{(z'+\lambda)^{1/2}}\,dz'. \quad \text{(A.8.15)}$$

The integral $J_{2i}^{(3)}$ here is similar to (Eqn. A.8.7). Substituting here $x = z' + \lambda$, and accounting for the change of the limits, $(0, \beta t)$ to $(\lambda, \lambda + \beta t)$ yields

$$J_{2i}^{(3)} = e^{\lambda}\int_{\lambda}^{\lambda+\beta t} x^{-1/2}e^{-x}\,dx = e^{\lambda}\left[\int_{\lambda}^{\infty} x^{-1/2}e^{-x}\,dx + \int_{\infty}^{\lambda+\beta t} x^{-1/2}e^{-x}\,dx\right]$$

$$= e^{\lambda}\left[\Gamma\left(\frac{1}{2},\lambda\right) - \Gamma\left(\frac{1}{2},\lambda+\beta t\right)\right], \quad \text{(A.8.16)}$$

where $\Gamma(\alpha, x)$ is again the incomplete gamma function. Substituting (Eqn. A.8.16) into (Eqn. A.8.15) and into (Eqn. A.8.14), we obtain

$$J_{0i}^{(3)} = \xi_{dep}^2 (\beta B_i)^{-1/2} e^{\beta t} e^{\lambda} \left[\Gamma\left(\frac{1}{2}, \lambda\right) - \Gamma\left(\frac{1}{2}, \lambda + \beta t\right) \right]. \tag{A.8.17}$$

It is more convenient in many cases to use the error function $\Phi(x) = erf(x)$, defined as

$$erf(x) \equiv \Phi(x) = \frac{2}{\sqrt{\pi}} \int_0^x e^{-x'^2} dx', \tag{A.8.17a}$$

instead of incomplete gamma functions, for which coding and finding asymptotics can be easier. This can be done using the relations (Gradshteyn and Ryzhik, 1994, Chapter 8, (Eqns. 8.359) and (8.356) therein)

$$\Gamma(1/2, \lambda) = \sqrt{\pi}[1 - erf(\sqrt{\lambda})], \tag{A.8.18}$$

$$\Gamma(\alpha + 1, \lambda) = \alpha \Gamma(\alpha, \lambda) + \lambda^{\alpha} e^{-\lambda}. \tag{A.8.19}$$

Using these two relations, the $\Gamma(3/2, \lambda)$ in Ψ_1 can be transformed as

$$\Gamma\left(\frac{3}{2}, \lambda\right) = \frac{1}{2} \Gamma\left(\frac{1}{2}, \lambda\right) + \lambda^{1/2} e^{-\lambda} = \frac{\sqrt{\pi}}{2}[1 - erf(\sqrt{\lambda})] + \lambda^{1/2} e^{-\lambda}. \tag{A.8.20}$$

Collecting all three integrals $J_{0i}^{(k)}$ yields the integral J_{0i} in (Eqn. 8.10.33):

$$J_{0i}(t) = \sum_k^3 J_{0i}^{(k)} = e^{\beta t} \Psi, \qquad \Psi = \Psi_1 + \Psi_2 + \Psi_3, \tag{A.8.21}$$

$$\Psi_1 = e^{\lambda} B_i^{1/2} \beta^{-3/2} \left[\Gamma\left(\frac{3}{2}, \lambda\right) - \Gamma\left(\frac{3}{2}, \lambda + \beta t\right) \right] \tag{A.8.22}$$

$$= e^{\lambda} B_i^{1/2} \beta^{-3/2} \{ (\sqrt{\pi}/2)[\Phi(\sqrt{\lambda + \beta t}) - \Phi(\sqrt{\lambda})]$$
$$+ e^{-\lambda}[\lambda^{1/2} - (\lambda + \beta t)^{1/2} e^{-\beta t)}]\}, \tag{A.8.23}$$

$$\Psi_2 = 2\xi_{dep} \beta^{-1}(e^{-\beta t} - 1), \tag{A.8.24}$$

$$\Psi_3 = e^{\lambda} \xi_{dep}^2 (\beta B_i)^{-1/2} \left[\Gamma\left(\frac{1}{2}, \lambda\right) - \Gamma\left(\frac{1}{2}, \lambda + \beta t\right) \right] \tag{A.8.25}$$

$$= e^{\lambda} \xi_{dep}^2 (\beta B_i)^{-1/2} \sqrt{\pi} [\Phi\sqrt{\lambda + \beta t} - \Phi\sqrt{\lambda}]. \tag{A.8.26}$$

These expressions are used in Section 8.10 for evaluation of the deposition integral J_{dep} in the parameterization of homogeneous nucleation and in Section 10.1 for the parameterization of heterogeneous nucleation.

The asymptotic expansion of $\Phi(\sqrt{\lambda})$ at large $\lambda >> 1$ is (Gradshteyn and Ryzhik, 1994, (Eqn. 8.254) therein)

$$\Phi(\sqrt{\lambda}) = erf(\sqrt{\lambda}) = 1 - \frac{1}{\sqrt{\pi}} \lambda^{-1/2} e^{-\lambda} \left(1 - \frac{1}{2\lambda} + \frac{3}{4\lambda^2}\right). \tag{A.8.27}$$

It follows from this equation and (Eqn. A.8.17a) that

$$\lim_{x\to\infty}\Phi(x)=1,\quad \int_0^{\infty} e^{-x'^2}\,dx'=\frac{\sqrt{\pi}}{2} \tag{A.8.28}$$

The other limit at small argument $x \ll 1$ with account taken for the first term is (Gradshteyn and Ryzhik, 1994, (Eqn. 8.253)):

$$\lim_{x\to 0}\Phi(x)=\frac{2}{\sqrt{\pi}}x\exp(-x^2) \tag{A.8.29}$$

The incomplete gamma function is related to the gamma function as

$$\Gamma(\mu,0)=\Gamma(\mu). \tag{A.8.30}$$

The last function has the property

$$\Gamma(1/2)=\sqrt{\pi} \tag{A.8.31}$$

These asymptotic properties of $\Phi(x)$ and $\Gamma(\mu)$ are used in Sections 8.10 and 10.1 for evaluation of the asymptotic behavior of the solutions to the supersaturation equations and parameterizations of homogeneous and heterogeneous ice nucleation processes.

9

Heterogeneous Nucleation of Drops and Ice Crystals

9.1. Introduction

It was shown in the previous chapter that homogeneous nucleation of water drops from vapor becomes effective only at very high supersaturations and homogenous nucleation of ice crystals occurs only at temperatures lower than −35 to −38 °C. Nucleation of droplets from vapor occurs at lower supersaturations on the surface of a foreign substrate that has the property of "*wettability*." Ice formation occurs at substantially higher temperatures, a few degrees below 0 °C, when the ice crystals form on the surface of specific foreign particles termed "*ice forming nuclei*" (IFN) or simply "*ice nuclei*" (IN). Nucleation of drops and crystals on the surfaces of foreign substances is called "*heterogeneous nucleation*."

Formation of the ice phase in clouds at temperatures warmer than −38 °C and small supersaturations via heterogeneous nucleation have been reviewed extensively in previous cloud physics textbooks—e.g., Fletcher (1962), Dufour and Defay (1963), Defay, Dufour, and Bellemans (1966), Young (1993), Pruppacher and Klett (1997), Kashchiev (2000). Heterogeneous nucleation also can act at temperatures colder than −38 °C if appropriative ice nucleating aerosols are present, although the relative importance of heterogeneous and homogeneous nucleation at these colder temperatures is less clear (e.g., Sassen and Dodd, 1988, 1989; DeMott et al., 1998, DeMott , 2002; Lin et al., 2002; Khvorostyanov and Sassen, 1998a,b,c, 2002; Gierens, 2003; Kärcher and Lohmann, 2003; Khvorostyanov et al., 2006; Barahona and Nenes, 2009; Hoose and Möhler, 2012). Heterogeneous ice nucleation dominates in polar clouds and multilayer frontal and cyclonic cloud systems (frequently observed to be mixed phase), and also in deep convective clouds and anvil cirrus, when large amounts of mixed aerosol particles can be brought into the middle and upper troposphere and may serve as ice nuclei.

This chapter describes heterogeneous nucleation processes. Heterogeneous nucleation of drops on insoluble cloud condensation nuclei is considered in Section 9.2, and the rest of this chapter is devoted to heterogeneous ice nucleation.

9.2. Nucleation of Drops by Vapor Deposition on Water-Insoluble Particles

We considered in Chapters 6 and 7 the nucleation of water drops on soluble or mixed aerosol particles, which is governed by Köhler's equation. That was an example of heterogeneous nucleation because of the presence of foreign substances, salt, and insoluble fractions. Another example of heterogeneous process is the nucleation of drops on the surface of solid insoluble aerosol particles, which also can serve as CCN. Consider a liquid germ forming on a planar surface, which has the shape of a spherical segment (cap) or a lens shown in Fig. 9.1.

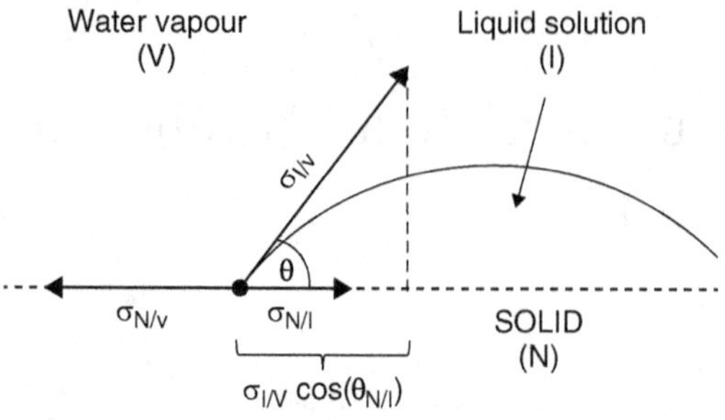

Figure 9.1. Conditions of mechanical equilibrium for a liquid drop on a horizontal solid substrate surface surrounded by humid air. The surface tensions are: σ_{Nv} at the solid–vapor interface (arrow to the left), σ_{Nl} at the solid–liquid interface (arrow to the right), and σ_{lv} at the liquid–vapor interface. Surface tension σ_{Nv} is balanced by the sum of the surface tension σ_{Nl} and the projection $\sigma_{lv}\cos(\theta_{Nl})$ of σ_{lv} onto the horizontal surface, $\sigma_{Nv} = \sigma_{Nl} + \sigma_{lv}\cos(\theta_{Nl})$.

Then, the germ is bounded by a spherical surface formed by the liquid, which is in contact with environmental vapor, and an underlying plane surface of insoluble solid, on which it forms. The geometry of the spherical cap is characterized by the two parameters: radius r of the sphere and a contact angle θ_{Nl} between the liquid germ and the solid surface.

The contact angle is determined by the properties of the liquid and solid surfaces or by the surface tensions at the three interfaces: solid–liquid (σ_{Nl}), liquid–vapor (σ_{lv}), and solid–vapor (σ_{Nv}) (the subscripts N, l, and v denote solid substrate, liquid, and vapor). The cosine of the contact angle $m_{lv} = \cos(\theta_{Nl})$ is determined by Young's relation for the balance of forces at the boundary of the liquid spherical cap and solid surface:

$$\sigma_{lv}\cos\theta_{Nl} + \sigma_{Nl} = \sigma_{Nv}, \tag{9.2.1}$$

where the left-hand side represents the force directed inside the cap, and the right-hand side represents the force directed outside of the cap. This gives

$$m_{lv} = \cos\theta_{Nl} = (\sigma_{Nv} - \sigma_{Nl})/\sigma_{lv}. \tag{9.2.2}$$

The parameter m_{lv} is also called the "*wetting coefficient*," the "*wettability coefficient*," the "*compatibility parameter*," or the "*contact parameter*" because it characterizes how "*wettable*" is this surface by this liquid, or how "*compatible*" are the liquid and surface. This parameter depends on the three surface tensions in (Eqn. 9.2.2), denoted by three letters, N, l, and v. It is customary to use for m the subscripts denoting the two outer media—here liquid and vapor—hence, it is denoted as m_{lv}.

If θ_{Nl} is small, then $m_{lv} = \cos(\theta_{Nl})$ tends to 1, and the liquid germ covers a large area on the surface as a thin liquid film, and such a surface is *hydrophilic* for this liquid. In the opposite case, if $\theta_{Nl} \geq \pi/2$ and $m_{lv} < 0$, the germ's volume is greater than half of a sphere and its shape tends to a sphere. Such a surface is *hydrophobic* for this liquid. If $\theta_{Nl} \to -\pi/2$, then $m_{lv} \to -1$, and the germ has the shape of a sphere that is located on the surface but touches the surface only at one point. Such a surface is not wetted, and is called *completely hydrophobic or non-wettable*. We will see that the smaller the θ_{Nl}, the more hydrophilic the surface, and the easier it is to form condensation on such a surface. The almost spherical drops of dew on the leaves of some plants indicate a hydrophobic surface.

The radius of the liquid germ on the plane surface (the segment of a sphere with radius r cut by the planar base) is $r_b = r\sin(\theta_{Nl})$. The volume of this spherical liquid cap V_l, the surface area A_{lv} of the spherical surface and the surface area A_{lN} of the planar base are provided by stereometry (e.g., Korn and Korn, 1969):

$$V_l = \frac{\pi r^3}{3}(2+m_{lv})(1-m_{lv})^2, \tag{9.2.3}$$

$$A_{lv} = 2\pi r^2(1-m_{lv}), \tag{9.2.4}$$

$$A_{lN} = \pi r^2(1-m_{lv}^2). \tag{9.2.5}$$

Now we can proceed to derive the nucleation rate in a manner similar to that for homogeneous nucleation in Chapter 8. We consider nucleation of water droplets from the vapor on an insoluble surface with the contact angle θ_{Nw} and contact parameter $\cos(\theta_{Nw})$ (the liquid is water in this example and the subscript "l" is replaced further with "w" to distinguish from aqueous solutions and ice considered later). We assume that the concentration c_{gs} of g-mers (embryos containing g molecules) is described by the Boltzmann distribution:

$$c_{gs} = c_{1s}\exp(-\Delta F_{gs}/kT), \tag{9.2.6}$$

where c_{1s} is the concentration of single molecules adsorbed on the surface, and ΔF_{gs} is the energy of the germ formation. The surface nucleation rate $J_{s,het}$—i.e., the rate of the germ formation on a unit surface area per second [cm^{-2}s^{-1}]—can be evaluated similarly to that for homogeneous nucleation in Chapter 8:

$$J_{s,het} = c_{gs} j_g^{(+)} A_{gs} Z_s, \tag{9.2.7}$$

where $j_g^{(+)}$ is the molecular flux onto this g-mer that can be written using the equation of state for water vapor, $c_{1s} = e_v/kT$, and assuming that the condensation coefficient $\alpha_c = 1$ (see Chapter 5, and Section 8.2):

$$j_g^{(+)} = \frac{c_{1s}V_w}{4} = \frac{e_v}{(2\pi m_{w1}kT)^{1/2}}, \tag{9.2.8}$$

A_{gs} is the surface area of the germ, and Z_s is the Zeldovich factor, introduced in Section 8.2,

$$Z_s = \left(\frac{\Delta F_{gs}}{3g\pi kT}\right)^{1/2}. \tag{9.2.9}$$

The critical energy for a germ formation can be written similarly to that for homogeneous nucleation as in (Eqns. 8.1.2) and (8.1.8), which we write again as a sum of the volume and surface terms:

$$\Delta F_{gs} = -V_g\frac{\Delta\mu_{vw}}{v_w} + \sigma_{wg}A_{wg}, \tag{9.2.10}$$

where $\Delta\mu_{vw} = \mu_v - \mu_w = RT\ln(S_w)$ is the difference of molar chemical potentials (affinity) of the vapor and liquid water, $V_g = V_l$ is the volume of the germ defined by (Eqn. 9.2.3), and $v_w = M_w/\rho_w$ is the molar volume of water. The surface term in (Eqn. 9.2.10) can be evaluated by taking into account the germ geometry as

$$\sigma_{wg} A_{wg} = \sigma_{vw} A_{wv} + (\sigma_{Nw} - \sigma_{Nv}) A_{Nw}, \tag{9.2.11}$$

where the three surface tensions on the RHS relate to the three interfaces: water–vapor, substrate–water, and substrate–vapor as defined by their subscripts, A_{vw} and A_{Nw} are the surface areas of the spherical and planar surfaces. Substituting (Eqns. 9.2.4) and (9.2.5) for A_{vw} and A_{Nw} into (Eqn. 9.2.11), we obtain

$$\sigma_{wg} A_{wg} = \sigma_{vw} 2\pi r^2 (1 - m_{wv}) + (\sigma_{Nw} - \sigma_{Nv}) \pi r^2 (1 - m_{wv}^2). \tag{9.2.12}$$

Using the relation $(\sigma_{Nw} - \sigma_{Nv}) = -\sigma_{vw} m_{wv}$ that follows from (Eqn. 9.2.2), we have

$$\begin{aligned} \sigma_{wg} A_{wg} &= \pi r^2 \sigma_{vw} (1 - m_{wv})[2 - m_{Nw}(1 + m_{wv})] \\ &= \pi r^2 \sigma_{vw} (2 + m_{wv})(1 - m_{wv})^2. \end{aligned} \tag{9.2.13}$$

The germ energy (Eqn. 9.2.10) can be rewritten using (Eqn. 9.2.3) for V_w and (Eqn. 9.2.13) for the surface term as

$$\begin{aligned} \Delta F_{gs} &= \left(-\frac{4\pi r^3}{3} \frac{\Delta\mu_{vw}}{v_w} + 4\pi r^2 \sigma_{vw} \right) \frac{[(2 + m_{wv})(1 - m_{wv})^2]}{4}, \\ &= \left(-\frac{4\pi r^3}{3} \frac{\Delta\mu_{vw}}{v_w} + 4\pi r^2 \sigma_{vw} \right) f(m_{wv}), \end{aligned} \tag{9.2.14}$$

where we introduced the geometric factor for the planar surface following Volmer (1939)

$$f(m_{wv}) = \frac{(2 + m_{wv})(1 - m_{wv})^2}{4} = \frac{m_{wv}^3 - 3m_{wv} + 2}{4}. \tag{9.2.15}$$

Both of these forms of $f(m_{wv})$ are used in applications.

Substituting $\Delta\mu_{vw} = \mu_v - \mu_w = RT\ln(S_w)$ into (Eqn. 9.2.14) yields another form of ΔF_{gs}:

$$\Delta F_{gs} = \left(-\frac{4\pi r^3}{3} \frac{RT \ln(S_w)}{v_w} + 4\pi r^2 \sigma_{vw} \right) f(m_{wv}). \tag{9.2.16a}$$

The critical energy is determined from the condition of minimum $\partial \Delta F_{gs}(r_{cr})/\partial r = 0$, which yields from (Eqn. 9.2.14):

$$4\pi r_{cr} \left(-r_{cr} \frac{\Delta\mu_{vw}}{v_w} + 2\sigma_{vw}\right) = 0, \tag{9.2.16b}$$

and so we obtain the critical radius

$$r_{cr} = \frac{2\sigma_{vw} v_w}{\Delta\mu_{vw}} = \frac{2\sigma_{vw} v_w}{\mu_v - \mu_w} = \frac{2\sigma_{vw} M_w}{RT \rho_w \ln S_w}. \tag{9.2.17}$$

Note that these equations have the same form as (Eqns. 8.1.6) or (8.1.9) in Section 8.1 for r_{cr} via molar chemical potentials. Substituting this r_{cr} into (Eqn. 9.2.14), we obtain ΔF_{cr}:

$$\Delta F_{cr} = 4\pi r_{cr}^2 \left(-\frac{r_{cr}}{3}\frac{\Delta\mu_{vw}}{v_w} + \sigma_{vw} \right) f(m_{wv}). \tag{9.2.18a}$$

Substituting here the relation $-(r_{cr}/3)\Delta\mu_{vw}/v_w = -(2/3)\sigma_{vw}$ that follows from (Eqn. 9.2.16b), we obtain

$$\Delta F_{cr} = \frac{4\pi}{3}\sigma_{vw} r_{cr}^2 = \frac{\sigma_{vw} A_{cr}}{3} = \frac{16\pi}{3}\frac{M_w^2 \sigma_{vw}^3}{[RT\rho_w \ln S_w]^2} f(m_{wv}), \tag{9.2.18b}$$

where A_{cr} is the surface area of a sphere with radius r_{cr}. The second form of this equation has the same form as in Chapter 8. If the insoluble substrate is completely wettable (a hydrophilic surface), $m_{wv} = 1$, according to (Eqn. 9.2.15), $f(m_{wv}) = 0$, and $\Delta F_{gs} = 0$. Thus, there is no energy barrier for nucleation. In the opposite case of a completely hydrophobic surface, $m_{wv} = -1$, then (Eqn. 9.2.15) shows that $f(m_{wv}) = 1$, and the energy is the same as for homogeneous nucleation. In the intermediate cases, $0 < f(m_{wv}) < 1$, and the energy of heterogeneous nucleation is smaller than that for homogeneous nucleation.

Now, the nucleation rate of drops by condensation from vapor per unit area of an insoluble plane surface per unit time [$cm^{-2}s^{-1}$] is obtained from (Eqn. 9.2.7) with the use of (Eqns. 9.2.6) and (9.2.8):

$$J_{s,het} = \frac{c_{1s} e_v}{(2\pi m_{w1} kT)^{1/2}} \pi r_{cr}^2 Z_s \exp(-\Delta F_{cr}/kT), \tag{9.2.19}$$

where m_{w1} is the mass of a water molecule. The expression before the exponent is termed the *kinetic coefficient*. Fletcher (1958, 1962) estimated it as $\sim 10^{24}$ to 10^{27}. The average value of 10^{25} is used sometimes in approximate calculations of the nucleation rates.

Similar to the previous derivation for nucleation on an insoluble plane particle, Fletcher (1958, 1962) considered the spherical cap formed on the curved surface of an insoluble particle of radius r_N. The volume of this spherical cap and its two surface areas change and yield a new geometrical factor $f(m_{wv}, x)$, and the critical energy becomes

$$\Delta F_{cr} = \frac{16\pi}{3}\frac{M_w^2 \sigma_{vw}^3}{[RT\rho_w \ln S_w]^2} f(m_{wv}, x). \tag{9.2.20}$$

Here, m_{wv} is the same as before the contact parameter at the interface of the solid substrate and liquid water, $x = r_N/r_{cr}$, and $f(m_{wv}, x)$ is Fletcher's geometric factor that accounts for the radius r_N of insoluble particles, and has the form (see derivation in Fletcher, 1958; or in Pruppacher and Klett, 1997):

$$f(m_{wv}, x) = (1/2)\{1 + [(1 - m_{wv}x)/y]^3 + x^3(2 - 3\psi + \psi^3) + 3m_{wv}x^2(\psi - 1)\},$$

$$\psi = (x - m_{wv})/y, \qquad y = (1 - 2m_{wv}x + x^2)^{1/2}. \tag{9.2.21}$$

Consider the particular cases of small and large r_N. In the small particle limit, $x = r_N/r_{cr} << 1$ (which may correspond to a vanishing insoluble substrate, $r_N \to 0$), the value of the contact angle $\theta_{ls} = 180°$, $m_{wv} = -1$, and $f(m_{wv}, x) = 1$, and this case is similar to homogeneous nucleation. In the large particle limit $x = r_N/r_{cr} >> 1$, the geometrical factor (Eqn. 9.2.21) reduces to the expression (Eqn. 9.2.15) for planar substrate.

Fletcher's geometrical factor was derived for the case when a germ forms on the surface of convex aerosol particles. For the case of spherically concave insoluble substrate with a radius of curvature r_N, the geometric factor can be derived in a similar way:

$$f(m_{wv}, x) = (1/2)\{1 - [(1 + m_{wv}x)/y]^3 - x^3(2 - 3\psi + \psi^3) + 3m_{Nw}x^2(\psi - 1)\},$$
$$\psi = (x + m_{wv})/y, \qquad y = (1 + 2m_{wv}x + x^2)^{1/2}. \tag{9.2.22}$$

Calculations of the drops' nucleation on this concave substrate showed that it can be faster than on the convex substrate with Fletcher's geometric factor (Eqn. 9.2.21).

The nucleation rate just shown was defined per unit area of the insoluble particle. The nucleation rate $J_{p,het}$ per particle [s^{-1}] can be calculated in a first approximation multiplying (Eqn. 9.2.19) by the particle surface area $4\pi r_N^2$, or:

$$J_{p,het} = \frac{4\pi r_N^2 c_{1s} e_v}{(2\pi m_{w1} kT)^{1/2}} \pi r_{cr}^2 Z_s \exp(-\Delta F_{cr}/kT). \tag{9.2.23}$$

Fletcher (1958, 1962) estimated the average kinetic coefficient (the expression before the exponent) as $\sim 10^{26} r_N^2$.

More sophisticated models of the surface nucleation have been developed that consider adsorption, desorption, and surface diffusion of the water molecules on the substrate (e.g., Pruppacher and Klett, 1997; Hellmuth et al., 2012, 2013, and references therein). These models require knowledge of the adsorption and desorption energies, which are still rather uncertain. A detailed consideration of these models is beyond the scope of this book.

9.3. Modes of Ice Nucleation and Properties of Ice Nuclei

9.3.1. Modes of Ice Nucleation

Classical cloud physics distinguishes four basic modes of heterogeneous ice nucleation when ice forms on ice nuclei (IN): deposition mode, condensation-freezing mode, immersion mode, and contact mode. The deposition mode is defined as a process that occurs at ice supersaturation, and water vapor is directly deposited onto the surface of IN, either directly as an ice layer, or as a liquid film, which is subsequently transformed into ice at sufficiently low temperatures. The condensation-freezing mode is usually defined as a two-step process that begins at water supersaturation. An aerosol particle in this mode acts first as a CCN forming a drop (condensation stage), and then this drop freezes (freezing stage). The immersion mode is defined as a process when IN is immersed in a drop, presumably at temperatures above the freezing threshold, and freezes after some time at some temperature below the threshold temperature. The contact mode is defined as the process when an ice nucleus is originally outside of a drop, and then is brought to the drop surface by any of the scavenging mechanisms (Brownian diffusion, diffusiophoresis, thermophoresis), contacts the drop, and ice is nucleated at the moment of contact. Aerosol particles initiating ice in these four modes are termed deposition or sorption nuclei, condensation-freezing nuclei, immersion nuclei, and contact nuclei, although the same ice nucleus may nucleate ice in different modes.

More recent research illustrates ambiguity among these four modes of ice nucleation. In many cases, the difference between the condensation-freezing and immersion modes become unclear because both processes assume condensation with subsequent freezing on the immersed IN. While the term *condensation-freezing* is defined earlier as a formation of the droplet at *water supersaturation* at the first stage, the first stage often occurs at *water subsaturation*, when the submicron haze particles become deliquescent but do not activate into cloud drops, and then freeze at water subsaturation at the second stage before the formation of cloud drops as was considered in Chapter 8. This may happen, for example, in the formation of cirrus clouds or diamond dust at water subsaturation. To emphasize the difference between processes occurring at water *sub-* and *supersaturations*, instead of condensation freezing, we use the term "*deliquescence-heterogeneous freezing mode*," or *DF* or "*DhetF mode*," introduced in Khvorostyanov and Curry (2004a,b, 2005a). The DF or DhetF mode is defined as a process of heterogeneous ice nucleation when mixed aerosol particles (CCN) become deliquescent but do not transform into drops, and freeze as haze particles still at water subsaturation. In contrast, immersion freezing is understood as a process of the formation of cloud drops from immersed ice nuclei and subsequent freezing at water supersaturation.

9.3.2. Properties of Ice Nuclei

Ice nuclei represent a small subset of aerosol particles that can catalyze ice nucleation—i.e., they have the ability to nucleate ice under conditions when ice would not form without these particles. Numerous studies of IN properties are reviewed in a number of monographs and papers (e.g., Fletcher, 1962, 1970a; Mason, 1971; Hobbs, 1974; Young, 1993; Pruppacher and Klett, 1997; Lohmann and Diehl, 2006; Vali, 2008; DeMott et al., 2011; Hoose and Möhler, 2012; Ladino et al., 2013; and references therein). Despite significant efforts studying IN and the collection of a vast amount of data on IN, a complete physical understanding of their nature, properties, and the mechanisms of ice nucleation remain elusive, and various experimental data are controversial. As emphasized by Pruppacher and Klett (1997, Section 9.2), "Unfortunately, none of the presently available devices which count the fraction of aerosol particles acting as IN is capable of allowing four different modes of action, nor can they realistically simulate the time scale over which temperature and supersaturation vary in atmospheric clouds. Therefore, the IN concentrations quoted in the literature have to be treated with considerable caution." Here we provide a brief outline of some properties of IN that are essential for understanding the mechanisms of IN activity in heterogeneous nucleation and calculations of ice nucleation.

The Nature of IN

Among the various methods of studying IN, two techniques are used most frequently: direct physico-chemical analyses of the substances and residues contained in ice crystals, and laboratory tests of the ice nucleating ability of various substances. A wide array of substances has been found within snow crystals by means of physico-chemical analyses that allow identification of the nature of ice nuclei. Many substances have been tested in the laboratory for their ice nucleating ability, including: mineral dust particles and clays, soot of various origin from anthropogenic and natural emissions, aerosols of biological origin (bacteria, leaf litter, fungal spores, various pollens, and diatoms), solid ammonium sulfate, organic acids, humic-like substances, and metallic compounds. A large group of artificially formed aerosols like AgI, CuI, and PbI has been tested in laboratories and used in cloud seeding experiments for artificial cloud crystallization (see Chapter 4).

Hoose and Möhler (2012) performed a detailed review of more than 60 laboratory experiments with various kinds of mineral dust particles, about 20 experiments with soot, and about 40 experiments with bioaerosols. The experiments were conducted using various devices and under a wide range of conditions. This comparison indicated that common clay minerals such as kaolinite, montmorillonite, and illite may serve as good IN, while natural desert dusts are less effective as IN and may require higher ice supersaturations or colder temperatures for activation. The conclusions on soot as IN are less certain and there is a greater spread in the results depending on the sort and origin of soot, but it can be confirmed that soot particles are in general less effective ice nuclei than mineral dust, also requiring colder temperatures and higher supersaturations. The ice nucleating ability of bioaerosols was found to be a selective property. A small fraction of bacteria was able to nucleate ice at subzero temperatures. Abbatt et al. (2006) found that solid ammonium sulfate particles were able to nucleate ice both at water subsaturation in deposition mode, and in immersion mode in solution drops. A detailed review of the experimental research on contact nuclei was provided by Ladino et al. (2013).

Concentrations

Measured concentrations N_{IN} of IN vary significantly. They exhibit a general tendency to increase with decreasing temperature and increasing humidity. It was found that various IN may nucleate ice heterogeneously over a wide temperature range from a few degrees below 0 °C to very cold temperatures below −80 °C and possibly lower. The values of N_{IN} measured prior to the 1970s vary in the range 10^{-3} to $0.1\,L^{-1}$ at −10 °C, and in the range from a few hundredths to a few per liter at −20 °C (Pruppacher and Klett, 1997). Many of these measurements used static filter techniques. Newer devices—e.g., continuous flow diffusion chambers (CFDC)—showed N_{IN} by about one order of magnitude higher (Cooper, 1986; Rogers, 1982; Al Naimi and Saunders, 1985; Meyers et al., 1992). Some of these measurements were performed at rather high water supersaturation, up to 20–25%, so that many new small drops could form and freeze in the immersion mode. The other experiments were performed at positive ice supersaturation but at water subsaturation, so that deliquescence-freezing, contact, or deposition modes could act. Thus, separation of the modes and identifying their relative contributions and concentrations of various types of IN is not a simple task.

Partial solubility

The old concept adopted in cloud physics treated ice nuclei as highly insoluble particles. In the real atmosphere, most ice nuclei have appreciable fractions of soluble materials, which has been emphasized in the field projects of the 1990s and 2000s—e.g., SUCCESS (Subsonic Aircraft: Contrail and Cloud Effects Special Study, Special issue of the Geophysical Research Letters, v. 25, 1998); and FIRE-SHEBA (First International Regional Experiment – Surface Heat Budget of the Arctic; Curry et al., 2000; and Special issue of the Journal of Geophysical Research, v. 106, 2001), CRYSTAL-FACE (Cirrus Regional Study of Tropical Anvils and Cirrus Layers – Florida Area Cirrus Experiment case study) (Jensen et al., 2005; Fridlind et al., 2004; Czizco et al., 2004).

Detailed analysis of IN physicochemical properties has shown that IN are not necessarily totally insoluble (Chen et al., 1998), but a significant fraction of the IN may represent mixtures of insoluble substrates with sulfates or other soluble materials with soluble volume fractions of 0.2–0.9. Hence, the same aerosol particles may serve both as IN and CCN. The fraction of these mixed IN has been observed to be as high as 40% of the IN in the lower and middle troposphere, and up to 27% in the upper troposphere (Chen et al. 1998, Rogers et al. 1998; 2001). Note that the energy-dispersive X-ray analysis used in these experiments was not able to detect soluble mass fractions smaller than 10%; if it had been possible to make this threshold lower (e.g., down to 1%), the fractions of these mixed IN-CCN could be considerably higher.

These experiments indicated the necessity of modifying the traditional view of IN as completely insoluble particles, as well as the parameterizations of heterogeneous ice nucleation. DeMott et al. (1998) developed a model of heterogeneous ice nucleation by assuming that all heterogeneous IN are freezing nuclei, which are deliquescent mixed cloud condensation nuclei (CCN) with 50% insoluble and 50% soluble matter (sulfuric acid in the upper troposphere), and are contained within 10% of the CCN population with sizes above 0.1 μm; the freezing fraction of these IN-CCN was parameterized as a function of temperature. Khvorostyanov and Curry (2000, 2004a,b, 2005a, 2009b, hereafter KC00, KC04a,b, KC05, KC09), and Curry and Khvorostyanov (2012, hereafter CK12) extended classical nucleation theory and developed a theory of heterogeneous ice nucleation by freezing of CCN containing both soluble and insoluble fractions, with the possibility of ice nucleation at both water sub- and supersaturations.

It has been assumed in many studies that the dominant mode of crystal formation in cold cirrus might be the homogeneous freezing of haze particles in conditions of subsaturation with respect to water (e.g., Sassen and Dodd, 1988, 1989; Heymsfield and Sabin, 1989; DeMott et al., 1994). High ice supersaturations in cirrus and an important role of heterogeneous freezing in cirrus and contrail formation have been recognized recently and have been intensely studied in subsequent years (e.g., Jensen et al., 1994, 1996, 2001, 2005; DeMott et al., 1994, 1997, 1998; 2002; Kärcher et al., 1996, 2006; Khvorostyanov and Sassen, 1998a,b,c, 2002; Kärcher and Lohmann, 2003; Sassen and Benson, 2000; Lin et al., 2002; Gierens, 2003; Haag et al., 2003; Khvorostyanov et al., 2001, 2006; Fridlind et al., 2004; Barahona and Nenes, 2009; Krämer et al., 2009; Spichtinger and Gierens, 2009; Hoose and Möhler, 2012), although significant differences still exist in the description of this process owing to the absence of adequate theories. The parameterizations and theories of heterogeneous freezing on partially soluble IN are discussed in the subsequent sections of this chapter.

Size

Laboratory and field experiments show that the efficiency of IN increases with increasing size of aerosol particles. Since ice nucleation occurs on the surface of aerosol particles, the catalyzing effect of IN increases with the surface area—i.e., with the particle size. The insoluble fraction of aerosol particles generally increases with their size, making them more efficient IN. Many substances known as good ice nucleators, like silicates, sand, or clay particles, are contained mostly in the large particle range. There is evidence that the IN efficiency rapidly decreases at small sizes below ~0.01 μm; however, the exact limit of the nucleating ability is uncertain. Thus, the nucleation ability of IN increases with size; however, the number of aerosol particles decreases with size. As a result of these two counteracting tendencies, some types of aerosol can be the most efficient IN in certain size ranges. For example, analysis of snow crystals in some experiments showed that the sizes of IN ranged between 0.1 and 15 μm with a mode between 0.5 and 5 μm (e.g., Pruppacher and Klett, 1997).

Chemical structure

Studies of the chemical composition of IN show that those aerosol particles having chemical bonds at the surface similar to the bond structure in the ice lattice (O-H-O) may be more effective ice nuclei. Many organic particles have similar hydrogen bond structures, therefore they may be efficient ice nucleants. These include some bacteria, leaf litter, cholesterol, phloroglucinol, and many others. Several organic substances were found to nucleate ice at temperatures warmer than −10 °C, and up to −1 to −2 °C. It was shown that the ice nucleating ability of bioaerosols is caused by the presence on the surface of these substances of crystalline structures similar to ice.

Crystallographic structure

Experimental studies of IN show that the ice nucleating properties of IN depend on their crystallographic structure. Many substances with crystallographic structure close or similar to ice structure appeared to be better ice nucleators than the substances with different structures. This, in particular, explains the good ice nucleability of artificial ice nuclei used for cloud seeding—e.g., AgI, PbI, CuS, and others that have been successfully used in weather modification programs since the 1940s as described in Chapter 4. The parameters of their crystalline lattice are similar to that of ice.

As was mentioned in Section 3.6, if the crystallographic differences between ice and the substrate are sufficiently small, both ice and substrate lattices may deform elastically and then join coherently. This misfit between the ice and substrate lattices causes an elastic strain; that is, an additional energy that hampers ice nucleation on the surface of the substrate. In general, this strain is characterized by the multicomponent tensor of elastic strain (Landau and Lifshitz, v.5, 1958b). In applications for nucleation problems, it is customary to use a simpler approach and to describe the effect of strain by the relative misfit between the ice and substrate lattices—i.e., by the relative difference between the lengths of ice and substrate lattice unit cells. A more detailed description of this effect will be given in Sections 9.5 and 9.7.

9.4. Empirical Parameterizations of Heterogeneous Ice Nucleation

For the past four decades, ice nucleation processes and the number concentration N_{IN} of atmospheric aerosol particles that initiate ice phase (ice nuclei, IN) have been parameterized based on experimental data mostly as two independent functions: of temperature and of supersaturation over ice or over water. There have been some attempts at representation of experimentally derived values of IN concentration simultaneously as a function of temperature and supersaturation.

Several authors developed parameterizations of ice nuclei concentrations $N_{IN}(T)$ or ice particle concentrations $N_i(T)$ as empirical functions of temperature T. Fletcher (1962) suggested a parameterization of $N_{IN}(T)$ as an empirical exponential function of the temperature T:

$$N_i(T) = A_F \exp[B_F(T_0 - T)], \qquad (9.4.1)$$

with the parameters $A_F = 10^{-5}\,\text{L}^{-1}$ and $B_F = 0.6\,(°\text{C})^{-1}$, $T_0 = 0\,°\text{C}$ and T in °C, although these parameters could substantially vary around these values (Pruppacher and Klett, 1997). Huffman and Vali (1973) and Huffman (1973) offered a power law by ice supersaturation s_i of

$$N_i(S_i) = C_s s_i^{b_s}, \tag{9.4.2}$$

with N_i in L^{-1}, s_i in %, and C_s and b_s being the constants for a given air mass. It was found later that $3 < b_s < 8$, and there was greater uncertainty in C_s.

Berezinsky and Stepanov (1986) offered a similar parameterization consisting of an exponential function by the temperature:

$$N_{IN}(T) = N_a 10^{k_{BS}(T-T_0)}, \qquad k_{BS} = 0.14 - 0.055\log(d_2), \tag{9.4.3}$$

where N_a is the concentration of aerosols of a given size, d_2 is the mean square diameter of aerosol in μm, and $\log(x)$ is the decimal logarithm of x. Thus, the segregation of N_{IN} by size was accounted for. The dependence on ice supersaturation was found similar to Huffman's (1973) power law, (Eqn. 9.4.2), but with additional dependence of parameters on temperature. Berezinsky and Stepanov (1986) also presented the simultaneous graphical dependences of $N_{IN}(T, s_i)$ on both T and s_i that showed the comparable roles of T and s_i on N_{IN}, so that the IN concentration was a function of both parameters as $N_{IN}(T, s_i)$.

Cooper (1986) derived a temperature-dependent parameterization that refined Fletcher's function as

$$N_i(T) = A_{TC} 10^{B_{TC}(T_0-T)}, \quad \text{or} \quad N_i(T) = A_{TC} e^{0.31(T_0-T)}, \tag{9.4.4}$$

where T is in degrees K, $B_{TC} = 0.135\,(\text{K})^{-1}$, and $A_{TC} = 4.47 \times 10^{-3}\,\text{L}^{-1}$. Cooper's parameterization has a smaller T-slope than Fletcher's (Eqn. 9.4.1), produces greater N_i at warmer $T > -22\,°\text{C}$, and more realistic smaller IN at lower T that are closer to observations. Another modification of Fletcher's parameterization was suggested by Sassen (1992) to account for the depletion of IN with height z and to make it more realistic at low T:

$$N_i(T) = A_{T0}\exp(-A_z z)\exp[B_T(T_0 - T)], \tag{9.4.5}$$

where $A_{T0} = 10^{-5}\,\text{L}^{-1}$ is the measured concentration of IN at the earth's surface, $A_z = 0.75\,\text{km}^{-1}$ determines a decrease of IN with height, and B_T is the same as in Fletcher's (Eqn. 9.4.1).

Both supersaturation and temperature dependencies were parameterized in a single equation by Cotton et al. (1986) who combined Fletcher's T-dependence and Huffman's s_i-dependence

$$N_i(T) = A_T (s_i/s_{i0})^{b_s} \exp[B_T(T_0 - T)], \tag{9.4.6}$$

where A_T and B_T are the same as in Fletcher's (Eqn. 9.4.1), s_i is the fractional ice supersaturation, and s_{i0} is the fractional ice supersaturation at water saturation at T.

Meyers et al. (1992) noted that the combined T-s_i parameterization by Cotton et al. (1986) underepredicts IN concentrations because it was constructed from the parameterizations (Eqns. 9.4.1) and (9.4.2) that were based on static filter and other measurements prior to use of continuous flow diffusion chambers (CFDC) and gave lower (up to an order of magnitude) IN concentrations than were obtained later with CFDC. Meyers et al. (1992) reanalyzed the data and suggested for the combined condensation-freezing and deposition modes another empirical parameterization of the condensation-freezing (CF) nucleation as a supersaturation-dependent only function

$$N_i(s_i) = \exp(a_M + b_M s_i), \tag{9.4.7}$$

with N_i in L^{-1}, s_i in %, $a_M = -0.639$, and $b_M = 0.1296$. This was suggested to be valid at $-20 < T < -7\,°C$, and $2 < s_i < 25\%$, but (Eqn. 9.4.7) was used later in several works outside of these regions (e.g., Comstock et al. (2008) used it for cirrus clouds at $T < -40\,°C$). Although the temperature dependence was present in the original data, Meyers et al. (1992) averaged it and retained only the supersaturation dependence. A similar s_i-dependent parameterization for deposition nucleation on dust particles was suggested by Möhler et al. (2006) based on measurements in a large expansion chamber of $84\,m^3$. Cotton et al. (2003) modified (Eqn. 9.4.7) to include the prognostic variable N_{IN}:

$$N_i(s_i) = N_{IN} \exp(b_M s_i), \tag{9.4.7a}$$

with N_i in L^{-1}, s_i in %, and where $T < -5\,°C$. A comparison with MDC92 (Eqn. 9.4.7) indicates that this could imply that $N_{IN} = \exp(a_M)$; however, (Eqn. 9.4.7a) assumes the variable N_{IN} can be deduced from the continuous flow diffusion chamber data.

In contrast, the contact nuclei N_{ac} were parameterized in Meyers et al. (1992) as a temperature power law. The supersaturation dependence was smoothed and disregarded although many data indicated such dependence in contact nucleation caused by the physical mechanisms of diffusiophoresis and thermophoresis that are governed by the supersaturation field (Young, 1993; Pruppacher and Klett, 1997; Seinfeld and Pandis, 1998). Contact nucleation will be considered in Chapter 10.

These parameterizations were revised again for the low temperature range by DeMott et al. (1998), who assumed that IN are mixed aerosol particles with radii over 0.1 μm, consisting of 50% soluble and 50% insoluble fractions—i.e., similar to cloud nuclei (CN) or to cloud condensation nuclei (CCN). The IN were measured simultaneously with CN using CFDC onboard of aircraft at low temperatures, and IN were parameterized in the temperature range $-25 < T < -38\,°C$ as the freezing fraction of CN:

$$N_{IN} = N_{CN} F_{IN/CN}; \tag{9.4.8}$$

$$F_{IN/CN} = a_D (T_0 - T)^{b_D}, \tag{9.4.9}$$

with $a_D = 1.3 \times 10^{-22}$; $b_D = 11.75$. This parameterization was suggested for the combined deposition and condensation-freezing modes. The water supersaturation s_w varied in these measurements in the range from subsaturation, when deliquescent submicron mixed haze particles (unactivated CCN) could be present, to ~18% when mixed CCN could transform into the drops and freezing could proceed in the immersion mode (DeMott et al., 1998, Fig. 1).

Note that these high water supersaturations $s_w = 1$–18% lead to very rapid activation of the cloud condensation nuclei (CCN) down to very small sizes < 0.01 μm (see Chapter 6 and Fig. 6.5). This activation takes from a few tenths of a second to a few seconds (see Chapters 6 and 7); that is, it is much faster than heterogeneous ice nucleation that may take 10–30 min (Pruppacher and Klett, 1997; Lin et al., 2002; Khvorostyanov and Curry, 2005a; Eidhammer et al., 2009; see Section 9.14 here), so that almost all mixed aerosol particles at $s_w > 0$ may have converted into cloud drops prior to freezing. Thus, the term "condensation" was understood literally but the condensation-freezing mode was not isolated and was actually mixed with the immersion mode at $s_w > 0$. This feature is pertinent to all freezing experiments where positive s_w is achieved.

Thus, Meyers et al. (1992) parameterized IN as a function of s_i only at $T > -20\,°C$, and DeMott et al. (1998) parameterized them as a function of T only at $T < -25\,°C$. We do not expect that there is an abrupt change of the mechanism of IN nucleation at $-20\,°C$, and a construction of a unified parameterization with both dependencies remained a problem.

Another empirical parameterization for the immersion mode with ice nuclei of soot, mineral dust, and biological nuclei were suggested by Diehl and Wurzler (2004) based on laboratory experiments and by generalizing Bigg's (1953) conception of the median freezing temperature. Diehl and Wurzler (2004) suggested a parameterization for the nucleation rate $J_{het}(T)$ and freezing rate dN_f/dt of

$$J_{het}(T) = a_{DW} B_{h,i} \exp(a_{DW} T_s) \tag{9.4.10a}$$

$$-\frac{dN_f}{dt} = \frac{dN_u}{dt} = N_u J_{het} V_d \frac{dT}{dt} = N_u a_{DW} B_{h,i} V_d \exp(a_{DW} T_s) \frac{dT}{dt}, \tag{9.4.10b}$$

where N_f and N_u are the numbers of frozen and unfrozen drops, respectively, $T_s = T_0 - T$ is supercooling, V_d is the droplet volume, and $B_{h,i}$ and a_{DW} are the empirical constants. The values of $B_{h,i}$ varied from $2.91 \times 10^{-9}\,\mathrm{cm}^{-3}$ for soot particles, to $6.15 \times 10^{-8}\,\mathrm{cm}^{-3}$ for kaolinite, $3.23 \times 10^{-5}\,\mathrm{cm}^{-3}$ for montmorillonite, and $6.19 \times 10^{-5}\,\mathrm{cm}^{-3}$ for illite, and were much greater for biological particles that appeared to be effective IN with $B_{h,i} = 1.01 \times 10^{-2}\,\mathrm{cm}^{-3}$ for pollen, $4.38 \times 10^{-1}\,\mathrm{cm}^{-3}$ for leaf litter, and $6.19 \times 10\,\mathrm{cm}^{-3}$ for bacteria. This choice of the parameters was based on fitting to laboratory measurements as described in Diehl and Wurzler (2004). The freezing point depression due to the presence of solutes was accounted for using the parameterization by Koop et al. (2000) for homogeneous nucleation, assuming that the depressions are similar for homogeneous and heterogeneous nucleation processes. This parameterization by Diehl and Wurzler (2004) was tested in the GCM ECHAM4 (Lohmann and Diehl, 2006) and in a mesoscale model COSMO (Zubler et al., 2011) and showed reasonable performance with realistic N_c.

Subsequent field experiments stimulated further revisions of the previous parameterizations. Measurements made during the Mixed-Phase Arctic Cloud Experiment (MPACE) in autumn 2004 (Verlinde et al., 2007) and subsequent numerical simulations showed that MDC92 parameterization (Eqn. 9.4.7) overestimates IN concentration and leads to cloud glaciation that is too rapid (Prenni et al., 2007). A modification of MDC92 was suggested by Prenni et al. (2007), so that the functional form was the same as in (Eqn. 9.4.7), but the coefficients were different: $a_M = -1.488$, $b_M = 0.0187$. As with MDC92, this parameterization by Prenni et al. (2007) also did not contain dependencies on temperature and aerosol concentration, had very weak s_i-dependence, but gave substantially smaller IN concentrations (0.2–$0.4\,\mathrm{L}^{-1}$) that were comparable to measured values in MPACE that permitted the existence of the mixed cloud state for several hours.

The next modification of the MDC92 parameterization was performed by Philips, DeMott and Andronache (2008, hereafter PDA08), who developed a new empirical parameterization using MDC92 as a basis. PDA08 extended this parameterization for various T- and s_i-ranges and generalized the parameterization to account for the three types of freezing aerosol (dust and metallic compounds, black carbon, and insoluble organics). This parameterization included appropriate scaling and integration over the surface areas of these aerosols, so that the concentration $N_{c,x}$ of IN of the x-th kind is

$$N_{c,x} = \int_{\log[0.1\mu m]}^{\infty} \{(1-\exp[-\mu_x(D_x, S_i, T)]\} \frac{dn_x}{d\log D_x} d\log D_x, \qquad (9.4.11)$$

where x denotes any of the 3 aerosol types, n_x is the aerosol mixing ratio, μ_x is the average activated IN per aerosol of diameter D_x, and μ_x is proportional to $N_c(s_i)$ from (Eqn. 9.4.7) multiplied by some coefficients. For a low freezing fraction, which often takes place, $N_{c,x} \sim \mu_x \sim N_i$ (PDA08).

Eidhammer et al. (2009, hereafter EDK09) compared in the parcel modeling the ice schems (PDA08; Diehl and Wurzler (2004), hereafter DW04, and Khvorostyanov and Curry (2000, 2004a,b), hereafter, KC00-04) and showed that PDA08 parameterization produces much lower crystal concentrations than the parameterizations DW04 and KC00-04. Curry and Khvorostyanov (2012) analyzed the results of simulations in EDK09 with these three ice schemes, and compared simulated in EDK09 crystal concentrations, LWC, IWC, and cloud phase state with the GCMs parameterizations and climatological data on cloud phase state. It was shown that the PDA08 parameterization is much less effective in ice nucleation and cloud glaciation than the parameterizations DW04, KC00-04 and other theoretical schemes and disagrees with climatological data on cloud phase state. Nucleation of a sensible number of crystals and transformation into a really mixed state occurs in PDA08 scheme only when homogeneous freezing of the large drops begins at $T \le -35\,°C$ (see Section 9.17 and Figs. 9.35 and 9.36). This indicates that the PDA08 scheme may require modification and it might be necessary to increase the coefficients or surface areas in this parameterization so that it would be able to produce mixed-phase clouds at $T < -15\,°C$ or $-20\,°C$.

DeMott et al. (2010) proposed a new parameterization for immersion and condensation freezing as a simple power law function by temperature:

$$N_c = a_D(-T_c)^{b_D}(N_{a,05})^{c_M T_c + d_M}, \qquad (9.4.12)$$

where N_c is in L^{-1}, T_c is the temperature in degrees Celsius, and $Na_{,05}$ is the concentration (in cm^{-3}) of aerosol particles larger than 0.5 μm. The first set of coefficients was: $a_D = 1.1968 \times 10^{-5}$, $b_D = 3.6434$, $c_D = -0.0167$, and $d_D = 0.2877$; the subsequent refined set was slightly changed to $a_D = 5.94 \times 10^{-5}$, $b_D = 3.334$, $c_D = 0.0264$, $d_D = 0.0033$. In contrast to MDC92 or PDA08, this parameterization includes only temperature and does not depend on supersaturation. Connolly et al. (2009) suggested a new scheme of heterogeneous nucleation based on a new concept of the active sites on the surface of IN. Niemand et al. (2012) developed a particle-surface-area-based parameterization of immersion freezing on desert dust particles. Development of similar new empirical or semi-empirical schemes is continuing.

This brief overview shows that all empirical parameterizations still require refinements and more experimental data—both laboratory and field—are needed.

9.5. Nucleation of Crystals in the Deposition Mode on Water-Insoluble Particles

The process of ice nucleation in the deposition mode is similar to the drop nucleation on an insoluble surface considered in Section 9.2, but replacing all the water properties with ice properties and replacing the water saturation ratio S_w with the ice saturation ratio S_i. However, an additional effect arises for the ice deposition. Since the parameters of the ice crystalline lattice are in general different from the lattice of an insoluble substrate, this is called the *lattice misfit*. Thus, the ice lattice should adjust to the substrate lattice. This causes either deformation of the ice lattice, *the elastic misfit strain*, or can cause dislocations at the ice–substrate interface. The misfit strain causes an increase in the bulk free energy, and dislocations increase the interfacial energy (see Section 3.6).

The elastic misfit strain can be characterized in terms of the relative misfit between the ice and substrate

$$\delta_{mis} = (a_n - a_i)/a_i, \tag{9.5.1}$$

where a_n is the substrate lattice vector and a_i is the length of an ice unit cell. If the misfit is small—that is, the substrate lattice is similar to the ice lattice—this is called the "*epitaxy*," and such substrates are expected to be good ice nucleants. For example, AgI, PbI, and CuI have lattice parameters sufficiently close to those of ice, have good nucleating ability, and are therefore used as the crystallizing agents for cloud seeding in the weather modification projects. If the misfit strain is small—i.e., the condition of epitaxy is fulfilled—then the number of dislocations may be small, and the elastic strain ε is equal to the misfit δ_{mis}. If the misfit is large, it may be assumed that the ice lattice is not accommodated to the substrate and the elastic strain is zero. The elastic strain causes the appearance of the additional energy in an ice embryo:

$$\Delta F_{el} = V_l C_\varepsilon \varepsilon^2, \tag{9.5.2}$$

where V_l is the particle volume. An estimate for the strain coefficient is $C_\varepsilon \approx 1.7 \times 10^{11}$ erg cm^{-3} at 0 °C (Turnbull and Vonnegut, 1952), and ε is measured in %.

The ice nucleation rate in the deposition mode can be written similarly to (Eqn. 9.2.7) for drop nucleation

$$J_{s,het} = c_{gs} j_g^{(+)} A_{gs} Z_s, \tag{9.5.3}$$

where $j_g^{(+)}$ is the molecular flux onto this g-mer, A_{gs} is the surface area of the germ, Z_s is the Zeldovich factor, and c_{gs} is the concentration of g-mers that is assumed to be described by the Boltzmann distribution using critical energies ΔF_{cr} similar to (Eqn. 9.2.6).

The germ critical energy and critical radius can be calculated in various ways. We present a derivation similar to that in Section 9.2 with additional accounting for the elastic strain energy. Consider formation of an ice embryo on the surface of an insoluble particle with radius r_N. We can write again the free energy as a sum of the volume and surface terms, similar to (Eqn. 9.2.16a), but replacing $S_w \to S_i$, and adding the term with the elastic strain energy ΔF_{el} from (Eqn. 9.5.2) multiplied by $f(m_{iv}, x)$ for the cap geometry:

$$\Delta F_{gs} = f(m_{iv}, x)\left[-\frac{4\pi r^3}{3}\left(\frac{RT\ln(S_i)}{v_{i\mu}} - C_\varepsilon \varepsilon^2\right) + 4\pi r^2 \sigma_{iv}\right], \tag{9.5.4}$$

where $v_{i\mu} = M_w/\rho_i$ is the molar volume of ice, ρ_i is the ice density, and m_{iv} is the cosine of the contact angle at the ice-vapor interface:

$$m_{iv} = \cos\theta_{iv} = (\sigma_{Nv} - \sigma_{Ni})/\sigma_{iv}, \tag{9.5.5}$$

determined by the three surface tensions: substrate–ice (σ_{Ni}), ice–vapor (σ_{iv}), and substrate–vapor (σ_{Nv}) (the subscripts N, i, and v here mean the corresponding substances: solid substrate, ice, and vapor). The shape factor $f(m_{iv}, x)$ in (Eqn. 9.5.4) is defined by (Eqn. 9.2.21) with the replacement of $m_{wv} \to m_{iv}$, which is defined by (Eqn. 9.5.5).

The critical energy of an ice germ is determined from the condition of minimum $\partial \Delta F_{gs}(r_{cr})/\partial r = 0$, which yields from (Eqn. 9.5.4)

$$r_{cr}\left(-\frac{RT\ln S_i}{v_{i\mu}} + C_\varepsilon \varepsilon^2\right) + 2\sigma_{iv} = 0, \tag{9.5.6}$$

and we obtain the critical radius

$$\begin{aligned} r_{cr} &= \frac{2\sigma_{iv}}{RT\ln S_i/v_{i\mu} - C_\varepsilon \varepsilon^2} \\ &= \frac{2\sigma_{iv} M_w}{RT\rho_i \ln S_i - M_w C_\varepsilon \varepsilon^2} = \frac{2\sigma_{iv}}{R_v T \ln S_i - C_\varepsilon \varepsilon^2}, \end{aligned} \tag{9.5.7}$$

As in Section 9.2, r_{cr} can be also written via the affinity $A_{\propto,vi} = \Delta\mu_{vi} = \mu_v - \mu_i$:

$$r_{cr} = \frac{2\sigma_{iv} v_{i\mu}}{\mu_v - \mu_i} \tag{9.5.8a}$$

Comparison with (Eqn. 9.5.7) shows that

$$\mu_v - \mu_i = RT\ln S_i - C_\varepsilon \varepsilon^2 v_{i\mu} \tag{9.5.8b}$$

Thus, the affinity (Eqn. 8.1.7a) for vapor–water $A_{\infty,vw} = \Delta\mu_{vw}$ is generalized for vapor–ice with accounting for the energy of the elastic misfit strain. The critical energy of a convex spherical ice cap with radius r_{cr} formed on the surface of an insoluble substrate with radius r_N is obtained from (Eqns. 9.5.4) and (9.5.7) similar to (Eqn. 9.2.20):

$$\begin{aligned} \Delta F_{cr} &= \frac{4\pi}{3}\sigma_{iv} r_{cr}^2 = \frac{\sigma_{iv} A_{cr}}{3} = \frac{16\pi}{3}\frac{M_w^2 \sigma_{iv}^3 f(m_{iv}, x)}{[RT\rho_i \ln S_i - M_w C_\varepsilon \varepsilon^2]^2} \\ &= \frac{16\pi}{3}\frac{\sigma_{iv}^3 f(m_{iv}, x)}{[R_v T\rho_i \ln S_i - C_\varepsilon \varepsilon^2]^2}, \end{aligned} \tag{9.5.9}$$

where A_{cr} is the surface area of a spherical germ with radius r_{cr}. Here, m_{iv} is the contact parameter of the angle between the ice germ and substrate, and the shape factor $f(m_{iv}, x)$ for a convex spherical ice cap with radius r_{cr} on the surface of an insoluble substrate with radius r_N is defined similar to (Eqn. 9.2.21) for liquid germs replacing $m_{wv} \to m_{iv}$:

$$\begin{aligned} f(m_{iv}, x) &= (1/2)\{1 + [(1 - m_{iv}x)/y]^3 + x^3(2 - 3\psi + \psi^3) + 3m_{iv}x^2(\psi - 1)\}, \\ \psi &= (x - m_{iv})/y, \qquad y = (1 - 2m_{iv}x + x^2)^{1/2}, \end{aligned} \tag{9.5.10}$$

and $x = r_N/r_{cr}$. The shape factor of the concave ice germs that may form in cavities or capillaries of the substrate is also defined by (Eqn. 9.2.22) with the replacement $m_{wv} \to m_{iv}$.

Now, the nucleation rate of crystals from vapor by the deposition mode per unit area of an insoluble surface per unit time [$cm^{-2} s^{-1}$] can be written similar to (Eqn. 9.2.19) for the drop nucleation on an insoluble substrate:

$$J_{sdep} = \frac{c_{1s} e_v}{(2\pi m_{w1} kT)^{1/2}} \pi r_{cr}^2 Z_s \exp(-\Delta F_{cr}/kT), \tag{9.5.11}$$

where m_{w1} is the mass of a water molecule, c_{1s} is the concentration of single molecules adsorbed on the surface, and r_{cr} and ΔF_{cr} are the critical radius and energy defined in (Eqns. 9.5.7) and (9.5.9). The nucleation rate J_{dep} per particle [s^{-1}] can be calculated to a first approximation similar to (Eqn. 9.2.23) by multiplying by the particle surface area $4\pi r_N^2$—i.e., $J_{dep} = 4\pi r N^2 Jsde_p$, or

$$J_{dep} = C_{dep} \exp(-\Delta F_{cr}/kT), \tag{9.5.12}$$

$$C_{dep} = \frac{4\pi^2 r_N^2 c_{1s} e_v}{(2\pi m_{w1} kT)^{1/2}} \pi r_{cr}^2 Z_s = \frac{4\pi^2 r_N^2 N_{Av} c_{1s} e_v}{(2\pi M_w RT)^{1/2}} \pi r_{cr}^2 Z_s, \tag{9.5.12a}$$

where N_{Av} is the Avogadro number. Fletcher (1958, 1962) estimated the average kinetic coefficient (the expression before the exponent) as $C_{dep} \sim 10^{26} r_N^2$. This expression along with (Eqn. 9.5.9) for ΔF_{cr} and (Eqns. 9.2.21) or (9.2.22) for $f(m_{iv}, x)$ can be used for calculations of the ice nucleation rates in the deposition mode.

The nucleation rates by the deposition mode become noticeable when the deposition contact parameters m_{iv} are sufficiently close to 1, or the contact angles are small (e.g., Fletcher, 1962; Young, 1993).

9.6. Ice Nucleation by Deliquescence-Freezing and Immersion

Consider a three-phase system consisting of an ice germ of radius r_{cr} that forms on a curved insoluble substrate with radius r_N inside an aqueous solution drop with radius r_d. The rate of heterogeneous germ formation in a supercooled droplet of water or solution per unit area per unit time, $J_{S,fr}$ [$cm^{-2} s^{-1}$], can be calculated similarly to that for homogeneous nucleation in Chapter 8 or to drop nucleation in Section 9.2, using the equation of balance for $J_{S,fr}$:

$$J_{S,fr} = c_{gs} j_g^{(+)} A_{gs} Z_s, \tag{9.6.1}$$

where c_{gs} is the concentration of g-mers or ice germs on the surface of an ice nucleus, $j_g^{(+)}$ is the diffusive flux of the molecules across the liquid–ice boundary onto this g-mer (see Chapters 5, 8), A_{gs} is the IN surface area adsorbing the molecular flux from the liquid, and Z_s is the Zeldovich factor. We assume that the Boltzmann distribution is valid for c_{gs}:

$$c_{gs} = c_{1s} \exp(-\Delta F_{cr}/kT), \tag{9.6.2}$$

where c_{1s} is the concentration of water molecules adsorbed on 1 cm^{-2} of a surface, and ΔF_{cr} is the energy of the germ formation.

The molecular flux to the surface of IN is determined by the potential barrier due to the activation energy ΔF_{act} and is assumed to be the same as in the case of homogeneous nucleation (Eqn. 8.3.2):

$$j_g^{(+)} = \frac{N_{cont} kT}{h} \exp(-\Delta F_{act}/kT), \tag{9.6.3}$$

where k and h are Boltzmann's and Planck's constants, ΔF_{act} is the activation energy at the solution–ice interface (or water–ice interface in the case of freezing of pure water), and N_{cont} is the concentration of molecules in contact with a unit area of ice. Substituting (Eqns. 9.6.2) and (9.6.3) into (Eqn. 9.6.1) we obtain

$$J_{S,fr}(T, r_N) = B_{het} \exp\left[-\frac{\Delta F_{act}}{kT} - \frac{\Delta F_{cr}}{kT}\right], \quad (9.6.4)$$

where

$$B_{het} = \frac{kT}{h} Z_s N_{cont} A_{gs} c_{1s}. \quad (9.6.5)$$

As in the case of homogeneous nucleation in Chapter 8, the nucleation rate is determined by the two energy barriers formed by the activation energy ΔF_{act} and the critical energy ΔF_{cr}. An estimate shows that $Z_s \sim 10^{-1}$ (i.e., accounting for the reverse flux of the molecules from an ice germ to the liquid decreases the nucleation rate by an order of magnitude), and that the product $N_{cont} A_{gs}$, ~ the number of molecules in contact with the ice germ, is ~10. Thus, the product $Z_s N_{cont} A_{gs} \sim 1$ and B_{het} can be approximately simplified as

$$B_{het} \approx \frac{kT}{h} c_{1s}. \quad (9.6.6)$$

The nucleation rate $J_{f,het}$ per particle [s^{-1} particle^{-1}] with surface area $A_{rN} = 4\pi r_N^2$, where r_N is the radius of an insoluble fraction of an aerosol particle that serves as IN, can be approximately evaluated by multiplying $J_{S,fr}$ by A_{rN}:

$$J_{f,het}(T, r_N) = C_{het} \exp\left[-\frac{\Delta F_{act}}{kT} - \frac{\Delta F_{cr}}{kT}\right], \quad (9.6.7)$$

where C_{het} is the kinetic coefficient

$$C_{het} \approx \frac{kT}{h} c_{1s} 4\pi r_N^2. \quad (9.6.8)$$

According to Fletcher (1962) and Young (1993), $c_{1s} \sim 10^{15}\ \mathrm{cm}^{-2}$ and $C_{het} \sim 10^{28} r_N^2$. Fletcher (1962, p. 45) emphasized that variation of C_{het} even by a few orders of magnitude has little effect on the final result because of the primary effect on the nucleation rates of the exponential term, whose variations are much greater.

Application of (Eqn. 9.6.7) for calculation of the nucleation rate $J_{f,het}$ requires knowledge of the critical energy ΔF_{cr} that is related to the ice germ critical radius r_{cr}. The equations for ΔF_{cr} and r_{cr} are derived in the next section.

9.7. Critical Radius and Energy of Heterogeneous Freezing

9.7.1. Basic Dependencies of Heterogeneous Freezing

A review in Section 9.4 of the empirical parameterizations showed that these methods attempted to describe the two major dependencies of the heterogeneous freezing process on the temperature and humidity or saturation ratio. These parameterizations mostly considered one of these dependencies, but not both together, except for a few attempts at empirical parameterizations simultaneously as a function of temperature and supersaturation. This section is devoted to derivation from the theory of both the T- and S_w dependencies that can justify these empirical parameterizations and provide the quantitative values of the empirical coefficients.

Earlier formulations of the classical nucleation theory for freezing (J. J. Thomson, 1888) described the temperature dependence of the critical radius r_{cr}, energy ΔF_{cr}, but did not contain a humidity dependence. Fletcher (1962) gave a formulation for r_{cr} and ΔF_{cr} in terms of ratio of the saturated humidities over water and ice, and this is equivalent to the T-dependence, as will be shown later. Dufour and Defay (1963) and Defay, Prigogine, and Bellemans (1966) derived from the classical nucleation theory the expressions for r_{cr} and ΔF_{cr}, that take into account the solution effect equivalent to the humidity dependence, but this was expressed via the osmotic potential or water activity that may require rather complicated calculations for an ensemble of polydisperse drops in a cloud model. Fukuta and Schaller (1982) developed a theory of condensation-freezing that proceeds in two steps: a) condensation on the surface of an insoluble nucleus at water supersaturation as described in Section 9.2 here; and b) the subsequent freezing of the water film formed on this insoluble nucleus following Fletcher's (1962) formulation taking into account the T-dependence. A similar model was suggested by Young (1993) for the immersion-freezing mode. This approach can be suitable for description of heterogeneous ice nucleation on insoluble IN at water supersaturation, in particular on artificial IN used in cloud seeding.

However, this treatment did not allow for calculations of ice nucleation by the freezing of solution drops at water supersaturation, haze solution particles (deliquescent CCN) at water subsaturation and the freezing of interstitial deliquescent CCN in a cloud. This requires a generalization of the classical theory, which is described in the next sections. As was discussed in Section 9.3, most ice nuclei in the real atmosphere have appreciable fractions of soluble materials (e.g., Young, 1993), which has been emphasized in recent field projects as discussed in Section 9.4. Hence, the same aerosol particles may serve both as IN and CCN.

The experiments and data on IN properties described in Section 9.3 indicate the necessity of modifying the traditional view of IN as completely insoluble particles, as well as the parameterizations and theories of heterogeneous ice nucleation. DeMott et al. (1998) and a few others developed empirical parameterizations of heterogeneous nucleation based on the conception of IN as deliquescent mixed cloud condensation nuclei (CCN) with 50% insoluble and 50% soluble matter (model of Colorado State University or CSU model, see Section 9.4).

Khvorostyanov and Curry (2000, 2004a,b, 2005a, 2009b, hereafter, KC00-09, or KC scheme), based on the modification of the classical nucleation theory, developed a theory of heterogeneous ice nucleation by freezing of the CCN and cloud drops containing both soluble and insoluble fractions, with the possibility of ice nucleation both at water sub- and supersaturations. The theory of homogeneous freezing of fully soluble CCN described in Chapter 8 was generalized in KC00-09 for the case of mixed CCN, extending the concept of freezing to temperatures as high as a few degrees below 0 °C. This theory is described in this and following sections.

Equations for the critical radius and activation energy show that nucleation becomes possible due to the catalyzing effect of insoluble substrate within CCN that counteracts the freezing point depression by the solution effect, significantly lowering the energy barrier for freezing and permitting freezing at smaller supercooling than that for homogeneous freezing. These equations show that ice nucleation may occur with noticeable and significant rates on the surface of the insoluble substance embedded in a solution drop that has formed on a dry CCN.

Therefore, the basic premises of this theory are: a) the same deliquescent hygroscopic aerosol (haze particles) that serves as cloud condensation nuclei (CCN) may also serve as ice nuclei (IN) under freezing conditions; b) in contrast to the homogeneous nucleation case, these haze particles contain an insoluble substrate (which is typical of CCN); c) in contrast to drop activation on CCN, where the aerosol soluble fraction determines the activity of a nucleus, heterogeneous ice nucleation is determined also by the insoluble fraction of CCN, which is a nucleation catalyzer; d) heterogeneous ice nucleation may occur on these CCN-IN; it will be shown that this process may take place both at water supersaturations and subsaturations; e) a similar approach is applicable for freezing of cloud drops in the immersion mode, and, as will be shown, leads to the similar equations for the critical radius, energy, and nucleation rates.

In the next few sections, we derive and analyze the analytical dependence of the critical radius, energy, and nucleation rate on the temperature, saturation ratio, misfit strain, pressure, and finite radius of freezing/melting particles. We consider three particular cases under variable external pressure and solution concentration: 1) volume heterogeneous freezing of aqueous solution drops (Section 9.7); 2) quasi-heterogeneous surface freezing of solutions; and 3) quasi-heterogeneous surface melting of ice crystals; the last two cases are considered in Section 9.8. The analysis of these relations is performed in Section 9.9.

9.7.2. Volume Heterogeneous Freezing

Heterogeneous volume freezing may be a characteristic of the deliquescence-heterogeneous-freezing (DHF) or immersion freezing modes of the several possible modes of ice nucleation, whereby the freezing does not initiate on the drop surface. As discussed in Chapter 8, the difference between the freezing of unactivated small solution drops (haze particles) and larger activated cloud drops is determined by the difference in the equilibrium states of these particles. As shown in Chapter 3, the small solution drops or unactivated interstitial CCN are close to the Köhler equilibrium, while larger cloud drops are close to the Kelvin equilibrium. Ice nucleation inside the volumes of both droplet types and both types of equilibrium is considered in this section.

To describe the first type of freezing, consider a three-phase system consisting of a solution or water drop with radius r_{dr} in humid air and an ice germ with radius r_{cr} that forms inside the drop on the surface of a curved insoluble foreign substrate with radius r_N. The ice germ is in equilibrium with the drop, which itself is in equilibrium with the environmental moist air; these conditions define the critical ice germ radius r_{cr}. Such a configuration was previously considered, e.g., by Fletcher (1962), Dufour and Defay (1963), Defay, Prigogine, and Bellemans (1966), Young (1993), Pruppacher and Klett (1997), and the expressions for the critical radii were derived under constant external pressure, often neglecting the effects of misfit strain, or without explicit direct analytical description of the solution effects. Here we consider variable pressure along with variable temperature, droplet size, and solution concentration that is accounted for via the water saturation ratio.

The derivation of r_{cr} and ΔF_{cr} shown next mostly follows the works by Khvorostyanov and Curry (2000, 2004a,b), where the dependencies of r_{cr} and ΔF_{cr} on temperature T, saturation ratio S_w, droplet size r_d, and pressure p were derived. The derivation is based on the entropy equation (Eqn. 3.6.4) of Section 3.6 for this system at equilibrium. In the context of (Eqn. 3.6.4), phase 1 is the liquid solution or water and phase 2 is ice, the subscript "k0,1" means pure water, "1" means solution, "2" means ice, and "e" means environmental moist air. Then, $\rho_{k0,1}$ and $\rho_{k0,1}$ are the densities of water ρ_w and ice ρ_i, the molar melting latent heat $\hat{L}_{12} = M_w L_m$, where L_m is the specific melting latent heat, and the activities of water and ice are $a_{k,1} = a_w$, $a_{k,2} = a_i$. Following observations (e.g., Dash et al., 1995; Pruppacher and Klett, 1997; Curry and Webster, 1999), as in the case with homogeneous freezing in Section 8.6, we assume that the retention coefficient is zero, and then $a_i = 1$ and $\ln(a_i) = 0$.

If an ice germ is approximated by a spherical cap at the surface of the insoluble substrate, the internal pressures inside a crystal, p_i, and inside a liquid solution drop, p_s, can be expressed in terms of external pressure p using conditions of mechanical equilibrium:

$$dp_i = dp_s + d\left(\frac{2\sigma_{is}}{r_{cr}}\right), \qquad dp_s = dp + d\left(\frac{2\sigma_{sa}}{r_{dr}}\right), \tag{9.7.1}$$

where subscripts i and s refer to ice and solution respectively, σ_{is} and σ_{sa} are the surface tensions at the ice–solution and solution–air interfaces. Equation (9.7.1) describes the equilibrium between an ice germ and liquid solution drop, and the equilibrium between the liquid drop and environmental air. Then, using (Eqn. 9.7.1) and multiplying the entropy equation (Eqn. 3.6.4) by T, we can rewrite it as

$$\begin{aligned} &-\frac{L_m}{T}dT - \frac{A_\rho}{\rho_i}dp - \frac{A_\rho}{\rho_i}d\left(\frac{2\sigma_{sa}}{r_{dr}}\right) - \frac{1}{\rho_i}d\left(\frac{2\sigma_{is}}{r_{cr}}\right) \\ &-d\left(\frac{C_\varepsilon \varepsilon^2}{\rho_i}\right) + \frac{RT}{M_w}d\ln a_w = 0 \end{aligned} \tag{9.7.2}$$

where we introduced the density function $A_\rho = 1 - \rho_i/\rho_w$, and a_w is the water activity defined in Section 3.5 by (Eqn. 3.5.16) for a solution with an insoluble substance.

The number of components in this thermodynamic system is $c = 4$ (water, solute, insoluble substance, air), and the number of thermodynamic degrees of freedom (or independent variables) would be $N_w = c + 1 = 5$ according to the phase rule (Eqn. 3.3.12) for the curved substances. If the mass of solute m_s is constant in the case considered here, we have an additional constraint and $N_w = 4$; if m_s is variable due to transfer processes (e.g., thermo- or diffusiophoresis transfer considered later), then $N_w = 5$. The number of variables in (Eqn. 9.7.2) is 6 (T, r_{cr}, r_{dr}, p, ε, a_w), and only 4 or 5 of them are independent according to this phase rule. We can consider the critical radius r_{cr} of a crystal as a function of 3 or 4 variables, and integrate (Eqn. 9.7.2) for several various possible combinations of variables, this would yield several analytical dependencies of the critical radius r_{cr} on various variables. To shorten and generalize calculations, as in Chapter 8, we will apply a method used in the theory of differential equations and integrate (Eqn. 9.7.2), temporarily considering all variables as independent. A selection of the possible combinations of independent variables can be done in the final result.

We integrate (Eqn. 9.7.2) from the triple point temperature T_0 to T, and from p_0 to p with the boundary conditions, $a_w = 1$, $r_{cr} = r_{dr} = \infty$, $p = p_0$, and $\varepsilon = 0$ at $T = T_0$, to a_w, r_{cr}, r_{dr}, p, ε, and T, and obtain after rearranging the terms:

$$L_m^{ef} \ln\frac{T_0}{T} = \frac{2\sigma_{is}}{\rho_i r_{cr}} + \frac{2\sigma_{sa}}{\rho_i r_{dr}} + \frac{A_\rho}{\rho_i}\Delta p + \frac{C_\varepsilon \varepsilon^2}{\rho_i} - \frac{RT}{M_w}\left(\frac{A_K}{r_{dr}} + \ln a_w\right), \tag{9.7.3}$$

where $\Delta p = p - p_0$, and $A_K = 2M_w\sigma_{sa}/\rho_w RT$ is the Köhler parameter, and we introduced the effective average melting heat, defined by (Eqn. 8.6.8). The water activity a_w is often assumed to be equal to the water saturation ratio S_w (Tabazadeh et al., 1997; Koop et al., 2000) or is calculated from the Köhler equation (e.g., DeMott et al., 1994).

As was discussed in Section 8.6 in the case with homogeneous freezing, an unactivated solution drop at subsaturation $S_w < 1$ grows or evaporates rapidly until the Köhler's equilibrium size and S_w are reached. The estimates in Section 8.6 showed that the characteristic time of adjustment to equilibrium varies from 10^{-4} s for $r_{dr} = 10^{-2}$ µm to a few seconds for r_{dr} of a few microns, and this equilibrium exists in most cases with moderate to high vertical velocities, even with rapid turbulent fluctuations. We also assume that an unactivated solution particle is in Köhler equilibrium before freezing. Therefore, the expression in the parentheses in the last term of (Eqn. 9.7.3), $A_K/r_{dr} + \ln a_w$, can be replaced with $\ln S_w$ according to the Köhler equation, and this $\ln S_w$ can be substituted into the last term in (Eqn. 9.7.3). For cloud drops of a few microns growing in a rising parcel at slight supersaturation, the effects of both curvature and solution are relatively small, and also approximately $\ln a_w \approx \ln S_w$.

Note that each term on the right-hand side of (Eqn. 9.7.3) is positive ($-\ln S_w > 0$ at $S_w < 1$), thus, $\ln(T_0/T) \geq 0$ and $T \leq T_0$—i.e., (Eqn. 9.7.3) describes lowering of the bulk triple point by all of the above factors: r_{cr}, r_{dr}, p, ε, $-\ln S_w$. Solving (Eqn. 9.7.3) for r_{cr}, we obtain

$$r_{cr} = \frac{2\sigma_{is}}{\rho_i L_m^{ef}\left[\ln\left(\frac{T_0}{T}S_w^{G_n}\right) - A_p\Delta p - \frac{A_f}{r_{dr}} - \frac{C_\varepsilon \varepsilon^2}{\rho_i L_m^{ef}}\right]}, \tag{9.7.4}$$

with the same $G_n = RT/(M_w L_m^{ef})$, $A_f = 2\sigma_{sa}/(\rho_i L_m^{ef})$, and $A_p = \Delta p/(\rho_w \rho_i L_m^{ef})$ as in (Eqn. 8.6.15), or in the more extended form

$$r_{cr} = \frac{2\sigma_{is}}{\rho_i L_m^{ef}\left[\ln\left(\frac{T_0}{T}\right) + \frac{RT}{M_w L_m^{ef}}\ln S_w - \frac{\Delta p \Delta p}{\rho_w \rho_i L_m^{ef}} - \frac{2\sigma_{sa}}{\rho_i L_m^{ef} r_{dr}} - \frac{C_\varepsilon \varepsilon^2}{\rho_i L_m^{ef}}\right]}. \tag{9.7.4a}$$

Note that the last three terms in the denominator caused by the excess pressure Δp, finite droplet radius r_{dr}, and misfit ε are negative—in other words, all these factors lead to an increase in r_{cr} and in ΔF_{cr}, and to a decrease in the ice nucleation rate. (Eqn. 9.7.4) can be written in slightly different forms that are convenient in some applications,

$$r_{cr} = \frac{2\sigma_{is}}{\rho_i L_m^{ef}\left[\ln\left(\frac{T_0}{T}S_w^{G_n}\exp(-H_{v,fr})\right)\right]} \tag{9.7.5a}$$

$$r_{cr} = \frac{2\sigma_{is}}{\rho_i L_m^{ef}\left[\ln\left(\frac{T_0}{T}\right) + G_n \ln S_w - H_{v,fr}\right]}, \tag{9.7.5b}$$

These expressions are similar to (Eqns. 8.6.14b) and (8.6.17b) for the critical radius of homogeneous freezing, with $\Delta p = \rho_w - \rho_i$, but the dimensionless function $H_{v,fr}$ is modified compared to (Eqn. 8.6.16) by addition of the term with misfit ε.

$$H_{v,fr} = A_p \Delta p + \frac{A_f}{r_{dr}} + C_\varepsilon \varepsilon^2 = \frac{\Delta\rho\Delta p}{\rho_w \rho_i L_m^{ef}} + \frac{2\sigma_{sa}}{\rho_i L_m^{ef} r_{dr}} + \frac{C_\varepsilon \varepsilon^2}{\rho_i L_m^{ef}}. \tag{9.7.6}$$

Equations (9.7.4), (9.7.4a), and (9.7.5a,b) account for the effects of temperature, water saturation ratio, pressure, elastic misfit strain, and the finite size of a freezing drop on r_{cr} (taking into account the number of independent variables). These expressions generalize the corresponding equations for the homogeneous freezing of unactivated solution drops derived in Chapter 8 and are valid for both heterogeneous ($\varepsilon \neq 0$) and homogeneous ($\varepsilon = 0$) freezing.

As in Sections 9.2 and 9.5, r_{cr} can be also written via the affinity $A_{\infty,li} = \Delta\mu_{li} = \mu_l - \mu_i$:

$$r_{cr} = \frac{2\sigma_{is} v_{i\mu}}{\mu_l - \mu_i} = \frac{2\sigma_{is} M_w}{\rho_i (\mu_l - \mu_i)}, \tag{9.7.6a}$$

where $v_{i\mu} = M_w/\rho_i$ is the molar volume, and $\mu_l = \mu_{mstab}$ and $\mu_i = \mu_{stab}$ are the molar chemical potentials of the metastable liquid phase and stable ice phase. Comparison with (Eqn. 9.7.5a) shows that

$$\mu_l - \mu_i = M_w L_m^{ef} \left[\ln\left(\frac{T_0}{T} S_w^{G_n} \exp(-H_{v,fr}) \right) \right]$$

$$= M_w L_m^{ef} \left[\ln\left(\frac{T_0}{T} \right) + \frac{RT}{M_w L_m^{ef}} \ln S_w - \frac{\Delta\rho\Delta p}{\rho_w \rho_i L_m^{ef}} - \frac{2\sigma_{sa}}{\rho_i L_m^{ef} r_{dr}} - \frac{C_\varepsilon \varepsilon^2}{\rho_i L_m^{ef}} \right]. \tag{9.7.6b}$$

Thus, the affinity for solution–ice $A_{\infty,li} = \Delta\mu_{li}$ is generalized for vapor–ice with an accounting for the energy of the elastic misfit strain, the solution effect, final drop radius, and external pressure; the last four terms in (Eqn. 9.7.6c) cause a decrease in the difference of chemical potentials, counteracting the freezing process. The quantity $\Delta\mu$ determines the transition from the metastable liquid to the stable ice state, $\mu_l > \mu_i$. The equality $\mu_l = \mu_i$ determines the equilibrium melting or freezing curves considered in Sections 9.11 and 9.12.

We have considered in this section the nucleation of ice crystals inside unactivated submicron solution droplets near Köhler's equilibrium. As mentioned in Section 8.6.4, the process of ice nucleation inside cloud drops differs because cloud drops have much larger radii and the solution inside the drops is very dilute. Therefore, the Köhler equilibrium is not applicable to such drops. The critical radii and energies of ice crystals nucleating inside cloud drops can be derived similarly to the derivation in Section 8.6.4 for homogeneous freezing.

We begin with the general equilibrium equation (Eqn. 3.6.4), assume again Kelvin's equilibrium of cloud drops with the environment's humid air, and replace in (Eqn. 3.6.4) the term $d\ln(a_w)$ with the term describing the contribution of the water vapor activity, $d\ln(e_v/p)$. Thus, (Eqn. 8.6.39) is modified for heterogeneous freezing, taking into account the misfit strain ε as

$$-L_m(T)\frac{dT}{T} + \frac{\Delta\rho}{\rho_w \rho_i} dp - \frac{\Delta\rho}{\rho_w \rho_i} d\left(\frac{2\sigma_{sa}}{r_{dr}} \right)$$

$$-\frac{1}{\rho_i} d\left(\frac{2\sigma_{is}}{r_{cr}} \right) - d\left(\frac{C_\varepsilon \varepsilon^2}{\rho_i} \right) + \frac{RT}{M_w} d\ln\left(\frac{e_v}{p} \right) = 0. \tag{9.7.6c}$$

Integration of (Eqn. 9.7.6c) from the initial conditions $T = T_0$, $r_{d0} = \infty$ (plane surface), $r_{cr} = \infty$, $e_v = e_{ws}$ (saturated vapor pressure over a plane water surface), $p = p_0$ to the state (T, r_d, r_{cr}, e_v, p) yields

$$L_m^{ef}(T)\ln\frac{T_0}{T} - \frac{\Delta\rho\Delta p}{\rho_w\rho_i} - \frac{2\Delta\rho\sigma_{sa}}{\rho_w\rho_i}\frac{1}{r_d} - \frac{2\sigma_{is}}{\rho_i r_{cr}} - \frac{C_\varepsilon\varepsilon^2}{\rho_i} + \frac{R\bar{T}}{M_w}\ln\left(S_w\frac{p_0}{p}\right) = 0, \quad (9.7.6d)$$

where $\Delta p = p - p_0$, $\bar{T}$ is the mean temperature in the range (T_0, T), and $S_w = e_v/e_{ws}$ is the water saturation ratio. This modification of the Kelvin equilibrium equation accounts for the ice crystal and insoluble substrate inside the drop. The assumptions $T = \text{const}$, $p = \text{const}$, $\varepsilon = 0$, and the exclusion of all the quantities related to ice, again would yield the Kelvin equation. Solving the modified (Eqn. 9.7.6d) relative to r_{cr}, we obtain

$$r_{cr} = \frac{2\sigma_{is}}{\rho_i L_m^{ef}\left[\ln\left(\frac{T_0}{T}S_w^{G_n}\frac{p_0}{p}\right) - H_{v,fr}\right]}, \quad (9.7.6e)$$

where $H_{v,fr}$ is defined in (Eqn. 9.7.6) and is the same as for the case of haze solution drops.

This equation is similar to (Eqn. 9.7.4a) for the r_{cr} of an unactivated solution drop with an additional factor (p_0/p) under the ln sign. An estimate of the terms in (Eqn. 9.7.6e) at $r_d \sim 10-30\,\mu\text{m}$, $T \sim 258$ K using L_m^{ef} from Fig. 8.14a and σ_{sa} from Section 4.4 shows that the second term in the denominator of (Eqn. 9.7.6e) under typical conditions is much smaller than the first term with ln and can be neglected. Also neglecting for simplicity the term with the misfit ε, (Eqn. 9.7.6e) is simplified to

$$r_{cr} = \frac{2\sigma_{is}}{\rho_i L_m^{ef}\left[\ln\left(\frac{T_0}{T}S_w^{G_n}\frac{p_0}{p}\right)\right]}. \quad (9.7.6f)$$

If droplet freezing occurs with a substantial pressure decrease, the factor $p_0/p > 1$ in (Eqn. 9.7.6f) leads to a smaller r_{cr} and ΔF_{cr} and accelerates droplet freezing. In processes without strong changes of pressure, $p \approx p_0$, this equation is further simplified as

$$r_{cr} = \frac{2\sigma_{is}}{\rho_i L_m^{ef}\left[\ln\left(\frac{T_0}{T}S_w^{G_n}\right)\right]}. \quad (9.7.6g)$$

As in Chapter 8, this equation coincides with the simplified versions of r_{cr} in (Eqns. 9.7.5a) and (9.7.5b), neglecting the effects of pressure, droplet curvature, and ε ($H_{v,fr} = 0$).

Thus, the effects of temperature and water saturation ratio are similar for both haze particles and activated cloud drops. As in the case with homogeneous nucleation in Section 8.6.4, (Eqn. 9.7.6g) allows introduction of "effective supercooling"

$$\Delta T_{ef} = \Delta T + \delta_s\Delta T, \qquad \delta_s\Delta T = T_0 G_n s_w, \quad (9.7.6h)$$

where ΔT is supercooling. Thus, positive supersaturation increases effective supercooling by $\delta_s \Delta T = T_0 G_n s_w$. Since $G_n \sim 0.4$ at $T \approx -10$ to $-20\,°\mathrm{C}$, the value $T_0 G_n \sim 10^2$; thus, $\delta_s \Delta T$ is equal to water supersaturation (positive or negative) measured in percent. If $s_w > 0$, this leads to an increase in the effective supercooling by $\delta_s \Delta T \approx s_w$ (%), and accelerates freezing in the immersion mode. At subsaturation, $s_w < 0$, as in the cloud evaporation layer, $\delta_s \Delta T < 0$, this hampers the freezing of cloud drops, or may even prevent it if $|\delta_s \Delta T| > \Delta T$.

9.7.3. Particular Cases of Critical Radius

As mentioned earlier, there are five variables in (Eqns. 9.7.4) and (9.7.5a), (9.7.5b), and (9.7.5c), $(T, S_w, p, \varepsilon, r_{dr})$, and not all of these can be independent. If the soluble mass m_s can vary, then $N_w = 5$, all five variables in (Eqns. 9.7.4) and (9.7.5) can be independent. If $m_s = \text{const}$, then the variance $N_w = 4$ and we can compose $C_5^4 = (5{\cdot}4{\cdot}3{\cdot}2)/(1{\cdot}2{\cdot}3{\cdot}4) = 5$ possible combinations of four variables of the total five by the exclusion of one variable. Thus, we can write five expressions for r_{cr}. These are $r_{cr}(T, S_w, p, \varepsilon,)$, $r_{cr}(T, S_w, p, r_{dr})$, $r_{cr}(T, S_w, \varepsilon, r_{dr})$, $r_{cr}(T, p, \varepsilon, r_{dr})$, and $r_{cr}(S_w, p, \varepsilon, r_{dr})$. If one of five variables is absent in (Eqns. 9.7.4) and (9.7.5), the corresponding term should be set equal to zero. Then, we can study the simultaneous dependences on the other four variables, and establish their equivalence, as was done for pressure Δp and solution (S_w) effects in Chapter 8. If additional conditions are imposed on the system, e.g., $T = \text{const}$ for isothermal process, or $p = \text{const}$ for isobaric process, then the corresponding terms should also be excluded from (Eqns. 9.7.4) and (9.7.5).

The expressions (Eqns. 9.7.4) and (9.7.5) can be compared with previously derived expressions by considering particular cases. For the simplest case with constant pressure ($\Delta p = 0$), no misfit strain ($\varepsilon = 0$), and a bulk solution ($1/r_{dr} = 0$), we have $H_{v,fr} = 0$. If additionally $S_w = 1$ (pure water and no external sources of supersaturation), (Eqns. 9.7.4) and (9.7.5) are reduced to J. J. Thomson's (1888) classical equation

$$r_{cr} = \frac{2\sigma_{iw}}{\rho_i L_m^{ef} \ln(T_0/T)} \tag{9.7.7}$$

where σ_{iw} is the surface tension at the ice–water interface. This equation is often used to describe the triple point lowering in polycrystalline crystals, and melting point shifts due to varying crystal size, the *Gibbs-Thomson effect* (Dufour and Defay, 1963; Defay et al., 1966; Dash et al., 1995; Johari, 1998).

Equation (9.7.7) is often written as

$$r_{cr} = 2\sigma_{iw}/[N_i kT \ln(e_{ws}/e_{is})] \tag{9.7.7a}$$

where N_i is the concentration of molecules per unit volume in ice, e_{ws} and e_{is} are the saturated vapor pressures over water and ice (e.g., Young, 1993, and Fletcher, 1962). Equations (9.7.7) and (9.7.7a) are equivalent expressions for ΔF_{cr}. Their equivalence can be shown using the Clausius–Clapeyron equation for e_{ws} and e_{is} (see Section 3.7):

$$d(\ln e_{ws})dT = L_e/R_v T^2, \qquad d(\ln e_{is})dT = L_s/R_v T^2, \tag{9.7.7b}$$

where L_e and L_s are the specific heats of evaporation and deposition, and R_v is the gas constant for water vapor. Subtracting these equations, then integrating from T to T_0, and using the relations $e_{ws} = e_{is}$ at $T = T_0$, and $L_s - L_e = L_m$, we obtain

$$\ln(e_{ws}/e_{is}) \approx L_m \Delta T/(R_v T_0^2), \tag{9.7.7c}$$

where $\Delta T = T_0 - T$. Substituting (Eqn. 9.7.7c) into (Eqn. 9.7.7a) and using the relation $N_i = N_{Av}/v_{i\mu} = N_{Av}\rho_i/M_w$ with N_{Av} and $v_{i\mu}$ being the Avogadro number and molar ice volume, the denominator in (Eqn. 9.7.7a) can be rewritten as

$$N_i kT \ln(e_{ws}/e_{is}) \approx \rho_i L_m \Delta T/T_0 \approx \rho_i L_m \ln(T_0/T). \tag{9.7.7d}$$

The second equation is valid for a not very large supercooling, when $\Delta T << T_0$ and $\ln(T_0/T) = \ln[T_0/(T_0 - \Delta T)] \approx \Delta T/T_0$. Substitution of (Eqn. 9.7.7d) into the denominator of (Eqn. 9.7.7a) yields

$$r_{cr} = 2\sigma_{iw}/[\rho_i L_m \ln(T_0/T)], \tag{9.7.7e}$$

which coincides with (Eqn. 9.7.7). Thus, (Eqns. 9.7.7) and (9.7.7a) for r_{cr} are equivalent for $\Delta T << T_0$. Note that although (Eqn. 9.7.7a) contains two vapor pressures, they both are saturated and depend only on the temperature but not on the saturation ratio; therefore, (Eqn. 9.7.7a) does not contain any information on ambient humidity or solution effects.

If misfit strain is included, (Eqn. 9.7.4) becomes

$$r_{cr} = \frac{2\sigma_{iw}}{\rho_i L_m^{ef} \ln(T_0/T) - C_\varepsilon \varepsilon^2}. \tag{9.7.8}$$

The equations equivalent to this generalization of Thomson's equation for r_{cr} for heterogeneous freezing of pure water were considered by Fletcher (1962) and Young (1993). In the presence of a solvent or external source of supersaturation, $S_w \neq 1$, but still for $\varepsilon = 0$ and $\Delta p = 0$, we have from (Eqn. 9.7.5) another expression derived by Khvorostyanov and Sassen (1998), and Khvorostyanov and Curry (2000):

$$r_{cr} = \frac{2\sigma_{is}}{\rho_i L_m^{ef} \ln[(T_0/T) S_w^{G_n}]}. \tag{9.7.9}$$

Similar equations were considered in Chapter 8 for the case of homogeneous nucleation. In the earlier works (e.g., Dufour and Defay, 1963; Pruppacher and Klett, 1997), r_{cr} was expressed in a different form using water activity a_w or molar fraction x_s, or osmotic potential Φ_s and molality. The use of S_w instead of a_w, Φ_s, or x_s is more convenient in applications such as cloud models where S_w is a readily determined environmental variable. Note that although a_w and S_w may be close in many cases at $S_w < 1$, they may differ at $S_w \sim 1$ due to the curvature effects, thus leading to different values of r_{cr}. The Kelvin parameter $A_k \sim 10^{-7}$ cm, and the curvature correction for a small particle with $r_{dr} \sim 10^{-6}$ cm, is $A_k/r_{dr} \sim 0.1$. In Köhler's equilibrium, reached very rapidly for a small droplet, $S_w \approx a_w + A_k/r_{dr}$. Thus, $S_w \approx a_w + 0.1$ in this case—i.e., the saturation ratio is greater by 0.1, which is equivalent to 10% in relative humidity. For example, S_w can be 1.05, and $a_w = 0.95$; therefore, using a_w instead of S_w can give a noticeably different r_{cr} and may lead to the error in calculations. Thus, the "*saturation ratio based criterion*" seems to be preferable compared to the "*water activity based criterion*," as was discussed in Chapter 8 for homogeneous nucleation, especially for small drops.

In the presence of a solvent or an external source of supersaturation, $S_w \neq 1$, but still for $\Delta p = 0$, we have from (Eqn. 9.7.5) the following equation:

$$r_{cr} = \frac{2\sigma_{is}}{\rho_i L_m^{ef} \ln[(T_0/T)S_w^{G_n}] - C_\varepsilon \varepsilon^2}. \tag{9.7.10}$$

Taking into account the solution effects, $S_w < 1$, and the external pressure, $\Delta p > 0$, but with the bulk solution, $r_{dr} = \infty$, (Eqn. 9.7.5) for the critical radius has the form

$$r_{cr} = \frac{2\sigma_{is}}{\rho_i L_m^{ef} \left[\ln\left(\frac{T_0}{T} S_w^{G_n}\right) - A_p \Delta p + \frac{C_\varepsilon \varepsilon^2}{\rho_i L_m^{ef}} \right]}. \tag{9.7.11}$$

All of these particular cases enable the establishment of the equivalence between any two variables. For example, (Eqn. 9.7.4) allows the quantitative relation between the solution and pressure to be determined. The pressure dependence in (Eqns. 9.7.4) and (9.7.5) has not been previously accounted for in this form. It will be shown in the following text that this allows for a simple evaluation of the pressure effects on the freezing and melting points. In the case $\varepsilon = 0$, $\Delta p = 0$, $r_{dr} = \infty$, and $T \to T_0$, (Eqns. 9.7.4) and (9.7.5) convert into

$$r_{cr} = \frac{2M_w \sigma_{is}}{RT \rho_i \ln S_w}, \tag{9.7.12}$$

and resemble Kelvin's expression for nucleation of a crystal from the vapor, except that it contains σ_{is} instead of σ_{iv} for nucleation from the vapor. Since $\sigma_{is} \sim (1/4)\sigma_{iv}$, (Eqn. 9.7.12) shows that r_{cr} for freezing is four times smaller; therefore ice nucleation by water freezing is energetically much more favorable than direct nucleation from the vapor.

So, according to the correspondence principle, (Eqns. 9.7.4), (9.7.4a), (9.7.5a), (9.7.5b), and (9.7.6e) generalize and unify all of these particular cases for the critical germ radius and are used in the following text to generalize the critical energy of freezing and to establish equivalence between pressure, solution, and temperature effects on freezing and melting.

9.7.4. Critical Energy of Volume Freezing

As discussed in Chapter 8, the free energy of germ formation $\Delta F_{cr}(T, S_w)$ in classical nucleation theory is written as a sum of the positive surface term, $\Delta F_{surf} \sim r^2$, and a negative volume term $\Delta F_V \sim r^3$, caused by the difference of the free energies of aqueous solution F_{as} and ice F_i—i.e., $\Delta F_V = (F_{as} - F_i)(4/3)\pi r^3 \rho_i$:

$$\Delta F_{cr}(r) = -\Delta F_V + \Delta F_{surf}. \tag{9.7.13}$$

Similar to Section 9.2, the critical energy ΔF_{cr} of volume freezing is obtained from the condition $\partial \Delta F(r)/\partial r = 0$. Then, $(F_{as} - F_i)$ is expressed via r_{cr}, and is substituted into (Eqn. 9.7.13) to yield for a crystal with the geometry of a spherical cap:

$$\Delta F_{cr} = \frac{4}{3}\pi \sigma_{is} r_{cr}^2 f(m_{is}, x), \tag{9.7.14}$$

where $f(m_{is}, x)$ is a geometrical factor defined in Section 9.2. Substituting (Eqn. 9.7.5a) for r_{cr} into the last expression, we obtain the critical free energy for volume freezing as a function of temperature, solution concentration (supersaturation), pressure, and misfit strain:

$$\Delta F_{cr} = \frac{16\pi}{3} \frac{\sigma_{is}^3 f(m_{is}, x)}{\left\{\rho_i L_m^{ef} \ln\left[\frac{T_0}{T} S_w^{G_n} \exp(-H_{v,fr})\right]\right\}^2}, \tag{9.7.15}$$

where $H_{v,fr}$ is defined in (Eqn. 9.7.6). The geometric factor $f(m_{is}, x)$ in (Eqns. 9.7.14) and (9.7.15) arises from the geometry of the spherical ice cap freezing on an insoluble substrate core with radius r_N. As in Section 9.2, this factor can be expressed similar to (Eqn. 9.2.15) for the planar surface of the substrate, or following Fletcher (1958), similar to (Eqns. 9.2.21) or (9.2.22) for the curved surface of the substrate is expressed via the ratio $x = r_N/r_{cr}$ and the contact parameter $m_{is} = \cos(\theta_{iN})$.

For volume freezing, the parameters and geometric factor were defined by Fletcher (1958) differently than in Section 9.2 for the drops nucleation from vapor; now $m_{is} = cos\,\theta_{iN} = (\sigma_{Ns} - \sigma_{Ni})/\sigma_{is}$, where σ_{Ns} and σ_{Ni} are the surface tensions at the solution–substrate and ice–substrate interfaces. For the convex surface of the insoluble core inside the solution drop, according to Fletcher (1958),

$$f(m_{is}, x) = (1/2)\{1 + [(1 - m_{is}x)/y]^3 + x^3(2 - 3\psi + \psi^3) + 3m_{is}x^2(\psi - 1)\},$$
$$\psi = (x - m_{is})/y, \qquad y = (1 - 2m_{is}x + x^2)^{1/2}. \tag{9.7.16}$$

It is shown in the next sections that the typical values of $r_{cr} \sim 10^{-3}$ to $10^{-2}\,\mu$m, while the major contribution to nucleation rates comes from the particles with $r_N \sim 0.05$–$1\,\mu$m and larger. Thus, the case with $x = r_N/r_{cr} >> 1$ is practically the most important. In this large particle limit, $x >> 1$, the geometrical factor (Eqn. 9.7.16) reduces to Volmer's expression for planar substrate,

$$f(m_{is}, x) \approx f(m_{is}) = \frac{(2 + m_{is})(1 - m_{is})^2}{4} = \frac{m_{is}^3 - 3m_{is} + 2}{4}. \tag{9.7.16a}$$

This expression does not depend on r_N. Thus, the r_N-dependence in the geometric factor can be neglected to the first approximation in many cases, which simplifies derivation for the polydisperse ensemble of particles. For example, typical values of $f(m_{is} = 0.5) = 0.156$ and $f(m_{is} = 0.36) = 0.242$ characterize critical energy reduction due to cap geometry and favors heterogeneous nucleation in the large particle limit as compared to the case of homogeneous nucleation.

For the concave insoluble core inside the solution drop, the geometric factor is similar to (Eqn. 9.2.22) for the drop nucleation from vapor on an insoluble substance, but now with the different contact angle and parameter m_{is}:

$$f(m_{is}, x) = (1/2)\{1 - [(1 + m_{is}x)/y]^3 - x^3(2 - 3\psi + \psi^3) + 3m_{is}x^2(\psi - 1)\},$$
$$\psi = (x + m_{is})/y, \qquad y = (1 + 2m_{is}x + x^2)^{1/2}. \tag{9.7.16b}$$

The geometric factor $f(m_{is}, x) < 1$, and thus the critical energy of heterogeneous nucleation is smaller than the energy of homogeneous nucleation, and so heterogeneous freezing begins at much higher temperatures (generally below −10 °C) than homogeneous freezing (below −38 °C). For small contact angles (good wettability of the surface), m_{is} is close to 1, and the geometric factor $f(m_{is}, x)$ is small, which favors heterogeneous nucleation.

For not very high pressure ($\Delta p \approx 0$), and the freezing of bulk solutions or large drops ($r_d = \infty$), the factor $H_{v,fr}$ becomes $C_\varepsilon \varepsilon^2/\rho_i L_m^{ef}$ and (Eqn. 9.7.15) reduces to the expression (Khvorostyanov and Curry, 2000, 2004a,b)

$$\Delta F_{cr} = \frac{(16\pi/3)\sigma_{is}^3 f(m_{is},x)}{\left[\rho_i L_m^{ef} \ln\left(\frac{T_0}{T} S_w^{G_n}\right) - C_\varepsilon \varepsilon^2\right]^2}. \quad (9.7.17)$$

For a particular case $S_w = 1$ (pure water) and neglecting misfit strain, $\varepsilon = 0$, this is reduced to

$$\Delta F_{cr} = \frac{16\pi}{3} \frac{\sigma_{is}^3 f(m_{is},x)}{[\rho_i L_m^{ef} \ln(T_0/T)]^2}, \quad (9.7.17a)$$

which corresponds to the classical expressions case of J. J. Thomson (1888) with r_{cr} in (Eqn. 9.7.7).

9.7.5. Modification of Critical Energy with Active Sites

Fletcher (1969) has further modified the classical expression for the critical energy using a simple model for the active sites on the surface of nucleating insoluble substrate. He assumed that the contact parameter on the active sites is $m_{is} = 1$, and $m_{is} < 1$ on the rest of the surface. If the relative area of active sites is α on an insoluble core with radius r_N, then the active sites occupy a surface area of αr_N^2. Then, the surface energy defined as in (Eqn. 9.2.11) should be modified as

$$\Delta F_{surf} = \sigma_{is} A_{is} + (\sigma_{Ni} - \sigma_{Ns}) A_{Ni} = \sigma_{is} A_{is} - m_{is} \sigma_{is} A_{Ni}, \quad (9.7.18)$$

where σ_{is}, σ_{iN}, and σ_{Ns} are the surface tensions at the interfaces ice–solution, ice–insoluble substrate, and solution–substrate (substrate is denoted by "N"). Transition from the first to the second equation here is made using a definition $m_{is} = (\sigma_{iN} - \sigma_{Ns})/\sigma_{is}$. Fletcher's (1969) active sites modification is performed by replacement of the surface energy:

$$-m_{is} \sigma_{is} A_{Ni} = -m_{is} \sigma_{is} (A_{Ni} - \alpha r_N^2) + 1 \times \sigma_{is} \alpha r_N^2, \quad (9.7.19)$$

where "1" on the right-hand side means $m_{is} = 1$ on the active site. Substituting this expression into (Eqn. 9.7.18), and then substituting ΔF_{surf} into (Eqn. 9.7.13), we obtain the total (volume plus surface) energy:

$$\begin{aligned} \Delta F_{cr} &= \Delta F_V + \Delta F_{surf} = \Delta F_V + \sigma_{is} A_{is} - m_{is} \sigma_{is} (A_{Ni} - \alpha r_N^2) - \sigma_{is} \alpha r_N^2, \\ &= \Delta F_V + \Delta F_{surf,0} - \Delta F_a^\alpha, \end{aligned} \quad (9.7.20)$$

where we defined the surface energy $\Delta F_{surf,0}$ without the active sites ($\alpha = 0$), and the term ΔF_a^α:

$$\Delta F_{surf,0} = \sigma_{is} (A_{is} - m_{is} A_{Ni}), \quad (9.7.20a)$$

$$\Delta F_a^\alpha = \alpha r_N^2 \sigma_{is} (1 - m_{is}) = \alpha r_N^2 \sigma_{is} \delta m_{is}, \quad \delta m_{is} = 1 - m_{is}. \quad (9.7.20b)$$

The term ΔF_a^α, introduced by Fletcher (1969), is the energy decrease due to existence of the active sites with surface fraction α on the surface of an insoluble core and with the contact parameter $m_{is} = 1$—i.e., with ideal wettability. The term ΔF_a^α in (Eqn. 9.7.20b) does not include the geometrical parameters of the germ, thus minimization of the total energy $\Delta F_{cr}(r)$ by r does not involve the term with α, and the term in energy without active sites is as before. It can be easily shown that if the surface areas in (Eqn. 9.7.18) for ΔF_{surf} are defined similar to (Eqn. 9.2.4) and (Eqn. 9.2.5) for a planar surface, $A_{Ni} = \pi r^2(1 - m_{is}^2)$, $A_{is} = 2\pi r^2(1 - m_{is})$, and the volume of the ice cap is as in (Eqn. 9.2.3), then the sum $(\Delta F_V + \Delta F_{surf,0})$ in (Eqn. 9.7.20) is proportional to Volmer's (1939) geometric factor (Eqn. 9.7.16a). For the cap geometry, this factor has Fletcher's form. Therefore, the critical energy with active sites can be written as a sum of the previously derived term and a term describing the active sites:

$$\Delta F_{cr} = \Delta F_{cr}^{\alpha 0} - \Delta F_a^{\alpha} = \Delta F_{cr}^{\alpha 0} - \alpha r_N^2 \sigma_{is}(1 - m_{is}), \tag{9.7.21a}$$

where $\Delta F_{cr}^{\alpha 0}$ is the critical energy without active sites ($\alpha = 0$) described by (Eqn. 9.7.15):

$$\Delta F_{cr}^{\alpha 0} = \frac{4}{3}\pi \sigma_{is} r_{cr}^2 f(m_{is}, x) = \frac{(16\pi/3)\sigma_{is}^3 f(m_{is}, x)}{\left\{\rho_i L_m^{ef} \ln\left[\frac{T_0}{T} S_w^{G_n} \exp(-H_{v,fr})\right]\right\}^2}, \tag{9.7.21b}$$

$$\Delta F_{cr}^{\alpha 0} = \frac{(16\pi/3)\sigma_{is}^3 f(m_{is}, x)}{\left\{\rho_i L_m^{ef}\left[\ln\left(\frac{T_0}{T} S_w^{G_n}\right) - H_{v,fr}\right]\right\}^2}, \tag{9.7.21c}$$

and ΔF_a^{α} is the decrease in the critical energy ΔF_{cr} due to active sites described by (Eqn. 9.7.20b). Thus, ΔF_{cr} that takes into account the active sites can be written now as

$$\Delta F_{cr} = \frac{(16\pi/3)\sigma_{is}^3 f(m_{is}, x)}{\left\{\rho_i L_m^{ef} \ln\left[\frac{T_0}{T} S_w^{G_n} \exp(-H_{v,fr})\right]\right\}^2} - \alpha r_N^2 \sigma_{is}(1 - m_{is}). \tag{9.7.21d}$$

Various forms of ΔF_{cr} are used in applications. In particular, for not very high pressure ($\Delta p = 0$), and for the freezing of bulk solutions or large drops ($r_d = \infty$):

$$\Delta F_{cr} = \frac{(16\pi/3)\sigma_{is}^3 f(m_{is}, x)}{\left[\rho_i L_m^{ef} \ln\left(\frac{T_0}{T} S_w^{G_n}\right) - C_\varepsilon \varepsilon^2\right]^2} - \alpha r_N^2 \sigma_{is}(1 - m_{is}). \tag{9.7.22}$$

It is clear that a polydisperse population of ice nuclei cannot be characterized by a single value of the active site area α. Curry and Khvorostyanov (2012) performed a further generalization of the active site concept and introduced the temperature-dependent $\alpha(T)$. This model and its effects will be considered in Section 9.17. Averaging over the active sites is considered in the following text.

Fletcher's concept of the active sites can be further generalized in the following way. The term ΔF_a^{α} in (Eqn. 9.7.20b) describes the decrease of the free energy if only one type of the active sites with "ideal" wettability, $m_{is} = 1$, is present. In reality, it can be expected that several types of active sites can be present on each IN. These sites may have different contact parameters m_{is}^k that satisfy the relation $m_{is} < m_{is}^k \le 1$ (Hobbs, 1974)—that is, are greater than the main contact parameter m_{is} characterizing the major IN surface, thereby ensuring greater wettability and ice nucleability, but are smaller than 1. Thus, (Eqn. 9.7.20b) for $\Delta F_a{}^{\alpha}$ can be further generalized for several types of active sites as

$$\Delta F_a^{\alpha} = \sum_k \Delta F_a^{\alpha k} = \sum_k f_{\alpha}^k \alpha_a^k r_N^2 \sigma_{is} \delta m_{is}^k, \qquad \delta m_{is}^k = m_{is}^k - m_{is}, \tag{9.7.23}$$

where m_{is}^k and α_a^k are the contact parameter and surface fraction of the active sites of the k-th kind and f_{α}^k is their density. It is assumed that $\delta m_{is}^k > 0$, thus $\Delta F_a{}^{\alpha} > 0$, and this term $-\Delta F_a{}^{\alpha} < 0$ in (Eqn. 9.7.21a) causes a decrease in ΔF_{cr} and favors ice nucleation. Then, (Eqn. 9.7.21a) for ΔF_{cr} can be generalized in the same way as before—i.e., considering the sum of the volume and surface energies and finding their minimum by radius.

Since the geometry of the ice cap is not involved in introducing these sites (Fletcher, 1969), the major first term of the RHS in (Eqns. 9.7.21b) or (9.7.22) remains the same and the expression for ΔF_{cr} can be generalized as

$$\Delta F_{cr} = \frac{(16\pi/3)\sigma_{is}^3 f(m_{is},x)}{\left\{\rho_i L_m^{ef} \ln\left[\frac{T_0}{T} S_w^{G_n} \exp(-H_{v,fr})\right]\right\}^2} - \sum_k f_{\alpha}^k \alpha_a^k r_N^2 \sigma_{is} \delta m_{is}^k. \tag{9.7.24}$$

Note that this generalization of classical nucleation theory with respect to the active sites introduced here is analogous to the concept developed in Connolly et al. (2009) where the active sites were introduced without invoking classical nucleation theory and their properties were studied experimentally. The parameters of the active sites from this study can be compared to (Eqn. 9.7.24).

The freezing nucleation rates $J_{S,fr}$ and $J_{f,het}$ can be now evaluated using (Eqns. 9.6.4) and (9.6.7) with the kinetic coefficients B_{het} and C_{het} defined by (Eqns. 9.6.6) and (9.6.8), and with ΔF_{cr} derived in this section. Using (Eqn. 9.7.24) for ΔF_{cr}, the nucleation rate $J_{f,het}$ (Eqn. 9.6.7) can be rewritten as

$$J_{f,het}(T,r_N) = \frac{kT}{h} c_{1s} 4\pi r_N^2 \exp\left[-\frac{\Delta F_{act}}{kT}\right]$$
$$\times \exp\left[-\frac{1}{kT}\frac{(16\pi/3)\sigma_{is}^3 f(m_{is},x)}{\left\{\rho_i L_m^{ef} \ln\left[(T_0/T) S_w^{G_n} \exp(-H_{v,fr})\right]\right\}^2} + \frac{1}{kT}\sum_k f_{\alpha}^k \alpha_a^k r_N^2 \sigma_{is} \delta m_{is}^k\right]. \tag{9.7.25}$$

Using a somewhat simplified (Eqn. 9.7.21d) for ΔF_{cr} with only one kind of active sites, the nucleation rate (Eqn. 9.6.7) can be rewritten as

$$J_{f,het}(T,r_N) = \frac{kT}{h} c_{1s} 4\pi r_N^2 \exp\left[-\frac{\Delta F_{act}}{kT}\right]$$
$$\times \exp\left[-\frac{1}{kT}\frac{(16\pi/3)\sigma_{is}^3 f(m_{is},x)}{\left\{\rho_i L_m^{ef} \ln\left[(T_0/T)\, S_w^{G_n} \exp(-H_{v,fr})\right]\right\}^2} + \frac{\alpha r_N^2 \sigma_{is}(1-m_{is})}{kT}\right], \tag{9.7.26}$$

with $H_{v,fr}$ defined in (Eqn. 9.7.6). These nucleation rates include the effects of the temperature, the solution concentration or saturation ratio S_w, the external pressure, misfit strain, the finite size of a freezing drop, and active sites.

9.8. Properties of the Deliquescence-Freezing Mode

9.8.1. Critical Radius, Energy, and Nucleation Rate

Properties of the critical ice germs radii, energies, and nucleation rates derived in Section 9.7 are illustrated in this section. In the calculations described here, we use the temperature and solution concentration parameterizations of ρ_i, ρ_w, L_m, σ_{sa}, and σ_{is} from Pruppacher and Klett (1997) and Tabazadeh et al. (1997) (Section 4.4), and $\Delta F_{act}(T)$ from Khvorostyanov and Curry (2004b) (Section 8.8.3), the contact parameter $m_{is} = 0.5$ for the deliquescence-heterogeneous-freezing (DHF) mode, the external pressure $\Delta p = 0$, and the misfit strain $\varepsilon = 0$ in most calculations. If these parameters are different in some cases, their values are indicated. Figs. 9.2–9.5 present r_{cr} and ΔF_{cr} calculated with (Eqns. 9.7.4), (9.7.5), and (9.7.15) in two forms: as 2D fields on a T-S_w diagram, and as functions of temperature for several humidities.

Fig. 9.2 shows isolines of the critical radius r_{cr} for the deliquescence-freezing mode as a function of the temperature T and saturation ratio S_w on the two-dimensional T-S_w diagram. Note that since r_{cr} does not depend on $f(m_{is},x)$, here r_{cr} is the same as for homogeneous nucleation, as described in Chapter 8. The isolines of r_{cr} extend from the lower right-hand corner to the upper left-hand corner, since the temperature increase of 50 °C is approximately compensated by the humidity increase of 50%. Values of r_{cr} range from ~10^{-7} cm at S_w ~ 1.3, T ~ −55 °C to 10^{-6} cm at warmer T or lower S_w. Thus, r_{cr} increases rapidly toward the left lower corner while the denominator in (Eqn. 9.7.4) for r_{cr} goes to zero, and so r_{cr} tends to infinity. In the left lower corner of the T-S_w diagram with too low S_w and supercooling ΔT, nucleation is prohibited since the denominator in (Eqn. 9.7.4) and r_{cr} become negative.

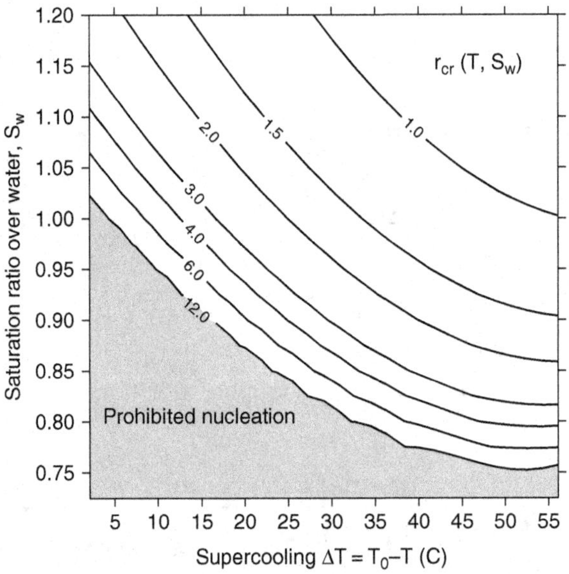

Figure 9.2. Critical radius r_{cr} (10^{-7} cm) for the freezing mode as a function of two variables: supercooling $\Delta T = T_0 - T$ and saturation ratio over water S_w (the T-S_w diagram) at $\Delta p = 0$ and $\varepsilon = 0$. r_{cr} is the same as for homogeneous nucleation. From Khvorostyanov and Curry (2004b), *J. Atmos. Sci.*, 61, © American Meteorological Society. Used with permission.

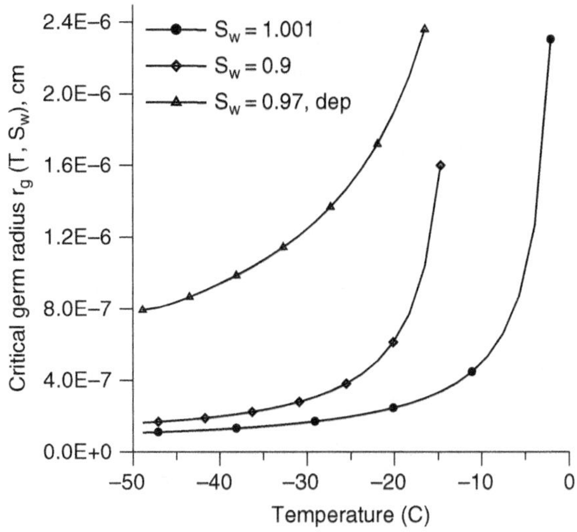

Figure 9.3. Critical germ radius r_{cr} for the freezing mode as a function of temperature with the contact parameter $m = 0.5$ for the saturation ratios over water, $S_w= 1.001$ (circles), $S_w= 0.9$ (diamonds), and for the deposition mode with $m = 0.95$ and $S_w= 0.97$ (triangles). From Khvorostyanov and Curry (2004b), *J. Atmos. Sci.*, 61, © American Meteorological Society. Used with permission.

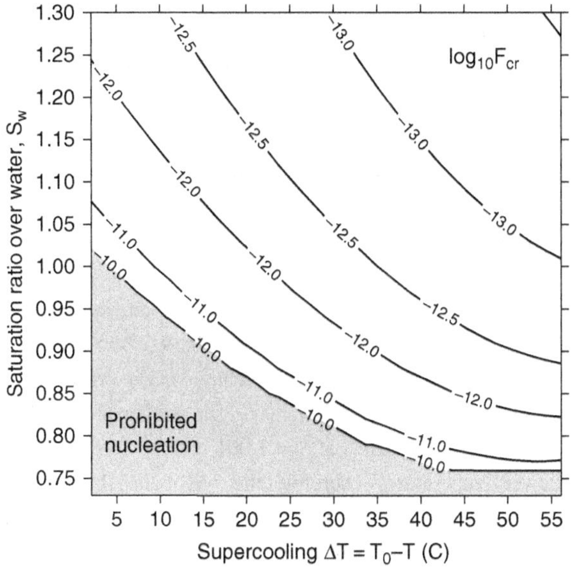

Figure 9.4. Logarithm of the critical energy F_{cr} (erg) for the freezing mode as a function of two variables: supercooling $\Delta T = T_0 - T$ and the saturation ratio over water S_w calculated with contact parameter $m = 0.5$ and $r_N= 0.26\,\mu$m. From Khvorostyanov and Curry (2004b), *J. Atmos. Sci.*, 61, © American Meteorological Society. Used with permission.

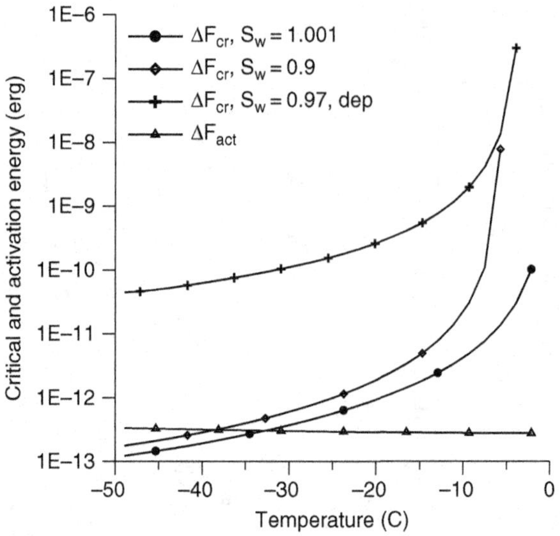

Figure 9.5. Critical germ energy F_{cr} for the deliquescence-freezing mode as a function of temperature with the contact parameter $m = 0.5$ for the saturation ratios over water, $S_w = 1.001$ (circles), $S_w = 0.95$ (diamonds), and for the deposition mode with $m = 0.95$ and $S_w = 0.97$ (crosses). Activation energy ΔF_{act} is presented for comparison (triangles). From Khvorostyanov and Curry (2004b), *J. Atmos. Sci.*, 61, © American Meteorological Society. Used with permission.

Fig. 9.3 shows an order of magnitude decrease of r_{cr} over the temperature range from −3 to −10 °C at $S_w = 1.001$ (supersaturation $s_w = 0.01\%$) typical of conditions in a mixed-phase cloud without strong glaciation and from −15 to −30 °C at $S_w = 0.9$, typical for colder crystalline clouds or diamond dust formation (e.g., Ohtake et al., 1982; Curry, 1983, 1995; Curry et al., 1990, 1993, 1996, 2000; Girard and Blanchet, 2001; Gultepe et al., 2003; Khvorostyanov et al., 2003). The value of r_{cr} may increase with decreasing T if S_w decreases rapidly with T. The critical radius for deposition is greater than for deliquescence-freezing for all values of T at $S_w = 0.97$.

The 2D field of the critical energy ΔF_{cr} on the T-S_w diagram (Fig. 9.4) is similar to the field of r_{cr}. Nucleation is prohibited in the lower left-hand portion of the diagram because the critical radius becomes negative here (Fig. 9.2). The T-curves for ΔF_{cr} (Fig. 9.5) show that lowering humidity by 10% leads to a decrease in ΔF_{cr} by 0.5–1 orders of magnitude. The critical energy of deposition is much greater than for deliquescence-freezing, indicating a smaller probability of deposition. The activation energy ΔF_{act} is smaller than ΔF_{cr} at $S_w = 1.001$ and $S_w = 0.95$ to $T > -35\ldots-38$ °C, and ΔF_{act} exceeds ΔF_{cr} at lower temperatures. Therefore, the role of ΔF_{act} in ice nucleation increases for very cold clouds. Since several different parameterizations were suggested for ΔF_{act} (Section 8.3.3), special studies are desirable to clarify their role relative to ΔF_{cr} in the nucleation rates.

Fig. 9.6 shows the logarithm of the freezing nucleation rate $J_{f,het}$ per particle (s^{-1}) on the T-S_w diagram, calculated with (Eqn. 9.6.7) and ΔF_{cr} with (Eqn. 9.7.17) using contact parameter $m_{is} = 0.5$ and $r_N = 0.26\,\mu$m. This figure shows very high gradients of $J_{f,het}$, varying by 10–15 orders of magnitude over the temperature range of 5 °C or a humidity range of 5%. The bold isoline "0" corresponds to

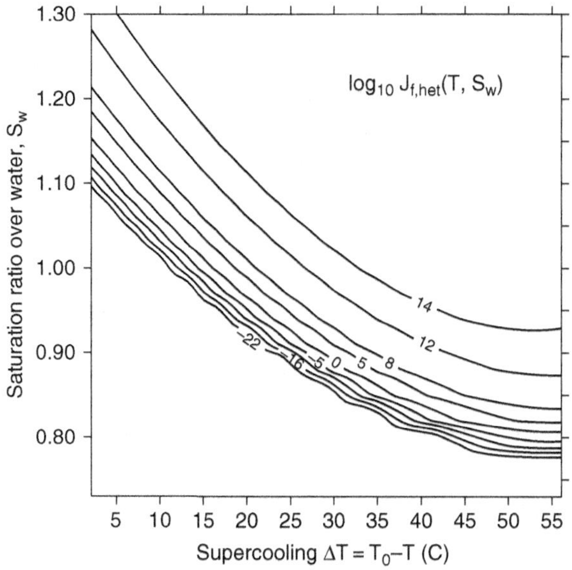

Figure 9.6. Logarithm of the freezing nucleation rate $\log_{10} J_{f,het}$ (s^{-1}) as a function of two variables: supercooling $\Delta T = T_0 - T$ and the saturation ratio over water S_w calculated with contact parameter $m = 0.5$ and $r_N = 0.26\,\mu m$. The bold isoline "0" corresponds to the rate $J_{f,het} = 1\ s^{-1}$. From Khvorostyanov and Curry (2004b), *J. Atmos. Sci.*, 61, © American Meteorological Society. Used with permission.

the rate $J_{f,het} = 1\ s^{-1}$ per particle and indicates the values of T, S_w where the nucleation rate becomes significant.

Fig. 9.7 illustrates the temperature dependence of $J_{f,het}$ for various humidities. Even relatively small variations in humidity and its variation with T lead to significant changes in the nucleation rate. At a typical cloud saturation ratio of 1.001 (supersaturation 0.1%), $J_{f,het}$ reaches values of 10^{-5} to $1\ s^{-1}$ per particle at $T = -13$ to $-15\,°C$. An increase in S_w to 1.05 ($s_w = 5\%$) causes the shift to higher T by 5 °C, so that ice nucleation may occur with a noticeable rate at T of about −10 °C or warmer, which is in agreement with observations by Hobbs and Rangno (1990) and Rangno and Hobbs (1991). It will be shown in Section 9.13 that an increase of m_{is} to 0.55–0.65 or the presence of active sites with a surface fraction $\alpha \sim 2 \times 10^{-5}$ causes noticeable nucleation rates at warmer temperatures −5 to −9 °C, and even at a low s_w of 0.1–0.2%. At temperatures below −20 °C, ice nucleation occurs at subsaturations. The nucleation rates and final crystal concentrations are discussed in more detail in Section 9.14 based on parcel model simulations.

9.8.2. Separation of the Temperature, Supersaturation, and Aerosol Dependencies of the Critical Energy and Nucleation Rate

The dependencies of ΔF_{cr} and $J_{f,het}$ on the temperature T and supersaturation s_w can be isolated and determined analytically, which is useful in using the functional relationships and for developing parameterizations similar to those described in Section 8.10 for homogeneous freezing. Ice nucleation

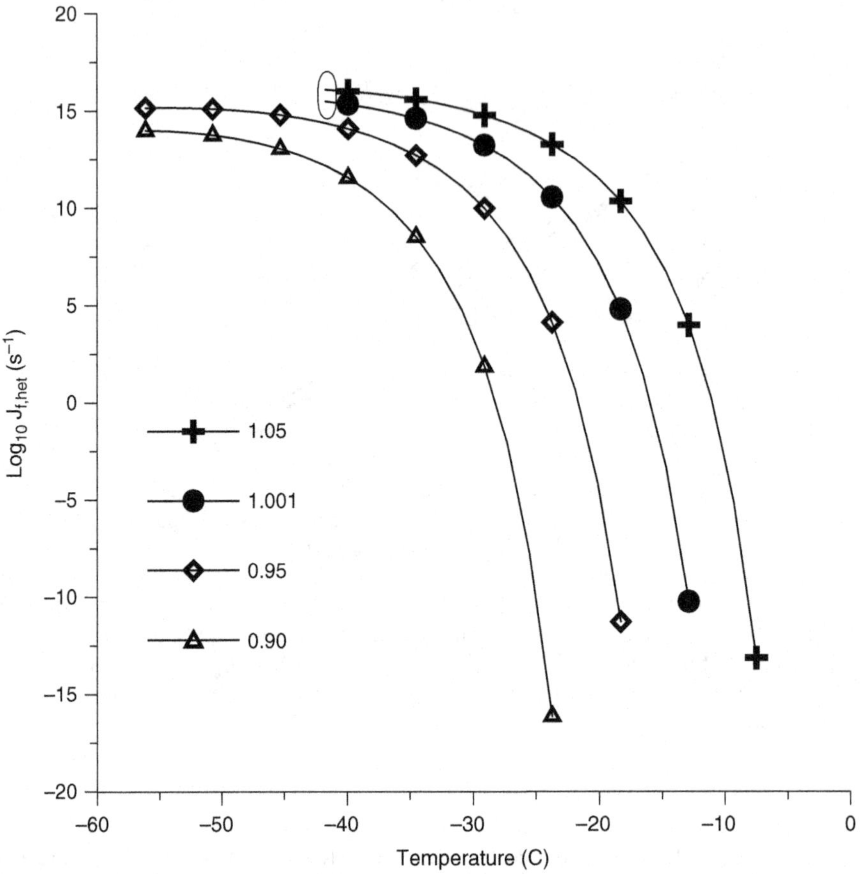

Figure 9.7. Freezing nucleation rate $J_{f,het}$ (s^{-1}) for the CCN freezing mode as a function of the temperature with the contact parameter $m = 0.5$ for the saturation ratios over water, $S_w = 1.05$ (crosses), $S_w = 1.001$ (circles), $S_w = 0.95$ (diamonds), and $S_w = 0.90$ (triangles). The threshold of the homogeneous freezing of drops with $r_{dr} \sim 0.2$–$5\,\mu$m at $T_c \sim -42\ldots-38\,°$C is denoted by the ellipse.

via the deliquescence-freezing mode occurs usually at the water saturation ratio S_w close to 1—i.e., at $|s_w| << 1$. Using the relations $S_w = 1 + s_w$ and $\ln(1 + s_w) \approx s_w$ for $|s_w| << 1$, and expanding the denominator of r_{cr} in (Eqn. 9.7.5b) and retaining only the term of the first order (this approximation is discussed in the following text), we obtain

$$r_{cr} = \frac{r_{cr,0}}{\left[1 + \frac{G_n}{\ln(T/T_0)}\ln(1+s_w) - \frac{H_{v,fr}}{\ln(T/T_0)}\right]},$$

$$\approx \frac{r_{cr,0}}{\left[1 + \frac{G_n s_w}{\ln(T/T_0)} - \frac{H_{v,fr}}{\ln(T/T_0)}\right]}, \tag{9.8.1a}$$

where $r_{cr,0}$ is the critical radius at zero water supersaturation, $s_w = 0$, with $\varepsilon = 0$ and $r_{dr} = \infty$,

$$r_{cr,0} = \frac{2\sigma_{is}}{\rho_i L_m^{ef} \ln(T/T_0)}. \tag{9.8.1b}$$

For convenience, we denote in the following text the fraction of active sites with subscript "a" as α_a. Substitution of (Eqn. 9.8.1a) into (Eqn. 9.7.21a,b) yields

$$\Delta F_{cr}(T, s_w, r_{dr}, \varepsilon, \alpha_a) = \frac{\Delta F_{cr,0}^{\alpha 0}(T)}{\left[1 + \dfrac{G_n s_w}{\ln(T_0/T)} - \dfrac{H_{v,fr}}{\ln(T/T_0)}\right]^2} - \Delta F_a^{\alpha}, \tag{9.8.2}$$

where $\Delta F_{cr,0}^{\alpha 0}$ is the critical energy at zero water supersaturation, $s_w = 0$, with $\varepsilon = 0$, $\alpha_a = 0$, and $r_{dr} = \infty$—i.e., as defined in the classical nucleation theory (see Section 9.7):

$$\Delta F_{cr,0}^{\alpha 0}(T) = \frac{4\pi}{3}\sigma_{is} r_{cr,0}^2 f(m_{is}, x)$$

$$= \frac{16\pi}{3}\frac{\sigma_{is}^3 f(m_{is}, x)}{[\rho_i L_m^{ef} \ln(T/T_0)]^2} \approx \frac{16\pi}{3}\frac{\sigma_{is}^3 f(m_{is})}{[\rho_i L_m^{ef} \ln(T/T_0)]^2}, \tag{9.8.3}$$

where the last equation accounts for the approximation (Eqn. 9.7.16a) for $f(m_{is}, x) \approx f(m_{is})$. For further transformation of (Eqn. 9.8.2), consider two examples of heterogeneous nucleation that occur via the freezing of interstitial deliquescent aerosols and deliquescent haze particles at $T < -10\,°\mathrm{C}$. For supersaturations typical in the clouds without extremely high updrafts and sufficiently far from the cloud boundaries, $s_w \sim 1 \times 10^{-4}$ to 1×10^{-3} (or 10^{-2} to 10^{-1} %). The parameter $G_n \sim 0.4$ (see Fig. 8.14c); thus, the term $G_n s_w \sim 0.4 \times 10^{-4}$ to 0.4×10^{-3}. At $T < -10\,°\mathrm{C}$, when heterogeneous nucleation becomes noticeable, $\ln(T_0/T) > 0.037$, and $X = G_n s_w/\ln(T_0/T) << 1$.

Consider another example, heterogeneous ice nucleation by freezing of haze particles at water subsaturation, $s_w < 0$, in cloudless atmosphere at low temperatures—e.g., in cirrus at $T = 223$ K. Then, $\ln(T_0/T) \approx 0.2$, the critical RHW $\approx 85\%$ (see Section 9.12)—i.e., $s_w = -0.15$. With $G_n \sim 0.4$, the term $G_n s_w/\ln(T_0/T) \approx -0.4 \times (-0.15)/0.2 = -0.3$. We can assume again $|X| = |G_n s_w/\ln(T_0/T)| << 1$ because the term of the second order in expansion by $|X|$ contributes less than 9%. A similar assumption can be made on the term $H_{v,fr}/\ln(T_0/T)$. Effects of the freezing drop radius r_{dr} and of the misfit strain ε can be considered as corrections to the main expressions for ΔF_{cr} in the physically reasonable variations of r_{dr} and ε (Young, 1993; Khvorostyanov and Curry, 2004a,b; see Section 9.11). The effects of pressure excess become noticeable only at very high pressures ~103 atm (Kanno and Angell, 1977; Mishima and Stanley, 1998; Koop et al., 2000; Khvorostyanov and Curry, 2004a; Curry and Khvorostyanov, 2012) that can be created in the laboratory experiments but are not met under usual atmospheric conditions. Therefore, we can omit here the terms with Δp. These terms can be easily included in the final equations for analysis of the experiments at high pressures.

Thus, we can expand the denominator in (Eqn. 9.8.2) by the power series of $X = G_n s_w/\ln(T_0/T)$, keeping only the first term in the series (the second term gives us a contribution of a few percent, or less than 1%). Thus, we obtain from (Eqn. 9.8.2)

$$\Delta F_{cr}(T, s_w) \approx \Delta F_{cr,0}^{\alpha 0}(T)\left[1 - (\kappa_s \Delta F_{cr,0}^{\alpha 0})s_w + \frac{2H_{v,fr}}{\ln(T_0/T)}\right] - \Delta F_a^{\alpha},$$

$$\approx \Delta F_{cr,0}^{\alpha 0}(T) - (\kappa_s \Delta F_{cr,0}^{\alpha 0})s_w + \Delta F_{cr}^{\varepsilon} - \Delta F_a^{\alpha}, \tag{9.8.4}$$

where

$$\kappa_s = \frac{2G_n}{\ln(T_0/T)} = \frac{2RT}{M_w L_m^{ef} \ln(T_0/T)}, \qquad \Delta F_{cr}^{\varepsilon} = \frac{2H_{v,fr}(\varepsilon, r_{dr})\Delta F_{cr,0}^{\alpha 0}}{\ln(T_0/T)}. \tag{9.8.5}$$

The four terms on the RHS of (Eqn. 9.8.4) describe, respectively, the effects of temperature, water saturation ratio, misfit strain and drop radius, and active sites. The range of s_w for the applicability of (Eqn. 9.8.4) can be estimated as follows. For not very large supercoolings $\Delta T = T_0 - T < 40$–60 K, and a simplified case, neglecting ε, r_{dr}, and Δp, and using the approximate relation $\ln(T_0/T) \approx \Delta T/T$, (Eqn. 9.8.4) can be further reduced to

$$\Delta F_{cr}(T, s_w) \approx \Delta F_{cr,0}^{\alpha 0}(T)\left[1 - 2G_n \frac{s_w}{\Delta T/T}\right]. \tag{9.8.6a}$$

Here the critical energy $\Delta F_{cr}(T,s_w)$ is separated into two factors: $\Delta F_{cr,0}^{\alpha 0}(T)$ is the critical energy at zero water supersaturation, $s_w = 0$ or ($S_w = 1$), and depends only on T, and the second term in the brackets describes the dependence on supersaturation. A quantitative limitation on supersaturation s_w follows from the condition that keeping only the first term in expansion by X is valid when $X \approx G_n T|s_w|/\Delta T << 1$ or, with $G_n \approx 0.4$:

$$s_w \ll \frac{\Delta T}{G_n T} \approx 2.5\frac{\Delta T}{T}. \tag{9.8.6b}$$

This criterion is usually satisfied for most situations with heterogeneous crystal nucleation. For example, at temperatures $T = 263$ and 253 K, or supercoolings $\Delta T = 10$ K and 20 K, the limitations are: $|s_w| < 0.095$ (9.5%) and $|s_w| < 0.20$ (20%), respectively. The supersaturations are much lower in mixed clouds with freezing drops, and subsaturations are smaller than 5–10% in the nucleation of crystals via the freezing of haze particles without a liquid phase. Thus, the expansion (Eqn. 9.8.4) is valid under a sufficiently wide range of cloud conditions, for most cloud types, which allows a great simplification of the nucleation rates.

For parameterization of the concentration of nucleated crystals, similar to that for homogeneous nucleation in Khvorostyanov and Curry (2012) (see Section 8.10), we need to separate the three dependencies of the nucleation rate per particle $J_{f,het}(T, s_w, r_N)$: on the temperature T, water supersaturation s_w, and aerosol properties, in particular, on the radius r_N of the insoluble core inside a freezing particle. Generalizing a separation method, developed in Khvorostyanov and Curry (2004b) and substituting (Eqn. 9.8.4) into (Eqn. 9.6.7) for $J_{f,het}$, we obtain after simple transformations

$$\begin{aligned} J_{f,het}(T, s_w, r_N) &\approx J_{f,het}^{(0,\alpha 0)}(T, \bar{r}_N)\frac{r_N^2}{\bar{r}_N^2}\exp\left(\frac{\kappa_s \Delta F_{cr,0}^{\alpha 0}}{kT} s_w\right)\exp\left[\frac{\Delta F_a^{\alpha}(r_N)}{kT}\right] \\ &\approx J_{f,het}^{(0,\alpha 0)}(T, \bar{r}_N) J_{act}^{\alpha}(T, r_N) J_{het}^{s}(T, s_w, m_{is}), \end{aligned} \tag{9.8.7}$$

where

$$J_{f,het}^{(0,\alpha 0)}(T, \bar{r}_N) = C_{het}(\bar{r}_N)\exp\left[-\frac{\Delta F_{act}(T)}{kT} - \frac{\Delta F_{cr,0}^{\alpha 0}}{kT} + \frac{\Delta F_{cr}^{\varepsilon}}{kT}\right], \tag{9.8.8}$$

$$J_{act}^{\alpha}(T, r_N) = [(r_N^2/\bar{r}_N^2)\exp(z_\alpha r_N^2)], \tag{9.8.9a}$$

$$z_\alpha = \frac{\Delta F_a}{kT} = \frac{\alpha_a \sigma_{is}(1-m_{is})}{kT}, \tag{9.8.9b}$$

$$J^s_{het}(T, s_w, m_{is}) = \exp(u_{shet} s_w), \tag{9.8.10}$$

$$u_{shet}(T, m_{is}) = \frac{\kappa_s \Delta F^{\alpha 0}_{cr,0}}{kT} = \frac{\Delta F^{\alpha 0}_{cr,0}(T, m_{is})}{kT} \cdot \frac{2G_n(T)}{\ln(T_0/T)} = \frac{2R}{kM_w L^{ef}_m} \frac{\Delta F^{\alpha 0}_{cr,0}(T, m_{is})}{\ln(T_0/T)}, \tag{9.8.11}$$

and $\bar{r}_N$ is the average of r_N over the aerosol population. If there are several active sites on the surface of an IN, $J^\alpha_{act}(T, r_N)$ is generalized using the expression for the energy of active sites (Eqn. 9.7.23). Then

$$J^\alpha_{act}(T, r_N) = (r_N^2/\bar{r}_N^2)\sum_k \exp(z^k_\alpha r_N^2), \tag{9.8.12a}$$

$$z^k_\alpha = \frac{1}{kT}\sum_k \Delta F^{\alpha k}_a = \frac{1}{kT}\sum_k f^k_\alpha \alpha^k_a r_N^2 \sigma_{is} \delta m^k_{is}, \tag{9.8.12b}$$

where $\delta m^k_{is} = m^k_{is} - m_{is}$, and f^k_α is the density of the active sites with the surface area α^k_a.

Thus, the nucleation rate is presented in (Eqn. 9.8.7) as a product of the three factors that describe respectively the major effects of T, r_N, and s_w. The first factor, $J^{(0,\alpha 0)}_{f,het}(T, \bar{r}_N)$, is the nucleation rate at $s_w = 0$, $\alpha_a = 0$ (superscript "$\alpha 0$"), and $r_N = \bar{r}_N$, evaluated with $C_{het}(\bar{r}_N)$—that is, does not depend on r_N. The factor J^α_{act} depends on z_α and describes the effect of the active sites area α_a and r_N. For example, at $T \le -20\,°\mathrm{C}$, when diamond dust or Ac may form, and with $m_{is} = 0.5$, the value of $z_\alpha \sim 0.7 \times 10^{10}\,\mathrm{cm}^{-2}$. Thus, $z_\alpha r_N^2 \sim 0.7$ for small aerosol particles with $r_N = 0.1\,\mu\mathrm{m}$ but grows to $z_\alpha r_N^2 \sim 70$ for larger particles with $r_N = 1\,\mu\mathrm{m}$, which results in a great enhancement of the nucleation rate by the factor $\exp(z_\alpha r_N^2)$ in (Eqn. 9.8.7) due to the active sites. The factors $J^s_{het}(T, s_w, m_{is})$ and $u_{s,het}(T, m_{is})$ depend on the supersaturation, temperature, and contact parameter and describe the effects due to variations of humidity or supersaturation s_w.

Equation (9.8.7) can be rewritten in two additional forms:

$$J_{f,het}(T, s_w, r_N, \varepsilon) \approx J^{(0)}_{f,het}(T, r_N, \varepsilon)\exp(u_{shet} s_w), \tag{9.8.13}$$

$$J_{f,het}(T, s_w, r_N, \varepsilon) = J^{(0)}_{f,het}(T, r_N, \varepsilon)[b_{het}(T)]^{s_w}, \tag{9.8.14}$$

where

$$J^{(0)}_{f,het}(T, r_N, \varepsilon) \approx J^{(0,\alpha 0)}_{f,het}(T, \bar{r}_N, \varepsilon) J^\alpha_{act}(T, r_N), \tag{9.8.15}$$

$$b_{het}(T) = \exp(u_{shet}) \approx \exp\left(\frac{\Delta F^{\alpha 0}_{cr,0}(T, m_{is})}{kT} \cdot \frac{2G_n(T)}{\ln(T_0/T)}\right). \tag{9.8.16}$$

The factor $J^{(0)}_{f,het}(T, r_N, \varepsilon)$ is the nucleation rate at water saturation, $s_w = 0$ or $S_w = 1$, which describes the effects of temperature, active sites, and misfit strain, but does not depend on supersaturation. The second multipliers in (Eqns. 9.8.13) and (9.8.14) describe the effects of s_w variations, and these equations show that the supersaturation dependence of the nucleation rate can be presented in different forms, either as an exponential function or as a power law. The power law (Eqn. 9.8.14) could be considered an analog of the humidity parameterization of Huffman (1973), although the functional dependence is somewhat different here. The exponent by s_w in (Eqn. 9.8.13) resembles the exponential parameterization by the ice supersaturation s_i of the crystal concentration of Meyers et al. (1992), and will be derived in this form in Section 10.1.

Equations (9.8.7)–(9.8.16) describe analytically the supersaturation dependence of the heterogeneous nucleation rate and can be used for development of parameterizations of this quantity as was done for homogeneous nucleation in Section 8.10 and will be done for heterogeneous freezing in Section 10.1. Equations (9.8.7) and (9.8.13) show that $J_{f,het} \sim \exp(u_{sher}s_w)$, and increases exponentially with s_w, which can be recalculated to the exponential dependence on ice supersaturation s_i, resembling empirical parameterizations of Meyers et al. (1992) and Rogers et al. (1998), but with additional T-dependence. It will be illustrated in more detail for the crystal concentrations in the following text. This exponential variation may explain very fast crystal nucleation rates in clouds at and after the stage of intensive coagulation among the drops, and the decrease of their concentration when supersaturation may increase above its normal value of 0.01–0.1% and produce high N_c, as was hypothesized by Hobbs and Rangno (1990) and Rangno and Hobbs (1991, 2001), and was discussed in Khvorostyanov and Curry (2005a). Other reasons for possible high N_c are discussed in the following text.

9.8.3. Separation of Insoluble Fractions between Activated Drops and Unactivated CCN

During drop activation, separation (or redistribution) of insoluble fractions occurs between activated drops and unactivated CCN. The soluble substance of CCN before activation can be characterized by the mass fraction ε_m, and so the insoluble fraction is $\varepsilon_N = 1 - \varepsilon_m$. As described in Chapters 6 and 7, the critical saturation ratio (or supersaturation $s_{w,cr}$) for a drop activation depends on the CCN dry radius r_{d0} and its soluble fraction ε_m. For any given r_{d0}, the critical saturation ratio increases with decreasing ε_m. So, when the maximum saturation ratio $S_{w,max}$ is reached in a cooling cloud parcel during drop activation, all CCN of the same radius r_{d0} with $\varepsilon_m > \varepsilon_{m,cr}$ will be activated, and all CCN with $\varepsilon_m < \varepsilon_{m,cr}$ ($\varepsilon_N > \varepsilon_{N,cr}$) will remain as deliquescent but unactivated interstitial CCN (e.g., Hänel, 1976). This leads to the natural separation of the insoluble substance during drop activation: its fraction ε_N will be substantially greater in these non-activated CCN than in the cloud drops, and the insoluble fraction in unactivated CCN increases with increasing r_{d0}. Note also that the radii of the dry particle r_{d0} and its insoluble fraction r_N are related as $r_N \sim r_{d0}\varepsilon_N^{1/3}$, so that r_N grows with increasing r_{d0} at constant ε_N, which may explain the higher ice nucleating activity of the larger CCN.

Since the heterogeneous nucleation rate is $J_{f,het} \sim r_N^2$, this accumulation of the insoluble substrate in deliquescent interstitial CCN will increase their integral surface area and ice nucleating ability relative to the activated drops. Thus, separation of insoluble fractions during drop activation may explain several effects in ice nucleation if these interstitial CCN serve as IN: a) higher insolubility of these particles as compared to the average CCN; b) higher ice nucleating efficiency of larger CCN that contain larger amounts of insoluble material; c) higher efficiency of the CCN deliquescence-freezing mode than that of the immersion mode; and d) higher efficiency of the contact mode compared to the immersion mode if these interstitial CCN collide with the drops and serve as contact nuclei. Note that the dilution m_w/m_d of these CCN, with m_w being the mass of accumulated water, is high enough even for small soluble fractions, and salt concentration is low at saturation ratios close to critical (e.g., Tables 6.2, 6.3 in Pruppacher and Klett, 1997, and Chapters 6 and 7 here). Thus, the freezing point depression due to the solution is small, and the catalyzing freezing action of insoluble substrates can overwhelm the effects of the freezing point depression.

A quantitative description of this separation and subsequent heterogeneous ice nucleation requires knowledge of the CCN distributions by dry radii and insoluble (or soluble) fractions. Unfortunately, information on the distribution of insoluble fractions among CCN of the same size is very scarce, since typical measurements of soluble fractions cover some size range but not single particles (Pruppacher and Klett, 1997). Thus, we can only hypothesize a plausible size distribution function of these interstitial deliquescent aerosols.

9.8.4. Characteristic Relaxation Times of CCN Size and Solution Concentration

The characteristic relaxation times of CCN size and the solution concentration illuminate the mechanism for the impact of supersaturation on an ice germ. As discussed in Section 9.7, the radius of a deliquescent CCN adjusts to equilibrium as predicted by the Kohler curves (Eqn. 3.9.8). Otherwise, supersaturation or subsaturation $s_w = S_w - 1$ occurs around a CCN droplet and it grows or evaporates until its radius r_d reaches such a value that the equilibrium condition in (Eqn. 3.9.8) is satisfied. The relaxation time τ_{eq}, during which a deliquescent CCN reaches equilibrium both at sub- and supersaturation, can be estimated from the diffusion equation as $\tau_{eq} \approx 0.4 \times 10^6 \times r_d^2/|S_w - 1|$ with r_d in cm. Both deliquescent CCN at $S_w < 1$ and interstitial CCN at $S_w > 1$ reach equilibrium very rapidly (generally much less than 1 s) in response to variations of external supersaturation.

Another timescale τ_{dif} characterizes the time during which salt concentration within a CCN drop reaches equilibrium due to diffusion processes and becomes homogeneous. It can be estimated as $\tau_{dif} \sim 0.14\, r_d^2/D_c$, where $D_c \sim 10^{-5}\,\mathrm{cm^2\,s^{-1}}$ is the coefficient of mass (salt) diffusion within a drop (Sedunov, 1974). This yields an estimate $\tau_{dif} \sim 10^{-6}$ s and 10^{-4} s for drops of radii 0.1 μm and 1 μm, respectively. The characteristic time of mean supersaturation variations τ_{sup} in natural clouds without vigorous updrafts is typically 10–10^3 s, so that the hierarchy of the timescales is $\tau_{sup} >> \tau_{eq} \geq \tau_{dif}$—i.e., τ_{sup} is much greater than both τ_{eq} and τ_{dif}.

These estimations illustrate the main mechanism by which variations of external supersaturation s_w influence an ice germ that is embedded in a solution drop and is not in direct contact with external supersaturation field: variations of supersaturation are immediately followed by the much faster adjustment of the CCN size and homogenization of solution concentration in the entire drops' volume, including the vicinity of the insoluble substrate and ice germ. The growth of the saturation ratio S_w (or s_w) is accompanied by the rapid swelling of a deliquescent CCN and a decrease in solution concentration, which causes a decrease in $r_{cr}(T, S_w)$ as described by (Eqns. 9.7.9) and (9.7.10). Vice versa, a decrease of S_w is followed by the shrinking of a CCN, an increase in solution concentration, and growth of $r_{cr}(T, S_w)$. Thereby, the link is established between varying S_w (or s_w) and r_{cr} inside CCN. An ice germ embedded in a CCN drop and that is "remote" to the external supersaturation is nonetheless influenced by it very quickly via the adjustment of the drop radius to supersaturation. Note that the same mechanism of impact of s_w on "remote" r_{cr} in a solution drop acts in homogeneous freezing as described by the equations for r_{cr} derived in Chapter 8.

The characteristic freezing time τ_{fr} under conditions considered here is typically a few tens to a few hundred seconds (see Section 9.14); thus, $\tau_{fr} >> \tau_{eq}$, and $\tau_{fr} >> \tau_{dif}$, and this rapid adjustment of r_{cr} to varying supersaturation also occurs during CCN freezing. For drops of a few microns, the impact of supersaturation on r_{cr} and the freezing point depression should be smaller because of the smaller contribution of the term with S_w in the denominator of r_{cr}, see Section 9.7.

9.9. Surface Freezing and Melting

9.9.1. Surface Freezing

There are several possible geometrical configurations of an ice germ forming at the surface of a solution drop. In Section 9.7, we considered volume freezing, whereby an ice germ forms on the surface of an insoluble core located inside a solution drop. In this section, we consider an ice germ located at the interface between the humid air and solution drop, whereby a solid forms at the interface of its own melt and gas (vapor or humid air). This quasi-heterogeneous ice germ formation was briefly outlined by Defay et al (1966) and was referred to as "*pseudo-heterogeneous freezing*" (Tabazadeh et al., 2002a,b; Djikayev et al., 2002) and as "*quasi-heterogeneous freezing*" in Khvorostyanov and Curry (2004a).

Consider an ice germ represented by a spherical segment (lens or "tiny pancake ice") and bounded by two surfaces. An outer surface (in contact with air) has a radius r_{cr} and the inner surface (in contact with the solution drop) is determined by a droplet radius r_{dr}. A specific feature of surface freezing and melting of many substances is that the contact angle between a solid substance (e.g., ice) and its own melt (liquid water) is small but non-zero—that is, a solid surface is not completely wettable by its liquid. For example, the measured contact angles between ice and water were 0.5 to 2 degrees, and at the beginning of slow melting, melted water on the ice surface does not form the continuous film and does not spread infinitely over the surface, but rather forms many thin liquid droplet caps or lenses, and similar situations are observed for some metals and other substances (Ketcham and Hobbs, 1969; Knight, 1971; Hobbs, 1974; Elbaum and Schick, 1991; Elbaum, Lipson, and Dash, 1993; Dash et al., 1995; Laaksonen et al., 1995; Oxtoby, 2003; Rosenberg, 2005).

Thus, the case of surface freezing and melting represents a mixture of homogeneous and heterogeneous processes because it proceeds homogeneously without a foreign substance, and due to incomplete wettability the geometrical decrease of free energy is similar to those in heterogeneous freezing as described in Sections 9.2 and 9.7. These processes can be called "*quasi-heterogeneous freezing and melting*."

Because both the ice germ and solution drop are in mechanical equilibrium with the environmental air, it follows that the pressures inside the solution drop, p_s, and the ice crystal cap, p_i, with corresponding radii r_{dr} and r_{cr} are described by the Laplace equation:

$$dp_s = dp + d\left(\frac{2\sigma_{sa}}{r_{dr}}\right), \qquad dp_i = dp + d\left(\frac{2\sigma_{ia}}{r_{cr}}\right), \tag{9.9.1}$$

where σ_{ia} is the surface tension at the ice–air interface. For freezing, phase 1 is water, and phase 2 is ice. Substituting (Eqn. 9.9.1) into the entropy equation (Eqn. 3.6.4) with the same notations and assumptions as in Section 9.7 ($\rho_{k0,1} = \rho_w$, and $\rho_{k0,2} = \rho_i$, $v_w = M_w/\rho_w$, $v_i = M_w/\rho_i$, $\hat{L}_{12} = M_w L_m$, $a_{k,1} = a_w$, $a_{k,2} = a_i = 1$—i.e., $\ln(a_i) = 0$, but without the strain term ($\varepsilon = 0$) because we do not consider undissolved particulates here, and dividing by M_w, we obtain

$$-\frac{L_m}{T}dT + \left(\frac{1}{\rho_w} - \frac{1}{\rho_i}\right)dp + d\left(\frac{2\sigma_{sa}}{\rho_w r_{dr}}\right) - d\left(\frac{2\sigma_{ia}}{\rho_i r_{cr}}\right) + \frac{RT}{M_w}d\ln a_w = 0 \tag{9.9.2}$$

Integration of (Eqn. 9.9.2) with the same boundary conditions as in Section 9.7, $r_{cr} = r_{dr} = \infty$, $a_w = 1$, and $p = p_0$ at $T = T_0$ yields

$$L_m^{ef} \ln\frac{T_0}{T} = \frac{2\sigma_{ia}}{\rho_i r_{cr}} - \frac{2\sigma_{sa}}{\rho_w r_{dr}} + \left(\frac{1}{\rho_i} - \frac{1}{\rho_w}\right)\Delta p - \frac{RT}{M_w}\ln a_w \tag{9.9.3}$$

It is interesting to note that Dufour and Defay (1963) considered a different configuration, with the drop and crystal coexisting in air, and obtained a similar equation (except without the pressure term) with a more complicated expression for the solution term. One can see that (Eqn. 9.9.3) contains both positive and negative terms on the right-hand side. In particular, the terms with pressure and a_w are positive. This causes an increase of the term $\ln(T_0/T)$ on the right-hand side—i.e., lowering the freezing temperature T with increasing pressure and solution molality.

Equation (9.9.3) predicts situations with $T < T_0$, which is a physical state, and with $T > T_0$; that is, the existence of ice above the bulk triple point (the superheating of crystals), which is an unobserved state (Dufour and Defay, 1963). This problem was prescribed to limitations of the classical nucleation theory (Oxtoby, 2003), which makes predictions that are symmetric relative to T_0. This symmetry is absent in the density functional theory that avoids freezing point elevation (Oxtoby, 2003), although superheating above T_0 before melting is observed in some metals and other solids (e.g., Hobbs, 1974; Lifshitz and Pitaevskii, 1997). Now, replacing a_w with S_w, in (Eqn. 9.9.3), we obtain finally the critical radius r_{cr} of an ice germ at the solution–air interface:

$$r_{cr} = \frac{2\sigma_{ia}}{\rho_i L_m^{ef} \ln[(T_0/T)S_w^{G_n}\exp(-H_{s,fr})]}, \tag{9.9.4}$$

$$H_{s,fr} = \frac{\Delta\rho\Delta p}{\rho_i\rho_w L_m^{ef}} - \frac{2\sigma_{sa}}{r_{dr}\rho_w L_m^{ef}}. \tag{9.9.4a}$$

Note that (Eqn. 9.9.4) for surface freezing contains the surface tension σ_{ia} at the ice–air interface instead of σ_{is} for solution–ice as in Section 9.7 for volume freezing, which reflects the difference in freezing mechanisms.

When a solid is incompletely wettable by its own melt, the critical free energy of an ice germ nucleating at the liquid–air interface at surface quasi-heterogeneous freezing in the absence of the foreign substances can be written similar to (Eqns. 9.7.14) and (9.7.15). Using (Eqn. 9.9.4) for r_{cr}, the critical free energy becomes

$$\Delta F_{cr} = \frac{4}{3}\pi\sigma_{ia} r_{cr}^2 f(m_{is}, x) = \frac{16\pi}{3}\frac{\sigma_{ia}^3 f(m_{is}, x)}{\left\{\rho_i L_m^{ef}\ln\left[\frac{T_0}{T}S_w^G \exp\left(-H_{s,fr}\right)\right]\right\}^2}, \tag{9.9.5}$$

where $x = r_{dr}/r_{cr}$ for the cases of surface freezing, $m_{is} = cos\theta_{is}$ is the corresponding wettability parameter, θ_{is} is the contact angle at the interface of ice and its own liquid (the angle between the plane surface of the solution and the dome-like surface of the ice, ~0.5–2 degrees in the case of pure water according to Ketcham and Hobbs, 1969; Knight, 1971) and $f(m_{is}, x)$ is the geometric factor defined by (Eqn. 9.7.16). Since θ_{is} is small in this case, $f(m_{is}, x)$ is close to 1, making nucleation easier. However, if to assume that $\sigma_{ia} \approx \sigma_{iv}$, and recall that $\sigma_{iv} \sim 4\sigma_{is}$, one can see that $\Delta F_{cr} \sim \sigma_{iv}^3$ for surface freezing can be much greater than $\Delta F_{cr} \sim \sigma_{is}^3$ for volume freezing. Which kind of freezing, volume or surface, would dominate, is not clear due to uncertainty in the parameters θ_{is} and σ_{ia}. A solution of this question requires refined measurements of the parameters, comparison of the volume and surface freezing modes in isolation, and careful calculations.

9.9.2. Surface Melting

The numerous theories of surface melting include intergranular melting in polycrystalline crystals, melting at grain and vein junctions and evaluation of the lowering of the bulk triple points and melted quasi-liquid film thickness with an accounting for various mechanisms of premelting (the melting or formation of liquid and quasi-liquid layers on ice surface below 0 °C, which, in particular, explains why ice and snow are slippery and allow skating and skiing, and how this slipperness decreases and ends toward –40 °C), and the effects of crystal size and impurities (Defay et al., 1966; Hobbs, 1974; Dash, et al., 1995; Laaksonen et al., 1995; Fletcher, 1968; Ketcham and Hobbs, 1969; Knight, 1971; Elbaum and Schick, 1991; Elbaum et al., 1993; Johari, 1998; Wettlaufer, 1999; Wettlaufer and Dash, 2000; Oxtoby, 2003; Rempel et al., 2004; Rosenberg, 2005).

Here we consider one simple mechanism of surface melting: nucleation of a liquid drop at the crystal–gas interface when melting is incomplete—i.e., drops or liquid lenses occur instead of a liquid film. We consider a quasi-heterogeneous process of nucleation of a liquid germ at the surface of its own solid at small but finite contact angles. This type of surface melting has been observed in laboratory experiments (Hobbs, 1974; Dash, et al., 1995; Ketcham and Hobbs, 1969; Knight, 1971; Elbaum et al., 1993; Elbaum and Schick, 1991). This case is similar to surface freezing considered in the previous section except with reversed configuration of liquid and ice. To determine the droplet critical radius r_{dr} and the melting temperature T_m, we again begin with the entropy equation (Eqn. 3.6.4). For melting, phase 1 is ice, phase 2 is the liquid, and $-L_m = (h_{i0} - h_{w0})/M_w$. We consider a liquid germ that is located at the interface of the crystal and humid air and represents a spherical segment (lens) bounded by two surfaces. An outer surface (in contact with vapor or humid air) is a spherical segment with radius r_{dr} and the bottom surface (in contact with ice crystal) has a radius r_{cr}.

Using the conditions of mechanical equilibrium, we write for the pressures inside the liquid lens p_s and crystal p_i the Laplace equations:

$$dp_s = dp + d\left(\frac{2\sigma_{sa}}{r_{dr}}\right), \qquad dp_i = dp + d\left(\frac{2\sigma_{ia}}{r_{cr}}\right). \tag{9.9.6}$$

Substituting (Eqn. 9.9.6) into (Eqn. 3.6.4) and dividing by M_w, we obtain

$$\frac{L_m}{T}dT + \left(\frac{1}{\rho_i} - \frac{1}{\rho_w}\right)dp + d\left(\frac{2\sigma_{ia}}{\rho_i r_{cr}}\right) - d\left(\frac{2\sigma_{sa}}{\rho_w r_{dr}}\right) - \frac{RT}{M_w}d\ln a_w = 0. \tag{9.9.7}$$

Integration of (Eqn. 9.9.7) with the same boundary conditions $a_w = 1$, $r_{dr} = r_{cr} = \infty$, and $p = p_0$ at $T = T_0$, yields

$$L_m^{ef}\ln\frac{T_0}{T} = -\frac{2\sigma_{sa}}{\rho_w r_{dr}} + \frac{2\sigma_{ia}}{\rho_i r_{cr}} + \left(\frac{1}{\rho_i} - \frac{1}{\rho_w}\right)\Delta p - \frac{RT}{M_w}\ln a_w, \tag{9.9.8}$$

which formally coincides with (Eqn. 9.9.3), except that the unknown is here r_{dr}. The terms with pressure at $\Delta p > 0$ and the solution effects are positive, shifting the melting temperatures T_m below the bulk triple point as in the case with homogeneous freezing in Chapter 8.

At negative pressure ($\Delta p < 0$), e.g., in the case of internal melting of ice caused by strong radiative heating and the formation of "*Tyndall flowers*," when ice may exist in metastable equilibrium at up to +8 °C and $\Delta p \sim -10^3$ bar (Hobbs, 1974), (Eqn. 9.9.8) describes the competition of the negative pressure, which causes elevation of T_m (possibly to well above 0 °C) and the counteracting solution effects ($a_w < 1$) that tend to lower T_m. Again replacing a_w with S_w, we find from (Eqn. 9.9.8) the critical radius r_{dr} of a liquid germ:

$$r_{dr} = \frac{2\sigma_{sa}}{-\rho_w L_m^{ef} \ln[(T_0/T) S_w^{G_n} \exp(-H_m)]}, \tag{9.9.9}$$

$$H_m = \frac{\Delta\rho\Delta p}{\rho_w \rho_i L_m^{ef}} + \frac{2\sigma_{ia}}{r_{cr}\rho_i L_m^{ef}}. \tag{9.9.10}$$

Equation (9.9.9) differs from (Eqn. 9.9.4) for freezing by the opposite sign of the temperature term $\ln(T_0/T)$.

For physical analysis, (Eqn. 9.9.9) can be rewritten as

$$r_{dr} = \frac{2\sigma_{sa}}{\rho_w L_m^{ef}[-\ln(T_0/T_m) + H_m - \ln S_w^{G_n}]} \tag{9.9.10a}$$

The physical condition $r_{dr} > 0$ requires that the denominator should be positive, which yields

$$\ln(T_0/T_m) \le H_m - G_n \ln S_w. \tag{9.9.10b}$$

For the bulk ice ($r_{cr} = \infty$) and the absence of external pressure, $\Delta p = 0$, we have $H_m = 0$. If a cap of the liquid is pure water, then $S_w \approx a_w = 1$. Thus, $\ln(S_w) = 0$ and (Eqn. 9.9.10) reduces to

$$\ln(T_0/T_m) \le 0, \qquad \text{or} \qquad T_m \ge T_0. \tag{9.9.10c}$$

That is, melting at these conditions may occur only above the triple point T_0. Presence of the external pressure Δp and the final radius of a melting crystal r_{cr} cause a decrease in the melting point. If solute penetrates into melting liquid from the environment, then $S_w < 1$, and this causes an additional depression of the melting point, so that the relations can be reached $T_m < T_0$, $\ln(T_0/T_m) > 0$, and melting may occur below the triple point. Equations (9.9.9)–(9.9.10b) allow us to study the effects of each of these factors (Δp, r_{cr}, S_w) separately. The effects of variations of the saturation ratio or water activity on the melting point depression are described in more detail in Section 9.11 and 9.12.

The critical free energy for this type of surface melting can be written as for the solid/air interface and using (Eqn. 9.9.9) for r_{dr} as

$$\Delta F_{cr} = \frac{4}{3}\pi\sigma_{sa} r_{dr}^2 f(m_{is}, x) = \frac{16\pi}{3} \frac{\sigma_{sa}^3 f(m_{is}, x)}{\left\{-\rho_w L_m^{ef} \ln\left[\frac{T_0}{T} S_w^G \exp(-H_m)\right]\right\}^2}, \tag{9.9.11}$$

where $x = r_{cr}/r_{dr}$ for surface melting, θ_{is} is the contact angle at the interface of ice and its own liquid, $m_{is} = cos\theta_{is}$ is the wettability parameter and $f(m_{is})$ is the geometric factor defined by (Eqn. 9.7.16), but with these m_{is} and x specific for surface melting. Because the contact angle for surface melting is small, the geometric factor $f(m_{is}, x)$ is close to unity, which may favor quasi-heterogeneous surface melting.

9.10. Nucleation in a Polydisperse Aerosol

We have considered so far single aerosol particles characterized by the radius of an insoluble substance r_N. When we consider heterogeneous freezing of the size spectra, there can be two different situations at water subsaturation, $s_w < 0$, and water supersaturation, $s_w > 0$. As described in Chapters 6 and 7, the size spectrum of the original dry CCN, $f_a(r_a)$, transforms into the spectrum of the wet aerosol above the threshold of deliquescence of the soluble fraction and $s_w < 0$. These CCN usually also contain an insoluble fraction and may serve as IN. The freezing of these wet haze particles serving as IN may occur at water subsaturation, $s_w < 0$, in the deliquescence-freezing (DF) mode, as occurs at the formation of cirrus clouds, diamond dust, and similar cloudless conditions at $s_w < 0$ when mostly only submicron wet haze particles produce crystals.

When the relative humidity increases further and reaches 100%, some of the CCN are activated and cloud drops form. Then, the aerosol spectrum splits into 2 fractions: a) the interstitial aerosol of unactivated particles with the size spectrum $f_{a,int}(r_a)$, which governs freezing in the DF mode, and b) aerosol immersed in the activated cloud drops with the size spectrum $f_{a,im}(r_a)$, which governs freezing in immersion mode. These two fractions are separated by the *boundary radius* r_b. This radius is of key importance for ice nucleation since it determines aerosol concentration in each fraction and therefore the efficiency of each mode. At $s_w > 0$, the boundary radius can be expressed via the water supersaturation s_w as $r_b = 2A_k/3s_w$ (see Chapter 6). Thus, $s_w(t)$ defines the splitting of the original dry CCN-IN size spectrum into interstitial and immersed fractions. An important feature of CCN at $s_w > 0$ considered in Chapters 6 and 7 is that the equilibrium deliquescent interstitial CCN may coexist with cloud drops. At supersaturation $s_w > 0$, the droplets with radii $r_d > r_b$ are activated cloud drops, and those with $r_d < r_b$ are non-activated CCN in equilibrium with the supersaturated environment. Since $A_K \sim 10^{-3}\,\mu\text{m}$ at the considered temperatures, the values of r_b are $r_b \sim 1\,\mu\text{m}$, $0.1\,\mu\text{m}$, and $0.01\,\mu\text{m}$ at $s_w = 0.001$, 0.01, and 0.1 (0.1, 1, and 10%, respectively). The value of r_b determines the well-known gap between the CCN and the cloud drop size spectra (see Fig. 2.5 here). Further, we consider heterogeneous ice nucleation in both situations: at water subsaturation with nucleation of haze particles in the DF mode, and at water supersaturation when the CCN-IN spectrum is split into two fractions, interstiatial and immersed in the drops.

9.10.1. Freezing of Haze Particles at Water Subsaturation in the DF Mode

The probability of freezing of an aerosol particle containing an insoluble part with r_N is

$$P_{fr}(r_N, t) = 1 - \exp\left(-\int_0^t J_{f,het}(r_N, t')\,dt'\right), \tag{9.10.1}$$

where $J_{f,het}$ is defined in (Eqn. 9.6.7), with ΔF_{cr} defined in (Eqns. 9.7.15) or (9.7.24) for volume freezing. The number of particles dN_{df} nucleated from CCN by the deliquescence-freezing mode in the range $(r_N, r_N + dr_N)$ is

$$dN_{df}(r_N, t) = P_{fr}(r_N, t) f_a(r_N) dr_N, \tag{9.10.2}$$

and the total number of nucleated particles is obtained by integrating over the aerosol size spectrum:

$$N_{fr}(t) = \int_{r_{min}}^{r_{max}} dN_{df}(r_N, t) = \int_{r_{min}}^{r_{max}} P_{fr}(r_N, t) f_a(r_N) dr_N. \tag{9.10.3}$$

The average probability of freezing $P_{fr,p}$ in a polydisperse aerosol is then

$$P_{fr,p}(t) = \frac{N_{fr}(t)}{N_a} = \frac{1}{N_a} \int_{r_{min}}^{r_{max}} P_{fr}(r_N, t) f_a(r_N) dr_N. \tag{9.10.4}$$

The heterogeneous crystal nucleation rate ($cm^{-3}\,s^{-1}$) in a polydisperse aerosol can be calculated as:

$$R_{f,het} = \frac{dN_{fr}}{dt} = \int_{r_{min}}^{r_{max}} dr_N f_a(r_N) J_{f,het}(t) \exp\left(-\int_0^t J_{f,het}(t')dt'\right). \tag{9.10.5}$$

The preceding equations include the radius r_N of the insoluble fractions but not the radius of the haze particles r_a. The reason for this is that the freezing rate in (Eqn. 9.10.1) depends on the surface area $\sim r_N^2$ of the insoluble inclusion inside a partially soluble CCN with r_N, but not on the surface $\sim r_d^2$ of the entire CCN particle. The radius r_a increases during deliquescence of the CCN, and, as described in Chapter 6, can be expressed in equilibrium via the dry radius r_{d0} and some function of water saturation ratio $\psi(S_w)$ as $r_a(S_w) = r_{d0}\psi(S_w)$. So, the CCN hygroscopic growth can be described by S_w; however, the effect of S_w is already accounted for in expressions for r_{cr} and ΔF_{cr} given in Section 9.7. Unfortunately, not much is known about the size distributions and properties of insoluble substance in CCN. We can estimate $f_a(r_N)$ using data on the mass soluble fraction of CCN ε_N, which varies over the range 0.1–1.0 with an average value of 0.5 (Pruppacher and Klett, 1997, Chapter 8). As mentioned earlier, the radii r_{d0} and r_N are related via the insoluble fraction ε_N as $r_N \sim r_{d0}\varepsilon_N^{1/3}$, so that r_N and r_{d0} are quite comparable for the mean $\varepsilon_N = 0.5$ and especially for the high $\varepsilon_N = 0.9$ insoluble fractions: $r_N \sim 0.8r_{d0}$ for $\varepsilon_N = 0.5$ and $r_N \sim 0.97r_{d0}$ for $\varepsilon_N = 0.9$. This means that the modal radii and dispersions of the size spectra of the insoluble fraction $f_a(r_N)$ are rather close to the aerosol spectra. This fact is used later for averaging over r_N with $f_a(r_N)$.

Fig. 9.8 shows the subintegral function ($cm^{-3}\,s^{-1}$) for the freezing nucleation rate R_{fr} in (Eqn. 9.10.5) for various combinations of temperature and humidity. This figure illustrates that the main contribution to the freezing rate comes from the submicron aerosol mode, around 0.1–0.5 μm, and contributions from nucleation and giant modes are smaller. This finding is in agreement with experimental data on the radii of IN in the central particles of snow crystals that were typically in the range 0.05–7.5 μm with a mode r_m between 0.25 and 2.5 μm (Pruppacher and Klett, 1997), recent measurements of IN in the Arctic that found modal radii r_m of 0.18–0.23 μm (Rogers et al., 2001), and measurements in the central U.S., $r_m \approx 0.2$ μm (Chen et al., 1998; Rogers et al., 1998). We considered here the simplest case of the monomodal aerosol size spectrum and did not include the larger size fractions, which also may contribute to the freezing rate. A generalization for several

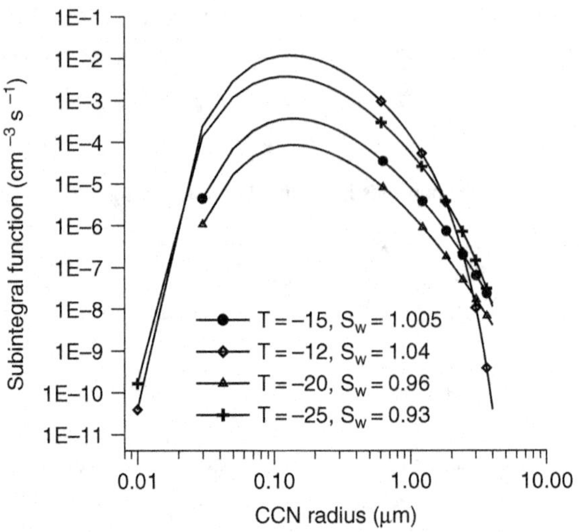

Figure 9.8. Subintegral function ($cm^{-3} s^{-1}$) for the freezing nucleation rate R_{fr} which shows the contributions of the various aerosol radii into the freezing rate for four cases: $T = -15\,°C$, $S_w = 1.005$ (circles); $T = -12\,°C$, $S_w = 1.04$ (diamonds); $T = -20\,°C$, $S_w = 0.96$ (triangles); $T = -25\,°C$, $S_w = 0.93$ (crosses). From Khvorostyanov and Curry (2004b), *J. Atmos. Sci.*, 61, © American Meteorological Society. Used with permission.

aerosol fractions can be easily made through summation over several fractions as shown later in the text. The addition of larger aerosol particles along with the variation of insoluble fractions over the size spectrum should affect heterogeneous freezing, since the insoluble fraction of CCN increases with size faster than the soluble fraction (e.g., Laktionov, 1972; Pruppacher and Klett, 1997), thereby increasing the catalyzing properties of large CCN in ice nucleation (e.g., Berezinsky and Stepanov, 1986).

The aerosol size $r_{N,max}(T,S_w)$ with maximum contribution to the nucleation rate, illustrated in Fig. 9.8, can be estimated analytically using this theory. First, we perform some simplifications. Since typical values of r_{cr} are 10^{-3}–$10^{-2}\,\mu m$, for most aerosol particles with $r_N \geq 0.1\,\mu m$, the parameter $x = r_N/r_{cr} >> 1$. Then the geometrical factor $f(m_{is},x)$ (Eqn. 9.7.16) equals its asymptotic value $f(m_{is},x) \approx (m_{is}^3 - 3m_{is} + 2)/4$ at $x \to \infty$. It does not depend on r_N; the nucleation rate per unit area, $J_{f,het}/r_N^2 \equiv y(T)$, also does not depend on r_N as shown in (Eqns. 9.6.7) and (9.6.8). Further, we assume that the aerosol is monodisperse with radius r_N. Then, the nucleation rate R_{fr} in (Eqn. 9.10.5) can be presented as $R_{fr} = y(T)\, r_N^2 \exp(-y(T)\, r_N^2 \Delta t)$ where Δt is the time interval. This is a Gaussian distribution relative to r_N, and a gamma distribution relative to $y(T)$. Maxima are determined from the conditions $dR_{fr}/dr_N = 0$ for r_N and $dR_{fr}/dT = 0$ for temperature. Both conditions lead to the equation

$$y(T,S_w) r_{N\max}^2 \Delta t = 1. \quad (9.10.6)$$

Thus, at some fixed T, we can calculate the radius of aerosol particles with the maximum freezing rate

$$r_{N\max}(T,S_w) = [y(T,S_w)\Delta t]^{-1/2} = r_N (J_{f,het}\Delta t)^{-1/2}. \quad (9.10.7)$$

This equation shows that ice nucleation initiates from the largest aerosol particles for small nucleation times Δt, and then the aerosol radius with maximum nucleation rate decreases with time. For example, for $y(T) \sim 10^8\,\text{cm}^{-2}\,\text{s}^{-1}$ (see Fig. 9.7), we have from (Eqn. 9.10.7) $r_N \sim 10\,\mu\text{m}$ at $\Delta t = 0.01$ s and $r_N \sim 0.1\,\mu\text{m}$ at $\Delta t = 100$ s.

A general picture of the aerosol radius $r_{N,max}(T,S_w)$ with maximum contribution to the freezing rate calculated with (Eqn. 9.10.7) and the same parameters as for $J_{f,het}$ in Fig. 9.7 is shown as a T-S_w diagram in Fig. 9.9. Aerosols in the size range r_N from 0.1 to a few tens of μm can preferably nucleate ice crystals in the rather narrow range of T, S_w. The dominant aerosol radius decreases rapidly at fixed T and increasing S_w, and more slowly at fixed S_w and decreasing T. This behavior predicted by (Eqn. 9.10.7) and illustrated in Fig. 9.9 is quite reasonable (the colder and more humid it is, the smaller the particles that are nucleated), and is in agreement with chamber experiments (e.g., Berezinsky and Stepanov, 1986; see Section 9.4 here), and may help identify experimentally the aerosol fractions with a maximum nucleation rate under various conditions.

If the aerosol size spectrum consists of several size fractions that also can include particles of different composition, then the concentration of nucleated crystals can be evaluated by summation over all fractions,

$$N_{fr}(t) = \sum_i \int_{r_{\min,i}}^{r_{\max,i}} dN_{df,i}(r_{Ni},t) = \sum_i \int_{r_{\min,i}}^{r_{\max,i}} P_{fr,i}(r_{Ni},t) f_{ai}(r_{Ni}) dr_{Ni}, \tag{9.10.8}$$

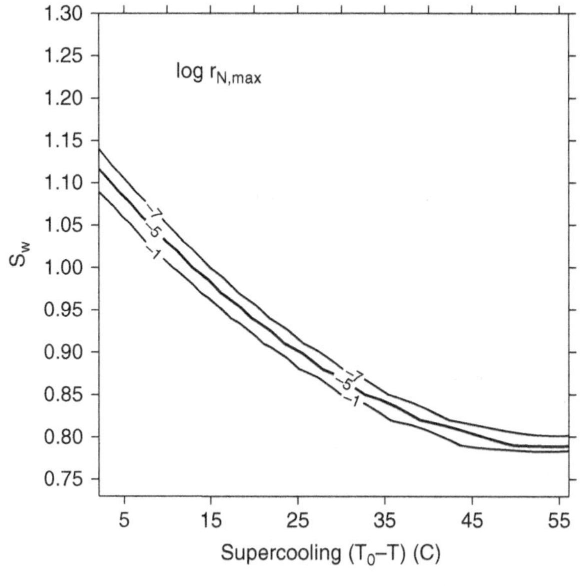

Figure 9.9. Logarithm of the radius $r_{N,max}$ (cm) of the insoluble aerosol fraction with the maximum heterogeneous freezing nucleation rate calculated with (Eqn. 9.10.7); log = −1 means $r_N = 10^{-1}$ cm or 1000 μm, log = −5 means $r_N = 10^{-5}$ cm or 0.1 μm (bold line), and log = −7 means $r_N = 10^{-7}$ cm. Aerosols in the size spectrum $r_N = 0.01$–100 μm can preferably nucleate ice crystals in the narrow corridor of T, S_w. From Khvorostyanov and Curry (2004b), *J. Atmos. Sci.*, 61, © American Meteorological Society. Used with permission.

where i denotes the fraction. The probability of freezing in a polydisperse aerosol is then

$$P_{fr,p}(t) = \frac{\sum_i N_{fr,i}(t)}{\sum_i N_{ai}} = \frac{1}{\sum_i N_{ai}} \sum_i \int_{r_{\min,i}}^{r_{\max,i}} P_{fr,i}(r_{Ni},t) f_{ai}(r_{Ni})\, dr_{Ni}. \tag{9.10.9}$$

The heterogeneous crystal nucleation rate ($cm^{-3}\,s^{-1}$) in a polydisperse aerosol with several fractions can be calculated as a sum over all fractions:

$$R_{f,het} = \sum_i \frac{dN_{fr,i}}{dt} = \sum_i \int_{r_{\min,i}}^{r_{\max,i}} dr_{Ni} f_{ai}(r_{Ni}) J_{fi,het}(t) \exp\left(-\int_0^t J_{fi,het}(t')dt'\right). \tag{9.10.10}$$

The values of $J_{fi,het}$ are usually small, as will be shown in Section 9.14, and the expressions for P_{fi}, N_{fr}, and $R_{f,het}$ can be simplified as shown in Section 9.14.3.

9.10.2. Simultaneous Freezing in the DF and Immersion Modes at Water Supersaturation

As described in Section 9.10.1, at $s_w > 0$, the aerosol spectrum splits into two fractions: the interstitial aerosol of unactivated particles with the size spectrum $f_{a,int}(r_a)$, and aerosol immersed in the activated cloud drops with the size spectrum $f_{a,im}(r_a)$. These two fractions govern freezing in the DF mode and in the immersion mode respectively. The corresponding concentrations of the interstitial and immersed in drops aerosol, $N_{a,int}$ and $N_{a,imm}$, can be evaluated by integration over the aerosol size spectrum $f_a(r_a)$, as

$$N_{a,\text{int}} = \int_{r_{\min}}^{r_b[s_w(t)]} f_a(r_a)\, dr_a, \qquad N_{a,imm} = \int_{r_b[s_w(t)]}^{r_{\max}} f_a(r_a)\, dr_a, \tag{9.10.11}$$

where r_{min} and r_{max} are the lower and upper boundaries of a given aerosol fraction, and r_b is the boundary radius between the fraction, $r_b = 2A_k/3s_w(t)$, which can change with time t along with evolving supersaturation. All the other relevant integrals like surface areas, probabilities of freezing, etc., can be evaluated in a similar way using the boundary $r_b(s_w)$. An important feature of this spectral splitting is that it depends on the history of a cloud parcel via $s_w(t)$. This means that different vertical velocities in the parcel, or more generally, different advective, convective, and radiative cooling rates in the models, or in the lab experiments, will lead to the different redistribution of the initial (input) dry aerosol between the interstitial and immersion modes. An accurate calculation would require introducing the distribution functions for aerosol $f_a(r_a, r_N)$ and for drops $f_d(r_d, r_N)$ and solving equations for them with tracking the amount of the insoluble fraction (r_N) in each haze particle and each drop. However, this approach would be too time-consuming for the present-day models. Thus, some simplifications and additional hypotheses are required. One reasonable hypothesis is: an assumption on the random distribution of the insoluble fraction (or r_N) inside haze particles and drops; an assumption discussed earlier that $f_a(r_a) \approx f_a(r_N)$, and introducing some average probabilities of freezing.

Then, the concentrations of haze particles N_{df} freezing in deliquescence-freezing (DF) mode and concentrations of drops N_{fimm} freezing in immersion mode can be calculated by integration over the corresponding ranges of radii. The number of particles dN_{df} nucleated in DF mode in the range of the aerosol radii $(r_a, r_a + dr_a)$ is

$$dN_{df}(r_a, r_N, t) = P_{fr}(r_a, r_N, t) f_a(r_a)\, dr_a, \qquad r_a < r_b. \tag{9.10.12}$$

The total number of particles nucleated in DF mode is obtained by integrating over the aerosol size spectrum from r_{min} to $r_b[s_w(t)]$:

$$N_{df}(t) = \int_{r_{\min}}^{r_b[s_w(t)]} dN_{df}(r_d, r_N, t) = \int_{r_{\min}}^{r_b[s_w(t)]} P_{fr}(r_a, r_N, t) f_a(r_a)\, dr_a. \qquad (9.10.13)$$

The average probability of freezing $\bar{P}_{df}$ of a polydisperse aerosol in DF mode is then

$$\bar{P}_{df}(t) = \frac{N_{df}(t)}{N_{a,\text{int}}} = \frac{1}{N_{a,\text{int}}} \int_{r_{\min}}^{r_b[(s_w(t)]} P_{fr}(r_a, r_N, t) f_a(r_a)\, dr_a, \qquad (9.10.14)$$

where $N_{a,int}$ is defined in (Eqn. 9.10.11). The heterogeneous crystal nucleation rate (cm^{-3} s^{-1}) in a polydisperse aerosol in DF mode can be calculated as:

$$R_{df} = \frac{dN_{df}}{dt} = \int_{r_{\min}}^{r_b[(s_w(t)]} dr_a f_a(r_a) J_{f,het}(r_N, t) \exp\left(-\int_0^t J_{f,het}(r_N, t')dt'\right). \qquad (9.10.15)$$

Similar relations could be written for the immersion mode, and the concentrations of the drops freezing in the range of drop radii (r_d, r_d+dr_d) could be defined as

$$dN_{f,imm}(r_d, r_N, t) = P_{fr}(r_d, r_N, t) f(r_d) dr_d, \qquad r_d > r_b. \qquad (9.10.16)$$

However, the problem becomes more complicated in this case since it is difficult to calculate the distribution of the aerosol particles inside the drops and to determine which r_N corresponds to which r_d. Thus, as discussed earlier, a reasonable approximation is to use the average over the aerosol spectrum the probability of freezing for the drops $\bar{P}_{imm}(t)$, defined similarly to (Eqn. 9.10.14):

$$\bar{P}_{imm}(t) = \frac{1}{N_{a,imm}} \int_{r_b[(s_w(t)]}^{r_{\max}} P_{fr}(r_N, t) f_a(r_a)\, dr_a, \qquad (9.10.17)$$

where $N_{a,imm}$ is defined in (Eqn. 9.10.11). The concentration of drops $dN_{f,imm}$ nucleated in the immersion mode in the range of radii (r_d, r_d +dr_d) is

$$dN_{f,imm}(r,t) = \bar{P}_{imm}(t) f(r_d) dr_d. \qquad (9.10.18)$$

The total amount of drops freezing in immersion mode is

$$N_{f,imm} = \int_{r_b[(s_w(t)]}^{r_{\max}} dN_{f,imm}(r,t) = \bar{P}_{imm}(t) \int_{r_b[(s_w(t)]}^{r_{\max}} f(r_d)\, dr_d = \bar{P}_{imm}(t) N_d. \qquad (9.10.19)$$

These equations can be used in the models to evaluate ice nucleation when the deliquescence-freezing and immersion modes act simultaneously. The parcel model simulations similar to those described in Section 9.14 showed a reasonable performance of this scheme and comparable contributions from the DF and immersion modes, although ice nucleation in the DF mode begins earlier still at subsaturation.

There is an important difference in calculations of the probabilities of freezing P_{fr} for interstitial aerosol and for drops in the DF and immersion modes at cloud evaporation or glaciation. When a cloud contains drops, the water supersaturation is usually very close to zero, $s_w \approx 0$. Therefore, the water activity in the solutions inside aerosol particles and inside drops is close to 1, $a_w \approx 1$. Thus, the nucleation rate $J_{f,het}$ defined in (Eqn. 9.6.7) should be calculated for both modes with the critical radius r_{cr} (Eqns. 9.7.5a) and (9.7.5b) and with ΔF_{cr} (Eqns. 9.7.15) or (9.7.24) at $s_w = 0$ ($S_w = 1$). At the stage of evaporation or intensive glaciation, when supersaturation in the cloud becomes negative, the fraction of interstitial aerosol rapidly adjusts to $s_w < 0$, the haze particles partially evaporate, their radii decrease, the water activity of the solution decreases, and the freezing point depression increases. This will hamper and eventually stop the freezing of aerosol in the DF mode. These effects are automatically accounted for by using (Eqns. 9.7.5a) and (9.7.5b) for r_{cr}, and (Eqns. 9.7.15) or (9.7.24) for ΔF_{cr} via $s_w < 0$ ($S_w < 1$).

Activated drops are very diluted and the water activity in the solution inside drops $a_w \approx 1$ even at the stage of evaporation at $s_w < 0$—i.e., $a_w \neq s_w$. Therefore, the calculation of r_{cr}, ΔF_{cr}, $J_{f,het}$, and P_{fr} for cloud drops should be performed using the corresponding equations derived in Section 9.7.2 from the the general equilibrium (Eqn. 3.6.4) for larger cloud drops (Eqns. 9.7.6e)–(9.7.6g). The reason for this is that, in contrast to unactivated CCN, activated drops do not obey Köhler's equation as unactivated solution drops but are close to Kelvin's equilibrium. Therefore the expressions derived for r_{cr}, ΔF_{cr} for solution drops and using water activity a_w are not applicable to activated drops. An attempt to modify the classical equations for r_{cr} valid for haze drops by introducing water activity $a_w < 0$ in cloud drops may lead to confusing results, as will be shown in Section 9.16. When a cloud evaporates or crystallizes at $s_w < 0$, the smallest drops evaporate first. Thus, the effects of subsaturation and the evaporation on drop freezing can be accounted by using r_{cr} for activated cloud drops as described in Section 9.7.6 and by changing the lower limit of integration in (Eqn. 9.10.19) from $r_b[s_w(t)]$ to the lower limit of the droplet size spectrum.

9.11. Critical Freezing and Melting Temperatures

9.11.1. General Equations

The critical or threshold temperatures and saturation ratios of heterogeneous freezing depending on the nucleation rates can be derived analogously to the derivation in Sections 8.7 and 8.8 for homogeneous nucleation. The critical freezing or melting temperatures depend on the mechanism of freezing or melting. The derivation in the following text follows Khvorostyanov and Curry (2004a,b, 2009b).

Volume Heterogeneous Freezing

For volume heterogeneous freezing of an ice crystal, the heterogeneous nucleation rate is described by (Eqn. 9.6.7). Solving equation (Eqn. 9.6.7) for $J_{f,het}$ relative to the critical energy ΔF_{cr}, we obtain

$$\Delta F_{cr} = -kT \ln \frac{J_{f,het}}{C_{het}} - \Delta F_{act}, \tag{9.11.1}$$

where the normalizing factor C_{het} for heterogeneous freezing on the surface of an insoluble core with radius r_N was defined in (Eqn. 9.6.8). The expression for ΔF_{cr} is given by (Eqns. 9.7.21) and (9.7.21b)

$$\Delta F_{cr} = \frac{(16\pi/3)\sigma_{is}^3 f(m_{is}, x)}{\left\{\rho_i L_m^{ef} \ln\left[\frac{T_0}{T} S_w^{G_n} \exp(-H_{v,fr})\right]\right\}^2} - \alpha r_N^2 \sigma_{is}(1 - m_{is}), \tag{9.11.2}$$

where $G_n = RT/(M_w L_m^{ef})$ and $H_{v,fr}$ was defined in (Eqn. 9.7.6). Equating (Eqns. 9.11.1) and (9.11.2), we obtain

$$\ln\left(\frac{T_0}{T} S_w^{G_n}\right) = H_{f,het} + H_{v,fr}, \tag{9.11.3}$$

where $H_{f,het}$ is

$$H_{f,het} = \frac{1}{\rho_i L_m^{ef}} \left[\frac{(-16\pi/3) f(m_{is}, x) \sigma_{is}^3}{kT \ln(J_{f,het}/C_{het}) + \Delta F_{act} - \alpha r_N^2 \sigma_{is}(1 - m_{is})} \right]^{1/2}. \tag{9.11.4}$$

Equation (9.11.3) is a key relation for the threshold T and S_w, that allows us to express the critical or threshold temperature of heterogeneous freezing $T_{f,het}$ as a function of the saturation ratio, which is done in this section, or, vice versa, to express the critical saturation ratio $S_{w,cr}$ as a function of the temperature, which is done in the next Section 9.12. Solving (Eqn. 9.11.3) relative to T, we obtain the critical or threshold temperature of volume heterogeneous freezing $T_{f,het}$:

$$T_{f,het} = T_0 S_w^{G_n} \exp(-H_{f,het} - H_{v,fr})]. \tag{9.11.5}$$

In expanded form, it becomes

$$T_{f,het}(J_{het}, S_w, r_{as}, \Delta p) = T_0 S_w^{G_n} \exp\left(-\frac{\Delta\rho\Delta p}{\rho_w \rho_i L_m^{ef}} - \frac{r_{fr}}{r_d} - \frac{C_\varepsilon \varepsilon^2}{\rho_i L_m^{ef}} \right)$$

$$\times \exp\left\{ -\frac{1}{L_m^{ef} \rho_i} \left[\frac{(-16\pi/3)\sigma_{is}^3 f(m_{is}, x)}{kT \ln(J_{f,het}/C_{het}) + \Delta F_{act} - \alpha r_N^2 \sigma_{is}(1 - m_{is})} \right]^{1/2} \right\}, \tag{9.11.6}$$

where, as in the case with homogeneous freezing in Chapter 8, the freezing curvature parameter $r_{fr} = 2\sigma_{sa}/\rho_i L_m^{ef}$, and r_d is the radius of the freezing particle. The parameters L_m^{ef}, ρ_i, C_{het}, σ_{is}, and ΔF_{act} are functions of temperature and should be evaluated at the same values of $T_{f,het}$ as on the left-hand side of (Eqn. 9.11.6). The solution to (Eqn. 9.11.6) requires an iteration procedure, but iterations converge rapidly, and one to two iterations are sufficient, as was illustrated in Section 8.7 for $T_{f,hom}$.

Fig. 9.10 provides an example of the heterogeneous freezing temperature $T_{f,het}(S_w)$ or $T_{f,het}(a_w)$ calculated with (Eqn. 9.11.6) for two nucleation rates, $J_{f,het} = 10^{-6}\,\mathrm{s}^{-1}$ and $1\,\mathrm{s}^{-1}$, and two radii of freezing droplets, $r_d = 5$ and $0.02\,\mu\mathrm{m}$. The four theoretical curves for $T_{f,het}(S_w)$ are located above the homogeneous freezing temperature $T_{f,hom}(S_w)$ due to the catalyzing action of the insoluble substrate, but below the melting curve T_m since heterogeneous freezing also requires some degree of supercooling. An increase in the nucleation rate $J_{f,het}$ by 6 orders of magnitude, from 10^{-6} to $1\,\mathrm{s}^{-1}$, relatively weakly influences $T_{f,het}(S_w)$, by 1–3 K in this example. A decrease of the drop size from $5\,\mu\mathrm{m}$ to $0.02\,\mu\mathrm{m}$ causes the shift of ΔT to colder freezing temperatures by about 10 K around $S_w \sim 1$ (weak solutions) and ΔT increases by 15–20 K in concentrated solutions at $S_w < 0.8$ and low $T < 220$ K.

The experimental points from different sources given here for comparison exhibit a wide scatter of the freezing temperatures for various substances and conditions of experiments performed with the drops of different sizes and cooling rates. A comparison of the experimental and theoretical results shows their general similarity and qualitative agreement. This indicates that classical nucleation

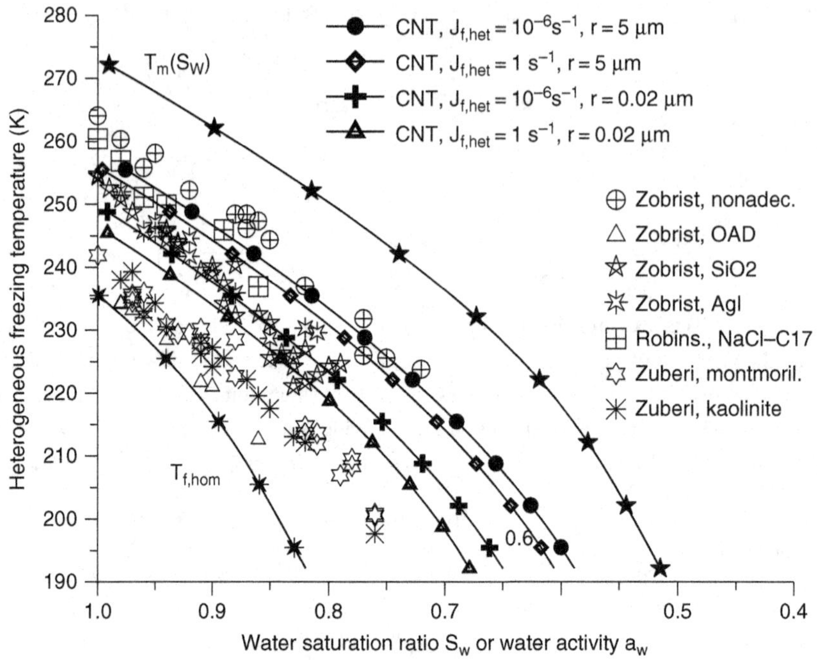

Figure 9.10. Heterogeneous freezing temperature $T_{f,het}(S_w)$ or $T_{f,het}(a_w)$ calculated with (Eqn. 9.11.6) based on the extended classical nucleation theory (CNT) with two nucleation rates, $J_{f,het} = 10^{-6}\,s^{-1}$, $J_{f,het} = 1\,s^{-1}$, and two radii of freezing droplets, $r_d = 5$ and 0.02 μm. The melting $T_m(S_w)$ and homogeneous freezing $T_{f,hom}(S_w)$ curves are given for comparison. The experimental data sets for the heterogeneous freezing data on $T_{f,het}(a_w)$ for various substances and authors: from Zuberi et al. (2002, montmorrilonite and kaolinite), Zobrist (2006, nonadecanol, oxalic acid dihydrate (OAD), SiO_2, and AgI), and from Cantrell and Robinson (2006, Robins., NaCl-C17).

theory allows us to describe the general features of heterogeneous freezing curves, $T_{f,het}(S_w)$. Although this comparison is shown just for illustration, and no special tuning of the parameters was done to reproduce any specific experimental results, the general agreement indicates that with the appropriate choice of the nucleation (cooling) rate, droplet size, activation energy, surface tension, and other parameters specific for a given substance, a better quantitative description can be reached.

For very slow processes, when $J_{f,het} \to 0$, and $\ln J_{f,het} \to -\infty$, and neglecting the term with r_{fr}/r_d, (Eqn. 9.11.6) is simplified and we obtain the threshold freezing temperature:

$$T_{f,het} = T_0 S_w^{G_n} \exp\left(-\frac{C_\varepsilon \varepsilon^2}{\rho_i L_m^{ef}(T_{f,het})} \right), \tag{9.11.7}$$

which for very slow processes tends toward the melting temperature $T_m(S_w)$ with correction for the misfit strain. Formally, this is a nonlinear equation for $T_{f,het}$. However, the temperature dependencies of $G_n(T)$ and $L_m^{ef}(T)$ are relatively weak in the region of interest ($G_n \approx 0.4$ at $T = -10\,°C$, see

Fig. 8.14c), and can be accounted for by iteration or neglected with a small error. It is convenient to rewrite (Eqn. 9.11.7) using the relation, $S_w = 1 + s_w$:

$$T_{f,het} = T_0(1+s_w)^{G_n} \exp\left(-\frac{C_\varepsilon \varepsilon^2}{\rho_i L_m^{ef}(T_{f,het})}\right). \tag{9.11.8}$$

The parameter $C\varepsilon^2/\rho_i L_m^{ef} << 1$ for $\varepsilon < 1$–2%, and at small $s_w << 1$, we can expand (Eqn. 9.11.8) into the Taylor series by s_w and by $C_\varepsilon \varepsilon^2/\rho_i L_m^{ef}$. Keeping only the first two terms in s_w and using the definition $G_n = RT/(M_w L_m^{ef})$, we have an expansion for $T_{f,het}$ similar to that obtained in Section 8.7 for the melting temperature, but with correction for the misfit strain effects:

$$T_{f,het} \approx [T_0 + G_n T_0 s_w + (1/2)G_n(G_n - 1)s_w^2](1 - C_\varepsilon \varepsilon^2/\rho_i L_m^{ef}) = T_0 + \Delta T_f, \tag{9.11.9}$$

$$\Delta T_f = \Delta T_{f1} + \Delta T_{f2} + \Delta T_\varepsilon, \tag{9.11.10}$$

$$\Delta T_{f1} = \frac{R_v T_0^2}{L_m^{ef}} s_w, \quad \Delta T_{f2} = \frac{R_v T_0^2}{2L_m^{ef}}\left(\frac{R_v T_0}{L_m^{ef}} - 1\right)s_w^2, \quad \Delta T_\varepsilon = -T_0 \frac{C_\varepsilon \varepsilon^2}{\rho_i L_m^{ef}}, \tag{9.11.11}$$

where R_v is the specific gas constant for water vapor. Equations (9.11.9)–(9.11.11) describe the depression or shift to colder temperatures of the freezing (melting) point temperature in a solution drop due to the solute effects, where ΔT_{f1} and ΔT_{f2} denote corrections of the first and second order by supersaturation or solution concentration respectively, and ΔT_ε is associated with the misfit strain effects of the insoluble substrate. It is easy to see that ΔT_{f1} is equivalent to the traditional formulation for the freezing point depression ΔT_f in the bulk water solution that is usually derived by equating the chemical potentials of ice and the solution (e.g., Pruppacher and Klett, 1997; Curry and Webster, 1999). The first term in expansion of ΔT_f by the mole fraction of a solute, $x_s = n_s/n_w$, can be written as (Curry and Webster, 1999, p. 122):

$$\Delta T_{f1} = -\frac{R_v T_0^2}{L_m} x_s. \tag{9.11.12}$$

In a dilute solution, $S_w \approx 1 - x_s$, or $x_s \approx -s_w$, so the expressions for ΔT_{f1} in both (Eqns. 9.11.11) and (9.11.12) are equivalent.

The freezing point depression is an important property of nucleating aerosols—their temperature threshold of nucleation. To estimate and compare the effects of salt ΔT_{f1} ("hygroscopic depression") and insoluble substrate ΔT_ε ("misfit depression"), note that $R_v T_0^2/L_m^{ef} \approx 103$ K and $\rho_i L_m^{ef} \approx 3.3 \times 10^9$ erg cm^{-3} near T_0, while $C\varepsilon^2 \sim 1.7 \times 10^7$ erg cm^{-3} at $\varepsilon = 1\%$ and $C\varepsilon^2 \sim 6.8 \times 10^7$ erg cm^{-3} at $\varepsilon = 2\%$. Since CCN serve in this freezing mode as IN, their equilibrium sizes and molalities can be easily calculated from the Köhler theory. The mole fraction in each haze particle is adjusted to the environmental relative humidity to determine the freezing point depression (see Sections 8.6 and 9.7). For instance, at $S_w = 0.9$ ($s_w = -0.1$ or the solute mole fraction $x_s \approx -s_w = 0.1$), we obtain $\Delta T_{f1} = 103$ K $\times (-0.1) = -10.3$ K (which well agrees with the point at $S_w = 0.9$ in Fig. 9.10), and $\Delta T_{f2} = -0.31$ K is much smaller. The misfit depression is $\Delta T_\varepsilon = -1.55$ °C at $\varepsilon = 1\%$ and $\Delta T_\varepsilon = -6.2$ °C at $\varepsilon = 2\%$. These simple estimates of ΔT_ε are in a good agreement with those in Young (1993). In addition, (Eqns. 9.11.10) and (9.11.11) allow a simple comparison of the freezing depression caused by the salts, induced supersaturation (subsaturation), and misfit strain.

The effects of supersaturation on the ice nucleation rate from CCN can also be characterized by introducing the "effective supercooling" ΔT_{ef} and the "effective temperature" of nucleation T_{ef} caused by supersaturation s_w (positive or negative), which can be defined using the denominator in (Eqn. 9.7.4) for r_{cr} and introducing T_{ef} with the relation

$$\ln(T_0/T_{ef}) = \ln[(T_0/T)S_w^{G_n}]. \tag{9.11.13}$$

Then, T_{ef} and ΔT_{ef} can be defined as

$$T_{ef} = TS_w^{-G_n} = T(1+s_w)^{-G_n} \approx T + \Delta T_{ef}, \tag{9.11.14}$$

$$\Delta T_{ef} = -TG_n s_w = -\frac{R_v T^2}{L_m^{ef}} s_w. \tag{9.11.15}$$

Since $G_n \sim 0.4$ at $T = -20$ to $0\,°C$, and $G_n T \sim 0.4 \cdot 263 \sim 100$ at $T \sim -10\,°C$, the value of $\Delta T_{ef} \approx -100 s_w$—i.e., ΔT_{ef} in °C is approximately equal to water supersaturation in %, but with the opposite sign. For example, the value of $\Delta T_{ef} \approx -5\,°C$ for $s_w = 0.05$ ($S_w = 1.05$), and $T_{ef} \approx -15\,°C$ for $T = -10\,°C$, which substantially increases the nucleation rate. This can be a quantitative formulation of the Hobbs–Rangno effect (1990–1991) for high crystal production at relatively warm temperatures and hypothesized high-water saturations of a few percent. Conversely, for $s_w = -0.05$ (the equivalent to solute concentration $x_s \approx 0.05$), the value of $\Delta T_{ef} \approx 5\,°C$ and $T_{ef} \approx -5\,°C$ for $T = -10\,°C$—that is, the effective temperature is higher by 5 °C at subsaturation caused by the dynamic factors of the presence of solute, with a corresponding suppression of nucleation. A limitation of the linear approximation is that if we consider solutions, they should be close to the ideal solutions, otherwise the freezing and melting point depressions should be evaluated with an accounting for the non-linear behavior of the freezing and melting curves as shown in Fig. 9.10.

Surface Quasi-Heterogeneous Freezing

When a solid is incompletely wettable by its own melt with the small but nonzero contact angle θ_{is}, as described in Section 9.9, surface freezing in the absence of the foreign substances can be considered as a quasi-heterogeneous nucleation of an ice germ at the drop surface. Then, the critical free energy of an ice germ at the liquid–air interface can be written as (Eqn. 9.9.5) in Section 9.9:

$$\Delta F_{cr} = \frac{16\pi}{3} \frac{\sigma_{ia}^3 f(m_{is}, x)}{\left\{\rho_i L_m^{ef} \ln\left[\frac{T_0}{T} S_w^{G_n} \exp\left(-H_{s,fr}\right)\right]\right\}^2}, \tag{9.11.16}$$

where $f(m_{is}, x)$ and $H_{s,fr}$ are defined by (Eqns. 9.7.16) and (9.9.4a), and m_{is} is the cosine of the contact angle θ_{is} between ice and its own liquid (water or aqueous solution). Equating (Eqn. 9.11.1) to (Eqn. 9.11.16) for ΔF_{cr} and solving for T, we obtain the surface freezing temperature:

$$T_{s,fr}(\Delta p) = T_0 S_w^{G_n(T)} \exp\left[-\frac{1}{\rho_w L_m^{ef}}\left(\frac{\Delta p \Delta \rho}{\rho_i} - \frac{2\sigma_{sa}}{r_d}\right)\right] \times$$

$$\exp\left\{-\frac{1}{L_m^{ef}\rho_i}\left[-\frac{16\pi}{3} \frac{\sigma_{ia}^3 f(m_{is}, x)}{kT \ln(J_{s,fr}/C_{s,fr}) + \Delta F_{act}}\right]^{1/2}\right\}, \tag{9.11.17}$$

where $J_{s,fr}$ and $C_{s,fr}$ are the nucleation rate and normalizing factor for surface freezing; their exact appearance is not important for us now for the reason explained in the following text.

Surface Quasi-Heterogeneous Melting

For incomplete wettability, a particular case of surface melting considered in Section 9.9 is a quasi-heterogeneous process of a liquid germ nucleation on the surface of a crystal. The critical free energy for the type of surface melting considered here can be written as for the solid/air interface (Defay et al., 1966) and using (Eqn. 9.9.11) as

$$\Delta F_{cr} = \frac{16\pi}{3} \frac{\sigma_{sa}^3 f(m_{is}, x)}{\left\{-\rho_w L_m^{ef} \ln\left[\frac{T_0}{T} S_w^{G_n} \exp(-H_m)\right]\right\}^2}, \qquad (9.11.18)$$

where $f(m_{is})$ and H_m are defined by (Eqns. 9.7.16) and (9.9.10), and m_{is} is again the cosine of the small contact angle between ice and its own liquid. Equating (Eqn. 9.11.1) to (Eqn. 9.11.18) for ΔF_{cr} and solving for T, we obtain the melting temperature:

$$T_m(\Delta p) = T_0 S_w^{G_n(T)} \exp\left[-\frac{1}{\rho_i L_m^{ef}}\left(\frac{\Delta p \Delta \rho}{\rho_w} + \frac{2\sigma_{ia}}{r_{cr}}\right)\right]$$
$$\exp\left\{\frac{1}{L_m^{ef}\rho_w}\left[-\frac{16\pi}{3}\frac{\sigma_{is}^3 f(m_{is}, x)}{kT\ln(J_{s,m}/C_{s,m}) + \Delta F_{act}}\right]^{1/2}\right\}, \qquad (9.11.19)$$

where $J_{s,m}$ and $C_{s,m}$ are the nucleation rate and normalizing factor for surface melting. For complete wettability $\theta_{is} = 0$, $m_{is} = 1$, $f(m_{is} = 1, x) = 0$, and $\Delta F_{cr} = 0$, there is no energy barrier for nucleation. For ice surface freezing or melting, θ_{is} has a nonzero yet very small value $\theta_{is} \sim 0.5$–$2°$ (Hobbs, 1974; Dash, et al., 1995; Ketcham and Hobbs, 1969; Knight, 1971; Elbaum et al., 1993; Elbaum and Schick, 1991; Rosenberg, 2005). Thus, $f(m_{is}) \to 0$, the arguments in the second exponents in (Eqns. 9.11.17) and (9.11.19) tend to zero, and the exponents tend to 1. Then, the expressions for the surface freezing and melting temperatures are simplified as

$$T_{s,fr}[f(m_{is}) \to 0] = T_0 S_w^{G_n(T)} \exp\left[-\frac{1}{\rho_w L_m^{ef}}\left(\frac{\Delta p \Delta \rho}{\rho_i} - \frac{2\sigma_{sa}}{r_d}\right)\right], \qquad (9.11.20)$$

$$T_m[f(m_{is}) \to 0] = T_0 S_w^{G_n(T)} \exp\left[-\frac{1}{\rho_i L_m^{ef}}\left(\frac{\Delta p \Delta \rho}{\rho_w} + \frac{2\sigma_{ia}}{r_{cr}}\right)\right]. \qquad (9.11.21)$$

These expressions do not depend on $J_{s,fr}$, $J_{s,m}$ owing to the small contact angle between water and ice. For substances when θ_{is} is not so small—e.g., some metals (Dash, et al., 1995; Elbaum et al., 1993; Elbaum and Schick, 1991), (Eqns. 9.11.17) and (9.11.19) should be used and the last exponent can make contributions lowering $T_{s,fr}$ and raising T_m. Equation (9.11.21) allows for a simple calculation of the melting point depression with increasing pressure. Differentiating (Eqn. 9.11.21) by p, we obtain

$$\frac{dT_m}{dp} = -\frac{\Delta\rho T_0}{\rho_i \rho_w L_m^{ef}} S_w^{G_n(T)} \exp\left[-\frac{1}{\rho_i L_m^{ef}}\left(\frac{\Delta p \Delta \rho}{\rho_w} + \frac{2\sigma_{ia}}{r_{cr}}\right)\right]. \qquad (9.11.22)$$

For a particular case of bulk pure ice ($r_{cr} \to \infty$, $S_w = 1$), (Eqn. 9.11.22) converts into the Clapeyron equation (Hobbs, 1974; Chapter 4 here), $T_0 = 273.15$ K, and yields $dT_m/dp = -1/138\ \mathrm{K\,atm^{-1}}$, close to the experimental values (Hobbs, 1974; Pruppacher and Klett, 1997; Dash et al., 1995).

Liquidus Curves

Equations (9.11.20) and (9.11.21) for $T_{s,fr}$, T_m at $\Delta p = 0$ and large r_d, r_{cr} or bulk solutions are reduced to

$$T_{s,fr}(\Delta p = 0) = T_m(\Delta p = 0) = T_0 S_w^{G_n(T)} = T_0 S_w^{RT/M_w L_m^{ef}} = T_0 a_w^{RT/M_w L_m^{ef}} . \quad (9.11.23)$$

This indicates the reversibility of slow surface freezing and melting. Equation (9.11.23) is the same as (Eqns. 8.7.12a) and (8.7.12b) for homogeneous nucleation considered in Section 8.7, because for very small contact angles $f(m_{is}) \to 0$, and heterogeneous nucleation becomes similar to the homogeneous process. These equations are nonlinear in the solution concentration, and are valid not only for dilute solutions but also for concentrated solutions. They describe with high accuracy the liquidus curves or freezing/melting point depressions over a wide range of solution concentration (for solutions without strong non-ideality), as was illustrated in Chapter 8 and in Fig. 9.10. For the particular case of a small solute mole fraction x_s (dilute solution), substituting $S_w \approx 1 - x_s$, and expanding (Eqn. 9.11.14) by x_s in the power series, $T_{s,fr}$ and T_m can be presented in a linear approximation as in (Eqn. 8.7.13), which is a known linear approximation for the freezing/melting point depression (e.g., Pruppacher and Klett, 1997; Curry and Webster, 1999). It was illustrated in the Figures in Chapter 8 and in Fig. 9.10 that (Eqn. 9.11.23) can describe with good accuracy a nonlinear decrease of the freezing or melting point to rather high values of solution concentrations.

9.12. Critical Saturation Ratios or Water Activities of Heterogeneous Freezing

The critical or threshold saturation ratios for homogeneous freezing were considered in Section 8.8. Threshold humidities and saturation ratios $S_{w,cr}^{het}$ for heterogeneous ice nucleation are much more variable due to the varying composition of aerosols that initiate freezing. The original parameterizations of the number density of ice crystals formed in clouds due to heterogeneous nucleation were formulated as functions of temperature (Fletcher, 1962; Cooper, 1986) or ice supersaturation (e.g., Huffman, 1973; Meyers et al., 1992) and usually did not assume any humidity thresholds for heterogeneous ice nucleation—i.e., nucleation in most parameterizations was allowed at any positive ice supersaturation. Later studies showed that there usually exists some critical saturation ratio that determines the onset of heterogeneous ice nucleation, and that the saturation ratios for heterogeneous nucleation are lower than those for homogeneous nucleation: in chamber and field experiments (e.g., Berezinsky and Stepanov, 1986; Sassen and Dodd, 1988, 1989; Young, 1993; Pruppacher and Klett, 1997; Rogers et al., 1998; Chen et al., 1998; Zuberi et al., 2002; DeMott et al., 2003; Shaw et al., 2005; Archuleta et al., 2005; Knopf and Koop, 2006; Abbatt et al., 2006; Möhler et al., 2006; Zobrist, 2006; Zobrist, et al., 2008; Cantrell and Robinson, 2006; Vali, 2008; Krämer et al., 2009; Spichtinger and Krämer, 2012; Hoose and Möhler, 2012); theoretical studies and numerical modeling (e.g., Khvorostyanov and Curry, 2000, 2004a,b, 2005a, 2009b; Kashchiev, 2000; Khvorostyanov and Sassen, 1998a,b, 2002; Sassen et al., 2002; Khvorostyanov, Curry et al., 2003; Kärcher and Lohmann, 2003; Gierens, 2003; Diehl and Wurzler, 2004; Morrison et al., 2005a,b; Barahona and Nenes, 2009).

A recent INCA field experiment (INterhemispheric differences in Cirrus properties from Anthropogenic emissions) was devoted to measurements of the properties of cirrus clouds and the upper troposphere and found the difference in critical saturation ratios in the Northern and Southern hemispheres. It has been hypothesized that the difference in critical saturation ratios between heterogeneous and homogeneous nucleation may cause global climatic effects as observed in the INCA experiment: homogeneous freezing in the cleaner southern hemisphere atmosphere versus a predominance of heterogeneous freezing in the more polluted northern hemisphere may be responsible for observations of lower average relative humidity of the upper troposphere with cirrus clouds in the northern hemisphere (Ovarlez et al., 2002; Ström et al., 2003; Haag et al., 2003; Gayet et al., 2004; Monier et al., 2006).

However, the exact values of the difference in critical humidities between the homogeneous and heterogeneous nucleation modes have not been established, nor has their dependence on various parameters such as the size of freezing particles, their physico-chemical properties, cooling rates, temperature, and other factors. Therefore, some recent parameterizations of the critical humidities for heterogeneous ice nucleation were based on some hypothetical but unverified relations, such as an assumption of constant difference in activities between the modes of homogeneous and heterogeneous nucleation, some parameterizations neglected factors such as dependence on the particle size, cooling rates, and chemical composition. Such assumptions introduce uncertainties and unknown errors in parameterizations. Therefore, parameterizations could be improved based on the theoretical evaluation of critical humidities for the two nucleation modes. In this section, the equations are derived for the critical saturation ratios of heterogeneous nucleation analogous to those derived in Chapter 8 for homogeneous nucleation, and the critical humidities of homogeneous and heterogeneous nucleation are compared.

9.12.1. General Equations

Derivation of the critical or threshold saturation ratios of heterogeneous nucleation in this section follows Khvorostyanov and Curry (2009b), is similar to the analogous derivation for homogeneous freezing in Section 8.8, and begins with (Eqn. 9.11.3). Solving (Eqn. 9.11.3) relative to S_w, we obtain the critical water saturation ratio:

$$S_{w,cr}^{het} = [(T/T_0)\exp(H_{f,het} + H_{v,fr})]^{1/G_n}. \tag{9.12.1}$$

Substituting $H_{f,het}$ from (Eqn. 9.11.4) and $H_{v,fr}$ from (Eqn. 9.7.6), we obtain an expanded form

$$S_{w,cr}^{het}(J_{het}, T, r_d, r_{fr}, \Delta p) = \left\{ \left(\frac{T}{T_0} \right) \exp\left(\frac{\Delta\rho\Delta p}{\rho_w \rho_i L_m^{ef}} + \frac{r_{fr}}{r_d} + \frac{C_\varepsilon \varepsilon^2}{\rho_i L_m^{ef}} \right) \right.$$
$$\left. \times \exp\left[\frac{1}{L_m^{ef}\rho_i} \left(\frac{-(16\pi/3) f(m_{is}, x)\sigma_{is}^3}{kT\ln(J_{f,het}/C_{het}) + \Delta F_{act} - \alpha r_N^2 \sigma_{is}(1 - m_{is})} \right)^{1/2} \right] \right\}^{1/G_n}. \tag{9.12.2}$$

Equation (9.12.2) expresses $S_{w,cr}^{het}$ as a function of atmospheric and aerosol variables T, r_d, r_{fr}, r_N, ε, m_{is}, p, α, and the nucleation rate $J_{f,het}$. Equation (9.12.2) depends also on a few fundamental parameters (L_m^{ef}, ρ_i, σ_{is}, ΔF_{act}), and accounts for the effects of active sites α and variable external pressure Δp that can be significant under certain circumstances as discussed in the following text. In applications for the freezing of bulk water, we should choose $r_d = \infty$, causing the term with r_{fr} to vanish.

The critical relative humidity over water is $RHW_{cr} = 100\,S_{w,cr}^{het}$, and the critical ice saturation ratio $S_{i,cr}^{het}$ and relative humidity over ice is RHI_{cr} are

$$S_{i,cr}^{het} = S_{w,cr}^{het}(e_{ws}/e_{is}), \qquad RHI_{cr} = RHW_{cr}(e_{ws}/e_{is}), \tag{9.12.3}$$

where $e_{ws}(T)$ and $e_{is}(T)$ are saturated vapor pressures over water and ice. Note that in this theory, the primary quantities are saturation ratios and humidities over water since they determine the process of ice nucleation, and the saturation ratios and humidities over ice are secondary quantities that determine the crystal growth but not the nucleation itself. The surface tension σ_{is} can be chosen in different ways (see Section 4.4). In the calculations in the following text it is chosen similar to Tabazadeh et al. (2000), Khvorostyanov and Curry, (2004a), and Archuleta et al. (2005), based on Antonoff's rule as described in Section 4.4.

9.12.2. Simplifications of Equations for the Heterogeneous Critical Saturation Ratio

Equations (9.12.1) and (9.12.2) can be used directly for evaluation of the critical S_w of heterogeneous nucleation and for comparison with homogeneous nucleation thresholds, as will be illustrated in the following text. Besides, they allow us to derive the two most popular empirical methods used over the last few decades for parameterizing the critical humidities in calculations of ice nucleation: the polynomial parameterizations of the critical S_w as a function of temperature and the shifted water activity method.

Equations (9.12.1) and (9.12.2) for the critical S_w of heterogeneous nucleation can be simplified by examining the temperature dependence of various terms in (Eqn. 9.12.1), shown in Fig. 8.14c of Chapter 8, where calculations were performed with $r_d = 0.05\,\mu\text{m}$, $\sigma_{sa} = 76\,\text{dyn}\,\text{cm}^{-2}$, $\alpha = 0$, $\varepsilon = 0$, $\Delta p = 0$ (so that $H_{v,fr}$ is the same for heterogeneous and homogeneous nucleation), and $J_{f,het} = 1\,\text{s}^{-1}$. Fig. 8.14c shows that G_n has a weak T dependence; $G_n \sim 0.4$ at 0°C and slightly increases toward colder T to $G_n \sim 0.6$ at 170 K. The $H_{f,het}$ increases in this range from 0.07 to 0.15. Both $H_{f,het} << 1$ and $H_{v,fr} << 1$; therefore, (Eqns. 9.12.1) and (9.12.2) can be simplified by expansion of the exponents into the power series

$$S_{w,cr}^{het}(T) \approx \left[\left(\frac{T}{T_0}\right)(1 + H_{f,het} + H_{v,fr})\right]^{1/G_n}$$

$$\approx \left(\frac{T}{T_0}\right)^{M_w L_m^{ef}/RT}\left(1 + \frac{H_{f,het}}{G_n} + \frac{H_{v,fr}}{G_n}\right). \tag{9.12.4a}$$

Using the definition of $H_{v,fr}$ (Eqn. 9.7.6), we further obtain from (Eqn. 9.12.4a)

$$S_{w,cr}^{het}(T) = \left(\frac{T}{T_0}\right)^{M_w L_m^{ef}/RT}(1 + \frac{1}{G_n}H_{f,het} + \Delta S_{w,rd} + \Delta S_{w,\varepsilon} + \Delta S_{w,p}), \tag{9.12.4b}$$

where the three corrections due to the curvature, misfit, and pressure are

$$\Delta S_{w,rd} = \frac{1}{G_n}\frac{r_{fr}}{r_d}, \qquad \Delta S_{w,\varepsilon} = \frac{1}{G_n}\frac{C_\varepsilon \varepsilon^2}{\rho_i L_m^{ef}}, \qquad \Delta S_{w,p} = \frac{1}{G_n}\frac{\Delta\rho\Delta p}{\rho_w \rho_i L_m^{ef}}. \tag{9.12.5}$$

Expressing G_n via thermodynamical parameters as before, we obtain

$$\Delta S_{w,rd} = \frac{\rho_w}{\rho_i}\frac{A_K}{r_d} \sim \frac{3.6\times 10^{-5}}{T r_d}, \qquad \Delta S_{w,\varepsilon} = \frac{R_v}{T}\frac{C_\varepsilon \varepsilon^2}{\rho_i}, \qquad \Delta S_{w,p} = \frac{R_v}{T}\frac{\Delta\rho\Delta p}{\rho_w \rho_i} \tag{9.12.6}$$

The expansion (Eqns. 9.12.4a) and (9.12.4b) is valid down to low temperatures ~ 200 K. If we express the temperature via supercooling ΔT as $T = T_0 - \Delta T$, and $\Delta T << T_0$, then the first T-factor in (Eqn. 9.12.4b) can also be expanded in the power series and we obtain

$$S_{w,cr}^{het}(T) \approx 1 + \Delta S_{w,T} + (1/G_n)H_{f,het} + (\Delta S_{w,rd} + \Delta S_{w,\varepsilon} + \Delta S_{w,p}), \tag{9.12.7}$$

$$\Delta S_{w,T} = -\frac{1}{G_n}\frac{\Delta T}{T} = -\frac{L_m^{ef}}{R_v}\frac{\Delta T}{T^2}. \tag{9.12.7a}$$

Each of the terms in (Eqn. 9.12.7) is a function of temperature, and this expression resembles previous empirical parameterizations of $S_{w,cr}^{\text{hom}}$, $S_{i,cr}^{\text{hom}}$ for homogeneous ice nucleation as polynomials by temperature (e.g., Sassen and Dodd, 1988, 1989; Heymsfield and Miloshevich, 1995, see Chapter 8) or polynomials by saturated vapor pressure e_{sw}, (Koop et al., 1998, 1999; Bertram et al., 2000), or similar parameterizations for heterogeneous nucleation (e.g., Zuberi et al., 2002; Archuleta et al., 2005). Thus, (Eqns. 9.12.1), (9.12.2), (9.12.4), and (9.12.7) serve as a physical justification for the previous parameterizations and as a basis for their improvement. However, these equations show that besides the T-dependence, there are several other dependencies not captured by the empirical parameterizations and may at least partially explain the differences in $S_{w,cr}^{het}$ determined by various laboratory and field experiments. In particular, the curvature correction $\Delta S_{w,rd}$ in (Eqn. 9.12.6) is similar to the slightly increased (by $\rho_w/\rho_i \sim 1.1$) Kelvin curvature effect ~ A_K/r_d on the saturated vapor pressure over the drop surface, but that was missing in previous ice nucleation parameterizations. The theory developed here indicates that this effect becomes noticeable for submicron particles, and neglecting it would lead to the same errors as neglecting the Kelvin effect on the vapor pressure for small particles. The similarity of the curvature effect on the critical humidity of freezing and of the Kelvin effect is not coincidental: It is an important consequence of the derivation of both these effects from the general entropy equation—for the Kelvin effect as described in Section 3.8 and for the freezing critical humidity in this section based on the critical radii and energies from Section 9.7.

Simple estimates of the effects of these 4 factors on variations ΔS_w of $S_{w,cr}^{het}$ can be done using (Eqn. 9.12.7), which shows that the temperature factor causes a decrease in $S_{w,cr}^{het}$, while each of the other 3 factors causes an increase in $S_{w,cr}^{het}$.

1. At $\Delta T = T_0 - T << T_0$, the temperature factor can be simplified as in (Eqn. 9.12.7a), and an estimate at $T = -15\,°\text{C}$ yields $\Delta S_{w,T} \sim -0.14$.
2. Assuming σ_{sa} ~76 dyn cm^{-1} (PK97, Fig. 5–2), we have $r_{fr} = 2\sigma_{sd}/\rho_i L_m^{ef} \sim 0.6 \times 10^{-7}$ cm at $T \sim 250$ K and $(\rho_w/\rho_i)A_K \sim 1.44 \times 10^{-7}$. If the size of aerosol particles is $r_d \sim 0.02\,\mu$m, then from (Eqn. 9.12.6) $\Delta S_{w,rd} \sim +0.075$, or $\Delta RHW_{cr} \sim 7.5\%$, a perceptible effect. The curvature effect decreases with increasing r_d: $\Delta RHW_{cr} \sim 1.5\%$ for $r_d \sim 0.1\,\mu$m and is negligible for $r_d \geq 0.5$–$1\,\mu$m.
3. The estimate from (Eqn. 9.12.6) of the misfit term is $\Delta S_{w,\varepsilon} \sim +0.04$ ($\Delta RHW_{cr} \sim 4\%$) for $\varepsilon = 1\%$, and $\Delta S_{w,\varepsilon} \sim +0.16$ ($\Delta RHW_{cr} \sim 16\%$) for $\varepsilon = 2\%$.
4. For $\Delta p \sim 1$ atm (10^6 dyn cm^{-2}), an estimate for the pressure term $\Delta S_{w,p}$ gives ~ 10^{-4}, which is negligible. However, for higher pressures of $\Delta p \sim 500$–1000 atm, as can exist in the atmospheres of the other planets, or are created in the laboratory devices (e.g., Kanno and Angell, 1977; Mishima and Stanley, 1998), we get $\Delta S_{w,p} \sim +0.05$ to +0.10 ($\Delta RHW_{cr} \sim 5$ to 10%) —i.e., a significant increase.

A difference of a few percent in RHW often may substantially alter the picture of cloud formation. For example, if RHW ≈ 100%, then a liquid cloud normally forms, but if RHW is just 3–5% lower and the temperature is cold enough (−15 to −20 °C), then a crystalline cloud may form in clear sky via the freezing of haze particles as diamond dust, and the liquid phase may not form or form later, which was found in lidar measurements and numerical modeling (e.g., Sassen et al., 2006; Sassen and Khvorostyanov, 2007, 2008).

9.12.3. Derivation from Classical Theory of the Water Activity Shift Method

Using the expansion of (Eqn. 9.12.4a) and (Eqn. 9.12.1) into the power series, and keeping only the terms of the first order, the critical saturation ratio can also be presented in the form

$$S_{w,cr}^{het}(T_{het}) \approx S_{w,cr}(T_m) + \delta S_{w,cr}^{het}(T), \tag{9.12.8a}$$

$$\delta S_{w,cr}^{het}(T) = \delta a_{w,het} = S_{w,cr}(T_m)\frac{(H_{f,het} + H_{v,fr})}{G_n}. \tag{9.12.8b}$$

Here, $S_{w,cr}(T_m) = (T/T_0)^{M_w L_m^{ef}/RT}$ is the same melting curve $a_w(T_m)$ or $S_w(T_m)$ discussed in Sections 8.7, 8.8, and 9.11, and

$$\delta S_{w,cr}^{het} = S_{w,cr}^{het}(T_{f,\mathrm{hom}}) - S_{w,cr}(T_m) \tag{9.12.8c}$$

is the difference or shift in the critical saturation ratios between the melting and heterogeneous freezing curves at the same temperature. It can be written in terms of activities as

$$\delta a_{w,het} = a_w(T_{f,het}) - a_w(T_m) = a_{w,het} - a_w(T_m). \tag{9.12.8d}$$

The quantity $\delta a_{w,het}$ is the difference or shift in the activities between the melting and heterogeneous freezing curves at the same temperature and is termed in the literature as *the water activity shift for heterogeneous freezing* by analogy with the *water activity shift for homogeneous freezing* considered in Section 8.8.

The $\delta a_{w,het}$ is usually introduced as an empirical quantity based on the measurements of the heterogeneous freezing temperatures and comparison with the melting curve $T_m(a_w)$, and characterizes the freezing point depression at heterogeneous freezing. Equations (9.12.8a)–(9.12.8d) allow expression of $\delta S_{w,cr}^{het}(T)$ and $\delta a_{w,het}$ in terms of the classical nucleation theory. Using the definition for $G_n = RT/(M_w L_m^{ef})$, (Eqn. 9.11.4) for $H_{f,het}$, and (Eqn. 9.7.6) for $H_{v,fr}$, the shifts in the saturation ratio or in the water activity can be presented in an expanded form:

$$\delta S_{w,cr}^{het}(T, J_{f,het}, \sigma_{is}, m_{is}, \alpha, \varepsilon, r_N, \Delta p) = \delta a_{w,het} = S_{w,cr}^{het}(T_m)\left(\frac{M_w L_m^{ef}}{RT}\right)$$
$$\times\left\{\frac{1}{\rho_i L_m^{ef}}\left[\frac{(-16\pi/3) f(m_{is}, x)\sigma_{is}^3}{kT\ln(J_{f,het}/C_{het}) + \Delta F_{act} - \alpha r_N^2 \sigma_{is}(1-m_{is})}\right]^{1/2}\right.$$
$$\left.+\frac{\Delta\rho\Delta p}{\rho_w \rho_i L_m^{ef}} + \frac{2\sigma_{sa}}{\rho_i L_m^{ef} r_{dr}} + \frac{C_\varepsilon \varepsilon^2}{\rho_i L_m^{ef}}\right\}. \tag{9.12.8e}$$

This equation shows that the water activity shift for heterogeneous freezing is not a universal quantity, but besides temperature, depends also on many other parameters: nucleation rate, surface tension, the contact angle of a substrate, misfit strain, the radius of a freezing drop, the shape factor and the size of the insoluble substrate, and several thermodynamic parameters like melting heat, water and ice densities, pressure, etc.

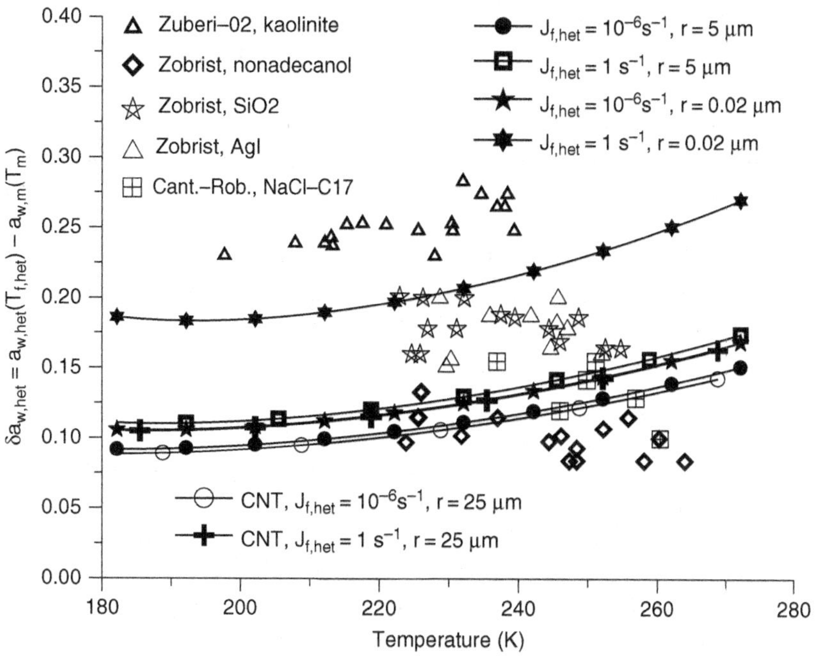

Figure 9.11. The shifts of water activity $\delta a_{w,het} = a_{w,het}(T_{f,het}) - a_w(T_m)$ calculated with equations of extended CNT described in Section 9.12.3 for two nucleation rates, $J_{f,het} = 10^{-6}\,s^{-1}$ and $J_{f,het} = 1\,s^{-1}$, and for three drop radii, 0.02 μm, 5 μm, and 25 μm. The five experimental data sets are from the heterogeneous freezing data on $T_{f,het}(a_{w,het})$ for various substances by various authors: from Zuberi et al. (2002, kaolinite), Zobrist et al. (2006, 2008), nonadecanol, SiO_2, and AgI), and from Cantrell and Robinson (2006, NaCl-C17). The shifts $\delta a_{w,het}$ were calculated using the experimental points from these data sets for freezing, $T_{f,het}(a_{w,het})$, and melting, $T_m(a_{w,m})$ as $\delta a_{w,het} = T_{f,het}(a_{w,het}) - T_m(a_{w,m})$ at $T_{f,het} = T_m$—i.e., as the difference in activities for each individual experimental point $T_{f,het}(a_{w,het})$ on the freezing curve and the corresponding temperature at the melting curve $T_m(a_{w,m})$.

Fig. 9.11 shows the shifts of saturation ratio $\delta S_{w,het}$ or water activity $\delta a_{w,het} = a_{w,het}(T_f) - a_w(T_m)$ calculated with (Eqn. 9.12.8e) for two nucleation rates, $J_{f,het} = 10^{-6}\,s^{-1}$ and $J_{f,het} = 1\,s^{-1}$, and for three drop radii, 0.02 μm, 5 μm, and 25 μm. These $\delta a_{w,het}$ calculated with classical nucleation theory are compared with five experimental data sets that were retrieved from the heterogeneous freezing data on $T_{f,het}(a_{w,het})$ for various substances. The experimental shifts $\delta a_{w,het}$ were calculated from these data sets using the experimental freezing points, $T_{f,het}(a_{w,het})$, and melting, $T_m(a_{w,m})$, as $\delta a_{w,het} = T_{f,het}(a_{w,het}) - T_m(a_{w,m})$ at $T_{f,het} = T_m$—i.e., as the difference in activities for each individual experimental point $T_{f,het}(a_{w,het})$ and the corresponding $T_m(a_{w,m})$ at the same temperature.

Fig. 9.11 shows that $\delta S_{w,het} = \delta a_{w,het}$ calculated with classical nucleation theory increases slightly with increasing temperature. For the drop radii of 5–25 μm, $\delta a_{w,het} \sim 0.1$–0.15, which is much lower than for homogeneous freezing $\delta a_{w,hom} \sim 0.3$–0.35 (see Sections 8.8), and agrees with most of the experimental data from Cantrell and Robinson (2006) and Zobrist et al. (2007, 2008) for four different IN, except for those from Zuberi et al. (2002) with kaolinite, which are located higher, near 0.25. An increase in $J_{f,het}$ from 10^{-6} to $1\,s^{-1}$ almost does not effect $\delta a_{w,het}$ for large drops 5–25 μm, but causes a noticeable increase in $\delta a_{w,het}$ up to 0.17–0.27 for small drops of 0.02 μm. As in the case with homogeneous freezing, the freezing of small submicron particles occurs at higher solution concentrations and at greater freezing point depressions. Thus, the laboratory results obtained for large drops should be applied with caution for submicron haze particles freezing in cirrus or similar clouds.

The experimental data on $\delta a_{w,het}$ for various substances and experiments exhibits a rather wide scatter, and (Eqn. 9.12.8a)–(9.12.8e) can explain at least a part of this scatter. The limitations of the shifted activity method are caused by the multi-parametric dependencies of $\delta a_{w,het}$. However, the general agreement of $\delta a_{w,het}$ calculated with classical nucleation theory suggests that the appropriate choice of parameters would allow more precise quantitative description of the measured water activity shifts for heterogeneous freezing.

9.12.4. Calculations of Critical Relative Humidities for Heterogeneous Nucleation

Equations (9.12.1) and (9.12.2) explain previously developed empirical parameterizations of the critical humidities. This is illustrated in this section with the results of calculations of critical values of RHW and RHI for heterogeneous nucleation performed with (Eqns. 9.12.2) and (9.12.3), using the relations RHW = $100\,S_{w,cr}^{het}$, RHI = $100\,S_{i,cr}^{het}$ and the following baseline set of parameters: $\sigma_{sa} = 76$ dyn cm^{-2}, $m_{is} = 0.5$ (close to typical values for surface soil, sand, and soot), $\alpha = 0$, $\varepsilon = 0$, $\Delta p = 0$, and two prescribed values of heterogeneous nucleation rate, $J_{f,het} = 10^{-6}\,s^{-1}$ and $1\,s^{-1}$. Additional calculations were performed where the values of ε, α, m_{is}, and r_d were varied, and sensitivity of RHW and RHI to these parameters was studied as described in the following text.

The parameter r_d in the baseline runs was chosen based on measurements in field and chamber experiments. In studies of heterogeneous freezing, measurements at aircraft altitudes found the modal radius of ice crystal residues of $r_d \approx 0.05$ μm in young cirrus clouds over Europe (Ström et al., 1997), of 0.1 μm in cirrus over Kansas in the SUCCESS campaign (Chen et al., 1998), and of 0.15–0.5 μm over the Florida area in the CRYSTAL-FACE campaign (Cziczo et al., 2004). Some chamber experiments on heterogeneous ice nucleation used nearly monodisperse core particles of comparable radii: 0.025, 0.05, and 0.1 μm (Archuleta et al., 2005). Measurements of aircraft engine soot as potential ice nuclei for contrail and cirrus formation found that the main soot fraction consisted of relatively hydrophilic particles with radii of 0.015–0.025 μm (Popovicheva et al., 2004). For homogeneous freezing, the Cirrus Parcel Model Comparison Project (CPMCP) (Lin et al., 2002) recommended the upper troposphere polydisperse aerosol of sulfuric acid with the similar modal radius $r_d = 0.02$ μm. The laboratory experiments used aqueous solution drops of larger radii ~ 0.2–5 μm (e.g., Pruppacher and Klett, 1997, Koop et al., 1998, 1999, 2000; Bertram et al., 2000; DeMott, 2002). Therefore, the value $r_d = 0.05$ μm was chosen for the baseline calculations to match the most typical value in this range; calculations and comparisons were performed also with the cases with 0.5 and 5 μm.

Critical values of RHW and RHI for heterogeneous ice nucleation calculated with the baseline set of parameters are shown in Fig. 9.12. RHW decreases from ~ 100–102.5% at $T \sim -18$ to $-20\,°C$ to ~ 65–70% at $-75\,°C$. RHI increases from ~ 120–125% to 137–145% over the same temperature range, especially rapidly at $T < -50\,°C$. Note the relatively weak dependence of RHI on the temperature at $T > -50\,°C$, which agrees with experimental data (e.g., Archuleta et al., 2005).

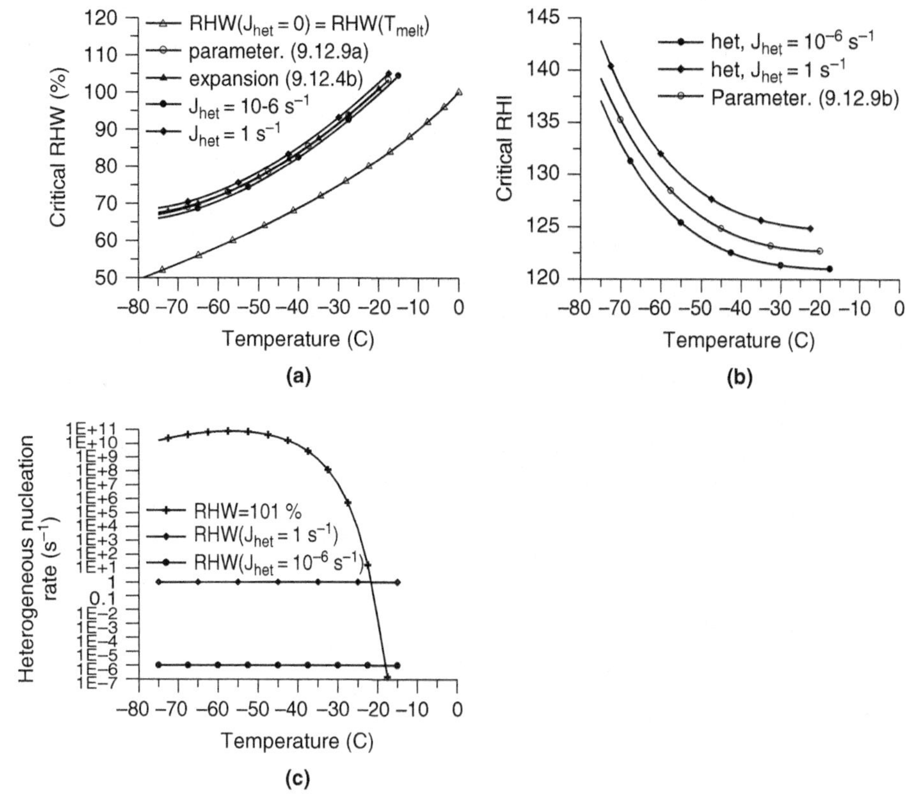

Figure 9.12. Critical RHW (a) and RHI (b) for heterogeneous ice nucleation calculated with the full (Eqn. 9.12.2) and indicated in the legend nucleated rates $J_{f,het} = 10^{-6}\,s^{-1}$ and $1\,s^{-1}$, and the parameters: $r_d = 0.05\,\mu m$, $m_{is} = 0.5$, $\alpha = 0$, $\varepsilon = 0$; compared with the power series expansion (Eqn. 9.12.4a,b) and with the parameterization (Eqns. 9.12.9a) and (9.12.9b); the lower curve in Fig. 9.12a (open triangles) shows RHW($J_{f,het} = 0$), which coincides with the melting curve T_m(RHW); (c) Verification of the solutions: heterogeneous nucleation rates $J_{f,het}(T,$ RHW) calculated with (Eqn. 9.6.7) using as input these RHW(T) from this figure, (a), corresponding to $J_{het} = 10^{-6}\,s^{-1}$ and $1\,s^{-1}$, and with RHW = const = 101%. From Khvorostyanov and Curry (2009b), *J. Geophys. Res.*, 114, D04207, reproduced with permission of John Wiley & Sons, Inc.

Fig. 9.12a shows that an increase in the nucleation rate by six orders of magnitude from $10^{-6}\,s^{-1}$ to $1\,s^{-1}$ is accompanied by a surprisingly small increase in critical RHW, 2–3%. However, the transition from $10^{-6}\,s^{-1}$ to the limit $J_{f,het} \to 0$ causes a substantial decrease in RHW, which merges with the melting curve RHW(T_m) (lower curve with open triangles in Fig. 9.12a), and is lower than the freezing RHW by 15–20%. The approximation (Eqn. 9.12.4b) shown in Fig. 9.12a,b (solid triangles) exhibits good accuracy, justifying the power series expansion.

We can parameterize these critical humidities as a polynomial of temperature T_c in °C, and account also for the curvature, misfit, and pressure corrections defined in (Eqns. 9.12.4a) and (9.12.4b)–(9.12.6):

$$S_{w,cr}^{het} = a_{sw} + b_{sw}T_c + c_{sw}T_c^2 + d_{sw}T_c^4 + (\Delta S_{w,rd} + \Delta S_{w,\varepsilon} + \Delta S_{w,p}), \quad (9.12.9a)$$

$$S_{i,cr}^{het} = S_{w,cr}^{het}(e_{ws}/e_{is}). \quad (9.12.9b)$$

The coefficients in (Eqn. 9.12.9a) are $a_{sw} = 1.21$, $b_{sw} = 1.23 \times 10^{-2}$, $c_{sw} = 9.6 \times 10^{-5}$, and $d_{sw} = 7.8 \times 10^{-10}$. Equation (9.12.9a) accounts for the newer parameterization of L_m^{ef}, allowing extension to lower T, and by accounting for the effects of finite drop size, r_{fr}/r_d, misfit strain, and pressure effects. The parameterized critical humidities RHW $= 100\,S_{w,cr}^{het}$ and RHI $= 100\,S_{i,cr}^{het}$ are also given in Fig. 9.12 (open circles) and illustrate sufficient accuracy of this parameterization for the chosen typical set of parameters.

It is shown later in this section that critical humidities depend on various aerosol properties and (Eqns. 9.12.9a) and (9.12.9b) may require corresponding corrections. However, variations of RHW_{cr} in this temperature range, ~ 30–35%, are much greater than variations with aerosol properties, which are typically a few percent. Therefore, these equations could be used as a first guess for simple parameterizations of the critical humidity of heterogeneous nucleation in the models that use some prescribed threshold value of RHI$_{cr}$. The value of this simple T-parameterization for heterogeneous nucleation is similar to those suggested in Sassen and Dodd (1989), Koop et al. (1999, 2000), Bertram et al. (2000) for homogeneous nucleation and to that in Zuberi et al. (2002) and Archuleta et al. (2005) for heterogeneous nucleation. A more precise evaluation of $S_{w,cr}^{het}$ that accounts for individual aerosol properties and nucleation rates can be based directly on (Eqns. 9.12.1), (9.12.2), or (9.12.7).

To illustrate the strong dependence of nucleation rates on RHW, Fig. 9.12c shows values of $J_{f,het}(T, S_w)$ calculated with (Eqn. 9.6.7) using the critical RHW values from Fig. 9.12a as input. If the humidities RHW are close to critical and vary with the temperature as shown in Fig. 9.12a for the corresponding $J_{f,het}(T, S_w)$, then $J_{f,het}(T, S_w)$ is constant with temperature. But if we assume RHW = 101%, or $S_w \approx 1$ at all T, as would be in the classical nucleation theory that accounts for only T-dependence, then $J_{f,het}(T)$ is by many orders of magnitude greater at $T < -20\,°C$, and is by many orders smaller at $T > 20\,°C$. The temperature dependence of the crystal concentrations calculated with such S_w-independent $J_{f,het}(T)$ would be substantially different than that calculated with a correct accounting for humidity.

In particular, Fig. 9.12c shows that correctly accounting for humidity RHW < 100% does not lead to an explosive increase in nucleation rates at low T. Thus, heterogeneous freezing can be an effective mode of ice nucleation in cirrus, which has been confirmed by numerical model simulations (e.g., Monier et al., 2006; Khvorostyanov et al., 2006; Comstock et al., 2008; Barahona and Nenes, 2009), and may explain the sufficiently high insoluble fractions in ice crystals residuals observed in cirrus in the field campaigns of SUCCESS (Chen et al., 1998) and CRYSTAL-FACE-2002 (Cziczo et al., 2004). If RHW is close to 100%, which is possible at not very low T, this rapid increase in $J_{f,het}(T, 101\%)$ at T below −20 to −30 °C is in agreement with chamber experiments on heterogeneous freezing that found threshold temperatures around −25 to −30 °C (e.g., Zuberi et al., 2002; Archuleta et al., 2005). However, $J_{f,het}(T$, RHW = 101%) rapidly decreases in the temperature range −20 °C to −15 °C (Fig. 9.12c), which explains the slower glaciation of natural liquid clouds at $T > -15\,°C$.

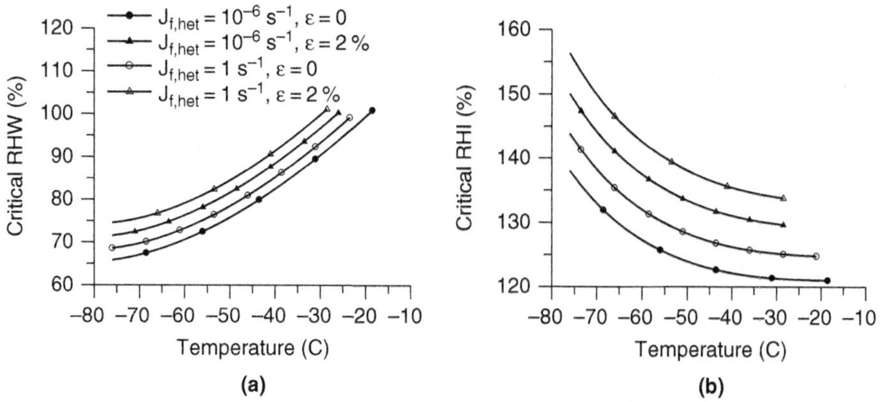

Figure 9.13. Dependence of the critical RHW and RHI for heterogeneous nucleation on the misfit strain $\varepsilon = 0$ and 2% at 2 nucleation rates, $J_{f,het} = 10^{-6}\,s^{-1}$ and $J_{f,het} = 1\,s^{-1}$; $r_d = 0.05\,\mu m$, and the other parameters are the same as in Fig. 9.12. From Khvorostyanov and Curry (2009b), *J. Geophys. Res.*, 114, D04207, reproduced with permission of John Wiley & Sons, Inc.

The dependence of the critical RHW and RHI for heterogeneous nucleation on the misfit strain ($\varepsilon = 0$ vs. 2%) is shown in Fig. 9.13, indicating that critical RHW and RHI increase by 7–8% and 8–13%, respectively, as ε increases from 0 to 2%. This is a significant increase that may inhibit ice nucleation if the crystalline lattice of the insoluble substrate substantially differs from the ice lattice.

Fig. 9.14 shows the effect of the active sites on the critical RHW and RHI for heterogeneous nucleation. When α increases from 0 to 10^{-3}, both humidities decrease by ~9–11%. Thus, the presence even of small active sites may substantially decrease the threshold RHW and favor heterogeneous nucleation. This effect is in agreement with parcel model simulations in (Khvorostyanov and Curry, 2005a) that showed that $\alpha \sim 10^{-4}$ may increase the threshold temperatures for heterogeneous ice nucleation above −10 °C (see Section 9.14.4).

Dependence of RHW and RHI for heterogeneous nucleation on the aerosol radius is illustrated in Fig. 9.15. When a particle radius grows 10 times from 0.05 μm to 0.5 μm, RHW decreases by 3–4% in this temperature range. When r_d further grows from 0.5 μm to 5 μm, a decrease in RHW is much less, ~0.8–1%, since the effect of the curvature correction decreases as $1/r_d$ according to (Eqn. 9.12.5). The total decrease is 4–5% in these radii range: 0.05 to 5 μm.

Fig. 9.16 illustrates the effect of the contact parameter m_{is} calculated with $r_d = 0.05\,\mu m$ and $J_{het} = 1\,s^{-1}$. When m_{is} decreases from 0.5 (the value used in most calculations here) to 0.2, RHW increases by 6–8%, or the freezing temperature decreases by about 10 °C; corresponding increases in RHI are 10–15%. When m_{is} decreases to −0.5, RHW further increases by 12–15% or the freezing temperatures decrease by almost 20 °C and RHI increases by ~20–30%. Thus, the effect of the wettability of the insoluble substrate is a significant factor.

The behavior of RHW and RHI at $m_{is} = 0.2$ in Fig. 9.16 is close to parameterizations suggested in Zuberi et al. (2002) and Archuleta et al. (2005) based on chamber experiments, the difference does not exceed 3–8% in the temperature range −30 to −70 °C. Note, however, that this comparison can be

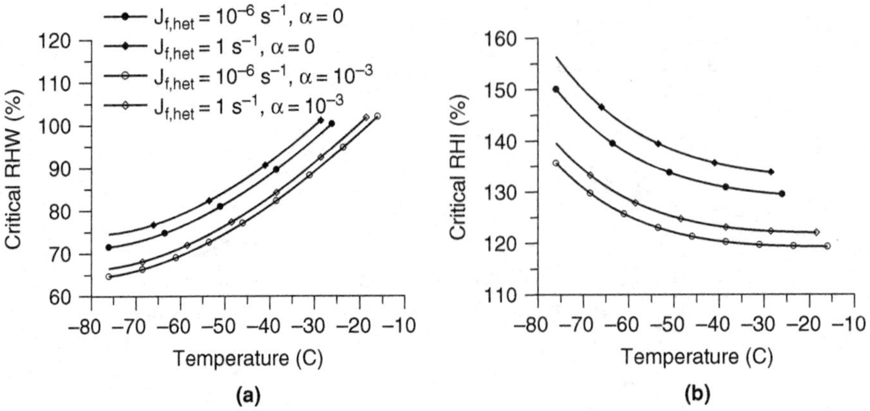

Figure 9.14. Effect of the active sites ($\alpha = 10^{-3}$ vs. $\alpha = 0$) on the critical RHW and RHI for heterogeneous nucleation at two nucleation rates, $J_{f,het} = 10^{-6}\,s^{-1}$, and $J_{het} = 1\,s^{-1}$, and $r_d = 0.05\,\mu m$. The other parameters are the same as in Figs. 9.12 and 9.13. From Khvorostyanov and Curry (2009b), *J. Geophys. Res.*, 114, D04207, reproduced with permission of John Wiley & Sons, Inc.

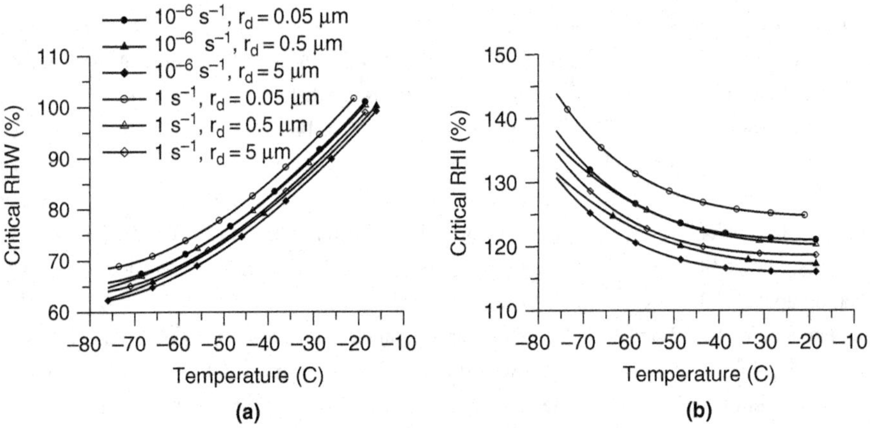

Figure 9.15. Dependence of RHW and RHI for heterogeneous nucleation on the aerosol radius r_d ($0.05\,\mu m$, $0.5\,\mu m$ and $5\,\mu m$) and two nucleation rates, $J_{f,het} = 10^{-6}\,s^{-1}$ and $1\,s^{-1}$. The other parameters are the same as in Fig. 9.12. From Khvorostyanov and Curry (2009b), *J. Geophys. Res.*, 114, D04207, reproduced with permission of John Wiley & Sons, Inc.

considered only as qualitative since many parameters in the chamber experiments are often uncertain (e.g., the contact parameter of the insoluble core, coverage, and properties of the soluble fraction, etc.), and if these parameters were specified with different values, the agreement could be best at different contact parameters.

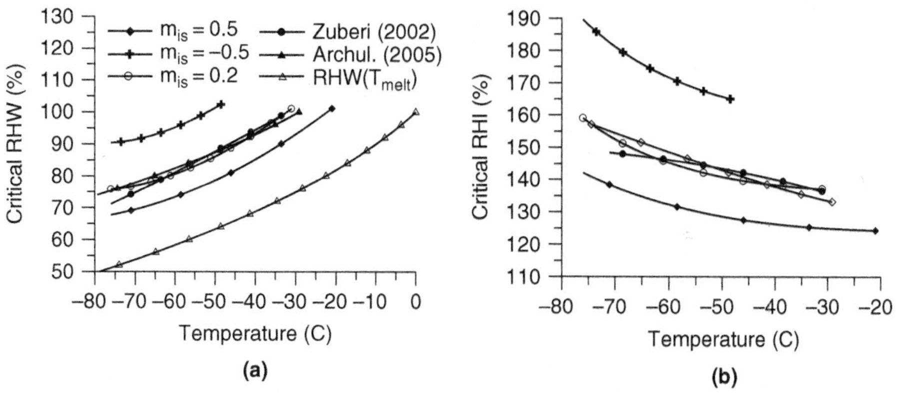

Figure 9.16. Effect of variations of the contact parameter m_{is} on the critical RHW (a) and RHI (b). Calculations are performed with $m_{is} = 0.5$, 0.2, and –0.5. A reasonably good agreement with experimental data from Zuberi et al. (2002) and Archuleta et al. (2005) is for calculations with $m_{is} = 0.2$. From Khvorostyanov and Curry (2009b), *J. Geophys. Res.*, 114, D04207, reproduced with permission of John Wiley & Sons, Inc.

9.12.5. Comparison of Critical Humidities for Heterogeneous and Homogeneous Nucleation

A principal issue for the mechanisms of cirrus formation is the difference between the critical humidities in homogeneous and heterogeneous modes of ice nucleation. Fig. 9.17 shows a comparison of RHW and RHI: homogeneous with $J_{hom}V = 1\ s^{-1}$, and heterogeneous with $J_{het} = 1\ s^{-1}$, and their differences, $\Delta RHW = RHW_{hom} - RHW_{het}$, and $\Delta RHI = RHI_{hom} - RHI_{het}$. Comparison with the benchmark data of Sassen and Dodd (1989, SD89) and Bertram et al. (2000, B00) indicates that both critical humidities of homogeneous nucleation at $r_d = 5\ \mu m$ are close to the parameterizations (SD89) and (B00). Both critical humidities of homogeneous nucleation are seen to be higher than the corresponding heterogeneous humidities with $\varepsilon = 0$ at all temperatures. At warmer T of −15 to −30 °C, the critical values of RHW of heterogeneous nucleation are ~22–30% lower than those for homogeneous nucleation, the RHI for heterogeneous nucleation is lower by 25–40%. This difference decreases to 15–25% toward colder $T = -75$ °C for RHW, and increases to 30–55% for RHI.

Over the range of the temperatures relevant for cirrus formation, the critical RHI of heterogeneous nucleation is often close to 120–130%, which is 20–30% lower than the critical ice relative humidity for homogeneous nucleation. These results provide an explanation for the lower RHI in the upper troposphere with cirrus clouds in the Northern Hemisphere (~130%) relative to the Southern Hemisphere (140–155%) found in the INCA field campaign, that were ascribed to the greater anthropogenic emissions and prevailing heterogeneous mode of cirrus formation in the Northern Hemisphere versus the cleaner Southern Hemisphere with the prevailing homogeneous mode (Ovarlez et al., 2001; Ström et al., 2003; Gayet et al., 2004; Monier et al., 2006).

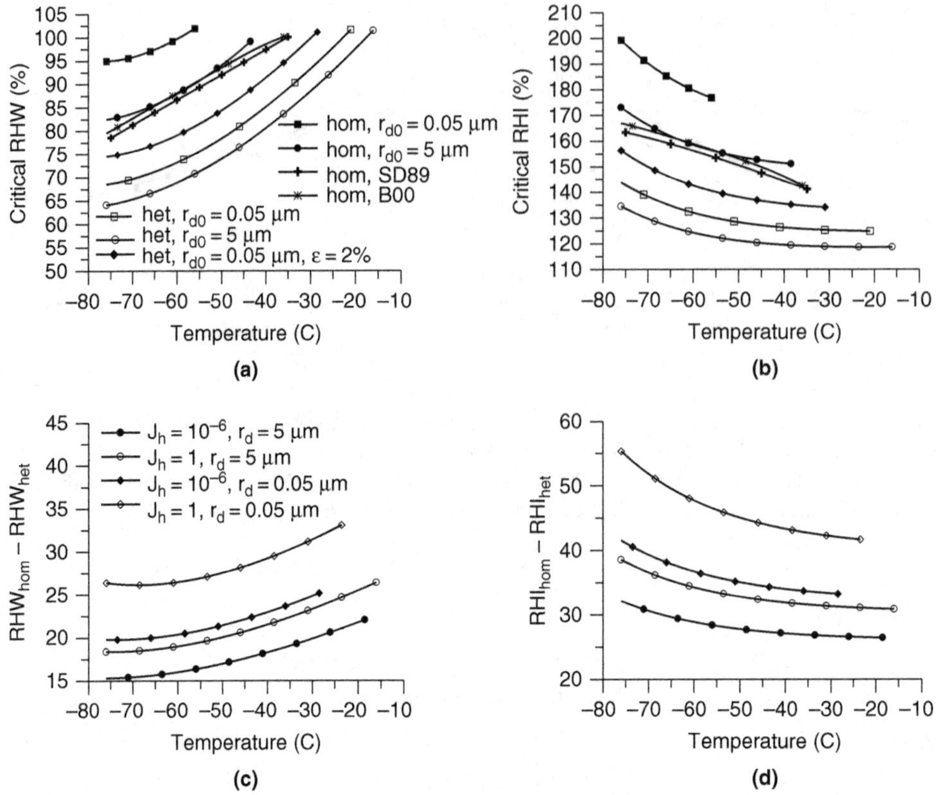

Figure 9.17. Comparison of *RHW* and *RHI* for ice nucleation modes. (a) and (b): homogeneous with $J_{f,hom}V = 1\ \mathrm{s}^{-1}$, and heterogeneous with $J_{f,het} = 1\ \mathrm{s}^{-1}$ for radii $r_d = 0.05$ and $5\ \mu\mathrm{m}$. These calculations are compared with parameterizations from Sassen and Dodd (1989, SD89) and Bertram et al. (2000, B00). The misfit strain is varied in the heterogeneous mode: $\varepsilon = 0$ and 2%. (c) The difference $RHW_{hom} - RHW_{het}$, and (d) the difference $RHI_{hom} - RHI_{het}$ for 2 radii ($r_d = 0.05$ and $5\ \mu\mathrm{m}$) and 2 nucleation rates ($J_{f,het}$ or $J_{f,hom}V = 10^{-6}\ \mathrm{s}^{-1}$ and $1\ \mathrm{s}^{-1}$). From Khvorostyanov and Curry (2009b), *J. Geophys. Res.*, 114, D04207, reproduced with permission of John Wiley & Sons, Inc.

Fig. 9.17 shows that although the heterogeneous critical humidities with misfit strain $\varepsilon = 2\%$ are higher than with $\varepsilon = 0$, they are still lower than humidities for homogeneous nucleation over the entire temperature range. Thus, heterogeneous ice nucleation may be favored over homogeneous nucleation if ice nuclei consisting of mixed aerosol particles containing soluble and insoluble fractions are present in the atmosphere and do not have a very large misfit strain (epitaxial to ice lattice).

An important implication of Fig. 9.17 is that the difference in both RHW and RHI, and therefore the shift in water activities between the homogeneous and heterogeneous nucleation processes, is not constant as hypothesized in some parameterizations based on the "shifted activity method" (e.g., Kärcher and Lohmann, 2003). Fig. 9.17 shows that these differences vary in the range 15–40% for the critical relative humidity over water and 25–55% for the critical relative humidity over ice, and depend substantially on the temperature, nucleation (cooling) rates, and physico-chemical properties of freezing aerosol (misfit strain, contact parameter, and others). Thus, parameterizations based on a "shifted activity method" should be corrected accordingly.

The equations derived here are presented both in complete form and in various simplified forms, including their expansions into power series and a simpler parameterization of the critical humidity of heterogeneous nucleation in the form of the temperature polynomial with corrections for the nucleation rate, curvature (finite size) of freezing drops, misfit strain, and external pressure. This parameterization can be applied to mixed aerosol with parameters typical for the submicron mode. In the simplified form, it is similar to the analogous parameterizations suggested in Sassen and Dodd (1989), Koop et al. (1999, 2000), Bertram et al. (2000) for homogeneous nucleation, and to Zuberi et al. (2002), Archuleta et al. (2005) for heterogeneous nucleation, but accounts also for the aforementioned four corrections that substantially change the critical humidities. Such polynomials with corrections could be used in GCMs with simple microphysics that cannot afford more sophisticated parameterizations.

However, the equations derived here for the critical water and ice saturation ratios are sufficiently simple even in complete form and can be directly used in many modern cloud and climate models. If a 3D temperature field and aerosol loading with its microphysics are specified in a model, these equations allow easy and fast calculation of the 3D fields of critical humidities for heterogeneous and homogeneous ice nucleation. Application of both of these ice nucleation schemes in the 1D (single column type) models with spectral bin and bulk double-moment microphysics (Khvorostyanov et al., 2001, 2006; Morrison et al., 2005a,b; Sassen and Khvorostyanov, 2008; Curry and Khvorostyanov, 2012), and in 2D and 3D mesoscale models with spectral bin microphysics of liquid and ice phases and detailed evaluation of supersaturation (Khvorostyanov and Sassen, 2002; Sassen et al., 2002; Khvorostyanov, Curry, et al., 2003) showed that the calculations require only very moderate resources of computer time and memory for simulations of mixed cloud formation and an evolution during several hours.

A comparison of the previous polynomial parameterizations for homogeneous and heterogeneous nucleation with the equations of extended nucleation theory described here showed that these parameterizations describe adequately the critical humidities on the average. However, RHW_{cr} and RHI_{cr} may substantially depend on the nucleation (cooling) rates, aerosol size, and other parameters that are now becoming available and usable in the advanced GCMs (e.g., Lohmann and Diehl, 2006; Yun and Penner, 2012; Morrison and Gettelman, 2008; Gettelman et al., 2008; Ghan et al., 2011), and in the chemical transport models, both global and regional (e.g., CHIMERE model, Menut et al., 2013). The use of the critical humidities based directly on nucleation theory may improve the accuracy of the models, and may help to refine the existing parameterizations taking into account the cooling rates, particle size, and other properties.

9.13. Parcel Model for a Mixed-Phase Cloud

The integral water supersaturation equation for a liquid-phase cloud was derived in Section 7.2, when droplets only are present. An analogous integral ice supersaturation equation for the crystalline clouds was derived in Section 8.5, and was used in Section 8.9.2 to study the kinetics of homogeneous nucleation. Heterogeneous ice freezing occurs at warmer temperatures than homogeneous freezing and usually in liquid or mixed but not fully glaciated clouds—that is, at small water supersaturation or small subsaturation, when both the drops and crystals may form. To study the kinetics of such mixed-phase clouds, we need a supersaturation equation for a mixed-phase cloud similar to that derived in Section 5.3, but taking into account the sources of the new drops and crystals.

In this section, a parcel model of a mixed cloud with spectral water and ice bin microphysics is formulated to simulate the evolution of cloud microphysical processes in adiabatic ascent. The microphysical formulation used here is similar to the Eulerian numerical 1D, 2D, and 3D spectral bin models used previously for simulation of mixed-phase Arctic clouds, orographic clouds, and other cloud types and their artificial seeding (Khvorostyanov, 1982, 1987, 1995; Kondratyev and Khvorostyanov, 1989; Kondratyev, Khvorostyanov, and Ovtchinnikov, 1990a,b; Khvorostyanov and Khairoutdinov, 1990; Khvorostyanov et al., 2001, 2003; Curry and Khvorostyanov, 2012), with some differences in equations owing to the Lagrangian approach in the parcel model, and newer schemes of drop and crystal nucleation described earlier. In the parcel model, vertical velocity is specified. The main thermodynamic equations are the prognostic equations for supersaturation and temperature. This system of equations includes terms that describe the phase transitions and is closed using the kinetic equations of condensation for the drop and ice crystal size distribution functions that account for nucleation and diffusional growth. The supersaturation and kinetic equations for particle size distributions are described in the following text.

9.13.1. Supersaturation Equation with Nucleation of Drops and Crystals

We begin with the general supersaturation equation derived in Section 5.3:

$$\frac{1}{(1+s_w)}\frac{ds_w}{dt} = c_{1w}w - \frac{\Gamma_1 I_{con}}{\rho_v} - \frac{\Gamma_{12} I_{dep}}{\rho_v}, \tag{9.13.1}$$

where c_{1w} was defined by (Eqn. 5.3.26b), and Γ_1 and Γ_{12} are the psychrometric corrections defined in (Eqn. 5.3.15). The sink terms I_{con} and I_{dep} include the integral supersaturation $y_w = \int_0^t s_w(t')dt'$. Therefore, it is convenient to rewrite (Eqn. 9.13.1) for y_w as

$$\frac{y_w'(t)}{[1+y_w'(t)]} = c_{1w}w - \frac{\Gamma_1 I_{con}}{\rho_v} - \frac{\Gamma_{12} I_{dep}}{\rho_v}, \tag{9.13.2}$$

where the prime means derivative by time. The vapor fluxes I_{con} and I_{dep} to the droplets and crystals (the sinks in the supersaturation equation) are the integrals of the mass growth rates that can be expressed using the equations of Chapter 5 for the growth rates dr_d/dt and dr_c/dt of the droplet and crystal radii over the corresponding size spectra of the drops $f_d(r_d, t)$ and crystals $f_c(r_c, t)$, where r_d and r_c are corresponding radii. The second term with I_{con} on the right-hand side of (Eqn. 9.13.1) describing the vapor flux on the activated drops was given by (Eqns. 7.2.12)–(7.2.14) with an assumption that the dry aerosol size spectrum is described by the algebraic size and activity spectrum equivalent to a lognormal distribution. Using the relation $\rho_{ws}/\rho_v = 1/(1+s_w) = (1+y_w')^{-1}$, the term with I_{con} in (Eqn. 9.13.1) can be written as

$$\frac{\Gamma_1}{\rho_v} I_{con}(t) = \frac{y_w'(t)}{[1+y_w'(t)]}[4\pi D_v N_a J_0(t)] = \frac{y_w'(t)[1+y_w'(t)]^{-1}}{\tau_{f,ac}(t)}, \tag{9.13.3}$$

where $\tau_{f,ac}(t) = [4\pi D_v N_a r_{act}(t)]^{-1}$ is the effective supersaturation relaxation time during the stage of drop activation defined by (Eqn. 7.2.13); the activation radius $r_{act}(t)$ and an integral $J_0(t)$ with the dimension of length are defined by (Eqn. 7.2.14) with $\mu = 2$. Substitution of $J_0(t)$ from (Eqn. 7.2.14) into (Eqn. 9.13.3) yields

$$\frac{\Gamma_1}{\rho_v} I_{con}(t) = \frac{y_w'(t)}{[1+y_w'(t)]} \left\{ 4\pi D_v N_a \frac{k_{s0}}{s_0^{k_{s0}}} \right.$$
$$\left. \times \int_0^t r_{d,ef}(t,t_0) \frac{[y_w'(t_0)]^{k_{s0}-1} y_w''(t_0)}{\{1+\eta_0 [y_w'(t_0)]^{k_{s0}}\}^2} dt_0 \right\}, \qquad (9.13.4)$$

where $r_{d,ef}$ is defined by (Eqn. 7.2.11b). The parameters of drop activation, the mean geometric supersaturation s_0, the supersaturation dispersion $\sigma_s = \sigma_d^{(1+\beta)}$, the index $k_{s0} = 4/(\sqrt{2\pi}\ln \sigma_s)$ and the parameter $\eta_0 = s_0^{-k_{s0}}$ are expressed via CCN microphysical properties in (Eqns. 6.6.35b), (6.6.35c), and (6.6.37b).

The term describing the vapor deposition rate on the crystals can be written similar to (Eqn. 8.5.5), as an integral of the crystal mass growth rate dm_c/dt over the size spectrum r_c:

$$I_{dep}(t) = 4\pi\rho_i \int_0^\infty \frac{dr_c(t,t_0)}{dt} r_c^2(t,t_0) f(r_c,t_0)\, dr_c, \qquad (9.13.5)$$

where $r_c(t, t_0)$ is the radius at the time t of the crystals nucleated at t_0. Substituting the crystal growth rate (dr_c/dt) from (Eqn. 8.5.6) yields

$$I_{dep}(t) = s_i(t) \frac{4\pi D_v \rho_{is}}{\Gamma_2} \int_0^\infty \frac{r_c^2(t,t_0)}{r_c(t,t_0)+\xi_{dep}} f_c(r_c,t)\, dr_c. \qquad (9.13.6)$$

Substituting $r_c(t, t_0)$ from (Eqn. 8.5.9) and using the equality $f_c(r_c)dr_c = R_{f,het}(t_0)dt_0$ similar to (Eqn. 8.5.12), we obtain an expression for I_{dep} via the freezing rate $R_{f,het}$ similar to (Eqn. 8.5.16):

$$I_{dep} = y_i' \frac{4\pi D_v \rho_{is}}{\Gamma_2} \int_0^t r_{c,ef}(t,t_0) R_{f,het}(t_0)\, dt_0, \qquad (9.13.7)$$

with $R_{f,het}$ defined in (Eqn. 9.10.5),

$$R_{fr} = \frac{dN_{fr}}{dt} = \int_{r_{\min}}^{r_{\max}} dr_N f_a(r_N) J_{f,het}(t) \exp\left(-\int_0^t J_{f,het}(t')\,dt'\right), \qquad (9.13.7a)$$

and the effective radius $r_{c,ef}(t, t_0)$ defined in (Eqn. 8.5.14). Note that here, as in the case with homogeneous nucleation, or drop nucleation, the integrand in expression (Eqn. 9.13.7) for I_{dep} represents the product of the two terms: the term $R_{f,het}(t_0)$ describes the "generation" of particles at a time t_0, and the term $r_c(t, t_0)$ describes their "propagation in time between t_0 and t," and plays the role of what is termed in physics "the propagator" or Green's function. This structure is similar to those arising in electrotechnique, the theory of radiation, or the theory or elementary particles (e.g., Landau and Lifshitz, v. 2, 3, 4). This analogy may suggest various solutions for this term used in the mentioned theories, in particular, the methods of Green function approximations or the perturbation theory.

Substitution of $R_{f,het}$ from (Eqn. 9.10.5) or (Eqn. 9.13.7a) into (Eqn. 9.13.7) yields the deposition flux expressed via heterogeneous freezing rate $J_{f,het}$, similar to (Eqn. 8.5.17) for homogeneous freezing,

$$I_{dep} = y_i' \frac{4\pi D_v \rho_{is}}{\Gamma_2}\left[\int_0^t r_{c,ef}(t,t_0)\int_{r_{\min}}^{r_{\max}} f_a(r_a)\right.$$
$$\left. \times J_{f,het}(t_0)\exp\left(-\int_0^t J_{f,het}(t')dt'\right)dr_a dt_0\right], \tag{9.13.8}$$

Using the relation $\rho_{is}/\rho_v = 1/(1+s_i) = (1+y_w')^{-1}$, the term $(\Gamma_{12}/\rho_v)I_{dep}$ in (Eqn. 9.13.2) can be now rewritten as

$$\frac{\Gamma_{12}}{\rho_v} I_{dep} = \frac{y_i'(t)}{[1+y_i'(t)]} 4\pi D_v \frac{\Gamma_{12}}{\Gamma_2}\left[\int_0^t r_{c,ef}(t,t_0)\int_{r_{\min}}^{r_{\max}} f_a(r_a)\right.$$
$$\left. \times J_{f,het}(t_0)\exp\left(-\int_0^t J_{f,het}(t')dt'\right)dr_a dt_0\right]. \tag{9.13.8a}$$

Substituting (Eqns. 9.13.4) and (9.13.8a) into the equation (Eqn. 9.13.2) and multiplying by $(1+y_w')$ yields

$$\frac{dy_w'}{dt} = c_{1w}[1+y_w'(t)]w$$
$$-y_w'(t)\left\{4\pi D_v N_a \frac{k_{s0}}{s_0^{k_{s0}}}\int_0^t r_{d,ef}(t,t_0)\frac{[y_w'(t_0)]^{k_{s0}-1} y_w''(t_0)}{\{1+\eta_0[y_w'(t_0)]^{k_{s0}}\}^2}dt_0\right\},$$
$$-y_i'(t)\frac{[1+y_w'(t)]}{[1+y_i'(t)]}4\pi D_v \frac{\Gamma_{12}}{\Gamma_2}\left\{\int_0^t dt_0 r_{c,ef}(t,t_0)\int_{r_{\min}}^{r_{\max}} dr_a f_a(r_a)\right.$$
$$\left. \times J_{f,het}(t_0)\exp\left(-\int_0^t J_{f,het}(t')dt'\right)\right\}. \tag{9.13.9}$$

This is an equation for the integral water supersaturation for a mixed-phase cloud that takes into account the possible drop activation and crystal nucleation by freezing of CCN or drops. It includes both integral supersaturations over water y_w and ice y_i, and can be written for y_w only using the relation (Eqn. 5.3.27) between s_w and s_i, which for the integral supersaturations becomes

$$y_i'\frac{1+y_w'}{1+y_i'} = y_w' + \frac{\rho_{ws}-\rho_{is}}{\rho_{ws}}. \tag{9.13.10}$$

Substitution into (Eqn. 9.13.9) yields an equation for y_w only that can be written in a more compact form using again I_{dep} as in (Eqn. 9.13.7) via $R_{f,het}$:

$$\frac{dy_w'}{dt} = c_{1w}[1+y_w'(t)]w$$
$$-y_w'(t)\left\{4\pi D_v N_a \frac{k_{s0}}{s_0^{k_{s0}}}\int_0^t r_{d,ef}(t,t_0)\frac{[y_w'(t_0)]^{k_{s0}-1} y_w''(t_0)}{\{1+\eta_0[y_w'(t_0)]^{k_{s0}}\}^2}dt_0\right\}$$
$$-y_w' 4\pi D_v \frac{\Gamma_{12}}{\Gamma_2}\int_0^t r_{c,ef}(t,t_0)R_{f,het}(t_0)dt_0$$
$$-\left(\frac{\rho_{ws}-\rho_{is}}{\rho_{ws}}\right)4\pi D_v \frac{\Gamma_{12}}{\Gamma_2}\int_0^t r_{c,ef}(t,t_0)R_{f,het}(t_0)dt_0. \tag{9.13.11}$$

This equation describes the vapor balance in a mixed-phase cloud and its four terms on the right-hand side have simple physical meanings. The first term with w describes supersaturation generation due to cooling in an updraft; the second term with $r_{d,ef}$ and the third term with $r_{c,ef}$ describe supersaturation depletion due to droplets and crystals absorption respectively, and the last term with $(\rho_{ws} - \rho_{is})$ describes the vapor flux from the drops to the crystals due to the difference of saturated humidities over water and ice (Wegener–Bergeron–Findeisen mechanism, WBF). The second term generalizes Twomey's (1959) equation for the drop activation with correct asymptotic and limited drop concentration at large s_w (as described in Chapters 6, 7), and the last two terms generalize this equation with an accounting for the supersaturation depletion by the crystals and WBF mechanism. Equation (9.13.11) can be used for analytical studies of the supersaturation y_w' and s_w and nucleated drop and crystal concentrations in a mixed-phase cloud, as was done in Chapter 7 for a liquid cloud. Equation (9.3.11) can be also used for numerical simulations in cloud models and advanced climate models. This equation can be simplified in many cases, as will be shown in the next sections.

Equation (9.13.11) is written in a closed form so that the source terms are included in the integrals I_{con}, I_{dep} with an accounting for the evolution of the size spectra of nucleated drops and crystals. An analog of (Eqn. 9.13.11) is (Eqn. 5.3.28a) rewritten for the integral supersaturation y_w' as

$$\frac{dy_w'(t)}{dt} = (1+y_w')c_{1w}w - y_w'(t)\left(\frac{1}{\tau_{fd}} + \frac{\Gamma_{12}}{\Gamma_2}\frac{1}{\tau_{fc}}\right) - \frac{\Gamma_{12}}{\Gamma_2}\frac{1}{\tau_{fc}}\frac{\rho_{ws}-\rho_{is}}{\rho_{ws}}, \tag{9.13.12}$$

where τ_{fd} and τ_{fc} are the droplet and crystal supersaturation relaxation times

$$\tau_{fd}^{-1} = 4\pi D_v \int_0^\infty \frac{r_d^2}{r_d + \xi_{con}} f_d(r_d, t)\, dr_d \approx (4\pi D N_d \bar{r}_d)^{-1}, \tag{9.13.13}$$

$$\tau_{fc}^{-1} = 4\pi D_v \int_0^\infty \frac{r_c^2}{r_c + \xi_{dep}} f_c(r_c, t)\, dr_c \approx (4\pi D N_c \bar{r}_c)^{-1}. \tag{9.13.14}$$

The times τ_{fd} and τ_{fc} are determined by the droplet and crystal size spectra, and the second equalities in (Eqns. 9.13.13), (9.13.14) are valid in the diffusion regime, $r_d << \xi_{con}$, $r_c << \xi_{dep}$. In (Eqns. 9.13.12)–(9.13.14), the source and sinks of the particles are described by the size spectra $f_d(r_d, t)$ and $f_c(r_c, t)$, and kinetic equations are needed to calculate $f_d(r_d, t)$ and $f_c(r_c, t)$ at every time moment.

The heat balance is calculated using the equation for the temperature T in a wet adiabatic process considered in Chapters 3 and 5 in the form

$$\frac{dT}{dt} = -\gamma_a w + \frac{L_e}{c_{pa}\rho_a} I_{con} + \frac{L_s}{c_{pa}\rho_a} I_{dep} \tag{9.13.15}$$

This equation is solved along with the integral ice supersaturation equation (Eqns. 9.13.11) or (9.13.12) that is required to describe droplet and crystal nucleation and growth.

9.13.2. Kinetic Equations for Droplet and Crystal Size Spectra with Particle Nucleation

Evolution of the droplet and crystal spectra is described using two kinetic equations for the $f_d(r_d, t)$ and $f_c(r_c, t)$ similar to that in Chapter 8,

$$\frac{\partial f_d}{\partial t} + \frac{\partial}{\partial r_d}\left(\frac{dr_d}{dt} f_d\right) = \psi_{fd}(r_d, t), \tag{9.13.16a}$$

$$\frac{\partial f_c}{\partial t} + \frac{\partial}{\partial r_c}\left(\frac{dr_c}{dt} f_c\right) = \psi_{fc}(r_c, t). \tag{9.13.16b}$$

As in Chapter 8, these equations are written for the simplest case, with an accounting for regular condensation and deposition only, and without taking into account aggregation and the effects of turbulence. As in Chapter 8, following the methodology recommended for the Cirrus Parcel Model Comparison Project (CPMCP, Lin et al., 2002), we deliberately exclude from consideration coagulation among the droplets and aggregation between the droplets and crystals, sedimentation, entrainment, turbulent exchange, etc. to isolate the effects directly related to nucleation processes. (Inclusion of these processes in multidimensional Eulerian models is described in Chapter 5, Sections 5.7 and 5.8). The source terms ψ_{fc} and ψ_{fd} on the right-hand side describe particle nucleation. The droplet nucleation term ψ_{fd} is calculated using the CCN activity spectrum $\varphi(s_w)$ which is the algebraic or modified power law, as described in Chapters 6 and 7:

$$\psi_{fd} = \varphi(s_w)\frac{ds_w}{dt}\delta(r_d - \Delta r_d), \tag{9.13.17}$$

where Δr_d is the first step by droplet radius, δ is Dirac's delta function, and its simplest finite difference approximation is $1/\Delta r_d$. If s_w reaches 0 and increases, then N_d increases by ΔN_d in one time step Δt, $\psi_{fd} = \Delta N_d/\Delta r_d/\Delta t$ and the new drops are placed in the first size bin Δr_d of $f_d(r_d)$.

The model includes both heterogeneous and homogeneous ice nucleation processes. The crystal source term for the heterogeneous deliquescent-freezing and immersion modes is calculated as $\psi_{fc} = \Delta N_{f,het}/\Delta r_c/\Delta t$ with the number $\Delta N_{f,het}$ of the crystals nucleated on frozen CCN-IN or drops in a time step Δt calculated as described in this chapter; Δr_c, is the first size step by the crystal radii (0.1–1 μm). Heterogeneous freezing rate $J_{f,het}$ is calculated based on Section 9.6 with Δr_{cr} and ΔF_{cr} from Section 9.7, and then $\Delta N_{f,het}$ is calculated following Section 9.10. The homogeneous nucleation rate $R_{f,hom}$ and $\Delta N_{f,hom}$ are calculated in a similar way, based on Chapter 8.

Both spectra include 30 points by radius: 10 steps by $\Delta r_{d,c}$ = 0.1–1 μm and the next 20 steps increasing logarithmically to the maximum r_{max}. The values of $\Delta r_{d,c}$ and r_{max} were varied in numerical experiments depending on the maximum size that the particles can reach, r_{max} = 350 μm at warmer temperatures −12 to −20 °C and 100 μm at low $T \sim -50$ °C. The experiments showed that this division allows coverage of both small and large size ranges without losing accuracy.

9.14. Parcel Model Simulations of Ice Nucleation Kinetics in Deliquescence-Freezing Mode

9.14.1. Introduction

The main disadvantage of many previous parameterizations of ice nucleation is that they are empirical, with parameters tuned to produce ice crystal concentrations N_c within some limited range of temperatures and supersaturation based on field or laboratory measurements. Such an approach has led to contradictions between parameterizations of the measured concentrations of ice nuclei, N_{IN}, and of the crystals, N_c, so that the "ice enhancement factor" $R_M = N_c/N_{IN}$, can reach values as high as 10^4–10^5 at $T = -5\,°C$ (e.g., Hobbs, 1969, 1974), which has produced several hypotheses on "ice multiplication" and stimulated their studies (e.g., Hallett and Mossop, 1974; Mossop and Hallett, 1974), or requirements for high water supersaturations (Hobbs and Rangno, 1990; Rangno and Hobbs, 1991).

Supersaturation dependence of ice nuclei concentration may produce a negative feedback and stop nucleation even in the presence of external sources of supersaturation generation; the temperature dependence of ice nucleation does not have this feature. If a source of cooling persists, the latent heat release of growing crystals may hamper nucleation but cannot stop it, commonly producing unrealistically high values of N_i, and artificial methods may be required to prevent unlimited ice nucleation when using a temperature-dependent parameterization (e.g., Spice et al., 1999). The heterogeneous freezing nucleation rate of classical nucleation theory describes the temperature dependence of the nucleation rate but does not contain supersaturation dependence (see Section 9.7). The theory of heterogeneous condensation-freezing by Fukuta and Schaller (1982) discussed in Section 9.7 showed that a combined temperature-supersaturation dependence occurs if heterogeneous freezing takes place as a two-step process: a) condensation on insoluble aerosol particles at positive water supersaturation; and b) subsequent freezing that depends on temperature. A similar model for the immersion-freezing mode was considered by Young (1993).

A theory of ice nucleation by heterogeneous freezing of deliquescent mixed CCN by Khvorostyanov and Curry (2000, 2004a,b; Section 9.7, hereafter KC00-04) allowed to describe simultaneously both the temperature and supersaturation dependencies of ice crystal nucleation by mixed CCN freezing. Heterogeneous ice crystal nucleation was examined in Section 9.7 for static conditions—i.e., for fixed temperature and supersaturation. However, nucleation rates and concentration of nucleated crystals cannot be explored and explained without a dynamical framework. Khvorostyanov and Curry (2005a) incorporated this KC00-04 theory of heterogeneous ice nucleation into a parcel model with explicit water and ice bin microphysics and simulated the process of ice nucleation under a wide range of transient thermodynamic conditions. The results of simulations in the work by Khvorostyanov and Curry (2005a, hereafter KC05a) illustrate the kinetics of heterogeneous freezing and are described in this section, similar to what has been done in Section 8.7 of Chapter 8 for homogeneous nucleation.

Simulations are conducted over the temperature range −4 to −60 °C, with vertical velocities varying from 1 to 100 cm s^{-1}, for varying initial relative humidities and aerosol characteristics. It is shown that the kinetics of heterogeneous ice nucleation exhibits a negative feedback regulated via water supersaturation. This feedback is similar to that found for homogeneous nucleation in works cited in Chapter 8, Section 8.7 (e.g., Lin et al., 2002) and Section 9.7, and is much stronger than the corresponding feedback for drop nucleation. This may partially explain discrepancies between observed ice nuclei concentrations and ice crystal concentrations. At the end of this section, the kinetics and the relative importance of heterogeneous versus homogeneous nucleation is examined for a variety of cloud conditions, and possible applications of this theory for cloud modeling are discussed.

9.14.2. Simulation Characteristics

To simulate the ice crystal nucleation process, the parcel model was run for 1–3 hours, varying the initial temperature (T_{0c}), relative humidity (RHW_0), vertical velocity (w), and aerosol characteristics. The only two additional parameters of this model of heterogeneous freezing are the relevant aerosol characteristics (see Section 9.7): the misfit strain ε and the wettability or contact parameter, m_{is}. Most runs were performed with $\varepsilon = 0$, and sensitivity to $\varepsilon = 1$–2% was tested in some experiments. The parameter m_{is} given in Table 5.2 in Pruppacher and Klett (1997) for surface soil ($m_{is} = 0.36$–0.42) and quartz ($m_{is} = 0.63$–0.72) might be representative for mixed CCN. Kärcher et al. (1996) measured the values $m_{is} = 0.44$–0.57 for soot that may constitute insoluble carbonaceous fraction of IN. Similar ranges of the contact parameter were reported in Young (1993), Marcolli et al. (2007), and many other studies, although smaller and even negative values of m_{is} were sometimes detected (e.g., Hung et al., 2003; Archuletta et al., 2005). The average values of $m_{is} = 0.50$ or 0.52 were taken in the simulations described in detail in the following text and the sensitivity to the lower and higher values of m_{is} in the range 0.12– 0.75 was also tested along with some negative $m_{is} < 0$ as in Section 9.12 for the critical humidities or saturation ratios. Numerical experiments showed that the values of m_{is} are mostly limited to this rather narrow range, otherwise freezing may occur too rapidly (for $m_{is} \geq 0.67$) or slowly (for small $m_{is} \leq 0.40$). This is caused by the exponential dependence of nucleation rate $J_{S,fr}$ on $\Delta F_{cr} \sim f(m_{is}, x)$, that was emphasized by Fletcher (1962) and discussed in Section 9.7. Models with lognormal and other distributions of the contact angles were also tested and showed that such ditribution can more smoothly describe ice nucleation in a wider temperature range (e.g., Marcolli et al., 2007; Lüönd et al., 2010; Hoose and Möhler, 2012). The results described in the following text and obtained with just one contact angle can be easily generalized for any statistical distributions of the contact angles.

The parcel model used here includes the option of isolating specific ice crystal nucleation modes, by turning off the other modes. In this section, we consider heterogeneous and homogeneous freezing of deliquescent CCN-IN and the other modes (deposition and contact) are considered in Chapter 10. The initial pressure p_0 is specified in most runs to be 800 hPa. The cloud condensation nuclei (CCN) concentration typical for continental clouds $N_a = 500\,\text{cm}^{-3}$ is chosen for all runs except for simulation of ice nucleation in maritime convective clouds, where N_a = 100–150 cm^{-3}, and for comparison of heterogeneous and homogeneous modes in cirrus clouds at low T_c and high altitudes, where $N_a = 200\,\text{cm}^{-3}$, as was recommended and used in CPMCP (Lin et al., 2002). Several experiments were performed with N_a = 50–100 cm^{-3} at $T_c = -20\,°\text{C}$ to $-30\,°\text{C}$ to test the possibility of mixed-phase cloud existence at these temperatures. The insoluble aerosol size spectrum over r_N is lognormal with modal radius $r_m = 0.02\,\mu\text{m}$ and dispersion $\sigma = 2.5$; at $s_w > 0$, it is limited by the boundary radius r_b as described in Section 9.8. More than 300 simulations were performed with variable combinations of T_{0c}, RHW_0, m_{is}, ε, w, aerosol microstructure, and the mode of freezing (heterogeneous deliquescent-freezing and/or homogeneous freezing).

The time steps were 0.01–0.2 s in the main program, but the time step can be divided further in the condensation subroutines as described in Section 5.8 to meet the stability condition $\Delta t \ll \min(\tau_d, \tau_c)$. The accuracy of the calculations was controlled by comparing the total number of crystals nucleated with those obtained by integration over the size spectrum of the grown crystals at the end of a parcel run. If the error exceeded 5% (especially at low temperatures), the time and radius steps were varied and several additional runs were performed until the error became less than 5%. The runs with vertical velocities w = 1–5 cm s^{-1} described in this section were conducted for a period of 3 hours, which was sufficient to trace all stages of supersaturation growth, nucleation, and subsequent supersaturation relaxation. In the runs with higher w = 30–100 cm s^{-1}, all the processes were faster, and it was sufficient to conduct simulations for a period of 0.5–1 h. The simulation results are described in the following text.

9.14.3. Kinetics of Ice Nucleation in the Mixed and Crystalline Clouds with Weak Updrafts

In this section, we describe the temporal evolution of the ice nucleation process. Figs. 9.18–9.21 show the results of the run with the initial $T_{0c} = -14\,°C$, $RHW_0 = 89\%$ ($s_w = -11\%$), vertical velocity $w = 2\,cm\,s^{-1}$ (which corresponds to a cooling rate of $-0.7\,°C\,h^{-1}$), and $m_{is} = 0.52$. The initial supersaturation over ice is $s_i = 11\%$. Figs. 9.18 and 9.19 illustrate that the temporal evolution can be separated into 3 stages. During the first stage (up to 115 minutes), the temperature decreases due to cooling and RHW increases (Fig. 9.18a,b), which causes a decrease in the critical radius r_{cr} and energy ΔF_{cr} defined in Section 9.7 (Fig. 9.18c,d). During the second stage, starting at $t = 115$ minutes, the nucleation rates reach their threshold values and grow very rapidly (Fig. 9.18e,f). At $t = 117$ minutes, RHW reaches and exceeds 100%, and drop activation occurs. Until $t = 145$ minutes, s_w slightly decreases because of the drop formation and growth, and r_{cr} and ΔF_{cr} decrease, while $J_{f,het}$ and dN_c/dt increase. Note that this occurs not at the threshold humidity RHW_{th}, but when RHW exceeds the threshold by ~13% (Fig. 9.18b), in agreement with estimations of the critical saturation ratio $S_{w,cr}$ given in Section 9.12. Ice nucleation rates are noticeable only for about 25 minutes, when the drops and crystals coexist (Fig. 9.19). During the third stage (starting nominally at $t = 145$ minutes), the rate of vapor absorption by the crystals increases sufficiently for complete evaporation of the drops by Wegener–Bergeron–Findeisen mechanism (Fig. 9.19d,f). Drops evaporate, RHW abruptly decreases, and r_{cr} and ΔF_{cr} increase rapidly, causing $J_{f,het}$ and dN_c/dt to decline below the threshold values, and hence ice nucleation stops.

Evolution of the integral characteristics over these 3 stages of nucleation is shown in Fig. 9.19. During the 1st stage, the initial subsaturation s_w decreases due to cooling and reaches -0.3% at $t = 115$ min (Fig. 9.19a) when the crystal formation starts (Figs. 9.19c,e,g), and crystal concentration N_c reaches $\sim 1\,L^{-1}$ in 5 minutes by $t = 120$ minutes. However, since the crystal supersaturation relaxation time is greater than 2 hours (Fig. 9.19b), the crystals cannot absorb all available vapor, and supersaturation generation is still faster than its absorption, so s_w continues to increase and becomes positive at $t \approx 120$ minutes. During the 2nd stage, water supersaturation becomes positive, droplets begin to form at $t = 118$ min (Fig. 9.19d,e,h). Their relaxation time is a few seconds (Fig. 9.19b), thus the supersaturation generation and absorption are balanced, s_w does not grow, and drops and crystals coexist for 30 min in a mixed cloud. Meanwhile, crystal nucleation is continuing, N_c reaches $\sim 8\,L^{-1}$, and the mean radius is $60\,\mu m$ by $t = 145$ min. The crystal capability of supersaturation absorption gradually increases during the second stage and finally exceeds the rate of supersaturation generation. The vapor flux is directed now from the droplets to the crystals, and due to the intensive Wegener–Bergeron–Findeisen process, the drops evaporate fast and disappear at $t = 150$ minutes, when s_w becomes negative. The crystal concentration continues to increase during the second stage (although at a slower rate) and reaches $17.6\,L^{-1}$at the end of the second stage. The third stage starts when both s_w and s_i begin to decrease. This stage is characterized by constant N_c, but the crystal radius and IWC still increase since s_i is positive.

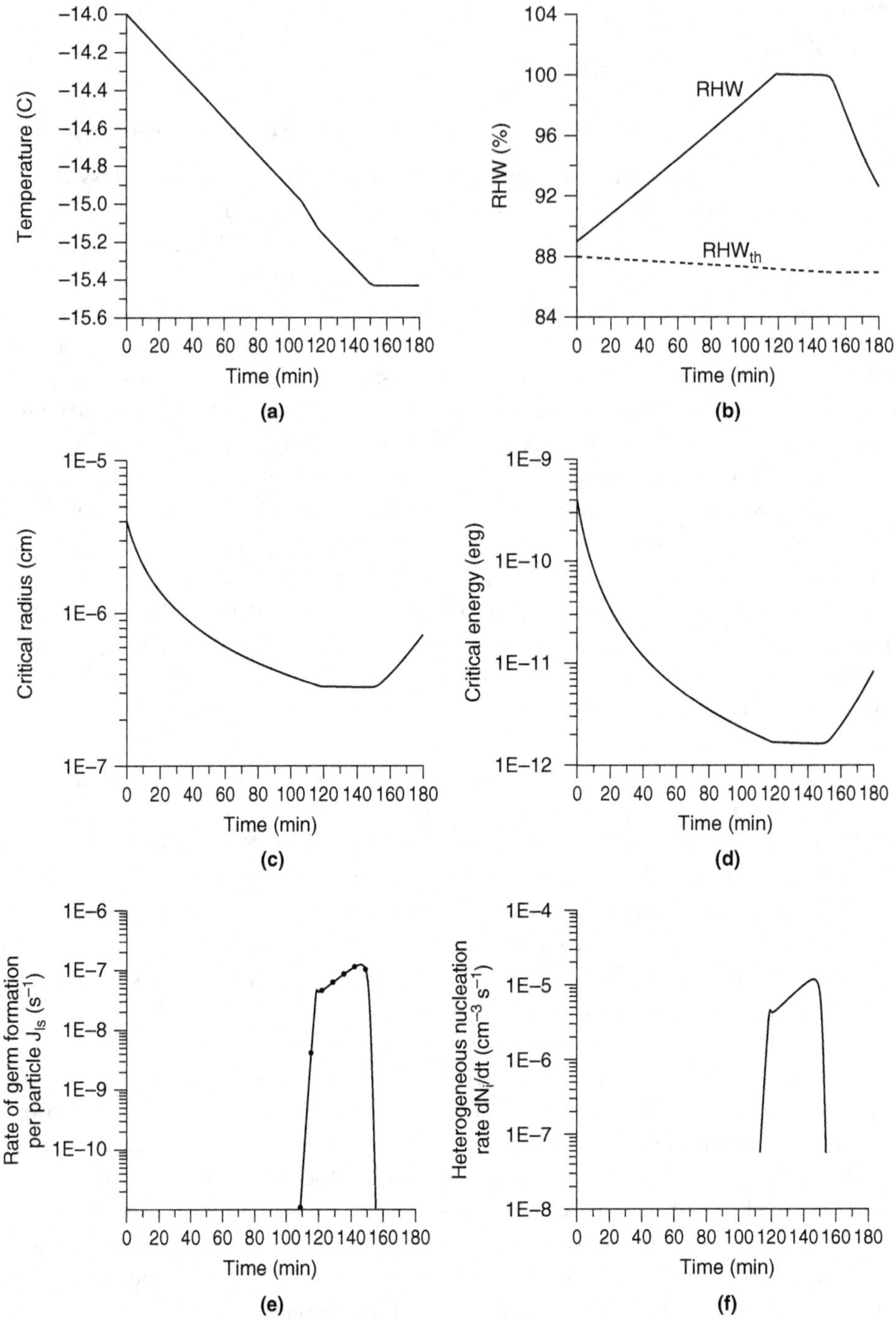

Figure 9.18. Temporal evolution of the characteristics of heterogeneous nucleation in the parcel at $T_0 = -14\,°C$; $RHW_0 = 89\%$, $w = 2\,cm\,s^{-1}$, $m_{is} = 0.52$. (a) Temperature, T, °C; (b) RHW and threshold humidity RHW_{th},%; (c) germ critical radius, r_{cr}, cm; (d) germ critical free energy, ΔF_{cr}, erg; (e) the rate of germ formation per particle $J_{f,het}$, s^{-1}; (f) polydisperse nucleation rate $R_{fr} = dN_{fr}/dt$, $cm^{-3}\,s^{-1}$. From Khvorostyanov and Curry (2005a), *J. Atmos. Sci.*, 62, © American Meteorological Society. Used with permission.

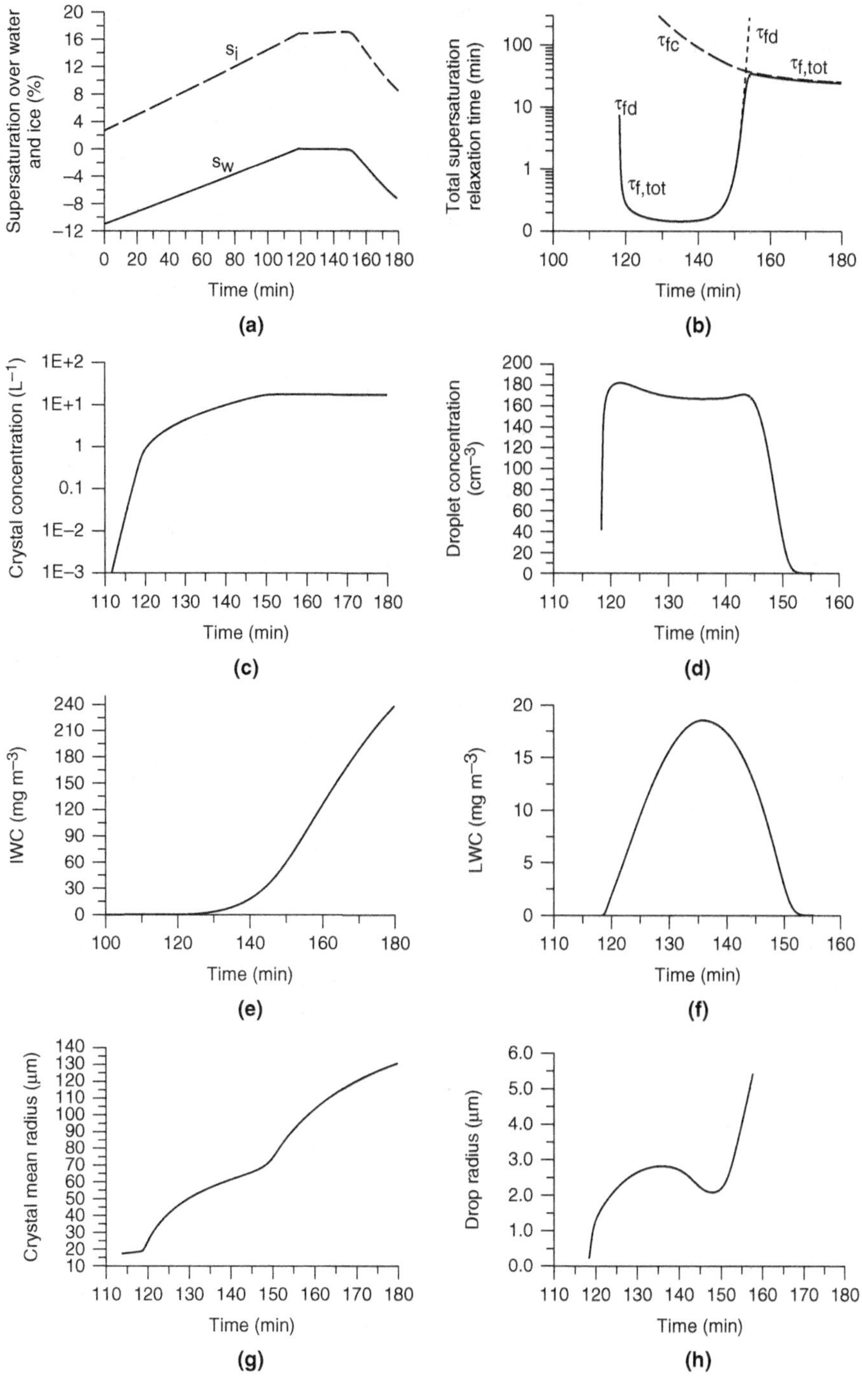

Figure 9.19. Time dependence of droplets and crystals microphysical properties with $T_0 = -14\,°C$, $RHW_0 = 89\%$, $w = 2\,cm\,s^{-1}$, $m = 0.52$. (a) Supersaturations over water and ice (%); (b) supersaturation relaxation times for drops τ_{fd}, crystals τ_{fc}, and total relaxation time $\tau_{f,tot}$ (min); (c) crystal concentration (L^{-1}); (d) droplet concentration (cm^{-3}); (e) IWC ($mg\,m^{-3}$); (f) LWC ($mg\,m^{-3}$); (g) crystal mean radius (μm); (h) droplet mean radius (μm). From Khvorostyanov and Curry (2005a), *J. Atmos. Sci.*, 62, © American Meteorological Society. Used with permission.

Thus, the modeled cloud is pure crystalline in the first stage (115–120 minutes), has a mixed-phase structure during the second stage for 30 minutes, and is again pure crystalline in the third stage. In this case, the cloud phase is governed by the relative humidity, which determines also the IN concentration (assuming that $N_{IN} = N_c$). As water saturation is reached, there is a competition between nucleation of the drops and crystals on the same population of the haze particles that serve at the same time as CCN and IN.

This simulation produced a value of $N_{dr} = 160\,\mathrm{cm}^{-3}$ out of a total $N_{CCN} = 500\,\mathrm{cm}^{-3}$ CCN. We can define "condensation efficiency" of CCN in producing droplets as $E_{dr} = N_{dr}/N_{CCN}$, and it is here $E_{dr} \approx 0.32$. The crystals are nucleating simultaneously, and only a very small fraction of the available CCN, $18\,\mathrm{L}^{-1}$ ($\sim 2 \times 10^{-2}\,\mathrm{cm}^{-3}$) was nucleated and would have been identified as IN in chamber experiments. A similar "freezing efficiency" of deliquescent CCN in producing crystals can be defined as $E_{cr} = N_c/N_{CCN} \sim N_{IN}/N_{CCN} = 2 \times 10^{-2}/500 = 40 \times 10^{-6}$. This estimate is close both conceptually and quantitatively to that obtained by Rogers et al. (2001) in recent measurements of IN in the Arctic ($N_{IN}/N_{CCN} \sim 20 \times 10^{-6}$), in SUCCESS (Chen et al., 1998; Rogers et al., 1998) and to the estimates of N_{IN}/N_{CCN} in the parcel model by DeMott et al. (1998).

Since the freezing efficiency E_{cr} of CCN is much less than the condensation efficiency E_{dr}, these results show that the present theory enables explanation at least of the observed differences between concentrations of the drops and crystals nucleated from the same population of deliquescent mixed CCN similar to the concepts derived from the field experiments (Hobbs and Rangno, 1990; Rangno and Hobbs, 1991; Pruppacher and Klett, 1997; DeMott et al., 1998). In numerical experiments here, ice nucleation on the CCN is prevented by a very strong humidity dependence of r_{cr}, ΔF_{cr}, $J_{f,het}$, and dN_c/dt. The dependence on humidity is an exponential function (see Section 9.8, Eqn. 9.8.11) that is much stronger than the power law for drop nucleation (Chapters 6 and 7), which is why crystal nucleation on CCN creates such a strong negative feedback via relative humidity and leads to such dramatic differences between the concentrations of drops and crystals.

The dissimilarity between the drops and crystals nucleation on CCN and the ratio of the final concentrations N_c and N_d in a mixed cloud can be illustrated from the analysis of the supersaturation (Eqns. 9.13.1) and (9.13.12) and Fig. 9.19. The second and third terms on the right-hand side in (Eqn. 9.13.1) are proportional to the vapor fluxes to the drops, I_{con}, and crystals, I_{dep}, and describe absorption by the drops and crystals. Note that according to (Eqns. 5.3.19), (9.13.13), and (9.13.14), $I_{con} \sim s_w/\tau_d \sim s_w N_d r_d$, and $I_{dep} \sim s_i/\tau_c \sim s_i N_c r_c$. Ice nucleation begins and proceeds at slow nucleation rates, and $I_{dep} << I_{con}$. Activation of drops occurs at $s_w \sim 0.1\%$, and during most of the time of liquid phase existence, supersaturation generation by w is balanced mostly by its absorption by the drops, by the flux I_{con}. After drop activation, $s_i \sim 20\%$, the newly nucleated crystals grow much faster than the drops and reach larger sizes, $r_i \sim 20\text{–}30 \cdot r_d$ by the end of cloud glaciation. At this time, supersaturation absorption by the crystals I_{dep} become comparable to I_{con} and then exceeds it, which causes the decrease of s_w and ceases ice nucleation (after 150 min in Fig. 9.19). Therefore, we can estimate the ratio of final concentrations N_c and N_d from the condition $I_{con} \approx I_{dep}$, or $s_w N_d r_d \sim s_i N_c r_c$, which yields in the preceding example $N_c/N_d \sim (s_w/s_i)(r_d/r_c) \sim (0.5 \times 10^{-2})(3 \times 10^{-2}) = 1.5 \times 10^{-4}$, an estimate close to the simulated $N_c/N_d \approx 20\,\mathrm{L}^{-1}/160\,\mathrm{cm}^{-3} = 1.3 \times 10^{-4}$.

Thus, the following factors determine the large difference between concentrations of crystals and drops formed on the same population of CCN: 1) crystal nucleation rate dN_c/dt is much slower than that for the drops; 2) ice supersaturation is ~10^2–10^3 times greater than water supersaturation in a mixed cloud and this dissimilarity determines the difference in growth rate, sizes, and supersaturation absorption by drops and crystals; 3) therefore, the absorption of supersaturation by crystals begins to exceed its generation by updrafts at much smaller concentrations, but at larger radii than that for drops; and 4) this causes a supersaturation decrease and a cease of ice nucleation at $N_c << N_d$.

Thus, in this parcel model without the precipitation of crystals, the mixed-phase cloud exists for half an hour and then full glaciation occurs. If the newly formed crystals were allowed to precipitate, the cloud mixed-phase stage could continue much longer. This was confirmed in simulations with a 3D model with spectral bin microphysics of a fog and cloud formed over the polar polynya in the Beaufort Sea observed during the SHEBA-FIRE field experiment in 1998 in the Arctic (Khvorostyanov, Curry, et al., 2003). The same heterogeneous ice nucleation scheme was applied at similar temperatures, and crystals nucleated with concentrations of 5–15 L^{-1}. However, crystal concentrations were depleted due to precipitation and wind transport, and mixed-phase cloud existed at least for several hours with very light precipitation a few tens of km away from the polynya. Simulation with a 1D model and this ice scheme of the cloud observed during the MPACE field campaign in the Arctic also showed the existence of the mixed-phase cloud without full glaciation for several hours. The microstructure and phase state of the simulated cloud were close to those observed (Curry and Khvorostyanov, 2012). Thus, this heterogeneous ice nucleation scheme allows us to explain the existence of the Arctic clouds in the mixed-phase state for a long time and its application in multidimensional models may provide new insights for phase transitions in clouds.

The equations of this ice nucleation theory allow significant simplifications. Note that the value of $J_{f,het}$ is very small ($<2 \times 10^{-7}\,\mathrm{s}^{-1}$) at all stages of the process (Fig. 9.18e), and the expressions introduced in Section 9.10 can be substantially simplified since in (Eqn. 9.10.1)

$$\exp\left(-\int_0^t J_{f,het}(r_N,t')\,dt'\right) \approx 1 - \int_0^t J_{f,het}(r_N,t')\,dt'. \qquad (9.14.1)$$

The probability of the freezing $P_{cf}(r_N, t)$ of an aerosol particle with r_N in (Eqn. 9.10.1) can be simplified as

$$P_{cf}(r_N,t) = 1 - \exp\left(-\int_0^t J_{f,het}(r_N,t')\,dt'\right) \approx \int_0^t J_{f,het}(r_N,t')\,dt'. \qquad (9.14.2)$$

The total number of crystals (Eqn. 9.10.3) is also simplified as

$$N_{fr}(t) = \int_{r_{\min}}^{r_{\max}} P_{cf}(r_N,t) f_a(r_N)\,dr_N \approx \int_{r_{\min}}^{r_{\max}} \int_0^t J_{f,het}(r_N,t') f_a(r_N)\,dt' dr_N. \qquad (9.14.3)$$

The crystal nucleation rate in a polydisperse aerosol (Eqn. 9.10.5) can be simplified as

$$R_{fr}(t) = \frac{dN_{fr}}{dt} = \int_{r_{\min}}^{r_{\max}} f_a(r_N) J_{f,het}(r_N,t)\,dr_N \qquad (9.14.4)$$

These simplifications can be very useful for analytical parameterizations similar to those developed in Section 8.10 for homogeneous nucleation and derived for heterogeneous nucleation in Section 10.1, and for numerical modeling, since they significantly shorten the number of arithmetic operations.

A key issue in heterogeneous nucleation is the contribution of the various regions of aerosol size spectra to the nucleation. Fig. 9.20 shows the subintegral functions of (Eqns. 9.10.5) and (9.10.3), or

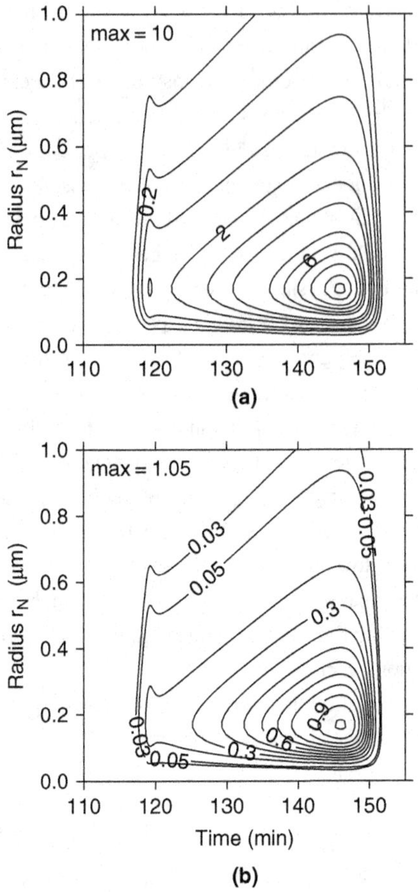

Figure 9.20. Characteristics of contributions of various aerosol size fractions into the polydisperse nucleation rate at various stages of nucleation. (a) Subintegral function (in $10^{-7}\,cm^{-3}\,s^{-1}$) in polydisperse nucleation rate $dN_{fr}/dt\ (r_N, t)$, (Eqns. 9.10.5) or (9.14.4); (b) Subintegral function in polydisperse concentration $N_f\,(r_N, t)$ of frozen aerosol (cm^{-4}), (Eqns. 9.10.3) and (9.14.3). Maxima of subintegral functions are indicated on top of Figures. From Khvorostyanov and Curry (2005a), *J. Atmos. Sci.*, 62, © American Meteorological Society. Used with permission.

(Eqns. 9.14.4) and (9.14.3) here, for the polydisperse nucleation rate $dN_c/dt(t)$ and crystal concentration $N_c(t)$, illustrating the size contributions to these quantities. The maximum contributions come from the region $r_a \sim 0.1$–$0.6\,\mu m$, in agreement with the available experimental data that show maximum occurrence of the IN radii ~0.2–2.5 μm (e.g., Pruppacher and Klett, 1997, p. 327), or 0.18–0.23 μm (Rogers et al., 1998, 2001; Chen et al. 1998). Note that we accounted here only for aerosol with $r_N < 4\,\mu m$. It would be interesting to estimate the contributions from larger particles. Berezinsky and Stepanov (1986) showed that these particles (up to 100 μm) might give contributions at higher T (in agreement with Section 9.10), and explained their action by the easier dissolution of the hygroscopic fraction of these IN. Accounting for such large IN requires knowledge of the distribution of soluble and insoluble fractions inside the large CCN.

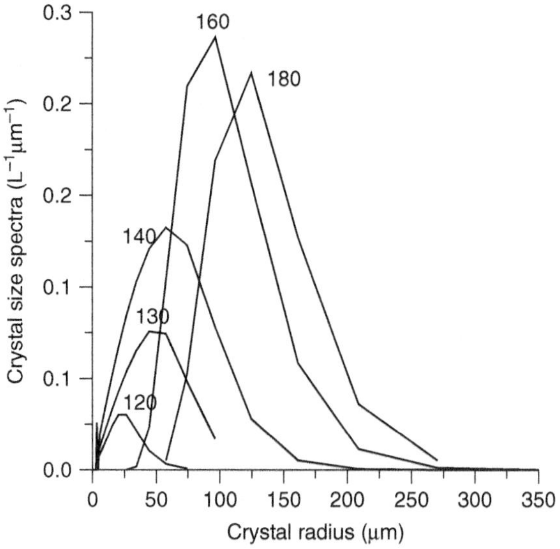

Figure 9.21. Temporal evolution of the crystal size spectra for the same conditions as in Figs. 9.18–9.19 ($T_0 = -14$ °C). The numbers at the curves indicate time in minutes. From Khvorostyanov and Curry (2005a), *J. Atmos. Sci.*, 62, © American Meteorological Society. Used with permission.

The temporal evolution of the ice crystal size spectra is shown in Fig. 9.21. The maximum crystal radius grows until $t = 150$ minutes, corresponding to increasing N_c (see Fig. 9.19). Nucleation ceases after 150 minutes, the small size fraction vanishes, the minimum and maximum radii at 180 minutes are 50 and 280 µm, and the sizes increase with time. The absolute dispersion of the size spectra σ_{abs} decreases due to the Maxwellian $1/r_i$ growth law, and the relative dispersions $\sigma_{abs}/r_{i,mean}$ decreases even faster because of the increasing mean radius $r_{i,mean}$. Thus, the crystal size spectra produced by a parcel model narrows with time similar to the droplet spectra in response to diffusional growth, as was found in the previous parcel model simulations (Howell, 1949; Mordy, 1959; Neiburger and Chien, 1960) and discussed in Section 5.5.2. In a real cloud, the spectral narrowing is overwhelmed by turbulent mixing, accretion, and other effects (e.g., Khvorostyanov and Curry, 1999c,d, 2008b,c; see Chapters 13 and 14).

When the initial temperature T_{c0} decreases further (from −18 to −57 °C), no drops occur with $m_{is} = 0.5$ and the cloud is entirely crystalline, even initially. The kinetics of this process at $T_{c0} = -18$ °C and $w = 2\,\mathrm{cm\,s^{-1}}$ is shown in Fig. 9.22. The crystals form via heterogeneous freezing of deliquescent CCN at water subsaturations similar to homogeneous freezing but at higher temperatures due to the catalyzing effect of the insoluble fraction. The maximum supersaturations over water, s_{wm}, and over ice, s_{im}, at which ice nucleation occurs are −3 and 20% respectively at $T_c = -19$ °C, and final $N_c = 27\,\mathrm{L^{-1}}$. This case illustrates how clear-sky crystal precipitation (diamond dust) may occur due to the heterogeneous freezing of deliquescent CCN even at relatively warm temperatures and humidities intermediate between water and ice saturation, such as have been observed in the Arctic, particularly near open leads with large relative humidity or are observed in winter at cold temperatures (Ohtake et al., 1982; Curry, 1983, 1995; Curry et al., 1990, 1996, 2000; Pinto et al., 2001; Girard and Curry, 2001; Girard and Blanchet, 2001; Gultepe et al., 2003; Khvorostyanov et al., 2003; Raddatz et al., 2011, 2012).

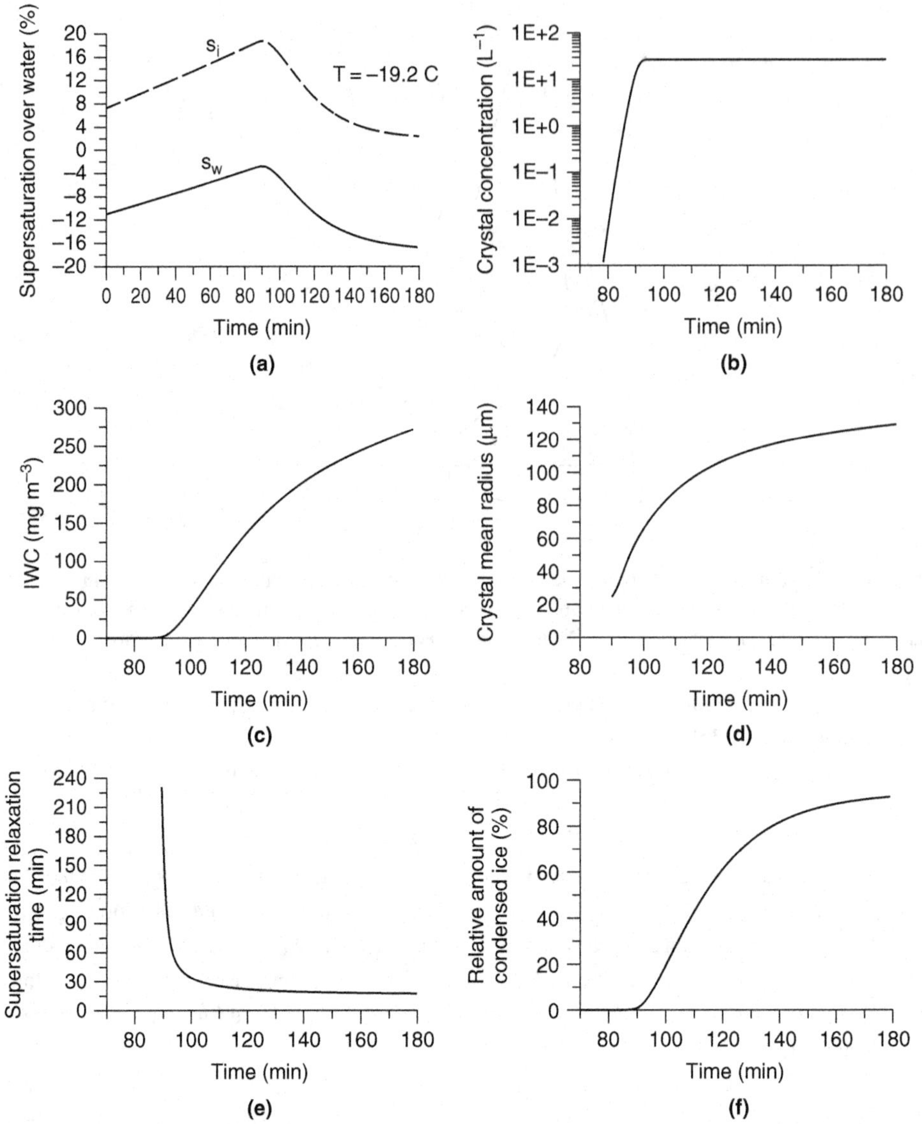

Figure 9.22. Time dependence of crystals microphysical properties with $T_0 = -18\,°C$, $RHW_0 = 89\%$, $w = 2\,cm\,s^{-1}$, $m = 0.52$. (a) Supersaturations over water, s_w, and ice, s_i,%; (b) crystal concentration, L^{-1}; (c) ice water content, mg m^{-3}; (d) crystal mean radius, μm; (e) crystal supersaturation relaxation time, min; (f) relative amount of condensed ice,%. From Khvorostyanov and Curry (2005a), *J. Atmos. Sci.*, 62, © American Meteorological Society. Used with permission.

A very cold case at $T_c = -54\,°C$ and $w = 2\,cm\,s^{-1}$ is presented in Fig. 9.23. The crystal formation occurs at $s_{wm} = -20.4\%$, final concentration is $110\,L^{-1}$, ice water content (IWC) increases slowly with time and is $15\,mg\,m^{-3}$ after 3 hours of simulation, with a mean crystal radius of 35 μm. This case may correspond to the initial stage of cirrus cloud development. Note that crystal concentration is not very high and matches those measured in cirrus (e.g., Heymsfield and McFarquhar, 2002), illustrating the possible use of this theory of ice nucleation for cirrus simulations.

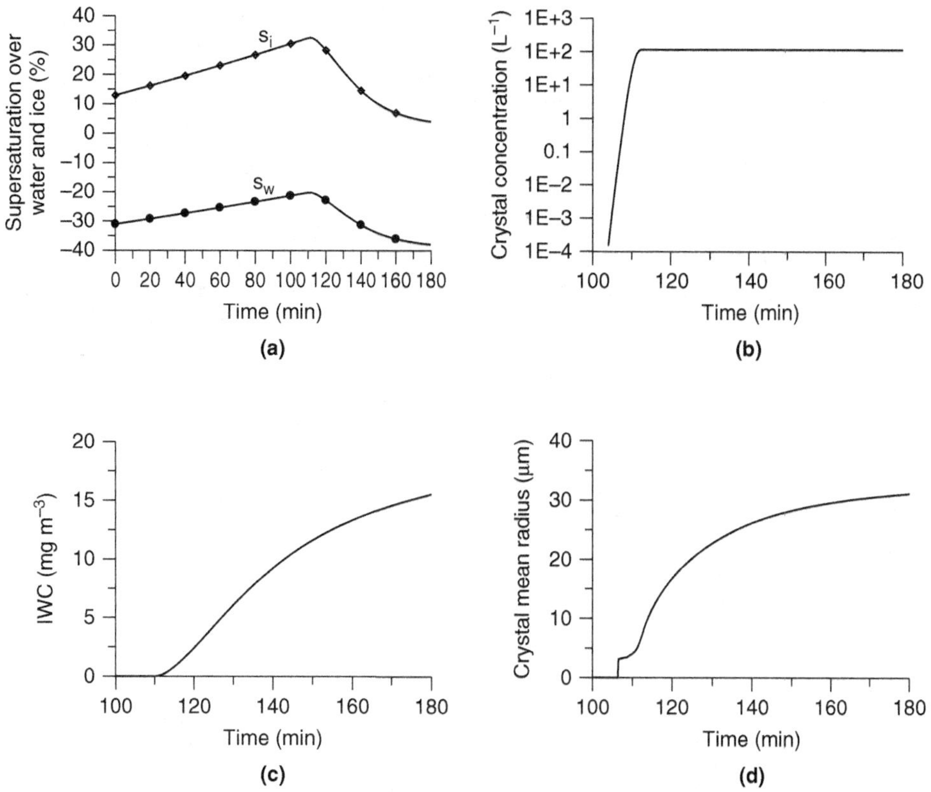

Figure 9.23. Time dependence of crystals microphysical properties with $T_0 = -54\,°C$ (as could be in cirrus), $RHW_0 = 69\%$, $w = 2\,cm\,s^{-1}$, and $m = 0.52$. (a) Supersaturations over water, s_w, and ice, s_i,%; (b) crystal concentration, L^{-1}; (c) ice water content, $mg\,m^{-3}$; (d) crystal mean radius, µm. From Khvorostyanov and Curry (2005a), *J. Atmos. Sci.*, 62, © American Meteorological Society. Used with permission.

Thus, this theory of heterogeneous crystal nucleation describes evolution of the cloud phase structure with lowering temperature and is in general agreement with observations and the climatological data on the cloud phase state, which show the occurrence of the ice phase and mixed clouds mostly at $T_c < -10\,°C$, and a substantial decrease (below 10%) of occurrence of the liquid phase in clouds below −20 °C (e.g., Borovikov et al., 1963; Pruppacher and Klett, 1997, Fig. 2.33 there; see also Section 9.17 and Fig. 9.36 here).

Fig. 9.24 illustrates the effect of the misfit strain ε at $T_{0c} = -14\,°C$, RHW = 93%, $w = 5\,cm\,s^{-1}$, and $m = 0.52$. With $\varepsilon = 0$, the drops form at $t \approx 30$ minutes, crystals form at $t = 30$–75 min with maximum $N_c = 63\,L^{-1}$, and they coexist in a mixed-phase cloud until 75 minutes, after which time the drops evaporate, the cloud completely glaciates, water supersaturation decreases, and ice nucleation ceases. With $\varepsilon = 1\%$, ice nucleation begins 15 min later and proceeds slowly until about 2 h. Then, the nucleation rate becomes comparable with the case $\varepsilon = 0$ and then exceeds it after $t = 2$ h with maximum $N_c = 180\,L^{-1}$ due to the lower temperature of nucleation, causing full crystallization by 150 min (recall that the parcel retains all particles, although in a real cloud precipitation would occur and the concentration would be much smaller). Because of the slower CCN freezing and slower

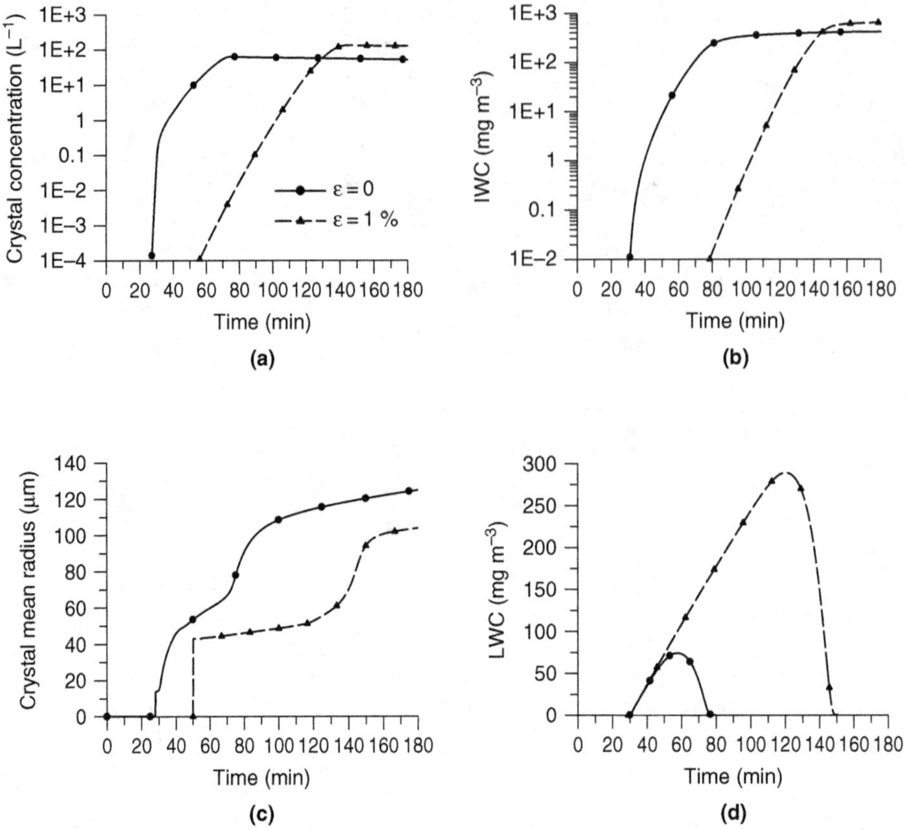

Figure 9.24. Comparison of the time evolution of cloud microphysical properties with $T_{0c} = -14\,°C$, $RHW_0 = 93\%$, $w = 5\,cm\,s^{-1}$, $m = 0.52$ in deliquescence-freezing mode with $\varepsilon = 0$ (circles) and $\varepsilon = 1\%$ (triangles). (a) crystal concentration, L^{-1}; (b) ice water content, $mg\,m^{-3}$; (c) crystal mean radius, µm; (d) liquid water content, mg m^{-3}. From Khvorostyanov and Curry (2005a), *J. Atmos. Sci.*, 62, © American Meteorological Society. Used with permission.

Wegener–Bergeron–Findeisen process in the run with $\varepsilon = 1\%$, maximum liquid water content is 4 times greater than in the case with $\varepsilon = 0$. The case with $\varepsilon = 2\%$ (not shown in Fig. 9.24) demonstrates the complete absence of crystallization, since the threshold supersaturation $s_{w,th}$ is not reached during simulation. This strong effect of the misfit strain indicates the necessity of laboratory measurements of the composition, chemical, and crystallographic properties of the aerosol particles.

9.14.4. Ice Nucleation Effects with Stronger Updrafts

Hobbs and Rangno (1990, hereafter HR90) and Rangno and Hobbs (1991, hereafter RH91) observed very rapid glaciation in maritime Cu clouds when crystal concentrations increased from $\leq 1\,L^{-1}$ to 300–$1100\,L^{-1}$ at warm temperatures of about -5 to $-10\,°C$, over a period of 9 to 15 minutes. Their analyses showed that the known mechanisms of ice multiplication could not explain the rapid rate of crystal production under the observed conditions. They hypothesized, that: 1) localized pockets of high water supersaturation of 5–10% may occur in a cloud after the stage of coalescence; 2) the less effective CCN, not activated at the first stage of cloud development, may be activated in these pockets and serve as condensation-freezing or deposition ice nuclei, causing rapid crystal nucleation. Rogers et al. (1994) performed simulations of such situations with a parcel model over a range of vertical velocities from 1–10 $\mathrm{m\,s^{-1}}$ and using parameterizations of IN concentration based on chamber experiments. A high water supersaturation and crystal concentration at $T_c = -15\,°\mathrm{C}$ and $w = 3\,\mathrm{m\,s^{-1}}$ was found just in one run of a total of 18, and the authors concluded that such a case may occur only under very specific conditions: with very low drop concentration (supermaritime CCN spectrum), high vertical velocities, and long duration adiabatic parcel ascent. Thus, this study raised the following issues: what is the probability of such high supersaturations in clouds; are such supersaturations required for high crystal concentrations; and if other possible mechanisms of high crystal concentrations and nucleation rates exist.

Khvorostyanov and Curry (2005a, hereafter KC05a) attempted to reproduce the rapid ice crystal nucleation observed by HR90 and RH91 using the CCN deliquescence-freezing model described here, which is in general agreement with the hypothesis from HR90 and RH91 on the role of CCN as IN. Since simulation of the whole life cycle of a convective cloud with an isolated parcel is a risky task, KC05a tried to reproduce a glaciation stage of this process under conditions similar to those observed. From the data in HR90 on cloud top growth and aircraft passes altitudes, the vertical velocity can be estimated as $w \sim 0.5$–$0.8\,\mathrm{m\,s^{-1}}$ at this stage. KC05a performed several simulations with $w = 0.5$–$1.0\,\mathrm{m\,s^{-1}}$.

Fig. 9.25 shows the results of the 1 hour simulation with average $w = 65\,\mathrm{cm\,s^{-1}}$, $RHW_0 = 82\%$, $T_{0c} = 0\,°\mathrm{C}$, $m_{is} = 0.5$, $\alpha = 2 \times 10^{-5}$ (α is the surface fraction of the active sites of IN, see Section 9.7), and maritime CCN concentration $N_a = 150\,\mathrm{cm^{-3}}$. With these parameters, condensation begins at 7 minutes and drop concentration is $92\,\mathrm{cm^{-3}}$ (Fig. 9.25c), similar to measured in HR90 during the aircraft passes 1–6 (90–120 $\mathrm{cm^{-3}}$). Crystal nucleation begins at 10 min ($T_c \approx -4\,°\mathrm{C}$), but concentration is low, ~0.1–0.3 $\mathrm{L^{-1}}$; nucleation accelerates at $t \sim 20$ min, and N_c increases from 1 $\mathrm{L^{-1}}$ to 290 $\mathrm{L^{-1}}$ from 20 to 33 min (Fig. 9.25d)—i.e., during 13 min in the temperature range −5.5 to −9 °C. Supersaturation generation is higher than its absorption by the drops and crystals until 37 min; LWC grows and reaches a maximum of 2.2 $\mathrm{g\,m^{-3}}$ (vs. measured 1.8–2.0 $\mathrm{g\,m^{-3}}$ in HR90 at this stage). After that time, absorption begins to prevail, the drops partially evaporate, LWC decreases to 1.3 $\mathrm{g\,m^{-3}}$ by 1 h (Fig. 9.25e) and IWC increases due to the WBF process (Fig. 9.25f). All of these characteristics (crystal concentrations and nucleation rates at this temperature range of −5 to −10 °C, droplet concentration, and LWC), along with their temporal evolution, are similar to those observed by HR90.

The effect of variations of the vertical velocity and α is as follows. Under the same initial conditions and different w, the general picture is similar but the maxima of N_c and the cloud phase state differ: $N_c = 436\,\mathrm{L^{-1}}$ with $w = 50\,\mathrm{cm\,s^{-1}}$, and the cloud is in the mixed phase. With $w = 1\,\mathrm{m\,s^{-1}}$, $N_c = 1700\,\mathrm{L^{-1}}$ and the cloud fully glaciates by 1 h. To study the sensitivity to the surface fraction of active sites α, we performed several runs with the same parameters but $\alpha = 0$. Rapid glaciation in these runs began later, at temperatures 4–6 °C lower, the final crystal concentrations were higher, up to 1–$4 \times 10^4\,\mathrm{L^{-1}}$, and were reached later at lower $T_c = -14$ to $-20\,°\mathrm{C}$. So, the agreement of the runs with $\alpha = 0$ with

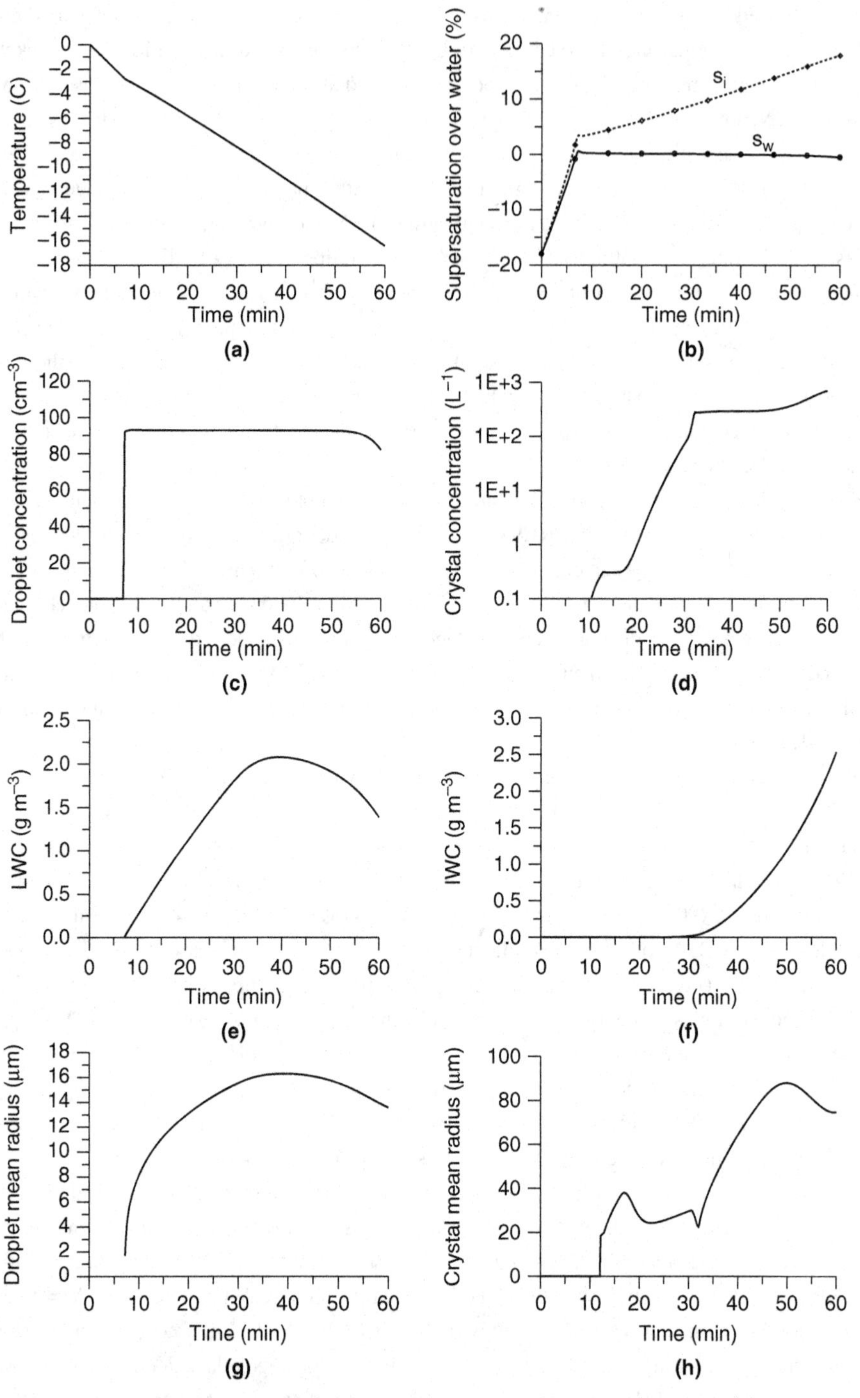

Figure 9.25. Kinetics of cloud glaciation at $w = 65\,\mathrm{cm\,s^{-1}}$ simulating the rapid crystal nucleation in small maritime Cu cloud similar to the case observed by Hobbs and Rangno (1990): evolution of the droplet and crystal microphysical properties with $w = 65\,\mathrm{cm\,s^{-1}}$, $RHW_0 = 82\%$, $T_{c0} = 0\,°\mathrm{C}$, $m_{is} = 0.5$, active sites fraction $\alpha = 2 \times 10^{-5}$. (a) Temperature, °C; (b) supersaturations over water, s_w, and ice, s_i,%; (c) droplet concentration N_d, $\mathrm{cm^{-3}}$; (d) crystal concentration N_c, $\mathrm{L^{-1}}$; (e) liquid water content, $\mathrm{g\,m^{-3}}$; (f) ice water content, $\mathrm{g\,m^{-3}}$; (g) droplet mean radius, µm; (h) crystal mean radius, µm. From Khvorostyanov and Curry (2005a), *J. Atmos. Sci.*, 62, © American Meteorological Society. Used with permission.

HR90 is worse, and this indicates that the active sites of CCN could play an important role in their freezing and rapid crystal formation in cumulus at $T_c = -5$ to -9 °C described in HR90. These simulations show that the action of the active site is the following: a) the CCN freezing shifts to the warmer temperatures; b) starts earlier; and c) ceases earlier at warmer T, which leads to the decrease in the final crystal concentration as compared to the case with $\alpha = 0$.

The value $\alpha = 2 \times 10^{-5}$ used here is from the range given by Fletcher (1969). A more detailed model of temperature-dependent $\alpha(T)$ is tested in Section 9.17. Reviews of numerous experimental studies of various types of aerosol active sites (morphological, chemical electrical, and others) is given in Fletcher (1962, 1969, 1970), Hobbs (1974), Young (1993), Pruppacher and Klett (1997, Chapters 5 and 9 therein), and in the recent papers by Connolly et al. (2009) and Hoose and Möhler (2012), along with the conclusion that little is still known about their properties. The essential effect of aerosol active sites on CCN freezing indicates the necessity of the detailed microscopic studies of the CCN surface properties after sampling in the chamber and field experiments.

Note that described above the high values of N_c ($>10^4\,L^{-1}$) simulated with $\alpha = 0$ at lower T are also not unrealistic. They were observed in Ac and Cu clouds even with the older techniques ($N_c > 10^4\,L^{-1}$, Fig. 2–42 in Pruppacher and Klett, 1997), and higher values were recently found in Arctic mixed clouds using the newer devices that can count small crystals (from a few hundred to > 4–$7 \times 10^4\,L^{-1}$, Lawson et al. 2001; Rangno and Hobbs, 2001) and post-convective anvils (e.g., Knollenberg et al. 1993). It was under conditions when ice multiplication mechanisms could not be effective, thus CCN freezing could also contribute to high N_c concentrations.

An important result of our simulations with $w = 0.5$–$1\,m\,s^{-1}$ is that water supersaturation never exceeds 1% and is 0.15 to 0.1% during the main crystal nucleation process in Fig. 9.25b, indicating that such high ice concentrations and nucleation rates as observed in HR90 and RH91 may occur in this theory without pockets of high water supersaturations, just via freezing of deliquescent mixed CCN acting as IN at small supersaturations $s_w < 0.2$–0.5% typical of mixed clouds.

9.14.5. Comparison with Homogeneous Nucleation Theory

Until the end of 1990s, absence of a heterogeneous nucleation theory suitable for low temperatures has resulted in the development and refinement of models of homogeneous nucleation (e.g., Sassen and Dodd, 1988, 1989; Jensen et al., 1994, 1998; DeMott et al., 1994; Pruppacher, 1995; Khvorostyanov and Sassen, 1998c, 2002; MacKenzie et al., 1998; Koop et al., 2000). Different models of heterogeneous nucleation were compared in the Idealized Cirrus Model Comparison Project (ICMCP, Starr et al., 2000), and Cirrus Parcel Model Comparison Project (CPMCP, Lin et al. 2002). All models used either homogeneous nucleation theory or various empirical parameterizations of heterogeneous nucleation. According to the theory presented here in Chapters 8 and 9, both homogeneous and heterogeneous nucleation processes are similar except that the nucleation energy barrier is lower in the presence of insoluble catalyzing substrate in the heterogeneous process. This conception is in agreement with the models employed in CPMCP for the heterogeneous freezing mode that also treat the freezing of deliquescent mixed CCN as discussed in Section 9.7 and in this section.

For a quantitative comparison of both the heterogeneous and homogeneous nucleation processes, Khvorostyanov and Curry (2005a, hereafter KC05a) performed 2 series of experiments using the parcel model with either homogeneous freezing of the deliquescent CCN described in Chapter 8, Section 8.9, or heterogeneous freezing described here in Sections 9.13, 9.14. Simulations were done under the conditions recommended for ICMCP and CPMCP; these two modes were switched on or off in the corresponding runs. The results from KC05a are described in the following text. The first series of simulations is for relatively "warm" cirrus at initial $T_{0c} = -40\,°C$, and the second series was for a cold cirrus at $T_{0c} = -60\,°C$; pressure was $p = 340\,\text{hPa}$, $w = 4\,\text{cm s}^{-1}$, $N_{CCN} = 200\,\text{cm}^{-3}$; contact parameter $m_{is} = 0.5$ for the heterogeneous case. The integration over the haze size spectrum was performed for the heterogeneous case as described in Section 9.10, and for the homogeneous case, as recommended by CPMCP (Lin et al. 2002), using the lognormal size spectrum with $r_{d0} = 0.02\,\mu\text{m}$ and $\sigma_d = 2.5$.

Simulations for the warm cirrus are illustrated in Figs. 9.26 and 9.27. For convenience of comparison, the initial RHW_0 was chosen after several preliminary runs in such a way that both nucleation processes occurred at approximately the same time, 30 minutes after the start of the simulation. This required $RHW_0 = 90\%$ for homogeneous nucleation and only 78% for heterogeneous nucleation. The nucleation reaches a significant rate at $RHW \approx 97.6\%$ for the homogeneous processes, and at 83% for heterogeneous processes (Fig. 9.26a). The critical radius is higher for heterogeneous nucleation because of lower humidity (Fig. 9.26b); however, the critical energy of heterogeneous nucleation is comparable and even a little smaller at $t = 35$ min than that for homogeneous nucleation (Fig. 9.26c). This is caused by the geometric factor $f(m_{is})$, which reduces the energy barrier for nucleation. The nucleation rates per particle [s^{-1}], homogeneous, $J_{hom} r_d^3$, and heterogeneous, $J_{f,het}(r_N, T)$ calculated with (Eqns. 8.6.21) and (9.7.26), respectively, as an example for a haze particle of 0.13 μm (Fig. 9.26d) and polydisperse nucleation rates (Fig. 9.26e) are comparable although the maxima for heterogeneous nucleation is ~4 times higher due to this small difference in ΔF_{cr}. The final crystal concentrations are comparable: $118\,\text{L}^{-1}$ in the heterogeneous case versus $56\,\text{L}^{-1}$ in the homogeneous case (Fig. 9.26f).

A comparison of the time evolution of the heterogeneous versus homogeneous processes at −40 °C is continued in Fig. 9.27. Both processes are similar in general (IWC, N_c, supersaturation relaxation time), although the lower ice supersaturations in the heterogeneous case lead to a smaller mean crystal radius and vapor excess. The percentage of condensed ice is higher with heterogeneous nucleation, so that the efficiency of deposition is higher in this case. The crystal size spectra 30 minutes after the main nucleation impulse (Fig. 9.28) have smaller maxima but are broader in the homogeneous nucleation process, since the difference in crystal ages is greater due to the longer nucleation time.

A comparison of the two ice nucleation modes in the colder case at $T_{0c} = -60\,°C$ is shown in Fig. 9.29. This time, both runs were started at the same $RHW_0 = 75\%$ in order to determine the difference in the two modes. As in the previous case, heterogeneous nucleation starts much earlier at $t = 28$ minutes at $RHW \approx 80\%$, and homogeneous nucleation begins 16 minutes later at $t = 44$ minutes at $RHW \approx 88\%$ (Fig. 9.29b). The impulse of nucleation in the heterogeneous case is again narrower in time ($\Delta t_{het} = 5$ minutes vs. $\Delta t_{hom} = 9$ minutes), with comparable maxima of the nucleation rate ($R_{max,het} = 2.2 \times 10^{-3}\,\text{cm}^{-3}\,\text{s}^{-1}$ vs. $R_{max,hom} = 2.9 \times 10^{-3}\,\text{cm}^{-3}\,\text{s}^{-1}$) and comparable $N_c = 188\,\text{L}^{-1}$ vs. $171\,\text{L}^{-1}$ for heterogeneous and homogeneous cases, respectively. The entire cirrus development proceeds similarly in both cases but is shifted by ~18 minutes later in the homogeneous case. The crystal size

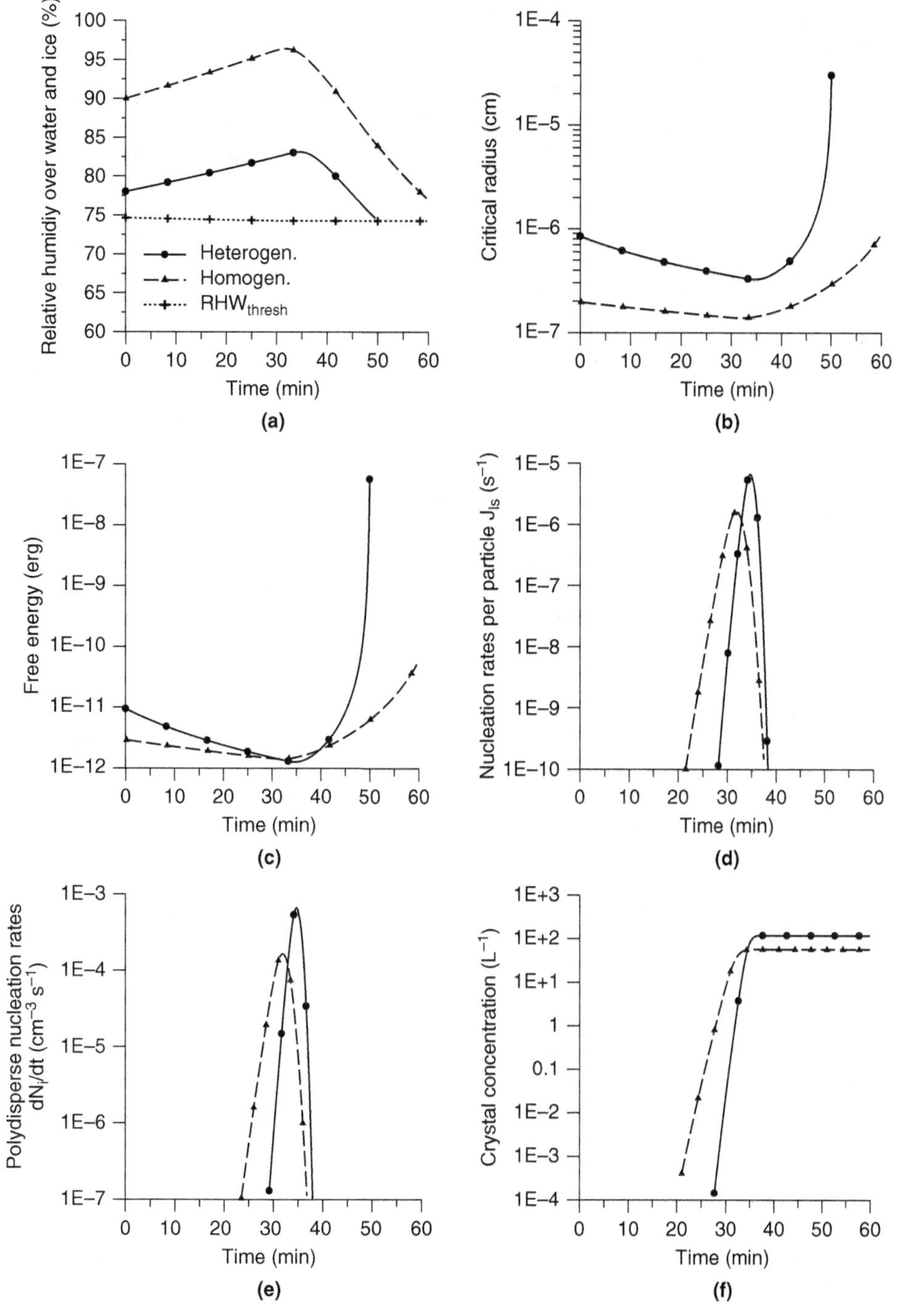

Figure 9.26. Comparison of primary properties of heterogeneous nucleation (solid circles, $m = 0.50$, $RHW_0 = 78\%$) and homogeneous nucleation (triangles, $RHW_0 = 90\%$) processes at $T_{0c} = -40\,°C$, $w = 4\,cm\,s^{-1}$, $p_0 = 340\,hPa$. (a) Relative humidity over water and threshold humidity RHW_{th}, % (crosses); (b) ice germ critical radius r_{cr}, cm; (c) free energy ΔF_{cr}, erg; (d) nucleation rates, homogeneous $J_{hom} r_d^3$, (Eqn. 8.3.5), and heterogeneous $J_{f,het}$, (Eqn. 9.6.7), (per particle, sec^{-1}); (e) polydisperse nucleation rates, $R_{f,hom}$ and $R_{f,het}$ in cm^{-3} s^{-1}, defined by (Eqns. 8.3.10) and (9.10.5); crystal concentration, L^{-1}. From Khvorostyanov and Curry (2005a), *J. Atmos. Sci.*, 62, © American Meteorological Society. Used with permission.

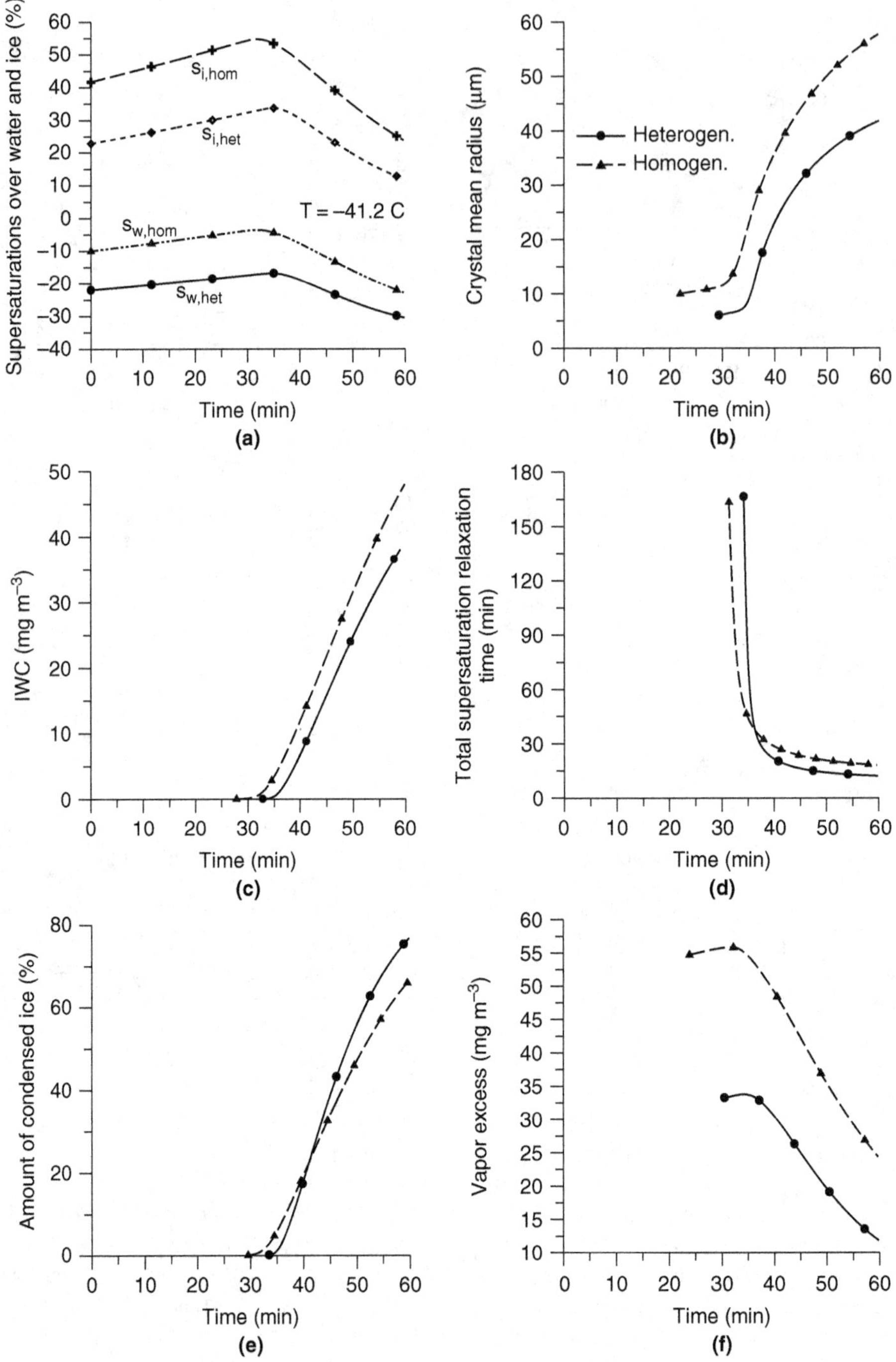

Figure 9.27. Comparison of crystal microphysical properties with heterogeneous (circles) and homogeneous (triangles) nucleation processes at $T_{0c} = -40\,°C$, $w = 4\,cm\,s^{-1}$ and the other parameters as in Fig. 9.26. (a) Supersaturations over water, s_w, and ice, s_i,%; (b) crystal mean radius, μm; (c) ice water content, $mg\,m^{-3}$; (d) crystal supersaturation relaxation time τ_{fc}, min; (e) relative amount of condensed ice, REL-ICE,%; (f) vapor excess, $mg\,m^{-3}$. From Khvorostyanov and Curry (2005a), *J. Atmos. Sci.*, 62, © American Meteorological Society. Used with permission.

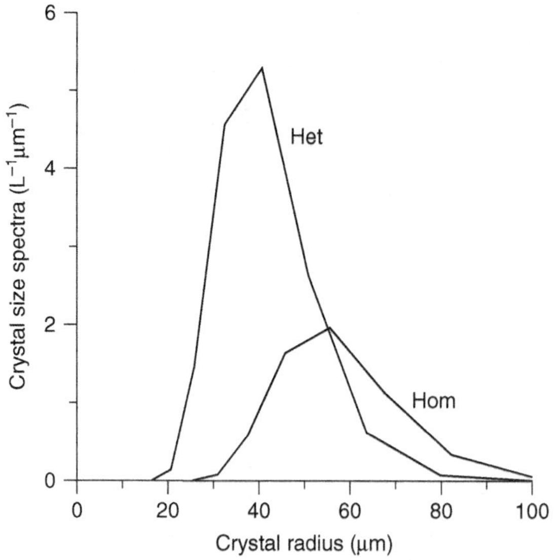

Figure 9.28. Comparison of the crystal size spectra at $T_0 = -40\,°C$ after $t = 1$ h with homogeneous and heterogeneous nucleation. From Khvorostyanov and Curry (2005a), *J. Atmos. Sci.*, 62, © American Meteorological Society. Used with permission.

spectra for both heterogeneous and homogeneous nucleation at $T = -60\,°C$ (Fig. 9.30) are also closer to each other than the spectra at warmer $T_{0c} = -40\,°C$ (Fig. 9.28). Thus the difference between the two ice nucleation modes decreases with lowering temperature. Note that crystal concentrations for both heterogeneous and homogeneous nucleation are quite comparable both at −40 °C and −60 °C, despite the fact that they were calculated with substantially different equations. This indicates that the heterogeneous theory can be used for the calculations of ice nucleation in cirrus.

Finally, we compare the results of KC05a with the other four models from the Cirrus Parcel Model Comparison Project (Lin et al., 2002). Table 9.1 shows that the results of KC05a for the

Table 9.1. *Comparison of the crystal concentrations, cm^{-3}, calculated here using this parcel model, heterogeneous nucleation theory KC00-05 and homogeneous theory from Khvorostyanov and Sassen (1998c, KS98), with the results from the parcel intercomparison project CPMCP (Lin et al., 2002). Capital letters mean the author of the model: L-Lin, S-Sassen, X- Xiahong Liu, D- DeMott. "W" means warm case (−40 °C), "C" means cold (−60 °C), "hom" means homogeneous nucleation, "het" means "all modes" (actually, heterogeneous case). The models are described in more detail in the text*

	Model				
Case	KC00-05, KS98	L	S	X	D
Ca (het)	0.188	0.0126	1.0655	–	0.4027
Ch (hom)	0.178	0.1380	0.1640	0.281	0.4740
Wa (het)	0.118	0.0054	0.0183	–	0.0064
Wh (hom)	0.056	0.0286	0.0275	0.044	0.0810

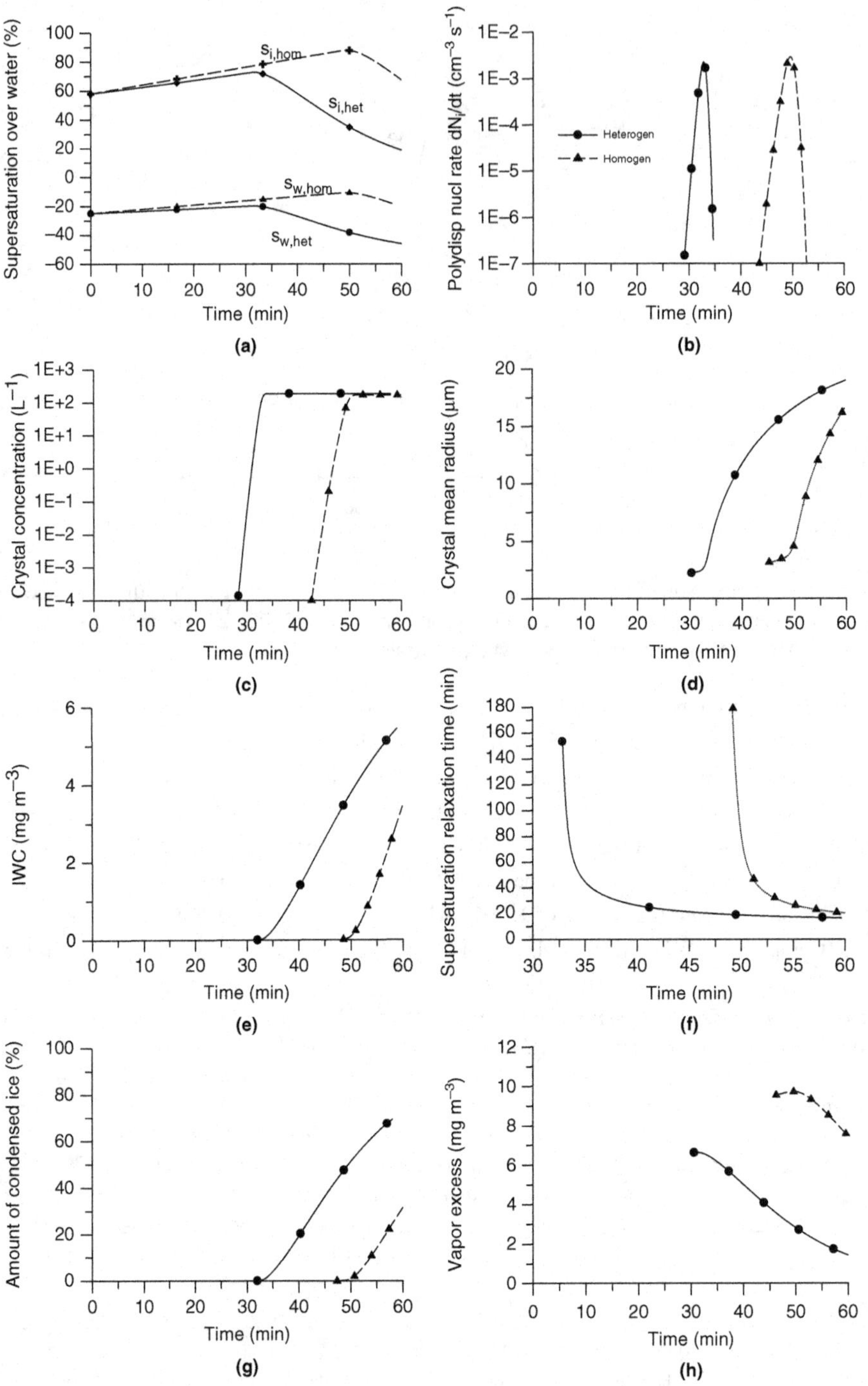

Figure 9.29. Comparison of crystal microphysical properties with heterogeneous (circles) and homogeneous (triangles) nucleation processes at $T_0 = -60\,°\mathrm{C}$, $w = 4\,\mathrm{cm\,s^{-1}}$ and the other parameters as in Fig. 9.26. (a) Supersaturations over water, $s_{w,hom}$, $s_{w,het}$, and ice, $s_{i,hom}$, $s_{i,het}$, (%); (b) polydisperse nucleation rates dN_c/dt ($\mathrm{cm^{-3}\,s^{-1}}$); (c) crystal concentration ($\mathrm{L^{-1}}$); (d) crystal mean radius, μm; (e) ice water content, $\mathrm{mg\,m^{-3}}$; (f) crystal supersaturation relaxation time τ_{fc}, min; (g) relative amount of condensed ice, REL-ICE,%; (h) vapor excess, $\mathrm{mg\,m^{-3}}$. From Khvorostyanov and Curry (2005a), *J. Atmos. Sci.*, 62, © American Meteorological Society. Used with permission.

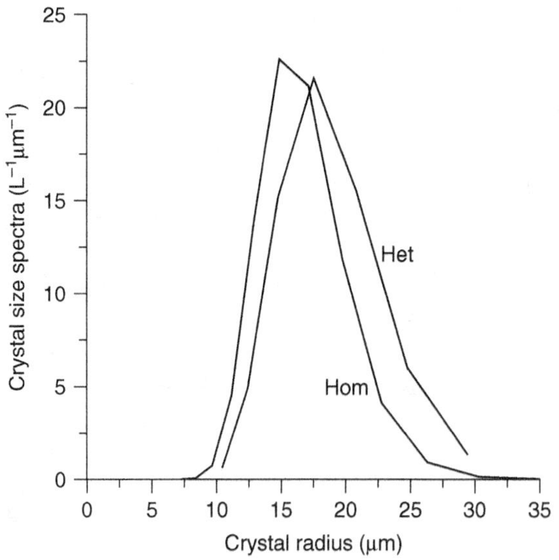

Figure 9.30. Comparison of the crystal size spectra at $T_0 = -60\,°\mathrm{C}$ after $t = 1$ h with homogeneous and heterogeneous nucleation. From Khvorostyanov and Curry (2005a), *J. Atmos. Sci.*, 62, © American Meteorological Society. Used with permission.

Warm homogeneous case (denoted Wh), $N_c = 0.056\,\mathrm{cm}^{-3}$, are in the middle of the range among the other models ($N_c = 0.0275$–$0.081\,\mathrm{cm}^{-3}$); the same is true for the Cold homogeneous case (Ch); $N_c = 0.178\,\mathrm{cm}^{-3}$ in KC05a is comparable to the others at $N_c = 0.138$–$0.474\,\mathrm{cm}^{-3}$. This agreement shows that the KC05a parcel model have sufficient accuracy and can be used for such simulations, and the theory of homogeneous ice nucleation is sufficiently developed for quantitative descriptions. With heterogeneous nucleation, the range of the other 4 models of the CPMCP is 0.0054–$0.0183\,\mathrm{cm}^{-3}$ for Wa (Warm all-mode) and 0.0126–$1.065\,\mathrm{cm}^{-3}$ for Ca (Cold all-mode, where mostly the heterogeneous nucleation mode acts). Heterogeneous nucleation results from KC05a are comparable with the homogeneous nucleation method, 0.118 and $0.188\,\mathrm{cm}^{-3}$ for warm and cold cases, since both processes are similar and proceed on the same CCN with comparable rates. However, heterogeneous nucleation requires lower humidity and starts earlier in a cooling air parcel.

9.15. Comparison of Simulated Crystal Concentrations with Experimental Data and Parameterizations

To illustrate the general dependence of the crystal concentration on the temperature and vertical velocities, the final crystal concentrations $N_c(T, w)$ (after nucleation has ceased) are plotted in Fig. 9.31. This figure includes simulations from KC05a based on several hundred runs of the parcel model with $w = 0.3$, 1, 2, 5, and $50\,\mathrm{cm\,s}^{-1}$. The Cooper's (1986) parameterization (Eqn. 9.4.4) is also plotted for comparison. Each solid symbol in Fig. 9.31 corresponds to a final value of N_c after a single run of the parcel model with the KC scheme. This figure shows a substantial variability of N_c that depends on the initial temperature T, vertical velocity w, contact parameter m_{is}, and the area α of the active sites.

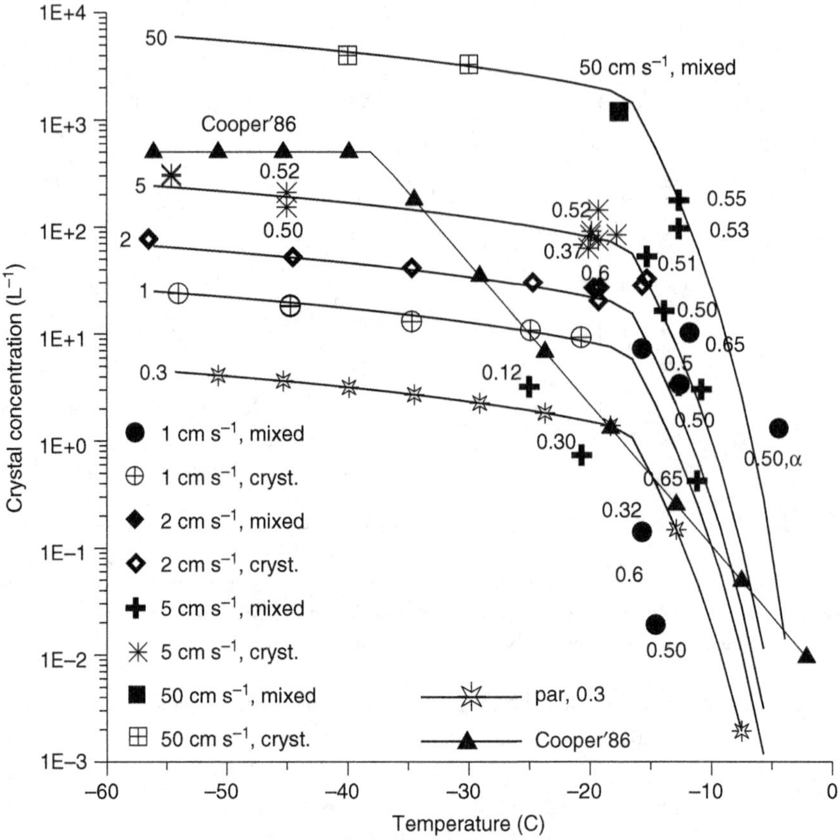

Figure 9.31. Temperature dependence of the crystal concentrations N_c calculated with $w = 1\,\mathrm{cm\,s^{-1}}$ (solid and open circles), $2\,\mathrm{cm\,s^{-1}}$ (solid and open diamonds), $5\,\mathrm{cm\,s^{-1}}$ (crosses and asterisks), and $50\,\mathrm{cm\,s^{-1}}$ (squares). Each symbol corresponds to a final N_c after a single run of the parcel model with the KC00-05 heterogeneous DF ice nucleation scheme. The values of the contact parameter $m_{is} = 0.52 = \text{const}$ along the continuous lines, the other values of m_{is} are shown near the points that are outside the lines; the symbol "α" denotes the runs with active sites fraction $\alpha = 2 \times 10^{-5}$, otherwise $\alpha = 0$. Solid symbols denote CCN freezing at $s_w > 0$ in the presence of drops in a mixed cloud, mostly at $T_c > -20\,°\mathrm{C}$, although mixed phase can be below $-20\,°\mathrm{C}$ and down to $-30\,°\mathrm{C}$ with lower $m_{is} = 0.12$–0.30. Open symbols denote ice nucleation at $s_w < 0$ in a crystalline cloud. The solid lines, along which the symbols are grouped, are parameterizations from KC05a of the simulation data as described in the text (Eqn. 9.15.1) plotted for $w = 0.3$, 1, 2, 3, 5, and $50\,\mathrm{cm\,s^{-1}}$. These lines are compared with Cooper's (1986) parameterization (triangles). These fits might be used as a simple parameterization of the average data in the Figure in cloud models and GCMs. From Curry and Khvorostyanov, Atmos. Chem. Phys. (2012), with changes.

The general tendency of the cloud phase state is the transition from the liquid state at $T_c \geq -10\,°\mathrm{C}$ to the mixed-phase state for $-10 < T_c < -20\,°\mathrm{C}$, and to the crystalline state at $T_c < -20\,°\mathrm{C}$. This is caused by the increasing crystal concentrations with decreasing temperature.

However, Fig. 9.31 shows that mixed-phase clouds may exist in these simulations down to $-30\,°\mathrm{C}$ at lower values of m_{is} (0.12–0.32), which produce low crystal concentrations of $N_c = 0.8$–$3\,\mathrm{L^{-1}}$ and do not lead to complete cloud glaciation. The same result was obtained with lower aerosol concentrations down to 50 cm^{-3} (not shown). This indicates that the KC scheme can be consistent with frequent observations of the mixed-phase Arctic clouds at low temperatures below −20 °C observed in the SHEBA-FIRE (Surface Heat Budget of the Arctic – First International Regional Experiment) of the International Satellite Cloud Climatology Project (ISCCP) and many other experiments (e.g., Curry, 1986; Curry et al., 1990, 1993, 1996, 2000; Gultepe et al., 2003; Lawson et al., 2001; Uttal et al., 2002; Intrieri et al. 2002; Korolev et al., 2003; Shupe et al., 2006; McFarquhar et al., 2007). This theory of heterogeneous nucleation indicates that observations of the mixed-phase strongly supercooled mixed Ac (Herman and Curry, 1984; Sassen and Campbell, 2001; Sassen and Khvorostyanov, 2007, 2008; Sassen and Wang, 2012) may also be caused by the specific chemical composition of CCN with low contact parameter m_{is} or low CCN concentration or low insoluble fractions.

The solid lines with the open symbols in Fig. 9.31 represent a parameterization of the simulation data as a function of two variables, T_c (temperature in °C) and w, found in KC05a and refined in Curry and Khvorostyanov (2012, CK12):

$$N_c(T,w) = C_g (T_{c0} - T_c)^{C_T} w^{C_w}, \tag{9.15.1}$$

where N_c is in L^{-1}, T_c is the temperature in Celsius, $T_{c0} = 0$ °C, $C_w = 1.41$; and there are two sets of the other constants: $C_g = 0.4 \times 10^{-8}$, $C_T = 8.0$, for $T_c > -15$ °C; and $C_g = 0.535$, $C_T = 1.05$ for $T_c < -15$ °C. Fig. 9.31 shows that (Eqn. 9.15.1) represents well the simulation data, the simulation points for each w lie along the T-curves calculated with (Eqn. 9.15.1), which therefore can be used as a simple parameterization in cloud models and GCMs. For example, Zhang et al. (2011) successfully used (Eqn. 9.15.1) in the Weather Research and Forecast (WRF) model incorporating the microphysics scheme from Morrison, Curry and Khvorostyanov (2005, hereafter, Morrison's microphysics) for simulations of dust effects on ice nucleation in the hurricane Helene-2005 development and found that dust aerosols nucleating ice in this mode may noticeably influence hurricane dynamics and cloud microphysics.

Fig. 9.31 shows that under relatively warm conditions ($T_c \geq -10$ °C) and in weak updrafts (w = 1–2 cm s^{-1}), the cloud is almost entirely liquid for m_{is} = 0.5–0.55. Ice crystals could probably form in this temperature range with values of $S_w \sim$ 1.03–1.05 (see Section 9.7), but do not form here since the growing droplets do not allow supersaturations higher than ~0.1–0.2%, which is insufficient for ice nucleation at these temperatures and contact parameters. At slightly higher values of m_{is} = 0.6–0.65 or in the presence of a small fraction of active sites, $\alpha = 2 \times 10^{-5}$, the probability of ice nucleation increases in this range. The simulations and Cooper's and Fletcher's empirical parameterizations described in Section 9.4 show increasing N_c with decreasing T.

It is interesting to note that although the dynamics in these simple parcel simulations is different from that in the continuous flow chambers, simulated values of N_c at -10 °C $> T_c > -20$ °C are comparable to the range of those measured in the chambers (e.g., Rogers, 1982, 1988; Al-Naimi and Saunders, 1985; Meyers et al., 1992). For a more detailed comparison with the chamber experiments, this theory of heterogeneous nucleation could be tested, e.g., in a 2D model that simulates a vertical plane in a chamber or in a 3D model that simulates the microphysical processes in a chamber with their spatial inhomogeneity.

The earlier T-parameterization by Fletcher (1962) was based on data available by the end of the 1950s, and was obtained for temperatures mostly above −25 °C. Its extrapolation down to very low temperatures led to very high values of N_c, which forced modelers to refuse use of heterogeneous nucleation for cold cirrus and altostratus and invoke homogeneous nucleation theory for these clouds as the main nucleation mode or introduce various modifications to Fletcher's parameterization for cold T_c (e.g., Sassen and Dodd, 1988, 1989; Heymsfield and Sabin, 1989; DeMott et al., 1994; Jensen et al., 1994, 1996; Khvorostyanov and Sassen, 1998b, 2002; Spice et al., 1999; Gu and Liou, 2000; Lin et al., 2002; Kärcher and Lohmann, 2002a,b). The theory presented here is in accord with numerous observations in recent decades (e.g., Meyers et al., 1992, Fig. 2 therein), and gives reasonable values of N_c at very cold temperatures that are in agreement with measurements in cirrus (see also the next section)—e.g., Gayet et al. (1996), Chen et al. (1998), Rogers et al. (1998), DeMott et al. (1998), Heymsfield and McFarquhar (2002), Lawson et al. (2001), and the other empirical parameterizations used in CPMCP for heterogeneous ice nucleation at cirrus temperatures (Lin et al., 2002). Hence, this theory of heterogeneous ice crystal nucleation can be used for simulation of cold altostratus, altocumulus, and cirrus clouds, serving as an alternative for homogeneous nucleation, or both modes could be included together. A comparison of the runs performed at colder temperatures (below −20 °C) showed that N_c increases by a factor ~3 when w increases from 1 to 2 cm s^{-1}, and by ~4 when w increases from 2 to 5 cm s^{-1}, indicating the nonlinear dependence of N_c on w, approximated in (Eqn. 9.15.1) by a power $C_w = 1.41$, similar to the parameterizations of homogeneous nucleation in Kärcher and Lohmann (2002a,b), Ren and MacKenzie (2005), and here in Chapter 8 with $C_w = 1.5$ (a more detailed parameterization and its coefficients for various regimes will be derived in Section 10.1). The KC heterogeneous ice nucleation scheme was used for simulations of cirrus clouds and cold Arctic clouds (Khvorostyanov et al., 2001, 2003, 2006; Khvorostyanov and Sassen, 2002; Sassen et al., 2002; Liu and Penner, 2005; Comstock et al., 2008) and for simulation of cold altocumulus clouds (Sassen and Khvorostyanov, 2007, 2008), and produced reasonable (not very high) crystal concentrations in all these simulations. Ervens and Feingold (2012) used the KC scheme with superimposed several kinds of random processes for simulation of the laboratory experiments on ice nucleation.

The KC parameterization (Eqn. 9.15.1) is compared in Fig. 9.32 to the experimental data from the six field campaigns described in Eidhammer et al. (2009): INSPECT1, INSPECT2, CRYSTAL-FACE, PACDEX, WISP, and MPACE (data kindly provided by Paul DeMott and Trude Eidhammer). Fig. 9.32 shows that the span of the KC parameterization curves in the range $w = 0.3$–5 cm s^{-1} encloses the majority of the field data—i.e., this ice nucleation scheme is in general agreement with the field experiments. The tendency of KC curves is in qualitative agreement with Cooper's (1986) parameterization used in the Morrison microphysics scheme currently employed in the CAM3 GCM (Morrison and Gettelman, 2008; Gettelman et al., 2008) but allows a greater $N_c(T, w)$ variability caused by the different cooling rates ($-\gamma w$).

The almost vertical curve in Fig. 9.32, marked "PDA-DHF," is from Phillips, DeMott, and Andronache (2008, hereafter, PDA08). This curve was intended in PDA08 to represent the T-dependence of the DHF mode in the KC theory . Therefore, it is labeled here "PDA-DHF." However, Figs. 9.31 and 9.32 here clearly illustrate that this "PDA-DHF" curve from PDA08 is completely different from the real T-dependences in the KC scheme shown in Figs. 9.31 and 9.32 and does not correspond to this scheme. Although this curve PDA-DHF was an erroneous attempt in PDA08 of representation of the DHF mode in the KC scheme, its discussion is useful because it illustrates the difficulties of the previous classical nucleation theory. The reasons of the discrepancy between the PDA-DHF curve and KC T-curves are: a) this curve was constructed in PDA08 without considering

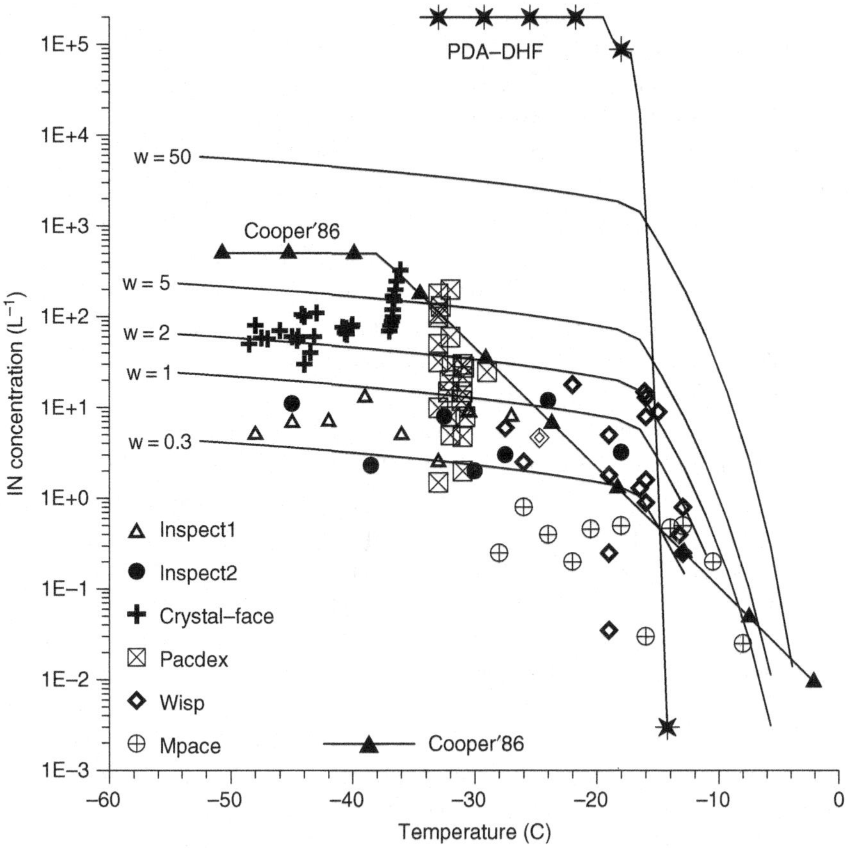

Figure 9.32. Parameterized with (Eqn. 9.15.1) parcel model simulations using the KC scheme with DF mode shown in Fig. 9.31 for the 5 values of $w = 0.3$, 1, 2, 5, and 50 cm s^{-1} (indicated near the left ends of the curves at $T \approx -56$ °C) are compared to the experimental data from the 6 field campaigns (INSPECT1, INSPECT2, CRYSTAL-FACE, PACDEX, WISP, MPACE) described in Eidhammer et al. (2009) and indicated in the legend with various symbols (data kindly provided by Paul DeMott and Trude Eidhammer). Cooper's (1986) parameterization is given for reference. The almost vertical curve labeled "PDA-DHF" is from Philips et al. (2008); it was calculated in Philips et al. (2008) without parcel simulations—i.e., with account for only T-dependence at $S_w = 1$; that is, neglecting the supersaturation dependence and its negative feedback. From Curry and Khvorostyanov, Atmos. Chem. Phys. (2012), with changes.

any time-dependent ice nucleation process (without parcel or any other simulations), just by calculating $N_c(T)$ from (Eqn. 9.10.3); b) the critical energy ΔF_{cr} and nucleation rates $J_{f,het}$ were calculated in PDA08 at S_w = 1 (or RHW = 100%), that is, neglecting the supersaturation dependence and its strong negative feedback that stops nucleation as was discussed in Section 9.14. Therefore, this curve essentially overestimates $N_c(T)$ and the slopes dN_c/dT.

This curve without S_w-dependence and without time evolution of the nucleation process actually represents the old T-dependence based on the classical equations (Eqn. 8.6.1) for r_{cr} and (Eqn. 8.6.2) for ΔF_{cr} by J. J. Thomson (1888) taking into account only the T-dependence (see Section 9.7). It has long been known that early formulations of the classical nucleation theory produce an unrealistic behavior of $N_c(T)$. This is well illustrated by this curve PDA-DHF, and this stimulated development of numerous empirical parameterizations, many of which, however, could not be extended to the colder temperatures. The use of the KC scheme which takes into account both T- and S_w-dependencies allows calculations of crystal concentrations in the wide T-range based on this extension of the classical theory. As Figs. 9.31 and 9.32 show, the KC scheme produces the results substantially different from the classical theory with only T-dependence represented by the curve PDA-DHF. It allows us to describe many nucleation events that begin at different initial conditions (T, S_w), pass different trajectories on the S_w-T phase plane, and span the wide range of final crystal concentrations nucleated at various conditions. The strong negative feedback due to S_w-dependence bounds N_c and produces much smoother $N_c(T)$ and parameterization that are in good general agreement with the data from six field campaigns as shown in Figs. 9.31 and 9.32.

A distinguishing feature of pristine N_c found in many experiments is its dependence on ice or water supersaturation discussed in Section 9.4 (e.g., Huffman, 1973; Rogers, 1982, 1988; Hussain and Saunders, 1984; Al-Naimi and Saunders, 1985; Berezinsky and Stepanov, 1986; Rogers, et al. 2001). These studies found that a power law (Eqn. 9.4.2) or exponential function (Eqn. 9.4.7) of s_i describe this dependence. Fig. 9.33 illustrates the ice supersaturation dependence $N_c(s_i)$ calculated in KC05a using this theory of CCN freezing; the same data as in Fig. 9.31, but grouped and plotted as the functions of s_i. Each solid triangle, circle, or diamond on the curves corresponds to a final N_c (after nucleation ceased) in a single run of the parcel model for 1–3 hours with w= 1, 2, and 5 cm s^{-1} plotted against the maximum value of s_i during the run. The symbols are connected by the fitting lines with the contact parameter m_{is} = 0.52 along the lines; the values of m_{is} are indicated near the points where it is different; the symbol "α" denotes the runs with $\alpha = 2 \times 10^{-5}$. Plotted here are also two previous empirical parameterizations described in Section 9.4, Meyers et al. (1992, MDC92, crosses) and Huffman's (1973) power law $N_c(s_i) = C_{iH} s_i^{b_H}$ (asterisks). Huffman found $3 < b_H < 8$, and C_{iH} was more uncertain. We have chosen here the values $C_{iH} = 10^{-5}\,\mathrm{L}^{-1}$ and $b_H = 4.9$ to match the lab data.

Fig. 9.33 shows that the simulated in KC05a with CNT values of N_c are in reasonable qualitative and quantitative agreement with the experimental points and both previous parameterizations. The main features of the simulated $N_c(s_i)$ are: *a*) increasing N_c with increasing s_i; and *b*) a marked decrease of the slopes $dN_c(s_i)/ds_i$ at s_i > 15–20%—i.e., some sort of "saturation" at higher s_i. This feature, convex dependence $N_c(s_i)$ with decreasing slopes, is similar to the measured water supersaturation dependence in the drop nucleation power law (e.g., Yum and Hudson, 2001; see Figs. 6.17 and 6.18 in Chapter 6) and to the ice supersaturation dependence for ice nucleation described by the empirical power law (Eqn. 9.4.2) and measured at constant temperatures (e.g., Rogers, 1982; Al-Naimi and Saunders, 1985; Berezinsky and Stepanov, 1986; Meyers et al., 1992), and is similar to the equivalent analytical dependence of homogeneous nucleation on w derived in Section 8.8. This functional similarity is caused by the analogous negative feedback with respect to supersaturation of both drop and crystal nucleation, although the latter is much stronger, as discussed earlier. Note again, a more detailed comparison with the laboratory data can be done using a more sophisticated model closer to the chamber conditions. However, Fig. 9.33 shows that even this simple parcel simulation with calculation of heterogeneous nucleation based on extended classical nucleation theory allows a reasonable description of the variety of experimental data.

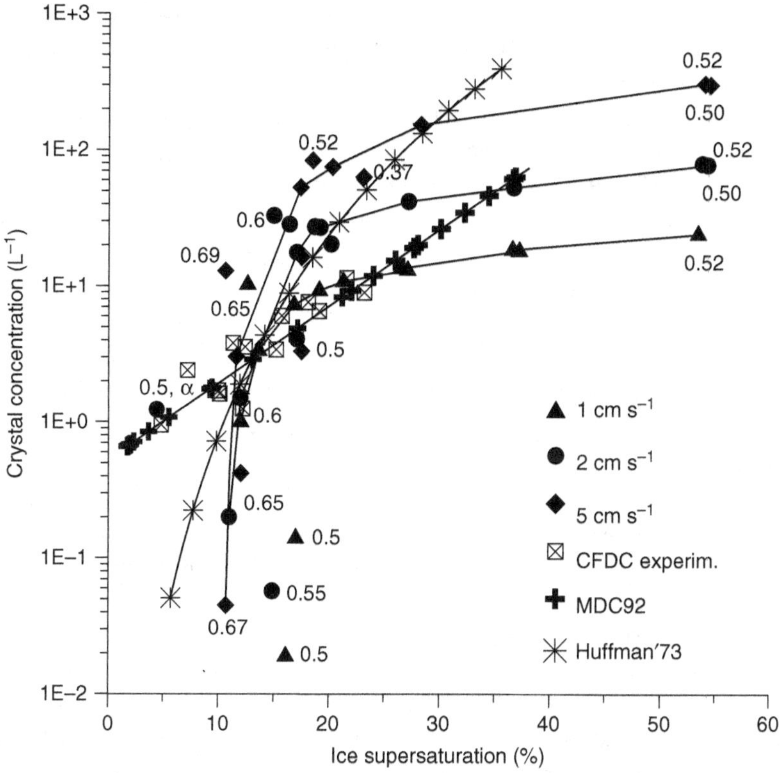

Figure 9.33. The simulated ice supersaturation dependence of the crystal concentration $N_c(s_i)$. The final N_c after a single run of the parcel model are denoted with solid triangles, circles, and diamonds for $w = 1$, 2, and 5 cm s^{-1}, respectively. The symbols are connected by the fitting lines, the contact parameter $m_{is} = 0.52$ along the lines; the values of m_{is} are indicated near the points where it is different; the symbol "α" denotes the runs with active sites $\alpha = 2 \times 10^{-5}$. Open squares (CFDC experim.) denote laboratory data from Rogers (1982) and Al Naimi and Saunders (1985). The parameterizations by Huffman (1973) and Meyers et al. (1992, MDC92) are marked with asterisks and crosses. From Curry and Khvorostyanov, Atmos. Chem. Phys. (2012), with changes.

9.16. Thermodynamic Constraints on Heterogeneous Ice Nucleation Schemes

Classical nucleation theory allows us to impose certain constraints on the empirical parameterizations of IN concentrations described in Section 9.4. Here we assess the range of validity of these parameterizations in the context of thermodynamic constraints based on the extended classical nucleation theory described in the previous sections. The critical radius r_{cr} of an ice germ in (Eqn. 9.7.5a) should be positive; thus, the denominator should be positive, yielding a condition for the threshold $S_{w,th}(T)$ for ice particle nucleation

$$S_w(T) \ge S_{w,th}^{het}(T) = \left(\frac{T}{T_0}\right)^{M_w L_m^{ef}/RT} \exp\left[\frac{M_w}{\rho_i RT}\left(C_\varepsilon \varepsilon^2 + \frac{2\sigma_{sa}}{r_a} + \frac{\Delta\rho\Delta p}{\rho_w}\right)\right]. \quad (9.16.1)$$

The notations were defined in Sections 9.7 and 9.12. Equation (9.16.1) represents a lower limit for S_w for infinitesimally small nucleation rates J_{het}, the more general equations for finite J_{het} derived in Section 9.12 predict somewhat higher $S_{w,th}^{het}$.

Equations (9.7.5a) and (9.16.1) show that the value $r_{cr} > 0$ if $S_w > S_{w,th}^{het}$, or $\delta S_{w,th} = S_w - S_{w,th}^{het} > 0$ at a given T, and only these states are thermodynamically allowed in the S_w-T domain. The denominator of r_{cr} of ice germs in (Eqn. 9.7.5a) (affinity $\Delta\mu = \mu_{mstab} - \mu_{stab}$, where μ_{mstab} and μ_{stab} are the chemical potentials of the metastable liquid and stable ice phases) becomes negative and the critical radii become negative, $r_{cr} < 0$, and unphysical in the S_w-T domain if $S_w < S_w^{het}$, or $\delta S_{w,th} < 0$ at a given T, that is, the relative humidity over water (RHW) or water saturation ratio S_w is smaller than its threshold value (see Section 9.12), or $\delta(RHW_{th}) = RHW - RHW_{th} = (S_w - S_{w,th}^{het}) \times 100\% < 0$.

As pointed out earlier, the condition $r_{cr} > 0$ or $\delta S_w > 0$ means that $\mu_{mstab} > \mu_{stab}$ (affinity $\Delta\mu > 0$), then such a transition $\mu_{mstab} \rightarrow \mu_{stab}$ (nucleation) is thermodynamically allowed. The reverse condition $r_{cr} < 0$ or $\Delta\mu < 0$ or $\Delta S_w < 0$ means that the transition is prohibited from the state with lower energy μ_{stab} to the state with higher energy μ_{mstab}, and an ice germ cannot form.

Note that the constraint (Eqn. 9.16.1) is sufficiently general, it follows from the entropy equation used for derivation of r_{cr} in (Eqn. 9.7.5a) that is based on classical thermodynamics.

The constraint (Eqn. 9.16.1) allows us to assess the range of thermodynamic validity of the IN parameterizations from Meyers et al. (1992, MDC92), Phillips et al. (2008, PDA08), and DeMott et al. (2010, DM10) on the S_w-T diagrams using values of N_c calculated with parameterizations MDC92, (Eqn. 9.4.7) here, and DM10, (Eqn. 9.4.12) here. Calculations with MDC92 and DM10 were performed over a wide range of values of s_i and T. For comparison with $S_{w,th}$ and ΔS_{th}, these quantities were recalculated for pairs of S_w and T values. Calculated with MDC92 and DM10 values of $N_c(S_i)$ and $N_c(T)$ were superimposed in Fig. 9.34 on the field of $\delta(RHW_{th}) = RHW - RHW_{th} = (S_w - S_{w,th}) \times 100\%$ in S_w-T coordinates calculated using (Eqn. 9.16.1).

Fig. 9.34 represents an S_w-T diagram over the domain $-30 < T_c < 0$°C and $0.7 < S_w < 1.0$. Superimposed here is the threshold difference $\delta(RHW_{th})$. The bold hatched line denotes the boundary $RHW - RHW_{th} = 0$ ($\delta(RHW_{th}) = 0$) and separates thermodynamically allowed and prohibited states. The states above this line (white field), as described by (Eqn. 9.7.6b) for $\Delta\mu$, correspond to the negative differences of $\delta S_{th} < 0$ or of chemical potentials $\Delta\mu = \mu_l - \mu_i = \mu_{mstab} - \mu_{stab} < 0$, i.e., to $\mu_{mstab} < \mu_{stab}$, and therefore to negative values of r_{cr}. The ice germs cannot be nucleated above this line in this S_w-T area, which corresponds to the reverse transition from the stable ice phase to metastable liquid phase and is thermodynamically prohibited. Only the states with $r_{cr} > 0$ or $\Delta\mu > 0$ below the bold hatched line $\delta(RHW_{th}) = 0$ (shaded fields) are thermodynamically allowed for heterogeneous ice nucleation by freezing. Fig. 9.34 shows that the allowed T-S_w domains are located in the right lower triangles below the temperature of −8 to −12 °C and at a water saturation ratio above 0.8 to 0.83, this area covering only about ⅛ of the entire domain is considered. We note here that ice nucleation in the MDC92, PDA08, DM10, and similar empirical schemes is allowed in the thermodynamically prohibited region, and certain IN concentrations are predicted here. The boundaries of the allowed domain depend on the size r_a of aerosol particles. When r_a increases from 0.05 μm, typical of the fine mode, to 1 μm, typical of the coarser mode, the allowed domain shifts to higher temperatures by about 5 °C.

It is interesting to note that the isolines of the MDC92 s_i-parameterization are in good correlation (almost parallel) with the isolines of $\delta(RHW_{th})$, indicating that MDC92 captures some basic features

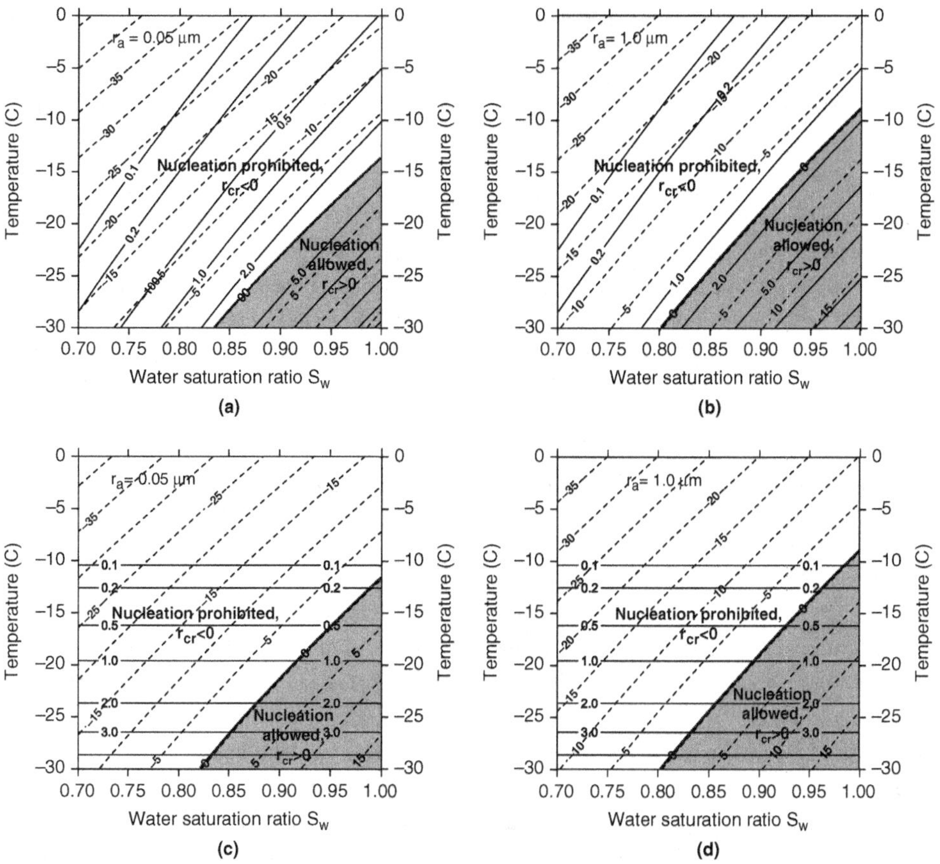

Figure 9.34. S_w-T diagrams of N_c calculated with MDC92 (a, b) and DM10 (c, d) parameterizations (solid lines) with superimposed threshold difference $\delta(RHW_{th}) = RHW - RHW_{th} = (S_w - S_{w,th}) \times 100\%$ calculated with KC scheme based on CNT (dashed lines), as described in the text. The bold and hatched line $\delta(RHW,_{th}) = 0$ or $RHW = RHW_{th}$ indicate the boundary between the thermodynamically allowed and prohibited states. The physical states and N_c above this line in the white field are thermodynamically prohibited and correspond to negative critical radii of ice germs, $r_{cr} < 0$. The states below the line $\delta(RHW,_{th}) = 0$ in the right lower corner are thermodynamically allowed and physical. From Curry and Khvorostyanov, Atmos. Chem. Phys. (2012), with changes.

of the nucleation process. However, the gradients dN_c/dS_w and dN_c/dT in MDC92 are noticeably smaller than predicted by the classical theory. This may be caused by the smoothing action of the mixtures of aerosols in the chamber experiments with different properties, while calculations with (Eqn. 9.16.1) are performed for only one aerosol type. The agreement of DM10 T-parameterization with classical theory is somewhat worse because they account only for the T-dependence but do not account for the humidity dependence.

Another thermodynamic constraint is imposed on the immersion freezing mode. In some works, it was hypothesized (without derivation) that the effect of humidity in the immersion mode can be described by introducing a new expression for the radius of the critical germ of immersion freezing in the form

$$r_{g,imm} = \frac{2v_{w1}\sigma_{iw}}{kT\ln\left(a_w \dfrac{e_{ws}}{e_{is}}\right)}, \tag{9.16.2}$$

where v_{w1} is the volume of a water molecule in ice, σ_{iw} is the surface tension between ice and water at the ice–water interface, e_{ws} and e_{is} are the saturation vapor pressures over water and ice, and a_w is the water activity inside a drop containing an ice germ. It was hypothesized in these works that this equation is based on the expression for r_{cr} from classical nucleation theory given in Chen et al. (2008, equation (5) and Section 2.2.2 therein) for the case of immersion freezing that can be written as

$$r_{g,imm} = \frac{2v_{w1}\sigma_{iw}}{kT\ln(e_{ws}/e_{is})}. \tag{9.16.3}$$

Using the relation $v_{w1} = 1/N_i$, where N_i is the concentrations of molecules per unit volume of ice, this equation can be rewritten as

$$r_{g,imm} = \frac{2\sigma_{iw}}{N_i kT\ln(e_{ws}/e_{is})}. \tag{9.16.4}$$

This equation exactly coincides with the classical equation (Eqn. 9.7.7a) of Section 9.7 given in Fletcher (1962), Young (1993), and other descriptions of CNT. Equations (9.16.3) and (9.16.4) are correct equations of CNT, and their equivalence to the J. J. Thomson's formulation (Eqn. 9.7.7) in terms of $\ln(T_0/T)$ was shown in Section 9.7. Chen et al. (2008) cited this equation as applicable for freezing in pure water—i.e., at water saturation, and it was stated further in Chen et al. (2008): "Note that for the freezing of solution, the parameter e_{ws} should be modified by the water activity of the solution." Some authors modified (Eqn. 9.16.3) for the case of water subsaturation (RHW < 100% or $S_w < 1$) simply multiplying the ratio e_{ws}/e_{is} by the water activity $a_w < 1$, and this equation got the form (Eqn. 9.16.2).

However, it can be shown that simple multiplication of the ratio e_{ws}/e_{is} by the water activity $a_w < 1$ is a risky trick and leads to a confusing result. It was shown in Section 3.10 and (Eqn. 3.10.24) that the water activity in equilibrium with ice $a_w^i(T)$ (as in a drop containing ice germ), is expressed via the ratio of saturated vapor pressures over ice and water, $a_w^i(T) = e_{is}(T)/e_{ws}(T)$, and it was illustrated in Fig. 8.14b that this equation gives a very good description of the melting curve for solution in agreement with the other methods. Substituting this $a_w^i(T)$ into (Eqn. 9.16.2), as has been done in several works, we obtain

$$r_{g,imm} = \frac{2v_{w1}\sigma_{iw}}{kT\ln\left(a_w^i \dfrac{e_{ws}}{e_{is}}\right)} = \frac{2v_{w1}\sigma_{iw}}{kT\ln\left(\dfrac{e_{is}}{e_{ws}}\dfrac{e_{ws}}{e_{is}}\right)} = \frac{2v_{w1}\sigma_{iw}}{kT\ln(1)} = \frac{2v_{w1}\sigma_{iw}}{0} = \infty. \tag{9.16.5}$$

Thus, (Eqn. 9.16.2) predicts an infinite germ radius, which is a confusing result. Therefore, generalization of (Eqn. 9.16.2) for solutions or water subsaturations by simple multiplication of the ratio e_{ws}/e_{is} by $a_w < 1$ leads to a wrong result. A more rigorous derivation is required for r_{cr}. Such a derivation for r_{cr} for cloud drop freezing in the immersion mode was presented in this chapter

(Eqns. 9.7.6c) to (9.7.6h), based on the general equilibrium (entropy) equation and Kelvin's equilibrium for cloud drops. The equations for r_{cr} for haze particles in the DF mode and for cloud drops in the immersion mode appeared to be similar. It was shown in parcel model simulations that the equations for r_{cr} and ΔF_{cr} can be applicable both in the deliquescence-freezing mode and in the immersion mode at water subsaturation ($S_w < 1$) and supersaturation ($S_w > 1$) (Section 9.14). However, thermodynamic constraints impose certain limitations on the physical domain in the (T-S_w) or (T-a_w) space, as illustrated in Fig. 9.34.

We do not present here similar thermodynamic analysis of the other existing parameterizations, but this can easily be done for any function $N_c(T)$ and $N_c(s_i)$. These thermodynamic limitations also should be accounted for when choosing and comparing the empirical and theoretical parameterizations of ice nucleation in the numerical models of various complexity as, for example, in Comstock et al. (2008) and Eidhammer et al. (2009). The empirical parameterizations should be applied with caution outside the ranges recommended for them and should not be applied outside of thermodynamically allowed conditions.

9.17. Evaluation of Ice Nucleation and Cloud Phase State Parameterizations

Another verification of the empirical versus theoretical ice nucleation parameterizations can be done by comparison of the cloud phase state predicted by various parameterizations with the climatology of the mixed-phase clouds as is observed and parameterized in the general circulation models (GCMs). Eidhammer et al. (2009; hereafter EDK09) compared three parameterizations of heterogeneous ice nucleation using a parcel model developed at Colorado State University (CSU). The model is based on the spectral bin microphysics for the mixed and ice states with various parameterizations of ice nucleation. The three ice nucleation schemes tested in EDK09 included the two empirical schemes described in Section 9.4, Diehl and Wurzler (2004, hereafter DW04), Phillips et al. (2008, hereafter PDA08), and the theoretical scheme based on the extended classical nucleation theory by Khvorostyanov and Curry (2000, 2004a,b, 2005a, hereafter KC). EDK09 compared mostly the crystal concentrations N_c produced by these schemes but not the ratio of the LWC to IWC—i.e., not the cloud mixed-phase state.

Comparing the results of simulations for the three parameterizations, EDK09 found that for small vertical velocities $w \sim 5\,\mathrm{cm\,s^{-1}}$, all three parameterizations yield similar results. For large w, only PDA08 compares well with typical observations of ice nucleation in CFDC, producing $N_c \sim 1$–$20\,\mathrm{L^{-1}}$, while the other two parameterizations (DW04 and KC) produce crystal concentrations significantly higher than PDA08. EDK09 concluded that the empirically derived "constraint" on the upper limit of N_c used in the PDA08 scheme should be used in cloud and climate model parameterizations.

Curry and Khvorostyanov (2012, hereafter CK12) compared further the PDA08 and KC schemes in order to understand the sources of the discrepancies between the two parameterizations (we note that the DW04 scheme performs comparably to KC and produces comparable values of N_c, LWC and IWC). The results of these comparisons are described in this section. CK12 carried out simulations using the parcel model described in KC05 (see Sections 9.13 and 9.14) and used the same composition of aerosol as in EDK09, and the KC deliquescence-heterogeneous freezing (DHetF) ice nucleation scheme (Section 9.7). The KC scheme and CNT in general were further extended in Curry and Khvorostyanov (2012) by generalizing the Fletcher's (1969) idea and including a temperature-dependent parameterization of the active site area α as a function of T,

$$\alpha(T) = \alpha_0(1 - T_c/T_v)\theta(T_{th} - T_c)\theta(T_c - T_v), \tag{9.17.1}$$

where $\theta(x)$ is the Heaviside function, $T_{th} = -5\,°\mathrm{C}$ is the threshold temperature of nucleation close to that assumed in EDK09, and $T_v = -20\,°\mathrm{C}$ is the scaling temperature that determined the rate of decrease of $\alpha(T)$. The value $\alpha_0 = 2 \times 10^{-5}$ was chosen because it has been successfully tested in KC05 to explain the fast glaciation of polar maritime cumulus observed in Hobbs and Rangno (1990) and Rangno and Hobbs (1991) without high water supersaturation (see Section 9.14). Equation (9.17.1) indicates that $\alpha(T)$ has a maximum $\alpha_0 = 2 \times 10^{-5}$ at warm T, decreases to 0 at $T_v = -20\,°\mathrm{C}$, and $\alpha(T) = 0$ at $T_c < T_v$. This parameterization accounts for the fact that the area of the active sites close to the structure of water ($m_{is} = 1$) that are favorable for nucleation increases toward 0 °C. We hypothesize that these sites can be formed by crystal defects, steps, or premelted sites. Their exact origin does not matter for now, but it is known that the number of such sites may increase toward 0 °C (Fletcher, 1970; Hobbs, 1974; Dash et al., 1995; Rosenberg, 2005).

Simulations were conducted under the following conditions: $w = 50\,\mathrm{cm\,s^{-1}}$, $RHW_0 = 96\%$, $T_0 = 10\,°\mathrm{C}$, and with the two models of active sites, $\alpha = 0$ and $\alpha(T)$, described by (Eqn. 9.17.1). The KC scheme was used with DHetF mode in three versions. The input data for these three runs were the following: #1) only one coarse aerosol fraction included as in EDK09, lognormal size spectrum, concentration $N_{IN,2} = 1\,\mathrm{cm^{-3}}$ ($1000\,\mathrm{L^{-1}}$), mean geometric radius $r_{d2} = 0.4\,\mu\mathrm{m}$, dispersion $\sigma_{d2} = 2$, and active site area $\alpha = 0$; #2) the same coarse aerosol fraction, $N_{IN,2} = 1\,\mathrm{cm^{-3}}$, but variable $\alpha(T)$ described by (Eqn. 9.17.1); #3) $\alpha = 0$, and including three IN lognormal fractions with equal concentrations $N_{IN,1} = N_{IN,2} = N_{IN,3} = 10\,\mathrm{L^{-1}} = 10^{-2}\,\mathrm{cm^{-3}}$—i.e., 100 times smaller than in runs #1 and #2 and in EDK09 for the KC scheme), with $r_{d1} = r_{d2} = r_{d3} = 0.4\,\mu\mathrm{m}$, but 3 different values of contact parameter, 0.85, 0.75, and 0.5 that can mimic a mixture of organic (bacteria or pollen), soot, and mineral IN. The IN concentration of $1\,\mathrm{cm^{-3}}$ in the runs #1 and #2 follows the choice in EDK09 for the KC scheme, although it is not clear why this very high concentration $N_{IN,2}$ was chosen in EDK09. This is an arbitrary choice, 2–3 orders of magnitude higher than typical IN concentrations in CFDC, and it is not related to any characteristic of the KC scheme. In run #3, the concentrations were chosen comparable to those measured in CFDCs and used in PDA08.

The results of simulations from EDK09 with ice scheme PDA08 and from the three simulations of our parcel model with the KC scheme are compared in Fig. 9.35. Due to high initial RHW, drop activation occurs in a few minutes (a bit earlier than in EDK09 due to a little higher RHW_0). The drop concentration N_d is $\sim 90\,\mathrm{cm^{-3}}$ in the EDK09 model and $160\,\mathrm{cm^{-3}}$ in the KC model, the difference associated with different drop activation methods, but this is unimportant. Values of N_d are constant in EDK09 simulations for 4 hours (Fig. 9.35c), and liquid water content (LWC) increases over this period due to drop growth down to $T = -34.5\,°\mathrm{C}$ (Fig. 9.35e). In the EDK09 model with the PDA08 ice scheme, noticeable heterogeneous crystal nucleation begins at about 75 min when $T < -3\,°\mathrm{C}$, the concentration N_c increases almost linearly and reaches $\sim 22\,\mathrm{L^{-1}}$ at $T \sim -32.5\,°\mathrm{C}$ at a height above 6 km, and time is 240 min (Fig. 9.35d). Thus, nucleation with the PDA08 scheme continues over almost 4 hours, much longer than in any other heterogeneous scheme (e.g., Sassen and Benson, 2000; Lin et al., 2002; Kärcher and Lohmann, 2003; KC05; Liu and Penner, 2005), and much longer than in CFDC experiments, 7–15 s, (Phillips et al., 2008) upon which the PDA08

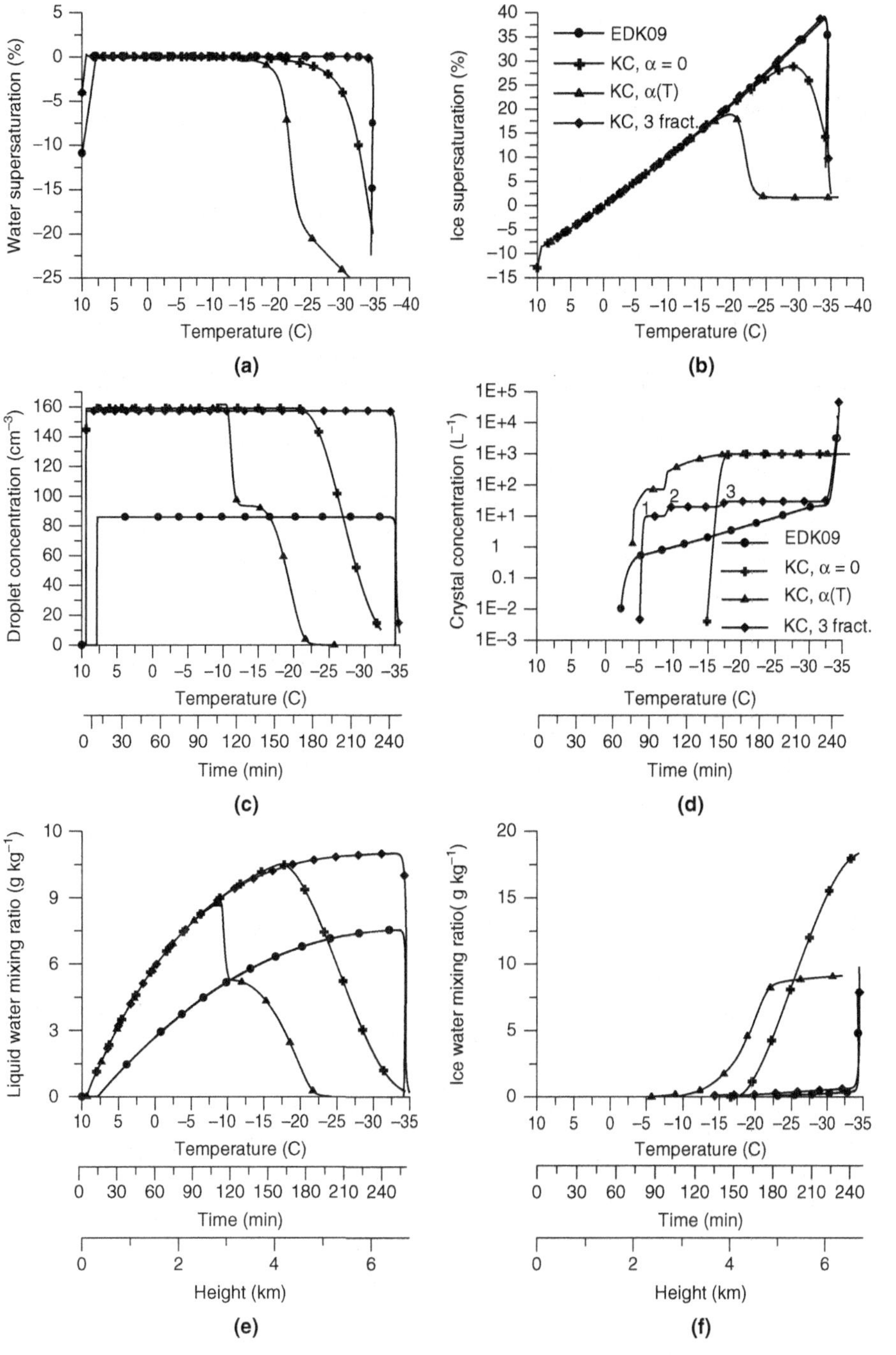

Figure 9.35. Comparison of the temperature and time dependencies of cloud microphysical properties in the parcel runs obtained in simulations by Eidhammer et al. (2009, EDK09) with parameterization of Phillips et al. (2008, PDA08, solid circles) and obtained in simulations in Curry and Khvorostyanov (2012) using the KC scheme with DHetF mode in three runs: #1) only 2nd aerosol mode included as in EDK09, $N_{d2} = 1\,\mathrm{cm}^{-3}$ ($1000\,\mathrm{L}^{-1}$), $\sigma_{d2} = 2$, $r_{d2} = 0.4\,\mu\mathrm{m}$, and active site area $\alpha = 0$ (crosses); #2) only the same 2nd aerosol mode and variable $\alpha(T)$ as a linear function of T as described in the text (triangles); #3) including three IN lognormal fractions with equal mean radii $r_{d1} = r_{d2} = r_{d3} = 0.4\,\mu\mathrm{m}$, equal concentrations $N_{IN,1} = N_{IN,2} = N_{IN,3} = 10\,\mathrm{L}^{-1} = 10^{-2}\,\mathrm{cm}^{-3}$ (i.e., 100 times smaller than in runs #1 and #2 and in EDK09 for the KC scheme), but with three different values of contact parameter for each fraction, $m_{is} = 0.85$, 0.75, and 0.5 (diamonds). The three nucleation impulses in N_c in the run #3 caused by nucleation of the IN fractions with different m_{is} are denoted by the numbers 1, 2, and 3 (Fig. 9.35d). The parameters: $w = 50\,\mathrm{cm\,s}^{-1}$, $RHW_0 = 96\%$, and $T_0 = 10\,°\mathrm{C}$. From Curry and Khvorostyanov, Atmos. Chem. Phys. (2012).

parameterization is based. Then, an abrupt increase in N_c occurs by almost 3 orders of magnitude to $1.6 \times 10^4 L^{-1}$, caused by homogeneous drop freezing, which begins in the EDK09 model at heights above 6 km, at $T \approx -34\,°C$, close to the homogeneous freezing threshold for the drops with radii of 18–20 μm. At temperatures warmer than −34 °C, the nucleated ice crystals do not influence N_d and LWC, and no signs of the Wegener–Bergeron–Findeisen process and crystallization are seen on the EDK09 curves. Instantaneous glaciation occurs due to drop homogeneous freezing only when the temperature falls below −35 °C; N_c rapidly increases by three orders of magnitude, and N_d and LWC abruptly drop to zero. Thus, the crystals heterogeneously formed in the PDA08 scheme are unable to produce any noticeable crystallization effect down to −34 °C, and the "constraints" imposed in the PDA08 scheme lead to a substantial underestimation of heterogeneous ice nucleation and ice phase.

In contrast, crystallization in the KC scheme in simulations #1 and #2 with $N_{IN,2} = 1\,cm^{-3}$ occurs much more smoothly with decreasing temperature, in the temperature range of ~20 °C. With $\alpha = 0$, crystal nucleation in the KC scheme begins at −15 °C and N_c reaches a maximum ~$10^3 L^{-1}$ within 2 °C. With the smooth function $\alpha(T)$ in simulation #2, crystal nucleation begins at about −5 °C, and ends at −17 °C, more smoothly than with $\alpha = 0$. Cloud glaciation with decreasing N_d and LWC begins at −15 °C with $\alpha = 0$ (at −7 °C with $\alpha(T)$, which better describes the effect of IN active at warmer temperatures) and ends at −35 °C with $\alpha = 0$ (at −23 °C with $\alpha(T)$), over the T-range of 16–20 °C and a time of 1 hour in both cases. (Note that the DW04 scheme, Fig. 1 in EDK09, is not shown here, but it performs similarly to the KC scheme in runs #1 and #2, and produces realistic crystallization and the cloud phase state.) In simulation #3 with 3 IN fractions and with the KC scheme, heterogeneous nucleation occurs in the three temperature ranges, near −5, −8 to −9, and −15 to −18 °C, corresponding to nucleation of each of 3 fractions, from the highest to lowest contact parameter. Each nucleation impulse produces values of N_c almost equal to the concentration in the corresponding fractions, ~$10 L^{-1}$, and the total is ~$30 L^{-1}$ at $T < -18\,°C$. Each nucleation impulse is located in a relatively narrow temperature range of 1–3 °C, but the total temperature range of ice nucleation stretches over 13 °C.

Fig. 9.35 shows that this nucleation picture and final crystal concentration with the KC scheme in simulation #3 are close to those produced in EDK09 with the PDA08 scheme. With this small final $N_c = 30 L^{-1}$, the KC scheme also does not produce glaciation down to the homogeneous freezing threshold of −34 °C, when rapid drop freeing and cloud crystallization occur. Thus, the conclusion in EDK09 that the KC and DW04 schemes produce very high crystal concentrations was caused by an arbitrary choice of very high ($1000 L^{-1}$) IN concentrations for these schemes in EDK09. The conclusion that nucleation in the KC scheme occurs in a very narrow temperature range was caused by the choice of just one IN fraction with "monodisperse" properties: contact angle, α, etc. A more realistic choice of IN properties produces nucleation with the KC scheme over a wide T-interval.

One of the major criteria for validity of ice nucleation parameterization is the cloud-phase state, an important characteristic in weather prediction and climate models. The phase state in clouds is characterized by the ratio of the liquid (LWC) to the total water (LWC + IWC) in mixed phase, $f_l = LWC/(LWC + IWC) \times 100\%$. Fig. 9.36 shows the observed climatology of f_l compiled from a few thousand aircraft measurements (Borovikov et al., 1963; reproduced in Pruppacher and Klett, 1997, PK97). In pure-liquid clouds at warm temperatures slightly below 0 °C, the observed climatological f_l is close to 100%, and then decreases with decreasing temperature (22% liquid at −15 °C) and tends to zero at $T < -30\,°C$—i.e., the clouds become purely crystalline.

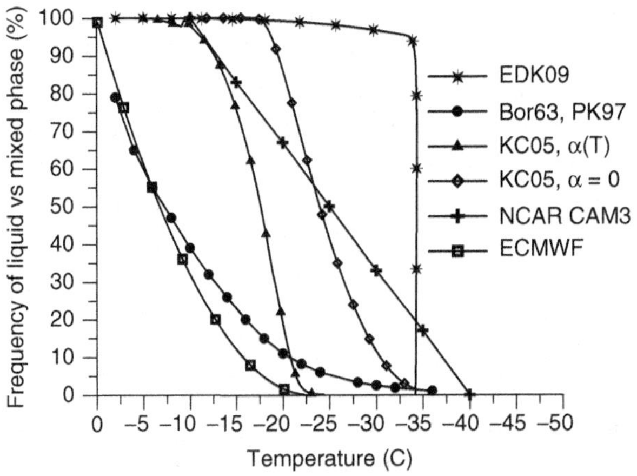

Figure 9.36. Frequency of liquid vs. mixed states. Climatological data after Borovikov et al. (1963, Bor63) (reproduced also in Fig. 2–33 in Pruppacher and Klett, 1997, PK97), compared to the characteristic of the liquid/mixed phase, the ratio f_l = LWC/(LWC + IWC), simulated in Curry and Khvorostyanov (2012) with the KC05 parcel model (Section 9.13) and heterogeneous KC ice scheme as shown in Fig. 9.35: with only 2nd aerosol mode included, $N_{d2} = 1\,\mathrm{cm}^{-3}$, $\sigma_{d2} = 2$, $r_{d2} = 0.4\,\mu\mathrm{m}$, and 2 values of active site areas: $\alpha = 0$ (diamonds), variable $\alpha(T)$ as described in the text (triangles), and simulated in the EDK09 parcel model with the PDA08 ice scheme with the same aerosol (asterisks). These simulations are compared to the corresponding T-partitioning of the liquid and ice phases in the climate models with single-moment microphysics: the NCAR CAM3 (Boville et al., 2006), (83% liquid at −15 °C) and ECMWF (ECMWF-2007), (12% liquid at −15 °C) as described in the text. From Curry and Khvorostyanov, Atmos. Chem. Phys. (2012).

Fig. 9.36 compares this climatological data with f_l calculated from the simulations data of EDK09 and from the three runs with the KC scheme shown in Fig. 9.35 earlier comparing two forms of $\alpha(T)$ and two input IN concentrations. These are also compared with the two parameterizations of f_l as a function of temperature in two general circulation models: the European Centre for Medium-Range Weather Forecasts (ECMWF-2007) and the National Center for Atmospheric Research Community Atmosphere Model 3 (NCAR CAM3), Boville et al. (2006). In ECMWF, the liquid fraction was chosen as $f_l = [(T - T_{ice})/(T_0 - T_{ice})]^2$, and $f_l = 0$ at $T < T_{ice}$, with $T_0 = 273.16$ and $T_{ice} = 250.16\,\mathrm{K}$ (12% liquid at −15 °C), (ECMWF-2007). In NCAR CAM3, the ice fraction was parameterized as $f_i(T) = (T - T_{max})/(T_{min} - T_{max})$ with $T_{max} = -10\,°\mathrm{C}$, $T_{min} = -40\,°\mathrm{C}$ (Boville et al., 2006). Then, f_l in percent can be written as $f_l(T) = 1 - f_i(T) = (T_{min} - T)/(T_{min} - T_{max}) \times 100$, and $f_l(T) = 0$ at $T < T_{min}$ (83% liquid at −15 °C). Fig. 9.36 shows that the ECMWF parameterization is very close to the climatological data of Borovikov et al. (1963), but ends at slightly warmer temperatures. The CAM3 parameterization has a slope close to the climatological data, but the curve CAM3 is displaced as a whole toward colder temperatures by about 10 °C, underestimating the ice phase at warm and medium temperatures (note that the T-limits in NCAR CAM2 were 0 and −20 °C (Boville et al., 2006), and $f_l(T)$ was closer to the ECMWF).

The $f_l(T)$ slopes in the KC scheme in simulations #1 and #2 with $N_{IN,2} = 1\,\mathrm{cm}^{-3}$ are steeper than the climatological ECMWF and CAM3 curves but are still comparable to them, and closer to CAM3.

A fraction of the ice phase increases in KC scheme at −16 °C with $\alpha = 0$ and at −7 °C with $\alpha(T)$; the threshold with $\alpha(T)$ is close to the threshold in CAM3. In general, the KC scheme in simulations #1 and #2 with $N_{IN,2} = 1\,\mathrm{cm}^{-3}$ may underestimate the ice phase at warm temperatures and overestimate the ice phase at cold T. However, there is a clear qualitative agreement of the KC scheme in simulations #1 and #2 with the climatological data and parameterizations ECMWF and CAM3, although a further smoothing of the KC curve over the wider T-range is desirable, which is discussed in the following text.

In contrast, the EDK09 simulations using the PDA08 parameterization, and simulations with the KC scheme in simulation #3 with low concentrations $N_{IN,1} = N_{IN,2} = N_{IN,1} = 10\,\mathrm{L}^{-1}$ are in sharp conflict with climatology. EDK09 and KC simulation #3 with low IN predict a more than 95% liquid phase down to −34.5 °C, where homogeneous nucleation begins to act in the drops with radii of ~20 μm. With homogeneous nucleation, the PDA08 and KC run #3 curves are very close and exhibit abrupt crystallization within a few tenths of a degree, and the curve $f_l(T)$ is actually vertical at $T_c \approx -35$ °C, which disagrees with climatological data and GCM parameterizations. The simulations in EDK09 show that the DW04 scheme produces ice crystals with concentrations similar to the KC scheme in simulations #1 and #2—i.e., with the limits $300\,\mathrm{L}^{-1}$ for dust and $1000\,\mathrm{L}^{-1}$ for soot, the corresponding limiting aerosol concentrations in the 2nd mode chosen in EDK09.

The low heterogeneous nucleation efficiency of the PDA08 scheme in ice production was somewhat masked in Fig. 9.35, where the characteristics of the liquid and ice phases were plotted separately, but it becomes clearer in Fig. 9.36, when considering the ratio of liquid to total water, $f_l(T)$. Fig. 9.36 illustrates that the increase in LWC during the parcel ascent is so rapid that the small amount of ice nucleated with the PDA08 scheme did not result in any noticeable crystal growth and liquid water depletion by the WBF mechanism. EDK09 argued that the KC and DW04 schemes produced crystal concentrations a few orders of magnitude greater and substantially overestimate ice production. However, Figs. 9.35 and 9.36 show that high crystal concentration is not a characteristic feature of the KC scheme. It was a result of choosing a high value of input IN concentration $N_{IN} \sim 1\,\mathrm{cm}^{-3}$ for the KC scheme in EDK09.

Fig. 9.36 shows that the KC scheme with higher values of $N_{IN} \sim 1\,\mathrm{cm}^{-3}$ is much closer to reality in reproducing the cloud phase state (and the DW04 scheme also), while the PDA08 scheme produces unrealistically high values of liquid water down to the threshold of homogeneous nucleation. It is not clear whether this is a consequence of the too low IN concentrations in the PDA08 scheme or a result of an unrealistic simulation with an isolated parcel model with high constant vertical velocities for a long time.

This question can be answered by running Eulerian models with a more realistic dynamic and physical framework with various ice schemes. To address this question, the KC scheme was tested with a more realistic 1D Eulerian model in simulations of the ARM Mixed Phase Arctic Cloud Experiment (MPACE) in autumn 2006 (Verlinde et al., 2007; Klein et al., 2009; Morrison et al., 2009). The results of simulation were described in Curry and Khvorostyanov (2012) and appeared to be similar to those observed in the MPACE mixed cloud with quite realistic crystal concentrations and cloud-phase states.

10

Parameterizations of Heterogeneous Ice Nucleation

Using the general equations derived in Chapter 9 based on extended classical nucleation theory, parameterizations are derived suitable for use in cloud and climate models with sufficiently large time steps. Heterogeneous freezing of haze particles and drops caused by supersaturation and temperature variations are considered in Sections 10.1 and 10.2, and similar parameterizations of the deposition nucleation is derived in Sections 10.3. In particular, it is shown that the functional forms of previous empirical or semi-empirical parameterizations described in Section 9.4 can be obtained and generalized from these theoretical parameterizations based on classical nucleation theory, and their empirical parameters can be expressed via the fundamental atmospheric parameters and aerosol properties.

In the next three sections, Sections 10.4–10.6, contact ice nucleation mode is described, its general properties, the three major mechanisms of aerosol scavenging by drops that may lead to contact nucleation, the collection rates, probabilities of freezing, and scavenging in polydisperse ensembles of drops and aerosols.

10.1. Analytical Parameterization of Heterogeneous Freezing Kinetics Based on Classical Nucleation Theory (CNT)

In this section, we consider the parameterization of heterogeneous ice nucleation by freezing, based on the generalization of the classical nucleation theory described in the previous sections of Chapter 9. Pure crystalline clouds are considered in detail, whereby ice crystals form by the freezing of haze particles without formation of the liquid phase. This mechanism of heterogeneous freezing competes with the homogeneous freezing of haze droplets in the formation of cirrus clouds (Jensen et al., 1994, 1998; DeMott et al., 1998; Khvorostyanov et al., 2001, 2006; Lin et al., 2002; Khvorostyanov and Sassen, 2002; Kärcher and Lohmann, 2003; Kärcher et al., 2006; Barahona and Nenes, 2009). This phenomenon also occurs at low altitudes in the polar regions and is called "clear sky ice crystal precipitation" or "diamond dust" (e.g., Ohtake and Huffman, 1969; Curry, 1983, 1995; Curry et al., 1990, 1996, 2000; Girard and Blanchet, 2001). Altocumulus clouds also may form via this pathway without the liquid phase (e.g., Sassen and Khvorostyanov, 2007, 2008; Sassen and Wang, 2012).

10.1.1. Nucleation Rates in a Polydisperse Aerosol

The nucleation rate per particle in a separable form was presented in (Eqn. 9.8.13):

$$J_{f,het}(T, s_{w,} r_N, t) \approx J_{f,het}^{(0)}(T, r_N, \varepsilon)\exp[u_{shet} s_w(t)], \qquad (10.1.1)$$

with $J^{(0)}_{f,het}$ defined in (Eqn. 9.8.15) and u_{shet} defined in (Eqn. 9.8.11). It was shown in Section 9.14 in the parcel model simulations of ice nucleation in the deliquescence-freezing mode that the freezing rate $J_{f,het}$ is usually small, and expressions for the probability of freezing and heterogeneous nucleation rate $R_{f,het}$ in a polydisperse aerosol given in Section 9.10 can be simplified as described by (Eqns. 9.14.2)–(9.14.4). Substituting (Eqn. 10.1.1) into (Eqn. 9.14.4) yields a simplified expression for the polydisperse nucleation rate

$$R_{f,het}(t) = \frac{dN_{fr}}{dt} \approx \int_{r_{\min}}^{r_{\max}} f_a(r_N) J_{f,het}(r_N, t)\, dr_N$$
$$= J^{(0,\alpha 0)}_{f,het}(T, \bar{r}_N) \bar{J}^{\alpha}_{act}(\alpha_a, m_{is}, \bar{r}_N) \exp[u_s(T) s_w(t)], \tag{10.1.2}$$

with $J^{(0,\alpha 0)}_{f,het}(T, \bar{r}_N)$ and $\bar{J}^{\alpha}_{act}(\alpha_a, m_{is}, \bar{r}_N)$ defined in (Eqns. 9.8.8) and (9.8.9a). The dependence on r_N vanishes after averaging over $f_a(r_N)$. In this section, we denote the active site fraction with α_a to distinguish from the deposition coefficient α_d. All other notations are the same as in Section 9.8. $\bar{J}^{\alpha}_{act}$ is the part of the nucleation rate that describes the effect of the active sites—i.e., $J^{\alpha}_{act}(T, r_N)$ defined in (Eqns. 9.8.9a) or (9.8.12a) and integrated over the size spectrum $f_a(r_N)$:

$$\bar{J}^{\alpha}_{act}(\alpha_a, m_{is}, \bar{r}_N) = \frac{1}{\bar{r}_N^2} \int_{r_{\min}}^{r_{\max}} f_a(r_N) r_N^2 \sum_k \exp(z_\alpha^k r_N^2)\, dr_N, \tag{10.1.3}$$

with the dimension cm^{-3}.

The integral in (Eqn. 10.1.3) depends on the specified size spectrum by r_N, and its analytical evaluation is possible for some types of spectra. The simplest estimate can be done for a monodisperse size spectrum such as the Dirac's delta function, $f_a(r_N) = N_a \delta(r - r_N)$, where N_a is the concentration of aerosol that can serve as IN, and for one type of active site with $m_{is} = 1$. Substituting this spectrum into (Eqn. 10.1.3), we obtain

$$\bar{J}^{\alpha}_{act}(\alpha_a, m_{is}, \bar{r}_N) = N_a \frac{\bar{r}_N^2}{\bar{r}_N^2} e^{z_\alpha \bar{r}_N^2} = N_a \exp\left[\frac{\alpha_a \sigma_{is}(1 - m_{is}) \bar{r}_N^2}{kT}\right]. \tag{10.1.4}$$

The delta-function type spectrum is infinitely narrow, and the integral over the other size spectrum of the finite width will contain some correction on the order of 1. We rewrite this integral with such a shape correction coefficient c_{sh},

$$\bar{J}^{\alpha}_{act}(\alpha_a, m_{is}, \bar{r}_N) \approx c_{sh} N_a \frac{\bar{r}_N^2}{\bar{r}_N^2} e^{z_\alpha \bar{r}_N^2} \approx c_{sh} N_a \exp\left[\frac{\alpha_a \sigma_{is}(1 - m_{is}) \bar{r}_N^2}{kT}\right]. \tag{10.1.5}$$

If there are several active sites with different α_a, m_{is}, this expression is easily generalized by summation over all sites with different z_α as in (Eqn. 10.1.3). Equations (10.1.1) and (10.1.2) are used in the following for parameterizations of the concentration and mean radius of the nucleated crystals.

10.1.2. Temporal Evolution of Supersaturation

The kinetics of the heterogeneous ice nucleation via deliquescence-freezing mode was described in Section 9.14 based on simulations using a parcel model with spectral bin microphysics. These simulations allow determination of the major features of the supersaturation evolution for developing an approximation suitable for analytical parameterizations. The nucleation rates described by (Eqns. 10.1.1 and (10.1.2) depend on T and s_w. Similar to the case of homogeneous freezing described in Section 8.10, the major time dependence is contained in $s_w(t)$ and is exponential, $\exp[u_{shet}s_w(t)]$. The time dependence of s_w and N_c is shown in more detail in Fig. 10.1, which is an enlarged fraction of Fig. 9.22 ($T_0 = -18\,°C$, $w = 2\,cm\,s^{-1}$) with a linear scale for N_c, which illustrates the difference in N_c at the time t_{max} of maximum supersaturation and after the end of nucleation. These Figures show that there are two branches in the supersaturation curves, with their growth (ascending branch) and decrease (descending branch). Noticeable nucleation begins at the time $t_{cr1} = 78.0$ min when N_c reaches $N_{c,cr1} = 10^{-3}\,L^{-1}$ and both water and ice supersaturations reach their first critical values, $s_{w,cr1} = -0.0385$ (−3.85%), $s_{i,cr1} = 0.176$ (17.6%).

The maximum values of both supersaturations, $s_{w,max} = -0.0281$ (−2.81%) and $s_{i,max} = 0.187$ (18.7%), occur at $t_{max} = 90.17$ min, when $N_c = 14.5\,L^{-1}$. However, in contrast to drop activation (Chapters 6, 7) and similar to homogeneous ice nucleation (Chapter 8), heterogeneous ice nucleation still does not cease at time t_{max} with maximum supersaturations s_{wmax} and s_{imax}, when $N_c(t_{max})$ is only about one half of the subsequent final value. Nucleation continues at $t > t_{max}$ with decreasing supersaturations, and reaches $27.1\,L^{-1}$ at the second critical time $t_{cr2} = 95$ min, when $s_{w,cr2} = -0.034$ (−3.4%) and $s_{i,cr2} = 0.181$ (18.1%). After that, the supersaturations fall below their critical values and nucleation ceases, although cooling continues. This analysis shows that the relative variation of s_w is $\delta s_w/s_w \sim 27\%$, while the change of s_i is much smaller, only $\delta s_i/s_i \sim 0.05\%$. Thus, as in the case with homogeneous nucleation, we can assume that $s_i \approx s_{i,cr}^{het} \approx$ const during nucleation.

The variation of s_w is $\delta s_w = s_{wmax} - s_{w,cr1} = -(3.85 \text{ to } 2.81) \times 10^{-2} = 1.04 \times 10^{-2}$. The value of u_{shet} is large, and even small variations in s_w lead to large variations in $\exp(u_{shet}s_w)$. For example, at $T = -20\,°C$, $u_{shet} \sim 4 \times 10^2$, and the product $u_{shet}\delta s_w \sim (4 \times 10^2)(1.04 \times 10^{-2}) \approx 4.2$. Thus, the factor $\exp(u_{shet}s_w)$ in

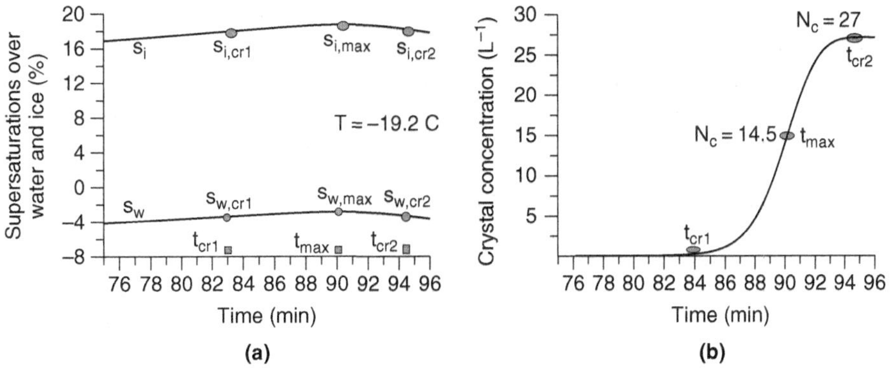

Figure 10.1. Enlarged fraction of Fig. 9.22 in the parcel run with heterogeneous ice nucleation, $w = 2\,cm\,s^{-1}$, and initial temperature $T_0 = -18\,°C$. (a) water (s_w) and ice (s_i) supersaturations, %; (b) crystal concentration N_c (L^{-1}). The subscript "$cr1$" means the first critical time when nucleation starts, "max" means the maxima of s_w, s_i, at the time t_{max}, and "$cr2$" means the second critical time, when nucleation ceases. The linear scale for N_c shows that only about half of the crystals are nucleated by the time t_{max} with $s_{i,w,max}$ and nucleation continues at $t > t_{max}$.

$R_{f,het}$ in (Eqns. 10.1.1) and (10.1.2) varies by $\exp(4.2) \approx 67$, while variations in $J_{f,het}^{(0,\alpha 0)}(T, \bar{r}_N)$ due to the temperature change are much smaller, as it was in the case with homogeneous nucleation in Chapter 8. Therefore, the time dependence of R_{fr} can be approximately accounted for via $s_w(t)$ at some average temperature or at $T = T_{cr,1}$.

We can therefore approximate differential s_i and integral y_i supersaturations as in Section 8.10 by the constant values $s_i = y_i' \approx s_{i,cr}^{het}$, and variations in $s_w = y_w'$ by the linear function $s_w(t) = y_w' \approx a_{1w}t + s_{w,cr}^{het}$. This is similar to a linear approximation for the drops activation described in Chapters 6 and 7 but with initial critical (threshold) values, which account for the specifics of ice nucleation described earlier. As discussed in Chapters 6–8, the parameter a_{1w} can be specified in various ways, which yield lower and upper limits of the solution. An approximation that gives a lower bound of the solution can be obtained with $a_{1w} = c_{1w}w$:

$$s_w(t) = y_w'(t) = s_{w,cr}^{het} + c_{1w}wt, \qquad y_w(t) = s_{w,cr}^{het}t + (c_{1w}w/2)t^2, \tag{10.1.6a}$$

$$s_i(t) = y_i'(t) \approx s_{i,cr}^{het}, \qquad y_i(t) = s_{i,cr}^{het}t. \tag{10.1.6b}$$

The difference between the lower and upper bounds is on the order of 5–20% or smaller, and the difference between the lower bound and mean values is even smaller, 3–10%. The analytical equations for the critical supersaturations $s_{w,cr}^{het}$ and $s_{i,cr}^{het}$ were derived in Section 9.12.

10.1.3. Heterogeneous Nucleation Rate Derived from CNT and Comparison with the Previous Parameterizations

Substitution of $s_w(t)$ from (Eqn. 10.1.6a) into the separable nucleation rate in (Eqn. 10.1.1) yields $J_{f,het}(T, s_w)$ as a function of time in the form

$$J_{f,het}[T, s_w(t), r_N, \varepsilon] \approx J_{f,het}^{(0)}(T, r_N, \varepsilon)\exp(u_{shet}s_{w,cr}^{het})\exp(u_{shet}c_{1w}wt). \tag{10.1.7}$$

Several previous parameterizations of heterogeneous ice nucleation were based on comparisons with the homogeneous nucleation rate $J_{f,hom}$ and modifications of parameterized $J_{f,hom}$ to make it suitable for description of heterogeneous nucleation. Kärcher and Lohmann (2003, hereafter KL03) and Kärcher et al. (2006) hypothesized that a relation between $J_{f,het}$ and $J_{f,hom}$ can be established based on the "shifted activity method" (see Sections 8.8 and 9.12). This relation from KL03 can be written as

$$\frac{J_{f,het}[T, \{X_k\}, a_w]}{J_{f,\mathrm{hom}}[T, a_w + \delta a_w]} = \Delta_l. \tag{10.1.8}$$

Here, a_w is the water activity identified with ambient relative humidity divided by 100; δa_w = const = 0.32 is an independent on T shift in water activity—i.e., the difference in the critical activities for freezing in these two modes of one particle with radius of 0.25 μm in 1 s, Δ_l is the parameter with the dimension of length that was estimated in KL03 as $\Delta_l \sim 30$ nm, and $\{X_k\}$ denotes the other parameters of heterogeneous freezing.

A direct and detailed comparison of $J_{f,het}$ with the homogeneous nucleation rate $J_{f,hom}$ can be done using (Eqn. 10.1.7) and an analogous expression (Eqn. 8.10.17a) for $J_{f,hom}(t)$ derived in Section 8.10, and also expressed in the form with separated temperature and saturation dependencies. Dividing

$J_{f,het}$ in (Eqn. 10.1.7) by $J_{f,hom}$ in (Eqn. 8.10.17a) and accounting for $\Delta F_{cr,0}^{0,\alpha 0} = \Delta F_{cr,0}^{\mathrm{hom}} f(m_{is})$ and $u_{shet} = u_{shom} f(m_{is})$ (see the notations in Section 8.10) yields

$$\frac{J_{f,het}[T, s_w(t), r_N]}{J_{f,\mathrm{hom}}[T, s_w(t), r_N]} = R_{J1} R_{J2} R_{J3}, \tag{10.1.9a}$$

where R_{J1}, R_{J2}, and R_{J3} are the ratios specified next.

$$R_{J1} = \frac{J_{f,het}^{(0)}(T, \bar{r}_N, \varepsilon)}{J_{f,\mathrm{hom}}^{(0)}(T,)} = \tilde{\Delta}_l \exp\left[-\frac{\Delta F_{cr,0}^{\mathrm{hom}}[f(m_{is}) - 1]}{kT} + \frac{\Delta F_{cr}^{\varepsilon}}{kT}\right], \tag{10.1.9b}$$

$$\tilde{\Delta}_l = \frac{C_{het}}{C_{\mathrm{hom}}} = \left(\frac{c_{1s}}{2N_{cont}} \frac{4\pi \bar{r}_N^2}{(\sigma_{is}/kT)^{1/2}}\right). \tag{10.1.9c}$$

Since $c_{1s} \sim N_{cont}$, an estimate at $T \sim 250$ K with $\sigma_{is} \sim 25\,\mathrm{erg\,cm^{-2}}$ gives $\tilde{\Delta}_l \sim (2.3\times10^{-7}\ \mathrm{cm^{-1}})\bar{r}_N^2$. The parameter $\tilde{\Delta}_l$ can be viewed as an analog of Δ_l in (Eqn. 10.1.8) from KL03, but now depends additionally on $\bar{r}_N$ and σ_{is}, and its dependence on T is more complicated. The factor R_{J2} is

$$\begin{aligned} R_{J2} &= \exp(u_{shet} s_{w,cr}^{het} - u_{s\mathrm{hom}} s_{w,cr}^{\mathrm{hom}}) \\ &= \exp[u_{s\mathrm{hom}} f(m_{is}) s_{w,cr}^{het} - u_{s\mathrm{hom}} s_{w,cr}^{\mathrm{hom}}], \end{aligned} \tag{10.1.9d}$$

where we used the relation between u_{shom} and u_{shet}. This factor can be viewed as an analog of the KL03 relation with the "shifted activity," but is it is more complicated. The shift in water activity (or critical humidity) hypothesized in KL03 would produce an exponent in (Eqn. 10.1.8) $J_{f,\mathrm{hom}} \sim \exp[u_{s\mathrm{hom}}(s_{w,cr}^{\mathrm{hom}} + \delta s_{w,cr})]$, while (Eqn. 10.1.7) shows that $J_{f,het} \sim \exp(u_{shet} s_{w,cr}^{het})$. Then, the equality $J_{f,het}(s_w) \sim J_{f,het}(s_w + \delta s_w)$ would be valid if $u_{s\mathrm{hom}}(s_{w,cr}^{\mathrm{hom}} + \delta s_{w,cr}) = u_{shet} s_{w,cr}^{het}$, or using again a relation $u_{s,het} = u_{shom} f(m_{is})$, we obtain an equation for $\delta s_{w,cr}$:

$$\delta s_{w,cr} = f(m_{is}) s_{w,cr}^{het} - s_{w,cr}^{\mathrm{hom}}. \tag{10.1.9e}$$

This equation shows that $\delta s_{w,cr}$ strongly depends on m_{is}; besides, $s_{w,cr}^{het}$, $s_{w,cr}^{\mathrm{hom}}$ themselves also depend on the temperature, IN properties (m_{is}, α, ε), and the chosen critical nucleation rate—i.e., on the cooling rate as described in Section 9.12 (see Fig. 9.17). Therefore $\delta s_{w,cr}$ cannot be constant, as was illustrated in Fig. 9.11. The next factor, R_{J3}, has the form

$$R_{J3} = \exp\{u_{s\mathrm{hom}}[f(m_{is}) - 1] c_{1w} wt\}, \tag{10.1.9f}$$

i.e., the ratio of $J_{f,het}(t)$ and $J_{f,hom}(t)$ is not constant but decreases with time, since $f(m_{is}) < 1$, the argument of exp is negative, and exp < 1.

Thus, although the ratio $J_{f,het}/J_{f,hom}$ bears some similarities with the KL03 relation (Eqn. 10.1.8), the ratio predicted by this extended CNT theory is more complicated, so that $J_{f,het}$ cannot be expressed by a simple relation via $J_{f,hom}$ and the shift in water activity or saturation ratios is not constant as was illustrated in Sections 8.8 and 9.12 both with calculations and experimental data. Fortunately, in the method developed here, $J_{f,het}$ does not need to build upon $J_{f,hom}$ and can be evaluated independently.

In some previous semi-empirical parameterizations, the exponential time dependence of $J_{f,het}$ was hypothesized and parameterized introducing some characteristic time scale $\tau_{nuc,het}$. The exponential time dependence is described by (Eqn. 10.1.7) and now, similar to the case of homogeneous nucleation considered in Section 8.10, we can derive an analytical expression for $\tau_{nuc,het}$. Figs. 9.18–9.28 and 10.1 show that the major time dependence of $J_{f,het}(t)$ is determined by $s_w(t)$, and the temperature change during nucleation is rather small; thus, the temperature dependence is determined near T_{cr}. Dividing $J_{f,het}(t)$ in (Eqn. 10.1.7) by $J_{f,het}(t_0)$ at some initial t_0 and assuming $T \approx$ const, we obtain the exponential time dependence of $J_{f,het}(t)$ in the form

$$J_{f,het}(t) = J_{f,het}(t_0)\exp[c_{1w} w u_{shet}(t - t_0)]. \tag{10.1.10}$$

This can be written in another form:

$$J_{f,het}(t) = J_{f,het}(t_0) e^{\beta_{het}(t-t_0)} = J_{f,het}(t_0) e^{(t-t_0)/\tau_{nuc,het}}. \tag{10.1.11}$$

We introduce the parameter β_{het}, which can be interpreted as a characteristic "nucleation frequency":

$$\beta_{het} = c_{1w} w u_{shet}, \tag{10.1.12}$$

and a characteristic "nucleation time" $\tau_{nuc,het}$ that can be expressed using (Eqn. 9.8.11) for u_{shet} and $G_n = RT/(M_w L_m^{ef})$, as

$$\begin{aligned} \tau_{nuc,het} &= \beta_{het}^{-1} = (c_{1w} w u_{shet})^{-1} = \left[c_{1w} w \frac{\Delta F_{cr,0}^{\alpha 0}(T, m_{is})}{kT} \frac{2G_n}{\ln(T_0/T)} \right]^{-1} \\ &= (c_{1w} w)^{-1} \frac{k L_m^{ef} \ln(T_0/T)}{2R_v \Delta F_{cr,0}^{\alpha 0}(T, m_{is})} = (c_{1w} w)^{-1} \frac{M_w L_m^{ef} \ln(T_0/T)}{2N_{Av} \Delta F_{cr,0}^{\alpha 0}(T, m_{is})}, \end{aligned} \tag{10.1.13}$$

where R_v is the gas constant for water vapor and N_{Av} is the Avogadro number. Equation (10.1.11) shows that the nucleation rate of the heterogeneous ice nucleation process can be presented as an exponential function of time with some scaling time $\tau_{nuc,het}$ given by (Eqn. 10.1.13). The time $\tau_{nuc,het}$ characterizes the average time scale of heterogeneous nucleation, although the full time of the nucleation process may be different as will be shown in the following.

It is useful to compare this time dependence with the case of homogeneous ice nucleation and with previous parameterizations. As discussed in Section 8.10, an exponential time dependence of the homogeneous nucleation rate $J_{f,hom}(t)$ similar to (Eqn. 8.10.18) with the nucleation time scale $\tau_{nuc,hom}$ was hypothesized by Ford (1998a,b). The $\tau_{nuc,hom}$ was parameterized as a function of temperature with auxiliary empirical relations and fitted to parcel model simulations in the parameterizations described in Section 8.10, and it was derived from CNT by Khvorostyanov and Curry (2012). Kärcher and Lohmann (2003), Kärcher et al. (2006), and Barahona and Nenes (2009) applied a similar exponential parameterization for the heterogeneous nucleation rate $J_{f,het}(t)$ assuming the same functional dependence of $\tau_{nuc,het}$ on $J_{f,het}(t)$ as in (Eqn. 9.6.11), with subsequent fitting to the results of the numerical parcel model. Equation (10.1.11) provides a justification for the hypothesized exponential dependence of the nucleation rates, while (Eqn. 10.1.13) provides a direct expression of $\tau_{nuc,het}$ derived from the first principles in terms of classical nucleation theory and its parameters.

Equations (8.10.19) and (10.1.13) show that both $\tau_{nuc,hom}$ and $\tau_{nuc,het}$ are inversely proportional to the corresponding critical energies of nucleation: $\Delta F_{cr,0}^{\alpha 0}(T, m_{is})$ defined in (Eqn. 9.8.3), and $\Delta F_{cr,0}^{\text{hom}}$

defined in (Eqn. 8.10.10), with $f(m_{is}) = 1$. From these two equations, we obtain a relationship between the time scales of the two nucleation modes:

$$\frac{\tau_{nuc,het}}{\tau_{nuc,\hom}} \approx \frac{\Delta F_{cr,0}^{\hom}}{\Delta F_{cr,0}^{\alpha 0}(T, m_{is})} \approx \frac{1}{f(m_{is})}. \qquad (10.1.14)$$

Since $f(m_{is}) < 1$, this suggests at first glance that the heterogeneous nucleation process lasts longer than the homogeneous nucleation. However, the times $\tau_{nuc,het}$ and $\tau_{nuc,hom}$ characterize only the time dependence of the nucleation rates $J_{f,het}$ and $J_{f,hom}$, but not the overall time dependence of the nucleation kinetics, which is governed by the supersaturation equation. Fig. 9.26 shows a comparison of the nucleation rates in these two modes simulated with a parcel model as described in Section 9.14 at cirrus conditions near −40 °C. The heterogeneous nucleation rate continues for about 10 min (from 28 to 38 min), and homogeneous nucleation continues for about 16 min (from 22 to 38 min). That is, homogeneous nucleation is slower under these conditions, which is determined by the kinetics of the process. The same slower homogeneous nucleation is seen in Fig. 9.29 at −60 °C. Thus, $\tau_{nuc,het}$ and $\tau_{nuc,hom}$ may be used for parameterization of the time dependence of the nucleation rates, but not for characterizing of the entire nucleation process, which requires solving the supersaturation equation or its parameterization.

Equation (10.1.14) also shows that a universal parameterization of $\tau_{nuc,het}$ based on the relations like (Eqn. 8.4.17) with $(\tau_{nuc,het})^{-1} \sim (dT/dt)$ is a much more difficult task than the parameterization of $\tau_{nuc,hom}$ because $\tau_{nuc,het}$ depends on the contact parameter m_{is}, and may significantly vary among various aerosol types. However, (Eqn. 10.1.13) provides a simple dependence on the contact parameter: $\tau_{nuc,het} \sim 1/f(m_{is})$.

For $t_0 = t_{cr}$, (Eqn. 10.1.10) can be rewritten using (Eqn. 10.1.6a) as

$$\ln\frac{J_{f,het}[s_w(t)]}{J_{f,het}[s_{w,cr}(t_{cr})]} = u_{shet}c_{1w}w(t - t_{cr}) = u_{shet}(T)[s_w(t) - s_{w,cr}]. \qquad (10.1.15)$$

This relation is expressed in terms of the water supersaturation. Some semi-empirical parameterizations hypothesize a similar relation in terms of the ice saturation ratio, which can be derived from (Eqn. 10.1.15). Using the relations in (Eqn. 8.10.17d) following from the Clausius–Clapeyron equation, $s_w = c_{iw}(s_i + 1) - 1$, $c_{iw} = \exp[-L_m/(T_0 - T)/R_v T_0 T]$, we express s_w in (Eqn. 10.1.15) via the ice saturation ratio $S_i = s_i + 1$ and obtain

$$\ln\frac{J_{f,het}[S_i(t)]}{J_{f,het}[S_{i,cr}(t_{cr})]} = u_{shet}(T)c_{iw}(T)[S_i(t) - S_{i,cr}]. \qquad (10.1.16)$$

This expression has the same form as hypothesized by Barahona and Nenes (2008) for homogeneous nucleation and Barahona and Nenes (2009) for heterogeneous nucleation, and their coefficient b_τ fitted with empirical data is expressed now using the extended classical nucleation theory as $b_\tau(T) = u_{shet}(T)c_{iw}(T)$.

10.1.4. Temporal Evolution of the Crystal Concentration

Substitution of (Eqn. 10.1.6a) for $s_w(t)$ into (Eqn. 10.1.2) for $R_{f,het}(t)$ and integration over time yields the crystal concentration $N_c(t)$ along the ascending branch of $s_w(t)$ shown in Fig. 10.1:

$$N_c(t) = \int_0^t R_{f,het}(t)dt = J_{f,het}^{(0,\alpha 0)}(T,\bar{r}_N)\bar{J}_{act}^{\alpha}(\alpha_a, m_{is}, \bar{r}_N) I_{sw}(t). \quad (10.1.17)$$

Here, $I_{sw}(t)$ is the integral over time, which is easily calculated using the linear approximation (Eqn. 10.1.6a) for $s_w(t)$ and assuming as in Section 8.10 that $T \approx$ const:

$$I_{sw}(t) = \int_0^t \exp[u_{shet} s_w(t)]dt \approx \beta_{het}^{-1} \exp(u_{shet} s_{w,cr}^{het})[\exp(\beta_{het} t) - 1]. \quad (10.1.18)$$

Substitution of $I_{sw}(t)$ into (Eqn. 10.1.17) yields

$$N_c(t) = \int_0^t R_{f,het}(t)dt \quad (10.1.19)$$

$$= J_{f,het}^{(0,\alpha 0)}(T,\bar{r}_N)\bar{J}_{act}^{\alpha}(\alpha_a, m_{is}, \bar{r}_N)\beta_{het}^{-1} \exp(u_{shet} s_{w,cr}^{het})[\exp(\beta_{het} t) - 1] \quad (10.1.20)$$

$$= J_{f,het}^{(0,\alpha 0)}(T,\bar{r}_N)\bar{J}_{act}^{\alpha}(\alpha_a, m_{is}, \bar{r}_N)\tau_{nuc,het}\{\exp[u_{shet} s_w(t)] - \exp(u_{shet} s_{w,cr}^{het})\}. \quad (10.1.21)$$

This is the basic parameterization of the number of ice crystals nucleated by heterogeneous freezing via the deliquescence-freezing (or condensation-freezing) that was sought, which can be applied for increasing $s_w(t)$.

An estimate at T = 250 K and w = 2 cm s^{-1} gives $u_{shet} \sim 4 \times 10^2$, and $c_{1w} \sim 10^{-5}$ cm^{-1}. Thus, $\beta_{het} = u_{shet} c_{1w} w \sim 8 \times 10^{-3}$ s^{-1}. At small times, $t << \tau_{nuc,het} = \beta_{het}^{-1} \sim 125$ sec, the value of $\beta_{het} t << 1$, the exponent can be expanded, $\exp(\beta_{het} t) \approx 1 + \beta_{het} t$, then $I_{sw}(t)$ and $N_c(t)$ are simplified as

$$N_c(t) = \int_0^t R_{f,het}(t)dt = t J_{f,het}^{(0,\alpha 0)}(T,\bar{r}_N)\bar{J}_{act}^{\alpha}(\alpha_a, m_{is}, \bar{r}_N)\exp(u_{shet} s_{w,cr}^{het}). \quad (10.1.22)$$

Using the approximation (Eqn. 10.1.5) for $\bar{J}_{act}^{\alpha}(\alpha_a, m_{is}, \bar{r}_N)$ with one kind of active site with $m_{is} = 1$, we obtain

$$N_c(t) = c_{sh} N_a t J_{f,het}^{(0,\alpha 0)}(T,\bar{r}_N)\exp\left[\frac{\alpha_a \sigma_{is}(1-m_{is})\bar{r}_N^2}{kT}\right]\exp(u_{shet} s_{w,cr}^{het}). \quad (10.1.23)$$

Without the active sites ($\alpha_a = 0$), the exponential term with α_a is equal to 1.

Thus, at small times $t << \beta_{het}^{-1} = \tau_{nuc,het}$, the crystal concentration $N_c(t)$ increases linearly with time, and is proportional to IN concentration N_a, and to the heterogeneous nucleation rate $J_{f\,het}^{(0,\alpha 0)}$ at zero water supersaturation, but is modulated (decreased) by the critical supersaturation $s_{w,cr}^{het} < 0$. At $s_{w,cr}^{het} = -3.5 \times 10^{-2}(-3.5\%)$, as in Fig. 10.1, and $u_{shet} \sim 4 \times 10^2$, the product $u_{shet}\, s_{w,cr}^{het} = -15.4$ and $\exp(u_{shet}\, s_{w,cr}^{het}) \approx \exp(-15.4) \sim 10^{-7}$. Hence, there is a substantial effect in the DF mode from the small subsaturation on the nucleation rate and on the concentration of nucleated crystals. As estimated earlier, an increase of s_w to -2.8×10^{-2} (−2.8%) at t_{max} in Fig. 10.1 causes an increase in the nucleation rate by a factor of exp(4.16) ~ 10^2.

At large times $t >> \beta_{het}^{-1} = \tau_{nuc,het}$ (125 sec in this example), $\exp(\beta_{het}\, t) >> 1$ and (Eqn. 10.1.20) yields

$$N_c(t) = J_{f,het}^{(0,\alpha 0)}(T,\overline{r}_N)\overline{J}_{act}^{\alpha}(\alpha_a, m_{is}, \overline{r}_N)\tau_{nuc,het}\exp[u_{shet}s_w(t)], \tag{10.1.24}$$

where $s_w(t)$ is a linear function (Eqn. 10.1.6a). With the approximation (Eqn. 10.1.5) for $\overline{J}_{act}^{\alpha}(\alpha_a, m_{is}, \overline{r}_N)$, it becomes

$$N_c(t) = c_{sh}\tau_{nuc,het}N_aJ_{f,het}^{(0,\alpha 0)}(T,\overline{r}_N)\exp\left[\frac{\alpha_a\sigma_{is}(1-m_{is})\overline{r}_N^2}{kT}\right]\exp[u_{shet}s_w(t)].$$

$$= c_{sh}\tau_{nuc,het}N_aJ_{f,het}^{(0,\alpha 0)}(T,\overline{r}_N)\exp\left[\frac{\alpha_a\sigma_{is}(1-m_{is})\overline{r}_N^2}{kT}\right]\times\exp[u_{shet}s_{w,cr}^{het}]\exp[u_{shet}c_{1w}wt]. \tag{10.1.25}$$

Since $s_w(t) \sim t$, the previous two equations show that $N_c(t)$ grows exponentially with time for $\tau_{nuc,het} < t < t_{max}$. This explains the linear growth with time of the $\log[N_c(t)] \sim t$ as plotted in log-linear scale figures of $N_c(t)$ simulated by the parcel model in Figs. 9.18–9.29.

10.1.5. Comparison of Crystal Concentrations with Empirical Parameterizations

These expressions provide a theoretical explanation for the empirical parameterization by Meyers et al. (1992, MDC92), $N_c(s_i) = \exp(a_M + b_M s_i)$, (Eqn. 9.4.7), and enables the expression of empirical parameters a_M, b_M via the parameters of classical nucleation theory and aerosol microphysical properties. We present $N_c(t)$ in (Eqn. 10.1.25) as

$$N_c(s_w) = \exp(\ln A_{DF} + u_{shet}s_w), \tag{10.1.26a}$$

$$A_{DF} = c_{sh}N_a\tau_{nuc,het}J_{f,het}^{(0,\alpha 0)}(T,\overline{r}_N)\overline{J}_{act}^{\alpha}(\alpha_a, m_{is}, \overline{r}_N). \tag{10.1.26b}$$

Similar to the derivation of (Eqn. 10.1.16), using (Eqn. 8.10.17d) we can replace s_w with s_i and obtain

$$N_c(s_i) = \exp[a_{DF}(T,\overline{r}_N, m_{is}, \alpha_a) + b_{DF}(T, m_{is})s_i]. \tag{10.1.27a}$$

Thus, we obtain expressions in a form similar to the MDC92 parameterization (Eqn. 9.4.7) with a_{DF}, b_{DF} as

$$a_{DF}(T,\overline{r}_N, m_{is}, \alpha_a) = \ln A_{DF} + u_{shet}(c_{iw} - 1)$$

$$= \ln[c_{sh}N_aJ_{f,het}^{(0,\alpha 0)}(T,\overline{r}_N)\tau_{nuc,het}] + \frac{\alpha_a\sigma_{is}(1-m_{is})\overline{r}_N^2}{kT} + u_{shet}(c_{iw} - 1), \tag{10.1.27b}$$

$$b_{DF}(T, m_{is}) = u_{shet}c_{iw} = \frac{2R}{kM_wL_m^{ef}}\frac{\Delta F_{cr,0}^{\alpha 0}(T, m_{is})}{\ln(T_0/T)}\exp\left[-\frac{L_m(T_0 - T)}{R_vT_0T}\right]. \tag{10.1.27c}$$

where the subscript "DF" means the deliquescence-freezing mode, and we substituted c_{iw} from (Eqn. 8.10.17d) and u_{shet} from (Eqn. 9.8.11). The coefficients a_{DF}, b_{DF} depend on the temperature, mean radius of IN, contact parameter, and fraction of active site α_a. Being physically based, this parameterization is sufficiently simple and can be applied in the models. The original experimental data shown in MDC92 and used for their parameterization exhibited both T- and s_i-dependencies. MDC92 averaged the data over the temperatures and recommended some fixed values for a_M, b_M. Equations (10.1.27a)–(10.1.27c) show that these coefficients are variable.

The dependence on the aerosol properties in a_{DF} contains a part with $\ln N_a$, and these equations can be rewritten as

$$N_c(s_i) = N_a \exp[\tilde{a}_{DF}(T,\bar{r}_N,m_{is},\alpha_a) + b_{DF}(T,m_{is})s_i], \tag{10.1.28a}$$

$$\tilde{a}_{DF}(T,\bar{r}_N,m_{is},\alpha_a) = \ln[c_{sh} J_{f,het}^{(0,\alpha 0)}(T,\bar{r}_N)\tau_{nuc,het}] + \frac{\alpha_a \sigma_{is}(1-m_{is})\bar{r}_N^2}{kT} + u_{shet}(c_{iw} - 1). \tag{10.1.28b}$$

Equation (10.1.28a) shows that $N_c(s_i)$ is proportional to IN concentration N_a, while the other coefficients contain explicit dependencies on T, α_a, and other aerosol parameters.

Field and laboratory experiments show that the concentration N_a of available aerosol particles that can serve as ice nuclei (IN) and nucleate ice heterogeneously in the DF mode may vary significantly in the atmosphere from a few per liter to a few hundred or thousands per liter. This indicates that there can be two different modes of DF ice nucleation that may occur at small or large values of N_a. These situations are considered next.

10.1.6. Parameterization for the Large-Scale Models. Case 1: Large N_a and Crystal Concentrations Limited by Kinetics

We assume that N_a is sufficiently large in this case, and the temporal behavior of supersaturations and crystal concentration is as shown in Fig. 10.1, so that $s_w(t)$ reaches a maximum at a time t_{max} and then decreases to the 2nd critical value t_{cr2} when nucleation ceases. Evaluation of the maximum supersaturations $s_{w,max}$ and $s_{i,max}$ and maximum $N_{c,max}$ can be done similar to Section 8.10. We use the equation for the integral ice supersaturation y_i' in the form similar to (Eqn. 8.5.15) for y_i' or to (Eqn. 9.13.2), but with $I_{con} = 0$ since we consider a pure crystalline cloud:

$$\frac{dy_i'}{dt} = c_{1i}w(1+y_i') - \frac{\Gamma_2}{\rho_{is}} I_{dep} = c_{1i}w(1+y_i') - \frac{s_i}{\tau_{fc}}. \tag{10.1.29}$$

The deposition integral I_{dep} defined by (Eqn. 8.5.8) is evaluated as in Section 8.10 using the conservation law, $dN_{fr} = f_c(r_c)dr_c = R_{f,het}(t_0)dt_0$:

$$\begin{aligned} I_{dep}(t) &= s_i \frac{4\pi D_v \rho_{is}}{\Gamma_2} \int_0^\infty \frac{r_c^2(t,t_0)}{r_c(t,t_0)+\xi_{dep}} f_c(r_c)dr_c. \\ &= s_i \frac{4\pi D_v \rho_{is}}{\Gamma_2} \int_0^t r_{c,ef}(t,t_0) R_{f,het}(t_0)dt_0. \end{aligned} \tag{10.1.30}$$

As in Section 8.10, $r_{c,ef}(t,t_0) = r_c(t,t_0)^2/[r_c(t,t_0) + \xi_{dep}]$, and $r_c(t,t_0)$ is a radius at time t of a crystal nucleated at time t_0. It is evaluated as described in Sections 8.5 and 8.10, by integrating the equations of the crystal growth rate, and is expressed by (Eqn. 8.5.9) via y_i'; $r_{c,ef}(t,t_0)$ is expressed by (Eqn. 8.10.29), in a form convenient for subsequent integration by time:

$$r_{c,ef}(t,t_0) = \frac{\{[(r_0+\xi_{dep})^2 + B_{ihet}(t-t_0)]^{1/2} - \xi_{dep}\}^2}{[(r_{c0}+\xi_{dep})^2 + B_{ihet}(t-t_0)]^{1/2}},$$
$$= r_{c,ef}^{(1)} + r_{c,ef}^{(2)} + r_{c,ef}^{(3)}, \qquad B_{ihet} = 2c_{i3}s_{i,cr}^{het}. \tag{10.1.31}$$

The three terms of $r_{c,ef}^{(k)}$ on the RHS of this equation are defined as in (Eqn. 8.10.30a)–(8.10.30c) but replacing $B_{ihom} \to B_{ihet}$. Substituting $R_{f,het}(t_0)$ from (Eqn. 10.1.2) and $r_{c,ef}(t, t_0)$ with B_{ihet} from (Eqn. 10.1.31) into (Eqn. 10.1.30) for I_{dep}, we can integrate over time using the method described in detail in Section 8.10 and Appendix A.8. Then, we obtain

$$I_{dep}(t) = s_i \frac{4\pi D_v \rho_{is}}{\Gamma_2} J_{f,het}^{(0,\alpha 0)} \bar{J}_{act}^{\alpha}(\alpha_a, m_{is}, \bar{r}_N) \exp(u_{shet} s_{i,cr}^{het}) J_{0i,het}(t), \tag{10.1.32}$$

where

$$J_{0i,het}(t) = \int_0^t r_{c,ef}(t,t_0) \exp(\beta_{het} t_0)\, dt_0 = \sum_{k=1}^{3} J_{0i,het}^{(k)}, \tag{10.1.33}$$

$$J_{0i,het}^{(k)} = \int_0^t r_{c,ef}^{(k)}(t,t_0) \exp(\beta_{het} t_0)\, dt_0. \tag{10.1.34}$$

The integrals $J_{0i,het}(t)$ and $J_{0i,het}^{(k)}$ for heterogeneous nucleation coincide with the corresponding integrals for the case of homogeneous nucleation evaluated in Khvorostyanov and Curry (2012) and in Section 8.10 (Appendix A.8 therein), although u_{shet}, β_{het}, and B_{ihet} differ due to different (smaller) critical energies $\Delta F_{cr,het}$ caused by the shape factor $f(m_{is})$, active sites, and lower $s_{i,cr}^{het}$. Thus, we can use the results for $J_{0i,het}^{(k)}$ and $J_{0i,het}(t)$ evaluated in Section 8.10 replacing $u_{shom} \to u_{shet}$, $\beta_{hom} \to \beta_{het}$, $B_{ihom} \to B_{ihet}$, and $s_{i,cr}^{\hom} \to s_{i,cr}^{het}$. Evaluating the integrals in (Eqn. 10.1.34) as described in Appendix A.8 and collecting all three integrals $J_{0i,het}^{(k)}$ yields

$$J_{0i,het} = \sum_{k}^{3} J_{0i,het}^{(k)} = e^{\beta_{het} t} \Psi_{het}, \tag{10.1.35}$$

$$\Psi_{het} = \Psi_{1,het} + \Psi_{2,het} + \Psi_{3,het}, \tag{10.1.36a}$$

$$\Psi_{1,het} = e^{\lambda_{het}} B_{ihet}^{1/2} \beta_{het}^{-3/2} \left[\Gamma\left(\frac{3}{2}, \lambda_{het}\right) - \Gamma\left(\frac{3}{2}, \lambda_{het} + \beta_{het} t\right) \right]$$
$$= e^{\lambda_{het}} B_{ihet}^{1/2} \beta_{het}^{-3/2} \{ (\sqrt{\pi}/2)\, [\Phi(\sqrt{\lambda_{het} + \beta_{het} t}) - \Phi(\sqrt{\lambda_{het}})]$$
$$+ e^{-\lambda_{het}} [\lambda_{het}^{1/2} - (\lambda_{het} + \beta_{het} t)^{1/2} e^{-\beta_{het} t)}]\}, \tag{10.1.36b}$$

$$\Psi_{2het} = 2\xi_{dep} \beta_{het}^{-1} (e^{-\beta_{het} t} - 1), \tag{10.1.36c}$$

$$\Psi_{3het} = e^{\lambda_{het}} \xi_{dep}^2 (\beta_{het} B_{ihet})^{-1/2} \left[\Gamma\left(\frac{1}{2}, \lambda_{het}\right) - \Gamma\left(\frac{1}{2}, \lambda_{het} + \beta_{het} t\right) \right]$$
$$= e^{\lambda} \xi_{dep}^2 (\beta_{het} B_{ihet})^{-1/2} \sqrt{\pi} [\Phi(\sqrt{\lambda_{het} + \beta_{het} t}) - \Phi(\sqrt{\lambda_{het}})]. \tag{10.1.36d}$$

The parameter λ_{het} here is defined similar to λ_{hom} in Section 8.10, (Eqn. 8.10.38), but with another critical parameters:

$$\lambda_{het} = \frac{(\xi_{dep} + r_0)^2 \beta_{het}}{B_{ihet}} = \frac{(u_{shet} c_{1w} w)(\xi_{dep} + r_0)^2}{2c_{i3} s_{i,cr}^{het}}. \tag{10.1.37}$$

Substituting (Eqn. 10.1.35) into (Eqn. 10.1.32), we obtain

$$\begin{aligned} I_{dep}(t) &= s_i \frac{4\pi D_v \rho_{is}}{\Gamma_2} J_{f,het}^{(0,\alpha 0)} \overline{J}_{act}^{\alpha} \exp(u_{shet} s_{w,cr}) \exp(\beta_{het} t) \Psi_{het} \\ &= s_i \frac{4\pi D_v \rho_{is}}{\Gamma_2} J_{f,het}^{(0,\alpha 0)} \overline{J}_{act}^{\alpha} \exp[u_{shet} s_w(t)] \Psi_{het}. \end{aligned} \tag{10.1.38}$$

Inserting (Eqn. 10.1.38) into (Eqn. 10.1.29) for y_i', and using the relation $\rho_v = \rho_{is}(1 + y_i')$, we obtain

$$\frac{dy_i'}{dt} = c_{1i} w(1 + y_i') - (4\pi D_v) s_i J_{f,het}^{(0,\alpha 0)}(\overline{r}_N) \overline{J}_{act}^{\alpha}(\alpha_a, m_{is}, \overline{r}_N) \exp[u_{shet} s_w(t)] \Psi_{het}. \tag{10.1.39}$$

At the point with maximum supersaturations, $s_w = s_{wmax}$ and $y_i' = s_i = s_{imax} \approx s_{i,cr}^{het}$ at $t = t_{max}$, we have the condition $ds_i/dt = dy_i'/dt = 0$, and (Eqn. 10.1.39) yields

$$\exp[u_{shet} s_{w\max}(t_{\max})] = c_{1i} w(1 + s_{i,cr}^{het})(s_{i,cr}^{het})^{-1} (4\pi D_v)^{-1} (J_{f,het}^{(0,\alpha 0)} \overline{J}_{act}^{\alpha})^{-1} \Psi_{het}^{-1}. \tag{10.1.40}$$

Figures with simulation results in Section 9.14 show that the nucleation process takes $t_{het} \sim$ 12–15 min, t_{max} is reached in 6–8 min, and the preceding estimates showed that $\tau_{nuc,het} = (\beta_{het})^{-1}$ is much smaller. Thus, we can assume the relation $\beta_{het} t_{max} >> 1$ for typical conditions and use (Eqn. 10.1.24) for $N_c(t_{max})$. After substitution of (Eqn. 10.1.40) into (Eqn. 10.1.24), the terms with nucleation rate $J_{f,het}^{(0,\alpha 0)}(T, \overline{r}_N)$ and active sites $\overline{J}_{act}^{\alpha}(\alpha_a, m_{is}, \overline{r}_N)$ vanish and we obtain a simple expression for $N_{chet}(t_{max})$:

$$N_{chet}(t_{\max}) = K_{gen}^{het}(1 + s_{i,cr}^{het})(s_{i,cr}^{het})^{-1} \Psi_{het}^{-1}, \tag{10.1.41a}$$

$$K_{gen}^{het} = (4\pi D_v)^{-1} u_{shet}^{-1} (c_{1i}/c_{1w}). \tag{10.1.41b}$$

This solution formally has the same analytical form as the corresponding solution for $N_{chom}(t_m)$ in homogeneous nucleation obtained in Section 8.10; however, the quantities u_{shet} and Ψ_{het} differ from the homogeneous freezing case by the presence of the geometric factor $f(m_{is})$. This significantly reduces the energy $\Delta F_{cr,0}$, parameter u_{shet} and all the other similar quantities relative to the case of homogeneous freezing, and shifts the nucleation process to higher temperatures.

Equations (10.1.41a) and (10.1.41b) along with the equations of Section 8.10 enable a simple estimate of the ratios of the crystal concentrations $N_{c,het}$ and $N_{c,hom}$ nucleated via heterogeneous and homogeneous mode. In the latter case, the expressions are the same with replacement of u_{shet} and Ψ_{het} with u_{shom} and Ψ_{hom} (Eqns. 8.10.51) and (8.10.52), and the ratio is

$$\frac{N_{chet}(t_{\max})}{N_{c\hom}(t_{\max})} = R_{ht/hm}^{(1)} R_{ht/hm}^{(2)}, \tag{10.1.42}$$

where

$$R_{ht/hm}^{(1)} = \frac{K_{gen}^{het}}{K_{gen}^{hom}} \frac{\Psi_{hom}}{\Psi_{het}} \approx \frac{u_{s\,hom}}{u_{shet}} \frac{\Psi_{hom}}{\Psi_{het}} \approx \frac{\Delta F_{cr,0}^{hom}}{\Delta F_{cr,0}^{0,\alpha 0}} \frac{\Psi_{hom}}{\Psi_{het}} \sim \frac{1}{f(m_{is})} \frac{\Psi_{hom}}{\Psi_{het}}, \qquad (10.1.43)$$

$$R_{ht/hm}^{(2)} = \frac{(1+s_{i,cr}^{het})}{(1+s_{i,cr}^{hom})} \frac{s_{i,cr}^{hom}}{s_{i,cr}^{het}}. \qquad (10.1.44)$$

Thus, the difference between the concentrations N_{chet} and N_{chom} is attributed to three factors that are easily calculated: geometric factor $f(m_{is})$, the difference in the critical supersaturations and in the Ψ-functions. It will be shown in the following text that this ratio is simplified for the limiting cases with the diffusion and kinetic regimes of crystal growth.

This method enables estimation of the supersaturation relaxation time τ_{fc} and the mean crystal radius at the time t_{max} by equating the two expressions for the second term I_{dep} with "relaxation" in the supersaturation (Eqn. 10.1.29), which yields

$$-(\Gamma_2/\rho_v)I_{dep} = -s_i\tau_{fc}^{-1}. \qquad (10.1.45)$$

Substituting (Eqn. 10.1.38) for I_{dep}, we obtain an expression for τ_{fc}^{-1}:

$$\tau_{fc}^{-1}(t) = (\Gamma_2/\rho_{is})(I_{dep}/s_i) \qquad (10.1.46a)$$

$$\approx (4\pi D_v)N_a J_{f,het}^{(0,\alpha 0)} \exp(z_\alpha \bar{r}_N^2)\exp[u_{shet}s_{w,cr}^{het}(t)]\exp(t/\tau_{het})\Psi_{het}(t). \qquad (10.1.46b)$$

This is an expression for τ_{fc} at a time $t \le t_{max}$, and it is significantly simplified at $t = t_{max}$. After substitution for $\exp[u_{shet}s_{w,cr}^{het}(t_{\max})]$ from (Eqn. 10.1.40) at $t = t_{max}$, most terms are canceled and we obtain simply

$$\tau_{fc}(t_{\max}) = [c_{1i}w(1+s_{i,cr}^{het})]^{-1}s_{i,cr}^{het}. \qquad (10.1.47)$$

Another expression for τ_{fc} can be obtained from (Eqn. 10.1.30) for I_{dep} and (Eqn. 10.1.46a):

$$\tau_{fc}^{-1} = \frac{\Gamma_2}{\rho_{is}}\frac{I_{dep}}{s_i} = 4\pi D_v \int_0^\infty \frac{r_c^2}{r_c+\xi_{dep}} f_c(r_c,t)dr_c. \qquad (10.1.48)$$

In the diffusion regime of crystal growth, when $r_c >> \xi_{dep}$, the integral is reduced to $N_c\bar{r}_c$, where $\bar{r}_c$ is the mean crystal radius, and we obtain as before

$$\tau_{fc}(t_{\max}) = [4\pi D_v N_{chet}(t_{\max})\bar{r}_c(t_{\max})]^{-1}. \qquad (10.1.49)$$

Equating expressions in (Eqns. 10.1.49) and (10.1.47), we obtain an expression for the mean crystal radius at $t = t_{max}$:

$$\bar{r}_c(t_{\max}) = [4\pi D_v N_{chet}(t_{\max})]^{-1}\tau_{fc}^{-1}$$
$$= [4\pi D_v N_{chet}(t_{\max})]^{-1}c_{1i}w(1+s_{i,cr}^{het})(s_{i,cr}^{het})^{-1}. \qquad (10.1.50)$$

Using values from Figs. 9.22 and 9.34, $N_{cm} \approx 15\,\text{L}^{-1}$ at $t = t_{max} \approx 90\,\text{min}$, $c_{1i} \sim 10^{-5}\,\text{cm}^{-1}$, $s_{i,cr}^{het} \sim 0.2$, and $w = 2\,\text{cm s}^{-1}$, an estimate from (Eqn. 10.1.50) gives $\bar{r}_c(t_m) \sim 26.6$ μm, which is in good agreement with Fig. 9.22d with $\bar{r}_c \approx 25$ μm at 90 min. Equation (10.1.50) can be further simplified by substituting (Eqn. 10.1.41a) and (10.1.41b) for $N_{chet}(t_{max})$:

$$\bar{r}_c(t_{\max}) = u_{shet} c_{1w} w \Psi_{het}^{-1}(t_{\max}) \approx \tau_{nuc.het} \Psi_{het}^{-1}(t_{\max}). \tag{10.1.51}$$

As in the case with homogeneous nucleation, heterogeneous freezing does not cease at t_{max}, but continues along the descending branch of s_w, when supersaturation absorption by the crystals exceeds its production by the cooling in updraft w, and nucleation ceases at time t_{cr2}. Calculations show that the final total $N_{c,tot} = N_c(t_{cr2})$ is proportional to $N_{chet}(t_{max})$ and can be evaluated in a simple manner following Section 8.10:

$$N_{chet}(t_{cr2}) \approx K_{cor}(w) N_{chet}(t_{\max}). \tag{10.1.52}$$

A comparison with parcel model simulations in Fig. 10.1b indicates that $K_{cor} \approx 2$ can be a suitable choice, since about half of the crystals nucleate at $t_{max} < t < t_{cr2}$. Fig. 9.22d shows that the value of the mean radius is also nearly doubled from t_{max} to t_{cr2}. Thus, an approximate relation can be used:

$$\bar{r}_c(t_{cr2}) \approx K_{cor}(w) \bar{r}_c(t_m). \tag{10.1.53}$$

Equations (10.1.41a) and (10.1.41b)–(10.1.53) can be used to parameterize the crystal concentration, mean radius, and ice water content in models with large time steps where the nucleation process is a substep process that occurs within one time step (e.g., Lohmann and Kärcher, 2002; Morrison and Gettelman, 2008; Gettelman et al., 2008; Sud et al., 2009; Hoose et al., 2010).

10.1.7. Diffusion Growth Limit

The limits of the general equations for heterogeneous ice nucleation are similar to those for homogeneous nucleation considered in detail in Section 8.10. The first practically important limit occurs at $\lambda_{het} << 1$. As seen from (Eqn. 10.1.37) and similar to the homogeneous freezing in Section 8.10, the condition $\lambda_{het} << 1$ implies small ξ_{dep} and r_0 that are typical of the diffusion regime of crystal growth with the deposition coefficient $\alpha_d \sim 1$ or $\alpha_d > 0.1$, not very large w and not very low T. In this case, we can neglect in (Eqns. 10.1.36a)–(10.1.36d) for Ψ_{het} all terms with ξ_{dep} and r_0. Using the estimates from earlier, we can then assume that $\beta_{het} t_{max} >> 1$, use the power series for $erf[(\lambda_{het}+\beta_{het}\, t_{max})^{1/2}]$ described in Section 8.10 and Appendix A.8 and neglect the terms with $\exp(-\beta_{het} t_{max})$. Then, Ψ_{het} is simplified in this diffusion regime as

$$\Psi_{dif,het} \approx (\sqrt{\pi}/2) e^{\lambda_{het}} B_{ihet}^{1/2} \beta_{het}^{-3/2} \approx (\pi/2)^{1/2} (c_{3i} s_{i,cr}^{het})^{1/2} u_{shet}^{-3/2} (c_{1w} w)^{-3/2}. \tag{10.1.54}$$

Substitution of this expression into (Eqns. 10.1.41a) and (10.1.41b) yields for the crystal concentration $N_{cm,dif}$ at t_{max} and $N_{c,dif}$ after the cease of nucleation at t_{cr2}:

$$N_{cm,dif}(t_{\max}) = K_{i,dif}^{het} (1 + s_{i,cr}^{het})(s_{i,cr}^{het})^{-3/2} (c_{1i} w)^{3/2}, \tag{10.1.55}$$

$$N_{c,dif}(t_{cr2}) \approx 2 N_{cm,dif}(t_{\max}), \tag{10.1.56}$$

where

$$K_{i,dif}^{het} = (2\pi D_v)^{-3/2}\left(\frac{\rho_i \Gamma_2}{\rho_{is}}\right)^{1/2} u_{shet}^{1/2}\left(\frac{c_{1w}}{c_{1i}}\right)^{1/2} \tag{10.1.57a}$$

$$= (2\pi D_v)^{-3/2}\left(\frac{\rho_i \Gamma_2}{\rho_{is}}\right)^{1/2}\left[\frac{2R}{kM_w L_m^{ef}}\frac{\Delta F_{cr,0}^{(\alpha 0)}(T)}{\ln(T_0/T)}\right]^{1/2}\left(\frac{c_{1w}}{c_{1i}}\right)^{1/2}. \tag{10.1.57b}$$

The dependence on the vertical velocities, $N_{c,het} \sim w^{3/2}$ in (Eqn. 10.1.55) derived here analytically is in good agreement with the parameterization $N_{c,het} \sim w^{1.47}$ found in Khvorostyanov and Curry (2005a) and Curry and Khvorostyanov (2012) as a numerical fit to several hundred runs of the parcel model and described by (Eqn. 9.15.1), which supports this analytical derivation. Note also that this dependence, $N_{chet} \sim w^{3/2}$, is the same as the dependence of the crystal concentration in homogeneous freezing, $N_{chom} \sim w^{3/2}$, found in Kärcher and Lohmann (2002a), Ren and MacKenzie (2995), in Khvorostyanov and Curry (2012) in the diffusion limit of crystal growth as described in Section 8.10.

A comparison of $N_{c,het}$ described by (Eqns. 10.1.55) and (10.1.56) with the homogeneously nucleated concentrations $N_{c,hom}$ described by (Eqns. 8.10.58) and (8.10.59) show that their ratio in the diffusion regime of crystal growth is simplified relative to (Eqns. 10.1.43) and (10.1.44) as

$$\frac{N_{cm,dif}^{het}(t_m)}{N_{cm,dif}^{\text{hom}}(t_m)} = R_{dif}^{\Delta F} R_{ht/hm}^{s}, \tag{10.1.58}$$

where

$$R_{dif}^{\Delta F} = \frac{K_{i,dif}^{het}}{K_{i,dif}^{\text{hom}}} \sim \left(\frac{\Delta F_{cr,0}^{\alpha 0}}{\Delta F_{cr,0}^{\text{hom}}}\right)^{1/2} \sim [f(m_{is})]^{1/2}, \tag{10.1.59}$$

$$R_{ht/hm}^{s} = \frac{(1+s_{i,cr}^{het})}{(1+s_{i,cr}^{\text{hom}})}\left(\frac{s_{i,cr}^{\text{hom}}}{s_{i,cr}^{het}}\right)^{3/2}. \tag{10.1.60}$$

Thus, the ratio of the Ψ-functions vanished and the difference of the concentrations $N_{c,het}$ and $N_{c,hom}$ is caused by two factors. The first factor, $R_{dif}^{\Delta F}$, arises from the difference in the critical energies of activation. Both concentrations are proportional to $N_{c,het} \sim (\Delta F_{cr,0}^{\alpha 0})^{1/2}$ and $N_{c,hom} \sim (\Delta F_{cr,0}^{\text{hom}})^{1/2}$, which yields the factor $N_{c,het}/N_{c,hom} \sim f(m_{is})^{1/2}$ and tends to produce smaller crystal concentrations in heterogeneous nucleation. The second factor $R_{ht/hm}^{s}$ arises from the difference in the critical supersaturations. Since $s_{i,cr}^{\text{hom}}$ is greater than $s_{i,cr}^{het}$, the ratio $(s_{i,cr}^{\text{hom}}/s_{i,cr}^{het})^{3/2} > 1$, and this leads to a tendency for $R_{ht/hm}^{s} > 1$ and greater $N_{c,het}$ than $N_{c,hom}$. This is in agreement with Fig. 9.26, which shows that $N_{c,het} > N_{c,hom}$, and (Eqns. 10.1.58)–(10.1.60) explain the reason of this. Which tendency will prevail depends on the contact parameter m_{is}, temperature, and other factors that determine critical supersaturations; as Fig. 9.29 shows, at −60 °C, $N_{c,het}$ only slightly exceeds $N_{c,hom}$. In any case, these equations allow a simple estimate of the ratio of the crystal concentrations that can be nucleated in via homogeneous or heterogeneous DF mode under similar conditions.

10.1.8. The Kinetic Growth Limit, and Small and Large Particle Limits

The other limit occurs at $\lambda_{het} >> 1$. Equation (10.1.37) shows that this condition means the kinetic regime with large ξ_{dep} (small α_d) or the large initial particle radius r_0 of freezing particles. As in Section 8.10, the kinetic limit for Ψ_{het} can be derived by expanding in (Eqns. 10.1.36a)–(10.1.36d) the functions $erf(\sqrt{\lambda})$ and $erf(\sqrt{\lambda+\beta t_{\max}})$ at $\lambda_{het} >> 1$, and neglecting the terms with $\exp(-\beta t_{max})$ and the terms $\lambda^{-3/2}$ compared to $\lambda^{-1/2}$. Collecting terms of the same order, we arrive at an equation that has the same form as (Eqn. 8.10.61) with replacement of β_{hom} and λ_{hom} with β_{het} and λ_{het}:

$$\Psi_{het,kin} = \beta_{het}^{-1/2}\lambda_{het}^{-1/2}[(1/2)B_{ihet}^{1/2}\beta_{het}^{-1} + \xi_{dep}^2 B_{ihet}^{-1/2}] + B_{ihet}^{1/2}\beta_{het}^{-3/2}\lambda_{het}^{1/2} - 2\xi_{dep}\beta_{het}^{-1}$$
$$= (r_0 + \xi_{dep})^{-1}[(1/2)B_{ihet}\beta_{het}^{-2} + \xi_{dep}^2\beta_{het}^{-1}] + (r_0 - \xi_{dep})\beta_{het}^{-1}. \quad (10.1.61)$$

This case contains two subcases: a) ξ_{dep} is large (small deposition coefficient α_d), but r_0 is small (small particle limit) and $\xi_{dep} >> r_0$; and b) r_0 is large (large particle limit) and $\xi_{dep} << r_0$, which may correspond to both the diffusion and kinetic regimes.

a. $\lambda_{het} >> 1$, $\xi_{dep} >> r_0$, kinetic regime, small particle limit

Under these conditions, r_0 can be neglected compared to ξ_{dep} and (Eqn. 10.1.61) is simplified as

$$\Psi_{het,kin,s} = \frac{B_{ihet}}{2\beta_{het}^2\xi_{dep}} = \frac{c_{i3}s_{i,cr}^{het}}{u_{shet}^2(c_{1w}w)^2}\frac{\alpha_d V_w}{4D_v}. \quad (10.1.62)$$

Substituting into (Eqns. 10.1.41a) and (10.1.41b) yields N_c at maximum t_{max} with s_{wmax}:

$$N_{cm,kin,s}^{het} = K_{kin,s}^{het}(1+s_{i,cr}^{het})(s_{i,cr}^{het})^{-2}(c_{1w}w)^2, \quad (10.1.63a)$$

$$K_{kin,s}^{het} = \frac{1}{(\pi D_v)}\frac{u_{shet}}{\alpha_d V_w}\left(\frac{\rho_i\Gamma_2}{\rho_{is}}\right)\frac{c_{1i}}{c_{1w}}. \quad (10.1.63b)$$

Thus, in this limit $N_{cm,kin,s}^{het} \sim w^2$, which is the same functional dependence described in Section 8.10 for homogeneous nucleation. The temperature dependence is determined by several factors: $N_{het,cm}$ ~ $[\rho_{is}(T)]^{-1}$, but the other factors, u_{shet}, V_w, Γ_2, c_{1i}, c_{1w}, and $s_{i,cr}^{het}$, are also the functions of T. Note also that the crystal concentration is inversely proportional to the deposition coefficient, $N_{cm,het} \sim \alpha_d^{-1}$; that is, the smaller α_d or the more polluted clouds, the greater nucleated crystal concentration. Small values of α_d were reported in some measurements, especially at low temperatures (e.g., Hobbs, 1974; Pruppacher and Klett, 1997) and Gierens et al. (2003) discussed possible reasons for small α_d in ice clouds down to 10^{-3}. The dependence $1/\alpha_d$ can be significant in these cases and may lead to enhanced crystal concentrations in polluted clouds with small α_d. This is also in agreement with the data from the INCA field experiment (Ovarlez et al., 2002; Ström et al., 2003; Haag et al., 2003; Gayet et al., 2004; Monier et al., 2006) that found greater ice crystal concentrations in cirrus in the more polluted northern hemisphere than in the cleaner southern hemisphere. This, in particular, could be caused by a small deposition coefficient in the heterogeneous ice nucleation mode in polluted areas. Equation (8.10.63) for N_{chom} derived in Section 8.10 in this limit has the same form as (Eqns. 10.1.63a) and (10.1.63b), but with u_{shom} instead of u_{shet}. This allows a simple estimate of the ratio N_{chet}/N_{chom} in this limit:

$$\frac{N_{cm,kin,s}^{het}}{N_{cm,kin,s}^{\hom}} = R_{het,kin,s}^{F} R_{het,kin,s}^{s}, \tag{10.1.64}$$

$$R_{het,kin,s}^{F} = \frac{u_{shet}}{u_{s\hom}} = \frac{\Delta F_{cr,0}^{0,\alpha 0}}{\Delta F_{cr,0}^{\hom}} = f(m_{is}), \tag{10.1.65}$$

$$R_{het,kin,s}^{s} = \frac{(1+s_{i,cr}^{het})}{(1+s_{i,cr}^{\hom})}\left(\frac{s_{i,cr}^{\hom}}{s_{i,cr}^{het}}\right)^2. \tag{10.1.66}$$

Here again, presence of the factor $f(m_{is})$ causes the tendency for higher crystal concentration with heterogeneous freezing, and the ratio of the critical supersaturations counteracts this tendency and leads to higher concentrations with homogeneous freezing, but the latter factor is stronger than in the diffusion growth limit.

b. $\lambda_{het} >> 1$, large initial $r_0 >> \xi_{dep}$, large particle limit

This limit is particularly important when heterogeneous nucleation in DF mode occurs on large particles with a large insoluble core (e.g., mineral dust particles with sizes of several microns to tens or hundreds of microns) covered by the shell of a soluble substance, which is deliquescent and forms a solution. Heterogeneous freezing may occur on the surface of the core inside this solution. If the soluble shell is thin, which is often observed (e.g., Levin et al., 1996; Wurzler et al., 2000), the solution in the shell may freeze rapidly after nucleation of a primary ice germ. Then, a large crystal may form consisting of an insoluble core covered by a thin ice layer whose radius r_0 is determined mostly by the insoluble core (e.g., dust particle) and may constitute several tens or hundreds of microns, so that $r_0 >> \xi_{dep}$. Neglecting ξ_{dep} compared to r_0, (Eqn. 10.1.61) can be further transformed as

$$\Psi_{het,kin,l} = \frac{r_0}{\beta_{het}}\left(\frac{B_{ihet}}{2\beta_{het} r_0^2} + 1\right) = \frac{r_0}{\beta_{het}}[(2\lambda_{het})^{-1} + 1] \approx r_0 \beta_{het}^{-1}. \tag{10.1.67}$$

The last equality takes into account that $\lambda_{het} >> 1$, so the first term in the parentheses is much smaller than the second one and can be neglected. Substituting $\Psi_{het,kin,l}$ into the general solution (Eqns. 10.1.41a) and (10.1.41b), we obtain

$$N_{cm,l}^{het} = (4\pi D_v r_{0het})^{-1}(1+s_{i,cr}^{het})(s_{i,cr}^{het})^{-1}(c_{1i} w), \tag{10.1.68}$$

where we denote r_0 as r_{0het} to emphasize the heterogeneous mode and to enable comparison with homogeneous freezing. Now, the dependence on w is linear, $N_{cm} \sim w$. This linear w-dependence is in agreement with predictions in Kärcher and Lohmann (2002a,b), Ren and MacKenzie (2005), and Khvorostyanov and Curry (2012) for homogeneous freezing described in Section 8.10. The term $\rho_{is}(T)$ is absent; thus, the temperature dependence is much weaker than in the previous cases, and is caused by the T-dependence of D_v, c_{1i}, and $s_{i,cr}^{het}$.

The inverse dependence on the size, $N_{cm,l}^{het} \sim 1/r_{0het}$, may seem unexpected because the nucleation rate per particle in (Eqns. 9.6.7) and (9.6.8) is proportional to its surface area, $J_{f,het} \sim r_N^2$ and one could expect that larger IN would produce higher N_c. However, (Eqn. 10.1.38) for I_{dep} clarifies this inverse dependence. The second term in the supersaturation (Eqn. 10.1.29) describes supersaturation relaxation and contains I_{dep}. The vapor sink term $I_{dep} \sim \Psi_{het} \sim r_{0het}$ in this limit; therefore, supersaturation relaxes faster with increasing r_{0het}, the maximum supersaturation is smaller (this is seen from (Eqn. 10.1.40) for this limit). Therefore, the number of nucleated crystals decreases with increasing r_{0het}. Thus, larger IN may nucleate smaller crystal concentrations due to kinetic limitations.

The expression (Eqn. 8.10.66) derived for homogeneous freezing in this limit in Section 8.10 has the same form with replacement $s_{i,cr}^{het} \to s_{i,cr}^{\text{hom}}$ and $r_{0het} \to r_{0hom}$. Thus, the ratio of the crystal concentrations in these modes under the same conditions is

$$\frac{N_{het,cm,l}}{N_{\text{hom},cm,l}} = \frac{r_{0\,\text{hom}}}{r_{0het}} \frac{(1+s_{i,cr}^{het})}{(1+s_{i,cr}^{\text{hom}})} \frac{s_{i,cr}^{\text{hom}}}{s_{i,cr}^{het}}. \tag{10.1.69}$$

The factor due to the difference of the critical humidities can be ~1.5–2 (see Fig. 9.17), which is much smaller than the ratio of the initial radii. Suppose that homogeneous freezing occurs on haze particles with a modal radius of ~0.02 μm. If heterogeneous freezing takes place on the surface of dust, kaolinite, or similar particles of coarse aerosol fractions with radii ~0.5 μm, then the ratio of the crystal concentrations that could potentially form via these two modes is $N_{chet}/N_{chom} \sim r_{0hom}/r_{0het} = 0.04$. If heterogeneous freezing via DF mode occurs on the particles with larger r_{0het} from the coarse mode then the ratio $N_{chet}/N_{chom} \sim r_{0hom}/r_{0het}$ would be smaller. This may explain lower crystal concentrations that form at low temperatures in cirrus in the kinetic regime of crystal growth via heterogeneous mode in the presence of ice nuclei than those formed via homogeneous freezing.

10.1.9. Parameterization for the Large-Scale Models. Case 2: Small IN Concentration N_a and Crystal Concentration Limited by N_a

If the value of N_a is small, then ice nucleation may cease before the maximum water supersaturation is reached. Then, all available IN can be nucleated and will determine the final crystal concentration. Such a situation was simulated in the parcel run #3 described in Section 9.17, with a 3-modal aerosol distribution and $N_a = 10\,\text{L}^{-1}$ in each mode, see Fig. 9.35. The time t_{het} of the nucleation process can be evaluated using (Eqn. 10.1.19), rewritten in the form:

$$N_c(t) = \int_0^t R_{f,het}(t)dt = N_a Q_{het}[\exp(t/\tau_{nuc,het}) - 1], \tag{10.1.70}$$

where we introduced the notation

$$Q_{het} = c_{sh}\tau_{het} J_{f,het}^{(0,\alpha 0)}(T,\bar{r}_N)\exp(z_\alpha \bar{r}_N^2)\exp(u_{shet} s_{w,cr}^{het}). \tag{10.1.71}$$

When the limit N_a of the available IN is reached, then $N_c = N_a$, and (Eqn. 10.1.70) yields

$$1 = Q_{het}[\exp(t_{het}/\tau_{nuc,het}) - 1]. \tag{10.1.72}$$

Solving for t_{het}, we obtain

$$t_{het} = \tau_{nuc,het} \ln(Q_{het}^{-1} + 1), \tag{10.1.73a}$$

or in expanded form,

$$t_{het} = \tau_{nuc,het} \ln\{[c_{sh}\tau_{nuc,het} J^{(0,\alpha 0)}_{f,het}(T,\bar{r}_N)\exp(z_\alpha \bar{r}_N^2)\exp(u_{shet}s_{w,cr})]^{-1} + 1\}. \quad (10.1.73b)$$

If any of the multipliers in (Eqn. 10.1.73b) is large enough (e.g., $J^{(0,\alpha 0)}_{f,het}$), then the nucleation process is rapid (for example, at large cooling rates), therefore $Q_{het} >> 1$ and $Q_{het}^{-1} << 1$, so the ln term in (Eqn. 10.1.73a) can be expanded and we obtain

$$t_{het} \approx \tau_{nuc,het} Q_{het}^{-1} \quad (10.1.74a)$$

$$= [c_{sh} J^{(0,\alpha 0)}_{f,het}(T,\bar{r}_N)\exp(z_\alpha \bar{r}_N^2)\exp(u_{shet}s_{w,cr})]^{-1}. \quad (10.1.74b)$$

That is, for rapid nucleation with limited N_a, the nucleation time does not depend on $\tau_{nuc,het}$ and is inversely proportional to the nucleation rate. In the opposite case of slow nucleation (small $J^{(0,\alpha 0)}_{f,het}$ and other factors in (Eqn. 10.1.73b)), we have the condition $Q_{het} << 1$, and $Q_{het}^{-1} >> 1$ then. (Eqns. 10.1.73a) and (10.1.73b) are transformed as

$$t_{het} = \tau_{nuc,het} \ln(Q_{het}^{-1}), \quad (10.1.75a)$$

$$= \tau_{nuc,het} \ln[c_{sh}\tau_{nuc,het} J^{(0,\alpha 0)}_{f,het}(T,\bar{r}_N)\exp(z_\alpha \bar{r}_N^2)\exp(u_{shet}s_{w,cr})]^{-1}, \quad (10.1.75b)$$

i.e., the nucleation time is proportional to the logarithm of the inverse of the nucleation rate. As mentioned earlier, the final crystal concentration in this case is equal to the concentration N_a of available IN. Calculations of these nucleation times for various conditions may help in analyses of the laboratory and field experiments on IN measurements—in particular, to determine if the time of processing in the experimental device is sufficient for the nucleation process if IN concentration is small.

10.2. Temperature Effects and Heterogeneous Freezing of Cloud Drops

The temperature effects in heterogeneous ice nucleation can be considered similar to homogeneous freezing in Section 8.11. According to (Eqn. 9.8.7), near water saturation ($s_w \approx 0$), the temperature variations of heterogeneous freezing are determined by changes in the nucleation rate $J^{(0,\alpha 0)}_{f,het}(T,\bar{r}_N)$ defined by (Eqn. 9.8.8), and in $J^{\alpha}_{act}(T,r_N)$ describing the effects of active sites defined by (Eqn. 9.8.9a):

$$J^{(0)}_{f,het}(T,r_N) \approx J^{(0,\alpha 0)}_{f,het}(T,\bar{r}_N)\bar{J}^{\alpha}_{act}(T,r_N), \quad (10.2.1)$$

while the factor in (Eqn. 9.8.7) describing s_w-effect is $J^{s}_{het}(T,s_w,m_{is}) = 1$ at $s_w = 0$. According to (Eqn. 9.8.4), the temperature effects in $J^{(0)}_{f,het}(T)$ are caused mostly by changes in the critical energy $\Delta F^{\alpha 0}_{cr,0}(T)$ and in ΔF_{act}, while T-variations in $\Delta F^{\varepsilon}_{cr}$ are smaller because of the small factor $H_{v,fr}$. The temperature dependence of the critical energy $\Delta F^{\alpha 0}_{cr,0}(T)$ and activation energy ΔF_{act} can be evaluated assuming, as in Section 8.11, that the temperature change $\Delta T << T$, and then expanding the terms $\sigma_{is}(T - \Delta T)$, $\rho_i(T - \Delta T)$, $L^{ef}_m(T - \Delta T)$, and $\ln[T/(T - \Delta T)]$ into the power series by $\Delta T/T$ to the first-order terms. Substituting these expansions into (Eqn. 9.8.3) for $\Delta F^{\alpha 0}_{cr,0}(T)$ yields

$$\Delta F^{\alpha 0}_{cr,0}(T - \Delta T) \approx \Delta F^{\alpha 0}_{cr,0}(T)[1 - \kappa_{Tn}(T)\Delta T(t)], \quad (10.2.2a)$$

where k_{Tn} is the same as in (Eqn. 8.11.5b),

$$\kappa_{Tn}(T) = 3\frac{\sigma'_{is}}{\sigma_{is}} - 2\left(\frac{\rho'_i}{\rho_i} + \frac{(L_m^{ef})'}{L_m^{ef}}\right) + \frac{2}{T\ln(T_0/T)}, \tag{10.2.2b}$$

and primes denote absolute values of the derivatives by T (i.e., the derivatives and ΔT are assumed to be positive for the quantities decreasing with T). The activation energy also can be presented in a similar form described by (Eqns. 8.11.6a) and (8.11.6b). Substituting these expressions into (Eqn. 9.8.8) yields

$$J_{f,het}^{(0,\alpha 0)}(T-\Delta T) = J_{f,het}^{(0,\alpha 0)}(T)\exp[(u_{Tact} + u_{Thet})\Delta T(t)], \tag{10.2.3a}$$

$$u_{Thet}(T) = \left(\kappa_{Tn}(T) - \frac{1}{T}\right)\frac{\Delta F_{cr,0}^{\alpha 0}(T)}{kT}, \tag{10.2.3b}$$

$$u_{Tact}(T) = \left(\kappa_{Ta}(T) - \frac{1}{T}\right)\frac{\Delta F_{act}(T)}{kT}. \tag{10.2.3c}$$

The quantity $\bar{J}_{act}^{\alpha}(T, \bar{r}_N)$ that describes the effect of active sites averaged over the size spectrum $f_a(r_N)$ is according to (Eqn. 10.1.5):

$$\bar{J}_{act}^{\alpha}(\alpha_a, m_{is}, \bar{r}_N) \sim \exp\left[\frac{\Delta F_a^{\alpha}(r_N)}{kT}\right] \approx c_{sh} N_a \exp\left[\frac{\alpha_a \sigma_{is}(1-m_{is})\bar{r}_N^2}{kT}\right]. \tag{10.2.4}$$

The energy of the active sites $F_a^{\alpha} = \alpha_a \sigma_{is}(1-m_{is})\bar{r}_N^2$ defined in (Eqn. 9.7.20b) can be presented in the form of expansion using (Eqn. 9.17.1) for the temperature-dependent active site fraction $\alpha_a(T)$ from Curry and Khvorostyanov (2012), and expanding $\alpha_a(T)$ and $\sigma_{is}(T)$ by $\Delta T/T$:

$$\Delta F_\alpha^{\alpha}(T-\Delta T) \approx \Delta F_\alpha^{\alpha}(T)[1 - \kappa_{Tas}(T)\Delta T(t)], \tag{10.2.5a}$$

$$\kappa_{Tas}(T) = \alpha'_a(T)/\alpha_a(T) + \sigma_{is}(T)/\sigma'_{is}(T). \tag{10.2.5b}$$

Substituting these expressions into (Eqn. 10.2.4), we obtain for the temperature effect of the active sites

$$\bar{J}_{act}^{\alpha}(T-\Delta T) = \bar{J}_{act}^{\alpha}(T)\exp[-u_{Tas}\Delta T(t)], \tag{10.2.6}$$

$$u_{Tas}(T) = \left(\kappa_{Tas}(T) - \frac{1}{T}\right)\frac{F_a^{\alpha}(T)}{kT}. \tag{10.2.7}$$

The parameters u_{Thet}, u_{Tact}, and u_{Tas} in [K^{-1}] determine changes in the corresponding expressions for nucleation rates at temperature decrease ΔT. Substituting these expressions into (Eqn. 10.2.1), we obtain the combined effect of the temperature variations ΔT in the critical and activation energies ΔF_{cr}, ΔF_{act}, and the energy of the active sites F_a^{α} on the nucleation rate $J_{f,het}^{(0)}$ in (Eqn. 9.8.15) (at supersaturation $s_w \approx 0$) that can be written similar to (Eqn. 8.11.7):

$$J_{f,het}^{(0)}(T-\Delta T, r_N) \approx J_{f,het}^{(0,\alpha 0)}(T, \bar{r}_N)\bar{J}_{act}^{\alpha}(T, \bar{r}_N)\exp[B_{Thet}\Delta T(t)], \tag{10.2.8}$$

$$B_{Thet} = u_{Ta} + u_{Thet} - u_{Tas}. \tag{10.2.9}$$

The parameter denoted B_{Thet} with the dimension [K^{-1}] describes an increase in $J_{f,het}$ with temperature decrease ΔT at $s_w \approx 0$. As in Section 8.11, the temperature decrease ΔT can be related to the time dependence via the temperature lapse rate γ and vertical velocity w using the temperature equations from Section 3.11, and its integration by time:

$$dT/dt = \dot{T} = -\gamma w, \qquad \Delta T(t) = -\dot{T}t = \gamma wt. \tag{10.2.10}$$

The minus sign in the second equation arises because we defined $\Delta T > 0$ as the temperature decreases and $dT/dt < 0$. Substituting this ΔT into (Eqn. 10.2.8), we obtain the time dependent $J^{(0)}_{f,het}$ and nucleation rate $R_{f,het}(t)$ at $s_w \approx 0$:

$$R_{f,het}(t) = \frac{dN_c}{dt} = J^{(0)}_{f,het}(T - \Delta T, \bar{r}_N),$$
$$\approx J^{(0)}_{f,het}(t_0, \bar{r}_N)\exp(\beta_{Thet}t) \approx J^{(0)}_{f,het}(t_0, \bar{r}_N)\exp(t/\tau_{Thet}), \tag{10.2.11}$$

where we introduce again the characteristic freezing frequency β_{Thet} and time τ_{Thet},

$$\beta_{Thet}(T) = B_{Thet}\gamma w = B_{Thet}(-dT/dt), \tag{10.2.12a}$$

$$\tau_{Thet} = [(B_{Thet})\gamma w]^{-1} = [(B_{Thet})(-dT/dt)]^{-1}. \tag{10.2.12b}$$

The first form of these equations with γw can be convenient in numerical models where γ and w are known, and the second form with $(-dT/dt)$ can be more convenient for analysis of laboratory experiments where dT/dt is known. Integration of (Eqn. 10.2.11) over time yields

$$N_c(t) = J^{(0)}_{f,het}(t_0, \bar{r}_N, t_0)\tau_{Thet}[\exp(\beta_{Thet}t) - 1]. \tag{10..2.13}$$

So, the time dependence of the crystal concentration due to temperature decreases is similar to that due to the supersaturation variations described in Section 10.1. If we consider freezing of deliquescent haze particles, then the temperature correction and the joint effect of the supersaturation and temperature variations can be simply accounted for by replacing in the equations of Section 10.1 for $J_{f,het}$ and $N_c(t)$ the frequency β_{shet} with $(\beta_{shet} + \beta_{Thet})$. If we consider the heterogeneous freezing of water drops near water saturation, $s_w \approx 0$, then only the terms with β_{Thet} are essential and the equations of this section can be used.

For small times or temperature changes, if $\beta_{Thet}t << 1$, the exponent in (Eqn. 10.2.13) can be expanded and (Eqn. 10.2.13) is reduced to

$$N_c(t) = tJ^{(0)}_{f,het}(t_0, \bar{r}_N, t_0) = tJ^{(0,\alpha 0)}_{f,het}(T, \bar{r}_N)\bar{J}^{\alpha}_{act}(T, \bar{r}_N), \tag{10.2.14}$$

where the time dependence of N_c is linear in this limit. For large times or temperature changes, if $\beta_{Thet}t >> 1$, the time dependence in (Eqn. 10.2.13) becomes exponential,

$$N_c(t) = J^{(0,\alpha 0)}_{f,het}(T, \bar{r}_N)\bar{J}^{\alpha}_{act}(T, \bar{r}_N)\tau_{Thet}\exp(\beta_{Thet}t). \tag{10.2.15}$$

Equations (10.2.14) and (10.2.15) are analogous to those derived in Section 8.11 for homogeneous nucleation and may explain the difference in the nucleation rates at small and large temperature changes ΔT. If a permanent and significant cooling rate is present, the rate of crystal formation is exponential according to (Eqn. 10.2.15). If a permanent source of cooling is absent, and there are only small random temperature fluctuations, the rate of crystal production is close to linear with time as in (Eqn. 10.2.14). Vali (1994, 2008) described the experiments on the heterogeneous freezing of drops with an alternating cooling rate and found exponential growth of $N_c(t)$ at a constant cooling rate and nearly constant $N_c(t)$ with a very weak increase in the absence of the regular cooling rate. Equations (10.2.14) and (10.2.15) may provide an explanation for these observations.

It follows from (Eqn. 10.2.12a) that

$$\beta_{Thet}\Delta t = B_{Thet}(-dT/dt)\Delta t = B_{Thet}\Delta T. \tag{10.2.16}$$

Using this relation, (Eqn. 10.2.15) can be rewritten for ice crystals nucleated during interval Δt as

$$\begin{aligned} N_c(\Delta t) &= J^{(0)}_{f,het}(t_0,\bar{r}_N,t_0)\tau_{Thet}\exp(B_{Thet}\Delta T) \\ &= J^{(0,\alpha 0)}_{f,het}(T,\bar{r}_N)\bar{J}^{\alpha}_{act}(T,\bar{r}_N)\frac{\exp(B_{Thet}\Delta T)}{B_{Thet}(-dT/dt)}, \end{aligned} \tag{10.2.17}$$

where we used (Eqn. 10.2.12b) for τ_{Thet}.

The theoretical exponential temperature dependence of N_c in (Eqn. 10.2.15) is similar to the corresponding dependence for homogeneous freezing in Section 8.11 and to the empirical parameterizations $N_c(T) \sim \exp(\beta_F \Delta T)$ for heterogeneous ice nucleation suggested by Fletcher (1962), Cooper (1986), Sassen (1992), Diehl and Wurzler (2004), and others described in Section 9.4. The parameter B_{Thet} [K^{-1}] is a theoretically derived analogue of the empirical parameter β_F introduced by Fletcher (1962) and others in similar parameterizations. The average value of the exponent in Fletcher's was $\beta_F \approx 0.6\,K^{-1}$, although its wide variations were found; Vali (1994) determined the average value to be $\sim 0.54\,K^{-1}$.

We can now compare the exponential T-dependence derived from the theory with the empirical parameterizations. An estimate is done at $-20\,°C$ with the parameterization (Eqn. 4.4.15a) for σ_{is} by Pruppacher and Klett (1997), (Eqn. 4.4.22) for $\rho_i(T)$, and (Eqn. 8.8.12a) for $L^{ef}_m(T)$, which gives $\sigma'_{is}(T)/\sigma_{is}(T) \sim 1.1 \times 10^{-2}\,K^{-1}$, $\rho'_i(T)/\rho_i(T) \sim 2.4 \times 10^{-3}\,K^{-1}$, $[L^{ef}_m(T)]'/L^{ef}_m(T) \sim 10^{-2}\,K^{-1}$, and $\ln(T_0/T) \approx 5 \times 10^{-2}\,K^{-1}$, which yields $k_{Tn} \sim 0.1\,K^{-1}$. Assuming $\Delta F_{cr} \sim 10^{-12}$ erg and $\delta F_{cr}/kT \sim 28.6$, we get $u_{Thet} \sim 2.75\,K^{-1}$. An estimate with (Eqn. 8.3.20) for $\Delta F_{act}(T)$ from Zobrist et al. (2007) gives $k_{Ta} \sim -1.2 \times 10^{-2}\,K^{-1}$, and $u_{Tact} \sim -0.14\,K^{-1}$. Using the model from Curry and Khvorostyanov (2012) of T-dependent active sites described by (Eqn. 9.17.1), $\alpha_a(T) = \alpha_0(1 - T_c/T_v)$ with $\alpha_0 = 2 \times 10^{-5}$, $T_c = T - T_0$, and $T_v = -25\,°C$, we estimate $\alpha_a \approx 0.4 \times 10^{-5}$ and $\alpha_a'/\alpha_a \approx 0.2\,K^{-1}$ at $T_c = -20\,°C$. This gives us $k_{Tas} \approx 0.214\,K^{-1}$. Assuming $r_N \sim 0.5\,\mu m$, $m_{is} = 0.5$, $\alpha_a \approx 0.4 \times 10^{-5}$, and the energy of active sites $F_a^{\alpha} \approx 0.2 \times 10^{-12}$ erg, then (Eqn. 10.2.7) gives $u_{Tas} \approx 1.2\,K^{-1}$. Finally, we obtain $B_{Thet} = u_{Thet} + u_{Tact} - u_{Tas} \approx 1.4\,K^{-1}$. This is comparable to but still higher than Fletcher's mean value $0.6\,K^{-1}$. If $\Delta F_{cr} \sim 0.7 \times 10^{-12}$ erg, which corresponds to a slightly smaller contact parameter $m_{is} < 0.5$ and shape factor $f(m_{is}, x)$, then with the same k_{Tn} and all the other parameters, we obtain $u_{Thet} \sim 1.92\,K^{-1}$ and $B_{Thet} = 0.58\,K^{-1}$, very close to the mean experimental value.

Thus, classical nucleation theory explains the exponential temperature dependence of ice nucleation by freezing, and the corresponding empirical parameter β_F can be expressed via the parameters of classical theory and fundamental constants. However, it should be noted that the parameter B_{Thet} itself is a function of temperature and may vary with T. Further, (Eqn. 10.2.15) shows that the actual temperature dependence of N_c is more complicated. Along with this exponential dependence, $N_c \sim \exp(B_{Thet}\Delta T)$, there is also temperature dependence of N_c in the denominator in $B_{Thet}(T)$. Another important fact follows from (Eqn. 10.2.15), that with the same temperature interval ΔT, the concentration of nucleated crystals is inversely proportional to the cooling rate dT/dt—i.e., the slower the cooling rate, the larger the number of crystals nucleating at the same temperature, which is in agreement with observations. A more detailed comparison of the theoretical description of the temperature dependence of heterogeneous ice nucleation with the experimental data would allow refinement of some parameters of classical nucleation theory.

10.3. Parameterization of Deposition Ice Nucleation Based on Classical Nucleation Theory

At sufficiently low temperatures (usually below −20 °C to −25 °C) and low humidities, liquid drops may be absent and ice nucleation occurs via direct deposition of water vapor on aerosol particles that serve as ice nuclei. Deposition nucleation has been studied in laboratory experiments by Möhler et al. (2006) and Connolly et al. (2009), among others. The critical energies and nucleation rates for the ice deposition process were derived in Section 9.5 and can be used in models with sufficiently small time steps. Parameterization of this process for large-scale models is quite similar to parameterization of homogeneous and heterogeneous freezing in Sections 8.10 and 10.1 and is based on several assumptions.

Since the critical radius r_{cr} of an ice germ for deposition is ~(1–3) × 10^{-2} µm (see Fig. 9.3), and the effective radii r_N of IN on which deposition occurs are usually > 0.1–0.5 µm, we assume again that $r_N >> r_{cr}$. Using this assumption, the geometric factor $f(m_{iv}, x)$ in (Eqn. 9.2.21) is reduced to $f(m_{iv})$ in (Eqn. 9.2.15) with replacement of $m_{wv} \rightarrow m_{iv}$ and does not depend on r_N. Numerical simulations of deposition nucleation with parcel models show that the temporal behavior of the ice supersaturation and crystal concentration is similar to those in the freezing process shown in Fig. 9.34. Nucleation begins at some critical supersaturation $s_{i,cr}^{dep}$, increases to maximum values s_{imax} at t_{max}, when $N_c = N_{cmax}$; then, supersaturation absorption begins to exceed supersaturation production, s_i decreases and reaches its second critical point, and then nucleation ceases.

As in the previous parameterizations of freezing in Sections 8.10 and 10.1, we approximate the growing branch of s_i at $t_{cr1} < t < t_{max}$ as a linear function, $s_i(t) = s_{i,cr}^{dep} + c_{1i}wt = s_{i,cr}^{dep} + \Delta s_i$, where $\Delta s_i = c_{1i}wt$, and c_{1i} is defined in (Eqn. 5.3.46a). The critical energy of a deposition ice germ formation can be written using (Eqn. 9.5.9) with the relation $S_i = 1 + s_i$, and incorporating these assumptions:

$$\Delta F_{cr} = \frac{16\pi}{3} \frac{\sigma_{iv}^3 f(m_{iv})}{[R_v T \rho_i \ln(1 + s_{i,cr}^{dep} + \Delta s_i) - C_\varepsilon \varepsilon^2]^2}. \tag{10.3.1}$$

Numerical simulations show that $\Delta s_i(t_{max}) << s_{i,cr}^{dep}$, and that the second term with ε in the denominator is much smaller than the first term with ln. Thus, we can expand the denominator and retain first-order terms using the relation $\ln(1 + x) \approx x$ and obtain

$$\Delta F_{cr}(s_i, t) = \Delta F_{dep}^{(0)}(s_{i,cr}^{dep})[1 - \kappa_{dep} \Delta s_i(t)], \tag{10.3.2}$$

where

$$\kappa_{dep} = 2\{(1 + s_{i,cr}^{dep})[\ln(1 + s_{i,cr}^{dep}) - M_w C_\varepsilon \varepsilon^2/(RT\rho_i)]\}^{-1}, \tag{10.3.3}$$

$$\Delta F_{dep}^{(0)} = \frac{16\pi}{3} \frac{\sigma_{iv}^3 f(m_{iv})}{[R_v T \rho_i \ln(1 + s_{i,cr}^{dep}) - C_\varepsilon \varepsilon^2]^2}. \tag{10.3.4}$$

That is, $\Delta F_{dep}^{(0)}$ is the critical energy at the beginning of nucleation. Substitution of (Eqn. 10.3.2) into (Eqn. 9.5.11) yields the deposition nucleation rate per unit area that can be written in various forms:

$$J_{sdep} = J_{sdep}^{(0)} \exp\left(\frac{\Delta F_{dep}^{(0)} \kappa_{dep} c_{1i} wt}{kT}\right) = J_{sdep}^{(0)} \exp\left(u_{sdep} c_{1i} wt\right),$$
$$= J_{sdep}^{(0)} \exp(\beta_{sdep} t) = J_{sdep}^{(0)} \exp(t/\tau_{sdep}), \quad (10.3.5)$$

where

$$u_{sdep} = \Delta F_{dep}^{(0)} \kappa_{dep} / kT, \quad (10.3.6)$$

$$J_{sdep}^{(0)}(r_N, s_{i,cr}^{dep}) = C_{sdep} \exp(-\Delta F_{dep}^{(0)}/kT), \qquad C_{sdep} = \frac{c_{1s} e_v \pi r_{cr}^2 Z_s}{(2\pi m_{w1} kT)^{1/2}}, \quad (10.3.7)$$

and we introduced the "deposition frequency" β_{dep}, and the "deposition nucleation time" τ_{sdep},

$$\beta_{sdep} = u_{sdep} c_{1i} w, \qquad \tau_{sdep} = \beta_{sdep}^{-1} = (u_{sdep} c_{1i} w)^{-1}. \quad (10.3.8)$$

The kinetic coefficient C_{sdep} depends on $s_i(t)$ and on time via r_{cr} as $C_{sdep} \sim r_{cr}^2$, but this power law time dependence of C_{sdep} is much weaker than the exponential time dependence and we assumed that $J_{sdep}^{(0)}(r_N, s_i) \approx J_{sdep}^{(0)}(r_N, s_{i,cr}^{dep})$.

The nucleation rate per particle $J_{dep} = 4\pi r_N^2 J_{sdep}$ in (Eqn. 9.5.12) can now be written as

$$J_{dep} = J_{dep}^{(0)}(r_N, s_{i,cr}^{dep}) \exp(\beta_{sdep} t), \quad (10.3.9)$$

$$J_{dep}^{(0)} = C_{dep} r_N^2 \exp(-\Delta F_{i,dep}^{(0)}/kT), \qquad C_{dep} = \frac{4\pi^2 r_{cr}^2 Z_s c_{1s} e_v}{(2\pi m_{w1} kT)^{1/2}}. \quad (10.3.10)$$

The polydisperse nucleation rate in an ensemble of particles is obtained by integration of J_{dep} over the size spectrum by r_N:

$$R_{dep} = \frac{dN_{dep}}{dt} = \exp(\beta_{sdep} t) \int_{r_{min}}^{r_{max}} dr_N f(r_N) J_{dep}^{(0)}(r_N, s_{i,cr}^{dep})$$
$$= N_a J_{dep}^{(0)}(\bar{r}_N, s_{i,cr}^{dep}) \exp(\beta_{sdep} t), \quad (10.3.11)$$

where $J_{dep}^{(0)}(\bar{r}_N, s_{i,cr}^{dep})$ depends on the mean radius $\bar{r}_N$,

$$J_{dep}^{(0)}(\bar{r}_N, s_{i,cr}^{dep}) = C_{dep}(s_{i,cr}^{dep}) \overline{r_N^2} \exp(-\Delta F_{i,dep}^{(0)}/kT). \quad (10.3.12)$$

Integration of (Eqn. 10.3.11) by time yields $N_{dep}(t)$:

$$N_{dep}(t) = \int_0^t R_{dep}(t')dt' = N_a J_{dep}^{(0)}(\bar{r}_N, s_{i,cr}^{dep}) \tau_{sdep} [\exp(\beta_{sdep} t) - 1] \quad (10.3.13a)$$

$$\approx N_a J_{dep}^{(0)}(\bar{r}_N, s_{i,cr}^{dep}) \tau_{sdep} \exp(\beta_{sdep} t), \quad (10.3.13b)$$

where the second equation holds if $\beta_{sdep}t >> 1$. This parameterization can be used in numerical models for evaluation of $N_{dep}(t)$ nucleating by deposition mode at a time t. Equation (10.3.13b) can be presented in an alternative form using the relations $\beta_{sdep}t = u_{sdep}c_{1i}wt = u_{sdep}\Delta s_i = u_{sdep}[s_i(t) - s_{i,cr}^{dep}]$:

$$N_{dep}(t) \approx N_a \exp\{a_{dep} + u_{sdep}[s_i(t) - s_{i,cr}^{dep}]\},$$

$$\approx N_a \exp[u_{sdep}s_i(t) + (a_{dep} - u_{sdep}s_{i,cr}^{dep})] \tag{10.3.13c}$$

$$a_{dep} = \ln[J_{dep}^{(0)}(\bar{r}_N, s_{i,cr}^{dep})\tau_{sdep}]. \tag{10.3.13d}$$

Equation (10.3.13c) resembles an empirical parameterization of deposition nucleation (Eqn. 9.4.7) suggested by Möhler et al. (2006) based on laboratory experiments, while the coefficients a_{dep} and u_{sdep} are expressed here via the aerosol properties and parameters of classical nucleation theory.

If the time step in a model is greater than the duration of nucleation, we need as before to calculate $N_{dep}(t_{max})$ with $s_i = s_{imax}$ and then after the end of nucleation. The supersaturation equation required for evaluation of $s_{imax}(t_{max})$ is the same as (Eqn. 10.1.29), but the deposition integral I_{dep} is now

$$I_{dep}(t) = s_i \frac{4\pi D_v \rho_{is}}{\Gamma_2} \int_0^t r_{c,ef}(t, t_0) R_{dep}(t_0)\, dt_0, \tag{10.3.14}$$

where $r_{c,ef}(t,t_0)$ is the same as defined in (Eqn. 10.1.31), but replacing $B_{ihet} \rightarrow B_{idep}$. We assume again that the crystal growth rate can be calculated with $s_i(t) \approx \text{const} \approx s_{i,cr}^{dep}$ because $\Delta s_i << s_{i,cr}^{dep}$, then the parameter that describes the crystal growth rate is $B_{idep} = 2c_{i3}s_{i,cr}^{dep}$. Substituting R_{dep} from (Eqn. 10.3.11) into (Eqn. 10.3.14) yields

$$I_{dep}(t) = s_i \frac{4\pi D_v \rho_{is}}{\Gamma_2} N_a J_{dep}^{(0)} J_{0i,dep}(t), \tag{10.3.15}$$

where, similar to Section 10.1,

$$J_{0i,dep}(t) = \int_0^t r_{c,ef}(t, t_0) \exp(\beta_{sdep}t_0)\, dt_0 = \sum_{k=1}^{3} J_{0i,dep}^{(k)}, \tag{10.3.16}$$

$$J_{0i,dep}^{(k)} = \int_0^t r_{c,ef}^{(k)}(t, t_0) \exp(\beta_{sdep}t_0)\, dt_0. \tag{10.3.17}$$

The integrals $J_{0i,dep}^{(k)}$ are evaluated as described in Section 8.10 and Appendix A.8, or in Section 10.1 and are expressed again via the Ψ-function but with different arguments

$$J_{0i,dep} = \sum_{k=1}^{3} J_{0i,dep}^{(k)} = e^{\beta_{sdep}t}\Psi_{dep}, \tag{10.3.18}$$

$$\Psi_{dep} = \Psi_{1,dep} + \Psi_{2,dep} + \Psi_{3,dep}. \tag{10.3.19}$$

The three components of the function Ψ_{dep} are expressed with analytical (Eqns. 10.1.36b)–(10.1.36d) with the parameter λ_{dep} defined similar to λ_{het} in Section 10.1, but with different critical parameters:

$$\lambda_{dep} = \frac{(\xi_{dep} + r_0)^2 \beta_{sdep}}{B_{idep}} = \frac{(u_{sdep}c_{1w}w)(\xi_{dep} + r_0)^2}{2c_{i3}s_{i,cr}^{dep}}. \tag{10.3.20}$$

Substitution of $J_{0i,dep}(t)$ from (Eqns. 10.3.17)–(10.3.18) into (Eqn. 10.3.14) yields

$$I_{dep}(t) = s_i \frac{4\pi D_v \rho_{is}}{\Gamma_2} N_a J_{dep}^{(0)} e^{\beta_{sdep}t} \Psi_{dep}(t). \tag{10.3.21}$$

Using again the supersaturation equation (10.1.29) and the condition $ds_i(t_{max})/dt = 0$ when $s_i = s_{imax}$, we obtain the integral $I_{dep}(t_{max})$:

$$I_{dep}(t_{\max}) = c_{1i}w(1+s_{i,cr}^{dep})\rho_{is}/\Gamma_2. \tag{10.3.22}$$

Equating (Eqn. 10.3.21) at $t = t_{max}$ and (Eqn. 10.3.22) with the approximation $s_{imax} \approx s_{i,cr}^{dep}$ yields

$$\exp(\beta_{sdep}t_{\max}) = c_{1i}w(1+s_{i,cr}^{dep})(s_{i,cr}^{dep})^{-1}(4\pi D_v)^{-1}(N_a J_{dep}^{(0)})^{-1}\Psi_{dep}^{-1}. \tag{10.3.23}$$

Parcel model simulations show that, as in the case with freezing in Sections 8.10 or 10.1, the duration of the nucleation process is usually much longer than τ_{sdep}, so that $\beta_{sdep}\, t_{max} >> 1$, and (Eqn. 10.3.13b) can be applied. Substituting (Eqn. 10.3.23) into (Eqn. 10.3.13b) yields

$$N_{cdep}(t_{\max}) = K_{dep}^{gen}(1+s_{i,cr}^{dep})(s_{i,cr}^{dep})^{-1}\Psi_{dep}^{-1}, \tag{10.3.24a}$$

$$K_{dep}^{gen} = (4\pi D_v)^{-1}u_{sdep}^{-1}. \tag{10.3.24b}$$

These equations are similar to those derived for freezing in Sections 8.10 or 10.1, with replacement of the corresponding parameters with u_{sdep}, $s_{i,cr}^{dep}$, and Ψ_{dep}. The mean crystal radius at t_{max} can be estimated as in Sections 8.10 and 10.1, equating the two expressions for the second term in the supersaturation equation, which yields as in (Eqn. 10.1.46a), $\tau_{fc}^{-1}(t) = (\Gamma_2/\rho_{is})(I_{dep}/s_i)$. Substituting here (Eqn. 10.3.22) for I_{dep} yields the supersaturation relaxation time at t_{max}:

$$\tau_{fc}(t_{\max}) = [c_{1i}w(1+s_{i,cr}^{dep})]^{-1}s_{i,cr}^{dep}. \tag{10.3.25}$$

Assuming the diffusion crystal growth rate, and using the expression $\tau_{fc}^{-1} \approx (4\pi D_v N_c \bar{r}_c)$, we obtain

$$\bar{r}_c(t_{\max}) = [4\pi D_v N_{cdep}(t_{\max})]^{-1}c_{1i}w(1+s_{i,cr}^{het})(s_{i,cr}^{het})^{-1}. \tag{10.3.26}$$

Numerical simulations with the parcel model indicate that N_{cdep} and the mean crystal radius after the end of nucleation are nearly doubled relative to t_{max}:

$$N_{cdep}(t_{cr2}) \approx 2N_{cdep}(t_{\max}), \qquad \bar{r}_c(t_{cr2}) \approx 2\bar{r}_c(t_{\max}). \tag{10.3.27}$$

The equations derived here can be used for parameterization of the crystal concentration and mean radius in the models with the large time steps where deposition nucleation is a substep process.

The asymptotics of N_{cdep} in various regimes are determined by the same asymptotic expressions of the Ψ-function as in Sections 8.10 and 10.1. The diffusion and kinetic limits for the deposition mode are determined by the values of the parameter λ_{dep} defined by (Eqn. 10.3.20).

The diffusion regime of crystal growth with the deposition coefficient $\alpha_d > 0.1$ implies $\lambda_{dep} << 1$. As is seen from (Eqn. 10.3.20) and that is similar to the homogeneous freezing in Section 8.10 or heterogeneous freezing in Section 10.1, the condition $\lambda_{dep} << 1$ implies small ξ_{dep} and r_0 that are typical of the diffusion regime of crystal growth with the deposition coefficient $\alpha_d \sim 1$ or $\alpha_d > 0.1$, not very large w and not very low T. Using the diffusion asymptotic Ψ_{dif} from (Eqn. 10.1.54) and replacing all the corresponding quantities with those for the deposition mode, we obtain N_{cdep} in the diffusion regime:

$$N_{cm,dif}^{dep}(t_{\max}) = K_{i,dif}^{dep}(1+s_{i,cr}^{dep})(s_{i,cr}^{dept})^{-3/2}(c_{1i}w)^{3/2}, \tag{10.3.28a}$$

$$K_{i,dif}^{dep} = (2\pi D_v)^{-3/2} \left(\frac{\rho_i \Gamma_2}{\rho_{is}} \right)^{1/2} u_{sdep}^{1/2} \left(\frac{c_{1w}}{c_{1i}} \right)^{1/2}. \quad (10.3.28b)$$

Another asymptotic limit with $\lambda_{dep} >> 1$, $\xi_{dep} >> r_0$, corresponds to the limit of the kinetic regime of crystal growth and small particles. Using the asymptotic value of $\Psi_{het,kin,s}$ from (Eqn. 10.1.62), and replacing all the corresponding quantities with those for the deposition mode, we obtain N_{cdep}:

$$N_{cm,kin,s}^{dep}(t_{\max}) = K_{kin,s}^{dep}(1+s_{i,cr}^{dep})(s_{i,cr}^{dep})^{-2}(c_{1w}w)^2, \quad (10.3.29a)$$

$$K_{kin,s}^{dep} = \frac{1}{(\pi D_v)} \frac{u_{sdep}}{\alpha_d V_w} \left(\frac{\rho_i \Gamma_2}{\rho_{is}} \right) \frac{c_{1i}}{c_{1w}}. \quad (10.3.29b)$$

The large particle asymptotic limit occurs for $\lambda_{dep} >> 1$ and large initial $r_0 >> \xi_{dep}$. This limit can be particularly important in the atmosphere when heterogeneous deposition nucleation occurs on large particles—e.g., on large mineral dust particles with $r_0 \geq 0.1$–$0.5\,\mu$m as often used in the lab experiments. In contrast to the deliquescence-freezing mode described in Section 10.1, these particles should not be covered by a soluble shell to prevent the freezing mode and initiate ice nucleation in the deposition mode. Substituting the limit (Eqn. 10.1.67) for the Ψ-function into the general (Eqns. 10.3.24a) and (10.3.24b), we obtain

$$N_{cm,l}^{dep}(t_{\max}) = (4\pi D_v)^{-1}(1+s_{i,cr}^{dep})(s_{i,cr}^{dep})^{-1} r_{0dep}^{-1}(c_{1i}w), \quad (10.3.30)$$

where we denote r_0 as r_{0dep} to emphasize the deposition mode and to relieve comparison with the freezing modes. Now, the dependence on w is linear, $N_{cm} \sim w$.

The final crystal concentrations after nucleation ceases can be approximately calculated for these limits, as for the general case, as $N_c \approx 2N_{cm}(t_{max})$.

10.4. General Properties and Empirical Parameterizations of Contact Nucleation

10.4.1. General Properties

In Chapter 9 and Sections 10.1–10.3, we considered three of the four modes of heterogeneous ice nucleation in clouds: deposition, deliquescence-freezing (or condensation-freezing), and immersion freezing. The last of the major heterogeneous modes, contact nucleation, is considered in the following sections. The term "contact nucleation" refers usually to a process when an aerosol particle (AP) collides with a supercooled drop and initiates freezing from outside (Young, 1974a,b, 1993; Pruppacher and Klett, 1997). Shaw et al. (2005) and Durant and Shaw (2005) suggested recently a more extended treatment. They observed freezing of evaporating drops, and found that some IN particles that did not cause freezing being fully immersed in the drop, initiated freezing when these IN reached and contacted the surface of an evaporating droplet from inside or were partially released from the drop. This led to an increase in ice nucleation temperatures. This mechanism was termed "*contact nucleation inside out*."

The contact mode has several distinguishing features. The temperature thresholds are usually higher in this mode than in the deposition and immersion modes and are comparable to those in the condensation-freezing mode (Young, 1974a,b; Cooper, 1995; Table 4.1 in Young, 1993; Shaw et al.,

2005, Durant and Shaw, 2005; Djikaev and Rukenshtein, 2008; Hoose and Möhler, 2012; Ladino et al., 2013). Therefore, this mode was invoked for possible explanation of intensive ice crystal formation in relatively warm clouds when ice multiplication mechanisms may be ineffective. In contrast to the other modes caused by the freezing of a single aerosol particle or cloud drop, contact nucleation is a more complicated phenomenon that includes two sub-processes: collection of aerosol particles by cloud drops and subsequent freezing. The primary three processes that cause the collection of aerosol particles by cloud drops and determine collection rates are Brownian diffusion, thermophoresis, and diffusiophoresis. Young (1974b, 1993) calculated these collection rates and showed that the net rate is higher at water subsaturation in clouds than at supersaturation. Due to these features, contact nucleation has often been invoked to explain fast cloud crystallization at relatively warm temperatures in the downdrafts near cloud edges, where mixing with the environment occurs, or in descending branches of the waves (e.g., Cooper and Vali, 1981; Hobbs and Rangno, 1985; Rangno and Hobbs, 1991; Cotton and Field, 2002).

The mechanism of contact nucleation is still poorly understood and several hypotheses have been suggested to explain the higher IN efficiency in this mode than in the others. Fletcher (1968, 1970b) considered the following two mechanisms. The first mechanism may act when a dry aerosol particle contains on its surface active sites (small areas of morphological, chemical, or electrical inhomogeneities stimulating ice germ formation). When such a particle is brought into contact with a supercooled drop and the etching of active sites is slower than nucleation of freezing, then this aerosol particle can be an effective contact nucleus. The second mechanism occurs when a small aerosol particle containing both hygroscopic and insoluble parts deliquesces and the solution is concentrated so that its freezing point can be depressed by 5–10 °C. The deliquesced particle does not freeze via condensation-freezing mode, but when it contacts a drop, the substantial dilution eliminates this depression and initiates freezing. Note that this mechanism includes the deliquescence-freezing heterogeneous mode considered in Chapter 9 as a part of contact nucleation.

Cooper (1974) based his hypothesized mechanism on an assumption that a dry aerosol particle in a saturated vapor possesses both deposition and immersion ice germs on its surface, with the critical germ radii r_{cr} proportional to the corresponding surface energies σ_{sv} at the ice–vapor interfaces and σ_{sl} at the ice–water interface as described in Sections 9.5 and 9.7. The value $\sigma_{sv} \approx 4$–$5\ \sigma_{sl}$, and r_{cr} is much smaller in the immersion mode. Thus, even if an aerosol particle is inactive in the deposition mode, it can be active in the contact mode when the immersion germ upon contact with a cloud drop meets "the native" liquid medium, growing above supercritical size and then freezing. Fukuta (1975) suggested another mechanism whereby, after contact of an aerosol particle with a drop, the water front propagates along the nucleus surface and creates local high interface energy zones that reduce the free energy and stimulate ice nucleation.

Djikaev and Ruckenstein (2008) compared efficiencies of the contact and immersion modes in the framework of a thermodynamic model. To determine if and how the surface of a liquid droplet can thermodynamically stimulate its heterogeneous crystallization, they examined crystal nucleation in the immersion and contact modes by deriving and comparing with each other the reversible works of formation of crystal nuclei in these cases. They found that the line tension of a three-phase contact gives rise to additional terms in the formation free energy of a crystal cluster and affects its Wulff (equilibrium) shape. The proposed model was applied to the heterogeneous nucleation of hexagonal ice crystals on generic macroscopic foreign particles in water droplets at $T = 253$ K. The results showed that the droplet surface does thermodynamically favor the contact mode over the immersion one.

In general, contact ice nucleation represents a two-stage process. The first stage consists of the formation of some number of contact ice nuclei from interstitial aerosol particles that could contact drops and potentially initiate freezing. The second stage consists of the scavenging of these potential IN by drops and freezing. The hypotheses by Fletcher (1968), Cooper (1974), and Fukuta (1975) dealt with ice embryos nucleation on a single aerosol particle and did not consider in detail the scavenging and accumulation of aerosol particles by the drops. Brock (1962), Waldman and Schmidt (1966), Slinn and Hales (1971), and Young (1974b) evaluated in detail aerosol fluxes, scavenging and collection rates with an emphasis on contact ice nucleation. The processes of aerosol scavenging by drops are considered in Section 10.5, and the combined processes of scavenging and freezing are considered in Section 10.6.

10.4.2. Empirical Parameterizations

Several empirical parameterizations of contact nuclei concentrations were developed and used in cloud and climate models. Young (1974a) suggested a parameterization of the concentration of active contact nuclei in terms of temperature T and surface concentration N_{cn0} as a power law:

$$N_{cn} = N_{cn0}(270.15 - T)^{1.3}, \tag{10.4.1}$$

where $N_{cn0} = 2 \times 10^2\,\mathrm{L}^{-1}$ is the number of active ice nuclei at sea level, decreasing to $10\,\mathrm{L}^{-1}$ at a height of 5000 m. This scheme was used by Young (1974a) in a cloud model for simulation of orographic clouds. Cotton et al. (1986) simulated in a cloud model with bulk microphysics both stages of contact nucleation—i.e., combining $N_{cn}(T)$ from (Eqn. 10.4.1) with evaluation of the total rate of scavenging of contact nuclei colliding with drops due to Brownian and phoretic aerosol fluxes, using a diagnosed supersaturation for phoretic fluxes, and then calculating the contact freezing rate. Meyers et al. (1992) concluded that Young's T-dependence (Eqn. 10.4.1) may overestimate the concentration of contact nuclei and suggested an exponential parameterization:

$$N_{cn} = N_{cn0} \exp[a_{cn} + b_{cn}(273.15 - T_c)], \tag{10.4.2}$$

with N_{cn} in L^{-1}, T_c in degrees Celsius, $N_{cn0} = 1\,\mathrm{L}^{-1}$, $a_{cn} = -2.80$, $b_{cn} = 0.262$, and where ice is not permitted to form at $T_c > -2\,°\mathrm{C}$. Equation (10.4.2) yields lower N_{cn} than (Eqn. 10.4.1), ~$0.2\,\mathrm{L}^{-1}$ at $-3\,°\mathrm{C}$ and ~$1\,\mathrm{L}^{-1}$ at $-10\,°\mathrm{C}$. These parameterizations may explain higher threshold temperatures for contact nucleation than for the condensation-freezing mode but remain insufficient to explain contact nucleation by some organic aerosol particles at $-3\,°\mathrm{C}$, with higher N_c observed by Fletcher (1972). The parameterization (Eqn. 10.4.2) coupled with the evaluation of aerosol collection rates was used in Meyers et al. (1992) to modify the Cotton et al. (1986) parameterization and simulate the orographic cloud system.

Similar approaches with use of submicron contact nuclei of various origin and sizes were applied to simulate the effect of contact nucleation on the convective clouds (e.g., Ovtchinnikov and Kogan, 2000; Ovtchinnikov et al., 2000), Arctic mixed-phase stratocumulus clouds (Morrison et al., 2005a,b;

Morrison and Pinto, 2005), and global cloudiness and indirect aerosol effects with the ECHAM4 climate model (Lohmann, 2002; Lohmann and Diehl, 2006). Sensitivity tests indicated that the assumed number and size of contact nuclei can have a large impact on the evolution and characteristics of mixed-phase clouds, particularly the partitioning of condensate between droplets and ice. However, there is currently no unified point of view on the properties and concentrations of the contact nuclei. Although aerosol fluxes to the drops substantially vary with the aerosol radius (Slinn and Hales, 1971; Young, 1974b, 1993; Pruppacher and Klett, 1997; see Section 10.5 here), models often assumed a monodisperse aerosol.

The effect of the contact nuclei size on the efficiency of freezing is also not clear. Some earlier experiments indicated an increase of the freezing probability with an increasing IN size (e.g., Young, 1993; Pruppacher and Klett, 1997). However, Hoose and Möhler (2012) analyzed the available recent data and found contradictory results. Rzesanke et al. (2011) found a clear increase of the contact freezing efficiency with particle size, although this effect was not observed or was much weaker in the studies by Ladino et al. (2011) and Bunker et al. (2012). Aerosol fluxes to the drops depend on the size of the aerosol particle; therefore, contact freezing should depend on the size of aerosol particles with maximum fluxes to the drops. The major mechanisms of aerosol scavenging that govern the fluxes of contact nuclei to or from the droplets, and their possible effects on the contact nucleation process are considered in the next sections.

10.5. Aerosol Scavenging by Drops

The collection scavenging rates of aerosol particles by a drop are analogous to the coagulation rates considered in Section 5.6. The scavenging rate is defined as the rate of accumulation of aerosol volume by a drop per unit time [$cm^3\ s^{-1}$]. As discussed in Section 10.4, the following three mechanisms govern the collection of aerosol particles by the drops: 1) Brownian diffusion; 2) thermophoresis; 3) diffusiophoresis. A brief description of these mechanisms is provided next.

10.5.1. Brownian Diffusion

It is a result of the random collisions of air molecules with AP that causes AP to move in the direction of the smallest concentration. The term *convective Brownian diffusion* is applied to the Brownian diffusion enhanced by the convection of the particles. In the case of the diffusion of AP toward the drops, this convective enhancement is caused by the drops' fall velocity and is described by the ventilation corrections. The collection or scavenging rates K_{CB} [$cm^3\ s^{-1}$] for convective Brownian diffusion can be written as:

$$K_{CB}(r_d, r_a) = 4\pi D_p(r_a) r_d \varphi_{v,br}(r_d, r_a). \tag{10.5.1}$$

Here, $D_p(r_a)$ is the Brownian diffusion coefficient of an aerosol particle:

$$D_p(r_a) = \frac{kT_\infty}{6\pi\eta_a r_a} C_{sl,r}(r_a), \tag{10.5.2}$$

where k is the Boltzmann constant, T_∞ is the temperature far from the drop (at "infinity"), η_a is the air dynamic viscosity, the first fraction in (Eqn. 10.5.2) is Einstein's coefficient of Brownian diffusion, and $C_{sl,r}(r_a)$ is Cunningham's slip-flow correction:

$$C_{sl,r}(r_a) = 1 + \alpha_{sl}(r_a) N_{Kn}(r_a). \tag{10.5.3}$$

Here, $N_{Kn}(r_a) = \lambda_a/r_a$ is the Knudsen number, λ_a is molecular free path, and

$$\alpha_{sl}(r_a) = A_{sl} + B_{sl} \exp[-C_{sl}/N_{Kn}(r_a)]. \tag{10.5.4}$$

The values of the parameters in (Eqn. 10.5.4) chosen here are $A_{sl} = 1.257$, $B_{sl} = 0.40$, $C_{sl} = 1.10$. These values are somewhat different from Waldman and Schmidt (1966) and Young (1974b) and are based on newer data (e.g., Seinfeld and Pandis, 1998, Pruppacher and Klett, 1997). The term λ_a is a function of temperature T and pressure p:

$$\lambda_a(T, p) = \lambda_{a0}(p_0/p)(T/T_0), \tag{10.5.5}$$

where $\lambda_{a0} = 0.66 \times 10^{-5}$ cm, $p_0 = 1013$ mb, and $T_0 = 273.15$ K. To determine the Brownian diffusion coefficient in (Eqn. 10.5.2), the term η_a is parameterized as a function of T following Pruppacher and Klett (1997):

$$\eta_a(T) = (1.718 + 0.0049T_c - 1.2T_c^2) \times 10^{-4}, \qquad T_c < 0^\circ\text{C}, \tag{10.5.6}$$

where T_c is temperature in Celsius, and η_a is in poise (1 poise = g cm^{-1} s^{-1}). The ventilation correction $\varphi_{v,br}$ in (Eqn. 10.5.1) arises from droplet sedimentation, and is parameterized following Slinn and Hales (1971), Young (1974b) as

$$\varphi_{v,p}(r_d, r_a) = 1 + 0.3 N_{\text{Re}}^{1/2}(r_d) N_{Sc,p}^{1/3}(r_a). \tag{10.5.7}$$

Here, $N_{Re}(r_d) = 2V_t(r_d) r_d \rho_d/\eta_a$ is the Reynolds number for the drops, ρ_a is the air density, $V_t(r_d) = C_{v,sl}(r_d) V_{t0}(r_d)$ is the terminal velocity of the drops, $V_{t0}(r_d)$ is the drop terminal velocity without slip correction parameterized analytically in Beard (1976, 1980), Böhm (1989, 1992), or Khvorostyanov and Curry (2002, 2005b) (see Chapter 12), and $C_{v,sl}(r_d)$ is the slip correction for V_t defined in (Eqn. 10.5.4) but with $N_{Kn}(r_d)$ for the drops. The $N_{Sc,p}(r_a) = \nu_a/D_p$ is the Schmidt number for aerosol particles, $\nu_a = \eta_a/\rho_a$ is the air kinematic viscosity.

10.5.2. Thermophoresis

It is the transport of aerosol particles in a temperature gradient in the suspending gas. A simplified explanation of this effect is that the aerosol particles in a temperature gradient are non-uniformly heated, and air molecules striking an aerosol particle from the warm side transfer a greater impulse to the aerosol particle than from the cold side, resulting in a flux of aerosol particles from warm to cold. The collection rates for this process are evaluated according to Brock (1962), Waldman and Schmidt (1966), and Slinn and Hales (1971) as:

$$K_{Th} = \frac{4\pi B_{th}(r_a) k_a r_d}{p}(T_\infty - T_s)\varphi_{v,th}(r_d), \tag{10.5.8}$$

where T_s and T_∞ are the temperatures on the drop surface, and far from the drop. The factor B_{th} is

$$B_{th}(r_a) = \frac{0.4(1+\alpha_{sl,r}N_{Kn})(k_a + 2.5k_p N_{Kn})}{(1+3N_{Kn})(k_p + 2k_a + 5k_p N_{Kn})}, \tag{10.5.9}$$

where k_a and k_p are thermal conductivities of air and aerosol particles, and the slip function $\alpha_{sl,r}(r_a)$ is defined by (Eqn. 10.5.4). The ventilation correction for this process is

$$\varphi_{v,th}(r_d) = 1 + 0.3N_{\mathrm{Re}}^{1/2}(r_d)N_{\mathrm{Pr}}^{1/3}, \tag{10.5.10}$$

where $N_{Pr} = \eta_a c_p / k_a$ is the Prandtl number, c_p is the isobaric specific heat capacity of the air. Slinn and Hales (1971) and Young (1974b, 1993) expressed the temperature difference $\Delta T = T_\infty - T_s$ via the corresponding difference of the vapor densities. It is more convenient, using the heat balance equation (Eqn. 5.1.28) to express ΔT directly via fractional water supersaturation $s_w = (\rho_{v\infty} - \rho_{ws})/\rho_{ws}$, where $\rho_{v\infty}$ and ρ_{ws} are the water vapor densities at infinity and on the drop surface (see Chapter 5):

$$T_\infty - T_s = -\frac{D_v L_e}{k_a}(\rho_{v\infty} - \rho_{ws}) = -\frac{D_v L_e \rho_{ws}}{k_a} s_w. \tag{10.5.11a}$$

Here D_v is the vapor diffusion coefficient and L_e is the latent heat of condensation.

The thermophoretic fluxes are often considered as in (Eqn. 10.5.8), which do not account for the kinetic correction. As we have seen in Section 5.1, this form corresponds to the diffusion regime of condensation. Since the thermophoretic flux is generated by and related to the condensation flux, which may occur in both the diffusion and kinetic regimes, a generalized parameterization is desirable for the thermophoretic flux. A generalized parameterization for both the diffusion and kinetic regimes is developed here based on the method of the boundary sphere as in Section 5.1. This leads to an effective diffusion coefficient D_v^* defined by (Eqn. 5.1.22a,b) and gives a kinetic correction φ_{kin} to the thermophoretic flux:

$$\varphi_{kin} = \frac{1}{1+\xi_{con}/r_d}, \qquad \xi_{con} = \frac{4D_v}{\alpha_c V_w}, \tag{10.5.11b}$$

where α_c is the condensation coefficient, and V_w is the molecular thermal velocity. Substituting (Eqns. 10.5.9)–(10.5.11b) into (Eqn. 10.5.8) we obtain the thermophoretic collection kernel:

$$K_{Th} = \frac{4\pi B_{th} r_d D_v L_e}{p}(-s_w)\rho_{ws}\varphi_{kin}\varphi_{v,th}(r_d). \tag{10.5.12}$$

This form of the thermophoretic collection kernel is convenient for cloud models where supersaturation is evaluated solving the supersaturation equations described in the previous chapters or where it is diagnosed as in Cotton et al. (1986). Equation (10.5.12) shows that $K_{Th} \sim (-s_w)$—i.e., the direction of the thermophoretic flux is opposite to the condensation flux. In the case of a growing drop, the thermophoretic aerosol flux is directed from the drop and prevents collection of aerosol particles by the drop, and in the case of an evaporating drop, the flux is directed to the drop, and increases aerosol collection rate.

10.5.3. Diffusiophoresis

It refers to the process of transport of aerosol particles by the hydrodynamic airflow, termed *Stefan flow*. For a growing drop, $s_w > 0$, the vapor pressure equals the saturated pressure e_{ws} at the drop surface and $e_{ws}(1 + s_w) > e_{ws}$ far from the drop. Thus, the water vapor pressure decreases toward the drop. Since the total pressure of the air–vapor mixture is roughly constant, the air pressure increases toward the drop and causes the diffusive counterflow of air molecules from the drop. Since there is no source of air molecules on the drop surface, this sets up a Stefan flow toward the drop when the drop is growing. For an evaporating drop, the directions of these fluxes are reversed. That is, the diffusiophoretic flux of aerosol particles is in the same direction as the vapor flux, toward a growing drop and away from an evaporating drop.

The collection rate K_{Df} for this process was derived in Waldman and Schmidt (1966), Slinn and Hales (1971), and Young (1974b) in terms of the vapor densities or pressures. Here we express K_{Df} in a more convenient form via water supersaturation s_w and also introduce the kinetic correction φ_{kin} that accounts for both the diffusion and kinetic regimes and intermediate cases:

$$\begin{aligned} K_{Df} &= 1.2\times 4\pi D_v r_d \frac{\rho_{v\infty} - \rho_{ws}}{\rho_a} \varphi_{kin}\varphi_{v,df}(r_d) \\ &= 1.2\times 4\pi D_v r_d s_w q_s \varphi_{kin}\varphi_{v,df}(r_d), \end{aligned} \tag{10.5.13}$$

where q_s is the saturated specific humidity. The ventilation correction for this case is

$$\varphi_{v,df}(r_d) = 1 + 0.3 N_{\text{Re}}^{1/2}(r_d) N_{Sc}^{1/3}, \tag{10.5.14}$$

where $N_{Sc} = \nu_a/D_v$ is the Schmidt number. The signs of K_{Df} and s_w are the same, showing that the directions of condensation and diffusiophoretic_flux coincide, while the directions of the thermophoretic and diffusiophoretic fluxes are opposite. The generalization here of the thermophoretic and diffusiophoretic fluxes accounting for the kinetic correction allows us to study the effects of variations of the condensation coefficient on contact nucleation and aerosol scavenging.

The total collection (or scavenging) rate K_{sc} can be evaluated as the sum of the preceding 3 processes:

$$K_{sc} = K_{Br} + K_{Th} + K_{Df}. \tag{10.5.15}$$

Fig. 10.2 shows the three relative collection rates—Brownian, diffusiophoretic (D), and thermophoretic (Th)—and the net rate (their sum) for a drop with a radius $r_d = 10\,\mu\text{m}$ at evaporation with $s_w = -2\%$ and condensation with $s_w = 0.1\%$. Calculations are performed at $T = -5\,°\text{C}$, $p = 600$ mb, and two values of the condensation coefficient: $\alpha_c = 0.04$ and 1. These Figures give the absolute values of all the collection rates. Diffusiophoresis D is negative at $s_w = -2\%$, and thermophoresis Th is negative at $s_w = 0.1\%$, their absolute values are denoted as |D| and |Th|, respectively. The phoretic fluxes with $\alpha_c = 1$ are 25–40% greater than with $\alpha_c = 0.04$. Thermophoresis here promotes aerosol collection at evaporation, while diffusiophoresis opposes it. The net collection rate is positive at evaporation in the considered range of aerosol radii $0.01 \le r_a \le 1\,\mu\text{m}$, but decreases more than 10 times with increasing aerosol size. At $s_w = 0.1\%$, which corresponds to the cloud volumes with condensation, the net collection rate decreases much faster with increasing aerosol size and becomes negative at $r_a \ge 0.4\,\mu\text{m}$

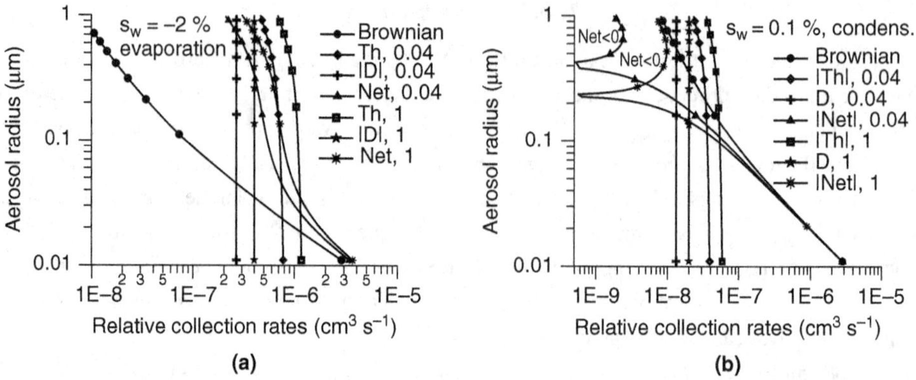

Figure 10.2. Relative collection rates. (a) Evaporation with $s_w = -2\%$; (b) condensation with $s_w = 0.1\%$. Calculations are performed at $T = -5\,°C$, $p = 600$ mb, and two values of the condensation coefficient, $\alpha_c = 0.04$ and 1. The components of aerosol fluxes are denoted as: Brownian, diffusiophoresis (D), and thermophoresis (Th). Net means the sum of these three fluxes. The digits near notations are the values of α_c (1 or 0.04). The absolute values of all the collection rates are given in these Figures. Diffusiophoresis D is negative at $s_w = -2\%$, and thermophoresis Th is negative at $s_w = 0.1\%$; their absolute values are denoted as |D| on the left panel and |Th| on the right panel. The net collection rate at condensation becomes negative at $r_a \geq 0.4\,\mu m$ with $\alpha_c = 0.04$ and at $r_a \geq 0.23\,\mu m$ with $\alpha_c = 1$ as indicated in the upper left corner of (b).

with $\alpha_c = 0.04$ and at $r_a \geq 0.23\,\mu m$ with $\alpha_c = 1$ as indicated in the upper left corner of Fig. 10.2b. Thus, collection of the larger particles is prohibited, and contact nucleation may be suppressed.

These features of scavenging led Young (1974b) and others to the conclusion that contact nucleation is favored in the evaporation regions of clouds. Figs. 10.3a,b represent the collection rates as isopleths for the range of aerosol radii r_a = 0–1 μm and drop radii r_d = 0–100 μm. Figs. 10.3a,b in general support the conclusion that the net collection rates are greater at evaporation. The net rates are negative at condensation at $r_d \leq 10\,\mu m$ for aerosol particles with $r_a \geq 0.2$–$0.4\,\mu m$, so that such particles cannot be collected by the small-size drop fraction. However, these figures show that the picture of aerosol scavenging may be more complicated and condensation regions may also contribute to contact nucleation for larger drops with $r_d \geq 10$–$20\,\mu m$. Fig. 10.3b shows that in the condensation regions, the net collection rates are positive except for a small range at $r_d \leq 10\,\mu m$, and increase with increasing drop radius. Collection rates are smaller at condensation, but since the probability of freezing is proportional to the squared radius of the aerosol particle, this could compensate for smaller collection rates at condensation and lead to comparable effects of aerosol in contact freezing in both the evaporating and condensing cloud portions.

Recent field studies during CRYSTAL-FACE-2002 of the size spectra even of relatively large dust aerosol transported from African dust storms to the Florida area showed that the submicron size fraction may be mostly responsible for cloud glaciation (e.g., Sassen et al., 2003; DeMott et al., 2003; Cziczo et al., 2004). Such dust particles could be captured by the clouds and initiate contact freezing, which may explain glaciation of rather warm Ac clouds at −5 to −9 °C (Sassen et al., 2003). Similar superposition of the submicron fractions of dust particles was used by Zhang et al. (2011) in simulations of the effects of dust as ice nuclei on tropical hurricane development. Although

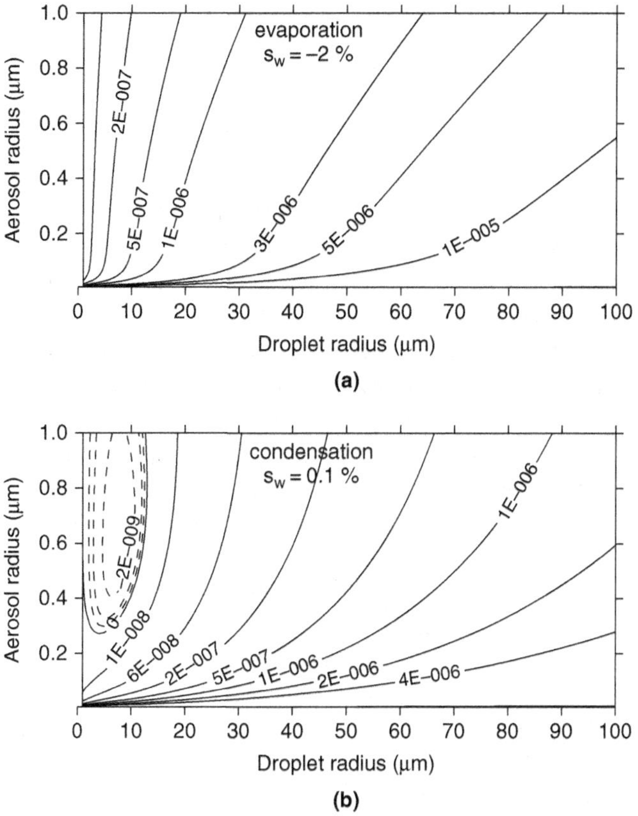

Figure 10.3. Net collection rates ($cm^3\ s^{-1}$) in the form of isopleths as the functions of the drop and aerosol radii r_d and r_a calculated with the same parameters as in Fig. 10.2 ($T = -5\,°C$, $p = 600$ mb) for evaporation with $s_w = -2\%$ (a) and condensation with $s_w = 0.1\%$ (b). The solid and dashed lines denote positive and negative values.

Figs. 10.2 and 10.3 are based on somewhat idealized choices of the aerosol and droplet size spectra and supersaturations, they illustrate the possibility of aerosol scavenging by droplets that may cause contact nucleation.

10.6. Freezing and Scavenging Rates

The number N_a of aerosol particles collected (scavenged) by a drop of a radius r_d can be calculated with the equation:

$$\frac{\partial N_a}{\partial t} = K_{sc}(r_d, r_a) n_{a\infty}(r_a), \qquad K_{sc} = K_{Br} + K_{Th} + K_{Df}, \qquad (10.6.1)$$

where $K_{sc}(r_d, r_a)$ is defined in Section 10.5 as the total collection (scavenging) kernel for aerosol particles with radius r_a by a drop of radius r_d, and $n_{a\infty}(r_a)$ is the concentration of aerosol particles of radius r_a far from the drop. When we consider a polydisperse ensemble of aerosol particles with the size spectrum $f_a(r_a)$, this equation should be generalized. If we denote $\delta N_a(r_d,r_a)$ as the number (dimensionless) of aerosol particles in the size range (r_a, $r_a + dr_a$) captured in unit time by a drop with r_d, then the collection rate can be written as

$$\frac{\partial \delta N_a(r_d,r_a)}{\partial t} = K_{sc}(r_d,r_a) f_a(r_a) dr_a, \tag{10.6.2}$$

and the number of particles captured by the drops in the radii interval (r_d, $r_d + dr_d$) in time Δt is $\delta N_a = (\partial \delta N_a/\partial t)\delta t$.

The aerosol particles that have not been frozen before being captured by a drop can be of various origins with different characteristics. For example, these can be haze solution particles with high concentrations of solute that depress the freezing point and suppress their freezing, or the insoluble particles without liquid film on their surface that could freeze. Another reason that could prevent their freezing before contact can be subsaturation $s_w < 0$ due to the factor $\exp[u_{shet} s_w(t)]$ as described by (Eqn. 10.1.1) with u_{shet} from (Eqn. 9.8.11). These can be interstitial aerosol particles or particles becoming interstitial due to the mixing of cloud volumes with the environment. The effect of different particles on contact freezing may be different depending on their nature, but the action of each type of such IN on contact freezing can be evaluated based on any hypothesis on their properties, and using the expressions for probabilities of freezing described in Chapter 9, Section 10.1, and evaluating collection rates as described in Section 10.5.

We illustrate these mechanisms with calculations of scavenging and freezing rates based on the hypothesis that a haze solution particle has not been frozen before the contacting drop due to solute effects or the freezing depression by subsaturation. When such a haze particle contacts a drop—e.g., in the slightly subsaturated portion of the cloud—rapid dilution of the solute occurs. Humidity around such contact IN becomes close to 100% on the drops' surfaces, favoring the initiation of freezing.

The probability of freezing of an aerosol particle with radius r_a in a time interval Δt is

$$P(r_a,\Delta t) = 1 - \exp[-J_{f,het}(r_a)\Delta t], \tag{10.6.3}$$

where $J_{f,het}(r_a)$ is the rate of ice germ formation per unit time per particle. Freezing depends on the temperature, humidity, and radius of the aerosol particle as specified in Section 9.6, and it can be assumed that $s_w \approx 0$ on the drop's surface. The probability of freezing in time Δt of a drop with r_d that captures $\delta N_a(r_d,r_a)$ of aerosol particles with similar properties is proportional to the number of captured particles and the probability of their freezing:

$$\hat{P}(r_d,r_a,\Delta t) = P(r_a,\Delta t)\delta N_a(r_d,r_a) = P(r_a,\Delta t) K_{sc}(r_d,r_a) f_a(r_a) dr_a \Delta t, \tag{10.6.4}$$

and the drop freezing rate is $\hat{P}(r_d,r_a,\Delta t)/\Delta t$. The corresponding freezing rate $\Psi_{cf}(r_d,\Delta t)$ (s^{-1}) of the drops that capture aerosol particles is the integral over the aerosol size spectrum:

$$\begin{aligned}\Psi_{cf}(r_d,\Delta t) &= \int_{r_a} P(r_a,\Delta t) K_{sc}(r_d,r_a) f_a(r_a) dr_a \\ &= \int_{r_a} P(r_a,\Delta t)[K_{Br}(r_d,r_a) + K_{Th}(r_d,r_a) + K_{Df}(r_d,r_a)] f_a(r_a) dr_a.\end{aligned} \tag{10.6.5}$$

The freezing rate of the drops in the size range (r_d, $r_d + dr_d$) is

$$\frac{df_{d,fr}(r_d)}{dt} = \Psi_{cf}(r_d,\Delta t) f_{d,uf}(r_d), \tag{10.6.6}$$

where $f_{d,uf}(r_d)$ is the size spectrum of unfrozen droplets. Depletion of the size spectrum of unfrozen drops is

$$\frac{\partial f_{d,uf}(r_d)}{\partial t} = -\Psi_{cf}(r_d, \Delta t) f_{d,uf}(r_d). \tag{10.6.7}$$

Now we introduce the characteristic contact freezing time $\tau_{c,fr}$ that describes the rate of drops freezing due to contact nucleation:

$$\tau_{c,fr} = \Psi_{cf}^{-1}(r_d, \Delta t). \tag{10.6.8}$$

This characteristic time is analogous to the supersaturation absorption time and characterizes an e-fold decrease in $f_{d,uf}(r_d)$ if Ψ_{cf} were constant.

Fig. 10.4 shows freezing coefficients $\Psi_{cf}(r_d)$ and times $\tau_{c,fr}(r_d)$ calculated with (Eqns. 10.6.5) and (10.6.8) due to Brownian diffusion, thermophoresis, and diffusiophoresis at evaporation with

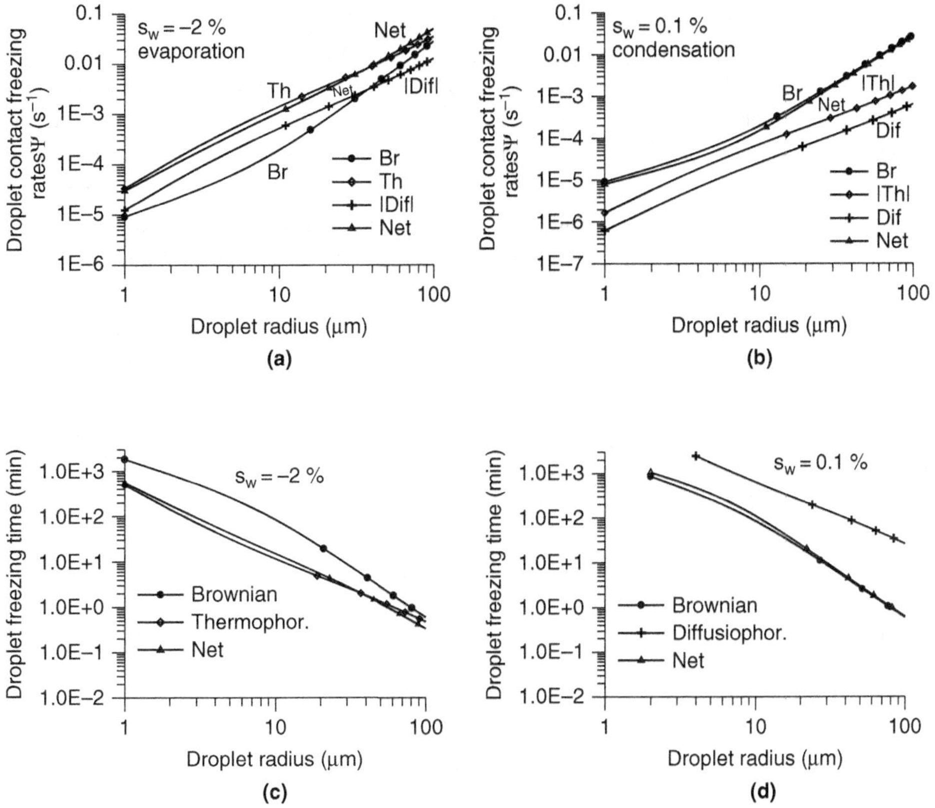

Figure 10.4. Freezing coefficients $\Psi_{cf}(r_d)$ and contact freezing times $\tau_{c,fr}(r_d)$ due to Brownian diffusion (Br), thermophoresis (Th), and diffusiophoresis (Dif) at evaporation with subsaturation −2% (left column, (a), (c) and growth with supersaturation 0.1% (right column, (b), (d). The integral in (Eqn. 10.6.5) of diffusiophoresis is negative at evaporation, its absolute value is denoted as |Dif| in Fig. 10.4a; the integral from thermophoresis is negative at condensation, and its absolute value is denoted as |Th| in Fig. 10.4b.

subsaturation −2% (left column) and condensation with supersaturation 0.1% (right column). Calculations are done using the same parameters as in Fig. 10.2: $T = -5\,°C$, $p = 600$ mb, and $\alpha_c = 0.04$. The aerosol size spectrum $f_a(r_a)$ includes 200 grid points by 0.05 μm from 0.01 to 1 μm and was chosen close to the continental type (Seinfeld and Pandis, 1998, Table 7.3) as a 3-modal lognormal spectrum with the mean geometric radii of 0.01, 0.058, and 0.9 μm, and dispersions 1.45, 1.65, and 2.45. The droplet spectrum included 100 points by 1 μm to 100 μm.

Fig. 10.4 shows that in the evaporation case with $s_w = -2\%$, the net freezing rates and times are determined mostly by the thermophoresis (positive here), with Brownian diffusion providing a minor contribution. Negative diffusiophoresis tends to decrease the freezing rate but its contribution is much smaller than that of thermophoresis. In the condensation case with $s_w = 0.1\%$, Brownian diffusion dominates the net rates, with diffusiophoresis and thermophoresis (negative here) playing minor roles. The freezing rates increase with increasing drop radius from $(1\text{–}3) \times 10^{-5}\,s^{-1}$ at $r_d = 1\,\mu m$ to $0.03\text{–}0.05\,s^{-1}$ at = 100 μm. The corresponding freezing times decrease with increasing radius, from ~600 s at evaporation and ~1000 s at condensation for $r_d = 1\,\mu m$ to 0.3 s ($s_w = -2\%$) and 0.6 s ($s_w = 0.1\%$) for $r_d = 100\,\mu m$. Thus, contact ice nucleation can be quite possible and rapid in the presence of moderately large drops, both in the subcloud layer at evaporation and in the growing cloud portion if sufficiently large particles have been formed by coagulation. This conclusion agrees with observations of Hobbs and Rangno (1990) and Rangno and Hobbs (1991), who observed rapid cloud glaciation after the stage of intensive coagulation in maritime convective clouds and concluded that contact nucleation could be one explanation for this effect. In interpreting Fig. 10.4, it should be emphasized that freezing may significantly vary with the concentrations and size spectra of the background aerosol. More general conclusions require a series of model simulations with varying aerosol properties and drop spectra.

The contact freezing rate initiated by the aerosol particle with radius r_a, or the rate of crystal production by this process, $[dN_c(r_a)/dt]_{cont}$, $[cm^{-3}\,s^{-1}]$, is the integral over the drop size spectrum

$$\left[\frac{dN_c(r_a)}{dt}\right]_{cont} = \int_0^\infty \Psi_{cf}(r_d, \Delta t) f_d(r_d)\, dr_d, \tag{10.6.9}$$

and the number of crystals formed due to contact nucleation by IN with r_a in a time step Δt is

$$\Delta N_{c,cont}(r_a, \Delta t) = \left(\frac{dN_c(r_a)}{dt}\right)_{cont} \Delta t = \int_0^\infty \Psi_{cf}(r_d, \Delta t) f_d(r_d)\, dr_d \Delta t. \tag{10.6.10}$$

The scavenging of aerosol particles by drops causes a decrease in aerosol concentration. The evolution of the aerosol size spectrum with an assumption of the complete capture of particles by the drops can be evaluated by integration of (Eqn. 10.6.2) over the drop size spectrum. This leads to a kinetic equation for $f_a(r_a)$:

$$\frac{\partial f_a(r_a)}{\partial t} = -\Lambda_{sc}(r_a) f_a(r_a), \tag{10.6.11}$$

where $\Lambda_{sc}(r_a)$ is the spectral aerosol scavenging coefficient,

$$\begin{aligned}\Lambda_{sc}(r_a) &= \int_0^\infty K_{sc}(r_d, r_a) f_d(r_d)\, dr_d \\ &= \int_0^\infty [K_{Br}(r_d, r_a) + K_{Th}(r_d, r_a) + K_{Df}(r_d, r_a)] f_d(r_d)\, dr_d.\end{aligned} \tag{10.6.12}$$

The inverse quantity

$$\tau_{sc}(r_a) = \Lambda_{sc}^{-1}(r_a), \tag{10.6.13}$$

is the characteristic scavenging time of aerosol with radius r_a.

Fig. 10.5 shows scavenging coefficients $\Lambda_{sc}(r_a)$ and times $\tau_{sc}(r_a)$ calculated using (Eqns. 10.6.12) and (10.6.13) due to Brownian diffusion, thermophoresis, and diffusiophoresis at evaporation with subsaturation −2% and condensation with supersaturation 0.1%. The aerosol size spectrum included 200 grid points by 0.005 μm from 0.001 to 1 μm. The droplet spectrum included 100 points by 1 μm to 100 μm and was approximated as a gamma distribution with the concentration 500 cm^{-3}, mean radius 10 μm, and an index of gamma distribution $p = 6$.

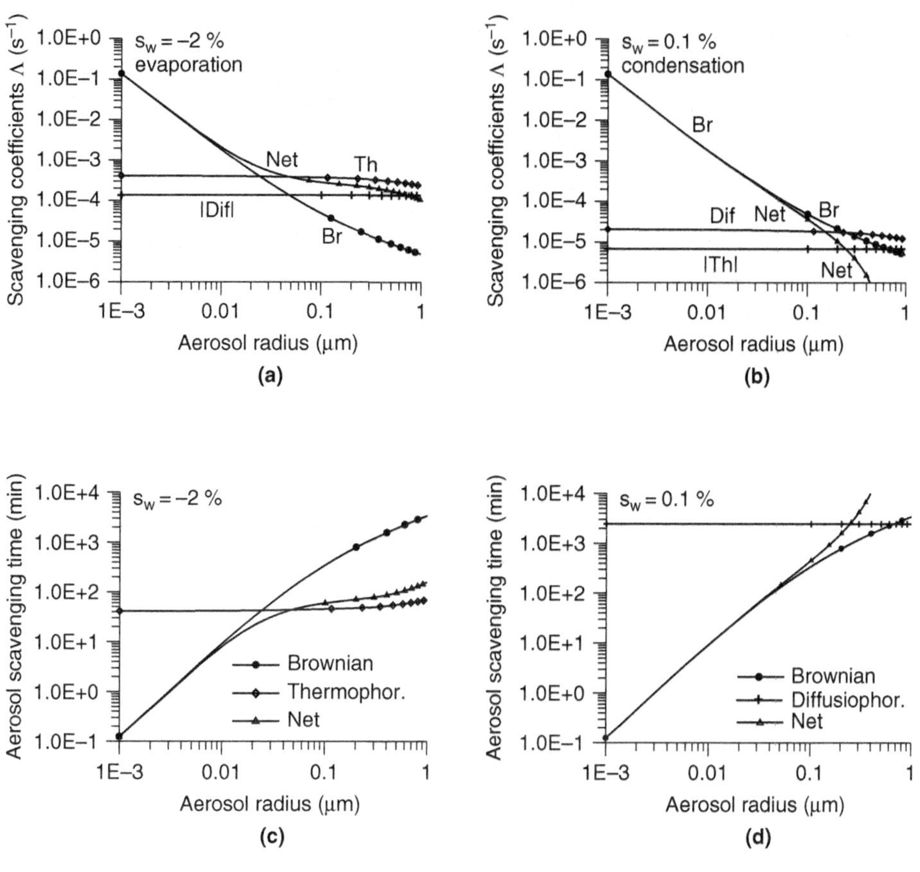

Figure 10.5. Scavenging coefficients $\Lambda_{sc}(r_a)$ and times $\tau_{sc}(r_a)$ due to Brownian diffusion (Br), thermophoresis, and diffusiophoresis at evaporation with subsaturation −2% (left column) and condensation with supersaturation 0.1% (right column). The integral in (Eqn. 10.6.12) of diffusiophoresis is negative at evaporation, and its absolute value is denoted as |Dif| in Fig. 10.5a. The integral from thermophoresis is negative at condensation, and its absolute value is denoted as |Th| in Fig. 10.5b.

Figs. 10.5a,c show that in the evaporation case, the net scavenging rates are determined mostly by Brownian diffusion for small aerosol particles with $r_a \leq 0.03$–$0.04\,\mu m$, and by thermophoresis at larger radii r_a. According to (Eqn. 10.5.2), the Brownian diffusion coefficient $D_p(r_a) \sim r_a^{-1}$, while thermophoresis in (Eqn. 10.5.12) very weakly depends on r_a. Therefore, $\Lambda_{sc}(r_a)$ strongly decreases from ~$0.1\,s^{-1}$ at $r_a = 10^{-3}\,\mu m$ to $4 \times 10^{-4}\,s^{-1}$ at $r_a = 0.04\,\mu m$, and is almost independent of radius at larger sizes. The net coefficient Λ_{sc} near $r_a \approx 1\,\mu m$ becomes slightly smaller than the thermophoretic integral due to the contribution of the negative diffusiophoresis. The scavenging time increases from 0.1 min at $r_a = 10^{-3}\,\mu m$ to 40 min at $r_a \approx 0.04\,\mu m$; weaker depends on the radius at larger sizes: It reaches ~60 min at $r_a = 0.1\,\mu m$ and 160 min at $r_a = 1\,\mu m$. Thus, scavenging by cloud of aerosol with radii larger than $0.1\,\mu m$ is a rather slow process in this example.

In the growth regime (Fig. 10.5b,d), the scavenging coefficient is determined mostly by Brownian diffusion in the range $r_a \approx 0.001$ to $0.25\,\mu m$, where it decreases from $0.1\,s^{-1}$ to less than $10^{-5}\,s^{-1}$. At $r_a > 0.3\,\mu m$, Brownian diffusion is small, the net coefficient $\Lambda_{sc}(r_a)$ is determined mostly by diffusiophoresis and thermophoresis, which have different signs, and their sum tends to zero at $r_a > 0.4\,\mu m$. Thus, $\Lambda_{sc}(r_a)$ becomes smaller $10^{-6}\,s^{-1}$, eventually becoming negative at $r_a > 0.56\,\mu m$, making scavenging absent. The corresponding scavenging time increases with increasing aerosol radius from ~0.1 min at $r_a = 0.001\,\mu m$ to 60 min at $r_a = 0.03\,\mu m$, and exceeds 7 hours at $r_a > 0.1\,\mu m$, making scavenging of such particles very slow in this example.

The total crystal contact nucleation rate $(dN_c/dt)_{cont}$ [$cm^{-3}\,s^{-1}$] includes integrals over both aerosol and drop size spectra and can be obtained by integration of (Eqn. 10.6.5) over the drop size spectrum or integration of (Eqn. 10.6.9) over the aerosol spectrum:

$$\left(\frac{dN_i}{dt}\right)_{cont} = \int_0^\infty P(r_a, \Delta t)\Lambda_{sc}(r_a) f_a(r_a)\, dr_a$$

$$= \int_0^\infty P(r_a, \Delta t) f_a(r_a)\, dr_a \int_0^\infty K_{sc}(r_d, r_a) f_d(r_d)\, dr_d. \tag{10.6.14}$$

The equations given allow calculations of the nucleation rates and concentrations of ice crystals formed at each time step in a model for the given collection rates described in Section 10.5 and freezing rates specified in Chapter 9 and Section 10.1.

11

Deliquescence and Efflorescence in Atmospheric Aerosols

11.1. Phenomena of Deliquescence and Efflorescence

In Chapters 6 and 7, we considered hygroscopic growth of aerosol particles. This growth can proceed only at sufficiently high relative humidities (RHW), exceeding some critical value. At very low humidities, atmospheric aerosol particles consisting of inorganic salts are solid. As RHW increases, the particles begin to absorb water on their surface and grow by adsorption, but this growth is usually insignificant.

Aerosol transformation at increase and decrease of the relative humidity is illustrated in Fig. 11.1. When RHW reaches some threshold value, specific for each substance, the salt particle begins to *deliquesce*, whereby liquid patterns or patches of solution form on the surface. As the humidity further increases, these patches merge and form a thin liquid film, which gradually thickens, and then the entire solid particle dissolves and transforms into a solution drop, and its radius abruptly grows. This is the point of deliquescence D in Fig. 11.1. If the solution is homogeneous and contains only one solute, this solid–liquid phase transition is similar to the surface melting considered in Section 9.9: It occurs as homogeneous or quasi-heterogeneous surface nucleation, usually at some critical relative humidity or within a very low range of humidities. This humidity is called *the deliquescence relative humidity* (DRH). For example, DRH is ≈75.3% for solutions of NaCl and ≈80% for ammonium sulfate at 298 K. If a less soluble salt or another insoluble foreign substance is contained in solution—e.g., mineral dust inclusions, the solid–liquid transition may start earlier and occurs as heterogeneous nucleation within some wider range of humidities as indicated by the curve with squares in Fig. 11.1 (Richardson and Snider, 1994; Tang, 1997; Oatis et al., 1998; Onasch et al., 1999, 2000). After deliquescence occurred, further hygroscopic growth is possible with increasing humidity as described in Chapters 6 and 7 and as indicated by the upward arrow in Fig. 11.1 at $S_w \geq 0.8$.

When relative humidity decreases, the reverse process of aerosol evaporation proceeds (downward arrow in Fig. 11.1). However, the solution drop does not transform back into a solid particle at the same DRH typical of deliquescence, but remains liquid down to substantially lower humidities. The solution is then supersaturated with respect to solute concentration or solute activity and exists in the metastable state. When decreasing humidity reaches another threshold value, called *efflorescence relative humidity* (ERH), a phase transition into solid state, or salt crystallization, occurs, which is called *efflorescence*. If a solution drop is homogeneous without admixture of any other substances or impurities, then *homogeneous efflorescence* occurs as the homogeneous nucleation of salt crystals at some critical ERH (point E in Fig. 11.1). For example, ERH was measured at ≈32–37% for bulk ammonium sulfate and ≈45% for NaCl at room temperatures for homogeneous efflorescence. The

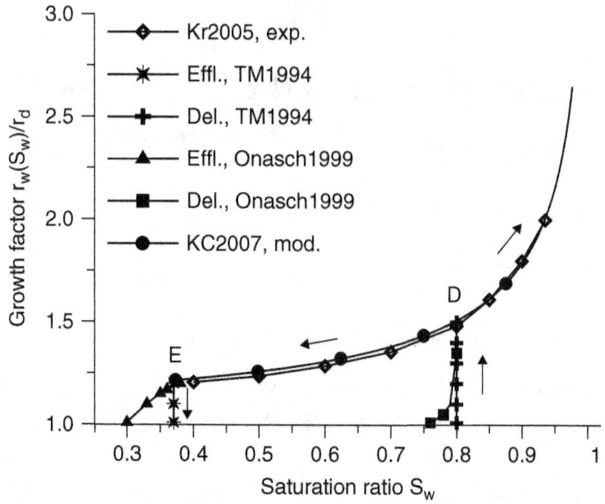

Figure 11.1. Illustration of aerosol deliquescence, hygroscopic growth, and efflorescence. The experimental data are from Kreidenweis et al. (2005, Kr2005) for hygroscopic growth, and adjustments from Tang and Munkelwitz (1994, TM1994) and from Onasch et al. (1999, Onasch1999) for deliquescence and efflorescence. Model calculations of hygroscopic growth are from Khvorostyanov and Curry (2007, KC2007) described in Chapter 6. The points D and E indicate the deliquescence and efflorescence points. The arrows indicate directions of the processes: deliquescence at point D, growth at $S_w \geq 0.8$, evaporation at $S_w \leq 0.8$, and efflorescence at point E.

solution can be characterized by the solute supersaturation $S_* = a_{s,ef}/a_{s,sat}$, defined as the ratio of the salt activity $a_{s,ef}$ at efflorescence to the saturated activity at deliquescence $a_{s,sat}$. At the point of homogeneous efflorescence, S_* may reach 20–30.

If there is a foreign insoluble or partially insoluble substrate inside this supersaturated solution, then *heterogeneous efflorescence* may occur as salt crystallization via the heterogeneous nucleation process, similar to heterogeneous ice nucleation (Section 9.7). For example, Richardson and Snider (1994) experimentally studied heterogeneous efflorescence using KCl as a catalyst in experiments on efflorescence of NaCl, CsCl, and KF that occurred in heterogeneous mode. Oatis et al. (1998) and Onasch et al. (2000) used $CaCO_3$ and $BaSO_4$ as heterogeneous catalysts of efflorescence of ammonium sulfate, $(NH_4)_2SO_4$, solutions. They found that ERH of heterogeneous efflorescence is 10–15% higher than at homogeneous efflorescence and the salt supersaturation S_* can be 40–70% lower than its homogeneous value. At the efflorescence point, ternary soluble salt mixtures may exhibit a step-wise salt crystallization at several critical values of ERH—e.g., the KCl-NaCl system (Tang and Munkelwitz, 1994; Oatis et al., 1998). This behavior is similar to the step-wise crystallization of seawater that contains several salts (Curry and Webster, 1999).

Thus, the process and critical humidity of the deliquescence-hygroscopic growth at increasing humidity are different from the process and critical humidity at evaporation-efflorescence at decreasing humidity. This phenomenon is called *humidity hysteresis*, by analogy with the phenomena of hysteresis in the magnetic and electrical fields. This hysteresis effect and the exact values of the DRH and ERH are important for cloud physics because the aerosol optical properties, the direct and indirect aerosol effects, and the processes of CCN activation or freezing are different in the liquid versus the solid states. In particular, if the environmental humidity exceeds these thresholds, cirrus clouds may form via homogeneous and heterogeneous freezing of solutions considered in the previous chapters, and when the salts are in the solid state at low humidities below ERH ice nucleation may proceed via deposition on the salt particles (Abbatt et al., 2006; Shilling et al., 2006).

The deliquescence and efflorescence relative humidities are temperature-dependent, and usually increase with decreasing temperature. Measurements usually indicate rather weak temperature dependence of deliquescence—e.g., for ammonium sulfate, DRH ≈ 80% at T = 25 °C, and increases slightly to 82–83% at −20 to −30 °C (Tang and Munkelwitz, 1993; Xu et al., 1998; Onasch et al., 1999). Tabazadeh and Toon (1998) suggested fitting DRH(T) with a polynomial of the inverse powers of the absolute temperature T:

$$DRH(T) = \exp\left(a_{t0} + \frac{a_{t1}}{T} + \frac{a_{t2}}{T^2} + \frac{a_{t3}}{T^3} \right). \qquad (11.1.1)$$

Onasch et al. (1999) determined the coefficients $a_{t0} = 4.9015$, $a_{t1} = -6.1612 \times 10^2$, $a_{t2} = -2.0656 \times 10^5$, and $a_{t3} = -2.0716 \times 10^7$. This fit illustrates the general tendency, but is valid only for ammonium sulfate for certain values of particle radius and environmental pressure.

Data obtained by various authors on the temperature dependencies of efflorescence critical humidity differ. Xu et al. (1998) found rather significant growth of ERH toward colder temperatures, while Onasch et al. (1999) and Czizco and Abbatt (1999) found much weaker dependence. It was found also that the ERH increases with decreasing size. This dependence becomes especially strong at particle radii smaller than 0.1–0.2 μm. Using experimental data on the deliquescence, efflorescence, and freezing of ammonium sulfate solutions, Xu et al. (1998) constructed a general phase diagram for the stable and metastable equilibriums and phase transitions for ammonium sulfate.

11.2. Theories and Models of Deliquescence and Efflorescence

Several theoretical models have been developed to explain the observed features of the deliquescence and efflorescence processes. Some of these models are described next.

Wexler and Seinfeld (1991) and Tang and Munkelwitz (1993) studied the temperature variations of deliquescence relative humidity DRH(T) using the Clausius–Clapeyron equations for solution and water vapor, combined with the empirical parameterizations of the salt solubilities. Xu et al. (1998) applied this model to explain measured properties of the saturated solutions and DRH(T). This model is described in more detail in the next section. Chen (1994) evaluated the positions of the deliquescence points on Köhler's curves in a simplified manner, assuming that the deliquescence point corresponds to a dry particle radius, prescribing a solubility limit for a soluble substance in a particle and using available data on solubility.

Classical homogeneous and heterogeneous nucleation theory has been applied to develop a quantitative description of deliquescence and efflorescence. Richardson and Snider (1994, hereafter RS1994 model) considered efflorescence as nucleation of a salt germ with the cap shape on the flat surface of a foreign substrate (catalyst) and derived the expressions for the critical radius r_{cr} and energy ΔF_{cr} of a salt germ:

$$r_{cr} = \frac{2\sigma_{cs} v_c}{kT \ln S_*}, \tag{11.2.1}$$

$$\Delta F_{cr} = \frac{16\pi\sigma_{cs}^3 v_c^2 f(m_{cs})}{3(kT \ln S_*)^2}, \tag{11.2.2}$$

where σ_{cs} is the surface tension at the interface of a salt crystal and solution, $S_* = a_{s*}/a_{s0}$ is the salt supersaturation, a_{s0} and a_{s*} are the solute activities at saturation and crystallization, k is the Boltzmann constant, v_c is the specific volume of the solid salt, $f(m_{cs})$ is the geometric factor, and m_{cs} is the cosine of the wetting angle between the salt and solid catalyst. Nucleation rates were evaluated with equations similar to those described in Chapters 8 and 9 as

$$J_{nuc} = A_s J_0 \exp(-\Delta F_{cr}/kT), \tag{11.2.3}$$

where A_s is the surface area of the solid inclusion in cases of heterogeneous nucleation, and $A_s = 1$ for homogeneous nucleation, and J_0 is the kinetic prefactor, which is different for homogeneous and heterogeneous nucleation. The kinetic prefactors were estimated as $J_0 \sim 10^{22}\,\mathrm{cm^{-2}\,s^{-1}}$ for heterogeneous nucleation and $J_0 \sim 7 \times 10^{31}\,\mathrm{cm^{-3}\,s^{-1}}$ for homogeneous nucleation (Newkirk and Turnbull, 1955; Oatis et al., 1998). Gao et al. (2006) estimated J_0 for NaCl as $2.8 \times 10^{32}\,\mathrm{cm^{-3}\,s^{-1}}$, applying the method of Richardson and Snyder (1994) and Onasch et al. (2000). The homogeneous nucleation in the absence of the foreign substrate was considered in a similar way by setting $m_{cs} = -1, f(m_{cs}) = 1$, and the nucleation rate was evaluated in a simplified manner as

$$J_{\mathrm{hom}} = J_{0,\mathrm{hom}} \exp(-\Delta F_{cr}/kT), \qquad J_{0,\mathrm{hom}} \approx (v_{jf}/v_c)\exp(-\Delta F_{act}/kT), \tag{11.2.4}$$

where $v_{jf} = (kT/h)$ is the jump frequency of the molecules' penetration through the potential barrier ΔF_{act}.

The critical supersaturation was calculated using measured water activities a_w of solutions at various concentrations and the Gibbs–Duhem relation derived in Section 3.1. As was shown in Section 3.6, for the two-component solutions consisting of water and solute, this relation can be written as

$$d \ln a_s = -(x_w/x_s) d \ln a_w. \tag{11.2.5}$$

Its integration for the process with varying solute concentration and corresponding changes in water activity from the deliquescence point with the water activity a_{wd} to the efflorescence point with a_{wef} yields

$$\ln S_{ef} = -\int_{a_{wd}}^{a_{wef}} \left(\frac{x_w}{x_s}\right) d\ln a_w \approx -\int_{a_{wd}}^{a_{wef}} \left(\frac{x_w}{x_s}\right) d\ln(RHW), \tag{11.2.6}$$

where $S_{ef} = (a_{sef}/a_{sd})$ is the supersaturation of the solution, a_{sef} and a_{sd} are the salt activities at the nucleation (efflorescence) point and at the deliquescence point with saturated salt concentration, x_w/x_s is the molar ratio of the water to solute, and the second equality here assumes that $a_w \approx$ RHW.

Application of this model allowed Richardson and Snider (1994) to explain the decrease in salt supersaturation and the increase in relative humidities in heterogeneous efflorescence in the presence of solid inclusions in the solution particles as compared to homogeneous nucleation. An important result of the RS1994 model was estimation of several fundamental properties of the salts that were examined (NaCl, CaCl, KF, and $NaNO_3$). The surface tensions σ_{cs} were found in the range 50–64 erg cm^{-2}, the activation energies were estimated as $\Delta F_{act} \sim 6\text{–}8\,kT$, the critical energies were found as $\Delta F_{cr} \sim 55\,kT$, the homogeneous nucleation rates $J_{hom} \sim 10^8\,\text{cm}^{-3}\,\text{s}^{-1}$, the kinetic coefficients were estimated as $J_0 \sim 7 \times 10^{31}\,\text{cm}^{-3}\,\text{s}^{-1}$ for homogeneous nucleation, and $J_0 \sim 10^{22}\,\text{cm}^{-2}\,\text{s}^{-1}$ for heterogeneous nucleation. These results and estimates served as a basis for subsequent works on salt crystallization.

Oatis et al. (1998) and Onasch et al. (2000) used this method for studies of homogeneous and heterogeneous efflorescence of ammonium sulfate, and found the ERH ≈ 37% and salt supersaturation $S_{ef} \approx 25\text{–}30$ for pure ammonium sulfate. In the presence of inclusions of $CaCO_3$ and $BaSO_4$, that served as catalyzers, the values of ERH increased to ≈45–48% and the critical supersaturations S_{ef} decreased to ≈14–17. Using classical nucleation theory for calculations of the nucleation rates along with Clegg et al.'s (1998) thermodynamic theory of solutions for calculations of salt supersaturation, Onasch et al. (2000) determined from the measurements of efflorescence the temperature dependence of the surface tension σ_{sc} at the interface between the solid ammonium crystal and solution: σ_{sc} decreased from 54–56 erg cm^{-2} at 298 K to 44 erg cm^{-2} at 210 K.

Russell and Ming (2002) used a simple thermodynamic equilibrium model to study deliquescence of the three soluble species (sodium chloride, ammonium sulfate, and a soluble organic compound) and showed that the role of the surface tension is to increase the deliquescence relative humidity, so that DRH for the very small particles (~15 nm) is 8–10% higher than that for the bulk substance. The two models of deliquescence were compared: deliquescence of a dry salt crystal, and a crystal wetted with several adsorbed liquid monolayers. The surface tension of 213 erg cm^{-2} was tested for the interface dry crystal–air, although there are large uncertainties in the measured data, with estimates ranging from 100 to 270 erg cm^{-2}. The values used for surface tensions at the solid–liquid and liquid–air interfaces were 29 ± 20 erg cm^{-2}, and 83 ± 2 erg cm^{-2}. It was emphasized in this work that the exact calculations for many species and mixtures are not possible since the precise values of the surface tensions have not been measured accurately; therefore, calculations were performed for the range of the surface tensions. However, these values tested by Russell and Ming (2002) served as a useful starting point in several subsequent models described next.

Gao et al. (2006, 2007), based on classical nucleation theory, elaborated a model of efflorescence for sodium chloride and ammonium sulfate at room temperature. This model considered efflorescence as homogeneous nucleation, similar to Richardson and Snider (1994) and Oatis et al. (1998), but with several modifications. For calculations of the critical germ radii and energies, the parameterizations of the solute density from Tang and Munkelwitz (1994) and of the solute–air surface tension from Korhonen et al. (1998) were used with a detailed accounting for the solution composition. The salt and water activities were calculated using the parameterizations from Tang and Munkelwitz (1994) and a more advanced parameterization from Ally et al. (2001). The last method can be more accurate for highly supersaturated solutions with molalities that can reach and exceed at efflorescence 35 mol kg^{-1}. The model developed by Ally et al. (2001) uses statistical mechanics of multilevel adsorption based on the Brunauer–Emmett–Teller (BET) adsorption isotherm and agrees well with experimental data even for highly supersaturated solutions with molalities greater than 35 mol kg^{-1}. The salt and water activities, a_s and a_w, are given in this model as

$$a_s = [(q_A\tilde{N}_s - X_{aw})/q_A\tilde{N}_s]^{q_A}, \tag{11.2.7}$$

$$a_w = (\tilde{N}_w - X_{aw})/\tilde{N}_w, \tag{11.2.8}$$

where $\tilde{N}_s$ and $\tilde{N}_w$ represent the salt and water moles in the solution, q_A is the number of adsorption sites per mole of the salt, and X_{aw} is the amount of adsorbed water and is determined from the quadratic equation

$$\frac{X_{aw}^2}{(q_A\tilde{N}_s - X_{aw})(\tilde{N}_w - X_{aw})} = \exp(-\varepsilon_A/kT) = c_A. \tag{11.2.9}$$

Here ε_A is the internal energy for a monolayer of water adsorbed onto the salt, and k is the Boltzmann constant. Ally et al. (2001) found the values for the aqueous solutions of these salts: for ammonium sulfate, $c_A = 2.073 \pm 0.293$, and $q_A = 2.47 \pm 0.476$; for NaCl solutions, $c_A = 3.813 \pm 0.2598$, and $q_A = 2.845 \pm 0.332$.

Combining Ally et al. model with classical theory, Gao et al. found the non-monotonic behavior of the calculated ERH with varying diameters D of the dry particles. When the size of ammonium sulfate particles decreases, the ERH first decreases from 39.1% at $D = 10\,\mu m$ to the minimum of 30% at $D \approx 30\,nm$, and then increases at smaller sizes, which is explained by the dominance of the Kelvin effect for the smallest particles. These results were in agreement with various measurements cited in Gao et al. (2006) and those earlier in this text. Similar behavior was found by Gao et al. (2007) for NaCl particles. For D larger than 70 nm, the ERH decreases with decreasing particle sizes from 48% at $D = 20\,\mu m$ to a minimum RHW = 44% at 60–70 nm. When the particles are smaller than 60–70 nm, the ERH for NaCl increases from this minimum to 57% for the particles with $D = 5\,nm$, in agreement with the experimental data by Hämeri et al. (2001) and Biskos et al. (2006a,b).

A more sophisticated model of the deliquescence with formation of the liquid films was developed by Shchekin, Shabayev, and Rusanov (2008a,b) and Shchekin and Shabayev (2010) (SSRH model) and further elaborated by Hellmuth et al. (2012, 2013). Instead of using the capillary approximation, their model was based on the concept of disjoining pressure. McGraw and Lewis (2009) promoted the theory of deliquescence considering the solution layer around the solid salt crystal and introducing a thin layer criterion (TLC) to define a limiting deliquescence relative humidity (DRH) for small particles. The definition of DRH requires: 1) equality of chemical potentials between salt in an undissolved core, and a thin adsorbed solution layer; and 2) equality of chemical potentials between water in the thin layer and vapor phase. The solution of this system of equations enabled determination of the DRH and ERH over a wide range of particle sizes (down to the nanoscale), as well as the construction of phase diagrams for this system.

This brief review gives a general outline of several theories of deliquescence and efflorescence that have been developed over the last two decades. Comparing these phenomena with homogeneous and heterogeneous freezing and melting considered in Chapters 8 and 9, we see the similarity of these phase transitions: the solid–liquid transition at deliquescence is similar to melting and salt crystallization at efflorescence is similar to freezing. This similarity motivates construction of the models of both deliquescence and efflorescence based on the general equilibrium (or entropy) (Eqn. 3.6.4) as was done in Chapters 8 and 9 for freezing and melting.

The approach developed here considers simultaneously the effects of temperature, particle size, the kinetic limitations due to various nucleation rates, and the effects of external pressure. The theory is similar to the theories of melting and freezing (Khvorostyanov and Curry, 2000, 2004a,b, 2009b) described in Chapters 8 and 9 and is called hereafter the KC theory of deliquescence and salt crystallization or the KC model for the convenience of comparison with experimental data and other models. Application of this model to the ice–solution and solid salt–solution transitions allows us to calculate various phase diagrams.

The KC theory is based on two simple hypotheses: 1) Deliquescence begins with formation of a liquid solution germ on the surface of a salt crystal (similar to surface melting). 2) Salt crystallization (efflorescence or precipitation) begins with formation of a salt germ in solution (similar to ice crystal nucleation in solution). In Section 11.3, the equations for the deliquescence are derived based on the entropy equation, and in Section 11.4 they are applied for calculations of the temperature dependencies of the dissolution heat, solubility, and deliquescence relative humidity for ammonium sulfate and NaCl. In Section 11.5, the equation for the eutectic point is derived and solved, and the phase diagram is constructed. In Section 11.6, the equations of efflorescence are derived based on the entropy equation. Finally, in Section 11.7, based on the calculations of the deliquescence and efflorescence in this chapter, and of the freezing–melting processes in Chapter 8, the general phase diagrams for an example of ammonium sulfate are constructed that include all four branches of stable and metastable equilibrium between the ice, solution, and the solid salt—that is, the deliquescence, efflorescence, and melting and freezing curves are calculated from the same theory.

11.3. A Model for Deliquescence of Salt Crystals Based on the Entropy Equation

We assume that deliquescence can be described as a nucleation process analogous to surface melting (Section 9.9) and begins with formation of a liquid solution germ on the surface of a crystal. This germ may have the shape of a spherical segment (lens) formed in the micro-cavity or defect of the salt crystal or may form as a thin liquid film around the crystal. Formation of the liquid lenses in the cavities or crystal defects may proceed earlier because the vapor pressure inside the cavities with negative radii of curvature is lower due to a negative Kelvin effect, which may cause vapor condensation and crystal deliquescence on such sites. Formation of the liquid lenses may proceed on the crystal surface even without crystal defects due to incomplete wettability of the crystal by the solution.

The liquid film stage may occur later, when many liquid lenses merge and form a continuous film as in McGraw and Lewis (2009), although this merging may proceed rapidly, depending on the rate of humidity increase. Some previous models of deliquescence with formation of the liquid films were briefly reviewed in the previous section. In the following text, we consider another model of the deliquescence process developed by the authors (the KC model), which is based on the classical nucleation theory and its extensions similar to our consideration of surface melting and freezing in Section 9.9. We suppose that a solution drop (liquid germ) has the shape of a spherical segment (lens) on the crystal surface or a liquid film around a solid crystal core bounded by the two surfaces. The outer surface between the solution and humid air has the radius of curvature r_d and surface tension σ_{sa}, and the inner surface (bottom) between the solution and crystal has a radius of curvature r_c (the radius of the solid crystal assumed to be spherical) and surface tension σ_{cs}.

Here we deal with the situation when liquid forms at the interface of its own solid and gas (vapor or humid air). As discussed in Section 9.9 for the water–ice interface, when a liquid lens nucleates, the contact angle θ_{cs} of such systems can be small but nonzero as in the case of water melting when wetting is incomplete—i.e., small liquid caps (spherical segments) occur on the surface of a salt crystal instead of a liquid film. Thus, at the first stage, deliquescence may occur as a quasi-heterogeneous process similar to surface melting. Hereafter, we use the term "drop" for a liquid solution germ, although this description is applicable also for the films.

The derivation is based on the general equilibrium (entropy) (Eqn. 3.6.4) of Section 3.6 for this system. For such a situation, in (Eqn. 3.6.4) phase 1 is salt in the crystal with molar enthalpy h_{c0} and pressure p_c, and phase 2 is salt in the solution drop with the molar enthalpy h_{s0} and pressure p_s. The molar enthalpy difference is $h_{10} - h_{20} = h_{c0} - h_{s0} = -\Delta h_s = -M_s L_c$. Here M_s is the salt molecular weight, and $-\Delta h_s$ and L_c are the molar and specific heats (enthalpies) of salt dissolution. The sign of L_c is defined earlier so that $L_c > 0$ if dissolution requires positive heat (as melting)—i.e., is an endothermic process as for most common salts, and $L_c < 0$ in the opposite case for salts with exothermic dissolution. For example, the molar dissolution heats Δh_s are positive, 1.88 kJ mol^{-1} for NaCl, 15.34 kJ mol^{-1} for KCl, and 6.32 kJ mol^{-1} for ammonium sulfate, but it is negative, –9.76 kJ mol^{-1}, for Na_2SO_4 (Seinfeld and Pandis, 1998). It is assumed that the activity of pure salt $a_c = 1$, thus, the last term in (Eqn. 3.6.4), $\ln(a_c/a_s) = -\ln(a_s)$. With these definitions, (Eqn. 3.6.4) can be written as

$$-\frac{(h_{c0} - h_{s0})}{T^2}dT + \frac{v_{c0}}{T}dp_c - \frac{v_{s0}}{T}dp_s - Rd\ln a_s = 0. \qquad (11.3.1)$$

Here v_{c0} and v_{s0} are the molar volumes of the solid crystal and solution, p_s and p_c are the internal pressures inside the solution drop and crystal, $a_s = \gamma_s x_s$ is the activity of the salt in solution, γ_s and x_s are the activity coefficient and molar fraction of the salt.

When the solution under consideration is an ionic solution, this equation should be somewhat modified. Defay et al. (1966) indicated that the term $\ln a_s$ should be replaced for an ionic solution as

$$\ln a_s = \ln(\gamma_s x_s) \rightarrow \nu \ln \tilde{a}_s, \qquad \tilde{a}_s = \gamma_{\pm}\alpha_s \hat{M}_s \qquad (11.3.2)$$

where $\nu = \nu_+ + \nu_-$ is the total effective number of ions, ν_+ and ν_- are the numbers of positive and negative ions, $\gamma_{\pm}$ is the mean ionic activity coefficient (the mean weighted activity coefficients of cations and anions), α_s is the degree of dissociation, and $\hat{M}_s$ is the stoichiometric molality of salt. The non-ideality of the solution can also be represented as $\ln(a_s) \rightarrow \nu\Phi_s \ln(\gamma_s x_s)$, where Φ_s is the osmotic potential.

If a salt crystal is approximated by a sphere (similar to an ice crystal in the cases of freezing considered earlier and to the previous deliquescence models reviewed in Section 11.1), the internal pressures inside a crystal, p_c, and inside a liquid solution drop, p_s, can be expressed in terms of external pressure p with use of Laplace's conditions of mechanical equilibrium:

$$dp_c = dp + d\left(\frac{2\sigma_{cs}}{r_c}\right), \qquad dp_s = dp + d\left(\frac{2\sigma_{sa}}{r_d}\right), \qquad (11.3.3)$$

where subscripts c and s refer to crystal and solution respectively, σ_{cs} and σ_{sa} are the surface tensions at the crystal–solution and solution–air interfaces, and we assume that $r_d << r_c$, and the curvature effect in dp_s due to r_c can be neglected in the second of (Eqns. 11.3.3). Equations (11.3.3) describe the mechanical equilibrium between a salt germ and liquid solution drop, and the equilibrium between the liquid drop and environmental air. Substituting (Eqn. 11.3.3), (Eqn. 11.3.2), and the enthalpy difference $h_{c0} - h_{s0} = -M_s L_c$ into (Eqn. 11.3.1), we obtain:

$$M_s L_c \frac{dT}{T^2} + \left(\frac{v_{c0}}{T} - \frac{v_{s0}}{T}\right) dp + \left(\frac{v_{c0}}{T}\right) d\left(\frac{2\sigma_{cs}}{r_c}\right) - \frac{v_{s0}}{T} d\left(\frac{2\sigma_{sa}}{r_d}\right) - \nu R d \ln \tilde{a}_s = 0. \qquad (11.3.4)$$

Now, we can integrate (Eqn. 11.3.4) in two slightly different ways, with and without multiplying first by T, which lead to the various forms of the final equation, each may be more convenient for comparison with the known equations. Before integration, we need to evaluate the number of thermodynamic degrees of freedom N_w (variance) which for a system including curved surfaces, according to Section 3.3 of Chapter 3, is related to the number of components c as $N_w = c + 1$. In our case here, the components are: 1) dry air; 2) water in liquid (solution) and gas (vapor around the drop) states; 3) salt in the dissolved and solid states. Hence, we have $c = 3$ components and $N_w = 3 + 1 = 4$ degrees of freedom. Equation (11.3.3) includes five variables, of which only four are independent and the fifth one is a function of these four. Thus, we have to fix any one variable in (Eqn. 11.3.3) when integrating, and this reduces (Eqn.11.3.4) to several particular cases: there are five different combinations that correspond to five different problems. If the salt is nonvolatile and its mass is constant, we have one more constraint, and the variance reduces by 1 to $N_w = 3$. Then, we have to fix two variables and have three independent variables. There are 10 combinations and 10 different situations. If the salt is volatile, its mass may change and we have $N_w = 4$. If there are chemical reactions, then N_w may be different.

Multiplying (Eqn. 11.3.4) by T and integrating by analogy with the case of aqueous solutions freezing–melting as in Section 9.7, from some reference point T_*, $1/r_d = 0$, $1/r_c = 0$, $p = p_*$, $\tilde{a}_{s*}$ to the values T, r_d, r_{cr}, p, $\tilde{a}_s$, we obtain:

$$-M_s L_c^{ef} \ln\frac{T_*}{T} + \Delta\bar{v}\Delta p + \bar{v}_{c0}\left(\frac{2\sigma_{cs}}{r_c}\right) - \bar{v}_{s0}\left(\frac{2\sigma_{sa}}{r_d}\right) - \nu R\bar{T} \ln \tilde{S}_s = 0, \qquad (11.3.5)$$

where we introduced the activities ratio

$$\tilde{S}_s = \tilde{a}_s / \tilde{a}_{s*}, \qquad (11.3.6)$$

$$L_c^{ef}(T) = \left(\int_T^{T_0} \frac{L_c(T)}{T} dT\right) \times \left(\int_T^{T_0} \frac{dT}{T}\right)^{-1} = \left(\int_T^{T_0} \frac{L_c(T)}{T} dT\right) \times \left[\ln\left(\frac{T_0}{T}\right)\right]^{-1}. \qquad (11.3.6a)$$

The symbols with asterisks in (Eqn. 11.3.5) correspond to the quantities in the reference state, L_c^{ef} is the specific dissolution heat averaged over the temperature range as was done in Chapters 8 and 9 for the melting heat, $\Delta v = v_{c0} - v_{s0}$ is the difference of the salt molar volumes in solid and solution, $\Delta p = p - p_*$ is the excess of external pressure, the bars over quantities denote mean values after integration.

If the reference point is chosen at solute saturation, $\tilde{a}_{s*}$ is the salt activity at saturation, and $\tilde{S}_s = \tilde{a}_s / \tilde{a}_{s*}$ is the solute supersaturation that can be smaller and greater than 1 since $\tilde{a}_s$ can be smaller than $\tilde{a}_{s*}$ in diluted solutions and much greater than $\tilde{a}_{s*}$ in supersaturated metastable solutions. Note that the ratio $\tilde{S}_s = \tilde{a}_s / \tilde{a}_{s*}$ under the logarithm sign in (Eqn. 11.3.5) is not very different from the activities ratio $S_s = a_s/a_{s*}$, which therefore can be used for approximate calculations as in Defay et al. (1966). The bars over the volumes are omitted in the following text for brevity; the factor ν (total number of ions) may be close to the van't Hoff factor i or to the product $\nu_{id}\Phi_s$, where ν_{id} is the number of ions in ideal solution and Φ_s is the osmotic potential. Thus, accounting for ν may substantially influence the results. For an isothermal (T = const) and isobaric (p = const) process, and large radius of liquid lens, $1/r_d = 0$, or small σ_{sa}, and assuming that $\tilde{S}_s = \tilde{a}_s/\tilde{a}_{s*} \approx x_s/x_{s*}$, (Eqn. 11.3.5) is simplified:

$$\frac{2\sigma_{cs}}{r_c} = \frac{R\bar{T}}{\bar{v}_{c0}} \ln \frac{x_s}{x_{s*}}, \tag{11.3.6b}$$

where x_s is the mole fraction of the solute in the crystal with radius r_c, and x_{s*} is the mole fraction of the solute in equilibrium with a large crystal or bulk phase ($r_c = \infty$). Equation (11.3.6b) is the Freundlich–Ostwald equation that describes the dependence of the solubility x_s/x_{s*} on the solution drop radius (Defay et al., 1996) and is considered in Section 11.6. Equation (11.3.5) generalizes (Eqn. 11.3.6b) taking into account the effects of the temperature, pressure, and non-ideality of the solution. Other generalizations of the Freundlich–Ostwald equation were derived by Shchekin, Shabayev, and Rusanov (2008a,b, 2010) based on the concept of disjoining pressure.

The critical radius r_d of the liquid germ of deliquescence on the crystal surface can be obtained from (Eqn. 11.3.5):

$$r_d = \frac{2\sigma_{sa} v_{s0}}{-M_s L_c^{ef} \ln\left(\dfrac{T_*}{T}\right) + \Delta v \Delta p + v_{c0} \dfrac{2\sigma_{cs}}{r_c} - \nu R\bar{T} \ln S_s}, \tag{11.3.7}$$

or in a more compact form

$$r_d = \frac{2\sigma_{sa} v_{s0}}{-M_s L_c^{ef} \ln\left(\dfrac{T_*}{T} S_s^{g_s}\right) + \Delta v \Delta p + v_{c0} \dfrac{2\sigma_{cs}}{r_c}}, \tag{11.3.8}$$

where we introduced a dimensionless parameter g_s analogous to G_n in ice nucleation

$$g_s = \frac{\nu R\bar{T}}{L_c^{ef} M_s}. \tag{11.3.9}$$

Dividing by M_s and introducing specific densities of salt in solution, $\rho_{s0} = M_s/v_{s0}$, and in the solid crystal, $\rho_{c0} = M_s/v_{c0}$, (Eqn. 11.3.8) can be rewritten in the form

$$r_d = \frac{2\sigma_{sa}}{-\rho_{s0} L_c^{ef} \ln\left(\dfrac{T_*}{T} S_s^{g_s}\right) + \dfrac{\Delta\rho_s \Delta p}{\rho_{c0}} + \dfrac{\rho_{s0}}{\rho_{c0}} \dfrac{2\sigma_{cs}}{r_c}}, \tag{11.3.10}$$

where $\Delta\rho_s = \rho_{s0} - \rho_{c0}$. It can be presented in a form similar to the case of surface melting, as in Section 9.7:

$$r_d = \frac{2\sigma_{sa} v_{s0}}{-M_s L_c^{ef} \ln\left[\dfrac{T_*}{T} S_s^{g_s} \exp(-H_{del})\right]}, \tag{11.3.11a}$$

$$r_d = \frac{2\sigma_{sa}}{-\rho_{s0} L_c^{ef} \ln\left[\frac{T_*}{T} S_s^{g_s} \exp(-H_{del})\right]}, \tag{11.3.11b}$$

where the term H_{del} describes the effects of pressure and finite crystal size

$$H_{del} = \frac{1}{L_c^{ef} M_s}\left(\Delta v \Delta p + v_{c0}\frac{2\sigma_{cs}}{r_c}\right) = \frac{1}{L_c^{ef} \rho_{c0}}\left(\frac{\Delta\rho_s \Delta p}{\rho_{s0}} + \frac{2\sigma_{cs}}{r_c}\right). \tag{11.3.12}$$

(Note that (Eqn. 11.3.11b) resembles (Eqn. 9.9.9) for surface melting.) Either of these two forms of (Eqns. 11.3.11a) and (11.3.11b) can be used when the data on the molar (molecular) volumes or specific densities are available. The form of these (Eqns. 11.3.11a) and (11.3.11b) enable establishing the equivalence of effects of various externals factors of deliquescence (T, r_c, Δp, a_s), following the previous chapters on freezing.

The critical energy of formation of the liquid solution germ for deliquescence can be written in terms of the classical nucleation theory similar to the surface melting in Section 9.9:

$$\begin{aligned}\Delta F_{cr,del} &= \frac{4}{3}\pi\sigma_{sa} r_d^2 f(m_{cs}, x) \\ &= \frac{16\pi}{3}\frac{\sigma_{sa}^3 v_{s0}^2 f(m_{cs}, x)}{\left\{L_c^{ef} M_s \ln\left[\frac{T_*}{T} S_s^{g_s} \exp\left(-H_{del}\right)\right]\right\}^2},\end{aligned} \tag{11.3.13}$$

where $f(m_{cs}, x)$ is Fletcher's geometric factor, and m_{cs} is the cosine of the contact angle between solid salt and its solution, $x = r_c/r_d$. The contact angle can be small but nonzero in the case of the incomplete wettability of the salt by solution as discussed in Section 9.9 for surface melting, thus slightly increasing the critical energy of deliquescence. Therefore, the shape factor $f(m_{cs}, x)$ occurs here as in the case of heterogeneous nucleation, although we consider homogeneous nucleation without foreign catalysts.

The nucleation rate now can be written by analogy with the surface melting (Section 9.7):

$$J_{del} = C_{del} \exp\left(-\frac{\Delta F_{cr,del}}{kT} - \frac{\Delta F_{a,del}}{kT}\right), \tag{11.3.14}$$

$$C_{del} \approx N_{cc}\left(\frac{kT}{h}\right) c_{1cs} 4\pi r_c^2. \tag{11.3.15}$$

Here C_{del} is the kinetic coefficient for deliquescence, $\Delta F_{a,del}$ is the activation energy for transition of the salt molecules from the solid to the liquid solution, N_{cc} is the number of salt molecules in crystal in contact with the unit area of solution, and c_{1cs} is the number of molecules adsorbed on the surface of the solid. If one considers the formation of a thin liquid solution film around a solid salt core instead of spherical caps or lenses, then the process can be treated as a homogeneous nucleation. In this case, we can assume $f(m_{cs}, x) = 1$, and the kinetic coefficient is as it was derived for homogeneous nucleation in Chapter 8:

$$C_{del} = 2N_{cc}\left(\frac{v_{c0}}{v_{s0}}\frac{kT}{h}\right)\left(\frac{\sigma_{cs}}{kT}\right)^{1/2} = 2N_{cc}\left(\frac{\rho_{s0}}{\rho_{c0}}\frac{kT}{h}\right)\left(\frac{\sigma_{cs}}{kT}\right)^{1/2}. \tag{11.3.16}$$

Equations (11.3.7) to (11.3.16) constitute the model for deliquescence and, as shown next, enable description of several essential features of this phenomenon.

11.4. Applications of the Deliquescence Model

Some useful relations can be obtained solving (Eqn. 11.3.14) relative to $\Delta F_{cr,del}$:

$$\Delta F_{cr,del} = -kT \ln \frac{J_{del}}{C_{del}} - \Delta F_{a,del}. \tag{11.4.1}$$

Equating (Eqn. 11.4.1) to (Eqn. 11.3.13), we obtain the temperature of deliquescence:

$$T_{del} = T_* \left(\frac{\tilde{a}_s}{\tilde{a}_{s*}} \right)^{g_s} \exp\left[-\frac{1}{M_s L_c^{ef}} \left(\Delta p \Delta v + v_{c0} \frac{2\sigma_{sc}}{r_c} \right) \right] \times$$

$$\exp\left\{ -\frac{1}{M_s L_c^{ef}} \left[-\frac{16\pi}{3} \frac{\sigma_{sa}^3 v_{s0}^2 f(m_{cs}, x)}{kT \ln(J_{del}/C_{del}) + \Delta F_{a,del}} \right]^{1/2} \right\}. \tag{11.4.2}$$

Note that this expression is similar in structure to the expressions for the critical temperatures of freezing and melting derived in Chapters 8 and 9. If $\tilde{a}_{s*}$ is the activity in a saturated solution, then $\tilde{a}_s/\tilde{a}_{s*} = S_{s*}$ is the solution supersaturation. This equation relates the temperature T_{del} and solute activity a_s at deliquescence, and can be used to express one as a function of another. If $L_c^{ef} > 0$ (dissolution requires positive heat), the arguments of both exponents are negative, the exponents are smaller than 1 and $T_{del} < T_*$. Thus, this equation shows that the following factors leads to a decrease of the deliquescence temperature: a) the possible existence of a small but finite contact angle θ_{cs} between the outer surface of the solution germ and the surface of the salt crystal; b) finite crystal radius r_c; the smaller the r_c, the greater the decrease in T_{del}; c) external pressure; and d) finite nucleation rate J_{del}. That is, the deliquescence temperature decreases with increasing external pressure (as can be in the atmospheres of the large planets, Curry and Webster, 1999) and with increasing nucleation rates.

In the case of water melting, the contact angle between liquid (water) and solid (ice) surfaces is 0.5–2 degrees although this angle may reach several degrees for the case of the melting of some metals. If the contact angle for solutions is small similar to water melting, then $m_{cs} = \cos\theta_{cs} \to 1, f(m_{cs}, x) \to 0$, and the second exponent on the right-hand side in (Eqn. 11.4.2) tends to 1. The same occurs when the nucleation rate J_{del} is very small due to a slow change of the external conditions. Then, $J_{del} \to 0$, $\ln(J_{del}) \to -\infty$, the denominator in the last exponent tends to zero, and the exponent tends to 1. Thus, (Eqn. 11.4.2) is simplified as

$$T_{del} = T_* \tilde{S}_s^{g_s} \exp\left[-\frac{1}{M_s L_c^{ef}} \left(\Delta p \Delta v + v_{c0} \frac{2\sigma_{sc}}{r_c} \right) \right]. \tag{11.4.3}$$

If we consider the process at constant pressure or at small pressure variations, $\Delta p \approx 0$, and for the bulk phase, $1/r_c = 0$, then the last exponent is also 1. Assuming also that $\tilde{S}_s = (\tilde{a}_s/\tilde{a}_{s*}) \sim (a_s/a_{s*})$, we obtain a simple relation:

$$T_{del}(a_s) = T_* S_s^{g_s} = T_* \left(\frac{\tilde{a}_s}{\tilde{a}_{s*}} \right)^{g_s} \approx T_* \left(\frac{a_s}{a_{s*}} \right)^{g_s} = T_* \left(\frac{\gamma_s x_s}{\gamma_{s*} x_{s*}} \right)^{g_s}. \tag{11.4.4}$$

This equation relates the deliquescence temperature and solution concentration—i.e., it describes the deliquescence curve on the phase T-x_s diagrams. It can be used in several ways.

11.4.1. The Temperature Dependence of Dissolution Heat

Using definition (Eqn. 11.3.9) of $g_s = \nu RT/(L_c^{ef}(T)M_s)$, dividing (Eqn. 11.4.4) by T_*, taking the logarithm and solving for $\Delta h_s = M_s L_c^{ef}$, we obtain

$$M_s L_c^{ef}(T) = \nu RT_{del}\frac{\ln S_s}{\ln(T/T_*)} \approx \nu RT_{del}\frac{\ln(\gamma_s x_s/\gamma_s x_{s*})}{\ln(T/T_*)}. \tag{11.4.5}$$

This equation can be used to evaluate the temperature dependence of the averaged dissolution heat using the data on solubility $x_s(T)$, with T_* sufficiently close to T_{del}. Such calculations can be done—e.g., using the data on solubility from Tang and Munkelwitz (1993) and Seinfeld and Pandis (1998) given in the polynomial form by temperature:

$$n_s(T) = A_s + B_s T + C_s T^2, \tag{11.4.6}$$

where n_s is the number of moles of solute per one mole of water. For example, the coefficients for the two most known and studied salts are the following. For NaCl, $n_s = 0.111$ at 298.15 K in saturated solution; $A_s = 0.1805$, $B_s = -5.310 \times 10^{-4}$, and $C_s = 9.965 \times 10^{-7}$. For ammonium sulfate, $(NH_4)_2SO_4$, $n_s = 0.104$ at 298.15 K in saturated solution; $A_s = 0.1149$, $B_s = -4.489 \times 10^{-4}$, and $C_s = 1.385 \times 10^{-8}$. Using (Eqn. 11.4.6), the mole fraction x_s can be calculated, substituted into (Eqn. 11.4.5), and assuming that γ_s/γ_{s*} only slightly varies, the temperature dependence of $M_s L_c^{ef}(T)$ can be estimated.

11.4.2. The Temperature Dependence of Solubility

If we consider the temperature as the variable and the solubility as its function, then (Eqn. 11.4.2) can be rewritten as:

$$\tilde{a}_s = \tilde{a}_{s*}\left(\frac{T_{del}}{T_*}\right)^{1/g_s}\exp\left[\frac{1}{M_s L_c^{ef}}\left(\Delta p\Delta v + v_{c0}\frac{2\sigma_{sc}}{r_c}\right)\right]\times$$
$$\exp\left\{\frac{1}{M_s L_c^{ef}}\left[-\frac{16\pi}{3}\frac{\sigma_{sa}^3 v_{s0}^2 f(m_{cs},x)}{kT\ln(J_{del}/C_{del}) + \Delta F_{a,del}}\right]^{1/2}\right\}. \tag{11.4.7}$$

This equation is an analog of the equations for the critical humidities of freezing derived in Chapters 8 and 9. It describes the temperature dependence of the salt activity (or solubility if $\gamma_s/\gamma_{s*} \sim 1$) with a simultaneous accounting for other effects (crystal curvature, external pressure, the finite nucleation rate, the value of dissolution heats, and activation energy $\Delta F_{a,del}$), whose number is determined by the variance of the system. Note that this equation is similar to that for the case of surface melting in Chapter 9. Consider again a particular case of (Eqn. 11.4.7) with p = const ($\Delta p = 0$), bulk solution, $1/r_d = 1/r_c = 0$, and small contact angle θ_{sc}—i.e., $f(m_{cs}, x) \to 0$. Then, (Eqn. 11.4.7) can be simplified. Assuming again that $\tilde{a}_s/\tilde{a}_{s*} \approx a_s/a_{s*}$, then (Eqn. 11.4.7) yields for the salt activity $a_s = \gamma_s x_s$:

$$a_s(T) = a_{s*}\left(\frac{T}{T_*}\right)^{1/g_s} = a_{s*}\left(\frac{T}{T_*}\right)^{\frac{L_c^{ef}M_s}{\nu RT}}. \tag{11.4.8}$$

If variations of γ_s are substantially smaller than those of x_s in the considered region, or for ideal solutions with $\gamma_s = 1$, (Eqn. 11.4.8) yields for the solubility:

$$x_s = x_{s*}(T/T_*)^{1/g_s} = x_{s*}(T/T_*)^{\frac{L_c^{ef} M_s}{\nu R\bar{T}}}. \tag{11.4.9}$$

Shown in Fig. 11.2a are the solubilities of NaCl and ammonium sulfate calculated with (Eqn. 11.4.9) (open symbols) using the parameters from Seinfeld and Pandis (1998): for NaCl, $\nu = 2$, $M_s L_c^{ef} = 1.88$ kJ mole^{-1} (approximately the same as $M_s L_c$ at $T = 298.15$ K), the reference saturated mole fraction $x_{s*} = 0.098$ mole mole^{-1} at $T = 273.15$ K; and for ammonium sulfate, $\nu\Phi_s = 2.6$, $M_s L_c^{ef} = 6.32$ kJ mole^{-1} (same as at $T = 298.15$ K), the reference saturated mole fraction $x_{s*} = 0.088$ mole mole^{-1} at $T = 273.15$ K. This figure illustrates a significant temperature dependence of ammonium sulfate solubility and a rather weak dependence for NaCl. This difference can be explained by the different values of the power indices g_s for these substances. An estimate with these parameters shows that at room temperature for NaCl, $g_s = 2.64$, and $1/g_s = L_c^{ef} M_s/(\nu R\bar{T}) = 0.38$; for ammonium sulfate, $g_s \approx 0.98$ and $1/g_s \approx 1.02$. Therefore, the variation of solubility, $\sim T^{1/g_s}$, is almost linear with temperature for ammonium sulfate and is smaller than a square root for NaCl.

A comparison with the polynomial parameterization (Eqn. 11.4.6) by Tang and Munkelwitz (1993) of the experimental data on solubility (see also Seinfeld and Pandis, 1998, p. 510) (solid symbols) shows a good accuracy of (Eqn. 11.4.9). This is quantitatively illustrated in Fig. 11.2b where the relative error of calculations of the mole fraction $x_s(T)$ with (Eqn. 11.4.9) is shown. One can see that the error of calculations with (Eqn. 11.4.9) relative to the empirical parameterization (Eqn. 11.4.6) does

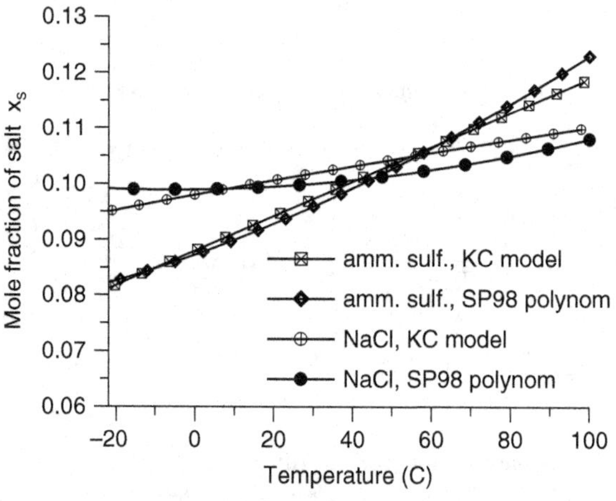

Figure 11.2. (a) Solubility of NaCl and ammonium sulfate calculated with (Eqn. 11.4.9) of the KC model (open symbols) using the parameters: for NaCl, $\nu = 2$, $M_s L_c^{ef} = 1.88$ kJ mole^{-1} (same as at $T = 298.15$ K), the reference saturated mole fraction $x_{s*} = 0.098$ mole mole^{-1} at $T = 273.15$ K; for ammonium sulfate: $\nu = 2.6$, $M_s L_c^{ef} = 6.32$ kJ mole^{-1} (same as at $T = 298.15$ K), and the reference saturated mole fraction $x_{s*} = 0.088$ mole mole^{-1} at $T = 273.15$ K. The results are compared to the parameterization with quadratic polynomials (Eqn. 11.4.6) from Tang and Munkelwitz (1993) and Seinfeld and Pandis (1998, p. 510), (solid symbols, SP98).

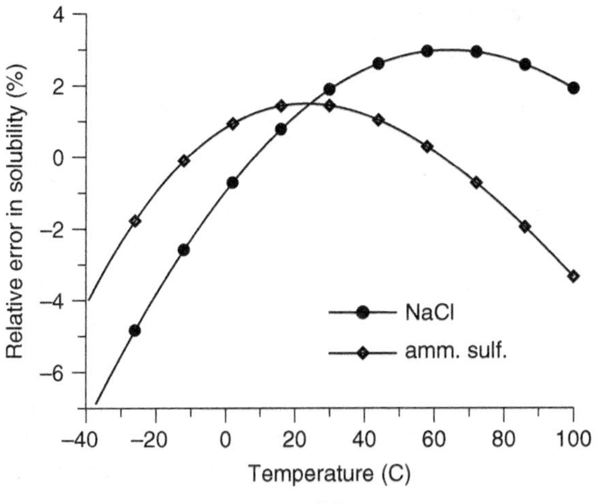

Figure 11.2. (b) The error of calculations of the mole fraction $x_s(T)$ of NaCl and ammonium sulfate for saturated solutions with (Eqn. 11.4.9) of the KC model relative to the polynomial parameterization (Eqn. 11.4.6) by Tang and Munkelwitz (1993).

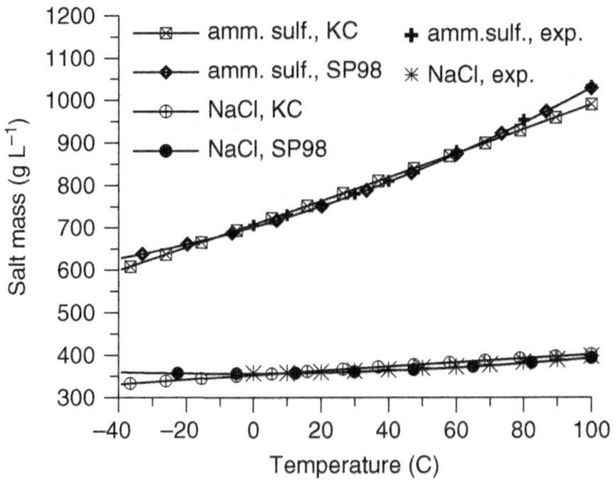

Figure 11.2. (c) The temperature dependence of the mass of salt (g) dissolved in 1 L of water calculated for ammonium sulfate and NaCl with (Eqn. 11.4.9) of the KC model (KC, open symbols) and compared to the parameterization from Tang and Munkelwitz (1993) and Seinfeld and Pandis (1998) (solid symbols, SP98), and to the experimental data from Tang and Munkelwitz (1993) and SP98 (crosses).

not exceed ±4%, and increases to −7% for NaCl near −40 °C. The mass of dissolved salt calculated with the KC model from the mole fraction (open symbols) is shown in Fig. 11.2c. A comparison with the parameterizations (diamonds and solid circles) and experimental data (crosses and asterisks) from Seinfeld and Pandis (1998) also shows a good accuracy of (Eqn. 11.4.9).

11.4.3. The Temperature Dependence of the Deliquescence Relative Humidity

When the deliquescence mole fraction x_s of the saturated salt is known, there are several ways to calculate the water activity a_w and deliquescence relative humidity, DRH. The simplest way would be that for the ideal solutions, where $a_w \approx x_w \approx 1 - x_s$. However, due to the strong non-ideality of solutions, this calculation can cause substantial errors. Another still sufficiently simple way is the following. The solute mole fraction x_s can be recalculated to the corresponding molality $\hat{M}$ and weight percent w, and then the water activity $a_w(T)$ can be easily calculated using the existing empirical parameterizations of a_w in terms of w or $\hat{M}$—e.g., Tang and Munkelwitz (1993, 1994, 1997); Chen (1994); and others. In the previous Section, we calculated the temperature dependence $x_s(T)$, which provides the temperature dependence $a_w(T)$ via w or $\hat{M}$. Since water activity in this case coincides with the environmental *RHW* at deliquescence, this yields the temperature dependence DRH(T).

We calculated DRH(T) for ammonium sulfate, using the parameterization for $a_w(w)$ from Tang and Munkelwitz (1994). Shown in Fig. 11.3 is the temperature dependence of the deliquescence water activity $a_w(T)$ or DRH(T)/100 for ammonium sulfate calculated with (Eqn. 11.4.9) of the KC model for $x_s(T)$. This calculation is compared to the calculation of DRH(T) from Tang and Munkelwitz (1993) and Seinfeld and Pandis (1998). Their derivation accounted for the fact that the enthalpy of solution Δh_{sol} is a sum of the enthalpy of condensation of water vapor $\Delta h_v = M_w L_e$ and the enthalpy of dissolution of n_s moles of solute, $n_s \Delta h_s = n_s M_s L_c$—that is, $\Delta h_{sol} = \Delta h_v - n_s \Delta h_s = M_w L_e - n_s M_s L_c$. Then, the Clausius–Clapeyron equation for solution and water vapor is a modification of this equation derived in Chapter 3:

$$\frac{d \ln e_{ws}}{dT} = \frac{\Delta h_{sol}}{RT^2} = \frac{M_w L_v}{RT^2} - n_s(T)\frac{M_s L_c}{RT^2}. \quad (11.4.10a)$$

The Clausius–Clapeyron equation for pure water vapor is

$$\frac{d \ln e_{ws,0}}{dT} = \frac{M_w L_v}{RT^2}. \quad (11.4.10b)$$

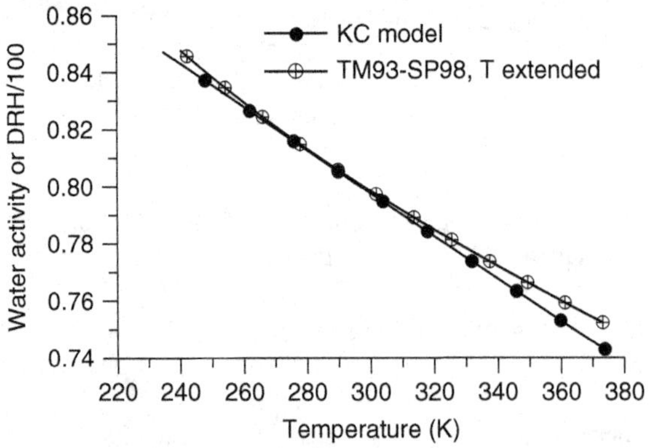

Figure 11.3. Temperature dependence of deliquescence water activity $a_w(T)$ or DRH(T)/100 for ammonium sulfate calculated with (Eqn. 11.4.9) of the KC model compared to the calculations with (Eqn. 11.4.12b) from Tang and Munkelwitz (1993) (see also SP98), based on the Clausius–Clapeyron equations at T = 0–50 °C, and extended here over a wider T-region.

Subtracting (Eqn. 11.4.10b) from (Eqn. 11.4.10a) and using the definition DRH = $(e_{ws}/e_{ws,0}) \times 100\%$, Tang and Munkelwitz (1993) obtained an equation for DRH

$$\frac{d\ln(e_{ws}/e_{ws,0})}{dT} = \frac{d\ln(DRH/100)}{dT} = -n_s(T)\frac{M_s L_c}{RT^2}. \tag{11.4.11}$$

Substituting (Eqn. 11.4.6) for $n_s(T)$ in (Eqn. 11.4.11), assuming that L_c does not depend on T and integrating from the reference point $T_* = 298\,\mathrm{K}$ to T yields

$$\ln\frac{DRH(T)}{DRH(T_*)} = \frac{M_s L_c}{R}\left[A_s\left(\frac{1}{T}-\frac{1}{T_*}\right) - B_s \ln T_* - C_s(T-T_*)\right], \tag{11.4.12a}$$

or

$$\frac{DRH(T)}{DRH(T_*)} = \exp\left\{\frac{M_s L_c}{R}\left[A_s\left(\frac{1}{T}-\frac{1}{T_*}\right) - B_s \ln T_* - C_s(T-T_*)\right]\right\}. \tag{11.4.12b}$$

Wexler and Seinfeld (1991) derived this equation for constant solubility—i.e., $B_s = C_s = 0$, Tang and Munkelwitz (1993, TM93 model) generalized it for variable $n_s(T)$ with nonzero B_s, C_s for the range $T = 0$–$50\,°\mathrm{C}$. The DRH(T)/100 calculated with TM93 model (Eqn. 11.4.12b) is compared in Fig. 11.3 with calculations using the KC model for a wider T-region from 230 K to 375 K. A comparison shows a good agreement of the KC model based on (Eqn. 11.4.9) with the TM93 model (Eqn. 11.4.12b) over this wide temperature range despite the different equations used, although a discrepancy somewhat increases toward the highest and lowest temperatures. Note that Wexler and Seinfeld (1991) and Tang and Munkelwitz (1993) derived DRH(T) based on the equilibrium of water in vapor and solution, and the T-dependence in the KC model is derived based on the equilibrium of salt in the crystal and solution. Both methods are complimentary, and as this comparison shows, are in good agreement.

11.5. Phase Diagram of the Solution and Evaluation of the Eutectic Point

Equation (11.4.4) for the deliquescence temperature is similar to the equation for the equilibrium melting temperature $T_m(S_w)$ (liquidus curve) derived in Section 8.7:

$$T_m(a_w) = T_0 a_w^{G_n} = T_0 a_w^{R\bar{T}/M_w L_m^{ef}} = T_0 S_w^{R\bar{T}/M_s L_m^{ef}}, \tag{11.5.1}$$

where $T_0 = 273.15$ is the triple point. Assuming that variations of the salt activity coefficients ratio (γ_s/γ_{s*}) in (Eqn. 11.4.4) is smaller than variations of the mole fraction ratio (x_s/x_{s*}), (Eqn. 11.4.4) can be rewritten as

$$T_{del}(a_s) \approx T_*\left(\frac{x_s}{x_{s*}}\right)^{g_s} = T_*\left(\frac{x_s}{x_{s*}}\right)^{\frac{\nu R\bar{T}}{L_c^{ef} M_s}}. \tag{11.5.2}$$

The equilibrium T_{del} calculated with (Eqn. 11.5.2) for ammonium sulfate is plotted in Fig. 11.4 as a function of the water saturation ratio S_w equal to the water activity a_w, which is calculated using a parameterization of Tang and Munkelwitz (1993), and is compared to the experimental data for the deliquescence temperature of ammonium sulfate from Tang and Munkelwitz (1993) and Seinfeld

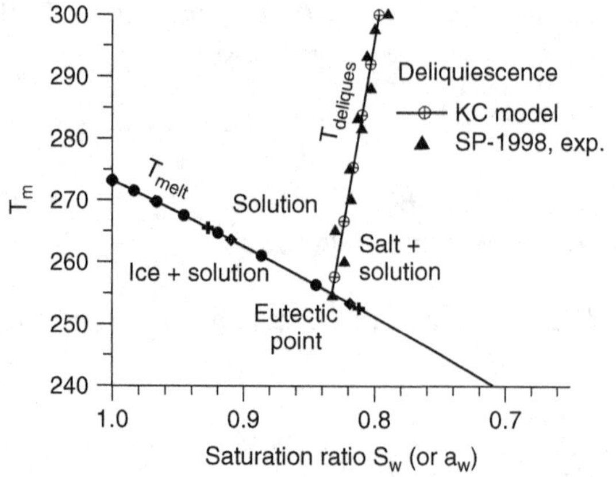

Figure 11.4. The phase diagram in $T - S_w$ coordinates of solution melting and deliquescence calculated with the KC model. Right curve: temperature of solute deliquescence $T_{del}(S_w)$ calculated with (Eqn. 11.5.2) as a function of the water saturation ratio S_w (equal to water activity a_w) for ammonium sulfate with $T_* = 0\,°C$ (open circles), compared to the experimental data for deliquescence from Seinfeld and Pandis (1998, SP-1998, exp.) and Tang and Munkelwitz (1993) (triangles). Left curve: calculated equilibrium melting temperatures $T_m(S_w)$ from Khvorostyanov and Curry (2004a) described in Chapter 8 or (Eqn. 11.5.1) here. Three curves for ammonium sulfate, NaCl, and sulfuric acid (circles, crosses, and diamonds) merge in one curve as the functions of S_w due to colligative properties. The intersection of these curves around $T \approx 255\,K$ and $S_w \approx 0.84$ is the eutectic point for ammonium sulfate.

and Pandis (1998, SP-1998, exp). A comparison shows again a good agreement, indicating the validity of the equations for deliquescence derived in Sections 11.3 and 11.4. Fig. 11.4 also shows the equilibrium melting temperature of ice T_m calculated with (Eqn. 11.5.1). As described in Chapter 8, the melting curves for the three indicated substances (NaCl, ammonium sulfate, and sulfuric acid) merge in the T–S_w coordinates due to colligative properties. These two curves for T_m and T_{del} constitute a phase diagram in coordinates T–S_w for ammonium sulfate, which is in good agreement with the known from experiments phase diagrams (e.g., Landau and Lifshitz, v.5, 1958b; Xu et al., 1998; Curry and Webster, 1999; Martin, 2000). Here, both branches of the diagram, the left curve for the ice–solution equilibrium and the right curve for the salt-solution equilibrium, are calculated from the same theory (extended CNT) using simple (Eqns. 11.5.1) and (11.5.2). Thus, these equations allow a simple description of both branches of the phase diagrams of solutions: the melting and deliquescence branches.

This phase diagram describes the evolution of the solution with changing composition. The uppermost point on the left with $S_w = 1$ corresponds to the melting point of pure water $T_m = 273.15\,K$. As the solution concentration increases (the water activity decreases), the melting temperature decreases along the curve $T_m(S_w)$. When a fraction of the solution freezes, almost pure ice forms, and solute is rejected into the solution making it increasingly concentrated. If a vertical trajectory on the T-S_w plane intersects the T_m curve, then with the same initial composition, the solution may exist above the T_m curve and a mixture of ice and solution may exist below T_m as indicated in Fig. 11.4. As the temperature falls, more and more ice forms leaving the solution increasingly saline. (When the solute is NaCl or contains other components of seawater, it is called *brine*, and similar diagrams are used for the salinity of seawater, Millero, 1978; Curry and Webster, 1999; Millero et al., 2008). When an increasing solute concentration reaches saturation, this point on the phase diagram is called *the eutectic point*, where all three phases (solution, ice, and salt) exist in equilibrium. The intersection of the melting and deliquescence curves around $T \approx 253.65\,\mathrm{K}$ (−19.35 °C) and $S_w = 0.83$ in Fig. 11.4 is the eutectic point for ammonium sulfate. The eutectic point for NaCl (the major component of seawater and many sea-generated aerosols) is at −21.2 °C. Further equilibrium at a higher solute concentration (or lower S_w) is possible along the deliquescence curve $T_{del}(S_w)$. The homogeneous solution may exist above $T_{del}(S_w)$, and a mixture of crystallized salt and solution exists below $T_{del}(S_w)$ but above the eutectic point. Below the eutectic point, the salts form various hydrates. The salt NaCl forms the bihydrate $NaCl \cdot 2H_2O$ (Curry and Webster, 1999), and ammonium sulfate forms thetrahydrate $(NH_4)_2SO_4 \cdot 4H_2O$ (Xu et al., 1998).

Fig. 11.4 shows that the slope of the deliquescence $T_{del}(S_w)$ branch is much steeper than that of the melting branch $T_m(S_w)$. This can be explained if we assume that the solution behavior is not very different from the ideal, and $x_s^{g_s} \approx (1-a_w)^{g_s}$. Then, (Eqns. 11.5.1) and (11.5.2) represent two power laws, but with the different indices. As was shown in Chapter 8, $G_n = R\overline{T}/(L_m^{ef} M_s) \approx 0.38$ at these values of T, and a calculation for ammonium sulfate gives $g_s = \nu R\overline{T}/(L_c^{ef} M_s) \approx 0.98$. Hence, the power index G_n in (Eqn. 11.5.1) for the T_m is much smaller than g_s, and the decrease in the equilibrium melting (freezing) temperatures with decreasing S_w (increasing solution concentration) on the left from the eutectic point on the phase T-S_w diagrams (liquidus curve) predicted by (Eqn. 11.5.1) is substantially slower than the temperature increase predicted by (Eqn. 11.5.2), in agreement with observations (see e.g., Fig. 4.8 in Curry and Webster, 1999; or Figs. 3–8 in Martin, 2000).

Equations (11.5.1) and (11.5.2) enable evaluation of the eutectic point—i.e., the mole fraction of solute at this point of intersection—for solutions without strong non-ideality. Equating (Eqns. 11.5.1) and (11.5.2), we obtain an equation for the eutectic point:

$$T_0 a_w^{R\overline{T}/M_s L_m^{ef}} = T_* (a_s/a_{s*})^{R\overline{T}/M_s L_c^{ef}}. \tag{11.5.3}$$

The water activity a_w can be approximately related to the molality $\hat{M}$ as in Tang and Munkelwitz (1993, 1994), Pruppacher and Klett (1997), (see Chapter 3 here):

$$a_w = 1 - B_s \hat{M}, \qquad B_s = 10^{-3} \nu \Phi_s M_w, \tag{11.5.4}$$

where ν is the number of ions in the solution, Φ_s is the osmotic potential, and M_w is the molecular weight of water. Note that the value $B_s = 0.034$ in (Eqn. 11.5.4) gives a very good approximation of the experimental data from the relation between a_w and $\hat{M}$ for ammonium sulfate. Molality can be expressed via the mole fraction x_s as $\hat{M} = \left(10^3/M_w\right) x_s/(1-x_s)$. Substituting this into (Eqn. 11.5.4), and taking into account that in saturated solutions $x_s \sim 0.1 << 1$ (Seinfeld and Pandis, 1998), and x_s can be neglected in the denominator, we obtain an approximate linear expression of a_w via x_s:

$$a_w = 1 - b_s x_s, \qquad b_s = 10^3 B_s / M_w = \nu \Phi_s. \tag{11.5.5}$$

Substituting this into (Eqn. 11.5.3), and assuming that $a_s/a_{s*} \approx x_s/x_{s*}$ (relatively small variations of the salt activity coefficients), we obtain:

$$T_0 (1 - b_s x_s)^{G_n} = T_* (x_s/x_{s*})^{g_s}. \tag{11.5.6}$$

Taking the power $1/g_s$ and expanding the left-hand side by the powers of small x_s, we obtain a linear equation for x_s. Its solution for the mole fraction $x_{s,eut}$ at the eutectic point is

$$x_{s,eut} = \frac{\lambda_{eu} x_{s*}}{1+\alpha_{eu} b_s \lambda_{eu} x_{s*}}, \tag{11.5.7}$$

where

$$\lambda_{eu} = \left(\frac{T_0}{T_*}\right)^{1/g_s} = \left(\frac{T_0}{T_*}\right)^{M_s L_c^{ef}/(\nu RT)}, \qquad \alpha_{eu} = \left(\frac{G_n}{g_s}\right) = \left(\frac{M_s L_c^{ef}}{\nu M_w L_m^{ef}}\right). \tag{11.5.8}$$

Equation (11.5.7) yields the coordinate of the eutectic point—i.e., an expression for the solute mole fraction $x_{s,eut}$ at this point via the mole fraction x_{s*} at a reference point $T = T_*$.

Consider the following application. Suppose we need to calculate the eutectic mole fraction for ammonium sulfate and choose the reference point $T_* = 298.15\,\text{K}$, then the reference mole fraction $x_{s*}(T_* = 298.15) = 0.0944\,\text{mol mol}^{-1}$ (Seinfeld and Pandis, 1998). The parameter b_s calculated from (Eqn. 11.5.5) with $B_s = 0.034$ for a saturated solution is $b_s \approx 1.83$; $\alpha_{eu} = G_n/g_s \approx 0.38/1.0 = 0.38$. Since $T_0 = 273.15\,\text{K}$, and $g_s \approx 1$, then $\lambda_{eu} \approx (273.15/298.15)^1 = 0.916$. Substituting these values into (Eqn. 11.5.7), we obtain $x_{s,eut} \approx 0.0816\,\text{mol mol}^{-1}$. Comparing this value to the experimental data $x_{s,eut} = 0.0828$ (Seinfeld and Pandis, 1998), we find that the relative error $\Delta x_{s,eut}/x_{s,eut}$ is 1.4%, which illustrates a good accuracy of (Eqn. 11.5.7). Calculations for the other salts can be done in the same way given the values of the required parameters. This simple estimate for the eutectic point may be helpful in constructing or constraining the phase diagrams for various substances.

11.6. A Model for the Efflorescence of Salt Crystals Based on the Entropy Equation

As discussed in Section 11.1, there are two types of efflorescence that proceed as homogeneous and heterogeneous nucleation. We consider first efflorescence as a process of homogeneous nucleation of a salt crystal in a solution supersaturated with respect to solute at decreasing humidity. The treatment is similar to homogeneous ice nucleation in Chapter 8 and is based again on the general equilibrium (or entropy) (Eqn. 3.6.4). The phase transition in this case is salt crystallization. Phase 1 is salt in solution, and phase 2 is salt in the crystal. With the same definitions as in Section 11.3 for deliquescence, (Eqn. 3.6.4) can be rewritten for efflorescence as

$$-\frac{(h_{s0} - h_{c0})}{T^2} dT + \frac{v_{s0}}{T} dp_c - \frac{v_{c0}}{T} dp_s + Rd \ln\left(\frac{a_s}{a_c}\right) = 0, \tag{11.6.1}$$

where h_{c0} and h_{s0} are the enthalpies of the solid and solution, p_c and p_s are pressures inside a crystal and inside a solution drop, v_{s0} and v_{c0} are the molar volumes of salt in a solution and a solid, and a_s and a_c are the activities of salt in a solution and a crystal. We assume again that the activity of the pure salt $a_c = 1$. As in the case with deliquescence, to account for the ionic dissolution and following Defay et al. (1966), we replace $\ln a_s = \ln(\gamma_s x_s) \to \nu \ln \tilde{a}_s$. The non-ideality of solutions can be accounted for by replacing $\nu \to \nu\Phi_s$. We assume that crystallization occurs with positive heat release, the enthalpies difference in (Eqn. 11.6.1) is $h_{s0} - h_{c0} = M_s L_c$, where notations are the same as in the previous section.

Pressures p_c and p_s inside a crystal of radius r_c and in a solution drop of radius r_s are related to the environmental pressure p in Laplace's equations:

$$p_s = p + \frac{2\sigma_{sa}}{r_d}, \qquad p_c = p_s + \frac{2\sigma_{cs}}{r_c}, \tag{11.6.2}$$

where σ_{cs} and σ_{sa} are the surface tensions at the crystal–solution and solution–environment interfaces, respectively. Substituting the enthalpies difference and Laplace's relations (Eqn. 11.6.2) into (Eqn. 11.6.1) and multiplying by T, we obtain:

$$-M_s L_c \frac{dT}{T} - \Delta v dp - \Delta v d\left(\frac{2\sigma_{sa}}{r_d}\right) - v_{c0} d\left(\frac{2\sigma_{cs}}{r_c}\right) + RTv d \ln \tilde{a}_s = 0, \tag{11.6.3}$$

where $\Delta v = v_{c0} - v_{s0}$. Integrating as before, from some reference point T_*, $1/r_d = 0$, $1/r_c = 0$, $p = p_*$, $\tilde{a}_{s*}$ to the current values T, r_d, r_{cr}, p, $\tilde{a}_s$, we obtain

$$M_s L_c^{ef} \ln\frac{T_*}{T} - \Delta\bar{v}\Delta p - \Delta\bar{v}\left(\frac{2\sigma_{sa}}{r_d}\right) - \bar{v}_{c0}\left(\frac{2\sigma_{cs}}{r_c}\right) + vR\bar{T}\ln\tilde{S}_s = 0, \tag{11.6.4}$$

where we introduced the activities ratio $\tilde{S}_s = \tilde{a}_s/\tilde{a}_{s*}$. As in the previous case with deliquescence, if we chose the reference point with saturated activity $\tilde{a}_{s*}$, then $\tilde{S}_s$ is the solution supersaturation. The tildes over a_s and S_s and the bars over volumes are omitted hereafter for brevity.

As in Section 11.3 with deliquescence, there are $c = 3$ components (salt, solvent, and air), and $N_w = 4$ degrees of freedom in this system. If the salt mass is constant, this imposes one more constraint, and $N_w = 3$. If we fix three variables in (Eqn. 11.6.3), T, p, and r_d, then the first three terms on the left-hand side in (Eqn. 11.6.4) vanish. If we additionally assume that variations of the activity coefficient γ_s are weaker than variations of the solution concentration x_s, or the solution is close to the ideal, so that $\tilde{a}_s/\tilde{a}_{s*} = \gamma_s x_s/\gamma_{s*} x_{s*} \approx x_s/x_{s*}$, and there is no ion dissociation, $v = 1$, then the remaining two terms in (Eqn. 11.6.4) yield

$$\left(\frac{2\sigma_{cs}}{r_c}\right) = \frac{R\bar{T}}{v_{c0}} \ln\left(\frac{x_s}{x_{s*}}\right). \tag{11.6.5}$$

This is again a known Freundlich–Ostwald equation (Defay et al., 1966) that describes the effect of the salt crystal radius curvature on its solubility and is similar to the Kelvin (Eqns. 3.8.5) or (3.8.6). If we assume that the surface tension of the crystal σ_{cs} does not depend on the crystal size, Freundlich–Ostwald (Eqn. 11.6.5) shows that the solution concentration in equilibrium with a small crystal is greater than the concentration of the bulk solution and increases as the crystal size decreases. That is, the solubility x_s of small crystals increases as their size decreases—i.e., the smaller crystals are more soluble than the larger ones. If crystals of various sizes are placed in a supersaturated solution, the smaller crystals will dissolute and the larger crystals will grow at their expense. A detailed theory of this process for a polydisperse salt crystal ensemble was developed by Lifshitz and Slezov (1958, 1961) (see also Lifshitz and Pitaevskii, 1997), and was called *coalescence* (with the other meaning than used in meteorology). Lifshitz and Slezov introduced in this problem of coalescence a kinetic equation for the diffusion growth of the salt crystal size spectrum. This kinetic equation was generalized and then applied by Buikov (1961, 1963, 1966a,b), Levin and Sedunov (1966) and others in cloud physics for the diffusion growth of the water drop and crystal size spectra as described in Chapters 5, 13, and 14. It is interesting to note that the kinetic equations for the diffusion growth of drops and crystals, the primary mechanism of initial cloud formation, were formulated only 40–50 years later than the Smoluchowski kinetic equation for coagulation that describes particles growth and precipitation formation at the later stages of cloud development.

The other consequence of (Eqn. 11.6.5) is that a small crystal of radius r_c cannot appear in a solution until its mole fraction $x_s(r_c)$ is greater than the equilibrium value x_{s*} for the bulk solution; that is, the solution should be supersaturated. This fact is substantial for explanation of the phenomenon of efflorescence considered further on. Equation (11.6.4) of this Khvorostyanov–Curry (KC) model is a generalization of the Freundlich–Ostwald (Eqn. 11.6.5) and can be rewritten as

$$\left(\frac{2\sigma_{cs}}{r_c}\right)=\frac{\nu R\overline{T}}{\overline{v}_{c0}}\ln\left(\frac{\tilde{a}_s}{\tilde{a}_{s*}}\right)+\frac{M_s L_c^{ef}}{\overline{v}_{c0}}\ln\frac{T_*}{T}-\frac{\Delta\overline{v}}{\overline{v}_{c0}}\left(\frac{2\sigma_{sa}}{r_d}\right)-\frac{\Delta\overline{v}}{\overline{v}_{c0}}\Delta p. \tag{11.6.6}$$

The terms of the right-hand side show that besides the effect of the crystal radius on solubility, this equation also describes the effects of the temperature T, the radius r_d of the liquid solution drop where a salt crystal nucleates, the external pressure Δp, the ion dissociation ν, and the non-ideality of solutions.

The critical radius r_c of the salt crystal germ in solution can be obtained from (Eqn. 11.6.4) as

$$r_{c,eff}=\frac{2\sigma_{cs}v_{c0}}{M_s L_c^{ef}\ln\left(\frac{T_*}{T}\right)-\Delta v\Delta p-\Delta v\frac{2\sigma_{sa}}{r_d}+\nu R\overline{T}\ln S_s}, \tag{11.6.7}$$

or, in a more compact form

$$r_{c,eff}=\frac{2\sigma_{cs}v_{c0}}{M_s L_c^{ef}\ln\left(\frac{T_*}{T}S_s^{g_s}\right)-\Delta v\Delta p-\Delta v\frac{2\sigma_{sa}}{r_d}}, \tag{11.6.8}$$

where $\Delta v = v_{c0} - v_{s0}$, and the parameter $g_s = \nu R\overline{T}/(L_c^{ef} M_s)$ is the same as in (Eqn. 11.3.9). More compact forms of r_c are

$$r_{c,eff}=\frac{2\sigma_{cs}v_{c0}}{M_s L_c^{ef}\ln\left[\frac{T_*}{T}S_s^{g_s}\exp(-H_{eff})\right]}, \tag{11.6.8a}$$

$$r_{c,eff}=\frac{2\sigma_{sa}}{\rho_{c0}L_c^{ef}\ln\left[\frac{T_*}{T}S_s^{g_s}\exp(-H_{eff})\right]}, \tag{11.6.8b}$$

where the term H_{eff} describes the effects of the pressure and finite size of the solution drop:

$$H_{eff}=\frac{\Delta v}{L_c^{ef}M_s}\left(\Delta p+\frac{2\sigma_{sa}}{r_d}\right)=\frac{\Delta\rho_s}{L_c^{ef}\rho_{c0}\rho_{s0}}\left(\Delta p+\frac{2\sigma_{sa}}{r_d}\right). \tag{11.6.9}$$

The critical energy $\Delta F_{cr,eff}$ of homogeneous nucleation of the efflorescence crystalline germ in solution can be derived in terms of the classical nucleation theory from the minimum of the sum of the volume and surface energies similar to the homogeneous freezing in Section 8.6, which yields

$$\Delta F_{cr,eff} = \frac{4}{3}\pi\sigma_{cs} r_{c,eff}^2 = \frac{16\pi}{3}\frac{v_{c0}^2\sigma_{cs}^3}{\left\{M_s L_c^{ef}\ln\left[\frac{T_*}{T}S_s^{g_s}\exp\left(-H_{eff}\right)\right]\right\}^2} \tag{11.6.10}$$

$$= \frac{16\pi}{3}\frac{\sigma_{cs}^3}{\left\{\rho_c L_c^{ef}\ln\left[\frac{T_*}{T}S_s^{g_s}\exp\left(-H_{eff}\right)\right]\right\}^2}. \tag{11.6.10a}$$

Equation (11.6.9) shows that the value of $H_{eff} > 0$. Therefore, this factor causes a decrease of the denominators in $r_{c,eff}$ and $\Delta F_{cr,eff}$, and an increase in $r_{c,eff}$ and $\Delta F_{cr,eff}$, and so hampers efflorescence, making it energetically more difficult. Thus, an increase of the external pressure and a decrease in the radius of the solution drop suppress efflorescence and shifts it to the lower humidities—i.e., ERH decreases with increasing pressure and decreasing drop radius.

The nucleation rate of homogeneous efflorescence [$\mathrm{cm^{-3}s^{-1}}$] can be written by analogy with homogeneous freezing described in Chapter 8:

$$J_{eff}^{\mathrm{hom}} = C_{eff}^{\mathrm{hom}}\exp\left(-\frac{\Delta F_{cr,eff}}{kT} - \frac{\Delta F_{act,eff}}{kT}\right), \tag{11.6.11}$$

$$C_{eff}^{\mathrm{hom}} = 2N_{cc}\left(\frac{v_{s0}}{v_{c0}}\frac{kT}{h}\right)\left(\frac{\sigma_{cs}}{kT}\right)^{1/2} = 2N_{cc}\left(\frac{\rho_{c0}}{\rho_{s0}}\frac{kT}{h}\right)\left(\frac{\sigma_{cs}}{kT}\right)^{1/2}, \tag{11.6.12}$$

where $\Delta F_{act,eff}$ is the activation energy barrier for efflorescence and all the quantities were defined in Section 11.3 and after (Eqn. 11.3.15).

As discussed in Section 11.2, if a foreign substrate (e.g., insoluble salt or a mineral dust particle) with a radius r_N is present inside a solution drop, then efflorescence may proceed via heterogeneous mode similar to heterogeneous freezing described in Chapter 9. The critical radius of a salt germ can be calculated with (Eqns. 11.6.8a), (11.6.8b), but the critical energy should be modified by multiplication by the geometric factor that arises due to the contact angle θ_{cs} between the salt and the foreign substance:

$$\Delta F_{cr,eff} = \frac{16\pi}{3}\frac{v_{c0}^2\sigma_{cs}^3 f(m_{cs},x)}{\left\{M_s L_c^{ef}\ln\left[\frac{T_*}{T}S_s^{g_s}\exp\left(-H_{eff}\right)\right]\right\}^2}, \tag{11.6.13}$$

where $m_{cs} = \cos(\theta_{cs})$, $x = r_N/r_{c,eff}$, and $f(m_{cs}, x)$ is described by (Eqns. 9.7.16), (9.7.16a), or (9.7.16b), depending on the geometry of efflorescence. The heterogeneous nucleation rate on the entire surface of a salt crystal [$\mathrm{s^{-1}}$] can be calculated similar to (Eqn. 11.6.11), but replacing C_{eff}^{hom} with the kinetic coefficient for heterogeneous nucleation:

$$J_{eff}^{het} = C_{eff}^{het}\exp\left(-\frac{\Delta F_{cr,eff}}{kT} - \frac{\Delta F_{act,eff}}{kT}\right), \tag{11.6.14}$$

$$C_{eff}^{het} \approx N_{cc}\left(\frac{kT}{h}\right)c_{1cs}4\pi r_{c,eff}^2, \tag{11.6.15}$$

where N_{cc} is the number of salt molecules in a crystal in contact with the unit area of a solution, c_{1cs} is the number of molecules adsorbed on the surface of the solid. The parameters N_{cc} and c_{1s} can be estimated using the parameters of crystalline lattice for a given salt.

11.7. Applications of the Efflorescence Model

11.7.1. The Temperature Dependence of Efflorescence

Some useful relations can be obtained by solving (Eqn. 11.6.11) relative to $\Delta F_{cr,eff}$ and equating to (Eqn. 11.6.10):

$$-kT\ln\frac{J_{eff}}{C_{eff}}-\Delta F_{a,eff}=\frac{16\pi}{3}\frac{v_{c0}^{2}\sigma_{cs}^{3}}{\left\{M_{s}L_{c}^{ef}\ln\left[\frac{T_{*}}{T}S_{s}^{g_{s}}\exp\left(-H_{eff}\right)\right]\right\}^{2}}. \tag{11.7.1}$$

Dividing the unity by both sides of (Eqn. 11.7.1), and then rearranging the terms and taking the square root, this equation is transformed into:

$$\ln\left[\frac{T_{*}}{T}S_{s}^{g_{s}}\exp(-H_{eff})\right]=\Pi_{eff}, \tag{11.7.2}$$

where

$$\Pi_{eff}=\frac{1}{M_{s}L_{c}^{ef}}\left(-\frac{16\pi}{3}\frac{v_{c0}^{2}\sigma_{cs}^{3}}{kT\ln(J_{eff}/C_{eff})+\Delta F_{a,eff}}\right)^{1/2}. \tag{11.7.3}$$

Equation (11.7.2) yields

$$\frac{T_{*}}{T}\left(\frac{a_{s}}{a_{s*}}\right)^{g_{s}}=\exp(H_{eff})\exp(\Pi_{eff}). \tag{11.7.4}$$

This equation expresses the salt activity a_s at efflorescence as a function of the temperature, or vice versa, the temperature of efflorescence as a function of a_s. Solving (Eqn. 11.7.4) for T, we obtain the temperature of efflorescence T_{eff} in the KC model:

$$T_{eff}=T_{*}\left(\frac{a_{s}}{a_{s*}}\right)^{g_{s}}\exp(-H_{eff})\exp(-\Pi_{eff}). \tag{11.7.5}$$

In expanded form, T_{eff} becomes

$$T_{eff}=T_{*}\left(\frac{a_{s}}{a_{s*}}\right)^{g_{s}}\exp\left[-\frac{\Delta v}{M_{s}L_{c}^{ef}}\left(\Delta p+\frac{2\sigma_{sa}}{r_{d}}\right)\right]$$
$$\times\exp\left\{-\frac{1}{M_{s}L_{c}^{ef}}\left[-\frac{16\pi}{3}\frac{\sigma_{cs}^{3}v_{c0}^{2}f(m_{cs},x)}{kT\ln(J_{eff}/C_{eff})+\Delta F_{act,eff}}\right]^{1/2}\right\}, \tag{11.7.6}$$

where the factor $f(m_{cs}, x)$ is added to account for the possible incomplete wettability by the solution of its own crystal or for the possible generalization for heterogeneous efflorescence. This equation can be applied for both homogeneous and heterogeneous efflorescence, using the corresponding expressions for J_{eff} and C_{eff} from Section 11.5. If efflorescence occurs as homogeneous nucleation with complete wettability, then the geometric factor $f(m_{cs}, x) = 1$.

11.7.2. The Solute Activity and Mole Fraction Dependence of Efflorescence

Solving (Eqn. 11.7.4) for a_s, we obtain an expression for the salt activity of efflorescence:

$$a_{s,eff} = \left(\frac{T}{T_*}\right)^{1/g_s} a_{s*}[\exp(H_{eff})\exp(\Pi_{eff})]^{1/g_s}. \tag{11.7.7}$$

In expanded form, substituting $1/g_s$, this becomes

$$a_{s,eff} = a_{s*}\left(\frac{T_{eff}}{T_*}\right)^{M_s L_c^{ef}/(\nu RT)} \exp\left[\frac{\Delta v}{\nu RT}\left(\Delta p + \frac{2\sigma_{sa}}{r_d}\right)\right]$$
$$\times \exp\left\{\frac{1}{\nu RT}\left[-\frac{16\pi}{3}\frac{v_{c0}^2\sigma_{cs}^3 f(m_{cs},x)}{kT\ln(J_{eff}/C_{eff})+\Delta F_{act,eff}}\right]^{1/2}\right\}. \tag{11.7.8}$$

This equation is simplified for the process at constant pressure ($\Delta p = 0$), and bulk solutions ($1/r_d = 0$):

$$a_{s,eff} = a_{s*}(T_{eff}/T_*)^{M_s L_c^{ef}/(\nu RT)}$$
$$\times \exp\left\{\frac{1}{\nu RT}\left[-\frac{16\pi}{3}\frac{v_{c0}^2\sigma_{cs}^3 f(m_{cs},x)}{kT\ln(J_{eff}/C_{eff})+\Delta F_{act,eff}}\right]^{1/2}\right\}. \tag{11.7.9}$$

If efflorescence proceeds via homogeneous nucleation without foreign substances, then $m_{cs} = -1$, and $f(m_{cs}, x) = 1$. If variations of the activity coefficients γ_s are smaller than variations of the mole fraction, then $a_{s,eff}/a_{s*} \approx x_{s,eff}/x_{s*}$, and (Eqn. 11.7.8) can be rewritten for the mole fraction as

$$x_{s,eff} = x_{s*}\left(\frac{T_{eff}}{T_*}\right)^{M_s L_c^{ef}/(\nu RT)} \exp\left[\frac{\Delta v}{\nu RT}\left(\Delta p + \frac{2\sigma_{sa}}{r_d}\right)\right]$$
$$\times \exp\left\{\frac{1}{\nu RT}\left[-\frac{16\pi}{3}\frac{v_{c0}^2\sigma_{cs}^3 f(m_{cs},x)}{kT\ln(J_{eff}/C_{eff})+\Delta F_{act,eff}}\right]^{1/2}\right\}. \tag{11.7.10}$$

Similar to the estimate of ice nucleation in Chapter 8, the nucleation rate of efflorescence J_{eff} can be determined as described in Section 11.2, from the particle volume V_c at efflorescence and the induction time t_{ind} for the nucleation, assuming that a single nucleation event occurs in a droplet at efflorescence:

$$J_{eff} = \frac{1}{t_{ind}V_c}. \tag{11.7.11}$$

As an example, the mole fraction x_s and weight percent w were calculated for homogeneous efflorescence for ammonium sulfate particles of 1 μm as the functions of temperature using (Eqn. 11.7.10) of this KC model. Calculations were performed with $f(m_{cs}, x) = 1$ (homogeneous nucleation) for the constant pressure ($\Delta p = 0$), and the values of the parameters close to those from Snider and Richardson (1994), Onasch et al. (2000), and others described in Section 11.2. The nucleation rate in (Eqn. 11.7.11) was calculated with $t_{ind} = 1$ s, and the volume V_c for the particles with a radius of 1 μm. The volume of a single-molecule crystal was taken as $v_{c0}^{(1)} = 12.4 \times 10^{-23}$ cm^3/molecule based on the sizes of the crystalline lattice of ammonium sulfate; thus, the molar volume used is $v_{c0} = v_{c0}^{(1)} N_{Av} = 12.4 \times 10^{-23}\,\text{cm}^3 \times 6.02 \times 10^{23}\,\text{mol}^{-1} = 74.4\,\text{cm}^3\,\text{mol}^{-1}$. The activation energies were taken from the cited works as $\Delta F_{act} = 6kT$. The most sensitive and still uncertain parameter is the surface tension σ_{cs} at the interface of the salt crystal surface and solution. Richardson and Snider (1994) found it in the range 50–64 erg cm^{-2}, and Onasch et al. (2000) estimated it as 55 erg cm^{-3} for homogeneous efflorescence at room temperatures. We selected values of σ_{cs} at $T = 298$ K, and then slightly fitted them to match the experimental data as clarified in the following text.

The mole fraction $x_s(T)$ calculated with (Eqn. 11.7.10) was recalculated to the weight percent w as described in Section 3.5. The results are shown in Fig. 11.5 as the temperature-composition phase diagram T-w. The laboratory data from Xu et al. (1998) and from Cziczo and Abbatt (1999, CA) are also shown. These two data sets are close in the region around $T = 300$ K, but differ at the other temperatures, variations in w are weaker in the Cziczo and Abbatt data set. The good agreement with both data sets was achieved using the two temperature-dependent models for the surface tension, and all the other parameters are the same. Thus, the two calculations with the KC model shown in Fig. 11.5 differ only by the temperature-dependent model of the surface tension σ_{cs}: calculations close to Cziczo and Abbatt (1999) (asterisks) are with $\sigma_{cs,1}(T) = [58 + (d\sigma_{cs}/dT)_1 \times (T - 298)]$ erg cm^{-2} with $(d\sigma_{cs}/dT)_1 = 0.07$ erg cm^{-2} K^{-1}, calculations close to Xu et al. (1998) (circles) are with

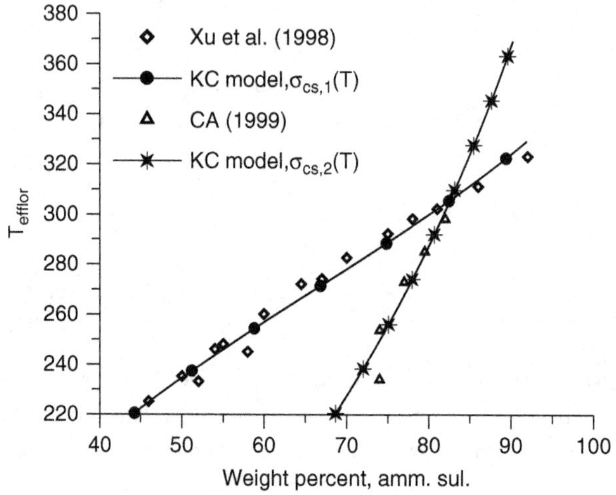

Figure 11.5. Measured and calculated with the KC model the temperature–composition phase diagrams for the efflorescence of ammonium sulfate. Experimental data on the efflorescence are from Xu et al. (1998, diamonds) obtained with the levitation technique and from Cziczo and Abbatt (1999, triangles) obtained with the flow tube technique. The corresponding calculations with the KC model described in the text differ only by the temperature-dependent model of the surface tension σ_{cs}: calculations close to Cziczo and Abbatt (1999) (asterisks) are with $\sigma_{cs,1}(T) = [58 + (d\sigma_{cs}/dT)_1 \times (T - 298)]$ erg cm^{-2} with $(d\sigma_{cs}/dT)_1 = 0.07$ erg cm^{-2} K^{-1}; calculations close to Xu et al. (1998) (circles) are with the $\sigma_{cs,2}(T) = [56 + (d\sigma_{cs}/dT)_2 \times (T - 298)]$ erg cm^{-2}, with $(d\sigma_{cs}/dT)_2 = 0.46$ erg cm^{-2} K^{-1}. The values at 298 K are close, and the slopes $(d\sigma_{cs}/dT)$ are positive in agreement with previous studies, but they significantly differ and may reflect the difference in the conditions of the experiments.

$\sigma_{cs,2}(T) = [56 + (d\sigma_{cs}/dT)_2 \times (T - 298)]\,\mathrm{erg\,cm^{-2}}$, with $(d\sigma_{cs}/dT)_2 = 0.46\,\mathrm{erg\,cm^{-2}\,K^{-1}}$. The values of $\sigma_{cs,2}(T)$ vary from $84.5\,\mathrm{erg\,cm^{-2}}$ at $T = 360\,\mathrm{K}$ to $20.1\,\mathrm{erg\,cm^{-2}}$ at $T = 220\,\mathrm{K}$; variations of $\sigma_{cs,1}(T)$ are much smaller; from 62.3 to $52.5\,\mathrm{erg\,cm^{-2}}$ in this T-range. The calculated $w(T)$ are close to both measured values near 300 K, which indicates the sufficient accuracy of this model. The composition in the case with $\sigma_{cs,1}(T)$ varies from $w = 82\%$ at 294 K to 73% at $T = 234\,\mathrm{K}$, which is also close to the average $w \approx 80\%$ determined by Onasch et al. (1999). The slopes $(d\sigma_{cs}/dT)$ are positive in agreement with previous studies, but significantly differ for the two experimental data sets and may reflect the difference in the conditions of experiments. There is no current agreement on the best measurements, and the model of the surface tension can be clarified when these data are refined.

The same calculations can be performed for heterogeneous efflorescence. According to Richardson and Snider (1994) and Onasch et al. (2000), the values of the contact angle for the tested less soluble salts were ~90°. Then, the contact parameter $m_{cs}(90°)$ is near zero, $m_{cs}(90°) \approx 0$, and Volmer's geometric factor in the equations of Sections 11.5 and 11.6 is described by (Eqn. 9.7.16a), $f(m_{cs}) = (m_{cs}^3 - 3m_{cs} + 2)/4 \approx 1/2$—that is, the critical energy is decreased twice. This value can be substituted into (Eqn. 11.7.10) and will yield the temperature or composition for the heterogeneous efflorescence. For very small particles, the more precise Fletcher's factor should be used, $f(m_{cs}, x)$ described by (Eqn. 9.7.16) in Chapter 9 that accounts for the droplet curvature. Since $f(m_{cs}, \mathrm{x}) < 1$, its presence in Equation (11.7.10) is equivalent to a decrease in the surface tension. As Fig. 11.5 shows, this is equivalent to a decrease in the solution concentration, in agreement with the cited experiments and models on the heterogeneous efflorescence.

Since $\Delta v = v_{c0} - v_{s0} > 0$ for ammonium sulfate and NaCl, (Eqn. 11.7.10) predicts an increase in x_s (solution supersaturation) with decreasing size r_d of the solution drops, in agreement with the model of Gao et al. (2006, 2007). This effect becomes especially large at very small r_d for nanoparticles since it is inversely proportional to r_d. The corresponding calculations can be easily performed with (Eqn. 11.7.10). This equation also predicts an increase in solution supersaturation with increasing pressure. Verification of this prediction requires experimental data on efflorescence at high pressures similar to the high-pressure experiments on water freezing and melting discussed in Chapters 4 and 8.

11.7.3. The Joint Phase Diagram

One of the most general and informative characteristics of solutions are the phase diagrams. Such diagrams are usually constructed based on various empirical data for the solutions (e.g., Curry and Webster, 1999; Martin, 2000, Landau and Lifshitz, v.5, 1958b). The results of Chapters 8, 9, and 11 show that these diagrams can be obtained theoretically. Collecting now the results of the calculations of solutions freezing and melting described in Chapter 8, and the calculations of the deliquescence and efflorescence in this chapter, we can plot the combined phase diagram for the stable and metastable equilibriums of solutions. Fig. 11.6 shows the phase diagram for ammonium sulfate in the $T - w$ coordinates composed of the four temperatures: melting T_m, freezing T_f, deliquescence T_{del}, and efflorescence T_{eff}—all calculated with this extension of the classical theory of ice and salt homogeneous nucleation described in Chapters 8 and 11, and compared with the experimental data for each curve.

Melting temperature T_m from Khvorostyanov and Curry (2004b, crosses) is calculated with (Eqn. 8.7.12b) or (Eqn. 11.5.1) and is compared with the laboratory data from Xu et al. (1998, stars)

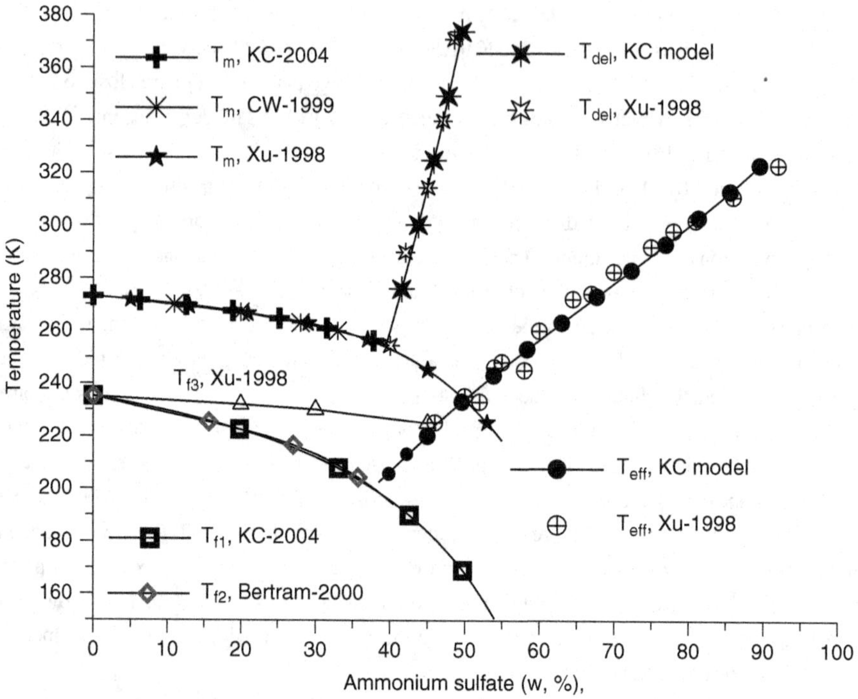

Figure 11.6. Phase diagram for ammonium sulfate in the $T-w$ coordinates composed of four temperatures, melting T_m, homogeneous freezing T_f, deliquescence T_{del}, and efflorescence T_{eff}, all calculated with this modification of the classical theory of ice and salt homogeneous nucleation described in Chapters 8 and 11 (the KC model), and compared with the experimental data for each curve. The melting temperature T_m from Khvorostyanov and Curry (2004b, crosses) calculated with (Eqns. 8.7.12b) or (11.4.1), compared with the lab data from Xu et al. (1998, stars) and with Curry and Webster (1999, asterisks); homogeneous freezing temperature T_{f1} (bold squares) from Khvorostyanov and Curry (2004a) calculated with (Eqn. 8.7.7) here and compared with the lab data T_{f2} from Bertram et al. (2000, open squares) and with the lab data T_{f3} from Xu et al. (1998, triangles); deliquescence temperature T_{del} (KC model) calculated with (Eqn. 11.4.2) here and compared with the experimental data from Xu et al. (1998); efflorescence temperature T_{eff} (solid circles, KC model) calculated with (Eqn. 11.7.6) here and compared with the experimental data from Xu et al. (1998, open circles).

and from Curry and Webster (1999, asterisks). This curve describes the stable equilibrium between ice and the solution. The homogeneous freezing temperature T_{f1} (bold squares) from Khvorostyanov and Curry (2004a) is here calculated with (Eqn. 8.7.7) and compared with the lab data T_{f2} from Bertram et al. (2000, open squares) and with the lab data T_{f3} from Xu et al. (1998, triangles). This curve describes the metastable equilibrium between ice and the solution. The difference between the melting and freezing temperatures describes *the melting–freezing hysteresis*.

The deliquescence temperature T_{del} is calculated with the KC model (Eqn. 11.4.2) here, and is compared with the experimental data from Xu et al. (1998). This curve describes the stable equilibrium between the salt and the solution. The efflorescence temperature T_{eff} (solid circles) is calculated with the KC model (Eqn. 11.7.6) here, and compared with the experimental data from Xu et al. (1998, open circles). This curve describes the metastable equilibrium between the salt and the solution. The difference between the deliquescence and efflorescence temperatures describes *the humidity hysteresis*.

As this figure shows, the melting and deliquescence temperatures from various sources agree well, but there is still some discrepancy among the experimental data from various sources for the freezing and efflorescence curves. In particular, the experimental efflorescence curve $T_{eff}(w)$ from Xu et al. (1998) could be replaced with that from Cziczo and Abbatt (1999) and with the corresponding calculated curve $T_{eff}(w)$ shown in Fig. 11.5 with the different model of the crystal surface tension. Thus, further experiments are needed to refine the exact positions for each curve.

However, a general comparison allows the following conclusions. 1) It is possible, using the general equilibrium (entropy) (Eqn. 3.6.4) of Section 3.6 and the extended classical nucleation theory described in Chapters 8 and 9 and in this chapter, to calculate all components of the phase diagrams for various solutions. Such calculations can be a useful tool for constraining and refining the experimental data. 2) The extended classical nucleation theory of the deliquescence and efflorescence developed here in Chapter 11, along with the corresponding theories of freezing and melting from Chapters, 8, 9, and 10, can be a reliable and precise enough tool for inclusion in numerical models of various complexities for description of the freezing–melting and deliquescence–efflorescence phenomena. This tool may help refine or replace many existing empirical parameterizations of these effects.

12

Terminal Velocities of Drops and Crystals

12.1. Review of Previous Theories and Parameterizations

Whereas a number of factors can contribute to inaccurate simulation of the gravitational fallout of cloud particles and hydrometeors (e.g., inaccuracies in parameterization of particle size distribution, habit and density), a key element in the parameterization of precipitation and gravitational settling of cloud particles and hydrometeors is the terminal velocity.

Probably the first known attempt to explain fall velocity was made by Aristotle in the 4th century B.C. Aristotle observed that heavy stones fall much faster than light leaves falling from trees and concluded that the fall velocity of a body is proportional to its weight. Aristotle did not clarify which velocity was meant: the initial velocity just after initiation of the fall, or the final constant velocity that the body can reach falling in a viscous media, which is now referred to as the *terminal velocity*.

Aristotle's conclusion was not challenged for almost 2000 years. In the 16th and 17th centuries, Galileo and then Newton performed experiments with falling bodies, discovering that the gravitational acceleration g is constant and universal for all bodies. Newton discovered later the gravitational law, which expresses g via the masses of two attracting bodies and the distance between them. That is, g for various planets is proportional to the planet mass and decreases with height above the surface. However, over distances comparable to the depths of the planetary atmospheres and clouds, variations of g are small and usually can be neglected. Assuming that g = const, Galileo and Newton derived the equation for the velocity of a falling body, $V = gt$, that does not depend on the body's mass without accounting for the resistance of the medium, but does depend on time t. It became clear that the fall velocities of heavy bodies measured over short times rather weakly depended on their weight, although there was some difference in the fall speeds for bodies of various shapes. This discovery of the gravitation law and universal gravitational acceleration g seemingly disproved Aristotle's theory.

Subsequently, studies of fall velocities branched into two major lines. One line of research concentrated on the fall velocities of the bodies falling in viscous media with a drag force. Another line of research was devoted to the free fall of the bodies in vacuum. The studies accounting for the drag force showed that the velocity of a body falling in a viscous medium reaches some limiting value, called the *terminal velocity* V_t. Thus, if a body falls at the terminal velocity, its acceleration is close to zero, and depends on the mass of the body. When a body falls in a gravitational field in vacuum without resistance, Galileo and Newton found that the acceleration of a falling body does not depend on its mass. This finding has important consequences. Einstein postulated later that free motion in a gravitational field is equivalent to the motion in a non-inertial coordinate system, when all the bodies move with the same acceleration. This postulate, termed the "*equivalence principle*", is one of the basic premises in the general relativistic theory formulated by Einstein since 1916 (e.g., Landau and Lifshitz, v. 2, 2005, chapters 10 and 11 therein.)

Another two centuries passed after Galileo's studies before Stokes derived and solved the hydrodynamic equations for a viscous fluid (the term fluid henceforth refers to any liquid or gaseous medium), now called the *Navier–Stokes equations.* Stokes found that the drag force F_D of a spherical body with radius r falling with the velocity V in a liquid or air with the dynamic viscosity η is determined by the equation $F_D = 6\pi\eta rV$ (e.g., Landau and Lifshitz, v.6, 1959). The motion of a falling body with density ρ_b is then determined by a balance of the gravitational force $F_{gr} = mg = (4/3)\pi r^3\rho_b$ and the drag force F_D (neglecting buoyancy, electrical, and other external forces that may act on the body). This balance of forces is written as

$$m\frac{dV}{dt} = F_{gr} - F_D = \frac{4}{3}\pi r^3 \rho_b g - 6\pi\eta rV. \tag{12.1.1}$$

Dividing both sides of (Eqn. 12.1.1) by m yields for the acceleration

$$\frac{dV}{dt} = g - \frac{V}{\tau_{v,rel}}, \qquad \tau_{v,rel} = \frac{2}{9}\frac{\rho_b}{\eta}r^2. \tag{12.1.2}$$

The parameter $\tau_{v,rel}$ is *the velocity relaxation time*, during which the falling body reaches a steady-state terminal velocity. The terminal velocity with the Stokes drag force is determined from (Eqn. 12.1.2) with the condition $dV/dt = 0$:

$$V_{St} = g\tau_{v,rel} = \frac{2}{9}\frac{g\rho_b}{\eta}r^2 = C_{St}r^2, \qquad C_{St} = \frac{2}{9}\frac{g\rho_b}{\eta} = \frac{2}{9}\frac{g\rho_b}{\nu\rho_F}, \qquad \nu = \frac{\eta}{\rho_F}, \tag{12.1.3}$$

where ν is the kinematic viscosity related to the dynamic viscosity η and the fluid density ρ_F. The velocity V_{St} is the *Stokes terminal velocity* for the sphere, the equation $V_t(r) = C_{St}r^2$ is *Stokes' law* and C_{st} is *the Stokes parameter.* Calculation with the standard conditions gives $C_{st} = 1.2 \times 10^6\,\mathrm{cm^{-1}\,s^{-1}}$. A general time-dependent solution of (Eqn. 12.1.2) is

$$V(t) = V_{St}(1 - e^{-t/\tau_{v,rel}}). \tag{12.1.4}$$

The limit at small times is determined by the expansion $1 - \exp(-t/\tau_{v,rel}) \approx t/\tau_{v,rel}$, and substitution into (Eqn. 12.1.4) with the use of (Eqn. 12.1.3), yields $V(t) \approx gt$. Thus, $v(t)$ tends to the free fall limit at $t << \tau_{v,rel}$; that is, the viscous drag force can be neglected at $t << \tau_{v,rel}$. The effect of the drag force becomes significant at $t \geq \tau_{v,rel}$, and $V(t)$ tends to V_{St} at $t >> \tau_{v,rel}$.

If a body begins motion in the horizontal direction with initial velocity V_0—e.g., in a turbulent fluctuation—its motion can be described by (Eqn. 12.1.2) with $g = 0$, that is

$$\frac{dV}{dt} = -\frac{V}{\tau_{v,rel}}, \quad \text{and} \quad V(t) = V_0 e^{-t/\tau_{v,rel}}. \tag{12.1.5}$$

This clarifies the meaning of the relaxation time $\tau_{v,rel}$: during this time the initial velocity decreases by e times. The values of $\tau_{v,rel}$ are small—e.g., at the normal conditions, $\tau_{v,rel}(r = 50\,\mu\mathrm{m}) \approx 0.03$ s, $\tau_{v,rel}(r = 100\,\mu\mathrm{m}) \approx 0.12$ s, and $\tau_{v,rel}(r = 200\,\mu\mathrm{m}) \approx 0.5$ s. Two conclusions follow from these estimates: 1) small droplets mostly follow the turbulent fluctuations of the air with very small inertia, allowing the turbulent transport of the drops to be calculated in the same manner as other conservative substances; and 2) differences in the velocities of the particles of various sizes may effect their collisions and accelerate the coagulation and accretion processes.

Returning to Stokes' law, we notice that since $V_t \sim r^2$, and the mass of a spherical body $m \sim r^3$, it follows from (Eqn. 12.1.3) that $V_t \sim m^{2/3}$; thus, Aristotle's intuitive guess $V_t \sim m$ was not so far from reality. However, Stokes' law is valid only for spherical bodies, and the size dependencies of the terminal fall velocities of nonspherical bodies (e.g., large drops and ice crystals) are more complex.

Precise parameterization of gravitational settling and fallout of cloud particles and hydrometeors is essential for accurate simulation of precipitation amounts, cloud dissipation, and cloud optical properties. Problems in parameterization of the fallout of cirrus cloud particles have been highlighted recently by the Intercomparison of the Cirrus Cloud and Parcel Models Projects performed within the Working Group 2 on Cirrus Clouds of the GEWEX Cloud System Study (GCSS; Randall et al., 2003). These intercomparison studies showed large differences in calculated ice crystal terminal velocities among similar models and illustrated the high sensitivity of the simulated cloud properties to the parameterization of fall velocity (Starr et al., 2000). The intercomparison studies stimulated a special activity on the sensitivity and comparison studies of the fall velocities. Recent numerical experiments with single column models (Petch et al., 1997) and with the ECMWF model (Stephens et al., 2000) showed that relatively small variations in parameterized ice crystal terminal velocities produce substantial differences in the simulated ice water paths, cloud boundaries, and cloud optical properties. This led to the conclusion that the discrepancies between the model simulated and the satellite-measured global radiative balance can be caused by inaccurate parameterization of fall velocities.

Numerous experimental studies have produced a wealth of data on terminal velocities, which are typically parameterized in the form of power laws by analogy with Stokes' law:

$$V_t = A_v D^{B_v} \tag{12.1.6}$$

where V_t is the terminal velocity, D is the particle diameter or maximum length, and the coefficients A_v, B_v are determined from best fits to the experimental data (e.g., Gunn and Kinzer, 1949; Litvinov, 1956; Bashkirova and Pershina, 1964; Jayaweera and Cottis, 1969; Locatelli and Hobbs, 1974; Beard, 1976, 1980; Heymsfield and Kajikawa, 1987; Mitchell, 1996). A universal dependence of V_t in the form (Eqn. 12.1.6) has not been found since the coefficients vary over the size spectrum; however, several fits to the experimental data have been given with A_v, B_v constant over some subranges of the size spectrum. For example, Rogers and Yau (1989), based on the data of Gunn and Kinzer (1949), gave the following approximation for liquid drops of radius r:

$$V_t = k_1 r^2, \ 0 < r < 40\,\mu\text{m}, \tag{12.1.7a}$$

$$V_t = k_2 r, \ 40\,\mu\text{m} < r < 600\,\mu\text{m}, \tag{12.1.7b}$$

$$V_t = k_3 r^{1/2}, \ r > 600\,\mu\text{m}, \tag{12.1.7c}$$

with $k_1 = 1.19 \cdot 10^6\,\text{cm}^{-1}\,\text{s}^{-1}$, $k_2 = 8 \cdot 10^3\,\text{s}^{-1}$, and $k_3 = 2.2 \cdot 10^3\,(\rho_{a0}/\rho_a)^{1/2}\,\text{cm}^{1/2}\,\text{s}^{-1}$. Similar parameterizations have been developed for ice crystals—for example, Starr and Cox (1985) found the best fit for crystals in cirrus clouds using five subregions of the size spectrum with different coefficients. Parameterizations of the power-law type (Eqns. 12.1.6) and (12.1.7) are used in many cloud models, general circulation models, and remote sensing techniques; however, attempts to derive these coefficients theoretically have been scarce.

Stokes' theory applies to spherical drops in a laminar viscous flow. Subsequent studies showed that a turbulent boundary layer and wake form around and behind the falling body, which influences the terminal velocity in a manner that is characterized by the Reynolds number $\mathrm{Re} = DV_t/\nu$, where D is the characteristic particle dimension and ν is the kinematic viscosity. The thickness of the turbulent boundary layer around the falling body can be estimated from the Navier–Stokes equations for the vertical velocity component u_i in a viscous incompressible fluid (Monin and Yaglom, 2007a)

$$\frac{\partial u_i}{\partial t} + u_\alpha \frac{\partial u_i}{\partial x_\alpha} = F_\alpha - \frac{1}{\rho_F}\frac{\partial p}{\partial x_i} + \nu \Delta u_i, \qquad (12.1.8)$$

where summation over three components is assumed by the repeated twice subscript α, F_α is an external force, and p is pressure. The thickness of the boundary layer around the body can be estimated by assuming steady-state ($\partial u_3/\partial t = 0$). Then, the balance of forces exists between the nonlinear second term on the left-hand side and the viscous force, the last term on the right-hand side. If the velocity of the flow away from the object is V, and the object's characteristic dimension is D, then an estimate for the nonlinear term is $u_3(\partial u_3/\partial x_3) \sim V^2/D$. If the depth of the boundary layer is δ, then the term $\partial^2 u_3/\partial z^2 \sim V/\delta^2$, and the viscous term is $\nu V/\delta^2$. Equating these two terms in the Navier–Stokes equation yields

$$\frac{V^2}{D} \sim \nu \frac{V}{\delta^2}, \quad \text{or,} \quad \delta \sim \frac{\nu D}{V} = \frac{D}{\mathrm{Re}^{1/2}}. \qquad (12.1.9)$$

Equation (12.1.9) provides a quantitative description of the terminal velocities where the boundary layer plays an important role (Section 12.2).

Numerous attempts have been made to extend Stokes' law for larger values of the Reynolds number and nonspherical drops and crystals. However, a general theory remained elusive because flow around a particle and its turbulent wake are complicated and evolve with increasing values of the Reynolds number. The nonsphericity of a hydrometeor further complicates the flow. Some sophisticated numerical models based on the solution of the Navier–Stokes equations have been applied to study the stream functions and vorticity of the turbulent flow around the objects of the simplest shapes: spheres, spheroids, and circular cylinders (for reviews, see Landau and Lifshitz, v.6, 1959; Monin and Yaglom, 2007a,b; Pruppacher and Klett, 1997). However, such models have not been successfully used to develop simple parameterizations like (Eqns. 12.1.6) and (12.1.7).

Substantial progress in finding a simple but general solution to the fall velocity problem was made by Abraham (1970). He showed that fluid motion around a rigid sphere can be divided into two regions: 1) a region close to the object where frictional effects are important; and 2) an outer region where friction may be neglected. The first regime corresponds to the viscous flow around the body with a maximum projected cross-sectional area A and the drag force $F_D(A) = (1/2)C_D\rho_F V_t A$, where C_D is the drag coefficient and ρ_F is the fluid density. Abraham (1970) suggested considering the second regime as the assembly of the body and the boundary layer with total projected area A_t so that the assembly moves in a potential flow with the drag force $F_{D0}(A_t) = (1/2)C_0\rho_F V_t^2 A_t$, where C_0 is the drag coefficient for the potential flow around the assembly without friction. Matching these two regimes, Abraham found a general functional dependence of the drag on Reynolds number.

The next significant step was made by Beard (1980) and Bohm (1989; 1992) who used Abraham's model of the flow around a falling particle to determine a general analytical relation between the

Reynolds number, Re, and Best (or Davies) number, $X = C_D\mathrm{Re}^2$. Then, inverting the definition of Re, Beard (1980) and Bohm (1989, 1992) obtained the general expressions for V_t as a function of X for both the drops and crystals in algebraic form. Mitchell (1996) extended this work, assumed a power-law representation for the X-Re relationship for four different regimes of X over the range $0.01 < X < 10^8$, and found the power-law coefficients for the fall velocities of nonspherical ice crystals using experimental mass-dimension (m-D) and area-dimension (A-D) relations. Mitchell's formulation led to a convenient power-law representation of V_t (D), although discontinuities are introduced at the matching points.

The studies of Beard (1980), Bohm (1989, 1992), and Mitchell (1996) were extended by Khvorostyanov and Curry (2002; hereafter KC02), who improved upon the discontinuous nature of parameterization and derived a power law–type representation for the Re-X relation and for fall velocities with coefficients as analytical continuous functions of X or the particle diameter D for the entire size range. KC02 applied these equations for evaluation of the fall speeds and found good agreement with experimental data for droplets and several crystal habits.

Mitchell and Heymsfield (2005; hereafter MH05) modified the Khvorostyanov and Curry (2002) scheme. They used the equations for the power law representation of V_t from KC02, and added an empirical correction term to the Re-X relation for turbulent effects with two new empirical constants. The scheme MH05 (this modification of the KC02 scheme in MH05 was called in some subsequent works the "MHKC scheme") showed a better accuracy for large crystal aggregates, but it had a few deficiencies: 1) presence of the two new unknown empirical constants limited its applicability since it required additional tuning; 2) the term added to the X-Re relation to account for the turbulent corrections was too smoothed over the wide range of sizes, from the smallest to the largest, while the experimental data show that the turbulent effect was localized in a rather narrow region of sizes and Re numbers as shown here in the following text; and 3) the drag and fall velocities in the MH05 scheme have incorrect asymptotics and can become negative at large particle dimensions.

Khvorostyanov and Curry (2005b; hereafter KC05) developed further their previous scheme KC02. KC05 did not contain any empirical elements of the MH05 scheme and improved KC02 by several modifications: 1) Instead of using the empirical corrections as in the MH05 scheme, KC05 analytically derived the turbulent corrections to the Reynolds number and to the coefficients of the power law; 2) The constants for crystals recommended by Böhm (1989, 1992) were used in calculations; and 3) The temperature and pressure corrections to the terminal velocities were analytically derived, which improved the accuracy in applications in wide regions of heights and temperatures. No further empirical corrections were needed, and the fall speeds of both droplets and crystals (including large aggregates and hail) could be evaluated with the KC05 scheme with a sufficiently high accuracy.

This chapter, mostly based on the works by Khvorostyanov and Curry (2002, 2005b), presents a unified treatment of cloud particle fall velocities for both liquid and crystalline cloud particles over the wide size range observed in the atmosphere. The representation is formulated in terms of the Best (or Davies) number, X, and the Reynolds number, Re. The fall velocities are represented as generalized power laws. The coefficients of the power laws for the Re-X relation and for the fall velocities are found as the continuous analytical functions of X or particle diameters over the entire hydrometeor size range. The turbulent corrections for the drag coefficients and terminal velocities are derived.

Analytical asymptotic solutions are obtained for these coefficients for the two regimes that represent large and small particles and correspond to potential and aerodynamic flows, respectively. The analytical temperature and pressure corrections for the wide size ranges are derived and compared with the previous parameterizations. The expressions for the Re-X relation and drag coefficients are applied for spherical and nonspherical drops and several crystal habits with special attention paid to the turbulent corrections. The accuracy of this scheme for the fall velocities is illustrated by the comparison with experimental data and previous formulations for small drops, large nonspherical drops, and various ice crystal habits, from small crystals to aggregates and hail. In these calculations, published mass-dimension and area-dimension relationships are used. Some applications are considered, including parameterizations for the bulk and bin cloud models, climate models, remote sensing, the atmospheres of other planets, and other fluids.

12.2. Basic Equations for Fall Velocities

The drag force F_D around a rigid sphere with radius r and projected area A is obtained by matching the drag forces for these two regions following Abraham (1970), as described in the previous section:

$$F_D = (1/2) C_D \rho_F V_t^2 A = F_{D0} = (1/2) C_0 \rho_F V_t^2 A_t, \tag{12.2.1}$$

where notations are the same as in Section 12.1. The total projected area A_t for a sphere of radius r according to Abraham (1970) is related to the sphere of radius r and the boundary layer depth δ:

$$A_t = \pi (r+\delta)^2 = \pi r^2 (1+\delta/r)^2. \tag{12.2.2}$$

Introducing Reynolds number:

$$\mathrm{Re} = V_t D/\nu = V_t D \rho_F/\eta, \tag{12.2.2a}$$

where $D = 2r$ is the sphere diameter, ν is the fluid kinematic viscosity (related to the dynamic viscosity η and fluid density ρ_F as $\nu = \eta/\rho_F$), and the depth δ is determined from boundary-layer theory using the estimate (Eqn. 12.1.9) through

$$\frac{\delta}{r} = \frac{\delta_0}{\mathrm{Re}^{1/2}}. \tag{12.2.3}$$

Substitution of (Eqns. 12.2.2) and (12.2.3) into (Eqn. 12.2.1), gives the drag coefficient as in Abraham (1970):

$$C_D = C_0 \left(1 + \delta_0/\mathrm{Re}^{1/2}\right)^2. \tag{12.2.4}$$

The constants C_0 and δ_0 were determined by Abraham to be $C_0 = 0.292$ and $\delta_0 = 9.06$ for the drops. Böhm (1989, 1992) determined the better constant for ice particles, $C_0 = 0.6$ and $\delta_0 = 5.83$. Using these values in (Eqn. 12.2.4) for the drops, the limiting value of C_D at low Re is determined to be $C_D = C_0\delta_0^2/\mathrm{Re} = 24/Re$, which is the well-known expression for drag in the Stokes regime (Landau and Lifshitz, v.6, 1959). These values of C_0 and δ_0 give an intermediate solution at Re > 1 between the Stokes solution, which underestimates the drag, and Oseen's solution to the Navier–Stokes equations, which overestimates C_D, and provides excellent agreement with experimental data.

The equation of motion of a falling body is determined by the forces of gravitation $F_{gr} = mg = \rho_b \mathrm{v}_b g$, buoyancy $F_b = \rho_F \mathrm{v}_b g$, drag F_D, and the electrical force $F_{el} = qE$,

$$m\frac{d\vec{v}}{dt} = \vec{F}_{gr} + \vec{F}_b + \vec{F}_D + \vec{F}_{el}, \tag{12.2.5a}$$

where ρ_b is the body density, v_b is its volume, g is the acceleration of gravity, q is the electrical charge of the particles, and E is the electrical field assumed directed upward. The terminal velocity V_t of a falling body in the steady state $dV/dt = 0$, as described by (Eqn. 12.2.5a), follows from (Eqn. 12.2.5a) by projection to the vertical direction. The force F_{gr} is directed downward, and the other three forces are upward. Thus, taking into account the signs, V_t is obtained by equating the drag force F_D to the difference of the gravitational force F_{gr} and the sum of the buoyancy force F_b and electrical force F_{el},

$$mg - F_b - F_{el} = (\rho_b - \rho_F)\mathrm{v}_b\, g - qE = F_D = (1/2) C_D\, \rho_F\, V_t^2 A, \tag{12.2.5b}$$

Solving for V_t, we obtain

$$V_t = \left[\frac{2\,|\,mg - F_b - F_{el}\,|}{\rho_F A C_D}\right]^{1/2} = \left[\frac{|\,2g\,\mathrm{v}_b(\rho_b - \rho_F) - 2qE\,|}{\rho_F A C_D}\right]^{1/2}$$

$$= \left[\frac{2g\,\mathrm{v}_b\,|\,\rho_b - \rho_F\,|}{\rho_F A C_D}\right]^{1/2} |\,1 - K_{el}\,|^{1/2}. \tag{12.2.6a}$$

K_{el} *is* the ratio of the electrical force to the difference of the gravitation and buoyancy force,

$$K_{el} = \frac{qE}{|\,\rho_b - \rho_F\,|\,\mathrm{v}_b g}. \tag{12.2.6b}$$

The notation |...| refers to the absolute value, since here we consider the positive differences. Equation (12.2.6a) shows that the effect of the electrical forces can be accounted for by multiplying by the factor $|1 - K_{el}|^{1/2}$, which is equal to 1 in the absence of the electrical charge. Hereafter, we omit this factor for the sake of simplicity, keeping in mind that it can be accounted for in the final equation for V_t. Then the expression for V_t becomes

$$V_t = \left[\frac{2g\,\mathrm{v}_b\,|\,\rho_b - \rho_F\,|}{\rho_F A C_D}\right]^{1/2}. \tag{12.2.6c}$$

The sign |...| of absolute value means that we consider falling bodies if they are denser than the fluid ($\rho_b > \rho_F$) and rising bodies in the opposite case $\rho_b < \rho_F$. For hydrometeors in the air, $\rho_b >> \rho_F$, and the term with ρ_F under the absolute value sign can be neglected, however, if $\rho_b/\rho_F \sim 1$, as, for example, for sand particles in the ocean or for dropsondes or radiosondes in the air or for falling objects in laboratory devices with dense liquids, then ρ_F should be retained.

The dependence of C_D on V_t complicates the solution of (Eqn. 12.2.6c). The problem is usually solved by introducing the Davies or Best number, X:

$$X = C_D\,\mathrm{Re}^2 \tag{12.2.7a}$$

$$= \frac{2(mg - F_b)\rho_F D^2}{A\eta^2} = \frac{2\mathrm{v}_b(\rho_b - \rho_F) g D^2}{A\rho_F \nu^2}, \tag{12.2.7b}$$

where D is the maximum dimension of the body (particle), and we use (Eqn. 12.2.2a) for Re and (Eqn. 12.2.6c) for V_t. The drag coefficient C_D can then be determined as a function of Re and X, which is a parameter that depends only on physical variables. To derive a continuous Re-X relationship and hence a continuous expression for V_t, we proceed as follows. Multiplying (Eqn. 12.2.4) by Re^2, we obtain a relation

$$\mathrm{Re}^2 C_D = X = \mathrm{Re}^2 C_0 (1+\delta_0/\mathrm{Re}^{1/2})^2. \qquad (12.2.8a)$$

Taking the square root, and then rearranging the terms and dividing by $C_0^{1/2}$ yields

$$\mathrm{Re}+\delta_0 \mathrm{Re}^{1/2} - (X/C_0)^{1/2} = 0. \qquad (12.2.8b)$$

Introducing a new variable $z = \mathrm{Re}^{1/2}$ leads to a quadratic equation for z. Its positive root gives the solution for z, and finally for $\mathrm{Re}(X) = z^2$:

$$\mathrm{Re}(X) = (\delta_0^2/4)[(1+C_1 X^{1/2})^{1/2} - 1]^2, \qquad (12.2.8c)$$

where we have introduced the constant $C_1 = 4/(\delta_0^2 C_0^{1/2})$. This equation was given by Beard (1980) with some numerical coefficients, and Böhm (1989, 1992) refined the values of C_0 and δ_0 for drops and crystals.

The Beard–Böhm relation (Eqn. 12.2.8c) is an algebraic expression by X, not a power law, but it can be presented in the power law form by X:

$$\mathrm{Re}(X) = a_{\mathrm{Re}} X^{b_{\mathrm{Re}}}. \qquad (12.2.9)$$

We will now find a representation for the coefficients a_{Re} and b_{Re} as continuous smooth functions of X in the entire X range. Similar to the method used for the power law representation of the aerosol size spectra in Chapter 6, consider a continuous function $\varphi(X)$ of the argument X with continuous derivative $\varphi'(X)$, where the prime means derivative. This function can be represented in a power-law form by X with the power b and coefficient a, so the function and its derivative are

$$\varphi(X) = aX^b, \qquad \varphi'(X) = abX^{b-1}). \qquad (12.2.10a)$$

Solving these two equations, the coefficients a and b can be expressed via φ, φ' as

$$b(X) = X(\varphi'/\varphi), \qquad a(X) = \varphi/X^b. \qquad (12.2.10b)$$

For the terminal velocity application, $\varphi(X) = \mathrm{Re}(X)$, and we can write, using (Eqn. 12.2.8c),

$$\varphi'(X) = \mathrm{Re}'(X) = \frac{[(1+C_1 X^{1/2})^{1/2} - 1]}{2C_0^{1/2}(1+C_1 X^{1/2})^{1/2} X^{1/2}}, \qquad (12.2.11)$$

where we used the definition of C_1. Substituting $\varphi(X) = \mathrm{Re}(X)$ from (Eqn. 12.2.8c) and $\varphi'(X)$ from (Eqn. 12.2.11) into (12.2.10b), we obtain the following expressions for a_{Re}, b_{Re}:

$$b_{\mathrm{Re}}(X) = X\frac{\mathrm{Re}'}{\mathrm{Re}} = \frac{C_1 X^{1/2}}{2[(1+C_1 X^{1/2})^{1/2} - 1](1+C_1 X^{1/2})^{1/2}}, \qquad (12.2.12)$$

$$a_{\mathrm{Re}}(X) = \frac{\mathrm{Re}(X)}{X^{b_{\mathrm{Re}}}} = \frac{\delta_0^2}{4}\frac{[(1+C_1 X^{1/2})^{1/2} - 1]^2}{X^{b_{\mathrm{Re}}(X)}}. \qquad (12.2.13)$$

Equations (12.2.12) and (12.2.13) provide a power-law representation of Re(X) in (Eqn. 12.2.9) with a_{Re}, b_{Re} being continuous functions of X that are consistent with the algebraic Re-X relation (Eqn. 12.2.8c).

The drag coefficient can be determined using (Eqns. 12.2.7a), (12.2.7b), and (12.2.9) to be

$$C_D = X/Re^2 = a_{\mathrm{Re}}^{-2}\, X^{1-2b_{\mathrm{Re}}} = a_{\mathrm{Re}}^{-2}\left(\frac{2(mg - F_b)\rho_F D^2}{A\eta}\right)^{1-2b_{\mathrm{Re}}}. \tag{12.2.14}$$

Substitution of (Eqn. 12.2.14) into (Eqn. 12.2.6c) yields the following expression for the terminal velocity:

$$V_t = a_{\mathrm{Re}}\, \nu^{1-2b_{\mathrm{Re}}}\left(\frac{2(mg - F_b)}{\rho_F A}\right)^{b_{\mathrm{Re}}} D^{2b_{\mathrm{Re}}-1} \tag{12.2.15a}$$

$$= a_{\mathrm{Re}}\, \nu^{1-2b_{\mathrm{Re}}}\left[\frac{2\mathrm{v}_b\, g}{A}\left(\frac{\rho_b - \rho_F}{\rho_F}\right)\right]^{b_{\mathrm{Re}}} D^{2b_{\mathrm{Re}}-1}. \tag{12.2.15b}$$

It is assumed in the preceding formula that $\rho_b > \rho_F$—that is, the body is denser than the environment and falls down as the particles in the atmosphere. But the density difference in (Eqn. 12.2.15b) should be $(\rho_F - \rho_b)$ in the reverse cases, when $\rho_b < \rho_F$—that is, the body is lighter than the environment and lifts up, like a balloon in the atmosphere or a light buoy in the ocean. These cases are discussed in the following text. In many applications, objects can be approximated by ellipsoids of specified axis ratio $\xi(D)$ as described in Sections 5.1 and 5.2 (e.g., oblate large drops or some crystals), then $\mathrm{v}_b = (\pi/6)D^3\xi(D)$, $A = (\pi/4)D^2$, and $\mathrm{v}_b/A = (2/3)D\xi(D)$. Then,

$$V_t = a_{\mathrm{Re}}\, \nu^{1-2b_{\mathrm{Re}}}\left[\frac{4}{3} g\,\xi(D)\left(\frac{\rho_b}{\rho_F} - 1\right)\right]^{b_{\mathrm{Re}}} D^{3b_{\mathrm{Re}}-1}. \tag{12.2.16}$$

The equations derived in the preceding text are valid for various fluids and bodies. In applications for falling hydrometeors in the atmosphere of a planet, we consider air as the "fluid" and a drop or a crystal as the "body"; henceforth, we use ρ_a (the density of air) in place of ρ_F, and ρ_b denotes the density of drops or crystals. We can use relationships for the particle mass and cross-sectional area to be a function of diameter as

$$m = \alpha D^{\beta}, \qquad A = \gamma D^{\sigma}, \tag{12.2.17}$$

where α, β, γ, and σ vary with particle size, phase, and habit. Incorporating the relations (Eqn. 12.2.17) into (Eqn. 12.2.15a) and neglecting F_b compared to mg (since $\rho_a << \rho_b$), we obtain

$$V_t = A_v\, D^{B_v}, \tag{12.2.18}$$

where the coefficients are

$$A_v = a_{\mathrm{Re}}\, \nu^{1-2b_{\mathrm{Re}}}\left(\frac{2\alpha g}{\rho_a \gamma}\right)^{b_{\mathrm{Re}}}, \tag{12.2.19}$$

$$B_v = b_{\mathrm{Re}}(\beta - \sigma + 2) - 1. \tag{12.2.20}$$

Since the coefficients a_{Re}, b_{Re} are given in (Eqns. 12.2.12) and (12.2.13) as continuous functions of X, expressions (Eqns. 12.2.18)–(12.2.20) give a continuous power-law representation of the terminal

Table 12.1. *Coefficients of mass,* $m = \alpha D^{\beta}$, *and area,* $A = \gamma D^{\sigma}$, *power laws for drops and crystals used in calculations of* a_v, b_v *for the size* D. *The data are from Auer and Veal (1970), Locatelli and Hobbs (1974, LH74), Matson and Higgins (1980), Heymsfield and Kajikawa (1987, HK87), Mitchell (1996, M96), Pruppacher and Klett (1997), Heymsfield and Iaquinta (2000, HI00), and Heymsfield (2003, H03).* $C_0 = 0.292$, $\delta_0 = 9.06$ *for drops;* $C_0 = 0.6$. $\delta_0 = 5.83$ *for crystals.*

	Mass		Area		
Particle Type	α	β	γ	σ	Remark
Spherical drops	$(\pi/6)\rho_w = 0.524$	3	$\pi/4 = 0.785$	2	
Nonspherical drops	$(\pi/6)\rho_w \times \xi(D)$	3	$\pi/4 = 0.785$	2	
Hex. Plates (P1a)					
$15\,\mu m < D < 100\,\mu m$	0.00739	2.45	0.24	2.00	$\alpha = 0.0376$, $\beta = 3.31$,
$100\,\mu m < D < 3000\,\mu m$	0.00739	2.45	0.65	1.85	$A_v = 297$, $B_v = 0.86$ (HK87)
Hex. Columns					
$30\,\mu m < D < 100\,\mu m$	0.1677	2.91	0.684	2.00	
$100\,\mu m < D < 300\,\mu m$	0.00166	1.91	0.0696	1.50	
$D > 300\,m$	0.000907	1.74	0.0512	1.414	
Rimed long columns					
$200\,\mu m \le D \le 2400\,\mu m$	0.00145(old), 0.00125(new)	1.8	0.0512	1.414	
Crystal w/sector-like branches (P1b)					
$10\,\mu m < D < 40\,\mu m$	0.00614	2.42	0.24	1.85	
$40\,\mu m < D < 2000\,\mu m$	0.00142	2.02	0.55	1.97	
Broad-branched crystal (P1c)					
$10\,\mu m < D < 100\,\mu m$	0.00583	2.42	0.24	1.85	
$100\,\mu m < D < 1000\,\mu m$	0.000516	2.02	021	1.76	
Stellar crystal with broad arms (P1d)					
$10\,\mu m < D < 90\,\mu m$	0.00583	2.42	0.24	1.85	
$90\,\mu m < D < 1500\,\mu m$	0.00027	1.67	0.11	1.63	
Densely rimed dendrites (R2b)					
$1800\,\mu m < D < 4000\,\mu m$	0.030 (old), 0.015 (new)	2.3	0.21	1.76	$A_v = 62$, $B_v = 0.33$ (LH74)
Bullet rosettes, 5 branches					
$200\,\mu m < D < 1000\,\mu m$	0.00308	2.26	0.0869	1.57	$A_v = 2150$, $B_v = 1.225$ ($D < 0.06\,cm$), $A_v = 492$, $B_v = 0.7$ ($D > 0.06\,cm$), (HI00)
Aggregates of thin plates	0.00145	1.8	0.2285	1.88	
Crystal aggregates in tropical cirrus	0.0038	1.37	0.196	1.73	Tropical Rain Measurement Mission (TRMM)
Lump graupel (R4b)					
$500\,\mu m < D < 3000\,\mu m$	0.049	2.8	0.50	2.0	LH74, HK87
Hail	0.466	3.0	0.625	2.0	Matson and Higgins

velocity over the entire particle size range. The parameters of *m-D* and *A-γ* relations for the drops and several crystal shapes are given in Table 12.1, based on the works by Auer and Veal (1970), Locatelli and Hobbs (1974), Heymsfield and Kajikawa (1987), Mitchell (1996), Pruppacher and Klett (1997), and Heymsfield and Iaquinta (2000).

12.3. Turbulent Corrections

When Re approaches ~10^3, C_D begins to increase, which is caused by the effects of turbulence in the flow. Böhm (1992) suggested an analytical parameterization of C_D in this transitional regime, whereby the turbulent drag coefficient C_{Dt} is related to the C_{Dl} in a laminar flow using an interpolation function $\psi_{tur}(X)$:

$$C_{Dt}(X) = C_{Dl}(X)\psi_{tur}^{-1}(X), \tag{12.3.1}$$

$$\psi_{tur}(X) = \frac{1+(X/X_0)^k}{1+C_t(X/X_0)^k}. \tag{12.3.2}$$

Equation (12.3.2) shows that limiting values of ψ_{tur} are $\psi_{tur} = 1$ at small $X << X_0$, and $\psi_{tur} = C_t^{-1}$ at large $X >> X_0$. Bohm (1992) used fixed values $C_t = 1.6$, $k = 2$, $X_0 = 6.7 \times 10^6$ for drops, and $X_0 = 2.8 \times 10^6$ for crystals. Thus, $C_{Dt} \approx C_t C_{Dl} = 1.6\ C_{Dl}$ at large X or large Re. The value of $C_t = 1.6$ characterizes an observed in some experiments increase of 60% in the drag coefficient caused by turbulence (see review in Böhm, 1992), while Mitchell (1996) used $C_t = 1.3$ to better conform to the data from Knight and Heymsfield (1983).

Here we generalize Böhm's parameterization. First, substitute (Eqn. 12.3.1) into (Eqn. 12.2.7a) and solve relative to Re:

$$\mathrm{Re}_t(X) = X^{1/2} / C_{Dt}^{1/2} = \mathrm{Re}_l(X)\left[\psi_{tur}(X)\right]^{1/2}, \tag{12.3.3}$$

where Re_t is the turbulent Reynolds number and Re_l is the "laminar" Reynolds number described by the Beard–Böhm relation (Eqn. 12.2.8c). Following KC02, Re_l can be formulated as a power law described by (Eqns. 12.2.9)–(12.2.13) for Re(X) in Section 12.2, and then (Eqn. 12.3.3) provides an analytical representation of the Re-X relation corrected for turbulence. Substituting (Eqn. 12.3.3) into (Eqn. 12.2.12) for b_{Re}, we obtain the coefficient $b_{Re,t}$ corrected for turbulence:

$$b_{\mathrm{Re},t}(X) = X\,\mathrm{Re}_t'(X)/\mathrm{Re}_t(X) = b_{\mathrm{Re},l}(X) + \Delta b_{\mathrm{Re},t}(X), \tag{12.3.4}$$

where $b_{\mathrm{Re},l} = X\,\mathrm{Re}_l'/\mathrm{Re}_l$ is the value of b_{Re} for the laminar flow defined in (Eqn. 12.2.12) and $\Delta b_{Re,t}$ is the turbulent correction:

$$\Delta b_{\mathrm{Re},t} = \frac{X}{2}\frac{\psi_{tur}'}{\psi_{tur}} = -\frac{k(C_t-1)z^k}{2(1+z^k)(1+C_t z^k)}, \tag{12.3.5}$$

where $z = X/X_0$. Substituting (Eqn. 12.3.3) into (Eqn. 12.2.13) for a_{Re}, we find:

$$a_{\mathrm{Re},t}(X) = \frac{\mathrm{Re}_t(X)}{X^{b_{\mathrm{Re},t}}} = a_{\mathrm{Re},l}(X)\xi_t(X), \tag{12.3.6}$$

where $a_{Re,l}$ is the value for the laminar flow defined by (Eqn. 12.2.13) and ξ_t is its turbulent correction:

$$a_{\mathrm{Re},l} = \frac{\mathrm{Re}_l(X)}{X^{b_{\mathrm{Re},l}}}, \qquad \xi_t(X) = \frac{\left[\psi_{tur}(X)\right]^{1/2}}{X^{\Delta b_{\mathrm{Re},t}}}. \tag{12.3.7}$$

The dependencies of the corrections $\Delta b_{Re,t}$ and ξ_t on X and D are described in more detail later in Sections 12.4 and 12.6.

12.4. Asymptotic Values and Applications for Spherical and Nonspherical Particles

The analytical representation of a_{Re}, b_{Re}, A_v, and B_v as given earlier allows for a straightforward estimate of the asymptotic limits, which is useful for estimating the limiting behavior of the fall velocities. It is seen from (Eqns. 12.2.12) and (12.2.13) that the asymptotic values of a_{Re}, b_{Re} are reached at $C_1X^{1/2} << 1$ and $C_1X^{1/2} >> 1$. Thus, a scaling Best parameter X_{sc} can be introduced from the condition $C_1X_{sc}^{1/2} = 1$, or

$$X_{sc} = \frac{1}{C_1^2} = \frac{\delta_0^4 C_0}{16}, \tag{12.4.1}$$

corresponding to $X_{sc} \approx 123$ for drops (rigid spheres) and $X_{sc} \approx 43$ for crystals with values of C_0 and δ_0 cited in Section 12.2. The value of X_{sc} separates the two regimes for small and large particles or the regimes of *potential* ($X << X_{sc}$) and *aerodynamical* ($X >> X_{sc}$) flows. The scheme for V_t described earlier is based upon Abraham's (1970) theory of the boundary layer, which should be valid for Re >> 1. Beard (1980) showed that Abraham's theory can be extended to the small particle limit, which reduces to Stokes' law for drops and appears to be valid with high accuracy for Re << 1 also. The analytical scheme allows obtaining analytical asymptotics both at $X << X_{sc,}$ and $X >> X_{sc}$.

For $X << X_{sc}$, the asymptotic values for the laminar coefficients $a_{Re,l}$, $b_{Re,l}$ are obtained from (Eqns. 12.2.12) and (12.2.13) by expanding into the power series by X:

$$b_{\mathrm{Re},l}(X) = 1, \qquad a_{\mathrm{Re},l}(X) = 1/(C_0\,\delta_0^2). \tag{12.4.2a}$$

This yields $a_{Re,l} = 0.0417$ for drops, in a good agreement with Mitchell's (1996) numerical fit $a_{Re} = 0.0439$, $b_{Re} = 0.97$ in the range of smallest $X = 0.01$ to 10. Since the turbulent correction $\psi_{tur}(X) = 1$ at $X << X_{sc,}$, the limits of the turbulent coefficients $a_{Re,t}$, $b_{Re,t}$ are the same.

In the opposite limit, $X >> X_{sc}$, it follows from (Eqns. 12.2.12) and (12.2.13) that

$$b_{\mathrm{Re},l} = 1/2, \qquad a_{\mathrm{Re},l} = 1/C_0^{1/2}, \tag{12.4.2b}$$

resulting in $a_{Re,l} = 1.85$ for drops ($C_0 = 0.292$) and 1.29 for crystals ($C_0 = 0.6$).

The turbulent corrections influence the behavior of $a_{Re,t}$, $b_{Re,t}$ in the region of large X, which is determined by a major change of the function $\psi_{tur}(X)$ around X_0. With the choice $X_0 = 2.8 \times 10^6$ for crystals in (Eqn. 12.3.2) for ψ_{tur} as in Böhm (1992), evaluation of $\psi_{tur}(X)$ and of $a_{Re,t}(X)$, $b_{Re,t}(X)$ described later in Section 12.6 shows that this region extends from $X_{1t} \sim 3 \times 10^4$ (Re $\sim 2 \times 10^2$) to $X_{2t} \sim 10^8$ (Re $\sim 10^4$). It is seen from (Eqn. 12.3.5) that the turbulent correction $\Delta b_{Re,t}$ tends to zero at asymptotically large X as $1/z^k$, or $1/X^k \sim 1/X^2$ and (Eqn. 12.3.4) shows that turbulence does not effect $b_{Re,t}$ at very large $X > X_2t$ and its asymptotic value coincides with that for the laminar flow. Equations (12.3.5) and (12.3.7) show that since at sufficiently large $X > X_{2t}$, $\Delta b_{Re,t} \to 0$, then $\xi(X) \to [\psi_{tur}(X)]^{1/2} \to 1/C_t^{1/2}$, so that the asymptotic values of $a_{Re,t}$ and $b_{Re,t}$ are

$$b_{\mathrm{Re},t} = 1/2, \qquad a_{\mathrm{Re},t} = 1/(C_0^{1/2} C_t^{1/2}), \tag{12.4.3}$$

(with $C_t = 1.6$).

The limit of the fall velocities at large sizes is considered in the following text. Thus, the asymptotic relations between the coefficients with and without turbulent corrections are

$$b_{\mathrm{Re},t} = b_{\mathrm{Re},l}, \qquad a_{\mathrm{Re},t} = a_{\mathrm{Re},l}/C_t^{1/2}. \tag{12.4.4}$$

Using $C_t = 1.6$ (following Böhm, 1992), (Eqn. 12.4.3) gives the asymptotic values $a_{Re,t} = 1.46$ for drops ($C_0 = 0.292$) and 1.02 for crystals ($C_0 = 0.6$), respectively. With $C_t = 1.3$, as in Mitchell (1996), we obtain $a_{Re,t} = 1.57$ and 1.09 for drops and crystals, respectively. Our analytical limiting value $b_{Re,t} = 1/2$ in (Eqn. 12.4.3) gives the known hydrodynamic limit of the power index, 1/2, for V_t of the large particles, $V_t \sim D^{1/2}$ (Landau and Lifshitz, v.6, 1959), as well as our limit of $a_{Re,t} = 1.09$ with $C_t = 1.3$ for ice crystals is close to Mitchell's numerical fit $a_{Re} = 1.0865$. Thus, (Eqn. 12.4.3) can be used with a good accuracy to estimate the asymptotic regimes of the fall velocities. The Re-X relation (Eqn. 12.2.9), taking into account (Eqn. 12.4.3), reduces for $X >> X_{sc}$ and $C_t = 1.6$ to a simple square root law:

$$\mathrm{Re}(X) = \frac{1}{(C_0 C_t)^{1/2}} X^{1/2}, \qquad (12.4.5)$$

which yields $\mathrm{Re}(X) = 1.46X^{1/2}$ for drops ($C_0 = 0.292$) and $\mathrm{Re}(X) = 1.02\,X^{1/2}$ for crystals ($C_0 = 0.6$), respectively. So, at large $X > X_{2t}$, the turbulent correction with $C_t = 1.6$ according to (Eqn. 12.4.4) leads to a decrease by $1/C_t^{1/2} = 1/1.265$ in the coefficient $a_{Re,t}$ in the power law (Eqn. 12.2.13). According to (Eqns. 12.2.18) and (12.2.19), this results in a decrease of ~26% in V_t for large crystals and aggregates, yielding calculated values of Re and terminal velocities that are close to those observed, as will be shown in Section 12.6. Hence, these equations with turbulent corrections yield a description for the large aggregates and can be used up to very large $X > 10^{12}$ and $\mathrm{Re} > 10^6$, in particular for large hailstones, which is illustrated in Section 12.6.

Turbulence does not change the asymptotic index $b_{Re,t}$ which determines the asymptotic behavior of Re(X) in (Eqn. 12.2.9) and of $V_t(D)$ in (Eqn. 12.2.18). Note however, that the correction Δb_{Re} and the coefficient b_{Re} have minima at moderately large $X \sim (2\text{–}3) \times 10^6$, corresponding to $D \sim 0.3\text{–}2$ cm for the large aggregates and hail. The larger values of $X \geq X_{2t} \sim 10^8$ may not be reached by many crystal habits due to size limitations, and their asymptotic behavior is determined by the intermediate region of transition to the turbulent regime $X_{1t} < X < X_{2t}$, as illustrated in Section 12.6. Also, at $\mathrm{Re} > 10^5$ or $X > 10^{10}$, the asymptotic expression (Eqn. 12.4.5) may become invalid as the "drag crisis" is encountered, which is considered in Section 12.6.2.

Note that the factor $\nu^{1-2b_{Re}}$ in (Eqn. 12.2.19) for A_v along with the asymptotic value for b_{Re} describes the correct dependence of V_t on viscosity: for small particles, $b_{Re} = 1$, and $1\text{–}2b_{Re} = -1$, so $V_t \sim 1/\nu$ (viscous Stokes regime, described by (Eqn. 12.1.3)); while for the large drops, $b_{Re} = 0.5$, and $1\text{–}2b_{Re} = 0$, so V_t is independent of viscosity, as should true be in an aerodynamic regime.

We consider here the application for spherical particles, including drops, graupel, and hail. For simplicity, we examine primarily drops using the water density ρ_w, however, the same equations are valid for spherical graupel and hail after replacing ρ_w with the ice density ρ_i (which may depend also on D and crystal habit). For spherical particles, the mass and area are related to the diameter by $m = (\pi/6)\rho_w D^3$ and $A = (\pi D^2/4)$; thus from m-D and A-D relations (Eqn. 12.2.17) we find that $\alpha = (\pi/6)\rho_w$, $\beta = 3$, $\gamma = \pi/4$, $\sigma = 2$ (see Table 12.1). Substituting these values into (Eqns. 12.2.18)–(12.2.20), we obtain the following expression for spherical particles:

$$A_v = a_{\mathrm{Re}}\, \nu^{1-2b_{\mathrm{Re}}} \left(\frac{4}{3} \frac{\rho_w g}{\rho_a} \right)^{b_{\mathrm{Re}}}, \qquad B_v = 3b_{\mathrm{Re}} - 1. \qquad (12.4.6)$$

This expression can be evaluated for small and large drops using the scaling diameter from (Eqn. 12.4.1), which yields (under standard atmospheric conditions) $D_{sc} \approx 122\,\mu\text{m}$.

For small droplets ($D << D_{cr}$), using the asymptotic values given in (Eqn. 12.4.2a), $a_{Re} = 1/(C_0\delta_0^2)$ and $b_{Re} = 1$, we obtain V_t in terms of radius, r, or diameter $D = 2r$ as

$$V_t(r) = A_v D^{B_v} = A_{vr} r^{B_v}, \qquad A_{vr} = 2^{B_v} A_v, \tag{12.4.7}$$

$$A_v = \frac{4\rho_w g}{3\delta_0^2 C_0 \eta}, \qquad A_{vr} = \frac{16\rho_w g}{3\delta_0^2 C_0 \rho_a \nu}, \tag{12.4.8}$$

$$B_v = 3b_{\text{Re}} - 1 = 3 - 1 = 2. \tag{12.4.9}$$

So, with the power index $B_v = 2$, (Eqns. 12.4.7) and (12.4.8) yield Stokes law $V_t \sim r^2$ (Eqn. 12.1.3), and using Abraham's constants $\delta_0 = 9.06$, $C_0 = 0.292$ in (Eqn. 12.4.8) gives $A_{vr} = 1.2 \times 10^6\,\text{cm}^{-1}\,\text{s}^{-1} = C_{st}$—i.e., the Stokes constant for the drops. These expressions are used in Section 12.5 for verification of the temperature and pressure dependence of the fall speeds for small particles. Thus, these equations based on the Abraham's concept of the "assembly" of the body and its boundary layer generalize the classical Stokes' law as the limit of small spherical particles, in agreement with the correspondence principle.

For large droplets, ($D > D_{cr}$), $a_{Re} = 1.85$ and $b_{Re} = 0.5$ are evaluated from (Eqns. 12.4.2b) and (12.4.3), and V_t is expressed as in (Eqn. 12.4.7) with coefficients

$$A_{vr} = 2^{1/2} A_v = 2^{1/2} a_{\text{Re}} \left(\frac{4}{3} \frac{\rho_w g}{\rho_a} \right)^{1/2}, \qquad B_v = 1/2. \tag{12.4.10}$$

So, $V_t \sim r^{1/2}$, as it should be for spherical particles in aerodynamic regime when C_D does not depend on V_t (Landau and Lifshitz, v. 6, 1959). Evaluation of the coefficient with the laminar value $a_{\text{Re},l} = 1.85$ for drops gives $A_{vr} = 2.72 \times 10^3\,\text{cm}^{1/2}\,\text{s}^{-1}$, which is higher than the corresponding value of $2.2 \times 10^3\,\text{cm}^{1/2}\,\text{s}^{-1}$ given in Rogers and Yau (1989), based on experimental data (see Eqn. 12.1.7c). Introducing the turbulent correction according to (Eqn. 12.4.4), $a_{\text{Re},t} = a_{\text{Re},l}/C_t^{1/2} = 1.46$ yields smaller $A_{vr} = 2.16 \times 10^3\,\text{cm}^{1/2}\,\text{s}^{-1}$, which is very close to Rogers and Yau's value. Thus, the turbulent effect causes a decrease in the fall velocity of large drops by ≈25%.

Terminal velocities of large drops are also influenced by the effects of nonsphericity. The nonsphericity of the drops becomes substantial for the fall velocity at $D > 535\,\mu\text{m}$ (Beard, 1976, 1980; Böhm, 1989). Large falling drops have shapes that can be approximated by an oblate spheroid or a convex upward pancake with maximum diameter D oriented horizontally, perpendicular to the flow, and the smaller diameter D_s oriented vertically. Measurements and models described in Pruppacher and Klett (1997) show that the aspect ratio $\xi_f = D_s/D = 1$ for small drops and decreases with D. We approximate the aspect ratio with the interpolation formula:

$$\xi_f(D) = \exp(-D/\lambda_1) + \frac{1 - \exp(-D/\lambda_1)}{1 + (D/\lambda_2)}. \tag{12.4.11}$$

In Khvorostyanov and Curry (2002), the parameters of the drop shape were chosen as $\lambda_1 = \lambda_2 = 4.7\,\mu\text{m}$, this led to a slight underestimate of V_t by 2–4%. Subsequent tuning showed that a better choice with two shape parameters, λ_1 and λ_2, leads to a better agreement with experimental data of

Gunn and Kinzer (1949). The interpolation (Eqn. 12.4.11) has two limits, $\xi_f(D) \to 1$ at small $D << \lambda_1$, λ_2—i.e., ensures the spherical shape of the small drops, and $\xi_f(D) \to \lambda_2/D$ at large $D >> \lambda_1$, which ensures a decrease of the axes ratio $\xi_f(D)$ with increasing D. The parameters λ_1, λ_2 can be determined from the condition $\xi_f \approx 0.5$ at $D \sim 8.5$ mm, after which drop breakup occurs due to hydrodynamic instability. The values $\lambda_1 = 5.5$ mm and $\lambda_2 = 4.7$ mm ensure good agreement with the experimental data on V_t from Gunn and Kinzer (1949) and the calculations of Beard (1980), as shown in Fig. 2.4. The factor $\xi_f(D)$ in (Eqn. 12.4.11) is close to the nonsphericity factor by Beard and Chuang (1987) and the other parameterizations of $\xi_f(D)$ (Pruppacher and Klett, 1997). Considering the large drop as an oblate ellipsoid, we obtain from the m-D relation (Eqn. 12.2.17) $\alpha(D) = (\pi/6)\rho_w\xi_f(D)$, $\beta = 3$, $\gamma = \pi/4$, $\sigma = 2$. Using (Eqns. 12.2.18)–(12.2.20) for V_t and asymptotics at large D, $b_{Re} = 0.5$ from (Eqns. 12.4.2b) and (12.4.4) and $\xi_f(D) \sim \lambda_2/D$ from (Eqn. 12.4.11), we obtain for large D:

$$V_t(D) = A_v D^{B_v} \sim \alpha^{b_{Re}} D^{b_{Re}(\beta-\sigma+2)-1}$$
$$\sim [\xi_f(D)]^{b_{Re}} D^{b_{Re}(\beta-\sigma+2)-1} \sim D^{-0.5} D^{0.5} \sim D^0. \tag{12.4.12}$$

So, V_t is asymptotically independent on D with $\xi_f(D)$ as (Eqn. 12.4.11). This is in agreement with observations of Gunn and Kinzer (1949) and calculations from Beard (1976, 1980) that show very weak dependence of V_t on the drop size at $D > 4$–5 mm (see Fig. 12.4).

12.5. Corrections for Temperature and Pressure

The equations for V_t were derived earlier for fixed values of pressure p, and temperature T. As seen from (Eqn. 12.2.18)–(12.2.20), there are two kinds of dependencies of V_t on p and T: 1) directly via the kinematic viscosity ν_a or the dynamic viscosity $\eta = \nu\rho_a$ and the air density ρ_a in $A_v(p, T)$, and 2) via dependence of a_{Re}, b_{Re} on X, which also depends on p, T.

Consider first the direct η and ρ_a dependence in $A_v(p, T)$, which is defined in (Eqn. 12.2.19) as

$$V_t \sim A_v(p,T) \sim \nu^{(1-2b_{Re})} \rho_a^{-b_{Re}} \sim \eta^{(1-2b_{Re})} \rho_a^{(b_{Re}-1)}. \tag{12.5.1}$$

The value of η is almost independent of p, and its temperature dependence is $\eta(T) = \eta_0\varphi(T)$, where $\eta_0 = 1.718 \times 10^{-4}$ poise is the dynamic viscosity at 0 °C, and

$$\varphi_\eta(T) = 1 + 0.00285T_c - 6.9\times10^{-6}T_c^2, \qquad T_c < 0°\text{C} \tag{12.5.2a}$$
$$\varphi_\eta(T) = 1 + 0.00285T_c, \qquad T_c > 0°\text{C}. \tag{12.5.2b}$$

$T_c = T - 273.15$ is the temperature in °C (Beard, 1980; Pruppacher and Klett, 1997). Using these expressions with the equation of state $\rho_a = p/(R_aT)$, with R_a being the specific gas constant for the dry air, and substituting them into (Eqn. 12.5.1), we obtain an expression for the velocity V_t at an altitude with some p, T via its value V_{t0} at an altitude with $p = p_0$ and $T = T_0$:

$$V_t(p,T) = C_{pT}V_{t0}, \tag{12.5.3a}$$

$$C_{pT} = \left[\frac{\eta(T)}{\eta(T_0)}\right]^{1-2b_{Re}} \left(\frac{\rho_a(p_0,T_0)}{\rho_a(p,T)}\right)^{1-b_{Re}} \tag{12.5.3b}$$

$$= \left[\varphi_\eta (T) \right]^{1-2b_{\mathrm{Re}}} \left(\frac{p_0}{p} \frac{T}{T_0} \right)^{1-b_{\mathrm{Re}}}. \tag{12.5.3c}$$

The two asymptotic cases follow from (Eqns. 12.5.3c), (12.4.2a), and (12.4.3). For small sizes, the index $b_{Re} = 1$, and $1 - 2b_{Re} = -1$, $1 - b_{Re} = 0$, and thus the pressure or air density dependencies in (Eqns. 12.5.3b) and (12.5.3c) vanish, and V_t weakly depends on the temperature via the function $\varphi_\eta(T)$ as

$$C_{pT} = [\varphi_\eta (T)]^{-1} = [\eta(p,T)/\eta_0]^{-1}, \tag{12.5.4}$$

and its increase from 0 °C to −40 °C is ~12%. This is confirmed by comparison of (Eqn. 12.5.4) to the Stokes limit for the small particles (Eqn.12.1.3) or (Eqns. 12.4.7)–(12.4.9), and to this limit in Beard (1980) where also $V_t \sim \eta^{-1}$, which indicates the validity of this C_{pT} limit in the viscous regime.

For large sizes, $b_{Re} = 0.5$ according to (Eqn. 12.4.3), thus $1 - 2b_{Re} = 0$, and $1 - b_{Re} = 0.5$, then the dependence on $\varphi(T)$ or dynamic viscosity η in (Eqn. 12.5.3c) vanishes, and the p, T-dependencies are described by the square root law by ρ_a or by p, T as

$$V_t(p,T) = c_{pT\infty} V_{t0}, \qquad c_{pT\infty} = \left(\frac{\rho_a(p_0,T_0)}{\rho_a(p,T)} \right)^{1/2} = \left(\frac{p_0}{p} \frac{T}{T_0} \right)^{1/2}. \tag{12.5.5}$$

The dependence (Eqn. 12.5.5) can be obtained directly from (Eqn. 12.2.6c) assuming C_D = const and $\rho_b >> \rho_a$. Then in (Eqn. 12.2.6c), $V_t \sim \rho_a^{-1/2}$, or using the equation of state, $\rho_a = p/(R_a T)$, we come again to (Eqn. 12.5.5) which is typical of the aerodynamic limit and is used in many numerical models for precipitating the hydrometeor species (e.g., Kessler, 1969, Lin et al., 1983; Rutledge and Hobbs, 1983; Fowler et al., 1996).

A comparison of (Eqns. 12.5.4) and (12.5.5) shows that the temperature and pressure dependence of the fall speeds is substantially different for the small ($r \le$ ~30–40 μm or $D \le$ 60–80 μm) and large particles: The velocities of the smaller sizes increase with increasing height much more slowly than those of the large sizes. Hence, the difference in fall speeds between the small and large particles increases with height. This increase between the surface (1000 hPa and 0 °C)—and, e.g., a height with p = 300 hPa, T = −40 °C—is ~12% for small sizes (due to φ_η^{-1}) and 170% for the large sizes according to (Eqn. 12.5.5). Accounting for this effect will lead to acceleration of gravitational coagulation and accretion with growing altitude, which depends on the difference of fall speeds between large and small particles (see Section 5.6).

Variations of V_t with p and T associated with changes in X and a_{Re}, b_{Re} can be estimated using the relations:

$$\delta a_{\mathrm{Re}} \sim \frac{\partial a_{\mathrm{Re}}}{\partial X} \delta X, \qquad \delta b_{\mathrm{Re}} \sim \frac{\partial b_{\mathrm{Re}}}{\partial X} \delta X. \tag{12.5.6}$$

For estimates of X variations, (Eqn. 12.2.7a) and (12.2.7b) can be rewritten using m-D, A-D relations (Eqn. 12.2.17), and equations for $\rho_a(T)$ and $\eta(T)$. Using Equations (12.2.12) and (12.2.13) for a_{Re}, b_{Re}, we performed both an analytical and numerical evaluation of these variations of V_t, which are 1–3 orders of magnitude smaller than the "direct" dependence described earlier and therefore can be neglected.

So, (Eqns. 12.5.3a)–(12.5.3c) describe the dependence of the fall velocities of the hydrometeors on temperature and pressure with sufficient precision over a wide size range and yield correct limits in both viscous and aerodynamic regimes for small and precipitating fractions, respectively. The accuracy of the (Eqns. 12.5.3a)–(12.5.3c) for C_{pT} correction is illustrated in Section 12.6.7 by a comparison with Beard's (1980) parameterization and with the direct numerical calculations of V_t at various heights.

12.6. Results of Calculations

In this section, the fall velocity scheme described here is compared to experimental data and to other methods. The features of the Re-X relation, drag coefficient, and of the coefficients a_{Re} and b_{Re} are illustrated, and this method is applied for evaluation of fall velocities of drops and various crystal habits, which facilitates the interpretation of the physical basis for the accuracy of the fall velocity scheme.

12.6.1. Re-X Relation

Understanding the influence of turbulence on the drag coefficient is explored through calculations of the Re-X relation. Fig. 12.1 shows the Re-X relation calculated using (Eqn. 12.2.9) and continuous functions $a_{Re}(X)$, $b_{Re}(X)$ (Eqns. 12.2.12) and (12.2.13) for various cases, and illustrates the effect of turbulence. A comparison with the Böhm (1992, B92) and Mitchell and Heymsfield (2005, MH05) schemes is also shown. The uppermost curve with open triangles in Fig. 12.1a describes rigid spheres or drops without account for turbulence, calculated using the constants from Abraham (1970) $C_0 = 0.29$, $\delta_0 = 9.06$. The curve calculated for ice crystals without turbulence, using $C_0 = 0.6$, $\delta_0 = 5.83$ recommended by Böhm (1989, 1992) lies lower than the curve for drops and is almost indistinguishable from the corresponding curve in Böhm (1989, 1992), which indicates a good accuracy of the power law approximation (Eqns. 12.2.9), (12.2.12), and (12.2.13) of the Re-X relation. Calculations for both drops and crystals including the turbulent corrections (Eqns. 12.3.3)–(12.3.7) leads to a decrease in $\mathrm{Re}_t(X)$, asymptotically by a factor of $\psi^{1/2} = 1/C_t^{1/2} = 1/1.265$. The curve calculated for the crystals using the KC05 model with turbulence (crosses) is very close to the corresponding Re-X relation from Böhm (1992) (solid diamonds).

The enhanced Figs. 12.1b,c,d show the comparison for the crystals in the three regions of Re numbers. The curves Re(X) calculated with the model of Khvorostyanov and Curry, 2005, KC05) using (Eqns. 12.2.9), (12.2.12), and (12.2.13) are very close to Böhm's (1992, B92) model in all three regions, both with and without turbulent corrections. The Re(X) curve from Mitchell and Heymsfield (2005, MH05) lies below the KC05 and Böhm curves at small $\mathrm{Re} < 300$, causing underestimates of the fall velocities for the small particles (Fig. 12.1b). At the intermediate region $10^6 < \mathrm{Re} < 10^8$, the MH05 curve lies between the curves from KC05 and Böhm calculated with and without turbulence (Fig. 12.1c), which means that the MH05 scheme underestimates the turbulent effect and overestimates the Re(X) and fall velocities. This difference is caused by the different accounting for the turbulent effect in the Böhm and KC05 schemes (the transition function localized in the limited Re-range in agreement with observations) and in MH05 (the transition function smoothed over the entire Re-range). At higher $X > 10^8$ or $\mathrm{Re} > 8 \times 10^3$, the MH05 curve goes lower than the Böhm and KC05 curves with turbulence, overestimating its effect, then MH05 reaches a maximum near $\mathrm{Re} \sim 10^4$, and then diverges with the other curves and abruptly falls almost vertically to negative values $\mathrm{Re} < 0$ (Fig. 12.1d).

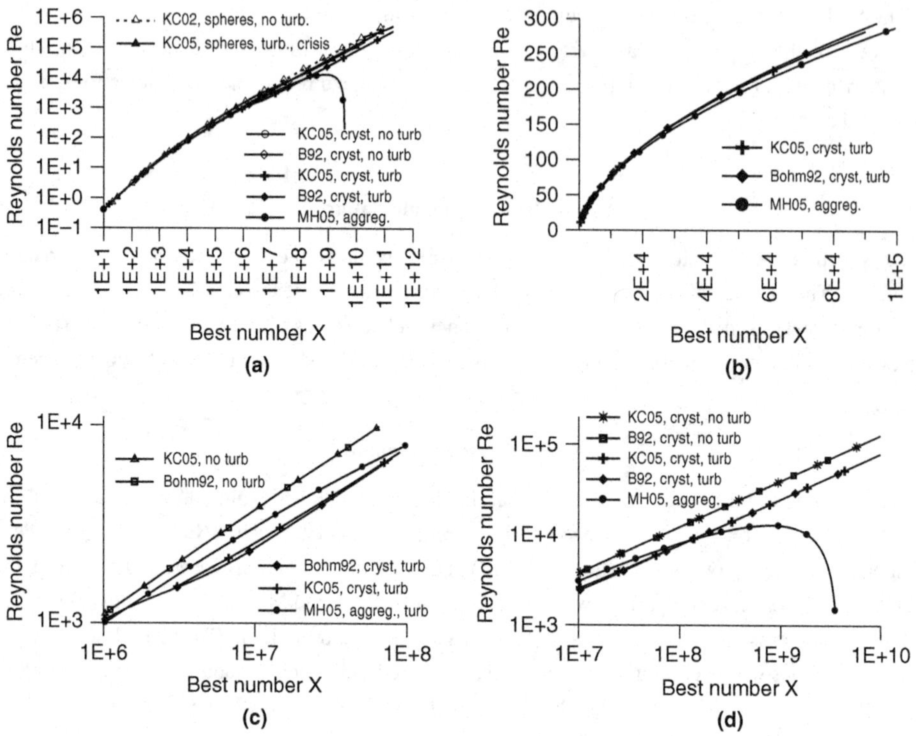

Figure 12.1. The Re-X relation calculated using various models. (a) KC02 with C_0, δ_0 for rigid spheres or drops without turbulent corrections (KC02, open triangles), for spheres with corrections for turbulence and drag crisis (KC05, solid triangles), with C_0, δ_0 for crystals without turbulent corrections (KC05, open circles) and with turbulent corrections (KC05, crosses). This is compared to Böhm (1992) without turbulent corrections (B92, open diamonds) and with turbulent corrections (B92, solid diamonds); and to Mitchell and Heymsfield for crystal aggregates (MH05, solid circles). (b), (c), and (d) are the enhanced fragments of (a) in various regions of X and Re. From Khvorostyanov and Curry (2005b) with changes, *J. Atmos. Sci.*, 62, © American Meteorological Society. Used with permission.

This comparison shows that the power law representation (Eqns. 12.2.9), (12.2.12), and (12.2.13) of the Re-X relation with the parameters C_0, δ_0 for drops and crystals found by fitting it to experimental data by Böhm (1989, 1992), and with the turbulent corrections (Eqns. 12.3.1)–(12.3.7), can be used for an accurate evaluation of crystal fall velocities.

12.6.2. The Drag Coefficient and "Crisis of Drag"

Fig. 12.2 shows calculations of the drag coefficient C_D for rigid spheres and crystals using (Eqn. 12.2.14), and Re(X) using (Eqns. 12.2.9), (12.2.12), and (12.2.13), with turbulent corrections (Eqns. 12.3.1)–(12.3.7) and without these corrections. The results are compared to the calculations of Böhm (1989, B89, and 1992, B92), showing good agreement. At small Re, the drag coefficient C_D decreases

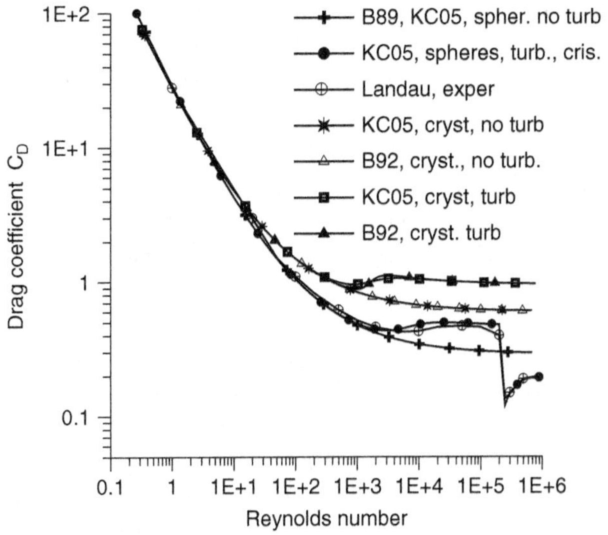

Figure 12.2. Drag coefficient C_D vs. Reynolds number calculated for the spheres and crystals by various methods. Rigid spheres, equations from this chapter based on the model of Khvorostyanov and Curry (2005b, KC05), and Böhm's (1989) for spheres (B89, KC05, crosses), (the curves B89 and KC05 are indistinguishable); equations from this chapter with turbulent corrections and corrections for drag crisis (KC05, solid spheres); experimental data for rigid spheres from Landau and Lifshitz (v. 6, 1959) (Landau, exper., open circles); equations from this chapter with parameters for crystals without turbulent corrections (KC05, asterisks), Böhm's equations without turbulence (B92, open triangles); KC05 model for crystals and turbulent corrections (KC05, cryst., turb., squares), Böhm's equations for crystals with turbulent corrections (B92, solid triangles). From Khvorostyanov and Curry (2005b), *J. Atmos. Sci.*, 62, © American Meteorological Society. Used with permission.

with increasing Re as 24/Re (Stokes' law). At Re > 10–50, this decrease in C_D continues but more slowly, leading to an asymptotic value for rigid spheres $C_D = C_0 = 0.29$ without the turbulent correction (crosses, the curves B92 and KC05 are indistinguishable). However, this behavior diverges at Re > 10^3 with the experimental data given for rigid spheres in Landau and Lifshitz (v.6, 1959) which show a minimum of C_D at Re ~ 2–5 × 10^3, then C_D increases to ~ 0.5 and then becomes almost constant till Re ~ 2 × 10^5. A similar increase of the measured C_D at Re > 2 × 10^2 to 10^3 was illustrated by Beard (1980) for the large drops, some idealized shapes (disk, oblate, and prolate spheroids), several crystal habits, and graupel.

These features are explained by the complex structure of the boundary layer and the turbulence around the falling body. At Re ≥ 10^3, the boundary layer surrounding the falling body is becoming turbulent and can be divided into three parts: the laminar boundary layer in the front part of the body, then the turbulent boundary layer farther downstream, and finally, the turbulent wake after the *separation line* (the roughly circular line in the plane perpendicular to the flow where the flow separates from the body forming the wake). It was noted in Landau and Lifshitz (1959) and Beard (1980) that the increase in C_D at Re > 10^3 and the relative maximum $C_D \approx 0.5$ between Re = 10^4 and

10^5 occurs because the separation line moves upstream, but it is not described by Abraham's theory which assumes a constant separation point. As Fig. 12.2 shows, the description is improved with turbulent corrections: C_D increases at $\mathrm{Re} > 2 \times 10^3$ to a larger value of $C_D = 0.29 \times 1.6 = 0.47$ (KC05, solid spheres) and merges with the experimental data. Thus, accounting for the turbulent corrections lets us describe this shift of the separation line upstream at $\mathrm{Re} \geq 10^3$, when the boundary layer becomes wider, increasing the drag.

An important and interesting feature of C_D for rigid spheres is an abrupt decrease at $\mathrm{Re}_{cris} \sim 2 \times 10^5$ by a factor of 4–5, which is referred to as the "crisis of drag" (Landau and Lifshitz, v. 6, 1959). At $\mathrm{Re}_{cris} \sim (2 \text{ to } 3) \times 10^5$, the separation line moves backward to the rear part of the sphere, and the turbulent wake behind the body abruptly becomes much narrower, causing the drag coefficient to decrease by a factor of 4–5 (Fig. 12.2). At $\mathrm{Re} > \mathrm{Re}_{cris}$, the separation line begins to move upstream to the front part of the sphere again, the boundary layer becomes thicker and the drag increases. KC05 found that a continuous parameterization of experimental data on C_D for rigid spheres up to $\mathrm{Re} > 10^6$ that takes into account turbulence provides the basis for the following formulation of the drag crisis. For $\mathrm{Re} < \mathrm{Re}_{cris} \sim 2 \times 10^5$, C_D is described by (Eqn. 12.2.7a) with Re(X) from (Eqns. 12.2.9), (12.2.12), and (12.2.13), the coefficients $C_0 = 0.292$, $\delta_0 = 9.06$ for the spheres, and turbulent corrections (Eqns. 12.3.1)–(12.3.7). For $\mathrm{Re} \geq \mathrm{Re}_{cris}$, C_D is described using these same equations, but with a stepwise decrease of C_0 from 0.29 to 0.12 at the crisis point $\mathrm{Re}_{cris}(X_{cris})$.

Fig. 12.2 shows satisfactory agreement between the values of C_D calculated with these equations using the following parameter values: $k = 2$ (parameter in the turbulent transition function $\psi_{tur}(X)$ (Eqn. 12.3.2), $X_{cris} = 4.8 \times 10^{10}$, $C_t = 1.6$ (curve KC05, solid circles) and the experimental data from Landau and Lifshitz (v.6, 1959, open circles) over the entire region of Re up to 10^6. The asymptotic value at $\mathrm{Re} \sim 10^6$ according to (Eqns. 12.3.1) and (12.3.2) is $C_D = 0.12 \times 1.6 = 0.19$, close to the value given in Landau and Lifshitz (v.6, 1959).

Calculations with the KC05 scheme but with $C_0 = 0.6$, $\delta_0 = 5.83$ for the crystals and the parameter $k = 1$ in (Eqn. 12.3.2) for $\psi_{tur}(X)$ shown in Fig. 12.2 yield greater values of C_D than for the drops (KC05, asterisks), which means greater drag force and smaller crystal fall velocities. The same increase at $\mathrm{Re} > 10^3$ with an accounting for the turbulent correction (solid diamonds) is caused by a displacement of the wake separation line forward to the front part of a crystal (against the flow), and C_D is again in good agreement with Böhm's (1992) method without and with turbulence (open and solid triangles), with the experimental data from Beard (1980) and with the field data for graupel from Heymsfield and Kajikawa (1987) that also exhibited an increase in C_D at the large Re. It is not clear whether the drag crisis regime can be achieved for ice crystals falling in the atmosphere. For the largest hailstones with $D \sim 10\,\mathrm{cm}$ and $V_t \sim 40\,\mathrm{m\,s^{-1}}$ (Matson and Huggins, 1980, Pruppacher and Klett, 1997), $\mathrm{Re} \sim 3 \times 10^5$ which exceeds Re_{cris} for drops. If the drag crisis occurs for crystals near $\mathrm{Re}_{cris} \sim 2 \times 10^5$ as for the spheres, this may mean an abrupt increase in V_t for large hailstones at $D \sim 10\,\mathrm{cm}$; however, if Re_{cris} for the crystals is greater, the drag crisis may not occur for these sizes. This possible occurrence should be verified experimentally.

12.6.3. Application to Drops

Fig. 12.3 compares the continuous coefficients A_v, B_v for the drops calculated using (Eqns. 12.2.19) and (12.2.20) with stepwise parameterization (Eqns. 12.1.7a), (12.1.7b), and (12.1.7c) from Rogers and Yau (1989).

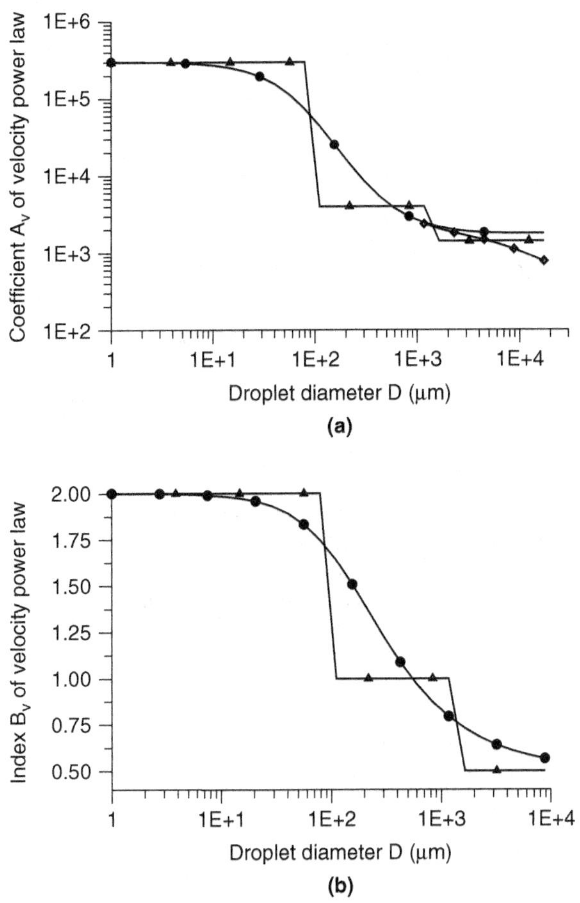

Figure 12.3. Comparison of the continuous coefficients A_ν calculated with the KC02-KC05 model, (Eqn. 12.2.19) (a), and B_ν (Eqn. 12.2.20) (b) of the velocity power law (Eqn. 12.2.18) (solid circles) with stepwise parameterization from Rogers and Yau (1989), (Eqns. 12.1.7a), (12.1.7b), and (12.1.7c) for the spherical drops (triangles) and oblate spheroid (curve with diamonds for A_ν). From Khvorostyanov and Curry (2002), *J. Atmos. Sci.*, 59, © American Meteorological Society. Used with permission.

Although calculations of A_ν, B_ν do not use any input parameters from the Rogers and Yau parameterization based on the experimental data, good agreement is seen.

Fig. 12.4 compares the fall velocities for liquid drops up to $D = 5\,\text{mm}$ calculated using the KC05 model (Eqns. 12.2.18)–(12.2.20) with turbulent corrections (Eqns. 12.3.3)–(12.3.7) and without them, and the experimental data from Gunn and Kinzer (1949). The values $\lambda_1 = 5.5\,\text{mm}$ and $\lambda_2 = 4.7\,\text{mm}$ were chosen in the nonsphericity factor $\xi(D)$ in (Eqn. 12.4.11). This figure shows that: a) both calculated curves with and without turbulent corrections are concave upward, which is caused by nonsphericity increasing with growing D described by $\xi(D)$ in (Eqn. 12.4.11); and b) the effect of turbulence becomes noticeable at $D \geq 1.5$–$2.0\,\text{mm}$, the drag especially increases at $D \geq 4\,\text{mm}$, causes the decrease of V_t as discussed earlier, and makes the calculated velocity close to the observed one, so that V_t at

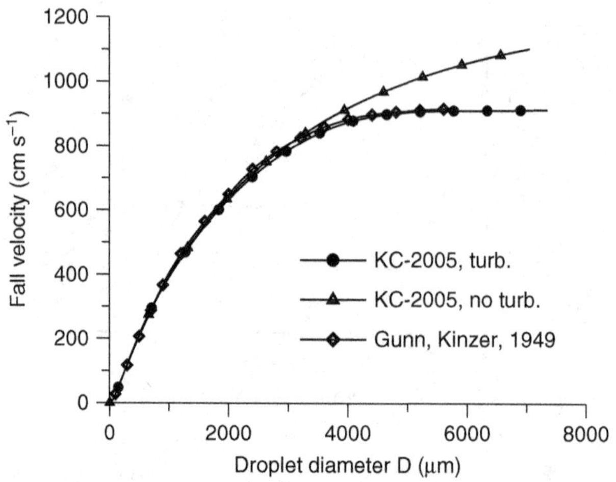

Figure 12.4. Drops' terminal velocities under standard atmospheric conditions calculated using Equation (12.2.16) for the drops with a_{Re}, b_{Re} from (Eqns. 12.2.12) and (12.2.13) and the newer two-parametric nonsphericity parameter (Eqn. 12.4.11). Calculations with turbulent correction and an accounting for nonsphericity (KC2005, turb., circles) practically coincides with the experimental curve from Gunn and Kinzer, 1949 (diamonds). The curve calculated without turbulent corrections (triangles) overestimates V_t at $D > 3500\,\mu$m.

normal conditions tends to the finite limit of about 9 m s^{-1}. At larger sizes, the drops become hydrodynamically unstable and breakup occurs (Pruppacher and Klett, 1997).

12.6.4. Turbulent Corrections and Their Application to Aggregates

The turbulent corrections are especially important for ice crystals that reach sizes of $D > 1$ cm, such as aggregates and hailstones. The coefficient $a_{Re}(X)$ and power index $b_{Re}(X)$ calculated with (Eqns. 12.2.12) and (12.2.13), and turbulent corrections a_{Re} (Eqn. 12.3.6) and b_{Re} (Eqn. 12.3.7) are shown in Fig. 12.5a,b as universal functions of X, independent of crystal type. The values $C_0 = 0.6$, $\delta_0 = 5.83$ were used, just as for crystals in Böhm (1992), Mitchell (1996, M96), MH05, and KC05. Although $C_t = 1.3$ was used in M96 to match observations from Heymsfield and Kajikawa (1987), we found that the value $C_t = 1.6$ recommended by Böhm (1992) (based on observations of a 60% increase in C_D by turbulence) yields better agreement for all considered crystal types. The coefficients $a_{Re,t}(X)$ and $b_{Re,t}(X)$ with turbulent corrections exhibit a maximum and minimum respectively in the X-range 10^5–10^8 as compared to their smooth behavior without turbulent corrections. This non-monotonic behavior is caused by the functions $\Delta b_{Re,t}(D)$ and $\xi_t(D)$ defined by (Eqns. 12.3.5) and (12.3.7) shown in Fig. 12.6a,b for the three crystal habits as the functions of D. The parameters α, β, γ, and σ in the m-D and A-D in calculations were taken from the sources indicated in Table 12.1.

Although these turbulent corrections are universal functions of X, they differ as functions of D due to different crystal habits (parameters α-σ). Fig. 12.6a shows that the correction $\Delta b_{Re,t}$ calculated with (Eqn. 12.3.5) has a minimum of –0.06 for all three crystal habits, centered at $D = 20$ mm for

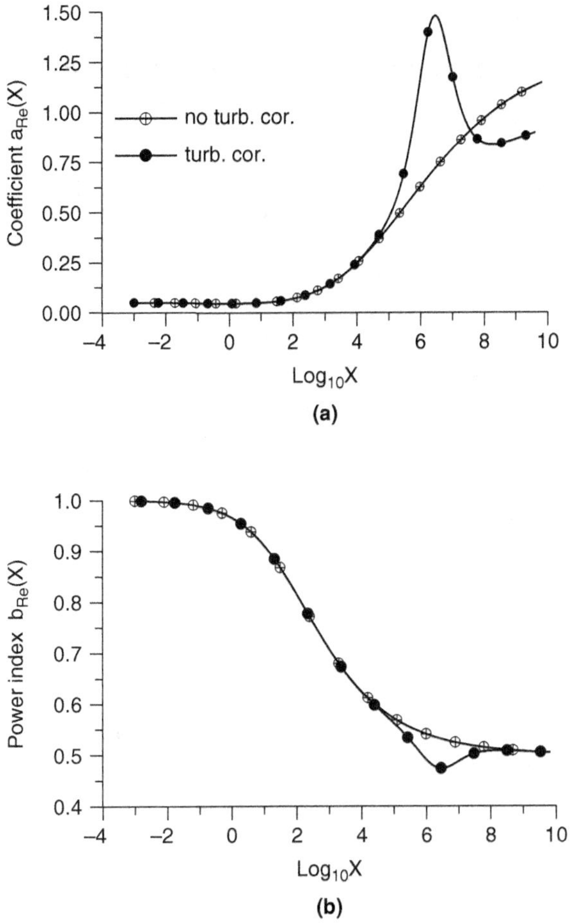

Figure 12.5. (a) Coefficient $a_{Re}(X)$ and (b) power index $b_{Re}(X)$ as functions of the Best parameter X with (solid circles) and without (open circles) turbulent corrections. From Khvorostyanov and Curry (2005b), *J. Atmos. Sci.*, 62, © American Meteorological Society. Used with permission.

aggregates of thin plates, at 3 mm for hail, and at 10 mm for crystals observed in the Tropical Rain Measurement Mission (TRMM) field campaign described in Heymsfield (2003). The location of the minimum is determined by the X-D relation and m-D, A-D relations: The denser a crystal, the smaller the value of D of minimum $\Delta b_{Re,t}$, as seen from the definition of X (Eqn. 12.2.7b). Asymptotically, $\Delta b_{Re,t}$ tends to zero at small and large D—i.e., turbulence does not affect the asymptotic values of the power index b_{Re}. The correction $\xi_t(D)$ begins with 1 at small D, reaches a maximum of 2.1 at the same locations as $\Delta b_{Re,t}$ for these habits, and tends to the asymptotic value $\psi_{tur}^{1/2} = 1/C_t^{1/2} = 0.79$ at large D according to (Eqn. 12.3.7), where $\Delta b_{Re,t} \to 0$ and $X^{\Delta b_{Re,t}} \to 1$. This explains why $a_{Re}(X)$ with turbulence tends in Fig. 12.5a to a smaller value than without turbulence. Since $V_t \sim a_{Re}$, this leads to a decrease in V_t at large sizes due to the turbulence effect that increases the drag force.

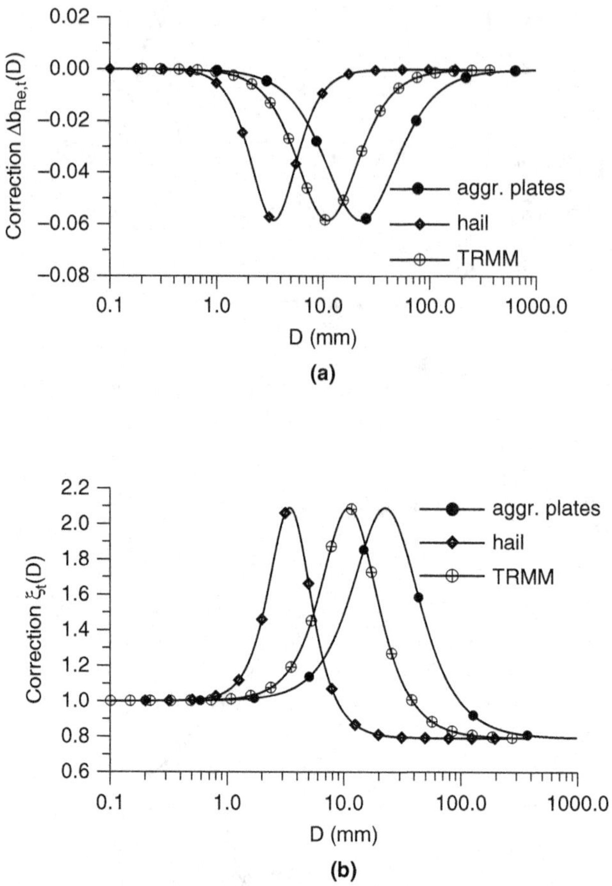

Figure 12.6. (a) correction $\Delta b_{Re,t}(D)$ to b_{Re} and (b) correction $\xi_t(D)$ to a_{Re} as a function of D for aggregates of thin plates (solid circles), hail (diamonds), and crystal aggregates observed in tropical cirrus during TRMM (open circles). From Khvorostyanov and Curry (2005b), *J. Atmos. Sci.*, 62, © American Meteorological Society. Used with permission.

The coefficients a_{Re} and b_{Re} are shown in Fig. 12.7 as functions of D with and without the turbulent corrections, and are compared to the corresponding calculations from MH05. Fig. 12.7 shows that all three curves for a_{Re} and b_{Re} are close at values less than $\text{Log}_{10}D \sim 3.7$ ($D \sim 0.5$ cm). At higher values, the curves KC05 without the turbulent correction exhibit a monotonic increase and decrease and tend to their limiting values, $a_{Re} \to 1$, and $b_{Re} \to 0.5$, as described in Section 12.4. The KC05 coefficients, with an accounting for turbulence, have local extrema in the range $D = 1$–10 cm: a_{Re} has a local maximum of 1.48 (~50% greater than the asymptotic value 1) and b_{Re} has a minimum of 0.47 (smaller than the asymptotic 0.5), both centered at $\log_{10}(D\,\mu\text{m}) \sim 4.4$ or $D \sim 2.5$ cm. This is caused by the behavior of corrections ξ_t and $\Delta b_{Re,t}$ shown in Fig. 12.6. The coefficients a_{Re} and b_{Re} in the KC05 scheme tend to their asymptotic values with a turbulence of 1.02 and 0.5 defined by (Eqn. 12.4.3).

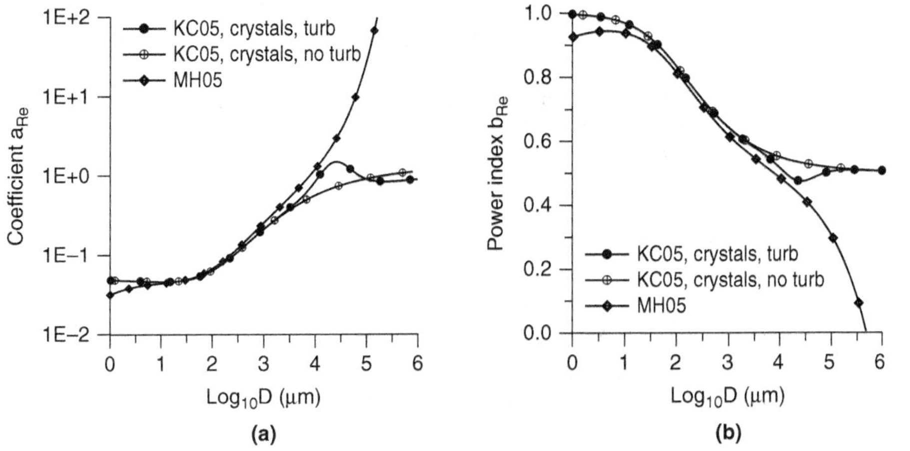

Figure 12.7. (a) Coefficient a_{Re} (Eqn. 12.2.13) of Re-*X* power law relation; and (b) power index b_{Re} (Eqn. 12.2.12) calculated for aggregates of thin plates as functions of *D* with turbulent correction (KC05, turb., solid circles), without correction (KC05, no turb., open circles) and compared to MH05 (diamonds). From Khvorostyanov and Curry (2005b), *J. Atmos. Sci.*, 62, © American Meteorological Society. Used with permission.

The coefficients in the MH05 scheme remain sufficiently close to the KC05 coefficients with turbulence until $\log_{10}D \sim 4.4$ ($D \sim 2.5$ cm), but at larger sizes, the coefficients in the MH05 scheme tend to much larger and smaller values, and do not tend to the correct asymptotic values.

These variations of a_{Re} and b_{Re} determine the behavior of the coefficients A_v, B_v of the velocity power law shown in Fig. 12.8a,b. Coefficients calculated with both schemes, the MH05 modification of the KC02 scheme, and the newer scheme KC05 are close in the range of $\text{Log}_{10}D$ (μm) from 2 to 4.5 (*D* from 100 μm to 3 cm), and then begin to diverge. The fall speeds (Fig. 12.8c) in both schemes are close up to $\text{Log}_{10}D = 4.5$ ($D = 3.1$ cm), and the V_t in the MH05 scheme decreases at larger $D > 3.1$ cm faster than in the KC05 scheme. Aggregates of this and larger sizes up to 10 cm were observed in frontal systems near 0 °C (Lawson et al., 1998). Calculated fall speeds are close in Fig. 12.8c to the field data on V_t by Locatelli and Hobbs (1974) for aggregates of unrimed radiating assemblages and aggregates of unrimed side planes.

An interesting feature of both the KC05 and MH05 schemes is that they predict a decrease in V_t at $D \geq 1$ cm. It is reached by subtracting an additional term from Böhm Re-*X* relation in the MH05 modification of the KC02 scheme and by introducing the turbulent correction in the KC05 scheme (Section 12.3). We used the values $\beta = 1.8$, $\sigma = 1.88$ from Table 12.1 for these aggregates. If the asymptotic value is used without the turbulent correction, $b_{Re} = 0.5$, then (Eqn. 12.2.20) gives $B_v = 0.5 \times (1.8 - 1.88 + 2) - 1 = -0.04$, predicting a decrease in V_t with increasing *D*. However, this asymptotic and negative value of B_v is attained at an unrealistically large *D*. The situation is changed by accounting for the turbulent correction: Due to negative Δb_{Re} (see Fig. 12.6a), B_v becomes negative at $D \sim 1$ to 10 cm, and V_t begins to decrease with *D* in this region. A minimum $b_{Re} = 0.47$ is reached at the more realistic value of $D = 2.5$ cm. The evaluation again from (Eqn. 12.2.20) with these β, σ gives $B_v = -0.1$ (Fig. 12.8b)—i.e., the fall speed decreases with size as $V_t \sim D^{-0.1}$ near this point. This

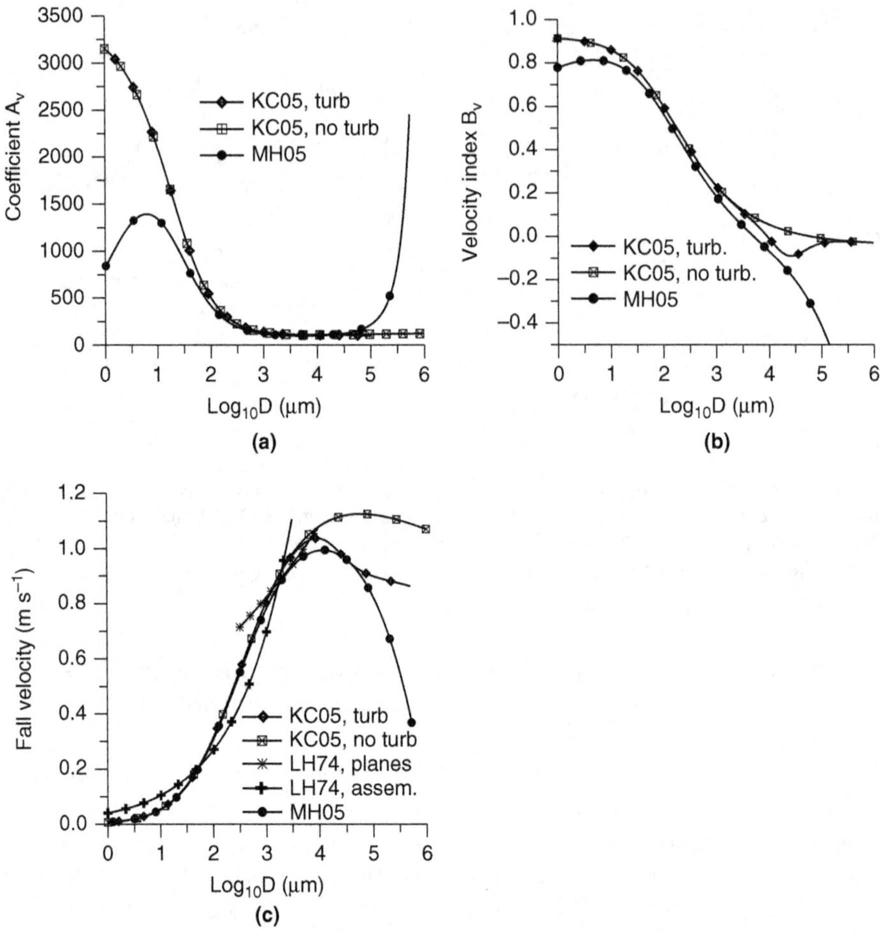

Figure 12.8. (a) Coefficient A_v; (b) power law index B_v; and (c) fall velocity V_t for unrimed aggregates of thin plates calculated with equations (Eqns. 12.2.18)–(12.2.20), C_0, δ_0 for crystals, with turbulent corrections (KC05, turb., solid diamonds), and without turbulent correction (KC05, no turb., open squares), compared to A_v, B_v calculated in MH05 with modified equations of KC02 (solid circles) and to the field data on V_t by Locatelli and Hobbs (1974, LH74) for aggregates of unrimed radiating assemblages (crosses) and aggregates of unrimed side planes (asterisks). From Khvorostyanov and Curry (2005b), *J. Atmos. Sci.*, 62, © American Meteorological Society. Used with permission.

decrease continues, but somewhat slower, at larger D (Fig. 12.8c). This decrease of fall speed with increasing aggregate size would lead to a suppressed gravitational aggregation. As shown in Fig. 12.8c, KC05 predicts at $D \geq 1$–$1.5\,\mathrm{cm}$ a decrease in fall speeds and a partial suppression of further aggregation.

Another comparison of the MH05 and KC05 schemes was performed for aggregates observed in tropical cirrus during the TRMM campaign (Heymsfield, 2003) using the same values of the parameters as in Fig. 12.6. The V_t shown in Fig. 12.9 were calculated with the KC05 scheme for

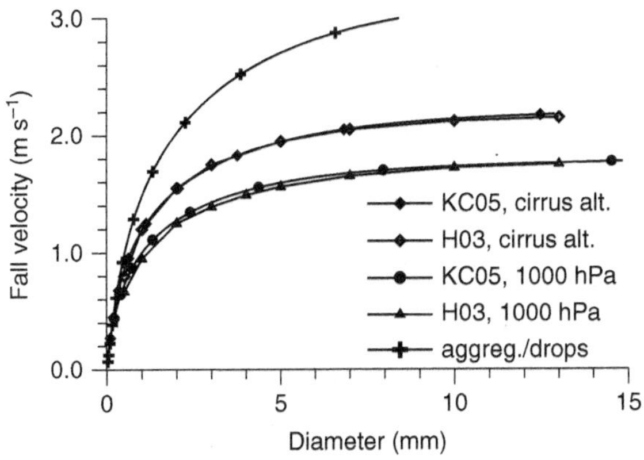

Figure 12.9. Fall velocities calculated with equations of Sections 12.2 and 12.3 and crystal parameters (α, β, γ, σ) from TRMM data by Heymsfield (2003) for the conditions of experiments at cirrus altitudes (solid diamonds, KC05) and recalculated for the surface level at p = 1000 hPa using the equations of Section 12.5 (solid circles, KC05). These are compared to the calculations from Heymsfield (2003) at corresponding altitudes (open diamonds and triangles, H03). The difference does not exceed 5% at 1000 hPa. The curve calculated with the KC02-05 scheme using the same α, β, γ, σ, but C_0, δ_0 for the drops (crosses, aggreg./drops) is given for illustration of the difference between the drops and crystals. From Khvorostyanov and Curry (2005b), *J. Atmos. Sci.*, 62, © American Meteorological Society. Used with permission.

the conditions of the experiment at high cirrus altitude, and recalculated to pressure p = 1000 hPa, using the C_{pT} correction (Eqn. 12.5.3c), and were compared to the corresponding data in Heymsfield (2003), where calculations were performed with an MH05 modification of the KC02 scheme. Fig. 12.9 exhibits good agreement between the two schemes, the difference not exceeding 5%. The curve calculated with the KC02 scheme using the same α, β, γ, σ but C_0, δ_0 for the drops (crosses) is shown for comparison. Its difference from the other two curves illustrates the effect of the particles' shape and the difference between the drops and crystals fall speeds. The fall speeds at these two heights are used in Section 12.6.7 for verification of the altitude correction (Eqn. 12.5.3c).

12.6.5. Other Crystal Habits

A comparison of the KC02–KC05 schemes with the other schemes and experimental data for various crystal habits is shown in Figs. 12.10–12.12. All calculations were performed with C_0, δ_0 for crystals and α-σ parameters from Table 12.1. Note that for crystal habits shown in Figs. 12.8–12.10, Re < 10^3 and the turbulent corrections are insignificant, so that actually the KC02 scheme could be used with C_0, δ_0 for crystals. A very good agreement of the KC02–KC05 schemes is seen with experimental data from Heymsfield and Kajikawa (1987) and Mitchell's (1996) calculations for hexagonal plates (Fig. 12.10), with field data from Locatelli and Hobbs (1974) and Mitchell's (1996) calculations for rimed columns and dendrites (Fig. 12.11), with experimental data from Heymsfield (1972) and Mitchell's (1996) calculations for hexagonal columns (Fig. 12.12), and with experimental parameterization from Brown (1970) and calculations from M96 for plates with sector branches (Fig. 12.12).

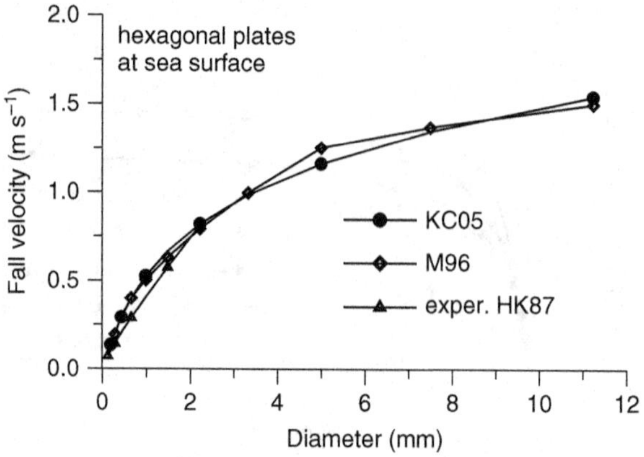

Figure 12.10. Fall velocities of hexagonal plates calculated with the equations of this chapter and parameters C_0, δ_0 for the crystals (KC05, solid circles) and by Mitchell (1996, diamonds, M96), compared up to 12 mm, and parameterization by Heymsfield and Kajikawa (1987) based on experimental data and extended to 2.2 mm (HK87, triangles). From Khvorostyanov and Curry (2005b), *J. Atmos. Sci.*, 62, © American Meteorological Society. Used with permission.

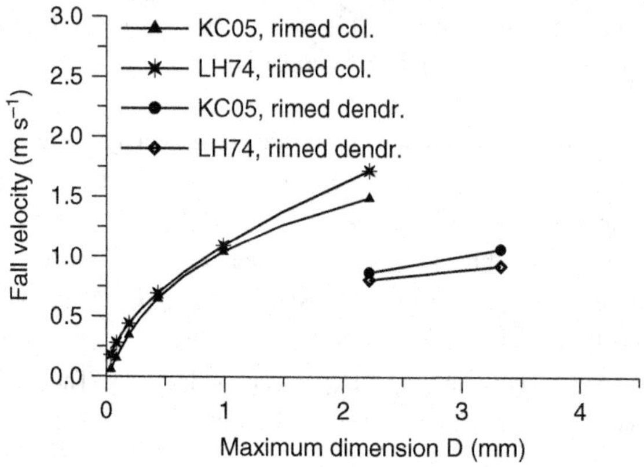

Figure 12.11. Fall velocities for densely rimed columns and dendrites calculated with equations of this chapter and C_0, δ_0 parameters for crystals (KC05, solid triangles and circles) compared to the field data by Locatelli and Hobbs (1974, LH74, asterisks and diamonds). From Khvorostyanov and Curry (2005b), *J. Atmos. Sci.*, 62, © American Meteorological Society. Used with permission.

Results for other dendritic crystals (plates with broad branches, stellars) are close to those for sector branches and to those calculated in M96 for these crystal types.

Thus, the method of fall velocity calculations based on the KC02–KC05 scheme can be applied for various crystal habits. The application of this method for snowflakes can be done with caution and may need modifications due to the high variability of snowflake properties, in particular habits and densities. Studies of snowflakes provide contradictory results—e.g., Magono and Nakamura (1965) predicted a decrease in snowflake fall velocity as $V_t \sim D^{-1/2}$, which was explained by the authors by a

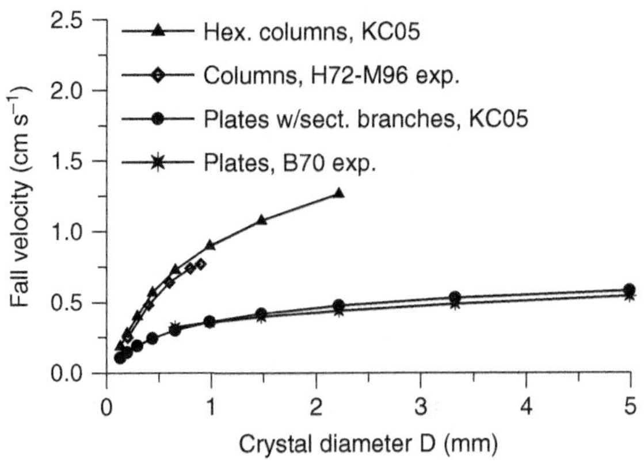

Figure 12.12. Fall velocities for hexagonal columns (solid triangles) and plates with sector branches (solid circles) calculated with equations of this chapter (KC05) and compared to experimental data for columns from Heymsfield (1972) and calculations of Mitchell (1996) (open diamonds, H72-M96) and the parameterization of Brown (1970) for plates (open circles, B70). From Khvorostyanov and Curry (2005b), *J. Atmos. Sci.,* 62, © American Meteorological Society. Used with permission.

significant density decrease with growing size. In contrast, Jiusto and Bosworth (1971) observed and parameterized a velocity increase with size as $V_t \sim D^{0.2}$. This uncertainty in snowflake properties and their high variability indicates a necessity of their further studies—in particular, parameterization of the *A-D*, *m-D*, and density-*D* relations. These relations can be incorporated into the method described here. The complicated shapes of snowflakes crystals may be parameterized using semi-empirical relations based on a combination of the theoretical approach and an empirical correlation analysis. Zawadski et al. (2010) and Szyrmer and Zawadski (2010) provided a review of the natural variability of snow terminal velocity, and parameterizations of the fall velocities of snowflakes, and found an average relationship between the mass of snowflakes and their terminal fall velocity.

12.6.6. Application to Hail

The final test of this scheme was performed with an application to hail. This crystal habit achieves the greatest sizes, $D \sim 10\,\text{cm}$, numbers Re $\sim 3 \times 10^5$, $X \sim 10^{11}$, and is especially sensitive to the turbulent corrections. Fig. 12.13 shows that calculations with the KC05 scheme and α-σ parameters from Table 12.1 are again in very close agreement with the experimental data for hailstones from Matson and Huggins (1980; hereafter MH80). The difference between MH80 and KC05 with turbulent corrections is less than 3.5% at $D = 50\,\text{mm}$ and 2% at $D = 100\,\text{mm}$. V_t without turbulent corrections is 26% higher at $D = 100\,\text{mm}$, which illustrates the effect of turbulence on a_{Re} described by (Eqn. 12.4.4): accounting for turbulence leads to a reduction in a_{Re} and thus in V_t by a factor of $1/C_t^{1/2} = 1/1.26$. Fig. 12.13 shows that our choice of $C_t = 1.6$ recommended by Böhm (1992) leads to agreement with the MH80 data. Note that the turbulent correction $\xi_t(D)$ for hail lies in the region of small $D \sim 2$–$5\,\text{mm}$ (see Fig. 12.6) and does not lead to a noticeable effect in fall speeds at large sizes in contrast to the aggregates (compare with Fig. 12.8c).

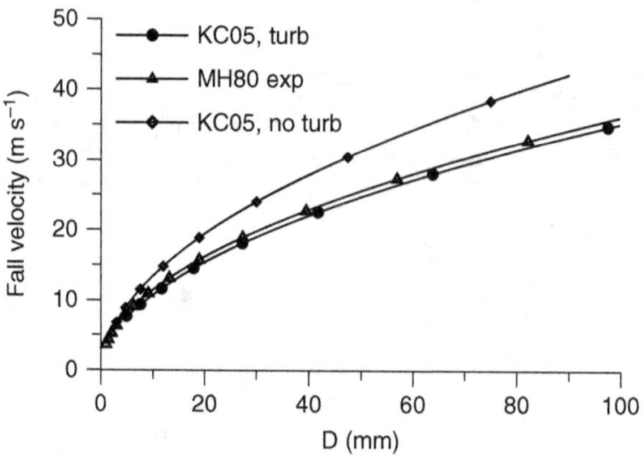

Figure 12.13. Terminal velocity of hailstones V_t calculated with equations from this chapter, δ_0 and C_0 for crystals, and $C_t = 1.6$ with turbulent correction (KC05 turb., solid circles) and without correction (diamonds, no turb.) compared to the experimental data from Matson and Huggins 1980 (MH80, triangles). The difference between KC05 with turbulence and MH80 is less than 3.5% at $D = 50$ mm and 2% at $D = 100$ mm. V_t with turbulent corrections is 26% smaller at $D = 100$ mm than without account for turbulence, upper curve), which illustrates the effect of turbulent corrections $(C_t)^{-1/2} = 1/1.26$. From Khvorostyanov and Curry (2005b), *J. Atmos. Sci.*, 62, © American Meteorological Society. Used with permission.

12.6.7. Altitude Correction Calculations

The altitude or pressure–temperature correction (Eqn. 12.5.3c) was verified by comparing with results from Beard (1980), who fitted slopes of the measured or calculated C_D(Re) curves or used similarity arguments, and then parameterized by constructing an empirical interpolation function C_{pT}(Re) linear by ln(Re) between the viscous C_{pT0} and aerodynamic $C_{pT\infty}$ limits, the same as (Eqns. 12.5.4) and (12.5.5), but with the stepwise constraints $C_{pT} = C_{pT0}$ at Re < 0.2 and $C_{pT} = C_{pT\infty}$ at Re > 10^3. We performed calculations of C_{pT} for the same altitude variation as in Beard (1980) between the reference level with p = 770 hPa, T_c = 0 °C and the top of the atmosphere with p = 232 hPa, T_c = −55 °C. Fig. 12.14a shows four curves of the C_{pT} correction, calculated with the KC05 scheme for crystals with and without turbulent correction for the spheres, and Beard's (1980) parameterization along with several experimental points from that work. It is seen that all four curves in the region $1 < \text{Re} < 5 \times 10^2$ lie close, the difference slightly increasing for smaller particles at Re ~ 0.1 and around large Re ~ 10^3. The latter local maximum (absent on the curve without turbulence) is caused by the turbulent correction in our scheme and allows explanation of an experimentally observed local increase in C_{pT} near Re ~ 10^3 (asterisks) that is not described by Beard's (1980) monotonous function. This agreement in the altitude correction indicates a validity of the KC05 scheme, which provides a theoretical justification and smoothing of Beard's parameterization.

Another test of (Eqn. 12.5.3c) for the C_{pT} correction was performed by comparing its values calculated with (Eqn. 12.5.3c) and $b_{Re,t}$ from (Eqn. 12.3.4) for crystal aggregates in tropical cirrus (see Table 12.1) with the direct calculation of the velocities ratio $C_{pT} = V_t(p,T)/V_t(p = 1000, T = -5\,°\text{C})$

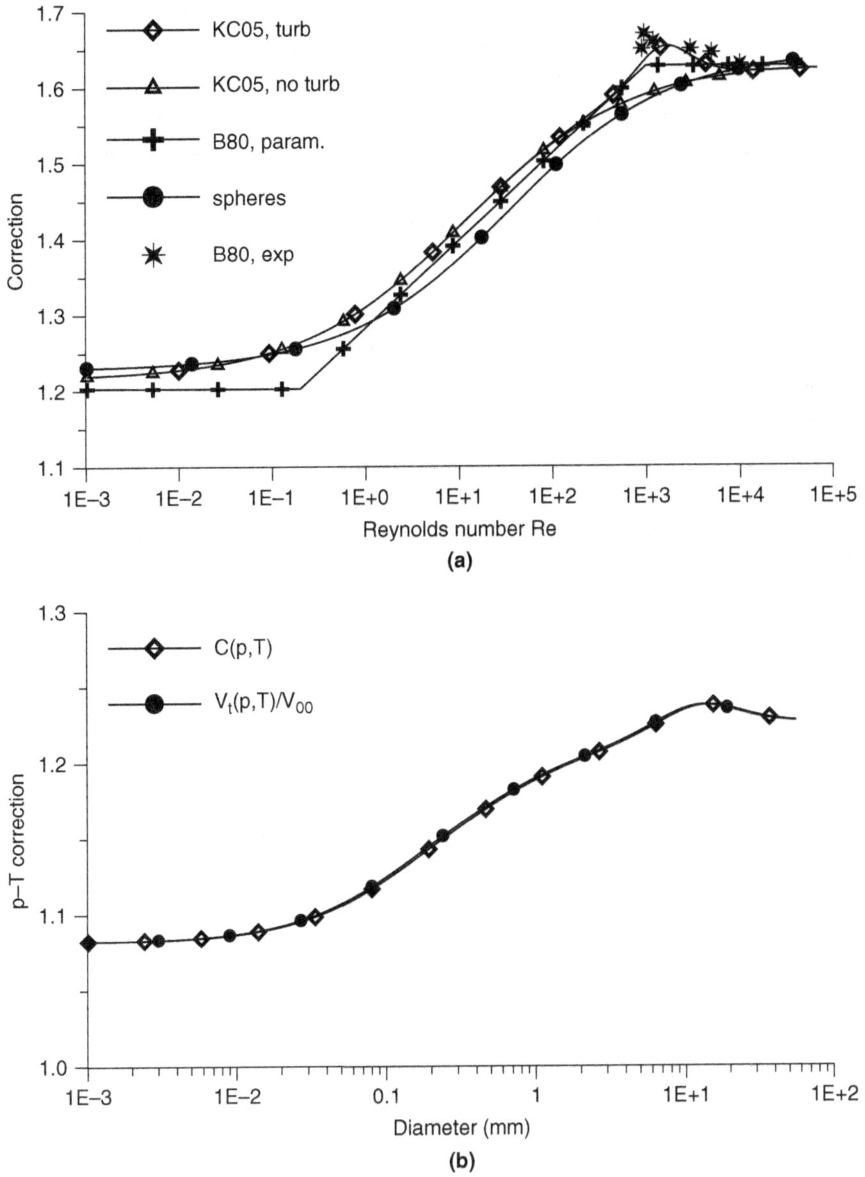

Figure 12.14. The altitude (pressure–temperature) correction C_{pT}. (a) Calculations using (Eqns. 12.5.3b) and (12.5.3c) for crystals with turbulence (KC05, turb., diamonds), and without turbulence (KC05, no turb., triangles); Beard's (1980) general parameterization (crosses, B80), calculations for spheres (solid circles) and some experimental points from Beard (1980), located higher than Beard's parameterization (B80, exp., asterisks); (b) correction $C(p,T)$ for tropical cirrus aggregates as in Fig. 12.9 calculated from (Eqn. 12.5.3c) and as the ratio $V_t(p,T)/V_t(p = 1000, T = -5\,°C)$. From Khvorostyanov and Curry (2005b), *J. Atmos. Sci.*, 62, © American Meteorological Society. Used with permission.

calculated with the equations of Section 12.2, and Section 12.3 (see Fig. 12.9). Fig. 12.14b shows that both methods give very close results over the entire D-range, where the C_{pT} correction significantly varies with D between the viscous limit (Eqn. 12.5.4) for small particles and the aerodynamic limit (Eqn. 12.5.5) for large particles. Fig. 12.14 shows that the difference in the fall velocities of the large and small particles increases with altitude by 30–70%, enhancing by this factor the rate of the gravitational coagulation and accretion. This effect can be accounted for in various models with use of the C_{pT} correction (Eqns. 12.5.3a), (12.5.3b), and (12.5.3c) in calculations of precipitation intensity.

12.7. Parameterizations for Large-Scale Models

The equations described earlier are directly applicable to cloud models with explicit microphysics that typically utilize the bin representation of the particle size spectra, with the particles in each bin falling at their own velocities. However, parameterization of these relations is desired for applications to larger-scale models and to remote sensing. Advanced general circulation models are using bulk cloud microphysical parameterizations (e.g., Fowler et al., 1996, hereafter FRR96). However, the long integration periods and the large number of the grid points impose strong limitations on the number of arithmetic operations per time step. Therefore, some simple parameterizations are desired for large-scale models.

Such a simple parameterization is easily developed using the continuous power law representation of the fall velocity. This parameterization is illustrated using the bulk microphysics parameterization from Fowler et al. (1996). The size spectrum of rain is approximated by Fowler et al. (1996) with the Marshall–Palmer distribution:

$$N_{DR} = N_{0R} \exp(-\lambda_R D_R), \qquad (12.7.1)$$

where D_R is the diameter of a raindrop, $N_{0R} = 8 \times 10^2 \text{cm}^{-4}$ is the intercept, $\lambda_R = (\pi\rho_w N_{0R}/\rho_a q_r)$ is the slope, and q_r is the rain water mixing ratio. The fall velocity of a raindrop was approximated in Fowler et al. (1996) by an expression based on the Gunn and Kinzer (1949) data,

$$V_R(D_R) = (c_{0R} + c_{1R}D_R + c_{2R}D_R^2 + c_{3R}D_R^3)(p/p_0)^{0.4} \qquad (12.7.2)$$

with $c_{0R} = -0.267$, $c_{1R} = 5.15 \times 10^3$, $c_{2R} = -1.0225 \times 10^6$, $c_{3R} = 7.55 \times 10^7$, and the average fall velocity is defined as the mass-weighted value,

$$\overline{V}_R = \frac{\int_0^\infty N_{0R}(D_R)M(D_R)V_R(D_R)dD_R}{\int_0^\infty N_{0R}(D_R)M(D_R)dD_R}. \qquad (12.7.3)$$

Substituting (Eqns. 12.7.1) and (12.7.2) into (Eqn. 12.7.3) and analytically evaluating the integrals yields

$$\overline{V}_R(D_R) = (b_{0R} + b_{1R}\lambda_R^{-1} + b_{2R}\lambda_R^{-2} + b_{3R}\lambda_R^{-3})(p/p_0)^{0.4}, \qquad (12.7.4)$$

with $b_{0R} = -0.267$, $b_{1R} = 206 \times 10^2$, $b_{2R} = -2.045 \times 10^7$, and $b_{3R} = -9.06 \times 10^9$. This equation is used in Fowler et al. (1996). It can be simplified and refined using the method described in this chapter. First, note that the subintegral functions in the numerator and denominator of (Eqn. 12.7.3) have pronounced maxima in the rather narrow range of sizes near the drop diameter $D_{max} \sim 2000$–$2300\,\mu\text{m}$ as shown in Fig. 12.15a. According to Section 12.2, $V_R\,(D)$ can be presented here by the power

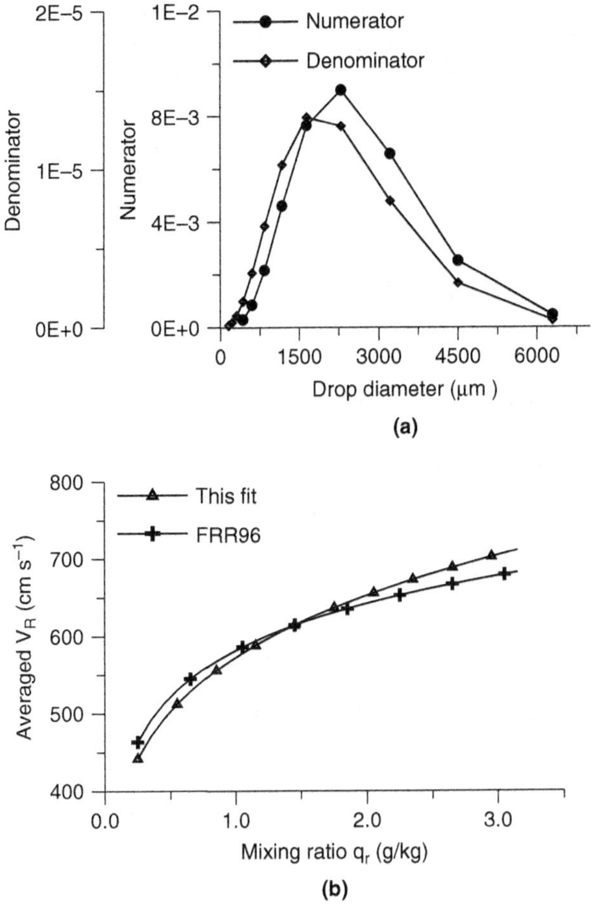

Figure 12.15. (a) Subintegral functions in the numerator (circles) and denominator (diamonds) of the average velocity $\overline{V}_R$ (Eqn. 12.7.3); (b) a comparison of $\overline{V}_R$ calculated with the previous equation (12.7.4) from Fowler et al. (1996) (crosses) and new (Eqn. 12.7.5) with $c_{vt} = 2.9$ (triangles) at surface pressure. From Khvorostyanov and Curry (2002), *J. Atmos. Sci.*, 59, © American Meteorological Society. Used with permission.

law $V_R(D_R) = A_v D_R^{B_v}$; the coefficients at the maximum D_{max} are easily evaluated with (Eqns. 12.2.19) and (12.2.20) or (Eqns. 12.4.7)–(12.4.11) and are $A_{vm} \sim 2 \times 10^3\,\text{cm}^{0.25}\,\text{s}^{-1}$ and $B_{vm} \sim 0.75$, as is seen in Fig. 12.3. These expressions for V_R, which are much simpler than (Eqn. 12.7.2), can be substituted into (Eqn. 12.7.3) now with these constants A_{vm}, B_{vm}, since the main contribution into the integrals comes from this region. The integral is easily evaluated with (Eqn. 12.7.1) analytically:

$$\overline{V}_R = c_{vt} A_{vm} \lambda_R^{-B_{vm}} = c_{vt} A_{vm} \left(\frac{\rho_a q_r}{\pi \rho_w N_{0R}} \right)^{B_{vm}/4}, \qquad (12.7.5)$$

where $c_{vt} = \Gamma(4.7)/\Gamma(4) = 3.7$, $\Gamma(x)$ is Euler's gamma function and $B_{vm}/4 = 0.188$. This expression is valid at standard atmospheric conditions. The temperature and pressure corrections can be calculated with (Eqns. 12.5.3a), (12.5.3b), and (12.5.3c); b_{Re} is evaluated using (Eqn. 12.2.20), $b_{Re} = (B_v + 1)/(\beta - \sigma + 2)$, which gives with this $B_{vm} \approx 0.75$ and $\beta = 3$, $\sigma = 2$ for spherical drops, $b_{Re} \approx 0.58$. Thus, in the large particle limit (Eqn. 12.5.5)

$$\bar{V}_R(p,T) = c_{pT}\,\bar{V}_R(p_0,T_0), \qquad c_{pT} = \left(\frac{p_0 T}{p T_0}\right)^{0.58}. \tag{12.7.6}$$

Note that this correction also accounts for the temperature dependence, in addition to Gunn and Kinzer's formulation (Eqn. 12.7.2), and the power index of p is almost 50% higher. A more precise T-p correction for various size fractions used in a model can be calculated directly with (Eqn. 12.5.3) without assuming the asymptotic large-size limit. In this example, the actual values of A_v, B_v vary slightly around the mentioned maxima in (Eqn. 12.7.3), and substitution of V_R with constant coefficients may cause an error. Hence, the coefficient c_{vt} arising from the gamma functions should be considered as a parameter that needs some tuning; the calculations show that the value $c_{vt} = 2.9$ is better than $c_v = 3.7$ in (Eqn. 12.7.5).

A comparison of V_R values calculated with (Eqn. 12.7.4) from Fowler et al. (1996) (crosses) and (Eqn. 12.7.5) with $c_{vt} = 2.9$ (triangles) for the surface pressure ($p_0 = p$) is given in Fig. 12.15b as the functions of the rain water mixing ratio q_r. Both equations are in excellent agreement, the maximum error is −4.7% at the lowest q_r and 4.1% at the highest q_r. Equations (12.7.5) and (12.7.6) are simpler and more universal for raindrop terminal velocity since the corrections for both temperature and pressure are accounted for by (Eqn. 12.7.6). The new parameterization (Eqns. 12.7.5) and (12.7.6) is also more economical since it contains fewer arithmetic operations, and can be recommended for use in cloud and climate models that use bulk microphysics parameterizations. The corresponding fall velocities for several ice crystal types accounted for in a model using a double-moment or another parameterization of the size spectra (e.g., the modified gamma distributions, and Marshall–Palmer or inverse power laws) can be easily calculated in a similar way.

12.8. Applications for Remote Sensing, Other Objects and Other Planets

Additional applications for parameterized fall velocities occur in remote sensing. In some algorithms for the measurements of vertical velocities with Doppler radars, power law approximations of V_t are used and the "statistical" or average relations between A_v and B_v are established based on numerous calculations of $A_v(D)$ and $B_v(D)$ (e.g., Matrosov and Heymsfield, 2000). The calculation of $V_t(D)$ based on the stepwise representation for X may create problems because the matching points by X, when the power law changes, correspond to different points in terms of diameter. This in turn may create problems when calculating the reflectivity-weighted or mass-weighted velocities by integration of the power law for V_t of individual particles with the size spectra. The method described in this chapter is free of such problems, A_v–B_v relations and averaged velocities can be easily calculated by integration using the continuous analytical form of $A_v(D)$ and $B_v(D)$ (Eqns. 12.2.18)–(12.2.20) for the various particles types. A comparison of the calculated values of A_v and B_v for hexagonal columns and bullet rosettes with the fit from Matrosov and Heymsfield (2000) for bullet rosettes was performed in Khvorostyanov and Curry (2002) and showed good agreement. Note that when evaluating the integrals for the reflectivity-weighted velocities, again the same analysis of the subintegral functions described earlier in this section can be performed. This may simplify the final expressions. Note that the integrals of the type (Eqn. 12.7.3) in remote sensing contain the higher powers of the particles' diameters, thus the maxima in the subintegral expressions with $D^n\exp(-\lambda_i D)$ should be sharper, the accuracy of using the constant values of A_v, B_v at the maximum location should be higher, which facilitates application of this power law representation for remote sensing.

In Sections 12.3–12.6, we considered falling drops and ice crystals in the air in Earth's atmosphere. However, the equations derived in Section 12.2, as was mentioned there, are general and can be applied to any fluid and body—e.g., to the clouds in the atmospheres of the other planets, or to the objects falling or rising in other hydrodynamical fluids by using the appropriate densities of the fluid ρ_F and of the body ρ_b, and the acceleration of gravity for a given planet. The examples of falling bodies include the clouds of aerosol particles in the air or interstitial aerosol in a cloud ($\rho_F = \rho_a$, $\rho_b = \rho_s$, with ρ_s being the density of the solid material); falling bodies in the cloud chambers or other laboratory devices filled with various gases and liquids—e.g., well-known laboratory experiments for measurements of the fall velocities in a vertical tube filled with glycerin, oil, water, or similar viscous liquids; and sinking clouds of aerosol particles in the ocean ($\rho_F = \rho_w$, $\rho_b = \rho_s$). In these cases, $\rho_s > \rho_F$ and the velocity in (Eqns. 12.2.15a), (12.2.15b), and (12.2.16) is directed downward.

Examples of rising bodies (with $\rho_b < \rho_F$) include the rising radiosonde balloon in the atmosphere ($\rho_F = \rho_a$, and ρ_b is the average density of the device), and the rising of small frazil ice particles formed in the supercooled river or ocean mixed layer ($\rho_F = \rho_w$, and $\rho_b = \rho_i < \rho_w$) (A description of the phenomenon of the frazil ice formation is described in Curry and Webster, 1999). An application of the Khvorostyanov and Curry (2002, 2005) scheme for evaluation of the frazil ice rising velocities in water was discussed by Morse and Richard (2009). The sizes of the ice particles of 1-3 mm and their rising velocities of 0.7-0.9 cm s^{-1} were measured and calculated by Morse and Richard, and the parameters δ_0 and C_0 specific the for frazil ice in this case were estimated. Some man-governed objects like spherical bathyspheres (an analogy of drops) and non-spherical bathyscaphs (an analogy of crystals) used to reach very deep layers in the ocean with very high water pressures can change their average density by controlled changing their volumes and may be falling or rising depending on the relation of ρ_F and ρ_b. Their velocities also can be described by this theory with an appropriate parameterization of the m–D and A–D relations.

These situations and many other cases can be considered with (Eqns. 12.2.15a), (12.2.15b), and (12.2.16) with the corresponding nonsphericity factor $\xi_f(D)$ when necessary. Equation (12.2.16) shows that these velocities can be characterized by a useful similarity criterion:

$$Y_{OF}(object/fluid) = v_F^{1-2b_{\mathrm{Re}}}\left(g\left|\frac{\rho_b}{\rho_F}-1\right|\right)^{b_{\mathrm{Re}}}, \tag{12.8.1}$$

where $|x|$ means the absolute value, and the subscripts "O" and "F" mean the object and fluid. This criterion Y_{OF} is independent of the particle dimension but does depend on the general properties of the fluid (ρ_F, v_F), the object (ρ_b), and the planet (acceleration of gravity g). From the definition (Eqns. 12.8.1) and (12.2.16), it follows that the terminal velocity for ellipsoids can be written as

$$V_t = a_{\mathrm{Re}}[(4/3)\xi_f(D)]^{b_{\mathrm{Re}}} D^{(3b_{\mathrm{Re}}-1)} Y_{OF}(object/fluid), \tag{12.8.2}$$

or $V_t \sim Y_{OF}$ for the particles of the same size and shape.

Thus, Y_{OF} allows us to establish simple similarity relations among the velocities of the falling or rising objects for various environments and planets. For example, consider sedimenting small aerosol particles with the density of $1.5\,\mathrm{g\,cm^{-3}}$ in the laboratory chamber designed for measurements of fall velocities and filled with water. Then, in (Eqn. 12.8.1) $\rho_b = \rho_s \approx 1.5\,\mathrm{g\,cm^{-3}}$, $\rho_F = \rho_w = 1\,\mathrm{g\,cm^{-3}}$, $b_{Re} = 1$, and according to (Eqn. 12.8.1), Y_{OF}(aerosol/water) = $(1/2)(g/\nu_w)$, where ν_w is the kinematic viscosity of water. For small drops in the air, Y_{OF}(drop/air) $\approx (g/\nu_a)(\rho_w/\rho_a)$. Using for $T = 5\,°\mathrm{C}$ the values $\nu_w = 0.015\,\mathrm{cm^2\,s^{-1}}$ and $\nu_a = 0.15\,\mathrm{cm^2\,s^{-1}}$, we find $(\nu_a/\nu_w) \sim 10$, and $(\rho_a/\rho_w) \sim 10^{-3}$, then Y_{OF}(aerosol/water)/Y_{OF}(drop/air) = $(1/2)(\nu_a/\nu_w)\cdot(\rho_a/\rho_w) \approx 0.5\cdot 10^{-2}$.

Thus, the fall speeds of the small aerosol particles in water are only a half percent of the value for the drops of the same radius in air, and measuring the fall velocities of aerosol particles with a given radius in this device filled with water and multiplying by ≈200 we can obtain the fall velocities of aerosol of the same size in the air. This similarity criterion also allows us to conclude that a cloud of hydrosol particles can be transported before their fallout by the ocean currents over the distances by two orders of magnitude greater than a cloud of an aerosol of the same size in the atmosphere. The other processes like convection or coagulation may change this relation, but the estimates of the relative role of sedimentation can be elucidated with the criterion Y_{OF}.

Other objects, fluids, and planets can be considered in the same way. For example, the difference in the fall velocities of the particles in the atmospheres of Earth and Mars is determined by the difference in their accelerations of gravity, g_E and g_M, and viscosities, ν_a and ν_M. For small particles with $b_{Re} = 1$, using (Eqn. 12.8.1) and the condition $\rho_b >> \rho_F$, we obtain an estimate V_t(Mars)/V_t(Earth) ~ $(g_M/g_E)(\nu_a/\nu_M)$. This relation is simplified for layers with close viscosities, $\nu_a \approx \nu_M$. Then, using the values $g_E = 9.81\,\mathrm{m\,s^{-2}}$ and $g_M = 3.37\,\mathrm{m\,s^{-2}}$ (e.g., Curry and Webster, 1999, Chapter 14 therein), we obtain V_t(Mars) = V_t(Earth) × (g_M/g_E) ~ $0.34V_t$(Earth). For large particles with $b_{Re} = 1/2$, we have from (Eqn. 12.8.1) V_t(Mars) = V_t(Earth) × $(g_M/g_E)^{1/2}$ ~ $0.59V_t$(Earth). The same comparison for Jupiter with the acceleration of gravity $g_J = 26.2\,\mathrm{m\,s^{-2}}$ gives for the layers with similar viscosities for the small particles V_t(Jupiter) ~ $2.67V_t$(Earth), and for the large particles, V_t(Jupiter) ~ $1.63V_t$(Earth). The vertical structures of the planetary atmospheres (temperature, pressure, and density, and chemical composition) are known for the planets of the solar system along with the probable cloud types at various heights (e.g., Curry and Webster, 1999, Chapter 14).

It is possible to create in the cloud chambers the conditions that correspond to various clouds on the planets and to measure the viscosities, densities, and particle fall velocities at the Earth gravity acceleration. Then, using the similarity criterion Y_{OF}, it is possible to estimate V_t of the particles on the other planets with their gravity accelerations and to calculate the temperature–pressure corrections C_{pT} for the fall velocities for various planets as in Section 12.3. These data could be used in studies of the fall velocities and numerical cloud simulations on various planets (e.g., Buikov et al., 1976; Lewis and Prinn, 1984; Ibragimov, 1990; Goody, 1995; Lewis, 1995; Curry and Webster, 1999).

13

Broad Size Spectra in Clouds and the Theory of Stochastic Condensation

13.1. Introduction

The characteristics and evolution of the cloud droplet and crystal size spectra determines cloud radiative properties and the formation of precipitation. Accounting for these processes in cloud-resolving models and climate models requires correct understanding and then parameterization of cloud microstructure and its dependence on predicted atmospheric parameters. During the past five decades, cloud physics and cloud optics have widely used empirical parameterizations of cloud particle size spectra such as the gamma distributions or exponential spectra described in Chapter 2.

These empirical spectra were used in the remote sensing of clouds and also in the modeling and parameterization of cloud properties and processes, both in the cloud-resolving models (CRM) and in some advanced general circulation models (GCMs). A new impetus for bulk microphysical models was given during the past two decades by development of the double-moment bulk schemes that include prognostic equations for the number concentrations of hydrometeors in addition to the mixing ratio, which allows a higher accuracy in predicting the cloud microphysical properties. The double moment schemes are used in cloud models (e.g., Ferrier, 1994; Harrington et al., 1995; Meyers et al., 1997; Cohard and Pinty, 2000; Girard and Curry, 2001; Seifert and Beheng, 2001, 2006; Morrison et al. 2005a,b; Milbrandt and Yau, 2005a,b). These schemes are also incorporated into advanced mesoscale weather prediction models such as the Weather Research and Forecast (WRF) model (e.g., Morrison et al. 2005a,b) and the weather prediction Consortium for Small-Scale Modeling (COSMO) model (e.g., Zubler et al., 2011). The most advanced general circulation models (GCMs) also began using such double-moment microphysics parameterizations that were included earlier only in cloud-resolving models (e.g., Sud and Lee, 2007; Morrison and Gettelman, 2008; Sud et al, 2009). These parameterizations are also based on using gamma distributions or exponential spectra with prescribed parameters.

A deficiency of empirically derived parameterizations of particle size spectra is that the parameters are generally unknown and are fixed to some prescribed values or parameterized based on experimental data. The index p of the gamma distributions (see Chapters 2–5) is of key importance because it determines the relative dispersion of the spectra, $\sigma_{rr} = (p + 1)^{-1/2}$, and thereby the cloud optical properties and the rates of water conversion and of precipitation formation. Experimental studies show wide variations in clouds of the relative dispersions—that is, of the indices p, which measured values varied from 1–2 to 25–30 (e.g., Austin et al., 1985; Curry, 1986; Mitchell, 1994; Pruppacher and Klett, 1997; Brenguier and Chaumat, 2001). As described in Chapter 2, the parameters of these distributions in the models have usually been prescribed from observations and may vary considerably from one model to another. Values of p have often been rather arbitrarily chosen between 0 and 10. Ideally, these parameters should be related to the local atmospheric conditions and evaluated from the cloud bulk model or climate model rather than prescribed uniformly over the globe.

The accuracy of calculations can be improved if the parameters of the size spectra are calculated with equations having a basis in theory. For example, the p-indices (and thereby spectral dispersions) were calculated using analytical expressions from Khvorostyanov and Curry (1999d) by Morrison et al. (2005a,b) and Morrison and Pinto (2005) in a cloud model with double-moment bulk microphysics scheme and by Sud and Lee (2007) and Sud et al. (2009) in the Goddard General Circulation Model. Calculated values of these p-indices showed substantial variations in time and space, resulting in variations in the onset of coalescence and in cloud optical properties.

In this section, we briefly review the major mechanisms and theories suggested for an explanation of the formation of the broad size spectra of cloud drops and crystals, with special emphasis on the kinetic equations of stochastic condensation that allow us to obtain analytical solutions suitable for parameterization of the size spectra in cloud and climate models.

13.1.1. Mechanisms and Theories of the Formation of Broad Size Spectra in Clouds

Understanding the formation of the size spectra and their parameterizations have been hampered by the contradiction between the theory of Maxwellian condensation and observations of size spectra in clouds. As was discussed in Chapter 5, the theory of regular condensation in a constant updraft has been shown to narrow the droplet spectrum with time. Since broad spectra are observed in clouds and are required to initiate collision and coalescence growth of precipitation, numerous attempts have been made to remove this contradiction. Various hypotheses and approaches have been developed to explain and describe broad spectra in clouds, including several attempts at analytical solutions to the kinetic equations for stochastic condensation.

The *inhomogeneous mixing theory* (Baker and Latham, 1979; Baker et al., 1980) considers penetration of the tongues or blobs of dry air into a cloud from the top or the sides with subsequent partial drop evaporation. This can be viewed as a two-stage process. At the first stage, the dry parcel and the neighboring saturated air parcels remain separated with the distinct interface surface. Complete evaporation occurs of all drops at the interfaces of the mixing parcels, but all drops remain unaffected at some distance from the interface. At the second stage, the interface breaks down and the two air parcels mix. This mixing may lead to a supersaturation difference in the mixed volumes and to spectral broadening, or even bimodal spectra. The essence of the interface was somewhat subtle in this model. Broadwell and Breidenthal (1982) extended this concept with *nongradient turbulent mixing*, whereby a source of supersaturation generation, instead of the vertical velocity, is determined by the rate at which the interface between the mixed volumes is formed. Baker et al. (1984) adapted this model of nongradient mixing in a shear layer to inhomogeneous mixing in clouds, which gave a better conceptual and mathematical basis for the model of inhomogeneous mixing.

An important feature of the inhomogeneous mixing model is that local values of supersaturation are much greater than in adiabatic or homogeneous mixing; thus, the largest drops grow faster. In the inhomogeneous mixing model, more droplets are completely evaporated, and the newly activated small droplets cannot compete for the water vapor as effectively as the previous larger drops. This leads to the local growth of the supersaturation in the vicinity of the older larger drops less effected by mixing, resulting in their more rapid growth.

Another model of turbulence effects on droplet spectral broadening, called *the theory of homogeneous mixing*, was suggested by Telford and Chai (1980) and Telford et al. (1984). This theory views a cloud as an ensemble of vertically coherent updrafts and downdrafts. When the entrained drier air near the cloud top mixes with the saturated cloud air, drops evaporate and the resulting mixture becomes denser, triggering its descent. Evaporation continues in this descending parcel, causing droplet sizes to decrease. The smaller drops in this parcel are replaced with the larger drops that are brought in this parcel from the neighboring undiluted parcels, providing a continuous source of the larger drops. When this downdraft becomes an updraft in turbulent motion and eventually reaches the cloud top, the drops in this parcel are larger than they were before the descent. Thus, according to Telford's hypothesis, recycling of the drops in turbulent down- and updrafts broadens the drop size spectra. Further field experiments with measurements in convective clouds did not support Telford's hypothesis, since the largest drops were found in the wettest and least diluted ascending cloud portions (e.g., Hill and Choularton, 1985; Paluch and Knight, 1986). Telford (1987) argued that this mechanism could operate in some other maritime cloud types.

Another similar mechanism suggested was *isobaric nongradient mixing* based on the Broadwell and Breidenthal (1982) model (e.g., Gerber, 1991). When two saturated cloud parcels at different temperatures are completely mixed, their temperature and humidity will be equal to average values, but due to the nonlinearity of the saturated humidity, the mixture will be supersaturated, this will accelerate drop growth in this domain and result eventually in spectral broadening.

Belyaev (1961, 1964) and Sedunov (1965) accounted for the fact that the fields of vapor, temperature, and supersaturation around a drop may experience random turbulent fluctuations. The authors considered Maxwellian drop growth in the random supersaturation field and obtained broad drop spectra in the form similar to Gaussian distribution. Other mechanisms have been proposed that include horizontal inhomogeneities in cloud condensation nuclei (CCN). Kabanov et al. (1970) assumed that the drop concentrations at a fixed cloud height is inhomogeneous due to the spatial variability of the cloud condensation nuclei and can be described by the analytical functions (gamma distribution or Gauss normal distribution). Using the mass balance equations, they derived the analytical size spectra, which had a finite spectral width, but were different from the observed gamma distributions. A similar approach was developed by Smirnov and Kabanov (1970) for the spatial inhomogeneity of the vertical velocity described by the Gauss normal distribution in agreement with the measurements by Warner (1969a,b).

Another hypothesis was based on observations of giant and ultragiant aerosol particles with sizes up to 100 μm (e.g., Ludwig and Robinson, 1970; Hobbs et al., 1978; Johnson, 1982) that could initiate formation of the precipitation particles. Smirnov and Sergeev (1973) performed calculations of the growth of cloud drops on the large hygroscopic aerosol particles described by the inverse power law. They showed that such particles may rapidly grow to the sizes of precipitation and that the size spectra of large drops formed on these CCN are also inverse power laws in good agreement with measurements in California stratus by Ludwig and Robinson (1970). Similar conclusions on the role of giant nuclei were made by Johnson (1982). In contrast, Woodcock et al. (1971) and Takahashi

(1976) found that giant salt nuclei do not give a substantial contribution to the process of precipitation formation.

A few other approaches were developed for derivation of the analytical form of the size spectra. Srivastava (1969) assumed that the cloud space is divided into a number of cells and the mass growth rate of each drop by diffusion is proportional to the volume of each cell. For these drops growing in the inhomogeneous supersaturation field, Srivastava derived the generalized gamma distribution (Eqn. 2.4.11) with $p = 2$ and $\lambda = p + 1 = 3$—i.e., the Weibull distribution $f(r) = c_N r^2 \exp(-r^3)$. Liu (1993, 1995), Liu et al. (1995), and Liu and Hallett (1998) derived the Weibull distribution for the drop size spectrum using Shannon's maximum entropy principle.

Srivastava (1989) introduced a concept of the microscopic and macroscopic supersaturation associated with the random distribution of drops. This important hypothesis explains different growth rates of various drops. It allows us to refine the kinetic equations and will be discussed in more detail in Section 13.2. Cooper (1989) explained spectral broadening by various drops' histories, so that the drops that come to some level in the cloud may have different lifetimes and therefore different sizes, which increases the spectral dispersion. Spectral broadening was also explained by averaging over the parcels with different ages (Considine and Curry, 1996) and various sedimentation of the cloud droplets (Baker et al., 1984; Considine and Curry, 1998). The stochastic nature of drop growth in the turbulent atmosphere was accounted for by Kulmala et al. (1997), who calculated cloud drop growth in the field of saturation ratio modeled as a stationary stochastic process that obeys the normal distribution with some mean value and a dispersion ~0.01 estimated from the measurements of the temperature and humidity fluctuations. The authors found acceleration of drop growth in this variable saturation ratio and appearance of bimodal spectra. Activation of new drops occurred in this work even when the mean supersaturation was slightly negative due to the effect of positive supersaturation dispersion.

Another mechanism to explain the broad cloud droplet spectra was based on observations of preferential concentration of the particles in a turbulent flow in some regions (e.g., Shaw et al., 1998; Vaillancourt and Yau, 2000; Shaw, 2003). This theory treats a turbulent flow as containing coherent structures similar to the "vortex tubes" that have various vorticities. The size of the tubes is about 10 times the Komogorov's scale, $l_K = (\nu_f^3/\varepsilon_d)^{1/4}$, where ν_f is the kinematic viscosity and ε_d is the dissipation rate of turbulent energy—i.e., $l_K \sim 0.5$–1.5 cm. The particles are ejected from the regions of higher vorticity by centrifugal forces and are accumulated in the regions with low vorticities between the cells. Thus, the initial spatially homogeneous particle distribution becomes inhomogeneous. This mechanism operates independent of entrainment and, therefore, can operate in adiabatic cloud cores. Shaw et al. (1998) applied this theory for cloud drops and showed that cloud droplets of sufficient size are not randomly dispersed in a cloud but are preferentially concentrated in regions of low vorticity in the turbulent flow field. Regions of high vorticity (low droplet concentration) develop higher supersaturation than regions of low vorticity (high droplet concentration). Therefore, on small spatial scales, cloud droplets are growing in a strongly fluctuating supersaturation field. These fluctuations in supersaturation exist independent of large-scale vertical velocity fluctuations. This breaks the link between fluctuations of the supersaturation and vertical velocity. Shaw et al. (1998) adjusted a parcel model for simulation of such turbulent vorticies with growing droplets and found that supersaturation in the regions with lower drop concentrations can be up to 2–3 times higher than predicted by classical theory. Therefore, droplets growing in regions of high vorticity experience enhanced growth rates, allowing some droplets to grow larger than predicted by the classic theory of condensational growth. This mechanism helps to account for two common observations in clouds: the presence of a large droplet tail in the droplet spectrum, important for the onset of collision-coalescence, and the possibility of new nucleation above the cloud base, allowing for the formation of a bimodal droplet spectrum. We note that this model of the vorticity cells with different supersaturations is conceptually similar to, and provides additional physical arguments, for Srivastava's (1969) concept of cloud microcells and Srivastava's (1989) concept of microscale supersaturation.

Another line of research of drop growth in the turbulent atmosphere is based on *direct numerical simulations* (DNSs). This is one of the most detailed and powerful models that studies 3D turbulence and directly solves the Navier–Stokes equation as well as conservation equations for heat and water vapor. The dynamical model is coupled with a droplet growth model that solves for the trajectories of several tens of thousands of droplets along with their growth as a function of conditions in their individual environments. Therefore, this method provides very detailed information on the simulated process, but it imposes limitations on the spatial scales bounded by a few tens of centimeters and on the time scales. Vaillancourtet al. (2002) performed a DNS for the condensation process and found significant fluctuations of the number density and supersaturation, but accounting for drop sedimentation tended to decrease these fluctuations. An unexpected result was that the drop size spectra were narrower than without turbulence. This finding contrasted other studies of turbulence effects and the origin of the effect is unclear. It could be a result of insufficient spatial resolution or an insufficient intensity of simulated turbulence.

A comprehensive series of DNS simulations was performed by Paoli and Shariff (2004, 2009) where the stochastic processes were modeled upon a set of *Langevin equations* for stochastic variables with various sources of fluctuations being superimposed on the mean fields. DNS simulations were performed with small spatial and temporal steps. The authors showed that the temperature and vapor fluctuations lead to supersaturation fluctuations, which are responsible for broadening the droplet size distribution in qualitative agreement with in situ measurements. Based on the results of their DNS, Paoli and Shariff formulated a set of Langevin stochastic equations for the droplet area, supersaturation, and temperature surrounding the droplets. One important result of this work obtained with their DNS was that there is a good correlation and scaling between the local fluctuations of the drop radius r' and supersaturation s', which can be written as

$$r'/\bar{r} \sim s'/\bar{s}, \qquad (13.1.1)$$

where $\bar{r}$ and $\bar{s}$ are mean values of the radius and supersaturation. This supports Srivastava's (1989) hypothesis on the microscale supersaturation.

The concepts of entrainment and mixing among the cloud parcels were further developed and quantified by introducing a supersaturation *Damköhler number* $Da = \tau_L/\tau_f$, where τ_L is a fluid time scale, and τ_f is the supersaturation relaxation time (e.g., Shaw, 2003; Kumar et al., 2012). The *Damköhler number* is a dimensionless parameter which relates the fluid time scale to the typical evaporation time scale, and can capture many aspects of the initial mixing process within the range of typical cloud parameters. Using the Damköhler number, the mixing process can be characterized by the limit of strongly homogeneous regime ($Da << 1$), when the supersaturation varies much slower than the fluid characteristics, and strongly inhomogeneous regime ($Da >> 1$), when the supersaturation relaxation time τ_f is much smaller than the fluid time τ_L. In the models of stochastic condensation considered in the following sections, the limits $Da << 1$ and $Da >> 1$ correspond to the high-frequency and low-frequency turbulence regimes respectively. The most general case corresponds to $Da \sim 1$, that is, when the fluid time and the supersaturation relaxation time are comparable. Such a case is considered later along with the both limits.

Some other possible mechanisms of spectral broadening, and of the effects of turbulence on the size spectra and warm rain initiation, are reviewed in Beard and Ochs (1993), Blyth (1993), Vaillancourt and Yau (2000), Shaw (2003), and Cotton et al. (2011).

13.1.2. Kinetic Equations of Stochastic Condensation

Simulation of the evolution of drop size spectra is enabled by kinetic equations that allow direct simulation of the spectra as well as the development of analytical solutions to the drop size spectra. Application of these equations in cloud models showed a fast progress over the last decades due to increasing computer power. Several versions of the kinetic equations of stochastic condensation have been developed and are reviewed in this section.

The *theory of stochastic condensation* attributes the presence of broad drop size spectra in clouds to the occurrence of condensation in a turbulent medium. Development of this theory and derivation of the kinetic equations of stochastic condensation began in the 1960s–1970s. These equations have been derived for two limiting cases: the low-frequency turbulence regime when the characteristic Lagrangian time scale of turbulent fluctuations τ_L (fluid time) is much greater than the supersaturation relaxation time τ_f—i.e., $\tau_L >> \tau_f$ (or $Da >> 1$), and the high-frequency turbulence regime when $\tau_L << \tau_f$ (or $Da << 1$).

As mentioned in Chapter 5, the method of kinetic equations for the condensation-deposition processes was introduced in cloud physics in the pioneering works of Buikov (1961, 1963, 1966a,b) as described by (Eqn. 5.5.4a). Representing all quantities as the sums of mean and fluctuating values, and applying Reynolds averaging procedure to (Eqn. 5.5.4a), Buikov arrived at the following equation:

$$\frac{\partial f}{\partial t}+\frac{\partial}{\partial x_i}\left[\left(u_i - v(r)\delta_{i3}\right)f\right]+\frac{\partial}{\partial r}(\dot{r}f)=\frac{\partial}{\partial x_i}k_{ij}\frac{\partial}{\partial x_j}f+\bar{J}, \qquad (13.1.2)$$

where the usual summation over the indices $i = 1, 2, 3$ is assumed and all quantities here are averaged over the turbulent ensemble. Buikov's first kinetic equation of stochastic condensation is analogous to the equations for the temperature or humidity, including terms with turbulent mixing described by the diffusion coefficient k_{ij} and advection terms $\partial(u_i f)/\partial x_i)$ with the wind velocity u_i. The equation incorporates the additional terms describing sedimentation $v(r)$ and condensational growth. The term $\partial(\dot{r}f)/\partial r$ for condensational growth has the form of the "advection in space of the radii," with the "velocity in radii space" $\dot{r}$. This simplified equation introduced by Buikov is equivalent to the high-frequency approximation, $\tau_L << \tau_f$.

Another version of the kinetic equation of condensation that correspond to the low-frequency approximation was derived by Levin and Sedunov (1966) and Sedunov (1974) who developed the theory of stochastic condensation that attributes the presence of broad drop size spectra in clouds to turbulent fluctuations. This stochastic method treats supersaturation and droplet growth rate as stochastic variables and replaces the usual operator of turbulent diffusion $k_z\partial/\partial z$ with the operator $k_z(\partial/\partial z + A_s\partial/\partial s)$ for a nonconservative substance s (the droplet surface), where A_s is proportional to the vertical velocity w.

The theory of stochastic condensation by Levin and Sedunov (1966) was criticized by Mazin and Smirnov (1969), who argued that since both dr/dt and A_s are proportional to w, the droplet radius will be a predetermined function of w. Therefore, the droplet radius will be a function of the height above the cloud base $z = w\Delta t$, which a cloud parcel reaches during the time Δt of its uplift, does not matter if the uplift is in a regular flow with the mean w_0 or in a turbulent flow with the sum of the mean and fluctuation, $w = w_0 + w'$. Thus, Mazin and Smirnov argued, the theory of stochastic condensation by Levin and Sedunov cannot explain the formation of the broad spectra. Similar criticisms were expressed by Warner (1969b) and Bartlett and Jonas (1972) who performed Lagrangian parcel model simulations of the drop growth in ascending motion with superimposed fluctuations of the vertical velocity but did not allow mixing among the parcels. In their simulations, the drop radius was highly correlated with height, and thus they obtained narrowing of the spectra and concluded that stochastic condensation cannot lead to spectral broadening. (These critics were answered in subsequent works, as will be described in Section 13.10.)

The equations of Levin and Sedunov (1966) and Sedunov (1974) were further refined by a number of authors. More rigorous and detailed derivations of the equation of stochastic condensation using Reynolds averaging and more careful evaluation of covariances were performed by Stepanov (1975, 1976) using the method of perturbation theory similar to that used in the plasma theory, which is valid for weak turbulence (fluctuations are smaller than the mean values), and by Voloshchuk and Sedunov (1977), who used the extended statistical theory of turbulence.

Although these low-frequency stochastic kinetic equations could explain some features of the drop size spectral dispersions in stratiform clouds, Manton (1979) argued that the time-dependent solutions of the Gaussian type to the stochastic equations of Levin and Sedunov (1966) lead to dispersions that tend to some constant values and are still narrower than those observed in convective clouds for the corresponding times. By hypothesizing a negative correlation between the fluctuations in mean droplet radius and vertical velocity, Manton (1979) derived a modified version of the kinetic equation that produced broader or bimodal size spectra with dispersions increasing for a longer time. In a subsequent discussion of Manton's theory (Merkulovich and Stepanov, 1981; Manton, 1981), it was clarified that the condition of mass balance imposes some additional limitations on the basic assumptions and analytical solutions to the kinetic equation. In particular, it was shown that the bimodal spectra predicted by this theory should be monomodal in many cases, and the dispersions should be smaller. Austin et al. (1985) showed that the basic assumption of Manton's theory on the negative correlation between the fluctuations in integral radius and vertical velocity is not observed in the continental cumuli, while Curry (1986) found some evidence of negative correlation in Arctic stratus. Thus, further verification or another hypothesis is required to break the link between supersaturation and vertical velocity, which appears to be the primary requirement for deriving sufficiently broad drop spectra using the stochastic kinetic equations.

The stochastic kinetic equations for condensation were generalized by Merkulovich and Stepanov (1977) and by Smirnov and Nadeykina (1986) to account for curvature and solution effects in droplet growth. The equations of Voloshchuk and Sedunov (1977) and Smirnov and Nadeykina (1986) have the form

$$\frac{\partial f}{\partial t}+\frac{\partial}{\partial x_i}\left[\left(u_i-v\delta_{i3}\right)f\right]+\frac{\partial}{\partial r}\frac{D_{eff}}{r}\left(a_1 w+A_{con}\right)f$$

$$=\left(\frac{\partial}{\partial x_i}+\delta_{i3}B_{con}\frac{\partial}{\partial r}\frac{D_{eff}}{r}\right)k_{ij}\left(\frac{\partial}{\partial x_j}+\delta_{j3}B_{con}\frac{\partial}{\partial r}\frac{D_{eff}}{r}\right)f, \tag{13.1.3}$$

where D_{eff} is an effective coefficient of vapor diffusion (corrected D_v), A_{con} is a thermodynamic function that includes the effects of the salts and fluctuations of vapor density, and

$$B_{con}=\frac{\rho_a}{\rho_w}\frac{c_p}{L}\tau_f(\gamma_a-\gamma_w). \tag{13.1.4}$$

The kinetic equations of the type (Eqn. 13.1.3) are valid in the low-frequency approximation—i.e., when $\tau_L >> \tau_f$. In this case, the fluctuations of supersaturation and droplet growth rate are highly correlated with the vertical velocity. The solutions to these low-frequency equations generalized for the exchange with environment were found for the quasi-steady state by Smirnov and Nadeykina (1986) in the diffusion and kinetic regimes of drop growth that differ by the condensation coefficient α_c and the kinetic correction (Eqn. 5.1.22a), $\xi_{con} = 4D_v/(\alpha_c V_w)$, to the drop growth rate. For diffusion regime, $r >> \xi_{con}$—that is, for $\alpha_c \sim 1$—the solution has the form of the generalized gamma distribution (Eqn. 2.4.11) but with a fixed index $p = 1$ and the indices related as $p = \lambda - 1$—in other words, it represents the Weibull distribution with the indices $p = 1$, $\lambda = 2$ (see Section 2.4, (Eqn. 2.4.20a)):

$$f(r)=c_{N1}r\exp(-\beta_1 r^2). \tag{13.1.5}$$

The relative dispersion of this spectrum is $\sigma_{rr} = 0.523$ according to (Eqn. 2.4.20c), does not depend on cloud properties, and is characteristic for broad size spectra. For the kinetic regime—that is, for $\alpha_c << 1$—Smirnov and Nadeykina (1986) found the solution for the small droplet fraction in the exponential form

$$f(r)=c_{N2}\xi_{con}\exp(-\beta_2\xi_{con}r), \tag{13.1.6}$$

with the relative spectral dispersion $\sigma_{rr} = 1$.

McGraw and Liu (2006) accounted for the random turbulent fluctuations of supersaturation and derived the kinetic equation of condensation in the form of the Fokker–Planck equation for the squared drop radius $A = r^2$:

$$\frac{\partial f(A)}{\partial t}=-v_{drift}\frac{\partial f(A)}{\partial A}+D_A\frac{\partial^2 f(A)}{\partial A^2}. \tag{13.1.7}$$

The drift velocity v_{drift} is proportional to supersaturation s and the diffusion coefficient D_A in the A-space was defined in this work as proportional to the supersaturation dispersion σ_{sA}:

$$v_{drift}=\left(\frac{dA}{dt}\right)=k_A s,\qquad k_A=\frac{2D_v\rho_{ws}}{\Gamma\rho_w},\qquad D_A=\frac{k_A^2\sigma_{sA}^2}{\gamma_\sigma}, \tag{13.1.8}$$

where γ_{σ}^{-1} is the correlation time of random supersaturation fluctuations. A stationary solution to this equation was obtained by McGraw and Liu in the form of the Weibull distribution. It had the form in terms of radii

$$f(A) = 2\pi N_d \left(\frac{N_d}{q_l} \right)^{2/3} r \exp\left[-\pi \left(\frac{N_d}{q_l} \right)^{2/3} r^2 \right], \qquad (13.1.9)$$

where N_d is the droplet concentrations and q_l is the dimensionless liquid water fraction.

An approach based on the Fokker–Planck equation was used by Jeffery et al. (2007) who derived an equation similar to Voloshchuk and Sedunov (1977) and found that it ensures a noticeable spectral broadening. Application of the Fokker–Planck equation will be discussed in more detail at the end of this chapter. Thus, the analytical solutions to the kinetic equations of stochastic condensation explained some features of the size spectra.

Since the supersaturation relaxation time $\tau_f \sim 1$–10 s in a developed cloud, and the Lagrangian turbulent time $\tau_L \sim 5$–10 minutes, the high-frequency approximation might be representative of rapidly changing conditions such as in an entrainment zone or during the initial phase of condensation, while the low-frequency approximation might be representative of a developed cloud that is quasi-steady. More detailed analysis of the turbulence in clouds showed that there are pronounced maxima in the turbulent power spectra of vertical velocity at the frequencies $\omega \sim 10^{-3}$–$10^{-2}\,\text{s}^{-1}$ and the long tails extend to $\omega \sim 10^{-1}\,\text{s}^{-1}$ and smaller frequencies (e.g., Curry et al., 1988; Sassen et al., 1989; Quante and Starr, 2002; Shaw, 2003; Monin and Yaglom, 2007b). As was shown in Chapter 5, the supersaturation relaxation (absorption) time in the cloud regions with smaller particle concentrations can be $\tau_f \sim 30$–60 s for liquid clouds and can exceed $\tau_f \sim 30$–180 min for crystalline clouds (Tables 5.1 and 5.2). Thus, the timescales τ_f and $\tau_L = 2\pi/\omega$ can be quite comparable over significant ranges of the frequencies of turbulent spectra, and their relation can change during cloud evolution. Hence, more general kinetic equations for arbitrary relations between τ_f and τ_L are required.

Khvorostyanov and Curry (1999c) performed the next step in development of this theory and derived a generalized kinetic equation of stochastic condensation. The two principal features that differentiate the new version of the kinetic equation from previous versions were the following: 1) consideration of supersaturation as a nonconservative substance with differentiation between the macroscale and microscale supersaturation similar to the Srivastava's (1989) concept; 2) consideration of the supersaturation fluctuations of various frequencies over the whole turbulent spectrum without assuming proportionality of supersaturation and vertical velocity, thus additional covariances and a turbulent diffusion coefficient tensor dependent on the supersaturation relaxation time, $k_{ij}(\tau_f)$, were introduced into the kinetic equation. Therefore, this version of the kinetic equation is applicable for arbitrary relationships between the supersaturation relaxation time τ_f and the time scale of turbulence τ_L or for arbitrary Damköhler numbers. These assumptions break the link between the fluctuations in supersaturation and vertical velocity. The differentiation between the macro- and microscale supersaturations produced the terms describing the diffusion in space of radii different from the previous formulations and allowed Khvorostyanov and Curry (1999d) to obtain the analytical solutions of the gamma distributions type with variable p-indices.

These equations were further generalized in Khvorostyanov and Curry (2008c,d) to include the effects of particle coagulation-accretion, breakup, and sedimentation and were applied to the small-size and large-scale fractions of cloud drops and crystals. This resulted in analytical solutions in the form of generalized gamma distributions, exponential functions, and inverse power laws.

In applying the kinetic equation for stochastic condensation in numerical cloud models, we are faced with the following dilemmas:

1. The small characteristic time scale for the condensation process and presence of cross-derivatives makes the numerical solution very computationally intensive. Despite development of the stochastic condensation theory, practically all spectral cloud models that use the kinetic equations of condensation-deposition, use them in the simplest form (Eqn. 13.1.2) suggested by Buikov. Thus, refinement of the kinetic equation and development of the numerical schemes for solution of the kinetic equations with cross-derivatives are required for the theory of stochastic condensation and the practical needs of modeling.
2. Analytical solutions to stochastic kinetic equations in the low-frequency regime have been obtained of the Weibull distribution (Eqn. 13.1.4) or of the Gaussian type, which does not agree very well with observations that are typically in better agreement with a gamma distribution, which is used in most cloud models with bulk parameterization of the microphysical processes. This suggests there are deficiencies in the formulation of the kinetic equations for stochastic condensation.

To address these concerns, a newer version of the kinetic equation for stochastic condensation for the small-size fraction of drops and crystals spectra is derived in this chapter. This equation is valid for arbitrary relative values of τ_L and τ_f that are suitable for numerical cloud models. Its analytical solutions are obtained under a number of assumptions following Khvorostyanov and Curry (1999d, 2008d). A model of the condensation process in a turbulent cloud is considered, and Reynolds procedure is applied to the regular and fluctuation parts of all the quantities, yielding an equation in terms of covariances, where the concept of macro- and microscale supersaturation is introduced. Evaluation of various covariances is performed by integration over the turbulent spectrum. Substitution of these covariances into the original kinetic equation yields the final kinetic equation of stochastic condensation, and its particular cases for low- and high-frequencies regimes are considered.

Using some assumptions and simplifications, and neglecting the diffusional growth of larger particles, analytical solutions of the gamma distribution type are obtained for the small–size fraction and their asymptotics are found. Additional solutions are obtained in the form of the generalized gamma distributions with account for the diffusional growth of large particles, sedimentation, and coagulation with the large–size fraction for the cases when such processes are important. An application of the solutions for liquid clouds is illustrated by comparison of the calculated spectral indices and size spectra with an example of a low St cloud similar to that observed in the Arctic (Curry, 1986). Comparison with some previous theories and observations of spectral broadening shows that previous criticisms of the stochastic condensation theory were inconsistent. An example of the calculation of the size spectra for a crystalline cloud similar to that observed during the SHEBA-FIRE field experiment in the Arctic in 1998 is described. Finally, based on the Chapman–Kolmogorov equation for stochastic processes, a general integral stochastic kinetic equation is formulated. Using this equation, the differential Fokker–Planck equation for stochastic condensation is derived, and it is shown that various versions of the kinetic equations considered in this chapter are particular cases of this Fokker–Planck equation.

13.2. Condensation in a Turbulent Cloud

13.2.1. Basic Equations

The kinetic equation of condensation can be written in the form described in Section 5.5 as

$$\frac{\partial f}{\partial t}+\frac{\partial}{\partial x_i}[(u_i - V_t(r))f]+\frac{\partial}{\partial r}(\dot{r}f) = J_s, \tag{13.2.1}$$

where r is the droplet radius, $\dot{r}$ is the droplet radius growth rate, x_i and u_i are the coordinates and velocity components, $V_t(r)$ is the droplet fall velocity, J_s describes droplet sources and sinks, and the usual summation convention over doubled indices is assumed, $i = 1, 2, 3$. We omit in this section the subscript "d" in all quantities for drops. Equation (13.2.1) is a continuity equation for the drop size spectra. The term $\partial(\dot{r}f)/\partial r$ in (Eqn. 13.2.1) represents the divergence of $f(r)$ due to condensation growth. The droplet growth rate $\dot{r}$ is determined from simplified (Eqn. 5.1.43) with $\xi_{con} = 0$ and $D_v^* = D_v$:

$$\dot{r} = \frac{D_v s_w \rho_{ws}}{\Gamma_1 \rho_w r}, \tag{13.2.2}$$

where all notations are the same as in Section 5.1. Equation (13.2.2) neglects for simplicity the solute (Raoult), kinetic (accomodation coefficient), and curvature (Kelvin) effects.

The fractional water supersaturation s_w in the parcel is described by (Eqn. 5.3.31a), and derived from the equations of heat balance and the continuity for water vapor:

$$\frac{ds_w}{dt} = -\frac{s_w}{\tau_f} + c_{1w} w_{ef}, \tag{13.2.3}$$

with c_{1w} defined in (Eqns. 5.3.26b) and (5.3.2c) as

$$c_{1w}(T) = \frac{c_p}{L_e}\frac{\rho_a}{\rho_{ws}}(\gamma_a - \gamma_w)\Gamma_1 = \left(\frac{dq_l}{dz}\right)_{ad}\Gamma_1\frac{\rho_a}{\rho_{ws}}, \tag{13.2.4}$$

where $\gamma_a = g/c_p$ is the dry adiabatic lapse rate, γ_w is the wet saturated over water adiabatic lapse rate (see Chapter 3), w_{ef} is an effective vertical velocity defined in the following text, τ_f is the supersaturation relaxation time (phase relaxation time) introduced in Section 5.1, related to the droplet concentration N and mean radius $\bar{r}$:

$$\tau_f = \left(4\pi D_v N\bar{r}\right)^{-1}, \qquad N = \int_0^\infty f(r)dr \qquad \bar{r} = \int_0^\infty rf(r)dr, \tag{13.2.5}$$

and all the other notations are the same as in Section 5.3.

The radiative effects on drop growth can be simply accounted for by replacing in the temperature equation the term $-\gamma_a w$:

$$-\gamma_a w + \left(\frac{\partial T}{\partial t}\right)_{rad} = -\gamma_a\left[w - \frac{1}{\gamma_a}\left(\frac{\partial T}{\partial t}\right)_{rad}\right] = -\gamma_a w_{ef}, \tag{13.2.6}$$

$$w_{ef} = w + w_{rad}, \tag{13.2.7}$$

$$w_{rad} = -\frac{1}{\gamma_a}\left(\frac{\partial T}{\partial t}\right)_{rad}. \tag{13.2.8}$$

Here, $(\partial T/\partial t)_{rad}$ is the radiative temperature change ($\mathrm{K\,s^{-1}}$) due to longwave and shortwave radiative transfer, and we have introduced an effective vertical velocity $w_{ef} = w + w_{rad}$, where w_{rad} is the radiative-effective vertical velocity related to $(\partial T/\partial t)_{rad}$. The advantage of introducing w_{rad} is that it allows direct comparison of the dynamical and radiative sources of supersaturation generation. The maximum values $(\partial T/\partial t)_{rad} \sim 10^{-3}\,\mathrm{K\,s^{-1}}$ ($-$ 3.6 $\mathrm{K\,s^{-1}K\,h^{-1}}$ or $-$ 86 $\mathrm{K\,day^{-1}}$) and $2 \times 10^{-3}\,\mathrm{K\,s^{-1}}$ ($-$7.2 $\mathrm{K\,h^{-1}}$) typical for the upper layer of the Arctic stratus (Curry, 1986) and other St–Sc clouds calculated with (Eqn. 13.2.8) correspond to $w_{rad} \sim 10 - 20\,\mathrm{cm\,s^{-1}}$ in the upper cloud layer. Thus, the magnitude of the radiative-effective velocity can be comparable to the turbulent updrafts and can be substantially larger than the synoptic-scale (mean) updraft velocity or the entrainment rate at the boundary layer top, providing the main source of condensation, instability, and turbulent kinetic energy production.

Equation (13.2.3) shows that the supersaturation change rate is determined by the sources of supersaturation (up- and downdraft and radiative cooling), the sink of supersaturation associated with condensation, and is inversely proportional to the phase relaxation time τ_f. If quasi-steady conditions are assumed in (Eqn. 13.2.3) ($ds_w/dt = 0$), the solution for equilibrium supersaturation, $s_{w,eq}$, is:

$$s_{w,eq} = c_{1w} w_{ef} \tau_f = \frac{c_p}{L}\frac{\rho_a}{\rho_{ws}}\frac{(\gamma_a - \gamma_w)}{4\pi D N \bar{r}}\Gamma_1 w_{ef}. \qquad (13.2.9)$$

13.2.2. Stochastic Equations

The kinetic (Eqn. 13.2.1), the droplet growth equation (Eqn. 13.2.2), and the supersaturation equation (Eqn. 13.2.3) close the system of equations required to determine the evolution of the drop size distribution if the velocity, temperature, and humidity fields are specified. To account for turbulence and other fluctuations, we represent all quantities as sums of mean (denoted by overbar) and fluctuating (denoted by prime) values:

$$u = \bar{u} + u', \qquad f = \bar{f} + f',$$
$$s_w = s_w + s_w', \qquad \dot{r} = \bar{\dot{r}} + \dot{r}' \qquad (13.2.10)$$

where the mean refers to an ensemble average. Substituting (Eqn. 13.2.10) into (Eqn. 13.2.1), neglecting the sedimentation for the small-size fraction for now, and averaging using the Reynolds procedure, we derive an equation for the drop size spectra that is averaged over the ensemble of realizations:

$$\frac{\partial \bar{f}}{\partial t} + \frac{\partial}{\partial x_i}\left(\bar{u}_i \bar{f}\right) + \frac{\partial}{\partial r}\left(\bar{\dot{r}}\,\bar{f}\right) = -\frac{\partial}{\partial x_i}\overline{u_i' f'} - \frac{\partial}{\partial r}\overline{\dot{r}' f'} + \bar{J}_s. \qquad (13.2.11)$$

Henceforth, we shall omit the bars over the mean values, other than for the covariance terms. The nucleation term $\bar{J}_s$ can be calculated by appropriate averaging of the CCN activity spectra as described in Chapters 6 and 7. Note that the averaging here can be understood also in the context of averaging over the grid scale of a Large-Eddy Simulation (LES) model, time, or grid-volume averaging, with the corresponding requirements on the averaging procedure.

In the case of the grid-volume average with the scale L_x, the average value of any variable is $\bar{\psi} = (1/L_x)\int_0^{L_x} \psi\, dx$. For typical LES models or a 1-Hz aircraft sampling distance, $L_x \sim$ 50–100 m. Averaging over this spatial scale implies filtering only of the highest frequencies of motion up to $\omega_{max} \sim 1\,\text{Hz}$, while the peak of the turbulent energy may lie near the smaller values of ω_{max} (e.g., Curry et al., 1988). Hence, the average quantities defined earlier still may contain turbulent fluctuations at longer wavelengths with $\omega < \omega_{max}$ and correlations still may exist between the average values and fluctuations denoted by primes. This should be kept in mind when defining the "mean" supersaturation $\bar{s}_w$, the "mean" vertical velocity $\bar{w}_{ef}$, and their relation. In this case, "mean" values implies grid averages.

The right-hand side of (Eqn. 13.2.11) is ascribed a stochastic meaning, since u and $\dot{r}$ (which depends on supersaturation fluctuations) vary randomly. We represent the fluctuations f' in a form analogous to the Prandtl mixing length concept with a generalization to the 4D phase space (x_j, r):

$$f' = -l'_j \frac{\partial f}{\partial x_j} - \frac{\partial}{\partial r} l'_r f. \tag{13.2.12}$$

The first term on the right-hand side of (Eqn. 13.2.12) corresponds to the mixing of a conservative passive scalar, where l'_j is the Prandtl mixing length along the x_j-axis. The second term arises from the nonconservativeness of f, and the Prandtl mixing length concept is extended to the r-dimension in the phase space following Voloshchuk and Sedunov (1977).

If $\tau_f << \tau_L$ (the low frequency regime), then the fluctuations of $\dot{r}'$ can be determined from the Maxwellian growth rate equation. Fluctuations in droplet growth rate can be determined as

$$\dot{r}' = \frac{D s'_w \rho_{ws}}{\Gamma_1 \rho_w r}. \tag{13.2.13}$$

The water supersaturation s_w in the presence of fluctuations is a stochastic variable. It can be described with (Eqn. 13.2.3), which has the typical form of the Langevin equation (Rodean, 1996; Lemons, 2002). The first term on the right-hand side describes the relaxation of s_w with time, while the second term describes the stochastic source of supersaturation, and w_{ef} should be treated as a stochastic variable. If the fluctuations s'_w are determined from (Eqn. 13.2.9), we have

$$s'_w = c_{1w} w'_{ef} \tau_f = \frac{c_p}{L} \Gamma_1 \frac{\rho_a}{\rho_{ws}} \frac{(\gamma_a - \gamma_w)}{4\pi D N \bar{r}} w'_{ef}. \tag{13.2.14}$$

Substituting (Eqns. 13.2.9), (13.2.12), (13.2.13), and (13.2.14) into (Eqn. 13.2.11) would yield the previous forms of the stochastic kinetic equation discussed in Section 13.1 that are valid in the low frequency regime, with the additional assumption that both mean supersaturation and its fluctuations are proportional to the corresponding vertical velocity. The assumption of low turbulence frequency is implicit in (Eqn. 13.2.14) because fluctuations of supersaturation are correlated exactly with the fluctuations of vertical velocity. Equations (13.2.13) and (13.2.14) are in conceptual agreement with the detailed DNS simulations of Paoli and Shariff (2004, 2009).

13.2.3. Supersaturation Fluctuations

As noted in Section 13.1, the basic assumption used previously in many derivations of the stochastic kinetic equation, that $\bar{s}_w$ and s'_w are both related to the respective components of the vertical velocity, leads to a size distribution of the Gaussian type or the Weibull distribution, which may be too narrow relative to observations. Thus, we hypothesize and seek alternative relationships to (Eqns. 13.2.13) and (13.2.14) that are not based on the assumption of proportionality of the fluctuations of growth rate and vertical velocity, and account for the different growth rates in fluctuations and in regular (low-frequency) growth.

Supersaturation fluctuations can be associated with turbulent fluctuations in temperature and humidity, which may or may not be associated with vertical velocity. There is a hierarchy of spatial scales of turbulent motions in the cloud, with the larger eddy sizes having a greater coherence with supersaturation. Additional supersaturation fluctuations may occur on the drop microscale that are unrelated to vertical velocity fluctuations, but might be induced by other turbulent fluctuations such as those caused by the mixing of the cloudy parcels with different properties that act to break the link between vertical velocity and supersaturation.

Suppose a supersaturation fluctuation s_w' arises in a cloudy parcel due to a turbulent fluctuation. Vapor begins to flow to the droplets and unless the equilibrium is attained in the volume V_L in the vicinity of each droplet, there will be a nonstationary vapor concentration field $\rho_{vi}(r)$. To calculate these vapor fields, we need to solve a complex system of many diffusion equations for $\rho_{vi}(r)$ from all drops that influence each other in order to calculate a self-consistent vapor field formed by the superposition of the overlapping fields from individual drops (Sedunov, 1974; Srivastava, 1989). The exact solution to this complicated diffusion problem has not been found; thus various approximate methods are used.

As described in Section 13.1, many models of condensation growth were developed over the last decades that may help to break the link between s_w' and w'. The common feature of several of them is that they predict the local growth of supersaturation fluctuations in proportion to the droplet radius, $s_w' \sim r$. Srivastava (1989) introduced the concept of the microscopic supersaturation fluctuations, which can vary from drop to drop due to randomness in the droplet spatial distribution. The main part of this microscopic supersaturation s_w' is proportional to the droplet radius.

The inhomogeneous mixing theory of Baker and Latham (1979), Baker et al. (1984), based on the chamber mixing experiments by Latham and Reed (1977), predicts the local increase of supersaturation in the vicinity of the larger drops that are less affected by mixing with entrained dry air, because the newly activated smaller droplets (that form later in the areas of complete evaporation) cannot compete as effectively for the available water vapor. Thus, the effective supersaturation in the inhomogeneous mixing theory also increases with the drop radius.

Further evidence for the proportionality $s_w' \sim r$ for small timescales comes from the detailed calculations of droplet growth that account for the kinetic correction (Fuchs, 1959, 1964; Fukuta and Walter, 1970) and the concept of the modified diffusion coefficient, $D_v^*(r)$, which accounts for this correction (see Chapter 5, Section 5.1). Calculations of droplet growth show that the magnitude of the kinetic correction depends on time and drop mass. Over short time periods comparable to τ_f (10–20 seconds), the kinetic correction causes a strong suppression of the growth of the smaller droplets. The value of $D_v^*(r) \sim r$ increases almost linearly with the radius for small droplets (Pruppacher and Klett, 1997, Table 13.1). Over periods much longer than ~20–60 s, the kinetic correction becomes negligible and Maxwellian growth occurs. Similar calculations of supersaturation during relaxation (~3–10 s) were performed by Khvorostyanov and Curry (1999c) taking into account the curvature and kinetic corrections with various accommodation coefficients (0.04–1), which also showed that the effective supersaturation, $s'_w(r, t)$ increases with radius such that it can be roughly approximated by a linear dependence, $s_w'(r, t) \sim \alpha(t)r$. The slope $\alpha(t)$ decreases with time; thus, the dependence of s'_w on r weakens and the usual Maxwellian growth takes place for the larger times. Thus, accounting for the kinetic and curvature corrections leads to the quasi-kinetic regime of growth for short time scales comparable to the phase relaxation time τ_f and the diffusion regime during long time scales.

Since we do not account explicitly here for the curvature and kinetic corrections, activation, and other effects that lead to the dependence $s'_w \sim r$ (i.e., suppression of the growth of smaller droplets), we consider an approximate model of condensational growth in supersaturation fluctuations. The average rate $\dot{s}'_{wi}$ of supersaturation absorption by the i-th drop from its vicinity $V_L \sim N^{-1}$ can be characterized by the local relaxation time as a scaling time, similar to the total relaxation time for the ensemble of drops (Eqn. 13.2.5). Then, $\dot{s}'_{wi}$ can be written approximately as

$$\tau_{fi} = (4\pi DNr_i)^{-1} \tag{13.2.15}$$

$$\dot{s}'_{wi} = s'_w / \tau_{fi} = s'_w (4\pi D_v N r_i), \tag{13.2.16}$$

where s'_w is the supersaturation fluctuation absorbed by all drops, so that the absorption rate of supersaturation fluctuation by an i-th drop increases with increasing radius. This model is conceptually close to the Srivastava's (1989) concept of microscale supersaturation. Now, the supersaturation s'_{wi} absorbed by the i-th drop during the entire period of relaxation in the ensemble of the drops is proportional to the total relaxation time τ_f. Using (Eqns. 13.2.16) and (13.2.5), we can determine

$$s'_{wi} = \dot{s}'_{wi}\tau_{fi} = s'_w \frac{\tau_f}{\tau_{fi}} = s'_w \frac{4\pi D_v N r_i}{4\pi D_v N \bar{r}} = s'_w \frac{r_i}{\bar{r}}. \tag{13.2.17}$$

So, the effective microscale supersaturation in (Eqn. 13.2.17) is approximated during relaxation as a linear function of drop radius. Incorporation of (Eqn. 13.2.17) into (Eqn. 13.2.13) for $\dot{r}_i'$ yields the desired result:

$$\dot{r}' = \frac{D_v s'_w \rho_{ws}}{\Gamma_1 \rho_w \bar{r}} = \phi_r s'_w, \qquad \phi_r = \frac{D_v \rho_{ws}}{\Gamma_1 \rho_w \bar{r}}. \tag{13.2.18}$$

The physical meaning of (Eqns. 13.2.16)–(13.2.18) is that due to suppressed growth of the smaller drops, the effective supersaturation is greater for the larger droplets, while the rate of their absorption is faster, so during relaxation the larger droplets absorb the larger fraction of the initial supersaturation fluctuation s'_w. Equation (13.2.18) shows that fluctuation of the growth rate described by this model is independent of radius, which is equivalent to the kinetic regime in fluctuations. Another important feature of (Eqn. 13.2.18) is that we do not assume here that $s'_w \sim w'$ as in previous theories, but simply consider s'_w as a stochastic variable.

Support for this model of microscale condensation is provided by several previous studies, as discussed earlier. After substituting into (Eqn. 13.2.13) the effective diffusivity $D_v^*(r) \sim r$ in approximation with kinetic correction for small drops (see Section 5.1), which can approximate the results by Fukuta and Walter (1970) and Pruppacher and Klett (1997, Table 13.1), one also arrives at a formulation (Eqn. 13.2.18) of $\dot{r}'$ that is independent of radius.

From (Eqn. 13.2.18), we can infer that the droplet spectrum during Δt in the kinetic regime would be displaced as a whole by $\Delta r = \phi_r s'_w \Delta t$, whereby the displacement is independent of radius and the shape of the spectrum is preserved. The possibility of such a spectral shift was noted by Baker et al. (1984) when analyzing the consequences of the Broadwell–Breidenthal (1982) mixing model.

Hence, this model of condensation in a turbulent medium is naturally separated by scales: 1) the usual Maxwellian diffusion regime controls condensational growth for longer time scales by using (Eqn. 13.2.2) for $\dot{r}$ (with the use of calculated values of supersaturation or equilibrium supersaturation $s_{w,eq}$ (Eqn. 13.2.9) for the steady state); and 2) droplet growth occurs in the quasi-kinetic regime (Eqns. 13.2.17) and (13.2.18) for small time scales in fluctuations. Therefore, fluctuations in supersaturation, s_w', and growth rate, $\dot{r}'$, have a stochastic meaning, but they are not proportional to the vertical velocity fluctuation w' as in the low-frequency approximation. This model therefore breaks the link between the fluctuations in s_w' (or growth rate) and w'.

By integrating (Eqn. 13.2.18) over the duration of a single fluctuation ($\sim \tau_L$), an expression for l_r' is derived that determines an analog of Prandtl's mixing length in radii space:

$$l_r' = \int_0^t \dot{r}' dt' = \phi_r \int_0^t s_w'(t') dt'. \tag{13.2.19}$$

As l_r' in (Eqn. 13.2.19) is independent of radius, (Eqn. 13.2.12) for f' then becomes

$$f' = -l_j' \frac{\partial f}{\partial x_j} - l_r' \frac{\partial f}{\partial r}. \tag{13.2.20}$$

Equations (13.2.17)–(13.2.20) allow evaluation of the covariances in the kinetic (Eqn. 13.2.11), which is done in the next section.

13.3. Evaluation of Correlation Functions

13.3.1. Expansions of Random Characteristics over the Turbulent Frequencies

In this section, we use concepts from the statistical theory of turbulence (Kolmogorov, 1941; Rodean, 1996; Monin and Yaglom, 2007b) to evaluate the correlation functions $\overline{u_i' f'}$ and $\overline{\dot{r}' f'}$ in (Eqn. 13.2.11). We assume that the velocity fluctuation vectors u_j', the supersaturation s_w', and mixing lengths l_j' are random quantities, which we represent as expansions in form of improper Fourier–Stieltjes integrals over the frequencies ω of the turbulent spectrum:

$$u_j'(t) = \int_{-\infty}^{\infty} e^{i\omega t}\, du_j'(\omega)$$

$$s_w'(t) = \int_{-\infty}^{\infty} e^{i\omega t}\, ds_w'(\omega)$$

$$l_j'(t) = \int_0^t u_j'(t_1)\, dt_1 = \int_0^t dt_1 \int_{-\infty}^{\infty} e^{i\omega t_1}\, du_j'(\omega), \tag{13.3.1}$$

where the symbol t_1 is used hereafter to denote the time variable of integration but not the fluctuation.

The complex amplitudes $du_j'(\omega)$, $ds_w'(\omega)$ have an analytic continuation into the region $\omega < 0$, with the amplitudes at $\omega > 0$ and $\omega < 0$ being the complex conjugates, $du_j'(-\omega) = du_j'^{*}(\omega)$, and are normalized to the Dirac delta-function, $\delta(\omega)$, and a spectral function of turbulence $F_{ij}(\omega)$ (Monin and Yaglom, 2007b):

$$\overline{du_i'(\omega)du_j'^*(\omega')} = F_{ij}(\omega)\delta(\omega - \omega')d\omega\, d\omega'$$

$$F_{ij}(\omega) = \frac{1}{2\pi}\int_0^{\infty} e^{-i\omega t} B_{ij}(t)\, dt, \tag{13.3.2}$$

where $F_{ij}(\omega)$ is related to the frequently used spectral density $E_{ij}(\omega)$ defined at positive $\omega > 0$ as $2F_{ij}(\omega) = E_{ij}(\omega)$. Assuming stationary, locally homogeneous, isotropic turbulence, we can also define (following Monin and Yaglom (2007b)) the velocity correlation function, B_{ij}, and the turbulence diffusion coefficient tensor, k_{ij}:

$$B_{ij}(t-t_1) = \int_{-\infty}^{\infty} e^{i\omega(t-t_1)} F_{ij}(\omega)d\omega = \int_0^{\infty} \cos[\omega(t-t_1)][2F_{ij}(\omega)]d\omega,$$

$$k_{ij} = \int_0^{\infty} B_{ij}(\tau)d\tau, \tag{13.3.3}$$

where $\tau = t - t_1$. The covariance $\overline{l_i'u_j'}$ is readily determined from (Eqn. 13.3.1) to be

$$\overline{l_i'u_j'} = \int_0^{t} dt_1 \int_{-\infty}^{\infty} e^{i\omega t} \int_{-\infty}^{\infty} e^{-i\omega' t_1} \overline{du_i'(\omega)du_i^*(\omega')} = \int_0^{t} dt_1 B_{ij}(t-t_1). \tag{13.3.4}$$

13.3.2. Supersaturation as a Nonconservative Variable

Determination of covariances involving supersaturation is complicated by the fact that supersaturation is nonconservative because of turbulent fluctuations of temperature and humidity, and because of phase changes when $s_w \neq 0$. Nonconservativeness of supersaturation requires introduction of nonconservative covariances and turbulence coefficients.

Separating fluctuations from the average quantities in the supersaturation (Eqn. 13.2.3), we can write

$$\frac{ds_w'}{dt} = -\frac{s_w'}{\tau_f} + c_{1w}u_3', \tag{13.3.5}$$

where the vertical velocity is denoted by u_3 for further convenience. By substituting (Eqn. 13.3.1) into (Eqn. 13.3.5), we can introduce a spectral analog of the supersaturation equation:

$$\int_{-\infty}^{\infty} e^{i\omega t}(i\omega)ds_w'(\omega) = -\frac{1}{\tau_f}\int_{-\infty}^{\infty} e^{i\omega t}\, ds_w'(\omega) + c_{1w}\int_{-\infty}^{\infty} e^{i\omega t}\, du_3'(\omega), \tag{13.3.6a}$$

$$ds_w'(\omega) = \frac{c_{1w}}{i(\omega - i\omega_p)} du_3'(\omega), \tag{13.3.6b}$$

where we introduced the inverse quantity $\omega_p = \tau_f^{-1} = 4\pi D_v N\overline{r}$ that can be referred to as the "*supersaturation relaxation frequency*," which is here the imaginary part of the frequency. The complex frequency in the denominator of (Eqn. 13.3.6b) with imaginary part $\omega_p > 0$ reflects the nonconservativeness of the supersaturation and is written in the form similar to that in electrodynamics for the wave propagation in the absorbing media (e.g., Landau and Lifshitz, 1966, v. 8; Lifshitz and Pitaevskii, 1997), describing relaxation of perturbation in a medium.

If we assume in (Eqn. 13.3.6b) that $\omega << \omega_p$ (low-frequency approximation), the amplitude $ds'_w(\omega)$ becomes proportional to the vertical velocity $du'(\omega)$ at all frequencies, and we come to the previous formulations of stochastic theory by Levin and Sedunov (1966), Voloshchuk and Sedunov (1977), and Manton (1979). Thus, our approach generalizes the previous theories by use of (Eqn. 13.3.6b), which breaks the links between supersaturation and vertical velocity. In deriving (Eqn. 13.3.6), we assumed for simplicity that $\tau_f = \omega_p^{-1}$ is constant; all subsequent manipulations can be readily generalized to allow for fluctuations of τ_f (e.g., Voloshchuk and Sedunov (1977) for the low-frequency regime; Cooper, 1989); however, as was noted in Section 13.2, the assumption of $\tau_f \approx$ const is a good approximation when considering supersaturation relaxation on the microscale over time scales $t \sim \tau_f$.

13.3.3. Covariances with Supersaturation

By incorporating (Eqns. 13.3.6b), (13.3.1), and (13.3.2), we obtain the following expression for the correlation function $\overline{u'_j S'}$:

$$\overline{s'_w(t)u'_j(t_1)} = \int_{-\infty}^{\infty} e^{i\omega t} \int_{-\infty}^{\infty} e^{-i\omega t_1} \frac{c_{1w}}{i(\omega - i\omega_p)} \overline{du'_3(\omega)du'_j{}^*(\omega')}$$
$$= \delta_{i3}(c_{1w}/\omega_p) B^n_{ij}(t-t_1,\tau_f), \qquad (13.3.7)$$

where δ_{ik} is Kronecker's symbol, we used the normalization condition (Eqn. 13.3.2) for the velocities amplitudes, and introduced the correlation function of velocity with supersaturation as a nonconservative substance (superscript "n") with a characteristic "nonconservativeness" time $\tau_f = \omega_p^{-1}$:

$$B^n_{ij}(t,\tau_f) = \int_{-\infty}^{\infty} e^{i\omega t} \frac{\omega_p}{i(\omega - i\omega_p)} F_{ij}(\omega)\, d\omega \qquad (13.3.8a)$$

$$= \int_0^{\infty} \frac{[\omega_p + \omega \tan \omega t]}{1+(\omega/\omega_p)^2} \cos \omega t [2F_{ij}(\omega)]\, d\omega. \qquad (13.3.8b)$$

The first term in the subintegral function of (Eqn. 13.3.8b) determines the nonconservative effects, and the rest coincides with the expansion of the conservative function (Eqn. 13.3.3). We can derive analogously the turbulence diffusion tensor coefficient for a nonconservative substance:

$$k^n_{ij}(\tau_f) = \int_0^{\infty} B^n_{ij}(\tau,\tau_f)\, d\tau. \qquad (13.3.9)$$

Following an approach analogous to the derivation of (Eqn. 13.3.7), we can obtain an expression for the supersaturation autocorrelation function:

$$\overline{s'_w(t)s'_w(t_1)} = \int_{-\infty}^{\infty} e^{i\omega t} \int_{-\infty}^{\infty} e^{-i\omega t_1} c_{1w}^2 \frac{\overline{du'_3(\omega)du'_3{}^*(\omega')}}{(i\omega + \omega_p)(-i\omega + \omega_p)}$$
$$= (c_{1w}/\omega_p)^2 \delta_{i3}\delta_{j3} B^{nn}_{ij}(t-t_1,\tau_f), \qquad (13.3.10)$$

where we use again (Eqn. 13.3.2) and introduce the autocorrelation function of a nonconservative substance, and the corresponding turbulence coefficient:

$$B_{ij}^{nn}(t,\tau_f) = \int_{-\infty}^{\infty} e^{i\omega t} \frac{\omega_p^2}{(\omega^2+\omega_p^2)} F_{ij}(\omega)\,d\omega = \int_0^{\infty} \frac{\cos\omega\, t}{1+(\omega/\omega_p)^2}[2F_{ij}(\omega)]\,d\omega$$

$$k_{ij}^{nn}(\tau_f) = \int_0^{\infty} B_{ij}^{nn}(t,\tau_f)\,dt. \tag{13.3.11}$$

The covariance of u_j' and l_r' is evaluated analogously using (Eqns. 13.2.19) and (13.3.7):

$$\overline{u_j'(t)l_r'(t_1)} = \phi_r \int_0^t dt_1 \overline{u_j'(t)s_w'(t_1)} = \phi_r \left(\frac{c_{1w}}{\omega_p}\right) \int_0^t dt_1 B_{ij}^n(t-t_1)\delta_{i3}$$

$$= \phi_r c_{1w}\tau_f \delta_{i3} \int_0^t dt_1 B_{ij}^n(t-t_1) = G_s\delta_{i3}\int_0^t dt_1 B_{ij}^n(t-t_1) \tag{13.3.11a}$$

where ϕ_r is defined in (Eqn. 13.2.18) and we have introduced the parameter G_s that plays a key role in the analytical solutions:

$$G_s = \phi_r c_{1w}\tau_f = \frac{D_v}{\rho_w \bar{r}} \frac{c_p}{L} \frac{\gamma_a - \gamma_w}{4\pi D N \bar{r}} \rho_a = \frac{\rho_a}{\rho_w} \frac{c_p}{L} \frac{\gamma_a-\gamma_w}{4\pi N \bar{r}^2}. \tag{13.3.12}$$

13.3.4. Covariances with the Drop Size Distribution Function

For the general case of arbitrary relations between τ_f and τ_L, we obtain for the correlation function $\overline{u_i'f'}$ in (Eqn. 13.2.11) using (Eqn. 13.2.20) for f':

$$\overline{u_i'f'} = -\overline{u_i'l_j'}\frac{\partial f}{\partial x_j} - -\overline{u_i'l_r'}\frac{\partial f}{\partial r}$$

$$= -\int_0^t dt_1 \left[B_{ij}(t-t_1)\frac{\partial f(t_1)}{\partial x_j} + B_{ij}^n(t-t_1,\tau_f)\delta_{j3}G_s \frac{\partial f(t_1)}{\partial r} \right], \tag{13.3.13}$$

where we used (Eqn. 13.3.4), for $\overline{u_i'l_j'}$ and (Eqn. 13.3.11a) for $\overline{u_i'l_r'}$. For $\overline{\dot{r}'f'}$, using (Eqn. 13.2.20) for f', we obtain

$$\overline{\dot{r}'f'} = \phi_r \overline{s_w'f'} = -\phi_r \int_0^t dt_1 \left[\overline{s_w'(t)u_j'(t_1)}\frac{\partial f(t_1)}{\partial x_j} + \phi_r \overline{s_w'(t)s_w'(t_1)}\frac{\partial f(t_1)}{\partial r} \right]$$

$$= -\int_0^t dt_1 \left[G_s\delta_{i3}B_{ij}^n(t-t_1,\tau_f)\frac{\partial f(t_1)}{\partial x_j} + G_s^2 \delta_{i3}\delta_{j3}B_{ij}^{nn}(t-t_1,\tau_f)\frac{\partial f(t_1)}{\partial r} \right], \tag{13.3.14}$$

where we used (Eqn. 13.3.7) for $\overline{s_w'(t)u_j'(t_1)}$ and (Eqn. 13.3.10) for $\overline{s_w'(t)s_w'(t_1)}$. Expressions (Eqns. 13.3.13) and (13.3.14) are integrated similar to Stepanov (1975, 1976) as is done in the statistical k-theory of turbulence (Monin and Yaglom, 2007a,b). The main contribution to the integrals comes from the region $t - t_1 \sim \tau_L$, because the correlation functions decrease rapidly beyond this interval. During the time period $t - t_1 \sim \tau_L$, the distribution function f varies much more slowly than the correlation functions, and we can remove $\partial f/\partial x_j$ and $\partial f/\partial r$ from the integrals and simplify (Eqns. 13.3.13) and (13.3.14) to

$$\overline{u_i'f'} \approx -\frac{\partial f}{\partial x_j}\int_0^t dt_1 B_{ij}(t-t_1) - \delta_{j3}G_s\frac{\partial f}{\partial r}\int_0^t dt_1 B_{ij}^n(t-t_1,\tau_f), \tag{13.3.15}$$

$$\overline{\dot{r}'f'} \approx -\delta_{i3}G_s\frac{\partial f}{\partial x_j}\int_0^t dt_1 B_{ij}^n(t-t_1,\tau_f) - \delta_{i3}\delta_{j3}G_s^2\frac{\partial f}{\partial r}\int_0^t dt_1 B_{ij}^{nn}(t-t_1,\tau_f). \tag{13.3.16}$$

The integrals in (Eqns.13.3.15) and (13.3.16) can then be evaluated from $t = 0$ to ∞ since the integrated functions decrease rapidly with time beyond τ_L. We can write (Eqns.13.3.15) and (13.3.16) by including the definitions of the turbulence coefficients (Eqns.13.3.3), (13.3.9) and (13.3.11) as

$$\overline{u_i'f'} = -k_{ij}\frac{\partial f}{\partial x_j} - \delta_{j3}\, k_{ij}^n(\tau_f)G_s\frac{\partial f}{\partial r} \tag{13.3.17}$$

$$\overline{\dot{r}'f'} = -\delta_{i3}k_{ij}^n(\tau_f)G_s\frac{\partial f}{\partial x_j} - \delta_{i3}\delta_{j3}k_{ij}^{nn}(\tau_f)G_s^2\frac{\partial f}{\partial r}. \tag{13.3.18}$$

For high-frequency turbulence, only the first term of the expression for $\overline{u_i'f'}$ on the right-hand side in (Eqn. 13.3.17) remains since the nonconservative turbulence coefficients vanish. In the high-frequency approximation $\omega >> \omega_p$ ($\tau_f >> \tau_L$), we find from (Eqns. 13.3.7)–(13.3.11) that $B_{ij}^n = 0$, $k_{ij}^n = 0$, $B_{ij}^{nn} = 0$, and $k_{ij}^{nn} = 0$. In the opposite case of low-frequency approximation $\omega << \omega_p$ ($\tau_L >> \tau_f$), the term $\omega_p/(i(\omega - i\omega_p))$ in (Eqn. 13.3.8a) tends to unity and the supersaturation-velocity correlation function reduces to (Eqn. 13.3.3), so, $B_{ij}^n = B_{ij}$, and $k_{ij}^n = k_{ij}$. It follows also from (Eqn. 13.3.11) that $B_{ij}^{nn} = B_{ij}$, $k_{ij}^{nn} = k_{ij}$; that is, nonconservative effects (dependence of the correlation functions and turbulence coefficients on the supersaturation relaxation time) vanish in the low-frequency regime.

In liquid clouds, $\tau_f \sim 1 - 10$ s (see Chapter 5), while $\omega \sim (0.2 - 1) \times 10^{-2}\,\mathrm{s}^{-1}$ and $\tau_L \sim 100$–600 s (e.g., Curry et al., 1988), which is closer to the low-frequency approximation; hence, the high-frequency kinetic equation used now in most cloud models with explicit spectral microphysics is oversimplified. The errors of neglecting the terms with k^n, k^{nn} are unknown and can be significant. In crystalline clouds like cirrus or diamond dust, for which this technique can also be applied as described in the following text, the concentration of particles can be 1–3 orders of magnitude less than in the liquid stratus, and the values of τ_f can be 10–180 min and greater (Khvorostyanov and Sassen, 1998a,b; 2002; Khvorostyanov et al., 2001, 2006), while $\tau_L \sim 5$–15 min. Thus, the values of τ_f and τ_L can be comparable or the situation can be closer to the high-frequency regime, which should influence k^n, k^{nn}. In particular, if the maximum in the turbulent spectrum is located near $\omega \sim \tau_L^{-1}$, and $\tau_f \sim \tau_L$, then (Eqns.13.3.8b) and (13.3.11) show that k^n, $k^{nn} \sim 0.5\,k$; these values will decrease with increasing τ_f. Equations (13.3.8) and (13.3.11) allow calculations of the nonconservative correlation functions and turbulent coefficients either with various theoretical models of the turbulence spectra or directly with the observed-in-clouds turbulence spectra (e.g., Curry et al., 1988; Sassen et al., 1989; Quante and Starr, 2002) with appropriate choices of limits in the integrals.

13.4. General Kinetic Equations of Stochastic Condensation

Substituting (Eqns. 13.3.17) and (13.3.18) into (Eqn. 13.2.11) and assuming that the droplet growth rate $\dot{r}$ is determined from (Eqn. 13.2.2), we obtain the general kinetic equation of stochastic condensation:

$$\frac{\partial f}{\partial t}+\frac{\partial}{\partial x_i}\left[\left(u_i - v(r)\delta_{i3}\right)f\right]+\frac{\partial}{\partial r}\left(\frac{b_{con}s_w}{r}f\right)$$
$$=\frac{\partial}{\partial x_i}k_{ij}\frac{\partial f}{\partial x_j}+\frac{\partial}{\partial x_i}\delta_{j3}k_{ij}^{n}G_s\frac{\partial f}{\partial r}+\delta_{\beta}k_{ij}^{n}G_s\frac{\partial^2 f}{\partial x_i \partial r}+\delta_{i3}\delta_{j3}k_{ij}^{nn}G_s^2\frac{\partial^2 f}{\partial r^2}+\overline{J}_s,$$
$$b_{con}=\frac{D_v\rho_{ws}}{\Gamma_1\rho_a} \qquad (13.4.1)$$

Recall, all the quantities on the left-hand side (f, u_i, w, s_w) indicate averaged values (in the sense discussed in Section 13.2.2), and all fluctuations are included on the right-hand side of the equation.

If the scale of averaging (e.g., over a grid box of a numerical model) is chosen to be sufficiently large so that the characteristic time of averaging is much larger than τ_f, then the mean supersaturation $\overline{s}_w$ can be equal to the equilibrium value $s_{w,eq}$ related in (Eqn. 13.2.9) to the "mean" vertical velocity $\overline{w}_{ef}$. Note, however, that when using this representation for $\overline{s}_w$, the value of $\overline{w}_{ef}$ should not be understood as simply the mean (synoptic-scale) vertical velocity.

The term $\overline{w}_{ef}$ represents an "effective" subgrid-averaged vertical velocity. Since the grid-box average still contains turbulent fluctuations of the larger scales (see Section 13.2), and correlations still exist between smaller-scale fluctuations (primes) and "mean" values (overbars), the effective subgrid average $\overline{w}_{ef}$ can be estimated from the relation $\overline{w}_{ef}\overline{f} \sim \overline{w'f'}$. Multiplying this relation by (4/3) $\pi\rho_w r^3$ and integrating over radii, we obtain a relation $\overline{w}_{ef} \sim \overline{w'q_L'}/q_L$, which allows estimation of $\overline{w}_{ef}$ from measurements of q_L and the covariance of the fluctuations of the vertical velocities and q_L—i.e., $\overline{w'q_L'}$.

Substituting the equilibrium value $s_{w,eq}$ (Eqn. 13.2.9) into (Eqn. 13.4.1), we obtain

$$\frac{\partial f}{\partial t}+\frac{\partial}{\partial x_i}\left[\left(u_i - v(r)\delta_{i3}\right)f\right]+\frac{\partial}{\partial r}\left(\frac{c_{con}}{r}w_{ef}f\right)$$
$$=\frac{\partial}{\partial x_i}k_{ij}\frac{\partial f}{\partial x_j}+\frac{\partial}{\partial x_i}\delta_{j3}k_{ij}^{n}G_s\frac{\partial f}{\partial r}+\delta_{i3}k_{ij}^{n}G\frac{\partial^2 f}{\partial x_i \partial r}+\delta_{i3}\delta_{j3}k_{ij}^{nn}G_s^2\frac{\partial^2 f}{\partial r^2}+\overline{J}_s,$$
$$c_{con}=\frac{D_v c_{1w}\rho_{ws}\tau_f}{\rho_w\Gamma_1}=D_v\frac{\rho_a}{\rho_w}\frac{c_p}{L}\frac{(\gamma_a-\gamma_w)}{4\pi D_v N\overline{r}}. \qquad (13.4.1a)$$

The representation (Eqn. 13.4.1a) with $s_{w,eq}$ may be suitable for models with grid boxes a few tens or hundreds of meters or larger and/or for processes where the mean supersaturation is close to the quasi-steady value, while the more complete form (Eqn. 13.4.1) with non-equilibrium s_w is preferable for models with a finer resolution (e.g., on the smaller scales of LES models), or for processes or cloud layers where supersaturation is not quasi-steady. Note that the minimum scales of averaging and conditions when $s_w \approx s_{w,eq}$ can probably only be determined with the use of the LES models with fine resolution after a detailed analysis of the entire supersaturation field.

To represent the kinetic equation in a more compact form and compare this with the previous equations, it is convenient to adopt a methodology from quantum field theory, functional analysis, and operational calculus that uses left and right operators. We introduce the left operator $\hat{H}_L$ and the right operator $\hat{H}_R$, such that

$$\hat{H}_L k_{ij} = k_{ij}^n, \quad k_{ij}^n = k_{ij}\hat{H}_R, \quad \hat{H}_L k_{ij}\hat{H}_R = k_{ij}^{nn}. \tag{13.4.2}$$

The left operator $\hat{H}_L$ stands on the left of a quantity and acts to the right; the right operator $\hat{H}_R$ stands on the right of a quantity and acts to the left. The operators convert the conservative tensor, k_{ij}, into a nonconservative tensor, k_{ij}^n, and commute with operators $\partial/\partial x_i$ and $\partial/\partial r$. $\hat{H}_{L,R}$ are the integral operators that convert the correlation function (Eqn. 13.3.3) into (Eqns. 13.3.8) and (13.3.11). Incorporating these operators into the right-hand side of (Eqns. 13.4.1 and 13.4.1a), we obtain

$$\frac{\partial f}{\partial t} + \frac{\partial}{\partial x_i}\left[\left(u_i - v(r)\delta_{i3}\right) f\right] + \frac{\partial}{\partial r}\left(\frac{b_{con} s_w}{r} f\right)$$
$$= \left(\frac{\partial}{\partial x_i} + \delta_{i3} G_s \hat{H}_L \frac{\partial}{\partial r}\right) k_{ij} \left(\frac{\partial}{\partial x_j} + \delta_{j3} G_s \hat{H}_R \frac{\partial}{\partial r}\right) f + \bar{J}_s. \tag{13.4.3a}$$

Equation (13.4.3a) does not use the condition $s_w \sim s_{w,eq}$, and if we assume that $s_w \sim s_{w,eq}$, we obtain

$$\frac{\partial f}{\partial t} + \frac{\partial}{\partial x_i}\left[\left(u_i - v(r)\delta_{i3}\right) f\right] + \frac{\partial}{\partial r}\left(\frac{c_{con}}{r} w f\right)$$
$$= \left(\frac{\partial}{\partial x_i} + \delta_{i3} G_s \hat{H}_L \frac{\partial}{\partial r}\right) k_{ij} \left(\frac{\partial}{\partial x_j} + \delta_{j3} G_s \hat{H}_R \frac{\partial}{\partial r}\right) f + \bar{J}_s. \tag{13.4.3b}$$

The assumption of equilibrium supersaturation $s_{w,eq}$ might be suitable for mesoscale models with appropriate parameterization of subgrid $\overline{w}_{ef}$.

For low-frequency turbulence $\hat{H}_{L,R} = 1$, whereas for high-frequency turbulence $\hat{H}_{L,R} = 0$. The high-frequency regime, for which generally $s_w \neq s_{w,eq}$, is thus represented by

$$\frac{\partial f}{\partial t} + \frac{\partial}{\partial x_i}\left[\left(u_i - v(r)\delta_{i3}\right) f\right] + \frac{\partial}{\partial r}\left(\frac{b_{con} s_w}{r} f\right) = \frac{\partial}{\partial x_i} k_{ij} \frac{\partial}{\partial x_j} f + \bar{J}_s. \tag{13.4.4}$$

This type of equation is used in most cloud models with explicit spectral bin microphysics and describes the evolution of the size spectra with broadening due to vertical and horizontal turbulent mixing between the cloud parcels or regions with different properties. The low-frequency regime with the assumption $s_w \approx s_{w,eq}$ is represented by

$$\frac{\partial f}{\partial t} + \frac{\partial}{\partial x_i}\left[\left(u_i - v\delta_{i3}\right) f\right] + \frac{\partial}{\partial r}\left(\frac{c_{con}}{r} w f\right)$$
$$= \left(\frac{\partial}{\partial x_i} + \delta_{i3} G_s \frac{\partial}{\partial r}\right) k_{ij} \left(\frac{\partial}{\partial x_j} + \delta_{j3} G_s \frac{\partial}{\partial r}\right) f + \bar{J}. \tag{13.4.5}$$

The term $(\partial/\partial r[(c_{con}/r)wf]$ describes "advection" in the space of radii with effective speed $(c_{con}/r)w$, inversely proportional to the radius. Thus, this "advection" is faster for the smaller droplets and this term alone causes narrowing of the size spectra. Analogous to the term $-k_{ij}\partial f/\partial x_i$, the turbulent flux of the droplets in the usual space, the term $-k_{ij}G_s\partial f/\partial r$ represents the turbulent flux in the space of radii. So, the term $k_{xr} = G_s k_{ij}$ is the effective diffusion coefficient in the combined space of coordinates–radii, the term

$$k_r = G_s^2 k_{ij} \tag{13.4.5a}$$

is the effective diffusion coefficient in the space of radii, and the parameter $G_s \sim 10^{-9}$ to 10^{-8} is the scaling factor between the diffusion coefficients in x_i and r spaces. Since the derivative $-\partial f/\partial r < 0$ for the radii smaller than the modal radius, $r < r_m$, the flux of droplets is directed along the gradient toward the smaller radii to the left of the mode, at $r < r_m$, and vice versa, $-\partial f/\partial r > 0$ at $r > r_m$, and the flux is directed toward the larger radii. So, this "diffusion" in the space of radii tends to smooth the gradients and broaden the size spectrum. The resulting shape of the size spectra is determined by the relative speed of the advection and diffusion in the space of radii and in the coordinate space. If the effects of advection dominate (e.g., in the vigorous updrafts but with weak turbulence), the spectra may narrow even in the presence of turbulence, which alone might be insufficient to produce the broad spectra.

When written using the left and right operators, the differences are clearer between the new equation and the earlier kinetic equations of stochastic condensation reviewed in Section 13.1. These earlier kinetic equations can be written using our notations after some simplifications as

$$\frac{\partial f}{\partial t} + \frac{\partial}{\partial x_i}\left[\left(u_i - v(r)\delta_{i3}\right)f\right] + \frac{\partial}{\partial r}\left(\frac{c_{con}}{r}wf\right)$$
$$= \left(\frac{\partial}{\partial x_i} + \delta_{i3}c_{con}\frac{\partial}{\partial r}\frac{1}{r}\right)k_{ij}\left(\frac{\partial}{\partial x_j} + \delta_{j3}c_{con}\frac{\partial}{\partial r}\frac{1}{r}\right)f. \tag{13.4.6}$$

Equation (13.4.3a) from Khvorostyanov and Curry (1999a, 2008a) does not assume the quasi-steady state supersaturation. Another version, (Eqn. 13.4.3b), assumes this approximation (although with another interpretation of $\overline{w}_{ef}$), but additionally accounts for the nonconservative character of supersaturation and can thus be used in both the low- and high-frequency regimes including the cases when $s_w \approx s_{w,eq}$, while the previous equations of the type (Eqn. 13.4.6) are applicable only in the low frequency regime ($\hat{H}_{L,R} = 1$) with $s_w \approx s_{w,eq}$. Additionally, by virtue of using the microscopic supersaturation fluctuation (Eqn. 13.2.17) instead of (Eqn. 13.2.14), the operator $G_s(\partial/\partial r)$ appears now in (Eqns. 13.4.3a) and (13.4.3b) and leads to solutions of gamma distribution, and exponential or inverse power law types (see Sections 13.5–13.7), while the operator $(\partial/\partial r)(c_{con}/r)$ appears in the previous (Eqn. 13.4.6) and leads to Gaussian solutions or the Weibull distribution, which decrease with radius faster than most observed size spectra.

In most cloud models with explicit spectral microphysics, the high-frequency form of the kinetic equation (13.4.4) is used and only a few attempts have been made to numerically solve the more complete kinetic equations in the low frequency approximation (e.g., Vasilyeva et al., 1984). Incorporation into cloud models of the more complete kinetic equations of stochastic condensation in the forms (Eqns. 13.4.3), (13.4.5), or (13.4.6) requires development of efficient economical numerical algorithms for solutions of equations with cross-derivatives by coordinates and radii and the small characteristic time and spatial scales that determine the condensation process.

13.5. Assumptions and Simplifications for Analytical Solutions

The basic equations developed in Sections 13.2–13.4 are applicable to pure liquid or pure crystalline clouds; mixed phase clouds are not explicitly addressed here. The kinetic equation of stochastic condensation derived in Section 13.4 can be easily generalized by adding the collection (aggregation) and breakup terms $(\partial f/\partial t)_{col}$ and $(\partial f/\partial t)_{br}$ as

$$\frac{\partial f}{\partial t}+\frac{\partial}{\partial x_i}\left[\left(u_i - v(r)\delta_{i3}\right)f\right]+\frac{\partial}{\partial r}\left(\dot{r}_{cond}f\right)$$
$$=\left(\frac{\partial}{\partial x_i}+\delta_{i3}G_s\hat{H}_L\frac{\partial}{\partial r}\right)k_{ij}\left(\frac{\partial}{\partial x_j}+\delta_{j3}G_s\hat{H}_R\frac{\partial}{\partial r}\right)f+\left(\frac{\partial f}{\partial t}\right)_{col}+\left(\frac{\partial f}{\partial t}\right)_{br}, \qquad (13.5.1)$$

where it is assumed that both liquid drops and ice crystals are represented by spheres of radius r. Hereafter in this section, we use the notations common for liquid and ice clouds—i.e., $\dot{r}_{cond}$ denotes the condensation growth rate of a drop or deposition growth of a crystal, and s denotes supersaturation over water or ice, unless specific notations are used for liquid or ice.

The turbulent coefficients k_{ij}^{n} and k_{ij}^{nn} involved in (Eqn. 13.5.1) and the correlation functions were derived in Sections 13.2–13.4 for pure liquid clouds. However, the equations for pure crystalline clouds have the same form as the expressions derived previously for pure liquid clouds. The derivation for pure crystalline clouds is the same, beginning with the kinetic equation of regular deposition for the crystal size spectrum $f_c(r_c)$ and the equation for a crystal growth rate dr_c/dt (Chapter 5). Then, the same consideration of fluctuations and correlation functions as in Sections 13.2 and 13.3 can be repeated for ice clouds, and we arrive at the same stochastic kinetic equation for the crystal size spectrum. The major difference between pure liquid and crystalline clouds is in the different values of supersaturation over water s_w, or ice s_i, in liquid or crystalline clouds, and in the supersaturation relaxation times $\tau_f = (4\pi D_v N\bar{r})^{-1}$, where N and $\bar{r}$ denote hereafter the drop or crystal number concentrations and their mean radii, and $\omega_p = \tau_f^{-1} = 4\pi D_v N\bar{r}$ is the supersaturation relaxation frequency introduced in Section 13.3. The times τ_f determine the rates of supersaturation absorption by drops or crystals and the degree of their "nonconservativeness" in interactions with supersaturation.

For the turbulent frequencies ω that provide the major contribution into the turbulent energy (e.g., Curry et al., 1988; Sassen et al., 1989; Quante and Starr, 2002; Shaw, 2003; Monin and Yaglom, 2007b), the expressions derived in Section 13.3 for the nonconservative turbulence coefficients show that $k_{ij}^{n}(\tau_f) \sim \omega_p^2 \sim \tau_f^{-2}$ and $k_{ij}^{nn}(\tau_f) \sim \omega_p^2 \sim \tau_f^{-2}$. That is, the nonconservative coefficients k_{ij}^{n}, k_{ij}^{nn} decrease with increasing τ_f and are generally smaller in crystalline clouds due to the greater τ_f than that in liquid clouds, which is caused by the different N, $\bar{r}$. The values of $\tau_f \sim 1-10\,\text{s}$ for liquid clouds with $N \sim 1-5\times 10^2\,\text{cm}^{-3}$ and $\bar{r} \sim 5 \sim 10\,\mu\text{m}$ (Section 5.3, Table 5.1). Calculations of τ_f in cirrus and mid-level ice clouds using spectral bin models show that it varies in the range 10^2–10^4 s for typical N and $\bar{r}$—e.g., Khvorostyanov and Sassen (1998a,b, 2002), Khvorostyanov, Curry et al. (2001, 2006), Sassen et al. (2002); Sassen and Khvorostyanov (2007); Table 5.2 in this book. Thus, the values of $k_{ij}^{n}(\tau_f)$, $k_{ij}^{nn}(\tau_f)$ are smaller for crystalline clouds than for liquid clouds, but estimation from the equations in Section 13.3 shows that values of $k_{ij}^{n}(\tau_f)$, $k_{ij}^{nn}(\tau_f)$ are comparable for both liquid and crystalline clouds as described in the following text.

Analytical solutions to the kinetic (Eqn. 13.5.1) can be obtained with the following additional assumptions and simplifications:

1. The cloud is in a quasi-steady state, $\partial f/\partial t = 0$. This does not imply a complete steady state, and the time dependence can be accounted for via the integral parameters in the solution.
2. The cloud is horizontally homogeneous, $\partial f/\partial x = \partial f/\partial y = 0$.
3. The size spectrum is divided into 2 size regions,

$$f(r) = f_s(r), \qquad r < r_0 \tag{13.5.2a}$$

$$f(r) = f_l(r), \qquad r > r_0 \tag{13.5.2b}$$

where $f_s(r)$ and $f_l(r)$ are the small-size and large-size fractions, and r_0 is the boundary radius between the two size fractions. The value of $r_0 \sim 30$–$50\,\mu m$ for the drops and ~30–80 μm (or maximum dimension $D_0 \sim 60$–$160\,\mu m$) for the crystals; r_0 separates the domain $r < r_0$ where accretion rates are small and particle growth is governed mostly by the condensation/deposition from the domain $r > r_0$ where collection rates become important and prevail (see e.g., Cotton et al., 2011, Fig. 4.1). The functions $f_s(r)$ and $f_l(r)$ correspond to the bulk categorization of the condensed phase into cloud water and rain for the liquid phase and cloud ice and snow for the crystalline phase.
4. The nonconservative turbulence coefficients are parameterized following Khvorostyanov and Curry (2008b,c) to be proportional to the conservative coefficient k as $k_{ij}^{n} = c_n k$, and $k_{ij}^{nn} = c_{nn} k$. In general, c_n, c_{nn} have values that are smaller than but close to unity; they are smaller in crystalline than in liquid clouds. A quantitative estimate of these nonconservative coefficients was made with the model spectrum of turbulence:

$$F_{ij}(\omega) = A_\omega \frac{\omega^{1/3}}{B_\omega + \omega^2}. \tag{13.5.3}$$

The maximum of (Eqn. 13.5.3) is found from the condition $dF_{ij}/d\omega = 0$, which yields $\omega_m = (B_\omega/5)^{1/2}$. It is easy to see that $F_{ij} \sim \omega^{-5/3}$ at $\omega >> \omega_m$. Thus, this spectrum satisfies the Kolmogorov–Obukhov asymptotic turbulence power law −5/3. The parameter B_ω was determined from the measured maxima in the turbulence spectrum in Curry et al. (1988) for liquid clouds, and in Sassen et al. (1989) in ice clouds, and the parameter A_ω was determined by normalization to the turbulence kinetic energy, which is the integral of (Eqn. 13.5.3) over the ω-spectrum. Evaluation with the spectrum (Eqn. 13.5.3) and these conditions showed that k^n and $k^{nn} \sim 0.2 - 0.7k$, depending on the cloud microstructure. This gives an estimate of c_n, $c_{nn} \sim 0.2 - 0.7$ and indicates that the nonconservative terms are quite comparable with the term $k(\partial^2 f/\partial x^2)$ and generally cannot be neglected, although it is done in almost all cloud models with spectral microphysics.
5. The coagulation term $(\partial f/\partial t)_{col}$ is described in detail in Section 14.2 of Chapter 14. We assume at this point that drop breakup does not influence the small fraction; a correction due to breakup can be added to the solution. For this analytical solution, the coagulation–accretion growth rate (both for the drop collision-coalescence and for the crystal aggregation with crystals) is considered in the continuous growth approximation (see Chapter 14 here; Pruppacher and Klett, 1997;

Seinfeld and Pandis, 1998), with an accounting for only the collision–coalescence between the particles of the different fractions of the spectrum, $f_s(r)$ and $f_l(r)$. The decrease in small fraction f_s is described in this approximation as a collection of the small-size fraction by the large-size fraction.

$$\left(\frac{\partial f_s}{\partial t}\right)_{col} = -I_{loss} = -\sigma_{col} f_s, \qquad r < r_0, \tag{13.5.4}$$

$$\sigma_{col} = \pi E_c \int_{r_0}^{\infty} r^2 v(r) f_l(r)\, dr, \tag{13.5.5}$$

where $v(r)$ is the particle fall velocity. These equations are derived in Section 14.2 from the Smoluchowski stochastic coagulation equation (see Section 5.6) along with the mass conservation at the transition between the small- and large-size fractions.

With these assumptions, the kinetic (Eqn. 13.5.1) can be written for f_s as

$$\frac{\partial}{\partial z}\left[(w - v(r)) f_s\right] + \frac{\partial}{\partial r}\left(\dot{r}_{cond} f_s\right) = k \frac{\partial^2 f_s}{\partial z^2}$$
$$+ c_n k G \left(\frac{\partial^2 f_s}{\partial r \partial z} + \frac{\partial^2 f_s}{\partial z \partial r}\right) + c_{nn} k G^2 \frac{\partial^2 f_s}{\partial r^2} - \sigma_{col} f_s. \tag{13.5.6}$$

All of the quantities on the left-hand side (f, w, $\dot{r}_{cond}$, s) indicate averaged values in the sense discussed in Sections 13.2–13.4, and all fluctuations are included on the right-hand side of the equation. The terms on the left-hand side describe convective and sedimentation fluxes and drop (crystal) condensation (depositional) growth/evaporation, and the terms on the right-hand side describe turbulent transport, the stochastic effects of condensation/deposition growth, and absorption by the large-size fraction.

To enable analytic solutions to (Eqn. 13.5.6), we make the following additional assumptions:

6. The vertical gradient of f_s can be parameterized as a separable function:

$$\frac{\partial f_s(r,z)}{\partial z} = \alpha_s(z) \varsigma_s(r) f_s(r,z). \tag{13.5.7a}$$

The functions $\alpha_s(z)$ and the $\zeta_s(r)$ can be different for each fraction, determining the magnitude of the gradient and its possible radius dependence. If $\zeta_s(r) = 1$, and

$$df_s/dz = \alpha_s f_s, \tag{13.5.7b}$$

it is easily shown from the definition of liquid (ice) water content that

$$\alpha_s(z) = (1/q_{ls})(dq_{ls}/dz). \tag{13.5.8}$$

7. We assume that terminal velocity $v(r)$ following Khvorostyanov and Curry (2002, 2005) can be parameterized in the form

$$v(r) = A_v(r) r^{B_v(r)}, \tag{13.5.9}$$

where $A_v(r)$ and $B_v(r)$ are continuous functions of particle size, that include consideration of various crystal habits and nonspherical drops and turbulent corrections to the flow around the particles. The derivation and details of this representation are given in Chapter 12.

8. The condensation or deposition growth/evaporation rate $\dot{r}_{cond}$ is described by the simplified Maxwell equation that we use in notations common for drops and crystals:

$$\dot{r}_{cond} = \frac{b_{con}s}{r}, \qquad b_{con} = \frac{D_v \rho_{vs}}{\Gamma_p \rho_w}, \tag{13.5.10}$$

where $s = (\rho_v - \rho_{vs})/\rho_{vs}$ is the fractional supersaturation over water (ice), ρ_v is the environmental vapor density, and ρ_{vs} (= ρ_{ws} or ρ_{is}) are the saturated over water (ice) vapor densities, ρ_w is the water (ice) density, and Γ_p (= Γ_1 or Γ_2) is the psychrometric correction associated with the latent heat of condensation.

Various assumptions on the form of $\zeta_s(r)$ lead to various analytical solutions, we consider the simplest case $\zeta_s(r) = 1$. Substituting (Eqns. 13.5.7b)–(13.5.10) into (Eqn. 13.5.6) allows elimination of z-derivatives and we obtain a differential equation that contains only derivatives by r:

$$c_{nn} k G_s^2 \frac{d^2 f_s}{dr^2} + \left[2c_n k G_s \alpha_s - \frac{b_{con}s}{r} \right] \frac{df_s}{dr}$$
$$+ \left[2c_n k G_s \alpha_s + \alpha_s^2 k + \frac{d(\alpha_s k)}{dz} - \alpha_s (w - v) - \frac{dw}{dz} - \sigma_{col} + \frac{b_{con}s}{r^2} \right] f_s = 0. \tag{13.5.11}$$

The quasi-equilibrium supersaturation s_{eq} can be written following (Eqn. 13.2.9):

$$s_{eq} = c_{1w} w_{ef} \tau_f = \frac{c_p}{L} \Gamma_p \frac{\rho_a}{\rho_{vs}} \frac{(\gamma_a - \gamma_w)}{4\pi D N \bar{r}} w_{ef}. \tag{13.5.12}$$

Then, the condensation term (Eqn. 13.5.10) can be rewritten as

$$\dot{r}_{cond} = \frac{b_{con} s_{eq}}{r} = \frac{c_{con} w_{ef}}{r}, \tag{13.5.13}$$

$$c_{con} = G_s \bar{r} = b_{con} c_{1w} \tau_f = D_v \frac{\rho_a}{\rho_w} \frac{c_p}{L} \frac{(\gamma_a - \gamma_w)}{4\pi D_v N \bar{r}}. \tag{13.5.14}$$

Analytical solutions of (Eqn. 13.5.11) are found in Sections 13.6 and 13.7.

13.6. Approximation Neglecting the Diffusional Growth of Larger Particles

In Sections 13.6 and 13.7, we obtain the analytical solutions of (Eqn. 13.5.11) for the small-size fraction following Khvorostyanov and Curry (1999d, 2008d), with some generalizations. In some situations, it can be assumed that the terms in (Eqn. 13.5.11) with diffusional growth can be neglected for the larger particles in the small-size fraction. These terms can be much smaller in the tails of the spectra than the other terms in (Eqn. 13.5.11) for small values of super- or subsaturation, or for sufficiently large gradients α_s in relatively thin clouds, or for sufficiently large values of G_s. This allows for neglect in (Eqn. 13.5.11) of the diffusional growth terms in the tails of the spectra since they decrease with radius faster than the other terms.

Equation (13.5.11) is similar in structure to the Schrödinger equation of quantum mechanics for the radial part of the wave function in the hydrogen atom. By analogy with this problem, following Landau and Lifsitz (v. 3, 1958), we obtain asymptotic solutions in the small and large radii limits, and then merge these two solutions to obtain the solution for all radii. In this section, we neglect the diffusional growth of larger particles, which is accounted for in Section 13.7.

13.6.1. Small Particle Solution

The analysis of asymptotic behavior of the individual terms in (Eqn. 13.5.11) shows that the most singular in r for small values of r are the 1st, 3rd, and last terms, and hence we retain in (Eqn. 13.5.11) only these terms. Dividing them by $c_{nn}kG_s^2$ and multiplying by r^2, we obtain

$$r^2\frac{d^2 f_s}{dr^2} - a_1 r\frac{df_s}{dr} + a_1 f_s = 0, \tag{13.6.1}$$

$$a_1 = \frac{b_{con}s}{c_{nn}kG_s^2}. \tag{13.6.2}$$

Equation (13.6.1) is a linear homogeneous differential Euler equation of the 2nd order and its solution can be obtained in the form of a power law:

$$f_s(r) \sim r^p. \tag{13.6.3}$$

Substitution of (Eqn. 13.6.3) into (Eqn. 13.6.1) yields

$$p = \frac{b_{con}s}{c_{nn}kG_s^2} = \frac{D_v\rho_{vs}s}{c_{nn}kG_s^2\Gamma_p\rho_w}. \tag{13.6.4}$$

As discussed in Section 13.2, (Eqn. 13.6.4) accounts in s for the effects of various possible sources of supersaturation (advective, convective, radiative, mixing among the parcels and with environment). In cloud layers with positive supersaturation, $s > 0$, then $p > 0$, and (Eqn. 13.6.1) has the form of the left branch of the gamma distribution, $f_s(r) \sim r^p$ (for r smaller than the modal radius). In evaporating cloud layers with negative supersaturation, $s < 0$, then $p < 0$, and the solutions have the form of inverse power laws, $f_s(r) \sim r^{-|p|}$. Thus, the index p of the gamma distribution or inverse power laws that is usually found from fitting experimental data is expressed now via the physical quantities.

13.6.2. Large Particle Solution

For the larger particles in the small-size fraction beyond the modal radius, at $r > r_m$ but $r < r_0$, with r_0 being the boundary between the small- and large-size fractions defined in Section 13.5, the terms with supersaturation and diffusional growth, proportional to r^{-1} and $\sim r^{-2}$, are smaller than the other terms, and we can eliminate in (Eqn. 13.5.11) the condensation growth terms. Then, (Eqn. 13.5.11) becomes

$$c_{nn}kG_s^2\frac{d^2 f_s}{dr^2} + 2c_n kG_s\alpha_s\frac{df_s}{dr}$$
$$+ [2c_n kG_s\alpha_s + \alpha_s^2 k(1-\mu_s^2)]f_s = 0, \tag{13.6.5}$$

where

$$\mu_s = \left[\frac{1}{\alpha_s^2 k}\left(\alpha_s(w-v) + \frac{dw}{dz} + \sigma_{col} - \frac{d\alpha_s k}{dz}\right)\right]^{1/2}. \tag{13.6.6}$$

We seek a solution as the exponential tail of the gamma distribution:

$$f_s(r) \propto \exp(-\beta_s r). \tag{13.6.7}$$

Substitution of (Eqn. 13.6.7) into (Eqn. 13.6.5) yields a quadratic equation for β_s:

$$c_{nn}kG_s^2\beta_s^2 - 2c_n kG_s\alpha_s\beta_s + k\alpha_s^2(1-\mu_s^2) = 0, \tag{13.6.8}$$

which has two solutions for β_s

$$\beta_{s,1,2} = \frac{c_n k\alpha_s \mp [c_n^2 k^2\alpha_s^2 - c_{nn}k^2\alpha_s^2(1-\mu_s^2)]^{1/2}}{c_{nn}kG_s}. \tag{13.6.9}$$

Hereafter in this section, the negative and positive signs relate to the first and second solutions, respectively. Equation (13.6.7) represents the exponential tail of the spectrum with the slopes β_s. Physical conditions require that $\beta_{s,1,2} > 0$, so that $f_s(r) \to 0$ at large r. The solutions (Eqn. 13.6.9) are simplified if the nonconservativeness of the k-components is neglected, $c_{nn} = c_n = 1$:

$$\beta_{s,1,2} = \frac{\alpha_s}{G_s}(1 \mp \mu_s). \tag{13.6.10}$$

Equation (13.6.7) represents the exponential tail of the spectrum and the slopes (Eqns. 13.6.9) and (13.6.10) account for the turbulence, coagulation/accretion, and vertical gradients of k, w, α_s.

Expression (Eqn. 13.6.6) for μ_s is also simplified if to neglect coagulation ($\sigma_{col} \approx 0$—i.e., if there are few large particles), sedimentation ($v = 0$, small particles), and the gradient $d(\alpha_s k)/dz$, and thus (Eqn. 13.6.6) yields

$$\mu_s = \left(\frac{w}{\alpha_s k} + \frac{1}{\alpha_s^2 k}\frac{dw}{dz}\right)^{1/2}. \tag{13.6.10a}$$

13.6.3. Merged Solution

The general solution to (Eqn. 13.5.11) neglecting the diffusional growth of large particles can be found by merging the two asymptotic solutions as

$$f_{1,2}(r) = c_{1,2}r^p \exp(-\beta_{s,1,2}r)\psi_{1,2}(r), \tag{13.6.11}$$

where $\psi_{1,2}(r)$ are some functions to be determined from the equation. The solutions for $\psi_{1,2}(r)$ are found under two simplifying assumptions. 1) We assume for simplicity $c_n = c_{nn} = 1$; a solution can be easily generalized for the other c_n and c_{nn}. 2) We assume $v =$ const, $\mu_s =$ const. The correction for fall velocity $v(r)$ can be introduced into the final solutions using the real fall speed $v(r)$. Substituting (Eqn. 13.6.11) into (Eqn. 13.5.11), applying (Eqns. 13.6.4) and (13.6.7), and performing simple but tiresome manipulations described in Appendix A.13, we obtain an equation for ψ, which for $\beta = \beta_{s1}$ is

$$r\frac{d^2\psi}{dr^2} + \left(p + \frac{2\alpha_s\mu_s}{G_s}r\right)\frac{d\psi}{dr} + \frac{p\alpha_s}{G_s}(1+\mu_s)\psi = 0. \tag{13.6.12}$$

Converting to the dimensionless variable $x = (2\alpha_s\mu_s/G_s)r$ and $dx = (2\alpha_s\mu_s/G_s)dr$, we obtain

$$x\frac{d^2\psi}{dx^2} + (p+x)\frac{d\psi}{dx} + \frac{p(\mu_s+1)}{2\mu_s}\psi = 0. \tag{13.6.13}$$

This is a confluent hypergeometric equation or Kummer equation, the solution to which, satisfying the conditions of finiteness, is the confluent hypergeometric function, $F(a, b; c)$ (Landau and Lifshitz, v. 3, 1958; Gradshteyn and Ryzhik, 1994):

$$\psi_1 = c_1 F\left(\frac{p}{2\mu_s}(\mu_s+1), p; -\frac{2\alpha\mu_s}{G_s}r\right). \tag{13.6.14}$$

The full solution for the droplet spectrum is then obtained from (Eqns. 13.6.11) and (13.6.14) to be

$$f_{s1}(r) = c_1 r^p \exp\left[-\frac{\alpha_s}{G_s}(1-\mu_s)r\right] F\left(\frac{p(\mu_s+1)}{2\mu_s}, p; -\frac{2\alpha_s\mu_s}{G_s}r\right), \tag{13.6.15}$$

where c_1 is a normalization factor.

Substituting (Eqn. 13.6.11) into (Eqn. 13.5.11) with $\beta = \beta_2$ from (Eqn. 13.6.10) yields another equation for ψ:

$$r\psi'' + \left(p - \frac{2\alpha_s\mu_s}{G_s}r\right)\psi' - \frac{p\alpha_s}{G_s}(\mu_s - 1)\psi = 0. \tag{13.6.16}$$

This is also a hypergeometric equation, with the solution

$$\psi_2 = c_2' F\left(\frac{p}{2\mu_s}(\mu_s - 1),\ p;\ \frac{2\alpha_s\mu_s}{G_s}r\right), \tag{13.6.17}$$

and the complete droplet size distribution function is obtained from (Eqns. 13.6.11) and (13.6.17),

$$f_{s2}(r) = c_2 r^p \exp\left[-\frac{\alpha_s}{G_s}(1+\mu_s)r\right] F\left(\frac{p(\mu_s - 1)}{2\mu_s},\ p;\ \frac{2\alpha_s\mu_s}{G_s}r\right). \tag{13.6.18}$$

The constants c_1 and c_2 for the distribution functions (Eqns. 13.6.15) and (13.6.18) can be determined using the integrals with Kummer functions (Landau and Lifshitz, v. 3, 1958; Gradshteyn and Ryzhik, 1994; see Appendix A.13). For solution (Eqn. 13.6.15), we obtain

$$c_1 = N\left\{\Gamma(p+1)\left[\frac{\alpha_s}{G_s}(1-\mu_s)\right]^{-(p+1)} F\left(\frac{p(1+\mu_s)}{2\mu_s},\ p+1,\ p;\ \frac{-2\mu_s}{1-\mu_s}\right)\right\}^{-1}, \tag{13.6.19}$$

where $\Gamma(x)$ is the Euler gamma function, and $F(\alpha, \beta, \gamma; x)$ is the Gauss hypergeometric function. For (Eqn. 13.6.18), we obtain

$$c_2 = N\left\{\Gamma(p+1)\left[\frac{\alpha_s}{G_s}(1+\mu_s)\right]^{-(p+1)} F\left(\frac{p(\mu_s - 1)}{2\mu_s}, p+1, p;\ \frac{2\mu_s}{1+\mu_s}\right)\right\}^{-1}. \tag{13.6.20}$$

Equations (13.6.19) and (13.6.20) define c_1 and c_2 for the distribution functions, normalized to the concentration N. From Appendix A.13, we can also determine the mean radius and water content as the 1st and 3rd moments. Alternatively, if the liquid water content is specified (say from a model prognostic equation), then the liquid or ice water content can be used to determine the constants c_1 and c_2. Expressions (Eqns. 13.6.15) and (3.6.18) are similar to gamma distributions with some modifications that generalize the gamma distributions. The left branch of these gamma distribution type spectra is described by the index p, which is now sign-variable, depending on supersaturation and allows for various sources of cooling. The right branch is a usually observed exponential tail.

Note that the existence of the two solutions can describe bimodal droplet spectra that are observed in clouds (e.g., Warner, 1969a) and in chamber mixing experiments (e.g., Baker et al., 1984). This possibility of bimodality of these solutions resembles Manton's (1979) theory. The existence, positions, and relative strengths of the two modes are determined by the values of α_s, μ_s, and G_s. The solution imposes some limitations on the relation among these parameters. In particular, the Kummer function can become negative or grow exponentially at large values of r for some combinations of α_s, μ_s, and G_s, and they should be properly chosen from the physical conditions.

13.6.4. Asymptotic Solutions

The derived expressions can be used for numerical calculations of the size spectra and their moments as described in this section. However, at least two asymptotic regimes exist where these solutions can be simplified. These asymptotic formulas can be obtained using the properties of the confluent hypergeometric function described in Appendix A.13. Using these expressions, we can consider two asymptotic regimes which simplify the solutions to (Eqns. 13.6.15) and (13.6.18). These regimes are determined by the term $x = 2\alpha_s\mu_s r/G_s$ which is the third argument of the confluent hypergeometric functions in (Eqns. 13.6.15 and 13.6.18).

First asymptotic regime, $x = 2\alpha_s\mu_s r/G_s << 1$. Since $F(a, b, x) \to 1$ at $x \to 0$ (see Appendix A.13), the Kummer function vanishes in solutions and we obtain from (Eqns. 3.6.15) and (3.6.18) the following two solutions:

$$f_{s1}(r) = c_1 r^p \exp[-\frac{\alpha_s}{G_s}(1-\mu_s)r], \tag{13.6.21}$$

$$f_{s2}(r) = c_2 r^p \exp[-\frac{\alpha_s}{G_s}(1+\mu_s)r]. \tag{13.6.22}$$

These expressions are the usual gamma distributions with index p. The condition $x = 2\alpha_s\mu_s r/G_s << 1$ implies that α_s should not exceed some value, $\alpha_s << G_s/2\mu_s r_{max}$, where r_{max} is the maximum radius for which the condition holds. To estimate this value of α_s, we assume $r_{max} = 10\,\mu\text{m}$, $N \sim 200\,\text{cm}^{-3}$, $\mu_s \sim 1$, and calculate $G_s \sim 6.4 \times 10^{-9}$. These values give an estimation of $\alpha_s < 0.3\,\text{km}^{-1}$, or for $q_L \sim 0.8\,\text{g m}^{-3}$, a vertical gradient $dq_L/dz < 0.25\,\text{g m}^{-3}\,\text{km}^{-1}$ (from Eqn. 13.5.8).

So the asymptotic formulae (Eqns. 13.6.21) and (13.6.22) can be valid in the center of the cloud layer near the maximum value of the liquid water content, or in any other region where the gradient in q_L is not very large. Note, however, that when α_s is very small, then $\mu_s >> 1$, and from the condition of finiteness, only the second solution (Eqn. 13.6.22) is valid. We call this regime the "internal asymptotic regime," as it is valid in the cloud interior far from the boundaries.

Second asymptotic regime, $x = 2\alpha_s\mu_s r/G_s >> 1$. Using the limit of the Kummer function $F(a, b, x) \to [\Gamma(b)/\Gamma(a)]\exp(x)x^{a-b}$ at $x >> 1$ (see Appendix A.13), we obtain from (Eqns. 13.6.15) and (13.6.18) the following two solutions:

$$f_1(r) = \tilde{c}_1 r^{\tilde{p}_1} \exp\left[-\frac{\alpha_s}{G_s}(1+\mu_s)r\right], \qquad \tilde{p}_1 = \frac{p}{2\mu_s}(\mu_s + 1), \tag{13.6.23}$$

$$f_2(r) = \tilde{c}_2 r^{\tilde{p}_2} \exp\left[-\frac{\alpha_s}{G_s}(1-\mu_s)r\right], \qquad \tilde{p}_2 = \frac{p}{2\mu_s}(\mu_s - 1). \tag{13.6.24}$$

This asymptotic condition should be valid for all radii, and we can rewrite it as $2\alpha_s\mu_s r_{min}/G_s >> 1$. This criterion is satisfied when α_s is large enough—i.e., when q_L is small and its vertical gradient is large. These conditions are typical near cloud boundaries. We can estimate the region of validity of this regime by approximating the q_L profile as a linear function of height: $q_L = q'_{L0} \times z$, where z is the height above the cloud base and q'_{L0} is a gradient of the liquid water content. Then, α_s can be simplified as $\alpha_s = (1/q_L)(dq_L/dz) = 1/z$. That is, α_s decreases rapidly with the height above the cloud base, and we can determine the following condition for z:

$$\frac{2\alpha_s\mu_s r_{\min}}{G_s} = \frac{2\mu_s r_{\min}}{G_s z} >> 1, \qquad z << \frac{2\mu_s r_{\min}}{G_s}. \tag{13.6.25}$$

Using the values $r_{min} = 1 - 1.5\,\mu m$, $\mu_s = 0.5 - 1$, $G_s = 1.3 \times 10^{-8}$, we obtain $z < 200$–$300\,m$, while for $G_s = 2.6 \times 10^{-8}$ we obtain $z < 100$–$150\,m$. So, this asymptotic regime can be valid within several hundred meters of the upper or lower boundary of the cloud (as long as these levels do not include a maxima in the liquid water content). We refer to this regime as the "boundary asymptotic regime" because it is valid near cloud boundaries.

The extent of applicability of the asymptotic solution in (Eqns. 13.6.23) and (13.6.24) is further illustrated by the following example. As the value of μ_s in (Eqn. 13.6.10a) should be real, the following condition should be satisfied by α_s:

$$\frac{w}{\alpha_s k} + \frac{1}{\alpha_s^2 k}\frac{dw}{dz} > 0, \qquad \text{or} \qquad \alpha_s > -\frac{1}{w}\frac{dw}{dz} \qquad \text{if } w > 0. \tag{13.6.26}$$

This condition allows α_s to be either positive or negative, depending on the vertical velocity profile in the cloud. A positive value of α_s indicates that q_L increases with the height in the cloud, while a negative value of α_s indicates that q_L decreases with the height in the cloud. The value of α_s would be zero at the maxima of q_L. The existence of vertical profiles of q_L that increase with height to a maximum value, then decrease with height above this maximum is commonly observed (see Sections 13.8 and 13.9) and may be caused by: 1) the corresponding vertical profiles of the vertical velocity in frontal stratiform or convective clouds; 2) the entrainment of dry air into the cloud through the cloud top or turbulent mixing with the dry air near the lower and upper boundaries with subsequent mixing in the cloud; and 3) the vertical profiles of radiative cooling near the cloud top in stratiform clouds.

The solution (Eqn. 13.6.23) is the usual gamma distribution, where $\tilde{p}_1 = (p/2\mu_s)(\mu_s + 1)$ is the effective index of the gamma distribution. The second solution (Eqn. 13.6.24) satisfies the condition of finiteness at large r only if $\mu_s < 1$ for $\alpha_s > 0$. For the solution (Eqn. 13.6.24), the effective index $\tilde{p}_2 = (p/2\mu_s)(\mu_s - 1)$ of the gamma distribution is negative. Size spectra of the form of the inverse power law have been observed for small ice crystals (e.g., Heymsfield and Platt, 1984; Ryan, 1996).

The inverse power law is not typically observed in water clouds for small drops but is observed for the large-size fraction (see Chapter 2 and the next sections). Calculations such as those presented in Section 13.9 show that the second solution (Eqn. 13.6.24) may exist in the lower cloud layer, but when added to (Eqn. 13.6.23), it influences the "tail" of the spectrum and does not produce the secondary mode. Thus, we examine primarily the first solution.

To calculate the drop size spectra in these asymptotic regimes from (Eqns. 13.6.21)–(13.6.24), we need to define the normalization constants c_2 and c_1. Using Appendix A.13, we can evaluate c_2 from the normalization to the concentration N. For (Eqn. 13.6.22) we have

$$c_2 = N\left[\Gamma(p+1)\right]^{-1}\left[\frac{\alpha_s}{G_s}(1+\mu_s)\right]^{(p+1)}. \qquad (13.6.27)$$

For the first solution f_1 of the "boundary" asymptotic solution (Eqn. 13.6.23), the normalization constant is

$$\tilde{c}_1 = N\left[\Gamma\left(\tilde{p}_1+1\right)\right]^{-1}\left[\frac{\alpha_s}{G_s}\left(1+\mu_s\right)\right]^{\tilde{p}_1+1}, \qquad (13.6.28)$$

where $\tilde{p}_1$ is defined in (Eqn. 13.6.23). Using (Eqns. 13.6.22) and (3.6.27) for the interior asymptotic solutions and (Eqns. 13.6.23) and (3.6.28) for the boundary asymptotic solutions, one can easily calculate the drop size spectra as shown in Section 13.9.

13.7. Solution Including the Diffusional Growth of Large Particles, Sedimentation, and Coagulation

The solution obtained in Section 13.6 assumed that the diffusion growth terms are small at sufficiently large r. However, if these terms are still significant in the tail of the small-size fraction, as it can be at large sub- and supersaturations, it is desirable to find another solution that accounts for these terms. Besides, it is desirable to estimate the effects of sedimentation and coagulation on the shapes of the spectra. In this section, we find a solution that takes into account these effects. As in Section 13.6, we again consider the two asymptotic cases at small and large r (again, within the small-size fraction, at $r < r_0$).

At small r, the same most singular terms (Eqns. 13.6.1) and (13.5.11) give the main contribution, and the solution is the same power law (Eqn. 13.6.3), $f_s \sim r^p$, with the index p defined in (Eqn. 13.6.4). At large r, if the diffusional growth terms are retained, and then (Eqn. 13.5.11) is a 2nd order differential equation with rather complicated variable coefficients, and an analytical solution is not easily obtained. Hence, we seek a simplification that will enable an analytical solution. At large r, it is reasonable to assume that the tail of the spectrum is smooth and the r-gradients of the spectrum are smaller than near the mode. Then, we can neglect in (Eqn. 13.5.11) the stochastic diffusion terms in radii space (the terms with G_s), and arrive at the following equation:

$$[w - v(r)]\alpha_s f_s + \frac{dw}{dz} f_s + b_{con} s \frac{d}{dr}\left(\frac{f_s}{r}\right)$$
$$= \left[\alpha_s^2 k + \alpha_s \frac{dk}{dz} + k\frac{d\alpha_s}{dz} - \sigma_{col}\right] f_s. \qquad (13.7.1)$$

A solution to this equation is found in Appendix B.13:

$$f_s(r) = \frac{r}{r_1} f_s(r_1) \exp\{-[\beta_{s2}(r) r^2 - \beta_{s2}(r_1) r_1^2]\}, \tag{13.7.2}$$

where r_1 is some reference point, and the slope β_{s2} is

$$\beta_{s2}(r) = \frac{1}{b_{con} s}\left[\frac{1}{2}\left(\alpha_s w + \frac{d(w - \alpha_s k)}{dz} + \sigma_{col} - \alpha_s^2 k\right) - \frac{\alpha_s v(r)}{B_v + 2}\right]. \tag{13.7.3}$$

This is the solution for the tail of the small fraction. Equation (13.7.2) can be written in a slightly different form that includes the terms with r_1 in the normalizing constant c_N:

$$f_s(r) = c_N r \exp[-\beta_{s2}(r) r^2]. \tag{13.7.4}$$

The merged solution can be constructed again as in Section 13.6 as an interpolation between the two asymptotic regimes at small and large r:

$$f_s(r) = c_N r^p \exp[-\beta_{s2}(r) r^2]\Phi(r). \tag{13.7.5}$$

Unfortunately, the resulting equation for Φ in this case is much more complicated than the confluent hypergeometric equation in Section 13.6 and its solutions cannot be easily reduced to Kummer functions. Thus, (Eqn. 13.7.5) can be used with some interpolation formulae for Φ—for example:

$$\Phi(r) = \exp(-r/r_{sc}) + (r/r_{sc})^{1-p} \tanh(r/r_{sc}), \tag{13.7.6}$$

where r_{sc} is a scaling radius comparable with $\bar{r}$. At $r << r_{sc}$, the first term tends to 1, the 2nd term tends to 0, and we obtain from (Eqn. 13.7.5) $f_s \sim r^p$ at small r since $\beta_{s2}(r)r^2 << 1$ in (Eqn. 13.7.5) and $\exp \to 1$. At $r >> r_{sc}$, the first term tends to 0, the 2nd term tends to (r/r_{sc}), and the solution (Eqn. 13.7.6) tends to the exponent (Eqn. 13.7.4). Thus, (Eqn. 13.7.5) ensures the correct limits at both small and large r.

An advantage of (Eqn. 13.7.5) is that the tail of the spectrum explicitly accounts for the diffusional growth process. This results in the inverse dependence of the slope β_{s2} (Eqn. 13.7.3) on supersaturation; that is, the slope becomes steeper as $|s|$ increases. This is physically justified since more vigorous condensation/evaporation should produce narrower spectra. At sufficiently large r, the 2nd term with $v(r)$ in (Eqn. 13.7.3) dominates, and the slope is

$$\beta_{s2} = -\frac{\alpha_s v(r)}{b_{con} s(B_v + 2)}. \tag{13.7.7}$$

Since f_s should decrease at large r, the slope β_{s2} should be positive. Therefore, the terms α_s and s should have opposite signs—i.e., $\alpha_s < 0$ and LWC (IWC) increases downward in the growth layer ($s > 0$, $\alpha_s < 0$), and decreases downward in the evaporation layer ($s < 0$, $\alpha_s > 0$), which is physically justified for this limit $v(r) >> w$. The argument in the exponent is $\beta_{s2} r^2 \sim r^2 v(r) \sim r^{B_v + 2}$—i.e., with $v(r) \sim r^2$ (Stokes regime, $B_v = 2$) or $v(r) \sim r$ (intermediate regime, $B_v = 1$) or $v(r) \sim r^{1/2}$ (aerodynamic regime, $B_v = 1/2$) (see Chapter 12), the tails of the spectra decrease as $\exp(-r^4)$, $\exp(-r^3)$, and $\exp(-r^{2.5})$, respectively. That is, the tail of the spectrum has the form assumed for the generalized gamma distributions, with the power of r in exponents greater than 1 (see Chapter 2), and a larger value of the fall velocity $v(r)$ is associated with a greater slope and shorter tail, which is consistent with increased precipitation from the tail.

When $v(r)$ is small relative to the other terms (e.g., $v(r) << |w|$), the exponent β_{s2} is determined by the 1st term in parentheses (Eqn. 13.7.3); neglecting for simplicity dw/dz, dk/dz, $d\alpha_s/dz$, we obtain

$$\beta_{s2} = \frac{1}{2b_{con}s}(\alpha_s w + \sigma_{col} - \alpha_s^2 k). \tag{13.7.8}$$

Now, the signs of s and of the expression in parentheses should coincide to obtain $\beta_{s2} > 0$. In particular, if $s > 0$, $w > 0$, and the term $\alpha_s w$ is greater than σ_{col} and $\alpha_s^2 k$ in (Eqn. 13.7.8), then $\alpha_s > 0$—i.e., q_{ls} increases with height, which is justified for this limit $w >> v(r)$. In this case, the slope becomes steeper (the spectrum narrows) when w increases (as in regular condensation), and when σ_{col} increases (faster absorption by the large fraction), and β_{s2} decreases when k increases (turbulence causes broadening of the spectra). Since β_{s2} does not depend on r for this case, the tail of the spectrum decreases as $\exp(-r^2)$.

The merged solution (Eqn. 13.7.5) differs from those found previously in the following ways. At small r, the solution is a power law with variable rather than fixed index p, allowing variable relative dispersions. Also, the slope of the tail is not constant but varies (generally increases) with r; accounting for the greater depletion of large particles with greater sedimentation rate. The slope also describes depletion of the small fraction due to accumulation by the large fraction via the term σ_{col}: the larger coagulation rate and σ_{col}, the greater the slopes and the faster the decrease with r. The tail in (Eqn. 13.7.4) at large r behaves as $\exp(-r^\lambda)$ with λ varying from 2 to 4 for various situations, in broad agreement with the range of values determined by previous analytical solutions and parameterizations as generalized gamma distributions (see Section 2.4.4).

13.8. Physical Interpretation of the Parameters

13.8.1. Various Forms of Solution Parameters

The dimensionless parameter G_s introduced in (Eqn. 13.3.12) in the process of averaging over the turbulence spectrum can be written in various forms:

$$G_s = \frac{\rho_a}{\rho_w}\frac{c_p}{L}\frac{\gamma_a - \gamma_w}{4\pi N\bar{r}^2} \approx \frac{c_f}{3}\frac{\rho_a c_p(\gamma_a - \gamma_w)}{L}\frac{\bar{r}}{q_{ls}}, \tag{13.8.1}$$

where c_p is the specific heat capacity, L is the latent heat of condensation or deposition, γ_a and γ_w are the dry and wet (water or ice) adiabatic lapse rates, ρ_a is the air density, N and $\bar{r}$ are the number concentration and mean radius of the particles, and q_{ls} is the liquid (ice) water content of the small fraction. The coefficient c_f occurs in (Eqn. 13.8.1) since we use a relation $q_{ls} = (4/3)\pi c_f \rho_w N\bar{r}^3$, or $(4\pi N\bar{r}^2)^{-1} = (c_f/3)\rho_w \bar{r}/q_{ls}$. The value of c_f depends on the shape of the spectra and is on the order of 1. For example, for gamma distributions with the index p considered in the following text, $c_f = (p+2)(p+3)/(p+1)^2$—i.e., $c_f/3 = 1$ for $p = 1$ and $c_f/3 = 0.5$ for $p = 6$. In the future, we assume in (Eqn. 13.8.1) that $c_f/3 \approx 1$, since its variations are several orders of magnitude smaller than variations of q_{ls}, N, and $\bar{r}$. Introducing the adiabatic liquid (ice) water mixing ratio $q_{ls,ad}(z) = (dq_{ls}/dz)_{ad}(z - z_{bot})$, where $(dq_{ls}/dz)_{ad}$ is the adiabatic vertical gradient of the mixing ratio (see Section 3.11):

$$\left(\frac{dq_{ls}}{dz}\right)_{ad} = \frac{c_p}{L}(\gamma_a - \gamma_w)\rho_a, \tag{13.8.2}$$

z_{bot} is the height of cloud bottom, $(z - z_{bot})$ is the altitude above z_{bot}, and γ_w hereafter denotes the wet adiabatic lapse rate in liquid or ice clouds (see Section 3.11). G_s in (Eqn. 13.8.1) can be also rewritten as

$$G_s = \frac{\rho_a}{\rho_w}\frac{1}{4\pi N\bar{r}^2}\left(\frac{dq_{ls}}{dz}\right)_{ad} = \frac{\bar{r}}{q_{ls}}\left(\frac{dq_{ls}}{dz}\right)_{ad} = \frac{\bar{r}}{(z-z_{bot})}\left(\frac{q_{ls}}{q_{ls,ad}}\right)^{-1} = \frac{\bar{r}}{\chi_{ad}\Delta z}, \tag{13.8.3}$$

where $\chi_{ad} = q_{ls}/q_{ls,ad}$ is the adiabatic liquid (ice) ratio, and $\Delta z = z - z_{bot}$ is the height above the cloud base. If the major source of supersaturation is uplift and/or radiative cooling, then p in (Eqn. 13.6.4) can be expressed via the "effective" vertical velocity w_{ef}, using (Eqns. 13.5.13) and (13.5.14):

$$p = \frac{b_{con}s}{c_{nn}kG_s^2} = \frac{c_{con}w_{ef}}{c_{nn}kG_s^2} = \frac{\bar{r}w_{ef}}{c_{nn}kG_s}. \tag{13.8.4}$$

Using (Eqns. 13.8.1)–(13.8.3) for G_s, the index p from (Eqn. 13.8.4) can also be written in the following forms

$$p = \frac{w_{ef}}{c_{nn}k}\frac{L}{c_p(\gamma_a - \gamma_w)}\frac{4\pi\rho_w N\bar{r}^3}{\rho_a}$$
$$\approx \frac{w_{ef}}{c_{nn}k}\frac{q_{ls}}{(dq_{ls,ad}/dz)} \approx \frac{w_{ef}}{c_{nn}k}\chi_{ad}\,\Delta z. \tag{13.8.5}$$

All of the parameters in (Eqn. 13.8.5) are available from bulk models, and the index p can be easily calculated. The first expression in (Eqn. 13.8.5) can be used in the double-moment models, when both q_{ls} and N are known, then $\bar{r}$ and p can be calculated (e.g., Morrison et al., 2005a,b). The 2nd and 3rd expressions in (Eqn. 13.8.5) can be used in single-moment models, when only q_{ls} is available. The relative dispersion of the spectra σ_r and parameter p are related as (Chapter 2)

$$\sigma_r = (p+1)^{-1/2}, \qquad p = 1/\sigma_r^2 - 1. \tag{13.8.6}$$

The term "effective velocities" here means subgrid velocities with addition of the "radiative-effective" velocities w_{rad} that allow direct comparison of the dynamical subgrid velocity w_{dyn} and w_{rad} that is determined by radiative cooling rates:

$$w_{ef} \approx w_{dyn} + w_{rad}, \quad w_{rad} = -(1/\gamma_a)(\partial T/\partial t)_{rad}, \tag{13.8.7}$$

where $(\partial T/\partial t)_{rad}$ is the radiative cooling rate caused by the total radiative flux divergence.

Since large-scale models with horizontal grid spacing ~ 100–400 km may produce w_{dyn} and $w_{ef} \le$ 1–2 cm s^{-1}, the problem of the subgrid parameterization of w_{ef} was addressed in several works and there were various definitions of the subgrid velocities in the models. Ghan et al. (1997) specified the subgrid velocities w_{dyn} in the general circulation model (GCM) via the turbulence coefficient k and mixing length l_{mix}:

$$w_{dyn} \approx k/l_{mix}. \tag{13.8.8}$$

Morrison et al. (2005a,b), and Morrison and Pinto (2005) developed a model with bulk double-moment microphysics based on gamma distributions. The p-indices in this model were not prescribed as some *a priori* fixed values but were calculated using (Eqns. 13.8.4) and (13.8.5), the subgrid velocities w_{sg} were specified similar to Ghan's et al. (1997) in (Eqn. 13.8.8). Sud and Lee (2007) and Sud et al. (2009) parameterized the particle size spectra in the form of a gamma distribution in the Goddard Space Flight Center (GSFC) GCM and also used the solutions (Eqns. 13.8.4) and (13.8.5) for the p-indices from Khvorostyanov and Curry (1999d) for evaluation of effective radius and cloud optical properties. Thus, the p-indices of the gamma distributions in these models were not constant but varied in space and time.

An alternative to these parameterizations of k, w_{dyn}, and l_{mix} was suggested by Dmitrieva-Arrago and Akimov (1998) in a GCM with prognostic equations for q_L and dew-point deficit. The subgrid value of w_{dyn} for stratiform clouds was defined in this GCM as the minimum velocity required to support all falling droplets with radii smaller than some threshold value—e.g., ~40 μm, close to the estimate of r_0 in Section 13.2. Assuming Stokes' velocities gives $w_{dyn} \approx 20\,\mathrm{cm\,s^{-1}}$, which is comparable to the estimation by Ghan et al. (1997) and Morrison et al. (2005a,b). Tables 13.1 and 13.2 show that with $k \sim 10\,\mathrm{m^2\,s^{-1}}$, this would give reasonable values of $p \sim 5$–10, so the method developed here can be used for evaluation of p-indices in such GCMs with parameterizations of subgrid w_{dyn} similar to Dmitrieva-Arrago and Akimov (1998). Ghan et al. (1993) derived an equation for w_{dyn} that also includes the effects of horizontal advection of heat and moisture. This equation was used in that work for evaluation of drop activation on CCN but can also be used for evaluation of the p-indices.

As noted in Section 13.4, another estimate of the effective velocity w_{ef} can be done from the relation $w_{ef} \sim \overline{w'q_L'}/q_L$ using the data gained in the field experiments. The values of the flux $\overline{w'q_L'}$ measured in arctic stratus presented in Curry (1986) are always positive (except near the cloud base). For the upper deck observed on 28 June 1980, the value of $\overline{w'q_L'}$ was 25–$55 \times 10^{-4}\,\mathrm{g\,m^{-2}\,s^{-1}}$ at $z = 818\,\mathrm{m}$ with corresponding $q_L \sim 0.12 - 0.2\,\mathrm{g\,m^{-3}}$. This gives an estimate $w_{ef} \sim 12 - 20\,\mathrm{cm\,s^{-1}}$. For the same deck, the covariances $\overline{w'w'} \sim 0.1 - 0.2\ \mathrm{m^2\ s^{-1}}$ (Curry et al., 1988)—i.e., $w' \sim (\overline{w'w'})^{1/2} \sim 33 - 45\,\mathrm{cm\ s^{-1}}$, which is larger than the previous estimate for w_{ef}. This is in agreement with the discussion of the covariances in Section 13.3, which showed that the correlation function $\overline{w'f'}$ or $\overline{w's'}$ due to the phase shift between them by τ_f should be smaller than the autocovariance $\overline{w'w'}$. Here, w_{ef} should be smaller than w', but still much greater than the synoptic-scale average $w_{syn} \sim 0.5$–$1\,\mathrm{cm\,s^{-1}}$.

The general (Eqn. 13.8.4) for p shows that the index $p \sim s$, and hence p can be positive or negative. These two cases are considered in the following text.

13.8.2. Solutions in the Form of Gamma Distributions

If $w_{ef} > 0$, then p is positive. Equations from Sections 13.6 and 13.7 show that the size spectrum of the small fraction is then the power law multiplied by the exponent, with some modification—i.e., is close to the gamma distribution. The expressions (Eqn. 13.8.5) for this case show that p depends on 5 factors: w, k, adiabatic ratio χ_{ad}, height Δz, and c_{nn}. Since $c_{nn} \sim 1$, we illustrate in the following text the role of the other factors.

Effect of parameters G_s and p. The equations for G_s and p indicate several features that are similar for both cloud drop and crystal size spectra. Equations (13.8.4) and (13.8.5) show that p increases and the relative dispersion σ_r decreases (the spectra become narrower) when positive w_{ef} increases, which is similar to the effect of regular condensation. The index p decreases (the spectra become broader) when k increases, which describes the effect of turbulence in stochastic condensation.

As discussed in Section 13.4, the effective diffusion coefficient in the space of radii is $k_r = G_s^2 k$. Equation (13.8.1) for G_s shows that: 1) since $G_s \sim \bar{r}$, the diffusion coefficient k_r is proportional to the squared mean radius $\bar{r}$—i.e., $k_r \sim \bar{r}^2$, which thereby is an analog of "mixing length in the space of radii," similar to the turbulent coefficient $k = l'^2(\partial u/\partial z)$—i.e., proportional to the squared mixing length l'; and 2) since $k_r \sim G_s$, and $G_s \sim 1/\chi_{ad}$, k_r at any given height z is inversely proportional to the liquid water ratio χ_{ad} (the ratio of q_L to the adiabatic value of $q_{L,ad}$).

Table 13.1 shows that the values of the parameter G_s calculated from (Eqn. 13.8.3) increase from 0.32×10^{-8} at $N = 400\,\text{cm}^{-3}$, $\bar{r} = 10\,\mu\text{m}$ to 5×10^{-8} at $N = 100\,\text{cm}^{-3}$, $\bar{r} = 5\,\mu\text{m}$. Table 13.2 shows that the values of p calculated from (Eqn. 13.8.5) for the "interior cloud" solution (Eqns. 3.6.21) and (3.6.22) range from 2 to 8 for $w_{ef} = 10\,\text{cm}\,\text{s}^{-1}$ in a cloud with mean radius $\bar{r} = 5\,\mu\text{m}$ and droplet concentration $N = 1 - 4 \times 10^2\,\text{cm}^{-3}$, typical for continental, summertime arctic stratus clouds (Curry, 1986) or some types of marine clouds near the coast (Noonkester, 1984). This value of p corresponds to dispersions of the spectra $\sigma_r \sim 0.33 - 0.58$ (Table 13.3), which is close to values observed in the central layers of such clouds. For $\bar{r} = 10\,\mu\text{m}$, typical of marine stratus clouds, the values of p increase; however, since typical drop concentrations are usually $N = 1 - 2 \times 10^2\,\text{cm}^{-3}$ for these clouds, values of p would not normally exceed $p = 8 - 16$ for this cloud type and would be less for smaller values of w_{ef}. Thus, from Table 13.2 we see that the spectra become narrower with increasing droplet concentration and mean radius. If the value of the turbulent coefficient is $k = 10\,\text{m}^2\,\text{s}^{-1}$ or the maximum effective vertical velocity is $5\,\text{cm}\,\text{s}^{-1}$, the values of p in Table 13.2 would be twice as small.

Table 13.1. *Parameter* G_s *(*10^{-8}*) for* $\gamma_w = 6\,K\,km^{-1}$ *and various values of droplet concentrations* N *and mean radii* $\bar{r}$.

	N, cm^{-3}		
$\bar{r}$, μm	10^2	$2 \cdot 10^2$	$4 \cdot 10^2$
5	5.1	2.6	1.27
10	1.3	0.64	0.32

Table 13.2. *Parameters* p *of the gamma distribution for* $k = 5\,m^2\,s^{-1}$, $\gamma_w = 6\,K\,km^{-1}$, *with various droplet concentrations* N, *mean radii* $\bar{r}$, *and vertical velocities* w_{ef}

	N, cm^{-3}		
$\bar{r}$, μm	10^2	$2 \cdot 10^2$	$4 \cdot 10^2$
		$w_{ef} = 1\,\text{cm}\,\text{s}^{-1}$	
5 μm	0.2	0.4	0.8
10 μm	1.6	3.2	6.4
		$w_{ef} = 10\,\text{cm}\,\text{s}^{-1}$	
5 μm	2	4	8
10 μm	16	32	64

Table 13.3. *Relative dispersions* $\sigma_r = (p+1)^{-1/2}$ *of the gamma distribution for various droplet concentrations* N, *mean radii* $\bar{r}$, *and* w_{ef} *for the same parameters used in Table 13.2*

	N, cm^{-3}		
$\bar{r}$, μm	10^2	2×10^2	4×10^2
		$w_{ef} = 1\,\text{cm s}^{-1}$	
5 μm	0.91	0.84	0.74
10 μm	0.62	0.49	0.37
		$w_{ef} = 10\,\text{cm s}^{-1}$	
5 μm	0.58	0.45	0.33
10 μm	0.24	0.17	0.12

These tables show that for $w_{ef} = 1 - 2\,\text{cm s}^{-1}$, as could be evaluated from the large-scale divergence (i.e., with the averaging over the large areas), the values of p are too small ($p < 1$ for $\bar{r} = 5\,\mu\text{m}$), and the dispersions are too large as compared to the local observations. The problem of correspondence of the local size spectra and those averaged over the large cloud areas has been widely discussed in the literature, and it was found that the scale of averaging influences the index p. The local (measured over the horizontal scales $L_x \sim 30$–$100\,\text{m}$) spectra are usually narrow, with $p = 6$, $\sigma_r = 0.38$ to $p = 10$, $\sigma_r = 0.30$ (e.g., Aleksandrov and Yudin, 1979; Curry, 1986; Liu and Hallett, 1998; Cotton et al., 2011).

This is illustrated in Table 13.4 with an example of the local drop spectra measured in low arctic stratus over small horizontal intervals (Curry, 1986), where $p = 5$–10 and $\sigma_r = 0.3$–0.4 except for the cloud base with wider spectra. However, the dispersions increase and the indices decrease with increasing scale of averaging, ($L_x \sim$ a few km) or with averaging over many clouds (e.g., Mason, 1971; Sedunov, 1974; Pruppacher and Klett, 1997). This effect of averaging is clearly seen from the data by Noonkester (1984) for marine stratocumulus shown in Table 13.4, which illustrates the increase in σ_r (artificial "broadening" of the spectra) with averaging over the scale 6.44 km ($\sigma_{r,all}$) as compared to the local spectra averaged over 0.529 km ($\sigma_{r,loc}$). Outside cloud top and bottom, the local $p = 6$–22, ($\sigma_r = 0.2$–0.4), and averaged over the scale 6.44 km $p \sim 0.13$–1.3 ($\sigma_r = 0.66$–0.94). This impact of averaging on the width of the size spectra is described by (Eqn. 13.8.5) since $p \sim w_{ef}$, which decreases with increasing scale of averaging. Note, however, that the cloud optical and radiative properties and the rate of precipitation formation are determined by the local ($L_x \sim 50$–$100\,\text{m}$) size spectra, hence the local values $p = 6$–10 with $\sigma_r = 0.3$–0.4 are representative and should be used for the parameterizations.

For convective clouds, typical values of vertical velocity are $w \sim 2.0 - 5.0\,\text{m s}^{-1}$; that is, 20–30 times larger than values used in Table 13.2 for stratiform clouds. Although the values of the turbulence coefficient for cumulus can be about 10–20 times larger than those for stratus clouds, this shows that the values of p should be larger and spectra narrower in cumulus, which is expected since the processes in the vigorous updrafts are closer to the regular condensation. In convective clouds, the factor $\chi_{ad}\Delta z$ can also influence the spectra as described in the following text.

An important feature of the solutions for p obtained here is that although (Eqn. 13.8.5) for p includes many parameters that vary by several orders of magnitude, the expression for p leads to a

Table 13.4. *Observed relative dispersions,* $\sigma_r = (p+1)^{-1/2} = (\sigma_{abs}/\bar{r})^{1/2}$, *and corresponding indices* p *in low stratus clouds*

	Arctic summertime stratus (Curry, 1986)			
Deck, date	Height, m	q_L, g m^{-3}	σ_r	p
#1, 20 June 1980	289 (top)	0.02	0.39	5.6
73.5 N,	241	0.42	0.30	10.1
159.3 W	183	0.35	0.31	9.4
	91	0.20	0.36	6.7
#2, 28 June 1980	1082 (top)	0.03	0.38	5.9
77.7 N,	940	0.30	0.33	8.2
155.2 W	819	0.10	0.40	5.3
	696 (base)	0.02	0.51	2.8

California coastal stratus (Noonkester, 1984)
$\sigma_{r,loc}$ and p_{loc} are averages over local paths of 529 m,
$\sigma_{r,all}$ and p_{all} are averages over all paths of 6.44 km.

Date	Height above cloud base, m	q_L, g m^{-3}	$\sigma_{r,loc}/\sigma_{r,all}$	p_{loc}/p_{all}
29 May 1981	274 (top)	0.288	0.42/0.67	4.7/1.2
	167	0.217	0.21/0.66	21.7/1.3
	18	0.034	0.57/0.89	2.1/0.3
18 August 1981	222 (top)	0.263	0.79/0.94	0.6/0.13
	121	0.199	0.37/0.78	6.3/0.6
	17	0.037	0.56/0.81	2.2/0.5

dimensionless quantity of order 1–10 for a wide variety of cloud types, and the relative dispersion is now related directly to the meteorological factors (w, γ_a, k), the properties of the cloud (N, $\bar{r}$), and fundamental atmospheric constants.

Effect of parameters χ_{ad} ***and*** Δz. The additional factor in (Eqn. 13.8.5) is the adiabatic ratio χ_{ad}, the index $p \sim \chi_{ad}\Delta z$. In convective clouds, χ_{ad} usually decreases with the height above the cloud base due to entrainment, but decreases with cloud width (e.g., Pruppacher and Klett, 1997, Figs. 2–22). A decrease in χ_{ad} may cause a decrease of p in (Eqn. 13.8.5) and a broadening of the spectra with height in the lower cloud layers. In the upper cloud half, χ_{ad} may tend to nearly constant values (e.g., Pruppacher and Klett, 1997), while the height-factor Δz causes a counter-effect in (Eqn. 13.8.7), causing an increase of p and spectral narrowing. Which effect will dominate depends on the product $\chi_{ad}\Delta z$. If χ_{ad} decreases with height faster than $1/\Delta z$, then $\chi_{ad}\cdot\Delta z$ decreases and the spectra broaden with height, as was observed in a convective cloud by Warner (1969a). It is this effect that is in contradiction with the theory of regular condensation that predicts narrowing of the spectra with height, and stimulated numerous attempts to explain the broad spectra, including the stochastic condensation theory as reviewed in the Introduction to Chapter 13.

However, later experimental studies showed that broadening with height is not a common feature of the droplet spectra, which may exhibit narrowing with height in cumulus clouds (e.g., Austin et al., 1985) and even in the nearly adiabatic cores of convective clouds (Brenguier and Chaumat, 2001). This feature may be explained by the factor $\chi_{ad}\Delta z$. If χ_{ad} in (Eqn. 13.8.5) decreases with height slower than $1/\Delta z$, then $\chi_{ad}\cdot\Delta z$ increases, p increases, and the spectra narrow with height, resembling the effect of regular condensation, as observed in a shallow convective cloud by Austin et al. (1985). The values of relative dispersion σ_r for a convective cloud described in Austin et al. (1985) varied from $\sigma_r = 0.2$–0.3 in the middle of the cloud (corresponding p calculated from (Eqn. 13.8.6) are 13–25) to $\sigma_r = 0.17$–0.21 ($p = 22$–35) near the cloud top. Thus, the spectra narrow with height in this case, in contrast to the convective cloud observed by Warner (1969a). If the product $\chi_{ad}\Delta z$ is close to but slightly lower than the adiabatic value (i. e., $\chi_{ad}\Delta z \approx$ const), such as in convective adiabatic cores due to lateral mixing, then the spectra can be narrow, but still wider than it would be in adiabatic ascent, similar to the situation observed by Brenguier and Chaumat (2001) in cumulus clouds.

Note that the entrainment processes in cumulus or near the tops of stratocumulus can be easily accounted for in this approach by addition into the RHS of the kinetic (Eqns. 13.4.3a), (13.4.3b), and (13.5.11) of the term $-\tau_m^{-1} f_s$, with τ_m being the characteristic mixing time. Then, the two solutions (Eqn. 13.6.10) for the slope $\beta_{s1,2}$ are modified:

$$\beta_{s,1,2} = \frac{\alpha_s}{G_s}\left[1 \pm \left(\frac{\tau_m^{-1}}{k\alpha_s^2} + \frac{w}{k\alpha_s} + \frac{1}{k\alpha_s^2}\frac{dw}{dz}\right)^{1/2}\right]. \tag{13.8.9}$$

The merged solution (Eqn. 13.6.11) becomes a superposition of two functions:

$$f_{1,2}(r) = c_1 r^p \exp(-\beta_{s,1} r)\psi_1(r) + c_2 r^p \exp(-\beta_{s,2} r)\psi_2(r), \tag{13.8.9a}$$

and can give the bimodal distributions. Equation (13.8.9) shows that with decreasing mixing time, τ_m^{-1} increases and leads to the larger separation of the modes and to broader spectra. Recall that one of Manton's (1979) results was that the size spectrum of each single mode was too narrow and only their bimodal superposition could ensure the dispersion growing with height at a rate comparable to the measured values (Warner, 1969a). A bimodal superposition of the gamma distributions considered here may also provide sufficient dispersion for the case of cumulus clouds.

In application of this parameterization for ice clouds, the difference of the crystal size spectra with the drop spectra arise due to the different values of N, $\bar{r}$, and lapse rates χ_w. In liquid clouds, $N \sim 1-5\times10^2\,\mathrm{cm}^{-3}$, $\bar{r} \sim 5-10\ \mu\mathrm{m}$, and according to (Eqn. 13.8.3) and Table 13.1, $G_s \sim (0.3-5)\times10^{-8}$. In crystalline cirrus clouds with $N \sim (1 - 5) \times 10^2\,\mathrm{L}^{-1}$ (~ 3 orders of magnitude smaller than droplet concentrations), $q_{ls} \sim 10 - 20\,\mathrm{mg\,m}^{-3}$, $\bar{r} \sim 5-10\,\mu\mathrm{m}$, and (Eqn. 13.8.3) yields $G_s \sim 10^{-7} - 10^{-5}$. Thus, both G_s and $\bar{r}$ are 1–2 orders greater in cirrus than in liquid clouds.

According to (Eqn. 13.8.4), $p \sim \bar{r}/G_s$, so that the increase in G_s is compensated by a similar growth in $\bar{r}$, and the indices p appear to be comparable in liquid and crystalline clouds, as illustrated in more detailed calculations in Section 13.10.

Temperature dependence. Equation (13.8.5) for p also shows that there is a temperature dependence of the drop and crystal size spectra. Liquid water clouds can occur over a large range of temperatures, particularly since supercooled liquid clouds are sometimes observed at temperatures as low as −35 °C and crystalline clouds are observed at −80 °C and lower (Borovikov et al., 1963; Pruppacher and Klett, 1997; see Chapter 4). According to (Eqn. 13.8.5), $p \sim (\gamma_a - \gamma_w)^{-1}$. Since $\sigma_r = (p+1)^{-1/2}$, then for the typical condition $p >> 1$, $\sigma_r \approx p^{-1/2} \sim (\gamma_a - \gamma_w)^{1/2}$. The wet adiabatic lapse rate γ_w tends to the dry adiabat γ_a with decreasing temperature; thus, p increases and σ_r decreases at low T. This explains observed narrower spectra in summertime colder liquid As-Ac clouds at $T \sim -20\,°C$, $P = 800\,hPa$ ($\gamma_a - \gamma_w = 1.7\,°C\,km^{-1}$) than in warmer Sc at $T \sim 0\,°C$ ($\gamma_a - \gamma_w = 4\,°C\,km^{-1}$) (e.g., Herman and Curry, 1984; Tsay and Jayaweera, 1984). This dependence is also strong in cirrus clouds at low temperatures for the wet adiabat γ_w over ice. For example, at a pressure $P = 200\,hPa$, the value $\gamma_a - \gamma_w = 4.14\,°C\,km^{-1}$ at −20 °C, $1.22\,°C\,km^{-1}$ at −40 °C, and $0.62\,°C\,km^{-1}$ at −50 °C. Thus, (Eqn. 13.8.5) predicts that the index p would increase by a factor of ~7 from −20 °C to −50 °C if N, $\bar{r}$, and the pressure were the same, and the crystal size spectra in cirrus should become narrower with decreasing T. This feature is clearly seen in the experimental data where the spectra are sorted by temperature or height (e.g., Heymsfield and Platt, 1984; Sassen et al., 1989; Platt, 1997; Poellot et al., 1999, Lawson et al., 2006). The theory described here provides a basis for the quantitative description of the temperature dependence of the drop and crystal size spectra.

This narrowing of the spectra with decreasing temperature causes slower droplet freezing, suppresses coagulation and accretion, and consequently precipitation—that is, it causes greater colloidal stability of the colder clouds. Vice versa, global atmospheric warming may lead to broader size spectra, intensification of coagulation, and precipitation. This may cause a decrease in cloud albedo—i.e., in global albedo, and accelerate warming. Thus, (Eqn. 13.8.5) predicts that a positive feedback may exist between warming and the broadening of the size spectra.

13.8.3. Solutions in the Form of Inverse Power Laws

Equation (13.8.4) predicts that the index p is negative in cloud layers with subsaturation, $s < 0$. Subsaturation may occur in a cloud due to downdrafts, advection of drier or warmer air, entrainment of dry ambient air, and radiative heating. Under these conditions, the smallest particles should have been evaporated to some boundary values r_* and the left branch of the size spectra at $r > r_*$ is described by the inverse power law

$$f_s(r) \sim (r/r_*)^p \sim (r_*/r)^{-|p|}. \qquad (13.8.10)$$

Such inverse power laws have been found by fitting experimental data from crystalline cirrus and frontal clouds in Heymsfield and Platt (1984), Platt (1997), Poellot et al. (1999), Ryan (2000), and in the data for the larger drops in liquid clouds (Okita, 1961; Borovikov et al., 1965; Nevzorov, 1967; Ludwig and Robinson, 1970), as described in Chapter 2. Here, this inverse power law is obtained as a solution to the kinetic equation and its index p is expressed in (Eqn. 13.8.5) via physical cloud parameters.

An estimation of p from (Eqn. 13.8.5) for cirrus clouds shows that at $T \sim -40\,°\text{C}$ with $k = 5 - 10\,\text{m}^2\,\text{s}^{-1}$ and $G_s \sim (1-2) \times 10^{-7}$, the parameters are $kG_s^2 \sim (0.5-1) \times 10^{-9}\,\text{cm}^2\,\text{s}^{-1}$, $b_{con} \sim 4 \times 10^{-8}\,\text{cm}^2\,\text{s}^{-1}$. Then, at subsaturation $s = -0.1$ (-10%), assuming $c_{nn} = 1$, we obtain $p = b_{con}s/kG_s^2 \sim -2$ to -4. The indices estimated here are quite comparable to those in Heymsfield and Platt (1984), Platt (1997), Poellot et al. (1999), and Ryan (2000) that give $p \sim -1$ to -8.

When size spectra are sorted by temperature, there is a distinct increase of the measured and fitted slopes $|p|$ with decreasing temperature. This can also be described using (Eqn. 13.8.5) with $w_{ef} < 0$. Then again, as in the case $p > 0$, the slopes in the crystalline clouds $|p| \sim (\gamma_a - \gamma_{is})^{-1}$, where γ_{is} is the lapse rate at the vapor–ice transition (see Section 3.11). Comparing the spectra at two temperatures, $-20\,°\text{C}$ (pressure $p_a = 600\,\text{hPa}$, $\gamma_a - \gamma_{is} = 2.14\,°\text{C}\,\text{km}^{-1}$) and at $-50\,°\text{C}$ ($p_a = 200\,\text{hPa}$, $\gamma_a - \gamma_{is} = 0.62\,°\text{C}\,\text{km}^{-1}$), we obtain that the slopes $|p|$ should be about $2.14/0.62 = 3.5$ times larger at colder temperatures if the other parameters are the same. A more precise evaluation for various p_a, T, N, and $\bar{r}$ can be performed using (Eqn. 13.8.5) instead of fitting to the measured spectra, as is usually done.

Subsaturated layers with negative p may constitute significant portions of crystalline clouds, particularly as falling particles enter the subsaturated regions below the cloud. Hall and Pruppacher (1976) showed that falling from cirrus clouds crystals can survive at subsaturation over 2–6 km, which was confirmed by later modeling studies of evolving cirrus (e.g., Jensen et al. 1994, 2001; Khvorostyanov and Sassen, 1998a,b, 2002; Sassen et al., 2002; Khvorostyanov, Curry et al. 2001). The life cycle of cirrus clouds usually begins with ice nucleation on haze particles that requires some threshold supersaturation (Chapters 8 and 9). After this threshold has been reached, the rate of vapor absorption by the nucleated crystals may exceed the rate of supersaturation generation, so that the mean supersaturation decreases, and the thickness of the cloud evaporation layer increases and can exceed the thickness of the growth layer. This process, governed by the crystal supersaturation relaxation time, is slow and may last up to several hours. This may cause a significant frequency of ice subsaturated (evaporation) layers in crystalline clouds and explain observations of the inverse power law spectra, based on the results of this section.

These features of the spectra discussed in this section and related to variations of the indices p are usually ignored and missed in models with parameterizations that use constant values of $p = 0$–3 for all spectra. This may reduce the accuracy of such models.

The effect of averaging is the same for ice clouds as that discussed earlier for liquid clouds. The measured and calculated σ_r increase and p decrease with increasing scales of averaging L_x. It should be emphasized that the local narrow spectra with $p \sim 6$–20, as all the cited experimental data indicate, are representative of the cloud processes and should be used in parameterizations, since these local spectra govern radiative transfer, coagulation-accretion, and precipitation. Averaging over large scales leads to an artificial broadening of the spectra, and unphysical lowering of the indices p to the values of 0–2; use of such broad spectra may cause artificial acceleration of the coagulation processes, unphysically faster precipitation formation and cloud dissipation, and errors in evaluation of the radiative scattering and absorption coefficients (Section 2.5).

The relationships described here are consistent with observations in clouds (e.g., Levin, 1954; Mason, 1971; Sedunov, 1974; Pruppacher and Klett, 1997; Cotton et al., 2011). Thus, (Eqns. 13.8.4) and (13.8.5) establish an analytical relationship of the index p, and hence drop spectra dispersion, with the meteorological characteristics of the atmosphere.

13.9. Applications of the Solution for Liquid Clouds

To illustrate the analytical solutions and evolution of the droplet spectra below the maximum in q_L, we consider the following characteristics typical of low stratus clouds and close to the arctic stratus observed on 28 June 1980 in the upper layer (Curry, 1986). The vertical profile of q_L is approximated by a parabolic profile of the form $q_L(z) = c_N q_{L,max}(z/H)^a(1 - z/H)^b$, where H is the cloud depth, $q_{L,max}$ is the maximum of LWC, $c_N = [a/(a + b)]^a\ [b/(a + b)]^b$; q_L and the parameter α_s are presented in Fig. 13.1. Cloud boundaries are at 700 m and 1100 m, and the maximum value $q_{L,max} = 0.4\,\mathrm{g\,m^{-3}}$ is located at ~1000 m. The vertical profile of the mean radius is chosen such that $\bar{r} = 2.5\ \mu$m at the cloud bottom and linearly increases to $\bar{r} = 6\ \mu$m at the cloud top. The vertical profile of the droplet concentration N calculated using these values of q_L and $\bar{r}$ is also presented in Fig. 13.1, and exhibits the fastest increase above the base with a much slower increase in mid cloud. Such profiles are also in agreement with the results of LES simulations of stratocumulus (e.g., Feingold et al., 1994), averaged over the domain.

To investigate the effect of vertical velocity, turbulence, and radiation on the drop size spectra, we examine four different cases, using the definition of effective vertical velocity $w_{ef} = w_{dyn} + w_{rad}$ described in Section 13.8. The four cases are: 1) w_{rad} is determined by accounting for the longwave cooling rate only, $w_{dyn} = 1\,\mathrm{cm\,s^{-1}}$, and $k = 5\,\mathrm{m^2\,s^{-1}}$; 2) w_{rad} is determined by accounting for longwave cooling only, $w_{dyn} = 10\,\mathrm{cm\,s^{-1}}$, and $k = 5\,\mathrm{m^2\,s^{-1}}$; 3) $w_{rad} = 0$, $w_{dyn} = 15\,\mathrm{cm\,s^{-1}}$, and $k = 5\,\mathrm{m^2\,s^{-1}}$; and 4)

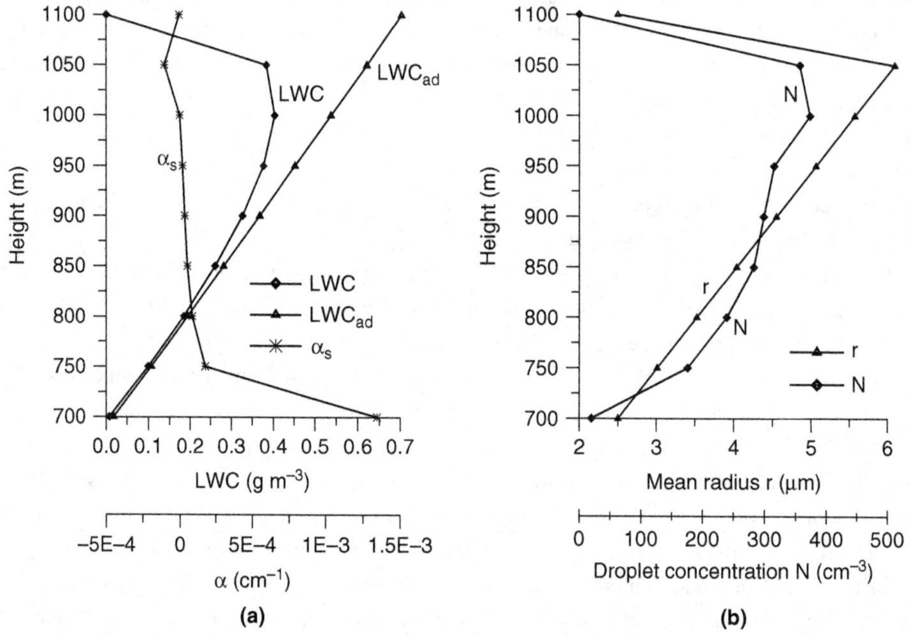

Figure 13.1. Vertical profiles for a prototype case of Arctic stratus clouds: (a) liquid water content (LWC), q_L, adiabatic liquid water content (LWC$_{ad}$), and parameter $\alpha_s = (1/q_L)(dq_L/dz)$; (b) droplet concentration, $N(z)$, and mean radius, $\bar{r}(z)$. From Khvorostyanov and Curry (1999d). *J. Atmos. Sci.*, 56, © American Meteorological Society. Used with permission.

w_{rad} is determined by accounting for longwave cooling only, $w_{dyn} = 10\,\text{cm s}^{-1}$, and $k = 10\,\text{m}^2\,\text{s}^{-1}$. These variations allow examination of the individual effects of w, k, and radiation on the drop size spectra.

Radiative heating rates are calculated as described in Khvorostyanov (1995), Khvorostyanov and Sassen (1998a,b; 2002), and Khvorostyanov et al. (2001) using the two-stream approximations for longwave (LW) and shortwave (SW) radiative transfer with parameterizations of the absorption and scattering coefficients depending on the droplet size spectra (see Section 2.5). The input variables were consistent with the observed arctic stratus clouds from Fig. 13.1. Calculated profiles of radiative heating rates presented in Fig. 13.2 show maximum longwave cooling of $-120\,\text{K day}^{-1}$ and solar heating of $\sim 40\,\text{K day}^{-1}$, which is in a good agreement with those determined from measurements of this cloud deck (Herman and Curry, 1984; Curry, 1986).

Corresponding radiative-effective vertical velocities (Fig. 13.3) calculated with (Eqn. 13.8.7) exhibit strong maxima at heights 50–100 m below the cloud top, equivalent to updrafts of $5\text{–}8\,\text{cm s}^{-1}$ for the longwave cooling and downdrafts of $-3\,\text{cm s}^{-1}$ for solar heating. These w_{rad} almost vanish in the lower half of the cloud, where the dynamic velocity plays the major role. The value of parameter α_s (Fig. 13.1) shows a strong maximum near the cloud base, sharply decreases upward in the lowest 50 m of the cloud, and then continues to decrease much more slowly until it reaches zero at a height of 1,000 m, and becomes negative above this level.

The values of parameter μ_s calculated from (Eqn. 13.6.10a) and shown in Fig. 13.4 exhibit a monotonic increase upward from about 0.2–0.6 near the cloud bottom to values $\mu_s = 4\text{–}5.5$ within 200–300 m above the cloud base, with values of μ_s significantly increasing above 900 m where radiative-effective velocities and their gradients increase.

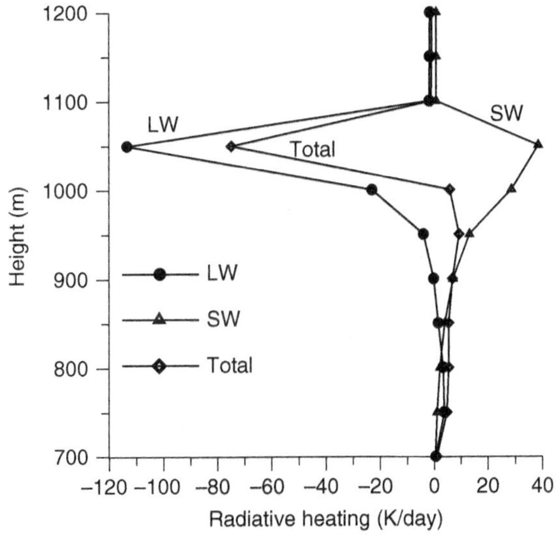

Figure 13.2. Vertical profiles of longwave (LW), shortwave (SW), and total (Total) radiative heating rates for the cloud shown in Fig. 13.1. From Khvorostyanov and Curry (1999d). *J. Atmos. Sci.*, 56, © American Meteorological Society. Used with permission.

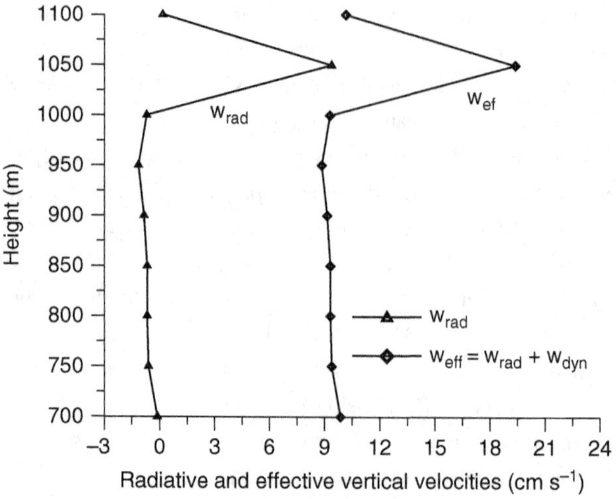

Figure 13.3. Vertical profiles of the "effective radiative" vertical velocities $w_{rad} = w_{long} + w_{short}$ calculated with (Eqn. 13.8.7) for the cloud shown in Figs. 13.1 and 13.2 that accounts for both longwave and shortwave cooling/ heating, and the total effective velocity $w_{eff} = w_{dyn} + w_{rad}$, where $w_{dyn} = 10\,\mathrm{cm\,s^{-1}}$. From Khvorostyanov and Curry (1999d). *J. Atmos. Sci.*, 56, © American Meteorological Society. Used with permission.

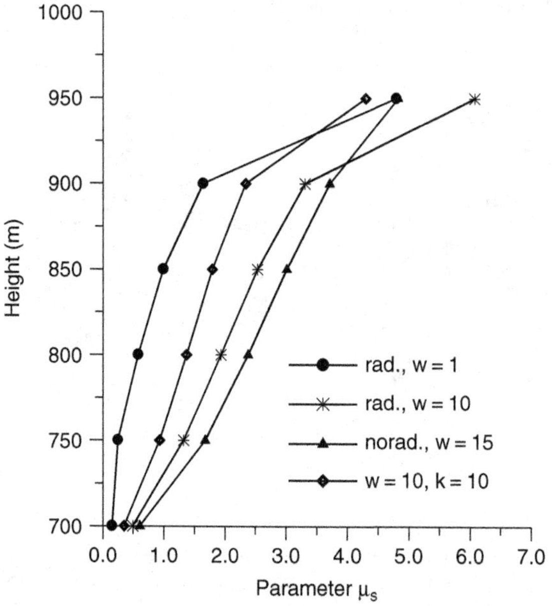

Figure 13.4. Vertical profiles of the parameter μ_s in the analytical solutions calculated with (Eqn. 13.6.10a) for four cases: 1) $w_{ef} = w_{dyn} + w_{rad}$; $w_{dyn} = 1\,\mathrm{cm\,s^{-1}}$, and w_{rad} calculated with (Eqn. 13.8.7) that accounts for longwave cooling only; $k = 5\,\mathrm{m^2\,s^{-1}}$; 2) same as case 1, but $w_{dyn} = 10\,\mathrm{cm\,s^{-1}}$; 3) same as case 1, but $w_{dyn} = 15\,\mathrm{cm\,s^{-1}}$ and $w_{rad} = 0$ (norad.); 4) same as case 2, but $k = 10\,\mathrm{m^2\,s^{-1}}$. From Khvorostyanov and Curry (1999d). *J. Atmos. Sci.*, 56, © American Meteorological Society. Used with permission.

The indices p evaluated with (Eqn. 13.8.4) for the general solutions (Fig. 13.5) for the case with $w_{dyn} = 1\,\text{cm}\,\text{s}^{-1}$ (synoptic-scale) do not exceed 1–2 in the whole cloud but reach 20 in the thin layer of maximum radiative cooling and maximum w_{rad} near 1050 m. The values of p for $w_{dyn} = 10\text{–}15\,\text{cm}\,\text{s}^{-1}$ are much greater and closer to the local observations. They also exhibit an increase from the cloud bottom with the values 5–15 at the mid-cloud depth, reach maximum values of 20–35 near the maximum q_L, and then decrease upward toward the cloud top. For the cases with $w_{rad} > 0$, this increase is especially rapid near the maximum q_L where w_{rad} is maximum.

The indices $\tilde{p}_1$ of the gamma distribution for the first solution (Eqn. 13.6.23) (Fig. 13.6) also increase upward from the cloud bottom but more slowly than the indices p. This is caused by the factor $(1 + \mu_s)/2\mu_s$ in $\tilde{p}_1$. As discussed in Section 13.6, the asymptotic solutions (Eqn. 13.6.23) with indices $\tilde{p}_1$ are valid only in the lower layer of ~100–200 m, while the larger indices p (Fig. 13.5) are more representative of the spectral dispersions in the upper half of the cloud. Smaller values of $\tilde{p}_1$ than p are in agreement with the measurements in clouds that show wider spectra near the cloud bottom than in the cloud interior (Table 13.4). It is seen from Figs. 13.5 and 13.6 that values of both p and $\tilde{p}_1$ increase with increasing vertical velocity, indicating smaller σ_r and narrower spectra, and decrease with increasing turbulence, indicating greater σ_r and wider spectra.

An example of the size spectra in the lower part of the cloud calculated with (Eqn. 13.6.23) is presented in Fig. 13.7. The shape and general behavior of the spectra are in reasonable agreement with those typically observed in clouds. The calculated size spectra can be compared to those measured on

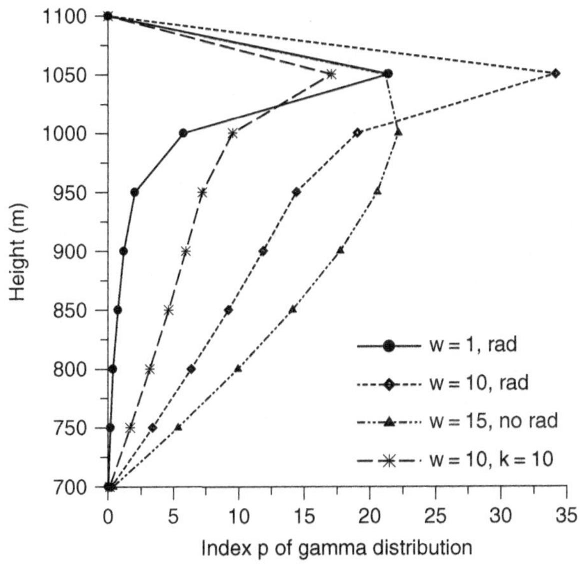

Figure 13.5. Vertical profiles of the indices p of gamma-distributions evaluated with (Eqn. 13.8.4) for the same four cases as in Fig. 13.4. Hereafter, the values of w_{dyn} and k are shown in the Figures; "*rad*" and "*no rad*" mean calculations with and without an accounting for the radiative velocities w_{rad}. From Khvorostyanov and Curry (1999d). *J. Atmos. Sci.*, 56, © American Meteorological Society. Used with permission.

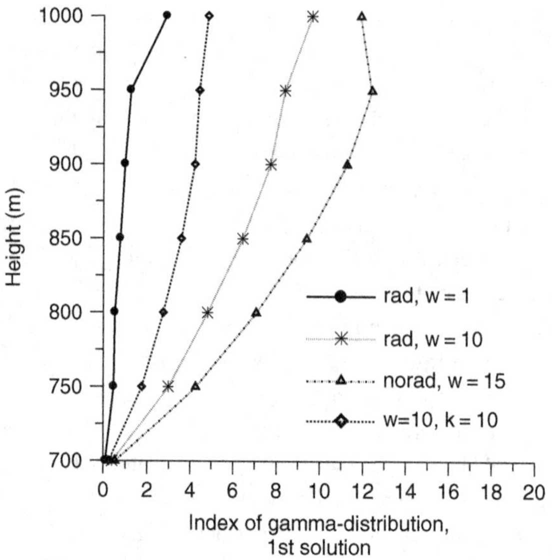

Figure 13.6. Vertical profiles of the indices $\tilde{p}_1$ of gamma-distributions for the first asymptotic solution (Eqn. 13.6.23) (2nd boundary asymptotic regime) for the same four cases as in Fig. 13.4. From Khvorostyanov and Curry (1999d). *J. Atmos. Sci.*, 56, © American Meteorological Society. Used with permission.

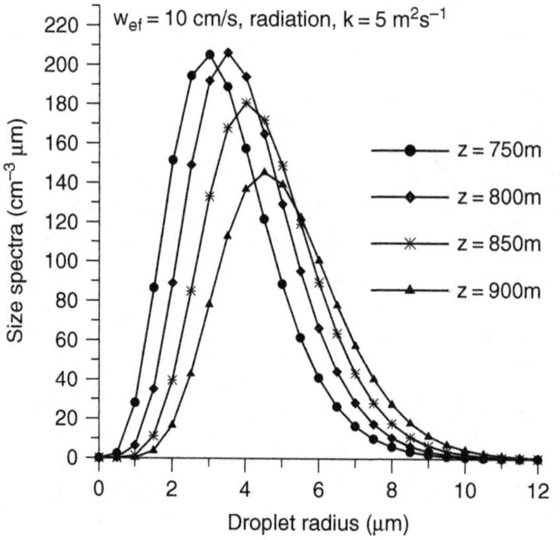

Figure 13.7. Evolution of the droplet size spectra above the cloud base calculated with the first asymptotic solution (Eqn. 13.6.23) of the 2nd asymptotic regime for the case with $w_{dyn} = 10\,\mathrm{cm\,s^{-1}}$, w_{rad} as in Fig. 13.2, $k = 5\,\mathrm{m^2\,s^{-1}}$, normalized to the concentration $N(z)$ shown in Fig. 13.1. From Khvorostyanov and Curry (1999d). *J. Atmos. Sci.*, 56, © American Meteorological Society. Used with permission.

28 June 1980 in stratus (Herman and Curry, 1984, Table 2). There were five aircraft passes through this cloud deck at various altitudes. Although the measured spectra are slightly different at various times, they exhibit some common features. The calculated spectra shown in Fig. 13.7 generally reproduce the shape of the measured spectra: a) the fast increase of $f(r)$ in the region from 0 to 1–2 μm caused by the high value $p \sim 5$–10; b) the slower exponential tail; c) the displacement of the maximum to the lager radii with height; d) the simultaneous decrease of the relative dispersion σ_r upward (see Table 13.4 for the measured σ_r and Figs. 13.8 and 13.9 for calculated σ_r). The feature of bimodality in some of the measured spectra is not reproduced in Fig. 13.7, as only one mode of the analytical solutions was accounted for. This bimodality may probably be caused by the mixing with the entrained dry air from above, and can be reproduced in this model by using a superposition of the solutions (Eqn. 13.8.9a) with the two exponents $\beta_{s,1,2}$ in (Eqn. 13.8.9) with appropriate mixing time τ_m.

13.10. Comparison with Previous Theories and Observations

The theory of stochastic condensation and the stochastic kinetic equations have been criticized by Warner (1969b), and Bartlett and Jonas (1972), who modeled the impact of turbulence on the size spectrum with random fluctuations of the vertical velocity in a cloud parcel using Lagrangian parcel models without exchange with environment. They found that turbulence modelled with such an isolated cloud parcel is unable to account for the observed drop size broadening. Criticisms by Mazin and Smirnov (1969) of the theory of stochastic condensation were also based on the concept of an adiabatic parcel ascent. This criticism was answered by Stepanov (1975), Voloschuk and Sedunov (1977), and Manton (1979), who noted that the kinetic equation under some conditions does not cause further broadening of the spectra with time if mixing among the parcels is absent. We consider in this section in more detail the correspondence between the approach based on the kinetic equation and the Lagrangian parcel models. It will be shown that there is no contradiction between these two approaches and that Warner (1969b) and Bartlett and Jonas (1972) considered a particular case of stochastic condensation.

As we see from the low-frequency form of the stochastic kinetic (Eqn. 13.4.5), the effect of turbulence vanishes when the right-hand side of the equation is zero, and (Eqn. 13.4.5) converts into the equation of regular condensation (Eqn. 13.2.1). The right-hand side of (Eqn. 13.4.5) can be zero if either

$$k_{ij} = 0 \tag{13.10.1}$$

or

$$\left(\frac{\partial}{\partial z} + G_s \frac{\partial}{\partial r}\right) f_s = 0 \tag{13.10.2}$$

Thus, the condition (Eqn. 13.10.2) can be regarded as equivalent to the condition of zero turbulence (Eqn. 13.10.1). Equation (13.10.2) can then be rewritten as

$$\frac{\partial f_s}{\partial z} = -G_s \frac{\partial f_s}{\partial r}. \tag{13.10.3}$$

Multiplying (Eqn. 13.10.3) by $(4\pi/3)\rho_w r^3$ and integrating, we obtain:

$$\frac{dq_L}{dz} = -\frac{4\pi}{3}\rho_w G_s \int_0^\infty r^3 \frac{\partial f}{\partial r} dr = 4\pi G_s \rho_w N \overline{r^2} \approx 4\pi G_s \rho_w N \bar{r}^2, \tag{13.10.4}$$

where q_L is the LWC. The last approximate relation is based on the fact that $\overline{r^2}$ is usually only slightly (by 10–20%) larger than $\bar{r}^2$. To interpret (Eqn. 13.10.4), we express G_s in terms of the vertical gradient of the adiabatic liquid water mixing ratio defined by (Eqn. 13.8.2) using the first of the expressions (Eqn. 13.8.3):

$$G_s = \frac{\rho_a}{\rho_w}\frac{1}{4\pi N \bar{r}^2}\left(\frac{dq_{ls}}{dz}\right)_{ad}, \tag{13.10.5}$$

where q_{ls} is the water mixing ratio. This equation shows that the parameter G_s relates the vertical gradient of the adiabatic liquid water mixing ratio to the effective area of the droplets per unit volume—that is, G_s characterizes the rate of absorption by the droplets of vapor excess released during adiabatic lifting. Substituting (Eqn. 13.10.5) for G_s into (Eqn. 13.10.4), we obtain the following relation:

$$\frac{dq_L}{dz} \approx 4\pi\rho_w N \bar{r}^2 \frac{\rho_a}{\rho_w}\frac{1}{4\pi N \bar{r}^2}\left(\frac{dq_{ls}}{dz}\right)_{ad} = \rho_a \left(\frac{dq_{ls}}{dz}\right)_{ad}. \tag{13.10.6}$$

Equation (13.10.6) shows that the condition (Eqn. 13.10.2) corresponds to the adiabatic gradient of the mixing ratio, or, as shown by (Eqns. 13.10.1) and (13.10.2), this is equivalent to $k_{ij} = 0$—i.e., to the absence of turbulence and mixing among the parcels. So, we see that turbulence does not influence the drop size spectrum if the vertical profile of q_L corresponds to the adiabatic value.

In the simulations made by Warner (1969b) and Bartlett and Jonas (1972) using Lagrangian parcel models, moist adiabatic ascents were assumed and adiabatic liquid water contents were derived. Vertical velocity fluctuations in these adiabatic parcel simulations led only to the reversible changes in the cloud properties, when the updraft causes an increase in the mean radius and q_L, and the downdraft removes the effects of a previous updraft. The major source of spectral broadening—vertical inhomogeneity and vertical turbulent exchange of the parcels with different spectra—is absent in the modeling with isolated parcels. Therefore, only a small broadening of the spectra was obtained in these parcel model simulations, probably due to the numerical diffusion.

As shown earlier, the kinetic equation of stochastic condensation leads in this wet adiabatic situation to the same results, absence of turbulent stochastic effects, and does not cause broadening of the spectra. Hence, there is no contradiction between the treatment of the stochastic condensation process as described here and the Lagrangian parcel models, which correspond to the adiabatic condition (Eqn. 13.10.2) and are equivalent to the absence of turbulence (Eqn. 13.10.1). These conditions along with the physical meaning of parameter G_s described by (Eqn. 13.10.5), mean that any vertical displacement in the cloud is accompanied by a change in the drop size spectrum. Displacements of parcels in regions of the clouds that have adiabatic liquid water profiles will not lead to the broadening of the drop spectra. The droplet radius will be a function of the height and the cloud will be in some equilibrium state. However, liquid water content in a cloud is typically much less than its adiabatic value, and the gradient dq_L/dz is different from the wet adiabatic profile (see Fig. 13.1) and the condition (Eqn. 13.10.6) is not satisfied. This departure of the cloud liquid water content from the adiabatic profile increasing upward will result in the evolution of the size spectra according to the kinetic equation of stochastic condensation. Since the adiabatic ratio $\chi_{ad} = q_L/q_{L,ad}$ usually decreases upward, the applicability of the parcel models without entrainment and with adiabatic gradients becomes more limited with increasing height above the cloud base.

Another good confirmation of the fact that the adiabatic LWC or IWC profiles in clouds effect the size spectra in a way equivalent to the absence of turbulence, $k = 0$, was described in Liu and Hallett (1998), who found very narrow size spectra in a lenticular cloud over the Sierra Nevada formed adiabatically in a laminar flow.

Considine and Curry (1996) used a simple statistical model to derive drop size spectra that have the shape of a modified gamma distribution. As such, the p-index is not directly comparable to the drop size spectra developed here. However, the parameters of Considine and Curry's drop size distribution are vertical velocity variance, temperature, pressure, the lapse rate in the cloud, and the lapse rate of a rising parcel. These parameters include essentially the same meteorological variables used in the drop size spectra developed here. Considine and Curry's results implied that if the lapse rate within the cloud is equal to the saturated adiabatic lapse rate, then the spectrum becomes infinitely narrow, approaching a monomodal distribution. This is consistent with the result that a cloud with adiabatic liquid water content will have a narrow spectrum and the spectrum will not be influenced by turbulent motions. Considine and Curry also found that the drop size spectra broaden as the vertical velocity variance increases, which is consistent with the results described here. This general agreement with Considine and Curry supports some of the assumptions used in their derivation. The present derivation provides a theoretical justification for some of these assumptions.

One of the most important quantities that characterizes the effect of turbulence on the size spectra is the relative dispersion $\sigma_r = (1 + p)^{-1/2}$. Its vertical profiles calculated with the values of $\tilde{p}_1$ from Fig. 13.6 are given in Fig. 13.8 for the four cases considered earlier. The dispersion for the cases

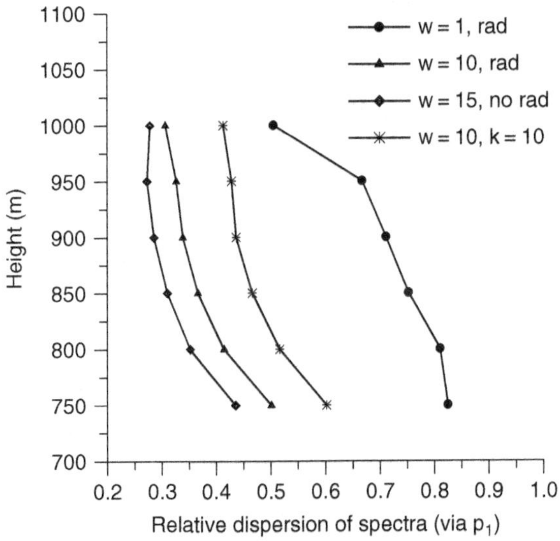

Figure 13.8. Vertical profiles of the relative dispersions of the size spectra for the first asymptotic solution (Eqn. 13.6.23) of the 2nd asymptotic regime for the same four cases as in Fig. 13.4 calculated with indices $\tilde{p}_1$ defined by (Eqn. 13.6.23) and shown in Fig. 13.6. From Khvorostyanov and Curry (1999d). *J. Atmos. Sci.*, 56, © American Meteorological Society. Used with permission.

with $w_{dyn} = 10 - 15\,\mathrm{cm\,s^{-1}}$ decreases from 0.45–0.6 above the cloud base to 0.3–0.4; that is, the size spectra in low stratus are narrowing with the height above the cloud base. These values are in good agreement with many observations, but the dispersions with small $w_{dyn} = 1\,\mathrm{cm\,s^{-1}}$ are 0.55–0.85, much larger than those observed locally. Note that the effect of the decrease of σ_r upward is opposite to that observed above the base in convective clouds where dispersions increase upward (Warner, 1969a); however, it is in good agreement with the observation in stratus clouds (Noonkester, 1984; Curry, 1986) and in the upper halves of some convective clouds (e.g., Austin et al., 1985). The data from Table 13.4 show that the relative dispersions have maxima near the cloud's top and bottom and a minimum in the middle of the cloud, near the maximum of q_L. The calculated values of σ_r are similar to the measured values shown in Table 13.4.

The indices $\tilde{p}_1$ corresponding to the asymptotic solution (Eqn. 13.6.23) characterize the breadth of the size spectra only in the lower layer, while the indices p can serve as a measure of the spectral breadth in the whole cloud. The dispersions calculated with the indices p from Fig. 13.5 are shown in Fig. 13.9. These dispersions exhibit minima of 0.2–0.3 near maximum values of q_L, and asymmetry of σ_r in vertical direction with larger dispersions near the base than in the upper layer, which is in agreement with observations shown in Table 13.4.

Hence, the analytical solutions derived in Section 13.6 provide an explanation of the observed dispersions in clouds along with their possible different behavior with height. We emphasize that these are only approximate estimations of the dispersions based on several simplifications. The more accurate evaluation of σ_r should be based on the numerical solution of the complete kinetic equation, with more realistic accounting for the profiles of $k(z)$, gradients $\partial f/\partial z$, and others.

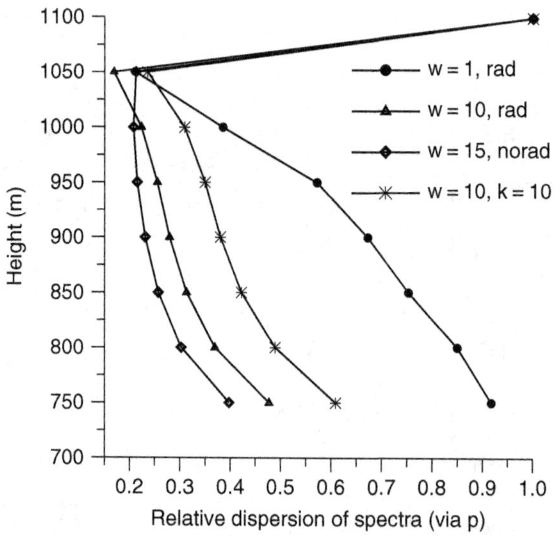

Figure 13.9. Vertical profiles of the relative dispersions of the size spectra for the same four cases as in Fig. 13.4 calculated with indices p defined by (Eqn. 13.8.4) and shown in Fig. 13.5. From Khvorostyanov and Curry (1999d). *J. Atmos. Sci.*, 56, © American Meteorological Society. Used with permission.

The technique described here can be used to study covariances of moments of the distribution function with various quantities, as was done based on observations by Curry (1986). The simplest is the covariance of liquid water content, q_L, with vertical velocity. Using (Eqn. 13.3.13) for $\overline{u_i' f'}$ in the low-frequency approximation ($k_{ij}^n = k_{ij}$) and the assumption of horizontal homogeneity (only the term $\partial/\partial z$ is kept), multiplying it by $(4\pi/3)\rho_w r^3$ and integrating over radii, we obtain

$$\overline{q_L' u_3'} = \frac{4\pi}{3}\rho_w \int_0^\infty r^3 \overline{u_3' f'} dr = \frac{4\pi}{3}\rho_w \int_0^\infty r^3 \left[-k\left(\frac{\partial f}{\partial z} + G_s \frac{\partial f}{\partial r} \right) \right] dr$$

$$= -k\frac{\partial q_L}{\partial z} - k\frac{4\pi}{3}\rho_w G_s \int_0^\infty r^3 \frac{\partial f}{\partial r} dr. \tag{13.10.7}$$

The last integral in (Eqn. 13.10.7) is evaluated by integration by parts and is equal to $-3N\overline{r^2} \approx -3N\bar{r}^2$. Including the first expression (Eqn. 13.8.3) for G_s, the covariance (Eqn. 13.10.7) is written as

$$\overline{q_L' u_3'} = k\left[\rho_a \left(\frac{dq_{ls}}{dz} \right)_{ad} - \left(\frac{dq_L}{dz} \right) \right]. \tag{13.10.8}$$

The expression (Eqn. 13.10.8) shows that covariance of q_L with vertical velocity should be positive for the cases when the vertical gradient of q_L is smaller than the adiabatic gradient of q_L, and is negative for the reverse situation. The first situation is met almost always, which causes a positive correlation. In particular, this positive correlation of q_L' and the vertical velocity fluctuations was described by Curry (1986) based on observations in arctic stratus (see Fig. 13.1), which means that vertical gradients of q_L were less than adiabatic, in accord with the present theory.

13.11. Calculation of Size Spectra for a Crystalline Cloud

Presented in this section are the results of calculations of the size spectra and their parameters for a crystalline cloud similar to those for liquid clouds described in Section 13.9. Calculations are performed using four different input profiles. The baseline profiles approximate a two-layer cloud of cirrus and nimbostratus observed on July 8, 1998, in the Surface Heat Budget of the Arctic Project (SHEBA) and the First International Satellite Cloud Climatology Project (ISCCP) Regional Experiment (FIRE) field campaign in the Arctic. The vertical profiles were chosen close to observations simulated with the spectral bin microphysics model in Khvorostyanov, Curry, et al. (2001). The cloud shown in Fig. 13.10 is pure crystalline, the small fraction exists in the region between 4 and 10 km with maximum IWC $q_L = 20\,\mathrm{mg\,m^{-3}}$ at 7 km, maximum crystal concentration $N_i \sim 90\,\mathrm{L^{-1}}$ at 9 km, and the mean crystal radius $\bar{r}$ increases downward from 30 μm near the cloud top to about 90 μm at the bottom. The following vertical profile of ice supersaturation is assumed: constant negative values $s = -15\%$ from 3 to 7.2 km, increasing from 0 at 7.5 km to a maximum of 12% at 9 km and then decreasing to −80% at 12 km (Fig. 13.10c). This design of s is consistent with observations (e.g., Jensen et al., 2001) and detailed simulations of this and similar clouds using a model with spectral bin microphysics and the supersaturation equation in Khvorostyanov, Curry, et al. (2001) and Khvorostyanov and Sassen (1998a,b, 2002), which showed that the maximum IWC in cirrus can be located near $s = 0$ since IWC increases downward in all the layers with $s > 0$ as crystals formed aloft grow and precipitate.

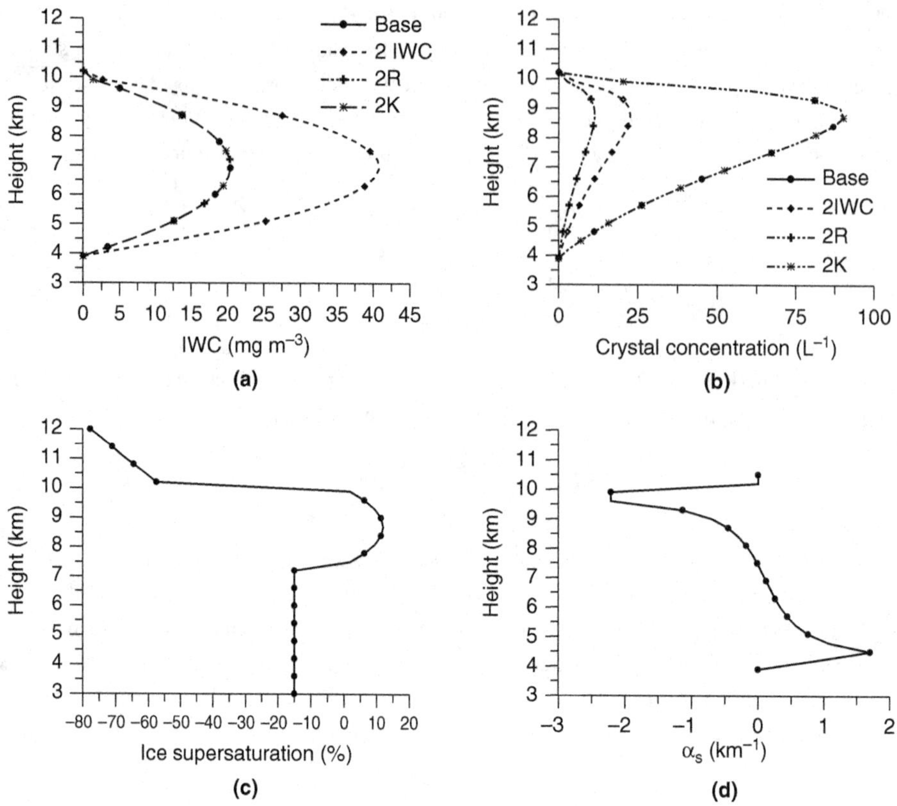

Figure 13.10. Vertical profiles of the input parameters for the four different cases with crystalline clouds indicated in the legend and described in the text (a) ice water content, mg m^{-3}; (b) crystal concentration, L^{-1}; (c) ice supersaturation, %; (d) the parameter α_s, km^{-1}. From Khvorostyanov and Curry (2008b). *J. Atmos. Sci.,* 65, © American Meteorological Society. Used with permission.

Maxima in s and N_i may coincide, reflecting the mechanism of ice nucleation governed by supersaturation (see Chapters 8–10). The values of α_s calculated from (Eqn. 13.5.8) are positive below the level of maximum IWC and are negative above, with extremes of about $\pm 2\,\text{km}^{-1}$ (Fig. 13.10d). We used the constant values of $w_{ef} = 3\,\text{cm s}^{-1}$ and $k = 5\,\text{m}^2\,\text{s}^{-1}$. These profiles are referred hereafter as the baseline (case 1) and denoted in the legends as "Base." To test the sensitivity of the results to the input profiles, the same calculations were performed with the following changes: doubled IWC and $\overline{r}$ (case 2); same IWC as in the baseline case but doubled $\overline{r}$ (case 3); and same conditions as in the baseline case but with doubled k (case 4).

Profiles of the parameter G_s calculated with (Eqn. 13.8.1) and these input data are shown in Fig. 13.11a. The values of G_s reach a maximum $\sim 10^{-5}$ near the cloud base, reach a minimum of $(1\text{–}2) \times 10^{-7}$ at a height of 9 km (coincident with maxima in N_i) and again increase upward. In case 3 (doubled $\overline{r}$), G_s is twice as large since $G_s \sim \overline{r}$. Note that the $\overline{r}$ dependence of G_s results in much greater values of G_s for ice clouds than for liquid clouds (compare with Section 13.9, Table 13.1). Cases 2 and 4 both show the same values of G_s as in the baseline case since in case 2 the increase in

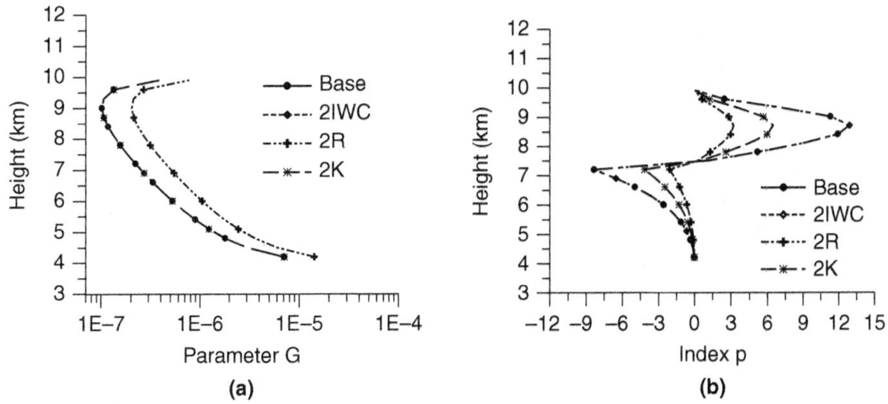

Figure 13.11. Profiles of the parameter G_s (a) and (b) the index p of the crystal size spectra. From Khvorostyanov and Curry (2008b). *J. Atmos. Sci.*, 65, © American Meteorological Society. Used with permission.

IWC is compensated in (Eqn. 13.8.1) by an increase in $\bar{r}$, and the same value in case 4 is because G_s does not depend on k.

Corresponding profiles of the index p calculated from the first of expressions (Eqn. 13.8.4) with $c_{nn} = 1$ are shown in Fig. 13.11b. In the baseline case, the values of p are negative, and increase from −9 to 0 in the cloud evaporation layer with $s < 0$ below 7 km since $p \sim s$. At the height ~ 4 km, $p \to 0$ since $p \sim N_i$ according to (Eqn. 13.8.5) and $N_i \to 0$. In the crystal growth layer with $s > 0$, values of p are positive, ranging from 0 to 13. These variations in p are caused by simultaneous variations in s and G_s. As discussed in Section 13.8, the values of p for this cirrus cloud are comparable to those in the liquid Arctic St cloud considered in Curry (1986), since larger G_s in Ci are compensated for in (Eqn. 13.8.4) by larger $\bar{r}$.

The values of p decrease by a factor of 2 with doubled k (case 4, asterisks) since $p \sim k^{-1}$; this causes an increase in relative dispersions σ_r according to (Eqn. 13.8.6) and reflects the effect of stochastic condensation as increasing turbulence broadens the spectra. The values of p decrease by a factor of 4 with doubled $\bar{r}$ (crosses) since $p \sim G_s^{-2}$, which is also an effect of stochastic condensation. As follows from the stochastic kinetic equations and discussed in Sections 13.4–13.8, the diffusion coefficient in radii-space is $k_r = c_{nn} G_s^2 k$, so that an increase in k causes an increase in k_r, which accelerates diffusion in radii space and leads to broader spectra.

Shown in Fig. 13.12 are the slopes β_s calculated from (Eqn. 13.7.3) in the upper layer with ice supersaturation at heights 7.5–8.7 km and in the lower layer at heights 4.8–6.0 km with $s < 0$. The common feature in both layers is a substantial increase of the slopes with increasing radius, which arises from the dominance of the second term with the fall velocity $v(r)$ in (Eqn. 13.7.3) that tends to the limit (Eqn. 13.7.7). In the upper layer, the slopes decrease downward causing spectral broadening, which agrees with observations. This is caused by dominance of the terms with α_s, and a decrease of $|\alpha_s|$ downward, toward the level with $s = 0$. Broadening of the spectra downward—i.e., with increasing temperature, also agrees with the observed temperature dependence of the spectra as discussed in Section 13.8. In the lower layer, the values of β_{s2} are much smaller due to smaller α_s (see Fig. 13.10d) and increase downward. The last feature would cause an incorrect narrowing of

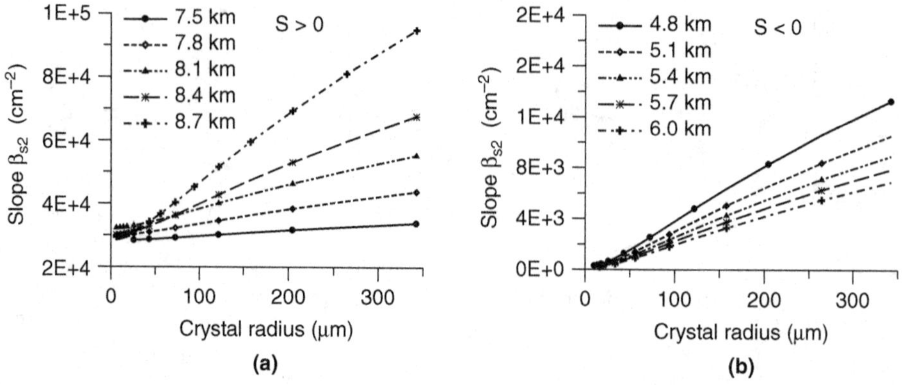

Figure 13.12. Slopes β_s of the crystal size spectra calculated from (Eqn. 13.7.3) at various altitudes in the layers with: (a) ice supersaturation and (b) subsaturation. The heights in km are indicated in the legends. From Khvorostyanov and Curry (2008b). *J. Atmos. Sci.*, 65, © American Meteorological Society. Used with permission.

the spectra downward, but it is overwhelmed by the effect of the negative indices p in this layer as described in the following text.

Fig. 13.13 depicts size spectra $f_s(r)$ calculated with input profiles from the baseline case. The left panels (*a, c*) show the tails of the spectra calculated from (Eqn. 13.7.4), and the right panels (*b, d*) show the composite spectra (Eqn. 13.7.5) with multiplication by $r^p\Phi(r)$ and $\Phi(r)$ from (Eqn. 13.7.6). One can see that in the upper ice supersaturated layer with $p > 0$, the spectra resemble typical gamma distributions. Addition of the factor $r^p\Phi(r)$ allows extension of the spectra to the small radii region but does not cause a substantial change of the spectra.

The situation is different in the lower ice subsaturated layer with $p < 0$. Without the factor $r^p\Phi(r)$, the small size region up to $100\,\mu$m in double log coordinates represents parallel straight lines with positive slopes caused by the factor r before the exponent in (Eqn. 13.7.4) (recall, the solution (Eqn. 13.7.4), $f_s \sim r\exp(-ar^2)$, is typical for some previous theories of stochastic condensation with $p = 1$). With multiplication by $r^p\Phi(r)$ in (Eqn. 13.7.5), the spectra become inverse power laws up to $r \sim 40$–$50\,\mu$m, the slopes (indices p) increase with height (i.e., with decreasing temperature), and the spectra show bimodality with secondary maxima at $r = 100$–$150\,\mu$m (Fig. 13.13d). Such bimodal behavior is often observed in crystalline clouds and parameterized with superposition of the two inverse power laws or gamma distributions that correspond to the two size fractions (e.g., Heymsfield and Platt, 1984; Mitchell et al. 1996; Platt, 1997, Ryan, 2000). Equation (13.7.5) and Fig. 13.13d show that this bimodality may occur within the small-size fraction: the left branch (inverse power law) is a result of stochastic and regular condensation, while the tail is dominated by the interaction of regular condensation, turbulent and convective transport, and sedimentation. A comparison with experimental data shows that these calculated spectra well describe the typical features of the crystal spectra measured in cirrus clouds and addition of the large fraction may cause multimodality. These results illustrate applicability of this theory of stochastic condensation for crystalline clouds.

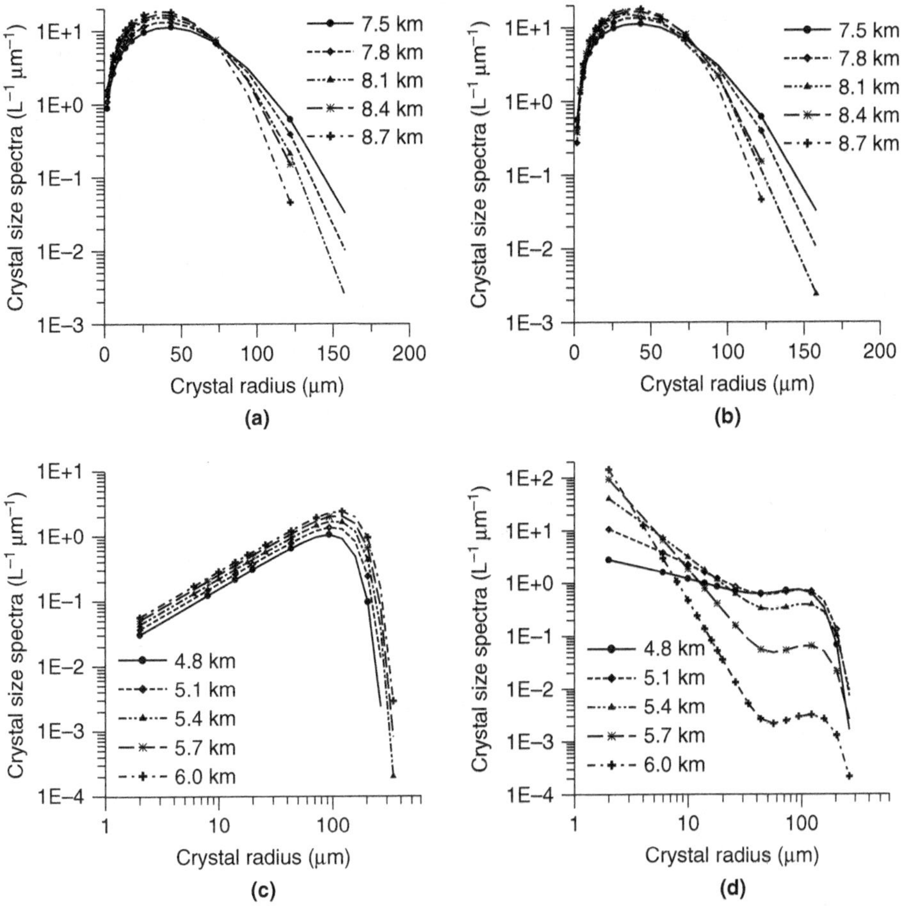

Figure 13.13. Crystal size spectra calculated for the baseline case at the heights indicated in the legends. (a) supersaturated layer using (Eqn. 13.7.4) for the tail; (b) supersaturated layer using (Eqn. 13.7.5) for the composite spectra with multiplication by $r^p\Phi(r)$; (c) subsaturated layer using (Eqn. 13.7.4) for the tail; (d) subsaturated layer using (Eqn. 13.7.5) for the composite spectra with multiplication by $r^p\Phi(r)$. The temperature falls by ~8 °C from the lower to the upper height of each layer. From Khvorostyanov and Curry (2008b). *J. Atmos. Sci.*, 65, © American Meteorological Society. Used with permission.

13.12. Derivation of the Generalized Stochastic Kinetic Equations from the Fokker–Planck Equation

13.12.1. Chapman–Kolmogorov and Fokker–Planck Equations

The kinetic equations of condensation and coagulation represent particular cases of the more general *Fokker–Planck equation*, which itself is a particular case of a more general *Chapman–Kolmogorov equation* (the history of these equations, detailed formulations, and bibliography can be found in Risken, 1989; van Kampen, 1992; and Rodean, 1996). In this section, we consider derivation of the generalized kinetic equation of condensation and coagulation that is a modification of the Chapman–Kolmogorov equation adapted for the size spectra, and derivation of the Fokker–Planck kinetic equations. This outlines the general approach to derivation of stochastic kinetic equations for the size spectra, allows us to establish relationships among various versions of the kinetic equations and possible ways to generalize them.

When random processes occur in a medium, the kinetic equation contains diffusion terms. In many fundamental derivations and applications, diffusion processes are described using the Fokker–Planck equation for the distribution function by various variables, which form a phase space, and the Fokker–Planck equation is a model of diffusion in this phase space (Rodean, 1996; Lifshitz and Pitaevskii, 1997). The diffusion terms are different for various processes and can be derived from this equation, as, for example, for Brownian diffusion (Levich, 1969), the mixing of gases and plasma theory (Lifshitz and Pitaevskii, 1997), and turbulent diffusion (Obukhov, 1959; van Kampen, 1992; Rodean, 1996). Here we apply the Fokker–Planck equation to derive a generalized stochastic kinetic equation for the cloud drops and crystals.

The Fokker–Planck equation can be derived from the more general Chapman–Kolmogorov equation, which is an integral equation that describes the evolution of the probability $P_p(y_1, t)$ of the system to be in a state y_1 at a time t if the probability of transition from y_1 to y_2 is $W(y_2, y_1)$ and the probability of a reverse transition from y_2 to y_1 is $W(y_1, y_2)$. The Chapman–Kolmogorov equation is applied to Markov processes, which is defined by two conditions (e.g., van Kampen, 1992; Rodean, 1996): 1) the present state y_1 can have its origin in any of the probable states y_2 in the past; 2) a process starting at time t_1 from a value y_1 reaches a value y_3 at time t_3 by passing any intermediate possible values of y_2 at time t_2. The Chapman–Kolmogorov equation can be written as (van Kampen, 1992; Rodean, 1996)

$$\frac{\partial P_p(y_1,t)}{\partial t} = \int [W(y_1|y_2)\, P_p(y_2,t) - W(y_2|y_1) P_p(y_1,t)] dy_2. \tag{13.12.1a}$$

We consider a spatially inhomogeneous cloud where the size distribution function, $f(\vec{x},r,t)$, depends on time, the coordinate vector with three components, $\vec{x}(x_\alpha)$, with $\alpha = 1, 2, 3$, and the particle radius r. Then, in the spirit of the Fokker–Planck equation, (x_α, r) are the parameters of a medium and form a 4-dimensional phase space (x_α, r). We consider the evolution of f due to the random processes that cause transitions $(\vec{x}-\vec{x}\,', r-r') \rightarrow (\vec{x}, r)$—that is, changes in (x_α, r) by some "jumps" (x_α', r'). The probability of transition, $W_{xr}(\vec{x}-\vec{x}\,', r-r' | \vec{x}\,', r')$, depends on the initial coordinates $(\vec{x}-\vec{x}\,', r-r')$ and on the values of the jumps $(\vec{x}\,', r')$.

The major assumptions usually made in derivations of the Fokker–Planck equation are: 1) the evolution of f is sufficiently slow—that is, its relaxation time τ_f is much greater than the characteristic time τ_{ran} of the random processes that cause these changes in x_α, r; 2) the probability $W_{xr}(\vec{x}-\vec{x}\,', r-r' | \vec{x}\,', r')$ decreases with increasing jumps $(x_\alpha - x_\alpha', r - r')$—i.e., the probability of a large jump $(x_\alpha - x_\alpha', r - r')$ is small, and the evolution of f proceeds via small jumps in (x_α, r) space. There are two equivalent types of derivation of the Fokker–Planck equation used in the literature: from the differential and integral forms of the Chapman–Kolmogorov equation. First, we present a brief derivation based on the differential form and then will illustrate application of the integral form.

Strictly speaking, the Chapman–Kolmogorov equation is defined for the probabilities of the states, but we will slightly modify it for the size distribution functions. If we consider evolution of the drop or crystal size spectra, then we can define the state y_1 as $(\vec{x},r)$ in the four-dimensional space. Since the size distribution function $f(\vec{x},r)$ is defined as proportional to the probability to find a particle in the radii range $(r, r + dr)$, and f is normalized to the particle concentration N, the probability $P_p(y_1, t)$ of this state is determined as, $P_p(\vec{x},r) = f(\vec{x},r)/N$. The state y_2 is $(\vec{x}-\vec{x}',r-r')$, its probability is $P_p(y_2,t) = f(\vec{x}-\vec{x}',r-r',t)/N$, and the probability $W(y_2|y_1)$ of the jumps $(\vec{x}',r')$ can be defined as the product of the probability of transition $W_{xr}(\vec{x}-\vec{x}',r-r'|\vec{x}',r')$ by the probability of the initial state. Since the probability of this state $(\vec{x}',r')$ is $\sim f(\vec{x}',r',t)$, we can write $W(y_2|y_1) = W_{xr}(\vec{x}-\vec{x}',r-r'|\vec{x}',r')f(\vec{x}',r',t)$. This probability determines the gain of the particles. The probability of transition from the state $(\vec{x}',r')$ into the state $(\vec{x},r)$ is $W(y_1|y_2) = W_{xr}(\vec{x},r|\vec{x}',r')f(\vec{x}',r',t)$, as the probability of the initial state is $f(\vec{x}',r',t)/N$. For Markov's process, the probabilities of a transition from an initial state $(\vec{x}-\vec{x}',r-r')$ to a final state $(\vec{x},r)$ should not depend on the intermediate states, therefore integration over $(d\vec{x}'dr')$ should be performed.

Then, we can write an analog of the Chapman–Kolmogorov (Eqn. 13.12.1a) in this 4-dimensional phase space as the stochastic kinetic equation for the size spectra,

$$\frac{\partial f(\vec{x},r,t)}{\partial t} = \int W_{xr}(\vec{x}-\vec{x}',r-r'|\vec{x}',r')f(\vec{x}-\vec{x}',r-r',t)f(\vec{x}',r',t)d\vec{x}'dr'$$
$$-f(\vec{x},r,t)\int W_{xr}(\vec{x},r|\vec{x}',r')f(\vec{x}',r',t)d\vec{x}'dr'. \tag{13.12.1b}$$

Here, the first and second terms on the right-hand side represent the gain and loss of the particles per unit time in an element of the 4D phase space, $W_{xr}(\vec{x}-\vec{x}',r-r'|\vec{x}',r')$ determines the probability of transition between the points $(\vec{x}-\vec{x}',r-r')$ and $(\vec{x},r)$ with the jumps $(\vec{x}',r')$ in the 4D phase space (denoted by the subscript "xr"), and W_{xr}, according to assumption (2) earlier, sufficiently rapidly decreases with increasing distance between these points.

The differences of this generalized stochastic kinetic (Eqn. 13.12.1b) with the original Chapman–Kolmogorov (CK) (Eqn. 13.12.1a) are the following. The CK (Eqn. 13.12.1a) is linear by the probabilities of the states, and the kinetic (Eqn. 13.12.1b) is nonlinear by the function f. The probabilities of transition $W(y_2|y_1)$ in the CK equation are defined per unit time, [s^{-1}], while the probabilities W_{xr} of the transition in the kinetic equation are defined per unit time and per unit volume, [$cm^3 s^{-1}$]—i.e., their dimension coincides with the dimension of the coagulation kernels in the Smoluchowski-Müller equation.

The generalized stochastic kinetic (Eqn. 13.12.1b), which is a modification of the Chapman–Kolmogorov equation, has a simple form and clear physical meaning and contains in a generalized form a wealth of the physical situations with particle transformations. It is shown in this section that the integral kinetic (Eqn. 13.12.1b) can serve for derivation of the differential Fokker–Planck equation in various forms—in particular, of various stochastic condensation equations considered earlier in this chapter. This is achieved by specifying various transition probabilities W_{xr}. Besides, one can see that the kinetic (Eqn. 13.12.1b) describes evolution of the size spectra due to the balance between the gain and loss in a random process and therefore can be viewed as a generalization of the Smoluchowski–Müller coagulation equation considered in Chapter 5.

We begin derivation of the Fokker–Planck equation for the size spectra with the generalized stochastic kinetic (Eqn. 13.12.1b). With the assumption of small jumps in the 4-dimensional space (x_α, r), the subintegral expressions in (Eqn. 13.12.1b) can be expanded into the Taylor series by x'_α, r' from the zeroth through the second order:

$$
\begin{aligned}
&W_{xr}(\vec{x}-\vec{x}',r-r'\,|\,\vec{x}',r')f(\vec{x}-\vec{x}',r-r',t) \approx W_{xr}(\vec{x},r\,|\,\vec{x}',r')f(\vec{x},r,t) \\
&-x'_\alpha \frac{\partial}{\partial x_\alpha}\left[W_{xr}(\vec{x},r\,|\,\vec{x}',r')f(\vec{x},r,t)\right] - r'\frac{\partial}{\partial r}\left[W_{xr}(\vec{x},r\,|\,\vec{x}',r')f(\vec{x},r,t)\right] \\
&+\frac{1}{2}x'_\alpha x'_\beta \frac{\partial^2}{\partial x_\alpha \partial x_\beta}\left[W_{xr}(\vec{x},r\,|\,\vec{x}',r')f(\vec{x},r,t)\right] + \frac{1}{2}x'_\alpha r' \frac{\partial^2}{\partial x_\alpha \partial r}\left[W_{xr}(\vec{x},r\,|\,\vec{x}',r')f(\vec{x},r,t)\right] \\
&+\frac{1}{2}r' x'_\alpha \frac{\partial^2}{\partial r \partial x_\alpha}\left[W_{xr}(\vec{x},r\,|\,\vec{x}',r')f(\vec{x},r,t)\right] + \frac{1}{2}r'^2 \frac{\partial^2}{\partial r^2}\left[W_{xr}(\vec{x},r\,|\,\vec{x}',r')f(\vec{x},r,t)\right].
\end{aligned}
\tag{13.12.2}
$$

The summation rule is assumed over the doubled Greek subscripts, α, $\beta = 1, 2, 3$. Substituting the expansion (Eqn. 13.12.2) into (Eqn. 13.12.1b) and rearranging the terms yields a differential kinetic equation:

$$
\begin{aligned}
\frac{\partial f(\vec{x},r,t)}{\partial t} &= -\frac{\partial}{\partial x_\alpha}(v_\alpha f) - \frac{\partial}{\partial r}(\dot{r}f) + \frac{\partial^2}{\partial x_\alpha \partial x_\beta}(k_{\alpha\beta} f) \\
&+\frac{\partial^2}{\partial x_\alpha \partial r}(k_{\alpha r} f) + \frac{\partial^2}{\partial r \partial x_\alpha}(k_{r\alpha} f) + \frac{\partial^2}{\partial r^2}(k_{rr} f).
\end{aligned}
\tag{13.12.3}
$$

The arguments $(\vec{x},r)$ of f and of coefficients on the right-hand side are omitted for brevity. We introduced here the velocities v_α in the 3D coordinate space and the velocity $\dot{r}$ in the radii space:

$$v_\alpha = \int x'_\alpha W_{xr}(\vec{x},r\,|\,\vec{x}',r')d\vec{x}'dr' \tag{13.12.4}$$

$$\dot{r}(\vec{x},r) = \int r' W_{xr}(\vec{x},r\,|\,\vec{x}',r')d\vec{x}'dr', \tag{13.12.5}$$

and the turbulence coefficient tensor

$$k_{\alpha\beta}(\vec{x},r) = \frac{1}{2}\int x'_\alpha x'_\beta W_{xr}(\vec{x},r\,|\,\vec{x}',r')d\vec{x}'dr', \tag{13.12.6}$$

$$k_{\alpha r}(\vec{x},r) = \frac{1}{2}\int x'_\alpha r' W_{xr}(\vec{x},r\,|\,\vec{x}',r')d\vec{x}'dr', \tag{13.12.7}$$

$$k_{rr}(\vec{x},r) = \frac{1}{2}\int r'^2 W_{xr}(\vec{x},r\,|\,\vec{x}',r')d\vec{x}'dr'. \tag{13.12.8}$$

The components $k_{\alpha\beta}$ in (Eqn. 13.12.6) are the components of the spatial turbulent diffusion tensor, $k_{\alpha r}$ in (Eqn. 13.12.7) are the components of the mixed coordinate-radius tensor, and k_{rr} in (Eqn. 13.12.8) are the components of the diffusion coefficient in the space of radii. The kinetic (Eqn. 13.12.3) with velocities (Eqns. 13.12.4)–(13.12.5) and turbulent diffusion tensors (Eqns. 13.12.6)–(13.12.8) includes many various models of condensation or deposition in a turbulent atmosphere. The specifics and physics of every model are determined by the form of W_{xr}, as will be shown in the following text.

If particles are described by the mass distribution function, and probabilities W describe transitions between the states $(\vec{x}-\vec{x}', m-m')$ and $(\vec{x}', m')$ in the 4D phase space, then the Chapman–Kolmogorov equation can be written as in (Eqn. 13.12.3) replacing $r \to m$, and so the expansion into the power series and substitution into the Chapman–Kolmogorov equation yields

$$\frac{\partial f(\vec{x},m,t)}{\partial t} = -\frac{\partial}{\partial x_\alpha}(v_\alpha f) - \frac{\partial}{\partial r}(\dot{m} f) + \frac{\partial^2}{\partial x_\alpha \partial x_\beta}(k_{\alpha\beta} f)$$
$$+\frac{\partial^2}{\partial x_\alpha \partial m}(k_{\alpha m} f) + \frac{\partial^2}{\partial m \partial x_\alpha}(k_{m\alpha} f) + \frac{\partial^2}{\partial m^2}(k_{mm} f), \tag{13.12.9}$$

where the velocity in the m-space and components of the tensor of turbulence $k_{\alpha m}$, $k_{m\alpha}$, and k_{mm} are defined as in (Eqns. 13.12.7) and (13.12.8) with replacement of radii r with masses m.

If particles are described by the distribution function by the area $A_r = r^2$, a characteristic of the Maxwellian growth rate, and probabilities W_{xA} describe transitions between the states, then the Chapman–Kolmogorov equation can be written as in (Eqn. 13.12.3) replacing $r \to A_r$, and the Fokker–Planck equation becomes

$$\frac{\partial f(\vec{x},A_r,t)}{\partial t} = -\frac{\partial}{\partial x_\alpha}(v_\alpha f) - \frac{\partial}{\partial A_r}(\dot{A}_r f) + \frac{\partial^2}{\partial x_\alpha \partial x_\beta}(k_{\alpha\beta} f)$$
$$+\frac{\partial^2}{\partial x_\alpha \partial A_r}(k_{\alpha A} f) + \frac{\partial^2}{\partial A_r \partial x_\alpha}(k_{A\alpha} f) + \frac{\partial^2}{\partial A_r^2}(k_{AA} f). \tag{13.12.9a}$$

One can see that the form of (Eqns. 13.12.3), (13.12.9), and (13.12.9a) is similar to the previous stochastic kinetic equations of the Fokker-Planck type with drift and diffusion reviewed in Section 13.1.2 and to the equations derived in this chapter based on statistical and semi-empirical theories of turbulence. The two features of these equations can be noted. On one hand, (Eqns. 13.12.3), (13.12.9), and (13.12.9a) generalize the equations from the previous sections and contain other stochastic kinetic equations derived previously as particular cases. On the other hand, the generalized velocities and turbulence tensors (Eqns. 13.12.4)–(13.12.8) remain the formal expressions until the models of turbulence and particle growth are specified, and a concrete form of the probabilities W_{xr} is chosen. The meaning of the Fokker–Planck equations derived earlier is illustrated by considering the particular cases in the next sections.

13.12.2. Spatially Homogeneous Cloud

For a spatially homogeneous cloud, the probability of transition depends only on radii as $W_{rr}(r|r')$, the spatial derivatives are zero, $\partial/\partial x_\alpha = 0$, and the Fokker–Planck (Eqn. 13.12.3) is simplified as

$$\frac{\partial f(r,t)}{\partial t} = -\frac{\partial}{\partial r}(\dot{r} f) + \frac{\partial^2}{\partial r^2}(D_r f), \tag{13.12.10}$$

where $\dot{r}$ represents the drift velocity and the diffusion coefficient D_r is determined by (Eqn. 13.12.8):

$$D_r(r) = \frac{1}{2}\int r'^2 W_r(r\,|\,r')dr'. \tag{13.12.11}$$

If particles are described by the distribution function by the area $A_r = r^2$, then (Eqn. 13.12.9a) for the spatially homogeneous case is

$$\frac{\partial f(A_r,t)}{\partial t} = -\frac{\partial}{\partial A_r}(\dot{A}_r f) + \frac{\partial^2}{\partial A_r^2}(D_A f), \tag{13.12.11a}$$

where the diffusion coefficient D_A is determined by (Eqn. 13.12.8):

$$D_A(A_r) = \frac{1}{2}\int A_r'^2 W_A(A_r \,|\, A_r')dA_r'. \tag{13.12.11b}$$

Equation (13.12.11a) has the form of the Fokker–Planck equation for $f(A_r)$ introduced by McGraw and Liu (2006). The drift velocity $\dot{A}_r$ was defined by McGraw and Liu via the mean supersaturation $\overline{s}$, and the diffusion coefficient D_A was expressed via the supersaturation variance σ_s and the inverse time γ_s of supersaturation relaxation. This can be written in our notations as

$$\dot{A}_r = 2b_{con}\overline{s}, \qquad D_A = (2b_{con}\sigma_s)^2/\gamma_s, \tag{13.12.11c}$$

where $b_{con} = D_v\rho_{ws}/(\Gamma_1\rho_a)$.

Thus, the equation of McGraw and Liu is a particular case of (Eqn. 13.12.9a), which in turn is a particular case of the generalized kinetic (Eqn. 13.12.1a).

Introducing the density flux $j(r,t)$ in the space of radii:

$$j(r,t) = \dot{r}f(r,t) - \frac{\partial}{\partial r}\left[D_r(r)f(r,t)\right], \tag{13.12.12}$$

(Eqn. 13.12.10) can be rewritten as the continuity equation

$$\frac{\partial f}{\partial t} = -\frac{\partial j}{\partial r}. \tag{13.12.13}$$

Equations (13.12.10) and (13.12.13) account for the diffusion term in the space of radii and generalize an original kinetic (Eqn. 5.5.5) (Chapter 5) for the regular condensation introduced by Buikov (1961, 1963). The explicit form of the diffusion coefficient D_r can be found either by specifying the corresponding probability $W_r(r|r')$ or by expressing D_r via f and $\dot{r}$.

The second way of deriving D_r is briefly illustrated here for the case of Brownian diffusion. The concentration $c(x)$ of the particles with radius r moving along the x-axis in an external potential field $U(x)$ with the mean velocity v is described by the diffusion equation

$$\frac{\partial c}{\partial t} = -\frac{\partial j_B}{\partial x}, \qquad j_B = vc - D_B\frac{\partial c}{\partial x}, \tag{13.12.14a}$$

where D_B is the coefficient of Brownian diffusion defined in Chapter 10, and j_B is the density flux. For the steady state, $\partial c/\partial t = 0$, and $j_B = 0$, hence

$$D_B = vc/(\partial c/\partial x). \tag{13.12.14b}$$

It is known that the equilibrium concentration $c(x)$ in the potential field $U(x)$ is described by the Boltzmann distribution, $c(x) = c_0\exp(-U/kT)$, where T is the temperature and k is the Boltzmann constant, so that $\partial c/\partial x = c(x)(-\partial U/\partial x)/kT$. But $-\partial U/\partial x = F$, where F is the force acting on the particle in the field U. If a particle is moving in a viscous medium without acceleration, this force F is equal to the viscosity force, which for small particles is determined by the Stokes law, $F = 6\pi\eta rv$, where η is the viscosity of the medium (see Chapter 12). Using these relations, we obtain $(\partial c/\partial x)/c(x) = F/kT = 6\pi\eta v/kT$. Substituting this expression into (Eqn. 13.12.14b), we obtain

$$D_B = \frac{kT}{6\pi\eta r}, \tag{13.12.14c}$$

which is Einstein's equation for the Brownian diffusion coefficient. Thus, the diffusion coefficient D_B in the coordinate x-space was determined by expressing it via the particles' distribution function $c(x)$ and their velocity v.

We use a similar method to determine the diffusion coefficient in the space of radii. For the steady-state conditions, $\partial f/\partial t = 0$, the condition $j(r,t) = 0$ allows us to express the diffusion coefficient $D_r(r)$ via $\dot{r}$ and f as earlier. In our case, we have from (Eqn. 13.12.12) and the condition $j = 0$:

$$j(r,t) = \dot{r}f(r,t) - f(r,t)\frac{dD_r(r)}{dr} - D_r(r)\frac{df(r)}{dr} = 0, \tag{13.12.15}$$

which can be rewritten as a linear inhomogeneous differential equation for D_r:

$$\frac{dD_r(r)}{dr} + D_r(r)\frac{1}{f(r,t)}\frac{df(r)}{dr} - \dot{r} = 0. \tag{13.12.16}$$

A solution for D_r is

$$D_r(r) = \frac{1}{f(r)}\int_{r_0}^{r} \dot{r}f(r)\,dr + C\frac{f(r_0)}{f(r)}, \tag{13.12.17}$$

where r_0 is some reference point and C is the constant of integration. It follows from (Eqn. 13.12.17) that $C = D_r(r_0) \equiv D_{r0}$ for $r = r_0$, thus

$$D_r(r) = \frac{1}{f(r)}\int_{r_0}^{r} \dot{r}f(r)\,dr + D_{r0}\frac{f(r_0)}{f(r)}. \tag{13.12.18}$$

This equation expresses D_r via $f(r)$, $\dot{r}$, and a reference value D_{r0}. A specific form of D_r depends on the type of cloud process and the resulting model of the size spectrum $f(r)$. For example, if $f(r)$ is the gamma distribution with the index p and modal radius r_m,

$$f(r) = c_N r^p \exp(-pr/r_m), \tag{13.12.19}$$

and the growth rate is described by Maxwell's law (Eqn. 13.2.2), substitution of these equations into (Eqn. 13.12.18) yields

$$D_r(r) = b_{con}s\left(\frac{pr}{r_m}\right)^{-p}\exp\left(\frac{pr}{r_m}\right)\left[\Gamma\left(p,\frac{pr_0}{r_m}\right) - \Gamma\left(p,\frac{pr}{r_m}\right)\right] + D_{r0}\left(\frac{r_0}{r}\right)^p \exp\left(-\frac{p(r_0 - r)}{r_m}\right), \tag{13.12.20}$$

where $\Gamma(x,a)$ is the incomplete gamma function. When the mean supersaturation $s = 0$, the regular growth rate is absent, the first term on the right-hand side vanishes, and the second term in (Eqn. 13.12.20) shows that

$$D_r(r) = D_{r0}\frac{f(r_0)}{f(r)}, \tag{13.12.21}$$

in agreement with (Eqn. 13.12.18). That is, D_r is in antiphase with $f(r)$: $D_r(r)$ decreases with increasing r and $r < r_m$, reaches a minimum at the point of modal radius r_m, and then increases with r. If the size spectra are parameterized in the form of the inverse power laws, $f \sim r^{-\beta}$, then it follows from (Eqn. 13.12.21) that $D_r \sim (r/r_0)^{\beta}$ at $s = 0$—i.e., the diffusion coefficient in the radii space resembles Taylor's form $k_x \sim (x/x_0)^{\gamma}$ of the horizontal eddy diffusion coefficient used in calculations of the dispersion of pollutants (Rodean, 1996; Monin and Yaglom, 2007a,b).

13.12.3. Spatially Inhomogeneous Cloud

Note first that in this case the jumps x_α' and r' may be not completely independent, and r' can be related to x_α' in one or another way via the supersaturation equation, and the type of this dependence determines the type of the stochastic kinetic equation. As discussed in Section 13.2, r' is analogous to the mixing length l_r' in the space of radii and can be evaluated by integration over time of the corresponding velocity $\dot{r}'$ of radius fluctuation. The choice of $\dot{r}'$ is crucial; it determines the type of final equation, and this is the major point where the differences among the theories occur. Consider now two models of r' that lead to different forms of the kinetic equations.

In the first model, assume that $\dot{r}'$ is of the Maxwellian type (Eqn. 14.3.2), thus the radius fluctuation velocity $\dot{r}'$ is

$$\dot{r}' = b_{con}s'/r, \tag{13.12.22}$$

where $b_{con} = D_v\rho_{vs}/(\Gamma_p\rho_w)$, and s' is the supersaturation fluctuation—that is, the growth rate in fluctuations is the same as in the regular updrafts, the Maxwellian diffusion regime, $\dot{r}' \sim 1/r$. A model for r' in the radii space follows from (Eqn. 13.12.22). Its integration by time yields

$$r' = \int_0^t \dot{r}'(t')dt' = \int_0^t \frac{b_{con}s'(t')}{r(t')}dt' \approx \frac{b_{con}}{r}\int_0^t s'(t')dt'$$
$$= \frac{B_c}{r}\delta_{\beta 3}x_\beta, \qquad B_c = b_{con}c_{1w}\tau_f = D_v B_{con}, \tag{13.12.23}$$

where B_{con} was defined in (Eqn. 13.1.4), and we accounted for the fact that the fluctuations are small, $r' << r$, and r can be approximately removed out of the integral sign over the short duration of the fluctuation. Substituting r' from (Eqn. 13.12.23) into (Eqn. 13.12.7) for $k_{\alpha r}$ into (Eqn. 13.12.8) for k_{rr}, we obtain

$$k_{\alpha r} = \frac{B_c}{r}\delta_{\beta 3}\frac{1}{2}\int x_\alpha' x_\beta' W(\vec{x},r\,|\,\vec{x}',r')d\vec{x}'dr' = \frac{D_v B_{con}}{r}\delta_{\beta 3}k_{\alpha\beta}, \tag{13.12.24}$$

$$k_{rr} = \left(\frac{B_c}{r}\right)^2\delta_{\alpha 3}\delta_{\beta 3}\frac{1}{2}\int x_\alpha' x_\beta' W(\vec{x},r\,|\,\vec{x}',r')d\vec{x}'dr' = \left(\frac{D_v B_{con}}{r}\right)^2\delta_{\alpha 3}\delta_{\beta 3}k_{\alpha\beta}, \tag{13.12.25}$$

that is, the diffusion coefficient in the radii space is inversely proportional to the squared radius.

Substituting these expressions into the Fokker–Planck kinetic (Eqn. 13.12.3), and accounting for sedimentation and the regular growth rate (Eqn. 13.2.2), we obtain

$$\frac{\partial f}{\partial t}+\frac{\partial}{\partial x_\alpha}\left[\left(u_\alpha - v_r(r)\delta_{\alpha 3}\right)f\right]+\frac{\partial}{\partial r}\left(\frac{b_{con}s}{r}f\right)$$
$$=\left(\frac{\partial}{\partial x_\alpha}+\delta_{\alpha 3}B_{con}\frac{\partial}{\partial r}\frac{D_v}{r}\right)\left(\frac{\partial}{\partial x_\beta}+\delta_{\beta 3}B_{con}\frac{\partial}{\partial r}\frac{D_v}{r}\right)k_{\alpha\beta}f. \qquad (13.12.26)$$

This is a simplified version of the kinetic (Eqn. 13.1.3) by Levin-Sedunov (1966), Sedunov (1974), Voloshchuk-Sedunov (1977), and Smirnov-Nadeykina (1986) considered in Section 13.1, which is written similar to (Eqn. 13.1.3), as if D_v may depend on r. For the particular case with spatial homogeneity, $\partial/\partial x_\alpha = 0$, it becomes

$$\frac{\partial f}{\partial t}+\frac{\partial}{\partial r}\left(\frac{b_{con}s}{r}f\right)=\left(\frac{\partial}{\partial r}\frac{D_v B_{con}}{r}\right)^2 k_{\alpha\beta}f. \qquad (13.12.26a)$$

This equation for $f(r)$ can be transformed to the equation for the size distribution $f(A_r)$ by the area $A_r = r^2$ considered in Section 13.12.1. Using the conservation law, $f(r)dr = f(A_r)dA_r$, and the relation $(\partial/\partial r) = (2r)(\partial/\partial A_r)$, (Eqn. 13.12.26a) can be transformed to a form similar to the equation by McGraw and Liu (2006) with the drift and diffusion in radii space but with a different diffusion coefficient.

Thus, we began with the general Fokker–Planck equation derived from the generalized integral kinetic (Eqn. 13.12.1b) of the Chapman–Kolmogorov type, and using the Maxwellian model both for the regular growth rate and its fluctuations, arrived at the differential stochastic kinetic equations of the Fokker–Planck type. The analytical solutions to these equations have the Gaussian form in the non-steady case, or the form of the Weibull distribution in the quasi-steady state (see Section 13.1).

Another model was considered by Khvorostyanov and Curry (1999c, 2008c) and was described in Section 13.2, when the growth rate in the regular updrafts is Maxwellian diffusion (Eqn. 14.3.2), but $\dot{r}'$ in fluctuations is parameterized as

$$\dot{r}' = \phi_r s'_w, \qquad \phi_r = \frac{D_v \rho_{ws}}{\Gamma \rho_w \bar{r}}. \qquad (13.12.27)$$

that is, $\dot{r}'$ does not depend on r, and the growth rate in fluctuations proceed in the quasi-kinetic regime. This relation was justified in Section 13.2 by the hypothesized proportionality $s' \sim r$ for small timescales in fluctuations. Then r' in the radii space can be evaluated as

$$r' = \int_0^t \dot{r}'(t')dt' = \phi_r \int_0^t s'_w(t')dt' = G_s \int_0^t u_3'(t')dt' = G_s \delta_{\alpha 3} x'_\alpha, \qquad (13.12.28)$$

where $\delta_{\alpha\beta}$ is the Kronecker symbol and G_s is a dimensionless parameter introduced in (Eqn. 13.3.12). Substituting this r' into (Eqn. 13.12.7) for $k_{\alpha r}$ and in (Eqn. 13.12.8) for k_{rr}, we obtain

$$k_{\alpha r} = G_s \delta_{\beta 3}\frac{1}{2}\int x'_\alpha x'_\beta W(\vec{x}, r \mid \vec{x}', r')d\vec{x}'dr' = G_s \delta_{\beta 3} k_{\alpha\beta}, \qquad (13.12.29)$$

$$k_{rr} = G_s^2 \delta_{\alpha 3}\delta_{\beta 3}\frac{1}{2}\int x'_\alpha x'_\beta W(\vec{x}, r \mid \vec{x}', r')d\vec{x}'dr' = G_s^2 \delta_{\alpha 3}\delta_{\beta 3} k_{\alpha\beta}. \qquad (13.12.30)$$

Substituting further these coefficients and $\dot{r}$ from (Eqn. 13.2.2) into the general (Eqn. 13.12.3), we obtain

$$\frac{\partial f}{\partial t}+\frac{\partial}{\partial x_\alpha}\left[\left(u_\alpha - v_r(r)\delta_{\alpha 3}\right)f\right]+\frac{\partial}{\partial r}\left(\frac{b_{con}s}{r}f\right)$$
$$=\left(\frac{\partial}{\partial x_\alpha}+G_s\delta_{\alpha 3}\frac{\partial}{\partial r}\right)\left(\frac{\partial}{\partial x_\beta}+G_s\delta_{\beta 3}\frac{\partial}{\partial r}\right)k_{\alpha\beta}f, \quad (13.12.31)$$

where we added the term with the terminal velocity $v_r(r)$ and accounted for the symmetry of the turbulence tensor, $k_{\alpha\beta} = k_{\beta\alpha}$. Equation (13.12.31) is a simplified version of the Khvorostyanov–Curry (1999c, 2008c) equation for stochastic condensation derived in Sections 13.2–13.4, which is valid for the low-frequency turbulence—i.e., with assumption on the quasi-steady supersaturation fluctuations. Again, it is derived here from the Fokker–Planck equation with some specification of the drift velocity and diffusion coefficient. The analytical solutions to this equation in the form of generalized gamma distributions were found in Sections 13.6 and 13.7.

The other models can be considered in the same way based on the Fokker–Planck (Eqn. 13.12.3) with various drift velocities and diffusion coefficients. These derivations and their comparison to the previous works show equivalence of the Fokker–Planck approach to the methods of the semi-empirical theory of turbulence with ensemble averaging. Thus, the generalized kinetic (Eqn. 13.12.1b) derived from the Chapman–Kolmogorov equation) and the Fokker–Planck equations may serve as a general basis for derivation of the stochastic condensation equations, but some additional arguments are needed every time to specify the drift velocity and the diffusion coefficient in the space of radii, mass, area, or equivalent variables.

Appendix A.13. Derivation and Solution of the Kummer Equation in Section 13.6.3

Here, we present details of the derivation and solution of the Kummer equation for the interpolation function $\Phi(r)$ because of the potential use of this technique for bulk cloud models. It is convenient to rewrite the composite spectrum (Eqn. 13.6.11) with auxiliary function $\eta(r)$:

$$f_s(r) = \eta(r)\Phi(r), \tag{A.13.1}$$

$$\eta(r) = c_{1,2}r^p \exp(-\beta_s r). \tag{A.13.2}$$

The derivatives f_s' and f_s'' are then expressed as

$$f_s' = (p/r - \beta_s)\eta\Phi + \eta\Phi', \tag{A.13.3}$$

$$f_s'' = (p/r - \beta_s)^2\eta\Phi - (p/r^2)\eta\Phi + 2(p/r - \beta_s)\eta\Phi' + \eta\Phi''. \tag{A.13.4}$$

Substituting (Eqns. A.13.3) and (A.13.4) into (Eqn. 13.5.11), applying (Eqns. 13.6.4) and (13.6.7), and deleting $\eta(r)$ in the resulting equation yields

$$G_s^2 k\left[\left(\frac{p}{r} - \beta\right)^2 \Phi - \frac{G_s^2 kp}{r^2}\Phi + 2\left(\frac{p}{r} - \beta_s\right)\Phi' + \Phi''\right]$$
$$+\left(2G_s\alpha_s k - \frac{b_{con}s}{r}\right)\left[\left(\frac{p}{r} - \beta_s\right)\Phi + \Phi'\right]\left[\alpha_s^2 k(1-\mu_s^2) + \frac{b_{con}s}{r^2}\right] = 0.Z \tag{A.13.5}$$

Now collecting the terms with various derivatives of Φ, we obtain the equation

$$A_1\frac{d^2\Phi}{dr^2} + A_2\frac{d\Phi}{dr} + A_3\Phi = 0, \tag{A.13.6}$$

where

$$A_1 = G_s^2 k, \tag{A.13.7}$$

$$A_2 = \left(\frac{2G_s^2 kp}{r} - 2G_s^2 k\beta_s + 2G_s\alpha_s k - \frac{b_{con}s}{r}\right), \tag{A.13.8}$$

$$A_3 = \frac{G_s^2 kp^2}{r^2} - \frac{2G_s^2 kp\beta_s}{r} + G_s^2 k\beta_s^2$$
$$-\frac{G_s^2 kp}{r^2} + \frac{2G_s\alpha_s kp}{r} - \frac{b_{con}sp}{r^2} - 2G_s\alpha_s k\beta_s$$
$$+\frac{bs\beta_s}{r} + \alpha_s^2 k(1-\mu_s^2) + \frac{b_{con}s}{r^2}. \tag{A.13.9}$$

These coefficients are significantly simplified after substituting the expression $b_{con}s = G_s{}^2kp$ that follows from (Eqn. 13.6.4) (with $c_{nn} = c_n = 1$) and (Eqn. 13.6.10) for β_s. First, for $\beta_{s1} = (\alpha_s/G_s)(1 - \mu_s)$, we have

$$A_3 = -\frac{2G_s kp\alpha_s}{r} + \frac{2G_s kp\alpha_s\mu_s}{r} + k\alpha_s^2 - 2k\alpha_s^2\mu_s + k\alpha_s^2\mu_s^2 + \frac{2G_s kp\alpha_s}{r}$$
$$-2k\alpha_s^2 + 2k\alpha_s^2\mu_s + \frac{G_s k\alpha_s p}{r} - \frac{G_s k\alpha_s p\mu_s}{r} + k\alpha_s^2 - k\alpha_s^2\mu_s^2. \tag{A.13.10}$$

This equation contains twelve terms, but one can see that the following nine terms are mutually cancelled: 1st and 6th; 4th and 8th; 3rd, 7th, and 11th; 5th and 12th. The remaining three terms yield

$$A_3 = \frac{G_s k\alpha_s p}{r}(1+\mu_s). \tag{A.13.11}$$

The coefficient A_2 is also simplified after substitution of $b_{con}s$ via p and β_{s1}:

$$\begin{aligned} A_2 &= \frac{2G_s^2 kp}{r} - 2G_s^2 k\frac{\alpha_s}{G_s}(1-\mu_s) + 2G_s\alpha_s k - \frac{G_s^2 kp}{r} \\ &= G_s^2 k\left(\frac{p}{r} + \frac{2\alpha_s\mu_s}{G_s}\right). \end{aligned} \tag{A.13.12}$$

Substituting these reduced terms A_2, A_3 into (Eqn. A.13.6), dividing by G_s^2k, and multiplying by r, we obtain the equation for Φ:

$$r\Phi'' + \left(p + \frac{2\alpha_s\mu_s}{G_s}r\right)\Phi' + \frac{\alpha_s p}{G_s}(1+\mu_s) = 0. \tag{A.13.13}$$

A solution to this equation is the confluent hypergeometric (Kummer) function $F(a, b; x)$ (Landau and Lifshitz, 1958, v.3; Gradshteyn and Ryzhik, 1994):

$$\Phi(r) = c_1 F\left(\frac{p}{2\mu_s}(\mu_s+1), p; -\frac{2\alpha_s\mu_s}{G_s}r\right), \tag{A.13.14}$$

where c_1 is the normalization constant. Then, the full solution for f_s is

$$\begin{aligned} f_{s,1}(r) &= c_1 r^p \exp\left(-\frac{\alpha_s}{G_s}(1-\mu_s)\right), \\ &\times F\left(\frac{p}{2\mu_s}(\mu_s+1), p; -\frac{2\alpha_s\mu_s}{G_s}r\right). \end{aligned} \tag{A.13.15}$$

For the second slope β_{s2}, substituting $\beta_{s2} = (\alpha_s/G_s)(1+\mu)$ from (Eqn. 13.6.10) and $b_{con}s = G_s^2kp$ into (Eqn. A.13.9) for A_3 yields

$$\begin{aligned} A_3 = &-\frac{2G_s kp\alpha_s}{r} - \frac{2G_s kp\alpha_s\mu_s}{r} + k\alpha_s^2 + 2k\alpha_s^2\mu_s + k\alpha_s^2\mu_s^2 + \frac{2G_s kp\alpha_s}{r} \\ &- 2k\alpha_s^2 - 2k\alpha_s^2\mu_s + \frac{G_s k\alpha_s p}{r} + \frac{G_s k\alpha_s p\mu_s}{r} + k\alpha_s^2 - k\alpha_s^2\mu_s^2. \end{aligned} \tag{A.13.16}$$

Here also, of the total twelve terms, the following terms are mutually cancelled: 1st and 6th; 3rd, 7th, and 11th; 4th and 8th; 5th and 12th. The remaining three terms yield

$$A_3 = \frac{G_s k\alpha_s p}{r}(1-\mu_s). \tag{A.13.17}$$

The coefficient A_2 in (Eqn. A.13.8) is simplified after substitution of $b_{con}s$ via p and β_{s2}:

$$A_2 = \frac{2G_s^2kp}{r} - 2G_s^2k\frac{\alpha_s}{G}(1+\mu_s) + 2G_s\alpha_sk - \frac{G_s^2kp}{r}$$
$$= G_s^2k\left(\frac{p}{r} + \frac{2\alpha_s\mu_s}{G_s}\right). \quad \text{(A.13.18)}$$

Again substituting reduced A_2, A_3 into (Eqn. A.13.6), dividing by G_s^2k, and multiplying by r, we obtain the equation for Φ:

$$r\Phi'' + \left(p - \frac{2\alpha_s\mu_s}{G_s}r\right)\Phi' + \frac{\alpha_sp}{G_s}(1-\mu_s) = 0. \quad \text{(A.13.19)}$$

A solution to this equation is the confluent hypergeometric function F (Gradshteyn and Ryzhik, 1994):

$$\Phi(r) = c_2F\left(\frac{p}{2\mu_s}(\mu_s - 1), p; +\frac{2\alpha_s\mu_s}{G_s}r\right). \quad \text{(A.13.20)}$$

The full solution for f_s is

$$f_{s,2}(r) = c_2r^p\exp\left(-\frac{\alpha_s}{G_s}(1+\mu_s)\right),$$
$$\times F\left(\frac{p}{2\mu_s}(\mu_s - 1), p; +\frac{2\alpha_s\mu_s}{G_s}r\right). \quad \text{(A.13.21)}$$

The confluent hypergeometric functions $F(a, b; x)$ have the following properties. The function $F(a,b;x)$ can be defined as a power series:

$$F(a,b,x) = 1 + \frac{a}{b}\frac{x}{1!} + \frac{a(a+1)}{b(b+1)}\frac{x^2}{2!} + \ldots \quad \text{(A.13.22)}$$

This series is convenient for coding, rather rapidly converges, and usually does not require a large number of terms. A few useful limits follow from this definition.

$$F(0,b;x) = 1, \qquad \lim_{x\to 0}F(a,b,x) = 1, \quad F(a,a,x) = e^x. \quad \text{(A.13.23)}$$

For analysis of the asymptotic behavior of $F(a, b; x)$ at large x, another representation of this function is used as a contour integral over a contour C in the complex plain of the complex variable x (Landau and Lifshitz, 1958, v.3; Gradshteyn and Ryzhik, 1994):

$$F(a,b;x) = \frac{\Gamma(b)}{2\pi i}\int_C e^t(t-x)^{-a}t^{a-b}\,dt, \quad \text{(A.13.24)}$$

where $\Gamma(b)$ is the Euler gamma function and $i = \sqrt{-1}$. The subintegral function in (Eqn. A.13.24) has two singular points, at $x = t$ and $x = 0$. Deforming contour C in the complex plane in such a way that it transforms into two contours each going in a counterclockwise direction around these two singular points, dividing subintegral function by $(-x)^{-a}$, expanding it by the powers t/x, and integrating all terms, the following expression is obtained:

$$F(a,b;x) = \frac{\Gamma(b)}{\Gamma(b-a)}(-x)^{-a}\Upsilon(a,\, a-b+1;\, -x) + \frac{\Gamma(b)}{\Gamma(a)}e^{x}x^{a-b}\Upsilon(b-a,\, 1-a;\, -x), \tag{A.13.25}$$

where the function $\Upsilon(a,\, b;\, x)$ is a power series of the inverse powers of x:

$$\Upsilon(a,\, b;\, x) = 1 + \frac{ab}{1!x} + \frac{a(a+1)b(b+1)}{2!x^2} + \ldots \tag{A.13.26}$$

It is easy to see that $\Upsilon(a,b;x) \to 1$ when $x \to \infty$. Using this limit in (Eqn. A.13.25), we see that the first term tends to zero at $x \to \infty$ and $a > 0$, and the second term has the following limit

$$\lim_{x\to\infty} F(a,b,x) \approx \frac{\Gamma(b)}{\Gamma(a)}e^{x}x^{a-b} \,. \tag{A.13.27}$$

This limit is convenient for analysis of the asymptotic behavior of the size spectra at large radii.

The spectral ν-th moment of the distribution functions (e.g., number density, mean radius, liquid or ice water content) can be evaluated using the relations (Landau and Lifshitz, 1958, v.3; Gradshteyn and Ryzhik, 1994)

$$M^{(\nu)} = \int_0^\infty x^{\nu} e^{-\lambda x} F(a,b,\beta x)\,dx = \Gamma(\nu+1)\lambda^{-(\nu+1)} F\left(a, \nu+1, b; \frac{\beta}{\lambda}\right), \tag{A.13.28}$$

where $F(\alpha, \beta, \gamma; x)$ is the Gauss hypergeometric function. Using this equation, we obtain the following n-th moment for the 1st solution (Eqn. A.13.15) or (Eqn. 13.6.15):

$$M^{(n)} = \int_0^\infty r^n f_s(r)\,dr = c_1\Gamma(p+n+1)\left[\frac{\alpha_s}{G_s}(1-\mu_s)\right]^{-(p+n+1)} \times F\left(\frac{p}{2\mu_s}(1+\mu_s), p+n+1, p; -\frac{2\mu_s}{1-\mu_s}\right), \tag{A.13.29}$$

and the n-th moment for the 2nd solution (Eqn. A.13.21 or 13.6.18)

$$M^{(n)} = \int_0^\infty r^n f_s(r)\,dr = c_2\Gamma(p+n+1)\left[\frac{\alpha_s}{G_s}(1+\mu_s)\right]^{-(p+n+1)} \times F\left(\frac{p}{2\mu_s}(\mu_s - 1), p+n+1, p, \frac{2\mu_s}{1+\mu_s}\right). \tag{A.13.30}$$

In particular, for $n = 0$, $M^{(0)} = N$ we obtain normalization constants c_1, c_2. For the solution of (Eqn. A.13.15) or (Eqn. 13.6.15), we obtain

$$c_1 = N\left\{\Gamma(p+1)\left[\frac{\alpha_s}{G_s}(1-\mu_s)\right]^{-(p+1)} F\left(\frac{p}{2\mu_s}(1+\mu_s),\ p+1,\ p,\ -\frac{2\mu_s}{1-\mu_s}\right)\right\}^{-1}. \quad \text{(A.13.31)}$$

For (Eqn. A.13.21) or (Eqn. 13.6.18), we obtain

$$c_2 = N\left\{\Gamma(p+1)\left[\frac{\alpha_s}{G_s}(1+\mu_s)\right]^{-(p+1)} F\left(\frac{p}{2\mu_s}(\mu_s - 1), p+1, p, \frac{2\mu_s}{1+\mu_s}\right)\right\}^{-1}. \quad \text{(A.13.32)}$$

These relations can be used for evaluation of the LWC (IWC), number densities, mean radii, extinction coefficients, and other moments of the solutions given in Section 13.6.

Appendix B.13 Solutions of Kinetic Equations of Section 13.7, Taking into Account Diffusional Growth in the Tail

It is convenient to solve (Eqn. 13.7.1) by rewriting it as

$$\frac{d}{dr}\left(\frac{f_s}{r}\right) = [\xi_1 + \xi_2(r)]f_s, \tag{B.13.1}$$

where

$$\xi_1 = \frac{1}{b_{con}s}\left(\alpha_s^2 k + \alpha_s \frac{dk}{dz} + k\frac{d\alpha_s}{dz} - \alpha_s w - \frac{dw}{dz} - \sigma_{col}\right), \tag{B.13.2a}$$

$$\xi_2 = \alpha_s v(r)/b_{con}s. \tag{B.13.3a}$$

Introducing a new variable $\varphi_s = f_s(r)/r$, (Eqn. B.13.1) can be rewritten as

$$\frac{d\varphi_s}{dr} = [\xi_1 + \xi_2(r)]r\varphi_s. \tag{B.13.4}$$

Integration from some r_1 to r yields

$$\varphi_s(r) = \varphi_s(r_1)\exp(J_{s1} + J_{s2}), \tag{B.13.5}$$

where

$$J_{s1} = \int_{r_1}^{r} \xi_1 r\,dr = \frac{(r^2 - r_1^2)\xi_1}{2}, \tag{B.13.6}$$

$$J_{s2} = \int_{r_1}^{r} \xi_2(r) r\,dr = \frac{\xi_2(r)r^2 - \xi_2(r_1)r_1^2}{B_v + 2}. \tag{B.13.7}$$

When evaluating J_{s2}, we assume that the fall velocity $v(r)$ at this size range can be approximated by the power law (Eqn. 13.5.9) $v(r) = A_v r^{B_v}$ with constant A_v, B_v. Substituting these integrals into (Eqn. B.13.5), and again using $f_s = \varphi_s/r$, we obtain the solution for the larger portion of the small fraction $r < r_0$:

$$f_s(r) = \frac{r}{r_1} f_s(r_1)\exp[-\beta_{s2}(r)r^2 - \beta_{s2}(r_1)r_1^2], \tag{B.13.8}$$

where the slope β_{s2} is

$$\begin{aligned}\beta_{s2} &= -\left(\frac{\xi_1}{2} + \frac{\xi_2}{B_v + 2}\right) \\ &= \frac{1}{b_{con}s}\left(\frac{\alpha_s w + dw/dz + \sigma_{col} - \alpha_s^2 k - \alpha_s k' - k\alpha_s'}{2} - \frac{\alpha_s v(r)}{B_v + 2}\right),\end{aligned} \tag{B.13.9}$$

and the primes here denote derivatives by z. This is the solution for the tail of the small fraction expressed via its value at $r = r_1$.

14

Analytical Solutions to the Stochastic Kinetic Equation for Precipitating Clouds

14.1. Introduction

As described in Chapter 2, the most often used parameterizations of the size spectra $f_l(r)$ of precipitating cloud particles (e.g., rain, snow, graupel, hail) are the Marshall–Palmer (1948, hereafter MP) and Gunn–Marshall (1958) exponential distributions,

$$f_l(r) = N_0 \exp(-\beta_l r), \tag{14.1.1}$$

where β_l is the slope and N_0 is the intercept. These parameterizations are widely used and routinely incorporated into bulk cloud models and remote sensing techniques. More recently, three-parameter gamma distributions have been suggested as a better alternative for rain and snow size spectra (Ulbrich, 1983; Willis, 1984):

$$f_l(r) = c_N r^{p_l} \exp(-\beta_l r), \tag{14.1.2}$$

where r is the particle radius, p_l is the index of the gamma distribution (shape parameter), positive or negative, and c_N is the coefficient determined from the normalization to the concentration or mass density. The exponential MP distribution (Eqn. 14.1.1) is a particular case of (Eqn. 14.1.2) with $p_l = 0$, whereby (Eqn. 14.1.2) is a more general form that allows additional degrees of freedom. Equations (14.1.1) and (14.1.2) are often formulated in terms of diameters or maximum crystal lengths D, $f_l(D) \sim \exp(-\Lambda D)$, with the slope $\Lambda = \beta_l/2$.

Since the beginning of the 1960s, numerous theoretical works have been devoted to explanations of the observed size spectra of cloud particles by means of various analytical and numerical solutions of the kinetic equation of coagulation. Here, we briefly review some of these studies that illustrate the evolution of various models and concepts and create a basis for derivation of the analytical solutions to the kinetic equation similar to (Eqns. 14.1.1) and (14.1.2). Such solutions are obtained in this chapter, and the indices p_l and slopes β_l are expressed via the cloud and atmospheric properties.

Studies of the solutions to the coagulation equations have proceeded along four lines: a) analytical solutions; b) numerical solutions; c) refinements of the collision-coalescent efficiencies; and d) generalizations of the coagulation kinetic equation itself.

Earlier theoretical studies of the size spectra of precipitating particles were directed toward explaining the exponential shape of the MP spectra and the evolution of its parameters. Golovin (1963), Scott (1968), Lushnikov (1973, 1974), Voloshchuk and Sedunov (1975), Lushnikov and Smirnov (1975), Srivastava and Passarelli (1980), Lushnikov and Piskunov (1982), and Voloshchuk (1984) obtained analytical solutions to the kinetic equation of condensation and coagulation with idealized assumptions including homogeneous kernels of the coagulation integral and non-Maxwellian models for the condensation growth rate. Srivastava (1971) hypothesized that a balance exists between the collision-coalescence and spontaneous breakup of raindrops, which leads to the exponential MP spectra, but the derived slopes were distinctly steeper than the observed spectra. Passarelli (1978a,b) assumed that snow spectra are described by the MP spectrum and found an analytical expression for the slopes via integral moments by solving the stochastic collection equation without accounting for breakup.

Although the coagulation equation is often called "the stochastic collection equation," accounting for the stochastic effects of turbulence on the coagulation process is often incomplete. Extending the theory of stochastic condensation described in Chapter 13 to the coagulation processes, Voloshchuk (1977) derived a stochastic coagulation equation that is a generalization of Equation (5.6.5) for the turbulent atmosphere. Applying the Reynolds procedure of decomposition into the mean and fluctuating parts as described in Section 13.2, averaging, and using the non-local parameterization of the semi-empirical theory of turbulence, Voloshchuk (1977) derived an equation for the volume distribution function $f_v(v)$:

$$\frac{\partial f_v(v,t)}{\partial t} + u_i \frac{\partial f_v(v,t)}{\partial x_i} = \frac{\partial}{\partial x_i} k_{ij} \frac{\partial f_v(v,t)}{\partial x_j}$$
$$+ \frac{1}{2}\int_0^v K(v',v-v') f_v(v-v') f_v(v')\,dv' - \int_0^\infty K(v,v') f_v(v) f_v(v')\,dv'$$
$$+ \frac{1}{2}\int_0^v \lambda_{ij} K(v',v-v') \frac{\partial f_v(v-v')}{\partial x_i}\frac{\partial f_v(v')}{\partial x_j}\,dv' - \int_0^\infty \lambda_{ij} K(v,v') \frac{\partial f_v(v)}{\partial x_i}\frac{\partial f_v(v')}{\partial x_j}\,dv'. \quad (14.1.3)$$

Here, $\lambda_{ij} = L_i L_j$, and L_i is the effective scale of turbulence in the i-th direction, u_i is the wind component in i-th direction, summation over doubled indices is assumed, $i = 1, 2, 3$, and all quantities are averaged over an ensemble of turbulent realizations. The first term and the last two terms on the right-hand side in (Eqn. 14.1.3) are the new terms that arise due to the effects of turbulence on the motion and collision-coalescence of the particles in the turbulent medium. Voloshchuk's equation is more complete and general than the original Smoluchowski–Müller (Eqn. 5.6.5); however, it is more complicated and its detailed analysis has not been performed until now. The subsequent description in this chapter is based on (Eqn. 5.6.5).

Numerical solutions of the stochastic collection (Eqn. 5.6.5) with more realistic kernels based on accurate finite difference approximations (e.g., Berry and Reinhardt, 1974a,b) or on the method of moments (e.g., Bleck, 1970; Tzivion et al., 1987) allowed more detailed studies of the temporal evolution of the size spectra of precipitating particles and created the benchmarks for testing the analytical solutions that were developing in parallel with the numerical methods. Verlinde et al. (1990) obtained an analytical solution to the collection growth equation for the original size spectra described by gamma distributions (Eqn. 14.1.2). The solution was rather complicated, expressed via the complete Gauss hypergeometric function ${}_2F_1(a,b,c;\,x)$ and the Euler gamma function and required simplifications for application in the models. Passarelli's model with exponential spectra was further developed and generalized by a number of authors (e.g., Mitchell, 1994, Mitchell et al., 1996).

By the end of 1970s, it became clear that collisional rather than spontaneous breakup may be more important in restricting drop growth and formation of the observed raindrop exponential spectra.

Srivastava (1978) formulated a simplified model of collisional breakup with a fixed constant number of fragments as a variable parameter and developed a parameterization for raindrop spectra in the form of a general exponential, but with time varying Λ and N_0. Low and List (1982a,b, hereafter LL82) developed a complex empirical parameterization of the fragment distribution function for collisional drop breakup. The parameterization of LL82 has been used in many numerical solutions of the stochastic coalescence/breakup equation to explain the mechanism of formation of the MP spectra and their slopes (e.g., Feingold et al., 1988; Hu and Srivastava, 1995; Brown, 1991, 1997; McFarquhar, 2004; Seifert, 2005).

These numerical solutions produced somewhat different equilibrium size spectra but with common features: a small size region from ~200 µm to ~2 mm consisting of several peaks with shallow troughs between them and a region beginning at ~2–2.5 mm and comprising the MP exponential tail. McFarquhar (2004) refined the LL82 equations and emphasized that measurement and sampling problems impose uncertainties on the solutions, motivating more detailed laboratory studies and improved parameterizations. Bott (1998, 2000, 2001) developed a precise flux method for a numerical solution of the collection equation that ensures mass conservation and can serve as the benchmark for the other methods.

The numerical studies focused on analyzing the positions of the peaks and values of the slopes but usually did not attempt to approximate the entire rain spectrum using gamma distributions. Parameterization of the large particle size spectrum in the form of the gamma distribution (Eqn. 14.1.2) has been undertaken by many empirical studies that were directed toward determination of the three parameters of the spectra, and in particular the index p_l. Ulbrich (1983) found a correlation between the type of the rain and the index p_l; there was $p_l < 0$ for orographic rain indicating broad spectra, and $0 < p_l < 2$ for thunderstorm rain indicating a narrower spectra. For widespread and stratiform rain, p_l was more variable but mostly positive. Willis (1984) found the best value to be $p_l \approx 2.5$ for raindrops from two hurricanes.

More recently, another type of p_l dependence was suggested: a Λ–p_l relation, whereby Λ was expressed as a quadratic polynomial of p_l or vice versa (e.g., Zhang et al., 2001, 2003a,b; Brandes et al., 2003). The validity of this parameterization was tested in direct simulations of convective rain with cloud models using the LL82 kernel (e.g., Seifert, 2005). The Λ–p_l relation allows reduction of the number of independent parameters in (Eqn. 14.1.1) to two, but the general dependence of the index p_l on the rain type described by Ulbrich (1983) is still unclear. A similar relation was suggested by Heymsfield (2003) for crystalline clouds.

Reviews of various methods and approaches in calculations of the large-size spectra can be found in the books of Voloshchuk and Sedunov (1975), Voloshchuk (1984), Rogers and Yau (1989), Young (1993), Pruppacher and Klett (1997), Straka (2009), and Cotton et al. (2011).

Previous research has revealed some fundamental properties of the size spectra of precipitating particles, and has shed some light on the mechanisms of their formation. However, direct application of these findings in cloud models and remote sensing retrievals encounters the following problems: owing to the complexity of collision/breakup kernels, mostly numerical solutions of the stochastic coalescence/breakup kinetic equation have been obtained for realistic representations of the gravitational kernel. The numerical solutions require small time steps, are rather time consuming, and do not provide simple analytical parameterizations for the indices and slopes of the exponential and gamma distributions that are needed in cloud and climate models and remote sensing retrievals. Many previous analytical solutions were expressed in terms of special functions, were too complicated for direct analysis, and did not provide the indices and slopes of the size spectra.

In Chapter 13, gamma distributions were derived for the small-size fraction as the solutions of the kinetic equation of stochastic condensation. The objective of this chapter is to obtain sufficiently simple yet physically-based analytical solutions of the stochastic kinetic equation for precipitating cloud particles and to explain observed variations in the size spectra that can be used to parameterize the size spectra for modeling and remote sensing applications (following mostly Khvorostyanov and Curry (2008b,c), with some generalizations). The kinetic equation of coagulation in approximation of continuous collection is derived from the general coagulation equation, and the corresponding assumptions are discussed. The basic stochastic kinetic equation is simplified and reduced to a differential equation in the ordinary derivatives of the second order by particle radius. Using some assumptions and simplifications, general analytical solutions are obtained in the form that is similar to the gamma distributions and exponential MP spectra but that contain several modifications that better describe the observed size spectra. Four particular cases are considered: the size ranges where fall speed is a linear function of particle size (intermediate regime of fall speeds) or is proportional to the square root of particle size (aerodynamic regime of fall speeds, see Chapter 12); conditions where coagulation growth is dominant; and subcloud layers with an absence of the small-size fraction. The solutions are illustrated with calculations for a crystalline cloud; a physical interpretation is given for these solutions, and an explanation is provided of the observed features of the size spectra that allow improvement of existing parameterizations. The process of autoconversion is considered and the corrections to the analytical size spectra that arise due to this process are described. Finally, it is shown that the coagulation equation represents a particular case of the integral Chapman–Kolmogorov equation, as was mentioned in Chapter 13.

14.2. Derivation of Kinetic Equations in Continuous Collection Approximation

As described in Chapter 5, the stochastic collection equation is written as follows:

$$\left(\frac{\partial f(m)}{\partial t}\right)_{col,ls} = I_{gain} - I_{loss}, \tag{14.2.1}$$

$$I_{gain} = \frac{1}{2}\int_0^m K(m-m',m')f(m-m')f(m')dm', \tag{14.2.2a}$$

$$I_{loss} = f(m)\int_0^\infty K(m,m')f(m')dm'. \tag{14.2.2b}$$

Here m, m', are the masses of the particles, thus

$$K(m,m') = \pi E_c(r,r')(r+r')^2 \,|\, v(r) - v(r') \,|\tag{14.2.2c}$$

is the coagulation gravitational kernel, r and r' denote the radii of the drops (crystals) corresponding to m and m', $E_c(r,r')$ is the collection efficiency, and $v(r)$ is the terminal velocity.

In the following sections, the continuous collection approximation is assumed for $(\partial f_l/\partial t)_{col,ls}$ to simplify the analytical solutions. In this approximation, only the collision-coalescence between particles of the different size fractions of the spectrum, $f_s(r)$ and $f_l(r)$, is considered—i.e., small particles are collected by large particles, as in most models with bulk microphysics. The continuous collection approximation is commonly used for evaluation of the accretion rate of the large-size fraction as in Kessler's (1969) and many subsequent works. If it appears in formulation of the kinetic equations, the corresponding term $(\partial f_l/\partial t)_{col,ls}$ is often written without derivation by analogy with Maxwellian growth, with the growth rate of individual particles $(dr/dt)_{coag}$ or $(dm/dt)_{coag}$ defined in the continuous collection approximation (e.g., Cotton et al., 2011; Pruppacher and Klett, 1997). This approach is intuitively clear, but its accuracy and relation to the full Smoluchowski stochastic collection equation is not clear. Therefore, the derivation of the continuous collection approximation from the integral Smoluchowski collection equation and its simplifications are given in this section.

We represent the size spectrum as the sum of the two size fractions,

$$f(m) = f_s(m)\theta(m_0 - m) + f_l(m)\theta(m - m_0), \tag{14.2.3}$$

where $\theta(x)$ is the Heaviside step function, $\theta(x) = 1$ at $x > 0$ and $\theta(x) = 0$ at $x < 0$. Substituting this decomposition into I_{gain}, we have the integral that contains in the integrand four combinations of f_s and f_l:

$$I_{gain} = \frac{1}{2}\int_0^{m_0} K(m-m',m')[f_s(m-m')f_s(m') + f_s(m-m')f_l(m') + f_l(m-m')f_s(m') + f_l(m-m')f_l(m')]dm'. \tag{14.2.4}$$

It is assumed in the continuous collection approximation that only the small-size fraction f_s interacts with the large-size fraction f_l. Thus, by definition of this approximation, the integrals of the 1st term in the square brackets with $f_s(m - m')f_s(m')$ and of the 4th term with $f_l(m - m')f_l(m')$ of (Eqn. 14.2.4) vanish.

Now we proceed in the continuous collection approximation and account only for the 2nd and 3rd terms—i.e., interactions between the fractions. In the 3rd term, we introduce the new variable $m'' = m - m'$ that accounts for the symmetry of the kernel, $K(m - m'', m'') = K(m'', m - m'')$, the integral of the 3rd term becomes equal to the integral of the 2nd term, the coefficient 1/2 vanishes after summation of the two integrals, and the sum yields

$$I_{gain} = \int_0^{m_0} K(m-m',m')f_l(m-m')f_s(m')dm'. \tag{14.2.5}$$

The upper limit is determined by definition of $f_s(m')$. Substituting the sum (Eqn. 14.2.3) into the I_{loss} yields four integrals:

$$I_{loss,ls} = f_l(m)\theta(m - m_0)\int_0^{m_0} K(m,m')f_s(m')dm', \tag{14.2.6a}$$

$$I_{loss,sl} = f_s(m)\theta(m_0 - m)\int_0^{\infty} K(m,m')f_l(m')dm', \tag{14.2.6b}$$

$$I_{loss,ss} = f_s(m)\theta(m_0 - m)\int_0^{m_0} K(m,m')f_s(m')dm', \tag{14.2.6c}$$

$$I_{loss,ll} = f_l(m)\theta(m - m_0)\int_0^{\infty} K(m,m')f_l(m')dm'. \tag{14.2.6d}$$

If we consider the size spectra of the large fraction $f_l(m)$ at $m > m_0$, the 2nd integral, $I_{loss,sl}$, vanishes. The 3rd integral, $I_{loss,ss}$, describes the process of self-collection of the small fraction called autoconversion that transfers the particles from f_s into f_l. This integral is also assumed to be zero in the continuous collection approximation; the autoconversion will be considered in Section 14.6. The 4th integral, $I_{loss,ll}$, represents the self-collection of the large fraction. This is also a kind of autoconversion and would transfer the particles from f_l into a larger size category, if it were present. In the continuous collection approximation, $I_{loss,ll}$ is set to zero. Thus, only the 1st integral, $I_{loss,ls}$, is non-zero at $m > m_0$ in this approximation, which will be considered further and is denoted as I_{loss}.

We assume, some average value of $E_c(r,r')$ = const = E_c. Also, in the continuous growth approximation

$$r >> r', \qquad v(r) >> v(r'), \qquad K(m,m') \approx K(m,0) = \pi E_c r^2 v(r). \tag{14.2.7}$$

Now we expand the subintegral expression in (Eqn. 14.2.5) for I_{gain} into a Taylor power series by the small parameter m' up to the term of the first order:

$$K(m-m',m')f_l(m-m') \approx K(m,0)f_l(m) - \frac{\partial}{\partial m}[K(m,0)f_l(m)]m'. \tag{14.2.8}$$

This is a truncated first-order approximation of a more general *Kramers–Moyal series* considered in Section 14.7. Substitution of this expression into (Eqn. 14.2.5) and using (Eqn. 14.2.7) yields

$$I_{gain} = K(m,0)f_l(m)\int_0^{m_0} f_s(m')dm' - \frac{\partial}{\partial m}[K(m,0)f_l(m)]\int_0^{m_0} m'f_s(m')dm'$$

$$= f_l(m)[\pi E_c r^2 v(r) N_s] - \frac{\partial}{\partial m}[\pi E_c r^2 v(r) q_{ls} f_l(m)] \tag{14.2.9}$$

$$= f_l(m)\sigma_{col,s} - \frac{\partial}{\partial m}[\pi E_c r^2 v(r) q_{ls} f_l(m)]. \tag{14.2.9a}$$

Here, we used the normalization of $f_s(m)$:

$$\int_0^{m_0} f_s(m')dm' = N_s, \qquad \text{and} \qquad \int_0^{m_0} m'f_s(m')dm' = q_{ls}, \tag{14.2.10}$$

where N_s and q_{ls} are the number concentration and LWC (or IWC) of the small-size fraction, and we introduced a collection frequency

$$\sigma_{col,s} = K(m,0)\int_0^{m_0} f_s(m')dm' = [\pi E_c r^2 v(r) N_s], \tag{14.2.10a}$$

which characterizes the frequency of collections of a large drop of mass m or radius r with all small drops. Since $m >> m'$, the kernel $K(m, m') \approx K(m,0)$ in (Eqn. 14.2.6a) for I_{loss}, and can be removed outside the integral. Then, incorporating (Eqn. 14.2.7), I_{loss} becomes

$$I_{loss} = f_l(m)[\pi E_c r^2 v(r)] \int_0^{m_0} f_s(m')\,dm'$$
$$= f_l(m)[\pi E_c r^2 v(r) N_s] = f_l(m)\sigma_{col,s}, \quad (14.2.11)$$

where we again use the normalization (Eqn. 14.2.10). Now, comparison of (Eqns. 14.2.9) and (14.2.11) shows that the 1st term in I_{gain} (Eqn. 14.2.9) is equal to I_{loss} (Eqn. 14.2.11); thus, they exactly cancel in (Eqn. 14.2.1) for $(\partial f_l/\partial t)_{col,ls}$. Substituting I_{gain} from (Eqn. 14.2.9) and I_{loss} from (Eqn. 14.2.11) into (Eqn. 14.2.1) and cancelling these two terms yields

$$\left(\frac{\partial f_l(m)}{\partial t}\right)_{col,ls} = -\frac{\partial}{\partial m}[\dot{m}_{coag} f_l(m)], \quad (14.2.12)$$

where we denote

$$\dot{m}_{coag} \equiv \left(\frac{dm}{dt}\right)_{coag} = \pi E_c r^2 v(r) q_{ls} \quad (14.2.13)$$

as the coagulation mass growth rate.

Thus, the collection rate $(\partial f_l/\partial t)_{col,ls}$ in the continuous collection approximation (Eqn. 14.2.12) is derived directly from the Smoluchowski stochastic collection equation, and so this represents a first-order approximation to the full kinetic equation. Equation (14.2.13) can be rewritten for $f_l(r)$ in terms of radii using the relation $f_l(m)dm = f_l(r)dr$. Then, substituting into (Eqn. 14.2.12) yields

$$\left(\frac{\partial f_l(r)}{\partial t}\right)_{col,ls} = -\frac{dm}{dr}\frac{\partial}{\partial m}[\dot{m}_{coag}\frac{dr}{dm} f_l(r)], \quad (14.2.14)$$

or

$$\left(\frac{\partial f_l(r)}{\partial t}\right)_{col,ls} = -\frac{\partial}{\partial r}[\dot{r}_{coag} f_l(r)] = -\chi_c \frac{\partial}{\partial r}[v(r) f_l], \quad (14.2.15)$$

where $\dot{r}_{coag}$ is the coagulation (accretion) radius growth rate, and so

$$\dot{r}_{coag} = \dot{m}_{coag}\frac{dr}{dm} = \chi_c v(r), \qquad \chi_c = \frac{E_c q_{ls}}{4\rho_w}. \quad (14.2.16)$$

Now the coagulation terms in (Eqn. 14.2.1) (the continuous collection approximation) can be written in a simplified differential form using the Heaviside step function $\theta(x)$ as

$$\left(\frac{\partial f_{s,l}(r)}{\partial r}\right)_{col} = -\chi_c \frac{\partial}{\partial r}[v(r) f_l]\theta(r - r_0) - \sigma_{col} f_s(r)\theta(r_0 - r). \quad (14.2.17)$$

The first term is the reduced gain I_{gain} for the large-size fraction. The second term is the reduced loss I_{loss} of the small size fraction, and the coefficient σ_{col} is defined in the following text from mass conservation.

If q_{ll} is the liquid (ice) water content of the large-size fraction, then (Eqns. 14.2.15) and (14.2.17) yield a simple parameterization of the accretion rate, dq_{ll}/dt, that can be used in bulk microphysical models. This expression is obtained by multiplying (Eqn. 14.2.15) by $(4/3)\pi\rho_w r^3$, and integrating over the radius:

$$\frac{dq_{ll}}{dt} = \frac{4\pi\rho_w}{3}\int_0^\infty r^3\left(\frac{\partial f_l}{\partial t}\right)_{col} dr = -\frac{4\pi\rho_w}{3}\int_0^\infty r^3 \frac{\partial}{\partial r}(\dot{r}_{coag} f_l)\, dr. \tag{14.2.18}$$

Here the lower limit is extended from r_0 to 0 noting that $f_l(r) = 0$ in this region. Integrating by parts and using (Eqn. 14.2.16), we obtain the gain rate $(dq_{ll}/dt)_{col}$ of the large-size fraction due to coagulation with the small-size fraction

$$\left(\frac{dq_{ll}}{dt}\right)_{col} = 4\pi\rho_w \int_0^\infty (\dot{r}_{coag} f_l) r^2\, dr = q_{ls}\left(\pi E_c \int_0^\infty r^2 v(r) f_l\, dr\right) = \sigma_{col} q_{ls}, \tag{14.2.19}$$

where σ_{col} is the collection frequency of the large-size fraction:

$$\sigma_{col} = \pi E_c \int_{r_0}^\infty r^2 v(r) f_l(r)\, dr. \tag{14.2.20}$$

Here, σ_{col} is the same collection frequency of the small-size fraction by the large-size fraction as defined and used in Chapter 13.

If $f_l(r_l)$ is a sufficiently narrow spectrum peaked near $r = r_{dm}$, then approximately

$$\sigma_{col} \approx \pi E_c r_{dm}^2 v(r_{dm}) \int_{r_0}^\infty f_l(r)\, dr = \pi E_c r_{dm}^2 v(r_{dm}) N_l, \tag{14.2.20a}$$

where N_l is the concentration of large drops. Now, using (Eqn. 14.2.17) here for the loss of small particles, we obtain a coagulation kinetic equation for the small fraction

$$\left(\frac{\partial f_s}{\partial t}\right)_{col} = -I_{loss} = -\sigma_{col} f_s, \qquad r < r_0. \tag{14.2.21}$$

Multiplying by $(4/3)\pi\rho_w r^3$ and integrating by radii, we obtain the loss rate $(dql_s/dt)_{col}$ of the small-size fraction due to coagulation with the large-size fraction:

$$\left(\frac{dq_{ls}}{dt}\right)_{col} = \frac{4\pi}{3}\rho_w \int_0^{r_0} r^3 \left(\frac{\partial f_s}{\partial t}\right)_{col} dr$$

$$= -\sigma_{col}\int_0^{r_0} \frac{4\pi\rho_w}{3} r^3 f_s\, dr = -\sigma_{col} q_{ls}, \qquad r < r_0. \tag{14.2.22}$$

Comparison of (Eqns. 14.2.19) and (14.2.22) shows that if σ_{col} is defined by (Eqn. 14.2.20) for both small- and large-size fractions, then $(dq_{ls}/dt)_{col} = -(dq_{ll}/dt)_{col}$; that is, the loss of the mass of small-size fraction particles is equal in magnitude to the gain of the mass of large-size fraction particles. Hence, $(dq_{ls}/dt)_{col} + (dq_{ll}/dt)_{col} = 0$, and mass is conserved in the continuous collection approximation with this definition of σ_{col}.

14.3. Basic Equations and Assumptions for the Large-Size Fraction

The continuous collection approximation is the simplest approximation for description of the coagulation/accretion processes. A more detailed consideration of analytical solutions to the kinetic equation for the large-size fraction requires some simplifications and assumptions. As in Chapter 13, we consider either pure liquid or crystalline clouds; a generalization for mixed phase clouds can be done in a similar way considering liquid and crystalline size spectra and interactions between them. The entire size spectrum of the drops or crystals consists of two size fractions, small, $f_s(r)$, and large, $f_l(r)$, with the boundary radius at $r = r_0$ between the two size fractions, where a minimum is usually observed in the size spectra composed of both fractions. Based on this minimum, the value $r_0 \sim 30$–$50\,\mu$m can be assumed for the liquid phase (see e.g., Cotton et al., 2011), and ~30–80 μm (or maximum dimension $D_0 \sim 60$–$160\,\mu$m) for the crystals as illustrated in Section 14.5 (see Figs. 14.3, 14.4). In the categorization of the condensed phase used in the bulk models parameterizations, the function $f_s(r)$ corresponds to cloud water for the liquid phase and cloud ice for the crystalline phase, and the function $f_l(r)$ corresponds to rain for the liquid phase and snow (or graupel or hail) for the crystalline phase.

The same assumptions as in Chapter 13 for the small-size fraction are made here for the large-size fraction size distribution function f_l: 1) horizontal homogeneity or horizontal averaging over some scale, so that horizontal derivatives are zero (the solutions can be easily generalized for the nonzero horizontal derivatives in parametric form); 2) quasi-steady state, so that $\partial f_l/\partial t = 0$. With these assumptions, the kinetic (Eqn. 13.5.6) from Chapter 13 for f_s can also be written for the size distribution function of the large-size fraction f_l of the drops or crystals:

$$\frac{\partial}{\partial z}\left[(w - v(r)) f_l\right] + \frac{\partial}{\partial r}\left(\dot{r}_{cond} f_l\right) = k\frac{\partial^2 f_l}{\partial z^2}$$
$$+ c_n k G_s \left(\frac{\partial^2 f_l}{\partial r \partial z} + \frac{\partial^2 f_l}{\partial z \partial r}\right) + c_{nn} k G_s^2 \frac{\partial^2 f_l}{\partial r^2} + \left(\frac{\partial f}{\partial t}\right)_{col,ls}$$
$$+ \left(\frac{\partial f}{\partial t}\right)_{col,g} + \left(\frac{\partial f}{\partial t}\right)_{col,l} + \left(\frac{\partial f}{\partial t}\right)_{br,g} + \left(\frac{\partial f}{\partial t}\right)_{br,l}. \tag{14.3.1}$$

Here z is height, r is the particle radius, k is the coefficient of turbulent diffusion, w is the vertical velocity, $v(r)$ is the particle fall velocity, and $\dot{r}_{cond}$ is the condensation (deposition) growth rate; a dimensionless parameter G_s and c_n, c_{nn}, are the coefficients arising from the nonconservative turbulence coefficients k_{ij}^n, k_{ij}^{nn} defined in Sections 13.3 and 13.8. The first three terms on the right-hand side describe the turbulent transport and effects of stochastic condensation. The fourth term, $(\partial f_l/\partial t)_{col,ls}$, is collection gain in the large-size fraction due to coagulation (accretion) with the small-size fraction, and the terms $(\partial f_l/\partial t)_{col,g}$, $(\partial f_l/\partial t)_{col,l}$, $(\partial f_l/\partial t)_{br,g}$, and $(\partial f_l/\partial t)_{br,l}$ denote respectively collection gain, collection loss, breakup gain, and breakup loss due to interactions within the large-size fraction alone. Such decomposition of the collection terms is similar to Srivastava (1978).

A simplified Maxwellian growth rate is assumed for condensation/deposition:

$$\dot{r}_{cond} = \frac{b_{con} s}{r}, \qquad b_{con} = \frac{D_v \rho_{vs}}{\Gamma_p \rho_w}, \tag{14.3.2}$$

where all the notations are the same as in Chapter 13.

The derivation of continuous collection approximation from the integral Smoluchowski collection equation and simplifications are given in Section 14.2. It was shown there that: 1) the continuous collection approximation is a first-order approximation by the mass of the small particle to the full stochastic coagulation equation; and 2) the Smoluchowski coagulation equation allows the term

$(\partial f_l/\partial t)_{col,ls}$ for the collection gain to be written analogously to the condensation term (the 2nd term on the left-hand side in (Eqn. 14.3.1)):

$$\left(\frac{\partial f_l(r)}{\partial t}\right)_{col,ls} = -\frac{\partial}{\partial r}[\dot{r}_{coag} f_l(r)] = -\chi_c \frac{\partial}{\partial r}[v(r) f_l], \tag{14.3.3}$$

where $\dot{r}_{coag}$ is the coagulation (accretion) radius growth rate

$$\dot{r}_{coag} = \dot{m}_{coag}\frac{dr}{dm} = \chi_c v(r), \qquad \chi_c = \frac{E_c q_{ls}}{4\rho_w}. \tag{14.3.4}$$

Now, the integral Smoluchowski coagulation equation can be written in the continuous collection approximation in a simplified differential form for both small and large fractions as (Eqn. 14.2.17), where σ_{col} is the same collection rate as introduced in Chapter 13 and used there in calculations of small-size spectra. It was derived in Section 14.2, (Eqn. 14.2.20). This form of σ_{col} ensures mass conservation—i.e., the mass loss of the small fraction is equal to the mass gain of the large fraction. We can also introduce the inverse quantity:

$$\tau_{col} = \sigma_{col}^{-1} = \left(\pi E_c \int_{r_0}^{\infty} r^2 v(r) f_l(r)\, dr\right)^{-1}, \tag{14.3.5}$$

which is the characteristic coagulation (accretion) time of the *e*-folding decrease in q_{ls} or the corresponding increase in q_{ll} if the other processes are absent. Then,

$$q_{ls}(t) = q_{ls}(0)\exp(-t/\tau_{col}), \tag{14.3.6}$$

$$q_{ll}(t) = q_{ll}(0) + q_{ls}(0)[1-\exp(-t/\tau_{col})]. \tag{14.3.7}$$

The sum $q_{ls}(t) + q_{ll}(t) = q_{ls}(0) + q_{ll}(0)$, which ensures mass conservation. It is interesting to compare this time scale τ_{col} with the supersaturation relaxation time τ_f discussed in Chapters 5 and 13, and to express it via the sum of f_s and f_l:

$$\tau_f = \left(4\pi D_v \int_0^{\infty} r[f_s(r) + f_l(r)]\, dr\right)^{-1} = [4\pi D_v (N_s \bar{r}_s + N_l \bar{r}_l)]^{-1}, \tag{14.3.8}$$

where $\bar{r}_s$, $\bar{r}_l$ are the mean radii of the small and large-size fraction, and N_l is the number density of the large-size fraction. This time determines the condensation (evaporation) growth rate. The term $N_s \bar{r}_s$ is usually greater than $N_l \bar{r}_l$, thus the time τ_f at condensation is determined mostly by the small particles. In the absence of small particles at evaporation, τ_f is determined by the large fraction. The coagulation (accretion) time τ_{col} is determined by the large particles, as follows from (Eqn. 14.3.5).

If the cloud water content of the large-size fraction is sufficiently small, the last four terms in (Eqn. 14.3.1)—$(\partial f_l/\partial t)_{col,g}$, $(\partial f_l/\partial t)_{col,l}$, $(\partial f_l/\partial t)_{br,g}$, $(\partial f_l/\partial t)_{br,l}$—for collection and breakup are also small and can be neglected, which is usually done in some bulk cloud models. This situation corresponds to sufficiently small concentrations N_l of the large drops, so that collisions among large particles are not frequent. With increasing N_l, water content and rain or snow intensity, the error of this approximation may increase and these terms should be accounted for; this is especially important for convective clouds. Srivastava (1982) and Feingold et al. (1988) evaluated the coalescence and breakup terms analytically, but the collision and breakup kernels were specified to be constant (independent of radii), and the solution was expressed via Bessel functions that are difficult to analyze analytically.

We use a parameterization of coalescence and breakup based on an assumption that is justified using the work by Hu and Srivastava (1995). The detailed analysis of various terms in the coalescence-breakup kinetic equation [the last four terms in (Eqn. 14.3.1)] performed in Hu and Srivastava (1995) showed that these terms are approximately proportional to each other over the major radii range, and are mostly mutually compensated near the equilibrium steady state. We hypothesize that these terms are proportional to the collection (accretion) gain $(\partial f_l/\partial t)_{col,ls}$ and can be roughly parameterized by expressing via $(\partial f_l/\partial t)_{col,ls}$ with some proportionality coefficients $c_{cg} > 0$, $c_{cl} < 0$, $c_{bg} > 0$, $c_{bl} < 0$ for the corresponding processes (the subscripts "*c*" and "*b*" mean collection and breakup; "*g*" and "*l*" mean gain and loss). These coefficients are, in general, functions of *r* but are proportional to each other. Therefore, the sum of $(\partial f_l/\partial t)_{col,ls}$ and all four last terms in (Eqn. 14.3.1) can be expressed via $(\partial f_l/\partial t)_{col,ls}$ as

$$\left(\frac{\partial f}{\partial t}\right)_{col,g} + \left(\frac{\partial f}{\partial t}\right)_{col,l} + \left(\frac{\partial f}{\partial t}\right)_{br,g} + \left(\frac{\partial f}{\partial t}\right)_{br,l} \approx c_{cb}\left(\frac{\partial f}{\partial t}\right)_{col,ls}, \qquad (14.3.9)$$

where $c_{cb} = 1 + c_{cg} + c_{cl} + c_{bg} + c_{bl}$. In a box model used in Hu and Srivastava (1995), equilibrium among these four terms is reached after some sufficient time so that they are mutually compensated, forming equilibrium spectra; thus, $c_{cb} \to 1$. In our case, the balance also includes the vertical mass gradient, and diffusion and accretion growth, so that there can be only approximate compensation among these four terms, and c_{cb} can slightly differ from 1. That is, the net effect of collection and breakup gain and loss is to change the accretion growth rate $\dot{r}_{coag}$ of the large particles described by (Eqn. 14.3.4)—i.e., the collection efficiency E_c. Then, we can introduce the correction c_{cb} into parameter E_c and solve the equation taking into account only $(\partial f_l/\partial t)_{col,ls}$, as in (Eqn. 14.3.3). This corresponds to the approximation adopted in many bulk cloud models. It will be shown in the following text that this simple parameterization yields the functional *r*-dependence of the collection growth rate that is in good agreement with more precise calculations in Hu and Srivastava (1995).

Note that the assumption (Eqn. 14.3.9) is not mandatory for analytical solutions obtained in the following text, we just use the fact, based on Hu and Srivastava (1995), that the sum of these four terms caused by interactions within only the large fraction can be much smaller (due to mutual compensation) than the term $(\partial f_l/\partial t)_{col,ls}$ caused by interactions between the small and large fractions. When the numerical models provide new information about the relative values of the terms in the coagulation-breakup equation, this assumption and analytical solutions can be modified accordingly.

The same closure as in Chapter 13 is chosen for the vertical gradient of the size spectra:

$$\frac{\partial f_l(r,z)}{\partial z} = \alpha_l(z)\varsigma_l(r)f_l(r,z). \qquad (14.3.10)$$

In the simplest model $\zeta_l(r) = 1$, α_l can be related from (Eqn. 14.3.10) to the relative gradient of the LWC (IWC) of the large-size fraction:

$$\alpha_l(z) = (1/q_{ll})(dq_{ll}/dz). \qquad (14.3.11)$$

Substituting (Eqn. 14.3.10) into (Eqn. 14.3.1) allows elimination of z derivatives; also using (Eqns. 14.3.2) and (14.3.4), and assuming now $\zeta_l(r) = 1$, we have a differential equation that is a function only of r:

$$c_{nn} k G_c^2 \frac{d^2 f_l}{dr^2} + \left[-\frac{b_{con} s}{r} - \chi_c v(r) \right] \frac{df_l}{dr}$$
$$+ \left[\alpha_l^2 k + \frac{d(\alpha_l k)}{dz} - \alpha_l (w - v) - \frac{dw}{dz} + \frac{b_{con} s}{r^2} - \chi_c \frac{dv(r)}{dr} \right] f_l = 0. \tag{14.3.12}$$

This equation allows analytical solutions that are found in the next section.

14.4. Solutions for the Large-Size Fraction Taking into Account Diffusion Growth and Coagulation

We seek analytical solutions for the general case and then for four particular cases that include the following: the size range where fall speed is a linear function of particle size; where fall speed is proportional to the square root of particle size (see Chapter 12); conditions where coagulation growth is dominant; and in the subcloud layer where the small-size fraction is absent as it has been evaporated in the upper subsaturated layers.

14.4.1. General Solution

The solutions in Chapter 13 indicate that stochastic condensation influences predominantly the small-size fraction. Therefore, it is reasonable to assume that the contribution of the stochastic diffusion terms are much smaller for the large-size fraction, and the shape of the spectra is determined mostly by the balance between regular growth/evaporation, vertical transport, collection, and sedimentation. Then, neglecting in (Eqn. 14.3.12) the terms with G_s and the effects of stochastic condensation, and using (Eqns. 14.2.10) and (14.2.11), the kinetic equation can be written for the quasi-steady state as

$$b_{con} s \frac{d}{dr}\left(\frac{f_l}{r} \right) + \chi_c \frac{d}{dr}(f_l v(r)) + [w - v(r) - \alpha_l k] \alpha_l f_l = 0 \tag{14.4.1}$$

(we neglect here for simplicity the terms with dw/dz, dk/dz, and $d\alpha_l/dz$ that can be introduced into the final equations).

Introducing a new variable $\varphi_l = f_l/r$, and solving (Eqn. 14.4.1) relative to φ_l, we obtain

$$\frac{1}{\varphi_l} \frac{d\varphi_l}{dr} = -\psi_l(r), \tag{14.4.2}$$

where

$$\psi_l(r) = \beta_{l0} + \frac{u_l(r)}{g_l(r)} + \frac{1}{g_l(r)} \frac{dg_l(r)}{dr}, \tag{14.4.3}$$

$$\beta_{l0} = -\frac{\alpha_l}{\chi_c}, \tag{14.4.4}$$

$$g_l(r) = b_{con} s + \chi_c r v(r) = r \dot{r}_{tot}(r), \tag{14.4.5}$$

$$u_l(r) = -\beta_{l0} b_{con} s + \alpha_l r(w - \alpha_l k), \tag{14.4.6}$$

$$\dot{r}_{tot}(r) = \dot{r}_{cond}(r) + \dot{r}_{coag}(r), \tag{14.4.6a}$$

where $\dot{r}_{tot}(r)$ is the total radius growth rate. The integral of (Eqn. 14.4.2) is

$$\varphi_l(r) = \varphi_l(r_0)\exp(-J_{l0}), \qquad J_{l0} = \int_{r_0}^{r} \psi_l(r')dr', \tag{14.4.7}$$

where r_0 is the left boundary of the large-size fraction. Substituting (Eqns. 14.4.3)–(14.4.6) into (Eqn. 14.4.7), and integrating, we obtain the solution for $f_l(r)$:

$$\begin{aligned} f_l(r) &= f_l(r_0)\frac{r}{r_0}\frac{g_l(r_0)}{g_l(r)}\exp\left[-\beta_{l0}(r - r_0) - J_{l1}\right] \\ &= f_l(r_0)\frac{\dot{r}_{tot}(r_0)}{\dot{r}_{tot}(r)}\exp\left[-\beta_{l0}(r - r_0) - J_{l1}\right], \end{aligned} \tag{14.4.8}$$

where

$$J_{l1} = \int_{r_0}^{r} \frac{u_l(r')}{g_l(r')}dr'. \tag{14.4.9}$$

This is the general solution to the kinetic (Eqn. 14.4.1) for the large-size fraction at $r > r_0$ taking into account the condensation and continuous collection. For application in bulk microphysical models, the integral J_{l1} can be evaluated numerically at any value of w, k, s, α_l, and q_{ls}; the fall speeds $v(r)$ for the drops accounting for nonsphericity and for various crystal habits can be evaluated as continuous functions of r as described in Chapter 12.

In certain particular cases, the integral J_{l1} can be obtained analytically and the solutions are simplified if the quasi-power law for terminal velocity

$$v(r) = A_v r^{B_v} \tag{14.4.10}$$

is applicable. Substituting (Eqns. 14.4.5), (14.4.6), and (14.4.10) into (Eqn. 14.4.9) and assuming that A_v = const and B_v = const over some interval of radii (r_1, r_2), we obtain

$$J_{l1} = I_1 + I_2, \tag{14.4.11}$$

$$I_1 = -\beta_{l0} b_{con} s \int_{r_1}^{r_2} \frac{dr}{b_{con} s + \chi_c A_v r^{B_v+1}}, \tag{14.4.12a}$$

$$I_2 = \alpha_l (w - \alpha_l k) \int_{r_1}^{r_2} \frac{rdr}{b_{con} s + \chi_c A_v r^{B_v+1}}. \tag{14.4.12b}$$

Tabulated analytical expressions for these integrals exist only for very limited values of B_v (Gradshteyn and Ryzhik, 1994). Therefore, we will illustrate the general solution for three particular cases: 1) when $v(r)$ is a linear function of r (intermediate regime of fall speeds of moderately large particles) 2) when the condensation rate is much smaller than the collection rate and can be neglected; and 3) when $v(r)$ is proportional to $r^{1/2}$ (aerodynamic regime of fall speeds, see Chapter 12).

14.4.2. Particular Case: Fall Speed as a Linear Function of Particle Size

The linear regime for particle terminal velocity with $B_v = 1$; that is, $v(r) = A_v r$, is valid in the intermediate range of drop radius 60 to 600 µm (e.g., Rogers and Yau, 1989), for spherical ice particles, and for some crystal habits in the region r = 90–300 µm (see Chapter 12; Sections 12.1 and 12.6, (Eqn. 12.1.7b). The integrals I_1 and I_2 for the linear function $v(r)$ are evaluated in Appendix A.14. The integrals have different expressions for the cases of condensation ($s > 0$) and evaporation ($s < 0$) and can be expressed using the new parameters:

$$R_l = \left(\frac{b_{con}\,|s|}{\chi_c A_v}\right)^{1/2} = r\left(\frac{|\dot{r}_{cond}(r)|}{\dot{r}_{coag}(r)}\right)^{1/2}, \qquad (14.4.13a)$$

$$U_l = \beta_{l0}\frac{(\alpha_l k - w)}{2A_v}, \qquad Q_l = \frac{\beta_0 R_l}{2}, \qquad (14.4.13b)$$

where $|s|$ and $|\dot{r}_{cond}|$ denote the absolute values of supersaturation and condensation growth rates.

Substituting expressions for I_1 and I_2 at $s > 0$ from Appendix A.14 into (Eqns. 14.4.12a) and (14.4.2b) and then into (Eqn. 14.4.8) yields the following $f_l(r)$ in the condensation layer:

$$f_l(r) = f_l(r_0)\frac{\dot{r}_{tot}(r_0)}{\dot{r}_{tot}(r)}\left[\frac{(r_0/R_l)^2+1}{(r/R_l)^2+1}\right]^{U_l}$$
$$\times\exp\left[-\beta_{l0}(r-r_0) - \beta_{l0}R_l\left(\operatorname{arctg}\frac{r_0}{R_l} - \operatorname{arctg}\frac{r}{R_l}\right)\right], \qquad s>0. \qquad (14.4.14)$$

Using expressions for I_1 and I_2 from Appendix A.14 for the evaporation layer ($s < 0$) yields two solutions for the two size regions, $r < R_l$ and $r > R_l$:

$$f_l(r) = f_l(r_0)\frac{\dot{r}_{tot}(r_0)}{\dot{r}_{tot}(r)}\left[\frac{(1-r_0/R_l)(1+r/R_l)}{(1-r/R_l)(1+r_0/R_l)}\right]^{Q_l}$$
$$\times\left[\frac{(r_0/R_l)^2+1}{(r/R_l)^2+1}\right]^{U_l}\exp\left[-\beta_{l0}(r-r_0)\right],$$

$$\text{for} \quad s<0, \quad r<R_l. \qquad (14.4.15)$$

$$f_l(r) = f_l(r_0)\frac{\dot{r}_{tot}(r_0)}{\dot{r}_{tot}(r)}\left[\frac{(r_0/R_l+1)(r/R_l-1)}{(r/R_l+1)(r_0/R_l-1)}\right]^{Q_l}$$
$$\times\left[\frac{(r_0/R_l)^2+1}{(r/R_l)^2+1}\right]^{U_l}\exp\left[-\beta_{l0}(r-r_0)\right],$$

$$\text{for} \quad s<0, \quad r>R_l. \qquad (14.4.16)$$

The equations for $s < 0$ are not applicable at one point $r = R_l$, since then $\dot{r}_{coag} = -\dot{r}_{cond}$ and the total growth rate $\dot{r}_{tot} = 0$, in contradiction with the assumption made when deriving (Eqns. 14.4.7) and (14.4.8).

The physical meaning of the parameter R_l and of the solutions (Eqns. 14.4.14)–(14.4.16) is clear from (Eqn. 14.4.13a): $(r/R_l)^2 = \dot{r}_{coag}/|\dot{r}_{cond}|$ is the ratio of the collection rate to the condensation/evaporation rate. The conditions $r < R_l$ or $r > R_l$ mean that the collection rate is smaller or greater than the condensation/evaporation rate, $\dot{r}_{coag} < |\dot{r}_{cond}|$ for $r < R_l$ or $\dot{r}_{coag} > |\dot{r}_{cond}|$ for $r > R_l$. The boundary condition $r = R_l = (b_{con}|s|/\chi_c A_v)^{1/2}$ can be estimated for the following parameters: $T \sim 0\,°\mathrm{C}$, $\rho_{vs} \approx 5\times10^{-6}\,\mathrm{g\,cm^{-3}}$, super- or subsaturation $s = \pm0.1$ ($\pm10\%$), $q_{ls} = 0.1\,\mathrm{g\,m^{-3}}$, $E_c = 0.5$, $A_v \approx 8\times10^3\,\mathrm{s^{-1}}$ (see Chapter 12; Sections 12.1 and 12.4). Then, $\chi_c \sim 10^{-8}$, $b_{con}|s| \sim 10^{-7}\,\mathrm{cm^2\,s^{-1}}$, and $R_l \sim 350\,\mu\mathrm{m}$. For $|s| = 0.2$ (or $\pm20\%$), or $|s| = 0.4$ (or $\pm40\%$), as can be in crystalline clouds or in evaporation layers, R_l increases to ~500 and 700 μm. With $|s| = 10^{-3}$ ($\pm$ 0.1%) and $q_{ls} = 1\,\mathrm{g\,m^{-3}}$, as can be in convective clouds, $R_l \sim$ 12–15 μm at $T = 0$ to 10 °C. For conditions of cirrus, $T = -40$ to -50 °C, $p = 200$ hPa, $q_{ls} = 20\,\mathrm{mg\,m^{-3}}$, $|s| = 0.10$ ($\pm$ 10%), the value $R_l \sim 200\,\mu\mathrm{m}$. These estimates can be used to assess the asymptotic solutions.

At $r >> R_l$ (i.e., when $\dot{r}_{coag} >> |\dot{r}_{cond}|$ and coagulation–accretion growth prevails), these solutions tend to the asymptotic, which is the same for both solutions (Eqns. 14.4.14)–(14.4.16) at sub- and supersaturations:

$$f_l(r) = f_l(r_0)\left(\frac{r}{r_0}\right)^{p_l} \exp[-\beta_0(r - r_0)], \tag{14.4.17a}$$

$$p_l = -(2U_l + 1) = \beta_{l0}\frac{(w - \alpha_l k)}{A_v} - 1. \tag{14.4.17b}$$

Thus, the analytical solutions for the large-size fraction in both the growth and evaporation layers are gamma distributions similar to those suggested and analyzed by Ulbrich (1983), Willis (1984), Zhang et al. (2001, 2003a,b), Brandes (2003), Heymsfield (2003), and others.

If in (Eqn. 14.4.17b) the parameter $(2U_l + 1) < 0$, then the index p_l is positive and the spectra represent the typical gamma distributions. If $(2U_l + 1) > 0$, then the index p_l is negative—i.e., these generalized gamma spectra represent a product of the inverse power law and the exponential Marshall–Palmer spectrum. Which functional type dominates depends on the combination of the parameters. The condition $w > \alpha_l k$ usually takes place in (Eqn. 14.4.17b), and the term with w mostly determines p_l. As discussed in Chapter 13, the effective w decreases with increasing scales of spatial-temporal averaging. At sufficiently large scales, w becomes small, p_l tends to zero, and the size spectrum (Eqn. 14.4.17a) tends to the Marshall–Palmer distribution. This prediction of our model coincides with the observations (e.g., Joss and Gori, 1978), and statistical theories of Marshall–Palmer spectra (e.g., Liu, 1993). However, the local spectra can be narrower than the Marshall–Palmer, and better described by gamma distributions like (Eqn. 14.4.17a), as discussed in Section 14.1. Which type of spectrum is preferable and what is the relation between β_l and p_l is still the subject of discussion in the literature (see Section 14.5).

14.4.3. Particular Case: Coagulation Growth Rate Much Greater than Diffusion Growth Rate

For sufficiently large r and small sub- or supersaturations, when $\dot{r}_{coag} >> \dot{r}_{cond}$, the condensation rate can be neglected. Then, using the quasi-power law form (Eqn. 14.4.10) for the fall speed $v(r)$, assuming A_v and B_v are approximately constant in some region of r and $B_v < 1$, the integral J_{l1} in (Eqn. 14.4.9) can be evaluated as

$$J_{l1} = \beta_{l0}(\alpha_l k - w)\int_{r_0}^{r}\frac{dr}{v(r)} = \beta_{l0}\left(\frac{w - \alpha_l k}{1 - B_v}\right)\left(\frac{r}{v(r)} - \frac{r_0}{v(r_0)}\right). \tag{14.4.18}$$

Introducing a size-dependent slope $\beta_l(r)$,

$$\beta_l(r) = \beta_{0l}\left[1 + \frac{w - \alpha_l k}{(1 - B_v)v(r)}\right], \tag{14.4.19}$$

and substituting (Eqns. 14.4.18) and (14.4.19) into (Eqn. 14.4.8), we obtain

$$f_l(r) = f_l(r_0)\frac{v(r_0)}{v(r)}\exp\{-[\beta_l(r)r - \beta_l(r_0)r_0]\}. \tag{14.4.20}$$

Assuming that $B_v \approx$ const in the power law $v(r) \approx A_v r^{B_v}$ for the considered region of r, the ratio $v(r_0)/v(r) \approx (r/r_0)^{-B_v}$ and (Eqn. 14.4.20) becomes

$$f(r) = f(r_0)(r/r_0)^{-B_v}\exp[-\beta(r)(r - r_0)]. \tag{14.4.21}$$

Thus, we obtain again a generalized gamma distribution with the negative index $p_l = -B_v$. This is again a product of the inverse power law and MP-type size distributions. Which dependence dominates depends on the combination of the parameters. In many cloud types, $v(r) >> w$ and $v(r) >> \alpha_l k$ for the large-size fraction, then $\beta(r) \approx \beta_{l0}$ and

$$f(r) = c_N(r/r_0)^{-B_v}\exp(-\beta_{l0}r). \tag{14.4.22}$$

Since $B_v \le 0.5$ for large particles, the spectrum tends to the MP exponent; however, there is also an algebraic term that is a function of r.

14.4.4. Particular Case: Aerodynamic Regime for the Fall Speed of Large Particles

As discussed in Chapter 12 (Sections 12.4 and 12.6), for sufficiently large radii, the terminal velocity can be approximated by the law $v(r) = A_v r^{1/2}$ where A_v also includes a height correction $(\rho_{a0}/\rho_a)^{1/2}$ with ρ_{a0} and ρ_a being the air densities at the surface and at a given height. This regime is approximately valid for large drops, some ice particles with shapes close to spherical (graupel, hail), and some other crystals' habits. The integrals I_1 and I_2 in (Eqns. 14.4.12a) and (14.4.12b) for this case, with $B_v = 1/2$, are evaluated in Appendix B.14. Substituting them into J_{l1} in (Eqn. 14.4.8) yields for the condensation layer with $s > 0$:

$$\begin{aligned} f_l(r) = f_l(r_0)&\frac{\dot{r}_{tot}(r_0)}{\dot{r}_{tot}(r)}\left(\frac{x+1}{x_0+1}\right)^{2\Phi_{1l}}\left(\frac{x^2 - x + 1}{x_0^2 - x_0 + 1}\right)^{-\Phi_{1l}} \\ &\times\exp[-\beta_{l0}(r - r_0) - V_l(x - x_0)] \\ &\times\exp\left[\frac{2\beta_{l0}H_l}{\sqrt{3}}\arctan\frac{2x-1}{\sqrt{3}}\bigg|_{x_0}^{x} + \frac{V_l}{\sqrt{3}}\arctan\frac{x\sqrt{3}}{2-x}\bigg|_{x_0}^{x}\right], \end{aligned} \tag{14.4.23}$$

where the vertical bar with limits is used in the equation for brevity and has the same meaning as in the integrals (i.e., the function at the lower limit is subtracted from the function at the upper limit), and

$$H_l = \left(\frac{b_{con}\,|s|}{\chi_c A_v}\right)^{2/3}, \quad x = \left(\frac{r}{H_l}\right)^{1/2}, \quad x_0 = \left(\frac{r_0}{H_l}\right)^{1/2}, \tag{14.4.24a}$$

$$V_l = \frac{2\alpha_l(w-\alpha_l k)H_l^{1/2}}{\chi_c A_c}, \qquad \Phi_{1l} = \frac{V_l}{6} - \frac{\beta_{l0}H_l}{3}. \tag{14.4.24b}$$

Substitution of I_1 and I_2 for the evaporation layer with $s < 0$ from Appendix B.14 into (Eqn. 14.4.11) for J_{l1} and then into (Eqn. 14.4.8) yields

$$\begin{aligned} f_l(r) &= f_l(r_0)\frac{\dot{r}_{tot}(r_0)}{\dot{r}_{tot}(r)}\left(\frac{x-1}{x_0-1}\right)^{-2\Phi_{2l}}\left(\frac{x^2+x+1}{x_0^2+x_0+1}\right)^{\Phi_{2l}} \\ &\times \exp\left[-\beta_{l0}(r-r_0) - V_l(x-x_0)\right] \\ &\times \exp\left[\frac{2\beta_{l0}H_l}{\sqrt{3}}\arctan\frac{2x+1}{\sqrt{3}}\bigg|_{x_0}^{x} + \frac{V_l}{\sqrt{3}}\arctan\frac{x\sqrt{3}}{2+x}\bigg|_{x_0}^{x}\right], \end{aligned} \tag{14.4.25}$$

$$\Phi_{2l} = \frac{V_l}{6} + \frac{\beta_{l0}H_l}{3}. \tag{14.4.26}$$

The parameter H_l is the characteristic length that determines the onset of the asymptotic regime. An estimate of H_l from (Eqn. 14.4.24a) with $q_{ls} = 0.1\,\mathrm{g\,m^{-3}}$ (stratiform clouds) with b_{con} and ρ_{vs} at $T \sim 0$°C, $|s| = 0.1$ (± 10%, ice phase cloud), $A_v = 2.2 \times 10^3\ (\rho_0/\rho)^{1/2}\mathrm{cm^{1/2}\,s^{-1}}$ (Section 12.4) yields $H_l \approx 300\,\mu\mathrm{m}$. For the same q_{ls} but with $|s| = 10^{-3}$ (± 0.1%, liquid phase cloud), $H_l \approx 15\,\mu\mathrm{m}$; with $q_{ls} \sim 1\,\mathrm{g\,m^{-3}}$ (convective clouds), $H_l \sim 5\,\mu\mathrm{m}$. For cirrus at $T = -40$ to -50 °C, $p \sim 200$ mb, $\rho/\rho_0 \sim 0.4$, and $q_{ls} = 10$–20 mg m^{-3}, an estimate yields $H_l \sim 60$–$120\,\mu\mathrm{m}$. Thus, in all these cases, $r >> H_l$, and $x >> 1$ for the large-size fraction. We can estimate the asymptotics of various terms in (Eqn. 14.4.23) at $x >> 1$. The product of two brackets with $x = V_l(r/H_l)^{1/2}$ before the exponent tends at large x to a limit $x^{2\Phi_{1l}}x^{-2\Phi_{1l}} \to 1$. The arguments of arctan in the exponent tend to infinity; arctan tends to $\pm\pi/2$, which results in a change of the normalizing constant. The term $1/\dot{r}_{tot}(r) \to r/[rv(r)] = v(r)^{-1/2} \to r^{-1/2}$. Incorporating these limits, we find the asymptotic of $f_l(r)$ at $s > 0$:

$$f_l(r) \sim r^{-1/2}\exp[-\beta_{l0}r - V_l(r/H_l)^{1/2}]. \tag{14.4.27}$$

The same estimates for f_l (Eqn. 14.4.25) at $s < 0$ yield the same asymptotic. An estimate of V_l with $q_{ls} = 0.1\,\mathrm{g\,m^{-3}}$ and $w = 1\,\mathrm{cm\,s^{-1}}$ gives $V_l \sim 0.1$. At $r \sim 1$ mm, the value $V_l x \sim 0.3$, while with $\beta_{l0} \sim 40\,\mathrm{cm^{-1}}$, the value $\beta_{l0}r = 4$. Thus, the term $V_l(r/H_l)^{1/2}$ in the exponent of (Eqn. 14.4.27) can be neglected and $f_l(r) \sim r^{-1/2}\exp(-\beta_{l0}r)$—i.e., its asymptotic coincides with that in Section 14.4.3 and tends to the Marshall–Palmer spectrum with a correction $r^{-1/2}$. However, for greater convective updrafts of $w \sim 1\,\mathrm{m\,s^{-1}}$ and $\sim q_{ls} = 1\,\mathrm{g\,m^{-3}}$, an estimate yields $V_l \sim 1$—i.e., the value $V_l x \sim 3$ at $r \sim 1$ mm, which is comparable in magnitude to $\beta_{l0}r$. Then, the term $V_l x$ should be retained in the exponent, the slope becomes nonlinear, and the spectra may be slightly concave downward in log-linear coordinates—i.e., the slopes increase with r and the spectra decrease with r faster than for the MP spectrum. Thus, this analytical solution explains observations of such concave downward spectra by Willis (1984) in convective clouds.

14.4.5. Solutions for Subcloud Layers

The previous solutions were obtained for the case of large particles coexisting with cloud liquid/ice, that is, with the small-size fraction f_s. When the small-size fraction has been evaporated—e.g., falling through the subsaturated subcloud layer, or in the downdraft or near the cloud edge—and large drops exist without the cloud particles, another regime occurs, $\dot{r}_{coag} << |\dot{r}_{cond}|$. That is, the coagulation–accretion rate is negligible and the large fraction evaporates with rate $\dot{r}_{cond}$. To address this situation, we solve (Eqn. 14.4.1) taking into account diffusional growth (evaporation) but neglecting the accretion term

$$b_{con}s\frac{d}{dr}\left(\frac{f_l}{r}\right)+[w-v(r)-\alpha_l k]\alpha_l f_l=0. \tag{14.4.28}$$

This equation coincides with (Eqn. 13.7.1) for the tail of the small-size fraction, except that the loss term with σ_{col} is absent and we omitted for brevity the terms with dk/dz and $d\alpha_l/dz$ (which can be included in the final solution). Assuming again a power law for the terminal velocity with coefficients A_v, B_v, we can modify the solution (Eqn. 13.7.2) as:

$$f_l(r)=\frac{r}{r_{be}}f_l(r_{be})\exp\{-[\beta_{le}(r)r-\beta_{le}(r_{be})r_{be}]\}, \tag{14.4.29}$$

where r_{be} denotes the lower boundary of the evaporating large-size fraction, and the size-dependent slope $\beta_{le}(r)$ for the evaporation layer is

$$\beta_{le}(r)=\frac{r\alpha_l}{b_{con}s}\left(\frac{w-\alpha_l k}{2}-\frac{v(r)}{B_v+2}\right). \tag{14.4.30}$$

For large enough r and without vigorous vertical velocities, the second term in (Eqn. 14.4.30) dominates and the slope is similar to that obtained in (Eqn. 13.7.7),

$$\beta_{le}(r)=c_{l1}rv(r), \qquad \text{where} \qquad c_{l1}=-\frac{\alpha_l}{b_{con}s(B_v+2)}. \tag{14.4.31}$$

To ensure a decrease of f_l with r, β_{le} should be positive; thus, the signs of α_l and s should be different in the evaporation layer (i.e., $\alpha_l > 0$ and $dq_{ll}/dz > 0$ at $s < 0$) and LWC (IWC) should decrease downward in the subcloud layer at $s < 0$, which is characterized by evaporating precipitation.

The spectra in (Eqn. 14.4.29) are generalized gamma distributions, and the behavior of the slopes depends on the velocity power index. In particular, $\beta_{le} \sim r^2$ and $f_l \sim \exp(-c_{l1}A_v r^3)$ for $v(r) = A_v r$ and $\beta_{le} \sim r^{3/2}$, $f_l \sim \exp(-c_{l1}A_v r^{5/2})$ for $v(r) = A_v r^{1/2}$. In vigorous downdrafts that cause subsaturation, such that $|w| >> v(r)$, the asymptotic slope (Eqn. 14.4.30) becomes

$$\beta_{le}=c_{l2}r, \qquad c_{l2}=\frac{\alpha_l(w-\alpha_l k)}{2b_{con}s}. \tag{14.4.32}$$

Since both w and s are negative and $\alpha_l > 0$, then $-\alpha_l k < 0$. Therefore, $c_{l2} > 0$ and $\beta_{le} > 0$, thus f_l decreases with r and (Eqn. 14.4.32) ensures a correct asymptotic. Then, the spectrum behaves as $f_l \sim \exp(-c_{l2}r^2)$. Thus, the spectra of the precipitating particles in the evaporating subcloud layer fall off more rapidly with r than in cases with the presence of the small fraction, $q_{ls} > 0$, and become concave downward in log-linear coordinates, which can explain observations described by Ulbrich (1983), Willis (1984), and Zawadski and de Agostinho Antonio (1988).

14.5. Interpretation of Solutions

In this section, a physical interpretation is given for solutions for the large-size fraction obtained in Section 14.4 and they are illustrated with calculations for a crystalline cloud.

14.5.1. General Analysis of the Parameters

The MP spectra, $f_l \sim \exp(-\Lambda D)$, play a fundamental role in many cloud and climate models and remote sensing (especially radar) techniques. The values of $\Lambda = \beta_{l0}/2$ can be estimated using (Eqn. 14.4.4) for β_{l0} that can be rewritten for layers with $q_{ls} > 0$ by using (Eqn. 14.3.4) for χ_c:

$$\beta_{l0} = -\frac{\alpha_l}{\chi_c} = \frac{4(-\alpha_l)\rho_w}{E_c q_{ls}}. \tag{14.5.1}$$

Note that $\alpha_l < 0$ since $\alpha_l = (dq_{ll}/dz)/q_{ll}$ and f_l and q_l increase downward, which is expected for falling particles growing by collection. Under this condition, $\beta_{l0} > 0$ and $\exp(-\beta_{l0}r)$ decreases with r. Equation (14.5.1) reveals that β_{l0} decreases—i.e., the spectra stretch toward large sizes, when the following processes occur: 1) E_c or q_{ls} increase—i.e., the rate of mass transfer from the small to the large-size fraction increases; 2) ρ_w decreases and the mass gained by the large-size fraction is distributed over the larger range of volumes and radii; 3) $(-\alpha_l) = |\alpha_l|$ decreases—i.e., the gravitational, convective, and turbulent fluxes of f_l decrease, causing weaker outflow of f_l from a given cloud level.

As an example, we can estimate from (Eqn. 14.5.1) the slope β_{l0} for the snow size spectra using data from Passarelli (1978b). Taking $E_c \sim 0.5$–1, snowflake bulk density $\rho_w \sim 0.1$–$0.2\,\mathrm{g\,cm^{-3}}$, the water content of the small-size fraction as $q_{ls} \sim 0.1\,\mathrm{g\,m^{-3}}$, and the thickness of the layer ~ 0.5–2 km, we obtain $\beta_{l0} \approx 40$–$160\,\mathrm{cm^{-1}}$, or $\Lambda = \beta_{l0}/2 \approx 20$–$80\,\mathrm{cm^{-1}}$. This is within the range of values of $\Lambda = 10$–$100\,\mathrm{cm^{-1}}$ given by Platt (1997), Passarelli (1978b), Houze et al. (1979), and Ryan (2000).

Equation (14.5.1) provides an explanation for some observed peculiarities of β_{l0} in crystalline clouds. The analyses performed by Platt (1997), Houze et al. (1979), and Ryan (2000) show that β_{l0} increases by about an order of magnitude when temperature T decreases from 0 °C to −50 °C (Fig. 3 in Platt, 1997). The same analysis shows that the ice water content decreases in this temperature range, although somewhat faster (Fig. 4 in Platt 1997). According to (Eqn. 14.5.1), $\beta_{l0} \sim q_{ls}^{-1}$, and this may explain the observed increase in β_{l0} with decreasing T and q_{ls}. The slower increase with T by β_{l0} relative to the decrease in q_{ls} may be caused by decreasing cloud thickness at lower T—i.e., vertical gradients of IWC and α_l in the numerator of (Eqn. 14.5.1). This temperature dependence of β_{l0} or Λ may cause the height dependence observed by Passarelli (1978b), who measured $\Lambda \approx 65\,\mathrm{cm^{-1}}$ at $z = 3.35$ km ($T = -20$ °C) and $\Lambda \approx 24\,\mathrm{cm^{-1}}$ at $z = 2.55$ km ($T = -12$ °C). This increase in Λ (or β_{l0}) with decreasing height can also be a consequence of the $q_{ls}(T)$ dependence.

An interesting feature of the exponential MP spectra is that the range of slopes is similar for both liquid and crystalline particles. Marshall and Palmer (1948) related the slopes with the rainfall rate R_0 in $\mathrm{mm\,h^{-1}}$ with an equation $\Lambda = 41R_0^{-0.21}\,(\mathrm{cm^{-1}})$, yielding $\beta_{l0} = 2\Lambda = 130$ to $50\,\mathrm{cm^{-1}}$ for $R_0 = 0.1$ to $10\,\mathrm{mm\,h^{-1}}$. An estimate from (Eqn. 14.5.1) with $E_c \sim 0.5$, $\rho_w \sim 1\,\mathrm{g\,cm^{-3}}$, $q_{ls} \sim 1\,\mathrm{g\,m^{-3}}$, $\alpha_l \sim 0.5\,\mathrm{km^{-1}}$ yields $\beta_{l0} = 40\,\mathrm{cm^{-1}}$, which is in the middle of the range of β_{l0} values determined for the MP spectra, and hence (Eqn. 14.5.1) can be applicable to MP spectra in liquid clouds.

Considering convective rain and the equations in Section 14.4, the p_l-index (Eqn. 14.4.17b) with $v(r) = A_v r$ is applicable near the observed modal diameter ~0.2–0.4 mm in warm rain (e.g., Ulbrich, 1983; Willis, 1984). Using the relation $-\alpha_l = \beta_{l0}\chi$ from (Eqn. 14.4.4), the p_l-index (Eqn. 14.4.17b) can be written as

$$p_l = c_{\beta 1}\beta_{l0}^2 + c_{\beta 2}\beta_{l0} - 1, \qquad (14.5.2a)$$

$$c_{\beta 1} = \chi_c k / A_v, \qquad c_{\beta 2} = w / A_v. \qquad (14.5.2b)$$

Equation (14.5.2a) can be rewritten in terms of $\Lambda = \beta_{l0}/2$ with corresponding coefficients $c_{1\Lambda} = 4c_{\beta 1}$ and $c_{2\Lambda} = 2c_{\beta 2}$. Equation (14.5.2a) resembles the empirical fit found by Zhang et al. (2001) from radar and disdrometer data

$$p_l = c_{\Lambda 1Z}\Lambda^2 + c_{\Lambda 2Z}\Lambda - 1.957, \qquad (14.5.3)$$

$c_{\Lambda 1Z} = -1.6 \times 10^{-4}\,\text{cm}^{-2}$, $c_{L2Z} = 0.1213\ \text{cm}^{-1}$. A similar relation was suggested by Heymsfield (2003) for crystalline clouds based on aircraft measurements:

$$p_l = c_{\Lambda 1H}\Lambda^{c_{\Lambda 2H}} - 2, \qquad (14.5.4)$$

where $c_{\Lambda 1H} = 0.076\,\text{cm}^{-0.8}$, and $c_{\Lambda 2H} = 0.8$.

The estimates from (Eqns. 14.5.2a) and (14.5.2b) with typical cloud parameters show that the value of $c_{\beta 1}$ has a smaller magnitude and is opposite in sign to $c_{\Lambda 1Z}$; hence, this relation is determined mostly by the coefficient $c_{\beta 2}$. This is in agreement with (Eqn. 14.5.3), which predicts a nearly linear relation except for very high values Λ, and with (Eqn. 14.5.4) where the power of Λ is 0.8, and the relation is close to linear. Thus, (Eqns. 14.5.2a) and (14.5.2b) predict positive correlations between p_l, Λ, and vertical velocities, and is in agreement with experimental data and parameterizations. The increase in p_l and Λ with increasing w predicts narrower spectra in stronger updrafts and broader spectra in downdrafts, which is similar to the effect of stochastic condensation described in Chapter 13.

However, it should be emphasized that this analysis is just an illustration of possible applications of these analytical solutions and should be used with caution. The slope α_l in (Eqn. 14.5.1) and the p_l–β_{l0} relation (Eqns. 14.5.2a) and (14.5.2b) are based on solutions with the presence of q_{ls}. In vigorous downdrafts, in the subcloud layer or near the surface where $q_{ls} \sim 0$, the solutions from Section 14.4 can be used. Then, (Eqns. 14.4.29)–(14.4.32) show that the slopes $\beta_{l0}(r)$ or $\Lambda(r)$ can be expressed as polynomials of p_l (equal to 1 in this case). These equations and the asymptotic analysis show that $\Lambda(r)$ is a 2nd-order polynomial of p_l if $v(r) << |w|$ and the slope is (Eqn. 14.4.32). Such parameterization was suggested in Zhang et al. (2003a,b) and in Brandes et al. (2003), and a 3rd- or 2.5-order polynomial if $v(r) >> |w|$ and the slope is (Eqn. 14.4.31).

Unfortunately, data on vertical velocities, the turbulent coefficient, and presence of the small-size fraction are absent in the cited papers, which precludes a more detailed comparison. Verification of the relations (Eqns. 14.5.2a) and (14.5.2b) would require simultaneous measurements of w, q_{ls}, q_{ll}, the turbulent coefficient, and the size spectra. However, these analytical solutions are consistent with the general findings from experimental observations: since the slopes and indices are expressed through related quantities, this leads to the existence of the p_l–Λ relations. At the same time, solutions in Section 14.4 for various particular cases show that these relations cannot be universal, but should depend on the altitude and position of the measured spectra in the cloud or below the cloud base, and specifically the sign of w, values of k and α_l, and the presence of q_{ls}.

14.5.2. Example Calculations for a Crystalline Cloud

The properties of snow spectra in a crystalline cloud are illustrated here in more detail. We select a generic case, chosen for illustration to mimic the profiles in similar clouds simulated in Khvorostyanov, Curry, et al. (2001) and in Khvorostyanov and Sassen (2002) using the spectral bin 1D and 2D models. The profiles for this case of $q_{ll}(z)$ and $\alpha_l(z)$, calculated from (Eqn. 14.3.11), are shown in Fig. 14.1a,b, along with the IWC of the small-size fraction q_{ls} and ice supersaturation, which are the same as shown in Fig. 13.10. The temperature decreases from about −5 °C at the lower boundary to −60 ° at 12 km.

Shown in Fig. 14.2 are the vertical profiles of the slopes β_{l0} and Λ calculated for this case from (Eqn. 14.5.1). The generalized empirically derived slope Λ for crystalline clouds from Platt (1997) is shown in Fig. 14.2b for comparison. The calculated slopes increase with decreasing temperature, although not linearly as predicted by the generalized experimental Λ, but somewhat faster, especially above 7 km, since α_l and Λ are inversely proportional to q_{ls}, which decreases upward nonlinearly at these heights. However, the general agreement of the calculated and experimental curves is fairly good, both in magnitude and vertical gradients. This indicates that if a large ensemble of values of q_{ls} and q_{ll}, measured at various temperatures, are used to calculate α_l and Λ, then the results would converge to the experimental curve by Platt (1997).

An example of size spectra at ice subsaturation, $s_i < 0$, at the heights 4.8–6.0 km is shown in Fig. 14.3. The small-size fraction (Fig. 14.3a) was calculated with the generalized gamma distributions from Section 13.6. In the spectral region from 6 to about 40–50 μm, the spectra are almost linear in log-log coordinates, close to the inverse power laws with the indices increasing toward the cloud top (to colder temperatures). At r = 50–130 μm, the effect of the exponential tail dominates and the spectra have a maximum at $r \sim$ 100–130 μm that decreases with height.

The spectra of the large-size fraction (Fig. 14.3b) are calculated using (Eqns. 14.4.18)–(14.4.20). The spectra plotted in log-linear coordinates are nearly linear—i.e., close to the Marshall–Palmer exponents. The size-dependent slope $\beta_l(r)$ slightly decreases with r as predicted by the second term in (Eqn. 14.4.19), but the departure from linearity is small, since the slope is determined mostly by β_{l0}. The composite spectra obtained by matching the small- and large-size fractions at r_0 = 72 μm (Fig. 14.3c) are seen to be bimodal. The calculated composite spectra and the experimental spectrum from Platt (1997) shown in Fig. 14.3d are in good agreement, having minima and maxima at similar positions (note the difference in radii and diameter scales in horizontal axes). The experimental and calculated values of f_l can be compared using the relation 1 $\mathrm{m}^{-4} \approx 10^{-9}$ $\mathrm{L}^{-1}\,\mu\mathrm{m}^{-1}$; the maximum $\sim 10^{11}\,\mathrm{m}^{-4}$ in Fig. 14.3d corresponds to $\sim 10^{2}\,\mathrm{L}^{-1}\,\mu\mathrm{m}^{-1}$, which is comparable to the maximum in Fig. 14.3c. In these calculations, the first bimodality occurs within the small-size fraction. If the matching point was located at greater $r_0 \sim$ 120–150 μm, there would be a second region of bimodality at r_0 due to different slopes of the small- and large-size fractions; bimodality is often observed in this region (e.g., Mitchell 1994, Mitchell et al. 1996), and polymodal spectra are also often observed (Sassen et al., 1989; Poellot et al., 1999).

The spectra in the condensation layer 7.5–8.7 km with positive ice supersaturation, $s_i > 0$, are depicted in Fig. 14.4. The indices of the small-size fraction (Fig. 14.4a) are positive since $p \sim s_i$ (see (Eqn. 13.8.4)), and the spectra are monomodal gamma distributions with maxima at $r \sim$ 30–50 μm. The portion of the spectra from ~10 to 50–60 μm in log-log coordinates is almost linear—i.e., it

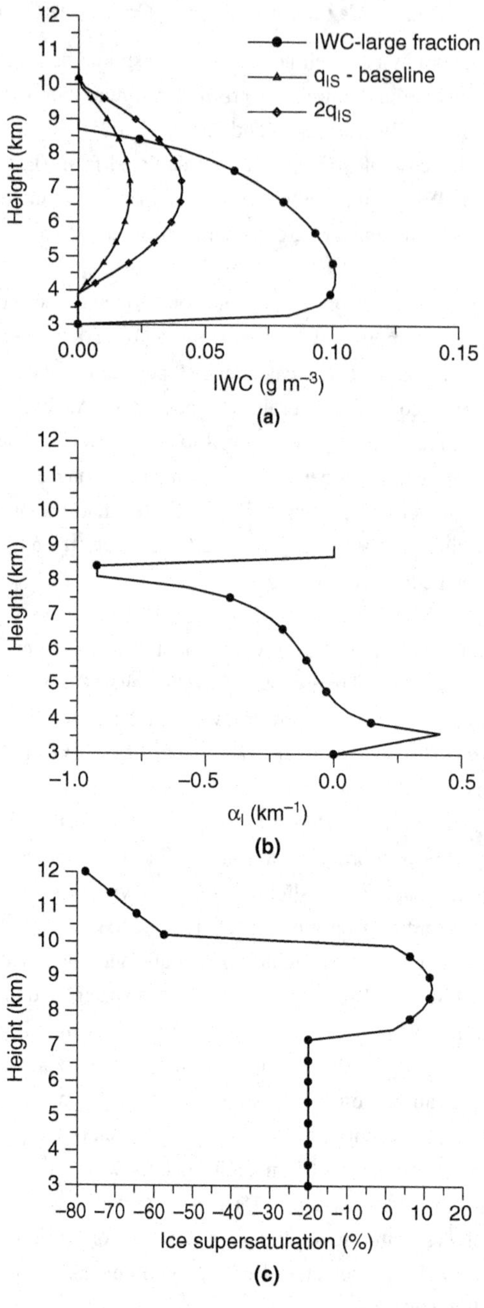

Figure 14.1. Profiles for calculations of the size spectra of the large-size fraction: (a) IWC of the large fraction (solid circles), of small fraction q_{ls} (triangles), and of the small fraction with doubled IWC, $2q_{ls}$ (diamonds); (b) parameter α_l; (c) ice supersaturation s. From Khvorostyanov and Curry (2008c). *J. Atmos. Sci.*, 65, © American Meteorological Society. Used with permission.

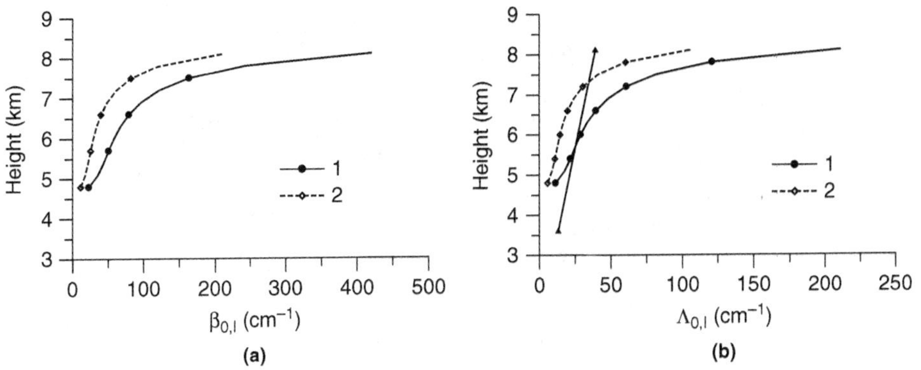

Figure 14.2. (a) Exponential slope β_{l0} and (b) Λ_{l0} of the size spectra of the large fraction calculated with the IWC of small fraction q_{ls} (circles) and $2q_{ls}$ (diamonds) shown in Fig. 14.1. The generalized experimental slope for crystalline clouds from Platt (1997) and Ryan (2000) recalculated from the temperature to the height with the average lapse rate 6.5 °K km^{-1} is plotted in (b) with triangles. From Khvorostyanov and Curry (2008c). *J. Atmos. Sci.*, 65, © American Meteorological Society. Used with permission.

obeys the power law with indices slightly increasing with height. The large-size fraction (Fig. 14.4b), as seen in log-linear coordinates, represents the MP distributions. The slopes are much steeper than in the lower layer and increase upward—i.e., the large-size spectra also become narrower at colder temperatures. These features are mostly due to the smaller q_{ls} and the dependence $\beta_{l0} \sim q_{ls}^{-1}$. The composite spectra matched at 90 μm (Fig. 14.4c) exhibit features of bimodality, but more weakly than in the lower layer. Now the bimodality occurs between the large- and small-size fractions rather than within the small-size fraction as shown in Fig. 14.3 at $s < 0$. For comparison, given in Fig. 14.4d are the average size spectra from Lawson et al. (2006) measured in cirrus clouds at three temperatures. One can see that the calculated spectra (Fig. 4c) are similar to the observed spectra. In particular, they exhibit a similar bimodality, become narrower at lower temperatures, and their bimodality decreases and vanishes with increasing height (decreasing temperature). The reason for this is the decrease with height of IWC of the small fraction, slowing down accretion, and diminishing the large fraction. This analysis is consistent with observations (Sassen et al., 1989; Mitchell, 1994; Platt, 1997; Ryan, 2000; Poellot et al., 1999) that bimodality is more pronounced in the lower layers.

Note that the spectra calculated at ice sub- and supersaturation are somewhat different. The experimental spectra, however, are usually presented without information about supersaturation, and may have been obtained from mixtures sampled in both sub- and supersaturated layers. This precludes a more detailed comparison at present and indicates that simultaneous measurements of the size spectra and supersaturation are desirable.

Fig. 14.5 shows the slopes $\beta_{le}(r)$ and size spectra calculated with (Eqns. 14.4.30) and (14.4.29) for the subsaturated subcloud layer where the small-size fraction has been evaporated. At small and large r, the behavior of $\beta_{le}(r)$ is determined by (Eqns. 14.4.32) and (14.4.31), respectively. The slopes rapidly increase with radius, but the rate of this growth $\beta_{le}(r)/dr$ decreases at large r, which is determined by the increasing contribution from the second term with $v(r)$ in (Eqn. 14.4.30). This results in

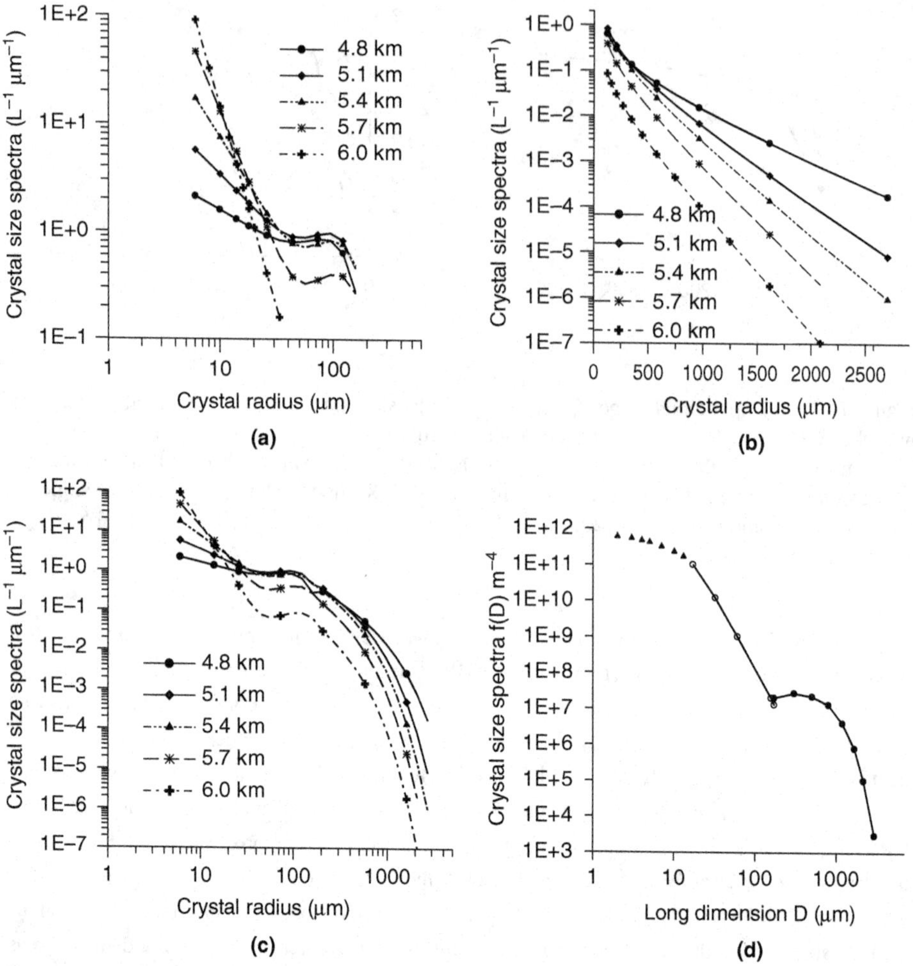

Figure 14.3. Crystal size spectra by radii r ($L^{-1}\,\mu m^{-1}$) at ice subsaturation ($s < 0$) at the heights in km indicated in the legends; temperature decreases from −14.9 °C at 4.8 km to −22.15 °C at 6.0 km. Size spectra: (a) for the small fraction; (b) large fraction; (c) composite; and (d) generalization by Platt (1997) of the observed crystal size spectra by long dimension D (m^{-4}) in the temperature interval of −25 °C to −30 °C. Note the different normalization of the spectra and axes scales in (a)–(c) by radius and (d) by diameter. From Khvorostyanov and Curry (2008c). *J. Atmos. Sci.*, 65, © American Meteorological Society. Used with permission.

a rapid decrease in $f_l(r)$ toward the larger values of r. This feature has been observed in liquid clouds (e.g., Willis 1984), and as Fig. 14.5 shows, can also be pertinent for subcloud layers of crystalline clouds. Since we consider an example with spherical particles and asymptotic $v(r) \sim r^{1/2}$ ($B_v = 1/2$), the asymptotic behavior of the spectra is $f_l \sim \exp(-c_{l1}A_v r^{5/2})$, as described in Section 14.4.5. For some crystal habits like aggregates or plates, the power B_v can be much smaller than 1/2 and closer to 0 (e.g., Mitchell, 1996; Khvorostyanov and Curry, 2002, 2005). Then, the decrease in f_l can be much slower and the tails in the subcloud layer can be much longer.

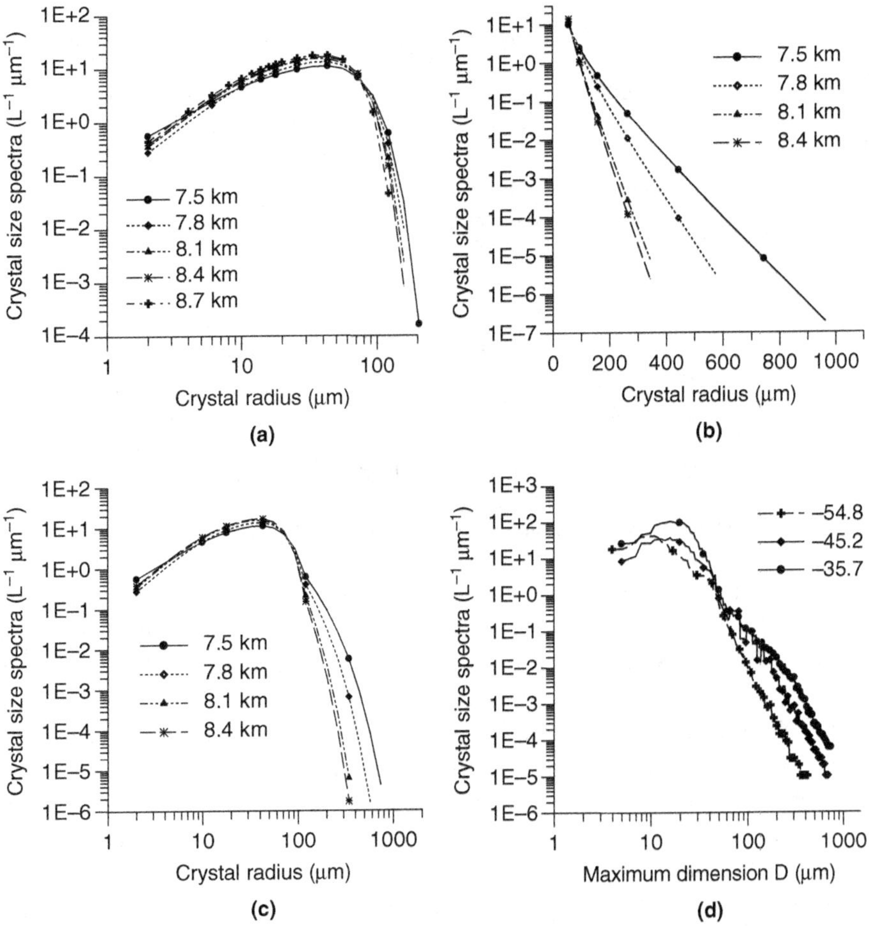

Figure 14.4. Crystal size spectra by radii ($L^{-1}\,\mu m^{-1}$) at ice supersaturation ($s > 0$) at the heights indicated in the legends. (a) small fraction extended to 200 μm; (b) large fraction; (c) composite spectra (merged the small-size and the large-size fractions); and (d) experimental spectra in cirrus clouds from Lawson et al. (2006) at 3 indicated temperatures. From Khvorostyanov and Curry (2008c). *J. Atmos. Sci.*, 65, © American Meteorological Society. Used with permission.

14.5.3. General Interpretation of the Solutions

The analytical solutions of the stochastic kinetic equations for the size spectra of the large fraction of liquid and crystalline precipitating particles were obtained in this chapter for various cases. The solutions are somewhat different in various situations but have common general features that allow us to explain and interpret observations and empirically derived expressions for rain and snow size spectra such as the Marshall–Palmer distribution, gamma distributions, or inverse power laws. This provides the basis for development of the more physical parameterizations of cloud microphysics in cloud and climate models. The major results can be summarized as follows.

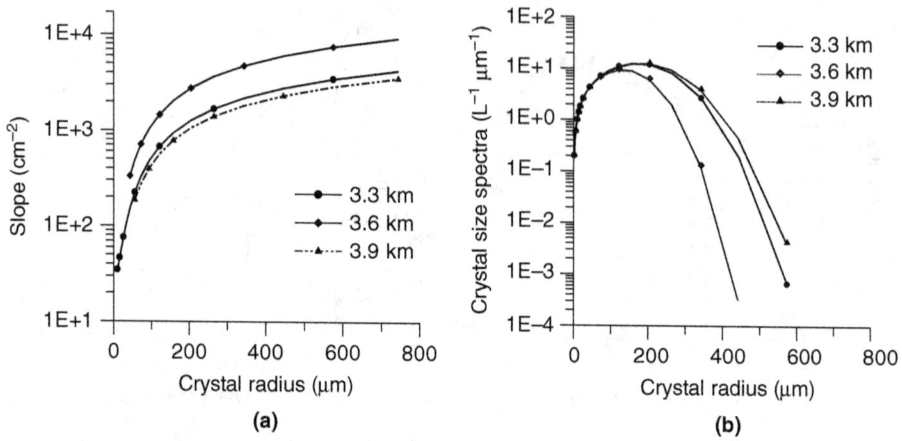

Figure 14.5. (a) Slopes and (b) size spectra in the subcloud layer with zero IWC of the small-size fraction q_{ls} at the heights (km) indicated in the legends. From Khvorostyanov and Curry (2008c). *J. Atmos. Sci.*, 65, © American Meteorological Society. Used with permission.

The general solution of the stochastic kinetic equations for the large-size fraction (precipitating particles) is described by the product of the two major terms: a) an exponential term, ~$\exp(-\beta_l r)$; and b) a prefactor before the exponent that is an algebraic function of radius like a power law or polynomial. The argument β_l of the exponent consists of a slope of the Marshall–Palmer type and an additional integral that depends on the condensation and accretion rates, vertical velocities, the turbulent coefficient, terminal velocity, and vertical gradient of the liquid (ice) water content. The algebraic function before the exponent is inversely proportional to the sum of the condensation and accretion rates and depends on the super- or subsaturation, terminal velocity, and collection efficiency.

Several particular cases were considered in this chapter: a) terminal velocity as a linear function of radius; b) terminal velocity as a square root function of radius; c) an accretion growth rate much greater than the condensation growth rate; and d) a subcloud evaporation layer with a very small or absent small-size fraction. The general solution is substantially simplified for these cases. The exponential part tends to the Marshall–Palmer exponent with the slope β_{l0}, but contains additional terms that make the slope radius-dependent and nonlinear, causing the spectra to decrease with radius faster than the MP exponent as observed in many experiments. This may influence the spectral moments (e.g., cause smaller radar reflectivity and precipitation rates than those evaluated with MP spectra) and change the relations between them.

The radius dependence of the algebraic prefactor before the exponent is weaker than that of the exponent, converts for sufficiently large radii to the power law, and the spectra also have the form of gamma distributions with the slope β_{l0} and index p_l. This index correlates with the supersaturation and in general is positive in the cloud growth layers, when the spectra tend to modified gamma distributions, and is negative in the evaporation layers, when the spectra tend to the inverse power laws at smaller sizes multiplied by the exponential tails. However, these gamma distributions are different from those obtained for the small-size fraction (cloud particles) described in Chapter 13.

A simple expression is derived for the slope β_{l0} via four parameters: β_{l0} is proportional to the relative gradient $\alpha_l = (dq_{ll}/dz)/q_{ll}$ of the liquid (ice) water content of the large-size fraction, to water or ice density, and is inversely proportional to the collection efficiency and the liquid (ice) water content q_{ls} of the small-size fraction. All of these parameters are available in cloud-scale and large-scale models, and these dependencies provide reasonable explanations for the observed features of β_{l0} with variations of each parameter. In particular, the inverse dependence, $\beta_{l0} \sim q_{ls}^{-1}$, provides an explanation of the observed strong inverse temperature dependence of β_{l0} since q_{ls} in general decreases with decreasing temperature. This explains experimentally observed narrowing of the size spectra with decreasing temperature.

Simple analytical expressions are also derived for the power indices p_l of the gamma distributions (shape parameters), which are expressed via the coefficients of the terminal velocity, the slopes β_{l0}, vertical velocity, and the turbulent coefficient. These features allow us to explain the observed shapes of the rain and snow size spectra (e.g., Ulbrich, 1983; Willis, 1984).

Based on these expressions for β_{l0} and p_l, a β_{l0}–p_l relation is derived for the case with a terminal velocity proportional to the radius as a 2nd-order polynomial. This relation is similar to the empirical parameterizations based on radar and disdrometer data (e.g., Zhang et al., 2001, 2003a,b; Brandes et al., 2003). The coefficients of this relation are expressed via vertical velocity, the turbulent coefficient, and the cloud liquid or ice water content.

These analytical solutions for the spectra of the large-size fraction and its parameters provide explanations for observed dependencies of the spectra on the temperature, turbulence, vertical velocities, liquid water or ice water content, and other cloud properties. The results are illustrated with calculations for a crystalline cloud. These analytical expressions can be used for parameterization of the size spectra and related quantities (e.g., coagulation–accretion rates, precipitation intensities, optical properties, radar reflectivities) in bulk cloud and climate models and in remote sensing techniques.

The solutions have been presented for liquid-only and ice-only size spectra. Treatment of mixed phase clouds would require simultaneous consideration of the small- and large-size fractions of the drop and crystal spectra with an accounting for their interactions (Wegener–Findeisen–Bergeron process and transitions among the fractions). The coexistence of drops and crystals would change the water and ice supersaturations in the three-phase system and the coagulation and accretion rates. However, it can be expected that the general functional shapes of the size spectra will be similar to those described earlier. The change of the spectra in a mixed-phase cloud will result in the corresponding change of the parameters of the spectra—in particular, of the p-indices proportional to sub- or supersaturation, and of slopes of the exponents that are expressed via the coagulation–accretion rates and LWC or IWC of the small fraction.

Many bulk parameterizations of cloud microphysics use schemes with the number of size categories larger than the two considered here: small- and large-size fractions. For example, the size spectrum of ice particles may be separated into ice, snow, graupel, and hail. The parameterization of the size spectra described in this chapter for two size fractions is easily generalized for several fractions. Then, several pairs of hydrometeors should be considered, and the parameterizations described earlier applied for each pair consisting of the smaller- and larger-size fractions.

14.6. Autoconversion and Corrections to the Analytical Solutions

As mentioned in Section 14.2, the autoconversion is a process of "self-collection" of the particles within the same fraction or size category, which transfers them into the next size category. The hypotheses on autoconversion and its parameterizations began developing in the 1960s along with the stochastic condensation theory considered in Chapter 13. While the stochastic condensation theory attempted to describe broadening of the originally narrow spectrum, the initiation of coagulation, and the formation of the embryonic precipitating particles, the parameterizations of autoconversion attempted to provide the equations for the rates P_{aut} of mass conversion from a smaller fraction into a larger fraction, usually without detailed consideration of the mechanism of this transition. Many features and quantitative characteristics of autoconversion are still unclear because this phenomenon is difficult to observe and it is not quite clear in general how to describe precisely gravitational coagulation among such comparable particles within a rather narrow size spectrum. One of the major problems in quantitative description of autoconversion is that interactions of the hydrometeors with comparable masses and fall velocities within the same size region make more questionable simplifications used in the other methods described in Section 14.2. Hereafter, we assume that the LWC (IWC) q_{ll} of the large size fraction is small and, in order to avoid duplication of autoconversion with the coagulation between the small- and large-size fractions, we will consider autoconversion only within the small-size fraction. The corresponding equations in the following text are formulated via q_{ls} rather than in terms of the total water content as in some other parameterizations.

A complete rigorous theory of autoconversion is still absent, and parameterizations of autoconversion have been developed based either on some intuitive considerations or on simulations with spectral bin models and the subsequent fitting of numerical results with some simple functions. One of the first and widely used parameterizations of autoconversion of cloud water q_{ls} into rainwater q_{ll} was suggested by Kessler (1969) as

$$P_{aut} = \frac{dq_{ll}}{dt} = K_{Kes} q_{ls} \theta(q_{ls} - q_{ls0}), \tag{14.6.1}$$

where $\theta(x)$ is the Heaviside step function, K_{Kes} is the conversion coefficient, and q_{ls0} is the threshold, from which autoconversion begins. Kessler assigned originally the values of $K_{Kes} = 10^{-3}\,\mathrm{s}^{-1}$ and $q_{ls0} = 5 \times 10^{-4}\,\mathrm{kg\,kg^{-1}}$ to $1 \times 10^{-3}\,\mathrm{kg\,kg^{-1}}$ for the deep convection. Presence of the threshold here implicitly accounts for the fact that autoconversion is considered as a coagulation process and begins only when the larger particles reach some minimum values and gravitational coagulation becomes noticeable.

Kessler's scheme was modified later by many authors who accounted for the fact that autoconversion is a function not only of the liquid water content but also of the drop concentration and size spectrum width (e.g., Berry, 1967; Berry and Reinhard, 1974a,b; Manton and Cotton, 1977; Tripoli and Cotton, 1980; Liou and Ou, 1989; Baker, 1993; Khairoutdinov and Kogan, 2000; Seifert and Beheng, 2001, 2006; a review of various schemes was given by Liu and Daum, 2004). Manton and Cotton (1977) (see also Tripoli and Cotton, 1980) suggested a modification of Kessler's scheme that can be written for the liquid water content as

$$\frac{dq_{ll}}{dt} = K_{MC} q_{ls} \theta(q_{ls} - q_{ls0}), \qquad q_{ls0} = \frac{4\pi}{3} r_{dm}^3 \rho_w N_d, \tag{14.6.2a}$$

$$K_{MC} = \pi r_d^2 E_c V_{dt}(r_d) N_d, \qquad r_d = (3q_{ls}/4\pi\rho_w N_d)^{1/3}. \tag{14.6.2b}$$

The threshold q_{ls0} is determined in this scheme by the threshold radius $r_{dm} = 10\,\mu\mathrm{m}$. The conversion coefficient K_{MC} is treated as the mean frequency of collisions among cloud droplets which become raindrops. In contrast to Kessler's parameterization, it is not constant but is expressed via the mean volume radius of cloud drops r_d related to the mean q_{ls}, the mean collision efficiency $E_c = 0.55$ assumed by Manton and Cotton, the terminal velocity of drops with radius r_d described by the Stokes' law $V_{dt} = c_s r_d^2$, where c_s is the Stokes constant (see Chapter 12), and the drop concentration N_d. Note that K_{MC} here resembles the collision efficiencies (Eqn. 14.2.10a) or collection rates (Eqns. 14.2.20) and (14.2.20a), the reason for this is discussed in the following text. After substitution of K_{MC} and r_d from (Eqn. 14.6.2b) into (Eqn. 14.6.2a), the autoconversion rate becomes

$$\frac{dq_{ll}}{dt} = K_{MC} q_{ls} = B_{aut} q_{ls}^{7/3} N_d^{-1/3} \theta(q_{ls} - q_{ls0}), \tag{14.6.3a}$$

$$B_{aut} = \pi c_s E_c \left(\frac{3}{4\pi\rho_w} \right)^{4/3}. \tag{14.6.3b}$$

Baker (1993) studied autoconversion in the marine boundary layer Sc clouds and, in order to account for the uncertainty in the self-collection rates within the same fraction, introduced an empirical multiplier γ_B. Thus, the autoconversion rate (Eqn. 14.6.3a) was modified as

$$\frac{dq_{ll}}{dt} = B_{aut} \gamma_B q_{ls}^{7/3} N_d^{-1/3} \theta(q_{ls} - q_{ls0}). \tag{14.6.4}$$

The coefficient γ_B was varying between 0.01 and 0.1 to decrease the autoconversion rate. Berry (1967), Khairoutdinov and Kogan (2000), and some other researchers used the results of numerical simulations with the spectral bin models and subsequent curve fits to refine the coefficients in relations like (Eqn. 14.6.4). The numerical simulations were usually performed for some specific cloud types. Therefore, these parameterizations are applicable for certain cloud types and may be different for others.

A comprehensive treatment of the autoconversion rates was performed by Liu and Daum (2004). They considered the mass growth rates in a polydisperse ensemble of drops, gave the derivations of several previous functional forms similar to (Eqn. 14.6.4) and suggested a newer parameterization. We will use an analogous but somewhat different way and will derive the autoconversion rate based on the coagulation kinetic equations of Section 14.2. Consider again the loss term in (Eqn. 14.2.2b) and represent the size spectrum as the sum (Eqn. 14.2.3) of the small and large fractions, $f = f_s + f_l$. Then, the term for self-collection $I_{loss,ss}$ of the small fraction that was omitted in (Eqn. 14.2.6) in the continuous growth approximation occurs now and accounts for autoconversion:

$$I_{loss,ss} = f_s(m) \int_0^{m_0} K(m,m') f_s(m') \, dm'. \tag{14.6.5}$$

It is convenient to rewrite this equation in terms of radii. Taking into account that loss of the small-size fraction $I_{loss,ss}$ due to self-collection is equal to the tendency $(\partial f_s/\partial t)_{aut}$, we obtain

$$\left(\frac{\partial f_s}{\partial t} \right)_{aut} = -I_{loss,ss} = -f_s(r) \int_0^{r_0} K(r,r') f_s(r') \, dr' = -K_{aut} f_s(r), \tag{14.6.6}$$

where we introduced the rate coefficient

$$K_{aut} = \int_0^{r_0} K(r,r') f_s(r') \, dr'. \tag{14.6.7}$$

It will be shown in the following text that this K_{aut} coincides with the autoconversion coefficient. Multiplying this equation by the drop mass $m(r) = (4/3)\pi\rho_w r^3$, and integrating over r yields

$$\left(\frac{dq_{ls}}{dt}\right)_{aut} = \int_0^{r_0} \left(\frac{\partial f_s}{\partial t}\right)_{aut} m(r)dr$$
$$= -\int_0^{r_0} f_s(r)m(r)dr \int_0^{r_0} K(r,r')f_s(r')dr'. \quad (14.6.8)$$

The first integral on the right-hand side is q_{ls} and we obtain the rate of the mass loss for the small fraction due to autoconversion:

$$\left(\frac{dq_{ls}}{dt}\right)_{aut} = -K_{aut}q_{ls}. \quad (14.6.9)$$

From the mass balance, we have a relation between the autoconversion gain rate $(dq_{ll}/dt)_{aut}$ in the large fraction, and the autoconversion loss $(dq_{ls}/dt)_{aut}$ in the small fraction:

$$\left(\frac{dq_{ll}}{dt}\right)_{aut} = -\left(\frac{dq_{ls}}{dt}\right)_{aut} = K_{aut}q_{ls}. \quad (14.6.10)$$

This is the usual autoconversion equation and K_{aut} is the autoconversion coefficient.

In can be shown that (Eqn. 14.6.10) with the general expression for K_{aut} in (Eqn. 14.6.7) allows us to obtain various forms of the autoconversion equation. For example, substituting into (Eqn. 14.6.7) the gravitational collection kernel from (Eqn. 14.2.2c), $K(r, r') = E_c(r, r')\pi(r' + r)^2)|v(r') - v(r)|$, we obtain

$$K_{aut} = \pi \int_0^{r_0} E_c(r,r')(r+r')^2 \,|\, v_t(r) - v_t(r') \,|\, f_s(r')dr'. \quad (14.6.11)$$

Applying now the mean-value theorem to this integral (similar to Chapter 7 for drop activation) and removing $K(r, r')$ out of the integral at some value $r' = r_M$, we obtain

$$K_{aut} = \pi E_c(r,r_M)(r+r_M)^2 \,|\, v_t(r) - v_t(r_M) \,|\, \int_0^{r_0} f_s(r')dr'$$
$$= \pi E_c(r,r_M)(r+r_M)^2 \,|\, v_t(r) - v_t(r_M) \,|\, N_s. \quad (14.6.12)$$

Various versions of K_{aut} can be obtained by specifying various r_M—e.g., by tuning to the results of numerical models. A particular case is obtained assuming, following Liu and Daum, the approximations $r >> r_M$, $(r_M + r)^2 \approx r^2$, and $v_t(r) - v_t(r_M) \approx v_t(r)$, similar to the continuous growth approximation in Section 14.2 for the large and small drops. Thus, K_{aut} can be written as

$$K_{aut} = \pi E_c(r,r_M)r^2 v_t(r)N_s. \quad (14.6.13)$$

Equation (14.6.13) was derived by Liu and Daum (2004) and shows that the Manton and Cotton parameterization (Eqns. 14.6.2a) and (14.6.2b) can be derived from (Eqn. 14.6.7) with the gravitational kernel, and this form implicitly assumes a kind of continuous growth approximation, although within the same size fraction. Liu and Daum (2004) considered a more precise equation for K_{aut} using Long's (1974) approximation for the collection kernel, $K(r, r') \approx \kappa_L r^6$ valid at $r < 50\,\mu\text{m}$ with $\kappa_L \approx 1.9 \times 10^{11}$ cm^{-3} s^{-1},

r in cm, and $K(r, r')$ in cm^3 s^{-1}. Substitution of this $K(r, r')$ into (Eqn. 14.6.7) yields the newer parameterization by Liu and Daum (2004):

$$\left(\frac{dq_{ll}}{dt}\right)_{aut} = K_{aut}^{LD} q_{ls}\theta(r_6 - r_{6c}) \qquad K_{aut}^{LD} = \kappa_L N_s r_6^6, \tag{14.6.14}$$

where r_6 is the mean radius of the sixth moment of the spectrum and r_{6c} determines the threshold of autoconversion. If the spectrum is the gamma distribution, $f_s(r) = c_N r^p \exp(-\beta r)$ (see Chapter 2), then r_6 is

$$r_6 = \frac{1}{\beta}\left[\frac{\Gamma(p+7)}{\Gamma(p+1)}\right]^{1/6}. \tag{14.6.15}$$

Expressing the mean radius via q_{ls} and N_s as was done in the cited works and substituting into (Eqn. 14.6.4), Liu and Daum presented the autoconversion rate in the form, allowing comparison with the previous works:

$$\left(\frac{dq_{ll}}{dt}\right)_{aut} = c_{LD} q_{ls}\theta(r_6 - r_{6c}), \tag{14.6.16a}$$

$$\left(\frac{dq_{ll}}{dt}\right)_{aut} = \alpha_{LD} N_s^{-1/3} q_{ls}^{7/3}\theta(r_6 - r_{6c}) \tag{14.6.16b}$$

$$c_{LD} = \alpha_{LD} N_s^{-1/3} q_{ls}^{4/3}, \qquad \alpha_{LD} = \left(\frac{3}{4\pi\rho_w}\right)^2 \kappa_L \beta_6^6 \left(\frac{q_{ls}}{N_s}\right)^{2/3}, \tag{14.6.17}$$

where β_6 is related to the relative dispersion of the spectrum σ_{rr},

$$\beta_6 = \left[\frac{(1+3\sigma_{rr}^2)(1+4\sigma_{rr}^2)(1+5\sigma_{rr}^2)}{(1+\sigma_{rr}^2)(1+2\sigma_{rr}^2)}\right]. \tag{14.6.18}$$

Equation (14.6.16a) is a linearized by the q_{ls} form similar to (Eqns. 14.6.2a), (14.6.2b), and (14.6.16a) and is a non-linear form by the q_{ls} and N_s, which extends the Manton and Cotton parameterization (Eqn. 14.6.3a). Thus, (Eqns. 14.6.16)–(14.6.18) of Liu and Daum contain all the factors of the previous parameterizations and include additionally the spectral dispersion in the general form. It is clear in advance that the autoconversion rate should depend on the width of the spectra and should be different in clouds of various forms with different spectral dispersions. Accounting for the effect of the spectral dispersion on the autoconversion rate in Liu and Daum's parameterization allows greater flexibility and higher accuracy than in the previous parameterizations. Indeed, this was demonstrated in Liu and Daum (2004) through numerical calculations and by comparing various parameterizations.

The spectral dispersion σ_{rr}, and therefore β_6, are external parameters in the Liu and Daum parameterization that can be varied or taken from the measurements like those described in Section 13.8. However, they can be expressed via the cloud properties. Substituting the relation (Eqn. 2.4.33) between σ_{rr} and the index of gamma distribution p, $\sigma_{rr} = (p + 1)^{-1/2}$, the expression (Eqn. 14.6.18) can be rewritten via p as

$$\beta_6 = \left[\frac{(p+4)(p+5)(p+6)}{(p+1)(p+2)(p+3)}\right]. \tag{14.6.19}$$

Then, p can be evaluated using (Eqns. 13.8.4) and (13.8.5) analyzed in Sections 13.8–13.10, and β_6 can be related to the cloud and atmospheric properties, this allows us to express β_6, c_{LD}, α_{LD}, and $(dq_{ll}/dt)_{aut}$ analytically and to study the effects on the autoconversion rates of atmospheric turbulence, LWC and its gradients, temperature, lapse rates, etc., in various clouds.

The autoconversion kinetic (Eqn. 14.6.6) with various specifications of K_{aut} described earlier lets us introduce corrections to the analytical solutions for the size spectrum of the small fraction obtained in Section 13.7. The structure of the autoconversion term in (Eqn. 14.6.6) for the small-size fraction, $(\partial f_s/\partial t)_{aut} = -K_{aut}f_s$, is similar to the coagulation term in the general kinetic (Eqn. 13.7.1) for f_s, which is the last term in (Eqn. 13.7.1), $(\partial f_s/\partial t)_{col} = -\sigma_{col}f_s$. Thus, (Eqn. 13.7.1) can be further generalized to account for autoconversion by adding the term $-K_{aut}f_s$ on the right-hand side. The terms with collection and autoconversion are similar, and the solutions (Eqns. 13.7.2) and (13.7.5) can be modified by replacing $-\sigma_{col}f_s$ with the sum $-[\sigma_{col} + K_{aut}\,\theta(q_{ls} - q_{ls0})]f_s$. Thus, the solution for the tail of the spectrum at large r is again (Eqn. 13.7.2), but with the slope β_{s2} from (Eqn. 13.7.3) modified as

$$\beta_{s2}(r) = \frac{1}{b_{con}s}\left[\frac{1}{2}\left(\alpha_s w + \frac{d(w-\alpha_s k)}{dz} + \sigma_{col}^{aut} - \alpha_s^2 k\right) - \frac{\alpha_s v(r)}{B_v + 2}\right], \tag{14.6.20a}$$

$$\sigma_{col}^{aut} = \sigma_{col} + K_{aut}\theta(q_{ls} - q_{ls0}). \tag{14.6.20b}$$

The general solution is again (Eqn. 13.7.5) with this modified $\beta_{s2}(r)$. The physical interpretation of this solution is clear: The effect of autoconversion is similar to the effect of coagulation; autoconversion leads to an increase in the slope parameter and a faster decrease of f_s with r. Equations (14.6.16a) and (14.6.16b) allow quantitative evaluation of the effect of autoconversion on the spectra in various situations.

Equations like (Eqns. 14.6.3a) or (14.6.16b) that contain $q_{ls}^{7/3}$ are nonlinear relations relative to q_{ls}, which may make parameterizations of the size spectra more complicated. It is more convenient to use linearized forms like (Eqn. 14.6.16a), which can be done by splitting them into two time steps as is often done in numerical models:

$$\frac{dq_{ll}^{(n)}}{dt} = K_{aut}^{(n-1)} q_{ls}^{(n)} \theta(q_{ls}^{(n-1)} - q_{ls0}^{(n-1)}), \tag{14.6.21a}$$

$$\frac{dq_{ls}^{(n)}}{dt} = -K_{aut}^{(n-1)} q_{ls}^{(n)} \theta(q_{ls}^{(n-1)} - q_{ls0}^{(n-1)}), \tag{14.6.21b}$$

where the subscript $(n-1)$ means the quantities from the previous time step, and n means the quantities from the current time step. These quantities can be used in (Eqn. 14.6.16b), which allows us to account for the effect of autoconversion on the size spectra.

14.7. The Coagulation Equation as the Integral Chapman–Kolmogorov and Differential Fokker–Planck Equations

As mentioned in Section 13.12, the integral Smoluchowski–Müller coagulation equation considered in Sections 5.6 and 14.2 is similar in structure to the generalized integral kinetic (Eqn. 13.12.1b) or the Chapman–Kolmogorov (Eqn. 13.12.1a). This analogy lets us derive the integral coagulation equation from the generalized kinetic (Eqn. 13.12.1b) and then derive the differential analog of the coagulation equation in the form of the Fokker–Planck equation. The generalized stochastic kinetic (Eqn. 13.12.1b) derived in Section 13.12 from the Chapman–Kolmogorov equation can be written in the four-dimensional phase space in terms of masses instead of radii as

$$\frac{\partial f(\vec{x},m,t)}{\partial t} = \int W_{xm}(\vec{x}-\vec{x}',m-m' \mid \vec{x}',m') f(\vec{x}-\vec{x}',m-m',t) f(\vec{x}',m',t) d\vec{x}'m'$$
$$- f(\vec{x},m,t) \int W_{xm}(\vec{x},m \mid \vec{x}',m') f(\vec{x},m',t) d\vec{x}'dm' \qquad (14.7.1)$$

Here, as in (Eqns. 13.12.1a) and (13.12.1b), the first and second terms represent gain and loss of the particles per unit of time in an element of the 4D phase space $(\vec{x},m)$. Recall, the probabilities of transition W_{xm} in (Eqn. 14.7.1) are defined per unit time and per particle and have dimension [cm^3 s^{-1}]. Let us assume that: a) the stochastic jumps $(\vec{x}',m')$ in the 4D phase space are caused by the acts of coagulation, b) the corresponding probabilities of transition $W_{xr}(\vec{x}-\vec{x}',m-m' \mid \vec{x}',m')$ and $W_{xm}(\vec{x},m \mid \vec{x}',m')$ in the spatially homogeneous medium do not depend on the coordinates x_α, and c) are described by the coagulation kernels $(1/2)K(m - m', m')$ and $K(m, m')$. Thus, (Eqn. 14.7.1) becomes

$$\left(\frac{\partial f(m)}{\partial t}\right)_{col,ls} = \frac{1}{2}\int_0^m K(m-m',m') f(m-m') f(m') dm'$$
$$- f(m) \int_0^\infty K(m,m') f(m') dm'. \qquad (14.7.2)$$

This is a familiar form of the coagulation equation where the first and second terms on the right-hand side represent the gain and loss of particles, I_{gain} and I_{loss}. Note that (Eqn. 14.7.2) is a particular case of (Eqn. 14.2.1) for the spatially homogeneous case, and a more general equation can be derived from (Eqn. 14.7.1) that would also include coordinate dependence. The coagulation kernel can be specified for various processes, as was described for gravitational coagulation in Sections 5.5 and 14.2, for Brownian coagulation as described in Chapter 10, or for other types of coagulation.

As was shown in Section 13.12, a differential analog of this integral equation can be derived by expansion of the kernels into the power series. We assume again, as in Section 14.2, that $f(m)$ can be represented as a sum of the small- and large-size fractions, $f(m) = f_l(m) + f_s(m)$. Assuming in the spirit of the Chapman–Kolmogorov equation that the stochastic "jumps" m' are small, the subintegral expression in (Eqn. 14.7.2) for I_{gain} can be expanded into the Taylor power series by the small parameter m':

$$K(m-m',m') f_l(m-m') = \sum_{n=0}^{\infty} \frac{(-1)^n}{n!} \frac{\partial^n}{\partial m^n} [K(m,0) f_l(m)] m'^n. \qquad (14.7.3)$$

The expansion with an infinite number of terms is called *the Kramers–Moyal series*. Such infinite expansion is mathematically equivalent to the original integral (Eqn. 14.7.2) but evaluation of the higher moments is difficult and the expansion is usually truncated at the first few terms. The continuous growth approximation used in many cloud models was derived in Section 14.2 by truncating this expansion at the first-order term, which yields the drift velocity in the m- or r-spaces but does not describe the diffusion in these spaces. Accounting for the next terms may improve the accuracy of the equation. Assuming as in Section 14.2 that $m >> m'$, expanding the kernel as in (Eqn. 14.7.3) and truncating the series at the terms of the second order with $n = 2$, we have

$$K(m-m',m')f_l(m-m') \approx K(m,0)f_l(m) - \frac{\partial}{\partial m}[K(m,0)f_l(m)]m' + \frac{1}{2}\frac{\partial^2}{\partial m^2}[K(m,0)f_l(m)]m'^2. \tag{14.7.4}$$

Substitution of this expression into (Eqn. 14.2.5) yields the gain term

$$I_{gain} = K(m,0)f_l(m)N_s - \frac{\partial}{\partial m}[K(m,0)f_l(m)]q_{ls} + \frac{1}{2}\frac{\partial^2}{\partial m^2}[K(m,0)f_l(m)]M_s^{(2)}, \tag{14.7.5}$$

where N_s and q_{ls} are the concentration and liquid (ice) water content of the small fraction as in (Eqn. 14.2.10) and $M_s^{(2)}$ is the second moment of the size spectrum by mass m':

$$M_s^{(2)} = \int_0^{m_0} m'^2 f_s(m')dm'. \tag{14.7.6a}$$

If the particles of the small fraction have the shape close to spherical, then the mass $m = (4/3)\pi\rho_w r^3$. Substituting this relation into (Eqn. 14.7.6a) yields

$$M_s^{(2)} = \left(\frac{4\pi\rho_w}{3}\right)^2 \int_0^{r_0} r^6 f_s(r)dr = \left(\frac{4\pi\rho_w}{3}\right)^2 Z_{r6}, \tag{14.7.6b}$$

where Z_{r6} is the 6th moment of the distribution by radii, proportional to the radar reflectivity. Assuming that the spectrum can be approximated by the gamma distribution $f_s(r) \sim r^p\exp(-\beta r)$, and that the upper limit here can be extended to infinity, the moment Z_{r6} is expressed analytically as described in Section 2.4, (Eqn. 2.4.36), and is in these notations:

$$Z_{r6} = \langle r^6 \rangle = N_s \bar{r}^6 c_{r6},$$
$$c_{r6} = \frac{(p+6)(p+5)(p+4)(p+3)(p+2)}{(p+1)^5}, \tag{14.7.6c}$$

where $\bar{r}$ is the mean radius of the spectrum. The coefficient $c_{r6} = 5.65$ for $p = 6$ and $c_{r6} = 27.65$ for $p = 2$.

The loss term is in this approximation is

$$I_{loss} = K(m,0)f_l(m)N_s. \tag{14.7.7}$$

Substituting I_{gain} (Eqn. 14.7.5) and I_{loss} (Eqn. 14.7.7) into the kinetic (Eqn. 14.7.2) written for f_l, we notice that I_{loss} is cancelled with the first term in I_{gain}. The rest yields

$$\frac{\partial f_l}{\partial t} = I_{gain} - I_{loss} = -\frac{\partial}{\partial m}[K(m,0)q_{ls}f_l(m)] + \frac{1}{2}\frac{\partial^2}{\partial m^2}[K(m,0)f_l(m)]M_s^{(2)}. \tag{14.7.8}$$

Introducing now the drift velocity $\dot{m}_{drift}$ and the diffusion coefficient D_{dif} in the mass space:

$$\dot{m}_{drift} = K(m,0)q_{ls}, \qquad D_{dif} = (1/2)K(m,0)M_s^{(2)}, \tag{14.7.9}$$

and (Eqn. 14.7.8) can be rewritten as

$$\frac{\partial f_l(m)}{\partial t} = -\frac{\partial}{\partial m}[\dot{m}_{drift} f_l(m)] + \frac{\partial^2}{\partial m^2}[D_{dif} f_l(m)] \tag{14.7.10}$$

This is the kinetic equation of the Fokker–Planck type with drift and diffusion. It can be applied for various types of coagulation, for which approximations made here are applicable. This can be done by specifying different coagulation kernels $K(m, 0)$ and will result in different drift velocities and diffusion coefficients.

In the following text, we consider the gravitational coagulation. The gravitational kernel is defined by (Eqn. 14.2.7), $K(m, 0) = \pi E_c r^2 v(r)$. Substituting it into (Eqn. 14.7.5), we obtain

$$I_{gain} = f_l(m)[\pi E_c r^2 v(r) N_s] - \frac{\partial}{\partial m}[\pi E_c r^2 v(r) q_{ls} f_l(m)]$$

$$+\frac{1}{2}\frac{\partial^2}{\partial m^2}[\pi E_c r^2 v(r) f_l(m)] M_s^{(2)}. \tag{14.7.11}$$

The loss term I_{loss} in the same approximation is described by (Eqn. 14.2.11), $I_{loss} = f_l(m)[\pi E_c r^2 v(r) N_s]$. Substituting I_{gain} and I_{loss} into the kinetic (Eqn. 14.2.1) shows that the term I_{loss} is again mutually cancelled with the first term in I_{gain}, and the rest yields

$$\frac{\partial f_l}{\partial t} = I_{gain} - I_{loss} = -\frac{\partial}{\partial m}[\pi E_c r^2 v(r) q_{ls} f_l(m)]$$

$$+\frac{1}{2}\frac{\partial^2}{\partial m^2}[\pi E_c r^2 v(r) f_l(m)] M_s^{(2)} \tag{14.7.12}$$

Introducing the drift velocity $\dot{m}_{m,coag}$ and a diffusion coefficient $D_{m,coag}$ in the mass space,

$$\dot{m}_{m,coag} = K(m,0)q_{ls} \approx \pi E_c r^2 v(r) q_{ls}, \tag{14.7.13}$$

$$D_{m,coag} = (1/2)K(m,0)M_s^{(2)} \approx (\pi/2)E_c r^2 v(r) M_s^{(2)}, \tag{14.7.14}$$

and so (Eqn. 14.7.12) can be rewritten as

$$\frac{\partial f_l(m)}{\partial t} = -\frac{\partial}{\partial m}\left[\dot{m}_{m,coag} f_l(m)\right] + \frac{\partial^2}{\partial m^2}\left[D_{m,coag} f_l(m)\right]. \tag{14.7.15}$$

Thus, we derived a differential kinetic equation of the Fokker–Planck type for coagulation with drift and diffusion. The expression for $\dot{m}_{m,coag}$ is the same as in the continuous growth approximation (Eqn. 14.2.13), while the diffusion coefficient $D_{m,coag}$ is new. It arose from the third term in the expansion (Eqn. 14.7.4) of the coagulation kernel. It can be written in a more explicit form using (Eqns. 14.7.6b) and (14.7.6c) for $M_s^{(2)}$ and Z_{r6},

$$D_{m,coag} = \frac{8\pi^3}{9} E_c c_{r6} \rho_w^2 r^2 v(r) N_s \bar{r}^6. \tag{14.7.16}$$

In particular, if $v(r) \sim r^2$, then $D_{m,coag} \sim r^4 \sim m^{4/3}$; if $v(r) \sim r$, then $D_{m,coag} \sim r^3 \sim m$. The dimension of $D_{m,coag}$ in (Eqn. 14.7.16) is [$g^2 s^{-1}$] which ensures the correct dimension of the diffusion term in (Eqn. 14.7.15). The values of $D_{m,coag}$ can be calculated using (Eqn. 14.7.16) or estimated from the measurements of radar reflectivity and Z_{r6} of the small fraction. Equation (14.7.15) can be rewritten in the form of a continuity equation, as was done in Section 8.2 for the nucleation theory:

$$\frac{\partial f_l(m)}{\partial t} = -\frac{\partial j_{m,coag}}{\partial m}, \tag{14.7.17}$$

where the flux $j_{m,coag}$ in the m-space is

$$j_{m,coag} = \dot{m}_{m,coag} f_l(m - \frac{\partial}{\partial m}\left[D_{m,coag} f_l(m)\right] \tag{14.7.18a}$$

$$= \tilde{\dot{m}}_{m,coag} f_l(m) - D_{m,coag}\frac{\partial f_l(m)}{\partial m}. \tag{14.7.18b}$$

The effective drift velocity introduced in the second equation is

$$\tilde{\dot{m}}_{m,coag} = \dot{m}_{m,coag} - \frac{\partial D_{m,coag}}{\partial m}. \tag{14.7.19}$$

The previous continuous growth approximation (Eqns. 14.2.12) and (14.2.13) of Section 14.2 predicted monotonous growth of the particles of the large-size fraction with the drift velocity $\dot{m}_{m,coag}$. Equations (14.7.15)–(14.7.19) of the newer differential analog of the coagulation equation of the Fokker–Planck type provide the corrections to the continuous growth and predict a more complicated growth. According to (Eqn. 14.7.16), $D_{m,coag}$ is a growing function of m, thus $\partial D_{m,coag}/\partial m > 0$, and the effective drift velocity $\tilde{\dot{m}}_{m,coag}$ is smaller than $\dot{m}_{m,coag}$. This factor hampers particle growth by coagulation. The second term in Equation (14.7.18b) for the flux, $-D_{m,coag}(\partial f_l/\partial m) > 0$, is negative at masses smaller than the modal mass m_m, $m < m_m$, where $(\partial f_l/\partial m) > 0$, and is positive at $m > m_m$ where $(\partial f_l/\partial m) < 0$. Thus, presence of the diffusion in the mass space hampers particle growth at smaller masses with $m < m_m$ and accelerates their growth at large masses with $m > m_m$. Therefore, the diffusion leads to broadening of the size spectrum and accelerates coagulation.

Equation (14.7.15) can be rewritten in terms of radii. Using the conservation law $f_l(m)dm = f_l(r)dr$, we obtain

$$\frac{\partial f_l(r)}{\partial t} = -\frac{\partial}{\partial r}\left[\dot{m}_{m,coag}\frac{dr}{dm} f_l(m)\right] + \frac{\partial}{\partial r}\left\{\frac{dr}{dm}\frac{\partial}{\partial r}\left[D_{m,coag}\frac{dr}{dm} f_l(m)\right]\right\}. \tag{14.7.20}$$

Using the relation for spherical particles $dm/dr = 4\pi\rho_w r^2$, we can introduce the drift velocity $\dot{r}_{coag}$ (the radius growth rate) and diffusion coefficient $D_{r,coag}$ in the radii space:

$$\dot{r}_{coag} = \dot{m}_{m,coag}\frac{dr}{dm} = \frac{E_c v(r) q_{ls}}{4\rho_w}, \tag{14.7.21}$$

$$D_{r,coag} = D_{m,coag}\frac{dr}{dm} = \frac{2\pi^2}{9} E_c \rho_w v(r) Z_{r6} = \frac{2\pi^2}{9} E_c c_{r6} \rho_w v(r) N_s \bar{r}^6. \tag{14.7.22}$$

Thus, we can rewrite (Eqn. 14.7.20) as

$$\frac{\partial f_l(r)}{\partial t} = -\frac{\partial}{\partial r}[\dot{r}_{coag} f_l(r)] + \frac{\partial}{\partial r}\left\{\frac{1}{4\pi\rho_w r^2}\frac{\partial}{\partial r}[D_{r,coag} f_l(r)]\right\}. \qquad (14.7.23)$$

Equation (14.7.23) is the differential equation of the Fokker–Planck type for the size spectrum taking into account the drift and diffusion in the space of the radii and generalizes (Eqn. 14.2.15) of Section 14.2 for coagulation in the continuous growth approximation that accounts only for the drift but does not account for the diffusion. Equation (14.7.23) contains the term with the drift velocity which coincides with (Eqn. 14.2.16) for the radius growth rate $\dot{r}_{coag}$, and the new term with diffusion. As discussed earlier for this equation in terms of masses, the new term in (Eqn. 14.7.23) with diffusion in the radii space $D_{r,coag}$ also causes faster spectral broadening and accelerates coagulation as compared to the previous continuous growth approximation without diffusion.

Note finally that (Eqns. 14.7.15) and (14.7.23) of the Fokker–Planck type are the differential analogs of the integral coagulation equations (Eqns. 5.6.4) and (5.6.7) of Section 5.6 or (Eqns. 14.2.1)–(14.2.1a) and (14.2.1b) of Section 14.2 and of the integral Chapman–Komogorov (Eqn. 13.7.1a). A similar analogy exists in the atmospheric models. The majority of the models are based on using the differential equations of the Fokker–Planck type with advective and turbulent transport, where the drift velocity is the wind speed and the diffusion is described by the turbulent coefficient. They are used for routine calculations of the atmospheric processes, while only a few stochastic Lagrangian models of turbulent diffusion are based on using the integral equations of the Chapman–Kolmogorov type (e.g., Rodean, 1996; van Kampen, 1992; Monin and Yaglom, 2007a,b). It is known that solving the integral equations of coagulation is computationally very expensive while solving the differential equations is simpler and faster. Even the simplest approximation of continuous growth extensively used in cloud modeling beginning from the pioneering work of Kessler (1969), and applied in many models for parameterizing conversion rates, appeared to be successful (Pruppacher and Klett, 1997; Cotton et al., 2011). Accounting for the next terms in expansion of the coagulation kernel and generalization taking into account diffusion in the space of mass or radii described in this section represents the next step in developing the approach of differential equations. Testing the differential equations for coagulation described here versus full integral equations of coagulation and further development of this approach (e.g., by accounting for the next terms in the Kramers–Moyal expansion (Eqn. 14.7.3)) may allow us to develop fast and efficient methods for calculation of coagulation–accretion based on the differential equations of the Fokker–Planck type.

Appendix A.14 Evaluation of the Integrals in Section 14.4.2 for $v(r) = A_v r$

If $v(r) = A_v r$, then $B_v = 1$ and (Eqns. 14.4.12a) and (14.4.12b) for I_1, I_2 in Section 14.4.1 can be rewritten as

$$I_1 = -\beta_{l0} b_{con} s \int_{r_1}^{r_2} \frac{dr}{b_{con} s + \chi_c A_v r^2}, \qquad \text{(A.14.1)}$$

$$I_2 = \alpha_l (w - \alpha_l k) \int_{r_1}^{r_2} \frac{r dr}{b_{con} s + \chi_c A_v r^2}. \qquad \text{(A.14.2)}$$

We introduce the notation

$$R_l = \left(\frac{b_{con} \,|\, s \,|}{\chi_c A_v} \right)^{1/2} = r \left(\frac{|\, \dot{r}_{cond}(r) \,|}{\dot{r}_{coag}(r)} \right)^{1/2}, \qquad \text{(A.14.3a)}$$

$$U_l = \beta_{l0} \frac{(\alpha_l k - w)}{2 A_v}, \qquad Q_l = \frac{\beta_0 R_l}{2}, \qquad \text{(A.14.3b)}$$

where $|x|$ means the absolute value of x. Here, the dimensions in CGS units are $[A_v] = \text{s}^{-1}$, $[R_l] = \text{cm}$. Then, I_1 for $s > 0$ can be rewritten as

$$I_1 = -\beta_{l0} R_l \int_{x_0}^{x} \frac{dx}{1 + x^2}, \qquad \text{(A.14.4)}$$

where $x = r/R_l$, $x_0 = r_0/R_l$. This is the standard table integral, and its evaluation yields

$$I_1 = -\beta_{l0} R_l [\arctan(r/R_l) - \arctan(r_0/R_l)]. \qquad \text{(A.14.5)}$$

For $s < 0$, using the relation $s = -|s|$, we see that the terms in the denominator of (Eqn. A.14.1) have different signs thus I_1 can be rewritten as

$$I_1 = -\beta_{l0} R_l \int_{x_0}^{x} \frac{dx}{1 - x^2}. \qquad \text{(A.14.6)}$$

This is also the standard table integral that is different for $|x| < 1$ and $|x| > 1$—that is:

$$I_1 = -\frac{\beta_{l0} R_l}{2} \ln \frac{1 + x}{1 - x} \bigg|_{x_0}^{x}, \quad \text{for} \quad |\, x \,| < 1 \qquad \text{(A.14.7)}$$

$$I_1 = -\frac{\beta_{l0} R_l}{2} \ln \frac{x - 1}{x + 1} \bigg|_{x_0}^{x} \quad \text{for} \quad |\, x \,| > 1. \qquad \text{(A.14.8)}$$

For $r < R_l$, we have $|x| < 1$, and so (Eqn. A.14.7) is reduced to

$$I_1 = -Q_l \ln \left(\frac{1 + r/R_l}{1 - r/R_l} \right) \bigg|_{r_0}^{r}$$

$$= \ln \left[\frac{(1 - r/R_l)(1 + r_0/R_l)}{(1 - r_0/R_l)(1 + r/R_l)} \right]^{Q_l}, \quad s < 0, \quad r < R_l. \qquad \text{(A.14.9)}$$

For $r > R_l$, we have $|x| > 1$, and (Eqn. A.14.8) is reduced to

$$I_1 = -Q_l \ln\left(\frac{r/R_l - 1}{r/R_l + 1}\right)\Bigg|_{r_0}^{r}$$

$$= \ln\left[\frac{(r/R_l + 1)(r_0/R_l - 1)}{(r_0/R_l + 1)(r/R_l - 1)}\right]^{Q_l}, \qquad s < 0, \qquad r > R_l. \tag{A.14.10}$$

The integral I_2 in (Eqn. A.14.2) can be transformed for $s > 0$ as

$$I_2 = 2U_l \int_{x_0}^{x} \frac{xdx}{x^2 + 1}, \qquad x = \frac{r}{R_l}, \tag{A.14.11}$$

and (Eqn. A.14.11) is evaluated with the standard integral

$$I_2 = U_l \ln(1 + x^2)\Big|_{x_0}^{x} = \ln\left[\frac{1 + (r/R_l)^2}{1 + (r_0/R_l)^2}\right]^{U_l}, \qquad s > 0. \tag{A.14.12}$$

If $s < 0$, then $x < 0$ and I_2 can be transformed as

$$I_2 = 2U_l \int_{x_0}^{x} \frac{xdx}{x^2 - 1}, \qquad x = \frac{r}{R_l}. \tag{A.14.13}$$

This integral is also standard and so we finally get

$$I_2 = \ln\left[\frac{(r/R_l)^2 - 1}{(r_0/R_l)^2 + 1}\right]^{U_l}, \qquad s < 0. \tag{A.14.14}$$

This yields a factor in the size spectrum

$$f_l \sim \exp(-I_2) = \left[\frac{(r_0/R_l)^2 + 1}{(r/R_l)^2 - 1}\right]^{U_l} \tag{A.14.15}$$

These expressions for I_1 and I_2 and the corresponding factors in $f_l(r)$ are used in Section 14.4.2 for the size spectra of the large-size fraction.

Appendix B.14 Evaluation of the Integrals in Section 14.4.4 for $v(r) = A_v r^{1/2}$ (Aerodynamic Regime of Fall Velocities)

If $v(r) = A_v r^{1/2}$, then $B_v = 1/2$ and (Eqns. 14.4.12a) and (14.4.12b) for I_1, I_2 in Section 14.4.2 can be rewritten for $s > 0$ as

$$I_1 = -\beta_{l0} H_l I_{11}, \qquad I_{11} = \int_{t_0}^{t} \frac{dt}{1+t^{3/2}} \tag{B.14.1}$$

$$I_2 = \frac{V_l}{2} I_{21}, \qquad I_{21} = \int_{t_0}^{t} \frac{tdt}{1+t^{3/2}}, \tag{B.14.2}$$

where

$$t_0 = \frac{r_0}{H_l}, \qquad t = \frac{r}{H_l} \qquad H_l = \left(\frac{b_{con} \,|\, s \,|}{\chi_c A_v}\right)^{2/3},$$

$$V_l = \frac{2\alpha_l (w - \alpha_l k) H_l^{1/2}}{\chi_c A_v}. \tag{B.14.3}$$

Here, the dimensions in CGS units are $[A_v] = \mathrm{cm}^{1/2}\ \mathrm{s}^{-1}$, $[H_l] = \mathrm{cm}$, and V_l is dimensionless. Introducing a new variable $x = t^{1/2}$, the integral I_{11} can be transformed into that given in Gradshteyn and Ryzhik (1994):

$$I_{11} = \int_{t_0}^{t} \frac{dt}{1+t^{3/2}} = 2\int_{x_0}^{x} \frac{xdx}{1+x^3}$$
$$= 2\left[\frac{1}{6}\ln\frac{1-x+x^2}{(1+x)^2} + \frac{1}{\sqrt{3}}\arctan\frac{2x-1}{\sqrt{3}}\right]\Bigg|_{x_0}^{x}. \tag{B.14.4}$$

Then, finally

$$I_1 = 2\beta_{l0} H_l \left\{\ln\left[\frac{(1+x)^2(1-x_0+x_0^2)}{(1+x_0)^2(1-x+x^2)}\right]^{1/6} - \frac{1}{\sqrt{3}}\left[\arctan\frac{2x-1}{\sqrt{3}} - \arctan\frac{2x_0-1}{\sqrt{3}}\right]\right\}. \tag{B.14.5}$$

Here and in subsequent equations, $x = (r/H_l)^{1/2}$.

The integral I_{21} in (Eqn. B.14.2) can be evaluated with the same substitution, $x = t^{1/2}$:

$$I_{21} = \int_{t_0}^{t} \frac{tdt}{1+t^{3/2}} = 2\int_{x_0}^{x} \frac{x^3dx}{1+x^3}$$
$$= 2\left[x - \frac{1}{3}\ln\frac{1+x}{(1-x+x^2)^{1/2}} - \frac{1}{\sqrt{3}}\arctan\frac{x\sqrt{3}}{2-x}\right]\Bigg|_{x_0}^{x}. \tag{B.14.6}$$

The last integral is from Gradshteyn and Ryzhik (1994). Substitution into (Eqn. B.14.2) yields

$$I_2 = V_l \left\{x - \ln\left[\frac{(1+x)(1-x_0+x_0^2)^{1/2}}{(1+x_0)(1-x+x^2)^{1/2}}\right]^{1/3} - \frac{1}{\sqrt{3}}\left[\arctan\frac{x\sqrt{3}}{2-x} - \arctan\frac{x_0\sqrt{3}}{2-x_0}\right]\right\}. \tag{B.14.7}$$

For $s < 0$, I_1 can be transformed as

$$I_1 = -\beta_{l0} H_l I_{12}, \qquad I_{12} = \int_{t_0}^{t} \frac{dt}{1 - t^{3/2}}. \tag{B.14.8}$$

Substitution $x = t^{1/2}$ allows us to reduce I_{12} to that given in Gradshteyn and Ryzhik (1994):

$$\begin{aligned} I_{12} &= \int_{t_0}^{t} \frac{dt}{1 - t^{3/2}} = 2\int_{x_0}^{x} \frac{x dx}{1 - x^3} \\ &= 2\left[-\frac{1}{6} \ln \frac{(1-x)^2}{1+x+x^2} - \frac{1}{\sqrt{3}} \arctan \frac{2x+1}{\sqrt{3}} \right]\Bigg|_{x_0}^{x}. \end{aligned} \tag{B.14.9}$$

Then, finally

$$I_1 = 2\beta_{l0} H_l \left\{ \ln\left[\frac{(1-x)^2 (1+x_0+x_0^2)}{(1-x_0)^2 (1+x+x^2)} \right]^{1/6} - \frac{1}{\sqrt{3}} \left[\arctan \frac{2x+1}{\sqrt{3}} - \arctan \frac{2x_0+1}{\sqrt{3}} \right] \right\}. \tag{B.14.10}$$

For $s < 0$, I_2 can be transformed as

$$I_2 = -\frac{V_l}{2} I_{22}, \qquad I_{22} = \int_{t_0}^{t} \frac{t dt}{1 - t^{3/2}}. \tag{B.14.11}$$

Substitution $x = t^{1/2}$ yields the form given in Gradshteyn and Ryzhik (1994):

$$\begin{aligned} I_{22} &= 2\int_{x_0}^{x} \frac{x^3 dx}{1 - x^3} \\ &= 2\left[-x + \frac{1}{3} \ln \frac{(1+x+x^2)^{1/2}}{1-x} + \frac{1}{\sqrt{3}} \arctan \frac{x\sqrt{3}}{2+x} \right]\Bigg|_{x_0}^{x}. \end{aligned} \tag{B.14.12}$$

Substitution into (Eqn. B.14.11) yields

$$I_2 = V_l \left\{ x - \ln\left[\frac{(1+x+x^2)^{1/2} (1-x_0)}{(1+x_0+x_0^2)^{1/2} (1-x)} \right]^{1/3} - \frac{1}{\sqrt{3}} \left[\arctan \frac{x\sqrt{3}}{2+x} - \arctan \frac{x_0\sqrt{3}}{2+x_0} \right] \right\}. \tag{B.14.13}$$

These four integrals are used in Section 14.4.4 for evaluation of the size spectra f_l for $v(r) = A_v r^{1/2}$.

References

Abbatt, J. P. D., S. Benz, D. J. Cziczo, Z. Kanji, U. Lohmann, and O. Möhler (2006). Solid ammonium sulfate aerosols as ice nuclei: A pathway for cirrus cloud formation. *Science*, **313**, 1770–1773.

Abdul-Razzak, H., and S. J. Ghan (2000). A parameterization of aerosol activation: 2. Multiple aerosol types. *J. Geophys. Res.*, **105**, 6837–6844.

Abdul-Razzak, H., and S. J. Ghan (2004). Parameterization of the influence of organic surfactants on aerosol activation. *J. Geophys. Res.*, **109**, D03205, doi:10.1029/2003JD004043.

Abdul-Razzak, H., S. J. Ghan, and C. Rivera-Carpio (1998). A parameterization of aerosol activation: 1. Single aerosol type. *J. Geophys. Res.*, **103**, 6123–6131.

Abraham, F. (1970). Functional dependence of drag coefficient of a sphere on Reynolds number. *Phys. Fluids*, **13**, 2194–2195.

Ackerman, S., and G. L. Stephens (1987). The absorption of solar radiation by cloud droplets: An application of anomalous diffraction theory. *J. Atmos. Sci.*, **44**, 1574–1588.

Agee, E. M. (1985). Extratropical cloud-topped boundary layers over the oceans. *Rep. JSC/CAS Workshop on Modeling of CTBL*, Fort Collins, CO, USA, WMO/TD No 75, 67 pp.

Agee, E. M. (1987). Mesoscale cellular convection over the oceans. *Dyn. Atmos. Oceans*, **10**, 317–341.

Albrecht, B. (1989). Aerosols, cloud microphysics and fractional cloudiness. *Science*, **245**, 1227–1230.

Aleksandrov, E. L., and K. B. Yudin (1979). On the vertical profile of the cloud microstructure in stratiform clouds. *Sov. Meteorol. Hydrol.*, **12**, 47–61.

Ally, M. R., S. I. Clegg, J. Braunstein, and J. M. Simonson (2001). Activities and osmotic coefficients of tropospheric aerosols: $(NH_4)_2SO_4$ (aq) and NaCl (aq). *J. Chem. Thermodynamics*, **33**, 905–915.

Al-Naimi, R., and C. P. R. Saunders (1985). Measurements of natural deposition and condensation-freezing ice nuclei with a continuous flow chamber. *Atmos., Environ.*, **19**, 1871–1882.

Angell, C. A. (1982). In: Water: A comprehensive treatise, vol. 7. Ed.: Franks, F., Plenum, New York, 1–81.

Ångström, A. (1929). On the atmospheric transmission of sun radiation and on dust in the air. *Geogr. Ann.*, **11**, 156–166.

Ångström, A. (1964). The parameters of atmospheric turbidity. *Tellus*, **16**, 64–75.

Archuleta, C. M., P. J. DeMott, and S. M. Kreidenweis (2005). Ice nucleation by surrogates for atmospheric mineral dust and mineral dust/sulfate particles at cirrus temperatures. *Atmos. Chem. Phys.*, **5**, 2617–2634.

Atkins, P. W. (1982). Physical chemistry, 2nd ed., Oxford Univ. Press., 1095 pp.

Auer, A. H., Jr., and D. L. Veal (1970). The dimensions of ice crystals in natural clouds. *J. Atmos. Sci.*, **27**, 919–926.

Austin, P. H., M. B. Baker, A. M. Blyth, and J. B. Jensen (1985). Small-scale variability in warm continental cumulus clouds. *J. Atmos. Sci.*, **42**, 1123–1138.

Bakan, S. (1984). Note on cellular convection with nonisotropic eddies. *Tellus*, **36A**, 87–89.

Baker, M. B. (1993). Variability in concentrations of CCN in the marine cloud-top boundary layer. *Tellus, Ser. B*, **45**, 458–472.

Baker, M. B., and M. Baker (2004). A new look at homogeneous freezing of water. *Geophys. Res. Lett.*, **31**, L19102, doi:10.1029/2004GL020483.

Baker, M. B., and J. Latham (1979). The evolution of the droplet spectra and the rate of production of embryonic raindrops in small cumulus clouds. *J. Atmos. Sci.*, **36**, 1612–1615.

Baker, M. B., R. G. Corbin, and J. Latham (1980). The influence of entrainment on the evolution cloud droplet spectra: I. A model of inhomogeneous mixing. *Q. J. Roy. Meteor. Soc.*, **106**, 581–598.

Baker, M. B., R. E. Briedenthal, T. W. Choularton, and J. Latham (1984). The effects of turbulent mixing in clouds. *J. Atmos. Sci.*, **41**, 299–304.

Bakhanov, V. P., and M. V. Buikov (1985). Modeling artificial crystallization, precipitation formation and dispersal of supercooled stratiform clouds. *Review. Proc. Sov. Inst. Sci. Inf.*, ser. Meteorology, No. 6, 50 pp., Obninsk, Moscow Region (in Russian).

Barahona, D., and A. Nenes (2008). Parameterization of cirrus formation in large scale models: Homogenous nucleation. *J. Geophys. Res.*, **113**, doi:10.1029/2007JD009355.

Barahona, D., and A. Nenes (2009). Parameterizing the competition between homogeneous and heterogeneous freezing in cirrus cloud formation monodisperse ice nuclei. *Atmos. Chem. Phys.*, **9**, 369–381.

Bartlett J. T., and P. R. Jonas (1972). On the dispersions of the sizes of droplets growing by condensation in turbulent clouds. *Quart. J. Roy. Meteor. Soc.*, **98**, 150–164.

Barton, E. F., and W. F. Oliver (1936). The crystal structure of ice at low temperatures. *Proc. Roy. Soc. Lond., A* **153**, 166–172.

Bashkirova, T. A., and T. A. Pershina (1964). On the mass of snow crystals and their fall velocity. *Proc. Main Geophys. Observ.*, **165**, 83–100.

Bauer, S. E. and D. Koch (2005), Impact of heterogeneous sulfate formation at mineral dust surfaces on aerosol loads and radiative forcing in the Goddard Institute for Space Studies general circulation model. *J. Geophys. Res.*, **110**, http://dx.doi.org/10.1029/2005JD005870.

Beard, K. V. (1976). Terminal velocity and shape of cloud and precipitation drops aloft. *J. Atmos. Sci.*, **33**, 851–864.

Beard, K. V. (1980). The effect of altitude and electrical force on the terminal velocity of hydrometeors. *J. Atmos. Sci.*, **37**, 1363–1374.

Beard, K. V., and C. Chuang (1987). A new model for the equilibrium shape of raindrops. *J. Atmos. Sci.*, **44**, 1509–1524.

Beard K. V., and H. T. Ochs (1993). Warm-rain initiation: an overview of microphysical mechanisms. *J. Appl. Meteorol.*, **32**, 608–625.

Beard, K. V., and H. R. Pruppacher (1971). A wind tunnel investigation of the rate of evaporation of small water drops at terminal velocity in air. *J. Atmos. Sci.*, **28**, 1455–1464.

Becker, R., and W. Döring (1935). Kinetische Behandlung der Keimbildung in übersättigten Dämpfen. *Ann. Physik*, **24**, 719–752.

Belyaev, V. I. (1961). Size distribution of drops in a cloud during its condensation stage of development. *Izv. Acad. Sci. USSR, Geophys. Ser.*, **8**, 1209–1213.

Belyaev, V. I. (1964). The method of Lagrange in the kinetics of cloud processes. Leningrad, Hydrometeoizdat, 119 pp. (in Russian).

Berezinsky, N. A., and G. V. Stepanov (1986). Dependence of natural ice-forming nuclei concentration of different size on the temperature and supersaturation. *Izv. Acad Sci. USSR, Atmos. Oceanic Phys.*, **22**, 722–727.

Bergeron, T. (1935). On the physics of clouds and precipitation. *Proc. Vth Assembly General of the International Union of Geodesy and Geophysics*, Lisbon, Portugal, International Union of Geodesy and Geophysics, 156–180.

Berjulev, G. P., A. A. Chernikov, B. G. Danelyan, V. I. Khvorostyanov, Y. A. Seregin, G. R. Toroyan, and M. V. Vlasyuk (1989). Field study and numerical simulation of an orographic cloud system: Natural evolution and seeding. *Proc. IVth International Conf. on Weather Modification*, Beijing, China, 128–132.

Berlyand, T. G., and L. A. Strokina (1980). Global distribution of total cloud amount. Hydrometeoizdat, Leningrad, 71 pp. (in Russian) (English translation by S. Warren).

Berry, E. X. (1967). Cloud drop growth by coalescence. *J. Atmos. Sci.*, **24**, 688–701.

Berry, E. X., and R. L. Reinhardt (1974a). An analysis of cloud drop growth by collection: Part I. Double distributions. *J. Atmos. Sci.*, **31**, 1814–1824.

Berry, E. X., and R. L. Reinhardt (1974b). An analysis of cloud drop growth by collection: Part II. Single initial distributions. *J. Atmos. Sci.*, **31**, 1825–1831.

Bertram, A. K., D. D. Patterson, and J. J. Sloan (1996). Mechanisms and temperatures for the freezing of sulfuric acid aerosols measured by FTIR spectroscopy. *J. Phys. Chem.*, **100**, 2376–2383.

Bertram, A. K., T. Koop, L. T. Molina, and M. J. Molina (2000). Ice formation in $(NH_4)_2SO_4$-H_2O particles. *J. Phys. Chem. A*, **104**, 584–588.

Bigg, E. K. (1953). The supercooling of water. *Proc. Phys. Soc., B* **66**, 688–703.

Bigg, E. K., and C. Leck (2001a). Cloud-active particles over the central Arctic Ocean. *J. Geophys. Res.*, **106** (D23), 32,155–32,166.

Bigg, E. K., and C. Leck (2001b). Properties of the aerosol over the central Arctic Ocean. *J. Geophys. Res.*, **106** (D23), 32,101–32,109.

Bigg, E. K., J. L. Brownscombe, and W. Thompson (1969). Fog modification with long-chain alcohols. *J. Appl. Met.*, **8**, 75–82.

Biskos, G., A. Malinowski, L. M. Russell, P. R. Buseck, and S. T. Martin (2006a). Nanosize effect on the deliquescence and the efflorescence of sodium chloride particles. *Aeros. Sci. Technol.*, **40**, 97–106.

Biskos, G., L. M. Russell, P. R. Buseck, and S. T. Martin (2006b). Nanosize effect on the hygroscopic growth factor of aerosol particles. *Geophys. Res. Lett.*, **33**, (1–4), L07, 801.

Bleck, R. (1970). A fast, approximative method for integrating the stochastic coalescence equation. *J. Geophys. Res.*, **75**, 5165–5171.

Blüthgen, J. (1966). Allgemeine klimageographie, Walter de Gruyter, Berlin, v. 2, 710 pp.

Bogdan, A., and M. J. Molina (2009). Why does large relative humidity with respect to ice persist in cirrus ice clouds? *J. Phys. Chem. A*, **113**, 14,123–14,130.

Bogdan, A., M. J. Molina, M. Kulmala, H. Tenhu, and T. Loerting (2013). Solution coating around ice particles of incipient cirrus clouds. *Proc. Natl. Acad. Sci.*, www.pnas.org/cgi/doi/10.1073/pnas.1304471110.

Blyth, A. (1993). Entrainment in cumulus clouds. *J. Appl. Meteorol.* **32**, 626–641.

Böhm, J. P. (1989). A general equation for the terminal fall speed of solid hydrometeors. *J. Atmos. Sci.*, **46**, 2419–2427.

Böhm, J. P. (1992). A general hydrodynamic theory for mixed-phase microphysics. Part I: Drag and fall speeds of hydrometeors. *Atmos. Res.*, **27**, 253–274.

Borick, S. S., P. G. Debenedetti, and S. Sastry (1995). A lattice model of network-forming fluids with orientation-dependent bonding: Equilibrium, stability, and implications for the phase behavior of supercooled water. *J. Phys. Chem.*, **99**, 3781–3792.

Born, M. (1963). Atomic physics, Blackie and Son, Ltd., London-Glasgow, 493 pp.

Borovikov, A. M., I. P. Mazin, and A. N. Nevzorov (1965). Some features of distribution of the large particles in clouds of various forms. *Izv. Acad. Sci. USSR, Atmos. Oceanic Phys.*, **1** (3), 357–369.

Borovikov, A. M., I. I. Gaivoronsky, E. G. Zak, V. V. Kostarev, I. P. Mazin, V. E. Minervin, A. Kh. Khrgian, and S. M. Shmeter (1963). The physics of clouds. Leningrad, Hydrometeoizdat, 1961, Transl. by Israel Program Scientif. Translation, U.S. Dept. Commerce, Washington, DC, 459 pp. 1963.

Bott, A. (1998). A flux method for the numerical solution of the stochastic collection equation. *J. Atmos. Sci.*, **55**, 2284–2293.

Bott, A. (2000). A flux method for the numerical solution of the stochastic collection equation: Extension to two-dimensional particle distributions. *J. Atmos. Sci.*, **57**, 284–294.

Bott, A. (2001). A new method for the solution of the stochastic collection equation in cloud models with spectral aerosol and cloud drop microphysics. *Atmos. Res.*, 59–60, 361–372.

Boville, B. A., P. J. Rasch, J. J. Hack, and J. R. McCaa (2006). Representation of clouds and precipitation processes in the Community Atmosphere Model version 3 (CAM3). *J. Climate*, **19**, 2184–2198.

Braham, R. R. (1976). CCN spectra in c-k space. *J. Atmos. Sci.*, **33**, 343–346.

Brandes, E. A., G. Zhang, and J. Vivekanandan (2003). An evaluation of a drop distribution-based polarimetric radar rainfall estimator. *J. Appl. Meteor.*, **42**, 652–660.

Brechtel, F. J., and S. M. Kreidenweis (2000a). Predicting particle critical supersaturation from hygroscopic growth measurements in the humidified TDMA. Part I: Theory and sensitivity studies. *J. Atmos. Sci.*, **57**, 1854–1871.

Brechtel, F. J., and S. M. Kreidenweis (2000b). Predicting particle critical supersaturation from hygroscopic growth measurements in the humidified TDMA. Part II: Laboratory and ambient studies. *J. Atmos. Sci.*, **57**, 1872–1887.

Brenguier, J.-L., and L. Chaumat (2001). Droplet spectra broadening in cumulus clouds. Part I: broadening in adiabatic cores. *J. Atmos. Sci.*, **58**, 628–641.

Bretherton, C. S., M. K. MacVean, P. Bechtold, A. Chlond, W. R. Cotton, et al. (1999a). An intercomparison of radiatively-driven entrainment and turbulence in a smoke cloud, as simulated by different numerical models. *Quart. J. Roy. Meteor. Soc.*, **125**, 391–423.

Bretherton, C. S., S. K. Krueger, P. Bechtold, E. van Meijgaard, B. Stevens, and J. Teixeira (1999b). A GCSS boundary layer model intercomparison study of the first ASTEX Lagrangian experiment. *Bound.-Layer Meteor.*, **93**, 341–380.

Bridgman, P. W. (1912). Water, in the liquid and five solid forms, under pressure. *Proc. Amer. Acad. Arts Sci.*, **47**, 441–558.

Broadwell, J. E., and R. E. Briedenthal (1982). A simple model of mixing and chemical reaction in a turbulent shear layer. *J. Fluid. Mech.*, **125**, 397–410.

Brock, J. R. (1962). On the theory of thermal forces acting on aerosol particles. *J. Colloid Sci.*, **17**, 768–780.

Brown, A. J., and E. Whalley (1966). Preliminary investigation of the phase boundaries between ice VI and VII and ice VI and VIII. *J. Chem. Phys.*, **45**, 4360–4361.

Brown, P. S. (1991). Parameterization of the evolving drop-size distribution based on analytic solution of the linearized coalescence breakup equation. *J. Atmos. Sci.*, **48**, 200–210.

Brown, P. S. (1997). Mass conservation considerations in analytic representation of rain drop fragment distribution. *J. Atmos. Sci.*, **54**, 1675–1687.

Brown, R. A. (1974). Analytical methods in planetary boundary layer modeling. Adam Higler, London, 150 pp.

Brown, S. R. (1970). Terminal velocities of ice crystals. *M. S. Thesis*, Dept. of Atmospheric Sciences, Fort Collins, Colorado, 52 pp.

Brüggeller, P., and E. Mayer (1980). Complete vitrification in pure water and dilute aqueous solutions. *Nature*, **288**, 569–571.

Brümmer, B. (1985). Structure, dynamics and energetics of boundary layer rolls from KonTur aircraft observations. *Contr.Atm.Phys.*, **58**, 237–254.

Brümmer, B., B. Rump, and G. Kruspe (1992). A cold air outbreak near Spitsbergen in springtime: Boundary-layer modification and cloud development. *Boundary Layer Meteorol.*, **61**, 13–46.

Bryant, F. D., and P. Latimer (1969). Optical efficiencies of large particles of arbitrary shape and orientation. *J. Colloid Interface Sci.*, **30**, 291–304.

Buikov, M. V. (1961). Kinetics of distillation in a polydisperse fog. *Izvestia Acad. Sci. USSR, Ser. Geophys.*, **7**, 1058–1065.

Buikov, M. V. (1963). A method of the kinetic equations in the theory of clouds. *Proc. All-Union Meteorol. Conf.*, Leningrad, **5**, 122–128.

Buikov, M. V. (1966a). Kinetics of heterogeneous condensation at adiabatic cooling. Part 1: Diffusion regime of droplet growth. *Colloid J.*, **28** (5), 628–634.

Buikov, M. V. (1966b). Kinetics of heterogeneous condensation at adiabatic cooling. Part 2: Kinetic regime of droplet growth. *Colloid J.*, **28** (5), 635–639.

Buikov, M. V., and V. I. Khvorostyanov (1976). Numerical simulation of radiation fog and stratus clouds formation with account for interaction among dynamic, radiative and microphysical processes. *Proc. 6th Intern. Conf. Cloud Phys.,* Boulder, CO, USA, 392–395.

Buikov, M. V., and V. I. Khvorostyanov (1977). Formation and evolution of radiative fog and stratus clouds in the boundary layer of the atmosphere with explicit account for microphysical processes. *Izv. Acad. Sci. USSR, Atmos. Oceanic Phys.*, **13**(4), 356–370.

Buikov, M. V., and V. I. Khvorostyanov (1979). Dispersal of fogs with surface-active reagents. A review. *Soviet Meteorol. Hydrol.* **5**, 28–35.

Buikov, M. V., and A. M. Pirnach (1973). A numerical model of a two-phase stratiform cloud with explicit account for microstructure. *Izv. Acad. Sci. USSR, Atmos. Oceanic Phys.*, **9** (5), 486–499.

Buikov, M. V., and A. M. Pirnach (1975). Numerical modeling of microphysical processes of precipitation formation in stratiform mixed-phase clouds with a 1D spectral bin microphysical model. *Izv. Acad. Sci. USSR, Atmos. Oceanic Phys.*, **11**(5), 469–480.

Buikov, M. V., K. Y. Ibragimov, A. M. Pirnach, and L. P. Sorokina (1976). A study of the two-phase stratiform clouds in the atmosphere of Jupiter. *Astrophys. J.*, **53**, 596–602.

Bunker, K. W., S. China, C. Mazzoleni, A. Kostinski, and W. Cantrell (2012). Measurements of ice nucleation by mineral dusts in the contact mode, *Atmos. Chem. Phys. Discuss.*, **12**, 20,291–20,309, doi:10.5194/acpd-12-20291-2012.

Butorin, G. T., and K. P. Skripov (1972). Crystallizaton of supercooled water. *Soviet Phys. Crystallogr.*, **17**, 322–326.

Callen, H. B. (1960). Thermodynamics: An introduction to the physical theories of equilibrium thermostatics and irreversible thermodynamics. J. Wiley and Sons, Inc., New York, 376 pp.

Cantrell, W., and C. Robinson (2006). Heterogeneous freezing of ammonium sulfate and sodium chloride solutions by long chain alcohols. *Geophys. Res. Lett.*, **33**, L07802, doi:10.1029/2005GL024945.

Charlson, R. J., J. H. Seinfeld, A. Nenes, M. Kulmala, A. Laaksonen, and M. C. Faccini (2001). Reshaping the theory of cloud formation, *Science*, **292**, 20205–2026.

Chen, J.-P. (1994). Theory of deliquescence and modified Köhler curves, *J. Atmos. Sci.*, **51**, 3505–3516.

Chen, J.-P., A. Hazra, and Z. Levin (2008). Parameterizing ice nucleation rates using contact angle and activation energy derived from laboratory data, *Atmos. Chem. Phys.*, **8**, 7431–7449.

Chen Y., P. J. DeMott, S. M. Kreidenweis, D. C. Rogers, and E. Sherman (2000). Ice formation by sulfate and sulfur acid aerosols under upper-tropospheric conditions, *J. Atmos. Sci.*, **57**, 3752–3766.

Chen, Y., S. M. Kreidenweis, L. M. McInnes, D. C. Rogers, and P. J. DeMott (1998). Single particle analyses of ice nucleating aerosols in the upper troposphere and lower stratosphere. *Geophys. Res. Lett.*, **25**, 1391–1394.

Chlond, A. (1992). Three-dimensional simulation of cloud street development during a cold air outbreak. *Boundary Layer Meteorol.*, **58**, 161–200.

Chuang, P. (2003). Measurement of the timescale of hygroscopic growth for atmospheric aerosols. *J. Geophys. Res.*, **108**, 4282, doi:10.1029/2002JD002757.

Chukin, V. V., and A. S. Platonova (2012). Model of crystallization of supercooled droplets of aqueous solution. *Proc. Intl. Conf. Clouds and Precipitation,* Leipzig, August 2012.

Chukin, V. V., E. A. Pavlenko, and A. S. Platonova (2010). Homogeneous ice nucleation rate in supercooled droplets of aqueous solutions. *Rus. Meteorol. Hydrol.*, **35**, No. 8, 524–529.

Chylek, P., and J. D. Klett (1991a). Extinction cross sections of non-spherical particles in the anomalous diffraction approximation. *J. Opt. Soc. Am.*, **8**, 274–281.

Chylek, P., and J. D. Klett (1991b). Absorption and scattering of electromagnetic radiation by prismatic columns: Anomalous diffraction approximation. *J. Opt. Soc. Am.*, **8**, 1713–1720.

Chylek, P. and Videen (1994). Longwave radiative properties of polydispersed hexagonal ice crystals. *J. Atmos. Sci.*, **51**, 175–190.

Chylek, P., and J. G. D. Wong (1998), Erroneous use of the modified Köhler equation in cloud and aerosol physics applications, *J. Atmos. Sci.*, **55**, 1473–1477.

Cirrus (2002). Eds.: Lynch, D., Sassen, K., Starr, D. O'C., and Stephens, G., Oxford University Press, New York, 480 pp.

Clark, T. (1974). A study of cloud phase parameterization using the gamma distribution. *J. Atmos. Sci.*, **31**, 142–155.

Clark, T. L. (1976). Use of log-normal distributions for numerical calculations of condensation and collection. *J. Atmos. Sci.*, **33**, 810–821.

Clegg, S. L., and P. Brimblecombe (1995). Application of a multicomponent thermodynamic model to activities and thermal properties of 0–40 mol kg−1 aqueous sulfuric acid from <200 to 328 K, *J. Chem. Eng. Ref. Data*, **40**, 43–64.

Clegg, S. L., P. Brimblecombe, and A. S. Wexler (1998). A thermodynamic model of the system H+-NH4+-SO42--NO3--H2O at tropospheric temperatures. *J. Phys. Chem.*, **102**, 2137–2154.

Clegg, S. L., S. S. Ho, C. K. Chan, and P. Brimblecombe (1995). Thermodynamic properties of aqueous $(NH)_2SO_4$ to high supersaturation as a function of temperature. *J. Chem. Eng. Data*, **40**, 1079–1090.

Coakley, J. A., Jr., et al. (1987). Effect of ship-stack effluents on cloud reflectivity. *Science*, **237**, 1020–1022.

Cohard, J.-M., and J.-P. Pinty (2000). A comprehensive two-moment warm microphysical bulk scheme. I: Description and tests. *Q. J. Roy. Meteorol. Soc.*, **126**, 1815–1842.

Cohard, J.-M., J.-P. Pinty, and C. Bedos (1998). Extending Twomey's analytical estimate of nucleated cloud droplet concentrations from CCN spectra. *J. Atmos. Sci.*, **55**, 3348–3357.

Cohard, J.-M., J.-P. Pinty, and K. Suhre (2000). On the parameterization of activation spectra from cloud condensation nuclei microphysical properties. *J. Geophys. Res.*, **105** (D9), 11,753–11,766.

Collins, W. D., P. J. Rasch, B. A. Boville, J. J. Hack, J. R. McCaa, D. L. Williamson, B. P. Briegleb, C. M. Bitz, S. J. Lin, and M. H. Zhang (2006). The formulation and atmospheric simulation of the Community Atmosphere Model version 3 (CAM3), *J. Clim.*, **19** (11), 2144–2161, doi:10.1175/JCLI3760.1.

Comstock, J. M., T. P. Ackerman, and D. D. Turner (2004). Evidence of high ice supersaturation in cirrus clouds using ARM Raman lidar measurements, *Geophys. Res. Lett.*, **31**, L10106, doi:10.1029/2004GL019539.

Comstock, J. M., R.-F. Lin, D. O'C. Starr, and P. Yang (2008). Understanding ice supersaturation, particle growth, and number concentration in cirrus clouds. *J. Geophys. Res.*, **113**, D23211, doi: 10.1029/2008JD010332.

Connolly, P. J., O. Möhler, P. R. Field, H. Saathoff, R. Burgess, T. Choularton, and M. Gallagher (2009). Studies of heterogeneous freezing by three different desert dust samples, *Atmos. Chem. Phys.*, **9**, 2805–2824, doi:10.5194/acp-9-2805-2009.

Considine, G. and J. A. Curry (1996). A statistical model of drop size spectra for stratocumulus clouds. *Quart. J. Roy. Meteor. Soc.*, **122**, 611–634.

Considine, G. and J. A. Curry (1998). Role of entrainment and droplet sedimentation on the microphysical structure in stratus and stratocumulus clouds. *Quart. J. Roy. Meteorol. Soc.*, **24**, 123–150.

Cooper, W. A. (1974). A possible mechanism for contact nucleation. *J. Atmos. Sci.*, **31**, 1832–1837.

Cooper, W. A. (1986). Ice initiation in natural clouds. In: Precipitation enhancement: A scientific challenge, *Meteor. Monogr.*, **21**, *Amer. Meteor. Soc.*, Boston, 29–32.

Cooper, W. A. (1989). Effects of variable droplet growth histories on droplet size distributions. Part I: Theory. *J. Atmos. Sci.*, **46**, 1301–1311.

Cooper, W. A. (1995). Ice formation in wave clouds: Observed enhancement during evaporation. In "Proc. Conf. on Cloud Physics." pp. 147–152. *Amer. Met. Soc, Dallas.*

Cooper, W. A., and G. Vali (1981). The origin of ice in mountain cap clouds. *J. Atmos. Sci.*, **38**, 1244–1259.

Cotton, R. J., and P. R. Field (2002). Ice nucleation characteristics of an isolated wave cloud. *Quart. J. Roy. Meteorol. Soc.*, **128**, 2417–2437.

Cotton, W. R., and R. A. Pielke (2007). Human impacts on weather and climate, 2nd ed. Cambridge Univ. Press, 315 pp.

Cotton, W. R., G. Bryan, and S. van den Heever (2011). *Storm and Cloud Dynamics*, Intern. Geophys. Ser., v. 99, Academic Press, Elsevier Publishers, The Netherlands, 809 pp.

Cotton, W. R., G. J. Tripoli, R. M. Rauber, E. A. Mulvihill (1986). Numerical simulation of the effects of varying ice nucleation rates and aggregation process on orographic snowfall. *J. Climate Appl. Meteorol.*, **25**, 1658–1680.

Cotton, W. R., R. A. Pielke Sr., R. L. Walko, G. E. Liston, C. J. Tremback, H. Jiang, R. L. McAnelly, J. Y. Harrington, M. E. Nicholls, G. G. Carrio, and J. P. McFadden (2003). RAMS 2001: Current status and future directions. *Meteor. Atmos. Phys.*, **82**, 5–29.

CTBL (1985). Report of the JSC/CAS Workshop on modeling of Cloud-Topped Boundary Layer. Fort Collins, CO, USA, 22–26 April 1985, 96 pp.

Curry, J. A. (1983). On the formation of continental polar air. *J. Atmos. Sci.*, **40**, 2278–2292.

Curry, J. A. (1986). Interactions among turbulence, radiation and microphysics in Arctic stratus clouds. *J. Atmos. Sci.*, **43**, 525–538.

Curry, J. A. (1995). Interactions among aerosols, clouds, and climate of the Arctic Ocean. *Sci. Total Environ.*, **160**, 777–791.

Curry, J. A., and G. F. Herman (1985). Infrared radiative properties of summertime Arctic stratus clouds. *J. Clim. Appl. Meteorol.*, **24**, 525–538.

Curry J. A., and V. I. Khvorostyanov (2012). Assessment of some parameterizations of heterogeneous ice nucleation in cloud and climate models. *Atmos. Chem. Phys.*, **12**, 1151–1172, www.atmos-chem-phys.net/12/1151/2012/, doi:10.5194/acp-12-1151-2012.

Curry, J. A., and P. J. Webster (1999), Thermodynamics of atmospheres and oceans, Academic Press, London, 467 pp.

Curry, J. A., E. E. Ebert, and G. F. Herman (1988). Mean and turbulent structure of the summertime Arctic cloudy boundary layer. *Quart. J. Roy. Meteor. Soc.*, **114**, 715–746.

Curry, J. A., J. Schramm, and E. E. Ebert (1993). Impact of clouds on the surface radiation budget of the Arctic Ocean. *Meteor. and Atmos. Phys*, **57**, 197–217.

Curry, J. A., W. B. Rossow, D. Randall, and J. L. Schramm (1996). Overview of Arctic cloud and radiation properties. *J. Clim.*, **9**, 1731–1764.

Curry, J. A., F. G. Meyer, L. F. Radke, C. A. Brock, and E. E. Ebert (1990). The occurrence and characteristics of lower tropospheric ice crystals in the Arctic. *Int. J. Climatol.*, **10**, 749–764.

Curry, J. A., P. V. Hobbs, M. D. King, D. A. Randall, P. Minnis, et al. (2000). FIRE Arctic clouds experiment, *Bull. Amer. Meteor. Soc.*, **81**, 5-29.

Cziczo, D. J., and J. P. D. Abbatt (1999). Deliquescence, efflorescence and supercooling of ammonium sulfate aerosols at low temperature: Implications for cirrus cloud formation and aerosol phase in the atmosphere. *J. Geophys. Res.*, **104**, 13,781–13,790.

Cziczo, D. J., D. M. Murphy, P. K. Hudson, and D. S. Thomson (2004). Single particle measurements of the chemical composition of cirrus ice residue during CRYSTAL-FACE, *J. Geophys. Res.*, **109**, (D4), D04201, 10.1029/2003JD004032.

Dash, J. G., H. Fu, and J. S. Wettlaufer (1995). The premelting of ice and its environmental consequences. *Rep. Progr. Phys.*, **58**, 115–167.

Debenedetti, P. K. (2003). Supercooled and glassy water. *J. Phys. Condens. Matter*, **15**, R1669–1726.

Debye, P. (1912). Zur theorie der spezifischen wärmen. *Annaln. Phys.*, **39**, 789–839.

Defay, R., I. Prigogine, A. Bellemans, and D. Everett (1966). Surface tension and absorption, 432 pp., Wiley, New York.

Deirmendijan, D. (1969). Electromagnetic scattering on spherical polydispersions. Elsevier, 291 pp.

DeMott, P. J. (2002). Laboratory studies of cirrus cloud processes. In: Cirrus. Eds.: Lynch, D., Sassen, K., Starr, D. O'C., and Stephens, G., Oxford University Press, 102–135.

DeMott, P. J., and D. C. Rogers (1990). Freezing nucleation rates of dilute solution droplets measured between –30 and –40°C in laboratory simulations of natural clouds. *J. Atmos. Sci.*, **47**, 1056–1064.

DeMott, P. J., M. P. Meyers, and W. R. Cotton (1994). Parameterization and impact of ice initiation processes relevant to numerical model simulation of cirrus clouds, *J. Atmos. Sci.*, **51**, 77–90.

DeMott, P. J., D. C. Rogers, and S. M. Kreidenweis (1997). The susceptibility of ice formation in upper tropospheric clouds to insoluble aerosol components. *J. Geophys. Res.*, **102**, 19,575–19,584.

DeMott, P. J., D. C. Rogers, S. M. Kreidenweis, Y. Chen, C. H. Twohy, D. Baumgardner, A. J. Heymsfield, and K. R. Chan (1998). The role of heterogeneous freezing nucleation in upper tropospheric clouds: Inferences from SUCCESS. *Geophys. Res. Lett.*, **25**, 1387–1390.

DeMott, P. J., D. J. Cziczo, A. J. Prenni, D. M. Murphy, S. M. Kreidenweis, D. S. Thomson, R. Borys, and D. C. Rogers (2003). Measurements of the concentration and composition of nuclei for cirrus formation, *Proc. Natl. Acad. Sci.*, **100**, 14,655–14,660.

DeMott, P. J., A. J. Prenni, X. Liu, S. M. Kreidenweis, M. D. Petters, C. H. Twohy, M. S. Richardson, T. Eidhammer, and D. C. Rogers (2010). Predicting global atmospheric ice nuclei distributions and their impacts on climate, *Proc. Natl. Acad. Sci. USA*, **107**, 11,217–11,222.

Denbigh, K. (1981). The principles of chemical equilibrium, Cambridge University Press, 494 pp.

Dennis, A. S. (1980). Weather modification by cloud seeding. Academic Press, New York, 274 pp.

Deryagin, B. V. and Y. S. Kurgin (1972). The theory of passivation of the droplet condensation growth by the vapor of cetyl alcohol. *Colloid. J.*, **35**, 26–42.

Deryagin, B. V., V. A. Fedoseev, and L. A. Rosentsveig (1966). A study of the adsorption of the cetyl alcohol vapor and its impact on evaporation of water drops. *Docl. Acad. Sci. USSR*, **167** (3), 616–620.

Dick, W. D., P. Saxena, and P. H. McMurry (2000). Estimation of water uptake by organic compounds in submicron aerosols measured during the Southeastern Aerosol and Visibility Study, *J. Geophys. Res.*, **105** (D1), 1471–1479.

Diehl, K., and S. Wurzler (2004). Heterogeneous drop freezing in the immersion mode: Model calculations considering soluble and insoluble particles in drops. *J. Atmos. Sci.*, **61**, 2063–2072.

Djikaev, Y. S. and E. Ruckenstein (2008). Thermodynamics of heterogeneous crystal nucleation in contact and immersion modes. *J. Phys. Chem. A*, **112**, 11,677–11,687.

Djikaev, Y. S., A. Tabazadeh, P. Hamill, and H. Reiss (2002). Thermodynamic conditions for the surface-stimulated crystallization of atmospheric droplets. *J. Phys. Chem., A*106, 10,247–10,253.

Dmitrieva-Arrago, L. R, and I. V. Akimov (1998). A method for calculation of nonconvective precipitation on the basis of liquid water content forecast with account for cloud microphysics. *Sov. Meteorol. Hydrol.*, **11**, 47–59.

Dorsey, N. E. (1968). Properties of ordinary water-substance. Hafner Publish. Co., 637 pp.

Dufour, L., and R. Defay (1963). Thermodynamics of clouds. Academic Press, New York, 255 pp.

Durant, A. J., and R. A. Shaw (2005). Evaporating freezing by contact nucleation inside-out. *Geophys. Res. Lett.*, **32**, L20814, doi: 10.1029/2005GL025175.

Eadie, W. J. (1971). A molecular theory of the homogeneous nucleation of ice from supercooled water. *PhD Dissertation*, University of Chicago, Cloud Physics Lab, Tech. Note 40, 117 pp.

Ebert, E. E., and J. A. Curry (1992). A parameterization of ice cloud optical properties for climate models. *J. Geophys. Res.*, **97**, 3831–3836.

ECMWF-2007: European Centre for Medium Range Weather Forecast (ECMWF), (2007). "IFS documentation cycle 31rl, Part IV: Physical processes." 155 pp., http://www.ecmwf.int/research/ifsdocs/CY31rl/index.html.

Eidhammer, T., P. J. DeMott, and S. M. Kreidenweis (2009). A comparison of heterogeneous ice nucleation parameterization using a parcel model framework. *J. Geophys. Res.*, **114**, D06202, doi: 10.1029/2008JD011095.

Einstein, A. (1906). Die plancksche theorie der Strahlung und die spezifischen Wärme. *Annaln. Phys.*, **22**, 180–190.

Elbaum, M., and M. Schick (1991). Application of the theory of dispersion forces to the surface melting of ice. *Phys. Rev. Lett.* **66**, 1713–1716.

Elbaum, M., S. G. Lipson, and J. D. Dash (1993). Optical study of surface melting on ice. *J. Cryst. Growth*, **129**, 491–505.

Ervens, B., and G. Feingold (2012). On the representation of immersion and condensation freezing in cloud models using different nucleation schemes. *Atmos. Chem. Phys.*, **12**, 5807–5826.

Facchini, M. C., M. Mircea, S. Fuzzi, and R. Charlson (1999). Cloud albedo enhancement by surface-active organic solutes in growing droplets. *Nature*, **401**, 257–259.

Falkovich, A. H., E. Ganor, Z. Levin, P. Formenti, and Y. Rudich (2001). Chemical and mineralogical analysis of individual mineral dust particles, *J. Geophys. Res.*, **106** (D16), 18,029–18,036.

Fan, J., S. Ghan, M. Ovchinnikov, X. Liu, P. J. Rasch, and A. Korolev (2011). Representation of Arctic mixed-phase clouds and the Wegener-Bergeron-Findeisen process in climate models: Perspectives from a cloud-resolving study. *J. Geophys. Res.*, **116**, D00T07, doi:10.1029/2010JD015375.

Farkas, L. (1927). Keimbildungsgeschwindigkeit in übersättigten Dämpfen. *Z. Phyik Chem.*, **A125**, 236–242.

Feingold, G., and P. Chuang (2002). Analysis of the influence of film-forming compounds on droplet growth: Implications for cloud microphysical processes and climate. *J. Atmos. Sci.*, **59**, 2006–2018.

Feingold, G., and Z. Levin (1986). The lognormal fit to raindrop spectra from frontal convective clouds in Israel. *J. Clim. Appl. Meteor.*, **25**, 1346–1363.

Feingold, G., S. Tzivion, and Z. Levin (1988). Evolution of raindrop spectra. Part I: Solution to the collection/breakup equation using the method of moments. *J. Atmos. Sci.*, **45**, 3387–3399.

Feingold G., B. Stevens, W. R. Cotton, and R. L. Walko (1994). An explicit cloud microphysics/LES model designed to simulate the Twomey effect. *Atmos. Res.*, **33**, 207–233.

Feistel, R. (2012). TEOS-10: A new international oceanographic standard for seawater, ice, fluid water and humid air. *Intern. J. Thermophysics*, **33**, doi: 10.1007/s10765-010-0901-y.

Feistel, R. and E. Hagen (1995). On the Gibbs thermodynamic potential of seawater. *Progr. Oceanogr.*, **36**, 249–327.

Feistel, R. and E. Hagen (1998). A Gibbs thermodynamic potential of sea ice. *Cold Reg. Sci. Technol.*, **28**, 83–142.

Feistel, R. and E. Hagen (1999). Corrigendum to "A Gibbs thermodynamic potential of sea ice." *Cold Reg. Sci. Technol.*, **29**, 173–176.

Feistel, R., and W. Wagner (2005). High-pressure thermodynamic Gibbs functions of ice and sea ice. *J. Marine Res.*, **63**, 95–139.

Feistel, R., and W. Wagner (2006). A new equation of state for H_2O ice Ih. *J. Phys. Chem. Ref. Data*, **35**, 1021–1047, doi: 10.1063/1.2183 324.

Feistel, R., and W. Wagner (2007). Sublimation pressure and sublimation enthalpy of H_2O ice I_h between 0 and 273.16 K. *Geochimica et Cosmochimica Acta*, **71**, 36–45, doi: 10.1016/j.gca.2006.08.034.

Feistel, R., D. G. Wright, D. R. Jackett, K. Miyagawa, J. H. Reissmann, W. Wagner, U. Overhoff, C. Guder, A. Feistel, and G. M. Marion (2010b). Numerical implementation and oceanographic application of the thermodynamic potentials of liquid water, water vapour, ice, seawater and humid air – Part 1: Background and equations. *Ocean Sci.*, **6**, 633–677.

Feistel, R., D. G. Wright, H.-J. Kretzschmar, E. Hagen, S. Herrmann, and R. Span (2010a). Thermodynamic properties of sea air. *Ocean Sci.*, **6**, 91–141, www.ocean–sci.net/6/91/2010/.

Feistel, R., D. G. Wright, K. Miyagawa, A. H. Harvey, J. Hruby, D. R. Jackett, T. J. McDougall, and W. Wagner (2008). Mutually consistent thermodynamic potentials for fluid water, ice and seawater: A new standard for oceanography. *Ocean Sci.*, 275–291.

Ferrier, B. S. (1994). A double-moment multiple-phase four-class bulk ice scheme. Part I: Description. *J. Atmos. Sci.*, **51**, 249–280.

Findeisen, W. (1938). Kolloid-Meteorologische Vorgänge bei Neiderschlags-bildung. *Meteor. Z.*, **55**, 121–133.

FIRE-SHEBA (2001). First international regional experiment: Surface heat budget of the Arctic. FIRE Arctic Clouds Experiment (FIRE), *J. Geophys. Res.*, **106**, Special section, Ed.: Curry, J. A., 14,985–15,376.

Fitzgerald, J. W. (1975). Approximation formulas for the equilibrium size of an aerosol particle as function of its dry size and composition and ambient relative humidity. *J. Appl. Meteorol.*, **14**, 1044–1049.

Fitzgerald J. W., W. A. Hoppel, and M. A. Vietty (1982). The size and scattering coefficient of urban aerosol particles at Washington D.C. as a function of relative humidity. *J. Atmos. Sci.*, **39**, 1838–1852.

Fletcher, N. H. (1958). Size effects in heterogeneous nucleation. *J. Chem. Phys.*, **29**, 572–576.

Fletcher, N. H. (1962). The physics of rainclouds. Cambridge University Press, Cambridge, UK, 390 pp.

Fletcher, N. H. (1968). Ice nucleation behavior of silver iodide smokes containing a soluble component. *J. Atmos. Sci.*, **25**, 1058–1060.

Fletcher, N. H. (1969). Active sites and ice nucleation, *J. Atmos. Sci.*, **26**, 1266–1271.

Fletcher, N. H. (1970a). The chemical physics of ice. Cambridge University Press, 265 pp.

Fletcher, N. H. (1970b). On contact nucleation. *J. Atmos. Sci.*, **27**, 1098–1099.

Fletcher, A. N. (1972). High-temperature contact nucleation of supercooled water by organic aerosols. *J. Appl. Meteorol.*, **11**, 988–993.

Flossmann, A. I., and W. Wobrock (2010). A review of our understanding of the aerosol–cloud interaction from the perspective of a bin resolved cloud scale modeling. *Atmos. Res.*, **97**, 478–497.

Flossman, A. I., W. D. Hall, and H. P. Pruppacher (1985). A theoretical study of the wet removal of atmospheric pollutants. Part I: The redistribution of aerosol particles captured through nucleation and impaction scavenging by growing cloud droplets. *J. Atmos. Sci.*, **42**, 583–606.

Flubacher, P., A. J. Leadbetter, and J. A. Morrison (1960). Heat capacity of ice at low temperatures. *J. Chem. Phys.*, **33**, 1751–1755.

Ford, I. J. (1998a). How aircraft nucleate ice particles: A simple model. *J. Aerosol Sci.*, **29**, S1117, doi:10.1016/S0021-8502(98)90741-8.

Ford, I. J. (1998b). Ice nucleation in jet aircraft exhaust plumes. In: Air pollution research report 68: Pollution from aircraft emissions in the North Atlantic flight corridor (POLINAT2). Ed.: Schumann, U., Report EUR 18877, European Commission, 1998, 269–287.

Fornea, A. P., S. D. Brooks, J. B. Dooley, and A. Saha (2009). Heterogeneous freezing of ice on atmospheric aerosols containing ash, soot, and soil. *J. Geophys. Res.*, **114**, D13201, doi:10.1029/2009JD011958.

Fountoukis, C, and A. Nenes (2005). Continued development of a cloud droplet formation parameterization for global climate models. *J. Geophys. Res.*, **110**, D11212, doi:10.1029/2004JD005591.

Fowler, L. D., D. A. Randall, and S. A. Rutledge (1996). Liquid and ice cloud microphysics in the CSU general circulation model. Part I: Model description and simulated microphysical processes. *J. Climate*, **9**, 489–529.

Frenkel, Ya. I. (1946). Kinetic theory of liquids. Oxford University Press, 592 pp.

Fridlind, A., A. S. Ackerman, E. J. Jensen, A. J. Heymsfield, M. R. Poellot, et al. (2004). Evidence for the predominance of midtropospheric aerosols as subtropical anvil cloud nuclei. *Science*, **304**, 718–722.

Fu, Q. (1996). An accurate parameterization of the solar radiative properties of cirrus clouds for climate models. *J. Climate*, **9**, 2058–2082.

Fu, Q., and K. N. Liou (1993). Parameterization of the radiative properties of cirrus clouds. *J. Atmos. Sci.*, **50**, 2008–2025.

Fu, Q., P. Yang, and W. B. Sun (1998). An accurate parameterization of the infrared radiative properties of cirrus clouds for climate models. *J. Climate*, **11**, 2223–2237.

Fuchs, N. A. (1959). Evaporation and droplet growth in gaseous media, Pergamon, 242 pp.

Fuchs, N. A. (1964). The mechanics of aerosols. Pergamon, New York, 234 pp.

Fukuta, N. (1975). A study of the mechanism of contact ice nucleation. *J. Atmos. Sci.*, **32**, 1597–1603.

Fukuta, N., and R. C. Schaller (1982). Ice nucleation by aerosol particles: Theory of condensation-freezing nucleation. *J. Atmos. Sci.*, **39**, 648–655.

Fukuta, N., and L. A. Walter (1970). Kinetics of hydrometeors growth from a vapor-spherical model. *J. Atmos. Sci.*, **27**, 1160–1172.

Gaivoronsly, I. I., and Y. A. Seregin (1962). Experiments on cloud dispersal over large areas. *Proc. Centr. Aerolog. Obs.*, **44**, 15–27 (in Russian).

Gao, R. S., P. J. Popp, D. W. Fahey, T. P. Marcy, et al. (2004). Evidence that nitric acid increases relative humidity in low-temperature cirrus clouds. *Science*, **303**, 516–520.

Gao, Y., S. B. Chen, and L. E. Yu (2006). Efflorescence relative humidity for ammonium sulfate particles. *J. Phys. Chem.*, **110**, 7602–7608.

Gao, Y., S. B. Chen, and L. E. Yu (2007). Efflorescence relative humidity of airborne sodium chloride particles: A theoretical investigation. *Atmos. Environ.*, **41**, 2019–2023, doi: 10.1016/j.atmosenv.2006.12.014.

Gayet, J.-F., J. Ovarlez, V. Shcherbakov, J. Strom, U. Schumann, A. Minikin, F. Auriol, A. Petzold, and M. Monier (2004). Cirrus cloud microphysical and optical properties at southern and northern midlatitudes during the INCA experiment. *J. Geophys. Res.*, **109**, D20206, doi:10.1029/2004JD004803.

Gerber, H. (1991). Supersaturation and droplet spectral evolution in fog. *J. Atmos. Sci.* **48**, 2569–2588.

Gettelman, A., H. Morrison, and S. J. Ghan (2008). A new two-moment bulk stratiform cloud microphysics scheme in the community atmosphere model, Version 3 (CAM3). Part II: Single-Column and Global Results. *J. Clim.*, **21**, 3660–3679.

Ghan, S., C. Chuang, and J. Penner (1993). A parameterization of cloud droplet nucleation. Part 1, Single aerosol species. *Atmos. Res.*, **30**, 197–222.

Ghan, S., C. Chuang, R. Easter, and J. Penner (1995). A parameterization of cloud droplet nucleation. Part 2, Multiple aerosol types. *Atmos. Res.*, **36**, 39–54.

Ghan, S., L. Leung, R. Easter, and H. Abdul-Razzak (1997). Prediction of cloud droplet number in a general circulation model. *J. Geophys. Res.*, **102**, 777–794.

Ghan, S. J., H. Abdul-Razzak, A. Nenes, Y. Ming, X. Liu, M. Ovchinnikov, B. Shipway, N. Meskhidze, J. Xu, and X. Shi (2011). Droplet nucleation: Physically-based parameterizations and comparative evaluation. *J. Adv. Model. Earth Syst.*, **3**, M10001, doi:10.1029/2011MS000074.

Giauque, W. F. and J. W. Stout (1936). The entropy of water and the third law of thermodynamics. The heat capacity of ice from 15 to 273 K. *J. Amer. Chem. Soc.*, **58**, 1144–1150.

Gierens, K. M. (2003). On the transition between heterogeneous and homogeneous freezing. *Atmos. Chem. Phys.*, **3**, 437–446.

Gierens, K. M., M. Monier, and J.-F. Gayet (2003). The deposition coefficient and its role in cirrus clouds. *J. Geophys. Res.*, **108** (D2), 4069, doi:10.1029/2001JD001558.

Girard, E., and J. A. Curry (2001). Simulation of Arctic low-level clouds observed during the FIRE Arctic Clouds Experiment using a new bulk microphysics scheme. *J. Geophys. Res.*, **106**, 15,139–15,154.

Girard, E., and J.-P. Blanchet (2001). Simulation of Arctic diamond dust, ice fog, and thin stratus using an explicit aerosol-cloud model. *J. Atmos. Sci.*, **58**, 1199–1221.

Glasstone, S., K. J. Laidler, and H. Eyring (1941). The theory of rate processes. McGraw-Hill, New York, 354 pp.

Golovin, A. M. (1963). On the kinetic equation for coagulating cloud droplets with allowance for condensation. *Izv. Acad. Sci. USSR, Ser. Geophys.*, **10**, 949–953.

Goody, R. M. (1995). Principles of atmospheric physics and chemistry. Oxford University Press, New York, 324 pp.

Grabowski, W. W. and H. Morrison (2008). Toward the mitigation of spurious cloud-edge supersaturation in cloud models. *Mon. Weather Rev.*, **136**, 1224–1234.

Gradshteyn, I. S., and I. M. Ryzhik (1994). Tables of integrals, series, and products, 5th ed., Ed.: Jeffery, A., Academic Press, 1204 pp.

Grassl, H. (1991). Possible climatic effects of contrails and additional water vapor. In: Air traffic and the environment—Background, tendencies and potential global atmospheric effects. Ed.: Schumann, U., Springer-Verlag, 124–137.

Gu, Y., and K. N. Liou (2000). Interactions of radiation, microphysics, and turbulence in the evolution of cirrus clouds. *J. Atmos. Sci.*, **57**, 2463–2479.

Gultepe I., G. A. Isaac, A. Williams, D. Marcotte, and K. B. Strawbridge (2003). Turbulent heat fluxes over leads and polynyas and their effect on Arctic clouds during FIRE-ACE: aircraft observations for April 1998. *Atmos. Ocean*, **41**, 15–34.

Gunn, R., and G. D. Kinzer (1949). The terminal velocity of fall for water droplets in stagnant air. *J. Meteorol.*, **6**, 243–248.

Gunn, R., and J. S. Marshall (1958). The distribution with size of aggregates snowflakes. *J. Meteorol.*, **15**, 452–461.

Gutzow, I., and J. W. P. Schmelzer (1995). The vitreous state. Thermodynamics, structure, rheology, and crystallization. Springer-Verlag, Berlin, Heidelberg.

Gutzow, I. S., and J. W. P. Schmelzer (2011). Glasses and the third law of thermodynamics. In *Glasses and the glass tradition*, edited by Schmelzer, J. W. P. and Gutzow, I. S., Wiley-VCH, Weinheim, p. 357–378.

Haag, W., B. Kärcher, J. Ström, A. Minikin, U. Lohmann, J. Ovarlez, and A. Stohl (2003). Freezing thresholds and cirrus cloud formation mechanisms inferred from in situ measurements of relative humidity. *Atmos. Chem. Phys.*, **3**, 1791–1806.

Haar, L., J. S. Gallagher, and G. S. Kell (1982). The anatomy of the thermodynamic surface of water: The formulation and comparison with data. *Proc. 8th Symp. Thermophys. Properties.* Ed.: J. V. Sengers. *The Amer. Soc. Mechan. Engineers*, vol. 2, New York, pp. 298–300.

Haar, L., J. S. Gallagher, and G. S. Kell (1984). NBS/NRC steam tables: Thermodynamic and transport properties and computer programs for vapor and liquid states of water in SI units. Hemisphere, Washington, and McGraw-Hill, New York, 271–276.

Hagen, D. E., R. J. Anderson, and J. L. Kassner, Jr. (1981). Homogeneous condensation-freezing nucleation rate measurement for small water droplets in an expansion clouds chamber. *J. Atmos. Sci.*, **38**, 1236–1243.

Hahn, C. J., and S. G. Warren (2007). A gridded climatology of clouds over land (1971–96) and ocean (1954–97) from surface observations worldwide. Report, Numeric Data Product NDP-026E, Carbon Dioxide Information Analysis Center, Oak Ridge National Laboratory, Oak Ridge, Tennessee, doi: 10.3334/CDIAC/cli.ndp026,

Hahn, C. J., S. G. Warren, J. London, and R. L. Jenne (1988). Climatological data for clouds over the globe from surface observations. NDP-026, Carbon Dioxide Information Analysis Center (CDIAC), Oak Ridge National Laboratory, Oak Ridge, TN. [Also available from Data Support Section, National Center for Atmospheric Research (NCAR), Boulder, CO.]

Hall, W. D. (1980). A detailed microphysical model within a two-dimensional dynamic framework: Model description and preliminary results. *J. Atmos. Sci.*, **37**, 2486–2506.

Hall, W. D. and H. R. Pruppacher (1976). The survival of ice particles falling from cirrus clouds in subsaturated air. *J. Atmos. Sci.*, **33**, 1995–2006.

Hallett, J., and Mossop, S. C. (1974). Production of secondary ice particles during the riming process. *Nature* (London) **249**, 26–28.

Hämeri, K., A. Laaksonen, M. Vakeva, and T. Suni (2001). Hygroscopic growth of ultrafine sodium chloride particles. *J. Geophys. Res., 106*, 20,749–20,757.

Hämeri, K., M. Vakeva, H.-C. Hansson, and A. Laaksonen (2000). Hygroscopic growth of ultrafine ammonium sulphate aerosol measured using an ultrafine tandem differential mobility analyzer. *J. Geophys. Res.*, 105, 22,231–22,242.

Han, Q., W. B. Rossow, J. Chou, and R. Welch (1998). Global variation of column droplet concentration in low-level clouds. *Geophys. Res. Lett.*, **25**, 1419–1422.

Hänel, G. (1976). The properties of atmospheric aerosol particles as functions of the relative humidity at thermodynamic equilibrium with the surrounding moist air, *Adv. Geophys.*, **19**, 73–188.

Hare, D. E., and C. M. Sörensen (1987). The density of supercooled water. II. Bulk samples cooled to the homogeneous nucleation limit. *J. Chem. Phys.*, **87**, 4840–4850.

Harrington, J. Y., M. P. Meyers, R. L. Walko, and W. R. Cotton (1995). Parameterization of ice crystal conversion process due to vapor deposition for mesoscale models using double-moment basis functions. Part I: Basic formulation and parcel model results. *J. Atmos. Sci.*, **52**, 4344–4366.

Hegg, D. A., and P. V. Hobbs (1992). Cloud condensation nuclei in the marine atmosphere: A review. In: Nucleation and atmospheric aerosols. Eds.: Fukuta N., and Wagner, P. E., A. Deepack Publishing, 181–192.

Hegg, D. A., D. S. Covert, M. J. Rood, and P. V. Hobbs (1996). Measurements of aerosol optical properties in marine air. *J. Geophys. Res.*, **101**(D8), 12,893–12,903.

Heide, H.-G. (1984). Observations of ice layers. *Ultramicroscopy*, **14**, 271–278.

Hellmuth, O., V. I. Khvorostyanov, J. A. Curry, A. K. Shchekin, J. W. P. Schmelzer, and V. G. Baidakov (2012). Review on the phenomenology and mechanism of atmospheric ice formation: Selected questions of interest. In: Nucleation theory and applications. Eds.: Schmelzer, J. W. P., Röpke, G., and Priezzhev, V. B., Joint Institute for Nuclear Research, Bogoliubov Lab. *Theor. Phys.*, Dubna, ISBN 978-5-9530-0301-8.

Hellmuth, O., V. I. Khvorostyanov, J. A. Curry, A. K. Shchekin, J. W. P. Schmelzer, R. Feistel, Y. S. Djikaev, and V. G. Baidakov (2013). Selected aspects of atmospheric ice and salt crystallization. In: Nucleation theory and applications: Special issues. Review series on selected topics of atmospheric sol formation, vol. 1. Eds.: Schmelzer, J. W. P., and Hellmuth, O., Joint Inst. Nuclear Res., Bogoliubov Lab. *Theor. Phys.*, Dubna, 548 pp.

Herman G. F., and J. A. Curry (1984). Observational and theoretical studies of solar radiation in Arctic stratus clouds. *J. Clim. Appl. Met.*, **23**, 5–24.

Herman, G. F., and R. M. Goody (1976). Formation and persistence of summertime Arctic stratus clouds. *J. Atmos Sci.*, **33**, 1537–1553.

Heymsfield, A. J. (1972). Ice crystals terminal velocities. *J. Atmos. Sci.*, **29**, 1348–1357.

Heymsfield, A. J. (2003). Properties of tropical and midlatitude ice cloud particle ensembles. Part 1: Median mass diameters and terminal velocities. *J. Atmos. Sci.*, **60**, 2573–2591.

Heymsfield, A. J., and J. Iaquinta (2000). Cirrus crystals terminal velocities. *J. Atmos. Sci.*, **57**, 916–938.

Heymsfield, A. J., and M. Kajikawa (1987). An improved approach to calculating terminal velocities of plate-like crystals and graupel. *J. Atmos. Sci.*, **44**, 1088–1099.

Heymsfield, A. J., and G. M. McFarquhar (2002). Mid-latitude and tropical cirrus. In: Cirrus. Eds.: Lynch, D., Sassen, K., Starr, D. O'C., and Stephens, G., Oxford University Press, 78–101.

Heymsfield, A. J., and L. M. Miloshevich (1993). Homogeneous ice nucleation and supercooled liquid water in orographic wave clouds, *J. Atmos. Sci.*, **50**, 2335–2353.

Heymsfield, A. J., and L. M. Miloshevich (1995). Relative humidity and temperature influences on cirrus formation and evolution: Observations from wave clouds and FIRE-II. *J. Atmos. Sci.*, **52**, 4302–4303.

Heymsfield, A. J., and C. M. R. Platt (1984). A parameterization of the particle size spectrum of ice clouds in terms of the ambient temperature and the ice water content. *J. Atmos. Sci.*, **41**, 846–855.

Heymsfield, A. J., and R. M. Sabin (1989). Cirrus crystal nucleation by homogeneous freezing of solution droplets. *J. Atmos. Sci.*, **46**, 2252–2264.

Heymsfield, A. J., L. M. Miloshevich, C. Twohy, G. Sachse, and S. Oltmans (1998). Upper-tropospheric relative humidity observations and implications for cirrus ice nucleation, *Geophys. Res. Lett.*, **25**, 1343–1346, doi:10.1029/98GL01089.

Hicks, I., and G. Vali (1973). Ice nucleation in clouds by liquefied propane spray. *J. Appl. Met.*, **12**, 1247–1258.

Hill, T. A., and T. W. Choularton (1985). An airborne study of the microphysical structure of cumulus clouds. *Q. J. Roy. Meteor. Soc.*, **111**, 517–544.

Hobbs, P. V. (1969). Ice multiplication in clouds. *J. Atmos. Sci.*, **26**, 315–318.

Hobbs, P. V. (1974). Ice physics. Clarendon Press, Oxford, 837 pp.

Hobbs, P. V., and A. L. Rangno (1985). Ice particle concentrations in clouds. *J. Atmos. Sci.*, **42**, 2523–2549.

Hobbs, P. V., and A. L. Rangno (1990). Rapid development of high ice particle concentrations in small polar maritime cumuliform cloud. *J. Atmos. Sci.*, **47**, 2710–2722.

Hobbs, P. V., M. K. Politovich, D. A. Bowdle, and L. F. Radke (1978). Airborne studies of atmospheric aerosol in the High Plains and the structure of natural and artificially seeded clouds in eastern Montana. *Rep. No. XIII, Dep. Atmos. Sci.*, Univ. of Washington, 125 pp.

Holten, V., C. E. Bertrand, M. A. Anisimov, and J. V. Sengers (2011). Thermodynamic modeling of supercooled water. *Tech. Rep., Intern. Assoc. for the Properties of Water and Steam* (IAPWS) (September 2011), Inst. Phys. Sci. Technol., Dept. Chem. Biomolec. Eng., Univ. Maryland, College Park, MD 20742, USA, 43 pp.

Holten, V., C. E. Bertrand, M. A. Anisimov, and J. V. Sengers (2012). Thermodynamics of supercooled water, *J. Chem. Phys.*, **136**, 094507, http://dx.doi.org/10.1063/1.3690497.

Hoose, C., and O. Möhler (2012). Heterogeneous ice nucleation on atmospheric aerosols: A review of results from laboratory experiments. *Atmos. Chem. Phys.*, **12**, 9817–9854.

Hoose, C., J. E. Kristjansson, J.-P. Chen, and A. Hazra (2010). A classical-theory-based parameterization of heterogeneous ice nucleation by mineral dust, soot and biological particles in a global climate model. *J. Atmos. Sci.*, **67**, 2483–2503.

Houze, R. A., P. V. Hobbs, P. H. Herzegh, and D. B. Parsons (1979). Size distribution of precipitating particles in frontal clouds. *J. Atmos. Sci.*, **36**, 156–162.

Howell, W. E. (1949). The growth of cloud drops in uniformly cooled air. *J. Meteorol.*, **54**, 134–149.

Hu, Z., and R. C. Srivastava (1995). Evolution of raindrop size distribution by coalescence, breakup, and evaporation: Theory and observations. *J. Atmos. Sci.*, **52**, 1761–1783.

Huang, J., and L. S. Bartell (1995). Kinetics of homogeneous nucleation in the freezing of large water clusters. *J. Phys. Chem.*, **99**, 3924–3931.

Hudson, J. G. (1984). Cloud condensation nuclei measurements within clouds. *J. Clim. Appl. Meteorol.*, **23**, 42–51.

Huffman, P. J. (1973). Supersaturation spectra of AgI and natural ice nuclei. *J. Appl. Meteorol.*, **12**, 1080–1082.

Huffman, P. J., and G. Vali (1973). The effect of vapor depletion on ice nucleus measurements with membrane filters. *J. Appl. Meteorol.*, **12**, 1018–1024.

Hung, H., A. Malinowski, and S. T. Martin (2002). Ice nucleation kinetics of aerosols containing aqueous and solid ammonium sulfate particles. *J. Phys. Chem. A*, **106**, 293–306.

Hung, H., A. Malinowski, and S. T. Martin (2003). Kinetics of heterogeneous ice nucleation on the surfaces of mineral dust cores inserted into aqueous ammonium sulfate particles. *J. Phys. Chem. A*, **107**, 1296–1306.

Hussain, K., and C. P. R. Saunders (1984). Ice nucleus measurement with a continuous flow chamber, *Quart. J. Roy. Meteor. Soc.*, **110**, 75–84.

Hyland, R. W., and A. Wexler (1983). Formulations for the thermodynamic properties of the saturated phases of H_2O from 173.15 K to 473.15 K. *Trans. Amer. Soc. Heating, Refrigerating and Air-Conditioning Engineers (ASHRAE)*, **89** (2A), Atlanta, GA, USA, 500–519.

IAPWS (2009a). Revised release on the equation of state 2006 for H_2O Ice I_h. *Tech. Rep., The International Association for the Properties of Water and Steam*, Doorwerth, The Netherlands, September 2009, http://www.iapws.org/relguide/Ice-Rev2009.pdf.

IAPWS (2009b). Revised release on the IAPWS formulation 1995 for the thermodynamic properties of ordinary water substance for general and scientific use. *Tech. Rep., The International Association for the Properties of Water and Steam*, Doorwerth, The Netherlands, September 2009, http://www.iapws.org/relguide/IAPWS-95.htm.

IAPWS (2011). Revised release on the pressure along the melting and sublimation curves of ordinary water substance. *Tech. Rep., The International Association for the Properties of Water and Steam*, Pilsen, Czech Republic, September 2011,http://www.iapws.org/relguide/MeltSub.htm.

IAPWS (2012). Guideline on a low-temperature extension of the IAPWS-95 formulation for water vapor. *Tech. Rep., The International Association for the Properties of Water and Steam*, Boulder, CO, USA, September/October 2012.

Ibragimov, K. Y. (1990). *Numerical modeling of stratiform cloudiness in the atmospheres of the giant planets*. Izd. Nauka, Alma-Ata, Kazakhstan, 239 pp. (in Russian).

Intrieri, J. M., M. D. Shupe, T. Uttal, and B. J. McCarty (2002). Arctic cloud statistics from radar and lidar at SHEBA. *J. Geophys., Res.*, **107**, 8030–8039.

IOC, SCOR, and IAPSO (2010). The international thermodynamic equation of seawater—2010: Calculation and use of thermodynamic properties. T. J. McDougall, R. Feistel, D. G. Wright, R. Pawlowicz, F. J. Millero, D. R. Jackett, B. A. King, G. M. Marion, S. Seitz, P. Spitzer, C. T. A. Chen. Tech. Rep., Intergovernmental Oceanographic Commission, Manuals and Guides No. 56, UNESCO, 196 pp., Paris 2010, http://www.teos-10.org.

IPCC (2007). Contribution of Working Group I to the Fourth Assessment Report of the Intergovernmental Panel on Climate Change, 2007. Eds.: Solomon, S., Qin, D., Manning, M., Chen, Z., Marquis, M., Averyt, K. B., Tignor, M., and Miller, H. L., Cambridge University Press, Cambridge, UK, and New York.

Jaenicke, R. (1988). Aerosol physics and chemistry. In: Numerical data and functional relationships in science and technology. Eds.: Fischer, G., Landolt-Börnstein New Series, vol. 4b, Springer, New York, pp. 391–457.

Jayaweera, L. O. L. F., and R. E. Cottis (1969). Fall velocities of plate-like and column ice crystals. *Quart. J. Roy. Meteor. Soc.*, **95**, 703–709.

Jeffery, C. A., and P. H. Austin (1997). Homogeneous nucleation of supercooled water: Results from a new equation of state. *J. Ceophys. Res.*, **102** (D21), 25,269–25,279.

Jeffery, C. A., and P. H. Austin (1999). A new analytic equation of state for liquid water. *J. Chem. Phys.*, **110** (1), 484–496.

Jeffery, C. A., J. M. Reisner, and M. Andrejczuk (2007). Another look at stochastic condensation for subgrid cloud modeling: Adiabatic evolution and effects. *J. Atmos. Sci.*, **64**, 3953–3973.

Jeffreys, H. (1918). Some problems of evaporation. *Phil. Mag.*, **35**, 270–280.

Jensen, E. J., O. B. Toon, D. L. Westphal, S. Kinne, and A. J. Heymsfield (1994). Microphysical modeling of cirrus, 1. Comparison with 1986 FIRE IFO measurements. *J. Geophys. Res.*, **99**, 10,421–10,442.

Jensen, E. J., O. B. Toon, A. Tabazadeh, G. W. Sachsse, B. E. Anderson, et al. (1998). Ice nucleation processes in upper tropospheric wave-clouds observed during SUCCESS. *Geophys. Res. Lett.*, **25** (9), 1363–1366.

Jensen, E. J., O. B. Toon, H. B. Selkirk, J. D. Spinhirne, and M. R. Schoeberl (1996). On the formation and persistence of subvisual cirrus near the tropical tropopause. *J. Geophys. Res.*, **101**, 21,361–21,375.

Jensen, E., L. Pfister, T. Bui, A. Weinheimer, E. Weinstock, J. Smith, J. Pittman, D. Baumgardner, P. Lawson, and M. J. McGill (2005). Formation of a tropopause cirrus layer observed over Florida during CRYSTAL-FACE. *J. Geophys. Res.*, **110**, D03208, doi:10.1029/2004JD004671.

Jensen, E. J., O. B. Toon, S. A. Vay, J. Ovarlez, R. May, T. P. Bui, C. H. Twohey, B. W. Gandrud, R. F. Pueschel, and U. Schumann (2001). Prevalence of ice-supersaturated regions in the upper troposphere: implications for optically thin ice cloud formation. *J. Geophys. Res.*, **106**, 17,253–17,266.

Ji, Q., and G. E. Shaw (1998). On supersaturation spectrum and size distribution of cloud condensation nuclei, *Geophys. Res. Lett.*, **25**, 1903–1906.

Jiusto, J. E., and G. E. Bosworth (1971). Fall velocity of snowflakes. *J. Appl. Meteorol.*, **10**, 1352–1354.

Jiusto, J. E., and G. G. Lala (1981). CCN-supersaturation spectra slopes (k). *J. Rech. Atmos.*, **15**, 303–311.

Johari, G. P. (1998). An interpretation for the thermodynamic features of ice $I_h \leftrightarrow$ ice XI transformations. *J. Chem. Phys.*, **109** (21), 9543–9548.

Johari, G. P., G. Fleissner, A. Hallbrucker, and E. Mayer (1994). Thermodynamic continuity between glassy and normal water. *J. Phys. Chem.*, **98**, 4719–4725.

Johnson, D. B. (1982). The role of giant and ultragiant aerosol particles in warm rain initiation. *J. Atmos. Sci.*, **39**, 448–460.

Jonas, P. R. (1972). The collision efficiency of small drops. *Quart. J. Roy. Meteor. Soc.*, **98**, 681–683.

Joss, J., and E. G. Gori (1978). Shapes of raindrop size distributions. *J. Appl. Meteorol*, **17**, 1054–1061.

Junge, C. E. (1952). Die Constitution der Atmospherischen Aerosols. *Ann. Meteorol.*, **1**, 128–135.

Junge, C. E. (1963). Air chemistry and radioactivity. Academic Press, New York and London, 424 pp.

Kabanov, A. S, I. P. Mazin, and V. I. Smirnov (1970). Effect of spatial inhomogeneity of nucleating droplets on their size spectrum in a cloud. *Izv. Acad. Sci. USSR, Atmos. Oceanic Phys.*, **6**, 265–277.

Kachurin, L. G. (1978). Physical foundations of artificial modification of the atmospheric processes. Leningrad, Hydrometeoizdat, 456 pp. (in Russian).

Kanno, H., and C. A. Angell (1977). Homogeneous nucleation and glass formation in aqueous alkali halide solutions at high pressures. *J. Phys. Chem.*, **81**(26), 2639–2643.

Kärcher, B., and U. Lohmann (2002a). A parameterization of cirrus cloud formation: Homogeneous freezing of supercooled aerosols. *J. Geophys. Res.*, **107**(D2), 4010, doi:10.1029/2001JD000470.

Kärcher, B., and U. Lohmann (2002b). A parameterization of cirrus cloud formation: Homogeneous freezing including effects or aerosol size. *J. Geophys. Res.*, **107**(D23), 4698, doi:10.1029/2001JD001429.

Kärcher, B., and U. Lohmann (2003). A parameterization of cirrus cloud formation: Heterogeneous freezing, *J. Geophys. Res.*, **108**, 4402, doi:10.1029/2002JD003220.

Kärcher, B., J. Hendricks, and U. Lohmann (2006). Physically based parameterization of cirrus cloud formation for use in global atmospheric models. *J. Geophys. Res.*, **111**, D01205, doi: 01210.01029/02005JD006219.

Kärcher, B., T. Peter, U. M. Biermann, and U. Schumann (1996). The initial composition of jet condensation trails. *J. Atmos. Sci.*, **53**, 3066–3083.

Kashchiev, D. (2000). Nucleation: Basic theory with applications. Batterworth-Heineman, Oxford, 512 pp.

Kashchiev, D., A. Borissova, R. B. Hammond, and K. J. Roberts (2010). Effect of cooling rate on the critical undercooling for crystallization. *J. Cryst. Growth*, **312**, 698–704.

Kasten, F. (1969). Visibility forecast in the phase of pre-condensation. *Tellus*, **21**(5), 631–635.

Kell, G. S. (1975). Density, thermal expansivity, and compressibility of liquid water from 0° to 150°C: Correlations and tables for atmospheric pressure and saturation reviewed and expressed on 1968 temperature scale. *J. Chem. Eng. Data*, **20**, 97–112.

Ketcham, W. M., and P. V. Hobbs (1969). An experimental determination of the surface energies of ice. *Phil. Mag.*, **19**, 1161.

Kessler, E. (1969). On the distribution and continuity of water substance in atmospheric circulation. *Meteor. Monogr.*, **10** (No. 32), *Amer. Meteor. Soc.*, 84 pp.

Khain, A. P., A. Pokrovsky, M. Pinsky, A. Seifert, and V. Phillips (2004). Simulation of effects of atmospheric aerosols on deep turbulent convective clouds using a spectral microphysics mixed-phase cumulus cloud model. Part I: Model description and possible applications. *J. Atmos. Sci.*, **61**, 2963–2982.

Khairoutdinov, M. F., and V. I. Khvorostyanov (1991). Modeling of artificial dispersal of orographic cloudiness by seeding from aircraft. *Atmos. Optics*, **4** (10), 650–676.

Khairoutdinov, M. F., and Y. L. Kogan (2000). A new cloud physics parameterization in a large-eddy simulation model of marine stratocumulus. *Mon. Wea. Rev.*, **128**, 229–243.

Khairoutdinov, M. F., and D. Randall (2003). Cloud resolving modeling of the ARM summer 1997 IOP: Model formulation, results, uncertainties, and sensitivities. *J. Atmos. Sci.*, **60** (4), 607–625, doi:10.1175/1520-0469.

Khvorostyanov, V. I. (1982). A two-dimensional time-dependent microphysical model of advective-radiative fog and low clouds. *Sov. Meteorol. Hydrol.*, **No. 7**, 16–28.

Khvorostyanov, V. I. (1984). Modeling artificial crystallization and dispersal of supercooled fogs. *Sov. Meteorol. Hydrol.*, No. 3, 35–45.

Khvorostyanov, V. I. (1987). Three-dimensional microphysical model of cloud crystallization after seeding with dry ice. *Sov. Meteorol. Hydrol.*, No. 4, 29–37.

Khvorostyanov, V. I. (1995). Mesoscale processes of cloud formation, cloud-radiation interaction and their modelling with explicit cloud microphysics. *Atmos. Res.* **39**, 1–67.

Khvorostyanov, V. I., A. P. Khain, and E. A. Kogteva (1989). A two-dimensional time-dependent microphysical model of the three-phase convective cloud and evaluation of seeding with a crystallizing agent. *Sov. Meteorol. Hydrol.*, No. 5, 33–45.

Khvorostyanov, V. I., and J. A. Curry (1999a). A simple analytical model of aerosol properties with account for hygroscopic growth. Part I: Equilibrium size spectra and CCN activity spectra. *J. Geophys. Res.*, **104** (D2), 2163–2174.

Khvorostyanov, V. I., and J. A. Curry (1999b). A simple analytical model of aerosol properties with account for hygroscopic growth. Part II: Scattering and absorption coefficients. *J. Geophys. Res.*, **104** (D2), 2175–2184.

Khvorostyanov, V. I., and J. A. Curry (1999c). Toward the theory of stochastic condensation in clouds. Part I: A general kinetic equation. *J. Atmos. Sci.*, **56**, 3985–3996.

Khvorostyanov, V. I., and J. A. Curry (1999d). Toward the theory of stochastic condensation in clouds. Part II. Analytical solutions of gamma distribution type. *J. Atmos. Sci.*, **56**, 3997–4013.

Khvorostyanov, V. I, and J. A. Curry (2000). A new theory of heterogeneous nucleation for application in cloud and climate models, *Geophys. Res. Lett.*, **27**, 4081–4084.

Khvorostyanov, V. I., and J. A. Curry (2002). Terminal velocities of droplets and crystals: Power laws with continuous parameters over the size spectrum. *J. Atmos. Sci.*, **59**, 1872–1884.

Khvorostyanov, V. I., and J. A. Curry (2004a). Thermodynamic theory of freezing and melting of water and aqueous solutions. *J. Phys. Chem. A*, **108** (50), 11,073–11,085.

Khvorostyanov, V. I., and J. A. Curry (2004b). The theory of ice nucleation by heterogeneous freezing of deliquescent mixed CCN. Part 1: Critical radius, energy and nucleation rate. *J. Atmos. Sci.*, **61**, 2676–2691.

Khvorostyanov, V. I., and J. A. Curry (2005a). The theory of ice nucleation by heterogeneous freezing of deliquescent mixed CCN. Part 2: Parcel model simulation. *J. Atmos. Sci.*, **62**, 261–285.

Khvorostyanov, V. I., and J. A. Curry (2005b). Fall velocities of hydrometeors in the atmosphere: Refinements to a continuous analytical power law. *J. Atmos. Sci.*, **62** (12), 4343–4357.

Khvorostyanov, V. I., and J. A. Curry (2006). Aerosol size spectra and CCN activity spectra: Reconciling the lognormal, algebraic and power laws. *J. Geophys. Res.*, **111**, D12202, doi:10.1029/2005JD006532.

Khvorostyanov, V. I., and Curry, J. A. (2007). Refinements to the Köhler's theory of aerosol equilibrium radii, size spectra, and droplet activation: Effects of humidity and insoluble fraction. *J. Geophys. Res.*, **112** (D5), D05206, http://dx.doi.org/10.1029/2006JD007672

Khvorostyanov, V. I., and J. A. Curry (2008a). Kinetics of cloud drop formation and its parameterization for cloud and climate models. *J. Atmos. Sci.*, **65**, 2784–2802.

Khvorostyanov, V. I., and J. A. Curry (2008b). Analytical solutions to the stochastic kinetic equation for liquid and ice particle size spectra. Part I: Small-size fraction. *J. Atmos. Sci.*, **65**, 2025–2043.

Khvorostyanov, V. I., and J. A. Curry (2008c). Analytical solutions to the stochastic kinetic equation for liquid and ice particle size spectra. Part II: Large-size fraction in precipitating clouds. *J. Atmos. Sci.*, **65**, 2044–2063.

Khvorostyanov, V. I., and J. A. Curry (2009a). Parameterization of cloud drop activation based on analytical asymptotic solutions to the supersaturation equation. *J. Atmos. Sci.*, **66**, 1905–1925.

Khvorostyanov, V. I., and J. A. Curry (2009b). Critical humidities of homogeneous and heterogeneous ice nucleation: Inferences from extended classical nucleation theory. *J. Geophys. Res.*, **114**, D04207, http://dx.doi.org/10.1029/2008JD011197.

Khvorostyanov, V. I., and J. A. Curry (2012). Parameterization of homogeneous ice nucleation for cloud and climate models based on classical nucleation theory. *Atmos. Chem. Phys.*, **12**, 9275–9302, 2012, www.atmos-chem-phys.net/12/9275/2012/, doi:10.5194/acp-12-9275-2012.

Khvorostyanov, V. I., J. A. Curry, J. O. Pinto, M. Shupe, B. Baker, and K. Sassen (2001). Modeling with explicit spectral water and ice microphysics of a two-layer cloud system of altostratus and cirrus observed during the FIRE Arctic Clouds Experiment. *J. Geophys. Res.*, **106**, 15,099–15,112.

Khvorostyanov, V. I., J. A. Curry, I. Gultepe, and K. Strawbridge (2003). A springtime cloud over the Beaufort Sea polynya: Three-dimensional simulation with explicit spectral microphysics and comparison with observations. *J. Geophys. Res.*, **108** (D9), 4296, doi:10.1029/2001JD001489.

Khvorostyanov, V. I., H. Morrison, J. A. Curry, D. Baumgardner, and P. Lawson (2006). High supersaturation and modes of ice nucleation in thin tropopause cirrus: Simulation of the 13 July 2002 Cirrus Regional Study of Tropical Anvils and Cirrus Layers case. *J. Geophys. Res.*, **111** (No. D2), D02201, http://dx.doi.org/10.1029/2004JD005235.

Khvorostyanov, V. I., and M. F. Khairoutdinov (1990). Modeling precipitation enhancement over extended mountainous country by aircraft seeding of orographic cloudiness. *Sov. Meteorol. Hydrol.*, No. 11, 43–54.

Khvorostyanov, V. I., and K. Sassen (1998a). Cirrus cloud simulation using explicit microphysics and radiation: Part I: Model description, *J. Atmos. Sci.*, **55**, 1808–1821.

Khvorostyanov, V. I., and K. Sassen (1998b). Cirrus cloud simulation using explicit microphysics and radiation: Part II: Microphysics, vapor and ice mass budgets, and optical and radiative properties. *J. Atmos. Sci.*, **55**, 1822–1845.

Khvorostyanov, V. I., and K. Sassen (1998c). Towards the theory of homogeneous nucleation and its parameterization for cloud models. *Geophys. Res. Lett.*, **25** (16), 3155–3158.

Khvorostyanov, V. I., and K. Sassen (2002). Microphysical processes in cirrus and their impact on radiation: A mesoscale modeling perspective. In: Cirrus. Eds.: Lynch, D., Sassen, K., Starr, D. O'C., and Stephens, G., Oxford University Press, 397–432.

Kimizuka, N., and T. Suzuki (2007). Supercooling behavior in aqueous solutions. *J. Phys. Chem. B*, **111**, 2268–2273.

Kiselev, S. B. (2001). Physical limit of stability in supercooled liquids. *Int. J. Thermophys.*, **22** (5), 1421–1433.

Kiselev, S. B., and J. F. Ely (2002). Parametric crossover model and physical limit of stability in supercooled water. *J. Chem. Phys.*, **116** (13), 5657–5665.

Klein, S. A., R. B. McCoy, H. Morrison, A. S. Ackerman, et al. (2009). Intercomparison of model simulations of mixed-phase clouds observed during the ARM Mixed-Phase Arctic Cloud Experiment. I. Single-layer cloud. *Q. J. Roy. Meteor. Soc.*, **135**, 979–1002.

Klotz, I. M., and R. M. Rosenberg (1972). Chemical thermodynamics, basic theory and methods, 3rd ed. Benjamin/Cummings: Menlo Park, CA, USA.

Klug, D. D. (2002). Dense ice in detail, *Nature*, **420**, 749–751.

Knight, C. A. (1971). Experiments on contact angle of water on ice. *Philos. Mag.* **23**, 153–165.

Knight, N. C., and A. J. Heymsfield (1983). Measurement and interpretation of hailstone density and terminal velocity. *J. Atmos. Sci.*, **40**, 1510–1516.

Knopf, D. A., and T. Koop (2006). Heterogeneous nucleation of ice on surrogates of mineral dust. *J. Geophys. Res.*, **111**, D12201, doi:10.1029/2005JD006894.

Köhler, H. (1921). Zur Kondensation des Wasserdampfes in der Atmosphäre, *Geophys. Publ.*, **2**, 3–15.

Köhler, H. (1936). The nucleus in and the growth of hygroscopic droplets, *Trans. Farad. Soc.*, **32**, 1152–1161.

Kolmogorov, A. N. (1941). Local structure of turbulence in an incompressible viscous fluid at very high Reynolds numbers. *Dokl. Acad. Sci. USSR*, **30**, 301–305. Reprinted in *Sov. Phys. Usp.*, 10, 734–736, 1968, and *Proc. Roy. Soc.*, London, A, 434, 9–13, 1991.

Kondratyev, K. Ya. (1969). Radiation in the atmosphere. Academic Press, New York, 912 pp.

Kondratyev, K. Ya., and V. I. Khvorostyanov (1989). Modeling of cloud formation due to air-sea interaction over the oceanic hydrological front. *Boundary Layer Meteorol.*, **46**, 229–249.

Kondratyev, K. Ya., M. V. Ovtchinnikov, and V. I. Khvorostyanov (1990a). Mesoscale model of mixed-phase cloud development with account for the interaction among optical, radiative and microphysical processes. *Atmos. Optics*, **3** (6), 639–646.

Kondratyev, K. Ya., M. V. Ovtchinnikov, and V. I., Khvorostyanov (1990b). Modeling the evolution of the optical, radiative and microphysical properties of the atmosphere after crystallization of cloudiness. Part I: Complete dispersal of the clouds. *Atmos. Optics*, **3** (No 6), 647–654.

Koop, T. (2004). Homogeneous ice nucleation in water and aqueous solutions. *Z. Phys. Chem.*, **218**, 1231–1258.

Koop, T., and B. Zobrist (2009). Parameterizations for ice nucleation in biological and atmospheric systems. *Phys. Chem. Chem. Phys.*, **11**, 10,839–10,850, doi:10.1039/b914289d.

Koop, T., A. K. Bertram, L. T. Molina, and M. J. Molina (1999). Phase transitions in aqueous NH4HSO4 solutions. *J. Phys. Chem. A.*, **103**, 9042–9048.

Koop, T., B. P. Luo, A. Tsias, and T. Peter (2000). Water activity as the determinant for homogeneous ice nucleation in aqueous solutions. *Nature*, **406**, 611–614.

Koop, T., H. P. Ng, L. T. Molina, and M. J. Molina (1998). A new optical technique to study aerosol phase transitions: The nucleation of ice from H_2SO_4 aerosols. *J. Phys. Chem. A*, **102**, 8924–8931.

Korhonen, P., A. Laaksonen, E. Batris, and Y. Viisanen (1998). Thermodynamics for highly concentrated water-ammonium sulfate solutions. *J. Aerosol Sci.*, **29** (Suppl. 1), S379–S380.

Korn, G. A., and T. M. Korn (1968). Mathematical handbook for scientists and engineers. McGraw-Hill Co., New York, 831 pp.

Korolev, A. V., G. A. Isaac, S. G. Cober, J. W. Strapp, and J. Hallett (2003). Microphysical characterization of mixed-phase clouds. *Q. J. Roy. Meteor. Soc.*, **129**, 39–65.

Kovetz, A., and B. Olund (1969). The effect of coalescence and condensation on rain formation in a cloud of finite vertical extent. *J. Atmos. Sci.*, **26**, 1060–1065.

Kotchenruther, R. A., P. V. Hobbs, and D. A. Hegg (1999). Humidification factors for atmospheric aerosols off the mid-Atlantic coast of the United States. *J. Geophys. Res.*, **104**, 2239–2251.

Krakovskaia, S. V., and A. M. Pirnach (2004). A theoretical study of the microphysical structure of mixed stratiform frontal clouds and their precipitation. *Atmos. Res.*, 47–48, 491–503.

Krämer, B., M. Schwell, O. Hubner, H, Vortisch, T. Leisner, E. Ruhl, H Baumgartel, and L. Woste (1996). Homogeneous ice nucleation observed in single levitated micro droplets. *Ber. Bunsenges Phys. Chem.*, **100**, 1911–1914.

Krämer, M., C. Schiller, A. Afchine, R. Bauer, I. Gensch, A. Mangold, S. Schlicht, N. Spelten, N. Sitnikov, S. Borrmann, M. de Reus, and P. Spichtinger (2009). Ice supersaturations and cirrus cloud crystal numbers. *Atmos. Chem. Phys.*, **9**, 3505–3522, doi:10.5194/acp-9-3505-2009.

Krasnovskaya, L. I. (1964). Physical basics of artificial cloud modification with cooling agents. *Proc. Centr. Aerolog. Obs.*, **58**, 79 pp. (in Russian).

Krasnovskaya, L. I., Y. A. Seregin, and V. I. Khvorostyanov (1987). State-of-the-art in studies of artificial seeding of supercooled clouds and fogs with cooling agents. In: Problems of cloud physics, weather modification. Leningrad, Hydrometeoidat, 1987, 50–64 (in Russian).

Kreidenweis, S. M., K. Koehler, P. J. DeMott, A. P. Prenni, C. Carrico, and B. Ervens (2005). Water activity and activation diameters from hygroscopicity data—Part I: Theory and application to inorganic salts. *Atmos. Chem. Physics*, **5**, 1357–1370.

Kulmala, M., A. Laaksonen, P. Korhonen, T. Vesala, T. Ahonen, and J. C. Barrett (1993). The effect of atmospheric nitric acid vapor on cloud condensation nucleus activation. *J. Geophys. Res.*, **98**, 22,949–22,958.

Kulmala, M., U. Rannik, E. Zapadinsky, and C. Clement (1997). The effect of saturation fluctuations on droplet growth. *J. Aerosol Sci.*, **28**, 1395–1409.

Kumar, B., F. Janetzko, J. Schumacher, and R. A Shaw (2012). Extreme responses of a coupled scalar–particle system during turbulent mixing. *New Journal of Physics*, **14**, 115,020–115,041, doi:10.1088/1367-2630/14/11/115020.

Kuo, J.-L., J. V. Coe, S. J. Singer, Y. B. Band, and L. Ojamäe (2001). On the use of graph invariants for efficiently generating hydrogen bond topologies and predicting physical properties of water clusters and ice. *J. Chem. Phys.*, **114**, 2527–2540.

Kuo, J.-L., M. L., Klein, S. J., Singer, and L. Ojamäe (2004). Ice I_h – Ice XI phase transition. A quantum mechanical study. In: Abstracts of papers, 227th ACS National Meeting, Anaheim, CA, USA, March 28–April 1, 2004, PHYS-463, Amer. Chem. Soc.

Laaksonen, A., V. Talanquer, and D. Oxtoby (1995). Nucleation: Measurements, theory and atmospheric applications. *Annu. Rev. Phys. Chem.*, **46**, 489–524.

Laaksonen, A., P. Korhonen, M. Kulmala, and R. J. Charlson (1998). Modification of the Köhler equation to include soluble trace gases and slightly soluble substances. *J. Atmos. Sci.*, **55**, 853–862.

Ladino, L., O. Stetzer, F. Lüönd, A. Welti, and U. Lohmann (2011). Contact freezing experiments of kaolinite particles with cloud droplets. *J. Geophys. Res.*, **116**, D22202, doi:10.1029/2011JD015727.

Ladino Moreno, L. A., O. Stetzer, and U. Lohmann (2013). Contact freezing: A review of experimental studies. *Atmos. Chem. Phys.*, **13**, 9745–9769, www.atmos-chem-phys.net/13/9745/2013/, doi:10.5194/acp-13-9745-2013.

Laktionov, A. G. (1972). Fraction of soluble in water substances in the particles of atmospheric aerosol. *Izv. Acad. Sci. USSR, Atmos. Oceanic Phys.*, **8**, 389–395.

Landau, L. D., and E. M. Lifshitz (1958a). Quantum mechanics. Non-relativistic theory. Course of theoretical physics, v. 3. Addison-Wesley, 526 pp.

Landau, L. D., and E. M. Lifshitz (1958b). Statistical physics, Part I. Course of theoretical physics, v. 5. Pergamon Press, 544 pp.

Landau, L. M., and E. M. Lifshitz (1959). Fluid mechanics. Course of theoretical physics, v. 6. 1st ed., Pergamon Press, 536 pp.

Landau, L. D., and E. M. Lifshitz (1966). Electrodynamics of continuous media. Course of theoretical physics, v. 8. Addison-Wesley, 417 pp.

Landau, L. D., and E. M. Lifshitz (2005). The classical theory of fields. Course of theoretical physics, v. 2. Pergamon Press, 428 pp.

Langham, E. J., and B. J. Mason (1958). The heterogeneous and homogeneous nucleation in supercooled water. *Proc. Roy. Soc.*, **A247**, 493–504.

Larson, B. H., and B. D. Swanson (2006). Experimental investigation of the homogeneous freezing of aqueous ammonium sulfate droplets. *J. Phys. Chem. A*, **110** (5), 1907–1916.

Latham, J., and R. L. Reed (1977). Laboratory studies of the effects of mixing on the evolution of cloud droplet spectra. *Quart. J. Roy. Meteor. Soc.*, **103**, 297–306.

Lawson, R. P., R. E. Stewart, and L. J. Agnus (1998). Observations and numerical simulations of the origin and development of very large snowflakes. *J. Atmos., Sci.*, **55**, 3209–3229.

Lawson, R. P., B. A. Baker, C. G. Schmitt, and T. L. Jensen (2001). An overview of microphysical properties of Arctic clouds observed in May and July 1998 during FIRE ACE. *J. Geophys. Res.*, **106** (D14), 14,989–15,014.

Lawson, R. P., B. Baker, B. Pilson, and Q. Mo (2006). In situ observations of the microphysical properties of wave, cirrus, and anvil clouds. Part II: Cirrus clouds. *J. Atmos. Sci.*, **63**, 3186–3203.

Leberman, R., and A. K. Soper (1995). Effect of high salt concentrations on water structure. *Nature*, **378**, 364–366.

Leck, K., E. D. Nilsson, E. K. Bigg, and L. Bäcklin (2001). Atmospheric program on the Arctic Ocean Expedition 1996 (AOE-96): An overview of scientific goals, experimental approach, and instruments. *J. Geophys. Res.*, **106** (D23), 32,051–32,067.

Lemons, D. S. (2002). An introduction to stochastic processes in physics. The Johns Hopkins University Press, 110 pp.

Levich, V. G. (1969). The course of theoretical physics, v. 1. Moscow, "Nauka," 910 pp. (in Russian).

Levin, L. M. (1954). On the size distributions functions of the cloud and rain droplets. *Dokl. Acad. Sci. USSR*, **44**, 1045–1049.

Levin, L. M., and Y. S. Sedunov (1966a). A theoretical model of condensation nuclei: The mechanism of cloud formation in clouds. *J. Rech. Atmos.*, **2** (2–3), 416–424.

Levin, L. M., and Y. S. Sedunov (1966b). Stochastic condensation of drops and kinetics of cloud spectrum formation. *J. Rech. Atmos.*, **2**, 425–432.

Levin, Z., E. Ganor, and V. Gladstein (1996). The effects of desert particles coated with sulfate on rain formation in the eastern Mediterranean. *J. Appl. Meteorol.*, **35**, 1511–1523.

Lewis, J. S. (1995). Physics and chemistry of the solar system. Academic Press, San Diego, CA, USA, 556 pp.

Lewis, J. S., and R. E. Prinn (1984). Planets and their atmospheres: Origins and evolutions. Academic Press, New York, 470 pp.

Li, Zh., A. L. Williams, and M. J. Rood (1998). Influence of soluble surfactant properties on the activation of aerosol particles containing inorganic solute. *J. Atmos. Sci.*, **55**, 1859–1866.

Lifshitz, E. M., and L. P. Pitaevskii (1997). Physical kinetics. Course of theoretical physics by Landau and Lifshitz, v. 10. Butterworth Heinemann, 452 pp.

Lifshitz, I. M., and V. V. Slezov (1958). On the theory of diffusional decay of supersaturated solid solutions. *J. Exper. Theor. Phys.*, **35**, 479–492.

Lifshitz, I. M., and V. V. Slezov (1961). On the kinetics of precipitation of supersaturated solid solutions. *J. Phys. Chem. Solids*, **19**, 35–50.

Lilly, D. K. (1968). Models of cloud-topped mixed layers under a strong inversion. *Quart. J. Roy. Met. Soc.*, (4), 292–309.

Lin, R.-F., D. O'C. Starr, P. J. DeMott, R. Cotton, K. Sassen, E. Jensen, B. Kärcher, and X. Liu (2002). Cirrus parcel model comparison project. Phase 1: The critical components to simulate cirrus initiation explicitly. *J. Atmos. Sci.*, **59**, 2305–2329.

Lin, Y. L., R. D. Farley, and H. D. Orville (1983). Bulk parameterization of the snow field in a cloud model. *J. Climate Appl. Meteorol.*, **22**, 1065–1092.

Liou, K. N. (1980). An introduction to atmospheric radiation. Academic Press, 384 pp.

Liou, K. N. (1992). Radiation and cloud processes in the atmosphere. Oxford University Press, 487 pp.

Liou, K. N., and S. C. Ou (1989). The role of cloud microphysical processes in climate: An assessment from a one-dimensional perspective. *J. Geophys. Res.*, **94D**, 8599–8607.

Liou, K.-N., S. C. Ou, and G. Koenig (1991). An investigation of the climatic effect of contrail cirrus. In: Air traffic and the environment—Background, tendencies, and potential global atmospheric effects. Ed.: Schumann, U., Springer-Verlag, 154–169.

Litvinov, I. V. (1956). Determination of falling velocity of snow particles. *Izv. Acad. Sci. USSR, Ser. Geophys.*, **7**, 853–856.

Liu, X., and J. E. Penner (2005). Ice nucleation parameterization for global models. *Meteorol. Z.*, **14**, 499–514.

Liu, Y. (1993). Statistical theory of the Marshall-Palmer distribution of raindrops. *Atmos. Environ.*, **27A**, 15–19.

Liu, Y. (1995). On the generalized theory of atmospheric particles systems. *Adv. Atmos. Sci.*, **12**, 419–438.

Liu, Y., and J. Hallett (1998). On size distribution of cloud drops growing by condensation: A new conceptual model. *J. Atmos. Sci.*, **55**, 527–536.

Liu, Y., and P. H. Daum (2004). Parameterization of the autoconversion Process. Part I: Analytical formulation of the Kessler-type parameterizations. *J. Atmos. Sci.*, **61**, 1539–1548.

Liu, Y., Y. Laiguang, Y. Weinong, and L. Feng (1995). On the size distribution of cloud droplets. *Atmos. Res.*, **35**, 201–216.

Locatelli, J. D., and P. V. Hobbs (1974). Fall speeds and masses of solid precipitation particles. *J. Geophys. Res.*, **79**, 2185–2197.

Lohmann, U. (2002). Possible aerosol effects on ice clouds via contact nucleation. *J. Atmos. Sci.*, **59**, 647–656.

Lohmann, U., and B. Kärcher (2002). First interactive simulations of cirrus clouds formed by homogeneous freezing in the ECHAM GCM, *J. Geophys. Res.*, **107**, doi:10.1029/2001JD000 767.

Lohmann, U., and J. Feichter (2005). Global indirect aerosol effects: A review. *Atmos. Chem. Phys.*, **5**, 715–737, www.atmos-chem-phys.org/acp/5/715/.

Lohmann, U., J. Feichter, C. C. Chuang, and J. E. Penner (1999). Predicting the number of cloud droplets in the ECHAM GCM. *J. Geophys. Res.*, **104**, 9169–9198.

Lohmann U., and K. Diehl (2006). Sensitivity studies of the importance of dust ice nuclei for the indirect aerosol effect on stratiform mixed-phase clouds. *J Atmos Sci.*, **63**, 968–982.

London, J. (1957). A study of the atmospheric heat balance, Final report. Contract AF19(122)-165. Dept. Meteorol. and Oceanogr., New York Univ. (ASTIA 117227, Air Force Geophysical Laboratory, Hanscom AFB), 99 pp.

Long, A. (1974). Solutions to the droplet collection equation for polynomial kernels. *J. Atmos. Sci.*, **31**, 1040–1052.

Low, R. D. H. (1969). A generalized equation for the solution effect in droplet growth. *J. Atmos. Sci.*, **26**, 608–612.

Low, T. B., and R. List (1982a). Collision, coalescence, and breakup of raindrops. Part I: Experimentally established coalescence efficiencies and fragment size distributions in breakup. *J. Atmos. Sci.*, **39**, 1591–1606.

Low, T. B., and R. List (1982b). Collision, coalescence, and breakup of raindrops. Part II: Parameterization of fragment size distributions. *J. Atmos. Sci.*, **39**, 1607–1618.

Lübken, F.-J., J. Lautenbach, J. Höffner, M. Rapp, and M. Zecha (2009). First continuous temperature measurements within polar mesosphere summer echoes. *J. Atmos. Solar-Terr. Phys.*, **71**, 453–463, doi: 10.1016/j.jastp.2008.06.001.

Ludwig, F. L., and E. Robinson (1970). Observations of aerosols and droplets in California stratus. *Tellus*, **22** (1), 78–89.

Lüönd, F., O. Stetzer, A. Welti, and U. Lohmann (2010). Experimental study on the ice nucleation ability of size selected kaolinite particles in the immersion mode. *J. Geophys. Res.*, **115**, D14201, doi:10.1029/2009JD012959.

Lushnikov, A. A. (1973). Evolution of coagulating systems. *J. Coll. Interface Sci.*, **45** (3), 549–556.

Lushnikov, A. A. (1974). Evolution of coagulating systems. II. Asymptotic size distributions and analytical properties of generating functions. *J. Coll. Interface Sci.*, **48** (3), 400–409.

Lushnikov, A. A., and V. N. Piskunov (1982). Three new exactly solvable models in the theory of coagulation. *Dokl. Acad. Sci. USSR*, **247**, 132–136.

Lushnikov, A. A., and V. I. Smirnov (1975). Stationary coagulation and size distribution of atmospheric aerosol particles. *Izv. Acad. Sci. USSR, Atmos. Oceanic Phys.*, **11**, 139–151.

MacKenzie, A. R., A. Laaksonen, E. Batris, and M. Kulmala (1998). The Turnbull correlation and the freezing of stratospheric aerosol droplets. *J. Geophys. Res. D*, **103**, 10875–10884.

Magono, C., and C. Lee (1966). Meteorological classification of natural snow crystals. *J. Fac. Sci. Hokkaido Univ.*, Ser. VII, **2**, 321–335.

Magono, C., and T. Nakamura (1965). Aerodynamic studies of falling snowflakes. *J. Meteor. Soc. Japan*, Ser. 2, **43**, 139–147.

Malkin, T. L., B. J. Murray, A. V. Brukhno, J. Anwar, and C. G. Salzmann (2012). Structure of ice crystallized from supercooled water. *Proc. Nat. Acad. Sci.*, www.pnas.org/lookup/suppl/, doi:10.1073/pnas.1113059109.

Manton, M. J. (1979). On the broadening of a droplet distribution by turbulence near cloud base. *Quart. J. Roy. Meteor. Soc.*, **105**, 899–914.

Manton, M. J. (1981). Reply to the comments on the paper "On the broadening of a droplet distribution by turbulence near cloud base." *Quart. J. Roy. Meteor. Soc.*, **112**, 977–978.

Manton, M. J., and W. R. Cotton (1977). Formulation of approximate equations for modeling moist deep convection on the mesoscale. *Atmospheric Science Paper No. 266*, Colorado State University, 62 pp.

Marchuk, G. I., K. Ya. Kondratyev, V. V. Kozoderov, and V. I. Khvorostyanov (1986). Clouds and climate. Leningrad, Hydrometeoizdat, 512 pp.

Marcolli, C., S. Gedamke, T. Peter, and B. Zobrist (2007). Efficiency of immersion mode ice nucleation on surrogates of mineral dust. *Atmos. Chem. Phys.*, **7**, 5081–5091.

Marshall, J. S., and W. M. K. Palmer (1948). The distribution of raindrops with size. *J. Meteor.*, **5**, 165–166.

Martin, S. T. (2000). Phase transitions of aqueous atmospheric particles. *Chem. Rev.*, **100**, 3403–3453.

Mason, B. J. (1971). The physics of clouds. Oxford Univ. Press, Clarendon, London, 481 pp.

Mason, P. J. (1985). A numerical study of cloud streets in the planetary boundary layer. *Boundary Layer Meteorol.*, **32**, 281–304.

Mason, P. J. (1989). Large-eddy simulation of the convective atmospheric boundary layer. *J. Atmos. Sci.*, **46**, 1492–1516.

Matrosov, S. Y. (1997). Variability of microphysical parameters in high-altitude ice clouds: Results of the remote sensing method. *J. Appl. Met.*, **36** (6), 633–648.

Matrosov, S. Y., and A. J. Heymsfield (2000). Use of Doppler radar to assess ice cloud particle fall velocity-size relations for remote sensing and climate studies. *J. Geophys. Res.*, **105**, 22,427–22,436.

Matson, R. J., and A. W. Huggins (1980). The direct measurement of the sizes, shapes and kinematics of falling hailstones. *J. Atmos., Sci.*, **37**, 797–816.

Matveev, L. T. (1984). Cloud dynamics. D. Reidel Publish Co., 212 pp.

Maxwell, J. C. (1890). Theory of the wet bulb thermometer. In: The scientific papers of James Clerk Maxwell, vol. 2. Dover Publisher, New York, 636–640 pp.

Mayer, E., and A. Hallbrucker (1987). Cubic ice from liquid water. *Nature*, **325**, 601–602.

Mazin, I. P., and V. I. Smirnov (1969). To the theory of formation of the size spectrum of cloud droplets at stochastic condensation. *Proc. Centr. Aerolog. Observ.*, **89**, 92–94.

McFarquhar, G. M. (2004). A new representation of collision-induced breakup of raindrops and its implications for the shapes of raindrop size distributions. *J. Atmos. Sci.*, **61**, 777–794.

McFarquhar, G., G. Zhang, M. R. Poellot, G. L. Kok, R. McCoy, T. Tooman, A. M. Fridlind, and A. J. Heymsfield (2007). Ice properties of single-layer stratocumulus clouds during the Mixed-Phase Arctic Cloud Experiment: 1. Observations, *J. Geophys. Res.*, **112**, D24202, doi:10.1029/2007JD008633.

McFiggans, G., P. Artaxo, U. Baltensperger, H. Coe, M. C. Facchini, G. Feingold, S. Fuzzi, M. Gysel, A. Laaksonen, U. Lohmann, T. F. Mentel, D. M. Murphy, C. D. O'Dowd, J. R. Snider, and E. Weingartner (2006). The effect of physical and chemical aerosol properties on warm cloud droplet activation. *Atmos. Chem. Phys.*, **6**, 2593–2649, www.atmos-chem-phys.net/6/2593/2006/.

McGraw, R., and E. Lewis (2009). Deliquescence and efflorescence of small particles. *J. Chem. Phys.*, **131**, 194705(1)–194705(14), doi:10.1063/1.3251056.

McGraw, R., and Y. Liu (2006). Brownian drift-diffusion model for evolution of droplet size distributions in turbulent clouds. *Geophys. Res. Lett.*, **33**, L03802, doi: 10.1029/2005GL023545.

Menut, L., B. Bessagnet, D. Khvorostyanov, M. Beekmann, A. Colette, I. Coll, G. Curci, G. Foret, A. Hodzic, S. Mailler, F. Meleux, J.-L. Monge, I. Pison, S. Turquety, M. Valari, R. Vautard, and M. G. Vivanco (2013). Regional atmospheric composition modeling with CHIMERE. *Geosci. Model Dev. Discuss.*, **6**, 203–329. www.geosci-model-dev-discuss.net/6/203/2013/, doi:10.5194/gmdd-6-203-2013.

Merkulovich, V. M., and A. S. Stepanov (1977). Hygroscopicity effects and surface tension forces during condensational growth of cloud droplets in the presence of turbulence. *Izv. Akad. Sci. USSR, Atmos. Oceanic Phys.*, **13**, 163–171.

Merkulovich, V. M., and A. S. Stepanov (1981). Comments on the paper "On the broadening of a droplet distribution by turbulence near cloud base" by M. J. Manton (Q. J., 1979, 105, 899–914). *Quart. J. Roy. Meteor. Soc.*, 112, 976–977.

Meyers, M. P., P. J. DeMott, and W. R. Cotton (1992). New primary ice-nucleation parameterizations in an explicit cloud model. *J. Appl. Meteor.*, **31**, 708–721.

Meyers, M. P., R. L. Walko, J. Y. Harrington, and W. R. Cotton (1997). New RAMS cloud microphysics parameterization. Part II: The two-moment scheme. *Atmos. Res.*, **45**, 3–39.

Milbrandt, J. A., and M. K. Yau (2005a). A multimoment microphysics parameterization. Part I: Analysis of the role of the spectral shape parameter. *J. Atmos. Sci.* **62**, 3051–3064.

Milbrandt, J. A., and M. K. Yau (2005b). A multimoment bulk microphysics parameterization. Part II: A proposed three-moment closure and scheme description. *J. Atmos. Sci.* **62**, 3065–3081.

Millero, F. J. (1978). Freezing point of seawater. In: Eighth Report of the Joint Panel on Oceanographic Tables and Standards. UNESCO Tech. Pap. Mar. Sci., No. 28, Annex 6, UNESCO, Paris.

Millero, F. J., R. Feistel, D. G. Wright, and T. J. McDougall (2008). The composition of standard seawater and the definition of the reference-composition salinity scale. *Deep Sea Res.* **55** (1), 50–72.

Ming, Y., V. Ramaswamy, L. J. Donner, and V. T. J. Phillips (2006). A new parameterization of cloud droplet activation applicable to general circulation models. *J. Atmos. Sci.*, **63**, 1348–1356.

Mishchenko, M. I. (1991). Light-scattering by randomly oriented axially symmetrical particles. *J. Optical Soc. Amer. A—Optics Image Sci. Vision*, **8**, 871–882.

Mishchenko, M. I., L. D. Travis, and D. W. Mackowski (1996). T-matrix computations of light scattering by nonspherical particles: A review. *J. Quant. Spectrosc. Radiat. Transfer*, **55**, 535–575.

Mishima, O. (1996). Relationship between melting and amorphization of ice. *Nature*, **384**, 546–549.

Mishima, O., and H. E. Stanley (1998). The relationship between liquid, supercooled and glassy water. *Nature*, **396**, 329–335.

Mishima, O., L. D. Calvert, and E. Whalley (1984). Melting ice I at 77 K and 10 kbar: A new method of making amorphous solids. *Nature*, **310**, 393–395.

Mishima, O., L. D. Calvert, and E. Whalley (1985). An apparently first-order transition between two amorphous phases of ice induced by pressure. *Nature*, **314**, 76–78.

Mitchell, D. L. (1994). A model predicting the evolution of ice particle size spectra and radiative properties of cirrus clouds. Part I: Microphysics. *J. Atmos. Sci.*, **51**, 797–816.

Mitchell, D. L. (1996). Use of mass- and area-dimensional power laws for determining precipitation particle terminal velocities. *J. Atmos. Sci.*, **53**, 1710–1723.

Mitchell, D. L., and W. P. Arnott (1994). A model predicting the evolution of ice particle size spectra and radiative properties of cirrus clouds. Part II: Dependence of absorption and extintion on ice crystal morphology. *J. Atmos. Sci.*, **51**, 817–832.

Mitchell, D. L., and A. J. Heymsfield (2005). Refinements in the treatment of ice particle terminal velocities: Highlighting aggregates. *J. Atmos. Sci.*, **62**, 1637–1644.

Mitchell, D. L., A. J. Baran, W. P. Arnott, and C. Schmitt (2006). Testing and comparing the modified anomalous diffraction approximation. *J. Atmos. Sci.*, **63**, 2948–2962.

Mitchell, D. L., S. K. Chai, Y. Liu, A. J. Heymsfield, and Y. Dong (1996). Modeling cirrus clouds. Part I: Treatment of bimodal spectra and case study analysis. *J. Atmos. Sci.*, **53**, 2952–2966.

Miyata, K., H. Kanno, K. Tomizawa, and Y. Yoshimura (2001). Supercooling of aqueous solutions of alkali chlorides and acetates. *Bull. Chem. Soc. Jpn.*, **74**, 1629–1633.

Miyata, K., H. Kanno, Y. Niino, and K. Tomozawa (2002). Cationic and anionic effects on the homogeneous nucleation of ice in aqueous halide solutions. *Chem. Phys. Lett.*, **354**, 51–55.

Moeng, C. H., and A. Arakawa (1980). A numerical study of a marine subtropical stratus cloud layer and its stability. *J. Atmos. Sci.*, **37**, 2661–2676.

Möhler, O., P. R. Field, P. Connolly, S. Benz, H. Saathoff, M. Schnaiter, R. Wagner, R. Cotton, M. Krämer, A. Mangold, and A. J. Heymsfield, (2006). Efficiency of the deposition mode ice nucleation on mineral dust particles. *Atmos. Chem. Phys.*, **6**, 3007–3021.

Mokhov, I. I., P. F. Demchenko, A. V. Eliseev, V. Ch. Khon, and D. V. Khvorostyanov (2002). Estimations of global and regional climate changes during the 19th–21st centuries on the basis of the IAP RAS Model with consideration for anthropogenic forcing. *Izvestia Rus. Acad. Sci., Atmos. Oceanic Phys.*, **38** (5), 629–642.

Monier, M., W. Wobrock, J.-F. Gayet, and A. Flossmann (2006). Development of a detailed microphysics cirrus model tracking aerosol particle's histories for interpretation of the recent INCA campaign. *J. Atmos. Sci.*, **63**, 504–525.

Monin, A. S., and A. M. Yaglom (2007a). Statistical fluid mechanics. Mechanics of turbulence, v. 1. Ed.: **Lumley, J. L.** Dover Publications Inc., Mineola, NY, USA, 769 pp.

Monin, A. S., and A. M. Yaglom (2007b). Statistical fluid mechanics. Mechanics of turbulence, v. 2. Ed.: **Lumley, J. L.** Dover Publications Inc., Mineola, NY, USA, 874 pp.

Mordy, W. A. (1959). Computations of the growth by condensation of a population of cloud droplets. *Tellus*, **11**, 16–44.

Morrison, H., and A. Gettelman (2008). A new two-moment bulk stratiform cloud microphysics scheme in the community atmosphere model, version 3(CAM3). Part I: Description and numerical tests. *J. Clim.*, **21**, 3642–3659, doi:10.1175/2008JCLI2105.1.

Morrison, H., J. A. Curry, and V. I. Khvorostyanov (2005a). A new double-moment microphysics parameterization for application in cloud and climate models, Part 1: Description. *J. Atmos. Sci.*, **62**, 1665–1677.

Morrison, H., J. A. Curry, M. D. Shupe, and P. Zuidema (2005b). A new double-moment microphysics scheme for application in cloud and climate models, Part 2: Single-column modeling of arctic clouds. *J. Atmos. Sci.*, **62**, 1678–1693.

Morrison, H., and J. Pinto (2005). Mesoscale modeling of springtime Arctic mixed-phase stratiform clouds using a new two-moment bulk microphysics scheme. *J. Atmos. Sci.*, **62**, 3683–3704.

Morrison, H., R. B. McCoy, S. A. Klein, et al. (2009). Intercomparison of model simulations of mixed-phase clouds observed during the ARM Mixed-Phase Arctic Cloud Experiment. II. Multilayer cloud. *Quart. J. Roy. Meteor. Soc.*, **135**, 1003–1019.

Morse, B., and M. Richard (2009). A field study of suspended frazil ice particles. *Cold Regions Sci. Technol.*, **55**, 86–102.

Mossop, S. C. (1955). The freezing of supercooled water. *Proc. Phys. Soc.*, **68**, 193–208.

Mossop, S. C., and J. Hallett (1974). Ice crystal concentration in cumulus clouds: Influence of the drop spectrum. *Science*, **186**, 632–633.

Moynihan, C. T. (1997). Two species / nonideal solution model for amorphous / amorphous phase transitions. *Mater. Res. Soc., Symp. Proc.*, **455**, 411–425.

Müller, H. (1928). Zur allgemeinen theory der raschen Koagulation. *Kolloid-Chem. Beib.*, Bd. **27**, 223–250.

Murphy, D. M. (2003). Dehydration in cold clouds is enhanced by a transition from cubic to hexagonal ice. *Geophys. Res. Lett.*, **30** (23), 2230, doi:10.1029/2003GL018566.

Murphy, D. M., and T. Koop (2005). Review of the vapor pressures of ice and supercooled water for atmospheric applications. *Quart. J. Roy. Meteorol. Soc.*, **131**, 1539–1565.

Murray, B. J., and A. K. Bertram (2006). Formation and stability of cubic ice in water droplets, *Phys. Chem. Chem. Phys*, **8**, 186–192.

Murray B. J., and E. J. Jensen (2010). Homogeneous nucleation of amorphous solid water particles in the upper mesosphere. *J. Atmos. Solar-Terrestrial Phys.*, **72**, 51–61.

Murray, B. J., D. A., Knopf, and A. K. Bertram (2005). The formation of cubic ice under conditions relevant to Earth's atmosphere. *Nature*, **434**, 202–205.

Murray, B. J., S. L. Broadley, T. W. Wilson, S. J. Bull, R. H. Wills, H. K. Christenson, and E. J. Murray (2010). Kinetics of the homogeneous freezing of water. *Phys. Chem. Chem. Phys.*, **12** (35), 10,380–10,387.

Myhre, C. E. L., C. J. Nielsen, and O. W. Saastad (1998). Density and surface tension of aqueous H_2SO_4 at low temperature. *J. Chem. Eng. Data*, **43**, 617–622.

Nagle, J. F. (1966). Lattice statistics of hydrogen-bonded crystals. I. The residual entropy of ice. *J. Math. Phys.*, **7**, 1484–1491.

Neiburger, M., and C. W. Chien (1960). Computations of the growth of cloud drops by condensation using an electronic digital computer. *Geophys. Monogr.*, No. 5, Amer. Geophys. Union, 191–208.

Nenes, A., and J. H. Seinfeld (2003). Parameterization of cloud droplet formation in global climate models. *J. Geophys. Res.*, **108**, 4415, doi:10.1029/2002JD002911.

Nenes, A., R. J. Charlson, M. C. Facchini, M. Kulmala, A. Laaksonen, and J. H. Seinfeld (2002). Can chemical effects on cloud droplet number rival the first indirect effect? *Geophys. Res. Lett.*, **29**, 1848, doi:10.1029/2002GL015295.

Nevzorov, A. N. (1967). Distribution of large drops in liquid stratiform clouds. *Proc. Centr. Aerolog. Observ.*, **79**, 57–69.

Newkirk, J. B., and D. Turnbull (1955). Nucleation of ammonium sulfate crystals from aqueous solutions. *J. Appl. Phys.*, **26**, 579–583.

Niemand, M., O. Möhler, B. Vogel, H. Vogel, C. Hoose, P. Connolly, H. Klein, H. Bingemer, J. Skrotzki, and T. Leisner (2012). A particle-surface-area-based parameterization of immersion freezing on mineral dust particles. *J. Atmos. Sci.*, doi:10.1175/JAS-D-11-0249.1.

Noonkester, V. R. (1984). Droplet spectra observed in marine stratus cloud layers. *J. Atmos. Sci.*, **41**, 829–845.

Oatis, S., D. Imre, R. McGraw, and J. Xu (1998). Heterogeneous nucleation of a common atmospheric aerosol: Ammonium sulfate. *Geophys. Res. Lett.*, **25**, 4469–4472.

Obukhov, A. M. (1959). Description of turbulence in terms of Lagrangian variables. *Adv. Geophys.*, **6**, 113–116.

Ohtake, T., and P. T. Huffman (1969). Visual range in ice fog. *J. Appl. Meteorol.*, **8**, 499–505.

Ohtake, T., K. O. L. F. Jayaweera, and K.-I. Sakurai (1982). Observation of ice crystal formation in lower Arctic atmosphere. *J. Atmos. Sci.*, **39**, 2898–2904.

Okita, T. (1961). Size distribution of large droplets in precipitating clouds. *Tellus*, **13**, 456–467.

Onasch, T. B., R. McGraw, and D. Imre (2000). Temperature-dependent heterogeneous efflorescence of mixed ammonium sulfate / calcium carbonate particles. *J. Phys. Chem. A*, **104**, 10,797–10,806.

Onasch, T. B., R. L. Siefert, S. D. Brooks, A. J. Prenni, B. Murray, M. A. Wilson, and M. A. Tolbert (1999). Infrared spectroscopic study of the deliquescence and efflorescence of ammonium sulfate aerosol as a function of temperature. *J. Geophys. Res.*, **104**, 21,317–21,326.

Orville, H. D., and J.-M. Chen (1982). Effects of cloud seeding. Latent heat of fusion and condensate loading on cloud dynamics and precipitation evolution: A numerical study. *J. Atmos. Sci.*, **39**, 2807–2827.

Ovarlez, J., J.-F. Gayet, K. Gierens, J. Strom, H. Ovarlez, F. Auriol, R. Busen, and U. Schumann (2002). Water vapour measurements inside cirrus clouds in northern and southern hemispheres during INCA. *Geophys. Res. Lett.*, **29**, 1813, 10.1029/2001GL014440.

Ovtchinnikov, M. V., and Y. L. Kogan (2000). An investigation of ice production mechanisms in small cumuliform clouds using a 3D model with explicit microphysics. Part I: Model description. *J. Atmos. Sci.*, **57**, 2989–3003.

Ovtchinnikov, M. V., Y. L. Kogan, and A. M. Blyth (2000). An investigation of ice production mechanisms in small cumuliform clouds using a 3D model with explicit microphysics. Part II: Case study of New Mexico cumulus clouds. *J. Atmos. Sci.*, **57**, 3004–3020.

Oxtoby, D. W. (2003). Crystal nucleation in simple and complex fluids. *Phil. Trans. Roy. Soc. London, A*, **361**, 419–428, doi: 10.1098/rsta.2002.1145.

Paltridge, G. W., and C. M. R. Platt (1976). Radiative processes in meteorology and climatology. Elsevier, 318 pp.

Paluch, I. R., and C. A. Knight (1986). Does mixing promote droplet growth? *J. Atmos. Sci.*, **43**, 1994–1998.

Paoli, R., and K. Shariff (2004). Direct numerical simulation of turbulent condensation in clouds. *Annual Res. Briefs, Center for Turbul. Res.*, NASA Ames Res. Center, 305–316.

Paoli, R., and K. Shariff (2009). Turbulent condensation of droplets: Direct simulation and a stochastic model. *J. Atmos. Sci.*, **66**, 723–740.

Passarelli, R. E. (1978a). An approximate analytical model of the vapor deposition and aggregation growth of snowflakes. *J. Atmos. Sci.*, **35**, 118–124.

Passarelli, R. E. (1978b). Theoretical and observational study of snow-sized spectra and snowflake aggregation efficiencies. *J. Atmos. Sci.*, **35**, 882–889.

Pauling, L. (1935). The structure and entropy of ice and of other crystals with some randomness of atomic arrangement. *J. Amer. Chem. Soc.*, **57**, 2680–2684.

Pawlowicz, R., T. McDougall, R. Feistel, and R. Tailleux (2012). Preface. An historical perspective on the development of the Thermodynamic Equation of Seawater – 2010. *Ocean Sci.*, **8**, 161–174, www.ocean–sci.net/8/161/2012/.

Penner, J. E., D. J. Bergmann, J. J. Walton, D. Kinnison, M. J. Prather, D. Rotman, C. Price, K. E. Pickering, and S. L. Baughcum (1998). An evaluation of upper troposphere NOx with two models. *J. Geophys. Res.*, **103** (D17), 22,097–22,113, doi:10.1029/98JD01565.

Petch, J. C., G. C. Craig, and K. P. Shine (1997). A comparison of two bulk microphysical schemes and their effects on radiative transfer using a single-column model. *Quart. J. Roy. Meteor. Soc.*, **123**, 1561–1580.

Petrenko, V. F., and R. W. Whitworth (1999). Physics of ice. Oxford University Press, 373 pp.

Pinto, J. O., J. A. Curry, and J. M. Intriery (2001). Cloud-aerosol interactions during autumn over Beaufort sea. *J. Geophys. Res.*, **106** (D14), 15,077–15,097.

Pirnach, A. M., and S. V. Krakovskaya (1994). Numerical studies of dynamics and cloud microphysics of the frontal rainbands. *Atmos. Res.*, **33**, 333–365.

Pirnach, A. M., and S. V. Krakovskaya (1998). Theoretical study of the microphysical structure of mixed stratiform frontal clouds and their precipitation. *Atmos. Res.*, **47–48**, 491–503.

Platt, C. M. R. (1997). A parameterization of the visible extinction coefficient in terms of the ice/water content. *J. Atmos. Sci.*, **54**, 2083–2098.

Poellot, M. R., K. A. Hilburn, W. P. Arnott, and K. Sassen (1999). In situ observation of cirrus clouds from the 1994 ARM RCS IOP. *Ninth ARM Science Team Meeting Proceedings*, San Antonio, Texas, March 22–26, 1999. Sponsored by the U.S. Department of Energy.

Ponyatovsky, E. G., V. V. Senitsyn, and T. A. Pozdnyakova (1994). Second critical point and low-temperature anomalies in the physical properties of water. *J.P.T. Lett.*, **60**, 360–364.

Ponyatovsky, E. G., V. V. Senitsyn, and T. A. Pozdnyakova (1998). The metastable T-P phase diagram and anomalous thermodynamic properties of supercooled water. *J. Chem. Phys.*, **109** (6), 2413–2422.

Poole, P. H., F. Sciortino, U. Essmann, and H. E. Stanley (1992). Phase behavior of metastable water. *Nature*, **360**, 324–328.

Poole, P. H., F. Sciortino, T. Grande, H. E. Stanley, and C. A. Angell (1994). Effect of hydrogen bonds on the thermodynamic behavior of liquid water. *Phys. Rev. Lett.*, **73** (12), 1632–1635.

Popovicheva, O. B., N. M. Persiantseva, E. E. Lukhovitskaya, N. K. Shonija, N. A. Zubareva, B. Demirdjian, D. Ferry, and J. Suzanne (2004). Aircraft engine soot as contrail nuclei. *Geophys. Res. Lett.*, **31**, L11104, doi: 10.1029/2003/GL018888.

Prenni, A. J., J. Y. Harrington, M. Tjernström, P. J. DeMott, A. Avramov, C. N. Long, S. M. Kreidenweis, P. Q. Olsson, and J. Verlinde (2007). Can ice-nucleating aerosols affect Arctic seasonal climate? *Bull. Am. Meteorol. Soc.* **88**, 541–550.

Pruppacher, H. R., and J. D. Klett (1997). Microphysics of clouds and precipitation, 2nd ed. Kluwer Academic Publishers: Boston, MA, USA, 954 pp.

Quante, M., and D. O'C. Starr (2002). Dynamical processes in cirrus clouds. In: Cirrus. Eds.: Lynch, D., Sassen, K., Starr, D. O'C., and Stephens, G., Oxford University Press, pp. 346–374.

Raddatz, R. L., M. G. Asplin, L. Candlish, and D. G. Barber (2011). General characteristics of the atmospheric boundary layer in a flaw lead polynya region for winter and spring. *Boundary Layer Meteorol.*, **138**, 321–335. doi:10.1007/s10546-010-9557-1

Raddatz, R. L., R. J. Galley, and D. G. Barber (2012). Linking the atmospheric boundary layer to the Amundsen Gulf sea-ice cover: A mesoscale to synoptic-scale perspective from winter to summer 2008. *Boundary Layer Meteorol.*, **142**, 123–148. doi:10.1007/s10546-011-9669-2

Räisänen, P., A. Bogdan, K. Sassen, M. Kulmala, and M. J. Molina (2006). Impact of H_2SO_4/H_2O coating and ice crystal size on radiative properties of sub-visible cirrus. *Atmos. Chem. Phys.*, **6**, 4659–4667.

Randall, D. A. (1980). Conditional instability of the first kind upside-down. *J.Atmos.Sci.*, **37**, 125–139.

Randall, D. A., S. Krueger, C. S. Bretherton, J. A. Curry, P. Duynkerke, M. Moncrieff, B. Ryan, D. Starr, M. Miller, W. Rossow, G. Tselioudis, and B. Wielicki (2003). Confronting models with data: The GEWEX Cloud Systems Study. *Bull. Am. Meteorol. Soc.*, **84**, 455–469.

Rangno, A. L., and P. V. Hobbs (1991). Ice particle concentrations and precipitation development in small polar maritime cumuliform clouds. *Quart. J. Roy. Meteor. Soc.*, **117**, 207–241.

Rangno, A. L., and P. V. Hobbs (2001). Ice particles in stratiform clouds in the Arctic and possible mechanisms for the production of high ice concentrations. *J. Geophys. Res.*, **106** (D14), 15,065–15,075.

Rasmussen, D. H. (1982). Thermodynamic and nucleation phenomena: A set of experimental observations. *J. Cryst. Growth*, **56**, 56–66.

Rasmussen, D. H., and A. P. Mackenzie (1972). Effect of solute on ice-solution interfacial free energy; calculation from measure homogeneous nucleation temperatures. In: Water structure at the water polymer interface. Ed.: Jellinek, H. H. G., Plenum Press, New York, pp. 126–145.

Rempel, A. W., J. S. Wettlaufer, and M. G. Worster (2004). Premelting dynamics in a continuum model of frost heave. *J. Fluid Mech.*, **498**, 227–244.

Ren, C., and A. R. MacKenzie (2005). Cirrus parameterisation and the role of ice nuclei. *Quart. J. Roy. Meteorol. Soc.*, **131**, 1585–1605.

Ren, C., and A. R. MacKenzie (2007). Closed-form approximations to the error and complementary error functions and their applications in atmospheric science. *Atmos. Sci. Let.*, **8** (3), 70–73, doi: 10.1002/asl.154.

Richardson, C. B., and T. D. Snider (1994). A study of heterogeneous nucleation in aqueous solutions. *Langmuir*, **10**, 2462–2465.

Risken, H. (1989). The Fokker–Planck Equation, 2nd ed., Springer-Verlag, 472 pp.

Rissler, J., A. Vestin, E. Swietlicki, G. Frisch, J. Zhou, P. Artaxo, and M. O. Andreae (2006). Size distribution and hygroscopic properties of aerosol particles from dry-season biomass burning in Amazonia. *Atmos. Chem. and Physics*, **6**, 471–491.

Rissman, T. A., A. Nenes, and J. H. Seinfeld (2004). Chemical amplification (or dampening) of the Twomey effect: Conditions derived from droplet activation theory. *J. Atmos. Sci.*, **61**, 919–930.

Robinson, R. A., and R. H. Stokes (1970). Electrolyte solutions, 2nd ed., Butterworths, 559 pp.

Rodean, H. C. (1996). Stochastic Lagrangian models of turbulent diffusion. *Meteorological Monographs, Amer. Meteorol. Soc.*, v. 26 (48), Boston, MA, USA, 84 pp.

Rogers, D. C. (1982). Field and laboratory studies of ice nucleation in winter orographic clouds. *PhD dissertation.* Dept. Atmospheric Science, Univ. of Wyoming, Laramie, WY, USA, 161 pp.

Rogers, D. C. (1988). Development of a continuous flow thermal gradient diffusion chamber for ice nucleation studies. *Atmos. Res.*, **22**, 149–181.

Rogers, D. C., P. J. DeMott, and L. O. Grant (1994). Concerning primary ice concentration and water supersaturations in the atmosphere. *Atmos. Res.*, **33**, 151–168.

Rogers, D. C., P. J. DeMott, and S. M. Kreidenweis (2001). Airborne measurements of tropospheric ice-nucleating aerosol particles in the Arctic spring. *J. Geophys. Res.*, **106**, 15,053–15,063.

Rogers, D. C., P. J. DeMott, S. M. Kreidenweis, and Y. Chen (1998). Measurements of ice nucleating aerosols during SUCCESS. *Geophys. Res. Lett.*, **25**, 1383–1386.

Rogers, P. S. Z., and K. S. Pitzer (1982). Volumetric properties of aqueous NaCl solutions. *J. Phys. Chem. Ref. Data*, **11**, 15–81.

Rogers, R. R., and M. K. Yau (1989). A short course in cloud physics, 3rd ed., Pergamon Press, 293 pp.

Rosenberg, R. (2005). Why is ice slippery? *Phys. Today*, **58** (12), 50–57.

Rosenfeld, D., Y. Rudich, and R. Lahav (2001). Desert dust suppressing precipitation: A possible desertification feedback loop. *Proc. Natl. Acad. Sci.*, **98**, 5975–5980.

Rossow, W. B., and R. A. Schiffer (1999). Advances in understanding clouds from ISCCP. *Bull. Amer. Meteorol. Soc.*, **80**, 2261–2287.

Rotman, D. A., et al. (2004). IMPACT, the LLNL 3-D global atmospheric chemical transport model for the combined troposphere and stratosphere: Model description and analysis of ozone and other trace gases. *J. Geophys. Res.*, **109**, D04303, doi:10.1029/2002JD003155.

Russell, L. M. and Y. Ming (2002). Deliquescence of small particles. *J. Chem. Phys.*, **116** (1), 311–321, doi: 10.1063/1.1420 727.

Rutledge, S. A., and Hobbs P. V. (1983). The mesoscale and microscale structure and organization of clouds and precipitations in midlatitude cyclones. VIII. A model for the "Seeder-feeder" process in warm frontal rainbands. *J. Atmos. Sci.*, **40**, 1185–1206.

Ryan, B. F. (1996). On the global variation of precipitating layer clouds. *Bull. Amer. Meteor. Soc.*, **77**, 53–70.

Rzesanke, D., D. Duft, and T. Leisner (2011). Laboratory experiments on the microphysics of electrified cloud droplets. In: Climate and weather of the Sun–Earth system (CAWSES): Highlights from a priority program. Ed.: Lübken, F.-J., Springer, Dordrecht, The Netherlands.

Sassen, K. (1980). Remote sensing of planar ice crystal fall attitudes. *J. Meteorol. Soc. Japan*, **58**, 422–429.

Sassen, K. (1992). Ice nuclei availability in the higher tropospheric: Implications of a remote sensing cloud phase climatology. In: Nucleation and atmospheric aerosols. Eds.: Fukuta, N., and Wagner, P., Deepak Publishing, pp. 287–290.

Sassen, K. (1997). Contrail-Cirrus and their potential for regional climate change. *Bull. Amer. Meteorol. Soc*, **78** (9), 1885–1903.

Sassen, K., and S. Benson (2000). Ice nucleation in cirrus clouds, A model study of the homogeneous and heterogeneous nucleation modes. *Geophys. Res. Lett.*, **27**, 521–524.

Sassen, K., and J. R. Campbell (2001). A midlatitude cirrus cloud climatology from the Facility for Atmospheric Remote Sensing. Part I: Macrophysical and synoptic properties. *J. Atmos. Sci.*, **58**, 481–496.

Sassen, K., and G. C. Dodd (1988). Homogeneous nucleation rate for highly supercooled cirrus cloud droplets. *J. Atmos. Sci.*, **45**, 1357–1369.

Sassen, K., and G. C. Dodd (1989). Haze particle nucleation simulation in cirrus clouds, and application for numerical and lidar studies. *J. Atmos. Sci.*, **46**, 3005–3014.

Sassen, K., and V. I. Khvorostyanov (2007). Microphysical and radiative properties of mixed-phase altocumulus: A model evaluation of glaciation effects. *Atmos. Res.*, **84** (4), 390–398.

Sassen, K., and V. I. Khvorostyanov (2008). Cloud effects from boreal forest fire smoke: evidence for ice nucleation from polarization lidar data and cloud model simulations. *Environ. Res. Lett.* **3**, 025006, doi: 10.1088/1748-9326/3/2/025006.

Sassen K., and Z. Wang (2012). The clouds of the middle troposphere: Composition, radiative impact, and global distribution. *Surv. Geophys.*, doi: 10.1007/s10712-011-9163-x.

Sassen, K., D. O'C. Starr, and T. Uttal (1989). Mesoscale and microscale sructure of cirrus clouds: Three case studies. *J. Atmos. Sci.*, **46**, 371–396.

Sassen, K., P. Cobb, J. Zhu, and V. Khvorostyanov (2006). Polarization lidar studies of Alaskan forest fire smoke and indirect effect of clouds. Summer Intern. Laser Radar Conf., Japan, 2006.

Sassen, K., P. J. DeMott, J. M. Prospero, and M. R. Poellot (2003). Saharan dust storms and indirect aerosol effects on clouds: CRYSTAL-FACE results. *Geophys. Res. Lett.*, **30**, 4714, doi:10.1029/2003GL017371.

Sassen, K., Z. Wang, V. I. Khvorostyanov, G. L. Stephens, and A. Bennedetti (2002). Cirrus cloud ice water content radar algorithm evaluation using an explicit cloud microphysical model. *J. Appl. Meteorol.*, **41**, 620–628.

Sastry, S. (2002). Sculpting ice out of water. *Nature*, **416**, 376–377.

Saul, A., and W. Wagner (1989). A fundamental equation for water covering the range from the melting line to 1273 K at pressures up to 25000 MPa. *J. Phys. Chem. Ref. Data*, **18**, 1537–1564.

Schaeffer, V. J. (1949). The formation of the ice crystals in the laboratory and the atmosphere. *Chem. Rev.*, **44**, 291–320.

Schiffer, R. A., and W. B. Rossow (1983). The International Satellite Cloud Climatology Project (ISCCP): The first project of the World Climate Research Programme. *Bull. Amer. Meteor. Soc.*, **64**, 779–784.

Schumann, U. (1994). On the effect of emissions from aircraft engines on the state of the atmosphere. *Ann. Geophys.*, **12**, 365–384.

Scott, W. T. (1968). Analytic studies of cloud droplet coalescence. *J. Atmos. Sci.*, **25**, 54–65.

Sedunov, Y. S. (1965). The fine structure of the clouds and its role in formation of the cloud droplet spectra. *Izv. Acad. Sci. USSR, Atmos. Oceanic Phys.*, **1**, 416–421.

Sedunov, Y. S. (1967). Kinetics of initial stage of condensation in clouds. *Izvestia Acad. Sci. USSR, Atmos Oceanic Phys.*, **3**, 34–46.

Sedunov, Y. S. (1974). Physics of drop formation in the atmosphere. Wiley, New York, 234 pp.

Seifert, A. (2005). On the shape-slope relation of drop size distributions in convective rain. *J. Appl. Meteorol.* **44**, 1146–1151.

Seifert, A., and K. D. Beheng (2001). A double-moment parameterization for simulating autoconversion, accretion and selfcollection. *Atmos. Res.*, 59–60, 265–281.

Seifert, A., and K. D. Beheng (2006). A two-moment cloud microphysics parameterization for mixed-phase clouds. Part 1: Model description, *Meteorol. Atmos. Phys.*, **92**, 45–66.

Seinfeld, J. H., and S. N. Pandis (1998). Atmospheric chemistry and physics, Wiley, New York, 1326 pp.

Serpolay, R. (1969). Perfectionnement d'une technique d'ensemencement et recherches de procedes de modification des brouillards. *Météorologie*, **9**, 45–54.

Shaw, R. A. (2003). Particle-turbulence interactions in atmospheric clouds. *Annu. Rev. Fluid Mech.*, **35**, 183–227.

Shaw, R. A., A. J. Durant, and Y. Mi (2005). Heterogeneous surface crystallization observed in undercooled water. *J. Phys. Chem.*, **109 B**, 9865–9868.

Shaw, R. A., W. C. Reade, L. R. Collins, and J. Verlinde (1998). Preferential concentration of cloud droplets by turbulence: Effects on the early evolution of cumulus cloud droplet spectra. *J. Atmos. Sci.*, **55**, 1965–1976.

Shchekin, A. K., and A. I. Rusanov (2008a). Generalization of the Gibbs-Kelvin-Köhler and Oswald-Freundlich equations for a liquid film on a soluble nanoparticle. *J. Chem. Phys.*, **129**, 154116, doi: 10.1063/1.2996590.

Shchekin, A. K., and I. V. Shabaev (2010). Activation barriers for the complete dissolution of condensation nucleus and its reverse crystallization in droplets in the undersaturated solvent vapor. *Colloid J.*, **72**, 432–439, doi: 10.1134/S1061 933X1003 018X, 2010.

Shchekin, A. K., I. V. Shabayev, and A. I. Rusanov (2008b). Thermodynamics of droplet formation around a soluble condensation nucleus in the atmosphere of a solvent vapor. *J. Chem. Phys.*, **129**, 214111, doi: 10.1063/1.3021 078.

Shifrin, K. S. (1955). On the calculation of radiative properties of clouds. *Proc. Main Geophys. Observatory*, Leningrad, **46** (108), 5–33 (in Russian).

Shifrin, K. S., and V. Y. Perelman (1960). Kinetics of distillation in a mixed cloud. *Proc. Acad. Sci. USSR*, **132**, 1148–1151.

Shilling, J. E., T. J. Fortin, and M. A. Tolbert (2006). Depositional ice nucleation on crystalline organic and inorganic solids. *J. Geophys. Res.*, **111**, D12 204, doi:10.1029/2005JD006664.

Shipway, B. J., and S. J. Abel (2010). Analytical estimation of cloud droplet nucleation based on an underlying aerosol population. *Atmos. Res.*, **96**, 344–355.

Shulmann, M. L., M. L. Jacobson, R. J. Charlson, R. E. Synovec, and T. E. Young (1996). Dissolution behavior and surface tension effects of organic compounds in nucleating cloud droplets. *Geophys. Res. Lett.*, **23**, 277–280.

Shupe M. D., S. Y. Matrosov, and T. Uttal (2006). Arctic mixed-phase cloud properties derived from surface-based sensors at SHEBA. *J. Atmos. Sci.*, **63**, 697–711.

Silverman, B. A., and A. I. Weinstein (1973). Fog modification – A technology assessment. *Air Force Surveys in Geophysics*, No. 2, AFC-0159, 126 pp.

Singer, S. J., J.-L. Kuo, T. K. Hirsch, C. Knight, L. Ojamäe, and M. L. Klein (2005). Hydrogen-bond topology and the ice VII-VIII and ice Ih-XI proton-ordering phase transitions. *Phys. Rev. Lett.*, **94**, 135,701–135,705.

Slingo, A. (1989). A GCM parameterization for the shortwave radiative properties of water clouds. *J Atmos. Sci.*, **46**, 1419–1427.

Slingo, A., and H. M. Schecker (1982). On the shortwave radiative properties of stratiform water clouds. *Quart. J. Roy. Meteor. Soc.*, **108**, 407–426.

Slinn, W. G. N., and J. M. Hales (1971). A reevaluation of the role of thermophoresis as a mechanism of in- and below-cloud scavenging. *J. Atmos. Sci.*, **28**, 1465–1471.

Smirnov, V. I. (1978). On the equilibrium sizes and size spectra of aerosol particles in a humid atmosphere. *Izv. Acad. Sci. USSR, Atmos. Oceanic Phys.*, **14**, 1102–1106.

Smirnov, V. I., and A. S. Kabanov (1970). Effect of horizontal inhomogeneity of activated cloud drops on their size spectrum. *Izv. Acad. Sci. USSR, Atmos. Oceanic Phys.*, **6**, 1262–1275.

Smirnov, V. I., and L. A. Nadeykina (1986). On the theory of the drop size spectrum formed by condensation in turbulent cloud. *Izv. Acad. Sci. USSR, Atmos. Oceanic Phys.*, **22**, 478–487.

Smirnov, V. I., and B. N. Sergeev (1973). Size distribution of large cloud drops grown on hygroscopic cloud condensation nuclei. *Izv. Acad. Sci. USSR, Atmos. Oceanic Phys.*, **9**, 1288–1300.

Smoluchowski, M. (1916). Drei Vortäge über Diffusion, Brounische Bewegung und Koagulation von Kolloidteilchen. *Phys. Zeits.*, **Bd. 17**, 557–585.

Snider, J. R., S. Guibert, J.-L. Brenguier, and J.-P. Putaud (2003). Aerosol activation in marine stratocumulus clouds: Köhler and parcel theory closure studies. *J. Geophys. Res.*, **108** (D15), 8629, doi: 10.1029/2002JD0026692.

Song, Y., and E. A. Mason (1990a). Analytical equation of state for molecular fluids: Kihara model for rodlike molecules. *Phys. Rev. A*, **42**, 4743–4748.

Song, Y., and E. A. Mason (1990b). Analytical equation of state for molecular fluids: Comparison with experimental data, *Phys. Rev. A*, **42**, 4749–4758.

Song, Y., and E. Mason (1989). Statistical-mechanical theory of a new analytical equation of state. *J. Chem. Phys.*, **91**, 7840.

Speedy, R. J. (1982). Stability-limit conjecture: An interpretation of the properties of water. *J. Phys. Chem.*, **86**, 982–991.

Speedy, R. J. (1996). Two waters and no ice please. *Nature*, **380**, 289–290.

Speedy, R. J., and C. A. Angell (1976). Isothermal compressibility of supercooled water and evidence for a thermodynamic singularity at –45°C. *J. Chem. Phys.*, **65**, 851–858.

Spice, A., D. W. Johnson, P. R. A. Brown, A. G. Darlison, and C. P. R. Saunders (1999). Primary ice nucleation in orographic cirrus cloud: A numerical simulation of the microphysics. *Quart. J. Roy. Meteor. Soc.*, **125**, 1637–1667.

Spichtinger, P., and K. M. Gierens (2009). Modelling of cirrus clouds – Part 2: Competition of different nucleation mechanisms. *Atmos. Chem. Phys.*, **9**, 2319–2334.

Spichtinger, P., and M. Krämer (2012). Tropical tropopause ice clouds: A dynamic approach to the mystery of low crystal numbers. *Atmos. Chem. Phys. Discuss.*, **12**, 28,109–28,153.

Squires, P. (1958). The microstructure and colloidal stability of warm clouds, II, The causes of the variations of microstructure. *Tellus*, **10**, 262–271.

Srivastava, R. C. (1969). Note on the theory of growth of cloud drops by condensation. *J. Atmos. Sci.*, **26**, 776–780.

Srivastava, R. C. (1971). Size distribution of raindrops generated by their breakup and coalescence. *J. Atmos. Sci.*, **28**, 410–415.

Srivastava, R. C. (1978). Parameterization of raindrop size distribution. *J. Atmos. Sci.*, **35**, 108–117.

Srivastava, R. C. (1982). A simple model of particle coalescence and breakup. *J. Atmos. Sci.*, **39**, 1317–1322.

Srivastava, R. C. (1989). Growth of cloud drops by condensation: A criticism of currently accepted theory and a new approach. *J. Atmos. Sci.*, **46**, 869–887.

Srivastava, R. C., and R. E. Passarelli (1980). Analytical solutions to simple models of condensation and coalescence. *J. Atmos. Sci.*, **37**, 612–621.

Stanley, H. E., P. Kumar, L. Xu, Z. Yan, M. G. Mazza, S. V. Buldyrev, S.-H. Chen, and F. Mallamace (2007). The puzzling unsolved mysteries of liquid water: Some recent progress. In: *Proc. Pan American Sci. Inst. (PASI) Conference "Disorder and Complexity,"* Mar del Plata, Argentina, 11–20 December 2006, *Physica, A*, **386**, 729–743.

Starr, D. O'C., et al. (2000). Comparison of cirrus cloud models: A Project of the GEWEX Cloud System Study (GCSS) Working Group on Cirrus Cloud Systems. *Proc. Int. Cloud Phys. Conf.*, Reno, NV, USA, August 2000, 1–4.

Starr, D. O'C., and S. K. Cox (1985). Cirrus clouds. Part I: A cirrus cloud model. *J. Atmos. Sci.*, **42**, 2663–2681.

Stepanov, A. S. (1975). Condensational growth of cloud droplets in a turbulized atmosphere. *Izv. Akad. Sci. USSR, Atmos. Oceanic Phys.*, **11**, 27–42.

Stepanov, A. S. (1976). On the effect of turbulence on the size spectrum of cloud droplets at condensation. *Izv. Acad. Sci. USSR, Atmos. Oceanic Phys.*, **12**, 281–292.

Stephens, G. L. (1983). The influence of radiative transfer on the mass and heat budgets of ice crystals falling in the atmosphere. *J. Atmos. Sci.*, **40**, 1729–1739.

Stephens, G. L. (1984). The parameterization of radiation for numerical weather prediction and climate models. Review. *Mon. Wea. Rev.*, **112**, 826–867.

Stephens, G. L. (2005). Cloud feedbacks in the climate system: A critical review. *J. Clim.* **18**, 237–273.

Stephens, G. L., D. G. Vane, and S. J. Walter (2000). The CLOUDSAT mission: A new dimension to space-based observations of cloud in the coming millennium. *Workshop on Cloud Processes and Cloud Feedback in Large-Scale Models*, 9–13 November 1999, Reading, UK. Report WCRP-110, WMO/TD-No 993, Geneva, 2000.

Stephens, G. L., S. C. Tsay, P. W. Stackhouse, and P. J. Flatau (1990). The relevance of the microphysical and radiative properties of cirrus clouds to climate and climatic feedback. *J. Atmos Sci.*, **47**, 1742–1753.

Stöckel, P., I. M. Weidinger, H. Baumgärtel, and T. Leisner (2005). Rates of homogeneous ice nucleation in levitated H_2O and D_2O droplets. *J. Phys. Chem. A*, **109**, 2540–2546.

Straka, J. M. (2009). Cloud and precipitation microphysics. Cambridge Univ. Press, 392 pp.

Ström, J., B. Strauss, T. Anderson, F. Schröder, J. Heintzenberg, and P. Wendling (1997). In situ observations of the microphysical properties of young cirrus clouds. *J. Atmos. Sci.*, **54**, 2542–2553.

Ström, J., M. Seifert, B. Kärcher, J. Ovarlez, A. Minikin, J.-F. Gayet, R. Krejci, A. Petzold, F. Auriol, W. Haag, R. Busen, U. Schumann, and H. C. Hansson (2003). Cirrus cloud occurrence as function of ambient relative humidity: A comparison of observations obtained during INCA experiment. *Atmos. Chem. Phys.*, **3**, 1807–1816.

Sud, Y. C., and D. Lee (2007). Parameterization of aerosol indirect effect to complement McRAS cloud scheme and its evaluation with the 3-year ARM-SGP analyzed data for single column models. *Atmos. Res.*, **86**, 105–125.

Sud, Y. C., E. Wilcox, W. K.-M. Lau, G. K. Walker, X.-H. Liu, A. Nenes, D. Lee, K.-M. Kim, Y. Zhou, and P. S. Bhattacharjee (2009). Sensitivity of boreal-summer circulation and precipitation to atmospheric aerosols in selected regions – Part 1: Africa and India. *Ann. Geophys.*, **27**, 3989–4007, www.ann-geophys.net/27/3989/2009/.

Swanson, B. D. (2009). How well does water activity determine homogeneous ice nucleation temperature in aqueous sulfuric acid and ammonium sulfate droplets? *J. Atmos. Sci.*, **66**, 741–754.

Swietlicki, E., et al. (1999). A closure study of submicrometer aerosol particle behavior. *Atmos. Res.*, **50**, 205–240.

Swietlicki, E., et al. (2000). Hygroscopic properties of aerosol particles in the north-eastern Atlantic during ACE-2. *Tellus, Ser. B*, **52**, 201–227.

Szyrmer, W., and I. Zawadski (2010). Snow studies. Part II: Average relationship between mass of snowflakes and their terminal fall velocity. *J. Atmos. Sci.*, **67**, 3319–3335.

Tabazadeh, A., and O. B. Toon (1998). The role of ammoniated aerosols in cirrus cloud nucleation. *Geophys. Res. Lett.*, **25**, 1379–1382.

Tabazadeh, A., E. J. Jensen, and O. B. Toon (1997). A model description for cirrus cloud nucleation from homogeneous freezing of sulfate aerosols. *J. Geophys. Res.*, **102** (D20), 23,845–23,850.

Tabazadeh, A., S. T. Martin, and J.-S. Lin (2000). The effect of particle size and nitric acid uptake on the homogeneous freezing of aqueous sulfuric acid particles. *Geophys. Res. Lett.*, **27**, 1111–1114.

Tabazadeh, A., Y. S. Djikaev, and H. Reiss (2002b). Surface crystallization of supercooled water in clouds. *Proc. Natl. Acad. Sci. (Geophysics)*, **99** (25), 15873–15878.

Tabazadeh, A., Y. S. Djikaev, P. Hamill, and H. Reiss (2002a). Laboratory evidence for surface nucleation of solid polar stratospheric cloud particles. *J. Phys. Chem. A*, **106**, 10,238–10,246.

Takahashi, T. (1976). Hail in axisymmetric cloud model. *J. Atmos. Sci.*, **33**, 1579–1601.

Takano, Y., and K.-N. Liou (1989). Solar radiative transfer in cirrus clouds, I, Single scattering and optical properties of hexagonal ice crystals. *J. Atmos. Sci.*, **46**, 3–19.

Tammann, G. (1900). Ueber die Grenzen des festen Zustandes IV. *Annaln. Phys.*, **2**, 1–31.

Tanaka, H. (1996). A self-consistent phase diagram for supercooled water. *Nature*, **380**, 328–330.

Tang, I. N. (1997). Thermodynamic and optical properties of mixed-salt aerosols of atmospheric importance. *J. Geophys. Res.*, **102** (D2), 1883–1893.

Tang, I. N., and H. R. Munkelwitz (1993). Composition and temperature dependence of the deliquescence properties of hygroscopic aerosols. *Atmos. Environ.*, **27A**, 467–473.

Tang, I. N., and H. R. Munkelwitz (1994). Water activities, densities, surface and refractive indices of aqueous sulfates and sodium nitrate droplets of atmospheric importance. *J. Geophys. Res.*, **99**, 18,801–18,808.

Tang, I. N., A. C. Tridicao, and K. H. Fung (1997). Thermodynamic and optical properties of sea salt aerosols. *J. Geophys. Res.*, **102** (D19), 23,269–23,275.

Tao, W.-K., J.-P. Chen, Z. Li, C. Wang, and C. Zhang (2012). Impact of aerosols on convective clouds and precipitation. *Rev. Geophys.*, **50**, RG2001, doi:10.1029/2011RG000369.

Tejero, C. F., and M. Baus (1998). Liquid polymorphism of simple fluids within a van der Waals theory. *Phys. Rev. E*, **57** (4), 4821–4823.

Telford, J. W. (1987). Comment on "Does mixing promote cloud droplet growth?" *J. Atmos. Sci.*, **44**, 2352–2354.

Telford, J. W., and S. K. Chai (1980). A new aspect of condensation theory. *Pure Appl. Geophys.*, **118**, 720–742.

Telford, J. W., T. S. Keck, and S. K. Chai (1984). Entrainment at cloud tops and the droplet spectra. *J. Atmos. Sci.*, **41**, 3170–3179.

Thomson, J. J. (1888). Application of dynamics to physics and chemistry, 1st ed., Cambridge University Press, 163 pp.

Tillner-Roth, R. (1998). Fundamental equations of state, Shaker Verlag, 126 pp.

Tomasi, C., E. Caroli, and V. Vitale (1983). Study of the relationship between Ångström's wavelength exponent and Junge particle size distribution exponent. *J. Clim. Appl. Meteorol.*, **22**, 1707–1716.

Tomasi, C., V. Vitale, A. Lupi, C. Di Carmine, et al. (2007). Aerosol in polar regions: A historical overview based on optical depth and in situ observations. *J. Geophys. Res.*, **112**, D16205.

Tripoli, G. J., and W. R. Cotton (1980). A numerical investigation of several factors contributing to the observed variable intensity of deep convection over South Florida. *J. Appl. Meteor.*, **19**, 1037–1063.

Truskett, T. M., P. G. Debenedetti, S. Sastry, and S. Torquato (1999). A single-bond approach to orientation-dependent interactions and its implications for liquid water. *J. Chem. Phys.*, **111**, 2647–2656.

Tsay, S. C., and K. Jayaweera (1984). Physical characteristics of Arctic stratus clouds. *J. Clim. Appl. Met.*, **23**, 584–596.

Tse, J. S. (1992). Mechanical instability in ice I_h: A mechanism for pressure-induced amorphization. *J. Chem. Phys.*, **96** (7), 5482–5487.

Tse, J. S., D. D. Klug, C. A. Tulk, I. Swainson, E. C. Svensson, C.-K. Loong, V. Shpakov, V. R. Belosludov, R. V. Belosludov, and Y. Kawazoe (1999). The mechanism for pressure-induced amorphization of ice I_h. *Nature*, **400**, 647–649.

Turnbull, D., and J. C. Fisher (1949). Rate of nucleation in condensed systems. *J. Chem. Phys.*, **17** (1), 71–73.

Turnbull, D., and B. Vonnegut (1952). Nucleation catalysis. *Industr. Eng. Chem.*, **44**, 1292–1298.

Twomey, S. (1959). The nuclei of natural cloud formation. II. The supersaturation in natural clouds and the variation of cloud droplet concentration. *Geoph. Pura Appl.*, **43**, 243–249.

Twomey, S. A. (1977a). The influence of pollution on the shortwave albedo of clouds. *J. Atmos. Sci.*, **34**, 1149–1152.

Twomey, S. (1977b). Atmospheric aerosols. Elsevier, New York, 302 pp.

Twomey, S., and T. A. Wojciechowski (1969). Observations of the geographical variations of cloud nuclei. *J. Atmos. Sci.*, **26**, 684–696.

Tzivion, S., G. Feingold, and Z. Levin (1987). An efficient numerical solution to the stochastic collection equation. *J. Atmos. Sci.*, **44**, 3139–3149.

Ulbrich, C. W. (1983). Natural variations in the analytical form of the raindrop size distribution. *J. Climate Appl. Meteor., 22*, 1764–1775.

Uttal, T., J. A. Curry, M. G. McPhee, et al. (2002). Surface heat budget of the Arctic Ocean. *Bull. Am. Meteorol. Soc.*, **83**, 255–275.

UNESCO (1981). *Tenth Report of the Joint Panel on Oceanographic Tables and Standards.* UNESCO Technical Papers in Marine Sci., No. 26, Paris, 126 pp.

Vaillancourt, P. A., and M. K. Yau (2000). Review of particle–turbulence interactions and consequences for cloud physics. *Bull. Amer. Meteor. Soc.*, **81**, 285–298.

Vaillancourt, P. A., M. K. Yau, P. Bartello, and W. W. Grabowski (2002). Microscopic approach to cloud droplet growth by condensation. Part II. Turbulence, clustering, and condensational growth. *J. Atmos. Sci.*, **59**, 3421–3435.

Vali, G. (1974). Contact ice nucleation by natural and artificial aerosols. In: "Conf. on Cloud Physics," Amer. Meteorol. Soc, Tucson., pp. 34–37.

Vali, G. (1976). Contact-freezing nucleation measured by the DFC instrument. In: "Third International Workshop on Ice Nucleus Measurements." Laramie, University of Wyoming, WY, USA, pp. 159–178.

Vali, G. (1985). Atmospheric ice nucleation: A review. *J. Rech. Atmos.*, **19**, 105–115.

Vali, G. (1994). Freezing rate due to heterogeneous nucleation. *J. Atmos. Sci.*, **51**, 1843–1856.

Vali, G. (2008). Repeatability and randomness in heterogeneous freezing nucleation. *Atmos. Chem. Phys.*, **8**, 5017–5031.

van de Hulst, H. C. (1957). Light scattering by small particles. Dover, 470 pp.

van Kampen, N. G. (1992). Stochastic processes in physics and chemistry, 2nd ed. North-Holland Publishing Company, 465 pp.

Vasilyeva, K. I., V. M. Merkulovich, and A. S. Stepanov (1984). Behavior of the droplet spectrum in a turbulized two-phase cloud. *Sov. Meteorol. Hydrol.*, **11**, 31–42.

Veltishchev, N. F., V. D. Zhupanov, and Yu. B. Pavlyukov (2011). Short-range forecast of heavy precipitation and strong wind using the convection-allowing WRF models. *Russian Meteorol. Hydrol.*, **36** (1), 47–58.

Verlinde, J., P. J. Flatau, and W. R. Cotton (1990). Analytical solutions to the collection growth equation: Comparison with approximate methods and application to cloud microphysics parameterization schemes. *J. Atmos. Sci.*, **47**, 2871–2880.

Verlinde, J., J. Y. Harrington, G. M. McFarquhar, V. T. Yannuzzi, et al. (2007). Overview of the Mixed-Phase Arctic Cloud Experiment (MPACE). *Bull. Am. Meteorol. Soc.*, **88**, 205–201.

Vertsner, V. N., and G. S. Zhdanov (1966). Electron-microscope study of the low-temperature forms of ice. *Sov. Phys. Crystallogr.*, **10**, 597–602.

Vignati, E., J. Wilson, and P. Stier (2004). An efficient size-resolved aerosol microphysics module for large-scale aerosol transport models. *J. Geophys. Res.*, D22 202.

Vlasiuk, M. P., N. G. Mukiy, Yu. A. Seregin, V. I. Khvorostyanov, A. A. Chernikov, L. V. Yaroshevich (1994). Progress in the development and the use of nitrogen technology for artificial dispersal of supercooled fogs at airports. *Proc. 6th WMO Scientif. Conf. Weather Modification*, Italy, May, WMO/TD-596, pp. 665–668

Volmer, M. (1939). Kinetic der phasenbildung. Steinkopf, Drezden and Leipzig, 325 pp.

Volmer, M., and A. Weber (1926). Keimbildung in übersättigten Gebilden. *Z. Phys. Chem.*, **A119**, 277–301.

Voloshchuk, V. M. (1977). The kinetic equation of stochastic coagulation. *Sov. Meteorol. Hydrol.*, No. 5, 3–12.

Voloshchuk, V. M. (1984). The kinetic theory of coagulation. Hydrometeoizdat, Leningrad, 283 pp. (in Russian).

Voloshchuk, V. M., and Y. S. Sedunov (1975). The processes of coagulation in the disperse systems. Leningrad, Hydrometeoizdat, 320 pp. (in Russian).

Voloshchuk, V. M., and Y. S. Sedunov (1977). The kinetic equation for the evolution of the droplet spectrum in the turbulent medium at the condensation stage of cloud development. *Sov. Meteorol. Hydrol.*, **3**, 1–10.

von der Emde, K., and U. Wacker (1993). Comments on the relationship between aerosol size spectra, equilibrium drop size spectra and CCN spectra. *Contrib. Atmos., Phys.*, **66**, 157–162.

Vonnegut, B. (1947). The nucleation of ice by silver iodide. *J. Appl. Phys.*, **18**, 593–595.

Wagner, W., and A. Pruβ (2002). The IAPWS formulation 1995 for the thermodynamic properties of ordinary water substance for general and scientific use. *J. Phys. Chem. Ref. Data*, **31**, 387–535.

Wagner, W., A. Saul, and A. Pruβ (1994). International equations for the pressure along the melting and along the sublimation curve of ordinary water substance. *J. Phys. Chem. Ref. Data*, **23** (3), 515–527.

Wagner, W., T. Riethmann, R. Feistel, and A. H. Harvey (2011). New equations for the sublimation pressure and melting pressure of H_2O ice I_h. *J. Phys. Chem. Ref. Data*, **40**, 043103, doi:10.1063/1.3657937.

Waldman, L., and K. H. Schmidt (1966). Thermophoresis and diffusionphoresis of aerosols. Academic Press, New York, 468 pp.

Wang, P. K., and W. Ji (1992). Preprints Cloud Phys. Conf. Montreal, McGill Univ., pp. 76–80.

WMO (1975). International cloud atlas, vol. I. Manual on the observations of clouds and other meteors. World Meteorological Organization (WMO), Geneva.

WMO (1987). International cloud atlas, vol. II. Plates. World Meteorological Organization (WMO), No. 407, Geneva.

Warner, J. (1969a). The microstructure of cumulus cloud. Part I: General features of the droplet spectrum. *J. Atmos. Sci.*, **26**, 1049–1059.

Warner, J. (1969b). The microstructure of cumulus cloud. Part II. The effect on droplet size distribution of the cloud nucleus spectrum and vertical velocity. *J. Atmos. Sci.*, **26**, 1272–1282.

Warren, S. G. (1984). Optical constants of ice from ultraviolet to the microwave. *Appl. Opt.*, **23**, 1206–1225.

Warren, S. G., C. J. Hahn, and J. London (1985). Simultaneous occurrence of different cloud types. *J. Clim. Appl. Meteor.* **24** (7), 658–667.

Warren, S. G., R. Eastman, and C. J. Hahn (2013). Cloud climatology. Encyclopedia of atmospheric sciences. Oxford University Press, 137–147.

Wegener, A. (1911). Thermodynamik der atmosphäre. J. A. Barth, 331 pp.

Weingartner, E., H. Burtscher, and U. Baltensperger (1997). Hygroscopic properties of carbon and diesel soot particles. *Atmos. Env.*, **31**, 3211–3227.

Wettlaufer, J. S. (1999). Impurity effects in the premelting of ice. *Phys. Rev. Lett.*, **82**, 2516–2523.

Wettlaufer, J. S., and J. G. Dash (2000). Melting below zero. *Scientific American*, **282** (2), 50–53.

Wexler, A. (1976). Vapor pressure formulation for water in range 0 to 100 °C: A revision. *J. Res. Nat. Bur. Stand.*, **80A**, 775–785.

Wexler, A. S., and J. H. Seinfeld (1991). Second-generation inorganic aerosol model. *Atmos. Environ.*, **25A**, 2731–2748.

Whalley, E. (1969). Structural problems of ice. In: Physics of ice. Eds.: Riehl, N., Bullemer, B., and Engelgardt, H., Plenum, New York, pp. 19–43.

Whalley, E., and D. W. Davidson (1965). Entropy changes in the phase transitions of ice. *J. Chem. Phys.*, **43**, 2148–2149.

Whalley, E., J. B. R. Heath, and D. W. Davidson (1968). Ice IX: An antiferroelectric phase related to ice III. *J. Chem. Phys.*, **48**, 2362–2370.

Whitby, K. (1978). The physical characteristics of sulfur aerosols. *Atmos. Environ.*, **12**, 135–159.

Willis, P. T. (1984). Functional fits to some observed drop size distributions and parameterization of rain. *J. Atmos. Sci.*, **41**, 1648–1661.

Woodcock, A. H., R. A. Duce, and J. L. Moyers (1971). Salt particles and raindrops in Hawaii. *J. Atmos. Sci.*, **28**, 1252–1257.

Wright, D. G., R. Feistel, J. H. Reissmann, K. Miyagawa, D. R. Jackett, W. Wagner, U. Overhoff, C. Guder, A. Feistel, and G. M. Marion (2010). Numerical implementation and oceanographic application of the thermodynamic potentials of liquid water, water vapour, ice, seawater and humid air – Part 2: The library routines. *Ocean Sci.*, **6**, 695–718, doi: 10.5194/os–6–695–2010.

Wurzler, S., T. G. Reisin, and Z. Levin (2000). Modification of mineral dust particles by cloud processing and subsequent effect on drop size distributions. *J. Geophys. Res.*, **105** (D4), 4501–4512.

Xu, J., D. Imre, R. McGraw, and I. Tang (1998). Ammonium sulfate: Equilibrium and metastability phase diagrams from 40 to −50 °C. *J. Phys. Chem., B*, **102**, 7462–7469.

Xue, L., A. Teller, R. Rasmussen, I. Geresdi, and Z. Pan (2010). Effects of aerosol solubility and regeneration on warm-phase orographic clouds and precipitation simulated by a detailed bin microphysical scheme. *J. Atmos. Sci.*, **67**, 3336–3354.

Yang, P., B.-C. Gao, B. A. Baum, Y. X. Hu, W. J. Wiscombe, S.-C. Tsay, D. M. Winker, and S. L. Nasiri (2001). Radiative properties of cirrus clouds in the infrared (8–13 μm) spectral region. *J. Quant. Spectrosc. Radiat. Transfer*, **70**, 473–504.

Young, K. C. (1974a). A numerical simulation of wintertime, orographic precipitation. Part I: Description of model microphysics and numerical techniques. *J. Atmos. Sci.*, **31**, 1735–1748.

Young, K. C. (1974b). The role of contact nucleation in ice phase initiation in clouds. *J. Atmos. Sci.*, **31**, 768–776.

Young, K. C. (1993). Microphysical processes in clouds. Oxford University Press, 427 pp.

Young, K. C., and A. J. Warren (1992). A reexamination of the derivation of the equilibrium supersaturation curve for soluble particles. *J. Atmos. Sci.*, **49**, 1138–1143.

Yum, S. S., and J. G. Hudson (2001). Vertical distributions of cloud condensation nuclei spectra over the springtime Arctic Ocean. *J. Geophys. Res.*, **106** (D14), 15,045–15,052.

Yun, Y., and J. E. Penner (2012). Global model comparison of heterogeneous ice nucleation parameterizations in mixed phase clouds. *J. Geophys. Res.*, **117**, D07203, doi:10.1029/2011JD016506.

Zawadski, I., and de M. A. de Agostinho (1988). Equilibrium raindrop size distributions in tropical rain. *J. Atmos. Sci.*, **45**, 3452–3459.

Zawadski., I., E. Jung, and G. W. Lee (2010). Snow studies. Part I: A study of natural variability of snow terminal velocity. *J. Atmos. Sci.*, **67**, 1591–1604.

Zeldovich, J. B. (1942). Toward the theory of formation of a new phase. Cavitation. *J. Exper. Theor. Phys.*, **12**, 525.

Zhang, G., J. Vivekanandan, and E. A. Brandes (2001). A method for estimating rain rate and drop size distribution from polarimetric radar measurements. *IEEE Trans. Geosci. Remote Sens.*, **39**, 830–841.

Zhang, G., J. Vivekanandan, and E. A. Brandes (2003a). Constrained gamma drop size distribution model for polarimetric radar rain estimation: Justification and development. *Preprints, 31st Int. Conf. on Radar Meteorology*, Seattle, WA, USA, *Amer. Meteor. Soc.*, 206–225.

Zhang, G., J. Vivekanandan, E. A. Brandes, R. Meneghini, and T. Kozu (2003b). The shape-slope relation in observed gamma raindrop size distributions: Statistical error or useful information? *J. Atmos. Oceanic Technol.*, **20**, 1106–1119.

Zhang, H., I. N. Sokolik, and J. A. Curry (2011). Impact of dust aerosols on Hurricane Helene's early development through the deliquescent heterogeneous freezing mode. *Atmos. Chem. Phys. Discuss.*, **11**, 14,339–14,381, doi:10.5194/acpd-11-14339-2011.

Zhou, J., E. Swietlicki, O. H. Berg, P. A. Aalto, K. Hämeri, E. D. Nilsson, and C. Leck (2001). Hygroscopic properties of aerosol particles over the central Arctic Ocean during summer. *J. Geophys. Res.*, **106** (D23), 32,111–32,123.

Zobrist, B. (2006). Heterogeneous ice nucleation in upper tropospheric aerosols. *Dr. Dissertation*, Swiss Federal Inst. Technology, Zurich, 148 pp.

Zobrist, B., C. Marcolli, T. Peter, and T. Koop (2008). Heterogeneous ice nucleation in aqueous solutions: The role of water activity. *J. Phys. Chem., A*, **112**, 3965–3975.

Zobrist, B., T. Koop, B. P. Luo, C. Marcolli, and T. Peter (2007). Heterogeneous ice nucleation rate coefficient of water droplets coated by a nonadecanol monolayer. *J. Phys. Chem. C*, **111**, 2149–2155, doi:10.1021/jp066080w.

Zuberi, B., A. K. Bertram, C. A. Cassa, L. T. Molina, and M. J. Molina (2002). Heterogeneous nucleation of ice in $(NH_4)_2SO_4$-H_2O particles with mineral dust immersions. *Geophys. Res. Lett*, **29** (10), 1504, doi:10.1029/2001GL014281.

Zubler, E. M., U. Lohmann, D. Lüthi, and C. Schär (2011). Statistical analysis of aerosol effects on simulated mixed-phase clouds and precipitation in the Alps. *J. Atmos. Sci.*, **68,** 1474–1492.

Notations

A_1, A_2 coefficients in the commonly used supersaturation equation that describe supersaturation generation and absorption

$A_f \equiv r_{fr} = \dfrac{2\sigma_{sa}}{\rho_i L_m^{ef}}$ curvature parameter for freezing (scaling freezing radius) that characterizes the effect of the drop radius on freezing

$A_g = 4\pi r_g^2$ surface area of an embryo of new phase with radius r_g containing g particles

$A_k = \dfrac{2M_w \sigma_{sa}}{RT\rho_w}$ Kelvin's curvature parameter

A_K backscatter cross sections for spherical particles at radar wavelengths

$A_p = \dfrac{\Delta\rho}{\rho_w \rho_i L_m^{ef}}$ coefficient that characterizes pressure effect on freezing; $\Delta\rho = \rho_w - \rho_i$

a_m parameter describing the pressure reduction in a non-ideal gas

a_v activity of water vapor

a_w activity of water in solution

$\tilde{a}_s$ activity of solute in solution

$\tilde{a}_{s*}$ activity of solute in saturated solution

$a_w^i(T)$ water activity in bulk solution in equilibrium with ice

$\delta a_{w,het} = a_w(T_{f,het}) - a_w(T_m)$ difference or shift in the activities between the melting and heterogeneous freezing curves at the same temperature

$\delta a_{w,\text{hom}} = a_w(T_{f,\text{hom}}) - a_w(T_m)$ difference or shift in the activities between the melting and homogeneous freezing curves at the same temperature

$B = \dfrac{3\nu\Phi_s m_s M_w}{4\pi M_s \rho_w} = br_d^{2(1+\beta)}$ Raoult's term in the Köhler equation (the activity of a CCN) and its parameterization via particle radius r_d

B_{ij} autocorrelation velocity function in turbulence diffusion

$B_{ij}^{nn}(t,\tau_f)$ autocorrelation velocity function of a nonconservative substance

$b = (\nu\Phi_s)\varepsilon_v \dfrac{\rho_s}{\rho_w}\dfrac{M_w}{M_s}$ parameter describing the volume-proportional soluble fraction in a CCN for $\beta = 1/2$

$b = r_{ad1}\varepsilon_{v0}(\nu\Phi_s)\dfrac{\rho_s}{\rho_w}\dfrac{M_w}{M_s}$ parameter describing the surface-proportional soluble fraction in a CCN for $\beta = 0$

$b_{con} = \dfrac{D_v \rho_{vs}}{\Gamma_p \rho_w}$ parameter in the particle growth rate equation in stochastic condensation theory

b_m parameter describing the excluded volume in a non-ideal gas

C_D drag coefficient

C_e electrical capacitance of a conductor

C_V heat capacity of a body at constant volume

C_P heat capacity of a body at constant pressure

C_{het} prefactor (kinetic coefficient) in the nucleation rate of heterogeneous freezing

C_{hom} prefactor (kinetic coefficient) in the nucleation rate of homogeneous freezing

C_{del} kinetic coefficient for deliquescence

$C_\varepsilon \approx 1.7 \times 10^{11}\,\mathrm{erg\,cm^{-3}}$ Turnbull–Vonnegut constant for the misfit strain energy density

$C_\varepsilon \varepsilon^2$ increase in activation energy density [$\mathrm{erg\,cm^{-3}}$] of nucleation due to misfit strain ε

$C_{sl,r}(r_a)$ Cunningham's slip-flow correction

$C_{st} = (2/9) g \rho_b r^2/\eta$ Stokes terminal velocity

c number of components in a thermodynamic system

$c_{1w}(T) = \left(\frac{L_e}{c_p T} \frac{M_w}{M_a} - 1 \right) \frac{g}{R_a T}$ coefficient in the generation term in the water supersaturation equation

$c_{1i}(T) = \left(\frac{L_s}{c_p T} \frac{M_w}{M_a} - 1 \right) \frac{g}{R_a T}$ coefficient in the generation term in the ice supersaturation equation

$c_{3w} = \frac{D_v \rho_{ws}}{\rho_w \Gamma_1}$ coefficient in the equation for droplet radius growth rate

$c_{3i} = \zeta_i \frac{D_v \rho_{is}}{\rho_i \Gamma_i}$ coefficient in the equation for crystal axis or radius growth rate

$c_{wi} = \exp\left[\frac{L_m (T_0 - T)}{R_v T_0 T} \right]$ coefficient in the relation between water s_w and ice s_i supersaturations

$c_{pi,m}$ isobaric molar heat capacity of ice

c_{pi} specific heat capacity of ice

c_v heat capacity at constant volume per mole or unit mass

$c_v = \rho_v/m_w$ molecular concentration

c_p heat capacity at constant pressure per mole or unit mass

c_i heat capacity of ice

c_w heat capacity of water

$c_g = c_{sat,w} \exp(-\Delta F_g/kT)$ Boltzmann'sdistribution of g-mers

$c_{sat,w}$ concentration of the monomers (water molecules) in a water-saturated environment

$c_{con} = G_s \bar{r} = b_{con} c_{1w} \tau_f$ parameter in the stochastic condensation theory

$D_{2,ext}$, $D_{3,ext}$, $D_{4,ext}$ the terms in the equation for the extinction cross section

$D_D(z)$ Debye's function

D_v the vapor diffusion coefficient

D_v^* effective vapor diffusion coefficient

$D_r = \frac{kT}{6\pi\eta_a r}$ Einstein's coefficient of Brownian diffusion

$D(g)$ diffusion coefficient of g-mers in the g-space in nucleation theory

$(dq_l/dz)_{ad}$ adiabatic gradient of liquid water mixing ratio or liquid water content

$(dq_i/dz)_{ad}$ adiabatic gradient of the ice water mixing ratio or ice water content

$dpdq$ the differential element of the phase space

E internal energy

$E_{kin}(p_m)$ kinetic energy of a particle depending on momentum p_m

$E(R, r)$ particles collision efficiency

E_n quantum energy at the n-th level

$E_{ij}(\omega)$ spectral density of turbulent energy

$\tilde{e}$ internal energy per mole or unit mass

e_v pressure of water vapor

e_{ws} pressure of water vapor saturated over water

e_{is} pressure of water vapor saturated over ice

erf(z) error function

F Helmholtz free energy of the system

$F_D = 6\pi\eta r V$ drag force of a spherical body with radius r falling with the velocity V in a medium with the dynamic viscosity η

$F_{gr} = mg$ gravitational force acting on a body in the gravitational field

F_{id} Helmholtz free energy F of an ideal gas

F_{EOS} free energy of a simple liquid described by the equation of state of van der Waals theory

$2F_{hb}$ free energy of the hydrogen bonds

$\Delta F_{a,del}$ activation energy for transition of the salt molecules from the solid to the liquid solution at deliquescence

ΔF_{act} activation energy for transition of the water molecules through a potential barrier at nucleation

$\Delta F_{cr,del}$ critical energy of deliquescence

$\Delta F_{cr,eff}$ critical energy of efflorescence

ΔF_{cr} critical energy of nucleation in the processes of condensation, deposition, freezing

$F_{ij}(\omega)$ spectral function of turbulence at frequency ω

$F(a, b; x)$ confluent hypergeometric (Kummer) function

$F(\alpha, \beta, \gamma, x)$ Gauss hypergeometric function

f_h specific Helmholtz free energy

$f(\bar{x}, r, \xi_i)$ distribution function in the phase space of (n + 4) of its variables

$f(m)$ size distribution function of the cloud particles by masses

$f(r)$ size distribution function of the cloud particles by radii r

$f_a(r_a) = C_{Na} r_a^{-\nu}$ inverse power law for the size spectra of aerosol or large drops and crystals

$f(r) = c_N r^p \exp(-\beta r^{\lambda})$ generalized gamma distributions

$f(r) = c_N r^p \exp(-\beta r)$ gamma distributions

$f_s(r)$ small-size distribution function

$f_l(r)$ large-size distribution function

f_v. ventilation coefficient in the drop and crystal growth rates

G Gibbs free energy

$G_n = RT/(M_w L_m^{ef})$ non-dimensional power index describing the effect of S_w in the theories of homogeneous and heterogeneous freezing

$G_s = \dfrac{D_v}{\rho_w \bar{r}} \dfrac{c_p}{L} \dfrac{\gamma_d - \gamma_w}{4\pi D N \bar{r}} \rho_a$ dimensionless parameter of the stochastic condensation theory

g Gibbs free energy per mole or unit mass

g acceleration of gravity

$g_s = \nu R\bar{T}/(L_c^{ef} M_s)$ dimensionless parameter of the theories deliquescence and efflorescence

H enthalpy or heat function

h enthalpy per mole or unit mass

h Planck's constant

h_w, h_i molar enthalpies of water and ice

$h(x, y)$ height of surface relief

$-\Delta h_s$ molar heat (enthalpies) of salt dissolution

I_{con} vapor flux to (from) the drops at condensation (evaporation), the source or sink term in the supersaturation equation

I_{dep} vapor flux to (from) the crystals at deposition (sublimation), the source or sink term in the supersaturation equation

$I_{con}^{eq} = \dfrac{\rho_{ws}}{\Gamma_1 \tau_{fd}} s_{w,liq}^{eq}$ equilibrium condensation rate in a liquid cloud

$I_{dep}^{eq} = \dfrac{\rho_{is}}{\Gamma_2 \tau_{fc}} s_{i,ice}^{eq}$ equilibrium deposition rate in a crystalline cloud

IWC ice water content

i van't Hoff's factor

$J_{c,dif}$ diffusional vapor flux to the drop surface

J_{kin}^{+}, J_{kin}^{-} kinetic condensation and evaporation molecular fluxes to and from the drop

J_{heat} heat flux from the drop or crystal surface

$J_{cond,hom}$ homogeneous nucleation rate of the drops from supersaturated vapor

J_{dep} deposition nucleation freezing rate

$J_{f,hom}$ homogeneous freezing nucleation rate

$J_{f,het}$ heterogeneous freezing nucleation rate

$j_h = -k_a \nabla T$ heat flux density around a drop or crystal

$j_g^{(+)}$ flux of the monomers to the germ (g-mer) of a new phase

K_b ebullioscopic constant

K_f cryoscopic constant

K_{CB} collection or scavenging rates [$cm^3 s^{-1}$] for convective Brownian diffusion

K_{Th} collection rates for thermophoresis

K_{Df} collection rates for diffusiophoresis

K_{sc} total aerosol collection or scavenging rate

K_{ik} coagulation collection kernel

$K(m, m')$ collection kernel in the mass space

$Kn = \lambda/r$ Knudsen number

K_{ext}, K_{sc}, *and* K_{abs} factors of radiation extinction, scattering, and absorption

$k = R/N_{Av} = 1.38 \times 10^{-16}\,\mathrm{erg\,K^{-1}}$ Boltzmann's constant

k_a thermal conductivity coefficient of humid air [cal/(cm sec °C)]

$k_f = C_e/r_c$ dimensionless crystal shape factor

$k_{a0} = 4/(\sqrt{2\pi}\sigma_d)$ index of the algebraic size spectra

$k_{s0} = \frac{4}{\sqrt{2\pi}\ln\sigma_s} = \frac{4}{\sqrt{2\pi}(1+\beta)\ln\sigma_d}$ index of the algebraic activity spectra

$\kappa_T = k_a/(\rho_a c_p)$ thermal diffusivity coefficient of air $[\mathrm{cm^2\,s^{-1}}]$

k_v the mean transfer coefficient in the ventilation corrections

k_w the rational activity coefficient in solutions

k_x, k_y, k_z horizontal and vertical components of the turbulent coefficient

k_{ij} turbulence diffusion tensor coefficient

$k_{ij}^{nn}(\tau_f)$ turbulence tensor coefficient for a non-conservative substance

L_c specific heat (enthalpy) of salt dissolution

$L_c^{ef}(T)$ specific salt dissolution heat averaged over the temperature

L_e, L_v specific latent heat of evaporation or condensation, $[\mathrm{cal\,g^{-1}}]$

$\hat{L}_e = M_w L_e$ molar latent heat of evaporation or condensation

L_m specific melting heat, $[\mathrm{cal\,g^{-1}}]$

$L_m^{ef}(T)$ specific melting heat averaged over the temperature

$\hat{L}_m = M_w L_m$ molar melting heat

L_s specific latent heat of sublimation or deposition, $[\mathrm{cal\,g^{-1}}]$

$\hat{L}_s = M_w L_s$ molar latent heat of sublimation or deposition

L_{vis} visibility or visual range in a fog or cloud

LWC liquid water content

l'_r mixing length in radii space

l_{mix}, l'_j Prandtl's mixing length

$\hat{M}$ molality, the number of moles of salt dissolved in 1000 g of water

$M^{(n)}$ n-th moment of the size spectrum

$M_a \approx 28.96\,\mathrm{g\,mole^{-1}}$ molecular weight of the dry air

M_s molecular weight of the soluble fraction (solute) in solution

$M_w \approx 18.015\,\mathrm{g\,mole^{-1}}$ molecular weight of water

m mass of any molecule

$m_{w1} = M_w/N_{Av} \approx 3 \times 10^{-23}\,g$ mass of a water molecule

$m_a = M_a/N_{Av} \approx 4.8 \times 10^{-23}\,g$ mass of the air molecule

m_d mass of a dry aerosol particle or mass of a droplet

$\dot{m}_{dif} = dm_d/dt$ diffusional drop mass growth rate

m_s, m_w the masses of the soluble fraction and water in solution

N total number of particles in the system

N particle concentration (number of particles in a unit volume; $1\,\mathrm{cm^3}$ or 1 L or $1\,\mathrm{m^3}$)

$N_{Av} = 6.023 \times 10^{23}$ molecules $\mathrm{mole^{-1}}$ Avogadro's number

N_d, N_c droplet and crystal concentrations or number densities, the numbers of particles in a unit volume of a cloud

N_{cc} number of salt molecules in a salt crystal in contact with the unit area of solution

N_w number of thermodynamic degrees of freedom in a system

$N_l = N_{Av}/v_w = 1/v_{w1} \approx 3.34 \times 10^{22}\,\mathrm{cm^{-3}}$ concentration of molecules in liquid water

$N_I = N_{Av}/v_i = 1/v_{i1} \approx 3.06 \times 10^{22}\,\mathrm{cm^{-3}}$ concentration of molecules in ice

$N_{dr,l}^{(1)}$, $N_{dr,l}^{(2)}$, $N_{dr,l}^{(3)}$, $N_{dr,l}^{(4)}$ concentrations of drops activated on CCN in the 1st, 2nd, 3rd, and 4th limits of the lower bound

$N_{dr,u}^{(1)}$, $N_{dr,u}^{(2)}$, $N_{dr,u}^{(3)}$, $N_{dr,u}^{(4)}$ concentrations of drops activated on CCN in the 1st, 2nd, 3rd, and 4th limits of the upper bound

n_k number of moles of the k-th chemical component in the system

n_k average number of particles in each energy state ε_k in statistical distributions

$\bar{n}_k$ average number of particles in each k-state

$\bar{n}_k = \exp\left(\frac{\mu - \varepsilon_k}{kT}\right)$ Boltzmann's distribution

$\overline{n}_k = \dfrac{1}{\exp[(\varepsilon_k - \mu)/kT - 1]}$ Bose–Einstein distribution

$\overline{n}_k = \dfrac{1}{\exp[(\varepsilon_k - \mu)/kT + 1]}$ Fermi–Dirac distribution

$n_\lambda = m_{i\lambda} - i\kappa_{i\lambda}$ complex radiation refractive index at the wavelength λ for water ($i = 1$) and ice ($i = 2$); $m_{i\lambda}$ and $\kappa_{i\lambda}$ are its real and imaginary parts

n_{solv} number of moles of solvent in solution

P precipitation rate

$\mathrm{Pe} = Du/\kappa_T$ Peclet number

$\mathrm{Pr} = \eta/(\rho_a\kappa_T)$ Prandtl number

p pressure

p generalized coordinates in statistical energy distributions

p spectral index or the shape parameter of the gamma distribution

p_{EOS} pressure determined from the van der Waals equation of state

p_{hb} pressure exerted by the hydrogen bonds

p_x, p_y, p_z components of the momentum

$p = \dfrac{b_{con}s}{c_{nn}kG_s^2} = \dfrac{\overline{r}w_{ef}}{c_{nn}kG_s}$ index of gamma distribution obtained in stochastic condensation theory

Q heat in the thermodynamic processes

Q_e electrical charge of a charged conductor

$Q_w(t) = \dfrac{2D_v\rho_a}{\rho_w\Gamma_w}\int_{t_j}^{t}\Delta_w(t')dt'$ normalized integral specific water supersaturation

$Q_i(t) = \dfrac{2D_v\rho_a\zeta_i}{\rho_i\Gamma_i}\int_{t_j}^{t}\Delta_i(t')dt'$ normalized integral specific ice supersaturation

q heat per mole or unit mass

q generalized coordinates in statistical energy distributions

q_l liquid water mixing ratio; liquid water content

q_i ice water mixing ratio; ice water content

q_v specific humidity

q_{ws}, q_{is} saturated over water and ice specific humidities

$q_{sup,w} = q_v - q_{ws}$ specific supersaturation over water

$q_{sup,i} = q_v - q_{is}$ specific supersaturation over ice

$\left(\dfrac{dq_l}{dz}\right)_{ad} = \dfrac{c_p}{L_e}(\gamma_a - \gamma_w)\rho_a$ adiabatic vertical gradient of the water mixing ratio

$\left(\dfrac{dq_i}{dz}\right)_{ad} = \dfrac{c_p}{L_s}(\gamma_a - \gamma_{is})\rho_a$ adiabatic vertical gradient of the ice mixing ratio

q_{ls} LWC or IWC of the small-size fraction

q_{ll} LWC or IWC of the large-size fraction

$R = kN_{Av} = 8.3144 \times 10^7\,\mathrm{erg\,mole^{-1}\,K^{-1}} = 1.9858\,\mathrm{cal\,mole^{-1}\,K^{-1}}$ universal gas constant

$Re = uD/\nu_a$ Reynolds number; D is the particle diameter, u is the velocity

$R_a = R/M_a = 2.8706 \times 10^6\,\mathrm{erg\,g^{-1}\,K^{-1}}$ gas constant for the air

$R_v = R/M_w = 4.6150 \times 10^6\,\mathrm{erg\,g^{-1}\,K^{-1}}$ gas constant for the water vapor

$R_{rad} = (\partial T/\partial t)_{rad}$ radiative temperature change rate

RHW, RHI relative humidities over water and ice in percent

$R_{f,het}$ Polydisperse nucleation rate in heterogeneous freezing

$R_{f,hom}$ Polydisperse nucleation rate in homogeneous freezing

r radius of an aerosol or cloud particle

$r_b = r_d + \Delta_v$ radius of the boundary sphere

$r_b = 2A_k/3s_w$ boundary radius between deliquescent but unactivated interstitial CCN and drops

r_d radius of a dry aerosol particle or cloud drop

r_c radius of a crystal

r_{a0} mean geometric radius of the lognormal spectra

$\overline{r}_d, \overline{r}_c$ mean radius of the droplets or the effective (mean) radius of the crystals

$\dot{r}_{d,dif}, \dot{r}_d = dr_d/dt$ diffusional droplet radius growth rate

$r_{d,eff}, r_{c,eff}$ *effective radius*, the ratio of the 3rd to the 2nd moments of the spectrum

r_m modal radius of the size spectrum

$\mathrm{Sc} = \eta_a/(\rho_a D_v)$ Schmidt number

$\mathrm{Sh} = 2k_v r_d/D_v$ Sherwood number

S_w water saturation ratio

S_i ice saturation ratio

S_η entropy of the system

$S_{w,cr}^{het}$ critical or threshold water saturation ratio of heterogeneous nucleation

$S_{i,cr}^{het} = S_{w,cr}^{het}(e_{ws}/e_{is})$ critical or threshold ice saturation ratio of heterogeneous nucleation

$S_{w,cr}^{\hom}$ critical or threshold water saturation ratio of homogeneous nucleation

$S_{i,cr}^{\hom} = S_{w,cr}^{\hom}(e_{ws}/e_{is})$ critical or threshold ice saturation ratio of homogeneous nucleation

$S_{w,cr}^{dep}$ critical or threshold water saturation ratio of deposition

$S_{i,cr}^{dep} = S_{w,cr}^{dep}(e_{ws}/e_{is})$ critical or threshold ice saturation ratio of deposition

$S_{s*} = \tilde{a}_s/\tilde{a}_{s*}$ solution supersaturation

$\delta S_{w,cr}^{het} = S_{w,cr}^{het}(T_{f,het}) - S_{w,cr}^{m}(T_m)$ shift or difference in the critical saturation ratios between the melting and heterogeneous freezing curves at the same temperature

$\delta S_{w,cr}^{\hom} = S_{w,cr}^{\hom}(T_{f,\hom}) - S_{w,cr}^{m}(T_m)$ shift or difference in the critical saturation ratios between the melting and homogeneous freezing curves at the same temperature

$s_w = (\rho_v - \rho_{ws})/\rho_{ws}$ fractional water supersaturation

$s_w = (\rho_v - \rho_{is})/\rho_{is}$ fractional ice supersaturation

$s_{w,liq}^{eq} = c_{1w} w \tau_{fd}$ equilibrium supersaturation in a liquid cloud

$s_{i,ice}^{eq} = \dfrac{c_{1i} w \tau_{fc}}{1 - c_{1i} w \tau_{fc}}$ equilibrium supersaturation in an ice cloud

$s_{w,mix}^{eq}$ equilibrium supersaturation in a mixed phase cloud

s_η molar entropy

$s_{ml}^{(1)}, s_{ml}^{(2)}, s_{ml}^{(3)}, s_{ml}^{(4)}$ 1st, 2nd, 3rd, and 4th limits of the lower bound of maximum supersaturation at CCN activation into drops

$s_{mu}^{(1)}, s_{mu}^{(2)}, s_{mu}^{(3)}, s_{mu}^{(4)}$ 1st, 2nd, 3rd, and 4th limits of the upper bound of maximum supersaturation at CCN activation into drops

T temperature

T_0, T_{tr} triple point temperature

$T_{cr} = 647.096\,\mathrm{K}$ critical temperature of water's first critical point

$T_{f,het}$ critical or threshold temperature of heterogeneous freezing

$T_{f,hom}$ critical or threshold temperature of homogeneous freezing

T_{dep} critical or threshold temperature of deposition

T_g temperature of glassy transition

T_m melting temperature

$T_r = (T - T_0)/T_0$ the reduced temperature

T_s spinodal temperature

T_{del} temperature of deliquescence

T_{eff} temperature of efflorescence

T_{eut} temperature of the eutectic point

$\Delta T = T_0 - T_\infty$ psychrometric temperature difference between the drop surface and environment

δT_f, ΔT_f freezing point depression

δT_m, ΔT_m melting point depression

$(\partial T/\partial t)_{rad}$ the radiative cooling rate caused by the total radiative flux divergence

t time

$t_{ml}^{(1)}, t_{ml}^{(2)}, t_{ml}^{(3)}, t_{ml}^{(4)}$ 1st, 2nd, 3rd, and 4th limits of the lower bound of time of CCN activation into drops

$t_{mu}^{(1)}, t_{mu}^{(2)}, t_{mu}^{(3)}, t_{mu}^{(4)}$ 1st, 2nd, 3rd, and 4th limits of the upper bound of time of CCN activation into drops

U potential energy of the molecular interactions in a non-ideal gas

$U_{pot}(q)$ potential energy of a particle depending on coordinate q

V volume of thermodynamic system

V_d volume of the dry aerosol particle or of the drop

V_t terminal velocity of a body in a medium

V_w volume of solvent (water) in aqueous solution or in an aerosol particle

$V_w = \left(\dfrac{8RT}{\pi M_w} \right)^{1/2}$ mean velocity of the water vapor molecules

$V_g(g)$ effective drift velocity of g-mers in g-space in nucleation theory

v volume per mole or unit mass

$v_w = M_w/\rho_w \approx 18\,\mathrm{cm^3\,g^{-1}}$ molar volume of water

$v_i = M_w/\rho_w \approx 19.63\,\mathrm{cm^3\,g^{-1}}$ molar volume of ice

$v_{w1} = v_w/N_{Av} \approx 3 \times 10^{-23}\,\mathrm{cm^3}$ volume of a single molecule in liquid water

$v_{i1} = v_i/N_{Av} \approx 3.27 \times 10^{-23}\,\mathrm{cm^3}$ volume of a single molecule in ice

v_{c0} *and* v_{s0} molar volumes of the solid crystal and solution at deliquescence and efflorescence

v_x, v_y, v_z components of the velocity

v_k terminal velocity of the k-th substance

W expansion work done on the system

$W_{xr}(\vec{x} - \vec{x}', r - r' | \vec{x}', r')$ probability of transition in the 4D phase space in a random process from the point $(\vec{x} - \vec{x}', r - r')$ to the point $(\vec{x}, r)$ due to a random jump $(\vec{x}', r')$

w expansion work per mole or unit mass

w weight concentration or weight percent of solute in solution

w_{nN} micro-canonic distribution

dw_p *and* dw_q energy distributions by momentum and coordinates

$dw_v(v_x, v_y v_z)$ Maxwell's differential velocity distribution

w_v water vapor mixing ratio

w_{ws}, w_{is} saturated over water and ice water vapor mixing ratio

$w_{rad} = -(1/\gamma_a)(\partial T/\partial t)_{rad}$ "radiative-effective" velocities

x_k mole fraction for the k-th substance

x_s, x_w mole fractions of solute and water in an aqueous solution

$x = 2\pi r/\lambda$ size parameter in radiation calculations

$y_w(t) = \int_{t_0}^{t} s_w(t')dt'$ integral fractional water supersaturation

$y_i(t) = \int_{t_0}^{t} s_i(t')dt'$ integral fractional ice supersaturation

Z statistical sum or the canonical partition function

Z_R, Z_D radar reflectivity factor

$Z = \left(\dfrac{\Delta F_g}{3\pi k T g^2} \right)^{1/2}$ Zeldovich factor for a g-mer in nucleation theory

α_c condensation coefficient

α_d deposition coefficient

α_{Li} spectral absorption coefficients of longwave (LW) radiation at wavelength λ

α_p coefficient of water thermal expansion

$\alpha_{sl}(r)$ Cunningham's slip-flow correction

α_T thermal accommodation coefficient

α_{k1}, $\alpha_{up}^{(1)}$, $\alpha_{low,up}^{(1)}$, $\alpha_{up}^{(2)}$, $\alpha_{low,up}^{(2)}$, $\alpha_{up}^{(3)}$, $\alpha_{low,up}^{(3)}$, $\alpha_{up}^{(4)}$, $\alpha_{low,up}^{(4)}$ coefficients of relations between the lower and upper bounds of t_m, s_m, and $N_{dr,m}$ in drop activation (Chapter 7)

$B(x, y)$ Euler's beta function

$\beta = 1/2$ parameter describing the volume-proportional soluble fraction in a CCN

$\beta = 0$ parameter describing the surface-proportional soluble fraction in a CCN

$\beta_T = 4\kappa_T/(\alpha_T V_w)$ kinetic correction to the growth rate in thermal conductivity

$\Gamma(x)$ Euler's gamma function

$\Gamma(x, \alpha)$ incomplete Euler's gamma function

$\Gamma_1 = 1 + \dfrac{L_e^2}{R_v c_p T^2} \dfrac{\rho_v}{\rho_a}$, $\Gamma_{12} = 1 + \dfrac{L_e L_s}{R_v c_p T^2} \dfrac{\rho_v}{\rho_a}$, $\Gamma_2 = 1 + \dfrac{L_s^2}{R_v c_p T^2} \dfrac{\rho_v}{\rho_a}$ psychrometric coefficients in the water and ice supersaturation equations and in the radii growth rates of drops and crystals

$d\Gamma$ the number of states in the differential element of the phase space

γ_a dry adiabatic lapse rate

γ_w wet adiabatic lapse rate in liquid clouds

γ_{is} wet adiabatic lapse rate in ice clouds

$\Delta_w = q_v - q_{ws}$ specific supersaturation over water

$\Delta_i = q_v - q_{is}$ specific supersaturation over ice

Δ_v thickness of the boundary sphere with kinetic molecular flux comparable to λ_a.

Δ_T thickness of the boundary sphere with kinetic molecular heat flux near the drop

$\delta_w = \rho_v - \rho_{ws}$ absolute supersaturation over water

$\delta_i = \rho_v - \rho_{is}$ absolute supersaturation over ice

ε elastic strain caused by the misfit between ice and substrate lattices [%]

$\varepsilon_{cw} = I_{con}/\rho_a$ specific condensation growth rate

$\varepsilon_{ci} = I_{dep}/\rho_a$ specific deposition growth rate

ε_f specific freezing rate

ε_m specific melting rate

ε_i cloud emissivity

ε_m mass soluble fraction of a cloud condensation nucleus

ε_v volume soluble fraction of a cloud condensation nucleus

ε_k quantum energy state

ε_k' internal energy of a molecule

$\zeta_i = k_{0i}\dfrac{k_{fi}}{\xi_f^n}$ factor describing the effect of the shape on the crystal growth rate

$\eta_a = \nu_a \rho_a$ dynamic viscosity of the air

θ potential temperature

$\theta_D = h\nu_m/k$ Debye's temperature

$<cos\theta>$ radiation asymmetry parameter

κ_T isothermal compressibility of water

κ_{vk} isobaric compressibility of water

$\Lambda_{sc}(r_a)$ spectral aerosol scavenging coefficient

λ wavelength of radiation

$\lambda_a(T, p) = \lambda_{a0}(p_0/p)(T/T_0)$ free molecular path

$\lambda_f \sim 0.1\ \mu m$ molecular mean free path length

μ_k chemical potential of the k -th chemical component in the system

$\mu_w^i(T)$, $\mu_w^0(T)$ chemical potentials of water in pure ice and of liquid water

ν frequency of radiation

ν the number of ions of a salt or acid in solution

ν_a kinematic viscosity of the air

ν_m maximum frequency of oscillators in Debye's energy spectrum of a solid

$\xi_{con} = 4D_v/(\alpha_c V_w)$ kinetic correction to the drop radius condensation growth rate

$\xi_{dep} = 4D_v/(\alpha_d V_w)$ kinetic correction to the crystal radius or axis deposition growth rate

$\xi_f = b/r_c$ crystal axes ratio (minor to major)

ρ the gas or liquid density

ρ_a air density

$\rho_{cr} = 322\ kg\ m^{-3}$ the critical density at water's first critical point

ρ_e density of the electrical charge

$\rho_{v\infty}$ vapor density at infinite distance from the drop

ρ_s'' solution density

ρ_{ws}, ρ_{is} densities of water vapor saturated over water and ice

ρ_i density of ice

ρ_w density of water

$d\rho_\varepsilon(\nu, T)$ energy density of radiation

σ_a dispersion of the lognormal spectrum

σ_{abs} absorption cross section

$\tilde{\sigma}_{ext,k}$, $\tilde{\sigma}_{sc,k}$, $\tilde{\alpha}_{abs,k}$ volume extinction, scattering, and absorption coefficients of the k-th substance in a cloud

$\sigma_{ext,k}$, $\sigma_{sc,k}$, $\sigma_{abs,k}$ mass extinction, scattering, and absorption coefficients of the k-th substance in a cloud

$\sigma_{\lambda Li}^s$, $\alpha_{\lambda Li}^s$ spectral scattering coefficients of shortwave (SW) radiation at wavelength λ

σ_{is} surface tension at the ice–solution interface

σ_{iv} surface tension at the ice–vapor interface

σ_{iw} surface tension at the ice–water interface

σ_{sa} surface tension at the solution–air interface

σ_{vw} surface tension at the vapor–water interface

σ_{wa} surface tension at the water–air interface

σ_{cs} surface tension at the crystal–solution interface at deliquescence or efflorescence

σ_{ra} absolute spectral dispersion by radii

σ_{rr} relative spectral dispersion by radii

σ_{col} collection frequency of the large-size fraction in the coagulation equation

τ_{cr} cloud crystallization time

τ_{fd}, τ_{fc} supersaturation absorption or relaxation times of drops and crystals; $\tau_{fd} \approx (4\pi D_v N_d \bar{r}_d)^{-1}$, $\tau_{fc} \approx (4\pi D_v N_c \bar{r}_c)^{-1}$ in the diffusion regime

$\tau_{f,mix} = [\tau_{fd}^{-1} + (\Gamma_{12}/\Gamma_2)\tau_{fc}^{-1}]^{-1}$ "effective" supersaturation relaxation time in a mixed cloud

$\tau_{c,fr} = \Psi_{cf}^{-1}(r_d, \Delta t)$ characteristic contact freezing time that describes the rate of drops freezing due to contact nucleation

$\tau_{sc}(r_a) = \Lambda_{sc}^{-1}(r_a)$ characteristic scavenging time of aerosol with radius r_a

$\tau_{ext,k}$ optical thickness of the k-th substance in a cloud

$\tau_{v,rel} = 2\rho_b r^2/(9\eta)$ velocity relaxation time

$\tau_{col} = (\sigma_{col})^{-1}$ characteristic coagulation (accretion) time of the e-folding decrease in q_{ls} or corresponding increase in q_{ll}

Φ_e electrostatic potential around a charged conductor

Φ_s molal or practical osmotic potential

$\Phi(x) = \mathrm{erf}(x)$ error function

φ number of bulk phases in a thermodynamic system

$\varphi_s(s_w)$ supersaturation activity spectrum of CCN

$\varphi(\delta_r, \tau_r)$ dimensionless Helmholtz free energy in the equation of state for liquid water

$\chi_{ad} = q_{ls}/q_{ls,ad}$ adiabatic liquid (ice) ratio

$\psi_{cf}(r_d, \Delta t)$ contact freezing rate (s^{-1}) of the drops that capture aerosol particles

ψ number of curved interfaces in a thermodynamic system

$\psi_h(s_w)$ supersaturation activity spectrum of IN

Ω Landau thermodynamic potential

ω frequency of turbulent fluctuation

ω_λ single-scattering albedo

Index

Printed in the United States
by Baker & Taylor Publisher Services